Methods and Applications of Statistics in the Life and Health Sciences

WILEY SERIES IN METHODS AND APPLICATIONS OF STATISTICS

Advisory Editor

N. Balakrishnan
McMaster University, Canada

The *Wiley Series in Methods and Applications of Statistics* is a unique grouping of research that features classic contributions from Wiley's *Encyclopedia of Statistical Sciences, Second Edition (ESS, 2e)* alongside newly written articles that explore various problems of interest and their intrinsic connection to statistics. The goal of this collection is to encompass an encyclopedic scope of coverage within individual books that unify the most important and interesting applications of statistics within a specific field of study. Each book in the series successfully upholds the goals of *ESS, 2e* by combining established literature and newly developed contributions written by leading academics, researchers, and practitioners in a comprehensive and accessible format. The result is a succinct reference that unveils modern, cutting-edge approaches to acquiring, analyzing, and presenting data across diverse subject areas.

WILEY SERIES IN METHODS AND APPLICATIONS OF STATISTICS

Balakrishnan · *Methods and Applications of Statistics in the Life and Health Sciences*

Forthcoming volumes:

Balakrishnan · *Methods and Applications of Statistics in Business, Finance, and Management Science*

Balakrishnan · *Methods and Applications of Statistics in Engineering, Quality Control, and the Physical Sciences*

Balakrishnan · *Methods and Applications of Statistics in the Social and Behavioral Sciences*

Balakrishnan · *Methods and Applications of Statistics in the Atmospheric and Earth Sciences*

Methods and Applications of Statistics in the Life and Health Sciences

N. Balakrishnan
McMaster University
Department of Statistics
Hamilton, Ontario, Canada

SOUTH UNIVERSITY LIBRARY
COLUMBIA, SC 29203

A JOHN WILEY & SONS, INC., PUBLICATION

Copyright © 2010 by John Wiley & Sons, Inc. All rights reserved.

Published by John Wiley & Sons, Inc., Hoboken, New Jersey. All rights reserved.
Published simultaneously in Canada.

No part of this publication may be reproduced, stored in a retrieval system, or transmitted in any form or by any means, electronic, mechanical, photocopying, recording, scanning, or otherwise, except as permitted under Section 107 or 108 of the 1976 United States Copyright Act, without either the prior written permission of the Publisher, or authorization through payment of the appropriate per-copy fee to the Copyright Clearance Center, Inc., 222 Rosewood Drive, Danvers, MA 01923, (978) 750-8400, fax (978) 750-4470, or on the web at www.copyright.com. Requests to the Publisher for permission should be addressed to the Permissions Department, John Wiley & Sons, Inc., 111 River Street, Hoboken, NJ 07030, (201) 748-6011, fax (201) 748-6008, or online at http://www.wiley.com/go/permission.

Limit of Liability/Disclaimer of Warranty: While the publisher and author have used their best efforts in preparing this book, they make no representations or warranties with respect to the accuracy or completeness of the contents of this book and specifically disclaim any implied warranties of merchantability or fitness for a particular purpose. No warranty may be created or extended by sales representatives or written sales materials. The advice and strategies contained herein may not be suitable for your situation. You should consult with a professional where appropriate. Neither the publisher nor author shall be liable for any loss of profit or any other commercial damages, including but not limited to special, incidental, consequential, or other damages.

For general information on our other products and services or for technical support, please contact our Customer Care Department within the United States at (800) 762-2974, outside the United States at (317) 572-3993 or fax (317) 572-4002.

Wiley also publishes its books in a variety of electronic formats. Some content that appears in print may not be available in electronic format. For information about Wiley products, visit our web site at www.wiley.com.

Library of Congress Cataloging-in-Publication Data:

Balakrishnan, N., 1956–
  Methods and applications of statistics in the life and health sciences / N. Balakrishnan.
    p. cm.
  Includes index.
  Preface: "This is the first of a series of handbooks on Methods and Applications of Statistics."
  ISBN 978-0-470-40509-3 (cloth)
  1. Medical statistics—Handbooks, manuals, etc. 2. Medicine—Research—Statistical methods—Handbooks, manuals, etc. 3. Life sciences—Statistical methods—Handbooks, manuals, etc. I. Title.
  R853.S7B35 2009
  610.72'7—dc22
                                                           2009015926

Printed in the United States of America.

10 9 8 7 6 5 4 3 2 1

Preface

People now are living longer than ever before. Many, despite the inevitable aging, lead active and healthy lives. This has resulted in an aging population bringing into limelight a host of key health-related issues such as population growth, disease detection and treatment, genetic research and possible cures for diseases, development of drugs, clinical trials and evaluation of the efficacy of new drugs and treatments, screening and prevention methods, assessment of rehabilitation and recovery, nutrition and physical exercise, alternative therapy and medication, and quality of life. Of course, over and above all these issues, there is a growing concern over medical care and the financial cost involved in health care and management.

These issues do place the onus on health agencies and organizations (public and private alike) to compile and process pertinent information in order to facilitate timely actions and health policies for improving the life and health of people. In this regard, statistical methods and applications play a vital role while addressing all the issues listed above. This is what provided the impetus for the volume to consolidate and present various statistical methods, techniques and applications, old and new, that are useful in handling a wide array of problems of interest in life and health sciences. It is my sincere hope that this would serve as a useful handbook for practitioners working in life and health sciences and also serve as a reference guide for statistical researchers and students.

It is important to mention that the recently revised edition of *Encyclopedia of Statistical Sciences* served as a basis for this handbook. While many pertinent entries from the *Encyclopedia* have been included here, a number of them have also been updated to reflect recent developments on their topics. Several new articles detailing modern advances on statistical methods in health sciences have also been included.

A volume of this size and nature cannot be successfully completed without the cooperation and support of the contributing authors, and my sincere thanks and gratitude go to all of them. Thanks are also due to Mr. Steve Quigley and Ms. Jacqueline Palmieri (of John Wiley & Sons, Inc.) for their keen interest in this project from day one, and their support and constant encouragement (and, of course, occasional nudges) throughout the course of this project. Careful and diligent work of Mrs. Debbie Iscoe and Ms. Emily Iscoe in the typesetting and production stages of this volume is gratefully acknowledged. Partial financial support of the Natural Sciences and Engineering Research Council of Canada also assisted the preparation of this handbook, and this support is much appreciated.

This is the first in a series of handbooks on *methods and applications of statistics*. While the present handbook has focused on life and health sciences, the forthcoming handbooks will cover such diverse disciplines as

Business, finance and management science,
Engineering, quality control and physical sciences,
Behavioral and social sciences, and
Environmental and earth sciences.

It is my sincere hope that this series of handbooks will become a basic reference resource for those involved in these fields of research!

PROF. N. BALAKRISHNAN
McMASTER UNIVERSITY
Hamilton, Canada
April 2009

Contributors

Colin G. G. Aitken, School of Mathematics, University of Edinburgh JCMB, Edinburgh, Scotland,
C.G.G.Aitken@ed.ac.uk

Ingrid A. Amara, Quintiles, Inc., Chapel Hill, NC

Christopher Amos, University of Texas MD Anderson Cancer Center, Houston, TX,
camos@mail.manderson.org

Per Kragh Andersen, Department of Biostatistics, University of Copenhagen, Copenhagen, Denmark,
pka@biostat.ku.dk

Charles Anello, Biostatistics Chief, Sylvester Comprehensive Cancer Center, Miller School of Medicine, University of Miami, Miami, FL,
canello@med.miami.edu

J. R. Ashford, Department of Mathematical Statistics and Operational Research, University of Exeter, Exeter, Devon, England

Stuart G. Baker, National Cancer Institute, Bethesda, MD,
sb16i@nih.gov

N. Balakrishnan, Department of Mathematics and Statistics, McMaster University, Hamilton, Ontario, Canada,
bala@mcmaster.ca

Donald A. Berry, Department of Biostatistics, University of Texas, MD Anderson Cancer Center, Houston, TX,
dberry@mdanderson.org

Ørnulf Borgan, Department of Mathematics, University of Oslo, Oslo, Norway,
borgan@math.uio.no

Larry L. Brant, NIH Biomedical Research Center, National Institute on Aging, Intramural Research Program, Baltimore, MD,
brantl@grc.nia.nih.gov

Germaine M. Buck Louis, *Eunice Kennedy Shriver* National Institute of Child Health and Human Development, National Institutes of Health, DHHS, Bethesda, MD,
louisg@mail.nih.gov

Chao W. Chen, Washington, DC

Chi Wan Chen, Center for Drug Evaluation and Research, FDA, Rockville, MD

Jie Chen, Department of Mathematics and Statistics, University of Missouri—Kansas City, Kansas City, MO,
chenj@umkc.edu

Shu L. Cheuk, LSUHSC School of Dentistry, Department of General Den-

tistry, New Orleans, LA,
scheuk@lsuhsc.edu

Pankaj K. Choudhary, Department of Mathematical Sciences, University of Texas at Dallas, Richardson, TX,
pankaj@utdallas.edu

Shein-Chung Chow, Duke Clinical Research Institute, Durham, NC,
sheinchung.chow@duke.edu

P. R. Cox

Noel Cressie, Department of Statistics, The Ohio State University, Columbus, OH,
ncressie@stat.ohio-state.edu

Susmita Datta, Public Health and Information Sciences, University of Louisville, Louisville, KY,
susmita.datta@louisville.edu

Jennifer Davidson

James A. Deddens, Department of Mathematics, University of Cincinnati, Cincinnati, OH,
james.deddens@uc.edu

Klaus Dietz, Director, Department of Medical Biometry, Eberhard-Karls-University of Tübingen, Tübingen, Germany,
klaus.dietz@uni-tuebingen.de

Alexei Dmitrienko, Lilly Research Labratories, Indianapolis, IN,
dmitrienko_alex@lilly.com

A. Donner, Department of Epidemiology and Biostatistics,
University of Western Ontario, London, Ontario, Canada,
allan.donner@schulich.uwo.ca

Satya D. Dubey, Food and Drug Administration, Bethesda, MD,
satya.dubey@comcast.net

Charles W. Dunnett, Department of Mathematics and Statistics, McMaster University, Hamilton, Ontario, Canada

Janet D. Elashoff, BMDP Statistics Software, Los Angeles, CA

W. J. Ewens, Department of Biology, University of Pennsylvania, Philadelphia, PA,
wewens@sas.upenn.edu

Vern T. Farewell, MRC Biostatistics Unit, Institute of Public Health, Cambridge, UK,
Vern.Farewell@mrc-bsu.cam.ac.uk

Mitchell H. Gail, Senior Investigator, Biostatistics Branch, Division of Cancer Epidemiology and Genetics, National Cancer Institute, Rockville, MD,
gailm@mail.nih.gov

Thomas A. Gerds, Institute of Medical Biometry and Medical Informatics, University of Freiburg, Freiburg, Germany,
gerds@fdm.uni-freiburg.de

Peter C. Gøtzsche, Nordic Cochrane Centre, Rigshospitalet, Kbenhavn, Denmark,
p.s.gotzsche@cochrane.dk

Joseph Glaz, Department of Statistics, University of Connecticut, Storrs, CT,
joseph.glaz@uconn.edu

Bernhard G. Greenberg, University of North Carolina, Chapel Hill, NC

Raymond S. Greenberg, Medical University of South Carolina, Charleston, SC,
greenber@musc.edu

Rudy Guerra, Department of Statistics, Rice University, Houston, TX,
rguerra@rice.edu

Larry V. Hedges, Institute for Policy Research, Northwestern University, Evanston, IL,
l-hedges@northwestern.edu

Jason C. Hsu, Ohio State University, Columbus, OH, *hsu.1@osu.edu*

Valen E. Johnson, The University of Texas, Graduate School of Biomedical Sciences at Houston, Houston, TX, *vejohnson@mdanderson.org*

Byron C. Jones, Department of Biobehavioral Health, The Pennsylvania State University, University Park, PA, *bcj1@psu.edu*

Richard Kay, *richard.kay@rkstatistics.com*

Niels Keiding, Institute of Public Health, University of Copenhagen, Copenhagen, Denmark, *N.Keiding@biostat.ku.dk*

Michael G. Kenward, London School of Hygiene and Tropical Medicine, London, UK, *Mike.Kenward@lshtm.ac.uk*

Sungduk Kim, *Eunice Kennedy Shriver* National Institute of Child Health and Human Development, Bethesda, MD, *kims2@mail.nih.gov*

N. Krishnan Namboodiri, Professor Emeritus, Department of Sociology, Ohio State University, Columbus, OH, *namboodiri.2@sociology.osu.edu*

Neil Klar, Department of Epidemiology and Biostatistics, University of Western Ontario, London, Ontario, Canada, *neil.klar@schulich.uwo.ca*

Gary G. Koch, School of Public Health, University of North Carolina, Chapel Hill, NC, *bcl@bios.unc.edu*

Andrew B. Lawson, Department of Biostatistics, Bioinformatics & Epidemiology, College of Medicine, Medical University of South Carolina, Charleston, SC, *lawsonab@musc.edu*

Emmanuel Lesaffre, Department of Biostatistics, Erasmus MC, Rotterdam, The Netherlands and Interuniversity Institute for Biostatistics and Statistical Bioinformatics, Katholieke Universiteit Leuven and Hasselt University, Diepenbeek, Belgium, *e.lesaffre@erasmusmc.nl*

Qizhai Li, Academy of Mathematics and Systems Science, Chinese Academy of Sciences, Beijing, People's Republic of China and Biostatistics Branch, National Cancer Institute, Bethesda, MD

Tsae-Yun Daphne Lin, Center for Drug Evaluation and Research, FDA, Rockville, MD, *daphne.lin@fda.hhs.gov*

Roderick J. Little, Department of Biostatistics-SPH, University of Michigan, Ann Arbor, MI, *rlittle@umich.edu*

Aiyi Liu, Biostatistics and Bioinformatics Branch, National Institutes of Health, Rockville, MD, *liua@mail.nih.gov*

Chunling (Catherine) Liu, Biostatistics and Bioinformatics Branch, National Institutes of Health, Rockville, MD, *liuc3@mail.nih.gov*

K. G. Manton, Center for Demographic Studies, Duke University, Medical Center, Durham, NC, *kgm@cds.duke.edu*

K. V. Mardia, Department of Statistics, The University of Leeds, Leeds, West

Yorkshire, UK,
K.V.Mardia@leeds.ac.uk

Ian W. McKeague, Department of Biostatistics, Columbia University, New York, NY,
im2131@columbia.edu

S. M. McKinlay, New England Research Institutes, Watertown, MA,
smckinlay@neriscience.com

Don McNeil, Faculty of Business and Economics, Macquarie University, Sydney, New South Wales, Australia,
dmcneil@efs.mq.edu.au

Carolina Meier-Hirmer, Freiburg Center for Data Analysis and Modeling, Freiburg, Germany,
cmh@fdm.uni-freiburg.de

Lee J. Melton III, Mayo Clinic, Rochester, MN,
melton.j@mayo.edu

Mounir Mesbah, LSTA, University of Pierre et Marie Curie, Paris 6, France,
mounir.mesbah@upmc.fr

David Moher, Children's Hospital of Eastern Ontario, University of Ottawa, Ottawa, Ontario, Canada

Geert Molenberghs, Center for Statistics, Hasselt University, Diepenbeek, Belgium,
geert.molenberghs@uhasselt.be

James E. Mosimann, Division of Computer Research and Technology, National Institutes of Health, Bethesda, MD

David M. Murray, Department of Psychology, University of Memphis, Memphis, TN,
d.murray@mail.psyc.memphis.edu

Joseph I. Naus, Department of Statistics, Rutgers University, Piscataway, NJ,
naus@stat.rutgers.edu

David Oakes, Department of Biostatistics and Computational Biology, University of Rochester,
Rochester, NY,
oakes@bst.rochester.edu

J. Keith Ord, McDonough School of Business, Georgetown University, Washington, DC,
ordk@georgetown.edu

Arthur V. Peterson Jr., Fred Hutchinson Cancer Research Center, Seattle, WA,
avpeters@fhcrc.org

German Rodriguez, Office of Population Research, Princeton University, Princeton, NJ,
grodri@princeton.edu

Kathryn Roeder, Carnegie Mellon University, Department of Statistics, Pittsburgh, PA,
roeder@cmu.edu

Bernard Rosner, Channing Laboratory, Harvard Medical School, Boston, MA,
stbar@channing.harvard.edu

Donald B. Rubin, Department of Statistics, Harvard University, Cambridge, MA,
rubin@stat.harvard.edu

Bruno Scarpa, Department of Statistical Sciences, University of Padua, Padua, Italy,
scarpa@stat.unipd.it

Nathaniel Schenker, National Center for Health Statistics, Centers for Disease Control and Prevention, Hyattsville, MD,
nschenker@cdc.gov

Stephen J. Senn, Department of Mathematics, University of Glasgow, Glasgow, UK,
stephen@stats.gla.ac.uk

Shan L. Sheng, NIH Biomedical Research Center, National Institute on Aging, Intramural Research Program, Baltimore, MD,
shengs@grc.nia.nih.gov

William K. Sieber Jr., National Institute for Occupational Safety and Health, Cincinnati, OH,
wsieber@cdc.gov

Richard Simon, Biometric Research Branch, National Cancer Institute, Rockville, MD,
rsimon@mail.nih.gov

K. Simonsen, Department of Statistics, Purdue University, West Lafayette, IN,
simonsen@stat.purdue.edu

Burton Singer, Office of Population Research, Princeton University, Princeton, NJ,
singer@princeton.edu

Cedric A. B. Smith, Galton Laboratory, University College London, London, UK

Edward J. Stanek III, Department of Biostatistics and Epidemiology, University of Massachusetts, Amherst, MA,
stanek@schoolph.umass.edu

Donna F. Stroup,
donnafstroup@comcast.net

Jianguo Sun, Department of Statistics, University of Missouri, Columbia, MO,
sunj@missouri.edu

Rajeshwari Sundaram, Eunice Kennedy Shriver National Institute of Child Health and Human Development, National Institutes of Health, DHHS, Bethesda, MD,
sundaramr2@mail.nih.gov

Wai-Y. Tan, Department of Mathematical Sciences, The University of Memphis, Memphis, TN,
waitan@memphis.edu

Man-Lai Tang, Department of Mathematics, Hong Kong Baptist University, Hong Kong, People's Republic of China,
mltang@math.hkbu.edu.hk

Elizabeth Thompson, Department of Statistics, University of Washington, Seattle, WA,
eathomp@u.washington.edu

D. M. Titterington, Department of Statistics, University of Glasgow, Glasgow, UK,
m.titterington@stats.gla.ac.uk

Robert K. Tsutakawa, Department of Statistics, University of Missouri, Columbia, MO,
Tsutakawar@Missouri.edu

B. W. Turnbull, School of Operations Research and Industrial Engineering, Cornell University, Ithaca, NY,
turnbull@orie.cornell.edu

Michael Væth, Department of Biostatistics, Aarhus University, Aarhus, Denmark,
vaeth@biostat.au.dk

Geert Verbeke, Interuniversity Institute for Biostatistics and Statistical Bioinformatics, Katholieke Universiteit Leuven and Hasselt University, Diepenbeek, Belgium,
geert.verbeke@med.kuleuven.be

Roger Weinberg

W. J. E. Wens

G. David Williamson, Agency for Toxic Substances and Disease Registry, Centers For Disease Control, Atlanta, GA,
dxw2@cdc.gov

M. E. Wise

M. A. Woodbury

Grace L. Yang, Department of Mathematics, University of Maryland, College Park, MD,
gly@math.umd.edu

Yaning Yang, Department of Statistics and Finance, University of Science and Technology of China, Hefei, People's Republic of China

P. A. Young, Wellcome Research Laboratories, Kent, England

Kai F. Yu, Biostatistics and Bioinformatics Branch, National Institutes of Health, Rockville, MD,
yukf@mail.nih.gov

Yong Zang, Department of Statistics and Actuarial Science, University of Hong Kong, Hong Kong

Hong Zhang, Department of Statistics and Finance, University of Science and Technology of China, Hefei, People's Republic of China and Biostatistics Branch, National Cancer Institute, Bethesda, MD

Gang Zheng, Office of Biostatistics Research, DPPS, National Heart, Lung and Blood Institute, Bethesda, MD,
zhengg@nhlbi.nih.gov

Contents

Preface		v
Contributors		vii
1	**Aalen's Additive Risk Model**	**1**
1.1	The Model	1
1.2	Model Fitting	2
1.3	Model Diagnostics	5
1.4	Related Models	5
1.5	Conclusion	5
2	**Aggregation**	**9**
2.1	Introduction	9
2.2	Aggregation as an Observable Phenomenon	9
2.3	Aggregation as a Statistical Method	10
2.4	A Perspective	12
3	**AIDS Stochastic Models**	**15**
3.1	Introduction	15
3.2	Stochastic Transmission Models of the HIV Epidemic in Populations	15
3.3	The Transmission Step: The Probability of HIV Transmission	17
3.4	The Progression Step: The Progression of HIV in HIV-Infected Individuals	17
3.5	Stochastic Modeling of HIV Transmission by Statistical Approach	18
3.6	The Stochastic Difference Equations and the Chain Multinomial Model	19
3.7	The Probability Distribution of the State Variables	21
3.8	The Staged Model of HIV Epidemic in Homosexual Populations	22
3.9	Stochastic Models of HIV Transmission in Complex Situations	23
3.10	State Space Models of the HIV Epidemic	25
3.11	A State Space Model for the HIV Epidemic in the San Francisco Homosexual Population	26
3.12	The Stochastic System Model	26
3.13	The Observation Model	27

		3.14	A General Bayesian Procedure for Estimating the Unknown Parameters and the State Variables	28

4 All-or-None Compliance   37
    4.1   Introduction . . . 37
    4.2   Examples . . . 37
    4.3   Randomized Consent Design   38
    4.4   The Paired Availability Design   39
    4.5   Estimating Efficacy   39

5 Ascertainment Sampling   43

6 Assessment Bias   47
    6.1   Lack of Blinding   47
    6.2   Harmless or False-Positive Cases of Disease   48
    6.3   Disease-Specific Mortality   48
    6.4   Composite Outcomes   49
    6.5   Competing Risks   49
    6.6   Timing of Outcomes   49
    6.7   Assessment of Harms   49

7 Bioavailability and Bioequivalence   53
    7.1   Definition   53
    7.2   History of Bioavailability   53
    7.3   Bioequivalence Measures   54
    7.4   Decision Rules   54
    7.5   Designs of Bioavailability Studies   55
    7.6   Statistical Methods for Average Bioavailability   55
    7.7   Sample Size Determination   57
    7.8   Current Issues   57

8 Cancer Stochastic Models   61
    8.1   Introduction   61
    8.2   A Brief History of Stochastic Modeling of Carcinogenesis   62
    8.3   Single-Pathway Models of Carcinogenesis   64
    8.4   Multiple-Pathways Models of Carcinogenesis   66
    8.5   The Mixed Models of Carcinogenesis   67
    8.6   Mathematical Analysis of Stochastic Models of Carcinogenesis by Markov Theories   68
    8.7   Mathematical Analysis of Stochastic Models of Carcinogenesis by Stochastic Differential Equations   71

9 Centralized Genomic Control: A Simple Approach Correcting for Population Structures in Case–Control Association Studies   81
    9.1   Introduction   81
    9.2   Methods   82
    9.3   Simulation Studies   84
    9.4   Simulation Results   88

9.5	Applications	90
9.6	Discussion	91

10 Change Point Methods in Genetics — 95
10.1	Introduction	95
10.2	A Brief Literature Review on Analysis of aCGH Data	97
10.3	A Brief Literature Review on Statistical Change Point Analysis	100
10.4	Statistical Change Point Methods Useful for aCGH Data Analysis	102
10.5	Concluding Remarks	111

11 Classical Biostatistics — 117
11.1	Introduction	117
11.2	Fields of Application or Areas of Concern	118
11.3	Future Trends	123
11.4	Education and Employment of Biostatisticians	127

12 Clinical Trials II — 131
12.1	Introduction	131
12.2	Types of Clinical Trials	131
12.3	Selection of Patients and Treatments	132
12.4	Endpoints	133
12.5	Randomization and Stratification	134
12.6	Sample Size	135
12.7	Large, Simple Trials	136
12.8	Monitoring of Interim Results	136
12.9	Subset Analysis	137
12.10	Meta-Analysis	138

13 Cluster Randomization — 143
13.1	Introduction	143
13.2	Examples of Cluster Randomization Trails	144
13.3	Principles of Experimental Design	145
13.4	Experimental and Quasi-Experimental Designs	146
13.5	The Effect of Failing to Replicate	147
13.6	Sample Size Estimation	148
13.7	Cluster-Level Analyses	149
13.8	Individual-Level Analysis	150
13.9	Incorporating Repeated Assessments	152
13.10	Study Reporting	153
13.11	Meta-Analysis	154

14 Cohort Analysis — 157
14.1	Definition	157
14.2	Methodology	157
14.3	Applications	161

15 Comparisons with a Control — 167
- 15.1 Introduction — 167
- 15.2 Some Examples — 168
- 15.3 Simultaneous Confidence Intervals — 168
- 15.4 Multiple Hypothesis Tests — 169
- 15.5 Other Results — 173
- 15.6 Other Approaches — 174

16 Competing Risks — 179
- 16.1 Observables — 179
- 16.2 Beyond Observables — 180
- 16.3 Estimation Assuming Independent $T_1, T_2, \ldots, T_M$ — 182
- 16.4 Estimation Assuming Dependent $T_1, T_2, \ldots, T_m$ — 184
- 16.5 Absolute Risks — 184
- 16.6 Markov Model and Extensions — 185

17 Countermatched Sampling — 189

18 Counting Processes — 193
- 18.1 Introduction — 193
- 18.2 Multivariate Counting Processes and Martingales — 193
- 18.3 Censoring — 194
- 18.4 Multiplicative Intensity Models — 194
- 18.5 Survival Analysis: The Kaplan–Meier Estimator — 197
- 18.6 Transitions in Nonhomogeneous Markov Processes — 198
- 18.7 Nonparametric Hypothesis Tests — 198
- 18.8 Regression Models and Random Heterogeneity — 199

19 Cox's Proportional Hazards Model — 203
- 19.1 The Model — 203
- 19.2 The Stratified Model — 204
- 19.3 Partial Likelihood — 204
- 19.4 Tied Data — 205
- 19.5 Parameter Estimation and Confidence Intervals — 207
- 19.6 Hypothesis Testing — 207
- 19.7 Estimating the Underlying Hazard Function — 208
- 19.8 Time-Dependent Covariates — 209
- 19.9 Model Checking — 211
- 19.10 Concluding Remarks — 212

20 Crossover Trials — 215
- 20.1 Introduction — 215
- 20.2 2×2 Crossover Trial — 217
- 20.3 Higher-Order Designs for Two Treatments — 217
- 20.4 Designs for Three or More Treatments — 217
- 20.5 Analysis of Continuous Data — 219
- 20.6 Analysis of Discrete Data — 220
- 20.7 Concluding Remarks — 222

21 Design and Analysis for Repeated Measurements — 225
- 21.1 Introduction — 225
- 21.2 Design — 226
- 21.3 Statistical Methods — 232

22 DNA Fingerprinting — 259
- 22.1 Introduction — 259
- 22.2 DNA Markers — 260
- 22.3 Summarizing the Evidence — 261
- 22.4 Independence and Population Substructure — 262
- 22.5 The Affinal Model (Same Subpopulation) — 263
- 22.6 The Cognate Model (Same Family) — 263
- 22.7 In Practice — 264

23 Epidemics — 269
- 23.1 Introduction — 269
- 23.2 History of Epidemic Models — 270
- 23.3 Infectious Disease Data — 271
- 23.4 The Basic Reproduction Number — 272
- 23.5 Efficacy of Vaccines and Effectiveness of Vaccination Programs — 273
- 23.6 AIDS — 274

24 Epidemiological Statistics I — 279
- 24.1 Introduction — 279
- 24.2 Role of the Statistician — 280
- 24.3 Computing — 281
- 24.4 Longitudinal and Cross-Sectional Data and Growth — 282
- 24.5 Small Risks Undergone by Millions of People: Human Radiation Hazards — 282
- 24.6 Human Radiation Hazards and Prospective or Forward-Looking Studies — 284
- 24.7 Epidemiology and Cigarette Smoking — 284
- 24.8 Unsolved Problems on the Effects of Cigarette Smoking — 286
- 24.9 Form of Age Incidence Curves in General — 288
- 24.10 An Example Where Age Must Be Allowed for: Side Effects from Adding Fluoride to Drinking Water — 288
- 24.11 Epidemiology and Heart Diseases — 291
- 24.12 Scope of Epidemiology in the 1980s — 293
- 24.13 Further Reading — 295

25 Epidemiological Statistics II — 299
- 25.1 Introduction — 299
- 25.2 Data Roles — 299
- 25.3 Concept Maps — 300
- 25.4 Data Types — 301
- 25.5 Disease Measurement — 302
- 25.6 Measuring Associations — 302
- 25.7 Types of Studies — 303
- 25.8 Clinical Trials — 304

25.9	Cohort Studies	305
25.10	Case–Control Studies	305
25.11	Cross-Sectional Studies	305
25.12	Bias	306
25.13	Sampling Variability	307
25.14	Matching	308
25.15	Statistical Significance versus Effect Importance	309
25.16	Basic Statistical Methods	309
25.17	Sample Size Determination	310
25.18	Adjusting for a Stratification Variable	311
25.19	Meta-Analysis	311
25.20	Survival Analysis	312
25.21	Logistic Regression	313
25.22	Poisson Regression	314
25.23	Proportional Hazards Regression	314
25.24	Correlated Outcomes	315

26 Event History Analysis — 319
26.1	Introduction	319
26.2	Recurrent Events	320
26.3	Parametric Estimation	320
26.4	Semiparametric Estimation	321
26.5	Multistate Models	321
26.6	Markov and Semi-Markov Models	322
26.7	Frailty Models	323
26.8	Goodness of Fit	324

27 FDA Statistical Programs: An Overview — 329

28 FDA Statistical Programs: Human Drugs — 335
28.1	Official Statistical Responsibilities	335
28.2	Role of Statistics	336
28.3	Statistical Guidelines	337
28.4	Selected Clinical Statistical Problems	337
28.5	Some Nonclinical Statistical Problems	339
28.6	Applied Statistical Research	339
28.7	Notes	340

29 Follow-Up — 343
29.1	Introduction	343
29.2	Concurrent Follow-Up	343
29.3	Historical Follow-Up	345

30 Frailty Models — 349

31 Framingham: An Evolving Longitudinal Study 353
31.1 Introduction 353
31.2 Description of the Framingham Population and Sample 354
31.3 Methodological Issues in the Analysis of Disease Processes in Longitudinal Population Studies 355
31.4 Adaptive Nature of Longitudinal Studies 356

32 Genetic Linkage 359

33 Group-Sequential Methods in Biomedical Research 365
33.1 Introduction 365
33.2 The Role of Brownian Motion for Large-Sample Statistics 366
33.3 Group-Sequential Testing of a Hypothesis 367
33.4 Terminal Analysis Following a Sequential Test 371

34 Group-Sequential Tests 377
34.1 Introduction 377
34.2 Two-Sided Tests 378
34.3 Unequal and Unpredictable Group Sizes 381
34.4 One-Sided Tests 381
34.5 Inference upon Termination 382
34.6 Repeated Confidence Intervals 382
34.7 Literature 383

35 Grouped Data in Survival Analysis 391
35.1 Introduction 391
35.2 Data Structure 392
35.3 Statistical Methods 393

36 Image Processing 397
36.1 Introduction 397
36.2 Hardware of an Imaging System 398
36.3 Processing of Image Data 399
36.4 Restoration and Noise Removal 402
36.5 Image Enhancement 404
36.6 Segmentation 405
36.7 Feature Extraction 409
36.8 Other Applications 411

37 Image Restoration and Reconstruction 415
37.1 Introduction 415
37.2 Image Restoration 415
37.3 Bayesian Restoration Models 417
37.4 Image Reconstruction 419
37.5 Summary 421

38 Imputation and Multiple Imputation — 425
- 38.1 Introduction — 425
- 38.2 Considerations for Creating Imputations — 426
- 38.3 The Underestimation of Uncertainty with Single Imputation — 427
- 38.4 Introduction to Multiple Imputation — 428
- 38.5 Analyzing a Multiply Imputed Dataset — 428
- 38.6 Reflecting Appropriate Variability and Choosing M — 429
- 38.7 Bayesian Iterative Simulation and Multiple Imputation — 431
- 38.8 More on Conditions for Validity of Inferences with Imputed Data — 432
- 38.9 Some Applications and Different Uses of Multiple Imputation — 433
- 38.10 Other Approaches to Estimating Variability with Imputed Data — 435
- 38.11 Summary and Conclusion — 436

39 Incomplete Data — 441
- 39.1 Introduction — 441
- 39.2 Common Incomplete-Data Problems — 441
- 39.3 Methods for Handling Incomplete Data — 444
- 39.4 General Data Patterns: The EM Algorithm — 447

40 Interval Censoring — 451
- 40.1 Censoring — 451
- 40.2 Classification and Examples — 452
- 40.3 Study Design and Statistical Modeling — 455
- 40.4 Statistical Analysis — 455
- 40.5 Worked Example — 457

41 Interrater Agreement — 461
- 41.1 Introduction — 461
- 41.2 Agreement Measures for Unreplicated Measurements — 462
- 41.3 Agreement Measures for Replicated Measurements — 468
- 41.4 Modeling Measurements and Inference on Agreement Measures — 471
- 41.5 Illustration — 473
- 41.6 Summary and Discussion — 475
- 41.7 Appendix — 476

42 Kaplan–Meier Estimator I — 481
- 42.1 Introduction — 481
- 42.2 Censored-Data Problem — 481
- 42.3 Kaplan–Meier Estimator — 482
- 42.4 Appropriateness — 485
- 42.5 Properties, Variance Estimators, Confidence Intervals and Confidence Bands — 485
- 42.6 Nonparametric Quantile Estimation Based on the Kaplan–Meier Estimator — 486

43 Kaplan–Meier Estimator II — 489
- 43.1 Introduction — 489
- 43.2 The Right-Censoring Model — 490
- 43.3 The Kaplan–Meier Estimator — 490
- 43.4 Model Identifiability — 490
- 43.5 The Inversion Formula — 491
- 43.6 Identifiability — 492
- 43.7 Finite-Sample Properties of the KM Estimator — 492
- 43.8 Consistency, Asymptotic Normality, and the Strong Law — 492
- 43.9 Asymptotic Optimality — 493
- 43.10 Counting Process Formulation and Martingales — 494
- 43.11 Confidence Intervals and Confidence Bands — 495
- 43.12 The Quantile Process of the KM Estimator — 495
- 43.13 Large Deviations — 495
- 43.14 Generalizations — 495

44 Landmark Data — 501
- 44.1 Introduction — 501
- 44.2 Shape Description — 502
- 44.3 Mean Shape — 503
- 44.4 Shape Variability — 504
- 44.5 Shape Comparisons and Deformation — 505
- 44.6 Shape Distributions — 507
- 44.7 Conditional Approach: the Complex Bingham Distribution — 508
- 44.8 Shapes in Higher Dimensions — 508
- 44.9 Distance-Based Method — 509
- 44.10 Shapes in Image Analysis — 509
- 44.11 Concluding Remarks — 510

45 Longitudinal Data Analysis — 515
- 45.1 Domain of Longitudinal Data Analysis — 515
- 45.2 Some History — 515
- 45.3 Designs and Their Implications for Analysis — 517
- 45.4 Analytical Strategies—Examples — 518
- 45.5 Brief Guide to Other Literature — 526

46 Meta-Analysis — 531

47 Missing Data: Sensitivity Analysis — 535
- 47.1 Introduction — 535
- 47.2 Sensitivity Analysis for Contingency Tables — 536
- 47.3 Selection Models and Local Influence — 538
- 47.4 Pattern-Mixture Modeling Approach — 539

xxii Contents

48 Multiple Testing in Clinical Trials — 547
- 48.1 Introduction — 547
- 48.2 Concepts of Error Rates — 548
- 48.3 Union–Intersection Testing — 549
- 48.4 Closed Testing — 550
- 48.5 Partition Testing — 552

49 Mutation Processes — 555

50 Nested Case-Control Sampling — 559
- 50.1 Introduction — 559
- 50.2 Nested Case–Control Studies and Other Case–Control Designs — 560
- 50.3 An Example — 560
- 50.4 Model and Data — 561
- 50.5 Estimation — 562
- 50.6 Relative Efficiency — 563
- 50.7 Extensions — 564

51 Observational Studies — 567
- 51.1 Introduction — 567
- 51.2 Prospective — 568
- 51.3 Retrospective — 569
- 51.4 Control of Variation — 569
- 51.5 Role of Observational Studies — 570

52 One- and Two-Armed Bandit Problems — 573
- 52.1 Introduction — 573
- 52.2 Finite-Horizon Bernoulli Bandits — 573
- 52.3 Geometric Discounting — 575
- 52.4 One-Armed Bandits — 576
- 52.5 Continuous Time — 576
- 52.6 Other Objectives — 577

53 Ophthalmology — 579
- 53.1 Introduction — 579
- 53.2 Clinical Trials — 579
- 53.3 Observational Studies — 580

54 Panel Count Data — 585
- 54.1 Introduction — 585
- 54.2 Nonparametric Estimation — 587
- 54.3 Nonparametric Comparison — 591
- 54.4 Regression Analysis — 594
- 54.5 Regression Analysis with Dependent Observation Process — 598
- 54.6 Discussion and Other Topics — 600

55 Planning and Analysis of Group-Randomized Trials   605
   55.1   Introduction 605
   55.2   The Research Question 605
   55.3   The Research Team 606
   55.4   The Research Design 606
   55.5   Potential Design Problems and Methods for Avoiding Them 607
   55.6   Potential Analytic Problems and Methods for Avoiding Them 608
   55.7   Variables of Interest and Their Measures 608
   55.8   The Intervention 609
   55.9   Power 611
   55.10 Summary 611

56 Predicting Preclinical Disease Using the Mixed-Effects Regression Model   613
   56.1   Introduction 613
   56.2   Single-Variable Prediction Model 617
   56.3   Predicting Alzheimer's Disease 618
   56.4   Predicting Prostate Cancer 621
   56.5   Multivariate Prediction Model 623
   56.6   Predicting Coronary Heart Disease 625
   56.7   Discussion 628

57 Predicting Random Effects in Group-Randomized Trials   635
   57.1   Introduction 635
   57.2   Background 635
   57.3   Methods 636
   57.4   Modeling Response for a Subject 637
   57.5   Random Assignment of Treatment and Sampling 638
   57.6   Intuitive Predictors of the Latent Value of a Realized Group 639
   57.7   Models and Approaches 640
   57.8   Henderson's Mixed-Model Equations 640
   57.9   Bayesian Estimation 641
   57.10 Superpopulation Model Predictors 641
   57.11 Random Permutation Model Predictors 642
   57.12 Practice and Extensions 643
   57.13 Discussion and Conclusions 644

58 Probabilistic and Statistical Models for Conception   647
   58.1   Introduction 647
   58.2   Biological Background 648
   58.3   Population Models 649
   58.4   Fecundability and Time to Pregnancy 654
   58.5   Day-Specific Probabilities of Conception 658
   58.6   Data 662
   58.7   Models for Markers 663
   58.8   Discussion 663

59 Probit Analysis — 669
- 59.1 Introduction — 669
- 59.2 Applications — 669
- 59.3 Framework for Quantal Response Models — 670
- 59.4 Relation between Quantitative and Quantal Responses — 670
- 59.5 The Tolerance Distribution: Probits and Logits — 671
- 59.6 Multivariate Tolerance Distributions — 672
- 59.7 Ordered Response Variables: Rankits — 672
- 59.8 Mixed Discrete–Continuous Response Variables: Tobits — 673
- 59.9 The Effects Function — 673
- 59.10 Estimation and Hypothesis Testing — 674

60 Prospective Studies — 677
- 60.1 Introduction — 677
- 60.2 The Selection of Subjects — 678
- 60.3 The Measurement of Effect — 679
- 60.4 Relative Merits of Prospective Studies — 680

61 Quality Assessment for Clinical Trials — 683

62 Repeated Measurements — 695
- 62.1 Introduction and Case Study — 695
- 62.2 Linear Models for Gaussian Data — 697
- 62.3 Models for Discrete Outcomes — 700
- 62.4 Design Considerations — 704
- 62.5 Concluding Remarks — 705

63 Reproduction Rates — 709

64 Retrospective Studies — 713
- 64.1 Introduction — 713
- 64.2 Types of Retrospective Studies — 713
- 64.3 The Frequency of Study Factors and Retrospective Research — 715
- 64.4 Other Considerations in Retrospective Research — 716

65 Sample Size Determination for Clinical Trials — 719
- 65.1 Introduction — 719
- 65.2 Sample Size for One-Sample Problem — 720
- 65.3 Sample Size for Two-Sample Problem — 721
- 65.4 Sample Size for Multiple-Sample Problem — 723
- 65.5 Sample Size for Multiple-Arm Dose–Response Trial Problem — 725
- 65.6 Sample Size for Multiple Regression Analysis — 727
- 65.7 Sample Size for Multiple Logistic Regression Analysis — 728
- 65.8 Discussion — 729

66 Scan Statistics — 733
- 66.1 Introduction — 733
- 66.2 History and Literature — 734
- 66.3 Higher-Dimensional Scans — 736
- 66.4 Scan Statistics with Nonuniform Background — 736
- 66.5 Scan Statistics with Variable-Size Windows — 737
- 66.6 The Conditional Probability P — 738
- 66.7 The Unconditional Probability P^* — 738
- 66.8 Sequence of Trials: Bernoulli Model — 738
- 66.9 Integer-Valued Random Variables — 739
- 66.10 The Circle — 739
- 66.11 Two-Dimensional Scan Statistics — 740
- 66.12 Multiple-Scan Statistics — 741
- 66.13 Spatial Scan Statistics — 741
- 66.14 Bioinformatics and Genetic Epidemiology — 742
- 66.15 Process Control and Monitoring — 742
- 66.16 Bayesian Scan Statistics — 742
- 66.17 Theoretical Methods — 743

67 Semiparametric Analysis of Competing-Risk Data — 749
- 67.1 Introduction and Notation — 749
- 67.2 Semiparametric Models — 751
- 67.3 Statistical Inference under the Additive Risks Model — 756
- 67.4 A Comprehensive Analysis of Malignant Melanoma Data — 758
- 67.5 Discussion — 765

68 Size and Shape Analysis — 769
- 68.1 Introduction — 769
- 68.2 Shape Vectors and Size Variables — 770
- 68.3 Size Variables Represent A Broad Class — 770
- 68.4 Regular Sequences of Size Variables — 771
- 68.5 Size Ratios of Regular Sequence Represent Shape — 771
- 68.6 Independence of Shape and Size; Isometry and Neutrality — 771
- 68.7 Isometry and Neutrality Characterize Distributions — 772
- 68.8 Diversity of Hypotheses for Isometry (Neutrality) — 773
- 68.9 Practical Considerations — 775
- 68.10 The Choice of a Size Variable — 775
- 68.11 Relationships to Other Concepts of Size and Shape — 777

69 Stability Study Designs — 781
- 69.1 Introduction — 781
- 69.2 Stability Study Designs — 782
- 69.3 Criteria for Design Comparison — 788
- 69.4 Stability Protocol — 791
- 69.5 Basic Design Considerations — 791
- 69.6 Conclusions — 793

70 Statistical Analysis of DNA Microarray Data — 795
- 70.1 Introduction — 795
- 70.2 cDNA versus Oligonucleotide Chips — 795
- 70.3 Preprocessing of Data — 796
- 70.4 Statistical Clustering — 796
- 70.5 Detection of Differentially Expressed Genes — 797
- 70.6 Regression Techniques — 797
- 70.7 Bootstrap — 797

71 Statistical Genetics — 801

72 Statistical Methods in Bioassay — 807
- 72.1 Introduction — 807
- 72.2 Estimation of Relative Potency from Quantitative Responses — 808
- 72.3 Quantal Response Models — 809
- 72.4 Estimation of Quantal Response Curves — 810
- 72.5 Estimation of Relative Potency from Quantal Responses — 812
- 72.6 Design of the Experiment — 812
- 72.7 Related Areas — 813

73 Statistical Modeling of Human Fecundity — 815
- 73.1 Introduction — 815
- 73.2 Statistical Models — 817
- 73.3 Analysis of a Prospective Pregnancy Study: New York Angler Cohort Study — 824
- 73.4 Biological Issues around Fecundity — 829
- 73.5 Further Extensions — 833

74 Statistical Quality of Life — 839
- 74.1 Introduction — 839
- 74.2 Measurement Models of Health-Related Quality of Life — 840
- 74.3 Validation of Health-Related Quality-of-Life Measurement Models — 847
- 74.4 Analysis of Quality of Life Variation between Groups — 853
- 74.5 Conclusion — 859

75 Statistics at CDC — 865
- 75.1 Introduction — 865
- 75.2 Public Health Data and Statistics at CDC/ATSDR — 866
- 75.3 Statistics and Research at CDC/ATSDR — 866
- 75.4 The CDC/ATSDR Statistical Advisory Group — 868
- 75.5 Future Directions for Statistics at CDC/ATSDR — 868

76 Statistics in Dentistry — 871

77 Statistics in Evolutionary Genetics — 873

78 Statistics in Forensic Science — 879
- 78.1 Introduction — 879
- 78.2 History — 879
- 78.3 Evaluation of Evidence — 881
- 78.4 Continuous Measurements — 882
- 78.5 Future — 885

79 Statistics in Human Genetics I — 887
- 79.1 Introduction — 887
- 79.2 Applications — 887
- 79.3 Genetic Mechanisms — 889
- 79.4 Literature — 891

80 Statistics in Human Genetics II — 893
- 80.1 Introduction — 893
- 80.2 Genomics — 893
- 80.3 Genetic Analysis of Traits — 894
- 80.4 Additional Topics — 901
- 80.5 Literature — 902

81 Statistics in Medical Diagnosis — 909
- 81.1 Introduction — 909
- 81.2 The Diagnosis Problem — 909
- 81.3 Relevant Statistical Methods — 910
- 81.4 Application and Acceptance in Practice — 913

82 Statistics in Medicine — 915
- 82.1 Introduction — 915
- 82.2 Key Features — 916
- 82.3 More Recent History — 917
- 82.4 Observational Plans — 917
- 82.5 Clinical Trials — 917
- 82.6 Stochastic Processes — 920
- 82.7 Meta-Analysis — 920
- 82.8 Bayesian Medical Statistics — 920
- 82.9 Current Research Interests — 921
- 82.10 The Practice of Statistics in Medicine — 921

83 Statistics in the Pharmaceutical Industry — 927
- 83.1 Introduction — 927
- 83.2 The Clinical Response to Treatment — 927
- 83.3 Quantal Responses — 929
- 83.4 Drug Interactions — 929
- 83.5 Biological Standardization (Bioassay) — 930
- 83.6 Stability of Products — 930
- 83.7 Quality Control — 931
- 83.8 Experiment Design — 931
- 83.9 Nonparametric Methods — 931

84 Statistics in Spatial Epidemiology — 933
- 84.1 Introduction — 933
- 84.2 Basic Definitions and Models — 934
- 84.3 Special Application Areas — 937

85 Stochastic Compartment Models — 939
- 85.1 Introduction — 939
- 85.2 Origin of Compartment Models in the Biological Sciences — 939
- 85.3 Linear Compartment Systems — 940
- 85.4 Estimation and Identification of Stochastic Compartment Model Parameters — 941
- 85.5 Model Building in Statistical Analysis — 943

86 Surrogate Markers — 945
- 86.1 Introduction — 945
- 86.2 Data from a Single Unit — 946

87 Survival Analysis — 953
- 87.1 Survival Data — 953
- 87.2 Nonparametric Survival Models — 954
- 87.3 Parametric Survival Models — 956
- 87.4 Regression Models — 957
- 87.5 Some Extensions — 960
- 87.6 An Example — 962

Index — 967

1 Aalen's Additive Risk Model

Ian W. McKeague

1.1 The Model

In medical statistics and survival analysis, it is important to assess the association between risk factors and disease occurrence or mortality. Underlying disease mechanisms are invariably complex, so the idea is to simplify the relationship between survival patterns and covariates in such a way that only essential features are brought out. Aalen's additive risk model [1] is one of three well-developed approaches to this problem; the others are the popular proportional hazards model introduced by D. R. Cox in 1972 [5], and the accelerated failure-time model, which is a linear regression model with unknown error distribution, introduced in the context of right-censored survival data by R. G. Miller in 1976 [18].

Aalen's model expresses the conditional hazard function $\lambda(t|\mathbf{z})$ of a survival time T as a linear function of a p-dimensional covariate vector $\mathbf{z}$

$$\lambda(t|\mathbf{z}) = \boldsymbol{\alpha}(t)'\mathbf{z} = \sum_{j=1}^{p} \alpha_j(t) z_j, \quad (1)$$

where $\boldsymbol{\alpha}(t)$ is a nonparametric p vector of regression functions [constrained by $\lambda(t|\mathbf{z}) \geq 0$] and $\mathbf{z} = (z_1, \ldots, z_p)'$. Some authors refer to (1) as the *linear hazard model*.

As a function of the covariates $z_1, \ldots, z_p$, the *additive* form of Aalen's model contrasts with the *multiplicative* form of Cox's model

$$\lambda(t|\mathbf{z}) = \lambda_0(t) \exp\{\boldsymbol{\beta}'\mathbf{z}\}$$
$$= \lambda_0(t) \prod_{j=1}^{p} \exp\{\beta_j z_j\},$$

where $\lambda_0(t)$ is a nonparametric baseline hazard function and $\boldsymbol{\beta}$ is a vector of regression parameters. Aalen's model has the feature that the influence of each covariate can vary separately and nonparametrically through time, unlike Cox's model or the accelerated failure-time model. This feature can be desirable in some applications, especially when there are a small number of covariates.

Consider the following simple example with three covariates: T is the age at which an individual contracts melanoma (if at all), $z_1 =$ indicator male, $z_2 =$ indicator female, and $z_3 =$ number of serious sunburns as a child. Then the corresponding regression functions, α_1, α_2, and α_3, can be interpreted as the (age-specific) background rates of melanoma for males and females and as the excess rate of melanoma due to serious sunburns is childhood, respectively.

Aalen's model is expected to provide a reasonable fit to data, since the first step of a Taylor series expansion of a general conditional hazard function about the zero of the covariate vector can be expressed in the form (1). It is somewhat more flexible than Cox's model and can be especially helpful for exploratory data analysis. A rough justification for the additive form can be given in terms of p independent competing risks, since the hazard function of the minimum of p independent random variables is the sum of their individual hazard functions.

It is generally sensible to include a nonparametric baseline function in the model, by augmenting $\mathbf{z}$ with a component that is set to 1. Also, it is often natural to center the covariates in some fashion, so the baseline can be interpreted as the "hazard" function for an "average" individual. In some cases, however, a baseline hazard is already implicit in the model and it is not necessary to center the covariates, as in the melanoma example above.

Aalen originally proposed his model in a counting process setting, which allows time-dependent covariates and general patterns of censorship, and which can be studied using powerful continuous-time martingale techniques. In a typical application the observed survival times are subject to right censorship, and it is customary to assume that the censoring time, C, say, is conditionally independent of T given $\mathbf{z}$. One observes $(X, \delta, \mathbf{z})$, where $X = T \wedge C$ and $\delta = I\{X = T\}$. Aalen's model (1) is now equivalent to specifying that the counting process $N(t) = I(X \leq t, \delta = 1)$, which indicates an uncensored failure by time t, has intensity process

$$\lambda(t) = \boldsymbol{\alpha}(t)' \mathbf{y}(t),$$

where $\mathbf{y}(t) = \mathbf{z} I\{X \geq t\}$ is a covariate process.

1.2 Model Fitting

To fit Aalen's model one first estimates the p vector of integrated regression functions $\mathbf{A}(t) = \int_0^t \boldsymbol{\alpha}(s)\, ds$. Denote by $(t_i, \delta_i, \mathbf{z}_i)$ the possibly right-censored failure time t_i, indicator of noncensorship δ_i, and covariate vector $\mathbf{z}_i$ for n individuals. Let $\mathbf{N} = (N_1, \ldots, N_n)'$ and $\mathbf{Z} = (\mathbf{y}_1, \ldots, \mathbf{y}_n)'$, where N_i is the counting process and $\mathbf{y}_i$ is the associated covariate process for individual i.

Aalen [1] introduced an ordinary least squares (OLS) estimator of $\mathbf{A}(t)$ given by

$$\hat{\mathbf{A}}(t) = \int_0^t (\mathbf{Z}'\mathbf{Z})^{-1} \mathbf{Z}' \, d\mathbf{N},$$

where the matrix inverse is assumed to exist; $\hat{\mathbf{A}}$ is a step function, constant between uncensored failures, and with jump

$$\boldsymbol{\Delta}_i = \left(\sum_{t_k \geq t_i} \mathbf{z}_k \mathbf{z}_k' \right)^{-1} \mathbf{z}_i \qquad (2)$$

at an uncensored failure time t_i. The matrix inverse exists unless there is collinearity between the covariates or there are insufficiently many individuals at risk at time t_i. A heuristic motivation for $\hat{\mathbf{A}}$ comes from applying the method of least squares to increments of the multivariate counting process $\mathbf{N}$. The estimator is consistent and asymptotically normal [14,9]. The covariance matrix of $\hat{\mathbf{A}}(t)$ can be estimated [1,2] by $\hat{\mathbf{V}}(t) = \sum_{t_i \leq t} \delta_i \boldsymbol{\Delta}_i \boldsymbol{\Delta}_i'$.

Plots of the components of $\hat{\mathbf{A}}(t)$ against t, known as *Aalen plots*, are a useful graphical diagnostic tool for studying time-varying covariate effects [2–4,7,9,12,13]. Mau [12] coined the term Aalen plots and made a strong case for their importance in survival analysis. Roughly constant slopes in the plots indicate periods when a covariate has a non-time-dependent regression coefficient; plateaus indicate times at which a covariate has no effect on the hazard. Interpretation of the plots is helped by the inclusion of pointwise or simultaneous confidence limits. An approximate pointwise $100(1-\alpha)\%$ confidence interval

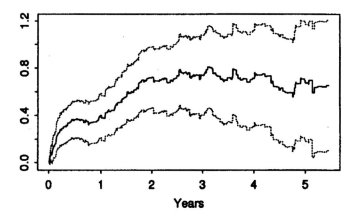

Figure 1: An Aalen plot with 95% confidence limits for the myelamatosis data.

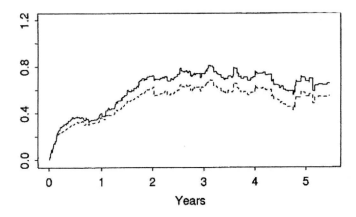

Figure 2: Comparison of Aalen plots of the WLS estimates (dashed line) and OLS (solid line) estimates for the myelamatosis data.

for the jth component of $\mathbf{A}(t)$ is given by

$$\hat{\mathbf{A}}_j(t) \pm z_{\alpha/2} \hat{\mathbf{V}}_{jj}(t)^{1/2},$$

where $z_{\alpha/2}$ is the upper $\alpha/2$ quantile of the standard normal distribution and $\hat{\mathbf{V}}_{jj}(t)$ is the jth entry on the diagonal of $\hat{\mathbf{V}}(t)$. To avoid wild fluctuations in the plots (which occur when the size of the risk set is small), estimation should be restricted to time intervals over which the matrix inverse in (2) is numerically stable.

Figure 1 shows an Aalen plot based on survival data for 495 myelamatosis patients [17]. The plot gives the estimated integrated regression function for one particular covariate, serum β_2-microglobulin, which was log-transformed to adjust for skewness. Pointwise 95% confidence limits are also shown. Serum β_2-microglobulin is seen to have a strong effect on survival during the first 2 years of follow-up.

The vector of regression functions $\boldsymbol{\alpha}$ can be estimated by smoothing the increments of $\hat{\mathbf{A}}$. One approach is to extend Ramlau-Hansen's kernel estimator [19] to the additive-risk-model setting [3,9,14]. For a kernel function K that integrates to 1 and some bandwidth $b > 0$,

$$\hat{\boldsymbol{\alpha}}(t) = b^{-1} \sum_{i=1}^{n} K\left(\frac{t - t_i}{b}\right) \boldsymbol{\Delta}_i$$

consistently estimates $\boldsymbol{\alpha}$ provided the bandwidth tends to zero at a suitable rate with increasing sample size. Plots of the regression function estimates in some real and simulated data examples have been given by Aalen [3].

Huffer and McKeague [9] introduced a weighted least-squares (WLS) estimator of $\mathbf{A}$; see Figure 2 for a comparison with the OLS estimator. The weights consistently estimate $[\lambda(t|\mathbf{z}_i)]^{-1}$ and are obtained by plugging $\hat{\boldsymbol{\alpha}}(t)$ and $\mathbf{z} = \mathbf{z}_i$ into (1). The WLS estimator is an approximate maximum-likelihood estimator and an approximate solution to the score equations [20]. It is consistent and asymptotically normal provided $\lambda(t|\mathbf{z})$ is bounded away from zero [9,14]. Furthermore, the WLS estimator is asymptotically efficient in the sense of having minimal asymptotic variance [6,20,4]. Simulation studies [9] show that significant variance reductions are possible using WLS compared with OLS estimators, especially in large samples where the weights are more stable. When there are no covariates ($p = 1$ and $\mathbf{z}_i = 1$), the OLS and WLS estimators reduce to the Nelson–Aalen estimator of the baseline cumulative hazard function. Simultaneous confidence bands for $\mathbf{A}$ based on OLS and WLS estimators, for continuous or grouped data, can be found in References 9, 15.

Tests of whether a specific covariate (say, the jth component of $\mathbf{z}$) has any effect on survival can be carried out within the Aalen model setting. The idea is to test the null hypothesis $H_0 : \mathbf{A}_j(t) = 0$ over the follow-up period. This can be done [2] using a test statistic of the form $\sum_{i=1}^{n} w(t_i) \boldsymbol{\Delta}_{ij}$ for a suitable weight function w. Kolmogorov–Smirnov-type tests are also available [9]; such tests are equivalent to checking whether the confidence band for the jth component of $\mathbf{A}$ contains the zero function.

To predict survival under Aalen's model, one estimates the conditional survival probability $P(T > t|\mathbf{z}) = \exp\{-\mathbf{A}(t)'\mathbf{z}\}$. This can be done using the product-limit estimator

$$\hat{P}(T > t|\mathbf{z}) = \prod_{t_i \leq t} (1 - \boldsymbol{\Delta}_i' \mathbf{z}),$$

or by plugging $\hat{\mathbf{A}}(t)$ into $P(T > t|\mathbf{z})$ in place of the unknown $\mathbf{A}(t)$. When there are no covariates, $\hat{P}(T > t|\mathbf{z})$ reduces to the Kaplan–Meier estimator of the survival function corresponding to the baseline hazard.

1.3 Model Diagnostics

Some goodness-of-fit checking procedures are available for additive risk models. Aalen [2,3] suggested making plots against t of sums of the martingale residual processes $\hat{M}_i(t) = \delta_i I(t_i \leq t) - \hat{\mathbf{A}}(t_i \wedge t)'\mathbf{z}_i$ over groups of individuals. If the model fits well, then the plots would be expected to fluctuate around the zero line. McKeague and Utikal [16] suggested the use of a standardized residual process plotted against t and $\mathbf{z}$, and developed a formal goodness-of-fit test for Aalen's model.

Outlier detection has been studied by Henderson and Oman [8], who considered the effects on $\hat{\mathbf{A}}(t)$ of deletion of an observation from a dataset. They show that unusual or influential observations can be detected quickly and easily. They note that Aalen's model has an advantage over Cox's model in this regard, because closed-form expressions for the estimators are available, leading to exact measures of the effects of case deletion.

Mau [12] noticed that Aalen plots are useful for diagnosing time-dependent covariate effects in the Cox model. To aid interpretation of the plots in that case, Henderson and Milner [7] suggested that an estimate of the shape of the curve expected under proportional hazards be included.

1.4 Related Models

More recently a number of variations on the additive structure of Aalen's model have been introduced. McKeague and Sasieni [17] considered a partly parametric additive risk model in which the influence of only a subset of the covariates varies nonparametrically over time, and that of the remaining covariates is constant

$$\lambda(t|\mathbf{x},\mathbf{z}) = \boldsymbol{\alpha}(t)'\mathbf{x} + \boldsymbol{\beta}'\mathbf{z}, \qquad (3)$$

where $\mathbf{x}$ and $\mathbf{z}$ are covariate vectors and $\boldsymbol{\alpha}(t)$ and $\boldsymbol{\beta}$ are unknown. This model may be more appropriate than (1) when there are a large number of covariates and it is known that the influence of only a few of the covariates is time-dependent. Lin and Ying [10] studied an additive analog of Cox's proportional hazards model that arises as a special case of (3):

$$\lambda(t|\mathbf{z}) = \alpha_0(t) + \boldsymbol{\beta}'\mathbf{z}. \qquad (4)$$

Efficient WLS-type estimators for fitting (3) and (4) have been developed.

A variation in the direction of Cox's proportional hazards model [5] has been studied by Sasieni [21,22]: the *proportional excess hazards* model

$$\lambda(t|\mathbf{x},\mathbf{z}) = \alpha_0(t|\mathbf{x}) + \lambda_0(t)\exp\{\boldsymbol{\beta}'\mathbf{z}\}, \qquad (5)$$

where $\alpha_0(t|\mathbf{x})$ is a known background hazard (available from national mortality statistics, say) and $\lambda_0(t)$ and $\boldsymbol{\beta}$ are unknown. A further variation in this direction is due to Lin and Ying [11], who considered an *additive–multiplicative hazards* model that includes

$$\lambda(t|\mathbf{x},\mathbf{z}) = \boldsymbol{\gamma}'\mathbf{x} + \lambda_0(t)\exp\{\boldsymbol{\beta}'\mathbf{z}\}, \qquad (6)$$

where $\boldsymbol{\gamma}$, $\boldsymbol{\beta}$, and $\lambda_0(t)$ are unknown. Finding efficient procedures for fitting the models (5) and (6) involves a combination of Cox partial likelihood techniques and the estimation of efficient weights similar to those needed for the standard additive risk model (1).

1.5 Conclusion

Despite the attractive features of Aalen's model as an alternative to Cox's model in many applications, it has received relatively little attention from practitioners or researchers. Cox's model has been perceived to be adequate for most applications, but it can lead to serious bias when the influence of covariates is time-dependent. Fitting separate Cox models over disjoint time intervals (years, say) is an ad hoc way around this problem. Aalen's model, however, provides a more effective approach. Interest in it, and especially in Aalen plots, is expected to increase in the future.

Acknowledgments. This research was partially supported by NSF Grant ATM-9417528.

References

1. Aalen, O. O. (1980). A model for nonparametric regression analysis of counting processes. *Lecture Notes Statist.*, **2**, 1–25. Springer-Verlag, New York. (Aalen originally proposed his model at a conference in Poland; this paper appeared in the conference proceedings.)

2. Aalen, O. O. (1989). A linear regression model for the analysis of life times. *Statist. Med.*, **8**, 907–925. (A readable introduction to additive risk models with an emphasis on graphical techniques. Results based on the Aalen and Cox models are compared.)

3. Aalen, O. O. (1993). Further results on the nonparametric linear regression model in survival analysis. *Statist. Med.*, **12**, 1569–1588. (Studies the use of martingale residuals for assessing goodness of fit of additive risk models.)

4. Andersen, P. K., Borgan, Ø., Gill, R. D., and Keiding, N. (1993). *Statistical Models Based on Counting Processes*. Springer-Verlag, New York. (A comprehensive and in-depth survey of the counting process approach to survival analysis, including the additive risk model and its associated estimators.)

5. Cox, D. R. (1972). Regression models and life tables (with discussion). *J. Roy. Statist. Soc. B*, **34**, 187–220.

6. Greenwood, P. E. and Wefelmeyer, W. (1991). Efficient estimating equations for nonparametric filtered models. In *Statistical Inference in Stochastic Processes*, N. U. Prabhu and I. V. Basawa, eds., pp. 107–141. Marcel Dekker, New York. (Shows that the weighted least-squares estimator is an approximate maximum-likelihood estimator and that this implies asymptotic efficiency.)

7. Henderson, R. and Milner, A. (1991). Aalen plots under proportional hazards. *Appl. Statist.*, **40**, 401–410. (Introduces a modification to Aalen plots designed to detect time-dependent covariate effects in Cox's proportional hazards model.)

8. Henderson, R. and Oman, P. (1993). Influence in linear hazard models. *Scand. J. Statist.*, **20**, 195–212. (Shows how to detect unusual or influential observations under Aalen's model.)

9. Huffer, F. W. and McKeague, I. W. (1991). Weighted least squares estimation for Aalen's additive risk model. *J. Am. Statist. Assoc.*, **86**, 114–129.

10. Lin, D. Y. and Ying, Z. (1994). Semiparametric analysis of the additive risk model. *Biometrika*, **81**, 61–71.

11. Lin, D. Y. and Ying, Z. (1995). Semiparametric analysis of general additive–multiplicative hazard models for counting processes. *Ann. Statist.*, **23**, 1712–1734.

12. Mau, J. (1986). On a graphical method for the detection of time-dependent effects of covariates in survival data. *Appl. Statist.*, **35**, 245–255. (Makes a strong case that Aalen plots can provide important information that might be missed when only Cox's proportional hazards model is applied.)

13. Mau, J. (1988). A comparison of counting process models for complicated life histories. *Appl. Stoch. Models Data Anal.*, **4**, 283–298. (Aalen's model is applied in the context of intermittent exposure in which individuals alternate between being at risk and not.)

14. McKeague, I. W. (1988). Asymptotic theory for weighted least squares estimators in Aalen's additive risk model.

In *Statistical Inference from Stochastic Processes*, N. U. Prabhu, ed.; *Contemp. Math.*, **80**, 139–152. American Mathematical Society, Providence. (Studies the asymptotic properties of estimators in the counting process version of Aalen's model.)

15. McKeague, I. W. (1988). A counting process approach to the regression analysis of grouped survival data. *Stoch. Process. Appl.*, **28**, 221–239. (Studies the asymptotic properties of grouped data based estimators for Aalen's model.)

16. McKeague, I. W. and Utikal, K. J. (1991). Goodness-of-fit tests for additive hazards and proportional hazards models. *Scand. J. Statist.*, **18**, 177–195.

17. McKeague, I. W. and Sasieni, P. D. (1994). A partly parametric additive risk model. *Biometrika*, **81**, 501–514.

18. Miller, R. G. (1976). Least squares regression with censored data. *Biometrika*, **63**, 449–464.

19. Ramlau-Hansen, H. (1983). Smoothing counting process intensities by means of kernel functions. *Ann. Statist.*, **11**, 453–466.

20. Sasieni, P. D. (1992). Information bounds for the additive and multiplicative intensity models. In *Survival Analysis: State of the Art*, J. P. Klein and P. K. Goel, eds., pp. 249–265. Kluwer, Dordrecht. (Shows asymptotic efficiency of the weighted least-squares estimators in the survival analysis setting.)

21. Sasieni, P. D. (1995). Efficiently weighted estimating equations with application to proportional excess hazards. *Lifetime Data Analysis*, **1**, 49–57.

22. Sasieni, P. D. (1996). Proportional excess hazards. *Biometrika*, **83**, 127–141.

2 Aggregation

J. Keith Ord

2.1 Introduction

Aggregation may be a phenomenon of direct interest, as in the study of biological populations, or it may reflect the necessary reduction of primary data to produce a usable statistical summary, as in the construction of index numbers. Since the two topics are quite distinct, we consider them separately.

2.2 Aggregation as an Observable Phenomenon

It is often true that events (individuals) cluster in time or space or both (e.g., larvae hatching from eggs laid in a mass, aftershocks of an earthquake). Thus if the random variable of interest is the number of events occurring in an interval of time (or in a selected area), the clustering is manifested in a greater probability of extreme events (large groups) than would be expected otherwise. Alternatively, individual members of a population may be in close proximity because of environmental conditions. In either case, the population is said to be aggregated.

A standard initial assumption (corresponding to the absence of aggregation) is that the random variable follows a Poisson distribution, and various indices have been proposed to detect departures from the Poisson process. These methods are based on data collected either as quadrat counts or as measurements of distance (or time) from randomly selected individuals (or points) to the nearest individual, known as *nearest-neighbor distances*. For example, the index of dispersion is defined as $I = s^2/m$, where m and s^2 denote the sample mean and variance. For the Poisson process, $E(I|m > 0) \doteq 1$; values of I significantly greater than 1 suggest aggregation; $I < 1$ is indicative of regular spacing of the individuals [5, Chap. 4]. Other measures, based on both quadrat counts and distances, are summarized in Pielou [15, Chaps. 8 and 10] and Cormack [6]. When different kinds of individual (e.g., species) have different aggregation patterns, this makes inferences about population characteristics such as diversity much more difficult.

If the Poisson process is used to describe parents (or centers), each parent may give rise to offspring (or satellites). If these clusters are independent but have identical size distributions, the resulting distribution for the total count is a (Poisson) randomly stopped sum distribution. If environmental heterogeneity is postulated, a compound distribution, usually based on the Poisson, is appropriate. For both classes of Poisson-based distributions, $I > 1$. These standard distributions lack an explicit spatial or temporal dimension, for which a dispersal mechanism must be incorporated. The resulting model, known as a *center–satellite process*, has three components: a Poisson process for locating the cluster center, a distribution to generate the number of satellites, and a dispersal distribution to describe displacements from the center. This class of processes was introduced by Neyman and Scott [14] and is mathematically equivalent to the class of doubly stochastic Poisson processes defined for heterogeneity (see Bartlett [3, Chap. 1]).

A more empirical approach to aggregation is that of Taylor [18], who suggests that the population mean, μ, and variance, σ^2, are related by the power law:

$$\sigma^2 = A\mu^b, \qquad A > 0, \quad b > 0.$$

It is argued that values of b greater than 1 (the Poisson value) reflect density dependence in the spatial pattern of individuals. Although this view has been contested (see the discussion in Taylor [18]), a substantial body of empirical evidence has been presented in its support [19].

Knox [12] developed a test to detect the clustering of individuals in space and time, which may be formulated as follows. Suppose that n individuals (e.g., cases of a disease) are observed in an area during a time period. If cases i and j are less than a specified critical distance from one another, set the indicator variable $w_{ij} = 1$; otherwise, set $w_{ij} = 0$. Similarly, if i and j occur within a specified time of one another, set $y_{ij} = 1$; otherwise, set $y_{ij} = 0$. Then the space–time interaction (STI) coefficient is

$$\text{STI} = \sum_{i \neq j} \sum w_{ij} y_{ij}.$$

For example, for a disease such as measles, we might consider cases within 1 mile of each other occurring 10 days or less apart (the length of the latent period). If n_S and n_T denote the number of adjacent pairs in space and time, respectively, and both are small relative to n, then the conditional distribution of STI given n_S, n_T, and n is approximately Poisson with expected value $n_S n_T / n$. The test has been extended to several spatial and temporal scales by Mantel [13]. For further details, see Cliff and Ord [5, Chaps. 1 and 2].

(*Editor's Addendum.* A population that has its individuals evenly distributed is frequently called *overdispersed*, while an aggregated population with its individuals clustered in groups can be described as *underdispersed.*)

2.3 Aggregation as a Statistical Method

Aggregation in this sense involves the compounding of primary data in order to express them in summary form. Also, such an exercise is necessary when a model is specified at the micro (or individual) level but the usable data refer to aggregates. Then the question that arises is whether the equations of the micro model can be combined in such a way as to be consistent with the macro (or aggregate) model to which the data refer.

We may wish to compound individual data records, such as consumers' expenditures, or to combine results over time and/or space. The different cases are described in turn.

2.3.1 Combining Individual Records

Consider a population of N individuals in which the ith individual ($i = 1, \ldots, N$) has response Y_i to input x_i of the form

$$Y_i = f(x_i, \beta_i) + \epsilon_i$$

where β_i denotes a (vector of) parameter(s) specific to the ith individual and

ϵ_i denotes a random-error term. For example, the equation may represent the consumer's level of expenditure on a commodity given its price. Then the total expenditure is $Y = \sum Y_i$ (summed over $i = 1, \ldots, N$), and the average input is $x = \sum x_i/N$.

In general, it is not possible to infer an exact relationship between Y and x from the micro relations. The few results available refer to the linear aggregation of linear equations. Theil [20] showed that when f denotes a linear function so that

$$Y_i = \alpha_i + \beta_i x_i + \epsilon_i,$$

perfect aggregation is possible, in that we may consider a macro relation of the form

$$Y = \alpha + \beta x^* + \epsilon,$$

where $\alpha = \sum \alpha_i$, $\beta = \sum \beta_i$, $x^* = \sum \beta_i x_i/\beta$, and $\epsilon = \sum \epsilon_i$; that is, we must use the weighted average x^* rather than the natural average x. Further, a different aggregation procedure is required for each regressor variable and for the same regressor variable with different response variables [2, Chap. 20; 20, Chap. 2]. If we use the natural average, the macro relation is

$$Y = \alpha + \beta x + N \operatorname{cov}(x_i, \beta_i) + \epsilon,$$

where the covariance is evaluated over the N members of the population and represents the aggregation bias. This bias is small, for example, when x is the price variable in a consumer demand equation, but may be much more substantial when x denotes consumers' income in such a relationship. When the micro relationship is a nonlinear function of x_i, the nonlinearity will generate a further source of aggregation bias. It must be concluded that exact aggregation is rarely possible, although the bias may be small in many cases. For further discussion and recent developments of the theory, see Ijiri [11].

2.3.2 Aggregation of Groups

Instead of forming a macro relation from a known group of individuals, we may wish to identify suitable groups from a finer classification of individuals. This is a necessary step in the construction of broad industrial classifications for use in input/output systems. Blin and Cohen [4] propose a method of cluster analysis for solving this problem.

2.3.3 Temporal Aggregation

Variates may be continuous or summed over a unit time interval, although the variate is recorded only as an aggregate over periods of r units duration. For a model that is linear in the regressor variables and has time-invariant parameters, aggregation is straightforward provided that there are no lagged variables. However, if

$$Y_t = \alpha + \beta x_{t-k} + \epsilon_t$$

for some $k > 0$ and k not a multiple of r, exact aggregation is not possible; any aggregated model will involve x values for two or more time periods [20, Chap. 4]. Such models are often formulated using distributed lags. Also, Granger and Morris [9] show that the autoregressive–moving average (ARMA) models are often appropriate in this case.

The aggregation of time series exhibiting positive autocorrelation tends to increase the values of the various test statistics and thereby give rise to overoptimistic assessments of the model [17]. However, Tiao and Wei [21] have shown that whereas aggregation can considerably reduce the efficiency of parameter estimators, it has much less effect on prediction efficiency. Indeed, it has been shown that there are circumstances where forecasts from aggregate relations may be more accurate than an aggregate of forecasts from micro relations [1,10].

The discussion so far has assumed that β does not vary over time. For a discussion of aggregation when there is a change of regime (time-varying parameters), see Goldfeld and Quandt [8, Chap. 4].

2.3.4 Spatial Aggregation

Many problems in spatial aggregation are similar to those of time series, but they are further compounded by the (sometimes necessary) use of areas of irregular shape and different size. Yule and Kendall [22] first showed how different aggregations of spatial units affect estimates of correlation. Cliff and Ord [5, Chap. 5] give a general review of methods for estimating the autocovariance and cross-correlation functions using a nested hierarchy of areal sampling units. The estimation of these functions for irregular areas depends on making rather restrictive assumptions about the nature of interaction between areas.

Cliff and Ord considered data from the *London Atlas*; a 24 × 24 lattice of squares of side 500 meters was laid over the greater London area and the percentage of land used for commercial (X), industrial (Y), office (Z), and other purposes was recorded for each square. The correlation between each X and Y for different combinations of grid squares were as follows:

	Size of Spatial Unit			
	1 × 1	2 × 2	4 × 4	8 × 8
Corr(X,Z)	0.19	0.36	0.67	0.71
Corr(Y,Z)	0.09	0.16	0.33	0.34

The correlation functions exhibit an element of mutual exclusion at the smallest levels, and the positive correlation for larger spatial units indicates the general effects of areas zoned for housing and nonhousing purposes. Here, as in time series, we must think in terms of a distance or time-dependent correlation function and not a unique "correlation" between variables.

2.4 A Perspective

Aggregation appears both as a phenomenon of interest in its own right and as a necessary evil in modeling complex processes. The center–satellite models have proved useful in astronomy [14], ecology [3,15], geography [5], and several other disciplines. At the present time, such processes offer a flexible tool for simulation work, although further work on the theory and analysis of such processes is desirable; the data analytic distance methods of Ripley [16] represent a useful step in the right direction. In epidemiology [12,13] and hydrology (models of storms, etc.) the development of clusters in both space and time is important, although relatively little work exists to date. The work of Taylor [18] represents a challenge to the theoretician, as useful models generating the empirical regularities observed by Taylor are still lacking.

Econometricians seem to be turning away from the view that an aggregate model is the sum of its parts and placing more emphasis on aggregated models per se. The nonlinearities of the aggregation procedure, combined with the complexities of the underlying processes [8], suggest that aggregated models with time-dependent parameters are likely to play an increasing role in economics and other social sciences.

Where the level of aggregation is open to choice [4], further work is needed to identify suitable procedures for combining finer units into coarser ones. Similar problems arise in quadrat sampling [15, p. 222].

The use of a sample to estimate the mean over an area or volume is of interest in the geosciences (e.g., drillings in an oil field). Estimators for such aggregates are based on a variant of generalized least squares known as "kriging"; see Delfiner and Delhomme [7] and the papers of Matheron cited therein for further details.

In all these areas, much remains to be discussed about the statistical properties of the estimators currently used, and there is still plenty of scope for the development of improved methods.

References

1. Aigner, D. J., and Goldfeld, S. M. (1974). *Econometrica*, **42**, 113–134.

2. Allen, R. G. D. (1959). *Mathematical Economics*. Macmillan, London. (Outlines the aggregation problem in econometric modeling and sets it in the context of other aspects of mathematical economics; written at an intermediate mathematical level.)

3. Bartlett, M. S. (1975). *The Statistical Analysis of Spatial Pattern*. Chapman & Hall, London. (A concise introduction to the theory of spatial point processes and lattice processes with a variety of applications in ecology.)

4. Blin, J. M. and Cohen, C. (1977). *Rev. Econ. Statist.*, **52**, 82–91. (Provides a review and references on earlier attempts to form viable aggregates as well as new suggestions.)

5. Cliff, A. D. and Ord, J. K. (1981). *Spatial Processes: Models, Inference and Applications*. Pion, London. (Discusses aggregation problems in the context of spatial patterns with examples drawn from ecology and geography; written at an intermediate mathematical level.)

6. Cormack, R. M. (1979). In *Spatial and Temporal Processes in Ecology*, R. M. Cormack and J. K. Ord, eds., pp. 151–211. International Co-operative Publishing House, Fairland, MD. (An up-to-date review of spatial interaction models with an extensive bibliography. Other papers in the volume cover related aspects.)

7. Delfiner, P. and Delhomme, J. P. (1975). In *Display and Analysis of Spatial Data*, J. C. Davis and J. C. McCullagh, eds., pp. 96–114. Wiley, New York. (This volume contains several other papers of general interest on spatial processes.)

8. Goldfeld, M. and Quandt, R. E. (1976). *Studies in Nonlinear Estimation*. Ballinger, Cambridge, MA.

9. Granger, C. W. J. and Morris, J. J. (1976). *J. Roy. Statist. Soc. A*, **139**, 246–257.

10. Grunfeld, Y. and Griliches, Z. (1960). *Rev. Econ. Statist.*, **42**, 1–13.

11. Ijiri, Y. (1971). *J. Am. Statist. Assoc.*, **66**, 766–782. (A broad review of aggregation for economic models, with an extensive bibliography.)

12. Knox, E. G. (1964). *Appl. Statist.*, **13**, 25–29.

13. Mantel, N. (1967). *Cancer Res.*, **27**, 209–220.

14. Neyman, J. and Scott, E. L. (1958). *J. Roy. Statist. Soc. B*, **20**, 1–43. (The seminal paper on clustering processes.)

15. Pielou, E. C. (1977). *Mathematical Ecology*. Wiley, New York. (Describes methods for the measurement of aggregation among individuals; extensive bibliography.)

16. Ripley (1977).

17. Rowe, R. D. (1976). *Int. Econ. Rev.*, **17**, 751–757.

18. Taylor, L. R. (1971). In *Statistical Ecology*, Vol. 1, G. P. Patil et al., eds., pp. 357–377. Pennsylvania State University Press, University Park, PA.

19. Taylor, L. R. and Taylor, R. A. J. (1977). *Nature (Lond.)*, **265**, 415–421.

20. Theil, H. (1954). *Linear Aggregates of Economic Relations*. North-Holland, Amsterdam. (The definitive work on aggregation for economic models.)

21. Tiao, G. C. and Wei, W. S. (1976). *Biometrika*, **63**, 513–524.

22. Yule, G. U. and Kendall, M. G. (1965). *An Introduction to the Theory of Statistics*. Charles Griffin, London.

3 AIDS Stochastic Models

Wai-Y. Tan

3.1 Introduction

AIDS is an infectious disease caused by a retrovirus called human immunodeficiency virus (HIV) [17]. The first AIDS case was diagnosed in Los Angeles, CA, USA in 1980 [14]. In a very short period, the AIDS epidemic has grown into dangerous proportions. For example, the World Health Organization (WHO) and the United Nation AIDS Program (UNAIDS) estimated that 5 million had acquired HIV in 2003, and about 40 million people are currently living with AIDS. To control AIDS, in the past 10 years, significant advances had been made in treating AIDS patients by antiviral drugs through cocktail treatment protocol [10]. However, it is still far from cure and the disease is still spreading, especially in Africa and Asia. For preventing the spread of HIV, for controlling AIDS, and for understanding the HIV epidemic, mathematical models that take into account the dynamic of the HIV epidemic and the HIV biology are definitely needed. From this perspective, many mathematical models have been developed [1,8,11,22,24–26,29,42,55]. Most of these models are deterministic models in which the state variables (i.e., the numbers of susceptible people, HIV-infected people, and AIDS cases) are assumed as deterministic functions, ignoring completely the random nature of these variables. Because the HIV epidemic is basically a stochastic process, many stochastic models have been developed [5–7,18,33–37,39–44,46–49,53–54,57–64, 66–68,73–76]. This is necessary because, as shown by Isham [23], Mode et al. [40,41], Tan [48,54], Tan and Tang [62], and Tan et al. [63], in some cases, the difference between the mean numbers of the stochastic models and the results of deterministic models could be very substantial; it follows that in these cases, the deterministic models would provide very poor approximation to the corresponding mean numbers of the stochastic models, leading to misleading and sometimes confusing results.

3.2 Stochastic Transmission Models of the HIV Epidemic in Populations

Stochastic transmission models for the spread of HIV in populations were first developed by Mode et al. [40,41] and by Tan and Hsu [59,60] in homosexual populations; see also References 5–7, 18, 47–49, 54, 57, 66–68, 73, and 74. These models have been extended to IV drug populations [19,53,62] and to heterosexual populations [39,46,75–76]. Many of these models have been summarized in books by Tan [55] and by Mode and Sleeman [42]. Some applications of these models have been illustrated in Reference 55.

To illustrate the basic procedures for deriving stochastic transmission models for the HIV epidemic, consider a large population at risk for AIDS. This population may be a population of homosexual men, or a population of IV drug users, or a population of single males, single females, and married couples, or a mixture of these populations. In the presence of HIV epidemic, then there are three types of people in the population: S people (susceptible people), I people (infective people), and A people (AIDS patients). S people are healthy people but can contract HIV to become I people through sexual contact and/or IV drug contact with I people or A people or through contact with HIV-contaminated blood. I people are people who have contracted HIV and can pass the HIV to S people through sexual contact or IV drug contact with S people. According to the 1993 AIDS case definition [15] by the Centers for Disease Control (CDC) at Atlanta, GA, USA, an I person will be classified as a clinical AIDS patient (A person) when this person develops at least one of the AIDS symptoms specified in Reference 15 and/or when his/her CD4$^{(+)}$ T-cell counts fall below 200/mm^3. Then, in this population, one is dealing with a high-dimensional stochastic process involving the numbers of S people, I people, and AIDS cases.

To develop realistic stochastic models for this process, it is necessary to incorporate many important risk variables and social and behavior factors into the model, and to account for the dynamics of the epidemic process. Important risk variables that have significant impacts on the HIV epidemic are age, race, sex, and sexual (or IV drug use) activity levels defined by the average number of sexual (or IV drug use) partners per unit time; the important social and behavior factors are the IV drug use, the mixing patterns between partners, and the condom use that may reduce the probability of transmission of the HIV viruses. To account for these important risk variables and for IV drug use, the population is further stratified into subpopulations.

Given that the population has been stratified into subpopulations by many risk factors, the stochastic modeling procedures of the HIV epidemic essentially boils down to two steps. The first step involves modeling the transmission of HIV from HIV carriers to susceptible people. This step would transform S people to I people ($S \longrightarrow I$). This step is the dynamic part of the HIV epidemic process and is influenced significantly by age, race, social behavior, sexual level, and many other factors. This step is referred to as the *transmission step*. The next step is the modeling of HIV progression until the development of clinical AIDS and death in people who have already contracted HIV. This step is basically the step transforming I people into A people ($I \longrightarrow A$) and is influenced significantly by the genetic makeup of the individual and by the person's infection duration that is defined by the time period elapsed since he/she first contracts the HIV. This step is referred to as the *HIV progression step*. By using these two steps, the numbers of S people, I people, and AIDS cases are generated stochastically at any time given the numbers at the previous time. These models have been referred by Tan and his associates [48,49,54,57,61,63,66–68,73–76] as chain multinomial models since the principle of random sampling dictates that aside from the distributions of recruitment and immigration, the conditional probability distributions of the numbers of S people, I people, and AIDS cases at any time, given the numbers at the previous time, are related to multinomial distributions.

3.3 The Transmission Step: The Probability of HIV Transmission

The major task in this step is to construct the probabilities of HIV transmission from infective people or AIDS cases to S people by taking into account the dynamic aspects of the HIV epidemic. These probabilities are functions of the mixing pattern that describes how people from different risk groups mix together and the probabilities of transmission of HIV from HIV carriers to susceptible people given contacts between these people. Let $S_i(t)$ and $I_i(u,t)(u = 0,\ldots,t)$ denote the numbers of S people and I people with infection duration u at time t in the ith risk group respectively. Let the time unit be a month and denoted by $p_i(t;S)$, the conditional probability that an S person in the ith risk group contracts HIV during the tth month given $\{S(t), I(u,t), u = 0,1,\ldots,t\}$. Let $\rho_{ij}(t)$ denote the probability that a person in the ith risk group selects a person in the jth risk group as a partner at the tth month and $\alpha_{ij}(u,t)$ the probability of HIV transmission from an infective person with infection duration u in the jth risk group to the susceptible person in the ith risk group given contacts between them during the tth month. Assume that because of the awareness of AIDS, there are no contacts between S people and AIDS cases and that there are n risk groups or subpopulations. Then $p_i(t;S)$ is given by

$$p_i(t;S) = 1 - \{1 - \psi_i(t)\}^{X_i(t)}, \quad (1)$$

where $X_i(t)$ is the number of partners of the S person in the ith risk group during the tth month and $\psi_i(t) = \sum_{j=1}^{n} \rho_{ij}(t)\{I_j(t)/T_j(t)\}\bar{\alpha}_{ij}(t)$ with $T_j(t) = S_j(t) + \sum_{u=0}^{t} I_j(u,t)$ and $\bar{\alpha}_{ij}(t) = \frac{1}{I_j(t)} \sum_{u=0}^{t} I_j(u,t)\alpha_{ij}(u,t)$ with $I_j(t) = \sum_{u=0}^{t} I_j(u,t)$.

If the $\alpha_{ij}(u,t)$ are small, then $\{1 - \psi_i(t)\}^{X_i(t)} \cong \{1 - X_i(t)\psi_i(t)\}$

so that

$$p_i(t;S) \cong X_i(t)\psi_i(t)$$
$$= X_i(t) \sum_{j=1}^{n} \rho_{i,j}(t) \frac{I_j(t)}{T_j(t)} \bar{\alpha}_{ij}(t). \quad (2)$$

Notice that in Equations (1) and (2), the $p_i(t;S)$ are functions of $\{S_i(t), I_i(u,t), u = 0,1,\ldots,t\}$ and hence in general are random variables. However, some computer simulation studies by Tan and Byers [57] have indicated that if the $S_i(t)$ are very large, one may practically assume $p_i(t;S)$ as deterministic functions of time t and the HIV dynamic.

3.4 The Progression Step: The Progression of HIV in HIV-Infected Individuals

The progression of HIV inside the human body involves interactions between $CD4^{(+)}T$ cells, $CD8^{(+)}T$ cells, free HIV, HIV antibodies, and other elements in the immune system, which will eventually lead to the development of AIDS symptoms as time increases. It is influenced significantly by the dynamics of the interactions between different types of cells and HIV in the immune system, treatment by antiviral drugs, and other risk factors that affect the speed of HIV progression. Thus, it is expected that the progression of I to A depends not only on the calendar time t but also on the infection duration u of the I people as well as the genetic makeup of the I people. This implies that the transition rate of $I \longrightarrow A$ at time t for I people with infection duration u is in general a function of both u and t; this rate will be denoted by $\gamma(u,t)$. Let T_{inc} denote the time from HIV infection to the onset of AIDS. Given $\gamma(u,t)$, the probability distribution of T_{inc} can readily be derived [55, Chap. 4].

In the AIDS literature, T_{inc} has been referred to as the *HIV incubation period* and the probability distribution of T_{inc} the *HIV incubation distribution*. These distributions have been derived by Bachetti [4], Longini et al. [33], and by Tan and his associates [50–53,57,61,64,65] under various conditions. These probability distributions together with many other distributions, which have been used in the literature have been tabulated and summarized in Reference 55, Chapter 4.

3.5 Stochastic Modeling of HIV Transmission by Statistical Approach

To develop stochastic models of HIV transmission, it is necessary to take into account the dynamic of the HIV epidemic to construct the probabilities $p_S(i,t) = \beta_i(t)\Delta t$ of HIV transmission. To avoid the dynamic aspect, statisticians assume $p_i(t;S)$ as deterministic functions of i and t and proceed to estimate these probabilities. This is a nonparametric procedure, which has ignored all information about the dynamics of the HIV epidemic; on the other hand, it has minimized the misclassification or misspecification of the dynamic of the HIV epidemic. In the literature, these probabilities are referred to as the infection incidence.

To illustrate the statistical approach, let T_I denote the time to infection of S people and $f_i(t)$ the probability density of T_I in the ith risk group. Then, $f_i(t) = \beta_i(t)\exp\{-\int_0^t \beta_i(x)dx\}$. The $f_i(t)$ have been referred by statisticians as the HIV infection distributions.

To model the HIV progression, let T_A denote the time to AIDS of S people and $h_i(t)$ the probability density of T_A in the ith risk group. Then $T_A = T_I + T_{inc}$, where T_{inc} is the HIV incubation period. If T_I and T_{inc} are independently distributed of each other, then

$$h_i(t) = \int_0^t f_i(x)g_i(x,t)dx, \qquad (3)$$

where $g_i(s,t)$ is the density of the HIV incubation distribution given HIV infection at time s in the ith risk group. Notice that since the transition rates of the infective stages are usually independent of the risk group, $g_i(s,t) = g(s,t)$ are independent of i. In what follows, it is thus assumed that $g_i(s,t) = g(s,t)$ unless otherwise stated.

Let ω_i denote the proportion of the ith risk group in the population. For an individual taken randomly from the population, the density of T_A is given by

$$\begin{aligned}
h(t) &= \sum_{i=1}^n \omega_i h_i(t) \\
&= \int_0^t \left\{\sum_{i=1}^n \omega_i f_i(x)g_i(x,t)\right\}dx \\
&= \int_0^t f(x)g(x,t)dx, \qquad (4)
\end{aligned}$$

where $\sum_{i=1}^n \omega_i f_i(t) = f(t)$ is the density of the HIV infection distribution for people taken randomly from the population.

Equation (4) is the basic equation for the backcalculation method. By using this equation and by interchanging the order of summation and integration, it can easily be shown that the probability that S person at time 0 taken randomly from the population will become an AIDS case for the first time during $(t_{j-1}, t_j]$ is

$$\begin{aligned}
&P(t_{j-1}, t_j) \\
&= \int_{t_{j-1}}^{t_j}\int_0^t f(u)g(u,t)\,du\,dt \\
&= \left\{\int_0^{t_j} - \int_0^{t_{j-1}}\right\}\int_0^t f(u)g(u,t)\,du\,dt \\
&= \int_0^{t_j} f(u)\left\{G(u,t_j) - G(u,t_{j-1})\right\}du, \\
&\hspace{7cm} (5)
\end{aligned}$$

where $G(u,t) = \int_u^t g(u,x)dx$ is the cumulative distribution function (cdf) of the HIV incubation period, given HIV infection at time u.

Equation (5) is the basic formula by means of which statisticians tried to estimate the HIV infection or the HIV incubation based on AIDS incidence data [8,55]. This has been illustrated in detail in References 8 and 55. There are two major difficulties in this approach, however. One difficulty is that the problem is not identifiable in the sense that one cannot estimate simultaneously the HIV infection distribution and the HIV incubation distribution. Thus, one has to assume the HIV incubation distribution as known if one wants to estimate the HIV infection distribution; similarly, one has to assume the HIV infection distribution as known if one wants to estimate the HIV incubation distribution. Another difficulty is that one has to assume that there are no immigration, no competing death, and no other disturbing factors for Equation (5) to hold; see Reference 55, Chapter 5. These difficulties can readily be resolved by introducing state space models; see References 55, 56, 73, and 74.

3.5.1 Stochastic Transmission Models of HIV Epidemic in Homosexual Populations

In the United States, Canada, Australia, New Zealand, and western European countries, AIDS cases have been found predominantly among the homosexual, bisexual, and intravenous drug user communities with only a small percentage of cases due to heterosexual contact. Thus, most of the stochastic models for HIV spread were first developed in homosexual populations [5–7, 40,41,47,49,57,59–60,62,66–68,73,74].

To illustrate how to develop stochastic models for the HIV epidemic, consider a large homosexual population at risk for AIDS that has been stratified into n risk groups of different sexual activity levels. Denote by $A_i(t)$ the number of new AIDS cases developed during the tth month in the ith risk group and let the time unit be a month. Then, one is entertaining a high-dimensional discrete-time stochastic process $\underset{\sim}{U}(t) = \{S_i(t), I_i(u,t), u = 0, \ldots, t, A_i(t), i = 1, \ldots, n\}$. To derive basic results for this process, a convenient approach is by way of stochastic equations. This is the approach proposed by Tan and his associates in modeling the HIV epidemic [49,54–57,63,64,66–68,73–76].

3.6 The Stochastic Difference Equations and the Chain Multinomial Model

To develop the stochastic model for the above process, let $\{\Lambda_i(S,t), \mu_i(S,t)\}$ and $\{\Lambda_i(u,t), \mu_i(u,t)\}$ denote the recruitment and immigration rate, and the death and migration rate of S people and $I(u)$ people at the tth month in the ith risk group, respectively. Then, given the probability $p_i(t;S)$ that the S person in the ith risk group would contract HIV during the tth month, one may readily obtain $\underset{\sim}{U}(t+1)$ from $\underset{\sim}{U}(t)$ by using multinomial distributions, for $t = 0, 1, \ldots$. This procedure provides stochastic difference equations for the numbers of S people, $I(u)$ people, and the number of new A people at time t. These models have been referred to by Tan [48,49,54,55] and by Tan and his coworkers [57,61,63,64, 66–68,73,74] as *chain multinomial models*.

To illustrate, let $R_i(S,t), F_i(S,t)$, and $D_i(S,t)$ denote respectively the number of recruitment and immigrants of S people, the number of $S \to I(0)$ and the total number of deaths of S people during the tth month in the ith risk group. Similarly, for $u = 0, 1, \ldots, t$, let $R_i(u,t)$,

$F_i(u,t)$ and $D_i(u,t)$ denote respectively the number of recruitment and immigrants of $I(u)$ people, the number of $I(u) \to A$, and the total number of deaths of $I(u)$ people during the tth month in the ith risk group. Then, the conditional probability distribution of $\{F_i(S,t), D_i(S,t)\}$ given $S_i(t)$ is multinomial with parameters $\{S_i(t), p_i(t;S), \mu_i(S,t)\}$ for all $i = 1,\ldots,n$; similarly, the conditional probability distribution of $\{F_i(u,t), D_i(u,t)\}$ given $I_i(u,t)$ is multinomial with parameters $\{I_i(u,t), \gamma_i(u,t), \mu_i(u,t)\}$, independently of $\{F_i(S,t), D_i(S,t)\}$ for all $\{i = 1,\ldots,n, u = 0,1,\ldots,t\}$. Assume that $E\{R_i(S,t)|S(t)\} = S_i(t)\Lambda_i(S,t)$ and $E\{R_i(u,t)|I_i(u,t)\} = I_i(u,t)\Lambda_i(u,t)$. Then, one has the following stochastic difference equations for $\{S_i(t), I_i(u,t), i = 1,\ldots,n\}$:

$$\begin{aligned}
S_i(t+1) &= S_i(t) + R_i(S,t) \\
&\quad - F_i(S,t) - D_i(S,t) \\
&= S_i(t)\{1 + \Lambda_i(S;t) - p_i(t;S) \\
&\quad - \mu_i(S,t)\} + \epsilon_i(S,t+1),
\end{aligned} \quad (6)$$

$$\begin{aligned}
I_i(0, t+1) &= F_i(S,t) = S_i(t)p_i(t;S) \\
&\quad + \epsilon_i(0, t+1),
\end{aligned} \quad (7)$$

$$\begin{aligned}
I_i(u+1, t+1) &= I_i(u,t) + R_i(u,t) \\
&\quad - F_i(u,t) - D_i(u,t) \\
&= I_i(u,t)\{1 + \Lambda_i(u,t) \\
&\quad - \gamma_i(u,t) - \mu_i(S,t)\} \\
&\quad + \epsilon_i(u+1, t+1),
\end{aligned} \quad (8)$$

$$\begin{aligned}
A_i(t+1) &= \sum_{u=0}^{t} F_i(u,t) \\
&= \sum_{u=0}^{t} I_i(u,t)\gamma_i(u,t) + \epsilon_i(A,t).
\end{aligned} \quad (9)$$

In Equations (6)–(9), the random noises $\{\epsilon_i(S,t), \epsilon_i(u,t), u = 0,1,\ldots,t, \epsilon_i(A,t)\}$ are derived by subtracting the conditional mean numbers from the corresponding random variables. It can easily be shown that these random noises have expected values 0 and are uncorrelated with the state variables.

Using these stochastic equations, one can readily study the stochastic behaviors of the HIV epidemic in homosexual populations and assess effects of various risk factors on the HIV epidemic and on some intervention procedures. Such attempts have been made by Tan and Hsu [59,60], Tan [48,54], and Tan et al. [63] by using some simplified models. For example, Tan and Hsu [59–60] have shown that the effects of intervention by decreasing sexual contact rates depend heavily on the initial number of infected people; when the initial number is small, say, 10, then the effect is quite significant. On the other hand, when the initial number is large, say, 10,000, then the effect of decreasing sexual contact rates is very small. The Monte Carlo studies by Tan and Hsu [59,60] have also revealed some effects of "regression on the mean" in the sense that the variances are linear functions of the expected numbers. Thus, although the variances are much larger than their respective mean numbers, effects of risk factors on the variance curves are quite similar to those of these risk factors on the mean numbers.

By using the above equations, one may also assess the usefulness of deterministic models in which $S_i(t), I_i(r,t), r = 1,\ldots,k, A_i(t)$ are assumed as deterministic functions of i and t, ignoring completely randomness of the HIV epidemic process. The system of equations defining the deterministic models is derived by ignoring the random noises from Equations (6)–(9). Thus, one may assume that the deterministic models are special cases of the stochastic models. However, the above equations for $S_i(t)$ and $I_i(0,t)$ are not the same as the equations for the mean numbers of $S_i(t)$ and $I_i(0,t)$, respectively. This follows from the observation that since the $p_S(i,t)$ are functions of $S_i(t)$ and $I_i(u,t), u = 1, \ldots, k$, $E[S_i(t)p_S(i,t)] \neq E[S_i(t)] \times E[p_S(i,t)]$. As shown in Reference 55, the equations for the mean numbers of $S_i(t)$ and $I_i(0,t)$ differ from the corresponding ones of the deterministic model in that the equations for the means of $S_i(t)$ and $I_i(0,t)$ contain additional terms involving covariances $\text{Cov}\{S_i(t), p_S(i,t)\}$ between $S_i(t)$ and $p_S(i,t)$. Thus, unless these covariances are negligible, results of the deterministic models of the HIV epidemic would in general be very different from the corresponding mean numbers of the stochastic models of the HIV epidemic. As shown by Isham [23], Mode et al. [40,41], Tan [48,54], Tan and Tang [62], and Tan et al. [63], the difference between results of deterministic models and the mean numbers of the stochastic models could be very substantial. The general picture appears to be that the stochastic variation would in general speed up the HIV epidemic. Further, the numbers of I people and A people computed by the deterministic model would underestimate the true numbers in the short run but overestimate the true numbers in the long run. These results imply that, in some cases, results of the deterministic model may lead to misleading and confusing results.

3.7 The Probability Distribution of the State Variables

Let $\boldsymbol{X} = \{\boldsymbol{X}(0), \ldots, \boldsymbol{X}(t_M)\}$ with $\boldsymbol{X}(t) = \{S_i(t), I_i(u,t), u = 0, 1, \ldots, t, i = 1, \ldots, n\}$. To estimate the unknown parameters and to assess stochastic behaviors of the HIV epidemic, it is of considerable interest to derive the probability density $P(\boldsymbol{X}|\Theta)$ of $\boldsymbol{X}$. By using multinomial distributions for $\{R_i(S,t), F_i(S,T)\}$ and for $\{R_i(u,t), F_i(u,T)\}$ as above, this probability density can readily be derived. Indeed, denoting by $g_{i,s}\{j;t|\boldsymbol{X}(t)\}$ and $g_{i,u}\{j;t|\boldsymbol{X}(t)\}$ the conditional densities of $R_i(S,t)$ given $\boldsymbol{X}(t)$ and of $R_i(u,t)$ given $B\boldsymbol{X}(t)$, respectively, one has

$$P(\boldsymbol{X}|\Theta)$$
$$= P\{\boldsymbol{X}(0)|\Theta\} \prod_{t=1}^{t_M} P\{\boldsymbol{X}(t)|\boldsymbol{X}(t-1)\}$$
$$= P\{\boldsymbol{X}(0)|\Theta\}$$
$$\times \prod_{t=1}^{t_M} \prod_{i=1}^{n} P\{S_i(t)|\boldsymbol{X}(t-1)\}$$
$$\times \left\{ \prod_{u=0}^{t} P[I_i(u,t)|\boldsymbol{X}(t-1)] \right\}. \quad (10)$$

In Equation (10), the $P\{S_i(t+1)|\boldsymbol{X}(t)\}$ and the $P\{I_i(u+1, t+1)|\boldsymbol{X}(t)\}$ are given respectively by

$$P\{S_i(t+1)|\boldsymbol{X}(t)\}$$
$$= \binom{S_i(t)}{I_i(0, t+1)}$$
$$\times [p_i(S,t)]^{I_i(0,t+1)} h_i(t|S), \quad (11)$$

$$P\{I_i(u+1, t+1)|\boldsymbol{X}(t)\}$$
$$= \sum_{r=0}^{I_i(u,t)} \binom{I_i(u,t)}{r}$$
$$\times [\gamma_i(u,t)]^r h_{i,r}(u,t|I)$$
$$\text{for } u = 0, 1, \ldots, t, \quad (12)$$

where the $h_i(t|S)$ and the $h_{i,r}(u,t|I)$ are

given respectively by

$$h_i(t|S) = \sum_{j=0}^{S_i(t+1)-S_i(t)+I_i(0,t+1)} g_{i,S}\{j,t|\boldsymbol{X}(t)\}$$
$$\times \binom{S_i(t)-I_i(0,t+1)}{a_{i,S}(j,t)}$$
$$\times [d_i(S,t)]^{a_{i,S}(j,t)}$$
$$\times \{1-p_i(S,t)-d_i(S,t)\}^{b_{i,S}(j,t)},$$

with

$$a_{i,S}(j,t) = S_i(t) - S_i(t+1) - I_i(0,t+1) + j$$

and

$$b_{i,S}(j,t) = S_i(t+1) - j,$$

and

$$h_{i,r}(u,t|I) = \sum_{j=0}^{I_i(u+1,t+1)-I_i(u,t)+r} g_{i,u}[j,t|\boldsymbol{X}(t)]$$
$$\times \binom{I_i(u,t)-r}{a_{i,u}(r,j,t)} [d_i(u,t)]^{a_{i,u}(r,j,t)}$$
$$\times \{1-\gamma_i(u,t)-d_i(u,t)\}^{b_{i,u}(r,j,t)},$$

with

$$a_{i,u}(r,j,t) = I_i(u,t) - I_i(u+1,t+1) - r + j$$

and

$$b_{i,u}(r,j,t) = I_i(u+1,t+1) + r - 2j.$$

In these equations, notice that aside from the immigration and recruitment, the distribution of $\boldsymbol{X}$ is basically a product of multinomial distributions. Hence, the above model has been referred to as a chain multinomial model; see References 48, 49, 54, 57, 61, 63, 64, 66–68, 73, and 74. Tan and Ye [74] have used the above distribution to estimate both the unknown parameters and the state variables via state space models.

3.8 The Staged Model of HIV Epidemic in Homosexual Populations

In the model presented above, the number of state variables $I_i(u,t)$ and hence the dimension of the state space increases as time increases; that is, if the infection duration is taken into account, then the size of the dimension of the state space increases as time increases. To simplify matters, an alternative approach is to partition the infective stage into a finite number of substages with stochastic transition between the substages and assume that within the substage, the effects of duration is the same. This is the approach proposed by Longini and his associates [33–37]. In the literature, such staging is usually achieved by using the number of CD4$^{(+)}$ T-cell counts per mm^3 blood. The staging system used by Satten and Longini [43,44] is I_1, CD4 counts $\geq 900/mm^3$; I_2, $900/mm^3 >$ CD4 counts $\geq 700/mm^3$; I_3, $700/mm^3 >$ CD4 counts $\geq 500/mm^3$; I_4, $500/mm^3 >$ CD4 counts $\geq 350/mm^3$; I_5, $350/mm^3 >$ CD4 counts $\geq 200/mm^3$; I_6, $200/mm^3 >$ CD4counts. (Because of the 1993 AIDS definition by CDC [15], the I_6 stage is merged with the AIDS stage (A stage).)

The staging of the infective people results in a staged model for the HIV epidemic. Comparing these staged models with the previous model, the following differences are observed:

1. Because of the infection duration, the nonstaged model is in general not Markov. On the other hand, if one assumes that the transition rates of the substages are independent of the time at which the substage were generated, the staged models are Markov.

2. For the nonstaged model, the number of different type of infectives always

increase nonstochastically as time increases. These are referred to as expanding models by Liu and Chen [32]. On the other hand, the number of substages of the infective stage in the staged model is a fixed number independent of time.

3. For the nonstaged model, the infective people increase its infection duration nonstochastically, always increasing by one time unit with each increase of one time unit. However, they transit directly to AIDS stochastically. On the other hand, for the staged model, the transition from one substage to another substage is stochastic and can either be forward or backward, or transit directly to AIDS. Because of the random transition between the infective substages, one would expect that the staging has introduced more randomness into the model than the nonstaged model.

Assuming Markov, one may use some standard results in Markov chain theory to study the HIV epidemic in staged models. This has been done by Longini and his associates [33–37]. Alternatively, by using exactly the same procedures as in the previous model, one may derive the stochastic equation for the state variables as well as the probability distributions of the state variables. By using these equations and the probability distributions, one can then study the stochastic behaviors and to assess effects of risk variables and the impact of some intervention procedures. This has been done by Tan and his associates [48,49,54,55,57,63,64,66–68]. By using the San Francisco homosexual population as an example, they have shown that the staged model gave similar results as the previous model and hence the same conclusions. These results indicates that the errors of approximation and the additional variations due to stochastic transition between the substages imposed by the staging system are in general quite small for the HIV epidemic in homosexual populations. On the other hand, because of the existence of a long asymptomatic infective period with low infectivity, it is expected that the staged model would provide a closer approximation to the real-world situations than the nonstaged model.

Another problem in the staged model is the impacts of measurement error as the CD4 T-cell counts are subject to considerable measurement error. To take this into account, Satten and Longini [44] have proposed a hidden Markov model by assuming the measurement errors as Gaussian variables. The calculations done by them did not reveal a significant impact of these errors on the HIV epidemic, however, indicating that the effects of measurement errors on the HIV epidemic of the staged model is not very significant.

3.9 Stochastic Models of HIV Transmission in Complex Situations

In Africa, Asia, and many South American countries, although homosexual contact and IV drug use may also be important avenues, most of the HIV epidemic are developed through heterosexual contacts and prostitutes [38]. The dynamics of HIV epidemic in these countries are therefore very different from those in the United States, Canada, and the western countries, where the major avenues of HIV transmission are homosexual contact and sharing needles and IV drug use. It has been documented that even in homosexual populations, race, age, and risk behaviors as well as many other risk variables would significantly affect the HIV epidemic [66,67,75,76]. To account for effects of many risk factors such as sex, race, and age, the above simple stochastic model has been extended into models under complex situations [19,39,46,53,66,67,75,76].

To develop stochastic models of HIV transmission in complex situations, the basic procedures are again the same two steps as described above:

1. Stratifying the population into subpopulations by sex, race, and risk factors, derive the probabilities of HIV transmission from infective people to susceptible people in each subpopulation. These probabilities usually involve interactions between people from different risk groups and the structure of the epidemic.

2. Develop steps for HIV progression within each subpopulation. It appears that because the dynamics of HIV epidemic are different under different situations, the first step to derive the probabilities of HIV transmission varies from population to population depending on different situations.

Given that the probabilities of HIV transmission from infective people to susceptible people have been derived for each subpopulation, the second step is similar to the progression step of the procedures described above. That is, the only major difference between different models lies in the derivation of the probabilities $p_i(t; S)$ for the ith subpopulation. Assuming that there are no sexual contacts with AIDS cases, the general form of $p_i(t; S)$ is

$$p_i(t; S) = X_i(t) \sum_j \rho_{ij}(t) \frac{I_j(t)}{T_j(t)} \bar{\alpha}_{ij}(t). \quad (13)$$

Notice that Equation (13) is exactly of the same form as Equation (2); yet, because of the different dynamics in different models, the $\rho_{ij}(t)$'s are very different between different models. In populations of IV drug users, HIV spread mainly through sharing IV needles in small parallel groups. In these populations, therefore, $\rho_{ij}(t)$ is derived by first forming small groups and then spread HIV by sharing needles between members within the group. This is the basic formulation by means of which Capasso et al. [9], Gani and Yakowitz [19], Kaplan [28], and Kaplan et al. [30] derived the probabilities of HIV infection of S people by infective people. Along this line, Tan [55, Chap. 4] has formulated a general procedure to derive this probability.

In populations stratified by race, sex, age, sexual activity levels, and risk behaviors and involving married couples, to derive $\rho_{ij}(t)$ one needs to take into account some realistic preference patterns. These realities include (1) People tend to mix more often with people of the same race, same age group, and same sexual activity level;(2) people with high sexual activity levels and/or old age tend to select sexual partners indiscriminately; (3) if the age difference between the two partners is less than 5 years, then age is not an important factor in selecting sexual partners; (4) race and sexual activity level may interact with each other to affect the selection of sexual partners; and (5) a happy marriage would reduce external marital relationship. Taking many of these factors into account and assuming that members of small populations select members from larger populations, Tan and Xiang [66,67] and Tan and Zhu [75,76] have proposed a selective mixing pattern through the construction of acceptance probabilities. Intuitively, this mixing can be expressed as a product of two probability measures: The first is the probability of selecting members from subpopulations with larger effective population size via acceptance probabilities; the second is the conditional probability of selecting members of infective people with different infection duration from the selected population.

By using the selective mixing pattern, Tan and Xiang [66,67] have developed stochastic models for the HIV epidemic in homosexual populations taking into account race, age, and sexual activity levels. Their results indicate that race and age affect the HIV epidemic mainly through the numbers of different sexual partners per partner per month and their interactions with the mixing pattern. Increasing the transition rates of infective people by race and/or by age seems to have some impact on the HIV progression but the effects are much smaller than those from the average numbers of different sexual partners per partner per month and the mixing patterns. Thus, the observed result that there is a much larger proportion of AIDS cases from black people than from white people in the US population [16] is a consequence of the following observations: (1) black people in general have larger number of sexual partners per unit time than do white people; and (2) there is a large proportion of restricted mixing pattern and mixing patterns other than proportional mixing while under these mixing patterns, black people appear to contract HIV much faster than white people.

By using selective mixing pattern, Tan and Zhu [75,76] have developed stochastic models involving single males, single females, married couples, and prostitutes. Their results indicate that the prostitute factor may be the main reason for the rapid growth of the HIV epidemic in some Asian countries such as Thailand and India, which have a large prostitute population. Their Monte Carlo studies also suggest that rapid growth of the HIV epidemic in some Asian countries may be arrested or controlled by promoting extensive use of condoms by prostitutes combined with a campaign of AIDS awareness in the younger and sexually active populations.

3.10 State Space Models of the HIV Epidemic

State space models of stochastic systems are stochastic models consisting of two submodels: the *stochastic system model*, which is the stochastic model of the system, and the *observation model*, which is a statistical model based on available observed data from the system. That is, the state space model adds one more dimension to the stochastic model and to the statistical model by combining both of these models into one model. This is a convenient and efficient approach to combine information from both stochastic models and statistical models. It takes into account the basic mechanisms of the system and the random variation of the system through its stochastic system model and incorporate all these into the observed data from the system; and it validates and upgrades the stochastic model through its observation model and the observed data of the system and the estimates of the state variables. It is advantageous over both the stochastic model and the statistical model when used alone since it combines information and advantages from both of these models. Specifically, one notes that (1) because of additional information, many of the identifiability problems in statistical analysis are nonexistent in state space models (see Refs. 73 and 74 for some examples); (2) it provides an optimal procedure to update the model by new data that may become available in the future—this is the smoothing step of the state space models [2, 12, and 20], (3) it provides an optimal procedure via Gibbs sampling to estimate simultaneously the unknown parameters and the state variables of interest [55, Chap. 6; 56, Chap. 9; 73; and 74]; and (4) it provides a general procedure to link molecular events to critical events in population and at the cellular level [58].

The state space model (Kalman filter model) was originally proposed by Kalman and his associates in the early 1960s for engineering control and communication [27]. Since then it has been successfully used as a powerful tool in aerospace research, satellite research, and military missile research. It has also been used by economists in econometrics research [21] and by mathematician and statisticians in time-series research [3] for solving many difficult problems that appear to be extremely difficult from other approaches. In 1995 and 2000, Wu and Tan [78,79] had attempted to apply the state space model and method to AIDS research. Since then many papers have been published to develop state space models for the HIV epidemic and the HIV pathogenesis; see References 13, 68–73, 77, and 78. Alternatively, by combining the Markov staged model with Gaussian measurement errors for the CD4 T-cell counts, Satten and Longini [44] have proposed a hidden Markov model for the HIV epidemic; however, it is shown by Tan [56] that this is a special case of the state space models.

Although Tan and Ye [74] have applied the state space model for the HIV epidemic in the Swiss population of IV drug users, to date, the state space models for HIV epidemic are developed primarily in homosexual populations. To illustrate how to develop state space models for the HIV epidemic, we will thus use the San Francisco homosexual population as an example, although the general results apply to other populations as well.

3.11 A State Space Model for the HIV Epidemic in the San Francisco Homosexual Population

Consider the San Francisco homosexual population, in which HIV spread primarily by sexual contact [16]. For this population, Tan and Ye [73] have developed a state space model for the HIV epidemic. For this state space model, the stochastic system model was represented by stochastic difference equations. The observation model of this state space model is based on the monthly AIDS incidence data (i.e., data of new AIDS cases developed during a month period). These data are available from the gofer network of CDC. This is a statistics model used by statistician through the backcalculation method. Combining these two models into a state space model, Tan and Ye [73] have developed a general Bayesian procedure to estimate the HIV infection, the HIV incubation, as well as the numbers of susceptible people, infective people, and AIDS cases. Notice that this is not possible by using the stochastic model alone or by using the statistic model alone because of the identifiability problem.

3.12 The Stochastic System Model

To develop a stochastic model for the San Francisco population, Tan and Ye [74] have made two simplifying assumptions: (1) By visualizing the infection incidence and hence the infection distribution as a mixture of several sexual levels, one sexual activity level may be assumed; and (2) because the population size of the city of San Francisco changes very little, for the S people and I people, it is assumed that the number of immigration and recruitment is about the same as the number of death and migration. Tan and Ye [74] have shown that these assumptions have little impacts on the probability distributions of the HIV infection and the HIV incubation. On the basis of these assumptions, then the state variables are $\underset{\sim}{U}(t) = \{S(t), I(u,t), u = 0, 1, \ldots, t, A(t)\}$. Assuming that the probability $p_S(t)$ of HIV infection of S people and the probability $\gamma(s,t) = \gamma(t-s)$ of $I(u) \to$ AIDS as deterministic functions of time, then the parameters are $\underset{\sim}{\theta}(t) = \{p_S(t), \gamma(u,t) = \gamma(t-u), u = 0, 1, \ldots, t\}'$. The densities of the HIV infection and the HIV incubation are given by $f_I(t) = p_S(t)\Pi_{i=0}^{t-1}[1 - p_S(i)] = G_S(t-1)p_S(t), t =$

$1, \ldots, \infty$ and $g(t) = \gamma(t)\Pi_{j=0}^{t-1}[1 - \gamma(j)] = G_I(t-1)\gamma(t), t = 1, \ldots, \infty$, respectively.

Let $F_S(t)$ be the number of S people who contract HIV during the tth month and $F_I(u,t)$ the number of $I(u,t) \to A$ during the tth month. Then

$$S(t+1) = S(t) - F_S(t), \quad (14)$$
$$I(0, t+1) = F_S(t), \quad (15)$$
$$I(u+1, t+1) = I(u,t) - F_I(u,t), \quad (16)$$
$$u = 0, \ldots, t,$$

where $F_S(t)|S(t) \sim B\{S(t), p_S(t)\}$ and $F_I(u,t)|I(ut) \sim B\{I(u,t), \gamma(u)\}, u = 0, 1, \ldots, t$.

Put $\Theta = \{\theta(t), t = 1, \ldots, t_M\}$, $\underset{\sim}{X}(t) = \{S(t), I(u,t), u = 0, 1, \ldots t\}$ and $\underset{\sim}{X} = \{\underset{\sim}{X}(1), \ldots, \underset{\sim}{X}(t_M)\}$, where t_M is the last timepoint. Then, the probability distribution of $\underset{\sim}{X}$ given Θ and given $\underset{\sim}{X}(0)$ is

$$P\{\underset{\sim}{X}|\underset{\sim}{X}(0)\}$$
$$= \prod_{j=0}^{t_M-1} P\{\underset{\sim}{X}(j+1)|\underset{\sim}{X}(j), \Theta\}$$

where

$$\Pr\{\underset{\sim}{X}(j+1)|\underset{\sim}{X}(j), \Theta\}$$
$$= \binom{S(t)}{I(0, t+1)}[p_S(t)]^{I(0,t+1)}$$
$$\times [1 - p_S(t)]^{S(t)-I(0,t+1)}$$
$$\times \prod_{u=0}^{t}\binom{I(u,t)}{I(u,t) - I(u+1, t+1)}$$
$$\times [\gamma(u)]^{I(u,t)-I(u+1,t+1)}$$
$$\times [1 - \gamma(u)]^{I(u+1,t+1)}. \quad (17)$$

Notice that the density here is a product of binomial densities and hence has been referred to as the chain binomial distribution.

For the HIV epidemic in the San Francisco homosexual population, Tan and Ye [73] have assumed January 1, 1970, as $t = 0$ since the first AIDS case in San Francisco appeared in 1981 and since the average incubation period for HIV is about 10 years. It is also assumed that, in 1970, there are no infective people but to start the HIV epidemic, some HIV were introduced into the population in 1970. Thus, one may take $I(0,0) = 36$ because this is the number of AIDS in San Francisco in 1981. Tan and Ye [73] have assumed the size of the San Francisco homosexual population in 1970 as 50,000 because with a 1% increase in population size per year by the US census survey [77], the estimate of the size of the San Francisco homosexual population is $58,048 = 50,000 \times (1.01)^{15}$ in 1985, which is very close to the estimate 58,500 of the size of the San Francisco homosexual population in 1985 by Lemp et al. [31].

3.13 The Observation Model

Let $y(j+1)$ be the observed AIDS incidence during the jth month and $A(t+1)$ the number of new AIDS cases developed during the tth month. Then the stochastic equation for the observation model is

$$y(j+1)$$
$$= A(j+1) + \xi(j+1)$$
$$= \sum_{u=0}^{j} F_I(u,j) + \xi(j+1)$$
$$= \sum_{u=0}^{j}[I(u,t) - I(u+1, t+1)]$$
$$\quad + \xi(t+1)$$
$$= \sum_{u=0}^{j} I(u,j)\gamma(u) + \epsilon_A(j+1)$$
$$\quad + \xi(j+1)$$
$$= \sum_{u=0}^{j} I(u,j)\gamma(u) + e(j+1), \quad (18)$$

where $\xi(t+1)$ is the random measurement error associated with observing $y(j+1)$ and $\epsilon_A(j+1) = [F_s(t) - S(t)p_s(t)] + \sum_{u=1}^{j}[F_I(u,t) - I(u,t)\gamma(u)]$.

Put $\mathbf{Y} = \{y(j), j = 1, \ldots, t_M\}$. Assuming that the $\xi(j)$ are independently distributed as normal with means 0 and variance σ_j^2, then the likelihood function $P\{\mathbf{Y}|\mathbf{X}, \Theta\} = L(\Theta|\mathbf{Y}, \mathbf{X})$ given the state variables is

$$P\{\mathbf{Y}|\mathbf{X}, \Theta\} \propto \prod_{j=1}^{t_M}$$
$$\times \left(\sigma_j^{-1} \exp\left\{ -\frac{1}{2\sigma_j^2}[y(j) - A(j)]^2 \right\} \right). \quad (19)$$

Notice that under the assumption that $\{p_s(t), \gamma(t)\}$ are deterministic functions of t,

$$E[S(t+1)]$$
$$= E[S(t)][1 - p_s(t)]$$
$$= E[S(t-1)] \prod_{i=t-1}^{t} \{1 - p_s(i)\}$$
$$= E[S(0)] \prod_{i=0}^{t} \{1 - p_s(i)\}$$
$$= E[S(0)]G_s(t), \quad (20)$$

$$E[F_s(t)] = E[S(t)]p_s(t)$$
$$= E[S(0)]G_s(t-1)p_s(t)$$
$$= E[S(0)]f_I(t), \quad (21)$$

$$E[I(u+1, t+1)]$$
$$= E[I(u,t)][1 - \gamma(u)]$$
$$= E[I(0, t-u)] \prod_{j=0}^{u}[1 - \gamma(j)]$$
$$= E[I(0, t-u)]G_I(u)$$
$$= E[S(0)]f_I(t-u)G_I(u). \quad (22)$$

Hence

$$E[I(u,t) - I(u+1, t+1)]$$
$$= E[S(0)]f_I(t-u)$$
$$\times \{G_I(u-1) - G_I(u)\}$$
$$= E[S(0)]f_I(t-u)G_I(u-1)\gamma(u)$$
$$= E[S(0)]f_I(t-u)g(u). \quad (23)$$

It follows that

$$\sum_{u=0}^{t} E[I(u,t) - I(u+1, t+1)]$$
$$= E[S(0)] \sum_{u=0}^{t} f_I(t-u)g(u)$$
$$= E[S(0)] \sum_{u=0}^{t} f_I(u)g(t-u), \quad (24)$$

so that

$$y(j+1)$$
$$= E[S(0)] \sum_{u=0}^{t} f_I(u)g(t-u)$$
$$+ e(j+1). \quad (25)$$

Notice that Equation (25) is the convolution formula used in the backcalculation method [4,55]. This implies that the backcalculation method is the observation model in the state space model. The backcalculation method is not identifiable because using Equation (25) alone and ignoring information from the stochastic system model, the information is not sufficient for estimating all the parameters.

3.14 A General Bayesian Procedure for Estimating the Unknown Parameters and the State Variables

By using the state space model, Tan and Ye [73,74] have developed a generalized

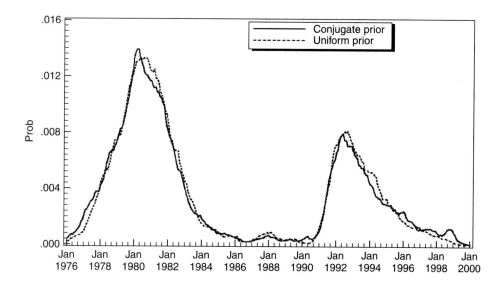

Figure 1: Plots of the estimated HIV infection distribution.

Bayesian approach to estimate the unknown parameters and the state variables. This approach will combine information from three sources: (1) previous information on and experience with the parameters in terms of the prior distribution of the parameters, (2) biological information via the stochastic system equations of the stochastic system, and (3) information from observed data via the statistical model from the system.

To illustrate the basic principle of this method, let $P(\Theta)$ be the prior distribution of Θ. Then, the joint distribution of $\{\Theta, X, Y\}$ is given by $P(\Theta, X, Y) = P(\Theta)P(X|\Theta)P(Y|X,\Theta)$. From this, the conditional distribution $P(X|\Theta, Y)$ of X given (Θ, Y) and the conditional posterior distribution $P(\Theta|X, Y)$ of Θ given (X, Y) are given respectively by

$$P(X|\Theta, Y) \propto P(X|\Theta)P(Y|X,\Theta) \quad (A)$$

$$P(\Theta|X, Y) \propto P(\Theta)P(X|\Theta)P(Y|X,\Theta) \quad (B)$$

Given these probability densities, one may use the multilevel Gibbs sampling method to derive estimates of Θ and X given Y [45]. This is a Monte Carlo sequential procedure alternating between two steps until convergence: (1) given $\{\Theta, Y\}$, one generates X by using $P(X|\Theta, Y)$ from (A) (these are the Kalman filter estimates); and (2) using the Kalman filter estimates of $\underset{\sim}{X}$ from (A) and given Y, one generates values of Θ by using $P(\Theta|X, Y)$ from (B). Iterating between these two steps until convergence, one then generates random samples from the conditional probability distribution $P(X|Y)$ independently of Θ, and from the posterior distribution $P(\Theta|Y)$ independently of X, respectively. This provides the Bayesian estimates of Θ given data and the Bayesian estimates of X given data, respectively. The proof of the convergence can be developed by using basic theory of stationary distributions in irreducible and aperiodic Markov chains; see Reference 56, Chapter 3.

Using the above approach, one can readily estimate simultaneously the numbers of S people, I people, and AIDS cases as well as the parameters $\{p_S(t), \gamma(t)\}$. With

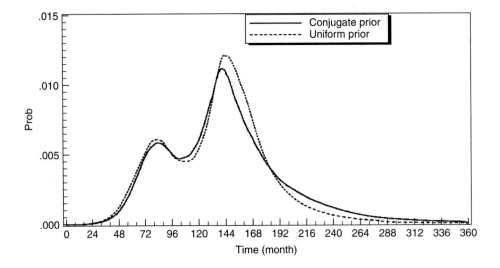

Figure 2: Plots of the estimated HIV incubation distribution.

the estimation of $\{p_S(t), \gamma(t)\}$, one may then estimate the HIV infection distribution $f_I(t)$ and the HIV incubation distribution $g(t)$. For the San Francisco homosexual population, the estimates of $f_I(t)$ and $g(t)$ are plotted in Figures 1 and 2. Given below are some basic findings by Tan and Ye [73]:

1. From Figure 1, the estimated density of the HIV infection clearly showed a mixture of distributions with two obvious peaks. The first peak (the higher peak) occurs around January 1980 and the second peak around March 1992. The mixture nature of this density implies that there are more than one sexual activity levels with a high proportion of restricted mixing (like-with-like) mixing. The second peak also implies a second wave of infection although the infection intensity is much smaller than the first wave.

2. From Figure 2, the estimated density of the HIV incubation distribution also appeared to be a mixture of distributions with two obvious peaks. The first peak is around 75 months after infection and is much lower than the second peak which occurs around 140 months after infection. This result suggests a multistage nature of the HIV incubation.

3. The Kalman filter estimates of the AIDS incidence by the Gibbs sampler are almost identical to the corresponding observed AIDS incidence respectively. This result indicates that the Kalman filter estimates can trace the observed numbers very closely if observed numbers are available.

4. Figuring a 1% increase in the population size of San Francisco yearly, Tan and Ye [73] have also estimated the number of S people and the I people in the San Francisco population. The estimates showed that the total number of S people before January 1978 were always above 50,000 and were between 31,000 and 32,000 during January 1983 and January 1993. The total number of people who do not have

AIDS were estimated around 50,000 before January 1993. It appeared that the total number of infected people reached a peak around the middle of 1985 and then decreased gradually to the lowest level around 1992. The estimates also showed that the number of infected people had two peaks, with the higher peak around the middle of 1985, the second peak around the year 2000 and with the lowest level around 1992.

Extending the above state space model to include immigration and death, Tan and Ye [74] have also analyzed the data of the Swiss population of homosexual men and IV drug users by applying the above generalized Bayesian method. The estimated density of the HIV infection in the Swiss homosexual population is quite similar to that of the San Francisco population except that in the Swiss population, the first peak appears about 6 months earlier and the second peak, about 2 years earlier. Similarly, the estimated density of the HIV incubation distribution in the Swiss population is a mixture of distributions with two obvious peaks. The higher peak occurs around 320 months after infection and the lower peak occurs around 232 months after infection. In the Swiss population, the estimates of the immigration and recruitment rates are about 10 times greater than those of the estimates of the death and retirement rates of the I people, suggesting that the size of the Swiss homosexual and bisexual populations is increasing with time. Another interesting point is that, in the Swiss population, the estimates of the death and retirement rates of infective people were much greater (at least 100 times greater) than those of S people, suggesting that HIV infection may have increased the death and retirement rates of HIV-infected people.

References

1. Anderson, R. M. and May, R. M. (1992). *Infectious Diseases of Humans: Dynamics and Control.* Oxford University Press, Oxford, UK.

2. Anderson, B. D. O. and Moore, J. B. (1979). *Optimal Filtering.* Prentice-Hall, Englewood Cliffs, NJ.

3. Aoki, M. (1990). *State Space Modeling of Time Series*, 2nd ed. Springer-Verlag, New York.

4. Bacchetti, P. R. (1990). Estimating the incubation period of AIDS comparing population infection and diagnosis pattern. *J. Am. Statist. Assoc.*, **85**, 1002–1008.

5. Billard, L. and Zhao, Z. (1991). Three-stage stochastic epidemic model: An application to AIDS. *Math. Biosci.*, **107**, 431–450.

6. Billard, L. and Zhao, Z. (1994). Multistage non-homogeneous Markov models for the acquired immune deficiency syndrome epidemic. *J. Roy. Statist. Soc. B*, **56**, 673–686.

7. Blanchard, P., Bolz, G. F., and Krüger, T. (1990). Modelling AIDS-epidemics or any veneral disease on random graphs. In *Stochastic Processes in Epidemic Theory*, Lecture Notes in Biomathematics, No. 86, J. P. Gabriel, C. Lefévre, and P. Picard, eds., pp. 104–117. Springer-Verlag, New York.

8. Brookmeyer, R. and Gail, M. H. (1994). *AIDS Epidemiology: A Quantitative Approach.* Oxford University Press, Oxford, UK.

9. Capasso, V., Di Somma, M., Villa, M., Nicolosi, A., and Sicurello, F. (1997). Multistage models of HIV transmission among injecting drug users via shared injection equipment. In *Advances in Mathematical Population Dynamics-Molecules, Cells and Man, Part II. Population Dynamics in Diseases in Man*, O. Arino, D. Axelrod, and M. Kimmel, eds., Chap. 8, pp. 511–528. World Scientific, Singapore.

10. Carpenter, C. C. J., Fischl, M. A., Hammer, M. D., Hirsch, M. S., Jacobsen, D. M., Katzenstein, D. A., Montaner, J. S. G., Richman, D. D., Saag, M. S., Schooley, R. T., Thompson, M. A., Vella, S., Yeni, P. G., and Volberding, P. A. (1996). Antiretroviral therapy for the HIV infection in 1996. *J. Am. Med. Assoc.*, **276**, 146–154.

11. Castillo-Chavez, C., ed. (1989). *Mathematical and Statistical Approaches to AIDS Epidemiology*, Lecture Notes in Biomathematics 83. Springer-Verlag, New York.

12. Catlin, D. E. (1989). *Estimation, Control and Discrete Kalman Filter*. Springer-Verlag, New York.

13. Cazelles, B. and Chau, N. P. (1997). Using the Kalman filter and dynamic models to assess the changing HIV/AIDS epidemic. *Math. Biosci.*, **140**, 131–154.

14. CDC. (1981). Pneumocystis pneumonia—Los Angeles. *MMWR (Morbidity and Mortality Weekly Report)*, **30**, 250–252.

15. CDC, (1992). Revised classification system for HIV infection and expanded surveillance case definition for AIDS among adolescents and adults. *MMWR*, **41**(RR-17), 1–19.

16. CDC. (1997). *Surveillance Report of HIV/AIDS*. Atlanta, GA, June 1997.

17. Coffin, J., Haase, J., Levy, J. A., Montagnier, L., Oroszlan, S., Teich, N., Temin, H., Toyoshima, K., Varmus, H., Vogt, P., and Weiss, R. (1986). Human immunodeficiency viruses. *Science*, **232**, 697.

18. Gani, J. (1990). Approaches to the modelling of AIDS". In *Stochastic Processes in Epidemic Theory*, Lecture Notes in Biomathematics No. 86, J. P. Gabriel, C. Leférre, and P. Picards, eds., pp. 145–154. Springer-Verlag, Berlin.

19. Gani, J. and Yakowitz, S. (1993). Modeling the spread of HIV among intravenous drug users. *IMA J. Math. Appl. Med. Biol.*, **10**, 51–65.

20. Gelb, A. (1974). *Applied Optimal Estimation*. MIT Press, Cambridge, MA.

21. Harvey, A. C. (1994). *Forcasting, Structural Time Series Models and the Kalman Filter*. Cambridge University Press, Cambridge, UK.

22. Hethcote, H. W. and Van Ark, J. M. (1992). *Modeling HIV Transmission and AIDS in the United States*, Lecture Notes in Biomathematics 95. Springer-Verlag, New York.

23. Isham, V. (1991). Assessing the variability of stochastic epidemics. *Math. Biosci.*, **107**, 209–224.

24. Isham, V. and Medley, G., eds. (1996). *Models for Infectious Human Diseases: Their Structure and Relation to Data*. Cambridge University Press, Cambridge, UK.

25. Jager, J. C. and Ruittenberg, E. J., eds. (1988). *Statistical Analysis and Mathematical Modelling of AIDS*. Oxford University Press, Oxford, U.K.

26. Jewell, N. P., Dietz, K., and Farewell, V. T. (1992). *AIDS Epidemiology: Methodological Issues*. Birkhäuser, Basel.

27. Kalman, R. E. (1960). A new approach to linear filter and prediction problems. *J. Basic Eng.*, **82**, 35–45.

28. Kaplan, E. H. (1989). Needles that kill: modeling human immuno-deficiency virus transmission via shared drug injection equipment in shooting galleries. *Rev. Infect. Dis.*, **11**, 289–298.

29. Kaplan, E. H. and Brandeau, M. L., eds. (1994). *Modeling the AIDS Epidemic*. Raven Press, New York.

30. Kaplan, E. H., Cramton, P. C., and Paltiel, A. D. (1989). Nonrandom mixing models of HIV transmission. In *Mathematical and Statistical Approaches to AIDS Epidemiology*, Lecture Notes in Biomathematics 83, C. Castillo-Chavez, ed., pp. 218–241. Springer-Verlag, Berlin.

31. Lemp, G. F., Payne, S. F., Rutherford, G. W., Hessol, N. A., Winkelstein, W.

Jr., Wiley, J. A., Moss, A. R., Chaisson, R. E., Chen, R. T., Feigal, D. W. Jr., Thomas, P. A., and Werdegar, D. (1990). Projections of AIDS morbidity and mortality in San Francisco. *J. Am. Med. Assoc. (JAMA)*, **263**, 1497–1501.

32. Liu, J. S. and Chen, R. (1998). Sequential Monte Carlo method for dynamic systems. *J. Am. Statist. Assoc.*, **93**, 1032–1044.

33. Longini Ira, M. Jr., Byers, R. H., Hessol, N. A., and Tan, W. Y. (1992). Estimation of the stage-specific numbers of HIV infections via a Markov model and back-calculation. *Statist. Med.*, **11**, 831–843.

34. Longini Ira, M. Jr., Clark, W. S., Byers, R. H., Ward, J. W., Darrow, W. W., Lemp, G. F., and Hethcote, H. W. (1989). Statistical analysis of the stages of HIV infection using a Markov model. *Statist. Med.*, **8**, 831–843.

35. Longini Ira, M. Jr., Clark, W. S., Gardner, L. I., and Brundage, J. F. (1991). The dynamics of CD4+ T-lymphocyte decline in HIV-infected individuals: A Markov modeling approach. *J. AIDS*, **4**, 1141–1147.

36. Longini Ira, M. Jr., Clark, W. S., and Karon, J. (1993). Effects of routine use of therapy in slowing the clinical course of human immunodeficiency virus (HIV) infection in a population based cohort. *Am. J. Epidemiol.*, **137**, 1229–1240.

37. Longini Ira, M. Jr., Clark, W. S., Satten, G. A., Byers, R. B., and Karon, J. M., (1996). Staged Markov models based on CD4$^{(+)}$ T-lymphocytes for the natural history of HIV infection. In *Models for Infectious Human Diseases: Their Structure and Relation to Data*, V. Isham and G. Medley, eds., pp. 439–459. Cambridge University Press, Cambridge, UK.

38. Mann, J. M. and Tarantola, D. J. M. (1998). HIV 1998: The global picture. *Sci. Am.*, **279**, 82–83.

39. Mode, C. J. (1991). A stochastic model for the development of an AIDS epidemic in a heterosexual population. *Math. Biosci.*, **107**, 491–520.

40. Mode, C. J., Gollwitzer, H. E., and Herrmann, N. (1988). A methodological study of a stochastic model of an AIDS epidemic. *Math. Biosci.*, **92**, 201–229.

41. Mode, C. J., Gollwitzer, H. E., Salsburc, M. A., and Sleeman, C. K. (1989). A methodological study of a nonlinear stochastic model of an AIDS epidemic with recruitment. *IMA J. Math. Appl. Med. Biol.*, **6**, 179–203.

42. Mode, C. J. and Sleeman, C. K. (2000). *Stochastic Processes in Epidemiology: HIV/AIDS, Other Infectious Diseases and Computers*. World Scientific, River Edge, NJ.

43. Satten, G. A. and Longini Ira, M. Jr. (1994). Estimation of incidence of HIV infection using cross-sectional marker survey. *Biometrics*, 50, 675–68.

44. Satten, G. A. and Longini Ira, M. Jr. (1996). Markov chain with measurement error: Estimating the "true" course of marker of the progression of human immunodeficiency virus disease. *Appl. Statist.*, **45**, 275–309.

45. Shephard, N. (1994). Partial non-Gaussian state space. *Biometrika*, **81**, 115–131.

46. Tan, W. Y. (1990). A Stochastic model for AIDS in complex situations. *Math. Comput. Model.*, **14**, 644–648.

47. Tan, W. Y. (1991). Stochastic models for the spread of AIDS and some simulation results. *Math. Comput. Model.*, **15**, 19–39.

48. Tan, W. Y. (1991). On the chain multinomial model of HIV epidemic in homosexual populations and effects of randomness of risk factors. In *Mathematical Population Dynamics 3*, O. Arino, D. E. Axelrod, and M. Kimmel, eds., pp. 331–353. Wuerz Publishing, Winnipeg, Manitoba, Canada

49. Tan, W. Y. (1993). The chain multinomial models of the HIV epidemiology in homosexual populations. *Math. Comput. Model.*, **18**, 29–72.

50. Tan, W. Y. (1994). On the HIV incubation distribution under non-Markovian

models. *Statist. Probab. Lett.*, **21**, 49–57.

51. Tan, W. Y. (1994). On first passage probability in Markov models and the HIV incubation distribution under drug treatment. *Math. Comput. Model.*, **19**, 53–66.

52. Tan, W. Y. (1995). On the HIV incubation distribution under AZT treatment. *Biometric. J.*, **37**, 318–338.

53. Tan, W. Y. (1995). A stochastic model for drug resistance and the HIV incubation distribution. *Statist. Probab. Lett.*, **25**, 289–299.

54. Tan, W. Y. (1995). On the chain multinomial model of HIV epidemic in homosexual populations and effects of randomness of risk factors. In *Mathematical Population Dynamics 3*, O. Arino, D. E. Axelrod, and M. Kimmel, eds., pp. 331–356. Wuerz Publishing, Winnipeg, Manitoba, Canada.

55. Tan, W. Y. (2000). *Stochastic Modeling of AIDS Epidemiology and HIV Pathogenesis*. World Scientific, River Edge, NJ.

56. Tan, W. Y. (2002). *Stochastic Models with Applications to Genetics, Cancers, AIDS and Other Biomedical Systems*. World Scientific, River Edge, NJ.

57. Tan, W. Y. and Byers, R. H. (1993). A stochastic model of the HIV epidemic and the HIV infection distribution in a homosexual population. *Math. Biosci.*, **113**, 115–143.

58. Tan, W. Y., Chen, C. W., and Wang, W. (2000). *A Generalized State Space Model of Carcinogenesis*. Paper presented at the 2000 International Biometric Conference at UC Berkeley, Calif., July 2–7, 2000.

59. Tan, W. Y. and Hsu, H. (1989). Some stochastic models of AIDS spread. *Statist. Med.*, **8**, 121–136.

60. Tan, W. Y. and Hsu, H. (1991). A stochastic model for the AIDS epidemic in a homosexual population. In *"Mathematical Population Dynamics*, eds. O. Arion, D. E. Axelrod, and M. Kimmel, eds., Chap. 24, pp. 347–368. Marcel Dekker, New York.

61. Tan, W. Y, Lee, S. R., and Tang, S. C. (1995). Characterization of HIV infection and seroconversion by a stochastic model of HIV epidemic. *Math. Biosci.*, **126**, 81–123.

62. Tan, W. Y. and Tang, S. C. (1993). Stochastic models of HIV epidemics in homosexual populations involving both sexual contact and IV drug use. *Math. Comput. Model.*, **17**, 31–57.

63. Tan, W. Y., Tang, S. C., and Lee, S. R. (1995). Effects of Randomness of risk factors on the HIV epidemic in homo-sexual populations. *SIAM J. Appl. Math.*, **55**, 1697–1723.

64. Tan, W. Y., Tang, S. C., and Lee, S. R. (1996). Characterization of the HIV incubation distributions and some comparative studies. *Statist. Med.*, **15**, 197–220.

65. Tan, W. Y., Tang, S. C., and Lee, S. R. (1998). Estimation of HIV seroconversion and effects of age in San Francisco homosexual populations. *J. Appl. Statist.*, **25**, 85–102.

66. Tan, W. Y. and Xiang, Z. H. (1996). A stochastic model for the HIV epidemic and effects of age and race. *Math. Comput. Model.*, **24**, 67–105.

67. Tan, W. Y. and Xiang, Z. H.(1997). A stochastic model for the HIV epidemic in homosexual populations and effects of age and race on the HIV infection. In *Advances in Mathematical Population Dynamics—Molecules, Cells and Man". Part II. Population Dynamics in Diseases in Man*, O. Arino, D. Axelrod, and M. Kimmel, eds., Chap. 8, pp. 425–452. World Scientific, River Edge, NJ.

68. Tan, W. Y. and Xiang, Z. H. (1998). State space models of the HIV epidemic in homosexual populations and some applications. *Math. Biosci.*, **152**, 29–61.

69. Tan, W. Y. and Xiang, Z. H. (1998). Estimating and predicting the numbers of T cells and free HIV by non-linear Kalman filter. In *Artificial Immune Systems and Their Applications*, D. DasGupta, ed., pp. 115–138. Springer-Verlag, Berlin.

70. Tan, W. Y. and Xiang, Z. H.(1998). State space models for the HIV pathogenesis. In *"Mathematical Models in Medicine and Health Sciences*, M. A. Horn, G. Simonett, and G. Webb, eds., pp. 351–368. Vanderbilt University Press, Nashville, TN.

71. Tan, W. Y. and Xiang, Z. H. (1999). Modeling the HIV epidemic with variable infection in homosexual populations by state space models. *J. Statist. Infer. Plan.*, **78**, 71–87.

72. Tan, W. Y. and Xiang, Z. H. (1999). A state space model of HIV pathogenesis under treatment by antiviral drugs in HIV-infected individuals. *Math. Biosci.*, **156**, 69–94.

73. Tan, W. Y. and Ye, Z. Z. (2000). Estimation of HIV infection and HIV incubation via state space models. *Math. Biosci.*, **167**, 31–50.

74. Tan, W. Y. and Ye, Z. Z. (2000). Simultaneous estimation of HIV infection, HIV incubation, immigration rates and death rates as well as the numbers of susceptible people, infected people and AIDS cases. *Commun. Stat.—Theory Methods*, **29**, 1059–1088.

75. Tan, W. Y. and Zhu, S. F. (1996). A stochastic model for the HIV infection by heterosexual transmission involving married couples and prostitutes. *Math. Comput. Model.*, **24**, 47–107.

76. Tan, W. Y. and Zhu, S. F. (1996). A stochastic model for the HIV epidemic and effects of age and race. *Math. Comput. Model.*, **24**, 67–105.

77. U.S. Bureau of the Census. (1987). *Statistical Abstract of the United States: 1980*, 108th ed. Washington, DC.

78. Wu, H. and Tan, W. Y. (1995). Modeling the HIV epidemic: A state space approach. *ASA 1995 Proc. Epidemiology Section*, pp. 66–71. ASA, Alexandria, VA.

79. Wu, H. and Tan, W. Y. (2000). Modeling the HIV epidemic: a state space approach. *Math. Comput. Model.*, **32**, 197–215.

4 All-or-None Compliance

Stuart G. Baker

4.1 Introduction

A frequent complication in the analysis of randomized trials of human subjects is noncompliance. Noncompliance can take many forms, such as switching from the assigned treatment to another treatment, not taking all the prescribed pills, or not attending all scheduled visits to receive treatment. The result, common to all forms, is that not every subjects in each arm of the randomized trial receives the same treatment. Meier [15] defines two causes of noncompliance: (1) a *selection* effect, in which selection for noncompliance may result in selection of subjects with certain risk of outcome; and (2) a *treatment* effect, in which noncompliance results from side effects or a change in health due to the treatment.

All-or-none compliance is a special type of noncompliance due only to a selection effect. Sometimes all-or-none compliance is used in a limited sense to mean the following situation: Subjects are randomized to receive treatment A or B, and subjects assigned to B may immediately switch to A. More generally, all-or-none compliance can be defined as follows: Subjects receive treatment A or B, but not both, with the fraction receiving A differing in the control and intervention groups. Because B is the treatment of interest and A is no treatment, placebo, or an old treatment, the fraction receiving B is greater in the intervention than in the control group.

All-or-none compliance is of particular interest because special methods have been developed to supplement the usual analysis by intent to treat. These methods make it possible to estimate the effect of treatment among subjects who would receive treatment if in the intervention but not the control group. This often leads to better extrapolations of the effect of treatment in a population who all receive the treatment.

4.2 Examples

We list a few types of all-or-none compliance.

1. Subjects are randomized to either A or B, but some assigned to B cannot receive it, so they receive A. Newcombe [16] described a study of the effect of type of analgesia on the amount needed in which patients in the control group received cryoanalgesia, and those in the intervention group received cervical epidural injection, unless administration was not possible, so they received cryoanalgesia. Sommer and Zeger [23] described a study

of the effect of vitamin supplement on mortality in which school children in the control group received nothing, and those in the intervention group received vitamin supplement, unless the distribution system failed, in which case they received nothing.

2. Subjects are randomized to either A or an offer of B; if subjects refuse the offer of B, they receive A. One of the earliest and largest studies involving this type of all-or-none compliance was the Health Insurance Plan of Greater New York (HIP) study [6,22], which began in 1963 and was designed to study the effect of screening for the early detection of breast cancer on breast-cancer mortality. Women in the control group were not offered screening and received none. Women in the intervention were offered screening; two-thirds accepted, while one-third refused. Bloom [4] described a study of the effect of job training on earnings in dislocated workers; those in the control group received no training, and some in the intervention group who were offered training did not participate. Bloom [4] also described a study of the effect of self-help care, on reversion to drug abuse among treated heroin addicts; those in the control group received no self-help care, and some in the intervention group who were offered self-help care did not participate. Zelen's single randomized consent design [25], described below, is another example.

3. Subjects in the control and intervention groups are offered A or B, with B encouraged only in the intervention group. Powers and Swinton [19] described a study of the effect of preparation on test performance in which subjects in the intervention group received extra encouragement to prepare for a test. Permutt and Hebel [18] described a study of the effect of smoking on birth weight in which pregnant women in the intervention group received encouragement to stop smoking. McDonald et al. [13] investigated the effect of vaccination on morbidity; subjects in the intervention group received additional vaccinations because their physicians were sent reminders to vaccinate.

4. Subjects in the control and intervention groups can receive A or B, but B is more likely to occur than A in the intervention group. Hearst et al. [8] described a "natural experiment" of the effects of military service on subsequent mortality; in this the draft lottery divided subjects into a control group exempt from the draft who were less likely to enter the military than an intervention group eligible for the draft.

Two designs with all-or-none compliance, the randomized consent design [25], and the paired availability design [3], are highlighted because they are alternatives in situations where standard randomized designs could not be easily implemented and because they are the only designs in which sample size and power have been investigated.

4.3 Randomized Consent Design

In a landmark series of papers Zelen [24–27] proposed the randomized consent design in order to make it easier for physicians to enter patients in a randomized clinical trial. Although the design avoids many ethical problems associated with informed consent, it raises other ethical issues [7].

In the single randomized consent design, patients, randomized to the control group receive treatment A and patients randomized to the intervention group are approached for consent to receive treatment B and told the risk, benefits, and treatment options; patients who do not agree to B receive A. In the double randomized consent design, patients in the control group are approached for consent to receive A, and receive B if they decline; and patients in the intervention group are approached for consent to receive B, and receive A if they decline.

To avoid bias in analyzing the results from the randomized consent design, Zelen [25] emphasized the need to compare outcomes in all subjects in the control group with all subjects in the intervention group. He compared the relative efficiency of his design with that of a usual randomized trial. Let p denote the probability of accepting the offered treatment, and assume it is the same for both groups in the double randomized consent design. For the single randomized consent design the relative efficiency is approximately p^2; for the double randomized consent design, it is approximately $(2p-1)^2$. Zelen [28] gave examples showing that easier patient accrual with the randomized consent design often more than offsets the loss in efficiency. Anbar [1], Brunner and Neumann [5], Matts and McHugh [12], and McHugh [14] also investigated the relative efficiency of the randomized consent design.

4.4 The Paired Availability Design

Extending Zelen's method as a way to reduce selection bias when randomization is not practical, Baker and Lindeman [3] developed the paired availability design; this involves two stages in which subjects may receive A or B, but B is more available in the second, later stage. This defines one pair, which should apply to a stable population in order to reduce the possibility of migration. To average the effect of random errors associated with each pair, the design is replicated in multiple pairs.

Baker and Lindeman [3] first proposed the paired availability design to study the effect of epidural analgesia on the rate of Caesarean section. The alternative of randomizing women to epidural analgesia or less effective pain relief is not practical because of difficulties in recruitment and in blinding obstetricians to the intervention. In their proposal, pregnant women in a geographic area served by only one hospital would receive treatment A (no epidural analgesia for labor) or B (epidural analgesia) before and after the increased availability of B in the hospital. The design would be replicated for multiple hospitals. Baker and Lindeman [3] provide a formula for computing the number of pairs required to achieve a given power.

4.5 Estimating Efficacy

With all-or-none compliance, one can estimate efficacy, the effect of receipt of treatment on endpoint, as opposed to effectiveness, the effect of group assignment on endpoint [6,15,16,21,23].

Let Y_g denote the observed effect in group g, where $g = C$ for control and $g = I$ for intervention. Alto let f_g denote the observed fraction who received A in group g. The simplest test statistic for estimating efficacy is $(Y_I - Y_C)/(f_I - f_C)$ [2–4,10,11,16–20]. This statistic estimates the benefit of receipt of B instead of A among those who would receive A if in the control group and B if in the intervention group. An underlying assumption is that no subject would receive B if in the control group or A if in the intervention group. Angrist et al. [2] showed how the statistic arises when using an approach based on instrumental variable or causal modeling. For a binary endpoint, Baker and Lindeman [3] showed that this statistic is a maximum-likelihood estimate. For the paired availability design they used this statistic as the basis of a permutation test.

When the endpoint is binary, an alternative test statistic is the estimated relative risk due to receipt of treatment [6,23]. Sommer and Zeger [23] derived the variance of this statistic.

References

1. Anbar, D. (1983). The relative efficiency of Zelen's prerandomization design for clinical trials. Biometrics, 39, 711–718.

2. Angrist, J. D., Imbens, G. W., and Rubin, D. R. (1996). Identification of causal effects using instrumental variables. J. Am. Statist. Assoc., 91, 444–472. (With discussion.)

3. Baker, S. G. and Lindeman, K. S. (1994). The paired availability design: A proposal for evaluating epidural analgesia during labor. Statist. Med., 13, 2269–2278.

4. Bloom, H. S. (1984). Accounting for no-shows in experimental evaluation designs. Evaluation Rev., 8, 225–246.

5. Brunner, D. and Neumann, M. (1995). On the mathematical basis of Zelen's prerandomization design. Methods Infer. Med., 24, 120–130.

6. Connor, R. Z., Prorok, P. C., and Weed, D. L. (1991). The case-control design and the assessment of the efficacy of cancer screening. J. Clin. Epidemiol., 44, 1215–1221.

7. Ellenberg, S. S. (1984). Randomization designs in comparative clinical trials. New Eng. J. Med., 310, 1404–1408.

8. Hearst, N., Newman, T. B., and Hulley, S. B. (1986). Delayed effects of the military draft on mortality. New Eng. J. Med., 314, 620–624.

9. Holland, P. (1988). Causal inference, path analysis, and recursive structural equations models. In Sociological Methodology, pp. 449–484. American Sociological Association, Washington, DC.

10. Ihm, P. (1991). Ein lineares modell für die Randomisierungspläne von Zelen (in German). In Medizinische Informatik und Statistik, Band 33, Therapiestudien, N. Victor, J. Dudeck, and E. P. Broszio, eds., pp. 176–184. Springer, Berlin.

11. Imbens, G. and Angrist, J. (1994). Identification and estimation of local average treatment effects. Econometrica, 62, 467–476.

12. Matts, J. and McHugh, R. (1987). Randomization and efficiency in Zelen's single-consent design. Biometrics, 43, 885–894.

13. McDonald, C. J., Hui, S. L., and Tierney, W. M. (1992). Effects of computer reminders for influenza vaccination on morbidity during influenza epidemics. Clin. Comput., 5, 304–312.

14. McHugh, R. (1984). Validity and treatment dilution in Zelen's single consent design. Statist. Med., 3, 215–218.

15. Meier, P. (1991). Comment on Compliance as an explanatory variable in clinical trials, by B. Efron and D. Feldman. J. Am. Statist. Assoc., 86, 19–22.

16. Newcombe, R. G. (1988). Explanatory and pragmatic estimates of the treatment effect when deviations from allocated treatment occur. Statist. Med., 7, 1179–1186.

17. Pearl, J. (1995). Causal inference from indirect experiments. Artif. Intell. Med. J., 7, 561–582.

18. Permutt, T. and Hebel J. R. (1989). Simultaneous-equation estimation in a clinical trial of the effect of smoking on birth weight. Biometrics, 45, 619–622.

19. Powers, D. E. and Swinton, S. S. (1984). Effects of self-study for coachable test item types, J. Educ. Psychol., 76, 266–278.

20. Robins, J. M. (1989). The analysis of randomized and nonrandomized AIDS treatment trials using a new approach to causal inference in longitudinal studies. In Health Service Research Methodology: A Focus on AIDS, L. Sechrest, H. Freeman, and A. Bailey, eds. NCHSR, US Public Health Service.

21. Schwartz, D. and Lellouch, Z. (1967). Explanatory and pragmatic attitudes in therapeutic trials. *J. Chron. Dis.*, **20**, 637–648.

22. Shapiro, S., Venet, W., Strax, P., and Venet, L. (1988). *Periodic Screening for Breast Cancer: The Health Insurance Plan Project and Its Sequelae, 1963–1986*. Johns Hopkins University Press, Baltimore.

23. Sommer, A. and Zeger, S. L. (1991). On estimating efficacy from clinical trials. *Statist. Med.*, **10**, 45–52.

24. Zelen, M. (1977). Statistical options in clinical trials. *Semin. Oncol.*, **4**, 441–446.

25. Zelen, M. (1979). A new design for randomized clinical trials. *New Eng. J. Med.*, **300**, 1242–1245.

26. Zelen, M. (1981). Alternatives to classic randomized trials. *Surg. Clin. N. Am.*, **61**, 1425–1432.

27. Zelen, M. (1982). Strategy and alternate designs in cancer clinical trials. *Cancer Treat. Rep.*, **66**, 1095–1100.

28. Zelen, M. (1990). Randomized consent designs for clinical trials: an update. *Statist. Med.*, **9**, 645–656.

5 Ascertainment Sampling

W. J. E. Wens

Ascertainment sampling is used frequently by scientists interested in establishing the genetic basis of some disease. Because most genetically based diseases are rare, simple random sampling will usually not provide sample sizes of affected individuals sufficiently large for a productive statistical analysis. Under ascertainment sampling, the entire family (or some other well-defined set of relatives) of a proband (i.e., an individual reporting with the disease) is sampled; we say that the family has been *ascertained through a proband*. If the disease does have a genetic basis, other family members have a much higher probability of being affected than individuals taken at random, so that by using data from such families we may obtain samples of a size sufficient to draw useful inferences. Additional benefits of sampling from families are that it yields linkage information not available from unrelated individuals and that an allocation of genetic and environmental effects can be attempted. Ascertainment sampling is used frequently by scientists interested in establishing the genetic basis of some disease. Because most genetically based diseases are rare, simple random sampling will usually not provide sample sizes of affected individuals sufficiently large for a productive statistical analysis. Under ascertainment sampling, the entire family (or some other well-defined set of relatives) of a proband (i.e., an individual reporting with the disease) is sampled; we say that the family has been *ascertained through a proband*. If the disease does have a genetic basis, other family members have a much higher probability of being affected than individuals taken at random, so that by using data from such families we may obtain samples of a size sufficient to draw useful inferences. Additional benefits of sampling from families are that it yields linkage information not available from unrelated individuals and that an allocation of genetic and environmental effects can be attempted.

The fact that only families with at least one affected individual can enter the sample implies that a conditional probability must be used in the data analysis, a fact recognized as early as 1912 by Weinberg [5]. However, the conditioning event is not that there is at least one affected individual in the family, but that the family is ascertained. These two are usually quite different, essentially because of the nature of the ascertainment process.

If the data in any family are denoted by D and the event that the family is ascertained by A, the ascertainment sampling likelihood contribution from this family is the conditional likelihood $P(DA)/P(A)$, where both numerator and denominator probabilities depend on the number of children in the family. The major problem in practice with ascertainment sampling is that the precise nature of the sampling procedure must be known in order to calculate both numerator and denominator probabilities. In practice this procedure is seldom well known, and this leads to potential biases in the estimation of genetic parameters, since while the numerator in the above probability can be written as the product $P(D)P(A|D)$ of genetic and ascertainment parameters, the denominator cannot, thus confounding estimation of the genetic parameters with the properties of the ascertainment process.

This is illustrated by considering two commonly discussed ascertainment sampling procedures. The first is that of *complete ascertainment*, arising for example from the use of a registry of families, and sampling only from those families in the registry with at least one affected child. Here the probability of ascertainment is independent of the number of affected children. The second procedure is that of *single ascertainment*; here the probability that a family is sampled is proportional to the number of affected children. There are many practical situations where this second form of sampling arises—for example, if we ascertain families by sampling all eighth-grade students in a certain city, a family with three affected children is essentially three times as likely to be ascertained as a family with only one affected child.

These two procedures require different ascertainment corrections. For example, if the children in a family are independently affected with a certain disease, each child having probability p of being affected, the probability of ascertainment of a family with s children under complete ascertainment is $1-(1-p)^s$ and is proportional to p under single ascertainment. The difference between the two likelihoods caused by these different denominators can lead to significant bias in estimation of p if one form of ascertainment is assumed when the other is appropriate.

In practice the description of the ascertainment process is usually far more difficult than this, since the actual form of sampling used is seldom clear-cut (and may well be neither complete nor single ascertainment). For example, age effects are often important (an older child is more likely to exhibit a disease than a younger child, and thus more likely to lead to ascertainment of the family), different population groups may have different social customs with respect to disease reporting, the relative role of parents and children in disease reporting is often not clear-cut (and depends on the age of onset of the disease), and the most frequent method of obtaining data (from families using a clinic in which a physician happens to be collecting disease data) may not be described well by any obvious sampling procedure.

Fisher [3] attempted to overcome these problems by introducing a model in which complete and single ascertainment are special cases. In his model it is assumed that any affected child is a proband with probability π and that children act independently with respect to reporting behavior. Here π is taken as an unknown parameter; the probability that a family with i affected children is ascertained is $1-(1-\pi)^i$, and the two respective limits $\pi = 1$ and $\pi \to 0$ correspond to complete and single ascertainment, respectively. However, the assumptions made in this model are often unrealistic; children in the same family will seldom act independently in reporting a disease, and the value of π will vary from family to family and will usually depend on the birth order of the child. Further, under the model, estimation of genetic parameters cannot be separated from estimation of π, so that any error in the ascertainment model will imply biases in the estimation of genetic parameters.

Given these difficulties, estimation of genetic parameters using data from an ascertainment sampling procedure can become a significant practical problem. An *ascertainment-assumption-free* approach which largely minimizes these difficulties is the following. No specific assumption is made about the probability $\alpha(i)$ of ascertaining a family having i affected children. [For complete ascertainment, $\alpha(i)$ is assumed to be independent of i and, for single ascertainment, to be proportional to i—but we now regard $\alpha(i)$ as an unknown parameter.] The denominator in the likelihood contribution from any ascertained family is thus $\sum_i p_i \alpha(i)$, where p_i, the probability that a family has i affected children, is a function only of genetic parameters. The numerator in the likelihood contribution is $P(D) = \alpha(i)$. The likelihood is now maximized jointly with respect to the genetic parameters and the $\alpha(i)$.

When this is done, it is found that estimation of the genetic parameters separates out from estimation of the $\alpha(i)$, and that the former can be estimated directly by using as the likelihood contribution $P(D)/P(i)$ from a family having i affected children. Estimation of ascertainment parameters is not necessary, and the procedure focuses entirely on genetic parameters, being unaffected by the nature of the ascertainment process.

More generally, the data D in any family can be written in the form $D = \{D_1, D_2\}$, where it is assumed that only D_1 affects the probability of ascertainment. Then the likelihood contribution used for such a family is $P(D_1, D_2)/P(D_1)$. This procedure [1] gives asymptotically unbiased parameter estimators no matter what the ascertainment process—all that is required is that the data D_1 that are "relevant to ascertainment" can be correctly defined.

These estimators have higher standard error than those arising if the true ascertainment procedure were known and used in the likelihood leading to the estimate, since when the true ascertainment procedure is known, this procedure conditions on more data than necessary, leaving less data available for estimation. The increase in standard error can be quantified using information concepts. In practice, the geneticist must choose between a procedure giving potentially biased estimators by using an incorrect ascertainment assumption and the increase in standard error in using the ascertainment-assumption-free method.

Conditioning not only on D_1 but on further parts of the data does not lead to bias in the estimation procedure, but will lead to increased standard errors of the estimate by an amount that can be again quantified by information concepts. Further conditioning of this type sometimes arises with continuous data. For such data the parallel with the dichotomy "affected/not affected" might be "blood pressure not exceeding T/blood pressure exceeding T," for some well-defined threshold T, so that only individuals having blood pressure exceeding T can be probands. To simplify the discussion, suppose that only the oldest child in any family can be a proband. Should the likelihood be conditioned by the probability element $f(x)$ of the observed blood pressure x of this child, or should it be conditioned by the probability $P(X \geq T)$ that his blood pressure exceeds T? The correct ascertainment correction is always the probability that the family is ascertained, so the latter probability is correct. In using the former, one conditions not only on the event that the family is ascertained, but on the further event that the blood pressure is x. Thus no bias arises in this case from using the probability element $f(x)$, but conditioning on further information (the actual value x) will increase the standard error of parameter estimates.

The above example is unrealistically simple. In more realistic cases conditioning on the observed value (or values) often introduces a bias, since when any affected child can be a proband, $f(x)/P(X \geq T)$ is a density function only under single ascertainment. Thus both to eliminate bias and to decrease standard errors, the appropriate conditioning event is that the family is ascertained.

A further range of problems frequently arises when large pedigrees, rather than families, are ascertained. For example if, as above, sampling is through all eighth-grade students in a certain city, and if any such student in the pedigree (affected or otherwise) is not observed, usually by being in a part of the pedigree remote from the proband(s), then bias in parameter estimation will, in general, occur. In theory this problem can be overcome by an exhaustive sampling procedure, but in practice this is seldom possible. This matter is discussed in detail in Reference 4. A general description of ascertainment sampling procedures is given in Reference 2.

References

1. Ewens, W. J. and Shute, N. C. E. (1986). A resolution of the ascertainment sampling problem I: Theory. *Theor. Popul. Biol.*, **30**, 388–412.
2. Ewens, W. J. (1991). Ascertainment biases and their resolution in biological surveys. In *Handbook of Statistics*, Vol. 8, C. R. Rao and R. Chakraborty, eds., pp. 29–61. North-Holland, New York.
3. Fisher, R. A. (1934). The effects of methods of ascertainment upon the estimation of frequencies. *Ann. Eugen.*, **6**, 13–25.
4. Vieland, V. J. and Hodge, S. E. (1994). Inherent intractability of the ascertainment problem for pedigree data: A general likelihood framework. *Am. J. Hum. Genet.*, **56**, 33–43.
5. Weinberg, W. (1912). Further contributions to the theory of heredity. Part 4: On methods and sources of error in studies on Mendelian ratios in man. (In German.) *Arch. Rassen- u. Gesellschaftsbiol.*, **9**, 165–174.

6. Assessment Bias

Peter C. Gøtzsche

6.1 Lack of Blinding

One of the most important and most obvious causes of assessment bias is lack of blinding. In empirical studies, lack of blinding has been shown to exaggerate the estimated effect by 14%, on average, measured as odds ratio [10]. These studies have dealt with a variety of outcomes, some of which are objective and would not be expected to be influenced by lack of blinding, for example, total mortality.

When patient-reported outcomes are assessed, lack of blinding can lead to far greater bias than the empirical average. An example of a highly subjective outcome is the duration of an episode of the common cold. A cold doesn't stop suddenly, and awareness of the treatment received could therefore bias the evaluation. In a placebo-controlled trial of vitamin C, the duration seemed to be shorter when active drug was given, but many participants had guessed that they received the vitamin because of its taste [12]. When the analysis was restricted to those who could not guess what they had received, the duration was not shorter in the active group.

Assessments by physicians are also vulnerable to bias. In a trial in multiple sclerosis, neurologists found an effect of the treatment when they assessed the effect openly but not when they assessed the effect blindly in the same patients [14].

Some outcomes can only be meaningfully evaluated by the patients, for example, pain and well-being. Unfortunately, blinding patients effectively can be very difficult, which is why active placebos are sometimes used. The idea behind an active placebo is that patients should experience side effects of a similar nature as when they receive the active drug, while it contains so little of a drug that it can hardly cause any therapeutic effect.

Since lack of blinding can lead to substantial bias, it is important in blinded trials to test whether the blinding has been compromised. Unfortunately, this is rarely done (Asbjørn Hróbjartsson, unpublished observations), and in many cases, double-blinding is little more than window dressing.

Some outcome assessments are not made until the analysis stage of the trial (see below). Blinding should, therefore, be used also during data analysis, and it should ideally be preserved until two versions of the manuscript—written under different assumptions, which of the treatments is experimental and which is control—have been approved by all the authors [8].

6.2 Harmless or False-Positive Cases of Disease

Assessment bias can occur if increased diagnostic activity leads to increased diagnosis of true but harmless cases of disease. Many stomach ulcers are silent, that is, they come and go and give no symptoms. Such cases could be detected more frequently in patients who receive a drug that causes unspecific discomfort in the stomach.

Similarly, if a drug causes diarrhea, this could lead to more digital, rectal examinations, and, therefore, also to the detection of more cases of prostatic cancer, most of which would be harmless, since many people die *with* prostatic cancer but rather few die *from* prostatic cancer.

Assessment bias can also be caused by differential detection of false-positive cases of disease. There is often considerable observer variation with common diagnostic tests. For gastroscopy, for example, a kappa value of 0.54 has been reported for the interobserver variation in the diagnosis of duodenal ulcers [5]. This usually means that there are rather high rates of both false-positive and false-negative findings. If treatment with a drug leads to more gastroscopies because ulcers are suspected, one would therefore expect to find more (false) ulcers in patients receiving that drug. A drug that causes unspecific, nonulcer discomfort in the stomach could, therefore, falsely be described as an ulcer-inducing drug.

The risk of bias can be reduced by limiting the analysis to serious cases that would almost always become known, for example, cases of severely bleeding ulcers requiring hospital admission or leading to death.

6.3 Disease-Specific Mortality

Disease-specific mortality is very often used as the main outcome in trials without any discussion of how reliable it is, even in trials of severely ill patients where it can be difficult to ascribe particular causes for the deaths with acceptable error.

Disease-specific mortality can be highly misleading if a treatment has adverse effects that increases mortality from other causes. It is only to be expected that aggressive treatments can have such effects. Complications to cancer treatment, for example, cause mortality that is often ascribed to other causes, although these deaths should have been added to the cancer deaths. A study found that deaths from other causes than cancer were 37% higher than expected and that most of this excess occurred shortly after diagnosis, suggesting that many of the deaths were attritutable to treatment [1].

The use of blinded endpoint committees can reduce the magnitude of misclassification bias, but cannot be expected to remove it. Radiotherapy for breast cancer, for example, continues to cause cardiovascular deaths even 20 years after treatment [2], and it is not possible to distinguish these deaths from cardiovascular deaths from other causes. Furthermore, to work in an unbiased way, death certificates and other important documents must have been completed and patients and documents selected for review without awareness of status, and it should not be possible to break the masking during any of these processes, including review of causes of death. This seems difficult to obtain, in particular, since those who prepare excerpts of the data should be kept blind to the research hypothesis [3].

Fungal infections in cancer patients with neutropenia after chemotherapy or bone marrow transplantation is another example of bias in severely ill patients. Not only is it difficult to establish with certainty that a patient has a fungal infection and what was the cause of death; there is also evidence that some of the drugs (azole antifungal agents) may increase the incidence of bacteriaemias [9]. In the largest placebo-controlled trial of fluconazole, more deaths were reported on drug than on placebo (55 vs. 46 deaths), but the authors also reported that fewer deaths were ascribed to acute systemic fungal infections (1 vs. 10 patients, $P = 0.01$) [6]. However, if this subgroup result is to be believed, it would mean that fluconazole increased mortality from other causes (54 vs 36 patients, $P = 0.04$).

Bias related to classification of deaths can also occur within the same disease. After publication of positive results from a trial in patients with myocardial infarction [16], researchers at the US Food and Drug Administration found that the cause-of-death classification was "hopelessly unreliable" [15]. Cardiac deaths were classified into three groups: sudden deaths, myocardial infarction, or other cardiac event. The errors in assigning cause of death, nearly all, favored the conclusion that sulfinpyrazone decreased sudden death, the major finding of the trial.

6.4 Composite Outcomes

Composite outcomes are vulnerable to bias when they contain a mix of objective and subjective components. A survey of trials with composite outcomes found that when they included clinician-driven outcomes, such as hospitalization and initiation of new antibiotics, in addition to objective outcomes such as death, it was twice as likely that the trial reported a statistically significant effect [4].

6.5 Competing Risks

Composite outcomes can also lead to bias because of competing risks [13], for example, if an outcome includes death as well as hospital admission. A patient who dies cannot later be admitted to hospital. This bias can also occur in trials with simple outcomes. If one of the outcomes is length of hospital stay, a treatment that increases mortality among the weakest patients who would have had long hospital stays may spuriously appear to be beneficial.

6.6 Timing of Outcomes

Timing of outcomes can have profound effects on the estimated result, and the selection of timepoints for reporting of the results is seldom made until the analysis stage of the trials, when possible treatment codes have been broken. A trial report of the antiarthritic drug, celecoxib, gave the impression that it was better tolerated than its comparators, but the published data referred to 6 months of follow-up, and not to 12 and 15 months, as planned, when there was little difference; in addition, the definition of the outcome had changed, compared to what was stated in the trial protocol [11].

Trials conducted in intensive care units are vulnerable to this type of bias. For example, the main outcome in such trials can be total mortality during the stay in the unit, but if the surviving patients die later, during their subsequent stay at the referring department, little may be gained by a proven mortality reduction while the patients were sedated. A more relevant outcome would be the fraction of patients who leave the hospital alive.

6.7 Assessment of Harms

Bias in assessment of harms is common. Even when elaborate, pretested forms have been used for registration of harms during a trial, and guidelines for their reporting have been given in the protocol, the conversion of these data into publishable bits of information can be difficult and often involves subjective judgments.

Particularly vulnerable to assessment bias is exclusion of reported effects because they are not felt to be important, or not felt to be related to the treatment. Trials that have been published more than once illustrate how subjective and biased assessment of harms can be. Both number of adverse effects and number of patients affected can vary from report to report, although no additional inclusion of patients or follow-up have occurred, and these reinterpretations or reclassifications sometimes change a nonsignificant difference into a significant difference in favor of the new treatment [7].

References

1. Brown, B. W., Brauner, C., and Minnotte, M. C. (1993). Noncancer deaths in white adult cancer patients. *J. Natl. Cancer Inst.*, **85**, 979–987.

2. Early Breast Cancer Trialists' Collaborative Group (2000). Favourable and unfavourable effects on long-term survival of radiotherapy for early breast cancer: an overview of the randomised trials. *Lancet*, **355**, 1757–1770.

3. Feinstein, A. R. (1985). *Clinical Epidemiology*. Saunders, Philadelphia, PA.

4. Freemantle, N., Calvert, M., Wood, J., Eastaugh, J., and Griffin, C. (2003). Composite outcomes in randomized trials: Greater precision but with greater uncertainty? *JAMA (J. Am. Med. Assoc.)*, **289**, 2554–2559.

5. Gjørup, T., Agner, E., Jensen, L. B, Jensen, A. M., and Møllmann, K. M. (1986). The endoscopic diagnosis of duodenal ulcer disease. A randomized clinical trial of bias and interobserver variation. *Scand. J. Gastroenterol.*, **21**, 561–567.

6. Goodman, J. L., Winston, D. J., Greenfield, R. A., Chandrasekar, P. H., Fox, B., Kaizer, H., et al. (1992). A controlled trial of fluconazole to prevent fungal infections in patients undergoing bone marrow transplantation. *New Engl. J. Med.*, **326**, 845–851.

7. Gøtzsche, P. C. (1989). Multiple publication in reports of drug trials. *Eur. J. Clin. Pharmacol.*, **36**, 429–432.

8. Gøtzsche, P. C. (1996). Blinding during data analysis and writing of manuscripts. *Control. Clin. Trials*, **17**, 285–290.

9. Gøtzsche, P. C. and Johansen, H. K. (2003). Routine versus selective antifungal administration for control of fungal infections in patients with cancer (Cochrane Review). *The Cochrane Library*, Issue 3. Update Software, Oxford.

10. Jüni, P., Altman, D. G., and Egger, M. (2001). Systematic reviews in health care: Assessing the quality of controlled clinical trials. *Br. Med. J.*, **323**, 42–46.

11. Jüni, P., Rutjes, A. W., and Dieppe, P. A. (2002). Are selective COX 2 inhibitors superior to traditional non steroidal anti-inflammatory drugs? *Br. Med. J.*, **324**, 1287–1288.

12. Karlowski, T. R., Chalmers, T. C., Frenkel, L. D., Kapikian, A. Z., Lewis, T. L., and Lynch, J. M. (1975). Ascorbic acid for the common cold: A prophylactic and therapeutic trial. *JAMA*, **231**, 1038–1042.

13. Lauer, M. S. and Topol, E. J. (2003). Clinical trials—multiple treatments, multiple end points, and multiple lessons. *JAMA*, **289**, 2575–2577.

14. Noseworthy, J. H., Ebers, G. C., Vandervoort, M. K., Farquhar, R. E., Yetisir, E., and Roberts, R. (1994). The impact of blinding on the results of a randomized, placebo-controlled multiple sclerosis clinical trial. *Neurology*, **44**, 16–20.

15. Temple, R. and Pledger, G. W. (1980). The FDA's critique of the anturane reinfarction trial. *New Engl. J. Med.*, **303**, 1488–1492.

16. The Anturane Reinfarction Trial Research Group. (1980). Sulfinpyrazone in the prevention of sudden death after myocardial infarction. *New Engl. J. Med.*, **302**, 250–256.

7 Bioavailability and Bioequivalence

Shein-Chung Chow

7.1 Definition

The *bioavailability* of a drug product is the rate and extent to which the active drug ingredient or therapeutic moiety is absorbed and becomes available at the site of drug action. A *comparative bioavailability study* refers to the comparison of bioavailabilities of different formulations of the same drug or different drug products. When two formulations of the same drug or different drug products are claimed to be *bioequivalent*, this means that they will provide the same therapeutic effect and that they are therapeutically equivalent. This assumption is usually referred to as the *fundamental bioequivalence assumption*. Under this assumption, the United States Food and Drug Administration (FDA) does not require a complete new-drug application (NDA) submission for approval of a generic drug product if the sponsor can provide evidence of bioequivalence between the generic drug product and the innovator through bioavailability studies.

7.2 History of Bioavailability

The study of drug absorption (e.g., sodium iodide) can be traced back to 1912 [2]. However, the concept of bioavailability and bioequivalence did not become a public issue until the late 1960s, when concern was raised that a generic drug product might not be as bioavailable as that manufactured by the innovator. These concerns rose from clinical observations in humans. The investigation was facilitated by techniques permitting the measurement of minute quantities of drugs and their metabolites in biologic fluids. In 1970, the FDA began to ask for evidence of biological availability in applications submitted for approval of certain new drugs. In 1974, a Drug Bioequivalence Study Panel was formed by the Office of Technology Assessment (OTA) to examine the relationship between the chemical and therapeutic equivalence of drug products. Based on the recommendations in the OTA report, the FDA published a set of regulations for submission of bioavailability data in certain new drug applications. These regulations became effective on July 1, 1977 and are currently codified in 21 Code of Federal Regulation (CFR) Part 320. In 1984, the FDA was authorized to approve generic drug products under the Drug Price Competition and Patent Term Restoration Act. In recent years, as more generic drug products become available, there is a concern that generic drug products may not be comparable in identity, strength, quality, or purity to the innovator drug product. To address this concern, the FDA conducted a hearing on Bioequivalence of Solid Oral Dosage Forms in Washington, D.C. in 1986. As a consequence of the hearing, a Bioequivalence Task Force (BTF) was formed to evaluate the current procedures adopted by the FDA for the assessment of bioequivalence between immediate solid oral dosage forms. A report from the BTF was released in January 1988. The FDA Division of Bioequivalence, Office of Generic Drugs, issued guidelines on statistical procedures for bioequivalence studies in 1992 [6], based on the BTF report recommendations.

7.3 Bioequivalence Measures

In bioavailability/bioequivalence studies, following the administration of a drug, the blood or plasma concentration–time curve is often used to study the absorption of the drug. The curve can be characterized by taking blood samples immediately prior to and at various time points after drug administration. The profile of the curve is then studied by means of several pharmacokinetic parameters such as the area under it (AUC), the maximum concentration ($C_{\max}$), the time to achieve maximum concentration ($t_{\max}$), the elimination half-life ($t_{1/2}$), and the rate constant (k_e). Further discussion of the blood or plasma concentration–time curve can be found in Gibaldi and Perrier [9]. The AUC, which is one of the primary pharmacokinetic parameters, is often used to measure the extent of absorption or total amount of drug absorbed in the body. Based on these pharmacokinetic parameters, bioequivalence may be assessed by means of the difference in means (or medians), the ratio of means (or medians), and the mean (or median) of individual subject ratios of the primary pharmacokinetic variables. The commonly used bioequivalence measures are the difference in means and the ratio of means.

7.4 Decision Rules

Between 1977 and 1980, the FDA proposed a number of decision rules for assessing bioequivalence in average bioavailability [14]. These decision rules include the 75/75 rule, the 80/20 rule, and the ±20 rule. The 75/75 rule claims bioequivalence if at least 75% of individual subject ratios [i.e., individual bioavailability of the generic (test) product relative to the innovator (reference) product] are within (75%, 125%) limits. The 80/20 rule concludes that there is bioequivalence if the test average is not statistically significantly different from the reference average and if there is at least 80% power for detection of a 20% difference of the reference average. The BTF does not recommend these two decision rules, because of their undesirable statistical properties. The ±20 rule suggests that two drug products are bioequivalent if the average bioavailability of the test product is within ±20% of that of the reference product with a certain assurance (say 90%). Most recently, the FDA guidance recommends a 80/125 rule for log-transformed data. The 80/125 rule claims bioequivalence if the ratio of the averages between the test product and the reference product falls within (80%, 125%) with 90% assurance. The ±20 rule for raw data and the 80/125 rule for log-transformed data are currently acceptable to the FDA for assessment of bioequivalence in average bioavailability. Based on the current practice of bioequivalence assessment, it is suggested that the 80/125 rule be applied to AUC and $C_{\max}$ for all drug products across all therapeutic areas.

7.5 Designs of Bioavailability Studies

The *Federal Register* [7] indicated that a bioavailability study (single-dose or multiple-dose) should be crossover in design. A crossover design is a modified randomized block design in which each block (i.e., subject) receives more than one formulation of a drug at different time periods. The most commonly used study design for assessment of bioequivalence is the two-sequence, two-period crossover design (Table 1), which is also known as the *standard crossover design*. For the standard crossover design, each subject is randomly assigned to either sequence 1 [reference–test (R-T)] or sequence 2 (T-R). In other words, subjects within sequence R-T (T-R) receive formulation R (T) during the first dosing period and formulation T (R) during the second dosing period. Usually the dosing periods are separated by a washout period of sufficient length for the drug received in the first period to be completely metabolized and/or excreted by the body.

In practice, when differential carryover effects are present, the standard crossover design may not be useful because the formulation effect is confounded with the carryover effect. In addition, the standard crossover design does not provide independent estimates of intrasubject variability for each formulation, because each subject receives each formulation only once. To overcome these drawbacks, Chow and Liu [3] recommend a higher-order crossover design be used. Table 1 lists some commonly used higher-order crossover designs, which include Balaam's design, the two-sequence dual design, and the optimal four-sequence design. As an alternative, Westlake [16] suggested the use of an incomplete block design. Note that crossover designs for comparing more than two formulations are much more complicated than those for comparing two formulations. In the interest of the balance property, Jones and Kenward [10] recommend a Williams design be used.

7.6 Statistical Methods for Average Bioavailability

Without loss of generality, consider the standard two-sequence, two-period crossover experiment. Let Y_{ijk} be the response [e.g., log(AUC)] of the ith subject in the kth sequence at the jth period, where $i = 1, \ldots, n_k$, $k = 1, 2$, and $j = 1, 2$. Under the assumption that there are no carryover effects, Y_{ijk} can be described by the statistical model

$$Y_{ijk} = \mu + S_{ik} + F_{(j,k)} + P_j + e_{ijk},$$

where μ is the overall mean, S_{ik} is the random effect of the ith subject in the kth sequence, P_j is the fixed effect of the jth period, $F_{(j,k)}$ is the direct fixed effect of the formulation in the kth sequence that is administered at the jth period, and e_{ijk} is the within-subject random error in observing Y_{ijk}.

The commonly used approach for assessing bioequivalence in average bioavailability is the method of the classical (shortest) confidence interval. Let μ_T and μ_R be the means of the test and reference formulations, respectively. Then, under normality assumptions, the classical $(1 - 2\alpha) \times 100\%$ confidence interval for $\mu_T - \mu_R$ is as follows:

$$\begin{aligned}
L &= (\overline{Y}_T - \overline{Y}_R) - t(\alpha, n_1 + n_2 - 2) \\
&\quad \times \hat{\sigma}_d \sqrt{\frac{1}{n_1} + \frac{1}{n_2}}, \\
U &= (\overline{Y}_T - \overline{Y}_R) + t(\alpha, n_1 + n_2 - 2) \\
&\quad \times \hat{\sigma}_d \sqrt{\frac{1}{n_1} + \frac{1}{n_2}},
\end{aligned}$$

Table 1: Crossover designs for two formulations.

Sequence	I	II	III	IV
Standard Crossover Design Period				
1	R	T	–	–
2	T	R	–	–
Balaam Design Period				
1	T	T	–	–
2	R	R	–	–
3	R	T	–	–
4	T	R	–	–
Two-Sequence Dual-Design Period				
1	T	R	R	–
2	R	T	T	–
Four-Sequence Optimal Design Period				
1	T	T	R	R
2	R	R	T	T
3	T	R	R	T
4	R	T	T	R

where $\overline{Y}_T$ and $\overline{Y}_R$ are the least-squares means for the test and reference formulations, $t(\alpha, n_1+n_2-2)$ is the upper αth critical value of a t distribution with n_1+n_2-2 degrees of freedom, and $\hat{\sigma}_d^2$ is given by

$$\hat{\sigma}_d^2 = \frac{1}{n_1+n_2-2}\sum_{k=1}^{2}\sum_{i=1}^{n_k}(d_{ik}-\overline{d}_{\cdot k})^2,$$

in which

$$d_{ik} = \frac{1}{2}(Y_{i2k}-Y_{i1k})$$

$$\overline{d}_{\cdot k} = \frac{1}{n_k}\sum_{i=1}^{n_k}d_{ik}.$$

According to the 80/125 rule, if the exponentiations of L and U are within (80%, 125%), then we conclude the two formulations are bioequivalent.

As an alternative, Schuirmann [15] proposed a procedure with two one-sided tests to evaluate whether the bioavailability of the test formulation is too high (safety) for one side and is too low (efficacy) for the other side. In it, we conclude that the two formulations are bioequivalent if

$$T_L = \frac{(\overline{Y}_T-\overline{Y}_R)-\theta_L}{\hat{\sigma}_d\sqrt{\frac{1}{n_1}+\frac{1}{n_2}}} > t(\alpha, n_1+n_2-2)$$

and

$$T_U = \frac{(\overline{Y}_T-\overline{Y}_R)-\theta_U}{\hat{\sigma}_d\sqrt{\frac{1}{n_1}+\frac{1}{n_2}}}$$
$$< -t(\alpha, n_1+n_2-2),$$

where $\theta_L = \log(0.8) = -0.2331$ and $\theta_U = \log(1.25) = 0.2331$ are bioequivalence limits.

In addition to the ± 20 rule (and the more recent 80/125 rule), several other methods have been proposed. These include the Westlake symmetric confidence interval [17], Chow and Shao's joint confidence region approach [5], an exact confidence interval based on Fieller's theorem

[8,12], Anderson and Hauck's test for interval hypotheses [1], a Bayesian approach for the highest posterior density (HPD) interval, and nonparametric methods. Some of these methods are actually operationally equivalent in the sense that they will reach the same decision on bioequivalence. More details can be found in Chow and Liu [2].

7.7 Sample Size Determination

For bioequivalence trials, a traditional approach for sample size determination is to conduct a power analysis based on the 80/20 decision rule. This approach, however, is based on point hypotheses rather than interval hypotheses and therefore may not be statistically valid. Phillips [13] provided a table of sample sizes based on power calculations of Schuirmann's two one-sided tests procedure using the bivariate noncentral t distribution. However, no formulas are provided. An approximate formula for sample-size calculations was provided in Liu and Chow [11]. Table 2 gives the total sample sizes needed to achieve a desired power for a standard crossover design for various combinations of $\theta = \mu_T - \mu_R$ and CV, where

$$\text{CV} = 100 \times \frac{\sqrt{2\hat{\sigma}_d^2}}{\mu_R}.$$

7.8 Current Issues

As more generic drug products become available, current regulatory requirements and unresolved scientific issues that may affect the identity, strength, quality, and purity of generic drug products have attracted much attention in recent years. For example, for regulatory requirements, the FDA has adopted the analysis for log-transformed data and the 80/125 rule. To allow for variability of bioavailability and to ensure drug exchangeability, the FDA is seeking alternative pharmacokinetic parameters, decision rules, and statistical methods for population and individual bioequivalence. Some unresolved scientific issues of particular interest include the impact of add-on subjects for dropouts, the use of female subjects in bioequivalence trials, in vitro dissolution as a surrogate for in vivo bioequivalence, postapproval bioequivalence, and international harmonization for bioequivalence requirements among the European Community, Japan, and the United States. A comprehensive overview of these issues can be found in Chow and Liu [4].

Table 2: Sample sizes for Schuirmann's procedures with two one-sided tests.

Power %	CV (%)	$\theta = \mu_T - \mu_R$			
		0%	5%	10%	15%
80	10	8	8	16	52
	12	8	10	20	74
	14	10	14	26	100
	16	14	16	34	126
	18	16	20	42	162
	20	20	24	52	200
	22	24	28	62	242
	24	28	34	74	288
	26	32	40	86	336
	28	36	46	100	390
	30	40	52	114	448
	32	46	58	128	508
	34	52	66	146	574
	36	58	74	162	644
	38	64	82	180	716
	40	70	90	200	794
90	10	10	10	20	70
	12	10	14	28	100
	14	14	18	36	136
	16	16	22	46	178
	18	20	28	58	224
	20	24	32	70	276
	22	28	40	86	334
	24	34	46	100	396
	26	40	54	118	466
	28	44	62	136	540
	30	52	70	156	618
	32	58	80	178	704
	34	66	90	200	794
	36	72	100	224	890
	38	80	112	250	992
	40	90	124	276	1098

Source: Liu and Chow [11].

References

1. Anderson, S. and Hauck, W. W. (1983). A new procedure for testing equivalence in comparative bioavailability and other clinical trials. *Commun. Statist. Theory Methods*, **12**, 2663–2692.

2. Chow, S. C. and Liu, J. P. (1992). *Design and Analysis of Bioavailability and Bioequivalence Studies*. Marcel Dekker, New York.

3. Chow, S. C. and Liu, J. P. (1992). On assessment of bioequivalence under a higher-order crossover design. *J. Biopharm. Statist.*, **2**, 239–256.

4. Chow, S. C. and Liu, J. P. (1995). Current issues in bioequivalence trials. *Drug. Inform. J.*, **29**, 795–804.

5. Chow, S. C. and Shao, J. (1990). An alternative approach for the assessment of bioequivalence between two formulations of a drug. *Biometrics J.*, **32**, 969–976.

6. FDA (1992). *Guidance on Statistical Procedures for Bioequivalence Studies Using a Standard Two-Treatment Crossover Design*. Division of Bioequivalence, Office of Generic Drugs, Food and Drug Administration, Rockville, MD, July 1.

7. *Federal Register* (1977). Vol. 42, No. 5, Section 320.26(b).

8. Fieller, E. (1954). Some problems in interval estimation. *J. Roy. Statist. Soc. B*, **16**, 175–185.

9. Gibaldi, M. and Perrier, D. (1982). *Pharmacokinetics*. Marcel Dekker, New York.

10. Jones, B. and Kenward, M. G. (1989). *Design and Analysis of Crossover Trials*. Chapman & Hall, London.

11. Liu, J. P. and Chow, S. C. (1992). Sample size determination for the two one-sided tests procedure in bioequivalence. *J. Pharmacokin. Biopharm.*, **20**, 101–104.

12. Locke, C. S. (1984). An exact confidence interval for untransformed data for the ratio of two formulation means. *J. Pharmacokin. Biopharm.*, **12**, 649–655.

13. Phillips, K. F. (1990). Power of the two one-sided tests procedure in bioequivalence. *J. Pharmacokin. Biopharm.*, **18**, 137–144.

14. Purich, E. (1980). Bioavailability/bioequivalency regulations: an FDA perspective. In *Drug Absorption and Disposition: Statistical Considerations*, K. S. Albert, ed., pp. 115–137. American Pharmaceutical Association, Academy of Pharmaceutical Sciences, Washington, DC.

15. Schuirmann, D. J. (1987). A comparison of the two one-sided tests procedure and the power approach for assessing the equivalence of average bioavailability. *J. Pharmacokin. Biopharm.*, **15**, 657–680.

16. Westlake, W. J. (1973). The design and analysis of comparative blood-level trials. In *Current Concepts in the Pharmaceutical Sciences*, J. Swarbrick, ed. Lea & Febiger, Philadelphia.

17. Westlake, W. J. (1976). Symmetrical confidence intervals for bioequivalence trials. *Biometrics*, **32**, 741–744.

8 Cancer Stochastic Models

Wai-Y. Tan and Chao W. Chen

8.1 Introduction

It is universally recognized that carcinogenesis is a multistage random process involving genetic changes and stochastic proliferation and differentiation of normal stem cells and genetically altered stem cells [61]. Studies by molecular biologists have confirmed that each cancer tumor develops from a single stem cell that has sustained a series of irreversible genetic changes. Stem cells are produced in the bone marrow and mature in the thymus; the matured stem cells move to the specific organ through the bloodstream. Stem cells are subject to stochastic proliferation and differentiation with differentiated cells replacing old cells of the organ. In normal individuals, there is a balance between proliferation and differentiation in stem cells and there are devices such as the DNA repair system and apoptosis in the body to protect against possible errors in the metabolism process. Thus, in normal individuals, the proliferation rate of stem cells equals to the differentiation rate of stem cells. If some genetic changes have occurred in a stem cell to increase the proliferation rate of the cell; then the proliferation rate (or birth rate) is greater than the differentiation rate (or death rate) in this genetically altered cell so that this type of genetically altered cells will accumulate; however, with high probability these genetically altered cells will eventually stop proliferating or be eliminated because of the existing protection devices unless more genetic changes have occurred in these cells to overcome the existing protection devices. Furthermore, since genetic changes are rare events, further genetic changes will occur in at least one of the genetically altered cells only if the number of these cells is very large. This may help explain why carcinogenesis is a multistage random process and why Poisson processes and stochastic birth–death processes are important components of this random process. This biological input has led to a new postulation that carcinogenesis may be considered as a microevolution process and that each cancer tumor is the outcome of growth of a most fitted genetically altered stem cell [6,26].

From the genetic viewpoint, carcinogenesis involves actions of oncogenes, suppressor genes, the mismatch repair (MMR) genes, the repair system, and the control of cell cycle [11,24,40,53]. Oncogenes are highly preserved dominant genes that regulate development and cell division. When these genes are activated or mutated, normal control of cell growth is unleashed, leading to the cascade of carcinogenesis. On the other hand, suppressor genes are recessive genes whose inactivation or mutation leads to uncontrolled growth. To date, about 200 oncogenes and 50 suppressor genes have been identified. The specific actions of these genes and its relationship with control of cell cycle have been discussed in detail in the book by Hesketh [24]. Specifically, some oncogenes such as the ras gene induce the cells entering into cell cycle through signal reception, signal transduction, and propagation; some oncogenes such as myc, jun, and fos serve as transcription factors to affect DNA synthesis during the S stage of the cell cycle while some other oncogenes such as bcl-2 serve as antiapoptosis agents. On the other hand, many of the suppressor genes such as the retinoblastoma (RB) gene control the checkpoints of the cell cycle. When a cell enters the cell division cycle, the RB gene protein forms a complex with E2F and some poked proteins; when the RB gene protein is phosphorylated or inactivated or mutated, E2F is unleashed to push the cell cycle from the $G1$ phase to the S phase. When the DNA is damaged and/or the cell proliferation is beyond control, the body then invokes the repair system and the apoptosis mechanism, which is controlled by the suppressor gene p53 and many other genes, to correct such aberrations; the inactivation or mutation of p53 or other relevant genes leads to the abrogation of apoptosis. The mutation and deletion of the MMR genes lead to microsatellite repeats and create a mutator phenotype, predisposing the affected cells to genetic instability and to increase the mutation rates of many relevant genes leading to the cascade of carcinogenesis [19,53,79]. All these mechanisms are controlled by many oncogenes, suppressor genes, and modifying or enhancing genes. It is the interaction of these genes and their interaction with the environment that creates the cancer phenotype. Specifically, one may liken carcinogenesis to the integrated circuit of electronics, in which transistors are replaced by proteins (e.g., kinases and phosphatases) and the electrons by phosphates and lipids, among others [20].

8.2 A Brief History of Stochastic Modeling of Carcinogenesis

Stochastic models of carcinogenesis were first proposed in the 1950s by Nording [52] and by Armitage and Doll [1,3] to assess effects of risk variables on cancer incidence. This model has been referred to as the multistage model. This model together with some other old carcinogenesis models have been reviewed by Whittemore and Keller [80] and Kalbfleish et al. [28]. The Armitage–Doll model has been widely used by statisticians to assess how exposure to carcinogens alters the cancer incidence rates and the distributions of time to tumor [4,5,12]. However, many results from molecular biology and molecular genetics have raised questions about the validity of this model as it has ignored stochastic proliferation of all intermediate initiated cells [14,32,46–48,61].

Biologically supported models of carcinogenesis was first proposed by Knudson [32], Moolgavkar and Venzen [48], and Moolgavkar and Knudson [47]. In the literature, this model was referred to as the Moolgavkar–Venzen–Knudson (MVK) two-stage model. It was first extended into nonhomogeneous cases by Tan and Gastardo [74] and analyzed by Tan and Brown [64] and by Moolgavkar et al. [46]. Because genetic mutations and cell proliferations occur during cell division, this model was further modified by Chen and Farland [7] and extended by Tan and Chen [66] to nonhomogeneous cases.

The MVK two-stage models and extensions of it, together with many other biologically supported models have been analyzed and discussed in detail in Reference

61. By merging initiation and promotion, alternative modeling approaches have been proposed by Klebanov et al. [30] for radiation carcinogenesis. Because many cancers involve many oncogenes and suppressor genes, extensions of the MVK model to stochastic models involving more than two stages have been developed by Chu [9], Chu et al. [10], and by Little and his colleagues [36–39]. Chu [9] has called his model the multievent model while Little [36] has called his model the *generalized MVK model*. Because of the difficulties in analyzing k-stage multievent models when $k > 2$, Herrero-Jimenez et al. [23] and Luebeck and Moolgavkar [42] have proposed a clonal expansion model by ignoring the stochastic proliferation and differentiation of the first $k-2$ stage of initiated cells. This model combines the Armitage–Doll model with the MVK two-stage model by assuming clonal expansion for the $(k-1)$-stage initiated cells in the k-stage Armitage–Doll model. Another important point is that all these models have ignored cancer progression by assuming that the last-stage-initiated cells grow instantaneously into malignant tumors as soon as they are produced. To account for cancer progression, Yang and Chen [82] and Tan and Chen [68] have further extended the two-stage model and the multievent model by postulating that cancer tumors develop from primary last-stage-initiated cells by clonal expansion. Tan and Chen [68] have called their model the *extended multievent model*, indicating that these models are extensions of the multievent models.

The multievent models and the two-stage models assume that cancer tumors develop from a single pathway through a multistage stochastic process. However, many biological data suggest that the same cancer can be derived by several different pathways [13,17–19,25,31,35,43,44,55,57,58]. To account for these mechanisms, Tan and Brown [63], Tan [61], and Tan and Chen [65] have developed multiple pathways models of carcinogenesis. Further extensions and applications of multiple pathways models have been given by Sherman and Portier [57], Mao et al. [43], Tan and Chen [69], and Tan et al. [70–72].

Two further extensions of stochastic models of carcinogenesis have been made by Tan [60,61], Tan and Chen [67], and Tan and Singh [76]. One is the mixed models of carcinogenesis [60,61,76] and the other is the multivariate stochastic models [67]. The mixed model of carcinogenesis arises because in the population, different individuals may develop cancer through different pathways.

To ease the problem of nonidentifiability and to provide a paradigm to combine information from different sources, Tan and Chen [68] and Tan et al. [70,72] have proposed stage space models (Kalman filter models) for multievent models and for multiple pathways models of carcinogenesis. Tan et al. [73] have applied the state space models to animal data to estimate the mutation rate of normal stem cells and the proliferation rate and the differentiation rate of initiated cells. Tan et al. [77] have applied the state space models to analyze the British physician data of lung cancer with smoking given in Reference 15. The state space models of a system are stochastic models that consist of two submodels: The stochastic model of the system and the statistical model based on available data from the system. Thus, these models provide a convenient and efficient approach to combine information from three sources [62,76,77]: (1) the mechanism of the system through the stochastic model of the system, (2) the information from the system though the observed data from the system, and (3) the previous knowledge about the system through the prior distribution of the parameters of the system 62,76,77.

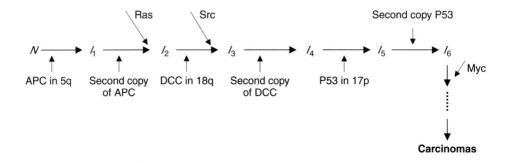

Figure 1: The APC–β-catenin–TCF–myc pathway of human colon cancer.

8.3 Single-Pathway Models of Carcinogenesis

The most general model for a single pathway is the extended k-stage ($k \geq 2$) multievent model proposed by Tan and Chen [68]. This is an extension of the multievent model first proposed by Chu [9] and studied by Tan [61] and Little [36,37]. It views carcinogenesis as the endpoint of k ($k \geq 2$) discrete, heritable, and irreversible events (mutations or genetic changes) with intermediate cells subjected to stochastic proliferation and differentiation.

Let N denote normal stem cells, T the cancer tumors, and I_j the jth stage initiated cells arising from the $(j-1)$th stage initiated cells ($j = 1, \ldots, k$) by mutation or some genetic changes. Then the model assumes $N \to I_1 \to I_2 \to \cdots \to I_k$ with the N cells and the I_j cells subject to stochastic proliferation (birth) and differentiation (death). It takes into account cancer progression by following Yang and Chen [82] to postulate that cancer tumors develop from primary I_k cells by clonal expansion (i.e., stochastic birth–death process), in which a primary I_k cell is an I_k cell that arises directly from an I_{k-1} cell.

As an example, consider the pathway involving the adenomatous polyposis cancer (APC) gene, β-catenin, the T-cell factor (Tcf) and the myc oncogene, referred to as the APC–β–catenin–Tcf–myc pathway for human colon cancer. This is a six-stage multievent model involving the suppressor genes in chromosomes 5q, 17p, and 18q [35,54,60]. A schematic presentation of this pathway is given in Figure 1. This is only one of the pathways for the colon cancer, although it is the major pathway that accounts for 80% of all colon cancers [35,54,59].

In Figure 1, the individual is in the first stage if one allele of the APC gene (a suppressor) in chromosome 5 has been mutated or deleted, the individual is in the second stage if the other allele of the APC gene has also been mutated or deleted; the individual who has sustained mutations or deletions of both copies of the APC gene is in the third stage if one copy of the deleted-in-colorectal cancer (DCC) gene in chromosome 18q has also been mutated or deleted, the individual who has sustained the two mutations or deletions of the APC gene and a mutation of the DCC gene is in the fourth malignant cancer stage if the second copy of the DCC gene in chromosome 18q has also been mutated or deleted, and the individual who has sustained mutations or deletions of both copies of the APC gene and mutations or deletions of both copies of the DCC gene is in the fifth stage if a gene in the p53 locus in chromosome 17p has mutated or deleted. In

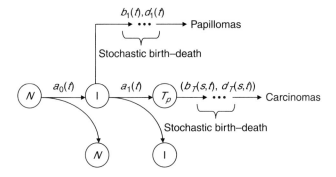

Figure 2: A two-stage (multi-event with $k = 2$) model for carcinogenesis in initiation-promotion experiments.

this model, the ras oncogene in chromosome 12q and the (MCC) gene in chromosome 5q are considered as promoter genes that promote cell proliferation of the initiated cells when these genes are mutated or activated.

The above example illustrates that carcinogenesis may involve a large number of cancer genes but only a few are stage-limiting genes, whereas other cancer genes may be dispensed with although these genes can enhance the cascade of carcinogenesis. In Reference 26, it has been noted that while mutation of a single gene may initiate the cascade of carcinogenesis in some cases such as retinoblastoma, the process of carcinogenesis would usually involve 5–10 genes.

8.3.1 The Extended Two-Stage Model

The extended two-stage model is an extension of the two-stage multievent model; the latter has also been referred to in the literature as the MVK two-stage clonal expansion model. This is the model used by Tan et al. [72] to describe the cascade of the mouse skin cancer in animal initiation–promotion experiments. In these experiments, normal stem cells are induced to mutate to I_1 cells. Primary I_1 cells grow into detectable papillomas through a stochastic birth and death process. The I_1 cells can further be induced to develop some more genetic changes by carcinogens to become I_2 cells. Primary I_2 cells grow into observable carcinomas through a stochastic birth–death process. This is described schematically in Figure 2.

The MVK two-stage model [61, Chap. 3] and its extensions differ from the two-stage model proposed by Armitage and Doll [2] in 1957, the two-stage model proposed by Kendall [29] in 1960, and the models proposed by Neyman and Scott [50,51] in two aspects: (1) the MVK two-stage model and extensions take into account different growth pattern of normal stem cells in different organs, and (2) these models assume that the first mutational event can occur either in germline cells or in somatic cells. These latter two features of the MVK two-stage model and extensions permit these models to fit most types of cancers whereas the two-stage models by Armitage and Doll [2], by Kendall [29], and by Neyman and Scott [50,51] can only fit some human cancers.

The MVK two-stage model has been proposed and used as a major model to assess risks of environmental agents [14,22,41,45, 46,61,78]. Dose–response curves based on the MVK two-stage model have been developed by Chen and Moini [8], and by Krewski and Murdoch [33]. They have used these dose–response curves to assess how a carcinogen alters cancer incidence through its effects on initiating mutations or on the rate of proliferation of initiated cells. If the carcinogen is a pure initiator, the dose-response curve for cancer incidence can be factorized as a product of a function of dose and a function of time and age; in these cases, the pattern of dose–response curves of the MVK model are quite similar to those of the Armitage–Doll multistage model. On the other hand, if the carcinogen is a promoter or a complete carcinogen, then the dose–response curves of the MVK model cannot be factorized, and it differs qualitatively from the Armitage–Doll model.

8.3.2 Some Clonal Expansion Models

In many cancers such as the colon cancer, while the initiation process requires changes or mutations of many genes, sustained increase of cell proliferation can only occur after accumulation of mutation or deletion of several genes. This is equivalent to stating that the cell proliferation rate (birth rate) would not be significantly greater than the cell differentiation or death rate unless the cell has sustained mutations or genetic changes of at least $k-1(k \geq 2)$ genes. Thus, for cells with mutations of less than $k-1$ genes, the expected number of cells with these genetic changes or mutations are only affected by the mutation rates of these genes and are independent of the birth rate and the death rate of the cells. In these cases, an approximation to the process of carcinogenesis is derived by ignoring the proliferation of the first $k-2$ intermediate cells. This appears to be a combination of the Armitage–Doll model and the two-stage MVK model with the first $k-2$ stage following the Armitage–Doll model but with the two-stage model following the MVK two-stage model. These are the models proposed by Herrero-Jimenez et al. [23] and by Luebeck and Moolgavkar [42].

8.4 Multiple-Pathways Models of Carcinogenesis

In many cancers, it has been observed that the same cancer may arise from different carcinogenic pathways [13,17,18,25,31,43,44,57,58]. This include skin cancers [18], liver cancers [13,17], and mammary gland [44] in animals, the melanoma development in skin cancer in human beings [25], and colon cancer in human beings [19,27,35,53,54,59,79].

To serve as an example, consider the colon cancer of human beings. For this cancer, genetic studies have indicated that there are at least three different pathways by means of which colon cancer is developed [19,27,35,53,54,59,79]. One pathway involves loss or mutation of the suppressor genes- the APC gene in chromosome 5q, the DCC gene in chromosome 18q, and the p53 gene in chromosome 17p. This pathway accounts for about 75–80% of all colon cancers and has been referred to as the "loss of heterozygosity" (LOH) pathway because it is often characterized by aneuploidy and/or loss of chromosome segments (chromosomal instability). Another pathway involves microsatellite MMR genes, hMLH1, hMSH2, hPMS1, hPMS2, hMSH6, and hMSH3. (Mostly hMLH1 and hMSH2.) This pathway accounts for about 10–15% of all colon cancers and appears mostly in the right colon. It has been referred to as the microsatellite instability (MSI) pathway or the *mutator phenotype pathway* because it is often characterized by the loss or mutations in the MMR genes creating a mutator phenotype to significantly increase the mutations rate of many critical genes. This pathway is also referred to as *microsatellite instability—high level* (MSI-H) pathway by Jass et al. [27] to reflect strong mutator effects. A third pathway is a mixture of LOH pathway and the MSI pathway and accounts for about 10% of all colon cancers. This pathway is driven by a mild mutator effects and the LOH force; in this pathway, colon cancer is induced by a mild mutator phenotype, but the progression involves the DCC gene and/or p53 gene. This pathway is also referred to as the *microsatellite instability-low level* (MSI-L) pathway by Jass et al. [27] to reflect mild mutator effects. Potter [54] has proposed two other pathways, but the frequency of these pathways are quite small and are negligible.

While biological evidences have long pointed to multiple pathways models of carcinogenesis, mathematical formulation of these models were first developed by Tan and Brown [63] and extended by Tan and Chen [65]. Further mathematical development of multiple pathways models have been given by Sherman and Portier [57] and by Tan et al. [70,71]. Analyzing p53-mediated data, Mao et al. [43] observed that a multiple-pathways model involving two pathways fitted the data much better than a single-pathway model. Using the multiple-pathway models involving one-stage and two-stage models, Tan and Chen [69] have assessed impacts of changing environments on risk assessment of environmental agents. They have observed that if there are initiation and promotions, then the two-stage model appears to be sufficient to account for the cancer incidence and the survival function. That is, one may ignore effects of one-stage disturbances if the mutation rate of the normal stem cells and the proliferation rate and differentiation rate of initiated cells in the two-stage models are affected by the changing environment.

8.5 The Mixed Models of Carcinogenesis

In the population, for the same type of cancer, different individual may involve different pathways or different number of stages [60,61]. These models have been referred by Tan [60,61] and Tan and Singh [75] as mixed models of carcinogenesis. These models are basic consequences of the observation that different individuals are subject to different environmental conditions and that the mutation of critical cancer genes can occur in either germline cells or in somatic cells. Depending on the environmental or genetic variation, several types of mixtures can therefore be discerned.

1. Foulds [18] noted that many cancers had developed by several independent pathways. Klein and Klein [31] and Spandidos and Wilkie [58] have provided evidence suggesting that the number of stages in carcinogenesis may depend on the cell's environmental conditions. In these situations, different individuals in the population may involve different pathways for the process of carcinogenesis.

2. Geneticists have confirmed that mutation of the critical cancer gene can occur either in germline cells or in somatic cells [32,61]. Thus, for a k-stage multievent model, if mutation of the first stage-limiting cancer gene occurs in a somatic cell, then starting at birth, carcinogenesis is a k-stage multievent model; on the other hand, if mutation of the first stage-limiting cancer gene occurs in a germline cell, then starting at birth, carcinogenesis is a $(k-1)$-stage multievent model. Thus, for retinoblastoma, one would

expect a mixed model of one-stage and two-stage models [61,75].

3. In many cancers, mixtures may arise from both environmental impacts and genetic variations. As an example, consider the colon cancer of human beings [19,27,35,53,54,59,79]. From the population perspective, this cancer is a mixture of at least five pathways: (1) the sporadic colon cancer by the LOH pathway in which all cells at the time of birth are normal stem cells (these cancers account for about 45% of all colon cancers), (2) the FAP (familial adenomatous polyposis) colon cancer is a special case of colon cancers by the LOH pathway in which the individual has already inherited a mutated APC gene (these cancers account for about 30% of all colon cancers), (3) the sporadic colon cancer by the MSI-H pathway in which all cells at the time of birth are normal stem cells (these cancers account for about 13% of all colon cancers), (4) the HNPCC (hereditary non-polyposis colon cancer) is a special case of colon cancers by the MSI-H pathway in which the individual has inherited a mutant MMR gene (mostly hMLH1 and hMSH2), (these cancers account for about 2% of all colon cancers), and (5) the sporadic colon cancer by the MSI-L pathway in which all cells at the time of birth are normal stem cells (these cancers account for about 10% of all colon cancers).

In Reference 61, Tan has provided evidence and examples for mixed models of one-stage models and two-stage models, mixed models of several different two-stage models and mixed models of different multiple pathways models. Tan and Singh [75] have analyzed and fitted the retinoblastoma data by using a mixed model of one-stage and two-stage models. As predicted, they found that the mixture model fitted much better than the single-pathway model.

8.6 Mathematical Analysis of Stochastic Models of Carcinogenesis by Markov Theories

In the literature, mathematical analyses of stochastic models of carcinogenesis were developed mainly by using Markov theories; see Reference 61. For the number of cancer tumors to be Markov, however, this would require the assumption that cancer tumor cells (i.e., the last-stage-initiated cells) grow instantaneously into cancer tumors as soon as they are produced; see References 61 and 68. In this approach, the first step is to derive the probability-generating function (PGF) of the numbers of intermediate initiated cells and cancer tumors through the Kolmogorov forward equation; using this PGF one then derives the incidence function (hazard function) of cancer tumors, the probability distribution of time to tumors, and the probability distribution of the number of tumors. There are several difficulties in this approach, however:

1. If the assumption of instantaneous growth of the last-stage-initiated cells into malignant tumors is violated [81], then the number of cancer tumors is not Markov since it depends on the time when the last-stage-initiated cell is generated. In these cases, the Markov theories are not applicable to cancer tumors [68].

2. The mathematical results become too complicated to be of much use if the number of stages is more than 2 and /or the number of pathways is more than 1, especially when the model is not time homogeneous [61,68].

3. The incidence function and the probability distribution of time to tumors are not identifiable for all parameters [21]; hence, not all parameters are estimable by using cancer incidence data unless some other data and some further external information about the parameters are available.

As an illustration, consider an extended k-stage multievent model. Let $I_j(t)$ ($j = 0, 1, \ldots, k$) denote the number of I_j cells at time t and $T(t)$ the number of cancer tumors at time t. Under the assumption that the I_k cells grow instantaneously into cancer tumors as soon as they are produced, then $I_k(t) = T(t)$ and the process $\{I_j(t), j = 1, \ldots, k-1, T(t)\}$ is Markov.

Let $\lambda(t)$ denote the incidence function of cancer tumors at time t given ($N(t_0) = N_0, I_i(t_0) = T(t_0) = 0, i = 1, \ldots, k-1$), $f(t_0, t)$ the probability density function (pdf) of the time to the onset of cancer tumors at time t given ($N(t_0) = N_0, I_i(t_0) = T(t_0) = 0, i = 1, \ldots, k-1$) and $P(j; t_0, t)$ the probability of $T(t) = j$ given ($N(t_0) = N_0, I_i(t_0) = T(t_0) = 0, i = 1, \ldots, k-1$). Denote by $\psi(y_j, j = 1, \ldots k : t_0, t) = \psi(t_0, t)$ the PGF of $\{I_j(t), j = 1, \ldots, k-1, T(t)\}$ given ($N(t_0) = N_0, I_i(t_0) = T(t_0) = 0, i = 1, \ldots, k$) and $\phi_i(y_j, j = i, \ldots, k : s, t) = \phi_i(s, t)$ the PGF of $\{I_j(t), j = i, \ldots, k-1, T(t)\}$ given one I_i cell at time s for $i = 1, \ldots, k-1$. Then, as shown in Reference 62, we obtain

$$\lambda(t) = \frac{-\psi'(1, \ldots, 1, 0; t_0, t)}{\psi(1, \ldots, 1, 0; t_0, t)}, \quad (1)$$

where

$$\psi'(1, \ldots, 1, 0; t_0, t) = \frac{d}{dt}\psi(1, \ldots, 1, 0; t_0, t);$$

$$f(t_0, t) = \lambda(t) \exp\left\{-\int_{t_0}^{t} \lambda(x) dx\right\}, \quad t > t_0, \quad (2)$$

and

$$P(j; t_0, t) = \frac{1}{j!}\left(\frac{d^j}{dz^j}\psi(t_0, t)\right)_{y_j=1, j=1, \ldots, k-1, y_k=0}. \quad (3)$$

From Equations (1)–(3), to derive the tumor incidence function and the probability distribution of cancer tumors, one would need to solve for $\psi_1(t_0, t)$. To derive this PGF, denote by $b_j(t)$ and $d_j(t)$ the birth rate and the death rate of the I_j cells at time t respectively and $\alpha_j(t)$ the transition rate or mutation rate of $I_j \to I_{j+1} j = 0, 1, \ldots, k-1$ at time t. Then the following results have been proved in References 61 and 68:

1. Assuming that the number of normal stem cells is very large (e.g., $N(t_0) = 10^8$), then $N(t) = I_0(t)$ is a deterministic function of t. In these cases, $\psi(t_0, t)$ is given by

$$\psi(t_0, t) = \exp\left\{\int_{t_0}^{t} N(x)\alpha_0(x) \right. $$
$$\left. \times [\phi_1(x, t) - 1] dx\right\}. \quad (4)$$

2. By using the Kolmogorov forward equation, it can readily be shown that the $\phi_i(t_0, t)$'s satisfy the following partial differential equations, respectively:

$$\frac{\partial}{\partial t}\phi_i(s, t)$$
$$= \sum_{j=i}^{k-1}\{(y_j - 1)[y_j b_j(t) - d_j(t)]$$
$$+ y_j(y_{j+1} - 1)\alpha_j(t)\}$$
$$\times \frac{\partial}{\partial y_j}\phi(s, t), \quad (5)$$

for $i = 1, \ldots, k-1$, where the initial condition is $\phi_i(s, s) = y_i$.

These above equations are difficult to solve. Hence, analytical solutions for the general cases are not available.

3. If $\{b_i(t) = b_i, d_i(t) = d_i, i = 1, \ldots, k-1, \alpha_j(t) = \alpha_j, j = 0, 1, \ldots, k-1\}$ are independent of time t, then the process is time-homogeneous. In these cases, $\phi_i(s, t) = \phi_i(t-s)$ and the $\phi_i(t)$'s satisfy the following system of Ricatti equations:

$$\begin{aligned}\frac{d}{dt}\phi_i(t) &= b_i[\phi_i(t)]^2 + \{\alpha_i\phi_{i+1}(t)\\&\quad -[b_i + d_i + \alpha_i]\}\phi_i(t)\\&\quad + d_i, i = 1, \ldots, k-1,\end{aligned} \quad (6)$$

with $\phi_k(t) = y_k$.

The initial condition is $\phi_i(0) = y_i, i = 1, \ldots, k-1$.

In these equations, the Ricatti equation for $\phi_{k-1}(t)$ is linear and the solution can easily be derived. This solution is given in Reference 61, Chap. 3. If $k > 2$, the Ricatti equation for $\phi_j(t)$ with $j < k-1$ is nonlinear and the solution is very difficult. Hence, general solution is not available if $k > 2$.

The above results indicate that the solution of $\phi_1(x, t)$ is not available in general cases; hence, it is possible only under some assumptions. Given below are some specific assumptions that have been made in the literature for solving $\phi_1(x, t)$.

1. Assume $k = 2$, then the model is the MVK two-stage model. In this case, when the model is time homogeneous, the solution of $\phi_1(t)$ is readily available and is given by

$$\begin{aligned}\phi_1(t) &= \{\beta_2(y_2 - \beta_1) + \beta_1(\beta_2 - y_2)\\&\quad \times \exp[b_I(\beta_2 - \beta_1)t]\}\\&\quad \times \{(y_2 - \beta_1) + (\beta_2 - y_2)\\&\quad \times \exp[b_I(\beta_2 - \beta_1)t]\}^{-1},\end{aligned} \quad (7)$$

where $\beta_2 > \beta_1$ are given by

$$2B_I\beta_i = (b_I + d_I + \alpha_1 - \alpha_1 y_2)\\\mp h(y_2), i = 1, 2$$

and

$$h(y_2) = \sqrt{\begin{aligned}\{(b_I + d_I + \alpha_1 - \alpha_1 y_2)^2\\-4b_I d_I\}\end{aligned}}.$$

This is the model used by Moolgavkar and his associates to analyze animal and human cancer data to estimate the mutation rate of $N \to I$ and the cell proliferation rate (the difference between birth rate and death rate) of I cells and to assess effects of some environmental agents on tumor incidence [14,22,41,42,48].

2. In most cancers such as colon cancer or lung cancer, $k > 2$ provides a more realistic description of the carcinogenesis process from the biological mechanisms. In these cases, the Ricatti equation for $\phi_j(t)$ with $j < k-1$ is nonlinear and the solution of $\phi_1(t)$ is very difficult. Thus, to derive results and to fit data by this approach, further assumption is needed. Herrero-Jimenez et al. [23] and Luebeck and Moolgavkar [42] assumed $b_j = d_j = 0$ for $j < k-1$. This is the clonal expansion model, which combines the Armitage–Doll model with the MVK model. Applying this model with $k = 2, 3, 4, 5$ stages, Luebeck and Moolgavkar [42] have fitted the NCI SEER data of colon cancer by further assuming $\alpha_0 = \alpha_1$ when $k = 2, 3, 4$ and by assuming $\{\alpha_0 = \alpha_1, \alpha_2 = \alpha_3\}$ when $k = 5$. Their analysis showed that the model with four stages were most appropriate, although all models appear to fit the data equally well. Notice also that even with the homogeneous clonal expansion model, only $k+1$ parameters are estimable; it appeared

that only the parameters $\{\alpha_i, i = 0, 1, \ldots, k-3\}$ and the parameter functions $\{g_1 = \alpha_{k-2}/b_{k-1}, \gamma_{k-1}, g_2 = b_{k-1} - d_{k-1}, \alpha_{k-1}/(1 - b_{k-1}/d_{k-1})\}$ are estimable by the traditional approaches.

8.7 Mathematical Analysis of Stochastic Models of Carcinogenesis by Stochastic Differential Equations

As shown above, for most of the stochastic models of carcinogenesis, the mathematics in the traditional Markov approach can easily become too complicated to be manageable and useful [62,69]; furthermore, in order to apply the Markov theories, it is necessary to assume that with probability $P = 1$ each cancer tumor cell grows instantaneously into a malignant cancer tumor [14,36,37,41,42,45–48,61], completely ignoring cancer progression. As shown by Yakovlev and Tsodikov [81], in some cases this assumption may lead to misleading results. To relax this assumption and to derive general results, Tan and Chen [68] have proposed alternative approaches by deriving stochastic differential equations for the state variables.

To serve as an example, consider a k-stage extended multievent model as above. Then the I_j cells arise from I_{j-1} cells by mutation or some genetic changes and cancer tumors (T) develop from primary I_k cells by following a stochastic birth–death process. Notice that the process $\boldsymbol{X}(t) = \{I_j(t), j = 1, \ldots, k-1\}$ is a high-dimensional Markov process. However, because the primary I_k cell can be generated at any time, s with $t_0 < s \leq t$, $T(t)$ is not Markov unless it is assumed that the I_k cells grow instantaneously into cancer tumors, in which case $I_k(t) = T(t)$ [61,68,81].

Using the method of PGFs, it is shown in References 62, 68, and 72 that given $\{I_{k-1}(s), s \leq t\}$, the conditional distribution of $T(t)$ is Poisson with intensity $\lambda(t)$, where $\lambda(t)$ is given by

$$\lambda(t) = \int_{t_0}^{t} I_{k-1}(x)\alpha_{k-1}(x) P_T(x,t) dx,$$

with $P_T(s,t)$ denoting the probability that a primary cancer tumor cell arising at time s will develop into a detectable cancer tumor by time t.

Assuming that a cancer tumor is detectable only if it contains at least N_T cancer tumor cells, Tan [62], Tan and Chen [68,72] have shown that $P_T(s,t)$ is given by

$$P_T(s,t) = \frac{1}{h(t-s) + g(t-s)} \times \left(\frac{g(t-s)}{h(t-s) + g(t-s)}\right)^{N_T - 1}, \tag{8}$$

where

$$\begin{aligned}
h(t-s) &= \exp\left\{-\int_s^t [b_k(y-s) - d_k(y-s)] dy\right\} \\
&= \exp\{-(\epsilon_T/\delta_T) \\
&\quad \times [1 - \exp(-\delta_T(t-s))]\},
\end{aligned} \tag{9}$$

and

$$\begin{aligned}
g(t-s) &= \int_s^t b_k(y-s) h(y-s) dy \\
&= (b_T/\epsilon_T)[1 - h(t-s)].
\end{aligned}$$

8.7.1 Stochastic Differential Equations for $\{I_i(t), i = 1, \ldots, k-1\}$

Let $\boldsymbol{X}(t) = \{I_j(t), j = 1, \ldots, k-1\}$. Then $\boldsymbol{X}(t + \Delta t)$ develops from $\boldsymbol{X}(t)$ through stochastic birth–death processes and mutation processes. $\boldsymbol{X}(t)$ is Markov although $T(t)$ is not. It can easily be shown that during $[t, t + \Delta t]$, to order of $o(\Delta t)$, the stochastic birth–death processes and the mutation processes are equivalent to the multinomial distributions and the Poisson distributions, respectively [62,68,72]. Thus, the transition from $\boldsymbol{X}(t)$ to $\boldsymbol{X}(t+1)$ is characterized by the following random variables:

- $B_j(t)$ = number of new I_j cells generated by stochastic cell proliferation and differentiation of I_j cells during $(t, t + \Delta t]$, $j = 1, \ldots, k - 1$,

- $D_j(t)$ = number of death of I_j cells during $(t, t + \Delta t]$, $j = 1, \ldots, k - 1$,

- $M_j(t)$ = number of transitions from I_j cells to I_{j+1} cells by mutation or some genetic changes during $(t, t + \Delta t]$, $j = 0, 1, \ldots, k - 1$.

Notice that the $M_k(t)$ cells are the primary cancer tumor cells generated during $(t, t + \Delta t]$.

Conditional on $\boldsymbol{X}(t)$, to order of $o(\Delta t)$ the above variables are basically multinomial variables and Poisson variables:

$$\{B_j(t), D_j(t)\} | I_j(t)$$
$$\sim ML\{I_j(t), b_j(t)\Delta t, d_j(t)\Delta t\},$$
$$j = 1, \ldots, k - 1, \quad (10)$$

and

$$M_j(t) | I_j(t) \sim \text{Poisson}\{I_j(t)\alpha_j(t)\Delta t\},$$
$$j = 0, 1, \ldots, k - 1, \quad (11)$$

independently of $\{B_j(t), D_j(t)\}$, $j = 1, \ldots, k - 1$.

This leads to the following stochastic difference equations for $I_i(t)$, $i = 1, \ldots, k - 1$:

$$\Delta I_j(t)$$
$$= I_j(t + \Delta t) - I_j(t) = M_{j-1}(t)$$
$$+ B_j(t) - D_j(t)$$
$$= \{I_{j-1}\alpha_{j-1} + I_j(t)\gamma_j(t)\}$$
$$\times \Delta t + \epsilon_j(t)\Delta t,$$
$$\text{for} \quad j = 1, \ldots, k - 1, \quad (12)$$

where $I_0(t) = N(t)$ and $\gamma_j(t) = b_j(t) - d_j(t)$.

In Equation (10), the random noises $\epsilon_j(t)$ are derived by subtracting the conditional expected values from the random variables:

$$\epsilon_j(t)\Delta t$$
$$= [M_j(t) - I_{j-1}(t)\alpha_{j-1}(t)\Delta t]$$
$$+ [B_j(t) - I_j(t)b_j(t)\Delta t]$$
$$- [D_j(t) - I_j(t)d_j(t)\Delta t],$$
$$j = 1, \ldots, k - 1. \quad (13)$$

It can easily be shown that the $\epsilon_j(t)$'s have expected value 0 and are uncorrelated with the state variables $I_j(t)$. The variances and covariances of the $\epsilon_j(t)$'s are easily obtained as, to order of $o(\Delta t)$:

$$Q_{jj}(t)$$
$$= \text{Var}[\epsilon_j(t)] = E\{I_{j-1}(t)\alpha_{j-1}(t)$$
$$+ I_j(t)[b_j(t) + d_j(t)]\},$$
$$\text{for } j = 1, \ldots, k - 1;$$
$$Q_{i,j}(t)$$
$$= \text{Cov}[e_i(t), e_j(t)] = 0, \text{if } i \neq j.$$
$$\quad (14)$$

8.7.2 The Probability Distribution of State Variables

Let $\Delta t \sim 1$ correspond to a small time interval such as 0.1 day and put $\boldsymbol{X} = \{\boldsymbol{X}(t), t = 0, 1, \ldots, t_M\}$. Then, by using the multinomial distribution for the numbers of birth and death of initiated cells and using the Poisson distribution for the numbers of mutations of initiated cells, the probability distribution $P(\boldsymbol{X}|\Theta)$ of the state variables $\boldsymbol{X}$ given the parameters Θ is

$$P(\boldsymbol{X}|\Theta) = P(\overset{X}{\sim}(0))\prod_{t=1}^{t_M}\prod_{i=1}^{k-1}P\{I_i(t)|\boldsymbol{X}(t),\Theta\}$$

where

$$\begin{aligned}&P\{I_r(t+1)|\boldsymbol{X}(t),\Theta\}\\&=\sum_{j=0}^{I_r(t+1)-I_r(t)}g_{r-1}(j,t)\\&\quad\times\sum_{i=0}^{I_r(t)}\binom{I_r(t)}{i}\binom{I_r(t)-i}{a_r(i,j;t)}\\&\quad\times[b_r(t)]^i[d_r(t)]^{a_r(i,j;t)}\\&\quad\times[1-b_r(t)-d_r(t)]^{I_r(t+1)-2i},\\&\quad r=1,\ldots,5,\end{aligned}\qquad(15)$$

with

$$\begin{aligned}a_0(i;t) &= I_0(t)-I_0(t+1)+i\\a_r(i,j;t) &= I_r(t)-I_r(t+1)+i+j,\\&\quad r=1,\ldots,k-1\end{aligned}$$

and for $i=0,1,\ldots,k-1$, the $g_i(j,t)$ is the density of a Poisson distribution with intensity $\lambda_i(t)=I_i(t)\alpha_i(t)$.

These distribution results have been used by Tan et al. [73] to estimate the unknown parameters and state variables in animal populations. Tan et al. [77] have also used these procedures to assess effects of smoking on lung cancer using the British physician data of lung cancer with smoking.

8.7.3 State Space Models of Carcinogenesis

State space models of carcinogenesis were first proposed by Tan and Chen [68] for multievent models. These models have then been extended to multiple pathways models [70–72] and applied to animal data to estimate the mutation rate of normal stem cells and the birth rate and death rate of initiated cells [71,72]. Tan et al. [77] have also used these models to assess effects of smoking on lung cancer using the British physician data of lung cancer with smoking given in Reference 15.

State space models of stochastic systems are stochastic models consisting of two submodels: the *stochastic system model*, is the stochastic model of the system; and the *observation model*, which is a statistical model based on available observed data from the system. For carcinogenesis, the stochastic system model is represented by a system of stochastic differential equations for the numbers of normal stem cells and initiated cells as described above; the observation model is represented by a statistical model based on available cancer data from the system. For human beings, the data available are usually the cancer incidence data that give the number of new cancer cases during some fixed time intervals such as 1 or 5 years; see the SEER data from NCI/NIH [56]. For animal carcinogenicity studies, the data available are usually the numbers of detectable preoplastic lesion per animal over time and/or the number of detectable cancer tumors per individual over time. For mouse initiation–promotion experiments on skin cancer, the preneoplastic lesion is the papillomas and the cancer tumor the carcinomas.

To serve as an example, consider an initiation–promotion experiment for skin cancer in mice; see References 16, 34, and 49. In these experiments, at time 0 similar mice are treated by an initiator for a very short period $(0,t_0]$ (normally a few days); these treated animals are then promoted during the period $(t_m,t_e]$ by a promoter. At some fixed times t_j after initiation, a fixed number (say, n_j at time t_j)

of treated animals are sacrificed and autopsies performed. The observed data are (1) the number of animals with detectable papillomas at time t_j among the sacrificed animals at $t_j, j = 1, \ldots, k$; and (2) given that the animal has detectable papillomas, the number of detectable papillomas per animal are then counted; see Reference 49. In some experiments, the number of animals with detectable carcinomas at time t_j among the sacrificed animals are also counted; given that the animal has detectable carcinomas, then the number of detectable carcinomas per animal at t_j are counted. Such data have been reported in Reference 49.

8.7.4 The Stochastic System Model

For the stochastic system model, assume a two-stage model as given in Figure 2 and assume that the numbers $N(t)$ of normal stem cells at time $t(t \geq 0)$ is very large so that $N(t)$ is a deterministic function of time t. Then the state variables are given by $\{M_{0,i}(t), I_i(t) t > 0\}$, where $M_{0,i}(t)$ is the number of cells initiated by the initiator at time t in the ith animal and $I_i(t)$ the number of the initiated cells (I cells) at time t in the ith animal. Let $b_I(t), d_I(t)$, and $\alpha_I(t)$ denote the birth rate, the death rate, and the mutation rate of the I cells at time t, respectively, and $\{B_i(t), D_i(t), M_i(t)\}$ the numbers of birth, death, and mutation of the I cell in the ith animal during $(t, t+\Delta t]$, respectively. Discretize the timescale by letting $\Delta t \sim 1$ corresponding to a small time interval such as 1 hour or 0.1 day. Then the stochastic system model is given by the following probability distributions and stochastic equations:

$$M_{0,i}(t) \sim \text{Poisson}\{\lambda(t)\}, \text{ independently}$$
$$\text{for } i = 1, \ldots, n, \quad (16)$$

where $\lambda(t) = N(t)\alpha_0(t)$ with $\alpha_0(t)$ being the mutation rate of normal stem cells;

$$\{B_i(t), D_i(t), M_i(t)\} | I_i(t) \sim ML\{I_i(t);$$
$$b_I(t), d_I(t), \alpha_I(t)\}$$
$$\text{independently for } i = 1, \ldots, n; \quad (17)$$

and

$$\Delta I_i(t)$$
$$= I_i(t+1) - I_i(t)$$
$$= \delta(t_0 - t)M_{0,i}(t) + B_i(t) - D_i(t)$$
$$= \delta(t_0 - t)\lambda(t) + \gamma_I(t)I_i(t) + \epsilon_i(t),$$
$$\text{with } I_i(0) = 0, \quad (18)$$

where $\delta(t)$ is defined by $\delta(t) = 1$ if $t \geq 0$ and $= 0$ if $t < 0$, $\gamma_I(t) = b_I(t) - d_I(t)$, and $\epsilon_i(t)$ the random noise that is the sum of residuals derived by subtracting the conditional mean numbers from the random variables.

Let $\overset{X}{\sim}_i(t) = \{M_{0,i}(t), I_i(t)\}$ if $t \leq t_0$, $\overset{X}{\sim}_i(t) = I_i(t)$ if $t > t_0$, and put $\boldsymbol{X}_i = \{\overset{X}{\sim}_i(t), t = 0, 1, \ldots, t_M\}$, where $t_M (t_M > t_0)$ is the termination time of the experiment. Denote by $g_P(i;t)$ the density of the Poisson distribution with intensity $\lambda(t)$. Using the distribution results given in Equations (12) and (13), the probability distribution of $\boldsymbol{X} = \{\boldsymbol{X}_i, i = 1, \ldots, n\}$ given the parameters $\Theta = \{\lambda(t), b_I(t), d_I(t), \alpha_I(t), \text{ all } t\}$ is

$$P\{\boldsymbol{X}|\Theta\}$$
$$= \prod_{i=1}^{n} P\{\boldsymbol{X}_i|\Theta\}$$
$$= \prod_{i=1}^{n} \{P[\overset{X}{\sim}_i(0)|\Theta] P[\boldsymbol{X}_i|\overset{X}{\sim}_i(0), \Theta]\}$$

$$P\{\boldsymbol{X}_i|\overset{X}{\sim}_i(0)\}$$
$$= \prod_{t=1}^{t_M} P\{\boldsymbol{X}_i(t)|\overset{X}{\sim}_i(t-1)\}, \quad (19)$$

where

$$P\{\boldsymbol{X}_i(t+1)|\overset{X}{\sim}_i(t)\}$$
$$= \{g_P[M_{0,i}(t);t]\}^{\delta(t_0-t)} \sum_{j=1}^{I_i(t)} \binom{I_i(t)}{j}$$
$$\times \binom{I_i(t)-j}{\eta_{i,1}(t)} [b_I(t)]^j [d_I(t)]^{\eta_{i,1}(t)}$$
$$\times [1-b_I(t)-d_I(t)]^{\eta_{i,2}(t)}$$
$$\times \left\{\prod_{k=1}^{2} \delta[\eta_{i,k}(t)]\right\},$$

$$\eta_{i,1}(t) = I_i(t) - I_i(t+1)$$
$$+\delta(t_0-t)M_{0,i}(t)+j,$$
$$\eta_{i,2}(t) = I_i(t+1)$$
$$-\delta(t_0-t)M_{0,i}(t)-2j.$$

8.7.5 The Observation Model

Assume that information about papillomas among the sacrificed animals has been collected. Let $Y_0(j)$ denote the number of animals with papillomas among the n_j sacrificed animals at time t_j and $Y_i(j)$ the number of papillomas per animal of the ith animals among the $Y_0(j)$ animals with papillomas. Let $P_I(s,t)$ denote the probability that an initiated cell arising from a normal stem cell at time s will develop into a detectable papillomas by time t. Since the number of papillomas are almost zero if there is no promotion, the observation model is specified by the following probability distributions (for proof, see Refs. 62 and 72):

$$Y_0(j)|n_j \sim \text{binomial}\{n_j, Q(j)\}, \quad (20)$$

independently for $j = 1, \ldots, k$, where

$$Q(j) = \left[1 - \exp\left(-\int_0^{t_0} \lambda(x)dx\right)\right]^{-1}$$
$$\times \left\{1 - \exp\left[-\int_0^{t_0} N(x)\alpha_0(x)\right.\right.$$
$$\left.\left. P_I(x,t_j)dx\right]\right\};$$

and for $Y_i(j) > 0$,

$$Y_i(j)|Y_0(j)$$
$$\sim \text{Poisson}$$
$$\left\{\int_0^{t_0} N(x)\alpha_0(x)P_I(x,t_j)dx\right\},$$
(21)

independently for $i = 1, \ldots, Y_0(j)$.

Assume that the preneoplastic lesion is detectable if it contains N_I initiated cells. Then the $P_I(s,t)$ in Equations (15) and (16) is given by

$$P_I(s,t) = \frac{1}{\xi(s,t)+\eta(s,t)}$$
$$\times \left(\frac{\eta(s,t)}{\xi(s,t)+\eta(s,t)}\right)^{N_I-1},$$
(22)

where

$$\xi(s,t) = \exp\left\{-\int_s^t [b_I(y)-d_I(y)]dy\right\}$$

and

$$\eta(s,t) = \int_s^t b_I(y)\xi(s,y)dy.$$

Tan et al. [72] have applied the above state space model to analyze the papillomas data from an animal initiation–promotion experiment for mice skin exposed to the emission of Nissan car in the US Environmental Protection Agency (EPA). Using this dataset, they have estimated the mutation rate from $N \to I_1$ and the time-dependent birth rate and death rate of I_1 cells as well as the numbers of the I_1 cells over time. Their results indicate that the emission of Nissan car is an initiator whereas the promotion effect of the emission is quite small.

Using the multievent model as the stochastic system model and using the cancer incidence data for constructing the observation model, Tan et al. [77] have developed a general state space model for carcinogenesis. They have applied this general state space model to the British physician data of lung cancer with smoking given in Reference 15. On the basis of this data set, they have estimated the mutation rate of $N \to I_1$ and the time-dependent birth rates and death rates of I_1 cells. Their results indicated that the tobacco nicotine is an initiator. If $t > 60$ years old, then the tobacco nicotine is also a promoter.

References

1. Armitage, P. and Doll, R. (1954). The age distribution of cancer and a multistage theory of carcinogenesis. *Br. J. Cancer*, **8**, 1–12.

2. Armitage, P. and Doll, R. (1957). A two-stage theory of carcinogenesis in relation to the age distribution of human cancer. *Br. J. Cancer*, **11**, 161–169.

3. Armitage, P. and Doll, R. (1961). Stochastic models for carcinogenesis. In *Proceedings of the 4th Berkeley Symposium on Mathematical Statistics and Probability: Biology and Problems of Health*, pp. 19–38. University of California Press, Berkeley.

4. Breslow, N. E. and Day, N. E. (1987). *Statistical Methods in Cancer Research, Volume II—The Design and Analysis of Cohort Studies*, International Agency for Research on Cancer, Lyon.

5. Brown, C. C. and Chu, K. C. (1983). Implications of multi-stage theory of carcinogenesis applied to occupational arsenic exposure. *J. Natl. Cancer Inst.*, **70**, 455–463.

6. Cahilll, D. P., Kinzler, K. W., Vogelstein, B., and Lengauer, C. (1999). Genetic instability and Darwinian selection in tumors. *Trends Cell Biol.*, **9**, M57–M60.

7. Chen, C. W. and Farland, W. (1991). Incorporating cell proliferation in quantitative cancer risk assessment: Approach, issues, and uncertainties. In *Chemically Induced Cell Proliferation: Implications for Risk Assessment*, B. Butterworth, T. Slaga, W. Farland, and M. McClain, eds., pp. 481–499. Wiley-Liss, New York.

8. Chen, C. W. and Moini, A. (1990). Cancer dose-response models incorporating clonal expansion. In *Scientific Issues in Quantitative Cancer Risk Assessment*, S. H. Moolgavkar ed., pp. 153–175. Birkhauser, Boston.

9. Chu, K. C. (1985). Multi-event model for carcinogenesis: A model for cancer causation and prevention. In *Carcinogenesis: A Comprehensive Survey, Volume 8: Cancer of the Respiratory Tract-Predisposing Factors*, M. J. Mass, D. G. Ksufman, J. M. Siegfied, V. E. Steel, and S. Nesnow, eds., pp. 411–421. Raven Press, New York.

10. Chu, K. C., Brown, C. C., Tarone, R. E., and Tan, W. Y. (1987). Differentiating between proposed mechanisms for tumor promotion in mouse skin using the multivent model for cancer. *J. Natl. Cancer Inst.*, **79**, 789–796.

11. Collins, K, Jacks, T., and Pavletich, N. P. (1997). The cell cycle and cancer. *Proc. Natl. Acad. Sci. (USA)*, **94**, 2776–2778.

12. Day, N. E. and Brown, C. C. (1980). Multistage models and primary prevention of cancer. *J. Natl. Cancer Inst.*, **64**, 977–989.

13. DeAngelo, A. (1996). *Dichloroacetic Acid Case Study*. Paper presented to Expert Panel to Evaluate EPA's Proposed Guidelines for Cancer Risk Assessment Using Chloroform and Dichloroacetate as Case Studies Workshop, Sept. 10–12, at ILSI Health and Environmental Sciences Institute, Washington, DC.

14. Dewanji, A., Moolgavkar, S. H., and Luebeck, E. G. (1991). Two-mutation

model for carcinogenesis: Joint analysis of premalignant and malignant lesions. *Math. Biosci.*, **104**, 97–109.

15. Doll, R. and Peto, R. (1978). Cigarette smoking and bronchial carcinoma: dose and time relationships among regular smokers lifelong non-smokers. *J. Epidemiol. Community Health*, **32**, 303–313.

16. DuBowski, A., Johnston, D. J., Rupp, T., Beltran, L. Couti, C. J., and DiGiovanni, J. (1998). Papillomas at high risk for malignant progression arising both early and late during two stage carcinogenesis in SENCAR mice. *Carcinogenesis*, **19**, 1141–1147.

17. Ferreira-Gonzalez, A., DeAngelo, A., Nasim, S., and Garrett, C. (1995). Ras oncogene activation during hepatocarcinogenesis in B6C3F1 male mice by dichloroacetic and trichloroacetic acids. *Carcinogenesis*, **16**, 495–500.

18. Foulds, L. (1975). *Neoplastic Development*, Vol. 2. Academic Press, New York.

19. Hawkins, N. J. and Ward, R. L. (2001). Sporadic colorectal cancers with microsatellite instability and their possible origin in hyperplastic polyps and serrated adenomas. *J. Natl. Cancer Inst.*, **93**, 1307–1313.

20. Hanahan, D. and Weinberg, R. A. (2000). The hallmarks of cancer. *Cell*, **100**, 57–70.

21. Hanin, L. G. and Yakovlev, A. Y. (1996). A nonidentifiability aspect of the two-stage model of carcinogenesis. *Risk Anal.*, **16**, 711–715.

22. Hazelton, W. D., Luebeck, E. G., Heidenreich, W. F., Peretzke, H. G., and Moolgavkar, S. H. (1999). Biologically-based analysis of the data for the Colorado plateau uranium miners cohort: Age, dose, dose-rate effects. *Radiat. Res.*, **152**, 339–351.

23. Herrero-Jimenez, P., Thilly, G., Southam, P. J., Mitchell, A., Morgenthaler, S., Furth, E. E., and Thilly, W. G. (1998). Mutation, cell kinetics and subpopulations at risk for colon cancer in the United States. *Mutat. Res.*, **400**, 553–578.

24. Hesketh, R. (1997). *The Oncogene and Tumor Suppressor Gene Facts Book*, 2nd ed. Academic Press, San Diego, CA.

25. Holman, L. D'Arcy, J., Armstrong, B. K., and Heenan, P. J. (1983). A theory of etiology and pathogenesis of human cutaneous malignant melanoma. *J. Natl. Cancer Inst.*, **71**, 651–656.

26. Hopkin, K. (1996). Tumor evolution: Survival of the fittest cells, *J. NIH Res.*, **8**, 37–41.

27. Jass, J. R., Biden, K. G., Cummings, M. C., Simms, L. A., Walsh, M., Schoch, E., Meltzer, S. J., Wright, C., Searle, J., Young, J., and Leggett, B. A. (1999). Characterization of a subtype of colorectal cancer combining features of the suppressor and mild mutator pathways. *J. Clin. Pathol.*, **52**, 455–460.

28. Kalbfleisch, J. D., Krewski, D., and Van Ryzin, J. (1983). Dose-response models for time-to-response toxicity data. *Can. J. Statist.*, **11**, 25–50.

29. Kendall, D. (1960). Birth-and-death processes, and the theory of carcinogenesis. *Biometrika*, **47**, 13–21.

30. Klebanov, L. B., Rachev, S. T., and Yakovlev, A. Y. (1993). A stochastic model of radiation carcinogenesis: latent time distributions and their properties. *Math. Biosci.*, **113**, 51–75.

31. Klein, G. and Klein, E. (1984). Oncogene activation and tumor progression. *Carcinogenesis*, **5**, 429–435.

32. Knudson, A. G. (1971). Mutation and cancer: statistical study of retinoblastima. *Proc. Natl. Acad. Sci. (USA)*, **68**, 820–823.

33. Krewski, D. R. and Murdoch, D. J. (1990). Cancer modeling with intermittent exposure. In *Scientific Issues in Quantitative Cancer Risk Assessment*, S. H. Moolgavkar ed., pp. 196–214. Birkhauser, Boston.

34. Kopp-Schneider, A. and Portier, C. J. (1992). Birth and death/differentiation rates of papillomas in mouse skin. *Carcinogenesis*, **13**, 973–978.

35. Laurent-Puig, P., Blons, H., and Cugnenc, P.-H. (1999). Sequence of molecular genetic events in colorectal tumorigenesis. *Eur. J. Cancer Prev.*, **8**, S39–S47.

36. Little, M. P. (1995). Are two mutations sufficient to cause cancer? Some generalizations of the two-mutation model of carcinogenesis of Moolgavkar, Venson and Knudson, and of the multistage model of Armitage and Doll. *Biometrics*, **51**, 1278–1291.

37. Little, M. P. (1996). Generalizations of the two- mutation and classical multistage models of carcinogenesis fitted to the Japanese atomic bomb survivor data. *J. Radiol. Prot.*, **16**, 7–24.

38. Little, M. P., Muirhead, C. R., Boice, J. D. Jr., and Kleinerman, R. A. (1995). Using multistage models to describe radiation-induced leukaemia. *J. Radiol. Prot.*, **15**, 315–334.

39. Little, M. P., Muirhead, C. R., and Stiller, C. A. (1996). Modelling lymphocytic leukaemia incidence in England and Wales using generalizations of the two-mutation model of carcinogenesis of Moolgavkar, Venzon and Knudson. *Statist. Med.*, **15**, 1003–1022.

40. Loeb, K. R. and Loeb, L. A. (2000). Significance of multiple mutations in cancer. *Carcinogenesis*, **21**, 379–385.

41. Luebeck, E. G., Heidenreich, W. F., Hazelton, W. D., and Moolgavkar, S. H. (2001). Analysis of a cohort of Chinese tin miners with arsenic, radon, cigarette and pipe smoke exposures using the biologically-based two stage clonal expansion model. *Radiat. Res.*, **156**, 78–94.

42. Luebeck, E. G. and Moolgavkar, S. H. (2002). Multistage carcinogenesis and colorectal cancer incidence in SEER. *Proc. Natl. Acad. Sci. (USA)*, **99**, 15095–15100.

43. Mao, J. H., Lindsay, K. A., Balmain, A., and Wheldon, T. E. (1998). Stochastic modelling of tumorigenesis in p53 deficient mice. *Br. J. Cancer*, **77**, 243–252.

44. Medina, D. (1988). The preneoplastic state in mouse mammary tumorigenesis. *Carcinogenesis*, **9**, 1113–1119.

45. Moolgavkar, S. H., Cross, F. T., Luebeck, G., and Dagle, G. (1990). A two-mutation model for radon-induced lung tumors in rats. *Radiat. Res.*, **121**, 28–37.

46. Moolgavkar, S. H., Dewanji, A., and Venzen, D. J. (1988). A stochastic two-stage for cancer risk assessment: The hazard function and the probability of tumor. *Risk Anal.*, **3**, 383–392.

47. Moolgavkar, S. H. and Knudson, A. G. (1981). Mutation and cancer: a model for human carcinogenesis. *J. Natl. Cancer Inst.*, **66**, 1037–1052.

48. Moolgavkar, S. H. and Venzen, D. J. (1979). Two-event models for carcinogenesis: incidence curve for childhood and adult tumors. *Math. Biosci.*, **47**, 55–77.

49. Nesnow, S., Triplett, L. L., and Slaga, T. J. (1985). Studies on the tumor initiating, tumor promoting, and tumor co- initiating properties of respiratory carcinogens. *Carcinogenesis*, **8**, 257–277.

50. Neyman, J. (1961). A two-step mutation theory of carcinogenesis. *Bull. Inst. Int. Statist.*, **38**, 123–135.

51. Neyman, J. and Scott, E. (1967). Statistical aspects of the problem of carcinogenesis. Proc. 5th Symp. on Mathematical Statistics and Probability, Vol. 4., pp. 745–776, Berkeley, CA.

52. Nording, C. O. (1953). A new theory on the cancer inducing mechanism. *Br. J. Cancer*, **7**, 68–72.

53. Peltomaki, P. (2001). Deficient DNA mismatch repair: A common etiologic factor for colon cancer. *Hum. Molec. Genet.*, **10**, 735–740.

54. Potter, J. D. (1999). Colorectal cancer: Molecules and population. *J. Natl. Cancer Inst.*, **91**, 916–932.

55. Richmond, R., DeAngelo, A., Potter, C., and Daniel, F. (1991). The role of nodules in dichloroacetic acid-induced hepatocarcinogenesis in B6C3F1 male mice. *Carcinogenesis*, **12**, 1383–1387.

56. Ries, L. A. G., Eisner, M. P., Kosary, C. L., Hankey, B. F., Miller, M. A., Clegg, L., and Edwards, B. K., eds. (2001). *SEER cancer statistic Review, 1973–1998*, National Cancer Institute, Bethesda, MD.

57. Sherman, C. D. and Portier, C. J. (1994). The multipath/multistage model of carcinogenesis. *Inf. Biomol. Epidemiol. Med. Biol.*, **25**, 250–254.

58. Spandido, D. A. and Wilkie, N. M. (1984). Malignant transformation of early passage rodent cells by a single mutated human oncogene H-ras-1 from T 24 bladder carcinoma line. *Nature*, **310**, 469–475.

59. Sparks, A. B., Morin, P. J., Vogelstein, B., and Kinzler, K. W. (1998). Mutational analysis of the APC/beta-catenin/Tcf pathway in colorectal cancer. *Cancer Res.*, **58**, 1130–1134.

60. Tan, W. Y. (1988). Some mixed models of carcinogenesis. *Math. Comput. Model.*, **10**, 765–773.

61. Tan, W. Y. (1991). *Stochastic Models of Carcinogenesis*. Marcel Dekker, New York.

62. Tan, W. Y. (2002). *Stochastic Models with Applications to Genetics, Cancers, AIDS and Other Biomedical Systems*. World Scientific, River Edge, NJ.

63. Tan, W. Y. and Brown, C. C. (1986). *A Stochastic Model of Carcinogenesis-Multiple Pathways Involving Two-Stage Models*. Paper presented at the Biometric Society (ENAR) Meeting. Atlanta, GA, March 17–19, 1986.

64. Tan, W. Y. and Brown, C. C. (1987). A nonhomogeneous two stages model of carcinogenesis. *Math. Comput. Model.*, **9**, 631–642.

65. Tan, W. Y. and Chen, C. W. (1991). A multiple pathways model of carcinogenesis involving one stage models and two-stage models. In *Mathematical Population Dynamics*, O. Arino, D. E. Axelrod, and M. Kimmel eds., Chap. 31, pp. 469–482. Marcel Dekker, New York.

66. Tan, W. Y. and Chen, C. W. (1995). A nonhomogeneous stochastic models of carcinogenesis and its applications to assess risk of environmental agents. In *Mathematical Population Dynamics 3*, O. Arino, D. E. Axelrod, and M. Kimmel eds., pp. 49–70. Wuerz Publishing, Winnepeg, Manitoba, Canada.

67. Tan, W. Y. and Chen, C. W. (1995). A bivariate stochastic model of carcinogenesis involving two cancer tumors. Paper presented at the 9th International Conference on Mathematical and Computer Modelling. University of California, Berkeley.

68. Tan, W. Y. and Chen, C. W. (1998). Stochastic modeling of carcinogenesis: Some new insight. *Math. Comput. Model.*, **28**, 49–71.

69. Tan, W. Y. and Chen, C. W. (2000). Assessing effects of changing environment by a multiple pathways model of carcinogenesis. *Math. Comput. Model.*, **32**, 229–250.

70. Tan, W. Y., Chen, C. W., and Wang, W. (1999). Some state space models of carcinogenesis. In *Proc. of 1999 Medical Science Simulation Conf.*, J. G. Anderson and M. Katzper eds., pp. 183–189. The Society for Computer Simulation, San Diego, CA.

71. Tan, W. Y., Chen, C. W. and Wang, W. (2000). Some multiple-pathways models of carcinogenesis and applications. In *Proceedings of 2000 Medical Science Simulation*, J. G. Anderson and M. Katzper ed., pp. 162–169. The Society for Computer Simulation, San Diego, CA.

72. Tan, W. Y., Chen, C. W., and Wang, W. (2001). Stochastic modeling of carcinogenesis by state space models: A new approach. *Math. Comput. Model.*, **33**, 1323–1345.

73. Tan, W. Y., Chen, C. W., and Zhu, J. H. (2002). *Estimation of Parameters*

in *Carcinogenesis Models via State Space Models*. Paper presented in person at the Eastern and Northern Biometric Society Meeting. Arlington, VA, March 15–17, 2002.

74. Tan, W. Y. and Gastardo, M. T. C. (1985). On the assessment of effects of environmental agents on cancer tumor development by a two-stage model of carcinogenesis. *Math. Biosci.*, **73**, 143–155.

75. Tan, W. Y. and Singh, K. P. (1990). A mixed model of carcinogenesis—with applications to retinoblastoma. *Math. Biosci.*, **98**, 201–211.

76. Tan, W. Y. and Ye, Z. Z. (2000). Estimation of HIV infection and HIV incubation via state space models. *Math. Biosci.*, **167**, 31–50.

77. Tan, W. Y., Zhang, L. J., and Chen, C. W. (2004). Stochastic modeling of carcinogenesis: state space models and estimation of parameters. *Discrete Continuous Dyn. Syst. Ser. B*, **4**, 297–322.

78. Thorslund, T. W., Brown, C. C., and Charnley, G. (1987). Biologically motivated cancer risk models. *Risk Anal.*, **7**, 109–119.

79. Ward, R., Meagher, A., Tomlinson, I., O'Connor, T., Norre, M., Wu, R., and Hawkins, N. (2001). Microsatellite instability and the clinicopathological features of sporadic colorectal cancer. *Gut*, **48**, 821–829.

80. Whittemore, A. S. and Keller, J. B. (1978). Quantitative theories of carcinogenesis, *SIAM Rev.*, **20**, 1–30.

81. Yakovlev, A. Y. and Tsodikov, A. D. (1996). *Stochastic Models of Tumor Latency and Their Biostatistical Applications*. World Scientific, River Edge, NJ.

82. Yang, G. L. and Chen, C. W. (1991). A stochastic two-stage carcinogenesis model: a new approach to computing the probability of observing tumor in animal bioassays. *Math. Biosci.*, **104**, 247–258.

9 Centralized Genomic Control: A Simple Approach Correcting for Population Structures in Case–Control Association Studies

Hong Zhang, Qizhai Li, Yong Zang, Yaning Yang, and Gang Zheng

9.1 Introduction

Although the case–control study is a popular design for detecting susceptibility genetic variants of a complex disease [15], it may yield spurious associations due to population structure. Two types of population structure have been discussed [3]. One is *population stratification* (PS), which occurs when genotype frequencies of a candidate marker vary across the subpopulations, the Hardy–Weinberg equilibrium (HWE) holds within each subpopulation, and the disease prevalence differs in the subpopulations. The other is *cryptic relatedness* (CR), which occurs when the genotypes of individuals within each subpopulation are correlated. Many statistical methods have been developed to correct for population structure. One is *structured association*, which infers membership probabilities for each sample using genomic data [13,14,17]. *Genomic control* (GC) is another approach; it adjusts for the variance distortion of the trend test using null loci that are not linked to the disease loci [1,4,6]. *Delta centralization* (DC), on the other hand, adjusts for the bias of the trend test using unlinked null loci [9,10]. Extensions of GC and DC also have been studied [5,21–23]. In genome-wide association studies, Price et al. [12] proposed a correction for PS based on principal-component analysis (PCA). It requires a large sample size and a large number of null loci to infer PS [20]. Alternatively, Epstein et al. [7] proposed a stratification method based on a predictor of the probability of the disease using substructure-informative loci. It can be applied to candidate gene studies.

The above approaches effectively correct for PS or CR, but they are not effective in correcting for both PS and CR simultaneously. Yu et al. [19] proposed a method of combining the structured association and GC to correct for both PS and CR. Zhu et al. [25] further developed a method to allow for familial correlation among individuals. Zheng et al. [24] demonstrate that, under PS, the trend test is biased and its variance is distorted. In the presence of CR, however, the trend test is not biased but its variance is distorted. Thus, under the null hypothesis of no association, the trend test does not asymptotically follow a standard normal distribution (or chi-square distribution with one degree of freedom). Intuitively, in the presence of population structure, it should suffice first to correct for the bias and variance distortion and then to apply the corrected trend test to test for association.

In this article, we integrate GC and DC to correct for both the variance distortion and the bias simultaneously. As illustrated in Zheng et al. [24], the bias and variance distortion can be removed effectively when the allele frequencies of the unlinked null loci roughly match the allele frequency of the candidate marker in the subpopulations. This matching, in practice, can be done only at the total population level using control samples. Therefore, such matching may not be achieved [24]. Our goals here include presenting the idea of integrating both GC and DC in one simple approach, conducting simulation studies with or without matching, studying matching probabilities using the HapMap data, and applying our results to two real datasets for comparison with other approaches.

9.2 Methods

9.2.1 Data and Trend Test

Throughout this chapter, we focus on a special genetic marker, *single-nucleotide polymorphism* (SNP), although the proposed method can be extended to other types of markers. The case–control data for a single SNP with alleles A and B consists of genotype counts (r_0, r_1, r_2) for genotypes (AA, AB, BB) in cases and (s_0, s_1, s_2) in controls, with $r = r_0 + r_1 + r_2$ cases and $s = s_0 + s_1 + s_2$ controls. The total number of samples is $n = r + s$. The counts for the three genotypes (AA, AB, BB) in the n samples are $n_0 = r_0 + s_0$, $n_1 = r_1 + s_1$ and $n_2 = r_2 + s_2$, respectively. The trend test [16]

$$\begin{aligned}Z = & \{n^{1/2}[(r_1/2 + r_2)/r \\ & - (n_1/2 + n_2)/n]\} \\ & \times \{[(s/r)\{(n_1/(4n) + n_2/n) \\ & - (n_1/(2n) + n_2/n)^2\}]^{1/2}\}^{-1},\end{aligned} \quad (1)$$

is often used to test for association. The trend test in (1) can be obtained as a score statistic in logistic regression model with a nuisance parameter and a covariate for the genotype, which is coded as 0, 1, and 2 for genotypes AA, AB, and BB, respectively. In the absence of population structure, it has an asymptotic standard normal distribution $N(0,1)$ under the null hypothesis (H_0) of no association. That is, Z^2 has an asymptotic chi-square distribution with 1 degree of freedom (dF) under H_0, denoted by $Z^2 \sim \chi_1^2$. The above trend test is asymptotically optimal under the additive genetic model (counting the number of the risk allele; here B is assumed to be the risk allele) [8,16]. In this chapter, we focus on the trend test for the additive model.

9.2.2 Impact of Population Structure

In the presence of PS, as shown by Zheng et al. [24], Z does not asymptotically follow $N(0,1)$ under H_0. In fact, under H_0, $Z \sim N(\delta, \sigma^2)$ in the presence of PS, where δ and σ^2 are the mean and variance of Z under H_0, respectively, where $\delta \neq 0$ or $\sigma^2 \neq 1$ or both in the presence of PS. In the presence of CR, however, $\delta = 0$ but $\sigma^2 \neq 1$ [24]. Hence, if we can estimate δ and σ^2, denoted by $\widehat{\delta}$ and $\widehat{\sigma}^2$, respectively, we can correct Z by using $Z^* = (Z - \widehat{\delta})/\widehat{\sigma}$ to test for association; Z^* has an asymptotic $N(0, 1)$ under H_0 in the presence of population structure. The GC [1,4–6,22,23] was designed to correct for σ while the DC [9,10] was designed to correct for normalized $\delta' = \delta/\sigma$. Other methods mentioned in the introduction were not proposed to correct for bias as for variance distortion.

9.2.3 Genomic Control, Delta Centralization, and Centralized Genomic Control

Suppose that $L > 1$ unlinked null loci are available. Let χ^2 be the 1-df Pearson chi-square test statistic (based on allele counts) for testing association between candidate gene and disease, and $\chi_1^2, \cdots, \chi_L^2$ be the analogies of χ^2 but for L null loci. The variance inflation parameter $\lambda = \sigma^2$ in χ^2 has two popular estimates. One is the robust estimate $\widehat{\lambda}_1$ the sample median of $\chi_1^2, \cdots, \chi_L^2$ divided by 0.456, the other one is the efficient estimate $\widehat{\lambda}_2$ the sample mean of $\chi_1^2, \cdots, \chi_L^2$. Genomic control (GC)-corrected test statistic is

$$T_{\text{GC1}} = \frac{\chi^2}{\widehat{\lambda}_1} \text{ or } T_{\text{GC2}} = \frac{\chi^2}{\widehat{\lambda}_2}. \quad (2)$$

The estimate $\widehat{\lambda}_1$ is valid when λ is constant across the null loci, which implies that Wright's measure of degree of genetic differentiation between subpopulations should be approximately constant across the loci. The other estimate $\widehat{\lambda}_2$ is efficient when the null loci have minor allele frequencies (MAFs) analogous to that of the candidate locus, but could be quite biased in the presence of outliers. In our simulation studies in the next two sections, we only consider the later correction.

Gorroochurn et al. [9] proposed to estimate $\delta' = \delta/\sigma$ by

$$\begin{aligned}\widehat{\delta}' &= \{(\widehat{P}_1 + 2\widehat{P}_0) - (\widehat{Q}_1 + 2\widehat{Q}_0)\} \\ &\times \left\{\left(\frac{1}{r} - \frac{1}{n}\right)\{(\widehat{Q}_1 + 4\widehat{Q}_0) \right. \\ &\left. - (\widehat{Q}_1 + 2\widehat{Q}_0)\}^2\right\}^{-1/2},\end{aligned}$$

where $\widehat{P}_i = \frac{1}{L}\sum_{j=1}^{L}\widehat{P}_i^{(j)}$, $\widehat{Q}_i = \frac{1}{L}\sum_{j=1}^{L}\widehat{Q}_i^{(j)}$, and $\widehat{P}_i^{(j)}$ and $\widehat{Q}_i^{(j)}$ are the null locus versions of r_i/r and s_i/s, respectively ($i = 0, 1$). The bias-corrected test statistic is

$$\begin{aligned}T_{\text{DC}} &= [\text{sign}\{n(r_1 + 2r_0) - r(n_1 + 2n_0)\} \\ &\times Z - \widehat{\delta}']^2. \quad (3)\end{aligned}$$

Here we propose a method, referred to as *centralized genomic control* (CGC), to correct for both the bias δ and the variance distortion σ simultaneously. Notice that Z has an asymptotic $N(\delta, \sigma^2)$ under H_0 in the presence of population structure (PS and/or CR). The results of Zheng et al. [24] indicate that δ and σ^2 are functions of the MAFs of the candidate SNP in the subpopulations. Let the trend test statistics on the null loci be $Z_1, \ldots, Z_L$. Then Z_l, $l = 1, \ldots, L$, can be regarded as an approximately random sample of size L drawn from the distribution $N(\delta_1, \sigma_1^2)$ in the presence of population structure.

Efficient estimates of δ_1 and σ_1^2 are given by

$$\begin{aligned}\widehat{\delta}_1 &= \bar{Z} = \frac{1}{L}\sum_{l=1}^{L} Z_l, \\ \widehat{\sigma}_1^2 &= \frac{1}{L-1}\sum_{l=1}^{L}(Z_l - \bar{Z})^2. \quad (4)\end{aligned}$$

Let the MAFs of the candidate SNP (null loci) in the jth subpopulation be p_{0j} (p_{1j}) for $j = 1, \ldots, J$. An ideal but unrealistic assumption is that

$$p_{1j} = p_{0j}, \text{ for } j = 1, \ldots, J, \quad (5)$$

under which $\delta_1 = \delta$ and $\sigma_1^2 = \sigma^2$. Thus, the estimates in (4) can be used to estimate δ and σ^2, respectively. Note that the condition $p_{1j} = p_{0j}$ for all j is sufficient but not necessary. When this condition holds, the corrected trend test can be written as

$$Z_{\text{CGC}}^2 = \frac{(Z - \hat{\delta}_1)^2}{\hat{\sigma}_1^2}, \quad (6)$$

which has an asymptotic F distribution with $(1, L-1)$ degrees of freedom, denoted by $Z_{\text{CGC}}^2 \sim F(1, L-1)$, under H_0 in the presence of population structure (PS and/or CR). We call Z_{CGC}^2 the CGC test. Note that, $F(1, L-1)$ converges to χ_1^2 as $L \to \infty$. Therefore, Z_{CGC}^2 can be used for small L, as shown by the real data analysis later.

9.2.4 Matching

In Gorroochurn et al. [9,10], the condition $p_{1j} = p_{0j}$ for all j in (5) was called *allele frequency matching*. This allele frequency matching guarantees $\delta = \delta_1$ and $\sigma^2 = \sigma_1^2$, so that the bias and variance distortion can be corrected directly as in (6). In reality, we do not estimate p_{0j} (the candidate SNP) or p_{1j} (the null loci). What we can estimate is the following

$$\text{MAF}_i = \sum_{j=1}^{J} \pi_j p_{ij}, \text{ for } i = 0, 1,$$

where π_j is the proportion of controls in the jth subpopulation. Thus, MAF_0 is the MAF for the candidate marker in the total population in controls while MAF_1 is the MAF for a null locus in the total population in controls. Both of them can be estimated using the observed case–control samples, denoted as $\widehat{\text{MAF}}_0$ and $\widehat{\text{MAF}}_1$. However, $\widehat{\text{MAF}}_0 = \widehat{\text{MAF}}_1$ does not imply that $p_{0j} = p_{1j}$ for all j unless $J = 1$, which means there is no population structure.

If $p_{0j} = p_{1j}$ for all j is referred to as allele frequency matching in *theory*, we refer to $\widehat{\text{MAF}}_0 = \widehat{\text{MAF}}_1$ as allele frequency matching in *practice*. The stronger the population structure, the less likely it is that $p_{0j} = p_{1j}$ holds for all j, given that $\widehat{\text{MAF}}_0 = \widehat{\text{MAF}}_1$ holds.

The original GC test [4] did not require the matching, but the original DC test [9,10] required such matching, or at least allele frequency matching in the total population. Zheng et al. [22,23] showed that the GC test needs allele frequency matching when the genetic model is not the additive model. Gorroochurn et al [9,10] studied the performance of matching with different windows. A 5% matching window means that the MAFs of null loci fall in $(p - 0.025, p + 0.025)$, where p is the estimated MAF of the candidate SNP in the total population in controls.

9.2.5 Choice of Null Loci

In genome-wide association studies, the choice of null loci was studied by Yu et al. [20], who suggested a panel of 12,898 SNPs selected across all the chromosomes based on their linkage disequilibrium and distributions of minor allele frequencies. For candidate gene studies, on the other hand, we may not have a large number of null loci from which to choose. The final choice of null loci would be limited to either the panel of Yu et al. [20] or the available ones subject to allele frequency matching with a given window, e.g., 5%, 10% or 20%.

9.3 Simulation Studies

Our first simulation study examines the performance of the GC test [4], the DC test [9], and the CGC test given in (6) in the presence of PS under three situations: (1) using Z without correction, (2) making corrections using the GC, DC and CGC tests without matching, and (3) making corrections using the GC, DC and CGC tests with matching based on a window 5%. Three different categories of PS are considered in our simulation: a random SNP and a differential SNP [12], and stratified sampling [4].

In the random SNP, an ancestral allele frequency of the candidate SNP is first generated from the uniform distribution, and a beta distribution is formed using this simulated allele frequency and the given Wright's coefficient of inbreeding, F. Then the allele frequency of the candidate SNP in each subpopulation is generated from the beta distribution. When $F = 0$, all the allele frequencies in the subpopulations are equal and PS vanishes. In the differential SNP, the allele frequencies in the subpopulations are not simulated but specified; for example, for $J = 2$ subpopulations, the allele frequencies in the two subpopulations are $p_1 = 0.2$ and $p_2 = 0.5$. The larger the difference between the two allele frequencies, the stronger the PS. In the stratified sampling, an ancestral allele frequency of the candidate SNP is fixed, say at 0.5 [4]. Then different Wright's coefficients of inbreeding are used for different subpopulations to generate an allele frequency for each subpopulation from different beta distributions. The numbers of cases R_j and controls S_j in the jth subpopulation are also prespecified. Detailed descriptions of three categories are given below.

Descriptions of random SNP, differential SNP, and stratified samplings. To evaluate the performance of the proposed CGC to control for type I error under the null hypothesis in the presence of PS, we conducted simulation studies and examined three tests GC, DC, and CGC. In the simulation studies, we used $L = 50$ null loci and a total of $r = 250$ cases and $s = 250$ controls. The type I error of the original trend test without any corrections was also estimated for comparison. Each simulation had 10,000 replications. We designed the simulation studies with three categories. Each category had allele frequency matching (Yes) with a 5% window and no allele frequency matching (No).

The first category was referred to as *random SNP* [15]. We specified the number of subpopulations J, $J = 2$ or 5, a single Wright's coefficient of inbreeding F, $F = 0, 0.005, 0.01$, or 0.05, and the proportion of samples and disease prevalence in the jth subpopulation (denoted by π_j and k_j, respectively, for $j = 1, \ldots, J$). In each replication: (1) We simulated an ancestral allele frequency p from the uniform distribution between 0.1 and 0.9, $U(0.1, 0.9)$, (2) we simulated the allele frequencies for the J subpopulations, denoted as p_j ($j = 1, \ldots, J$), independently from the beta distribution $\beta((1-F)p/F, (1-F)(1-p)/F)$, when $F \neq 0$; the mean and variance of this beta distribution are p and $p(1-p)F$, respectively (when $F = 0$, $p_j = p$ for all j); (3) Assuming HWE in each subpopulation, we calculated the genotype distributions for the jth subpopulation as $g_{0j} = \Pr(G_0) = (1-p_j)^2$, $g_{1j} = \Pr(G_1) = 2p_j(1-p_j)$, and $g_{2j} = \Pr(G_2) = p_j^2$, where (G_0, G_1, G_2) are three genotypes; (4) then genotype distributions for cases and controls can be written as $p_i^* = \Pr(G_i|\text{case}) = \sum_{j=1}^{J} \pi_j f_{ij} g_{ij} / \sum_{j=1}^{J} \pi_j k_j$ and $q_i^* = \Pr(G_i|\text{control}) = \sum_{j=1}^{J} \pi_j (1-f_{ij}) g_{ij} / \sum_{j=1}^{J} \pi_j (1-k_j)$, respectively, where f_{ij} is the penetrance for G_i in the jth subpopulation. Given the disease prevalence k_j and genotype relative risks (GRRs) λ_1 and λ_2, $f_{0j} = k_j/(g_{0j} + g_{1j}\lambda_1 + g_{2j}\lambda_2)$, $f_{1j} = f_{0j}\lambda_1$, and $f_{2j} = f_{0j}\lambda_2$. Under the null hypothesis, $\lambda_1 = \lambda_2 = 1$. (5) We simulated the genotype counts for r cases and s controls from the multinomial distributions $Mul(r; p_0^*, p_1^*, p_2^*)$ and $Mul(s; q_0^*, q_1^*, q_2^*)$,

Table 1: Random SNPs. The observed type I errors in the presence of PS (5 expected) when there is no correction (No) and corrections by the DC, GC, and CGC tests using $L = 50$ null loci. The number of subpopulations (SPs) is $J = 2$ or 5 with different Wright's coefficients of inbreeding F, various proportions of subpopulations, and different prevalence in the subpopulations. The matching window is 5%.

J	F	Match	No	DC	GC	CGC
2^a	0	No	4.89	5.08	5.05	5.15
		Yes		5.22	3.39	3.61
	0.005	No	14.87	14.87	4.56	4.87
		Yes		15.29	3.77	3.91
	0.01	No	21.55	21.86	4.87	5.21
		Yes		22.16	4.05	4.30
	0.05	No	50.07	50.18	4.69	4.91
		Yes		50.54	4.36	4.70
5^b	0	No	4.93	5.25	5.12	5.37
		Yes		5.37	4.52	4.84
	0.005	No	8.38	8.66	5.01	5.13
		Yes		8.51	4.30	4.52
	0.01	No	11.26	11.58	4.69	4.84
		Yes		11.72	4.43	4.60
	0.05	No	30.05	30.04	4.90	5.13
		Yes		30.56	4.88	5.02

[a] Proportions of SP = (0.6,0.4); prevalence = (0.01,0.2).
[b] Proportions of SP = (0.2,0.1,0.3,0.3,0.1); prevalence = (0.005,0.15,0.05,0.01,0.2).

respectively. Without allele frequency matching, procedures 1–5 for the candidate SNP were repeated independently for each of the null loci with $\lambda_1 = \lambda_2 = 1$. With allele frequency matching, procedures 2–5 were also repeated independently for each of the null loci, but with the ancestral allele frequency simulated from $U(p-d/2, p+d/2)$ truncated at 0.1 and 0.9, where p is the ancestral allele frequency simulated for the candidate SNP and $\lambda_1 = \lambda_2 = 1$. The second category is referred to as *differential SNP*, in which allele frequencies in the subpopulations ($p_j, j = 1, \ldots, J$) were prespecified rather than simulated from the beta distribution, so this approach is to compare the performance of different correction methods under various levels of PS. For this category, we considered $J = 2$ and specified two different values for p_1 and p_2. We fixed $p_1 = 0.2$ and varied p_2 from 0.2 to 0.8 with increments of 0.1. The PS becomes stronger when p_2 increases. Given p_1 and p_2, the genotype counts for the candidate SNP and null loci were simulated like the first category with allele frequency matching. When there was no allele frequency matching, p_1 and p_2 for each of the L null loci were randomly generated from $U(0.2, 0.8)$. The third category was referred to as *stratified sampling*. It is similar to the sampling considered by Devlin and Roeder [4]. We specified R_j cases and S_j controls from the jth subpopulation, with $J = 2$ subpopulations ($r = R_1 + R_2$ and $s = S_1 + S_2$). We also specified Wright's coefficients F_1 and F_2 ($F_1 \neq F_2$) for the two subpopulations. Then the ancestral allele frequency was fixed at $p = 0.3$ rather than simulated from the uniform distribution (Devlin and Roeder [4] used $p = 0.5$): (1) The allele frequency p_j was simulated for the jth subpopulation from the beta distribution, $\beta((1-F_j)p/F_j, (1-F_j)q/F_j)$ for $j = 1, 2$; (2) the genotype distributions for the jth subpopulation are given by $g_{0j} = \Pr(G_0) = (1-p_j)^2$, $g_{1j} = \Pr(G_1) = 2p_j(1-p_j)$, and $g_{2j} = \Pr(G_2) = p_j^2$. We calculated f_{ij} as for the first category. Then the genotype distributions of G_i for cases and controls are $p_{ij} = g_{ij}f_{ij}k_{ij}$ and $q_{ij} = g_{ij}(1-f_{ij})/(1-k_{ij})$ in the jth subpopulation. Under the null hypothesis, $p_{ij} = q_{ij} = g_{ij}$ for each j. Then, (3) for the candidate SNP, the genotype counts for cases and controls in the jth subpopulation were simulated from $Mul(R_j; p_{0j}, p_{1j}, p_{2j})$ and $Mul(S_j; q_{0j}, q_{1j}, q_{2j})$ for $j = 1, 2$. Without allele frequency matching, the null loci in each subpopulation were simulated independently using procedures 1–3 with $\lambda_1 = \lambda_2 = 1$. With allele frequency matching, we generated the matched allele frequencies first for the null loci. Then, the null loci in each subpopulation were simulated independently from procedures 1 and 2 with $\lambda_1 = \lambda_2 = 1$.

The second simulation study examines the genotype frequency matching probability using HapMap data [18], including 45 unrelated Han Chinese (CHB) individuals, and 45 unrelated Japanese individuals (JPT). The matching probability for a given SNP is defined as the proportion of the SNPs matching in the subpopulations among all the SNPs matching in the total population for a given window; see below for a detailed description. After the SNP rs numbers (IDs) in the HapMap data were matching with the 12,898 null SNPs given by Yu et al. [20] and SNPs with MAFs less than 0.01 were removed, 1870 SNPs remained for our analysis.

Allele frequency-matching probability. We first define the allele frequency matching probability of a SNP. For each SNP, we calculate its MAF, p, which forms a window $[p-d/2, p+d/2]$, where d is a prespecified window width for matching. We identify SNPs whose MAFs fall in this window. Then we calculate the proportion of these

SNPs whose MAFs in subpopulations also match those of the candidate SNP with the same window width. This provides an estimate of the matching probability for that candidate SNP. We then use HapMap data, including 45 unrelated Han Chinese individuals (CHB) and 45 unrelated Japanese individuals (JPT), to study the allele frequency matching probability. We use the 12,898 SNPs with identical SNP IDs given by Yu et al. [20] and remove SNPs with MAFs less than 0.01. Then the remaining 1870 SNPs are used in our analysis. We consider a population with two subpopulations: JPT+CHB. Note that our approach to estimating the matching probability is different from that of Zheng et al. [24], who did not identify all SNPs with matching to a given SNP. They predefined subintervals for MAFs. For a given SNP in a subinterval, they identified only SNPs from the same subinterval for matching. It is possible SNPs from nearby sub-intervals could have matching MAFs in the controls. Therefore, our estimation of matching probabilities is more accurate.

9.4 Simulation Results

9.4.1 Comparison among the GC, DC, and CGC Tests

In the simulation, we used 250 cases and 250 controls for a candidate SNP and $L = 50$ null loci. When allele frequency matching was considered, matching with a 5% window was used. In the random SNPs, we chose $J = 2$ or 5 subpopulations with different proportions of subpopulations and disease prevalence rates. Results are reported in Table 1, based on 10,000 replicates. Under the random SNP, when $F > 0$, the trend test without correction (NO) has inflated type I error rates. Correcting for the bias alone (DC) does not control type I error regardless of whether or not matching was used. On the other hand, GC and CGC have good control of type I error rates even without matching. After matching using a 5% window, both GC and CGC appear to be conservative, particularly GC. When F increases, the GC and CGC tests become less conservative. Overall, under the matching, CGC has better control of type I error than GC. For example, when $F = 0.005$ and $J = 2$, the observed type I error rates for GC and CGC are 3.77% and 3.91% (the expected is 5%), which become 4.30% and 4.51% when $J = 5$. Thus, the impact of the random SNP is reduced when J increases.

In the differential SNPs, we chose only $J = 2$ subpopulations. The results are presented in Table 2, which shows that without matching, neither test has good control of type I error rates. Under matching, both DC and CGC have good control of type I error rates, although the DC test appears to be conservative when p_2 increases. Table 2 also shows that the GC test does not control type I error under the differential SNP. In the stratified sampling, we also considered only $J = 2$ subpopulations. Results from four simulation studies are reported in Table 3, with two rows for each study. The first two rows are for a pairwise matching case–control study, in which one case is matched to one control in each subpopulation ($R_j = S_j$). Thus, when $F_1 \neq F_2$, the type I error rates are under control for each method, including the original trend test without corrections. When the level of PS increases, the original trend test and the DC test without matching have inflated type I error rates. Under matching, the DC test has good control of type I error rates unless the PS is very extreme, in which case the variance distortion has some impact. For example, in the last row, the observed type I error for the DC test with matching is 7.93%, which is much larger than the expected type I error rate. The GC test has good control of type

Table 2: Differential SNPs. The observed type I errors in the presence of PS (5 expected) when there is no correction (No) and corrections by the DC, GC, and CGC tests with $L = 50$ null loci. The number of subpopulations is $J = 2$ with an equal proportion of two subpopulations. The prevalence is $(0.15, 0.05)$ and the allele frequencies in the two subpopulations are (p_1, p_2), where $p_1 = 0.2$ and p_2 varies. The matching window is 5%.

		No Matching			With Matching		
p_2	No	DC	GC	CGC	DC	GC	CGC
0.2	4.74	4.86	0.00	0.00	4.90	4.22	4.33
0.3	17.54	19.54	0.04	0.04	4.99	3.28	4.82
0.4	48.70	63.84	1.09	3.16	5.10	1.01	5.08
0.5	77.01	84.99	1.72	3.92	4.80	0.07	4.75
0.6	93.06	97.94	9.59	27.84	4.87	0.01	4.93
0.7	98.34	99.06	45.94	54.66	4.53	0.00	4.66
0.8	99.60	99.19	59.21	46.29	4.56	0.00	5.03

Table 3: Stratified sampling. The observed type I errors in the presence of PS (5 expected) when there is no correction (No) and corrections by the DC, GC, and CGC tests with $L = 50$ null loci. The number of subpopulations (SP) is $J = 2$ with ancestral allele frequency $p = 0.3$. The numbers of cases and controls in the jth subpopulation are (R_j, S_j) with Wright's coefficients of inbreeding $(F_1, F_2) = (0.01, 0.05)$. The matching window is 5%.

(R_1, S_1)	(R_2, S_2)	Match	No	DC	GC	CGC
(150,150)	(100,100)	No	4.66	4.76	5.06	5.26
		Yes		4.74	4.75	4.98
(150,100)	(100,150)	No	11.48	11.91	4.89	5.03
		Yes		4.78	4.09	5.00
(200,50)	(50,200)	No	44.71	45.41	4.25	4.56
		Yes		5.34	1.77	3.89
(250,0)	(0,250)	No	73.16	73.49	3.87	4.21
		Yes		7.93	0.50	4.01

I error rates, even without matching, when the PS is moderate. However, GC appears to be very conservative when the level of PS increases or with matching. For example, the observed type I error rate is 1.77% for the GC test when $(R_1, S_1) = (200, 50)$ and $(R_2, S_2) = (50, 200)$. This is due to the bias effect, which is not corrected for in GC. The CGC test has good control of type I error rates overall. Even under the most extreme PS (the last two rows), regardless of matching, the CGC test is only slightly conservative while the GC test is very conservative and the DC test is anti-conservative.

In summary, in addition to its moderate conservativeness under strong PS, the CGC test can be applied without matching under the random SNP and stratified sampling, and with matching under the differential SNP. DC and GC can also be applied under certain situations with or without matching. However, the CGC test overall outperforms the DC and GC tests.

9.4.2 Matching Probabilities Using HapMap Data

Figure 1 plots the matching probabilities given PS based on HapMap data with the CHB and the JPT subpopulations mixed. Figure 1 shows that the matching probabilities increase with the matching window width. The matching probabilities are much higher with the 20% window (the second panel) than with the 10% window (the first panel). With the 20% matching window, the matching probabilities are usually greater than 0.6 when the MAFs of the candidate SNPs are either less than 0.1 or greater than 0.45. When the MAFs are between 0.1 and 0.45, most matching probabilities are between 0.5 and 0.7. In some cases, however, the matching probabilities could be below 0.20 when the MAFs are between 0.1 and 0.45. Note that in reality PS is much weaker than the one we created

in the simulation by mixing the CHB and JPT. The variance inflation factor (VIF) is often used in practice to measure the PS. When there is no PS, the VIF should be 1. When VIF is much larger than 1, a false-positive association could be obtained due to the PS. In the population that mixes the CHB and JPT, the VIF is 1.672, which is much larger than one could observed in practice, due to hidden PS [12]. Therefore, in practice, the matching probabilities are expected to be higher than those estimated in Figure 1.

9.5 Applications

9.5.1 Application to the Height Data

A spurious association between the candidate SNP rs4988235 in the *lactase gene* and the tall/short status resulting from the PS in the European American sample was reported by Campbell et al. [2]. Altogether 192 tall individuals and 176 short individuals were genotyped at the candidate SNP. In addition, 67 null ancestry-informative SNPs (AIMs) were genotyped. The p value of the trend test without adjusting for the PS is 0.0038, which indicates a significant association. Neither GC (p-value = 0.0038) nor DC (p value = 0.024) can completely correct for this confounding. When EIGENSTRAT (Price et al. [12]) is used, the p value is 0.003 when the top 10 principal components are included. The possible reason is that the number of null loci is too small [20].

As demonstrated by Epstein et al. [7], their method can effectively correct for PS in the height data using all 67 null loci (p value was 0.44). Using CGC with a matching window of 10%, we identified 7 null SNPs whose MAFs matched with that of SNP rs4988235 in the controls. Then we applied CGC and obtained the p value 0.069. When we used a matching window

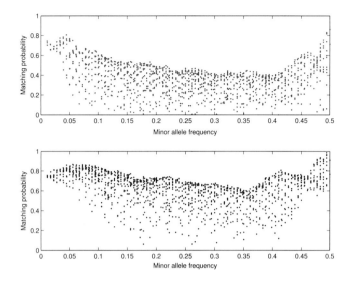

Figure 1: Matching probability based on HapMap data. The first panel has a 10% matching window and the second panel has a 20% matching window.

of 20%, 14 matching null SNPs were identified, and the obtained p value was 0.055; both showed that CGC successfully corrected for PS in the height data.

9.5.2 Application to Age-Related Macular Degeneration Data

Klein et al. [11] conducted a genome-wide association study for age-related macular degeneration (AMD) using 116,206 SNPs and 96 cases and 50 controls. We deleted the SNPs on the X chromosome and the SNPs on the autosomal chromosomes with call rates less than 0.85, missing heterogenuous, or MAFs less than 0.01. The remaining 99,650 SNPs were analyzed.

In order to infer the underlying population substructure, we used 1,870 SNPs after matching the 12,898 SNPs with identical SNP IDs [20] with those of the 99,650 SNPs and removing SNPs with MAF less then 0.01. To apply CGC, we used a 10% matching window. Although there is no

PS in the AMD data, we applied GC, DC, CGC, and EIGENSTRAT to correct for PS as if there were PS in the AMD data. The VIFs calculated, after these approaches were applied to each of the 99,650 SNPs, were equal to 1.01, 0.66, 1.01, and 1.17 for GC, DC, CGC, and EIGENSTRAT, respectively. When there is no PS, the GC and CGC tests each has VIF=1.01; which is very close to 1. The EIGENSTRAT has a relatively larger VIF, 1.17, due to a small sample size in the AMD dataset. The DC test, however, over corrected for PS.

9.6 Discussion

There have been some methods that could control for PS or CR or both in disease marker association analysis using case–control data. The existing methods that can correct for both PS and CR require extensive computation time, as using software STRUCTURE to assign populations. In this chapter we propose an alternative method that is as simple to implement as

GC or DC. We demonstrated that our approach controls for type I error due to PS at least as well as GC and DC, and perhaps better than them at the cost of requiring matching of null loci is some situations. We studied the probabilities of such matching using the HapMap data and found that the matching probabilities could vary depending on the minor allele frequency and the choice of matching windows. From our results, with a 10% matching window, the matching probabilities range from below 20% to over 80%. In practice, the matching probabilities are expected to be higher than those obtained from our simulation studies because mixing the Chinese and Japanese from the HapMap data created a much stronger PS (with the variance inflation factor 1.672) than one could observe in practice (the variance inflation factor is usually less than 1.2). Simulation results and applications to real data demonstrate the usefulness of our CGC approach.

Acknowledgments. We thank Drs. K. Yu and B. J. Stone for their valuable help. The work is partially supported by the Natural Science Foundation of China (10701067, 10671189) and the Knowledge Innovation Program of the Chinese Academy of Sciences.

References

1. Bacanu, S. A., Devlin, B., and Roeder, K. (2000). The power of genomic control. *Am. J. Hum. Genet.*, **66**, 1933–1944.

2. Campbell, C. D., Ogburn, E. L., Lunetta, K. L., Lyon, H. N., Freedman, M. L., Groop, L. C., Altshuler, D., Ardlie, K. G., and Hirschhorn, J. N. (2005). Demonstrating stratification in a European American population. *Nat. Genet.*, **37**, 868–872.

3. Crow, J. F. and Kimura, H. (1970). *An Introduction to Population Genetics Theory*. Burgess Publication Co., Minneapolis, MN.

4. Devlin, B. and Roeder, K. (1999). Genomic control for association studies. *Biometrics*, **55**, 997–1004.

5. Devlin, B., Bacanu, S. A., and Roeder, K. (2004). Genomic control to the extreme. *Nat. Genet.*, **36**, 1129–1130.

6. Devlin, B., Roeder, K., and Wasserman, L. (2001). Genomic control, a new approach to genetic-based association studies. *Theor. Popul. Biol.*, **60**, 155–166.

7. Epstein, M. P., Allen, A. S., and Satten, G. A. (2007). A simple and improved correction for population stratification in case-control studies. *Am. J Hum. Genet.*, **80**, 921–930.

8. Freidlin, B., Zheng, G., Li, Z., and Gastwirth, J. L. (2002). Trend tests for case-control studies of genetic markers: power, sample size and robustness. *Hum. Hered.*, **53**, 146–152.

9. Gorroochurn, P., Heiman, G.A., Hodge, S.E. and Greenberg, D.A. (2006). Centralizing the non-central chi-square: a new method to correct for population stratification in genetic case-control association studies. *Genet Epidemiol* **30**, 277-289.

10. Gorroochurn, P., Hodge, S. E., Heiman, G. A., and Greenberg, D. A. (2007). A unified approach for quantifying, testing and correcting population stratification in case-control association studies. *Hum. Hered.*, **64**, 149–159.

11. Klein, R. J., Zeiss, C., Chew, E. Y., Tsai, J. Y., Sackler, R. S., Haynes, C., Henning, A. K., SanGiovanni, J. P., Mane, S. M., Mayne, S. T., Bracken, M. B., Ferris, F. L., Ott, J., Barnstable, C., and Hoh, J. (2005). Complement factor H polymorphism in age-related macular degeneration. *Science*, **308**, 385–389.

12. Price, A. L., Patterson, N. J., Robert, M. P., Weinblatt, M. E., Shadick, N. A., and Reich, D. (2006). Principal components analysis corrects for stratification in genome-wide association studies. *Nat. Genet.*, **38**, 904–909.

13. Pritchard, J. K. and Rosenberg, N. A. (1999). Use of unlinked genetic markers

to detect population stratification in association studies. *Am. J. Hum. Genet.*, **65**, 220–228.

14. Pritchard, J. K., Stephens, M., Rosenberg, N. A., and Donnelly, P. (2000). Association mapping in structured populations. *Am. J. Hum. Genet.*, **67**, 170–181.

15. Risch, N. and Merikangas, K. (1996). The future of genetic studies of complex human diseases. *Science*, **273**, 1516–1517.

16. Sasieni, P. D. (1997). From genotypes to genes: doubling the sample size. *Biometrics*, **53**, 1253–1261.

17. Satten, G. A., Flanders, W. D., and Yang, Q. (2001). Account for unmeasured population substructure in case-control studies of genetic association using a novel latent-class model. *Am. J. Hum. Genet.*, **68**, 466–477.

18. The International HapMap Consortium (2003). The International HapMap Project. *Nature*, **426**, 789–796.

19. Yu, J., Pressoir, G., Briggs, W. H., Bi, I. V., Yamasaka, M., Doebley, J. F., McMuller, M. D., Gaut, B. S., Nielsen, D. M., Holland, J. B., Kresovich, S., and Buckler, E. S. (2006). A unified mixed-model method for association mapping that accounts for multiple levels of relatedness. *Nat. Genet.*, **38**, 203–208.

20. Yu, K., Wang, Z., Li, Q., Wacholder, S., Hunter, D. J., Hoover, R. N., Chanock, S., and Thomas, G. (2008). Population substructure and control selection in genome-wide association studies. *PLoS One*, **3**, e2551.

21. Zang, Y., Zhang, H., Yang, Y. N., and Zheng, G. (2007). Robust genomic control and robust delta centralization tests for case-control association studies. *Hum. Hered.*, **63**, 187–195.

22. Zheng, G., Freidlin, B., and Gastwirth, J. L. (2006). Robust genomic control for association studies. *Am. J. Hum. Genet.*, **78**, 350–356.

23. Zheng, G., Freidlin, B., Li, Z., and Gastwirth, J. L. (2005). Genomic control for association studies under various genetic models. *Biometrics*, **61**, 186–192.

24. Zheng, G., Li, Z., Gail, M. H., and Gastwirth, J. L. (2009). Impact of population substructure on trend tests for genetic case-control association studies. *Biometrics*, in press.

25. Zhu, X., Li, S., Cooper, R. S., and Elston, R.C. (2008). A unified association analysis approach for family and unrelated samples correcting for stratification. *Am. J. Hum. Genet.*, **82**, 352–365.

10. Change Point Methods in Genetics

Jie Chen

10.1 Introduction

In the last decade, fast developing biotechnologies have revolutionized the life science research. Among the many well-developed biotechnologies, microarray technology is a breakthrough biotechnology that makes it possible to quantify the expression patterns of thousands or tens of thousands of genes in various tissues, cell lines, and conditions simultaneously. Biologists, geneticists, and medical researchers now routinely use microarray technology in their specified research projects, thus resulting in voluminous numerical data related to expression of each gene encoded in the genome, the content of proteins and other classes of molecules in cells and tissues, and cellular responses to stimuli, treatments and environmental factors. The spontaneous genetic datasets have given statisticians a golden opportunity to contribute their wisdom to the modeling of such large dimensional genetic datasets, with the goal of making statistically significant and biologically meaningful inferences about the biological systems from which the datasets are acquired.

Statistical data analysis and statistical inference (estimation and hypothesis testing) are essential for interpreting the rapidly increasing databases of numerical measurements collected from biological processes using modern biotechnology such as microarray technology. In current life science research, there are many open statistical problems, including issues such as experimental design of gene chips, signal filtering, data analysis, and inference of cellular networks, and inference of genome-wide profiles, among others. Genome-wide studies describe the genomic landscape of disease, as well as molecular mechanisms of genome evolution and function. Thus, in addition to hypothesis testing, genomic studies are hypothesis-generating as well. While biologists can identify important biological problems, statisticians can develop statistical modeling and analytic methods to analyze the data produced by biological observations and experiments, and together they will be able to validate the results biologically and statistically. Therefore, statistical applications to specific genome-wide data resulting from various biological experiments are in great demand.

The microarray technology has not only provided powerful tools for life science research, it has also revolutionized the genetic research; process for medical diagnostics of cancer and other genetic diseases. To study the genetic reasons of tumor growth, cancer formation and genetic diseases such as down syndromes, researchers now are able to conduct genetic experiments, called DNA copy number experiments, for detecting DNA copy number variants (CNVs) using the microarray-based technology such as array comparative genomic hybridization (aCGH) technique ([28],[46]) or single nucleotide polymorphism (SNP) arrays [40]. The conventional comparative genomic hybridization (CGH) technique has proven to be useful in producing a map of DNA sequence copy number at chromosomal DNA locations [28]. The modification of conventional CGH using microarray is called *array CGH* (aCGH). This aCGH is a high-resolution, high-throughput technique for screening of DNA copy numbers within the entire genome of less than 1 Mb resolution, and is very useful for genome-wide studies of DNA copy number changes [46,40], or CNVs. The aCGH data contain rich information such as gene positions, log base 2 ratio intensities, and DNA copy numbers along the genome. The log ratio intensities along the biomarkers reflect DNA copy number variations and are valuable information resources for accurately identifying loci of DNA copy number variations along the chromosome or genome. Most recently, single-nucleotide polymorphism (SNP) arrays have improved the DNA copy number experiments by providing the gene contents in the range of less than 100 kb resolution [40]. The SNP array data are of higher throughput and have many applications in medical fields and genetics in addition to cancer research.

Biological and medical research reveals that some forms of cancer are caused by somatic or inherited mutations in oncogenes and tumor suppressor genes, [37] and cancer development and genetic disorders often result in chromosomal DNA copy number changes or CNVs. Consequently, identification of these loci where the DNA copy number changes or CNVs have taken place will (at least partially) facilitate the development of medical diagnostic tools and treatment regimes for cancer and other genetic diseases. However, due to the random noise inherited in the imaging and hybridization process in the DNA copy number experiments, identifying statistically significant CNVs or DNA copy number changes in aCGH data and in SNP array data is challenging.

It turns out that identifying loci of DNA copy number variations along the chromosome or genome can be viewed as a change point problem of detecting the changes presented in the sequence of log ratio intensities. Therefore, in this review chapter, the author will discuss the background of DNA copy number experiments, especially the experiments using the array comparative genomic hybridization (aCGH) technique, the resulting aCGH data, and the appropriate change point methods that can be used to analyze the aCGH data. Several applications of the discussed change point methods to the analysis of fibroblast cell line data, the breast tumor data, and the breast cancer data will be given in this review chapter as well.

Specifically, this review chapter is organized as follows. A recent literature review on various statistical and computational approaches to the analysis of aCGH data is given in Section 10.2. It is this author's opinion that statistical change point methods are very suitable for modeling aCGH data and hence a brief review on change point methods is given in Section 10.3. The formulation of the change point models for aCGH data and some examples of using change point methods to analyze aCGH data are given in Section 10.4. Finally, concluding remarks are given in Section 10.5.

10.2 A Brief Literature Review on Analysis of aCGH Data

10.2.1 aCGH Data

As a background, the aCGH data are introduced briefly first. In DNA copy number experiments such as aCGH copy number experiments, differentially labeled sample and reference DNA are hybridized to DNA microarrays [46,47], and the sample intensities of the test and reference samples are obtained by Myers [39]. As the reference sample is assumed or chosen to have no copy number changes, markers whose test sample intensities are significantly higher (or lower) than the reference sample intensities are corresponding to DNA copy number gains (or losses) in the test sample at those locations [42].

Concretely, the test sample intensity at locaus i on the genome is usually denoted by T_i and the corresponding reference sample intensity by R_i, and the normalized log base 2 ratio of the sample and reference intensities, $\log_2 T_i/R_i$, at the ith biomarker, is one of the default outputs after the DNA copy number experiment is conducted using aCGH technique. Here, $\log_2 T_i/R_i = 0$ indicates no DNA copy number change at locus i, $\log_2 T_i/R_i < 0$ reveals a deletion at locus i, and $\log_2 T_i/R_i > 0$ signifies a duplication in the test sample at that locus. Due to various random noise, which occurs largely during the experimental and image processing stages, the $\log_2 T_i/R_i$ becomes a random variable. Ideally, this random variable is assumed to follow a Gaussian distribution of mean 0 and constant variance σ^2. Then, deviations from the constant parameters (mcan and variance) presented in $\log_2 T_i/R_i$ data may indicate a copy number change. Hence, the key to identifying true DNA copy number changes becomes the problem of how to identify changes in the parameters of a normal distribution based on the observed sequence of $\log_2 T_i/R_i$. Figure 1 is the scatter plot of the log base 2 ratio intensities of the genome of the breast tumor cell line S0034 [56].

10.2.2 A Brief Survey of Methods used for Analysis of aCGH Data

It is evident from Figure 1 that there are changes in the log ratio intensities along the biomarkers on the genome of S0034, and these changes correspond to copy number changes [56]. How to use the log ratio intensities to accurately infer the loci of changes and how to estimate the copy number gains or losses for those segments of biomarkers, whose log ratio intensities have significantly deviated from 0, has become a challenging computational and statistical issue.

Hodgson et al. [22] studied the copy number alterations in mouse islet cascinimas using aCGH technique. They analyzed their aCGH data for DNA copy number changes using a finite Gaussian mixture model with three components (a "no change" component, a "loss" component, and a "gain" component). The parameters of their mixture Gaussian model include the proportion of each component, the mean, and the variance of each component. The estimation of these model parameters is carried out by two steps, the visual estimation and the least-squares estimation. However, the significance of their findings has not been accessed statistically.

Pollack et al. [48] studied DNA copy number change in breast tumors and breast cancer cell lines using pairwise t tests for classes of genes, correlation coefficients between DNA copy number and mRNA level, and linear regression model among the tumors based on their aCGH data. Weiss et al. [63] proposed using a threshold analysis method to identify DNA copy num-

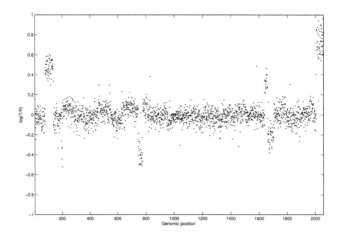

Figure 1: The genome of the breast tumor cell line S0034 [56].

ber changes based on aCGH data. However, there is a lack of statistical assessment of such thresholds. Lucito et al. [36] developed a method, *representational ologonucleotide microarray analysis* (ROMA), for the detection of copy number aberrations in cancer and normal humans. They profiled a primary breast cancer sample (CHTN159) and a breast cancer cell line compared with a normal male reference (SK-BR-3) using ROMA and analyzed their data using mean ratios for the detection of DNA copy number changes.

Autio et al. [1] provided a practical software solution to the analysis of DNA copy number data by introducing the Matlab toolbox: CGH-Plotter. They proposed a three stage procedure in the CGH-Plotter, namely, filtering, k-means clustering, and dynamic programming. Myers et al. [39] proposed a method called CHARM (chromosomal aberration region miner) to identify segmental aneuploidies in gene expression data and aCGH data. CHARM consists of three stages: an edge detection filter that identifies the potential starting or ending point of aberrations on chromosomes, an EM-based edge placement algorithm that statistically optimizes these potential starting and ending points, and a window significant test that determines whether the predicted aberrations are statistically significant.

Fridlyand et al. [17] used an unsupervised hidden Markov models approach to map the clones into the states which represent the underlying copy number of the group of clones and use this approach to identify possible copy number changes. Hupé et al. [25] used the adaptive weights smoothing procedure to estimate copy number gains and losses on simulated and some publicly available reference aCGH data. Their approach provided fine estimations on the changes by an optimal choice of the parameters in their algorithm. However, the choices of the optimal parameters were either empirical or through other extensive computational algorithm, thus this feature of their algorithm made it impractical for application.

More recently, Zhao et al. [66] used single-nucleotide polymorphism (SNP) array analysis to study copy number changes in primary human lung carcinoma specimens and cell lines. They applied the hidden Markov model to their SNP data and identified copy number changes that are vital to the development of lung cancer [17].

Picard et al. [45] proposed an adaptive criterion to estimate the DNA copy number changes in aCGH data. Price et al. [49] provided a dynamic programming solution for detecting CNVs in aCGH data. Both of these methods are alternatives to the estimation of DNA copy number changes, but there was lack of statistical assessment (in terms of p value) on how good those estimations were.

A Bayesian HMM approach was proposed for the analysis of aCGH data using extensive computational techniques [18]. Posterior inferences are made about the additions and deletions in DNA copy numbers and computational algorithm for calculating the posterior probabilities are developed [18]. A robust HMM approach, which is resistant to outliers presented in the aCGH data, was developed for the analysis of DNA copy number changes in aCGH data [53]. A continuous-index HMM approach was also proposed for identifying the CNVs in aCGH data, along with a Monte Carlo EM algorithm for the estimation of the parameters in the proposed model [58]. Korbel [29] proposed an approach, called BreakPtr, which is based on a bivariate HMM that accounts for both intensity ratios and sequence information for the detection of CNVs. Most recently, a computational approach using a Bayesian segmentation modeling of aCGH data was introduced, aiming at giving good estimation and confidence intervals of the DNA aberration regions [31].

Although the aforementioned studies employed some statistical estimation methods in analyzing aCGH and gene expression data for copy number changes, the validation of using those methods is not completely satisfactory. Lai et al. [32] compared 11 methods (including several methods mentioned above) and pointed out that such comparisons of methods are difficult due to possibly suboptimal choice of parameters in those methods. They nevertheless reveal general characteristics that are helpful to the biological investigator who needs to analyze copy number change data. Lai et al. [32] finally concluded that a statistical analysis of copy number change should essentially address two important points: (1) how to estimate the loci where DNA copy number has changed, and (2) how good the estimation is in terms of giving the probability of an observed significance (or simply the p-value) for the estimated locus of a change.

Reviewing the nature of the log ratio intensity data shown in Figure 1, it is evident that an appropriate statistical change point model [11] is most suitable for analyzing DNA copy number data. In fact, Linn et al. [35] employed a mean change point model to verify DNA copy number changes in dermatofibrosarcoma protuberans that were observed in their gene expression experiment. Olshen et al. [42] proposed a circular binary segmentation (CBS) method to identify DNA copy number changes in aCGH data base on the mean change point model proposed in Sen and Srivastava [52]. Their work demonstrated that by using a change point model, the estimation of DNA copy number changes in aCGH data could be improved. However, as pointed out in Lai it et al. [32], employing the right change point model is then the key factor for accurately identifying the loci of DNA copy numbers and further inferring the copy number gains and losses at those loci. In this regard, a brief literature review on various change point methods that can be used for modeling aCGH data is provided in the next section.

10.3 A Brief Literature Review on Statistical Change Point Analysis

Statistical change point analysis is an active topic in the statistics literature. In order for readers to quickly grasp the concept of change point analysis, a brief review is hereby provided.

Let $\{X_i\}$, $i = 1, \ldots, n$, be a sequence of independent random variables taken from a distribution with cumulative distribution function (cdf) F_i for $i = 1, \ldots, n$. A general multiple change points problem [11] is frequently viewed as the hypothesis testing of the null hypothesis:

$$H_0 : F_1 = F_2 = \cdots = F_n$$

versus the alternative hypothesis:

$$\begin{aligned}H_1 : F_1 &= \cdots = F_{k_1} \neq F_{k_1+1} \\ &= \cdots = F_{k_2} \neq F_{k_2+1} \\ &= \cdots = F_{k_q} \neq F_{k_q+1} \\ &= \cdots = F_n,\end{aligned}$$

where $1 < k_1 < k_2 < \cdots < k_q < n$, q is the unknown number of change points and $k_1, k_2, \ldots, k_q$ are the respective unknown change points' positions. If $F_i \in \{F(\theta_i)\}$, for all $\theta_i \in \mathbf{R}^p$, inference about multiple change points is formally characterized [11] by the following test of statistical hypothesis on the parameter (can be a vector), θ, that is, to test

$$H_0 : \theta_1 = \theta_2 = \cdots = \theta_n = \theta(\text{ unknown}) \quad (1)$$

versus the alternative:

$$\begin{aligned}H_1 : \theta_1 &= \cdots = \theta_{k_1} \neq \theta_{k_1+1} \\ &= \cdots = \theta_{k_2} \\ &\neq \cdots \neq \theta_{k_q-1} \\ &= \cdots = \theta_{k_q} \neq \theta_{k_q+1} \cdots = \theta_n,\end{aligned} \quad (2)$$

where $1 < k_1 < k_2 < \cdots < k_q < n$, q is the unknown number of change points, and $k_1, k_2, \ldots, k_q$ are the respective unknown change points' positions (loci). Apparently, there are two tasks in change point inference: (1) to find whether there is (are) a change point (change points); and (2) to estimate the number and the location(s) of the change point(s).

Historically, the single change point problem [with $q = 1$ in Eq. (2) above] was mostly studied in the literature, and most of the studies were on the detection of a single change point in the mean of a sequence of independent Gaussian random variables. Page [43] was the first to study a single change point model in the mean of a sequence of Gaussian random variables. Chernoff [16] derived a Bayesian estimator of the current mean for a priori uniform distribution on the whole real line using a quadratic loss function. Sen and Srivastava [52] derived the exact and asymptotic distribution of their test statistic for testing a single change in mean of a sequence of normal random variables. Hawkins [20] and Worsley [64] derived the null distribution for the case of known and unknown variances of a single change in the mean a normal model. Srivastava and Worsley [57] studied the multiple changes in the multivariate normal mean and approximated the null distribution of the likelihood ratio test statistic based on Bonferroni inequality.

Horváth [23] was one of the first authors who studied the single change point problem in both mean and variance for normal distributions. Inclán and Tiao [26] used the CUSUM method to detect and estimate multiple variance change points. Chen and Gupta [8] derived the asymptotic null distribution of the likelihood procedure [34] statistic for the simultaneous change in the mean vector and covariance of a sequence of normal random vectors.

Many works related to change point(s) problem have occurred in the literature. To name only a few, Hawkins [20] studied the shifts (change points) in the functions of multivariate location and covariance parameters, Joseph and Wolfson [27] studied the maximum-likelihood estimation in the multipath change point problem, Parzen [44] did a comparison change analysis from the nonparametric point of view; Yao [65] studied tests for change points with epidemic alternatives; Vlachonikolis and Vasdekis [61] studied a class of change-point models in covariance structures for growth curves and repeated measurements, Gupta and Chen [19] applied a mean vector multiple-change-point model to study literature and geology data, Chen and Gupta [10] studied the mean and change point model using information criterion and derived an unbiased Schwartz information criterion SIC for such a change point model, Chen and Gupta [12] provided a modified information-theoretic approach for detecting a change in the parameters of a normal model, Chen [6] studied the change point problem for in a failure rate model, Chen and Balakrishnan [7] studied a change point problem in a Gaussian model when an outlier is present, and Balakrishnan and Chen [2] studied the single change point for a sequence of extreme value observations.

For detecting multiple change points, an effective method [52,11] is the binary segmentation procedure (BSP) proposed by Vostrikova [62]. It searches the first significant change point in a sequence, then breaks the original sequence into two subsequences: one before the first significant change point (including the change point) and the other after the first significant change point. Thereafter, the procedure tests the two subsequences separately for a change point. One repeats the process until no further subsequences have change points. The collection of change point locations found at the end is denoted by $\{\widehat{k}_1, \widehat{k}_2, ..., \widehat{k}_q\}$, and the estimated total number of change points is then q. This BSP approach has been implemented in many multiple change point problems such as in Gupta and Chen [19] and Chen and Gupta [9], among others. With BSP, to test (1) versus (2), one can now simply test (1) against

$$H_1 : \theta_1 = \cdots = \theta_k \neq \theta_{k+1} = \cdots = \theta_n, \quad (3)$$

where k is the single-change-point position at each stage.

A more recent modification of the BSP is the circular binary segmentation (CBS) proposed in Olshen et al. [42]. The idea of CBS for the search of multiple change points in one parameter is to consider the sequence $X_1, X_2, \ldots, X_n$ to be spliced at the two ends to form a circle. The null hypothesis is that there are not changes in the parameter; and the alternative hypothesis is that the arc from $i+1$ to j and its complement arc have different parameters for $1 \leq i \leq j \leq n$. The CBS searches for one change point (or two change points if they are nearby the two ends) at a time, and it also inherits an edge effect. That is, if the i and j that correspond to a selected test statistic (extreme-value type) are such that either i is "close" to 1 or j is "close" to n, then there might be only one true change instead of the two changes suggested by the data [42].

It should be noted that when the BSP or CBS search scheme is used with a single change point hypothesis test, one is, in fact, testing multiple hypotheses without prior knowledge of how many tests will be conducted. Along this line, there is the problem of handling multiple testing or comparison procedure (MCP) for the hypothesis testing of multiple changes. No one in the literature has considered this issue because of its complexity. In traditional multiple testing/comparison problems, the total number of tests/comparisons is known and hence the traditional Bonferroni approach of controlling the familywise error rate (FWER) and the modern Benjamini–

Hochberg [3] approach of controlling the false discovery rate (FDR) can be readily used. For multiple change point hypothesis testing, when the BSP or the CBS search scheme is used, the number of changes is unknown, and hence a new method of controlling the FWER needs to be developed.

10.4 Statistical Change Point Methods Useful for aCGHData Analysis

10.4.1 The Circular Binary Segmentation (CBS) Method for the Mean Change Point Model

As mentioned earlier in this chapter, Olshen et al. [42] proposed a circular binary segmentation (CBS) method to identify DNA copy number changes in aCGH database on the mean change point model proposed in Sen and Srivastava [52]. This CBS method is mainly the combination of the likelihood ratio based test for testing no change in the mean against exactly one change in the mean with the BSP [62] for searching multiple change points in the mean, assuming that the variance is unchanged. The idea of CBS method [42] can be summarized as follows.

Let X_i denote the normalized $\log_2 T_i/R_i$ at the ith locus along the chromosome, then $\{X_i\}$ is considered as a sequence of normal random variables taken from $N(\mu_i, \sigma_i^2)$, respectively, for $i = 1, \ldots, n$. Consider any segment of the sequence of the log ratio intensities $\{X_i\}$ (assumed to follow normal distributions) to be spliced at the two ends to form a circle, the test statistic Z_c of the CBS is based on the modified likelihood ratio test and is specifically,

$$Z_c = \max_{1 \leq i < j \leq n} |Z_{ij}|, \quad (4)$$

where Z_{ij} is the likelihood ratio test statistic [52] for testing the hypothesis that the arc from i+1 to j and its complement have different means (that is, there is a change point in the mean of the assumed normal distribution for the X_i's) and is given by:

$$Z_{ij} = \left\{ \frac{1}{j-i} + \frac{1}{n-j+i} \right\}^{-1/2} \\ \times \left\{ \frac{S_j - S_i}{j-i} - \frac{S_n - S_j + S_i}{n-j+i} \right\}, \quad (5)$$

with

$$S_i = X_1 + X_2 + \cdots + X_i, \quad 1 \leq i < j \leq n.$$

Note that Z_c allows for both a single change ($j = n$) and the epidemic alternative ($j < n$). A change is claimed if the statistic exceeds an appropriate critical value at a given significant level based on the null distribution. However, the null distribution of the test statistic Z_c is not attainable so far in the literature of change point analysis. Then, as suggested in Olshen et al. [42], the critical value when the X_i's are normal need to be computed using Monte Carlo simulations or the approximation given by Siegmund [54] for the tail probability. Once the null hypothesis of no change is rejected the change point(s) is (are) estimated to be i (and j) such that $Z_c = |Z_{ij}|$ and the procedure is applied recursively to identify all the changes in the whole sequence of the log ratio intensities of a chromosome (usually of hundreds to thousands of observations). The CBS algorithm is written as an R package and is available from the R Project Website.

The influence of the CBS to analyses of aCGH data is tremendous, as it provided a statistical framework to the analysis of DNA copy number analysis. The p value given by the CBS for a specific locus being a change point; however, is obtained only by a permutation method, and the calculation of such a p value takes a long computation time when the sequence is long (which is the case for high-density array data). Hence the CBS method in Olshen et al. [42] has the slowest computational speed [45]. A recent result in Venkatraman and Olshen [60] has improved the computational speed of CBS. If there is an analytic formula for calculating the p value of the change point hypothesis, the computational speed will undoubtedly be faster and convenient.

For the application of the CBS method to the analysis of 15 fibroblast cell lines [56], the readers are referred to Olshen et al. [42].

10.4.2 Methods for the Mean and Variance Change Point Model for Identifying DNA Copy Number Changes or CNVs

In Linn et al. [35] and Olshen et al. [42] DNA copy number changes were viewed as mean changes (MCM) with a fixed variance in the distributions of the sequence $\{X_i\}$. As the aCGH technology may not guarantee the aCGH data to have a constant variance [22], it is more reasonable to analyze the DNA copy number changes using the mean-and-variance change model (MVCM) [15] in the distributions of the sequence $\{X_i\}$. Observing the following normalized log-ratio intensities obtained through aCGH experiments [36] on breast cancer cell line SK-BR-3 (see Fig. 2), it is evident that both mean and variance of the sequence have changed.

Let $\theta_i = (\mu_i, \sigma_i^2)'$, the multiple DNA copy number changes in the sequence of log ratio intensities can be defined as the hypothesis testing problem given in (1) and (2). Using BSP, we just need to focus on how to detect the single change (the most significant one) each time and repeat the searching scheme of BSP to get all the significant changes. Specifically, to identify a single DNA copy number change in the mean and variance parameters, we test the null hypothesis

$$\begin{aligned} H_0 : \mu_1 &= \mu_2 = \cdots = \mu_n = \mu \\ \sigma_1^2 &= \sigma_2^2 = \cdots = \sigma_n^2 = \sigma^2 \end{aligned} \quad (6)$$

versus the alternative:

$$\begin{aligned} H_1 : \mu_1 &= \cdots = \mu_k \neq \mu_{k+1} \\ &= \cdots = \mu_n \\ \sigma_1^2 &= \cdots = \sigma_k^2 \neq \sigma_{k+1}^2 \\ &= \cdots = \sigma_n^2, \end{aligned} \quad (7)$$

where μ and σ^2 are the unknown common mean and variance under the null hypothesis; and k, $1 < k < n$, is the unknown position of the single change at each single stage. Failure to the rejection of H_0 at a given significance level α indicates no change in the DNA copy number sequence and the search scheme stops at this stage; and the rejection of H_0 at a given significance level α indicates that a significant change in the DNA copy number sequence is found and the search scheme of BSP continues until no more significant changes are found.

To carry out the hypothesis testing of the null hypothesis (6), which claims no DNA copy number changes, versus the alternative hypothesis (7), the research hypothesis that there is a change in the mean and variance and hence a change in the DNA copy number, a test statistic is needed. Three commonly used approaches for testing (6) against (7) will be outline thereafter.

As pointed out in Chen and Wang [15], the advantage of using the MVCM model comes with the fact that MVCM leads to fewer change points than that of the mean change point model (MCM) as MCM tends to divide large segments into smaller pieces so that the homogenous variance assumption for all segments can be met [45].

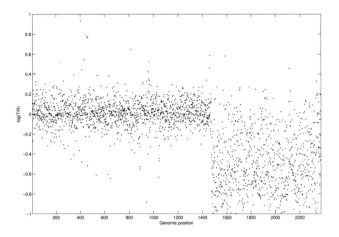

Figure 2: Chromosome 16 of the breast cancer cell line SK-BR-3 [36].

Therefore, the MVCM model has the potential to give fewer false positives than does MCM. Adding the variance component in the change point analysis will improve the estimation of the change point location even if just the mean shifts greatly. This is because in the MVCM model, the variances under the alternative hypothesis are estimated for each subsequence without pooling all subsequences (with possible different means) together, while in MCM the homogeneous variance under the alternative hypothesis is estimated by pooling all subsequences with different means together. Using either MVCM or MCM also depends on the biological experiment in which the scientists may have prior knowledge on whether there are potential variance changes. In that case, the MVCM model is proposed as an alternative to MCM when possible variance changes exist in the sequence.

The Likelihood-Ratio-Based Approach. The first method is the likelihood ratio procedure approach [34]. Under the null hypothesis (6), the natural logarithm of the maximum-likelihood function is easily obtained as

$$\log L_0(\widehat{\mu}, \widehat{\sigma}^2) = -\frac{n}{2}\log 2\pi - \frac{n}{2}\log \widehat{\sigma}^2 - \frac{n}{2}, \quad (8)$$

where $L_0(\widehat{\mu}, \widehat{\sigma}^2)$ is the maximum-likelihood function with respect to H_0, and $\widehat{\mu}$ and $\widehat{\sigma}^2$ are the maximum-likelihood estimators (MLEs) of μ and σ^2 under H_0, respectively, and are found to be

$$\widehat{\mu} = \overline{X} = \frac{1}{n}\sum_{i=1}^{n} X_i,$$

$$\widehat{\sigma}^2 = \frac{1}{n}\sum_{i=1}^{n}(X_i - \overline{X})^2. \quad (9)$$

Corresponding to H_1 given by (7), the natural logarithm of the maximum likelihood function is also easily obtained as

$$\log L_1(\widehat{\mu}_1, \widehat{\mu}_k, \widehat{\sigma}_1^2, \widehat{\sigma}_k^2)$$
$$= -\frac{n}{2}\log 2\pi - \frac{k}{2}\log \widehat{\sigma}_1^2$$
$$\quad -\frac{n-k}{2}\log \widehat{\sigma}_k^2 - \frac{n}{2}, \quad (10)$$

where $L_1(\widehat{\mu}_1, \widehat{\mu}_k, \widehat{\sigma}_1^2, \widehat{\sigma}_k^2)$ is the maximum-likelihood function under H_1, and $\widehat{\mu}_1$, $\widehat{\mu}_k$,

$\widehat{\sigma}_1^2$, and $\widehat{\sigma}_k^2$ are the MLEs of μ_1, μ_k, σ_1^2, and σ_k^2, respectively, satisfying the following equations:

$$\widehat{\mu}_1 = \overline{X}_k = \frac{1}{k}\sum_{i=1}^{k} X_i,$$

$$\widehat{\sigma}_1^2 = \frac{1}{k}\sum_{i=1}^{k}(X_i - \overline{X}_1)^2,$$

$$\widehat{\mu}_k = \overline{X}_{n-k} = \frac{1}{n-k}\sum_{i=k+1}^{n} X_i,$$

$$\widehat{\sigma}_k^2 = \frac{1}{n-k}\sum_{i=k+1}^{n}(X_i - \overline{X}_k)^2, \quad (11)$$

The likelihood-ratio-procedure-based approach is to derive a test statistic, $T(X_1, X_2, \ldots, X_n)$, based on

$$\delta_n = -2\log \Lambda_n = 2*[\log L_0(\widehat{\mu}, \widehat{\sigma}^2) - \log L_1(\widehat{\mu}_1, \widehat{\mu}_k, \widehat{\sigma}_1^2, \widehat{\sigma}_k^2)]. \quad (12)$$

The exact null distribution of δ_n still remains unknown so far in the literature, and only the asymptotic null distribution of δ_n is found in Horváth [23].

The Schwarz Information Criterion Approach. Chen and Wang [15] proposed to use the Schwarz information criterion (SIC) [51] approach to test (6) against (7). The SIC approach is a likelihood-based approach with the penalty term added for possible model overparameterization. It converts hypothesis testing into model selection process where the null hypothesis H_0 in (6) corresponds to a model of no change in the sequence and the alternative hypothesis H_1 in (7) corresponds to $n-3$ change point models with change located at $2, \ldots, n-2$, respectively. The SIC approach is also appealing to engineers and other practionners as information criterion is widely used in engineering fields. In general, SIC is defined as

$$SIC = -2\log L(\widehat{\theta}) + k\log n,$$

where $L(\widehat{\theta})$ is the maximum-likelihood function of a model, k is the number of parameters to be estimated, and n is the sample size. The information criterion principle for model selection is to choose the model whose SIC is the minimum (equivalent to the maximum likelihood being the maximum) as the best possible model. The SIC corresponding to the no change model specified by H_0, denoted by SIC(n) is obtained as [10]:

$$\begin{aligned}
\text{SIC}(n) &= -2\log L_0(\widehat{\mu}, \widehat{\sigma}^2) + 2\log n \\
&= n\log 2\pi + n\log \sum_{i=1}^{n}(X_i - \overline{X})^2 \\
&\quad + n + (2-n)\log n, \quad (13)
\end{aligned}$$

where $L_0(\widehat{\mu}, \widehat{\sigma}^2)$ is the maximum likelihood function with respect to H_0 given by (6), and $\widehat{\mu}$ and $\widehat{\sigma}^2$ are the maximum likelihood estimators (MLEs) of μ and σ^2 under H_0, respectively, and are expressed as in (9).

Corresponding to H_1, there are $n-3$ change point models, and the SIC for each model, denoted by SIC(k) for fixed k, $2 \leq k \leq n-2$, is obtained as

$$\begin{aligned}
\text{SIC}(k) &= -2\log L_1(\widehat{\mu}_1, \widehat{\mu}_k, \widehat{\sigma}_1^2, \widehat{\sigma}_k^2) \\
&\quad + 4\log n \\
&= n\log 2\pi + k\log \widehat{\sigma}_1^2 \\
&\quad + (n-k)\log \widehat{\sigma}_k^2 \\
&\quad + n + 4\log n, \quad (14)
\end{aligned}$$

where $L_1(\widehat{\mu}_1, \widehat{\mu}_k, \widehat{\sigma}_1^2, \widehat{\sigma}_k^2)$ is the maximum likelihood function under H_1, and $\widehat{\mu}_1$, $\widehat{\mu}_k$, $\widehat{\sigma}_1^2$, and $\widehat{\sigma}_k^2$ are the MLEs for μ_1, σ_1^2, μ_k, and σ_k^2, respectively, and are given earlier in (11).

According to the principle of information criterion, if

$$\text{SIC}(n) < \min_{2 \leq k \leq n-2} \text{SIC}(k),$$

then there is no change in the sequence, and if

$$\text{SIC}(n) > \min_{2 \leq k \leq n-2} \text{SIC}(k),$$

then there is a change in the sequence and the change point position k is estimated by $\hat{k}$ such that

$$\text{SIC}(\hat{k}) = \min_{2 \leq k \leq n-2} \text{SIC}(k). \quad (15)$$

Using this method, the change is easily located and the algorithm is fast in identifying change point estimates. However, due to random disturbance during the process of aCGH experiments, aCGH data does contain fluctuations which may not indicate a copy number change. To address this issue, Chen and Gupta [9] proposed to view the following quantity

$$\Delta_n = \min_{2 \leq k \leq n-2} [\text{SIC}(k) - \text{SIC}(n)] \quad (16)$$

as a statistic and hence use the null distribution of Δ_n to make formal inference decision regarding H_0 versus H_1. The exact null distribution of Δ_n still remains unknown so far. However, the asymptotic null distribution of a function of Δ_n is [23,10]:

$$\lim_{n \to \infty} P[a(\log n)(2 \log n - \Delta_n)^{1/2}$$
$$- b(\log n) \leq x]$$
$$= \exp(-2e^{-x}),$$

where $a(\log n) = (2 \log \log n)^{1/2}$, and $b(\log n) = 2 \log \log n + \log \log \log n$. Therefore, an approximate p-value for rejecting the null hypotheses (hence estimating the change location as $\hat{k}$) is obtained as

$$p \text{ value} = 1 - \exp\{-2 \exp[b(\log n)$$
$$- a(\log n) \lambda_n^{1/2}]\}, \quad (17)$$

where $\lambda_n = 2 \log n - \Delta_n$.

The applications of the SIC method to the detection of change point loci in the 15 fibroblast cell lines [56] and other known aCGH data are given in Chen and Wang [15]. There are comparisons of using the mean and variance change point model with the CBS method which is based on a mean change point model in Chen and Wang [15]. For an illustration of the application of this SIC approach, let's look at one example.

Snijders et al. [56] conducted aCGH copy number experiments on fifteen fibroblast cell lines, namely, GM03563, GM00143, GM05296, GM07408, GM01750, GM03134, GM13330, GM03576, GM01535, GM07081, GM02948, GM04435, GM10315, GM13031, and GM01524, and obtained aCGH data on the genome of all the cell lines. The samples were obtained from the NIGMS Human Genetics Cell Repository (Coriell Institute for Medical Research), and the experiments underwent several stages such as specimens preparation, genomic cloning, isolation of BAC/P1 DNA, preparation of BAC/P1 DNA representations by ligation-mediated PCR, preparation of DNA spotting solutions, array printing, DNA labeling, hybridization, imaging, and analysis. Finally, an aCGH dataset was obtained for each of the cell lines under study. The resulting aCGH data for each cell line are available at the fibroblast cell lines data Website [59]. Many researchers have analyzed these datasets using different computational and statistical approaches. The change point methods, CBS and MVCM, which were used for the analysis of the fibroblast aCGH data, were compared in Chen and Wang [15] in terms of the change loci identified, the sensitivity, and specificity of the two methods. Two applications of the SIC approach are presented below.

The first one is a chromosome-wide CNV search using SIC in MVCM on chromosome 7 of the fibroblast cell line GM07081. There are 157 genomic positions on which log base 2 ratio of intensities were recorded. The SIC values at all of the genomic locations were calculated according to expressions (13) and (14). The minimum SIC is occurred at location index of 69 with minSIC = −256.1510 and corresponding p value [according to Eq. (17)] of 0.0000. The graph of SIC values for this chromosome is given in the following Figure 3. Transferring back to the log ratio intensities, a scatterplot of the log ratio intensities of chromosome 7 of the fibroblast cell line GM07081 is provided as Figure 4 with the circle indicating the change point identified.

The second application is a genome-wide CNV search using SIC in MVCM on the cell line GM02948. It is found that the minimum SIC value of the whole sequence of 2048 log ratio intensity values occurs at 1457 locus with min(SIC) = −2968.601 and the p value of 8.353×10^{-9}. The BSP is applied to the searching process. For the subsequence containing the first through the 1457th observations, the search results in no significant CNV, and for the subsequence consisting of the 1458th through the 2048th log ratio intensity value, the minimum SIC occurs at the 1513th locus of the original sequence with min(SIC) = −931.4321 and the p-value of 4.866×10^{-11}. These two loci are circled in the scatterplot, Figure 5, of the genome of the fibroblast cell line GM02948.

A Bayesian Approach to the Identification of a Change in Both Mean and Variance. In the meantime, a Bayesian approach to the identification of a change in both mean and variance, or the Bayesian approach to MVCM, was just established [14]. The advantage of using a Bayesian approach towards the hypothesis testing problem of MVCM is that, a posterior distribution of the change locus gives the probability of the identified locus being a locus of DNA copy number aberration and makes the application very easy to understand in practical situations.

In the Bayesian framework, we assume the existence of a change, located at k, in the mean and a change, located at k, in the variance; and we derive a posterior probability distribution function, $\pi_1(k)$, for the location of the change introduced in the alternative hypothesis (7) above.

Several authors have used the Bayesian approach to study the problems of change point in mean or variance for a sequence of normal variables; (e.g., see Refs. 55, 24, and 38). Several testing methods— the L test based on the U statistic of Lehmann [33], a Bayesian B test, the R test derived from likelihood method, the C test based on the CUSUM of squares [5], and the LM-test based on Lagrange multiplier method—were compared with the Bayesian approach in the literature [4,41]. More recently, a Bayesian approach to model some gene expression that follows a smooth-and-abrupt change pattern was proposed and interesting results were found; see [13]. We elucidate our proposed Bayesian approach for the one change case in the mean and variance change point model as follows.

For the model specified in (7), it is assumed that the location, τ, of the change point is uniformly distributed, or the prior distribution of τ is taken to be

$$\pi_0(\tau) = \begin{cases} \frac{1}{n-3}, & \tau = 2, \ldots, n-2 \\ 0, & \text{otherwise} \end{cases}. \quad (18)$$

The prior distribution of the means μ and μ_1 is thought as constant, or simply

$$\pi_0(\mu, \mu_1 \mid \sigma^2, \sigma_1^2, \tau) \propto \text{constant}, \quad (19)$$

and the prior distribution of the variances σ^2 and σ_1^2 is taken as an improper prior as following:

$$\pi_0(\sigma^2, \sigma_1^2 \mid \tau) = \frac{1}{\sigma^2 \sigma_1^2}. \quad (20)$$

Then, under the normality assumption of the random sample X_i and the respective priors (18), (19), and (20), the posterior probability mass function of the

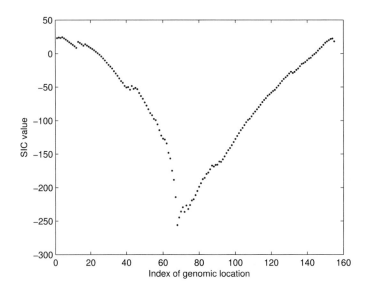

Figure 3: SIC values for every locus on chromosome 7 of the fibroblast cell line GM07081 [56].

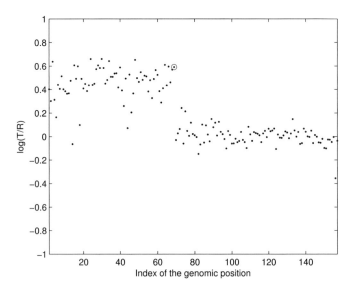

Figure 4: Chromosome 7 of the fibroblast cell line GM07081 [56].

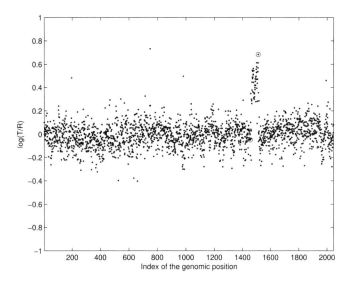

Figure 5: Genome of the fibroblast cell line GM02948 [56].

change point location τ is derived as

$$\pi_1(\tau) = \frac{\pi_1^*(\tau)}{\sum\limits_{t=2}^{n-2} \pi_1^*(t)}, \text{ for } \tau = 2, ..., n-2, \quad (21)$$

where

$$\begin{aligned}
&\pi_1^*(\tau) \\
&= \Gamma\left(\frac{\tau-1}{2}\right) \Gamma\left(\frac{n-\tau-1}{2}\right) \tau^{\frac{\tau}{2}-1} \\
&\times \left\{\tau \sum_{i=1}^{\tau} X_i^2 - \left(\sum_{i=1}^{\tau} X_i\right)^2\right\}^{-\frac{\tau-1}{2}} \\
&\times (n-\tau)^{\frac{n-\tau}{2}-1} \\
&\times \left\{(n-\tau) \sum_{i=\tau+1}^{n} X_i^2 - \left(\sum_{i=\tau+1}^{n} X_i\right)^2\right\}^{-\frac{n-\tau-1}{2}}.
\end{aligned} \quad (22)$$

Once the posterior probability mass function (21) is obtained, then the locus τ at which posterior probability mass function attains its maximum is the estimated locus where a CNV has occurred. As a quick application of this posterior probability mass function (21) for identifying a change point loci on the aCGH data, the fibroblast cell line aCGH data of Snijders et al. [56] are used again. For chromosome 1 of the fibroblast cell line GM13330 and chromosome 14 the fibroblast cell line GM01750, the posterior probability mass function values are calculated according to Equation (21). In Figures 6 and 7, the scatterplots of these chromosomes are given in the upper panels, and the corresponding posterior probability mass functions are plotted in the respective lower panels. It is very clear that the maximum posterior probability reveals the accurate change locus in each chromosome.

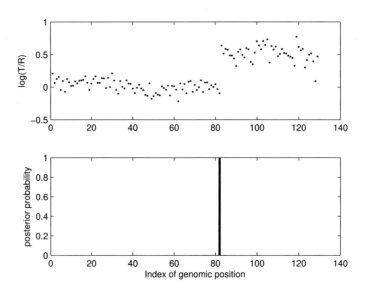

Figure 6: Chromosome 1 the fibroblast cell line GM13330 [56].

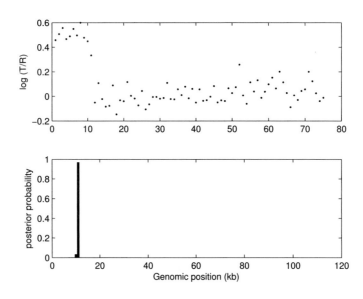

Figure 7: Chromosome 14 the fibroblast cell line GM01750 [56].

10.4.3 New Challenge: Methods for Data from SNP Arrays

Accompanying the fast development of biotechnology, single-nucleotide polymorphism (SNP) microarray chips have been developed for studying CNVs. The resulting SNP data are of a lot higher throughput than the aCGH data, in the sense that there are 100,000 to 500,000 SNPs, along with their genomic positions, log base 2 ratio intensities, CNV state calls, etc., contained in a typical SNP data set. For instance, in the newest Affymetrix SNP 6.0 array, there are about 900,000 SNPs and the resulting array data consists of 2.7 million measurements per sample [30] including SNP positions, intensities, CNV calls, etc. This type of spontaneous data information casts even more challenge in modeling SNP data for the searching of CNVs.

Very few SNP data based on Affymetrix SNP 6.0 arrays have been published so far. Korn et al. [30] give a comprehensive discussion is given, regarding the importance of accurately and competently studying both SNPs and CNVs (common and rare) to the understanding of the role of genetic variation in disease. It turns out that SNPs and CNVs may coexist throughout the genome, and so they may influence on each other's measurement, and may act individually or coherently to affect human genotypes [30]. The biological importance of discovering rare CNVs, which are unobserved previously, is emphasized and the challenge of how to discover rare CNVs is presented and is predicted to be much harder [30]. A four-stage algorithm, Birdsuite, was proposed [30] to sequentially assign DNA copy number across regions of common copy number polymorphisms (CNPs), to call genotypes of CNVs, to identify rare CNVs using a hidden HMM, and to generate an integrated sequence and copy number genotype at each locus along the genome. This is the first algorithm that integrates the discovery of SNPs, rare CNVs, genotyping, and association analysis together. It requires large-scale computations, and does not access the accurate detection of the boundaries of CNV regions with statistical significance. Therefore, it is predicted that the development of more change point methods for modeling SNP data is strongly in demand.

10.5 Concluding Remarks

In this review chapter, emphasis is on the parametric change point methods useful for genetics, especially in modeling the aCGH data generated from DNA copy number experiments. Cancer formation and genetic diseases are caused by somatic or inherited mutations in oncogenes and tumor suppressor genes, which in turn result in DNA copy number variations [37]. Moreover, DNA copy number variation, or CNV, of a DNA sequence or of a genome is a functionally significant event but yet has many biological aspects to be fully ascertained [50]. In the *Nature* article of Redon et al. [50], a general biological description of CNV is stated as

> Genetic variation in the human genome takes many forms, ranging from large, microscopically visible chromosome anomalies to single-nucleotide changes. Recently, multiple studies have discovered an abundance of submicroscopic copy number variation of DNA segments ranging from kilobases (kb) to megabases (Mb) in size. Deletions, insertions, duplications and complex multisite variants, collectively termed copy number variations (CNVs) or copy number polymorphisms (CNPs), are found in all humans and other mammals examined.

Therefore, DNA copy number experiments become widespread approaches for biologists and geneticists to study the causes of various types of cancer, tumor growth, genetic diseases, and population genetic variations for different ethnic groups. The resulting aCGH data from copy number experiments are of high throughput and the analysis of such aCGH data for inferring CNVs requires a statistical framework. In this chapter, brief literature reviews on the computational and statistical approaches for analyzing aCGH data and on statistical change point analysis are given. The formulation of aCGH data via change point models comes in naturally from the ground of statistical change point analysis.

With the new challenge inherited in the higher throughput SNP array data (e.g., the data obtained using Affymetrix SNP 6.0 arrays, or any other customized SNP arrays), the need for a powerful statistical change point analysis of the SNP data will be strongly in demand for genetic studies. From a statistical perspective, a fundamental study of the theoretic perspectives of the change point models will benefit the medical community in the analysis of copy number experimental data. Thus, research on proposing new change point models that can make full use of the rich data information in SNP data and development of a relevant method for detecting changes in such change point models will not only provide effective statistical methods to researchers who are facing change point problems in genetic studies but also advance the statistical change point analysis itself.

References

1. Autio, R., Hautaniemi, S., Kauraniemi, P., Yli-Harja, O., Astola, J., Wolf, M., and Kallioniemi (2003). A., CGH-Plotter: MATLAB toolbox for CGH-data analysis. *Bioinformatics*, **19**, 1714–1715.

2. Balakrishnan, N. and Chen, J. (2004). Detecting a change point in a sequence of extreme value observations. *J. Probab. Statist. Sci.*, **2**, 55–64.

3. Benjamini, Y. and Hochberg, Y. (1995). Controlling the false discovery rate: practical and powerful approach to multiple testing. *J. Roy. Statist. Soc.*, **57**, 289–300.

4. Bhatti, M. I. and Wang, J. (2000). On testing for a change-point in variance of normal distribution. *Biometr. J.*, **42**, 1021–1032.

5. Brown, R. L., Durbin, J., and Evans, J. M. (1975). Techniques for testing the constancy of regression relationships over time. *J. Roy. Statist. Soc.*, **37**, 149–163.

6. Chen, J. (2003). A note on change point analysis in a failure rate. *J. of Probab. Statist. Sci.*, **1**, 135–140.

7. Chen, J. and Balakrishnan, N. (2002). Locating a change point in a Gaussian model when an outlier is present. *J. Appl. Statist. Sci.*, **11**, 101–109.

8. Chen, J. and Gupta, A. K. (1995). Likelihood procedure for testing change point hypothesis for multivariate Gaussian model. *Random Operator. Stoch. Eq.*, **3**, 245–254.

9. Chen, J. and Gupta, A. K. (1997). Testing and locating variance change points with application to stock prices. *J. Am. Statist. Assoc.*, **92**, 739–747.

10. Chen, J. and Gupta, A. K. (1999). Change point analysis of a Gaussian model. *Statist. Papers*, **40**, 323–333.

11. Chen, J. and Gupta, A. K. (2000). *Parametric Statistical Change Point Analysis*, Birkhauser, Boston.

12. Chen, J. and Gupta, A. K. (2003). Information-theoretic approach for detecting change in the parameters of a normal model. *Math. Methods Statist.*, **12**, 116–130.

13. Chen, J. and Gupta, A. K. (2007). A Bayesian approach to the statistical analysis of a smooth-abrupt change point model, *Adv. Appl. Statist.*, **7**, 115–125.

14. Chen, J., Yiğiter, A., and Chang, K.-C. *A Bayesian Approach to Inference about a Change Point Model with Aplication to DNA Copy Number Experimental Data*, preprint.

15. Chen, J. and Wang, Y.-P. (2008). A statistical change point model approach for the detection of DNA copy number variations in array CGH data, *IEEE/ACM Trans. on Computational Biology and Bioinformatics*, Nov. 25 2008. IEEE Computer Society Digital Library, http://doi.ieeecomputersociety.org/10.1109/TCBB.2008.129.

16. Chernoff, H. and Zacks, S. (1964). Estimating the current mean of a normal distribution which is subject to change in time, *Ann. Math. Statist.*, **35**, 999–1018.

17. Fridlyand, J., Snijders, A. M., Pinkel, D., Albertson, D. G., and Jain, A. N. (2004). Hidden Markov models approach to the analysis of array CGH data. *J. Multivar. Anal.*, **90**, 132–153.

18. Guha, S., Li, Y., and Neuberg, D. (2006). *Bayesian Hidden Markov Modeling of Array CGH Data*. Harvard University Biostatistics Working Paper Series. Oct. 2006, Working Paper 24.

19. Gupta, A. K. and Chen, J. (1996). Detecting changes of mean in multidimensional normal sequences with application to literature and geology. *Comput. Statist.*, **11**, 211–221.

20. Hawkins, D. M. (1992). Detecting shifts in functions of multivariate location and covariance parameters. *J. Statist. Plan. Infer.*, **33**, 233–244.

21. Hawkins, D. M. (1997). Testing a sequence of observations for a shift in location. *J. Am. Statist. Assoc.*, **72**, 180–186.

22. Hodgson, G., Hager, J. H., Volik, S., Hariono, S., Wernick, M., Moore, D., Nowak, N., Albertson, D. G., Pinkel, D., Collins, C., Hanahan, D., and Gray J. W. (2001). Genome scanning with array CGH delineates regional alterations in mouse islet carcinomas. *Nat. Genet.*, **29**, 459–464.

23. Horváth, L. (1993). The maximum likelihood methods for testing changes in the parameters of normal observations. *Ann. Statist.*, **21**, 671–680.

24. Hsu, D. A. (1979). Detecting shifts of parameter in Gamma sequences with applications to stock price and air traffic flow analysis. *J. Am. Statist. Assoc.*, **74**, 31–40.

25. Hupé, P., Stransky, N., Thiery, J., Radvanyi, F., and Barillot, E. (2004). Analysis of array CGH data: From signal ratio to gain and loss of DNA regions. *Bioinformatics*, **20**, 3413–3422.

26. Inclán, C. and Tiao, G. C. (1994). Use of sums of squares for retrospective detection of changes of variance. *J. Am. Statist. Assoc.*, **89**, 913–923.

27. Joseph, L. and Wolfson, D. B. (1993). Maximum likelihood estimation in the multi-path change-point problem. *Ann. Inst. Statist. Math.*, **45**, 511–530.

28. Kallioniemi, A., Kallioniemi, O.-P., Sudar, D., Rutovitz, D., Gray, J. W., Waldman, F., and Pinkel D. (1992). Comparative genomic hybridization for molecular cytogenetic analysis of solid tumors. *Science*, **258**, 818–821.

29. Korbel, J. O., Eckehart Urban, A., Grubert, F., Du, J., Royce, T. E., Starr, P., Zhong, G., Emanuel, B. S., Weissman, S. M., Snyder, M., and Gerstein, M. B. (2007). Systematic prediction and validation of breakpoints associated with copy-number variants in the human genome, *Proc. of the Natl. Acad. Sci.*, **104**, 10110–10115.

30. Korn, J. M., Kuruvilla, F. G., McCarroll, S. A., Wysoker, A., Nemesh, J., Cawley, S., Hubbell, E., Veitch, J., Collins, P. J., Darvishi, K., Lee, C., Nizzari, M. M., Gabriel, S. B., Purcell, S., Daly, M. J., and Altshuler, D. (2008). Integrated genotype calling and association analysis of SNPs, common copy number polymorphisms and rare CNVs. *Nat. Gene.*, Sept. 7, 2008, doi:10.1038/ng.237.

31. Lai, T. L., Xing, H., and Zhang, N. (2008). Stochastic segmentation models

for array-based comparative genomic hybridization data analysis. *Biostatistics*, **9**, 290–307.

32. Lai, W. R., Mark, D., Johnson, M. D., Raju Kucherlapati, R., and Park, P. J. (2005). Comparative analysis of algorithms for identifying amplifications and deletions in array CGH data. *Bioinformatics*, **21**, 3763–3770.

33. Lehmann, E. L. (1951). Consistency and unbiasedness of certain nonparametric tests. *Ann. Math. Statist.*, **22**, 165–179.

34. Lehmann, E. L. (1986). *Testing Statistical Hypotheses*, 2nd ed. Wiley, New York.

35. Linn, S. C., West, R. B., Pollack, J. R., Zhu, S., Hernandez-Boussard, T., Nielsen, T. O., Rubin, B. P., Patel, R., Goldblum, J. R., Siegmund, D., Botstein, D., Brown, P. O., Gilks, C. B., and van de Rijn, M. (2003). Gene expression patterns and gene copy number changes in dermatofibrosarcoma protuberans. *Am. J. Pathol.*, **163**, 2383–2395.

36. Lucito, R., Healy, J., Alexander, J., Reiner, A., Esposito D, Chi, M., Rodgers, L., Brady, A., Sebat, J., Troge, J., West, J. A., Rostan, S., Nguyen, K. C., Powers, S., Ye, K. Q., Olshen, A., Venkatraman, E., Norton, L., and Wigler, M. (2003). Representational oligonucleotide microarray analysis: A high-resolution method to detect genome copy number variation. *Genome Res.*, **13**, 2291–2305.

37. Lucito, R., West, J., Reiner, A., Alexander, D., Esposito, D., Mishra, B., Powers, S., Norton, L., and Wigler, M. (2000). Detecting gene copy number fluctuations in tumor cells by microarray analysis of genomic representations. *Genome Res.*, **10**, 1726–1736.

38. Menzefricke, U. (1981). A Bayesian analysis of a change in the precision of a sequence of independent normal random variables at an unknown time point. *Appl. Statist.*, **30**, 141–146.

39. Myers, C. L., Dunham, M. J., Kung, S. Y., and Troyanskaya, O. G. (2004). Accurate detection of aneuploidies in array CGH and gene expression microarray data. *Bioinformatics*, **20**, 3533–3543.

40. Nannya, Y., Sanada, M., Nakazaki, K., Hosoya, N., Wang, L., Hangaishi, A., Kurokawa, M., Chiba, S., Bailey, D. K., Kennedy, G. C., and Ogawa, S. (2005). A robust algorithm for copy number detection using high-density oligonucleotide single nucleotide polymorphism genotyping arrays. *Cancer Res.*, **65**, 6071–6079.

41. Nyblom, J. (1989). Testing for constancy of parameter over time. *J. Am. Statist. Assoc.*, **84**, 223–230.

42. Olshen, A. B., Venkatraman, E. S., Lucito, R., and Wigler, M. (2004). Circular binary segmentation for the analysis of array-based DNA copy number data. *Biostatistics*, **5**, 557–572.

43. Page, E. S. (1955). A test for a change in a parameter occurring at an unknown point. *Biometrika*, **42**, 523–527.

44. Parzen, E. (1992). Comparison change analysis. In *Nonparametric Statistics and Related Topics*, A. K. Md. E. Saleh, ed., Elsevier.

45. Picard, F., Robin, S., Lavielle, M., Vaisse, C., and Daudin, J. (2005). A statistical approach for array CGH data analysis. *BMC Bioinform.*, **6**, 27.

46. Pinkel, D., Seagraves, R., Sudar, D., Clark, S., Poole, I., Kowbel, D., Collins, C., Kuo, W.-L., Chen, C., Zhai, Y., Zhai, Y, Dairkee, S., Ljjung, B.-M., Gray, J. W., and Albertson, D. (1998). High resolution analysis of DNA copy number variation using comparative genomic hybridization to microarrays. *Nat. Gen.*, **20**, 207–211.

47. Pollack, J. R., Perou, C. M., Alizadeh, A. A., Eisen, M. B., Pergamenschikov, A., Williams, C. F., Jeffrey, S. S., Botstein, D., and Brown, P.O. (1999). Genome-wide analysis of DNA copy-number changes using cDNA microarrays. *Nat. Gene.*, **23**, 41–46.

48. Pollack, J. R., Sorlie, T., Perou, C. M., Rees, C. A., Jeffrey, S. S., Lonning, P. E., Tibshirani, R., Botstein, D., Borresen-Dale, A. L., and Brown, P. O. (2002).

Microarray analysis reveals a major direct role of DNA copy number alteration in the transcriptional program of human breast tumors. *Proc. Natl. Acad. Sci*, **99**, 12963–12968.

49. Price, T. S., Regan, R., Mott, R., Hedman, A., Honey, B., Daniels, R. J., Smith, L., Greenfield, A., Tiganescu, A., Buckle, V., Ventress, N., Ayyub, H., Salhan, A., Pedraza-Diaz, S., Broxholme, J., Ragoussis, J., Higgs, D. R., Flint, J., and Knight, S. J. L. (2005). SW-ARRAY: A dynamic programming solution for the identification of copy-number changes in genomic DNA using array comparative genome hybridization data. *Nucleic Acids Res.*, **33**, 3455–3464.

50. Redon, R., Ishikawa, S., Fitch, K. R., Feuk, L., Perry, G. H., Andrews, T. D., Feigler, H., Shaperio, M. H., Carson, A. R., Chen, W., Cho, E. K., Dallaire, S., Freeman, J. L., González, J. R., Gratacòs, M., Huang, J., Kalaitzopoulos, D., Komura, D., MacDonald, J. R., Marshall, C. R., Mei, R., Montgomery, L., Nishimura, K., Okamura, K., Shen, F., Somerville, M. J., Tchinda, J., Valsesia, A., Woodwark, C., Yang, F., Zhang, J., Zerjal, T., Zhang, J., Armengol, L., Conrad, D. F., Estivill, X., Tyler-Smith, C., Carter, N. P., Aburatani, H., Lee, C., Jones, K. W., Scherer, S. W., and Hurles, M. E. (2006). Global variation in copy number in the human genome. *Nature*, **444**, 444–454.

51. Schwarz, G. (1978). Estimating the dimension of a model. *Ann. Statist.*, **6**, 461–464.

52. Sen, A. and Srivastava, M. S. (1975). On tests for detecting a change in mean. *Ann. Statist.*, **3**, 98–108.

53. Shah, S. P., Xuan, X., DeLeeuw, R. J., Khojasteh, M., Lam, W. L., Ng, R., and Murphy, K. P. (2006). Integrating copy number polymorphisms into array CGH analysis using a robust HMM. *Bioinformatics*, **22**, e431–e439.

54. Siegmund, D. (1986). Boundary crossing probabilities and statistical applications. *Ann. Statist.*, **14**, 361–404.

55. Smith, A. F. M. (1975). A Bayesian approach to inference about a change-point in a sequence of random variables. *Biometrika*, **62**, 407–416.

56. Snijders, A. M., Nowak, N., Segraves, R., Blackwood, S., Brown, N., Conroy, J., Hamilton, G., Hindle, A. K., Huey, B., Kimura, K., Law, S., Myambo, K., Palmer, J., Ylstra, B., Yue, J. P., Gray, J. W., Jain, A. N., Pinkel, D., and Alberston D. G. (2001). Assembly of microarrays for genome-wide measurement of DNA copy number. *Nat. Genet.*, **29**, 263–264.

57. Srivastava, M. S. and Worsley, K. J. (1986). Likelihood ratio tests for a change in the multivariate normal mean. *J. Am. Statist. Assoc.*, **81**, 199–204.

58. Stjernqvist, S., Rydén, T., Sköld, M., and Staaf, J. (2007). Continuous-index hidden Markov modelling of array CGH copy number data. *Bioinformatics*, **23**, 1006–1014.

59. The Fibroblast Cell Lines Data. Available at
http://www.nature.com/ng/journal/v29/n3/full/ng754.html.

60. Venkatraman, E. S. and Olshen, A. B. (2007). A faster circular binary segmentation algorithm for the analysis of array CGH data. *Bioinformatics*, **23**, 657–663.

61. Vlachonikolis, I. G. and Vasdekis, V. G. S. (1994). On a class of change-point models in covariance structures for growth curves and repeated measurements. *Commun. Statist.—Theor. Methods*, **23**, 1087–1102.

62. Vostrikova, L. Ju. (1981). Detecting "disorder" in multidimensional random processes. *Soviet Math. Dokl.*, **24**, 55–59.

63. Weiss, M. M., Snijders, A. M., Kuipers, E. J., Ylstra, B., Pinkel, D., Meuwissen, S. G. M., van Diest, P. J., Albertson, D. G., and Meijer, G. A. (2003). Determination of amplicon boundaries at 20q13.2 in tissue samples of human gastric adenocarcinomas by high-resolution microarray comparative genomic hybridization. *J. Pathol.*, **200**, 320–326.

64. Worsley, K. J. (1979). On the likelihood ratio test for a shift in location of normal populations. *J. Am. Statist. Assoc.*, **74**, 365–367.

65. Yao, Q. (1993). Tests for change-points with epidemic alternatives. *Biometrika*, **80**, 179–191.

66. Zhao, X., Weir, B. A., LaFramboise, T., Lin, M., Beroukhim, R., Garraway, L., Beheshti, J., Lee, J. C., Naoki, K., Richards, W. G., Sugarbaker, D., Chen, F., Rubin, M. A., Janne, P. A., Girard, L., Minna, J., Christiani, D., Li, C., Sellers, W. R., and Meyerson, M. (2005). Homozygous deletions and chromosome amplifications in human lung carcinomas revealed by single nucleotide polymorphism array analysis. *Cancer Res.*, **65**, 5561–5570.

11 Classical Biostatistics

Bernhard G. Greenberg

11.1 Introduction

Biostatistics is that branch of science that applies statistical methods to biological problems, the common prefix being derived from the Greek word *bios*, meaning life.

The first major applications started in the middle of the seventeenth century when Sir William Petty and John Graunt conceived new and creative methods to analyze the London Bills of Mortality. Petty and Graunt essentially invented the field of *vital statistics* by studying the reported christenings and causes of death, and proposing measures of what they called "political arithmetick." Graunt recognized problems of inference when there has been inaccurate reporting of causes of death; he created methods of estimating mortality rates by age when age was not even recorded on the death certificate; and he devised estimates of birth rates, as well as a method to estimate the population from birth rates and other ingenious techniques for interpreting the data in the records of christenings and burials. Sir William Petty developed an enumeration schedule for a population census, proposed a centralized statistical department in the government, conceived the idea of life expectancy before Halley developed the first actual life table, and proposed clever and original ideas on how to estimate population sizes. For further details, see the monograph by Greenwood [29].

Today, vital statistics is generally restricted by definition to the statistics of births, deaths, marriages, and divorces, and thus the term has a current connotation considerably more limited than "biostatistics," despite its derivation from the same root in its Latin form *vita*. Biometry or biometrics is another term closely identified with biostatistics but also more restricted in scope. The biostatistician must deal not only with biometrical techniques used in the design and analysis of experiments but also with some sociometric and psychometric procedures plus most of the methods used by demographers. Thus, the biostatistician works closely not only with the biological researcher but also with the epidemiologist, survey researcher, local community planner, state and national health policy analyst, and those government officials concerned with developing procedures for registering births, deaths, marriages, divorces, abortions, morbidity reports, the description of populations by sample surveys and census enumeration, and with health regulatory agencies.

11.2 Fields of Application or Areas of Concern

The biostatistician differs from the traditional statistician by being confronted by a wider range of problems dealing with all the phenomena that affect people's physical, social, and mental well being. These phenomena consist of our relationship to other human beings, to animals and microbes, to plants, and to the physical and chemical elements in the environment. In dealing with these problems the biostatistician encounters theoretical difficulties, such as analyzing autocorrelated data in time series, in addition to practical and applied problems, such as working with accountants and economists to calculate costs versus benefits in evaluating the efficiency of a health program.

This means that the biostatistician must have familiarity with the concepts, goals, and specialized techniques of numerous fields beyond what might be considered a standard knowledge of statistics and probability. Some of these fields and areas of concern are mentioned here briefly, and the remainder of this chapter discusses a few of them at greater length.

11.2.1 Statistical Genetics

After the early developments in vital statistics, the field of *statistical genetics* was the next area that benefited most from the new ideas emerging in statistics. Any discussion of biostatistics and biometry would be incomplete without the names of Charles Darwin (1809–1882), Francis Galton (1822–1911), Karl Pearson (1857–1936), and Ronald A. Fisher (1890–1962).

Galton was responsible for the use of the term "regression" when he observed that sons regressed linearly on their fathers with respect to stature. His thesis was to call the phenomenon a "regression to mediocrity" because children deviated less from the mean height of all children than the degree to which their fathers deviated from the mean height of all fathers. This bivariate normal distribution gave rise to the measurement of the association by the coefficient of (product–moment) correlation in 1897 by Karl Pearson and to many other contributions by him. He is also generally credited with the creation of the new discipline of biometry and established with Walter F. R. Weldon and C. B. Davenport, in consultation with Galton, a new journal called *Biometrika* to provide for study of these problems. The journal has been in continuous publication since 1901, and after an unsigned editorial presumably written by Pearson, had as its first paper an article entitled "Biometry" by Francis Galton. The journal is still a highly regarded source for communications in biometry. Fisher's major contributions were to genetics and statistical theory, and he published the genetical theory of natural selection in 1930. This landmark book, plus earlier and later publications, represented attempts by Fisher to give quantitative form to Darwin's views and a statistical theory of evolution.

For a history of early developments in statistical genetics, see Norton [48]. For biographical accounts of the statistical geneticists, see the appropriate entries in Kruskal and Tanur [35].

11.2.2 Bioassay

Bioassay techniques cover the use of special transformations such as probits and logits, as well as the application of regression to the estimation of dosages that are p percent effective within stated confidence limits. There are also problems in measuring relative potency, slope-ratio assays, and quantal responses vis-à-vis tolerance distributions. The reader interested in this subject is well advised to consult bioassay and a standard textbook such as

Finney [18].

11.2.3 Demography

A knowledge of *demography*, which includes traditional vital statistics, rates and ratios, life tables, competing risks, actuarial statistics, and census enumeration techniques, is necessary in biostatistics. In this category, many tabulations of data will consist of a time series of events or rates classified by age. For the appropriate analysis of such data, reference should be made to cohort analysis techniques collected in a monograph by Hastings and Berry [31]. For further details in this broad area, see Linder and Grove [37] and the book by Spiegelman [52].

11.2.4 Epidemiology

Some knowledge is required about the measurement of disease, including false-negative and false-positive results, so that sensitivity and specificity of a diagnostic test can be estimated, as well as survey results used to estimate the true incidence and prevalence of disease. It is necessary to have knowledge of epidemic theory and the use of deterministic and stochastic models [1]. Fundamental to this whole field of application is an understanding of causality and association [3,24,36].

In the case where clinical trials can be conducted, two groups of persons, one "treated" and the other "untreated," are observed over a period of time with respect to attack by or relief from the disease that is the object of the study. Here the biostatistician must know how to develop a protocol [17], how to randomize, use double-blind techniques, and combine multiple-response variables into a multivariate analysis. If several medical centers are involved, it is important to know how to operate a statistical coordinating center for collaborative clinical trials (see Refs. 20 and 23).

In situations where moral concerns prohibit a comparative experiment, such as in the study of whether exposure of a woman during pregnancy to infection by German measles (rubella) causes congenital malformations, it is necessary to know how to conduct retrospective case–control studies and measure the relative risk caused by exposure. In fact, with the sole exception of clinical trials, almost all the statistical research in epidemiology is retrospective in nature. That is, the research is ex post facto because the investigators seek to describe and analyze a series of events that are customarily a rather sudden, unusual, and significant increase in the incidence of disease.

The so-called case–control study is the most common procedure used to investigate an epidemic or unusual increase in disease. By this approach, a special group, frequently a 100% sample of available cases, is studied in detail to ascertain whether there were one or more common factors to which the members of the group were exposed. The exposure might be a drug, a food, or an environmental factor. A comparable group of noncases, frequently called *controls*, *compeers*, or *referents*, is also selected at random in order to determine whether its members had the same, less, or more exposure to the suspected factor(s).

In the typical design of such studies, the data are presented in a 2×2 contingency table of the following form, wherein a, b, c, and d are category frequencies.

Factor F	Cases	Compeers	Total
Exposed	a	b	$a+b$
Nonexposed	c	d	$c+d$
Total	$a+c$	$b+d$	N

If the proportion $a/(a+c)$ is significantly greater than $b/(b+d)$, one can safely assume that factor F is associated in some way with the occurrence of the event. The test of significance to validate this may be the common χ^2 with 1 degree of freedom.

Owing to the fact that the design is retrospective, the comparable groups are cases and compeers, *not* exposed and nonexposed. Thus one cannot calculate the rates of disease as simply $a/(a+b)$ and $c/(c+d)$ in order to divide the former by the latter to derive a measure of relative risk associated with factor F. Although other researchers in genetics had previously used a solution similar to his, it was Cornfield [10] who demonstrated clearly that an estimate of relative risk is obtainable from the ratio of cross-products, ad/bc. [If $a/(a+c)$ is designated as p_1, and $b/(b+d)$ is designated as p_2, the relative risk is equivalently estimated as $p_1(1-p_2)/p_2(1-p_1)$.] The ratio of cross-products is commonly referred to as the *odds ratio*, motivated by the comparison of exposed-to-nonexposed "odds" in the two groups, $a:c$ and $b:d$.

Cornfield clearly emphasized that the validity of such estimation is contingent upon the fulfillment of three assumptions:

1. The rate of disease in the community must be comparatively small, say in the order of magnitude of 0.001 or less, relative to both the proportion of exposed cases and the proportion of nonexposed persons in the nonattacked population.

2. The $(a+c)$ cases must represent a random, unbiased sample of all cases of the disease.

3. The $(b+d)$ controls must represent a random, unbiased sample of all non-cases of the disease.

In actual practice, fulfillment of the first assumption is usually easily attainable, and any minor deviation from it causes no serious distortion in the results. The remaining two assumptions, however, are extremely difficult, if not actually impossible, to satisfy. Failure can cause considerable bias in the results, and is the basis of disagreement among both biostatisticians and epidemiologists. For instance, the detection of cases, referred to as ascertainment by R. A. Fisher, may be biased because cases are selected in a large medical referral center which is not representative of all cases in the community. In addition to being certain that the cases have all been diagnosed properly, the biostatistician must check to ensure that the cases were not selected because of distinguishing attributes such as socioeconomic status, location with respect to the center, race, sex, medical care previously received, or even whether the cases had close relatives with a similar disease and sought special diagnostic attention. The controls are sometimes chosen to be persons in the same hospital with a different diagnosis, or neighbors in the community. The biostatistician has to determine whether they are comparable in such factors as age, race, sex, severity (stage or grade) of the disease, and many other variables that tend to confound a just and fair comparison on factor F alone.

Three statistical problems are mentioned as a result of this type of study.

1. How to select the cases and controls? The literature on this is voluminous, but a few references may be mentioned [9,32,33].

2. In selecting the compeers, is it worthwhile to try to pair one or more controls to each case on the basis of certain characteristics that influence the probability of disease so that factor F will be the primary residual influence? References to this are found in Cochran [8] and McKinlay [38,39], among others.

3. After selecting cases and controls, it is difficult to estimate the influence of all factors other than F. These other variables are referred to as *confounding variables*, and they need to

be adjusted or accounted for so as to enable a valid comparison to be made on factor F. Historically, the procedure of adjusted or standardized rates is one way of achieving this goal. Another common statistical procedure is the use of covariance analysis [22]. More elaborate statistical procedures for dealing with confounding may be found in Miettinen [42–45], Rothman [51], and many others.

An excellent introductory reference to this class of problems is the monograph of Fleiss [19] and a review paper by Walter [54].

11.2.5 Clinical Trials

In the case of *clinical trials*, a whole host of special problems arise for which the biostatistician has had to develop special techniques. One of these is the detection of unexpected and untoward rare effects of drugs with the consequent need to terminate a trial early. Moreover, when data are demonstrating a trend earlier than expected, there is also a need to end the accession of patients and to stop further treatment with what may be an inferior regimen. This means the biostatistician must be familiar with the problems of multiple examinations of data [40,41], multiple comparisons [46,53], and other adjustment procedures made necessary by the ex post facto dredging of data. An excellent pair of references on randomized clinical trials is the set of articles by Peto et al. [49,50] and one by Byar et al. [5]. For the ethical problems involved in conducting clinical trials, the reader is urged to read Gilbert et al. [21] and Courand [12].

11.2.6 Confidentiality and Privacy of Records

The biostatistician is continually confronted with the demands of *confidentiality* and *privacy* of records in dealing with health and medical records. This subject begins with the controversial area of what constitutes informed consent when a patient is a minor, ill, comatose, or otherwise incompetent. The use of informed consent and randomization may present a conflict as far as a patient is concerned, and other design techniques may have to be used to overcome this problem [55].

11.2.7 Safeguarding of Computer Data

Related to the problem of privacy and confidentiality of data is how to safeguard these attributes when data are stored in *computers*, or records are to be linked from multiple sources. The linkage of records regarding an individual or an event requires the knowledge of computers, computer languages, and programming, as well as means for protecting identification of information stored in computers [6,14,15]. Of course, knowledge of computers is necessary in general because the tabulation of large volumes of data and analysis by current techniques could not be carried out if the methods in vogue in the mid-1970s, such as punch-card machines and desk calculators, were still relied on to carry out the necessary mechanical procedures.

11.2.8 Nature of Human Surveys

The nature of the field of inquiry in *human surveys* involving personal knowledge, attitudes, and behavior confront the biostatistician with a challenge akin to that of the sociologist. The goal is to obtain cooperation and a truthful response when the question(s) to be asked may be highly personal, sensitive, or stigmatizing in nature. Biostatisticians and sociologists have developed techniques to maximize cooperation with incentives of various kinds, including even monetary payment, as well as devices to assure anonymity. The use of response cards, telephone surveys, and a technique known as *randomized response* have all shown promise in this field [27,28].

11.2.9 Censoring of Observations

The *censoring of observations* in statistics is not a matter related to the sensitivity of highly classified data but rather is the purposeful or accidental blurring of an actual value for other reasons. For instance, in studies of life expectancy subsequent to an event such as exposure to a carcinogenic agent, the limited amount of time available for observation may require the experimenter to terminate the experiment after 1 or 2 years. Those individuals who have survived up to that point have had their *actual* life expectancy "censored," and the appropriate analysis must consider such observations in estimating parameter values. Similarly, the speed of some reactions may be so rapid, or the recording of data so crude at the beginning of the observation period, that a certain number of early observations are censored initially. Thus, in deaths of infants under 1 day old, some countries do not record the actual number of hours lived under 24. For most of these problems the use of order statistics has been the preferred solution to the problem of calculating unbiased estimates of the distribution parameters. (See Refs. 16 and 26.)

11.2.10 Community Diagnosis

The biostatistician collaborates with health planning personnel to establish benchmarks that describe the health status of a community so as to earmark places where greater attention and/or funds should be directed. The first step in community diagnosis is to study the population in terms of its magnitude and distribution by attributes such as age, sex, ethnicity, occupation, residence, and other factors that are related either to health or to the ability to obtain needed health services. Special studies may also be made of the community itself with respect to special environmental factors (industries and occupations, climate, pollution, etc.) and availability of health facilities, resources, and personnel. This information is combined with vital statistics and morbidity data to ascertain health problems characteristic of the community. For example, an unusually high birth rate may signify the need for a program of family planning services or it may simply be a fact caused by the peculiar age and sex distribution of the population. An excessive rate of lung cancer may suggest the need for an antismoking campaign or an investigation as to whether air pollution or a special industry, like the manufacture of products containing asbestos, may be involved.

In making judgments regarding the health status of a community, the biostatistician must be aware of the many possible comparisons that might be drawn. The age-adjusted death rate in a county, for example, might be compared to that of an adjacent county, to the rate of the entire state or nation, to the lowest county rate in the area, to the rate of a county or counties similar in the composition of the population of the study county, or simply to the trend of the rates for that county over the past 10 to 20 years.

Finally, community diagnosis would be incomplete without an attempt to study the effectiveness of treatment efforts. The biostatistician collaborates with health service providers to evaluate the effectiveness of any program instituted to improve the status of the community. With the assistance of cost-accounting specialists, the benefits (and any undesirable effects) of the program are compared with the costs in terms of funds and personnel so as to balance the relative weights of these items from a societal point of view. There are many references for studying this aspect in greater detail, and two helpful ones are Caro [7] and Greenberg [25].

11.3 Future Trends

There are two areas in which the biostatistician has been playing a leading role lately, and these cut across many kinds of applications and problems. It is highly likely that considerable research in methodology will continue to be devoted to these two special areas of concern. The first of these areas might be called modeling.

11.3.1 Mathematical Models

The relationship between a set of independent variables and the dependent or response variable(s) is usually referred to as the *mathematical model*. This model may take the form of a standard multiple regression analysis with a single response variable, a surface, or multiple response variables as in multivariate analysis.

It is generally assumed that the technical specialist with substantive knowledge of the field of application (epidemiology, toxicology, pharmacology, radiology, genetics, etc.) will play a crucial role in determining the model or relationship between a set of independent variables and the response variable(s). In actual practice, however, the biostatistician is the one who finally selects the specific model that establishes this functional relationship and then attempts to measure the strength or influence of the independent variables therein. Moreover, the biostatistician is often expected to contribute strongly to the decision as to whether the relationship is a causal one or merely one of association and correlation.

For example, in the measurement of the carcinogenicity of a food additive or drug, questions may arise as to whether a substance can be judged harmful if it "accelerates" the appearance of a tumor even though it does not increase the incidence of the abnormal growth. In general, the answer to this question is in the affirmative provided that there can be unequivocally demonstrated a dosage–response relationship between the substance and the tumor—i.e., the more of the suspected compound that is given or exposed to the test animal, the greater is the probability or likelihood that the tumor will occur earlier.

In the case of bioassay procedures, the biostatistician may be called upon to decide which sigmoid or S-shaped curve to use in order to relate the dosage to the response. This problem is more than simply a decision between the integrated normal curve (or probit) vis-à-vis the logit function [i.e., $\log_e(p/(1-p))$, where $0 < p < 1$]. In the case of harmful or toxic substances or low-level irradiation (as explained in the subsequent section), it is a question as to whether even a regression relationship can be assumed for purposes of extrapolation.

In many datasets that arise in biostatistics the data are in the form of contingency tables arising from frequency counts in cells created as a result of an n-way cross-classification of the population. The variable being measured is frequently on a nominal scale, i.e., the categories are simply the set of names corresponding to an attribute such as occupation, place of residence, cause of death, religion, or sex. In a few instances the categories of classification may be ordered or ranked on a scale whose orientation is clear but whose spacings are not known. For example, socioeconomic class may be classified as low, medium, or high; or birth order may be determined as first, second, third, and so on. In those special cases where the number of possible categories for an attribute is limited to two, the data are referred to as *dichotomous*. Thus one can have dichotomous information on sex disaggregated by male, female; or data where simply the presence or absence of a factor is noted and a dummy variable is created such that 1 = yes and 0 = no.

If we consider one kind of contingency table encountered frequently in surveys, we may have as cell entries the annual mean number of physician visits, illnesses, or days of disability. The number of respondents in each cell may be known but the relationship of the attributes in the classification scheme to the dependent variable is not. The aim of the analysis is to study how the different subclasses of the categories relate to the response variable.

For instance, let us consider that the number of doctor visits for a given age group has been averaged in each cell, and the number of classification variables is three: location (L = urban, rural), sex (S = male, female), and highest level of education (E = no high school, high school, college, postcollege). This means that there are $2 \times 2 \times 4 = 16$ cells, and the response variable, R, can be a function of all 16 parameters.

Instead of having to contend with 16 parameters, a general linear model might consist of

$$R = \lambda + \lambda_L + \lambda_S + \lambda_E + \lambda_{LS} + \lambda_{LE} + \lambda_{SE} + \lambda_{LSE},$$

where λ = general mean,

$\lambda_L, \lambda_S, \lambda_E$
= main effects of location, sex, and education, respectively

$\lambda_{LS}, \lambda_{LE}, \lambda_{SE}$
= the first-order interactions, each the simultaneous effect of two factors shown

λ_{LSE}
= a second-order interaction, or the simultaneous effect of all three factors

In fitting the parameters in this model, it is highly unlikely that the general mean will itself provide an adequate fit. By adding the main effects, one can ascertain whether an adequate fit has been obtained. If it doesn't, one proceeds to the next level and introduces whichever of the first-order interactions seem necessary. Finally, and only if required, the highest-order interaction would be brought into the final model. This hierarchical approach is similar to that of a stepwise regression analysis and can be carried out either forward or backward.

Now, if the model is restructured so that one may consider the effects as a linear sum for the *logarithm* of R, then the model is referred to as multiplicative, because effects are being multiplied. Thus, if

$$\log_e R = \lambda + \lambda_L + \lambda_S + \lambda_E + \lambda_{LS} + \lambda_{LE} + \lambda_{SE} + \lambda_{LSE},$$

we can define a new set of parameters such that $\lambda = \log_e \lambda'$, whence

$$\log_e R = \log \lambda' + \log \lambda'_L + \log \lambda'_S + \log \lambda'_E + \log \lambda'_{LS} + \log \lambda'_{LE} + \log \lambda'_{SE} + \log \lambda'_{LSE}$$

or

$$\log_e R = \log(\lambda' \cdot \lambda'_L \cdot \lambda'_S \cdot \lambda'_E \cdot \lambda'_{LS} \\ \cdot \lambda'_{LE} \cdot \lambda'_{SE} \cdot \lambda'_{LSE}).$$

Taking antilogarithms results in a form that shows why the model is referred to as a multiplicative model:

$$R = \lambda' \cdot \lambda'_L \cdot \lambda'_S \cdot \lambda'_E \cdot \lambda'_{LS} \cdot \lambda'_{LE} \cdot \lambda'_{SE} \cdot \lambda'_{LSE}.$$

This form of relationship is referred to as a *loglinear model*, and the predicted response value must always be positive.

In the case where the response variable is dichotomous yes–no, the mean values in the cell entries are really proportions. In the case of proportions, one may also use a logit–linear model not only to assure that the predicted values of p will lie between zero and 1 but also to help obtain a better fit.

A frequently encountered multiway contingency table consists of s samples from s multinomial distributions having r categories of response. There are then counterparts to the linear models and loglinear models discussed earlier, but the problem of estimation of the parameters involves choices between ordinary least squares, weighted least squares, maximum likelihood, and minimum χ^{2*}. The complexity of these considerations is beyond the scope of this section, but References 2, 13, 30, and 47 fairly well summarize the state of the art.

11.3.2 Detection of Hazardous Substances

With the successful conquest of most of the infectious diseases that have plagued humans throughout history, health authorities have been concentrating recently upon two chronic diseases whose etiology is yet to be determined: cardiovascular disease and cancer. In both cases, there is no disagreement with the thesis that heredity exercises a determining influence, but the role of the environment in causing many cases is also unquestioned. Attempts to measure the harm that may be caused by potentially hazardous substances in the environment, principally with respect to these two diseases, represent the greatest challenge to the biostatistician today. The benefit to society when successes are obtained make this area of research rewarding emotionally and scientifically despite the exceptional complexities involved.

The number of factors included under the rubric of environment which have a human impact is probably infinite. In addition to food, air, and water, environment includes everything that is not included under the genetics label. Thus in addition to known or unknown chemical and physical substances, there are such variables as exercise, noise, tension, and stress, plus all the psychosocial elements that affect people.

When one starts to limit study to hazardous substances by themselves, such as cigarette smoking and exposure to asbestos, the interaction of these two factors and of these two with stress, sleep, use of alcohol, and psychosocial elements soon points to the impracticality of drawing too constraining a line around the hazardous substances. Thus, if one accepts the assertion that an overwhelming majority of cancers are environmentally induced, and that of these perhaps 5 to 10% are occupation-related, any good inquiry into the incriminating substance(s) in the workplace cannot overlook the personal characteristics and habits of the workers themselves.

This highlights the first two difficulties in measuring the health importance of hazardous substances: that the list of substances and important factors is substantially great if not infinite, and that these factors have interactions or synergistic reactions that may be more important than the main effects themselves. (The effect of cigarette smoking and asbestos exposure referred to a moment ago is a perfect example of how the two factors in combination are much more important than the addition of the two by themselves in promoting lung cancer.)

Some of the other difficulties in studying the hazardous substances is that they are frequently available only in low doses and administered over a long period of time. The low dosage creates many problems of its own. For example, there is the question of how reliable the measurement is, especially when many of the estimates of exposure have to be retrospective or ex post facto in nature. Even if prospective in time, a sampling of the air in an environment requires a suitable model of air circulation so as to know whether to collect samples at different locations, at different times of the day, indoors or outdoors, with or without sunlight and wind, and so on. Furthermore, total exposure over an n-hour period may be more or less important than a peak period of exposure during a short period of time.

Determination of the impact of low doses of hazardous substances is especially complex because of other peculiarities. Experiments on human beings are, of course, out of the question, so reliance must be placed on accidental exposure of human beings, long-term exposure retrospectively, or the effect upon animals. Since low doses are likely to cause small effects, the results are extremely difficult to detect and usually require large numbers of animals plus extensive and expensive examinations of many organs. Since the number of animals required might be prohibitive anyway, reliance is often placed upon artificially high doses. This confuses the picture further because one needs to know what model to use in projecting by extrapolation from high doses the effect at low dose levels, as well as what rationale for judging how valuable it is to measure the possible effects on human beings from data on the health of animals, usually mice and other rodents. The models used to extrapolate from high dose to low dose, especially in problems involving radiation, include the one-hit, two-hit, probit, and a variety of empirical models. This subject gets still more complicated when assumptions are made about threshold levels and rates of neutralization or detoxification [11].

There is here also the effect of confounding variables, which were referred to earlier, when discussing the selection of epidemiological cases and controls, as a special field of application.

There is also the entire question of risk versus benefit, which is an important consideration in determining public policy. As an example of how legislation in the United States which fails to consider both risk and benefit can adversely affect public policy, the Delaney Amendment to the Food Additive Amendments of 1958 [Public Law (PL) 85–959] can be cited. The Delaney clause stated that any food additive that is capable of producing cancer in animals or human beings is assumed harmful to human beings and must be banned by the Food and Drug Administration regardless of any benefit. Although the methodology for studying risks vis-à-vis benefits is in a primitive state, the present law does not encompass the possibility that the risk might be minuscule and the benefit substantial. Such has probably been the situation regarding the use of saccharin. Another drawback with the Delaney clause is that certain food additives that have a low carcinogenic effect, such as nitrites, might be found naturally in other foods. They have to be banned as a food additive, even though the benefits might outweigh the disadvantages and even though other foods already possess the particular substance.

This has been a brief overview of a most important area. Readers who are interested in this subject are urged to examine the technical report by Hunter and Crowley [34].

11.4 Education and Employment of Biostatisticians

What does it take to make a biostatistician? To this question there are a host of responses but no firm, positive answer. This is not surprising in view of the fact that two other quite different occupations that have been studied probably more than any other since the late 1960s—airplane piloting and medicine—still have a long way to go before most of the answers are known regarding the optimal selection of applicants and the most effective method of education.

Undoubtedly, one of the most important characteristics in the outstanding biostatistician is the individual himself or herself. Essential attributes are an inquisitive curiosity or "burning yearning for learning," a constancy of purpose, an ability to think quantitatively, an interest in applying statistical methods to biological problems, and probably a personality or mental disposition that encourages close working relationships with collaborators from many fields.

The field of biostatistics has many avenues of access for entry purposes. Although most persons probably enter from a traditional mathematical or mathematical–statistical background, many others have come from an original interest in biomedical fields, sociology, psychology, engineering, and computer sciences. There is no unique or assured pathway to biostatistics; the right person can approach it from whatever direction maximizes his or her own potential.

The most frequently used institutions for educating biostatisticians are the departments of biostatistics and statistics. These departments should be located in a university setting where there are, first, an academic health center or, at the very least, a medical school and hospital. There must also be strong units in the remainder of the university concerned with the teaching of graduate students in mathematics, probability and statistics, and computer sciences. More than the structural units in the university, however, there must be a pattern or tradition of close working relationships between the biostatistics faculty and those from the other entities. Training at the doctoral level will not be truly meaningful unless the biostatistical faculty are actively engaged in and publish work on applied as well as theoretical research concerning statistical methods.

Merely because the university has, say, an accredited school of public health with a department of biostatistics is no guarantee that it is a good one. (Moreover, biostatisticians can be, and have been, educated in statistical departments other than those found in schools of public health.) Students seeking training in biostatistics would be well advised to be certain that the teaching faculty are engaged in both applied and methodological research, and most important, that there is a close affiliation or working relationship with a medical unit of some kind. Unless such affiliation exists, it will be more difficult to get exposure to good experience involving clinical data, clinical trials, epidemiology, hospital and clinic studies, and health services research.

The doctoral training might consist of either the program for the traditional academic Ph.D. degree or a program that is professionally oriented, such as one directed to the Doctor of Hygiene or Doctor of Public Health degree. The Ph.D. is usually intended for those persons planning careers in an academic or research setting where emphasis is on developing statistical methodology to solve important biological and public health problems. The related doctoral dissertations are most often published in statistical journals.

The professional doctoral degree is usually intended for persons who plan careers in government or industry and whose emphasis is on service to persons seeking statistical advice. The nature of the doctoral dissertation is frequently the new application of statistical concepts to important public health or biological problems. What is novel in the dissertation is the application of a known statistical technique to solve an important health problem. This type of dissertation is usually published in biological and public health journals.

Training at the master's degree level for persons interested in beginning and intermediate-level positions in biostatistics is determined and evaluated by approximately the same guidelines as the foregoing. The main difference, perhaps, is that the criterion of an active ongoing research program is not as stringent. Instead, emphasis should be placed on good teaching and other pedagogical processes that will enable students to learn practical techniques at the same time that a few of them are stimulated to pursue advanced training at the doctoral level.

Employers of biostatisticians have ranged from local, state, and federal government to industry and academic institutions. Each employer will require a different set of areas of knowledge over and above the general field of statistics and probability. For example, one type of government biostatistician may be required to be an expert in registration of vital statistics, demography, survey research, and the special problems associated with confidentiality. The person working as a biostatistician in a pharmaceutical firm may require special training in bioassay techniques, mathematical modeling, and those aspects of clinical trials related to the study of new drugs and their toxicity, dosage, and effectiveness, in order to collaborate in preparing applications for a new drug to be approved by a regulatory agency such as the Food and Drug Administration, U.S. Public Health Service.

It is difficult to know even approximately how many persons there are in the United States who would classify themselves as biostatisticians. There is no one professional organization designed for affiliation of persons who are primarily biostatisticians. A rough guess would be that one-fifth of the 15,000 statisticians listed in the 1978 Directory published by the American Statistical Association are strongly interested in biostatistical problems, and, of these, about one-half would be individuals who classify themselves primarily as biostatisticians. An international Biometric Society was established in 1947, and in 1950 it assumed responsibility for publication of the journal *Biometrics*, which had had its inception as the *Biometrics Bulletin* in 1945 under the aegis of the American Statistical Association.

The reader interested in further readings about biostatistics is recommended to consult especially the journals *Biometrika* and *Biometrics*.

References

The publications cited in this chapter are listed in alphabetical order. For convenience, they have been classified into one of seven categories shown at the end of the reference and coded according to the following scheme:

(A) historical and biographical
(B) epidemiology: models and causality
(C) epidemiology: relative risk
(D) epidemiology: clinical trials
(E) demography and community diagnosis
(F) surveys, privacy and confidentiality
(G) general biometry

1. Bailey, N. T. J. (1957). *The Mathematical Theory of Epidemics*. Charles Griffin, London/Hafner, New York. (B)

2. Bishop, Y. M. M., Fienberg, S. E., and Holland, P. W. (1975). *Discrete Multivariate Analysis*. MIT Press, Cambridge, MA. (G)

3. Blalock, H. C., Jr. (1964). *Causal Inference in Nonexperimental Research*. University of North Carolina Press, Chapel Hill, NC. (B)

4. Box, J. F. (1978). *R. A. Fisher: The Life of a Scientist*. Wiley, New York. (A)

5. Byar, D. P., Simon, R. M., Friedewald, W. T., Schlesselman, J. J., DeMets, D. L., Ellenberg, J. H., Gail, M. H., and Ware, J. H. (1976). *New Engl. J. Med.*, **295**, 74–80. (D)

6. Campbell, D. T., Baruch, R. F., Schwartz, R. D., and Steinberg, J. (1974). *Confidentiality—Preserving Modes of Access to Files and to Interfile Exchange for Useful Statistical Analysis*. Report of the National Research Council Committee on Federal Agency Evaluation Research. (F)

7. Caro, F. G., ed. (1971). *Readings in Evaluation Research*. Russell Sage Foundation, New York. (E)

8. Cochran, W. G. (1953). *Am. J. Public Health*, **43**, 684–691. (C)

9. Cochran, W. G. (1965). *J. Roy. Statist. Soc. A*, **128**, 234–255. (C)

10. Cornfield, J. (1951). *J. Natl. Cancer Inst.*, **11**, 1269–1275. (C)

11. Cornfield, J. (1977). *Science*, **198**, 693–699. (G)

12. Courand, A. (1977). *Science*, **198**, 699–705. (D)

13. Cox, D. R. (1970). *The Analysis of Binary Data*. Methuen, London. (G)

14. Dalenius, T. (1974). *Statist. Tidskr.*, **3**, 213–225. (F)

15. Dalenius, T. (1977). *J. Statist. Plan. Infer.*, **1**, 73–86. (F)

16. David, H. A. (1981). *Order Statistics*, 2nd ed. Wiley, New York. (G)

17. Ederer, F. (1979). *Am. Statist.*, **33**, 116–119. (D)

18. Finney, D. J. (1964). *Statistical Method in Biological Assay*, 2nd ed. Hafner, New York. (G)

19. Fleiss, J. L. (1973). *Statistical Methods for Rates and Proportions*. Wiley, New York. (C)

20. George, S. L. (1976). *Proceedings of the 9th International Biomedical Conference, (Boston)*, **1**, 227–244. (D)

21. Gilbert, J. P., McPeek, B., and Mosteller, F. (1977). *Science*, **198**, 684–689. (D)

22. Greenberg, B. G. (1953). *Am. J. Public Health*, **43**, 692–699. (C)

23. Greenberg, B. G. (1959). *Am. Statist.*, **13**(3), 13–17, 28. (D)

24. Greenberg, B. G. (1969). *J. Am. Statist. Assoc.*, **64**, 739–758. (B)

25. Greenberg, B. G. (1974). *Medikon*, **6/7**, 32–35. (E)

26. Greenberg, B. G. and Sarhan, A. E. (1959). *Am. J. Public Health*, **49**, 634–643. (G)

27. Greenberg, B. G. and Sirken, M. (1977). *Validity Problems, Advances in Health Survey Research Methods*. National Center of Health Services Research, Research Proceedings Series, DHEW Publication No. (HRA) 77-3154, pp. 24–31. (F)

28. Greenberg, B. G. and Abernathy, J. R., and Horvitz, D. G. (1970). *Milbank Mem. Fund Quart.*, **48**, 39–55. (F)

29. Greenwood, M. (1948). *Medical Statistics from Graunt to Farr*. Cambridge University Press, Cambridge, UK. (A)

30. Grizzle, J. E., Starmer, C. F., and Koch, G. G. (1969). *Biometrics*, **25**, 489–503. (G)

31. Hastings, D. W. and Berry, L. G., eds. (1979). *Cohort Analysis: A Collection of Interdisciplinary Readings*. Scripps Foundation for Research in Population Problems, Oxford, OH. (E)

32. Horwitz, R. I. and Feinstein, A. R. (1978). *New Engl. J. Med.*, **299**, 1089–1094. (C)

33. Hulka, B. S., Hogue, C. J. R., and Greenberg, B. G. (1978). *Am. J. Epidemiol.*, **107**, 267–276. (C)

34. Hunter, W. G., and Crowley, J. J. (1979). *Hazardous Substances, the Environment and Public Health: A Statistical Overview.* Wisconsin Clinical Cancer Center Technical Report No. 4, University of Wisconsin, Madison, Wis. (G)

35. Kruskal, W. and Tanur, J. M., eds. (1978). *International Encyclopedia of Statistics.* Free Press, New York. (A)

36. Lave, L. B. and Seskin, E. P. (1979). *Am. Sci.*, **67**, 178–186. (B)

37. Linder, F. E. and Grove, R. D. (1947). *Vital Statistics Rates in the United States, 1900–1940.* U.S. Government Printing Office, Washington, DC (see especially Chaps. 3 and 4.) (E)

38. McKinlay, S. M. (1975). *J. Am. Statist. Assoc.*, **70**, 859–864. (C)

39. McKinlay, S. M. (1977). *Biometrics*, **33**, 725–735. (C)

40. McPherson, K. (1974). *New Engl. J. Med.*, **290**, 501–502. (D)

41. McPherson, C. K. and Armitage, P. (1971). *J. Roy. Statist. Soc. A*, **134**, 15–25. (D)

42. Miettinen, O. S. (1970). *Biometrics*, **26**, 75–86. (C)

43. Miettinen, O. S. (1970). *Am. J. Epidemiol.*, **91**, 111–118. (C)

44. Miettinen, O. S. (1972). *Am. J. Epidemiol.*, **96**, 168–172. (C)

45. Miettinen, O. S. (1974). *Am. J. Epidemiol.*, **100**, 350–353. (C)

46. Miller, R. G. (1966). *Simultaneous Statistical Inference.* McGraw-Hill, New York. (G)

47. Nelder, J. A. and Wedderburn, R. W. M. (1972). *J. Roy. Statist. Soc. A*, **135**, 370–384. (G)

48. Norton, B. J. (1976). *Proceedings of the 9th International Biomedical Conference (Boston)*, Vol. 1, pp. 357–376. (A)

49. Peto, R., Pike, M. C., Armitage, P., Breslow, N. E., Cox, D. R., Howard, S. V., Mantel, N., McPherson, K., Peto, J., and Smith, P. G. (1976). *Br. J. Cancer*, **34**, 585–612. (D)

50. Peto, R., Pike, M. C., Armitage, P., Breslow, N. E., Cox, D. R., Howard, S. V., Mantel, N., McPherson, K., Peto, J., and Smith, P. G. (1977). *Br. J. Cancer*, **35**, 1–39. (D)

51. Rothman, K. J. (1976). *Am. J. Epidemiol.*, **103**, 506–511. (C)

52. Spiegelman, M. (1968). *Introduction to Demography*, Harvard University Press, Cambridge, MA. (E)

53. Tukey, J. W. (1977). *Science*, **198**, 679–684. (D)

54. Walter, S. D. (1976). *Biometrics*, **32**, 829–849. (C)

55. Zelen, M. (1979). *New Engl. J. Med.*, **300**, 1242–1245. (D)

12 Clinical Trials II

R. Simon

12.1 Introduction

Clinical trials are a major success story for our society. They represent one of the few areas in which the effects of new technology or new programs on humans are studied using modern statistical principles of experimental design. Although the most fundamental statistical principles—randomization, replication, and unbiased measurement—are similar to those in other areas of experimentation, there are many complexities to clinical trials resulting from the use of human subjects. These complexities have stimulated the development of important new statistical methods.

12.2 Types of Clinical Trials

Clinical trials may be categorized into those that attempt to identify promising treatments and those that attempt to determine whether such treatments provide meaningful benefit to the subjects. These two objectives generally require very different types of trials. Many of the controversies that arise in the design and interpretation of clinical trials are related to these differences in objectives. Schwartz and Lellouch [39] have used the terms "explanatory" and "pragmatic" to distinguish trials whose objectives are to provide information about biological effects from those aimed at determining subject benefit. The terms "phase 2" and "phase 3" are also sometimes used to indicate the same distinction.

Today many clinical trials do not actually involve administering treatments to patients. For example, there are prevention trials, disease-screening trials, and trials of diagnostic methods. We will continue to refer to "clinical trials" of "treatments" for "patients," but broader meanings of these terms will be assumed. There are many similarities in the design of these various types of trials, but there are also important differences. For example, prevention trials tend to involve large numbers of subjects because the event rate of the disease is often quite low. Such trials may require long-term interventions, so that the disease process may be influenced long before the disease becomes clinically evident. Prevention trials may involve lifestyle interventions which require substantial efforts for achieving good subject compliance. Imaging modalities are improving at a more rapid rate than are treatments or prevention agents in many areas. Because of the expense of new technology, there is an increased need for well-designed prospective trials of diagnostic technologies. This area, and particularly developments in receiver operating curve (ROC) analysis, have been reviewed by Begg [2].

12.3 Selection of Patients and Treatments

The guiding principle of clinical trials is to ask an important question and get a reliable answer. The former generally means asking a question that has the potential for influencing medical or public health practice, using a control group that is widely accepted, using an experimental treatment that is widely applicable, using an endpoint that is a direct measure of patient benefit, and studying a group of patients that is broadly representative. Physicians tend to view clinical trials from the explanatory perspective, preferring extensive characterization of each patient and narrow eligibility criteria in an attempt to achieve a homogeneous selection. The treatment may subsequently be used for a broader selection of patients, however. Consequently, and because eligibility restrictions are often somewhat arbitrary, there has been a movement toward broader eligibility criteria in many clinical trials. For Medical Research Council–sponsored trials in the UK the concept of very broad eligibility criteria utilizing the "uncertainty principle" has become popular. This principle replaces a list of eligibility criteria specified by the trial organizers with the judgments of the participating physicians. If the patient has the disease in question and the physician believes that both treatments being studied are appropriate and is uncertain which is preferable, then the patient is eligible for randomization.

A run-in period is sometimes used to screen subjects for participation. This is most often done in prevention trials. After the patient is determined to be eligible for the trial, but before randomization, the patient is given a trial period of some medication, which may be a placebo. If during the run-in period the patient has complied with taking the medication and still wishes to participate, then the patient is randomized. Patient dropouts and noncompliance after randomization have major deleterious effects on the power of a trial. The purpose of the run-in period is to avoid such events [27].

The selection of treatments for study is a key factor in the design of a clinical trial. In many cases the choice will be obvious. Sometimes, however, it is difficult to obtain agreement among participating physicians to randomize among the treatments that represent the most medically important study. For example, among a group of radiotherapists there may be reluctance to have a control group not involving radiotherapy.

Usually the selection and refinement of the experimental treatment is accomplished in phase 2 trials. Phase 2 trials often employ short-term biological endpoints because the focus is on selecting the most promising treatment for evaluation in a phase 3 trial rather than on establishing the medical value of that treatment. Phase 2 trials may involve randomization among treatment variants or may be single treatment group studies of the biological effects of a given treatment variant.

In some cases a single experimental treatment may not have been identified at the time that the phase 3 clinical trial is initiated. Several designs have been introduced to simultaneously screen experimental treatments and test against a control those treatments selected as most promising [37,44].

Factorial designs are being more widely used. With a 2×2 factorial design, patients are randomized between the two levels of intervention A and also between the two levels of intervention B. For example. subjects on the Physicians Health Study are randomized to daily aspirin or placebo and to daily beta carotene or placebo [48].

Often, the primary endpoints will be different for the two factors. In the physician's health study the primary endpoint for the aspirin randomization was cardiovascular events, whereas the primary endpoint for the beta carotene randomization was cancer events.

If the effect of intervention A is the same (possibly zero) at each of the levels of randomization B, and vice versa, then no interaction exists and the two treatment questions can be answered with the same sample size as would be required for a single-factor study. If interactions do exist, then the factorial trial addresses primarily the question of the average effect of intervention A (averaged over the levels of randomization B) and the average effect of intervention B [3]. To address the effect of intervention A at each level of factor B, one would require twice the sample size in order to have the same statistical power. To attain the same statistical power for testing the statistical significance of the interaction would require 4 times the sample size if the size of interaction to be detected is the same magnitude as the main effect. Consequently, a factorial trial designed under the assumption of no interaction will not have good statistical power for testing that assumption. A similar situation exists for the two-period crossover design [12]. Unless it can be assumed that interactions are unlikely or unless the average effects are themselves important, factorial designs may provide ambiguous results.

When there are different endpoints for the different factors, the assumption of independence may be warranted. Factorial designs are efficient for the screening of treatments, most of which will be ineffective so long as the administration of one treatment does not interfere with the administration of others. In this context, high-order 2^p factorial designs may be useful. Although factorial designs may give ambiguous results concerning interactions, subsequent clinical trials can be designed to address the interaction hypotheses. Chen and Simon [4] have developed new methods for designing multirandomization group clinical trials in which the treatment groups are structurally related but not necessarily in a factorial manner.

12.4 Endpoints

Endpoints refer to the outcome measures used to evaluate a treatment. The definition of a clinically meaningfully endpoint is crucial to the design of a clinical trial. The endpoint determines what claims can be made for a new treatment and influences the nature of patient examinations, costs, data collection, and data analysis. For example, oncologists often think of tumor shrinkage as a measure of anticancer effect and measure the activity of a treatment for metastatic cancer by its response rate. But "partial response," which represents a 50% shrinkage of the tumor, often does not result in palliation of symptoms, improved quality of life, or prolongation of life. For studies of patients with metastatic disease, survival and direct measures of palliation are more appropriate endpoints.

It is common to have multiple types of toxicity monitored in a clinical trial. This seldom results in difficulty, because toxicities are often of high enough prevalence to be clearly associated with the treatment. Having several efficacy endpoints can be more problematic, however [50]. If combined inference for all the efficacy endpoints is medically meaningful, then a serious multiplicity problem can be avoided. One simple approach is to compute a significance test of the null hypothesis for each of the endpoints and then combine these test statistics into an overall test of the composite null hypothesis that treatment does not affect any of the endpoints. Many other methods for performing a combined analysis of endpoints have been proposed [29,32,45]. Combined analysis uses the endpoints to reinforce each other and can have good power for alternatives where the treatment effect is consistent in direction for the individual endpoints.

Making separate inferences on each of many efficacy endpoints carries greater risk of error (either false positives or false negatives). There should generally not be many primary efficacy endpoints for a phase 3 trial. By the time a treatment is ready for phase 3 evaluation, the hypotheses concerning treatment effects of real patient benefit should be rather well defined. If there are few (e.g., 1–3) efficacy endpoints, then planning the trial to have 90% power for detecting medically significant effects with a separate two-sided 5% significance level for each endpoint should generally be acceptable.

There has been considerable work done in the decade 1985–1994 on the topic of surrogate endpoints [9,33,53]. The motivation for using a surrogate endpoint is to facilitate trial design and permit earlier conclusions. If, for example, a drop in CD4 count were a valid surrogate for survival in patients with AIDS, then clinical trials would not be influenced by variability in treatments received following the CD4 drop. There is also often pressure to make a new treatment available to patients in the control group after disease progression. Unless time to progression is established as a valid surrogate of survival, however, this may make it impossible to evaluate whether the introduction of the new drug has affected survival of the patients. Unfortunately, it can be very difficult to establish an endpoint as a valid surrogate of the endpoint of real interest, and the validity only holds in the context of specific treatments.

12.5 Randomization and Stratification

The history of medicine contains numerous examples of useless and harmful treatments that persisted in practice for many years. For example, insulin coma was widely used for 25 years as a treatment for schizophrenia before it was finally subjected to a randomized clinical trial in 1957 that showed it to be dangerous and ineffective [24]. The randomized clinical trial has given modern medicine a tool for reliability determining whether a treatment provides real benefit to patients. If a treatment cures a large proportion of patients with a previously incurable disease, then the results will be apparent without randomization. This was the case for advanced Hodgkin's disease, but such examples are rare. Generally, treatment benefits are small compared with the variations in prognosis among patients, and randomization is essential. Many physicians have accepted the notion that nonrandomized trials are appropriate for determining what treatments are sufficiently promising to warrant evaluation in randomized clinical trials, but that definitive conclusions require randomized trials. If treatment effects are really large, then few patients will be required before significant differences are indicated by interim analyses. The importance of randomization in Bayesian analyses was described by Rubin [35].

Stratified randomization is a class of methods for ensuring a greater balance of the treatment groups with regard to potentially important baseline covariates than might be achieved by pure randomization. Physicians like to see the treatment groups balanced with regard to potential prognostic factors. They do not accept the "closurization principle" that states "first you randomize and then you close your eyes" [34]. They trust evidence of balance and "comparability" more than they do complex covariate adjustments. Consequently, stratified randomization has been widely practiced. Adaptive stratification methods have also been developed that ensure good balance marginally with regard to many covariates [30, "Imbalance Functions"]. Although the increase in power compared to analytic covariate adjustment is often small, good marginal balance of covariates is better ensured, and there appear to be no serious analytical difficulties [18,21,40]. Standard methods of analysis are generally satisfactory. For nondeterministic adaptive stratification designs, valid analysis can always be accomplished by using a randomization test implemented by simulating the reassignment of treatments to patients using the adaptive stratification procedure. Stratification also serves to prespecify the major covariates and so may be used to limit the extent of subsequent subset analysis.

12.6 Sample Size

Physicians often wish to do their own clinical trial rather than participating in a multicenter or intergroup clinical trial. Doing their own trial permits them to pursue their own ideas and to get greater recognition in subsequent publications. Such physicians are often wildly optimistic about the size of treatment effects that will be obtained and unrealistic about the number of patients that their institution can place on a clinical trial. Consequently, a large number of inadequate-sized clinical trials are initiated, and many of the studies that are published as "negative" are really just indeterminate, giving results consistent with both significant effects and no effects. An even more serious result is that small positive studies may represent the outlying false positives among the many initiated trials [41]. Consequently, independent assessment of the likely size of treatment effects and accrual potential is important. The former can be obtained by review of the literature of similar trials or elicitation from independent investigators, and the latter by review of accrual to previous studies.

The conventional approach to sample size determination involves specified statistical power for achieving statistically significant rejection of the null hypothesis under a specified alternative hypothesis representing the smallest medically important treatment effect. The actual calculations depend on the nature of the data. For survival data the most commonly used formulae are based on the assumption of proportional hazards. In this case the required total number of events to observe is approximately

$$E = 4(z_{1-\alpha/2} + z_{1-\beta})^2/(\log\theta)^2,$$

where z_p denotes the pth percentile of the standard normal distribution and θ denotes the hazard ratio to be detected with power $1-\beta$ at a two-sided significance level α. Since the power depends on the number of events observed, it is best to target a specified number of events at the time of the final analysis. The number of patients to be accrued, the duration of accrual and the duration of follow-up after conclusion of accrual can then be estimated and adjusted according to the parametric distributions expected and the event rates observed for the control group. George and Desu [16] and Rubinstein et al. [36] derived the above formula under the assumption of exponential failure distributions. Schoenfield [38] derived it for general proportional hazard alternatives. Sposto and Sather [47] studied cure models. Others have incorporated other factors such as noncompliance [22,23]. The above expression is based on a two-sample model in which the survival distributions are proportional hazard alternatives. It is also possible to define a model in which for each covariate stratum the survival curves for the two treatments are proportional-hazard alternatives; the same equation results for that model. For binary-response endpoints, the method of Ury and Fleiss [51] is widely used as an accurate approximation to the power function of Fisher's exact test. More recently, "exact" unconditional methods have been developed by Suissa and Shuster [49].

The usual approach of testing the null hypothesis has been misleading for the development and reporting of therapeutic equivalence or positive control trials [7]. In a *therapeutic equivalence trial*, the new treatment will be accepted if it appears equivalent to the control treatment with regard to the efficacy endpoint. For cancer trials this is usually because the control treatment is considered effective and the new treatment has some advantage in terms of toxicity, morbidity, cost, or convenience. In general, *positive control trials* are conducted when the effectiveness of a new treatment is to be evaluated and it is not ethically appropriate to use a placebo or no-treatment group. For such trials the medical decisionmaking structure is different than usual because failure to reject the null hypothesis may result in adoption of the new treatment. When the disease is life-threatening, only very small reductions in treatment effectiveness are usually acceptable in exchange for reductions in side effects; hence small values of θ and β are necessary. Often somewhat larger values of α are acceptable. This is sometimes not recognized in the planning and analysis of such trials, however. The planning of therapeutic equivalence and positive control trials has received increasing attention since the 1990s. The use of confidence intervals in the planning and reporting of such studies [7] seems particularly appropriate. In reporting the results of a significance test, the important information is hidden in often unstated statistical power. Also, statistical power takes no account of the results actually obtained.

12.7 Large, Simple Trials

Reliable detection of treatments that cause a 15–20% reduction in the annual hazard of death requires very large clinical trials. Yet a reduction of this amount for a widely applicable treatment of a common disease can represent a saving of tens of thousands of lives each year [54]. To do very large clinical trials requires broad participation of community physicians, and to facilitate this the trial should be well integrated with the usual procedures for the care of patients and should not require more work or expense than usual care. This approach has several implications. It means that the paperwork requirements must be kept to a minimum and that the tests used for workup and follow-up procedures should not be rigidly standardized. Because intensive monitoring of follow-up is not required, mortality is used as the endpoint. In addition, the eligibility criteria are made broad and left largely to the discretion of physicians entering patients, so that the results are widely applicable. The multinational ISIS-2 trial comparing streptokinase, aspirin, both, or neither randomized 17,187 patients with suspected acute myocardial infarction, and subsequent trials for this disease have been even larger [19].

12.8 Monitoring of Interim Results

Because clinical trials involve human subjects, interim monitoring of efficacy and safety are necessary. The mistakes possible from the naive interpretation of repeated analyses of accumulating data are well known to statisticians. For example, if a clinical trial of two equivalent treatments is analyzed repeatedly over the periods of accrual and follow-up, then the probability that a $P < 0.05$ will be found in at least one interim or final analysis may exceed 25%. Sequential designs have been developed in order to control the type I error with interim monitoring. For multicenter trials, group sequential methods are popular [26,28,31]. These accommodate a limited number of interim analyses and are more practical where data collection is logistically complex. Bayesian versions of group sequential designs have also been proposed [14]. It has become more widely recognized that clinical trials are not essentially decision processes; they are vehicles to provide information to a broad audience who will be making their own decisions. Consequently most current Bayesian methods are oriented to quantifying evidence based on posterior distributions but do not attempt to define loss functions. The concepts of "skeptical" and "enthusiastic" prior distributions for treatment effects have been introduced in recognition of the diverse consumers of clinical trial reports [46].

A second popular approach to interim monitoring is the stochastic curtailment or conditional power approach developed by Lan et al. [25]. Consider a trial designed to provide power $1-\beta$ for rejecting the null hypothesis at a significance level α in favor of a specified alternative. At various points during the trial the probability of rejecting the null hypothesis at the end conditional on the data already accrued is computed under the original alternative hypothesis. If this probability is less than $1-\gamma$, then the trial is terminated and the null hypothesis accepted. Similarly, the probability of accepting the null hypothesis at the end of the trial is computed, conditional on the results at the time of interim analysis. This is computed under the null hypothesis. If this probability is less than $1-\gamma'$, then the trial is terminated and the null hypothesis is rejected. Even with continuous interim monitoring of this type, the type I and II error rates are bounded by α/γ' and β/γ respectively. These upper bounds are conservative if intermittent rather than continuous monitoring is used. The bounds are valid even for continuous monitoring, however, and hence the method is useful for monitoring trials without predesigned sequential plans. Values of γ and γ' of 0.8 provide conservative monitoring with relatively little effect on the error probabilities.

A third approach to interim monitoring is the use of repeated confidence intervals [20]. This method is particularly appropriate for therapeutic equivalence or positive control trials [7]. Continuous monitoring designs are also used, primarily for single-institution trials [52].

In addition to a very powerful body of statistical methodology for interpreting accumulating data, major multicenter clinical trials have adopted data-monitoring committees to review the accumulating data and make recommendations about whether accrual to the study should be terminated, the protocol changed, or the results released. During the trial, interim efficacy results are kept from the physicians entering patients. This approach is designed to protect both the patients and the study. The patients are protected because decisions to terminate, change, or continue the study are made by persons with no vested professional or financial interest in the results of the trial. The study is protected from inappropriate early termination resulting from physicians entering patients getting nervous about unreliable interim trends or losing interest because of the lack of such trends.

12.9 Subset Analysis

No two patients are exactly alike, and physicians must make treatment recommendations for individual patients. Because treatment of the many for the benefit of the few is problematic, there is often interest in determining which types of patients actually benefit from a treatment. For example, a major NIH-sponsored randomized clinical trial compared zidovudine and zalcitabine with each other and with their combination for treating advanced HIV disease [10]. Although no differences were found for the patient groups overall, attention focused on the subset of patients with higher CD4 lymphocyte counts. The dangers of subset analysis are well known, however. If you test true null hypotheses in each of k disjoint subsets of patients, the probability of rejecting at least one null hypothesis by chance alone at the α level is $1 - (1 - \alpha)^k$. If $\alpha = 0.05$ and $k = 10$, then the probability of a type I error is about 0.40. If the subsets are not disjoint but are defined by the levels of $k/2$ binary covariates considered one at a time, the results are similar [13]. Several approaches to evaluating subset effects have been proposed. These include the traditional requirement of establishing a significant treatment by subset interaction before analyzing the subsets separately [42]. Gail and Simon [15] have developed tests for qualitative interactions. A qualitative interaction indicates that one treatment is preferable for some subsets and the other treatment is preferable for other subsets. Bayesian methods for subset analysis have also been developed [6].

One particular type of subset analysis that often arises is that involving center effects. Main effects and even interactions involving centers should be expected in multicenter clinical trials. Having a broad basis for generalization of conclusions is a strength of multicenter clinical trials, and one should not expect such trials to provide statistically powerful evidence for addressing center effects. When there are few centers involved, standard fixed-effect models can be used to evaluate whether there is evidence of large center effects or interactions (with treatment effects) not explained by baseline patient covariates [11]. If such effects are detected, further investigation of their causes should be pursued. When there are many centers involved, fixed-effect models are unlikely to be useful. Mixed models, treating patient covariates as fixed effects and center as a random effect, can be employed to assess the extent of intercenter variability. These are very similar to the empirical Bayes [5] and Bayesian [17] models that have been developed for examining disjoint subsets. The robustness of the overall conclusion to omission of data from individual centers can also be studied.

12.10 Meta-Analysis

One of the best examples of a meta-analysis of randomized clinical trials was the evaluation of tamoxifen or chemotherapy for the treatment of women with primary breast cancer [8]. This illustrated most of the desirable qualities of a meta-analysis. A meta-analysis identifies all relevant randomized trials and attempts to use the combined data from all the trials to obtain a more reliable answer than that obtainable from any one trial. Only randomized trials are included, because the biases of nonrandomized comparisons are too great relative to the size of treatment effect of interest. Attention is generally not limited to published trials, because there is a publication bias toward the reporting by investigators and acceptance by journals of positive results [1]. The best meta-analyses are based on obtaining individual patient data on all randomized patients on each trial rather than relying on published summaries. Published analyses often exclude some randomized patients and thereby introduce potential bias. With individual patient data, subsets can be examined in uniform ways across studies. Meta-analyses often use mortality as the endpoint, because it is objective and because individual studies may have already provided adequate answers for shorter-term endpoints. Meta-analysis has two broad objectives. The first is to address questions that individual trials were too small to address reliably. The second is to review all the evidence systematically and overcome the problems of publication bias that tend to focus attention on the positive trials.

Meta-analyses have important limitations, however. Many meta-analyses are based on published reports and hence do not avoid publication bias or the other biases of the original publications. A more inherent limitation, however, is that the therapeutic question addressed by a meta-analysis must usually be a general one, because there are often differences between the individual trials. Meta-analyses are seldom successful in deriving reliable information from a mass of varied and inadequate trials. Good meta-analyses require good large randomized clinical trials.

References

1. Begg, C. B. and Berlin, J. A. (1989). Publication bias and dissemination of clinical research, *J. Natl. Cancer Inst.*, **81**, 107–115.
2. Begg, C. B. (1991). Advances in statistical methodology for diagnostic medicine in the 1980's. *Statist. Med.*, **10**, 1887–1895.
3. Brittain, E. and Wittes, J. (1989). Factorial designs in clinical trials: the effects of noncompliance and subadditivity. *Statist. Med.*, **8**, 161–171.
4. Chen, T. T. and Simon, R. (1994). A multiple decision procedure in clinical trials. *Statist. Med.*, **13**, 431–446.
5. Davis, C. E. and Leffingwell, D. P. (1990). Empirical Bayesian estimates of subgroup effects in clinical trials. *Controlled Clin. Trials*, **11**, 37–42.
6. Dixon, D. O. and Simon, R. (1991). Bayesian subset analysis. *Biometrics*, **47**, 871–882.
7. Durrleman, S. and Simon, R. (1990). Planning and monitoring of equivalence studies. *Biometrics*, **46**, 329–336.
8. Early Breast Trialists Collaborative Group (1992). Systemic treatment of early breast cancer by hormonal, cytotoxic or immune therapy. *Lancet*, **339**, 1, 15, 71–85.
9. Ellenberg, S. S. and Hamilton, J. M. (1989). Surrogate endpoints in clinical trials: cancer. *Statist. Med.*, **8**, 405–414.
10. Fischl, M. A., Stanley, K., Collier, A. C., et al. (1995). Combination and monotherapy with zidovudine and zalcitabine in patients with advanced HIV disease. *Ann. Intern. Med.*, **122**, 24–32.
11. Fleiss, J. L. (1986). Analysis of data from multiclinic trials. *Controlled Clin. Trials*, **7**, 267–275.
12. Fleiss, J. L. (1989). A critique of recent research on the two-treatment crossover design. *Controlled Clin. Trials*, **10**, 237–243.
13. Fleming, T. R. and Watelet, L. F. (1989). Approaches to monitoring clinical trials. *J. Natl. Cancer Inst.*, **81**, 188–193.
14. Freedman, L. S. and Spiegelhalter, D. S. (1989). Comparison of Bayesian with group sequential methods for monitoring clinical trials. *Controlled Clin. Trials*, **10**, 357–367.
15. Gail, M. and Simon, R. (1985). Testing for qualitative interactions between treatment effects and patient subsets. *Biometrics*, **41**, 361–372.
16. George, S. L. and Desu, M. M. (1974). Planning the size and duration of a clinical trial studying the time to some critical event. *J. Chron. Dis.*, **27**, 15–24.
17. Gray, R. J. (1994). A Bayesian analysis of institutional effects in a multicenter cancer clinical trial. *Biometrics*, **50**, 244–253.
18. Halperin, J. and Brown, B. W. (1986). Sequential treatment allocation procedures in clinical trials-with particular attention to the analysis of results for the biased coin design. *Statist. Med.*, **5**, 211–230.

19. ISIS-2 Collaborative Group (1988). Randomized trial of intravenous streptokinase, oral aspirin, both, or neither among 17187 cases of suspected acute myocardial infarction. *Lancet*, Aug. 13, 1988, pp. 349–359.
20. Jennison, J. and Turnbull, B. (1989). The repeated confidence interval approach (with discussion). *J. Roy. Statist. Soc. B*, **51**, 305–362.
21. Kalish, L. A. and Begg, C. B. (1985). Treatment allocation methods: a review. *Statist. Med.*, **4**, 129–144.
22. Lachin, J. M. and Foulkes, M. A. (1986). Evaluation of sample size and power for analyses of survival with allowance for nonuniform patient entry, losses to follow-up, non-compliance and stratification. *Biometrics*, **42**, 507–519.
23. Lakatos, E. (1988). Sample sizes based on the log-rank statistic in complex clinical trials. *Biometrics*, **44**, 229–242.
24. Lambert, E. C. (1978). *Modern Medical Mistakes*. Indiana University Press, Bloomington, IN.
25. Lan, K. K. G., Simon, R., and Halperin, M. (1982). Stochastically curtailed tests in long-term clinical trials. *Commun. Statist. Seq. Anal.*, **1**, 207–219.
26. Lan, K. K. G. and DeMets, D. (1983). Discrete sequential boundaries for clinical trials. *Biometrika*, **70**, 659–663.
27. Lang, J. M., Buring, J. E., Rosner, B., Cook, N., and Hennekens, C. H. (1991). Estimating the effect of the run-in on the power of the Physicians' Health Study. *Statist. Med.*, **10**, 1585–1593.
28. O'Brien, P. O. and Fleming, T. R. (1979). A multiple testing procedure for clinical trials. *Biometrics*, **35**, 549–556.
29. O'Brien, P. O. (1984). Procedures for comparing samples with multiple endpoints. *Biometrics*, **40**, 1079–1087.
30. Pocock, S. J. and Simon, R. (1975). Sequential treatment assignment with balancing for prognostic factors in the controlled clinical trial. *Biometrics*, **31**, 103–115.
31. Pocock, S. J. (1982). Interim analysis for randomized clinical trials: The group sequential approach. *Biometrics*, **38**, 153–162.
32. Pocock, S. J., Geller, N. L., and Tsiatis, A. A. (1987). The analysis of multiple endpoints in clinical trials. *Biometrics*, **43**, 487–498.
33. Prentice, R. L. (1989). Surrogate endpoints in clinical trials: Definition and operational criteria. *Statist. Med.*, **8**, 431–440.
34. Royall, R. M. (1976). Current advances in sampling theory: implications for human observational studies. *Am. J. Epidemiol.*, **104**, 463–474.
35. Rubin, D. B. (1978). Bayesian inference for causal effects: The role of randomization. *Ann. Statist.*, **6**, 34–58.
36. Rubinstein, L. V., Gail, M. H., and Santner, T. J. (1981). Planning the duration of a comparative clinical trial with loss to follow-up and a period of continued observation. *J. Chron. Dis.* **34**, 469–479.
37. Schaid, D. J., Wieand, S., and Therneau, T. M. (1990). Optimal two-stage screening designs for survival comparisons. *Biometrika*, **77**, 507–513.
38. Schoenfeld, D. A. (1983). Sample size formula for the proportional hazards regression model. *Biometrics*, **39**, 499–503.
39. Schwartz, D. and Lellouch J. (1967). Explanatory and pragmatic attitudes in therapeutic trials. *J. Chron. Dis.*, **20**, 637–648.
40. Simon, R. (1979). Restricted randomization designs in clinical trials. *Biometrics*, **35**, 503–512.
41. Simon, R. (1982). Randomized clinical trials and research strategy. *Cancer Treatment Rep.*, **66**, 1083–1087.
42. Simon, R. (1982). Patient subsets and variation in therapeutic efficacy. *Br. J. Clin. Pharmacol.*, **14**, 473–482.
43. Simon, R. (1991). A decade of progress in statistical methodology for clinical trials. *Statist. Med.*, **10**, 1789–1817.

44. Simon, R., Thall, P. F., and Ellenberg, S. S. (1994). New designs for the selection of treatments to be tested in randomized clinical trials. *Statist. Med.*, **13**, 417–429.

45. Simon, R. (1995). Problems of multiplicity in clinical trials. *J. Statist. Plan. Infer.* **42**, 209–221.

46. Spiegelhalter, D. J., Freedman, L. S., and Parmar, M. K. B. (1994). Bayesian approaches to randomized trials. *J. Roy. Statist. Soc. A*,

47. Sposto, R. and Sather, H. N. (1985). Determining the duration of comparative clinical trials while allowing for cure. *J. Chron. Dis.*, **38**, 683–690.

48. Stampfer, M. J., Buring, J. E., Willett, W., Rosner, B., Eberlein, K., and Hennekens, C. H. (1985). The 2×2 factorial design: Its application to a randomized trial of aspirin and carotene in U.S. physicians. *Statist. Med.* **4**, 111–116.

49. Suissa, S. and Shuster, J. J. (1985). Exact unconditional sample sizes for the 2×2 binomial trial. *J. Roy. Statist. Soc. A*, **148**, 317–327.

50. Tukey, J. W. (1977). Some thoughts on clinical trials, especially problems of multiplicity. *Science*, **198**, 679–784.

51. Ury, H. and Fleiss, J. (1980). On approximate sample sizes for comparing two independent proportions with the use of Yates' correction. *Biometrics*, **36**, 347–352.

52. Whitehead, J. (1982). *The Design and Analysis of Sequential Clinical Trials*. Ellis Harwood, Chichester, UK.

53. Wittes, J., Lakatos, E., and Probstfield, (1989). Surrogate endpoints in clinical trials: Cardiovascular diseases. *Statist. Med.*, **8**, 415–426.

54. Yusuf, S., Collins, R., and Peto, R. (1984). Why do we need large, simple, randomized trials *Statist. Med.*, **3**, 409–420.

55. Yusuf, S., Simon, R., and Ellenberg, S. S., eds. (1987). Proceedings of "Methodologic issues in overviews of randomized clinical trials." *Statist. Med.*, **6**, 217–403.

13 Cluster Randomization

Neil Klar and A. Donner

13.1 Introduction

Randomized trials in which the unit of randomization is a community, worksite, school, or family are becoming increasingly common for the evaluation of lifestyle interventions for the prevention of disease. This form of treatment assignment is referred to as *cluster randomization* or *group randomization*. Reasons for adopting cluster randomization are diverse, but include administrative convenience, a desire to reduce the effect of treatment contamination, and the need to avoid ethical issues that might otherwise arise.

Data from cluster randomization trials are characterized by between-cluster variation. This is equivalent to saying that responses of cluster members tend to be correlated. Dependences among cluster members typical of such designs must be considered when determining sample size and in the subsequent data analyses. Failure to adjust standard statistical methods for within-cluster dependences will result in underpowered studies with spuriously elevated type I errors.

These statistical features of cluster randomization were not brought to wide attention in the health research community until the now famous article by Cornfield [4]. However, the 1980s saw a dramatic increase in the development of methods for analyzing correlated outcome data, in general [1], and methods for the design and analysis of cluster randomized trials, in particular [8,14]. Books summarizing this research have also appeared [6,18] and new statistical methods are in constant development.

Several published trials, which we review in the Section 13.2, will be used to illustrate the key features of cluster randomization. Principles of experimental design, including the benefits of random assignment and the importance of replication are discussed in Sections 13.3, 13.4, and 13.5, respectively, while issues of sample size estimation are considered in Section 13.6. Methods of analysis at the cluster level and at the individual level are discussed in Sections 13.7 and 13.8, respectively while designs involving repeated assessments are considered in Section 13.9. In Section 13.10, we provide recommendations for trial reporting and, in Section 13.11, we conclude the paper by considering issues arising in meta-analyses that may include one or more cluster randomization trials. Readers interested in a more detailed discussion might wish to consult reference 6 from which much of this article was abstracted.

13.2 Examples of Cluster Randomization Trails

1. A group of public health researchers in Montreal [21] conducted a household randomized trial to evaluate the risk of gastrointestinal disease due to consumption of drinking water. Participating households were randomly assigned to receive an in-home water filtration unit or were assigned to a control group that used tap water. Households were the natural randomization unit in this trial for assessing the effectiveness of the water filtration unit. There were 299 households (1206 individuals) assigned to the filtered water group and 308 households (1201 individuals) assigned to the tap water group. The annual incidence of gastrointestinal illness was analyzed using an extension of Poisson regression that adjusted for the within-household correlation in the outcome variable. On the basis of these analyses, investigators concluded that approximately 35% of the reported gastrointestinal illnesses among control group subjects were preventable.

2. The National Cancer Institute of the United States funded the Community Intervention Trial for Smoking Cessation (COMMIT) that investigated whether a community-level, 4-year intervention would increase quit rates of cigarette smokers [3]. Communities were selected as the natural experimental unit since investigators assumed that interventions offered at this level would reach the greatest number of smokers and possibly change the overall environment, thus making smoking less socially acceptable. Random-digit dialing was used to identify approximately 550 heavy smokers and 550 light to moderate smokers in each community. Eleven matched pairs of communities were enrolled in this study with one community in each pair randomly assigned to the experimental intervention with the remaining community serving as a control. Matching factors included geographic location, community size, and general sociodemographic factors. Each community had some latitude in developing smoking cessation activities, which included mass media campaigns and programs offered by health care providers or through worksites. These activities were designed to increase quit rates of heavy smokers, which, in theory, should then also benefit light to moderate smokers whose tobacco use tends to be easier to change. The effect of the intervention was assessed at the community level by calculating the difference in community-specific quit rates for each pair. Hypothesis tests were then constructed by applying a permutation test to the 11 matched-pair difference scores, an analytic approach that accounts for the between-community variability in smoking quit rates as well as for the matching. Further details concerning this cluster-level method of analysis are provided in Section 13.7. Unfortunately, while the experimental intervention offered by COMMIT significantly increased smoking quit rates among light to moderate smokers from about 28% to 31% ($p = 0.004$), no similar effect was identified among the cohort of heavy smokers.

3. Prenatal care in the developing world has attempted to mirror care that is offered in developed countries even though not all antenatal care interventions are known to be effective. The World Health Organization (WHO) prenatal care randomized trial [26] compared a new model of prenatal

care that emphasized health care interventions known to be effective with the standard model of prenatal care. The primary hypothesis in this equivalence trial was that the new model of prenatal health care would not adversely effect the health of women or of their babies. Participating clinics, recruited from Argentina, Cuba, Saudi Arabia, and Thailand, were randomly assigned to an intervention group or control group separately within each country. Clinics were selected as the optimal unit of allocation in this trial for reasons of administrative and logistic convenience. This decision also reduced the risk of experimental contamination that could have arisen had individual women been randomized. However, random assignment of larger units (e.g., communities) would have needlessly reduced the number of available clusters, thus compromising study power. In the WHO study 27 clinics (12,568 women) were randomly assigned to the experimental arm, while 26 control group clinics (11,958 women) received standard prenatal care. The primary analyses examining low birthweight (<2500 g) as the principal endpoint were based on (a) extensions of the Mantel–Haenszel test statistic and (b) extensions of logistic regression, both of which were adjusted for between-cluster variability and which also took into account the stratification by country. The resulting odds ratio comparing experimental to control group women with respect to low birthweight was estimated as equal to 1.06 (95% confidence interval 0.97–1.15). On the basis of these results, and bolstered by similar findings from other study outcomes, the investigators concluded that the new prenatal care model does not adversely affect perinatal and maternal health.

13.3 Principles of Experimental Design

The science of experimental design as initially put forward by R. A. Fisher [22] was based on the principles of random assignment, stratification, and replication. These principles may be illustrated by considering, in more detail, the cluster randomization trials described above.

All three trials reviewed in Section 13.2 used random allocation in assigning clusters to an intervention group. They differed, however, in the experimental unit; random assignment was by household in the Montreal trial [21], by clinic in the WHO prenatal care trial [26], and by community in COMMIT [3]. The advantages of random allocation for cluster randomization trials are essentially the same as for clinical trials randomizing individuals. These include the assurance that selection bias has played no role in the assignment of clusters to different interventions, the balancing, in an average sense, of baseline characteristics in the different intervention groups, and formal justification for the application of statistical distribution theory to the analysis of results. A final compelling reason for randomized assignment is that the results are likely to have much more credibility in the scientific community, particularly if they are unexpected.

There are three designs that are most frequently adopted in cluster randomization trials:

1. *Completely randomized*, involving no prestratification or matching of clusters according to baseline characteristics;

2. *Matched-pair*, in which one of two clusters in a stratum are randomly assigned to each intervention;

3. *Stratified*, involving the assignment of two or more clusters to at least some combinations of stratum and intervention.

The completely randomized design is most appropriate for studies having large numbers of clusters such as the Montreal trial, which enrolled over 600 households. Matching or stratification is often considered for community intervention trials such as COMMIT [3] in which the numbers of clusters that can be enrolled are usually limited by economic or practical considerations. The main attraction of this design is its potential to provide very tight and explicit balancing of important prognostic factors, thereby improving statistical power.

The stratified design is an extension of the matched-pair design in which several clusters, rather than just one, are randomly assigned within strata to each of the intervention and control groups. This design, although selected for the WHO prenatal care trial [26], has been used much less frequently than either the matched-pair or completely randomized design. However, for many studies, it would seem to represent a sensible compromise between these two designs in that it provides at least some baseline control on factors thought to be related to outcome, while easing the practical difficulties of finding appropriate pairmatches.

13.4 Experimental and Quasi-Experimental Designs

Investigators may sometimes be hesitant to adopt a random allocation scheme because of ethical and/or practical concerns that arise. For example, the systematic allocation of geographically separated control and experimental clusters might be seen as necessary to alleviate concerns regarding experimental contamination. Nonrandomized designs may also seem easier to explain to officials, to gain broad public acceptance, and, more generally, to allow the study to be carried out in a simpler fashion, without the resistance to randomization that is often seen. In this case, well-designed quasi-experimental comparisons [24] will be preferable to obtaining no information at all on the effectiveness of an intervention. At a minimum, it seems clear that nonexperimental comparisons may generate hypotheses that can subsequently be tested in a more rigorous framework.

Several reviews of studies in the health sciences have made it clear that random assignment is, in fact, being increasingly adopted for the assessment of nontherapeutic interventions [6]. Similar inroads have been made in the social sciences, where successfully completed randomized trials help investigators to recognize that random assignment may well be possible under circumstances in which only quasi-experimental designs had been previously considered.

The availability of a limited number of clusters has been cited as a reason to avoid randomization, since it may leave considerable imbalance between intervention groups on important prognostic factors. This seems a questionable decision as there is no assurance that quasi-experimental designs will necessarily create acceptably balanced treatment groups with a limited number of available clusters. Moreover, a matched-pair or stratified design can usually provide acceptable levels of balance in this case.

Concern has been raised by some researchers at the relatively small number of community intervention trials, which have actually identified effective health promotion programs [16]. It is worth noting, however, that the dearth of effective programs identified using cluster randomization is largely confined to evaluations of behavioral interventions. There has been considerably more success in evaluating medical interventions as in trials assessing the effect of vitamin A on reducing childhood morbidity and mortality [11].

There is a long history of methodological research comparing the effectiveness of methods of treatment assignment in controlled clinical trials [15]. For example, Chalmers et al. [2] reviewed 145 reports of controlled clinical trials examining treatments for acute myocardial infarction published between 1946 and 1981. Differences in case fatality rates were found in only about 9% of blinded randomized trials, 24% of unblinded randomized studies, and in 58% of nonrandomized studies. These differences in the effects of treatment were interpreted by the authors as attributable to differences in patient participation (i.e., selection bias), which should give pause to investigators touting the benefits of nonrandomized comparisons as a substitute for cluster randomization. Murray [19], in addressing the ability of investigators to identify effective behavioral interventions, stated that

> The challenge is to create trials that are (a) sufficiently rigorous to address these issues, (b) powerful enough to provide an answer to the research question of interest, and (c) inexpensive enough to be practical. The question is not whether to conduct group-randomized trials, but how to do them well.

13.5 The Effect of Failing to Replicate

Some investigators have designed community intervention trials in which exactly one cluster has been assigned to the experimental group and one to the control group, either with or without the benefit of random assignment [6]. Such trials invariably result in interpretational difficulties caused by the total confounding of two sources of variation: (1) the variation in response due to the effect of intervention, and (2) the natural variation that exists between two communities (clusters) even in the absence of an intervention effect. These sources of confounding can be only be disentangled by enrolling multiple clusters per intervention group.

The natural between-household variation in infection rates may easily be estimated separately for subjects in the control and experimental arms of the completely randomized Montreal water filtration trial [21]. However, this task is more complicated for COMMIT [3], since the matched pair design adopted for this trial implies that there is only a single community in each combination of stratum and intervention. Inferences concerning the effect of intervention must, therefore, be constructed using the variation in treatment effect assessed across the 11 replicate pairs. Between-cluster variation is also directly estimable in stratified trials such as the WHO prenatal care trial [26] using, for example, the replicate clusters available per intervention group within each of the four participating countries.

The absence of replication, while not exclusive to community intervention trials, is most commonly seen in trials that enrol clusters of relatively large size. Unfortunately, the large numbers of study subjects in such trials may mislead some investigators into believing valid inferences can be constructed by falsely assuming individuals within a community provide independent responses.

A consequence of this naive approach is that the variance of the observed effect of intervention will invariably be underestimated. The degree of underestimation is a function of the variance inflation due to clustering (i.e., the design effect). In particular, the variance inflation factor is given by $1+(\bar{m}-1)\rho$, where $\bar{m}$ denotes average cluster size while ρ is a measure of intracluster correlation, interpretable as the standard Pearson correlation between any two responses in the same cluster. With the additional assumption that the intracluster correlation is nonnegative, ρ may also be interpreted as the proportion of overall variation in response that can be accounted for by the between-cluster variation.

As demonstrated in Section 13.6, even relatively small values of intracluster correlation, combined with large cluster sizes, can yield sizable degrees of variance inflation. Moreover, in the absence of cluster replication, positive effects of intervention artificially inflate estimates of between-cluster variation, thus invalidating the resulting estimates of ρ.

More attention to the effects of clustering when determining the trial sample size might help to eliminate designs that lack replication. Even so, investigators will still need to consider whether statistical inferences concerning the effect of intervention will be interpretable if only two, or a few, replicate clusters are allocated to each intervention group.

13.6 Sample Size Estimation

A quantitatively justified sample-size calculation is almost universally regarded as a fundamental design feature of a properly controlled clinical trial. Methodologic reviews of cluster randomization trials have consistently shown that only a small proportion of these studies have adopted a predetermined sample size based on formal considerations of statistical power [6].

There are several possible explanations for the difficulties investigators have faced in designing adequately powered studies. One obvious reason is that the required sample size formulas still tend to be relatively inaccessible, not being available, for example, in most standard texts or software packages. A second reason is that the proper use of these formulas requires some prior assessment of both the average cluster size and the intracluster correlation coefficient ρ.

The average cluster size for a trial may at times be directly determined by the selected interventions. For example, households were the natural unit of randomization in the Montreal water filtration study [21], where the average cluster size at entry was approximately four. When relatively large clusters are randomized, on the other hand, subsamples of individual cluster members may be selected to reduce costs. For example, the primary endpoint in the community intervention trial COMMIT [3] was the quit rate of approximately 550 heavy smokers selected from each cluster.

Difficulties in obtaining accurate estimates of intracluster correlation are slowly being addressed as more investigators begin publishing these values in their reports of trial results. Summary tables listing intracluster correlation coefficients and variance inflation factors from a wide range of cluster randomization trials and complex surveys are also beginning to appear [6]. In practice, estimates of ρ are almost always positive and tend to be larger in smaller clusters.

The estimated variance inflation factor may be directly applied to determine the required sample size for a completely randomized design. Let $Z_{\alpha/2}$ denote the two-sided critical value of the standard normal distribution corresponding to the error rate α and Z_β denote the critical value corresponding to β. Then, assuming the difference in sample means for the experimental and control groups, $\bar{Y}_1 - \bar{Y}_2$, can be regarded as approximately normally distributed, the number of subjects required per intervention group is given by

$$n = \frac{(Z_{\alpha/2} + Z_\beta)^2 (2\sigma^2)[1 + (\bar{m} - 1)\rho]}{(\mu_1 - \mu_2)^2},$$

where $\mu_1 - \mu_2$ denotes the magnitude of the difference to be detected and $\bar{m}$ denotes the average cluster size. Equivalently, the number of clusters required per group is given by $k = n/\bar{m}$. Formulas needed to estimate sample size for matched pair and stratified designs for a variety of study outcomes are provided by Donner and Klar [6].

Regardless of the study design, it is useful to conduct sensitivity analyses exploring the effect on sample size by varying values of the intracluster correlation, the number of clusters per intervention group, and the subsample size. The benefits of such a sensitivity analysis are illustrated by a report describing methodological considerations in the design of the WHO prenatal care trial [10,26]. The values for ρ, suggested from data obtained from a pilot study, varied from 0 to 0.002, while cluster sizes were allowed to vary between 300 and 600 patients per clinic. Consequently, the degree of variance inflation due to clustering could be as great as $1 + (600 - 1)0.002 = 2.2$. As is typical in such sensitivity analyses, power was seen to be much more sensitive to the number of clusters per intervention group than to the number of patients selected per clinic. Ultimately a total of 53 clinics were enrolled in this trial, with a minimum of 12 clinics per site and with each clinic enrolling approximately 450 women.

It must be emphasized that the effect of clustering depends on the joint influence of both $\bar{m}$ and ρ. Failure to appreciate this point has led to the occasional suggestion in the epidemiological literature that clustering may be detected or ruled out on the basis of testing the estimated value of ρ for statistical significance, that is, testing $H_0 : \rho = 0$ versus $H_A : \rho > 0$. The weakness of this approach is that observed values of ρ may be very small, particularly for data collected from the very large clusters typically recruited for community intervention trials. Therefore, the power of a test for detecting such values as statistically significant tends to be unacceptably low [6]. Yet small values of ρ, combined with large cluster sizes, can yield sizable values of the variance inflation factor, which can seriously disturb the validity of standard statistical procedures if unaccounted for in the analyses. Thus, we would recommend that investigators inherently assume the existence of intracluster correlation, a well-documented phenomenon, rather than attempting to rule it out using statistical testing procedures.

13.7 Cluster-Level Analyses

Many of the challenges of cluster randomization arise when inferences are intended to apply at the individual level while randomization is at the cluster level. If inferences were intended to apply at the cluster level, implying that an analysis at the cluster level would be most appropriate, the study could be regarded, at least with respect to sample size estimation and data analysis, as a standard clinical trial. For example, one of the secondary aims of the Community Intervention Trial for Smoking Cessation (COMMIT) was to compare the level of tobacco control activities in the experimental and control communities after the study ended [25]. The resulting analyses were then, naturally, conducted at the cluster (community) level.

Analyses are inevitably more complicated when data are available from individual study subjects. In this case, the investigator must account for the lack of statistical independence among observations within a cluster. An obvious method of simplifying the problem is to collapse the data in each cluster, followed by the construction of a meaningful summary measure, such as an average, which then serves as the unit of analysis. Standard statistical methods can then be directly applied to the collapsed measures. This removes the problem of nonindependence since the subsequent significance tests and confidence intervals would be based on the variation among cluster summary values rather than on variation among individuals.

An important special case arises in trials having a quantitative outcome variable when each cluster has a fixed number of subjects. In this case, the test statistic obtained using the analysis of variance is algebraically identical to the test statistic obtained using a cluster-level analysis [6]. Thus, the suggestion that is sometimes made that a cluster-level analysis intrinsically assumes $\rho = 1$ is misleading, since such an analysis can be efficiently conducted regardless of the value of ρ. It is important to note, however, that this equivalence between cluster-level and individual-level analyses, which holds exactly for quantitative outcome variables under balance, holds only approximately for other outcome variables (e.g., binary, time to event, count). A second implication of this algebraic identity is that the well-known ecological fallacy cannot arise in the case of cluster-level intention-to-treat analyses, since the assigned intervention is shared by all cluster members.

In practice, the number of subjects per cluster will tend to exhibit considerable variability, either by design or by subject attrition. Cluster-level analyses which give equal weight to all clusters may, therefore, be inefficient. However, it is important to note that appropriately weighted cluster-level analyses are asymptotically equivalent to individual-level analyses. On the other hand, if there are only a small number of clusters per intervention group, the resulting imprecision in the estimated weights might even result in a loss of power relative to an unweighted analysis. In this case it might, therefore, be preferable to consider exact statistical inferences constructed at the cluster level, as based on the randomization distribution for the selected experimental design (e.g., completely randomized, matched-pair, stratified). As noted in Section 13.2, COMMIT investigators [3] adopted this strategy, basing their primary analysis of tobacco quit rates on the permutation distribution of the difference in event rates within a matched pair study design. Using a two-stage regression approach, investigators were also able to adjust for important baseline imbalances on known prognostic variables.

13.8 Individual-Level Analysis

Standard methods of analysis applied to individually randomized trials have all been extended to allow for the effect of between-cluster sources of variation. These include extensions of contingency table methods (e.g., Pearson chi-square test, Mantel–Haenszel methods) and of two sample t-tests. More sophisticated extensions of multiple regression models have also been developed and are now available in standard statistical software.

We will focus primary attention, in this section, to methods for the analysis of binary outcome data, which arise more frequently in cluster randomization trials than continuous, count, or time-to-event data. We will also limit the discussion to data obtained from completely randomized and stratified designs. Methods of analysis for other study outcomes are considered in detail elsewhere [6] while some analytic challenges unique to matched pair trials are debated by Donner and Klar [6] and by Feng et al. [13].

We now consider analyses of data from the WHO Antenatal Care Trial [26] in which clinics were randomly assigned to experimental or control groups separately within each of the four participating sites (countries). An extension of the Mantel–Haenszel statistic adjusted for clustering was used to compare the risk of having a low-birthweight outcome for women assigned to either the new model of antenatal care or a standard model. For clusters of fixed size m, this statistic is equal to the standard Mantel–Haenszel statistic divided by the variance inflation factor $1+(m-1)\hat{\rho}$ where $\hat{\rho}$ is the sample estimate of ρ. Thus, failure to account for between-cluster variability (i.e., incorrectly assuming $\rho = 0$) will tend to falsely increase the type I error rate.

A key advantage of this approach is that the resulting statistic simplifies to the standard Mantel–Haenszel test statistic. Similar advantages are shared by most other individual-level test statistics.

Additional analyses reported in this trial used an extension of logistic regression which allowed adjustment for other potential baseline predictors of low birthweight including maternal education, maternal age, and nulliparity. These analyses allowed examination of the joint effects of individual-level and cluster-level predictors (i.e., intervention, strata).

Two frequently used extensions of logistic regression are the logistic–normal model and the generalized estimating equation (GEE) extension of this procedure [6]. The logistic-normal model assumes that the logit transform of the probability of having a low-birthweight outcome follows a normal distribution across clusters. The resulting likelihood ratio tests will have maximum power for detecting effects of intervention as statistically significant when parametric assumptions such as these are satisfied.

It may be difficult in practice to know whether the assumptions underlying the use of parametric models are reasonable. We, therefore, limit attention here to the GEE approach, which has the advantage of not requiring specification of a fully parametric distribution. Two distinct strategies are available to adjust for the effect of clustering using the GEE approach. The first can be said to be model-based, as it requires the specification of a working correlation matrix, which describes the pattern of correlation between responses of cluster members. For cluster randomization trials, the simplest assumption to make is that responses of cluster members are equally correlated, that is, to assume the correlation structure within clusters is exchangeable. The second strategy that may be used to adjust for the effect of clustering employs "robust variance estimators" that are constructed using between-cluster information. These estimators consistently estimate the true variance of estimated regression coefficients even if the working correlation matrix is misspecified. Moreover, provided there are a large number of clusters, inferences obtained using robust variance estimators will become equivalent to those obtained using the model-based strategy provided the working correlation matrix is correctly specified.

The examples we have considered involve only a single level of clustering. More sophisticated multilevel methods of analysis are available [5,6] that allow examination of effects at two more levels. For example, women participating in the WHO prenatal care trial might have been cared for by a specific physician within each clinic. Responses of women would then be clustered by physician nested within clinics, generating two levels of clustering. This additional structure could then be used to explore differential intervention effects across physicians, for example, to consider whether years of training was associated with relatively fewer women having low weight babies. While these analyses may enrich our understanding of the trial results, they are almost always exploratory in nature.

It is important to note that statistical inferences constructed using individual-level analyses are approximate, with their validity only assured in the presence of a large number of clusters. This requirement essentially flows from the difficulty in accurately estimating between-cluster sources of variation. Thus, the validity of statistical inferences constructed using individual-level analyses may be in question should there be fewer than 20 clusters enrolled. If a small number of clusters are enrolled, it may only be possible to consider cluster-level analyses by constructing statistical inferences based on the selected randomization distribution.

13.9 Incorporating Repeated Assessments

Investigators are often interested in considering the longitudinal effects of intervention as part of a cluster randomized trial. The choice here is between a cohort design that tracks the same individuals over time and a repeated cross-sectional design that tracks the same clusters over time but draws independent samples of individuals at each calendar point. In this section, we outline how the study objectives should determine the choice between these designs and how the resulting decision affects data collection, data analysis, and the interpretation of study results.

Cohort samples of subjects were included in each of the three trials presented in Section 13.2. For example, smoking quit rates were measured for subsamples of heavy and light to moderate smokers selected from each of the participating COMMIT communities [3]. Cohort members were followed and contacted annually during the 4-year intervention. The length of follow-up time was a consequence of the study objectives for which smoking quit rates were defined "as the fraction of cohort members who had achieved and maintained cessation for at least six months at the end of the trial." This outcome illustrates how cohort designs are best suited to measuring change within individual participants, implying that the unit of inference is most naturally directed at the individual level.

A desirable design feature of COMMIT was that subjects were selected prior to random assignment with the avoidance of any concerns regarding possible selection bias. This strategy was not available for the WHO prenatal care trial [26], as women could be identified only following their initial clinic visit, which for most women occurred after random assignment. Selection bias is unlikely, however, since all new patients from participating clinics were enrolled in the trial and birthweight data from singleton births were available for 92% of women from each intervention group.

A secondary objective of COMMIT was to determine if the intervention would decrease the prevalence of adult cigarette smoking [3]. This objective was achieved by conducting separate surveys prior to random assignment and following completion of the intervention. A principal attraction of such repeated cross-sectional surveys is that any concerns regarding the effects of possible attrition would be avoided. Of course, differential rates of participation in cross-sectional surveys conducted after random assignment can still compromise validity, since willingness to participate may be a consequence of the assigned intervention. Nonetheless, random samples of respondents at each assessment point will be more representative of the target population than a fixed cohort of smokers.

The final decision regarding the selection of a cohort or cross-sectional design should be based primarily on the study objectives and the associated unit of inference. However, it can still be informative to quantitatively evaluate the relative efficiency of the two designs [12]. Since repeated assessments are made on the same subjects, the cohort design tends to have greater power than a design involving repeated cross-sectional surveys. Note, however, that, in practice, subject attrition may eliminate these potential gains in power.

The number and timing of assessments made after baseline should be determined by the anticipated temporal responses in each intervention group. For example, it might be reasonable to expect different linear trends over time across intervention groups in a community randomized trial of smoking cessation if the effects of intervention were expected to diffuse slowly through each community. Alternatively, the effects of intervention might diffuse rapidly but be transient, requiring a more careful determination of assessment times in order to ensure that important effects are not missed.

The methods of analysis presented in Sections 13.7 and 13.8 assumed the presence of only a single assessment following random assignment. Extensions of these methods to cluster randomization trials having longitudinal outcome measures are beginning to appear [20,23].

13.10 Study Reporting

Reporting standards for randomized clinical trials have now been widely disseminated [17]. Many of the principles that apply to trials randomizing individuals also apply to trials randomizing intact clusters. These include a carefully posed justification for the trial, a clear statement of the study objectives, a detailed description of the planned intervention, and an accurate accounting of all subjects randomized to the trial. Unambiguous inclusion–exclusion criteria must also be formulated, although perhaps separately for cluster-level and individual-level characteristics. There are, however, some unique aspects of cluster randomization trials that require special attention at the reporting stage. We focus here on some of the most important of these.

The decreased statistical efficiency of cluster randomization relative to individual randomization can be substantial, depending on the sizes of the clusters randomized and the degree of intracluster correlation. Thus, unless it is obvious that there is no alternative, the reasons for randomizing clusters rather than individuals should be clearly stated. This information, accompanied by a clear description of the units randomized, can help a reader decide if the loss of precision due to cluster randomization is, in fact, justified.

Having decided to randomize clusters, investigators may still have considerable latitude in their choice of allocation unit. As different levels of statistical efficiency are associated with different cluster sizes, it would seem important to select the unit of randomization on a carefully considered basis. An unambiguous definition of the unit of randomization is also required. For example, a statement that "neighbourhoods" were randomized is clearly incomplete without a detailed description of this term in the context of the planned trial.

The clusters that participate in a trial may not be representative of the target population of clusters. Some indication of this lack of representativeness may be obtained by listing the number of clusters that met the eligibility criteria for the trial, but which declined to participate, along with a description of their characteristics.

A continuing difficulty with reports of cluster randomization trials is that justification for the sample size is all too often omitted. Investigators should clearly describe how the sample size for their trial was determined, with particular attention given to how clustering effects were adjusted for. This description should be in the context of the experimental design selected (e.g., completely randomized, matched pair, stratified).

It would also be beneficial to the research community if empirical estimates of ρ were routinely published (with an indication of whether the reported values have been adjusted for the effect of baseline covariates).

It should be further specified what provisions, if any, were made in the sample size calculations to account for potential loss to follow up. Since the factors leading to the loss to follow-up of individual members of a cluster may be very different from those leading to the loss of an entire cluster, both sets of factors must be considered here.

A large variety of methods, based on very different sets of assumptions, have been used to analyze data arising from cluster randomization trials. For example, possible choices for the analysis of binary outcomes include adjusted chi-square statistics, the method of generalized estimating equations (GEEs), and logistic-normal regression models. These methods are not as familiar as the standard procedures commonly used to analyze clinical trial data. This is partly because methodology for analyzing cluster randomization trials is in a state of rapid development, with virtually no standardization and a proliferation of associated software. Therefore, it is incumbent on authors to provide a clear statement of the statistical methods used, and accompanied, where it is not obvious, by an explanation of how the analysis adjusts for the effect of clustering. The software used to implement these analyses should also be reported.

13.11 Meta-Analysis

Meta-analyses involving the synthesis of evidence from cluster randomization trials raise methodologic issues beyond those raised by meta-analyses of individually randomized trials. Two of the more challenging of these issues are (1) the increased likelihood of study heterogeneity, and (2) difficulties in estimating design effects and selecting an optimal method of analysis [7].

These issues are illustrated in a meta-analysis examining the effect of vitamin A supplementation on child mortality [11]. This investigation considered trials of hospitalized children with measles as well as community-based trials of healthy children. Individual children were assigned to intervention in the four hospital-based trials, while allocation was by geographic area, village, or household in the eight community-based trials.

One of the community-based trials included only one geographic area per intervention group, each of which enrolled approximately 3000 children. On the other hand, there was an average of about two children from each cluster when allocation was by household. Thus, an important source of heterogeneity arose from the nature and size of the randomization units allocated in the different trials. This problem was dealt with by performing the meta-analysis separately for the individually randomized and cluster randomized trials.

It is straightforward to summarize results across trials when each study provides a common measure for the estimated effect of intervention (such as an odds ratio, for example) and a corresponding variance estimate that appropriately accounts for the clustering. Unfortunately, the information necessary for its application, in practice, is rarely available to meta-analysts.

One consequence of this difficulty is that investigators are sometimes forced to adopt ad hoc strategies when relying on published trial reports that fail to provide estimates of the variance inflation factor. For example, in the meta-analysis described above only four of the eight community-based trials reported that they accounted for clustering effects. The authors argued that increasing the variance of the summary odds ratio estimator computed over all eight trials by an arbitrary 30% was reasonable since the design effects ranged from 1.10 to 1.40 in those studies that did adjust for clustering effects.

Even when each trial provides an estimate of the design effect several different approaches could be used for conducting a meta-analysis. For example, a procedure commonly adopted for combining the results of individually randomized clinical trials with a binary outcome variable is the well-known Mantel–Haenszel test. The adjusted Mantel–Haenszel test [6] may be used to combine results of cluster randomized trials. Other possible approaches are discussed by Donner et al. [9].

References

1. Ashby, M., Neuhaus, J. M., Hauck, W. W., Bacchetti, P., Heilbron, D. C., Jewell, N. P., Segal, M. R., and Fusaro, R. E. (1992). An annotated bibliography of methods for analyzing correlated categorical data, *Statist. Med.*, **11**, 67–99.

2. Chalmers, T. C., Celano, P., Sacks, H. S., and Smith, Jr. H. (1983). Bias in treatment assignment in controlled clinical trials. *New Engl. J. Med.*, **309**, 1358–1361.

3. COMMIT Research Group. (1995). Community intervention trial for smoking cessation (COMMIT): i. cohort results from a four-year community intervention. *Am. J. Public Health*, **85**, 183–192.

4. Cornfield, J. (1978). Randomization by group: a formal analysis, *Am. J. Epidemiol.*, **108**, 100–102.

5. Diez-Roux, A. V. (2000). Multilevel analysis in public health research. *Annu. Rev. Public Health*, **21**, 171–192.

6. Donner, A., and Klar, N., (2000). *Design and Analysis of Cluster Randomization Trials in Health Research*. Arnold, London, UK..

7. Donner A., and Klar, N. (2002). Issues in the meta-analysis of cluster randomized trials. *Statist. Med.*, **21**, 2971–2980.

8. Donner, A., Birkett, N., and Buck, C. (1981).Randomization by cluster: sample size requirements and analysis. *Am. J. Epidemiol.*, **114**, 906–914.

9. Donner, A., Paiggio, G., and Villar, J. (2001). Statistical methods for the meta-analysis of cluster randomization trials. *Statist. Methods in Med. Res.*, **10**, 325–338.

10. Donner, A., Piaggio, G., Villar, J., Pinol, A., Al-Mazrou, Y., Ba'aqeel, H., Bakketeig, L., Belizan, J. M., Berendes, H., Carroli, G., Farnot, U., and Lumbiganon, P., for the WHO Antenatal Care Trial Research Group. (1998). Methodological considerations in the design of the WHO Antenatal care randomised controlled trial. *Paediatr. Perinat. Epidemiol.*, **12**(Suppl. 2), 59–74.

11. Fawzi, W. W., Chalmers, T. C., Herrera, M. G., and Mosteller, F. (1993). Vitamin A supplementation and child mortality, a meta-analysis. *J. Am. Med. Assoc. (JAMA)*, **269**, 898–903.

12. Feldman, H. A., and McKinlay, S. M. (1994). Cohort vs. cross-sectional design in large field trials: Precision, sample size, and a unifying model. *Statist. Med.*, **13**, 61–78.

13. Feng, Z., Diehr, P., Peterson, A., and McLerran, D. (2001). Selected statistical issues in group randomized trials. *Annu. Rev. Public Health*, **22**, 167–187.

14. Gillum, R. F., Williams, P. T., and Sondik, E. (1980). Some consideration for the planning of total-community prevention trials: when is sample size adequate?. *J. Community Health*, **5**, 270–278.

15. McKee, M., Britton, A., Black, N., McPherson, K., Sanderson, C., and Bain, C. (1999). Interpreting the evidence: choosing between randomised and non-randomised studies. *Br. J. Med.*, **319**, 312–315.

16. Merzel, C., and D'Affitti, J. (2003). Reconsidering community-based health promotion: Promise, performance, and potential. *Am J. Public Health*, **93**, 557–574.

17. Moher, D., Schulz, K. F., and Altman, D. G., for the CONSORT Group. (2001). The CONSORT statement: Revised recommendations for improving the quality of reports of parallel-group randomised trials. *Lancet*, **357**, 1191–1194.

18. Murray, D. M. (1998). *Design and Analysis of Group-randomized Trials*. Oxford University Press, Oxford, UK.

19. Murray, D. M. (2001). Efficacy and effectiveness trials in health promotion and disease prevention: Design and analysis of group-randomized trials, In *Integrating Behavioral and Social Sciences with Public Health*, N. Schneiderman, M.A. Speers, J. M. Silva, H. Tomes and J. H. Gentry, eds., Chap. 15. American Psychology Association, Washington, DC.

20. Murray, D. M., Hannan, P. J., Wolfinger, R. D., Baker, W. L., and Dwyer, J. H. (1998). Analysis of data from group-randomized trials with repeat observations on the same groups. *Statist. Med.*, **17**, 1581–1600.

21. Payment, P., Richardson, L., Siemiatycki, J., Dewar, R., Edwardes, M., and Franco, E. (1991). A randomized trial to evaluate the risk of gastrointestinal disease due to consumption of drinking water meeting microbiological standards. *Am. J. Public Health*, **81**, 703–708.

22. Preece, D. A. (1990). RA fisher and experimental design: A review. *Biometrics*, **46**, 925–935.

23. Sashegyi, A. I., Brown, K. S., and Farrell, P. J. (2000). Application of a generalized random effects regression model for cluster-correlated longitudinal data to a school-based smoking prevention trial. *Am. J. Epidemiol.*, **152**, 1192–1200.

24. Shadish, W. R., Cook, T. D., and Campbell, D. T. (2002). *Experimental and Quasi-Experimental Designs for Generalized Causal Inference*. Houghton-Mifflin, Boston, MA.

25. Thompson, B., Lichtenstein, E., Corbett, K., Nettekoven, L., and Feng, Z. (2000). Durability of tobacco control efforts in the 22 community intervention trial for smoking cessation (COMMIT) communities 2 years after the end of intervention. *Health Educ. Res.*, **15**, 353–366.

26. Villar, J., Ba'aqeel, H., Piaggio, G., Lumbiganon, P., Belizan J. M., Farnot, U., Al-Mazrou, Y., Carroli, G., Pinol, A., Donner, A., Langer, A., Nigenda, G., Mugford, M., Fox-Rushby, J., Hutton, G., Bergsjo, P., Bakketeig, L., and Berendes, H., for the WHO Antenatal Care Trial Research Group (2001).WHO antenatal care randomised trial for the evaluation of a new model of routine antenatal care. *Lancet*, **357**, 1551–1564.

14 Cohort Analysis

N.K. Namboodiri

14.1 Definition

Ancient Romans seem to have understood by the term "cohort" a military unit consisting of several hundred soldiers. Now, however, the dictionary definitions of the term include one's accomplice, associate, or supporter, or in the collective sense a band or a group. Demographers adapted the term to denote a group of persons in a geographically or otherwise defined population who experienced a given significant life event (e.g., birth, marriage, or graduation) during a given time interval (e.g., a calendar year or a decade). Thus persons born in the United States during the calendar year 1935 form the U.S. birth cohort 1935. Marriage cohorts, high school graduation cohorts, and others are similarly defined. It may be noted in passing that the concept can be extended to include nonhuman populations. For example, the housing starts in a country in a given year, the passenger cars manufactured by a company in a given period, and similar aggregates are also cohorts.

Cohort analysis is a quantitative research orientation that emphasizes intracohort and intercohort comparisons of aggregate measures of life experiences, classified by age (duration from the start of the life history). Cohort analysis has been used by demographers primarily in fertility studies, but it has also been applied to the analysis of mortality, nuptiality, migration, and other demographic phenomena. In recent years its field of application has been extended to include nondemographic topics, such as the effects of human aging and the dynamics of political behavior.

14.2 Methodology

Cohort analysis often use the presentation known as a Lexis diagram to locate in a two-dimensional space the life experiences of cohorts. Figure 1 is a Lexis diagram in the age–time plane for birth cohorts, with death as the relevant experience. Time (e.g., calendar year) is laid out on the horizontal axis and age on the vertical. Both axes use the same time unit (e.g., year). Each individual member of each cohort is represented by a life line inclined at 45° to either axis, starting on the horizontal axis at the moment of birth, and terminating at a point corresponding to the moment of death and the age at that moment. Events such as marriage, high school graduation, or entry into the labor force can be marked on the life lines by points corresponding to the time and age when each such event occurs.

Consider a particular cohort in Figure 1,

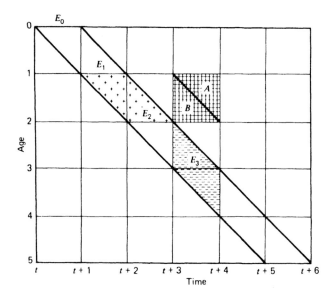

Figure 1: Lexis diagram showing birth cohorts in a time–age plane.

say the one with its origin in the time interval t to $t+1$. If E_0 is the number of births that occur during the period t to $t+1$, the cohort in question starts with E_0 life lines between t and $t+1$. Suppose that E_x of those born during the period survive to their xth birth day, $E_x - E_{x+1}$ dying between their xth and $(x+1)$st birth days. Then the diagram will show E_x of the original E_0 life lines reaching the horizontal at age x, and $E_x - E_{x+1}$ life lines terminating between the horizontals at x and $x+1$. The sequence of numbers $E_0, E_1, \ldots$ provide the full information regarding the experience of the cohort, as far as death or survival is concerned.

There are several ways in which the information in a Lexis diagram can be aggregated. One mode of aggregation is by age and cohort, i.e., by parallelograms of the type marked by +'s in Figure 1. An example is the sequence $(E_0 - E_1), (E_1 - E_2), \ldots$, referred to above. This mode of aggregation mixes adjacent periods, e.g., t to $t+1$ with $t+1$ to $t+2$.

Another mode of aggregation is by period and cohort, i.e., by parallelograms of the type marked by $-$'s in Figure 1. An example of this type of aggregation is a table that presents for each marriage cohort, i.e., persons who got married in a given year, the number of marriages that end in divorce, separation, or widowhood during a given year after marriage. This mode of aggregation mixes adjacent age groups.

A third form of aggregation is by age and period, i.e., by squares of the type crosshatched in Figure 1. A common example of this is the tabulation of deaths in a population by year of occurrence and age at last birth day of the deceased. This mode of aggregation mixes the experiences of adjacent cohorts, as can be seen by noting that the triangle marked A in the crosshatched square (Figure 1) belongs to one cohort and the one marked B to another.

Instead of squares or parallelograms, one may use lines to define aggregation units. The sequence $E_0, E_1, \ldots$ mentioned above is an example. This sequence represents, for a given birth cohort of initial size E_0, the numbers of persons who survive to the first second, ... birthdays or anniversaries. The unit of aggregation in this case is a period segment on the age (horizontal) line. Similarly, it is possible to use age segments on time (vertical) lines as units of aggregation. To give an example, suppose that a particular interview question was repeated in opinion surveys conducted on November 1 of each presidential election year in the United States. The date of each survey is thus 4 years after that of the preceding one. If the responses to the interview questions in the successive surveys are tabulated using 4-year age groups, the resulting aggregation is by age segments on time lines.

It should be noted that often a particular aggregation used may only be an approximation to one or another of the forms just described. For example, the field work of an opinion survey may have been completed over a short interval of time such as a week, rather than in a day or in one instant. In such situations, the aggregation involved will not be exactly by the age segments on a single time (vertical) line, but by those on a small strip of time interval. For practical purposes one often regards such aggregations as if age segments on a single time line had been used.

Data useful for cohort analysis may be obtained from any of a number of sources, including registration systems, censuses, and sample surveys. In nondemographic studies, the required data are drawn more often than not from sample surveys or censuses. Survey (census) data may be classified into three different types: data from reinterviewing a panel, retrospective data from a single survey (census), and comparable data from two or more independent surveys (censuses).

Whatever the source of data, a cohort analyst will have to first assemble the available information in a form that permits tracing the experience of each cohort as it ages. Often this is done by constructing a "standard" cohort table in which counts, rates, proportions, averages, or probabilities pertaining to the phenomenon taken for study are arranged by time and age, such that the width of the age intervals bears a 1:1, 1:2, 2:1 or similar relation to the intervals between the time points for which there are data. It is not infrequent, however, that the available data may be spaced irregularly in time or that the width of the age intervals used for aggregation for the data may not be the same as those used for the data collected at another point in time. In such situations it becomes difficult, if not impossible, to trace the experiences of cohorts as they age. A common strategy used by cohort analysts in such situations is to derive a table in the standard form from the available data, using techniques such as interpolation. This strategy has its own drawbacks, however. Consider, for example, using interpolation to estimate rates or probabilities for regular 5-year age groups from the data available for irregular age groups. The technique is based on certain assumptions, of unknown validity, concerning the underlying age pattern of the rates or probabilities. Inferences based on interpolated data are more likely to be a function of the assumptions underlying the interpolation technique than a reflection of any basic features of the phenomenon studied.

In analyzing cohort data, attention is very often focused on cohort, age, and period effects. It is logical to assume that each environmental circumstance or stimulus, such as a change in the economic conditions, evokes from each cohort an immediate response as well as delayed response. (These correspond to the "direct" and "carryover" effects referred to in the literature of experimental designs.) They vary according to the age or stage of development of the cohort at the moment when the stimulus is received, the reason for the age dependence being that the response to any stimulus is likely to vary according to the history of past experiences. To illustrate the point, consider a sequence of three decades: the first and the last being prosperous ones, whereas the middle decade represents a depression period. Imagine birth cohort A entering young adult ages (say, 20 to 29 years) during decade 1, cohort B entering the same age group in decade 2, and cohort C doing so in decade 3. These cohorts are likely to show different responses to the prosperity of decade 3: cohort A is unlikely to show any significant fertility response to the prosperity of the decade, having almost reached the end of its childbearing period by the beginning of the decade. The response of cohort B might very well be to show a tendency toward later marriage and later childbearing, because of a transfer of marriages and births from the depression decade 2 to the prosperous decade 3. Cohort C, on the other hand, since it enters young adult ages during a prosperous decade, might very well show a tendency toward early marriage and early age of childbearing. Thus the same environmental stimulus (prosperity in decade 3 in the illustration above) may evoke different "immediate" responses from different cohorts, depending upon the stage of the life history at which they receive the stimulus and on their life experiences until then. Similar comments apply also to "delayed" responses. Such intercohort differences in responses constitute a cohort effect. The period effect and age effect are defined in the same way as column and row effects are defined in the analysis of variance.

There are certain problems in identifying the cohort, period, and age effects from cohort tables. One source of the difficulty lies in the fact that as a (birth) cohort ages, its composition may change because of death or outmigration. Also, the survivors of a cohort may get mixed up with inmigrants. Those who die or migrate may differ from the rest of the cohort in regard to the phenomenon being studied, and if so, some of the observed intracohort and intercohort variations may be due to the compositional changes within cohorts, resulting from death or migration. Standardization techniques can be used to separate some of the compositional effects, but this could only be tentative, because it is impossible to know how the deceased or the outmigrants would have behaved in regard to the phenomenon being studied if they had survived or not migrated.

Another source of the difficulty in dealing with cohort, period, and age effects is that in a cohort table, each effect is to a certain degree confounded with either of the other two. To see this, consider Table 1, in which (Y_{ijk}) measures on a dependent variable are displayed by age for three successive, evenly spaced cross-sectional surveys. The dates of data collection, which are taken to represent instants rather than time intervals, define the periods. The interval between each two consecutive periods match the width of the age groups used for aggregation. The cohort membership is defined in terms of the age group reached in a given period. The youngest cohort in Table 1 is the one that reaches the first age group in period 3; the oldest is the one that reaches age group 3 in period 1; and so on. The third subscript of Y_{ijk} denotes cohort, the second the period, and the first the age group. It is easily seen that cohort membership $(k) = \text{period}(j) - \text{age}(i) + 3$, which is a perfect linear relationship. This situation resembles the one in the regression analysis when one regressor is a perfect linear function of two or more other regressors. In such situations, some of the effects (regression coefficients) are not estimable. This is known as the identification problem. Situations of this type arise in the analysis of social mobility, status inconsistency, and similar topics.

To see the nature of the identification

Table 1: Cohort table involving three periods and three age groups.

Age Group	Period 1	2	3
1	Y_{113}	Y_{122}	Y_{131}
2	Y_{214}	Y_{223}	Y_{232}
3	Y_{315}	Y_{324}	Y_{333}

problem just referred to, suppose for the data in Table 1 that information is available on the exact age of each respondent. One may then treat period, age, and cohort as continuous variables of time and write a cohort–age–period model, considering only main effects, thus:

$$E(Y) = \alpha + \sum_1^m \beta_r A^r + \sum_1^k \gamma_s C^s + \sum_1^2 \delta_1 P^t, \quad (1)$$

where A stands for age, C for cohort, and P for period. (The maximum degree of the polynomial in P is constrained to 2 because there are only three periods represented in the data. The degrees of the polynomials in A and C are similarly constrained by the available number of data points.)

Now, since by definition $C = P - A$, one may substitute $(A+C)$ for P in the model, thus getting

$$E(Y) = \alpha + (\beta_1 + \delta_1)A$$
$$+ (\beta_2 + \delta_2)A^2 + \sum_3^m \beta_r A^r$$
$$+ (\gamma_1 + \delta_1)C + (\gamma_2 + \delta_2)C^2$$
$$+ \sum_3^k \gamma_s C^s + 2\delta_2 AC. \quad (2)$$

Clearly, except for β_1 and γ_1, which are both indistinguishable from δ_1, all parameters in the foregoing model are estimable.

It follows that it is the linear components of the cohort, age, and period main effects that are indistinguishable; all higher-order components are estimable [2,10].

By restricting δ_1/δ_2, δ_1/β_1, or δ_1/γ_1 to equal a known constant, one can make the model (2) exactly identified. Obviously, except in very limited contexts, such restrictions may not be justifiable. For this reason it is not advisable to base one's interpretation of cohort data exclusively on the results given by formal models fitted to the data, unless the models, including the restrictions imposed on the parameters, have been carefully selected on the basis of substantive considerations. In this connection, a debate that has been going on for some time now between two camps of cohort analysts may be worth mentioning. The issue of the debate is whether it is appropriate to use additive age–period–cohort models for interpreting data of the type shown in Table 1. One camp (see, e.g., Glenn [4]) holds that there is likely to be age–period, age–cohort, or period–cohort interaction in cohort tables, and hence additive models, if applied to such tables, may suffer from specification errors, because of ignoring interactions. The opposite camp (see, e.g., Fienberg and Mason [2]) holds that all models are simplifications (abstractions) of reality; that whether a given simplification is acceptable in a given context is to be decided on substantive grounds; and that this should be decided on a case-by-case basis rather than by invoking a general principle such as "interactions should always be included in the model".

14.3 Applications

14.3.1 Fertility

As already mentioned, the cohort method has been used more extensively in fertility analysis than in any other demographic studies. The cohort analysis of fertility may employ birth cohorts or marriage cohorts or both in combination. Confining attention to birth cohorts, imagine that schedules of age-specific fertility rates are available for a number of consecutive years. A table of these rates arranged in columns for years (and ages in rows) can be summarized in two obvious ways: by column or diagonally. For each column the sum of the age-specific fertility rates gives the period total fertility rate, and the arithmetic mean of the age distribution of the birth rates gives the period mean age of fertility. The cohort schedule (the figures in the diagonal) can be summarized similarly to give the cohort total fertility rate and the cohort mean age of fertility. It can be shown that the time series of period total fertility rate diverges from that of the cohort total fertility rate to the extent that there is temporal change in the period or cohort mean age of fertility, and that the time series of the period mean age of fertility will diverge from the cohort mean age of fertility to the extent that there is a temporal shift in the period or cohort total fertility rate. To take a specific example, which is only slightly exaggerated, suppose that in a large population, for a long period of time, a constant age pattern of childbearing has been prevailing, with each cohort having on an average 3 children per woman by the end of the childbearing period. Suppose that the younger cohorts at time t suddenly change the timing of births in such a way that they bear their children earlier than usual, still having the same cohort total fertility rate as the older cohorts. Then an examination of the period total fertility rate for several successive years will show a rise in the birth rate. But such a rise does not really constitute increased (cohort) fertility. Here the divergence between the period and cohort total fertility rates is attributable to the temporal shift in the cohort mean age of fertility.

Suppose again that women vary the time at which they have children, delaying pregnancies during recessions and depressions and advancing them when times are prosperous, but with no change in the average number of children borne by the end of childbearing period. The period data would show considerable fluctuations, but they do not reflect variations in total cohort fertility.

Referring to Figure 1, it may be noted that the childbearing of women is represented by density along the life lines. If the net maternity function, which is the product of survivorship to a given age and reproduction rate at that age, is represented by $\varphi(a,t)$, a function of age and time, then the integral (or total) along the vertical strip will give the period net reproduction rate and that along the diagonal strip the cohort net reproduction rate. It is possible to relate the integral (or total) along the diagonal to that along the vertical (see Ryder [7]), thereby enabling one to translate cohort information into period information, and vice versa.

14.3.2 Mortality

The sequence of numbers $E_0, E_1, \ldots$ in Figure 1 shows that of E_0 births that occur in the time interval $(t, t+1)$, E_x survive to the xth birthday and $(E_x - E_{x+1})$ die between the xth and $(x+1)$st birthdays. From the sequence $(E_0, E_1, \ldots)$, mortality rates according to age in successive years can be computed using the formula $q_x = (E_x - E_{x+1})/E_x$. The life table completed on the basis of the q_x-values thus computed is known as the cohort or generation life table, since it reflects the actual mortality experience of a cohort as it ages in successive years. The cohort life table shows among other things e_x^0, the average length of life remaining after the xth birthday for those who survive to that birthday.

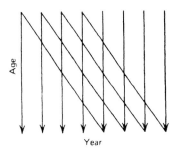

Figure 2: The life table.

The cohort life table is to be distinguished from the current life table, which is derived from the mortality rates for a single calendar year or period. In Figure 2, the life table based on the q_x-values in any vertical line is the current life table, whereas the one based on the q_x sequence on any diagonal is a cohort life table.

The current life table obviously is derived from a mixture of the mortality rates that different cohorts experience at different stages in their life history, one cohort during its first year of life, another during its second year, and so on. Consequently, given the usual downward trend in mortality over time, the current life table understates the average length of life of the newborn, and the average remaining life of those who survive to a given age, during the period of the life table.

That cohort analysis helps to avoid erroneous inferences from data is more dramatically illustrated by Frost's analysis of the age pattern of tuberculosis mortality in Massachusetts during the years 1880, 1910, and 1930 [3]. An examination of the age curve of tuberculosis mortality for the three years showed:

1. The curve for 1930 was consistently below that for 1910, which in turn was consistently below that for 1880.

2. For each year the curve started at a high level in infancy, dropped to a low level in childhood, then rose again to reach higher levels in adult ages.

3. The curve for 1930 peaked between ages 50 and 60, whereas that for 1910 peaked between ages 30 and 40, and that for 1880 between ages 20 and 30.

Earlier explanations for the apparent shift in the peak of the age curve of tuberculosis mortality were based on the notion that, over time, tuberculosis lost its killing power, shifting fatality to older ages at which the vital resistance of the body characteristically is low. When Frost examined the data in cohort form, however, he noticed that all the cohorts exhibited similarly shaped age curves of mortality, peaking at about the same age (20 to 29 years), each cohort having its age curve consistently above those of all younger cohorts. Frost concluded that the apparent shift in the peak age of tuberculosis mortality (from between 20 and 30 in 1880 to between 50 and 60 in 1930) was an artifact of the period method of organizing data. The period method, it may be noted, juxtaposes the experience of an older cohort at an older age with that of younger cohorts at younger ages. Consequently, given the consistently higher mortality of older cohorts at all ages, this method produces the artifact of the age curve peaking at higher ages for more recent periods.

Frost's work on tuberculosis mortality illustrates why the experiences of real cohorts rather than the age pattern observed in period data should be the basis for inferences regarding the age pattern of mortality or survival functions.

The difficulty with the cohort method in mortality studies, of course, is that 100 years or more must elapse before the death of the last survivor of a birth cohort. This difficulty can be partially overcome by confining attention to a part of the cohort experience, for example, mortality up to age 50, or by studying cohorts between two age points, say, between ages 20 and 50 [8].

14.3.3 Nuptiality

In the area of nuptiality, cohort analysis has frequently been applied to the study of the age patterns of first marriage. From cohort data, one may construct the nuptiality table along the lines of the usual life table or in the fashion of the double-decrement table, the latter if mortality is recognized as an additional factor responsible for the decrement of never-married persons in the cohort as it advances in age.

Mathematical functions, parallel to the survival functions in mortality studies, have been developed for the age pattern of cumulative proportions ever married [1,5].

The fluctuations in the number of marriages from period to period, with changing economic conditions despite stability in the likelihood of getting married eventually, have been interpreted in terms of temporary shifts in the time pattern of cohort nuptiality [6].

14.3.4 Migration

The cohort approach to migration studies has taken two forms: one in which the focus is on birth cohort and the other on migration cohort. In the former, migration is viewed as an event in the life history of the individual or cohort, similar to marriage or childbearing. When the focus is on migration cohorts, the emphasis is on duration of residence in a given area.

14.3.5 Effects of Aging

In the next few years, given the trend toward older age structures in almost all the developed populations, interest in the study of the effects of aging on responses to forces of change is likely to be on the increase. A basic issue for examination would be whether aging individuals tend to become so inflexible in their attitudes and behavior that they will be a formidable obstacle to social and economic change. The method appropriate for investigation of this issue involves separating cohort effect from age effect in intercohort differences in susceptibility to change.

14.3.6 Other Fields

The cohort approach has also been applied to the study of educational careers from the time of entry into school, of occupational careers from the time of entry into the labor force, and of morbidity history beginning with the first exposure to a condition; it has also been used in many other fields (see, e.g., United Nations [9]).

Underlying these and other applications of the cohort method is the hypothesis that in the macro-biography of any group, what happens today is contingent on what happened yesterday, and what happens tomorrow depends upon what happens today. It is the plausibility of this hypothesis on which rests the utility of the cohort method.

References

1. Coale, A. J. (1971). *Popul. Stud.*, **25**, 193–196.
2. Fienberg, S. E. and Mason, W. M. (1979). In *Sociological Methodology*, Karl F. Schuessler, ed., Jossey-Bass, San Francisco, cA, 1–67.
3. Frost, W. H. (1939). *Am. J. Hyg. A*, **30**, 91–96.
4. Glenn, N. D. (1977). *Cohort Analysis*. Sage Univ. Paper Ser. Quant. Appl. Social Sci.: Ser. No. 07–005. Sage Publications, Beverly Hills, CA. (Contains a useful set of references.)
5. Hernes, G. (1972). *Am. Sociol. Rev.*, **37**, 173–182.
6. Ryder, N. B. (1956). *Population (Paris)*, **11**, 29–46.

7. Ryder, N. B. (1964). *Demography,* **1,** 74–82.
8. Spiegelman, M. (1969). *Demography,* **6,** 117–127.
9. United Nations (1968). *Methods of Analysing Census Data on Economic Activities of the Population.* Ser. A. Popul. Stud. No. 43. United Nations, New York.
10. Winsborough, H. H. (1976). Discussion. *Proc. Social Statist. Sect. Part I,* American Statistical Association, Washington, DC.

15. Comparisons with a Control

Charles W. Dunnett

15.1 Introduction

Consider $k+1$ samples from normal distributions with a common variance σ^2, where one of the samples represents a control or standard, or otherwise specified group. The observations are denoted by y_{ij} ($i = 0, 1, \ldots, k$; $j = 1, \ldots, n_i$), the subscript 0 denoting the specified group. Denote the sample means by $\bar{y}_i$, assumed to be distributed independently as $N(\mu_i, \sigma^2/n_i)$. If the parameter σ^2 is unknown, let s^2 be an unbiased estimator, which is independent of the $\bar{y}_i$, with ν df and such that $\nu s^2/\sigma^2$ has a χ^2_ν distribution.

The problem is (1) to obtain confidence interval estimates of $\mu_i - \mu_0$ for $i = 1, \ldots, k$ simultaneously, with joint confidence coefficient $1 - \alpha$, or (2) to test the null hypotheses $H_i : \mu_i - \mu_0 = 0$ (or any other particular value) against the alternatives $A_i : \mu_i - \mu_0 \neq 0$ (or $H_i : \mu_i - \mu_0 \leq 0$ against $A_i : \mu_i - \mu_0 > 0$) for $i = 1, \ldots, k$ simultaneously, such that the *familywise* (or *experimentwise*) *error rate* (FWE) is $\leq \alpha$. Either of these formulations is referred to as *simultaneous inference* on the parameters $\theta_i = \mu_i - \mu_0$. Whether or not to formulate the problem as one of simultaneous inference depends on the nature of the inferences required concerning the values of the $\mu_i - \mu_0$.

A *multiple comparison procedure* (MCP) is a statistical procedure for making simultaneous inferences concerning the expected values of differences or other contrasts between treatment means. In most cases, the aim of the MCP is to *adjust* for the multiplicity of the inferences. The most widely accepted method of making this adjustment is to control the FWE, defined as the probability of making one or more type I errors, so that it is not greater than α under any null configuration of the parameters.

An alternative Bayesian approach to MCPs advocates on *adaptive* handling of the comparisons rather than *adjusting* for the multiplicity by controlling error rates; this is described later in this chapter. Another formulation is the use of a selection procedure to select the best treatment provided that it is better than the specified treatment, or to select a subset of the k treatments that are better than the specified treatment: see Chapter 5 in Bechhofer et al. [2].

The first to suggest an adjustment for multiplicity in comparing treatments with a control was Roessler [42]. His proposal was that each of the k tests be performed at a significance level $1 - (1 - \alpha)^k$ instead of α, and he provided a short table of adjusted critical values to use with Student's t test. This adjustment would be exactly right for achieving FWE $\leq \alpha$ if the k tests were independent, which, of course. they are not. Roessler's adjustment does control the FWE, but it overadjusts by making the critical values larger than they need be. The correct critical values, under the assumptions stated above, require the use of a multivariate t distribution.

15.2 Some Examples

Comparisons with a control occur frequently in medical trials and pharmacological experiments, where the specified group is often a placebo control or a group receiving a standard treatment, but they arise in other areas as well. For a nonmedical application, see Bitting et al. [6], who describe an investigation in which automobiles were evaluated for tailpipe emissions and four different deposit conditions were studied, one of which was designated as the control with which the other three were compared. For a biopharmaceutical application, see Bedotta et al. [4], who carried out an experiment on rats divided into 10 treatment groups, consisting of an untreated control group and a group receiving a thyroid hormone (T4), four other groups receiving various treatments, and an additional four groups that received one of the latter in combination with T4. Measures of cardiovascular response were observed, and two sets of comparisons of the treatments-vs.-control type were of interest for the purpose of elucidating the mechanisms of action of the four treatment compounds: (1) T4 and each of the four treatments groups vs. the untreated control group, and (2) the four treatments given in combination with T4 vs. T4 given alone.

In the medical area, comparisons between a new treatment and any standard therapies that may be available are required by the regulatory authorities in order to determine the safety and efficacy of the new treatment compared with the standard treatments, before it can be approved for therapeutic use. This is usually done by carrying out a series of clinical trials at various locations and on various patient groups, involving comparisons between the new treatment and the standards with respect to their efficacies and side effects. In this application, the new treatment represents the specified group.

It should be noted that not all treatment-vs.-control comparisons necessarily require adjustment for multiple inferences. Sometimes, several such comparisons are made in the same experiment for reasons of efficiency only. An example is given in Chapter 11 of Finney [24], in which the bioassay of several test substances compared with a common standard is described. The purpose of including them together in the same bioassay was that this was more efficient than carrying out separate bioassays for each one; hence the inference methods applied should be the same as if they had been run separately. An analogous situation arises in drug screening for detecting possible new drugs. Here, several potential candidates are usually tested together against a common placebo control or standard, in order to determine whether any are worth testing further in follow-up studies. Again there is no need to "adjust" for the presence of other treatments in the same experiment; see, for example, Redman and Dunnett [40].

15.3 Simultaneous Confidence Intervals

Two-sided simultaneous confidence interval estimates of $\mu_i - \mu_0$ for $i = 1, \ldots, k$

are given by

$$\bar{y}_i - \bar{y}_0 \pm hs\sqrt{1/n_i + 1/n_0}. \quad (1)$$

The corresponding one-sided intervals are given by

$$\bar{y}_i - \bar{y}_0 + gs\sqrt{1/n_i + 1/n_0} \quad \text{or}$$
$$\bar{y}_i - \bar{y}_0 - gs\sqrt{1/n_i + 1/n_0}, \quad (2)$$

depending on whether upper or lower limits are required. The constants h and g are determined so that the intervals in each case have a joint coverage probability of $1 - \alpha$.

Denote by T_i the random variable

$$\frac{\bar{y}_i - \bar{y}_0 - (\mu_i - \mu_0)}{s\sqrt{1/n_i + 1/n_0}},$$

which has a univariate Student's t distribution with ν df. Then h is chosen so that

$$\mathbb{P}(-h < T_1 < h, -h < T_2 < h, \ldots,$$
$$- h < T_k < h) = 1 - \alpha. \quad (3)$$

The joint distribution of $T_1, \ldots, T_k$ is k-variate central Student's t with ν df and correlation matrix $\boldsymbol{R} = \{\rho_{ij}\}$, where ρ_{ij} is the correlation coefficient between $\bar{y}_i - \bar{y}_0$ and $\bar{y}_j - \bar{y}_0$, given by $\rho_{ij} = \lambda_i \lambda_j$ with $\lambda_i = 1/\sqrt{1 + n_0/n_i}$. Thus $h = |t|^\alpha_{k,\nu,\boldsymbol{R}}$, the two-sided (equicoordinate) upper α-point of this multivariate t distribution. Similarly, $g = t^\alpha_{k,\nu,\boldsymbol{R}}$, the one-sided upper α-point of the multivariate t distribution, which is defined by

$$\mathbb{P}(T_1 < g, T_2 < g, \ldots, T_k < g) = 1 - \alpha. \quad (4)$$

The above results were originally obtained by Dunnett [17].

15.4 Multiple Hypothesis Tests

Here the simultaneous inference problem is to test the null hypothesis H_i vs. an alternative hypothesis A_i for $i = 1, \ldots, k$. The hypotheses are of the form

$$H_i : \mu_i - \mu_0 = 0 \quad \text{vs.} \quad A_i : \mu_i - \mu_0 \neq 0$$

or

$$H_i : \mu_i - \mu_0 \leq 0 \quad \text{vs.} \quad A_i : \mu_i - \mu_0 > 0,$$

depending on whether two-sided or one-sided alternatives are required. (The inequalities in the latter are reversed if the desired alternative is to show that μ_i is less than the control mean.) In either case, the test statistic for H_i is $t_i = (\bar{y}_i - \bar{y}_0)/s\sqrt{1/n_i + 1/n_0}$. The critical constants for the t_i are to be chosen so that FWE $\leq \alpha$.

15.4.1 Single-Step Methods

There are two fundamentally different single-step methods, and which is the appropriate one to use in a particular application depends on the nature of the problem.

Problem 15.4.1 Here, the simultaneous inference problem is testing the null hypothesis that all the individual H_i are true versus the alternative that at least one of them is false. The method used (denoted SS) is the hypothesis-testing analog of the confidence interval method in the preceding section. For two-sided alternatives, the hypothesis testing problem is written

$$H : \mu_0 = \mu_1 = \cdots = \mu_k$$
$$(\text{or } H : \mu_i = \mu_0 \text{ for all } i) \quad \text{vs.}$$
$$A : \mu_i \neq \mu_0 \text{ for at least one } i. \quad (5)$$

[Because A can be represented as the union $\bigcup(A_i)$ of the A_i and H as the intersection $\bigcap(H_i)$ of the H_i, this is called a *union–intersection* multiple testing problem.] The test statistic for H is $\max(|t_i|)$ and the critical constant c is chosen so that

$$\mathbb{P}(\max|T_i| < c) = \mathbb{P}(-c < T_1 < c, \ldots,$$
$$- c < T_k < c)$$
$$= 1 - \alpha. \quad (6)$$

The same multivariate t distribution is involved here as in simultaneous confidence intervals, and the same solution is obtained, namely, $c = h = |t|_{k,\nu,\mathbf{R}}^{\alpha}$. The test rejects H if $\max|t_i| \geq h$. In terms of the individual H_i, rejection of H means that the H_i associated with $\max|t_i|$ is rejected (as well as any other H_i whose test statistic also exceeds the same critical constant). For an example of the application of this test, see Dunnett [18].

For one-sided alternatives, the null hypothesis and its alternative are

$$H : \mu_i \leq \mu_0 \text{ for all } i \quad \text{vs.}$$
$$A : \mu_i > \mu_0 \text{ for at least one } i. \quad (7)$$

The test statistic is $\max t_i$, and the critical value is $c = g = t_{k,\nu,\mathbf{R}}^{\alpha}$, the same constant used in one-sided simultaneous confidence intervals. The test rejects H if $\max t_i \geq g$. If H is rejected, the H_i associated with $\max t_i$ is rejected (as well as any other H_i whose test statistic also exceeds g).

Problem 15.4.2 The simultaneous inference problem here is to test the overall null hypothesis that at least one of the individual H_i is true vs. the alternative that all of them are false. For two-sided alternatives, this problem is written

$$H : \mu_i = \mu_0 \text{ for at least one } i \quad \text{vs.}$$
$$A : \mu_i \neq \mu_0 \text{ for all } i. \quad (8)$$

(This is called an *intersection–union* multiple testing problem, because A can be considered as the intersection $\bigcap A_i$ of the A_i and H as the union $\bigcup H_i$ of the H_i.) The test statistic is $\min|t_i|$.

The correct critical constant c for testing H in (8) turns out to be the two-sided upper α point of univariate Student's t. This somewhat surprising result is due to Berger [5]. If $\min|t_i| \geq c = t_\nu^{\alpha/2}$, then H is rejected. This means that *all* the individual H_i can be rejected.

For one-sided alternatives, the test statistic is $\min t_i$, and its critical constant c is the one-sided upper α point of univariate Student's t. The test is to reject H if $\min t_i \geq c = t_\nu^{\alpha}$. If H is rejected, then *all* the individual H_i can be rejected.

The use of $\min|t_i|$ for testing H in (8) or $\min t_i$ in the case of a one-sided alternative is the basis for the MIN test developed by Laska and Meisner [36]. It does not require the normality and homogeneous variance assumptions, as the t-tests described above can be replaced by nonparametric methods, such as rank tests.

15.4.2 Stepwise Test Methods

A stepwise multiple testing procedure is more appropriate than a single-step procedure when the main interest is in the individual H_i rather than an overall H as specified above. A sequence of critical constants $c_1 < c_2 < \cdots < c_k$ is used for testing these hypotheses. First, the t statistics, along with the corresponding hypotheses, are renumbered so that $|t_1| \leq \cdots \leq |t_k|$ for two-sided tests, or $t_1 \leq \cdots \leq t_k$ for one-sided tests. There are two types of stepwise testing procedures: "stepdown" and "stepup."

Stepdown Method (SD). Step-down testing starts with H_k, the hypothesis corresponding to the most significant test statistic. The steps are as follows:

Step 1. If $|t_k| \geq c_k$, rejected H_k and proceed to H_{k-1}; otherwise, accept all H_i and stop testing.

Step 2. If H_k was rejected and $|t_{k-1}| \geq c_{k-1}$, reject H_{k-1} and proceed to H_{k-2}; otherwise, accept $H_1, \ldots, H_{k-1}$ and stop testing.

General Step. If $H_{i+1}, \ldots, H_k$ were rejected and $|t_i| \geq c_i$, reject H_i and proceed to H_{i-1}; otherwise, accept $H_1, \ldots, H_i$ and stop testing.

The values of the constants $c_1, \ldots, c_k$ for stepdown testing are given by

$$\mathbb{P}(-c_m < T_1 < c_m, \ldots, -c_m < T_m < c_m) = 1 - \alpha \quad (m = 1, \ldots, k). \quad (9)$$

$T_1, \ldots, T_m$ are the random variables associated with $t_1, \ldots, t_m$, the first m t-statistics in order of significance. For solving (9), they are jointly distributed as central m-variate Student's t with ν df and correlation matrix $\boldsymbol{R}_m$, which is the submatrix of the correlation matrix $\boldsymbol{R}$ for the entire set of t statistics obtained by deleting all except the first m rows and columns. Thus, $c_m = |t|_{m,\nu,\boldsymbol{R}_m}^\alpha$, the two-sided α point of m-variate Student's t. In particular, c_k is obtained from k-variate t and is the same constant h used in the two-sided confidence intervals, while c_1 is the corresponding two-sided value $t_\nu^{\alpha/2}$ for univariate Student's t.

For one-sided testing, similar considerations apply except $c_m = t_{m,\nu,\boldsymbol{R}_m}^\alpha$, the one-sided upper α point of m-variate t. In particular, c_k is identical with the constant g used in the one-sided confidence intervals, while c_1 is the corresponding one-sided value for univariate Student's t, which is t_ν^α.

Since the critical constant c_k used in the first step coincides with that used in the SS procedure, the SD procedure may be considered as a stepwise extension of SS. It is easy to see why SD, which uses the constants $c_1 < \ldots < c_k$, is superior to SS, which uses h (or g), since all the $c_i < h$ except for $c_k = h$ and thus SD tends to reject more of the individual hypotheses.

The first literature reference to the effect that a stepdown test could be used in treatments-vs.-control multiple comparisons was by Miller in the 1966 edition of his book [38, pp. 78,85,86]. Marcus et al. [37] provided the theoretical justification for it through their proposal of the *closure method* [30, p. 54] for constructing stepdown procedures that satisfy the requirement FWE $\leq \alpha$, citing the treatments-vs.-control stepdown testing method as an example. Dunnett and Tamhane [20] gave the procedure for unbalanced data (unequal sample sizes).

Stepup Method (SU). Stepup testing starts with H_1, the hypothesis corresponding to the least significant test statistic. The steps are as follows:

Step 1. If $|t_1| < c_1$, accept H_1 and proceed to H_2; otherwise, reject all the H_i and stop testing.

Step 2. If H_1 was accepted and $|t_2| < c_2$, accept H_2 and proceed to H_3; otherwise, reject $H_2, \ldots, H_k$ and stop testing.

General Step. If $H_1, \ldots, H_{i-1}$ were accepted and $|t_i| < c_i$, accept H_i and proceed to H_{i+1}; otherwise, reject $H_i, \ldots, H_k$ and stop testing.

The values of the constants $c_1, \ldots, c_k$ for stepup testing are determined by the equations

$$\mathbb{P}(-c_1 < T_{(1)} < c_1, \ldots, -c_m < T_{(m)} < c_m) = 1 - \alpha \quad \text{for } m = 1, \ldots, k, \quad (10)$$

where $T_{(1)}, \ldots, T_{(m)}$ are the ordered values of the random variables $T_1, \ldots, T_m$ associated with the first m t statistics in order of significance. Note the differences from (9). For $m = 1$, we obtain the same result as in (9), namely, $c_1 = t_\nu^{\alpha/2}$, the two-sided α point of Student's t. For $m > 1$, it is necessary to determine the values of $c_1, \ldots, c_{m-1}$, and then solve Equation (10) for c_m. For $m > 1$, c_m is slightly larger than the corresponding constant for stepdown testing.

For one-sided testing, similar considerations apply. Thus $c_1 = t_\nu^\alpha$, the one-sided upper α point of univariate Student's t, which is the same solution obtained for stepdown testing. For $m > 1$, the solution for c_m is slightly larger than the corresponding value for SD.

This SD method was developed by Dunnett and Tamhane [21,23]. It can be considered as a step-wise extension of Laska and Meisner's [36] MIN test, in that its first step coincides with MIN. It has the advantage that, in cases when the MIN test fails to be significant, it may still be possible to reject some of the H_i by proceeding stepwise to see whether some t-statistic other than t_1 exceeds the appropriate critical constant.

For an example of the use of both the SD and SU methods, and a discussion of when each is appropriate, see Dunnett and Tamhane [22].

Determining the Values of the Constants. Tables of the constants needed for confidence intervals or for stepdown testing are readily available for the equicorrelated case $\rho_{ij} \geq 0$: see Hochberg and Tamhane [30] or Bechhofer and Dunnett [1]. Equal correlations require the sample sizes in the k comparison groups to be approximately the same, although the size of the specified group may be different. For unequal sample sizes, a good approximation is to use the average correlation and interpolate in the tables. Alternatively, they can be computed exactly by using a computer program based on the algorithm of Dunnett [19]; this program is available on the World Wide Web from the statlib at Carnegie-Mellon University, using the address *http://lib.stat.cmu.edu/apstat*, or by email to *statlib@lib.stat.cmu.edu*, asking "send 251 from apstat."

Limited tables of the constants needed for step-up testing are given in Dunnett and Tamhane [21] for equal correlation, along with a description of the algorithm used to compute them. Dunnett and Tamhane [23] describe methods for determining the constants for SU when the correlations ρ_{ij} among the $\bar{y}_i - \bar{y}_0$ are unequal.

Normality and Variance Assumptions. All these tests based on multivariate t assume normally distributed data and homogeneous variances in the groups, assumptions that should be checked before using the methods. The normality assumption is usually not an issue for moderate or large samples, since the central-limit theorem ensures that normality applies approximately to the means. However, because the multivariate t distribution is based on a pooled variance estimate, homogeneous variances are important. If there is any doubt, the variances should not be pooled and separate variance estimates used instead. But then this involves making approximations in the determination of the c's. A method that can be used which is conservative is to employ Bonferroni critical values; this is equivalent to using the stepdown Bonferroni method of Holm [31] in place of SD, and the stepup Bonferroni method of Hochberg [29] in place of SU. Another approach is to use robust estimates of the means and variances, as described by Fung and Tam [27]. Chakraborti and Desu [10–12] have developed nonparametric methods, including methods to handle censored data. Fligner [25] and Fligner and Wolfe [26] have also developed distribution-free methods; see also Hothorn [33] and Hothorn and Lehmacher [34]. Rudolph [45] investigated the type I and type II error rates by simulation to compare the robustness of two nonparametric procedures and SS with respect to nonnormality and variance heterogeneity.

15.5 Other Results

15.5.1 Extensions to Treatments-vs.-Control Comparisons

The question is sometimes asked whether the widths of the confidence intervals given in (1) and (2) can be shortened by the use of a stepwise method. Bofinger [7] shows that the answer is a qualified *yes*: the one-sided SD multiple testing method can be followed by one-sided lower-confidence-interval estimates for the $\mu_i - \mu_0$ corresponding to the accepted hypotheses by taking $g = t^\alpha_{m,\nu,\mathbf{R}_m}$ in (2), where m is the number of accepted hypotheses. For the rejected hypotheses, on the other hand, the confidence limit can only be specified as ≥ 0. Thus, when $0 < m < k$, it provides sharper limits for the accepted hypotheses than the method using (2), but less sharp for the rejected hypotheses.

Cheung and Holland [13,14] have extended the SS and SD tests to the case of more than one group of treatments, each group containing a control, and the tests are done so that FWE $\leq \alpha$ overall. Shaffer [46] and Hoover [35] extended the problem of simultaneous confidence intervals for k treatments vs. a control to two or more controls. Bristol [9] and Hayter and Liu [28] have considered the problem of determining the sample sizes necessary for the SS procedure to achieve a specified power.

It has long been known that the optimum allocation of observations between the control and the treatments is to have approximately $\sqrt{k}$ times as many on the control as on each treatment: see "square root allocation rule" in Hochberg and Tamhane [30]. Spurrier and Nizam [47] have given tables of exact numbers for situations where the total number is a specified value $n \leq 50$.

Balanced treatment incomplete block (BTIB) designs, where one of the treatments is a control treatment, have been developed by Bechhofer and Tamhane (see [3] and other references cited there). The balanced nature of these designs ensures that, as in the BIB designs, the treatment comparisons of interest are equally correlated, with correlation coefficient ρ dependent on the design. The stepwise tests SD and SU can be applied to these designs.

Selection procedures for selecting treatments relative to a specified treatment group often are equivalent to related MCPs; see the contributed chapter by Horn and Vollandt [32].

15.5.2 Comparing Dose Levels with a Zero-Dose Control

A special case of treatments-vs.-control comparisons is when the treatments are ordered: for example, when the treatment groups correspond to different dose levels. The problem is to determine the lowest dose that differs significantly from that of the control. Williams [51,52] used maximum-likelihood estimates of the dose-level effects under the assumption that the dose response is monotonic, and determined the required critical constants for the multiple testing of the differences of these from the control.

More recent developments have been obtained by Tukey et al. [50] and Rom et al. [43], who proposed that various linear regressions be tested, and Ruberg [44], who proposed other contrasts to be tested. Tamhane et al. [48] provided a general framework for the methods (classified by type of contrast and type of stepwise testing method) and carried out extensive comparisons of the various methods, including some new methods, by simulation.

15.6 Other Approaches

15.6.1 Order-Restricted Methods: Multiple-Contrast Tests

Methods based on order-restricted inference assume that the null hypothesis and its alternative are

$$H : \mu_0 = \mu_1 = \ldots = \mu_k \quad \text{vs.}$$
$$A : \mu_i \geq \mu_o \text{ with } \mu_i > \mu_0$$
for at least one i.

Note the difference from the null hypothesis in (7); here, values $\mu_i < \mu_0$ are excluded from the null hypothesis, as there is assumed to be prior information that the μ_i cannot be less than μ_0. This imposes a restriction on the parameters, $\mu_0 \leq [\mu_1, \ldots, \mu_k]$, called a *simple tree* ordering.

Methods have been developed to handle inference problems where the parameters are subject to such order restrictions. The methods involve obtaining the ML estimates of the μ_i under the order restriction and then calculating the likelihood ratio test (LRT) statistic. The LRT of H under this particular order restriction has been described by Robertson and Wright [41]. The algorithm for determining the MLEs is the following: If $\bar{y}_0 \leq \bar{y}_i$ for all i, the estimates are simply the observed means. Otherwise, denote the ordered means (excluding the control) by $\bar{y}_{(1)} \leq \bar{y}_{(2)} \leq \ldots \leq \bar{y}_{(k)}$, and define A_j to be the weighted average of $\bar{y}_0, \bar{y}_{(1)}, \ldots, \bar{y}_{(j)}$, namely,

$$A_j = \frac{n_0 \bar{y}_0 + \sum_{i=1}^{j} n_{(i)} \bar{y}_{(i)}}{n_0 + \sum_{i=1}^{j} n_{(i)}}$$
$$j = 1, \ldots, k-1,$$

where $n_{(i)}$ is the sample size associated with $\bar{y}_{(i)}$. Define j to be the smallest integer $< k$ for which $A_j < \bar{y}_{(j+1)}$; if none exists, then define $j = k$.

Then the restricted MLE of μ_0 is $\bar{\mu}_0 = A_j$; that for μ_i is $\bar{\mu}_i = A_j$ if $\bar{y}_i$ is included in A_j, and is $\bar{\mu}_i = \bar{y}_j$ if it is not. The LRT statistic for H is

$$S_{01} = \frac{\sum_{i=0}^{k} n_i(\bar{\mu}_i - \hat{\mu})^2}{\sum_{i=0}^{k} n_i(\bar{y}_i - \hat{\mu})^2 + (N-k-1)s^2},$$

where $\hat{\mu} = \sum_{i=0}^{k} \sum_{j=1}^{n_i} y_{ij}/N$, $N = \sum_{i=0}^{k} n_i$ and $(N-k-1)s^2/\sigma^2$ has a χ^2_{N-k-1} distribution. Tables of critical points for S_{01} are given in Robertson and Wright [41].

Conaway et al. [16] developed an approximation to the LRT method, called a *circular-cone* test, which has similar power properties and is simpler to compute.

Mukerjee et al. [39] considered a class of *multiple-contrast tests*, defined by a scalar $r (0 \leq r \leq 1)$, as follows. For a specific r, the ith contrast S_i is the weighted average of two contrasts: $\tilde{y} - \bar{y}_0$, where $\tilde{y}$ is the mean of all y_{ij} in the k treatment groups, and $\bar{y}_i - \bar{y}_i^*$, where $\bar{y}_i^*$ is the mean of all y_{ij} in the k treatment groups with the exception of the ith group—the weights being proportional to r and $1-r$, respectively. The first of the two defining contrasts is a measure of how much the control mean differs from the mean of all the remaining observations, while the second is a measure of how much the mean for the ith group differs from the mean of the remaining observations excluding the control. A test statistic for S_i is $t_i = S_i/\text{SE}(S_i)$ (where SE = standard error), and the test is to reject H if $\max t_i \geq c$, where c is chosen to achieve significance level α. The solution is obtained as in (7) and is given by $c = t^\alpha_{k,\nu,\mathbf{R}}$, where $\mathbf{R}$ is the correlation matrix for the S_i.

Mukerjee et al. [39] computed the power of rejecting H as a function of r, and found that the power was approximately constant for values of $\sum(\mu_i - \mu_0)^2$ restricted to be constant if r is chosen so that the $S_i(r)$ are orthogonal (viz., $\mathbf{R} = \mathbf{I}$, the identity matrix). They recommended using this value of r and provided an explicit formula for it.

Cohen and Sackrowitz [15] have shown that all multiple-contrast tests, including the orthogonal contrast test and the test SS described earlier, fail to satisfy the conditions for being admissible under this order-restricted inference model.

15.6.2 A Bayesian Method: Duncan's k-Ratio Method

The k-ratio method of multiple comparisons, which has long been advocated by Duncan and his followers as an alternative to MCP methods which control error rates, has been extended to the treatments-vs.-control problem by Brant et al. [8]. This approach assumes normal priors for the control and for the other treatment groups and linear loss functions which are additive. The critical value c for testing any hypothesis, instead of being a function of the number of treatment comparisons as in the case of MCPs which control FWE, is a continuous function (subject to $c \geq 0$) of the observed values of two statistics: the t statistic for testing the difference between the mean of the k treatments and the control treatment (t_G), and the F statistic for testing the homogeneity of the k treatments, excluding the control (F_T). Note that the contrast tested by t_G coincides with the first of the two defining contrasts in the orthogonal contrasts method. If t_G and F_T are small, indicating that the observed treatment differences are small, then the method takes a more cautious approach in determining significant differences by setting a large value for c, and vice versa. In this way, it is *adaptive* to the amount of observed variation between the treatment groups. It has also been applied to treatments-vs.-control comparisons by Tamhane and Gopal [49].

References

1. Bechhofer, R. E. and Dunnett, C. W. (1988). Tables of percentage points of multivariate t distributions. *Selected Tables Math. Statist.*, **11**, 1–371.

2. Bechhofer, R. E., Santner, T. J., and Goldsman, D. M. (1995). *Design and Analysis of Experiments for Statistical Selection, Screening and Multiple Comparisons*. Wiley, New York.

3. Bechhofer, R. E. and Tamhane, A. C. (1985). Tables of admissible and optimal BTIB designs for comparing treatments with a control. *Selected Tables Math. Statist.*, **8**, 41–139.

4. Bedotta, J. E., Gay, R. G., Graham, S. D., Morkin, E., and Goldman, S. (1989). Cardiac hypertrophy induced by thyroid hormone is independent of loading conditions and beta adrenoceptor. *J. Pharm. Exp. Therap.*, **248**, 632–636.

5. Berger, R. L. (1982). Multiparameter hypothesis testing and acceptance sampling. *Technometrics*, **24**, 295–300.

6. Bitting, W. H., Firmstone, G. P., and Keller, C. T. (1994). *Effects of Combustion Chamber Deposits on Tailpipe Emissions*, Paper 940345, Society of Automotive Engineers Technical Paper Series, pp. 1–5.

7. Bofinger, E. (1987). Step down procedures for comparison with a control. *Austral. J. Statist.*, **29**, 348–364.

8. Brant, L. J., Duncan, D. B., and Dixon, D. O. (1992). k-ratio t tests for multiple comparisons involving several treatments and a control. *Statist. Med.*, **11**, 863–873.

9. Bristol, D. R. (1989). Designing clinical trials for two-sided multiple comparisons with a control. *Controlled Clin. Trials*, **10**, 142–152.

10. Chakraborti, S. and Desu, M. M. (1988). Generalizations of Mathisen's median test for comparing several treatments with a control. *Commun. Statist. Simul.*, **17**, 947–967.

11. Chakraborti, S. and Desu, M. M. (1990). Quantile tests for comparing several treatments with a control under unequal right-censoring. *Biomed. J.*, **32**, 697–706.

12. Chakraborti, S. and Desu, M. M. (1991). Linear rank tests for comparing several treatments with a control when data are subject to unequal patterns of censorship. *Statist. Neerd.*, **45**, 227–254.

13. Cheung, S. H. and Holland, B. (1991). Extension of Dunnett's multiple comparison procedure to the case of several groups. *Biometrics*, **47**, 21–32.

14. Cheung, S. H. and Holland, B. (1992). Extension of Dunnett's multiple comparison procedure with differing sample sizes to the case of several groups. *Comput. Statist. Data Anal.*, **14**, 165–182.

15. Cohen, A. and Sackrowitz, B. (1992). Improved tests for comparing treatments against a control and other one-sided problems. *J. Am. Statist. Assoc.*, **87**, 1137–1144.

16. Conaway, M., Pillars, C., Robertson, T., and Sconing, J. (1991). A circular-cone test for testing homogeneity against a simple tree order. *Can. J. Statist.*, **19**, 283–296.

17. Dunnett, C. W. (1955). A multiple comparison procedure for comparing several treatments with a control. *J. Am. Statist. Assoc.*, **50**, 1096–1121.

18. Dunnett, C. W. (1964). New tables for multiple comparisons with a control. *Biometrics*, **20**, 482–491.

19. Dunnett, C. W. (1989). Multivariate normal probability integrals with product correlation structure. Algorithm AS251, *Appl. Statist.*, **38**, 564–579. Correction note, **42**, 709.

20. Dunnett, C. W. and Tamhane, A. C. (1991). Step-down multiple tests for comparing treatments with a control in unbalanced one-way layouts. *Statist. Med.*, **10**, 939–947.

21. Dunnett, C. W. and Tamhane, A. C. (1992). A step-up multiple test procedure. *J. Amer. Statist. Ass.*, **87**, 162–170.

22. Dunnett, C. W. and Tamhane, A. C. (1992). Comparisons between a new drug and active and placebo controls in an efficacy clinical trial. *Statist. Med.*, **11**, 1057–1063.

23. Dunnett, C. W. and Tamhane, A. C. (1995). Step-up multiple testing of parameters with unequally correlated estimates. *Biometrics*, **51**, 217–227.

24. Finney, D. J. (1978). *Statistical Methods in Biological Assay*, 3rd ed. Griffin, London and High Wycombe, UK.

25. Fligner, M. A. (1984). A note on two-sided distribution-free treatment versus control multiple comparisons. *J. Am. Statist. Assoc.*, **79**, 208–211.

26. Fligner, M. A. and Wolfe, D. A. (1982). Distribution-free tests for comparing several treatments with a control. *Statist. Neerl.*, **36**, 119–127.

27. Fung, K. Y. and Tam, H. (1988). Robust confidence intervals for comparing several treatment groups to a control group. *Statistician*, **37**, 387–399.

28. Hayter, A. J. and Liu, W. (1992). A method of power assessment for tests comparing several treatments with a control. *Commun. Statist. Simul.*, **21**, 1871–1889.

29. Hochberg, Y. (1988). A sharper Bonferroni procedure for multiple tests of significance. *Biometrika*, **75**, 800–802.

30. Hochberg, Y. and Tamhane, A. C. (1987). *Multiple Comparison Procedures*. Wiley, New York.

31. Holm, S. (1979). A simple sequentially rejective multiple test procedure. *Scand. J. Statist.*, **6**, 65–70.

32. Horn, M. and Vollandt, R. (1993). Sharpening subset selection of treatments better than a control. In *Multiple Comparisons, Selection, and Applications in Biometry*, F. M. Hoppe, ed., pp. 381–389. Marcel Dekker, New York.

33. Hothorn, L. (1989). On the behaviour of Fligner—Wolfe—Trend test "Control versus k treatments" with application in toxicology. *Biomed. J.*, **31**, 767–780.

34. Hothorn, L. and Lehmacher, W. (1991). A simple testing procedure "Control versus k treatments" for one-sided ordered alternatives, with application in toxicology. *Biomed. J.*, **33**, 179–189.

35. Hoover, D. R. (1991). Simultaneous comparisons of multiple treatments to two (or more) controls. *Biom. J.*, **33**, 913–921.

36. Laska, E. M. and Meisner, M. J. (1989). Testing whether an identified treatment is best. *Biometrics*, **45**, 1139–1151.

37. Marcus, R., Peritz, E., and Gabriel, K. R. (1976). On closed testing procedures with special reference to ordered analysis of variance. *Biometrika*, **63**, 655–660.

38. Miller, R. G., Jr. (1981). *Simultaneous Statistical Inference*, 2nd ed. McGraw Hill, New York.

39. Mukerjee, H., Robertson, T., and Wright, F. T. (1987). Comparison of several treatments with a control using multiple contrasts. *J. Am. Statist. Assoc.*, **82**, 902–910.

40. Redman, C. E. and Dunnett, C. W. (1994). Screening compounds for clinically active drugs. In *Statistics in the Pharmaceutical Industry*, 2nd ed., Chap. 24, pp. 529–546. Marcel Dekker, New York.

41. Robertson, T. and Wright, F. T. (1987). One-sided comparisons for treatments with a control. *Can. J. Statist.*, **13**, 109–122.

42. Roessler, E. B. (1946). Testing the significance of observations compared with a control. *Proc. Am. Soc. Hort. Sci.*, **47**, 249–251.

43. Rom. D. M., Costello, R. J., and Connell, L. T. (1994). On closed test procedures for dose-response analysis. *Statist. Med.*, **13**, 1583–1596.

44. Ruberg, S. J. (1989). Contrasts for identifying the minimum effective dose. *J. Am. Statist. Assoc.*, **84**, 816–822.

45. Rudolph, P. E. (1988). Robustness of multiple comparison procedures: treatment versus control. *Biomed. J.*, **30**, 41–45.

46. Shaffer, J. P. (1977). Multiple comparisons emphasizing selected contrasts: an extension and generalization of Dunnett's procedure. *Biometrics*, **33**, 293–303.

47. Spurrier, J. D. and Nizam, A. (1990). Sample size allocation for simultaneous inference in comparison with control experiments. *J. Am. Statist. Assoc.*, **85**, 181–186.

48. Tamhane, A. C., Hochberg, Y., and Dunnett, C. W. (1996). Multiple test procedures for dose finding. *Biometrics*, **52**, 21–37.

49. Tamhane, A. C. and Gopal, G. V. S. (1993). A Bayesian approach to comparing treatments with a control. In: *Multiple Comparisons, Selection, and Applications in Biometry*, F. M. Hoppe, ed., pp. 267–292. Marcel Dekker, New York.

50. Tukey, J. W., Ciminera, J. L., and Heyse, J. F. (1985). Testing the statistical certainty of a response to increasing doses of a drug. *Biometrics*, **41**, 295–301.

51. Williams, D. A. (1971). A test for differences between treatment means when several dose levels are compared with a zero dose level. *Biometrics*, **27**, 103–117.

52. Williams, D. A. (1972). The comparison of several dose levels with a zero dose control. *Biometrics*, **28**, 519–531.

16 Competing Risks

Mitchell H. Gail

The term "competing risks" applies to problems in which an object is exposed to two or more causes of failure. Such problems arise in public health, demography, actuarial science, industrial reliability applications, and experiments in medical therapeutics. More than three decades before Edward Jenner published his observations on inoculation against smallpox, Bernoulli [7] and d'Alembert [14] estimated the effect that eliminating smallpox would have on population survival rates. This is a classic competing risk problem, in which individuals are subject to multiple causes of death, including smallpox. Other examples follow. The actuary charged with designing a disability insurance plan must take into account the competing risks of death and disability. In offering joint life annuities, which cease payment as soon as either party dies, an insurance company must consider the competing risks of death of the first and second parties. An epidemiologist trying to assess the benefit of reducing exposure to an environmental carcinogen must analyze not only the reduced incidence rate of cancer but also effects on other competing causes of death. These examples illustrate the pervasive importance of competing-risk problems.

The actuarial literature on competing risks has been reviewed by Seal [35]. Statisticians interested in reliability theory or survival analysis have rediscovered many actuarial methods, emphasized certain theoretical dilemmas in competing-risk theory, and introduced new estimation procedures. The statistical literature has been reviewed by Chiang [11], Gail [23], David and Moeschberger [15], Prentice et al. [34] and Tsiatis [39].

16.1 Observables

We suppose that an individual is subject to m causes of death, and that for each individual we observe the time of death T and the cause of death J. This process is described in terms of the "cause-specific hazard functions"

$$g_j(t) = \lim_{\Delta t \downarrow 0} \Pr[t \leq T < t + \Delta t, J = j | T \geq t]/\Delta t. \qquad (1)$$

The term "cause-specific hazard function" is used by Prentice et al. [34]. These functions are identical to the "force of transition" functions in Aalen's Markov formulation of the competing-risk problem [1] and to the functions $g_j(t)$ in Gail [23]. Indeed, the functions $g_j(t)$ are the "forces of mortality" which an actuary would estimate from a multiple-decrement table and were termed "decremental forces" by the English actuary Makeham [31]. Assuming the existence of the quantities $g_j(\cdot)$, which we assume throughout, the hazard

$$\lambda(t) = \lim_{\Delta t \downarrow 0} \Pr[t \leq T < t + \Delta t | T \geq t]/\Delta t \quad (2)$$

satisfies

$$\lambda(t) = \sum_{j=1}^{m} g_j(t). \quad (3)$$

Prentice et al. [34] emphasize that only probabilities expressible as functions of $\{g_j(\cdot)\}$ may be estimated from the observable data (T, J). We define a probability to be *observable* if it can be expressed as a function of the cause-specific hazards corresponding to the original observations with all m risks acting. For example,

$$\begin{aligned} S_T(t) &\equiv \Pr[T > t] \\ &= \exp\left[-\int_0^t \lambda(u)du\right], \end{aligned}$$

the probability of surviving beyond time t, is observable. The conditional probability of dying of cause j in the interval $(\tau_{\Delta-1}, \tau_i]$, given $T > \tau_{i-1}$, is computed as

$$\begin{aligned} &Q(j;i) \\ &= [S_T(\tau_{i-1})]^{-1} \int_{\tau-1}^{\tau} g_j(u) S_T(u)\, du. \end{aligned} \quad (4)$$

This probability is a function of $\{g_j(\cdot)\}$ and is, by definition, observable. The probabilities $Q(j;i)$ are termed "crude" probabilities by Chiang [11] because the events represented occur with all the m original risks acting.

The conditional probabilities $Q(j;i)$ are the basic parameters of a multiple-decrement life table. If the positive time axis is partitioned into intervals $(\tau_{i-1}, \tau_i]$, for $i = 1, 2, \ldots, I$, with $\tau_0 = 0$, then $Q(j;i)$ gives the conditional probability of dying of cause j in interval i given $T > \tau_{i-1}$. Such a partitioning arises naturally in actuarial problems in which only the time interval and cause of death are recorded, and multiple-decrement life tables also arise when exact times to death T are grouped. The conditional probability of surviving interval i, $\rho_i = 1 - \sum_{j=1}^{m} Q(j;i)$, may be used to calculate such quantities as $S_T(\tau_i) = \prod_{l=1}^{i} \rho_l$, which is the probability of surviving beyond τ_i, and $\sum_{i=1}^{k} S_T(\tau_{i-1})Q(j;i)$, which is probability of dying of risk j in the interval $(0, \tau_k]$. Such quantities are observable because they are functions of the observable probabilities $Q(j;i)$. The maximum likelihood estimate of $Q(j;i)$ is $\hat{Q}(j,i) = d_{ji}/n_i$, where d_{ji} is the number who die of risk j in time interval i, and n_i is the number alive at τ_{i-1}. The corresponding estimate of ρ_i is $\hat{\rho}_i = s_i/n_i$, where $s_i = n_i - \sum_j d_{ji}$ survive interval i. Maximum-likelihood estimates of other observable probabilities related to the multiple-decrement life table are obtained from $\hat{Q}(j;i)$ and $\hat{\rho}_i$.

16.2 Beyond Observables

Observable probabilities can be thought of as arising in the original observational setting with all risks acting. Many interesting probabilities arise when one tries to predict what would happen if one or more risks were eliminated. For example, Chiang [11] defines the "net" probability q_{ji} to be the conditional probability of dying in interval $(\tau_{i-1}, \tau_i]$ given $T > \tau_{i-1}$ when only risk j is acting, and he defines $Q_{ji \cdot \delta}$ to be the corresponding "partial crude" probability of dying of risk j when risk δ has been eliminated. Such probabilities are of great practical interest, yet they are not observable because they are not functions of $\{g_j(\cdot)\}$, as estimated from the original experiment. Such calculations depend on assumed models of the competing-risk problem.

Suppose that each individual has m failure times $T_1, T_2, \ldots, T_m$ corresponding to risks $1, 2, \ldots, m$. Only $T = \min_{1 \leq l \leq m} T_l$

and the risk j such that $T_j = T$ are observed. This is termed the latent-failure time model. To use this model to compute nonobservable probabilities, a joint survival distribution $S(t_1, \ldots, t_m) = P[T_1 > t_1, \ldots, T_m > t_m]$ must be assumed. The distribution of T is

$$S_T(t) = S(t, t, \ldots, t). \quad (5)$$

Whenever the following partial derivatives exist along the equiangular line $t_1 = t_2 = \cdots = t_m$, the cause-specific hazards are given by

$$g_j(t) = -\left(\frac{\partial \ln S(t_1, \ldots, t_m)}{\partial t_j}\right)_{t_1 = t_2 = \ldots t_m = t} \quad (6)$$

as in Gail [23]. From its definition in (2), the hazard $\lambda(t)$ satisfies

$$\lambda(t) = -d \ln S_T(t)/dt, \quad (7)$$

and (3) can be derived by applying the chain rule to $-\ln S(t_1, t_2, \ldots, t_m)$. Gail [23] points out that (3) may be invalid if $S(t_1, \ldots, t_m)$ is singular, so that $\Pr[T_i = T_j] > 0$ for some $i \neq j$.

Suppose that $T_1, T_2, \ldots, T_m$ are independent with marginal survival distributions $S_j(t) = \Pr[T_j > t]$ and corresponding marginal hazards $\lambda_j(t) = -d \ln S_j(t)/dt$. Under independence, $S(t_1, \ldots, t_m) = \prod S_j(t_j)$, from which it follows that

$$\lambda_j(t) = g_j(t) \quad \text{for } j = 1, 2, \ldots, J. \quad (8)$$

We note parenthetically that the converse is false. For suppose that the density of T_1 and T_2 is constant on the triangular region with vertices $(0, 0)$, $(1, 0)$, and $(0, 1)$. Then, from an example in Gail [23], $g_1 = g_2 = \lambda_1 = \lambda_2 = 1/(1-t)$, yet T_1 and T_2 are dependent. The relationships (8) allow one to compute nonobservable probabilities as follows. First, estimate $g_j(\cdot)$ and hence $\lambda_j(\cdot)$ from the original data. Second, invoke Makeham's assumption [31] that

the effect of eliminating cause J is to nullify $\lambda_j(\cdot)$ without altering other marginal hazards. Taking $m = 3$ risks for ease of exposition, the net probability of surviving exposure to risk 1 alone until time t is $S_1(t) = S(t, 0, 0) = \exp[-\int_0^t \lambda_1(u) du]$, and the partial crude probability of dying of risk 3 in $[0, t]$ with risk 1 eliminated is $\int_0^t \lambda_3(v) S_2(v) S_3(v) dv = \int_0^t \lambda_3(v) S(0, v, v) dv$.

Three important assumptions are made in the previous calculations:

A1. A structure for $S(t_1, \ldots, t_m)$ is assumed.

A2. It is assumed that the effect of eliminating a set of risks is known and expressible in terms of $S(t_1, \ldots, t_m)$.

A3. It is assumed that the experimental procedures used to eliminate a risk will only produce the changes specified in assumption A2 without otherwise altering $S(t_1, \ldots, t_m)$.

Makeham [31] questioned assumption A3 in his discussion of smallpox. It seems self-evident, for example, that a campaign to reduce lung cancer by banning cigarettes will have wide-ranging effects on other health hazards. We turn now to an examination of assumptions A1 and A2.

In the previous case of independence, the net survival distribution for risk 1 was taken to be the marginal distribution $S_1(t) = S(t, 0, 0)$. One might generalize this result to assume that the effect of eliminating risk j is to nullify the corresponding argument in $S(t_1, t_2, \ldots, t_m)$. This is an example of assumption A2 for modeling elimination of a risk. The implications of this viewpoint have been discussed by Gail [23], who shows how various observable and nonobservable probabilities may be computed given $S(t_1, \ldots, t_m)$ and given this version of assumption A2. These methods may be used whether

$T_1, T_2, \ldots, T_m$ are independent or dependent. Elandt-Johnson [17] gives an alternative preferable model for eliminating risks. For example, she asserts that the appropriate net survival distribution in the previous case is the limiting conditional distribution

$$\lim_{t_2, t_3 \to \infty} \Pr[T_1 > t | T_2 = t_2, T_3 = t_3]. \quad (9)$$

Another similar version of assumption A2 might be to take

$$\lim_{t_2, t_3 \to \infty} \Pr[T_1 > t | T_2 > t_2, T_3 > t_3] \quad (10)$$

as the net survival distribution from risk 1. If T_1, T_2, and T_3 are independent, the net survival distributions (9), (10), and $S_1(t) = S(t, 0, 0)$ are equivalent. Thus under independence it is reasonable to take the net survival distributions as corresponding marginal distributions of $S(t_1, \ldots, t_m)$ and, more generally, to model elimination of risk j by nullifying the jth argument in $S(t_1, t_2, \ldots, t_m)$. With dependent models, however, a good argument can be made for modeling elimination of risks as in (9) or (10). Some assumptions A2 defining the effect of elimination are inevitably required.

Although $S(t_1, \ldots, t_m)$ defines $\{g_j(\cdot)\}$ and hence all observable probabilities, the observables (T, J) do not uniquely define $S(t_1, t_2, \ldots, t_m)$. Suppose that the data are so numerous as to permit perfect estimation of $\{g_j(\cdot)\}$. Whatever the distribution of $T_1, T_2, \ldots, T_m$, define a new set of independent random variables $T_1^*, T_2^*, \ldots, T_m^*$ with marginal hazards $\lambda_j^*(t) \equiv g_j(t)$ and marginal distributions

$$\begin{aligned} S_j^*(t) &= \Pr[T_j^* > t] \\ &= \exp\left[-\int_0^t \lambda_j^*(u) du\right]. \end{aligned}$$

It is clear that the distribution $S^*(t_1, t_2, \ldots, t_m) = \prod_{j=1}^m S_j^*(t_j)$ has the same cause specific hazards $g_j^*(t) = \lambda_j^*(t) \equiv g_j(t)$ as the original distribution $S(t_1, t_2, \ldots, t_m)$. Thus even if the data are so complete as to specify $\{g_j(\cdot)\}$ exactly, they do not define $S(t_1, t_2, \ldots, t_m)$ uniquely. This conundrum was noted by Cox [12] and termed nonidentifiability by Tsiatis [38], who gave a more formal construction.

Nonidentifiability has two important practical implications. First, assumption A1 specifying the structure $S(t_1, t_2, \ldots, t_m)$ cannot be verified empirically. Second, one can estimate $\{g_j(\cdot)\}$ from any distribution $S(t_1, \ldots, t_m)$ by pretending that the independent process $S^*(t_1, \ldots, t_m)$ is operative and using methods for estimating the marginal hazards $\{\lambda_j^*(\cdot)\}$ corresponding to independent random variables $T_1^*, \ldots, T_m^*$. These estimation methods for independent processes are discussed in the next section.

It is apparent that estimates of nonobservable probabilities, which must depend on assumptions A1, A2, and A3, are suspect.

16.3 Estimation Assuming Independent $T_1, T_2, \ldots, T_M$

First we consider parametric estimation in which each marginal distribution $S_j(t; \theta_j) = \Pr[T_j > t]$ depends on parameters θ_j. Let $\{t_{ij}\}$ denote the times of death of the d_j individuals dying of cause j for $J = 1, 2, \ldots, m$ and $i = 1, 2, \ldots, d_j$, and let $\{t_l^*\}$ denote the s follow-up times of those who survive until follow-up ceases. Then the likelihood is given by

$$\left[\prod_{j=1}^m \prod_{i=1}^{d_j} \lambda_j(t_{ij}) S_T(t_{ij})\right] \prod_{l=1}^s S_T(t_l^*), \quad (11)$$

where $S_T(t)$ is obtained from (5) and $\lambda_j(t) = -d \ln S_j(t; \theta_j)/dt$. David and Moeschberger [15] give a detailed account for exponential, Weibull, normal, and Gompertz marginal survival distributions, and they give a similar likelihood for grouped data. Once $\{\theta_j\}$ have been estimated, estimates of observable and nonobservable probabilities may be obtained from $\hat{S}(t_1,\ldots,t_m) = \prod S_j(t_j;\hat{\theta}_j)$. The likelihood (11) assumes that the distribution of any right censoring that occurs because follow-up ceases is functionally independent of the parameters governing $S(t_1,\ldots,t_m)$.

Estimates based on the multiple-decrement life table are essentially nonparametric. The quantities $\hat{Q}(j;1) = d_{ji}/n_i$ and $\hat{\rho}_i = 1 - \sum_{j=1}^{m}\hat{Q}(j;i)$ defined in the preceding section may be used to estimate observable probabilities. To estimate the net conditional probability of surviving interval i with risk j alone present, $p_{ji} = S_j(\tau_i)/S_j(\tau_{i-1})$, we invoke the piecewise proportional hazards model

$$\lambda_j(t) = \omega_{ji}\lambda(t) \quad \text{for } t \in (\tau_{i-1},\tau_i), \quad (12)$$

where $\omega_{ji} \geq 0$ and $\sum_j \omega_{ji} = 1$. This assumption was used by Greville [26] and Chiang [11] and is less restrictive than the assumption that the marginal hazards are constant on time intervals introduced in the actuarial method. The relation (12) will hold for fine-enough intervals, provided that $\{\lambda_j(\cdot)\}$ are continuous. It follows from (12) that $\omega_{ji} = Q(j;i)/(1-\rho_i)$, which yields the estimate $\hat{\omega}_{ji} = d_{ji}/\sum_{j=1}^{m}d_{ji}$. Also, (12) implies that

$$p_{ji} = \rho_i^{\omega_{ji}}. \quad (13)$$

Hence the net conditional probability may be estimated from $\hat{p}_{ji} = (s_i/n_i)^{\hat{\omega}_{ji}}$. The corresponding estimate of $S_j(\tau_i)$ is $\prod_{i=1}^{i}\hat{p}_{ji}$, from which other competing risk calculations follow. Gail [23] shows that the actuarial estimate $p_{ji}^* = d_{ji}/(n_i - \sum_{l\neq j}d_{li}/2)$ is an excellent approximation to $\hat{p}_{ji}$.

A fully nonparametric approach to estimation of $S_j(t)$ under independence is outlined by Kaplan and Meier [28], who refine the actuarial method to produce a separate interval at the time of each death. They credit this approach to Böhmer [8]. The resulting product limit estimate of $S_j(t)$ is

$$S_j^{PL}(t) = \prod_r [1 - d_j(t_r)/n(t_r)], \quad (14)$$

where r indexes the distinct times $t_r \leq t$ at which deaths from cause j occur, $d_j(t_r)$ is the number of such deaths at t_r, and $n(t_r)$ is the number known to be at risk at t_r. The asymptotic theory has been studied by Breslow and Crowley [10], who proved that $S_j^{PL}(t)$ converges to a Gaussian process after normalization. Aalen [1,2] and Fleming [21,22] generalized these ideas to estimate the partial crude probabilities $p_j(t;A)$ of death from cause j in $[0,t]$ when an index set of risks $A \subset \{1,2,\ldots,m\}$ defines the only risks present. For the special case $A = \{j\}$, $P_j(t;A) = 1 - S_j(t)$. For the case $A = \{1,2,\ldots,m\}$, $P_i\{t,A\}$ is the crude probability given by (4) with $\tau_{i-1} = 0$ and $\tau_i = t$. Aalen's estimates $\hat{P}_j(t,A)$ are uniformly consistent, have bias tending to zero at an exponential rate, and tend jointly to a Gaussian process after normalization.

In this section we have emphasized estimation. Parametric and nonparametric methods for testing the equality of two or more net survival curves are found in the statistical literature on survival analysis. Suppose that each individual in a clinical trial has a death time T_1 and a censorship time T_2, indicating when his or her followup ends. If patients are assigned to treatments 1 or 2 with net survival curves $S_1^{(1)}(t)$ and $S_1^{(2)}(t)$, respectively, treatments are compared by testing the equality $S_1^{(1)}(t) = S_1^{(2)}(t)$. Such tests are made under the "random censorship" assumption that T_1 and T_2 are independent. Pertinent references are cited in Breslow [9].

To summarize, standard parametric methods and nonparametric extensions of the life-table method are available under the independence assumption. One can estimate both observable and nonobservable probabilities using these methods, and one can test equality of net survival curves from different treatment groups, but the unverifiable independence assumption is crucial.

16.4 Estimation Assuming Dependent $T_1, T_2, \ldots, T_m$

The nonparametric methods just described may be used to estimate observable probabilities even when $T_1, T_2, \ldots, T_m$ are dependent because the functions $g_j(t)$ may be regarded as marginal hazards $\lambda_j^*(t)$ from an independent process $T_1^*, \ldots, T_m^*$ as mentioned in Section 16.2. Thus crude probabilities such as $P_j(t, A)$ with $A = \{1, 2, \ldots, m\}$ may be estimated by pretending that the death times are independent and using the methods of Aalen [2] or related actuarial methods.

In contrast, nonobservable probabilities, such as net or partial crude probabilities, depend on a hypothesized structure $S(t_1, \ldots, t_m)$ and cannot be estimated as in the preceding section. Once this joint distribution has been estimated, the computation of nonobservable probabilities proceeds as in the second section. If a specific parametric form $S(t_1, \ldots, t_m; \theta)$ is posited, θ may be estimated from the likelihood (11) with $g_j(t_{ij})$ replacing $\lambda_j(t_{ij})$, provided that $S(t_1, \ldots, t_m)$ is absolutely continuous. David and Moeschberger [15] discuss the bivariate normal model and a bivariate exponential model proposed by Downton [16], as well as the bivariate exponential model of Marshall and Olkin [32], which is singular.

Peterson [33] obtained bounds on $S(t_1, t_2)$, $S(t, 0)$, and $S(0, t)$ when the observable probabilities $\Pr[T_1 > t, T_1 < T_2]$ and $\Pr[T_2 > t, T_2 < T_1]$ are known. These bounds were derived without special assumptions on the structure $S(t_1, t_2)$, but the bounds are too wide to be useful. Narrower bands, such as those in Slud and Rubinstein [37], require assumptions that cannot be tested with competing risk data.

16.5 Absolute Risks

The crude probability that a person alive at age a will die of a cause 1 in the interval $[a, a + \tau]$ is

$$Q_1(a, \tau) = \int_a^{a+\tau} g_1(t) \exp\left\{-\int_a^t \lambda(u)du\right\}dt.$$

This quantity is sometimes called the "absolute risk" or "cumulative risk" or the "cumulative incidence function". The absolute risk is observable [see Eq. (4)] and is of great interest in applications to disease prevention and to managing patients after the diagnosis of a serious chronic disease. For example, a woman may decide to take a drug like tamoxifen to prevent breast cancer if her absolute risk of breast cancer is high enough that the benefits of preventive intervention outweigh the risks. A 65-year old man just diagnosed with prostate cancer may decide that no treatment is necessary if his absolute risk of dying of prostate cancer is small because he will probably die of other causes first.

Gray [25] devised tests of the quality of several cumulative incidence functions, $Q_1(0, \tau | X = k)$ for $k = 1, 2, \ldots, K$ independent treatment groups, for example. More generally, it is of interest to model $Q_1(0, \tau | X)$ in terms of a set of covariates, X. For example, to model the absolute risk of breast cancer, X might include factors such as age at menache and number of relatives with breast cancer. One approach is to model the cause-specific hazards as $g_{j0}(t) \exp(\beta_j^T X_j)$, where X_j contains covariates associated with cause j and $g_{j0}(t)$ is the baseline cause-specific hazard corresponding to $X_j = 0$. This cause-specific model has several advantages. First, the effects β_j have a familiar interpretation as log relative risk parameters for the cause-specific hazards. Second, if cohort data are available and if any censoring apart from the K events of interest acts independently [39], standard survival methods can be used to estimate $\{\beta_j\}$ and the cumulative cause-specific baseline hazards, $\{\int_0^t g_{j0}(u)du\}$. Combining these estimates to estimate $Q_1(0,\tau|X)$ and computing the variance of $\hat{Q}_1(0,\tau|X)$ requires additional calculations [5]. Third, estimates $\hat{Q}_1(0,\tau|X)$ and corresponding methods of inference are available for case–cohort data [36], for case–control studies nested within a cohort [30], and for combining population-based case–control data with population registry data. See Gail [24] for discussion of these designs and of family-based designs for estimating absolute risk under the cause-specific hazards model. Fine and Gray [18] developed alternative models of the type $Q_1(0,\tau|X) = \psi^{-1}\{h_0(\tau) + \beta^T X\}$, where ψ is a known increasing function and h_0 is an arbitrary unknown invertible monotone increasing function. Estimation under this model requires cohort data and can be very complex in the presence of censoring. Moreover, if a component of $\hat{\beta}$ is positive, one does not know whether the effect of a unit increase in the corresponding component of X is to increase the cause-specific hazard of cause 1 or to decrease one or more of the competing cause-specific hazards.

16.6 Markov Model and Extensions

Aalen [1] modeled the classical competing risk problem as a continuous-time Markov process with one alive state $j = 0$ and m absorbing death states $\{1, 2, \ldots, m\}$. The only permissible transitions in this model are $0 \to j$ for $j = 1, 2, \ldots, m$, and the corresponding "forces of transition" are the functions $g_j(t)$ in (1). It is further supposed that elimination of risk j merely nullfies $g_j(t)$. This model of risk elimination, which was adopted by Makeham [31], is equivalent to the latent-failure-time model with independent latent failure times (we call this the ILFT model) and marginal hazards $\lambda_j(t) = g_j(t)$. To see this, let $p(t,j)$ be the probability of being in state j at time t. The governing differential equations for the Markov process are $dp(t,0)/dt = -\sum_{l=1}^m g_l(t)$ and $dp(t,j)/dt = p(t,0)g_j(t)$ for $j = 1, 2, \ldots, m$. Hence

$$p(t,0) = \exp\left[-\int_0^t \sum_{l=1}^m g_l(u)du\right]$$

and $p(t,j) = \int_0^t p(u,0)g_j(u)du$. In terms of the ILFT model with $\lambda_j(t) = g_j(t)$, these probabilities are $S_T(t)$ and $\int_0^t S_T(u)\lambda_j(u)du$, respectively. In the Markov formulation with $m = 3$ risks, the probability of being in state 0 at time t when risks 2 and 3 have been eliminated is seen to be

$$\exp\left[-\int_0^t g_1(u)du\right]$$

by solution of the first differential equation above with $g_2(t) = g_3(t) = 0$. This is, however, precisely equal to $S_1(t) = S(t,0,0) = \exp\left[-\int_0^t \lambda_1(u)du\right]$ in the ILFT model. Extensions of these arguments show that the Markov method for competing risks is entirely equivalent to the ILFT model for the classical multiple-decrement problem.

However, if one considers more general transition structures, one is led to a new class of problems that admit new analytical approaches because more data are available. For example, suppose that following cancer surgery a patient is in a cancer-free state (state 0), a state with recurrent cancer (state 1), or the death state (2), and suppose only the transitions $0 \to 1, 1 \to 2$, and $0 \to 2$ are possible. In a patient who follows the path $0 \to 1 \to 2$, one can observe transition times t_{01}, and t_{12} but not t_{02}. In one who dies directly from state 0, only t_{02} is observable. Such data allow us to answer such questions as: "Is the risk of death at time t higher in a patient with recurrent disease than in a patient in state 0?" The essential feature of these models is the inclusion of intermediate nonabsorbing states, such as state 1 above. The work of Fix and Neyman [20], Chiang [11], and Hoem [27] assumes that a stationary Markov process governs transitions among states. Nonparametric methods of Fleming [21,22], Aalen [3], and Aalen and Johansen [4] allow one to estimate transition probabilities even when the Markov process is not homogeneous. Berlin et al. [6] discuss a Markov model for analyzing animal carcinogenesis experiments. The nonabsorbing intermediate states in this model are defined by the presence or absence of certain diseases which can only be diagnosed at autopsy. Thus one does not know the state of an animal except at the time of its death. To surmount this difficulty, which is not present in the applications treated by Fix and Neyman [20] or Chiang [11], the experiment is designed to obtain additional data by sacrificing animals serially. These extensions of the simple Markov model for multiple-decrement life tables, and the results of Prentice et al. [34], Lagakos et al. [29], and Fine et al. [19] indicate the variety of methods that may be of use when the competing-risk structure is relaxed to include additional data.

References

1. Aalen, O. (1976). *Scand. J. Statist.*, **3**, 15–27.

2. Aalen, O. (1978). *Ann. Statist.*, **6**, 534–545.

3. Aalen, O. (1978). *Ann. Statist.*, **6**, 701–726.

4. Aalen, O. and Johansen, S. (1977). *An Empirical Transition Matrix for Nonhomogeneous Markov Chains Based on Censored Observations*. Preprint No. 6, Institute of Mathematical Statistics, University of Copenhagen.

5. Benichou, J. and Gail, M. H. (1990). *Biometrics*, **46**, 813–826.

6. Berlin, B., Brodsky, J., and Clifford, P. (1979). *J. Am. Statist. Assoc.*, **74**, 5–14.

7. Bernoulli, D. (1760). Essai d'une nouvelle analyse de la mortalité causeé par la petite vérole, et des avantages de l'inoculation pour le prévenir. *Historie avec les Mémoirs*, pp. 1–45. Académie Royale des Sciences, Paris.

8. Böhmer, P. E. (1912). Theorie der unabhängigen Wahrscheinlichkeiten. Rapport. *Mémories et Procès-verbaux de Septième Congrès International d'Actuaires*, Vol. 2, pp. 327–346. Amsterdam.

9. Breslow, N. (1970). *Biometrika*, **57**, 579–594.

10. Breslow, N. and Crowley, J. (1974). *Ann. Statist.*, **2**, 437–453.

11. Chiang, C. L. (1968). *Introduction to Stochastic Processes in Biostatistics*, Chap. 11. Wiley, New York.

12. Cox, D. R. (1959). *J. Roy. Statist. Soc. B*, **21**, 411–421.

13. Crowley, J. (1973). *Nonparametric Analysis of Censored Survival Data with Distribution Theory for the k-Sample Generalized Savage Statistic*. Ph.D. dissertation, University of Washington.

14. D'Alembert, J. L. R. (1761). Onzième mémoire: Sur l'application du calcul des probabilités à l'inoculation de la petite vérole. *Opusc. Math.*, **2**, 26–95.

15. David, H. A. and Moeschberger, M. L. (1978). *The Theory of Competing Risks*. Griffin's Statistical Monograph No. 39, Macmillan, New York.

16. Downton, F. (1970). *J. Roy. Statist. Soc. B*, **32**, 408–417.

17. Elandt-Johnson, R. C. (1976). *Scand. Actuarial J.*, **59**, 37–51.
18. Fine, J. P. and Gray, R. J.(1999). *J. Am. Statist. Assoc.*, **94**, 496–509.
19. Fine, J. P., Jiang, H., and Chappell, fnmR. (2001). *Biometrika*, **88**, 907–919.
20. Fix, E. and Neyman, J. (1951). *Hum. Biol.*, **23**, 205–241.
21. Fleming, T. R. (1978). *Ann. Statist.*, **6**, 1057–1070.
22. Fleming, T. R. (1978). *Ann. Statist.*, **6**, 1071–1079.
23. Gail, M. (1975). *Biometrics*, **31**, 209–222.
24. Gail, M. (2008). *Lifetime Data Anal.*, **14**, 18–36.
25. Gray, R. J. (1988). *Ann. Statist.*, **16**, 1141–1154.
26. Greville, T. N. E. (1948). *Rec. Am. Inst. Actuaries*, **37**, 283–294.
27. Hoem, J. M. (1971). *J. Roy. Statist. Soc. B*, **33**, 275–289.
28. Kaplan, E. L. and Meier, P. (1958). *J. Am. Statist. Assoc.*, **53**, 457–481.
29. Lagakos, S. W., Sommer, C. J., and Zelen, M. (1978). *Biometrika*, **65**, 311–317.
30. Langholz, B. and Borgan, O. (2003). *Biometrics*, **53**, 767–774.
31. Makeham, W. M. (1874). *J. Inst. Actuaries*, **18**, 317–322.
32. Marshall, A. W. and Olkin, I. (1967). *J. Am. Statist. Assoc.*, **62**, 3–44.
33. Peterson, A. V. (1976). *Proc. Natl. Acad. Sci. (USA)*, **73**, 11–13.
34. Prentice, R. L., Kalbfleisch, J. D., Peterson, A. V., Jr., Flournoy, N., Farewell, V. T., and Breslow, N. E. (1978). *Biometrics*, **34**, 541–554.
35. Seal, H. L. (1977). *Biometrika*, **64**, 429–439.
36. Self, S. G. and Prentice, R. L.(1988). *Ann. Statist.*, **16**, 64–81.
37. Slud, E. V. and Rubinstein, L. V. (1983). *Biometrika*, **70**, 643–649.
38. Tsiatis, A. (1975). *Proc. Natl. Acad. Sci. (USA)*, **72**, 20–22.
39. Tsiatis, A. (2005). Competing Risks. In Encyclopedia of Biostatistics, Wiley. Chichester, UK.

17 Countermatched Sampling

Ørnulf Borgan

Countermatching is a novel design for stratified sampling of controls in epidemiological case–control studies. It is a generalization of nested case–control sampling and will often give an efficiency gain over that classical design.

Countermatched sampling and the nested case–control design are closely related to Cox's regression model for failure-time data. This model relates the vector of covariates $\mathbf{x}_i(t) = (x_{i1}(t), \ldots, x_{ip}(t))'$ at time t for an individual i to its hazard rate function $\lambda_i(t)$ by the equation

$$\lambda_i(t) = \lambda_0(t) e^{\boldsymbol{\beta}' \mathbf{x}_i(t)}. \tag{1}$$

Here $\boldsymbol{\beta} = (\beta_1, \ldots, \beta_p)'$ is a vector of regression coefficients, while the baseline hazard rate function $\lambda_0(t)$ is left unspecified. Estimation in Cox's model is based on a partial likelihood that, at each failure time, compares the covariate values of the failing individual to those of all individuals at risk at the time of the failure. In large epidemiological cohort studies of a rare disease, Cox regression requires the collection of information on exposure variables and other covariates of interest for all individuals in the cohort, even though only a small fraction of these actually get diseased. This may be very expensive, or even logistically impossible.

Nested case–control studies, in which covariate information is needed only for each failing individual ("case") and a small number of controls selected from those at risk at the time of the failure, may give a substantial reduction in the resources required for a study. Moreover, as most of the statistical information is contained in the cases, a nested case–control study may still be sufficient to give reliable answers to the questions of main interest.

In the classical form of a case–control study nested within a cohort, the controls are selected by simple random sampling. Often some information is available for all cohort members, e.g., a surrogate measure of exposure, such as the type of work or duration of employment, may be available for everyone. Langholz and Borgan [3] have developed a stratified version of the simple nested case–control design which makes it possible to incorporate such information into the sampling process in order to obtain a more informative sample of controls. For this design, called *countermatching*, one applies the additional information on the cohort subjects to classify each individual at risk into one of L strata, say.

Then at each failure time t_j, one samples randomly without replacement m_l controls from the $n_l(t_j)$ at risk in stratum l, except for the case's stratum, where only $m_l - 1$ controls are sampled. The failing individual i_j is, however, included in the sampled risk set $\tilde{\mathcal{R}}(t_j)$, so this contains a total of m_l from each stratum $l = 1, 2, \ldots, L$. In particular, for $L = 2$ and $m_1 = m_2 = 1$, the single control is selected from the opposite stratum of the case. Thus countermatching is, as the name suggests, essentially the opposite of matching, where the case and its controls are from the same stratum.

Inference from countermatched data concerning β in (1) can be based on the partial likelihood

$$L(\beta) = \prod_{t_j} \frac{e^{\beta' x_{i_j}(t_j)} w_{i_j}(t_j)}{\sum_{k \in \tilde{\mathcal{R}}(t_j)} e^{\beta' x_k(t_j)} w_k(t_j)}, \quad (2)$$

using the usual large-sample likelihood methods [1,3]. Here $w_k(t_j) = n_l(t_j)/m_l$ if individual k belongs to stratum l at time t_j. The partial likelihood (2) is similar to Oakes' partial likelihood [6] for simple nested case–control data. But the contribution of each individual, including the case, has to be weighted by the reciprocal of the proportion sampled from the individual's stratum in order to compensate for the different sampling probabilities in the strata. The cumulative baseline hazard rate function $\Lambda_0(t) = \int_0^t \lambda_0(u) \, du$ can be estimated [1] by

$$\hat{\Lambda}_0(t) = \sum_{t_j \leq t} \frac{1}{\sum_{k \in \tilde{\mathcal{R}}(t_j)} e^{\hat{\beta}' x_k(t_j)} w_k(t_j)}, \quad (3)$$

where $\hat{\beta}$ is the maximum partial likelihood estimator maximizing (2). The estimator (3) is also similar to the one used for nested case–control data.

Countermatching may give an appreciable improvement in statistical efficiency for estimation of a regression coefficient of particular importance compared to simple nested case–control sampling. Intuitively this is achieved by increasing the variation in the covariate of interest within each sampled risk set. The efficiency gain has been documented both by calculations of asymptotic relative efficiency [3,4,5] and by Steenland and Deddens' study of a cohort of gold miners [7]. For the latter, a countermatched design (with stratification based on duration of exposure) using three controls per case had the same statistical efficiency for estimating the effect of exposure to crystalline silica as a simple nested case–control study using 10 controls. According to preliminary investigations by the author of this entry, a similar increase in efficiency is not seen for the estimator (3). One important reason for this is that, for estimation of the baseline hazard rate function, even a nested case–control study has quite high efficiency compared to the full cohort.

The idea of countermatching originated in the middle of the 1990s and is rather new at the time of writing (1997). It has therefore not yet been put into practical use. But it has attracted positive interest from researchers in epidemiology [2,7], and it is quite likely to be a useful design for future epidemiological studies.

References

1. Borgan, Ø., Goldstein, L., and Langholz, B. (1995). Methods for the analysis of sampled cohort data in the Cox proportional hazards model. *Ann. Statist.*, **23**, 1749–1778. (The paper uses marked point processes to describe a general framework for risk set sampling designs, including simple nested case–control sampling and countermatched sampling as special cases. Large-sample properties of the estimators of the regression coefficients and the cumulative baseline hazard rate function are studied using counting process and martingale theory.)

2. Cologne, J. B. (1997). Counterintuitive matching. *Epidemiology*, **8**, 227–229. (Invited editorial advocating the use of countermatching for an epidemiological audience.)

3. Langholz, B. and Borgan, Ø. (1995). Countermatching: A stratified nested case–control sampling method. *Biometrika*, **82**, 69–79. (The basic paper where the concept of countermatching was introduced and studied. Comparisons with simple nested case–control sampling are also provided.)

4. Langholz, B. and Clayton, D. (1994). Sampling strategies in nested case–control studies. *Environ. Health Perspect.*, **102**(Suppl. 8), 47–51. (A nontechnical paper that discusses a number of practical situations where countermatching may be useful. Comparisons with simple nested case–control sampling are also provided.)

5. Langholz, B. and Goldstein, L. (1996). Risk set sampling in epidemiologic cohort studies. *Statist. Sci.*, **11**, 35–53. (The paper reviews a broad variety of risk set sampling designs, including simple nested case–control sampling and countermatched sampling, and their suitability for different design and analysis problems from epidemiological research.)

6. Oakes, D. (1981). Survival times: Aspects of partial likelihood (with discussion). *Int. Statist. Rev.*, **49**, 235–264.

7. Steenland, K. and Deddens, J. A. (1997). Increased precision using countermatching in nested case–control studies. *Epidemiology*, **8**, 238–242. (An applied paper that compares countermatching with simple nested case–control sampling for a real dataset.)

18 Counting Processes

Neils Keiding

18.1 Introduction

A counting process N on the positive half line $[0, \infty)$ is a stochastic process $(N(t), t \in [0, \infty))$ with $N(0) \equiv 0$ and whose sample paths are (almost surely) step functions with steps $+1$. Probabilistically, a counting process is just one representation of a stochastic point process. The present entry focuses on counting-process-based statistical models for event-history analysis, using martingales, stochastic integrals, and product–integrals. The exposition here is based on the monograph [5], which besides the full mathematical regularity conditions also contains a large set of worked practical examples.

In event history analysis individuals are assumed to move between states. Simple cases include survival analysis, with the two states "alive" and "dead" and transition only possible from "alive" to "dead"; competing-risk models with several types of failure, corresponding to transitions from "alive" to "dead of cause i" for $i = 1, \ldots, j$; and illness–death or disability models, usually with three states; transitions are allowed back and forth between "healthy" and "diseased" and from each of these to "dead."

18.2 Multivariate Counting Processes and Martingales

The transitions between each pair of states as just described are counted by a multivariate counting process $\boldsymbol{N} = (\boldsymbol{N}(t) = ((N_1(t), \ldots, N_k(t)), t \in \mathcal{T})$ defined on a measurable space $(\Omega, \mathcal{F})$ where $\mathcal{T}$ is an interval of the form $[0, \tau]$ or $[0, \tau]$. Each component of $\boldsymbol{N}$ is a univariate counting process as defined above, and with probability one, no two components may jump simultaneously.

On $(\Omega, \mathcal{F})$ a filtration $(\mathcal{F}_t, t \in \mathcal{T})$ is given; this is specified as a family of σ-algebras that is both

Increasing: $s < t \Rightarrow \mathcal{F}_s \subset \mathcal{F}_t$;

Right-continuous:
$$\mathcal{F}_s = \bigcap_{t \geq s} \mathcal{F}_t \text{ for all } s.$$

The filtration is interpreted as recording the "history" of the process. The development in time of a multivariate counting process is assumed to be governed by its (random) intensity process $\boldsymbol{\lambda} = (\boldsymbol{\lambda}(t), t \in \mathcal{T})$, where $\boldsymbol{\lambda}(t) = (\lambda_1(t), \ldots, \lambda_k(t))$ and $\lambda_h(t)dt$ is, heuristically, the conditional probability of a jump of N_h in $[t, t + dt)$ given the "history" $\mathcal{F}_{t-}$ up to but not including t.

The mathematical rigorization of this formulation is based on noting that each component N_h is an increasing right-continuous process and therefore a local submartingale allowing a *compensator* $(\Lambda_h, t \in \mathcal{T})$. Here a compensator is a nondecreasing predictable process with $\Lambda_h(0) = 0$, where $M_h(t) = N_h(t) - \Lambda_h(t)$, for $t \in \mathcal{T}$, is a local martingale. The key property of a martingale is that $E[M_h(t)|\mathcal{F}_s] = M_h(s)$ for $s < t$. For our purposes here we can always replace "predictable" by "left-continuous."

So far this structure is very general, but we also assume that the compensator is absolutely continuous, so that there exists an intensity process $(\lambda_h(t), t \in \mathcal{T})$ with $\Lambda_h = \int_0^t \lambda_h(s) ds$. From the standard theory of martingales, the *predictable variation process* $\langle M \rangle$ of a local square-integrable martingale M is defined as the compensator of M^2, and the predictable covariation process $\langle M, M' \rangle$ of two local square-integrable martingales M and M' is defined as the compensator of MM'. In our situation we have $\langle M_h \rangle = \Lambda_h = \int \lambda_h$ and $\langle M_h, M_j \rangle = 0$ for $h \neq j$ such that M_h and M_j are orthogonal.

18.3 Censoring

Censored data are allowed for in the present framework by a predictable indicator process $(C_h(t), t \in \mathcal{T})$ that assumes the value 1 when the individual is under observation. The censored counting process is then

$$N_h^c(t) = \int_0^t C_h(s) dN_h(s),$$

which has intensity process $C_h(t)\lambda_h(t)$, because

$$\begin{aligned} M_h^c(t) &= N_h^c(t) - \int_0^t C_h(s)\lambda_h(s) ds \\ &= \int_0^t C_h(s)[dN_h(s) - \lambda_h(s)]ds \\ &= \int_0^t C_h(s) dM_h(s) \end{aligned}$$

is the stochastic integral of the predictable process C_h with respect to the local square-integrable martingale M_h, and hence is itself a local square-integrable martingale.

By suitable (sometimes somewhat delicate) choices of the filtration $(\mathcal{F}_t, t \in \mathcal{T})$, this approach extends to many previously investigated censoring patterns, and its flexibility is one of the main contributions of the counting process perspective to event-history analysis. Besides censoring, truncation may also be handled in this framework. In what follows, our discussion will extend in a straightforward way to counting processes for censored and truncated data, though without further specific reference.

18.4 Multiplicative Intensity Models

A key class of statistical models $\mathcal{P}$ for a multivariate counting process $\boldsymbol{N}$ on $(\Omega, \mathcal{F}, (\mathcal{F}_t))$ is the *multiplicative intensity model*, in which the $(P, (\mathcal{F}_t, t \in \mathcal{T}))$ intensity process $\boldsymbol{\lambda}$ is given by

$$\lambda_h(t) = \alpha_h(t) Y_h(t)$$

for $h = 1, \ldots, k$ and $P \in \mathcal{P}$ [1]. Here α_h is a nonnegative deterministic function depending on P, whereas Y_h is a predictable process not depending on P, to be interpreted as *observable*. In many event-history models α_h will be an individual transition intensity and Y_h will count the number at risk.

To motivate an estimator of $\mathbf{A} = (A_1(t), \ldots, A_k(t))$, where we define

$$A_h(t) = \int_0^t \alpha_h(s) ds$$

for $h = 1, \ldots, k$, interpret the local square-integrable martingale

$$M_h(t) = N_h(t) - \int_0^t \alpha_h(s) Y_h(s) ds$$

as noise, leading to the heuristic estimating equation

$$0 = dN_h(t) - \alpha_h(t) Y_h(t) dt$$

with the solution $\alpha_h(t) dt = dN_h(t)/Y_h(t)$. This suggests the *Nelson–Aalen estimator*

$$\hat{A}_h(t) = \int_0^t Y_h(s)^{-1} dN_h(s).$$

It is possible to develop an interpretation of $\hat{A}_h$ as a nonparametric maximum-likelihood estimator. An important property of the Nelson–Aalen estimator is its conceptual and technical simplicity: let $0 < T_1 < T_2 < \cdots$ be the jump times of N_h; then

$$\hat{A}_h(t) = \sum_{T_j \leq t} \frac{1}{Y(T_j)},$$

a simple sum. Formally, define $J_h(t) = I\{Y_h(t) > 0\}$ and

$$A_h^*(t) = \int_0^t \alpha_h(s) J_h(s) ds;$$

then since N_h may only jump when Y_h is positive, we also have (with $0/0$ taken as 0)

$$\hat{A}_h(t) = \int_0^t \frac{J_h(s)}{Y_h(s)} dN_h(s),$$

so that

$$\hat{A}_h(t) - A_h^*(t) = \int_0^t \frac{J_h(s)}{Y_h(s)} dM_h(s),$$

which is the stochastic integral of the predictable locally bounded process J_h/Y_h with respect to the local square-integrable martingale M_h, and hence itself a local square-integrable martingale. It follows that

$$E[\hat{A}_h(t)] = E[A_h^*(t)]$$
$$= \int_0^t \alpha_h(s) P[Y_h(s) > 0] ds,$$
$$t \in \mathcal{T},$$

so that the Nelson–Aalen estimator is in general biased downward with bias

$$E[\hat{A}_h(t) - A_h(t)]$$
$$= -\int_0^t \alpha_h(s) P[Y_h(s) = 0] ds.$$

The predictable variation process $\langle \hat{A}_h - A_h^* \rangle$ is given by

$$\langle \hat{A}_h - A_h^* \rangle(t) = \int_0^t \frac{J_h(s)}{Y_h(s)} \alpha_h(s) ds,$$

so that

$$\tilde{\sigma}_h^2(t) = E[\langle \hat{A}_h - A_h^* \rangle(t)]$$
$$= \int_0^t E[\frac{J_h(s)}{Y_h(s)}] \alpha_h(s) ds$$

which, because $\langle \hat{A}_h - A_h^* \rangle$ is the compensator of $(\hat{A}_h - A_h^*)^2$, is interpreted as

$$\tilde{\sigma}_h^2(t) = E[\{\hat{A}_h(t) - A_h^*(t)\}^2],$$

a "mean-squared-error function" of $\hat{A}_h$. This mean of the predictable variation process may be estimated by the (observable) *optional variation process*, as

$$\hat{\sigma}_h^2(t) \approx \int_0^t J_h(s) Y_h(s)^{-2} dN_h(s).$$

When there is only a small probability that $Y_h(s) = 0$ for some $s \leq t$, then $A_h^*(t)$ is almost the same as $E[\hat{A}_h(t)]$, and thus $\hat{\sigma}_h^2(t)$ will be a reasonable estimator of the variance of $\hat{A}_h(t)$.

Note that $\hat{\sigma}_h^2$, like $\hat{A}_h$ itself, is just a simple sum. Because of the orthogonality of the local square-integrable martingales $M_1, \ldots, M_k$ it furthermore follows that $\hat{A}_1 - A_1^*, \ldots, \hat{A}_k - A_k^*$ are also orthogonal.

In order to use the Nelson–Aalen nonparametric estimator, one needs to understand its large-sample properties. The framework for developing this is a sequence of counting processes $\boldsymbol{N}^{(n)} = (N_1^{(n)}, \ldots, N_k^{(n)}), n = 1, 2, \ldots$, each satisfying the multiplicative intensity model $\lambda_h^{(n)}(t) = \alpha_h(t) Y_h^{(n)}(t)$ with the same α_h for all n. Let $J_h^{(n)}(t) = I\{Y_h^{(n)}(t) > 0\}$. The consistency and asymptotic normality results below are based on Lenglart's inequality and martingale central limit theory, respectively. The uniform consistency result states the following theorem.

Theorem 18.4.1 *Let $t \in \mathcal{T}$, and assume that, as $n \to \infty$,*

$$\int_0^t \frac{J_h^{(n)}(s)}{Y_h^{(n)}(s)} \alpha_h(s) ds \xrightarrow{P} 0,$$

where $\xrightarrow{P}$ indicates convergence in probability, and assume

$$\int_0^t [1 - J_h^{(n)}(s)] \alpha_h(s) ds \xrightarrow{P} 0.$$

Then, as $n \to \infty$,

$$\sup_{s \in [0,t]} \|\hat{A}_h^{(n)}(s) - A_h(s)\| \xrightarrow{P} 0.$$

The second basic theorem is an asymptotic normality result

Theorem 18.4.2 *Let $t \in \mathcal{T}$, and assume that there exist a sequence of positive constants a_n, increasing to infinity as $n \to \infty$, and nonnegative functions y_h such that α_h/y_h is integrable over $[0, t]$ for $h = 1, 2, \ldots, k$. Let*

$$\sigma_h^2(s) = \int_0^s \frac{\alpha_h(u)}{y_h(u)} du, \quad h = 1, 2, \ldots, k,$$

and assume that

(A) *For each $s \in [0, t]$ and $h = 1, 2, \ldots, k$,*

$$a_n^2 \int_0^s \frac{J_h^{(n)}(u)}{Y_h^{(n)}(u)} \alpha_h(u) du \xrightarrow{P} \sigma_h^2(s)$$

as $n \to \infty$.

(B) *For $h = 1, 2, \ldots, k$ and all $\varepsilon > 0$,*

$$a_n^2 \int_0^t \frac{J_h^{(n)}(u)}{Y_h^{(n)}(u)} \alpha_h(u)$$

$$\times I\left\{\left|a_n \frac{J_h^{(n)}(u)}{Y_h^{(n)}(u)}\right| > \varepsilon\right\}$$

$$du \xrightarrow{P} 0 \text{ as } n \to \infty.$$

(C) *For $h = 1, 2, \ldots, k$,*

$$a_n \int_0^t [1 - J_h^{(n)}(u)] \alpha_h(u) du \xrightarrow{P} 0$$

as $n \to \infty$.

Then

$$a_n(\hat{\boldsymbol{A}}_{(n)} - \boldsymbol{A}) \xrightarrow{\mathcal{D}} \boldsymbol{U}$$
$$= (U_1, \ldots, U_k)$$

as $n \to \infty$,

where $\xrightarrow{\mathcal{D}}$ denotes convergence in law and $U_1, \ldots, U_k$ are independent Gaussian martingales with $U_h(0) = 0$ and $\text{Cov}[U_h(s_1), U_h(s_2)] = \sigma_h^2(s_1 \wedge s_2)$ (the $\wedge$ operator takes the minimum of s_1 and s_2). Also, for $h = 1, 2, \ldots, k$, we obtain

$$\sup_{s \in [0,t]} |a_n^2 \hat{\sigma}_h^2(s) - \sigma_h^2(s)| \xrightarrow{P} 0 \text{ as } n \to \infty.$$

These results may be used to generate approximate pointwise confidence intervals and simultaneous confidence bands.

Of course, the Nelson–Aalen estimator is concerned with the *integrated intensity* $A_h(t)$, a concept usually of rather less direct interest in applications than the intensity $\alpha_h(t)$ itself, for which, however, there is no similar canonical estimator. Indeed, various smoothing techniques (as known from nonparametric regression and density estimation) could be applied to the increments of $\hat{A}_h(t)$. Kernel smoothing [15] is particularly obvious in this context, because its properties may be studied by the same tools from martingale theory and stochastic integrals as above.

Let $K(t)$ be a kernel function, which we shall here take to be a bounded function vanishing outside $[-1, 1]$. For a given bandwidth b define the estimator

$$\hat{\alpha}_h(t) = b^{-1} \int_{\mathcal{T}} K(\frac{t-s}{b}) d\hat{A}_h(s).$$

Setting

$$\alpha_h^*(t)$$
$$= b^{-1} \int_{\mathcal{T}} K\left(\frac{t-s}{b}\right) dA_h^*(s)$$
$$= b^{-1} \int_{\mathcal{T}} K\left(\frac{t-s}{b}\right) J_h(s)\alpha_h(s) \, ds,$$

one may again arrive at a stochastic integral:

$$\hat{\alpha}_h(t) - \alpha_h^*(t)$$
$$= b^{-1} \int_{\mathcal{T}} K\left(\frac{t-s}{b}\right)$$
$$\times J_h(s)Y_h(s)^{-1} dM_h(s).$$

As usual in smoothing, $\hat{\alpha}_h(t)$ is in general not even approximately unbiased as an estimator of α_h, because α_h is a smoothed version of α_h. The statistical analysis of $\hat{\alpha}_h$ therefore entails a balance between the bias and the variability of the estimator, often expressed via a decomposition of the mean integrated squared error into two terms.

18.5 Survival Analysis: The Kaplan–Meier Estimator

An important example of the counting process model arises in the context of a sample of n i.i.d. nonnegative random variables with absolutely continuous distribution function F and survival function $S = 1 - F$, hazard rate $\alpha = F'/(1 - F) = -S'/S$, and integrated hazard $A(t) = \int_0^t \alpha(s) ds$. We do not observe $X_1, \ldots, X_n$, but only $(\tilde{X}_i, D_i), i = 1, \ldots, n$, where $\tilde{X}_i = X_i \wedge U_i$ and $D_i = I\{\tilde{X}_i = X_i\}$ (an indicator function to show whether the observation is actually censored) for some censoring times $U_i, \ldots, U_n$.

Under suitable assumptions on the joint distribution of X and U (e.g., that they are independent),

$$N(t) = \sum_{i=1}^n I\{\tilde{X}_i \leq t, D_i = 1\}$$

is a counting process with intensity process $\lambda(t) = \alpha(t)Y(t)$, where

$$Y(t) = \sum_{i=1}^n I\{\tilde{X}_i \geq t\},$$

the number at risk at time t.

In this situation the Nelson–Aalen estimator will yield an estimator of the integrated hazard A, whose increments may be smoothed, as just described, to obtain an estimate of the hazard α itself. However, often the interest is more focused on the survival function S. It turns out to be useful to express the general connection between S and A by the *product–integral* [10]

$$S(t) = \prod_{0 \leq s \leq t} [1 - dA(s)].$$

The product–integral may be defined in several ways. One attractive definition is based on a limit of finite products over finite partitions $0 < t_1 < t_2 < \cdots < t$:

$$\prod_{0 \leq s \leq t} [1 + dG(s)]$$
$$= \lim_{\max |t_i - t_{i-1}| \to 0} \prod [1 + G(t_i)$$
$$- G(t_{i-1})].$$

In particular, if G is continuous, $\prod(1 + dG) = e^G$, and if G is a step function with finitely many steps, $\prod(1 + dG) = \prod(1 + \Delta G)$, the ordinary finite product of $1 + \Delta G$, where ΔG is the step size.

The estimator of S obtained by plugging the Nelson–Aalen estimator $\hat{A}$ into the product–integral formula

$$\hat{S}(t) = \prod_{0 \leq s \leq t} [1 - d\hat{A}(s)]$$

$$= \prod_{\substack{\tilde{x}_j \leq t \\ D_j = 1}} \times \left(1 - \frac{1}{Y(\tilde{X}_j)}\right)$$

is identical to the well-known Kaplan–Meier estimator in survival analysis. Its statistical properties are very similar in nature to those of the Nelson–Aalen estimator and may be derived using either generalized δ methodology, or directly, again using martingales and stochastic integrals. In the latter case a key tool is the local square-integrable martingale [8]

$$\frac{\hat{S}(t)}{S^*(t)} - 1 = -\int_0^t \frac{\hat{S}(s-)J(s)}{S^*(s)Y(s)} dM(s),$$

where $S^*(t) = e^{-A^*(t)} = \prod_{0 \leq s \leq t}[1 - dA^*(s)]$.

18.6 Transitions in Nonhomogeneous Markov Processes

Consider a nonhomogeneous, time-continuous Markov process $X(t)$ on $\mathcal{T} = [0, \tau]$ or $[0, \tau)$ with finite state space $\{1, 2, \ldots, k\}$ having transition probabilities $P_{hj}(s, t)$ and transition intensities $\alpha_{hj}(t)$, where h and j are distinct states. For n conditionally (given the initial states) independent replications of this process, subject to quite general censoring patterns, the multivariate counting process $\mathbf{N} = (N_{hj}; h \neq j)$, with $N_{hj}(t)$ counting the number of observed direct transitions from h to j in $[0, t]$, has intensity process $\boldsymbol{\lambda} = (\lambda_{hj}, h \neq j)$ of the multiplicative form $\lambda_{hj}(t) = \alpha_{hj}(t)Y_h(t)$. Here $Y_h(t) \leq n$ is the number of sample paths observed to be in state h just before time t.

The previous theory specializes directly to yield Nelson–Aalen estimators $\hat{A}_{hj}$ of the integrated transition intensities A_{hj}. However, in practice, there will often be a need to combine the estimated transition intensities into a synthesis describing the net effect of the various transitions. The transition probabilities

$$P_{hj}(s, t) = P\{X(t) = j | X(s) = h\}$$

depend on the transition intensities α_{hj} through the Kolmogorov forward differential equations, whose solution may be represented as the matrix product–integral

$$\mathbf{P}(s, t) = \prod_{(s,t)}[\mathbf{I} + d\mathbf{A}(u)]$$

with $\mathbf{I}$ the identity matrix. Aalen and Johansen [3] used this relation to motivate the estimator

$$\hat{\mathbf{P}}(s, t) = \prod_{(s,t)}[\mathbf{I} + d\hat{\mathbf{A}}(u)],$$

which may be given a nonparametric maximum-likelihood interpretation. Our rather compact notation may not fully reveal that the estimator is really a simple finite product of elementary matrices.

As before, martingales and stochastic integrals are available to derive exact and asymptotic properties and to estimate covariance matrices.

18.7 Nonparametric Hypothesis Tests

A common hypothesis-testing problem is that of comparing two counting processes N_1 and N_2 with intensity processes $\alpha_1 Y_1$ and $\alpha_2 Y_2$. Under the null hypothesis that

$\alpha_1 = \alpha_2$, and for any predictable weight process L, the stochastic integral

$$Z(t) = \int_0^t L(s) d[\hat{A}_1(s) - \hat{A}_2(s)]$$

is a local square-integrable martingale with predictable variation process

$$\langle Z \rangle(t) = \int_0^t L^2(s) \left(\frac{1}{Y_1(s)} + \frac{1}{Y_2(s)} \right) \alpha(s) ds,$$

where α denotes the common value of α_1 and α_2. Approximating this α by $d(N_1 + N_2)/(Y_1 + Y_2)$ suggests the estimate

$$\hat{\sigma}^2(t) = \int_0^t L^2(s) [Y_1(s) Y_2(s)]^{-1} \\ \times d[N_1(s) + N_2(s)]$$

of $\text{Var} Z(t)$. An obvious test statistic is therefore $Z(t)/\hat{\sigma}(t)$ for some t (e.g., $t = \tau$), and the martingale–stochastic integral machinery is again available to generate exact and asymptotic results.

Special choices of L lead to specific two-sample censored-data rank tests. For example, $L = Y_1 Y_2 / (Y_1 + Y_2)$ yields the logrank test, $L = Y_1 Y_2$ yields the Gehan–Gilbert generalized Wilcoxon test, and if the two counting processes to be compared concern right-censored survival data, $L = Y_1 Y_2 \hat{S} / (Y_1 + Y_2 + 1)$ yields Prentice's generalized Wilcoxon test (here $\hat{S}$ is the Kaplan–Meier estimator based on the joint sample). Andersen et al. [4] showed how one- and k-sample linear nonparametric tests may be similarly derived and interpreted.

The test statistic process $Z(t)$ allows not only test statistics based on one fixed time t, but also the utilization of the complete test-statistic process. One such application is to maximal deviation (Kolmogorov–Smirnov type) or squared integrated deviation (Cramér–von Mises or Anderson–Darling type) statistics. Another application is to sequential-analysis hypothesis tests. In all cases the master asymptotic theorems based on martingale central limit theory provide sufficiently powerful approximations to Gaussian martingales for which relevant existing results may be used as approximations to the counting process models.

18.8 Regression Models and Random Heterogeneity

A very popular semiparametric analysis for survival data is based on Cox's regression model [7]. This assumes that the hazard $\alpha(t)$ of an individual with covariates $z = (z_1, \ldots, z_p)$ is

$$\alpha(t; z) = \alpha_0(t) e^{\beta' z},$$

where β is a vector of unknown regression coefficients and $\alpha_0(t)$ an unknown hazard function for individuals with $z = 0$. Andersen and Gill [6] generalized this model to the present counting process framework.

A different regression model was suggested by Aalen [2], as follows. Consider again a multivariate counting process $N = (N_1, \ldots, N_n)$ corresponding to n individuals, where N_i has intensity process

$$\lambda_i(t) = [\beta_0(t) + \beta_1(t) Z_{i1}(t) + \cdots \\ + \beta_p(t) Z_{ip}(t)] Y_i(t),$$

in which $Z_{ij}(t)$ are covariate processes, $\beta_i(t)$ are regression functions, and $Y_i(t)$ indicates whether individual i is at risk at time t. The model may be written in matrix form

$$\boldsymbol{N}(t) = \int_0^t \boldsymbol{Y}(u) \boldsymbol{\beta}(u) du + \boldsymbol{M}(t),$$

showing that it can be viewed as a matrix multiplicative intensity model. Here M is a vector of martingales, $\boldsymbol{\beta} = (\beta_0, \beta_1, \ldots, \beta_p)$, and $Y(t)$ is the $n \times (p+1)$ matrix with ith row, $i = 1, \ldots, n$, given by $\mathbf{Y}_i(t) = (1, Z_{i1}(t), \ldots, Z_{ip}(t))$.

The main interest is in deriving estimates of the integrated regression functions

$$B_j(t) = \int_0^t \beta_j(u)du$$

and their variances. Aalen proposed what could be interpreted as generalized Nelson–Aalen estimators

$$\hat{\boldsymbol{B}}(t) = \int_0^t J(u)\boldsymbol{Y}^-(u)d\boldsymbol{N}(u),$$

where $\boldsymbol{Y}^-(t)$ is a predictable generalized inverse of $Y(t)$. Exact and asymptotic properties of these estimators are available; the asymptotic results are primarily from Huffer and McKeague [12].

The Cox model describes individual heterogeneity in intensities through regression models relating the heterogeneity to relevant covariates registered for each individual. Another tool for modeling heterogeneity is to assume additional random variation, often to allow for positive statistical dependence between several transitions for the same individual (serial dependence) or the same transitions for several individuals within the same stratum, e.g., twins, litters, parent–offspring combinations, or other matched pairs (parallel dependence).

Borrowing a term originally proposed by Vaupel et al. [16] in demography, such models are often termed "frailty models"; this usage is common in the counting process context. The idea is to stay with multiplicative intensities but to add an *unobservable* random factor (the "frailty").

In the simplest situation, let $\boldsymbol{N} = (N_1, \ldots, N_n)$ be a multivariate counting process with intensity process $\boldsymbol{\lambda} = (\lambda_1, \ldots, \lambda_n)$ satisfying

$$\lambda_i(t) = Z_i Y_i(t)\alpha(t)$$

for some observable predictable process Y_i, an unknown deterministic baseline intensity function α, and unobservable random variables Z_i, independently drawn from some distribution. The most commonly used distribution has been the gamma, although Hougaard in a series of papers [11] has advocated other classes, notably the positive stable distributions.

Nonparametric estimation in this setting may naturally be performed using an EM algorithm approach (suggested by Gill [9] and elaborated on by Nielsen et al. [14]). For each given parameter δ of the gamma frailty distribution, the E-step predicts Z_i by its conditional expectation $\hat{Z}_i$ (under the current parameter values) given the data $\boldsymbol{N}$, $\boldsymbol{Y}$; the M step is to calculate the Nelson–Aalen estimator as if Z had been observed (and were equal to $\hat{\boldsymbol{Z}}$). The resulting profile likelihood is then maximized over δ.

Asymptotic properties of these estimators have been hard to derive, although Murphy [13] now seems to have opened the way. Yashin et al. [17] gave important comments on precise interpretations of the frailty variables, introducing the concept of "correlated frailty" in addition to the usual "shared frailty."

References

1. Aalen, O. O. (1978). Nonparametric inference for a family of counting processes. *Ann. Statist.* **6**, 701–726.

2. Aalen, O. O. (1980). A model for nonparametric regression analysis of counting processes. *Springer Lecture Notes Statist.*, **2**, 1–25.

3. Aalen, O. O. and Johansen, S. (1978). An empirical transition matrix for nonhomogeneous Markov chains based on censored observations. *Scand. J. Statist.*, **5**, 141–150.

4. Andersen, P. K., Borgan, Ø., Gill, R. D., and Keiding, N. (1982). Linear nonparametric tests for comparison of counting processes, with application to censored survival data (with discussion). *Int. Statist. Rev.*, **50**, 219–258. Amendment: **52**, 225 (1984).

5. Andersen, P. K., Borgan, Ø., Gill, R. D., and Keiding, N. (1993). *Statistical Models Based on Counting Processes.* Springer-Verlag, New York.

6. Andersen, P. K. and Gill, R. D. (1982). Cox's regression model for counting processes: a large sample study. *Ann. Statist.*, **10**, 1100–1120.

7. Cox, D. R. (1972). Regression models and life-tables (with discussion). *J. Roy. Statist. Soc. B*, **34**, 187–220.

8. Gill, R. D. (1980). *Censoring and Stochastic Integrals,* Mathematical Centre Tracts 124, Mathematisch Centrum, Amsterdam.

9. Gill, R. D. (1985). Discussion of the paper by D. Clayton and J. Cuzick. *J. Roy. Statist. Soc. A.*, **148**, 108–109.

10. Gill, R. D. and Johansen, S. (1990). A survey of product integration with a view towards application in survival analysis. *Ann. Statist.*, **18**, 1501–1555.

11. Hougaard, P. (1987). Modelling multivariate survival. *Scand. J. Statist.*, **14**, 291–304.

12. Huffer, F. W. and McKeague, I. W. (1991). Weighted least squares regression for Aalen's additive risk model. *J. Am. Statist. Assoc.*, **86**, 114–129.

13. Murphy, S. A. (1994). Consistency in a proportional hazards model incorporating a random effect. *Ann. Statist.*, **22**, 712–731.

14. Nielsen, G. G., Gill, R. D., Andersen, P. K., and Sørensen, T. I. A. (1992). A counting process approach to maximum likelihood estimation in frailty models. *Scand. J. Statist.*, **19**, 25–43.

15. Ramlau-Hansen, H. (1983). Smoothing counting process intensities by means of kernel functions. *Ann. Statist.*, **11**, 453–466.

16. Vaupel, J. W., Manton, K. G., and Stallard, E. (1979). The impact of heterogeneity in individual frailty on the dynamics of mortality. *Demography*, **16**, 439–454.

17. Yashin, A. I., Vaupel, J. W., and Iachine, I. A. (1995). Correlated individual frailty: an advantageous approach to survival analysis of bivariate data. *Math. Popul. Stud.*, **5**, 145–159.

19 Cox's Proportional Hazards Model

Richard Kay

19.1 The Model

In medical statistics methods for evaluating the dependence of survival or some other time to event variable T on treatment intervention and independent variables or covariates has received considerable attention. The proportional hazards model [4] is specified through the hazard function $\lambda(t; \underline{x})$ for a subject with treatment indicator and covariates $\underline{x} = (x_1, x_2, \ldots, x_p)'$ as follows:

$$\lambda(t; \underline{x}) = \lambda_0(t) \exp(\underline{\beta}' \underline{x}). \quad (1)$$

The function $\lambda_0(t)$ is termed the *underlying hazard function* and corresponds to the hazard function for a subject with $\underline{x} = \underline{0}$. If $x_1 = 0/1$ is the treatment indicator, then $\lambda_0(t)$ is the hazard function for a subject in the treatment group with $x_1 = 0$ and all covariate values equal to 0.

To aid interpretation it is sometimes useful to express the covariate values in terms of differences from the corresponding means. For example, if z_2 is the covariate age and $\bar{z}_2$ is the mean age, then defining $x_2 = z_2 - \bar{z}_2$ and so on for the remaining covariates allows $\lambda_0(t)$ to be interpreted as the hazard function for an "average" patient receiving treatment defined by $x_1 = 0$.

The focus for this discussion will be the clinical trial with patients randomized to receive one of two treatments. Time T will be assumed to be measured from the point of randomization. The model, however, is used extensively outside this framework, for example, in epidemiology and in observational studies where the requirement is to explore causality or the predictors of outcome where the outcome is in the form of a well-defined time-to-event variable.

If $x_1 = 0/1$, where $x_1 = 0$ corresponds to the control treatment and $x_1 = 1$ denotes the experimental treatment, then the ratio of the hazard rates (experimental/control) for a subject in each group with the same covariate values is e^{β_1}. This ratio is a constant, independent of time, and it is this property that gives the model its name. The effect of treatment is multiplicative on the hazard function. The covariates also affect the hazard function in the same way. For a binary covariate, for example sex with $x_2 = 0/1$ for male/female the ratio of the hazard rates (female/male) will be e^{β_2}. For a continuous covariate x_2, the ratio of the hazard rates for subjects with values x_{i2} and x_{j2} will be $e^{\beta_2(x_{i2} - x_{j2})}$.

Under the proportional hazards model the survivor function $pr(T > t)$ for a subject with treatment indicator and covariates $\underline{x} = (x_1, x_2, \ldots, x_p)'$ is given by

$$S(t; \underline{x}) = \exp\left\{-\int_0^t \lambda(u; \underline{x})du\right\}$$
$$= \{S_0(t)\}^{e^{\beta' \underline{x}}}$$

where $S_0(t) = \exp\left\{-\int_0^t \lambda_0(u)du\right\}$ is the *underlying survivor function*; the survivor function for a subject in treatment group $x_1 = 0$ with zero values for the covariates.

19.2 The Stratified Model

As outlined in the previous section the proportional hazards model makes the assumption that treatment and covariates affect the hazard function in a multiplicative way. This assumption of course may not always be satisfied and the *stratified proportional hazards model* [8, Sec. 4.4; 9] is a more general form of the model that allows departures from proportionality. Suppose that there is a factor, for example center or baseline disease severity, that is thought to affect the hazard function in a nonproportional way and further assume that this factor defines $c = 1, \ldots, C$ strata. The stratified form of the model is then

$$\lambda(t; \underline{x}*) = \lambda_{0c}(t)\exp(\beta' \underline{x}*) \quad (2)$$

for a subject in stratum $c = 1, \ldots, C$. Note here that $\underline{x}*$ is the vector incorporating the treatment indicator and the covariates but excluding the factor that defines the strata.

This more general model can be useful in several ways: (1) it allows factors to be used in the modelling process that do not satisfy the proportional hazards assumption, for example in a meta-analysis "study" can be used as the factor defining the strata, while (2) it can be useful for assessing proportionality in model checking, and (3) it can be used to provide Kaplan–Meier curves for treatment groups adjusted for baseline factors; in this case the treatment groups would form the strata.

19.3 Partial Likelihood

Usually time-to-event data of this kind will be subject to right censoring. For example, in a clinical trial with a fixed period of follow-up there will be subjects who reach the end of the follow-up period without suffering the event of interest. Alternatively there may be subjects who are lost to follow-up. If it can be assumed that the censoring and event mechanisms are independent, then the only information that we have regarding time to event for a patient who provides a censored observation equal to t is that time to event is $> t$.

It is inevitable that time-to-event data of the type being considered here will be subject to censoring so that time to event will not be known for all subjects in the study. It is the presence of censoring that has primarily led to the development of special methods for time-to-event data

The concept of *partial likelihood* was initially introduced by Cox [4] and subsequently formalized in Cox [5]. This methodology allows a likelihood function to be constructed as a function of the unknown coefficients $\underline{\beta}$ of $\underline{x}$ that does not involve the underlying hazard function $\lambda_0(t)$. Estimation of $\underline{\beta}$ can then proceed without the need to make specific assumptions about the form of $\lambda_0(t)$. The proportional hazards model can therefore be considered as a semiparametric model; the effect of the treatment indicator and the effects of the covariates on the hazard function are parameterized in a very specific way, but the underlying hazard is left unspecified. In one sense it is this feature associated with the method of estimation that has made the proportional hazards model so attractive.

To simplify the notation, assume in the remainder of this section that the event of interest is death. Let $t_{(1)} < t_{(2)} < \cdots < t_{(K)}$ be the ordered times at which deaths occur, and for the moment assume that there are no ties. Further, let $\underline{x}_{(j)}$ be the vector containing the treatment indicator and covariate values for the subject who dies at time $t_{(j)}$. Finally, let $R\{t_{(j)}\}$ denote the collection of subjects who are still alive and in the trial (the risk set) just prior to time $t_{(j)}$.

The partial likelihood is made up of K terms, one for each timepoint at which an event occurs. These terms correspond to the conditional probability that the subject identified by $\underline{x}_{(j)}$ dies at $t_{(j)}$, conditional on one subject in $R\{t_{(j)}\}$ dying at time $t_{(j)}$ These terms are constructed as follows:

$$\Pr\{\underline{x}_{(j)} \text{ dies at } t_{(j)} \mid \text{one subject}$$
$$\text{in } R\{t_{(j)}\} \text{ dies at } t_{(j)}\}$$
$$= [\Pr\{\underline{x}_{(j)} \text{ in } R\{t_{(j)}\} \text{ dies at } t_{(j)}\}]$$
$$\times [\Pr\{\text{one subject in } R\{t_{(j)}\}$$
$$\text{dies at } t_{(j)}\}]^{-1}$$
$$= \left[f(t_{(j)}; \underline{x}_{(j)}) \right.$$
$$\times \prod_{l^* \notin R\{t_{(j)}\} - \{\underline{x}_{(j)}\}} S(t_{(j)}; \underline{x}_{l^*})\right]$$
$$\times \left[\sum_{l \in R\{t_{(j)}\}} f(t_{(j)}; \underline{x}_l) \right.$$
$$\left. \times \prod_{l^* \notin R\{t_{(j)}\} - \{\underline{x}_{(j)}\}} S(t_{(j)}; \underline{x}_{l^*}) \right]^{-1}.$$

Here $R\{t_{(j)}\} - \{\underline{x}_k\}$ denotes the risk set at time $t_{(j)}$ with the subject identified by $\underline{x}_k$ for $k = (j)$ removed. The probabilities in the numerator and denominator have implicitly accounted for the fact that although one subject dies at $t_{(j)}$, all others survive beyond $t_{(j)}$.

Since $f(t; \underline{x})$ the probability density function for T and subject $\underline{x}$ is the product

$\lambda(t; \underline{x}) \times S(t; \underline{x})$, the above ratio reduces to

$$\frac{\lambda\left(t_{(j)}; \underline{x}_{(j)}\right)}{\sum_{l \in R\{t_{(j)}\}} \lambda\left(t_{(j)}; \underline{x}_l\right)} = \frac{e^{\underline{\beta}'\underline{x}_{(j)}}}{\sum_{l \in R\{t_{(j)}\}} e^{\underline{\beta}'\underline{x}_l}}. \quad (3)$$

The full partial likelihood is then

$$L\left(\underline{\beta}\right) = \prod_{j=1}^{K} \frac{e^{\underline{\beta}'\underline{x}_{(j)}}}{\sum_{l \in R\{t_{(j)}\}} e^{\underline{\beta}'\underline{x}_l}}. \quad (4)$$

Estimation of β then proceeds by maximizing the log partial likelihood function to produce *maximum (partial)-likelihood estimates* $\hat{\underline{\beta}}$ of β

It is of interest to note that the partial likelihood function does not involve time t and is based only on a ranking of the subjects in terms of when deaths occur. The subjects with censored values contribute to the likelihood through the risk sets that are constructed prior to that subject being censored.

For the stratified form of the proportional hazards model (2), the likelihood terms are constructed separately within each stratum giving a series of C terms $L_c(\underline{\beta})$. The full likelihood function is then a product of these terms.

19.4 Tied Data

It is inevitable, even with time to event measured on a continuous scale, that there will be ties. Assume that at time $t_{(j)}$ there are in fact d_j events occurring amongst the n_j subjects at risk at $t_{(j)}$. In reality these events will have occurred in a particular order and there are $d_j!$ such orderings. If E denotes a particular ordering $j_1, j_2, \ldots, jd_j$ of the $D\{t_{(j)}\}$ subjects who have events at

$t_{(j)}$ then

$$\Pr(E) = \frac{e^{\beta' x_{j_1}}}{\sum_{l \in R\{t_{(j)}\}} e^{\beta' x_l}}$$

$$\times \frac{e^{\beta' x_{j_2}}}{\sum_{l \in R\{t_{(j)}\}-\{j1\}} e^{\beta' x_l}}$$

$$\times \cdots \times$$

$$\times \frac{e^{\beta' x_{j d_j}}}{\sum_{l \in R\{t_{(j)}\}-\{j1, j2, \ldots, j(d_j-1)\}} e^{\beta' x_l}} \quad (5)$$

where $R\{t_{(j)}\}-\{j1\}$ for example is the risk set at time $t_{(j)}$ excluding patient j_1. The contribution to the partial likelihood function at time $t_{(j)}$ is then

$$\Pr\{E(t_{(j)})\}$$
$$= \Pr(E_1 \cup E_2 \cup \cdots \cup E_{d_j!})$$
$$= \Pr(E_1) + \Pr(E_2) + \cdots + \Pr(E_{d_j!}) \quad (6)$$

the sum of $d_j!$ terms of the form $\Pr(E)$ in equation (5) where each E_i corresponds to one of the possible orderings. The full partial likelihood is then the product of the $\Pr\{E(t_{(j)})\}$ terms over the distinct time points at which events occur. With substantial amounts of tied data this approach becomes computationally very difficult since at each tied event time the partial likelihood term involves consideration of the $d_j!$ orderings of the d_j subjects and several approximations exist

The numerator in each of the terms $\Pr\{E\}$ on the right-hand side of Equation (6) is the same and equal to

$$e^{\beta' x_{j_1}} \times e^{\beta' x_{j_2}} \times \cdots \times e^{\beta' x_{j d_j}} = e^{\beta' s_j}$$

where $s_j = x_{j_1} + x_{j_2} + \cdots + x_{j d_j}$ is the vector of covariate and treatment indicator values summed over the subjects who have events at time $t_{(j)}$ The approximations to the partial likelihood for tied data all derive from approximations for the denominator.

Peto [14] and Breslow [2] proposed as an approximation for $\Pr\{E(t_{(j)})\}$ the term

$$\frac{e^{\beta' s_j}}{\left\{\sum_{l \in R\{t_{(j)}\}} e^{\beta' x_l}\right\}^{d_j}}$$

while Efron [7] suggested

$$[e^{\beta' s_j}]$$

$$\times \left[\prod_{k=1}^{d_j} \left\{\sum_{l \in R\{t_{(j)}\}} e^{\beta' x_l} - \frac{(k-1)}{d_j}\right.\right.$$

$$\left.\left.\times \sum_{l \in D\{t_{(j)}\}} e^{\beta' x_l}\right\}\right]^{-1}$$

Cox [4], in his original development of the proportional hazards model, proposed a discrete argument to deal with ties. The term corresponding to the d_j events at time $t_{(j)}$ is constructed by considering the probability associated with the observed collection of subjects $D\{t_{(j)}\}$ who have events at $t_{(j)}$, conditional on some collection of d_j subjects having events at $t_{(j)}$ as follows:

$$\Pr\{\text{observed } d_j \text{ subjects} \in D\{t_{(j)}\}$$
$$\text{have events at } t_{(j)} | d_j \text{ subjects}$$
$$\in R\{t_{(j)}\} \text{ have events at } t_{(j)}\}$$
$$= [\Pr\{\text{observed } d_j \text{ subjects}$$
$$\in D\{t_{(j)}\} \text{ have events}$$
$$\text{at } t(j)\}]$$
$$\times [\Pr\{d_j \text{ subjects in } R\{t_{(j)}\}$$
$$\text{have events at } t_{(j)}\}]^{-1}$$
$$= \frac{e^{\beta' s_j}}{\sum e^{\beta' x_{l_1}} e^{\beta' x_{l_2}} \cdots e^{\beta' x_{l_{d_j}}}}$$

where the sum in the final denominator is over all possible subsets $\{l_1, l_2, \ldots, l_{d_j}\}$ of d_j subjects selected from the risk set $R\{t_{(j)}\}$

Allison [1, Chap. 5] suggests that in practice the Efron method seems to give results closest to the exact method using all possible orderings.

19.5 Parameter Estimation and Confidence Intervals

Maximisation of the partial likelihood provides estimated values $\hat{\beta}$ for the β coefficients. Cox [4] showed that the usual properties relating to maximum likelihood also hold for in the context of partial likelihood and maximum (partial)-likelihood estimators have the following properties: asymptotically normally distributed, asymptotically unbiased with variance–covariance matrix estimated by the inverse of the matrix of second partial derivatives $I(\underline{\beta})$, evaluated at the maximum (partial)-likelihood estimates $\hat{\underline{\beta}}$.

The diagonal elements of $I^{-1}(\hat{\underline{\beta}})$ provide the estimated variances of the individual components of $\hat{\underline{\beta}}$ while the off-diagonal provide the covariances. The standard error for a particular component $\hat{\beta}_j$ (labeled $\mathrm{SE}(\hat{\beta}_j)$) is then the square root of the jth diagonal element of $I^{-1}(\hat{\underline{\beta}})$ and an approximate 95% confidence interval for β_j is given by

$$\hat{\beta}_j \pm \{1.96 \times se(\hat{\beta}_j)\}.$$

A confidence interval for the corresponding hazard ratio, e^{β_j}, can then be obtained by taking antilogs of the confidence interval for β_j.

19.6 Hypothesis Testing

There are several, large sample procedures for testing hypotheses regarding model parameters; Wald's test, the likelihood ratio test and the score test. In practice it is Wald's test and the likelihood ratio test that are most useful although the score test in the model with just a single treatment term has links with the logrank test.

Consider the test of a simple hypothesis $H_0 : \beta_j = 0$ against the alternative $H_1 : \beta_j \neq 0$ relating to one of the coefficients in the proportional hazards model; of main interest usually is the coefficient of the treatment indicator.

Wald's test is based on comparing the ratio of the estimate $\hat{\beta}$ of β to its standard error with the standard normal distribution. The standard error is obtained through the estimated variance–covariance matrix so that the test statistic for Wald's test is

$$\frac{\hat{\beta}_j}{\mathrm{SE}(\hat{\beta}_j)}.$$

This test is sometimes presented through the square of this statistic, which is distributed as χ_1^2 under the null hypothesis.

The *likelihood ratio test* compares the maximized likelihood evaluated at $\hat{\underline{\beta}}$ with the maximized likelihood evaluated under the null hypothesis. Let $\hat{\underline{\beta}}$ denote the maximum (partial) likelihood estimate of β under the assumption that $\beta_j = 0$; this is in practice achieved by fitting the model with the x_j term omitted. The likelihood ratio statistic is given by $\frac{L(\hat{\underline{\beta}})}{L(\hat{\underline{\beta}})}$ and under the null hypothesis

$$-2\log\frac{L(\hat{\hat{\underline{\beta}}})}{L(\hat{\underline{\beta}})} \sim \chi_1^2.$$

For the score test, assume that there is only one term x in the model, for example, the treatment indicator, so that the null hypothesis is $H_0 : \beta = 0$. The first derivative of the log partial likelihood $U(\beta) = \frac{d\log L(\beta)}{d\beta}$ can be shown to reduce to

$$\sum_{j=1}^{K} x_{(j)} - \sum_{j=1}^{K} \left\{ \frac{\sum_{l \in R\{t_{(j)}\}} x_l e^{\beta x_l}}{\sum_{l \in R\{t_{(j)}\}} e^{\beta x_l}} \right\}.$$

The solution to $U(\beta) = 0$ is the maximum (partial) likelihood estimate. The second derivative of the log partial likelihood $I(\beta) = \frac{d^2 \log L(\beta)}{d\beta^2}$ is given by

$$-\sum_{j=1}^{K} \left\{ \sum_{l \in R\{j_{(j)}\}} e^{\beta x_l} \sum_{l \in R\{t_{(j)}\}} x_l^2 e^{\beta x_l} - \left(\sum_{l \in R\{t_{(j)}\}} x_l e^{\beta x_l} \right)^2 \right\}$$
$$\times \left\{ \left(\sum_{l \in R\{t_{(j)}\}} e^{\beta x_l} \right)^2 \right\}^{-1}.$$

The *score test* is based on the result that, asymptotically and under the null hypothesis, $\frac{\{U(0)\}^2}{I(0)} \sim \chi_1^2$.

In practice either the likelihood ratio test or Wald's test is usually used. The likelihood ratio test tends to be preferred in general; it reaches its asymptotic properties more rapidly than Wald's test. The construction of confidence intervals, as discussed earlier is mathematically linked to Wald's statistic. If Wald's test rejects the null hypothesis $\beta_j = 0$ at the two-sided 5% level then the 95% confidence interval will exclude zero and vice versa, so using Wald's test gives a direct link between the associated p value and the confidence interval, and this correspondence alone can be useful.

If the discrete method is used to handle ties, then the score test in the proportional hazards model containing a single treatment indicator term is mathematically equivalent to the logrank test [3].

19.7 Estimating the Underlying Hazard Function

The main focus of attention in the proportional hazards model is the estimation of the coefficients of the treatment indicator and covariates. It is also of interest, however, to obtain an "estimate" of the underlying hazard function $\lambda_0(t)$ and the corresponding underlying survivor function $S_0(t)$. This then provides a complete specification of both the hazard function $\lambda(t; \underline{x})$ and the survivor function $S(t; \underline{x})$ for any particular subject. There are several reasons why this is useful:

- *Prediction*—enables estimation of a complete survivor function for any specific individual in either of the two treatment groups

- *Goodness of fit*—allows comparison of the fitted survivor function for an "average" individual with the corresponding Kaplan–Meier curves for each of the two treatment groups

If T_0 denotes time to event for a patient with $\underline{x} = \underline{0}$, then let

$$\alpha_k = \Pr(T_0 > t_{(k)} | T_0 > t_{(k-1)})$$
$$= \frac{\Pr(T_0 > t_{(k)})}{\Pr(T_0 > t_{(k-1)})} = \frac{S_0(t_{(k)})}{S_0(t_{(k-1)})}.$$

Under the PH model

$$\Pr(T > t_{(k)} | T > t_{(k-1)})$$
$$= \frac{S(t_{(k)}; \underline{x})}{S(t_{(k-1)}; \underline{x})} = \alpha_k^{e^{\beta' \underline{x}}}$$

since $S(t; \underline{x}) = S_0(t)^{e^{\beta' \underline{x}}}$.

Assuming no ties, then at time $t = t_{(k)}$ the subject with covariates vector $\underline{x}_{(k)}$ has an event and the likelihood contribution at that time point can be loosely written as

$$\Pr(T = t_{(k)} | T > t_{(k-1)})$$
$$\times \prod \Pr(T > t_{(k)} | T > t_{(k-1)})$$
$$= [1 - \Pr(T > t_{(k)} | T > t_{(k-1)})]$$
$$\times \prod \Pr(T > t_{(k)} | T > t_{(k-1)})$$
$$= \left[1 - \alpha_k^{e^{\beta' \underline{x}_{(k)}}} \right] \prod \alpha_k^{e^{\beta' \underline{x}_l}} = \ell_k,$$

where the products are over $l \in R\{t_{(k)}\} - \{\underline{x}_{(k)}\}$. The full likelihood is then $\prod_{k=1}^{K} \ell_k$. Putting $\underline{\beta} = \underline{\hat{\beta}}$, taking logs, differentiating and equating to zero, leads to estimates of the α_k parameters as

$$\hat{\alpha}_k = \left[1 - \frac{e^{\underline{\hat{\beta}}'\underline{x}_{(k)}}}{\sum_{l \in R\{t_{(k)}\}} e^{\underline{\hat{\beta}}'\underline{x}_l}}\right]^{e^{-\underline{\hat{\beta}}'\underline{x}_{(k)}}}$$

and $\hat{S}_0(t) = \prod_{j=1}^{k} \hat{\alpha}_j$ for $t_{(k-1)} \le t < t_{(k)}$.

The predicted survivor function for subject $\underline{x}$ is then $\hat{S}(t;\underline{x}) = \hat{S}_0(t)^{e^{\underline{\hat{\beta}}'\underline{x}}}$

For ties, if we let $D\{t_{(k)}\}$ denote the collection of patients who have events at $t_{(k)}$, then the likelihood cannot be solved explicitly for the α_k parameters. The likelihood function is given by

$$\prod_{k=1}^{K} \left[\prod_{l \in D\{t_{(k)}\}} \left(1 - \alpha_k^{e^{\underline{\beta}'\underline{x}_l}}\right) \right.$$
$$\left. \times \prod_{l \in R\{t_{(k)}\} - D\{t_{(k)}\}} \alpha_k^{e^{\underline{\beta}'\underline{x}_l}} \right],$$

and the estimates $\hat{\alpha}_k$ are obtained as solutions to this expression.

Finally, in the absence of a covariates vector $\underline{x}$, the term involving α_k in the above likelihood reduces to

$$(1 - \alpha_k)^{d_k} \alpha_k^{n_k - d_k},$$

where d_k and n_k are respectively the number of subjects with events and at risk at time $t_{(k)}$.

Taking logs, differentiating and equating to zero gives $\hat{\alpha}_k = 1 - \frac{d_k}{n_k}$ and $\hat{S}_0(t)$ reduces to the Kaplan–Meier estimate of the survivor function.

19.8 Time-Dependent Covariates

Cox [4], in his original development of the proportional hazards model, allowed the possibility that the covariates included in the model could vary with time; the so-called *time-dependent covariates*.

On a technical point, in order to fit models that include covariates that are allowed to vary with time, there needs to be very detailed information available in relation to those covariates. In particular, because of the form of the partial likelihood, it is necessary to know the value $x_i(t)$ of the time-dependent covariate for subject i at all time points at which events occur until that subject either has an event or is censored. In effect this means that complete profiles are needed for the covariate "process" for every subject up to the time at which they leave the risk set. This condition of itself means that the range of applications for time-dependent covariates is somewhat limited.

A further issue limits their application even further in clinical trials. The value of a covariate that is allowed to vary with time may well be influenced, either directly or indirectly, by the treatment received. Including that covariate in the model may then bias the measure of treatment effect.

There are, however, particular formats for time-dependent covariates that can be of value in certain ways, and we will consider three areas of application:

- *Modeling*—it may be appropriate to allow movement away from proportional hazards, for example, to model a treatment effect that decays over time and this can be done through a time-dependent covariate.

- *Model checking*—including a time-dependent treatment effect models departures from the proportional hazards assumption, and this can be used in model checking.

- *Multistate models*—a time-dependent indicator variable could be used to flag the occurrence of an intermediate event of interest that could impact on the hazard function and this can be useful in modeling.

It should be pointed out here that outside of model checking, there are always concerns regarding the interplay between the covariate process, the underlying "event" process and the influence of treatment and all applications with time-dependent covariates should be very carefully thought through. The model-checking element in relation to time-dependent covariates will be covered in Section 19.9.3.

19.8.1 Modeling Nonproportional Hazards

We may expect clinically that the effect of treatment, for example, following several cycles of chemotherapy in oncology, may decrease over time, particularly in long term studies. This decay in the treatment effect can be built into the model through a time-dependent covariate:

$$\lambda(t; \underline{x}) = \lambda_0(t) \exp\{\beta x + \gamma x h(t) + \underline{\beta} *' \underline{x} *\},$$

where $h(t) = t$ or $\log t$ depending on the nature of the decay and $\underline{x}*$ contains the covariates.

So, for example, with $h(t) = t$, the hazard ratio is $\exp(\beta + \gamma t)$, and this varies with time according to the value of γ; for a treatment decay the value of γ should be negative. Including a time-dependent covariate in this way destroys the proportional hazards aspect of the model and depending on the value of γ the direction of the treatment difference may change. In Figure 1 the hazard ratio is greater than 1 for $t < \tau$ and less then 1 for $t > \tau$.

We also now have two parameters modelling the treatment effect and there is consequently no simple measure of that effect. A test of the overall treatment effect based on the composite null hypothesis $H_0 : \beta = \gamma = 0$ against the general alternative hypothesis can be undertaken through a likelihood ratio test. Statistical significance however does not necessarily imply an advantage for one of the treatments; it depends on what $\hat{\beta} + \hat{\gamma} t$ looks like.

A test of $H_0 : \gamma = 0$ provides a test of proportionality for the treatment effect and this is precisely what would be considered for model checking.

19.8.2 Modeling Intermediate Events

A particular example of modeling intermediate events through a time-dependent covariate that has received considerable attention is the Stanford Heart Transplantation Program. This particular case study was first developed from a modeling perspective by Turnbull et al. [15]. See also Crowley and Hu [6], Kalbfleisch and Prentice [8, Sec. 6.4.3], Collett [3, Sec. 8.4], and Machin et al. [13, Sec. 7.2]. Patients were entered into the program ($t = 0$) once the decision had been taken to seek a donor heart. Some patients died before a donor could be identified, while others received a new heart and died or were censored subsequent to that; the primary endpoint was survival time.

The time-dependent covariate was a binary indicator that took the value 0 prior to receipt of a new heart and 1 after that. Suppose that patient i receives a new heart at time τ_i; for those never receiving a new heart, $\tau_i = \infty$. The time-dependent covariate is then defined by

$$x_i(t) = \begin{cases} 0 & t < \tau_i \\ 1 & t \geq \tau_i \end{cases}.$$

The coefficient β then measures the impact of receiving a new heart, in terms of changing the hazard rate for death. A negative value for β tells us that receiving a new

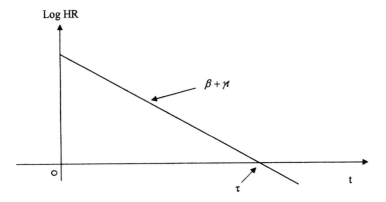

Figure 1: Linear time-dependent covariate.

heart reduces the hazard rate, a positive value for β is indicating that receiving a heart increases the hazard rate for death. Machin et al. [13] report that $\hat{\beta} = 0.992$ with an associated p value of 0.98 suggesting that transplantation has no effect on prolonging life. The model can also contain covariates measured at baseline that remain fixed and several were considered in the Heart Transplant Program, including age at acceptance into the program and year of acceptance. The analysis undertaken by Kalbfleisch and Prentice [8] identified year of acceptance as a key covariate in predicting survival possibly as a result of changing policy over time in terms of the kinds of patients admitted into the program.

19.9 Model Checking

Numerous methods have been provided for evaluating goodness of fit of the proportional hazards model. For an earlier review, see Kay [11]. Three methods will be presented here. For simplicity assume that we are considering a single treatment indicator in the model (no covariates). These methods in the main can be extended to include covariates if required.

19.9.1 Log–Log Plots

If $S_1(t)$ and $S_2(t)$ are the separate survivor functions then under the proportional hazards assumption they are related in the sense that $S_1(t) = [S_2(t)]^{e^\beta}$ or equivalently $\log[S_1(t)] = e^\beta \log[S_2(t)]$. Multiplying both sides of the equation by -1 and taking logs again gives

$$\log\{-\log[S_1(t)]\} = \beta + \log\{-\log[S_2(t)]\}.$$

If the proportional hazards assumption holds, then $\log\{-\log[S_1(t)]\}$ and $\log\{-\log[S_2(t)]\}$ should be separated by a constant. Estimating each of the individual survivor functions using the Kaplan–Meier method and plotting these functions against t will allow this assumption to be checked. Machin et al. [13, Sec. 5.1] provide an example of this methodology.

19.9.2 Time Axis Split

Consider a subdivision of the timescale and allow the β coefficient to take different values over these intervals: β_1 for

$0 < t \le \tau_1$ β_2 for $\tau_1 < t \le \tau_2$ and β_3 for $\tau_2 < t < \infty$.

The resulting partial likelihood function (4) is the product of three separate components:

$$L(\beta_1, \beta_2, \beta_3) = L_1(\beta_1) L_2(\beta_2) L_3(\beta_3),$$

where $L_1(\beta_1)$ is the partial likelihood function corresponding to considering all events in the dataset after τ_1 to be censored, $L_2(\beta_2)$ is the partial likelihood with all events and censored observations before τ_1 in the dataset removed and all events after τ_2 considered as censored observations and finally $L_3(\beta_3)$ is the partial likelihood with all events and censored observations, before τ_2 removed from the dataset.

Each of these three components can then be maximised separately to obtain the maximum (partial) likelihood estimates $\hat{\beta}_1, \hat{\beta}_2, \hat{\beta}_3$. If $L*(\beta)$ denotes the standard likelihood (single treatment coefficient) with maximum partial likelihood estimate $\hat{\beta}$, then under the assumption of proportional hazards,

$$-2\log\frac{L*(\hat{\beta})}{L(\hat{\beta}_1, \hat{\beta}_2, \hat{\beta}_3)} \sim \chi_3^2.$$

This test will frequently lack power and it may be prudent to consider $p < 0.10$ as indicative of a departure from proportional hazards.

The method can be extended to take in more than three intervals in an obvious way, although in practice three seems to work well and allows quadratic departures from a constant coefficient. Cutpoints can be defined by requiring equal numbers of events in the three intervals to give similar levels of information in the intervals. See Kay [12] for an application.

19.9.3 Time-Dependent Covariates and Other Methods

As discussed earlier a time-dependent covariate can be included in the model (Section 19.8) and a test constructed for the null hypothesis $H_0 : \gamma = 0$. Again using a significance level of 10% would be indicative of a departure from proportionality.

Residuals can also be defined for the proportional hazards model and inspected visually for model departures. Machin et al. [13, Sec. 5.5] provide more details and give an example.

19.10 Concluding Remarks

This short chapter has given details of the proportional hazards model, initially introduced by Cox [4], in the context of medical statistics and clinical trials. It is important to point out that the range of applications is much more widespread than this, encompassing epidemiology (time to onset of disease and its dependence on risk factors), sociology (duration of time in a job and its relationship with numerous factors at time of employment), economics (time to a stockmarket crash), engineering (the whole area of equipment failure and reliability), and many more.

The model is flexible and provides a broad framework within which dependences can be explored. Its power however comes with its ability to deal with censoring through the methods of estimation based around partial likelihood.

References

1. Allison, P. (1995). *Survival Analysis Using the SAS® System: A Practical Guide*. SAS Institute Inc., Cary, NC.

2. Breslow, N. E. (1974). Covariance analysis of censored survival data. *Biometrics*, **30**, 89–100.

3. Collett, D. (2003). *Modelling Survival Data in Medical Research, 2nd ed.* Chapman & Hall, London.

4. Cox, D. R. (1972). Regression models and life tables (with discussion). *J. Roy. Statist. Soc. B*, **74**, 187–220.

5. Cox, D. R. (1975). Partial likelihood. *Biometrika*, **62**, 269–276.

6. Crowley, J. and Hu, M. (1977). Covariance analysis of heart transplant survival data. *J. Am. Statist. Assoc.*, **72**, 27–36.

7. Efron, B. (1977). The efficiency of Cox's likelihood function for censored data. *J. Amer. Statist. Assoc.*, **72**, 557–565.

8. Kalbfleisch, J. D. and Prentice, R. L. (2002). *The Statistical Analysis of Failure Time Data, 2nd ed.* Wiley, New York.

9. Kay, R. (1977). Proportional hazard regression models and the analysis of censored survival data. *Appl. Statist.*, **26**, 227–237.

10. Kay, R. (1982). The analysis of transition times in multistate stochastic processes using proportional hazard regression models. *Commun. Statist. Theory Methods*, **11**, 1743–1756.

11. Kay, R. (1984). Goodness of fit methods for the proportional hazards regression model. *Revue Epidemiologie et de Sante Publique*, **32**, 185–198.

12. Kay, R. (1986). Treatment effects in competing-risks analysis of prostate cancer data. *Biometrics*, **42**, 203–211.

13. Machin, D., Cheung, Y. B., and Parmar, M. K. B. (2006). *Survival Analysis. A Practical Approach.* Wiley, Chichester, UK.

14. Peto, R. (1972). Contribution to the discussion of paper by D.R. Cox. *J. Roy. Statist. Soc. B*, **74**, 205–207.

15. Turnbull, B. W., Brown, B. W. Jr., and Hu, M. (1974). Survivorship analysis of heart transplant data. *J. Am. Statist. Assoc.*, **69**, 74–86.

20 Crossover Trials

Michael G. Kenward and Byron C. Jones

20.1 Introduction

The aim of medical research is to develop improved treatments or cures for diseases and medical ailments. Part of that research involves comparing the effects of alternative treatments with a view to recommending those that should be used in practice. The treatments are compared using properly controlled randomized clinical trials. In such trials the treatments are given either to healthy volunteers (in the early phase of development) or to patients (in the later phases of development). We will refer to patients, volunteers, or whomever is being compared in the trial as the *subjects*.

Table 1: Plan of 2×2 trial.

Group	Period 1	Period 2
1	A	B
2	B	A

Two types of design are used in these trials: the *parallel-group* design and the *crossover* design. In order to explain these we will consider trials for comparing two treatments, A and B. The latter might be different ingredients in an inhaler used to treat asthma attacks or two drugs used to relieve the pain of arthritis, for example.

In a *parallel-group trial* the subjects are randomly divided into two groups of equal size. Everyone in the first group gets A, and everyone in the second group gets B. The difference between the treatments is usually estimated by the difference between the group means.

In a crossover trial the subjects are also randomly divided into two groups of equal size. (In agriculture and dairy science, crossover trials often are referred to as *changeover trials*.) Now, however, each subject gets both treatments for an equal period of time. In the first group the subjects begin by getting A for the first period and then cross over to B for the second period. Each subject in the second group begins with B in the first period and then crosses over to A for the second period. The basic plan of this design is given in Table 1. This type of crossover trial uses two treatment sequences AB and BA and is usually referred to as the 2×2 trial.

The main advantage the crossover trial has over the parallel-group trial is that the two treatments are compared *within subjects* as opposed to *between subjects*. That is, the 2×2 trial provides two repeated measurements on each subject, and the difference between these is used to estimate the difference between A and B. In this way each subject "acts as his or her own control" and any variability between subjects is eliminated. As the variability within subjects is usually much smaller than that between subjects, a relatively precise estimate of the treatment difference is obtained.

In contrast, the treatment difference in the parallel-group trial is estimated by taking differences of measurements taken on different subjects, and so is based on between-subject variability. As a consequence the crossover trial requires far fewer subjects than a parallel-group trial to achieve equivalent power to detect a particular size of treatment difference. A detailed comparison of the two types of design is given by Grieve [18].

If $\bar{y}_{ij}$ is the mean for period j in group i, then for the 2×2 trial the within-subject estimator of the $A-B$ treatment difference is $D = \frac{1}{2}[(\bar{y}_{11} - \bar{y}_{12}) - (\bar{y}_{21} - \bar{y}_{22})]$.

Obviously, crossover trials are not suitable for treatments that effect a cure. A basic assumption is that subjects will be in the same state at the start of the second period as they were in at the start of the first period. Therefore, it is essential to ensure that the effect of the first treatment is not present at the start of the second period. One way of achieving this is to separate the two active periods by a *washout* period of sufficient length to ensure that the effects of the first treatment have disappeared by the start of the second period. Any effect of previous treatment allocation that affects the second period is a *carryover effect*.

If τ_A and τ_B denote the effects of treatments A and B, respectively, and λ_B and λ_B denote the carryover effects of treatments A and B, then in the presence of unequal carryover effects the expected value of D, the within-subjects estimator of the treatment difference, is $(\tau_A - \tau_B) - \frac{1}{2}(\lambda_A - \lambda_B)$, i.e., D is biased. If the carryover difference is of the same sign as the treatment difference, then D underestimates the true treatment difference. Therefore, if a significant treatment difference is detected, it is still appropriate to conclude that the treatments are different, because the trial has detected an even smaller difference between the treatments than anticipated at the planning stage.

In the basic 2×2 trial it is not possible to estimate the difference between the carryover effects using within-subject information. This is because the estimate is based on differences of subject totals. However, for designs with more than two periods or treatment sequences it is possible to estimate the carryover difference using within-subject information. Also, in the basic 2×2 trial the carryover difference is completely confounded with the group difference and the treatment-by-period interaction.

Both the parallel-group trial and the crossover trial are usually preceded by a *run-in* period, when subjects are acclimatized and, perhaps, are monitored for eligibility. Response measurements are usually taken during the run-in period of the parallel-group trial and during the run-in and washout periods of the 2×2 trial. The run-in and washout measurements can be used to estimate the carryover difference in the 2×2 trial, and the run-in measurements can be used to improve the precision of the parallel-group trial. However, even when run-in measurements are used, the parallel-group trial still falls short of the crossover trial as far as precision of estimation is concerned and so needs more subjects than the crossover trial to achieve comparable power.

In the general case of t treatments, a crossover trial consists of randomly dividing the subjects into s groups and assigning a different sequence of treatments to each group. The sequences are of length p, corresponding to the p periods of the trial, and some of the sequences must include at least one change of treatment. The choice of sequences depends on the number of treatments and periods and on the purposes of the trial. Examples of designs for $t = 3$ and $t = 4$ treatments are given later.

20.2 2×2 Crossover Trial

In the absence of run-in and washout measurements, a standard analysis for this design follows the two-stage approach of Grizzle [21] and Hills and Armitage [24]. In the first stage, a test, based on the subject totals, is done to determine if the carryover effects are equal. If they are not significantly different (usually at the 10% level), then in the second stage the within-subjects test for a treatment difference, based on D, is done (usually at the 5% level). If the first-stage test is significant, then the test for a treatment difference uses only the data collected in the first period of the trial, i.e., is based on $\bar{y}_{11} - \bar{y}_{21}$. This analysis has been criticized, particularly because the actual significance level can be much higher than the nominal level of 5% [14,46].

However, some improvement in the performance of this two-stage procedure is possible if measurements are available from the run-in and washout periods [25,28]. The best advice is to base the analysis on the assumption that carryover effects are either absent or equal for both treatments and to proceed directly to the test for a treatment difference that uses within-subject comparisons. This assumption would need to rely on prior knowledge or washout periods of adequate length. As noted above, the within-subject estimate of the treatment difference is biased downward if carryover effects are different, but the extent of the bias will not be great unless the carryover difference is large. The detailed views of a working group of the Biopharmaceutical Section of the American Statistical Association are given in Peace [42, Chap. 3]. Their view is that the 2×2 design is not the design of choice if carryover effects exist.

A Bayesian approach to the analysis is described in Grieve [16,19,20], and further discussion on analysis can be found in Jones and Kenward [26, Sec. 2.13] and Senn [46, Chap. 3].

Overall, if significant carryover effects are likely to occur, then the 2×2 design is best avoided if possible. A number of better designs for two treatments are mentioned below.

20.3 Higher-Order Designs for Two Treatments

The disadvantages of the 2×2 design can be overcome if more periods or sequences are used, given certain assumptions about the behavior of the carryover effects. The two-period design with four sequences AA, BB, AB, and BA enables the treatment difference to be estimated within subjects even if carryover effects are present and different.

However, in common with all two-period designs in general, the estimate of the treatment difference is inefficient, and it is better to use at least three periods. For three periods the recommended design has two sequences, ABB and BAA, and for four periods it has four sequences, $AABB$, $BBAA$, $ABBA$, and $BAAB$.

20.4 Designs for Three or More Treatments

There is a great variety of designs for three or more treatments. The choice of design will be determined by the purpose of the trial, the number of permissible periods or sequences, and other practical constraints. An important feature of these designs is that carryover differences can be estimated using within-subject information. In a variance-balanced design the variance of the estimated difference between any two treatments, allowing for subjects, periods, and any carryover effects, is the same whichever pair of treatments is considered. Plans of such designs for $t = 3$ and $t = 4$ and t periods are given below. For properties of balanced and nearly balanced designs [41].

A general introduction including a review, tables of designs, discussion of optimality, and choice of design are given in Jones and Kenward [26, Chap. 5]. More recent reviews are given in Afsarinejad [1] and Matthews [37].

Study of the optimality of crossover designs has generally concentrated on a model that has fixed effects for the subject, period, treatment, and carryover effects and independent within-subject errors that have constant variance. It is mostly assumed that if carryover effects can occur they are only of first order, i.e., last for only one period. Most results in the literature refer to universal optimality [27]; a useful review is given in Matthews [37].

Closely linked to optimality are the concepts of uniformity and balance. A design is *uniform* if each treatment occurs equally often in each period and each treatment is allocated equally often to each subject. A design is (combinatorially) *balanced* if every treatment follows every other treatment equally often. Balanced uniform designs are universally optimal for the estimation of treatment and carryover effects [22,23,6].

When $p = t$, balanced uniform designs are the Williams designs [50] and can be constructed easily using the algorithm given by Scheehe and Bross [48]. Examples of these designs are given in Tables 2 and 3 for $t = 3$ and $t = 4$, respectively.

If every treatment follows every other treatment, including itself, equally often, the design is *strongly balanced*, and strongly balanced uniform designs are universally optimal [6]. A simple way of gen-

Table 2: Balanced design for $t = 3, p = 3$.

Sequence No.	Treatment Sequence
1	ABC
2	ACB
3	BAC
4	BCA
5	CAB
6	CBA

Table 3: Balanced design for $t = 4, p = 4$.

Sequence No.	Treatment Sequence
1	ABCD
2	BDAC
3	CADB
4	DCBA

erating a strongly balanced design is to repeat the last period of a Williams design to give a design with $p = t + 1$. These designs are variance-balanced and have the additional property that the treatment and carryover effects are orthogonal.

For those combinations of t, s, and p for which a variance-balanced design does not exist, it may be possible to construct a partially balanced design. In such a design the variances of the estimated treatment differences are not all the same but do not vary much. This makes them attractive in practice, as they usually require fewer periods or sequences than a fully balanced design. Such designs are tabulated in Jones and Kenward [26] and in Ratkowsky et al. [45].

Another potentially important group of designs is where the treatments are made up of factorial combinations of two or more ingredients. For example, the four treatments A, B, C, and D might correspond to all possible combinations of two ingredients X and Y, where each ingredient can occur either at a high or low level. Here designs that are efficient for estimating the main effects or the interaction of X and Y

can be constructed [13].

20.5 Analysis of Continuous Data

Crossover data are examples of repeated measurements; that is, they consist of a set of short sequences of measurements. The observations from one subject will typically be correlated, which needs to be accommodated in the analysis.

Continuous crossover data are most commonly analyzed using a conventional factorial linear model and analysis of variance. The model will almost invariably include terms for period and treatment effects. Other terms may be included as required, such as first- and higher-order carryover effects, treatment-by-period interaction, and treatment-by-carryover interactions, although for some designs there may be aliasing among these, and the inclusion of more than a small number of such terms can seriously reduce the efficiency of the analysis. For all terms except carryover, the definition of appropriate factor levels is straightforward. Construction of factors for the latter is not obvious, because there are observations for which these effects cannot occur, for example those in the first period. One simple solution for this is to deliberately alias part of the carryover effect with period effects. For example, for a first-order carryover factor, levels follow treatment allocation in the preceding period, except for observations in the first period, when any factor level can be used provided it is the same in all sequences. After adjustment for the period term, this factor gives the correct sums of squares and degrees of freedom for the first-order carryover.

Within-subject dependence is normally allowed for by the inclusion of fixed subject effects in the linear model. Essentially, a randomized block analysis is used with subjects as blocks.

In the case of the 2×2 trial, this analysis reduces to a pair of t-tests, each comparing the two sequence groups [24]. For the treatment effect, the comparison is of the within-subject differences, and for the carryover-treatment-by-period interaction, it is of the subject totals. Baseline measurements contribute little to the efficiency of the direct treatment comparison, but may substantially increase that of the carryover. Analyses for higher-order two-treatment two-sequence designs can be expressed in terms of t tests in a similar way.

For designs in which treatment effects are not orthogonal to subjects (for example, when $t > p$, or generally when a carryover term is included), there exists some treatment information in the between-subject stratum; this is lost when fixed subject effects are used. It has been suggested that this between-subject (interblock) information should be recovered through the use of *random* subject effects. Restricted maximum likelihood (REML) is an appropriate tool for this. However, small, well-designed crossover trials are not ideal for the recovery of interblock information; most of the treatment information lies in the within-subject stratum, between-subject variability will typically be high, and the reliance on asymptotic estimates of precision means that standard errors of effects can be seriously underestimated.

The use of random subject effects implies a simple uniform covariance structure for the sequence of observations from a subject. A more general structure can be used, for example, to allow for a serial correlation pattern, but such analyses are not widely used, and there is little evidence as yet to suggest that these models are needed for routine crossover data. Such modeling may be important, however, when there are repeated measurements *within* treatment periods, and the multivariate linear model provides an appropriate framework for this

Table 4: Binary data from a 2 × 2 trial.

	Joint Response			
Sequence	(0,0)	(0,1)	(1,0)	(1,1)
AB	n_{11}	n_{12}	n_{13}	n_{14}
BA	n_{21}	n_{22}	n_{23}	n_{24}

Table 5: Contingency table for mainland–Gart test.

Sequence	(0,1)	(1,0)
AB	n_{12}	n_{13}
BA	n_{22}	n_{23}

setting.

Nonparametric methods of analysis for crossover data are not well developed, apart from the two-treatment two-sequence designs in which the t tests can be replaced by their nonparametric analogs. In other designs the occurrence of different treatment sequences precludes a straightforward application of an orthodox multivariate rank test, while the use of the rank transformation followed by a conventional analysis of variance is best avoided [8]. Senn [46] develops an ad hoc test based on a combination of two-sequence tests, and a general review is given in Tudor and Koch [49].

Table 6: Contingency table for Prescott's test.

Sequence	(0,1)	(0,0) or (1,1)	(1,0)
AB	n_{12}	$n_{11}+n_{14}$	n_{13}
BA	n_{22}	$n_{21}+n_{24}$	n_{23}

20.6 Analysis of Discrete Data

We consider a binary response first, coded 0/1. As with continuous data, the 2 × 2 trial forms a simple special case. Each subject then provides one of four categories of joint response: (0,0), (0,1), (1,0), and (1,1). Given n_i subjects on sequence i, the data from such a trial can be summarized in a 2 × 4 contingency table as in Table 4.

Two tests for treatment effect (assuming no carryover effect) are based on the entries in this table. The Mainland–Gart is the test for association in the 2 × 2 contingency table [35,15] as given in Table 5. This involves the data from only those subjects who make a *preference*. Prescott's test [43] introduces the pooled nonpreference data, and it is the test for *linear trend* in the 2 × 3 table given as Table 6. The test for carryover–treatment×period interaction is the test for association in the table involving only the *nonpreference* outcomes [2] as given in Table 7. Conventional chi-square, likelihood ratio, or conditional exact tests can be used with these tables.

These tests are quick and simple to use. Unfortunately, they do not generalize satisfactorily for higher-order designs and ordinal responses, and they are awkward to use when there are missing data. Recent developments in modeling dependent discrete data have made available a number of more flexible model-based approaches that are applicable to crossover data.

Generalized estimating equation (GEE) methods can be used to fit *marginal models* to binary data from any crossover design, whether data are complete or not [52,51,30,31] and have been extended for use with ordinal data [32,31].

A marginal, or *population-averaged*, model defines the outcome probabilities for any notional individual in the population under consideration for the given covariate values (treatment, period, and so on).

Table 7: Contingency table for Hills and Armitage's test.

Sequence	(0,0)	(1,1)
AB	n_{11}	n_{14}
BA	n_{21}	n_{24}

It is marginal with respect to the other periods and, provided a crossover trial is used to draw conclusions about constant as opposed to changing treatment conditions, can be regarded as the appropriate model from which to express conclusions of most direct clinical relevance. The simpler forms of GEE (GEE1) are comparatively straightforward to use, but may provide poor estimates of precision in small trials. Extended GEE methods (GEE2) are more complicated, but give better estimates of precision. The full likelihood for a marginal model cannot be expressed in closed form for $p > 2$, so likelihood-based analyses require considerably more elaborate computation (e.g., [32,3]).

In contrast to marginal models, *subject-specific models* include subject effect(s) that determine an individual's underlying outcome probabilities. Other effects, such as period and direct treatment, modify these subject-specific probabilities, and generally these effects will not have the same interpretation as their marginal-model analogues. Marginal representations of probabilities can be obtained from subject-specific models by taking expectations over the distribution of the subject effects, but only in special cases will the treatment–covariate structure of the subject-specific model be preserved.

In analyses using subject-specific models, subject effects cannot be treated as ordinary parameters as with continuous data, because the number of these effects increases at the same rate as the number of subjects. This implies that estimates of other effects will be inconsistent, a generalization of the well-known result for matched case–control studies. Two alternative approaches can be used: conditional likelihood and random subject effects.

If, for binary data, a logistic regression model is used, or, for categorical data, a model based on generalized (or adjacent-category) logits, then a *conditional* likelihood analysis can be used in which the subject effects are removed through conditioning on appropriate sufficient statistics [29]. In the binary case these statistics are the subject totals; in the categorical case, the subject joint outcomes *ignoring the order*. The application of this approach to binary data from the two-period two-treatment design produces the Mainland–Gart test. The conditional likelihood can be calculated directly, or the whole analysis can be formulated as a loglinear analysis for a contingency table of the form of the 2×4 table above, with appropriate extension for other designs and for categorical outcomes. One advantage of this approach is the availability of conditional exact tests when sample sizes are very small. The two main disadvantages are (1) the discarding of between-subject information in the process of conditioning, which precludes a population-averaged interpretation of the results, and (2) the use of generalized logits, which are not ideal for ordinal categorical outcomes.

If the subject effects are assumed to follow some distribution, typically the normal, then the likelihood for the model can be obtained through numerical integration [11,12].

In general such analyses are computationally intensive, but are usually manageable for crossover trials, for which sample sizes are commonly small. The inferences from such models are subject-specific, but population-averaged summary statistics, for example marginal probabilities, can be produced using integration. Numerical integration can be avoided through the use of an approximate or *hierarchical* likelihood in place of the full marginal likelihood [4,34]. However, the consistency of such procedures is not guaranteed for all sample configurations, and the small-sample properties of the resulting analyses for crossover data have not yet been explored.

20.7 Concluding Remarks

There is a large and diverse literature on the statistical aspects of crossover trials, which reflects their extensive use in medical research. There are at present three books on the subject [26]; [46]; [45] and several reviews. The literature is scattered over numerous journals and conference proceedings (e.g., [5,7,10,33,36]). A particularly useful review is given in *Statist. Methods Med. Res.*, **3**, (4) (1994). In addition to medicine, crossover trials are used in areas such as psychology [39], agriculture [40], and dairy science. An industrial example is given by Raghavarao [44].

References

1. Afsarinejad, K. (1990). Repeated measurements designs—a review. *Commun. Statist. Theory Methods*, **19**, 3985–4028.

2. Armitage, P. and Hills, M. (1982). The two-period cross-over trial. *Statistician*, **31**, 119–131.

3. Balagtas, C. C., Becker, M. P., and Lang, J. B. (1995). Marginal modelling of categorical data from crossover experiments. *Appl. Statist.*, **44**, 63–77.

4. Breslow, N. E. and Clayton, D. G. (1993). Approximate inference in generalized linear models. *J. Am. Statist. Assoc.*, **88**, 9–24.

5. Carriere, K. C. and Reinsel, G. C. (1992). Investigation of dual-balanced crossover designs for two treatments. *Biometrics*, **48**, 1157–1164.

6. Cheng, C.-S. and Wu, C-F. (1980). Balanced repeated measurements designs. *Ann. Statist.*, **6**, 1272–1283. Correction (1983), **11**, 349.

7. Chi, E. M. (1992). Analysis of cross-over trials when within-subject errors follow an AR(1) process. *Biometr. J.*, **34**, 359–365.

8. Clayton, D. and Hills, M. (1987). A two-period cross-over trial. In *The Statistical Consultant in Action*, D. J. Hand and B. S. Everitt, eds. Cambridge University Press.

9. Cochran, W. G., Autrey, K. M., and Cannon, C. Y. (1941). A double change-over design for dairy cattle feeding experiments. *J. Dairy Sci.*, **24**, 937–951.

10. Cornell, R. G. (1991). Non-parametric tests of dispersion for the two-period crossover design. *Commun. Statist. Theory Methods*, **20**, 1099–1106.

11. Anonymous (1985–1990). *EGRET: Epidemiological, Graphics, Estimation and Testing Package*. Statistics and Epidemiology Research Corp., Seattle, WA.

12. Ezzet, F. and Whitehead, J. (1991). A random effects model for ordinal responses from a cross-over trial. *Statist. Med.*, **10**, 901–907.

13. Fletcher, D. J., Lewis, S. M., and Matthews, J. N. S. (1990). Factorial designs for crossover clinical trials. *Statist. Med.*, **9**, 1121–1129.

14. Freeman, P. R. (1989). The performance of the two-stage analysis of two-treatment, two-period cross-over trials. *Statist. Med.*, **8**, 1421–1432.

15. Gart, J. J. (1969). An exact test for comparing matched proportions in crossover designs. *Biometrika*, **56**, 75–80.

16. Grieve, A. P. (1985). A Bayesian analysis of the two-period cross-over trial. *Biometrics*, **41**, 979–990. Correction (1986), **42**, 456.

17. Grieve, A. P. (1987). A note on the analysis of the two-period crossover design when period–treatment interaction is significant. *Biometr. J.*, **29**, 771–775.

18. Grieve, A. P. (1990). Crossover vs parallel designs. In *Statistics in Pharmaceutical Research*, D. A. Berry, ed. Marcel Dekker, New York.

19. Grieve, A. P. (1994). Extending a Bayesian analysis of the two-period crossover to allow for baseline measurements. *Statist. Med.*, **13**, 905–929.

20. Grieve, A. P. (1994). Bayesian analyses of two-treatment crossover studies. *Statist. Methods Med. Res.*, **4**, 407–429.

21. Grizzle, J. E. (1965). The two-period changeover design and its use in clinical trials. *Biometrics*, **21**, 467–480.

22. Hedayat, A. and Afsarinejad, K. (1978). Repeated measurements designs, I. In *A Survey of Statistical Design and Linear Models*, J. N. Srivastava, ed., pp. 229–242, North-Holland, Amsterdam.

23. Hedayat, A. and Afsarinejad, K. (1978). Repeated measurements designs, II. *Ann. Statist.*, **6**, 619–628.

24. Hills, M. and Armitage, P. (1979). The two-period cross-over clinical trial. *Br. J. Clin. Pharm.*, **8**, 7–20.

25. Jones, B. and Lewis, J. A. (1995). The case for cross-over trials in phase III. *Statist. Med.*, **14**, 1025–1038.

26. Jones, B. and Kenward, M. G. (1989). *Design and Analysis of Crossover Trials*. Chapman & Hall, London. (This text takes a broad view with emphasis on crossover trials used in medical research. Both theory and practice are covered in some detail. The analysis of repeated measurements both between and within periods is considered. Methods for analyzing binary and categorical data are described as well as methods for continuous data.)

27. Kiefer, J. (1975). Construction and optimality of generalized Youden designs. In *A Survey of Statistical Design and Linear Models*, J. N. Srivastava, ed., pp. 333–341, North-Holland, Amsterdam.

28. Kenward, M. G. and Jones, B. (1987). The analysis of data from 2 × 2 crossover trials with baseline measurements. *Statist. Med.*, **6**, 911–926.

29. Kenward, M. G. and Jones, B. (1991). The analysis of categorical data from cross-over trials using a latent variable model. *Statist. Med.*, **10**, 1607–1619.

30. Kenward, M. G. and Jones, B. (1992). Alternative approaches to the analysis of binary and categorical repeated measurements. *J. Biopharm. Statist.*, **2**, 137–170.

31. Kenward, M. G. and Jones, B. (1994). The analysis of binary and categorical data from crossover trials. *Statist. Methods Med. Res.* **3**, 325–344.

32. Kenward, M. G., Lesaffre, E., and Molenberghs, G. (1994). An application of maximum likelihood and generalized estimating equations to the analysis of ordinal data from a longitudinal study with cases missing at random. *Biometrics*, **50**, 945–953.

33. Laserre, V. (1991). Determination of optimal designs using linear models in crossover trials. *Statist. Med.*, **10**, 909–924.

34. Lee, Y. and Nelder, J. A. (1996). Hierarchical generalized linear models. *J. Roy. Statist. Soc. B*, **58**, 619–678.

35. Mainland, D. (1963). *Elementary Medical Statistics*, 2nd ed. Saunders, Philadelphia.

36. Matthews, J. N. S. (1990). The analysis of data from crossover designs: The efficiency of ordinary least squares. *Biometrics*, **46**, 689–696.

37. Matthews, J. N. S. (1994). Multi-period crossover designs. *Statist. Methods Med. Res.*, **4**, 383–405.

38. Molenberghs, G. and Lesaffre, E. (1994). Marginal modelling of ordinal data using a multivariate Plackett distribution. *J. Am. Statist. Assoc.*, **89**, 633–644.

39. Namboodiri, K. N. (1972). Experimental design in which each subject is used repeatedly. *Psychol. Bull.*, **77**, 54–64.

40. Patterson, H. D. (1950). The analysis of change-over trials. *J. R. Statist. Soc. B*, **13**, 256–271.

41. Patterson, H. D. (1982). Change over designs. In *Encyclopedia of Statistical Sciences*, S. Kotz, N. L. Johnson, and C. B. Read, eds., pp. 411–415, Wiley, New York.

42. Peace, K. E. (1990). *Statistical Issues in Drug Research and Development*. Marcel Dekker, New York.

43. Prescott, R. J. (1979). The comparison of success rates in cross-over trials in the presence of an order effect. *Appl. Statist.*, **30**, 9–15.

44. Raghavarao, D. (1989). Crossover designs in industry. In *Design and Analysis of Experiments, with Applications to Engineering and Physical Sciences*, S. Gosh, ed. Marcel Dekker, New York.

45. Ratkowsky, D. A., Evans, M. A., and Alldredge, J. R. (1993). *Crossover Experiments*. Marcel Dekker, New York. (The contents of this text are presented from the viewpoint of someone who wishes to use the SAS statistical analysis system to analyze crossover data. Two nonstandard features are the way designs are compared and the approach suggested for the analysis of categorical data. Extensive tables of designs, which would otherwise be scattered over the literature, are included.)

46. Senn, S. (1993). *Cross-over Trials in Clinical Research*. Wiley, Chichester. (This text is written mainly for biologists and physicians who want to analyze their own data. The approach is nontechnical, and explanations are given via worked examples that are medical in nature. A critical view is expressed on mathematical approaches to modeling carryover effects.)

47. Senn, S. (1994). The AB/AB crossover: Past, present and future? *Statist. Methods Med. Res.*, **4**, 303–324.

48. Scheehe, P. R. and Bross, I. D. J. (1961). Latin squares to balance for residual and other effects. *Biometrics*, **17**, 405–414.

49. Tudor, G. and Koch, G. G. (1994). Review of nonparametric methods for the analysis of crossover studies. *Statist. Methods Med. Res.*, **4**, 345–381.

50. Williams, E. J. (1949). Experimental designs balanced for the estimation of residual effects of treatments. *Austral. J. Sci. Res.*, **2**, 149–168.

51. Zeger, S. L. and Liang, K. -Y. (1992). An overview of models for the analysis of longitudinal data. *Statist. Med.*, **11**, 1825–1839.

52. Zhao, L. P. and Prentice, R. L. (1990). Correlated binary regression using a quadratic exponential model. *Biometrika*, **77**, 642–648.

21. Design and Analysis for Repeated Measurements

Gary G. Koch, J. D. Elashoff, and I. A. Amara

21.1 Introduction

A broad range of statistical investigations can be classified as repeated measurements studies. Their essential feature is that each subject is observed under two or more conditions. Four important classes of repeated measures studies are as follows:

1. *Split-plot experiments in agriculture.* An example is the evaluation of the effects of fertilizer and crop variety on crop yield, where fertilizer types are randomly assigned to fields and crop varieties are randomly assigned to plots within fields. The fields represent the "subjects" that are randomly assigned to levels of one factor (fertilizer type), and the plots correspond to the observational conditions that are randomly assigned to levels of a second factor (crop variety).

2. *Longitudinal studies.* For example, cattle are randomly assigned to one of three diets and their weight is measured every week for three months. The cattle are the subjects, and the successive time intervals correspond to the observational conditions.

3. *Change over design studies.* For example, each subject is randomly assigned either to the sequence group with treatment A first and B second or to the sequence group with treatment B first and then A. Responses to each successive treatment are measured for their corresponding periods of administration. The observational conditions are not only the successive time periods, but also the treatments, and possibly the immediately preceding treatments.

4. *Sources of variability studies.* An example is the study of the number of gastrin cells per unit length of rat stomach tissue via measurements for adjacent microscopic fields, in adjacent biopsies, from selected gastric sites, by two different observers. The rats constitute the subjects, and the sites, biopsies, fields, and observers can be either a fixed set or a random sample from a large population.

For studies like those described in 1–4, the subjects are primary sampling units randomly selected to represent various strata or randomly assigned to levels of a grouping factor; they are often called *experimen-*

tal units. The responses measured under the respective conditions constitute the observational units; because within each subject these constitute a profile of inherently multivariate data, their covariance structure plays an important role in the formulation of statistical methods for their analysis.

A useful strategy for a broad range of repeated measurements studies is to view the subjects as the units of analysis in the following two-stage procedure: certain measures of interest (e.g., sums for total response, differences between conditions, orthogonal contrasts between conditions) are first constructed from the observational unit data within each subject; then this information is analyzed across experimental units by parallel, but separate, univariate methods and/or by simultaneous multivariate methods. This general strategy can appropriately account for the measurement scale of the response (categorical, ordinal, interval) and the nature of the randomization in the study design.

21.2 Design

In this section, principles pertaining to the design of repeated measurements studies are discussed through representative examples. Separate attention is given to split-plot experiments, longitudinal studies, change over design studies, and sources of variability studies. The features that are shared by these different classes of designs are noted as well as those that characterize them; also, relative advantages and limitations are identified.

21.2.1 Split-Plot Experiments

The distinguishing aspect of split-plot experiments relative to other types of studies is the use of two or more stages of randomization. In the first stage, subjects are randomly allocated to treatments or randomly selected from strata; in the subsequent stages, conditions are randomly allocated within subjects. In the usual agricultural experiment, the subjects are whole plots or fields within which split plots are the observational units. Some other representative examples are as follows:

S1. The whole plots are six batches of synthetic fiber; these correspond to three batches from each of two different combinations of ingredients. The split plots are four subsamples from each batch; they are tested for pulling strength under four different temperature conditions.

S2. The whole plots are litters (or cages) of rats that are assigned to diets containing different amounts of fat, and the split plots are the different doses of carcinogen under which the tumor levels of individual animals are observed; or, the split plots could be the different times at which the rats were sacrificed for tumor evaluation.

S3. The whole plots are school systems that are assigned to one of two different sets of materials for a science unit, and the split plots are individual schools in which different teaching strategies are used. The response measure is student examination performance.

S4. Rats are paired on the basis of weight; within each pair, one rat receives an experimental diet and the other receives the same amount of a control diet (i.e., it is a pair-fed control). Seven days later, the rats are sacrificed and the liver of each rat is divided into three parts. For each section of a rat liver, the amount of iron absorbed is determined from a solution with one of three randomly assigned pH (acidity) levels and with randomly assigned temperature for the pair to which the animal belonged. Pairs of rats are whole plots, rats within pairs are split plots, and liver thirds within a rat are split split plots. Further discussion of this *split-split-plot experiment* is given in Koch [45].

S5. Three generations of animals with large litters are used to assess the effect of hormone and diet treatments on hormonal blood levels in the offspring. In the first generation, four females from the same litter are randomly assigned to four hormone doses. When these females have offspring, three females are selected from each litter and randomly assigned to be fed one of three different diets; also, the four animals receiving the same diet are maintained in the same cage. Thus, at the second generation, there is a block of 12 females originating from the same first generation litter and residing in three cages (for diet) with four animals per cage (for hormone dose). The third generation litters from these 12 females represent whole plots, and five females within each of them correspond to split plots to which five weekly intervals for time of sacrifice are assigned in order to measure hormonal blood levels. These 60 animals are raised in 12 cages with the five animals from the same litter sharing the same cage. In all, there are 10 blocks with the (4×3) (mother $\times$ cage) structure described here. This type of study is called a *split-block experiment*. It can be naturally applied in agriculture by assigning one type of treatment to rows of a field (e.g., planting method) and another type to the columns (e.g., fertilizer); row $\times$ column cells are whole plots within which a third factor (e.g., varieties of a crop) can be assigned to split plots; see Federer [22] for further discussion.

A broad range of potential research designs is thus available for split-plot experiments, and so relevant aspects of a particular situation can be taken into account. The treatment groups can be based on a single factor or on the cross-classification of two or more factors; they can be assigned to whole plots according to a completely randomized design, a randomized complete blocks design, or some type of incomplete blocks design. A similar statement applies to the nature of conditions and their assignment to split plots. When incomplete blocks structures are used for cross-classified treatment or condition factors, issues concerning the *confounding* of effects require careful attention for both design specification and analysis. Finally, the preceding considerations also apply to situations where whole plots are randomly selected from strata that correspond to the groups.

Split-plot experiments have two important advantages over other research designs without their nested structure. They can be less costly or more straightforward to implement when the treatment for each whole plot can be applied to its entirety (i.e., to all of its split plots simultaneously rather than separately). The second advantage is that split plots within the same whole plot are usually more homogeneous than those from different whole plots. As a result, comparisons between conditions and (treatment $\times$ condition) interaction effects (i.e., between treatment differences for comparisons of conditions) are estimated more precisely. The basic consideration here is that the variability of such estimates comes from within whole plot variability for split plots rather than overall (i.e., across whole plots) variability. Thus, it corresponds to the way in which the design of split-plot experiments enables the researcher to control the extent of variability influencing comparisons between conditions. Since whole plots are the units through which such control is applied, they are often said to serve as their own controls.

Three limitations of split-plot experiments should be noted. One is that differences between treatments applied to whole plots are estimated less precisely than differences between conditions applied to split plots and also less precisely than if they had been applied to the same number of independent split plots; this occurs because split plots within whole plots usually have a positive *intraclass correlation*. Second, greater cost or effort might be required for the administration of split-plot experiments in order to ensure that each condition only affects the split plots to which it was assigned. Contamination of condition effects to neighboring split plots needs to be negligible so that estimated comparisons among conditions and treatments are not biased to a potentially misleading extent. The third limitation of split-plot experiments is that complexities in their structure can make the analysis of their data relatively difficult.

The statistical literature for split-plot experiments is extensive; bibliographies have been published by Federer and Balaam [24], Hedayat and Afsarinejad [32], and Federer [23]. A useful basic reference is Snedecor and Cochran [82]. Some textbooks that discuss alternative designs are Allen and Cady [2], Bennett and Franklin [5], Cochran and Cox [12], Cox [17], Federer [21], Gill [27], Kempthorne [40], Myers [64], and Winer [92].

21.2.2 Longitudinal Studies

The primary way in which longitudinal studies differ from split-plot experiments is that the observational units for their subjects are systematically linked to the conditions rather than being randomly assigned. The usual dimension for such linkage is time, but it can also be location in space or different components of a concept, item, or process. The design of longitudinal studies for subjects specifies parallel groups, which can be based on either random allocation to treatments, random selection from strata, or both. Some representative examples are:

L1. Two treatments for chronic pain are randomly assigned to subjects, and the extent of pain relief is evaluated at weekly visits for 6 weeks. Alternatively, for studies of the rapidly occurring effects of treatments, heart rate, blood pressure, or gastric pH might be measured at more frequent intervals such as every hour or every 10 minutes.

L2. Boys and girls from a cohort of 1-year-olds are observed every 6 months for 5 years to assess their ability to perform a manual dexterity task (or measurements of height, weight, or physical fitness might be made).

L3. Two treatments for a dental problem are randomly assigned to children. The status of teeth on the upper and lower jaws is evaluated every three months for one year. Here, the eight conditions are determined by the site × time cross-classification.

L4. In a political survey, each subject is asked about the degree of trust in three political institutions (the Presidency, the Senate, and the Supreme Court). Subjects are drawn from different demographic groups. The three institutions constitute the conditions.

An important aspect of the design of longitudinal studies is the specification of the number of conditions and their nature (e.g., a set of time intervals). As the number of conditions is increased, the amount of information for the response is increased; but the cost or effort for its acquisition and management is also usually increased (a potential exception being studies with automated data collection and processing devices). Since conditions represent situations for comprehensively observing within-subject response across some dimension(s) of interest (e.g., time course) rather than an experimental factor, which can be assigned either within or among subjects, their cost is a necessary consequence of their scope. Also, there are no alternative designs that would provide the same information. Thus, conditions are specified to encompass the observational situations of greatest interest in a manner compatible with available resources.

In some cases, the number of conditions may be too extensive for all of them to be observed on each subject (e.g., more than 100 timepoints over a period exceeding 5 years), owing to cost or subject tolerance, say. One way of dealing with this problem is to define incomplete subsets that suitably cover the range of conditions and to assign them randomly to the subjects in the respective treatment groups. Such subsets can be formed in ways that appropriately account for what is feasible in a particular application. For example, when the conditions are based on time, overlapping subsets of timepoints (0–12, 6–18, 12–24, etc.) can be used. Studies with this structure are sometimes said to have a *panel design* with *rotating groups* of new subjects entering every six months and leaving after 12 months of participation. Similarly, when age is the principal dimension for conditions, subjects with consecutive initial ages (e.g., children in the range 6–17 years) can be subsequently observed for some response (e.g., height) at the same time intervals over some specified period (e.g., every 6 months over 3 years); in this way, longitudinal information is obtained for the entire age range under study (e.g., children 6–20 years old). Studies of this nature are said to have a linked cross-sectional or mixed longitudinal design; see Rao and Rao [72] and Woolson et al. [93].

The range of potential designs for longitudinal studies is very broad. Since the data for each subject form a multivariate profile, the relevant design principles are the same as for multivariate studies. Discussion and related references are given in Roy et al. [77]. For more specific consideration of the number of conditions to be used and the spacings between them, see Morrison [62] and Schlesselman [79]. When the number of conditions is small relative to the number of subjects and all conditions are observed on all subjects, data from longitudinal studies can often be satisfactorily analyzed by multivariate analysis of variance procedures like those described in the Methods section of this entry; also, see textbooks dealing with multivariate analysis such as Anderson [4], Bock [7], Gill [27], Morrison [63], Rao [71], and Timm [86]. For situations in which the number of conditions is large or that require methods specifically designed for longitudinal studies, refer to Cook and Ware [16], Dielman [19], Geisser [26], Grizzle and Allen [30], Laird and Ware [52], Nesselroade and Baltes [66], Rao [70], Snee et al. [83], and Ware [90]. The longitudinal studies described here do not include follow-up studies for the time until some event such as death or recurrence of disease.

21.2.3 Changeover Designs

Change over designs have similarities to both split-plot and longitudinal studies. Subjects are assigned at random to groups or selected from strata. As in longitudinal studies, the response of each subject is observed at several points in time (or locations in space); and as in split-plot experiments, a treatment condition is assigned to each time period. Subjects are randomly assigned to groups that receive alternative sequences of treatments. More than one sequence is needed to allow the separate estimation of time or location effects and treatment effects (since they are completely confounded in any single sequence). For this reason, sequences are often based on the rows of a *Latin square*.

Another important aspect of changeover studies is that the preceding or neighboring treatment can influence the response to the next treatment. Such factors are called *carryover effects* or *residual effects*. When their extent cannot be presumed negligible, adjustments for their presence become necessary. In such cases, the sequences in the design of a changeover study need to be constructed so as to allow the estimation of location effects, treatments effects and carryover effects. Alternatively, it may suggest that some other research design in which subjects receive only one treatment be used. Brown [10], Cochran and Cox [12], Constantine and Hedayat [15], Gill [27], Kershner and Federer [42], Hedayat and Afsarinejad [32,33], Koch et al. [51], Laycock and Seiden [53], and Wallenstein and Fisher [88] provide additional discussion concerning construction and properties.

Some examples that illustrate the repeated measurements nature of changeover studies are

C1. Information on recent smoking was obtained for each subject by two different methods; one was the subject's self-report to a direct question and the other was a biochemical determination based on carbon monoxide levels in the blood. Subjects were randomly assigned to one of two sequence groups; for one group, the self-report preceded the biochemical determination; and for the second group, the self-report followed the biochemical determination.

C2. Patients with an unfavorable skin condition on both sides of the body are randomly assigned to one of two groups; one group receives an active treatment for daily application on the right side of the body and placebo for the left side, and the other receives the active treatment for the left side and placebo for the right side. The extent of healing is evaluated separately for each side at weekly visits during a 4-week period.

C3. The relative potency of two drugs that influence cardiovascular function is assessed through a change over design. Volunteers are randomly assigned to one of two sequence groups. One group receives drug A during the first 6-week study period and drug B during the second, and the other group receives the opposite regimen. A 2-week washout period separates the two treatment periods. During each treatment period, three doses of the drug are tested with the drug dose being successively increased every 2 weeks. At the beginning of treatment and at the end of each 2-week dose interval, heart rate is measured before and after a treadmill exercise test. Additional discussion of this type of example is given in Elashoff [20].

C4. Subjects are randomly assigned to one of six groups based on the cross-classification of a treatment factor (alcohol vs. no alcohol) and a sequence factor for the assignment of three drugs; the sequence factor is based on the three rows of a 3×3 Latin square. At each of the three one-day study periods, performance on a task is measured in response to one of the three drugs. The study periods are separated by one-week washout intervals in which no drugs or alcohol are to be used.

For changeover studies like those in C1–C4, note that subjects are analogous to whole plots in split-plot experiments, and treatment sequences are analogous to their treatment groups. A straightforward extension of this structure is its application within different subpopulations of subjects. These subpopulations could correspond to different strata from which subjects were randomly selected or to one or more additional experimental factors that were randomly assigned to subjects (e.g., alcohol status in C4). The structure of conditions in change over studies can be extended by observing the response to each within-subject treatment at a longitudinal set of locations (e.g., several timepoints as in C2 and C3). All of the considerations outlined here imply that changeover studies can be designed in many different ways.

Since changeover designs have a structure similar to that of split-plot experiments, they have similar advantages and disadvantages. The advantages are potentially reduced cost through the need for fewer subjects (because more than one treatment is tested for each) and more precise estimates for within-subject treatment comparisons. The principal disadvantage is that research design, management, and analysis strategies to minimize or deal with carryover effects may be complex and expensive. The major issue is that carryover effects may induce substantial bias in treatment comparisons if they are ignored; adjustments to eliminate such bias may lead to use of less efficient estimation methods such as analyzing only the data from each subject's first treatment.

21.2.4 Sources of Variability Studies

Sources of variability studies are concerned with identifying the amount of variability between responses that is attributable to each component of the sampling or measurement process. Some representative examples are as follows:

V1. The variability in product performance is to be studied. On each of 10 randomly identified production days, two samples are obtained at randomly specified times. Each sample is divided into five subsamples, and each subsample is assigned to one of five evaluators who make two performance determinations. The same five evaluators are used on each of the 10 days.

V2. The variability between and within observers for ratings of severity of a periodontal condition is to be assessed. For each of a sample of patients, two photographs of gum status are obtained. All photographs are shown in random order to five observers for rating.

V3. The variability associated with households within clusters and with interviewers is to be assessed in a survey of a socioeconomic variable. A random sample of 288 clusters of eight households is randomly divided into three sets (A, B, C) of 96 clusters. Twenty-four interviewers are randomly assigned to each set of clusters. Each assignment in the following protocol is made at random. In set A of household clusters, each interviewer is assigned four clusters and obtains responses from all eight households in each cluster. In set B, 12 sets of two interviewers are formed and eight clusters are assigned to each pair; one member of the pair of interviewers is assigned to four households in each cluster, and the second to the other four households. In set C, six blocks of four interviewers are formed and each block is assigned 16 clusters; each interviewer in the block is assigned to two households in each cluster.

Studies of sources of variability can thus deal with fixed components of a measurement process such as the evaluators in V1, and random components such as days and samples in V1. Their structure can be straightforward or very complex. For designs where interest is focused on a fixed component of variability with few levels in a straightforward design, as in the case of the observers in V2, a satisfactory analysis can usually be undertaken with the procedures described in Section 20.3. See also, for example, Anderson and Bancroft [3], Kempthorne [41], Scheffé [78], and Searle [81].

21.2.5 Combination Designs

The previous discussion has dealt with split-plot experiments, longitudinal studies, changeover studies, and variability studies as separate types of repeated measurements studies; however, many research designs combine features across these types. Longitudinal data can be obtained for each split plot or for each successive treatment in a changeover study (see C2 and C3). Responses can be assessed by more than one observer for each split plot or subject. Repeated measurements studies can be designed to provide information for factors of direct interest (treatment effects or time patterns) and on background factors (place or observer effects); the latter are often called *lurking variables* so as to reflect their potential influence on the analysis of a response variable; see Joiner [39]. The examples presented here illustrate the wide variety of research designs involving repeated measurements; others are given in Federer [22], Gill [27], Koch et al. [46], and Monlezun et al. [61].

21.3 Statistical Methods

In this section, several strategies for the analysis of repeated measurements studies are described. They include

1. Univariate analysis of within-subject functions

2. Multivariate analysis for within-subject functions

3. Repeated measures analysis of variance

4. Nonparametric rank methods

5. Categorical data methods

The specification of these methods takes five important considerations into account: the measurement scale, the number of subjects, the role of randomization, the nature of the covariance structure of the observational units within subjects, and the potential influence of carryover effects. Basic methods for relatively straightforward situations are emphasized. Examples 21.3.1–21.3.4 provide insight on their application and are potentially helpful to read in parallel with the discussion of methods or prior to it.

21.3.1 Data Structure

A general framework for repeated measurements studies involves a set of subjects for whom data are obtained for each of d conditions; these conditions can be based on a single design factor or the cross-classification of two or more factors. For each subject, the data for each condition can include a set of N determinations for each of R response variables and M concomitant variables for which potential associations with the variation between conditions are to be taken into account. Also, the information for the respective subjects can include classifications for one or more study design factors by which the subjects are partitioned into G groups, as well as K background variables for which potential associations with the variation between groups are to be taken into account. The subsequent discussion is directed at situations with $N = 1$ determination for $R = 1$ response variable and M concomitant variables for the respective conditions within each subject; when $N > 1$, the methods for $N = 1$ are usually applicable to averages across the multiple determinations. When $R > 1$, the multivariate analogs of methods for a univariate response variable are usually applicable; see Reinsel [73] and Thomas [85].

Let $i = 1, 2, \ldots, G$ index the G groups and $j = 1, 2, \ldots, d$ index the d conditions. Let $l = 1, 2, \ldots, n_i$ index the n_i subjects within the ith group. Let Y_{ijl} denote the response of the lth subject in the ith group for the jth condition and $\mathbf{z}_{ijl} = (z_{ijl1}, z_{ijl2}, \ldots, z_{ijlM})'$ be the corresponding set of observed values for M concomitant variables. Finally, let $\mathbf{x}_{i*l} = (x_{i1l}, x_{i2l}, \ldots, x_{iKl})'$ denote the set of K background variables that represent fixed characteristics for the lth subject in the ith group. Interpretation of this notation can be clarified by considering a changeover design example like C4, where $i = 1, 2, 3, 4, 5, 6$ indexes the six groups of subjects with respect to alcohol status and drug sequence and $j = 1, 2, 3$ indexes the three periods during which they receive the three drugs; the background variables $\mathbf{x}_{i*l}$ for the lth subject represent characteristics such as age, sex, and previous medical history, for which the values are defined prior to the study (or apply to all conditions); and the concomitant variables $\mathbf{z}_{ijl}$ represent aspects of health status such as baseline performance at the time of administration of the jth drug, period effects, drug effects, or carryover effects; the response y_{ijl} represents performance after the administration of the jth drug. More specific illustrations are given in Examples 21.3.1 and 21.3.2

21.3.2 Univariate Analysis of Within-Subject Functions

For many repeated measurements studies, the questions of interest can be expressed in terms of some summary functions of within-subject responses to the conditions. These summary functions could be contrasts; the general form of a contrast for the lth subject in the ith group is $f_{il} = \Sigma_{j=1}^{d} a_j y_{ijl}$, where $\Sigma_{j=1}^{d} a_j = 0$. Examples are pairwise comparisons of conditions $\{(y_{ijl} - y_{ij'l})\}$; or when the conditions represent equally spaced doses or times, the contrasts could be orthogonal polynomials, e.g., for four doses, the linear contrast is $-3y_{i1l} - y_{i2l} + y_{i3l} + 3y_{i4l}$. Summary functions could also be the average response over conditions ($f_{il} = \bar{y}_{i \cdot l} = \Sigma_{j=1}^{d} y_{ijl}/d$) or nonlinear functions such as ratios.

Suppose that the variation of the $\{y_{ijl}\}$ across groups and conditions can be described by the additive model

$$E\{y_{ijl}\} = \mu + \xi_i + \theta_j + (\xi\theta)_{ij}, \quad (1)$$

where μ is a reference parameter for group 1 and condition 1, the $\{\xi_i\}$ are incremental effects for the respective groups, the $\{\theta_j\}$ are incremental effects for the respective conditions, and the $\{(\xi\theta)_{ij}\}$ are interaction effects between group and condition; also, $\xi_1 = 0, \theta_1 = 0$, the $\{(\xi\theta)_{1j} = 0\}$ and the $\{(\xi\theta)_{i1} = 0\}$ to avoid redundancy. To simplify the exposition, we shall concentrate the subsequent discussion in this section on a contrast function. Two cases are considered:

1. One group or multiple groups with no (group × condition) interaction

2. Multiple groups that possibly interact with conditions

Case 1: One Group or No Group-by-Condition Interaction. Here the contrast functions $\{f_{il}\}$ for the respective subjects have the same expected value

$$E\{f_{il}\} = \sum_{j=1}^{d} a_j \theta_j = \theta_f, \qquad (2)$$

since no (group × condition) interaction implies $\{(\xi\theta)_{ij} = 0\}$ for (1), and if there is only one group, the $\{(\xi\theta)_{ij}\}$ do not exist. The variances for the $\{f_{il}\}$ have the form

$$\text{Var}\{f_{il}\} = \sum_{j=1}^{d} \sum_{j'=1}^{d} a_j a_{j'} v_{i,jj'}, \qquad (3)$$

where $v_{i,jj'} = \text{Cov}\{y_{ijl}, y_{ij'l}\}$ denotes the (jj')th element of the covariance matrix $\mathbf{V}_i$ for the d responses of subjects in the ith group. The $\{f_{il}\}$ will have homogeneous variances

$$\text{Var}\{f_{il}\} = v_f \qquad (4)$$

if the elements of the $\{\mathbf{V}_i\}$ have the structure

$$v_{i,jj'} = v_{i,*} + v_{*,jj'}; \qquad (5)$$

here $v_{i,*}$ represents the between-subject variance component for the ith group and $v_{*,jj'}$ the within-subject covariance component for the jth and j'th conditions. However, it is often realistic to make the stronger assumption that the d responses for all subjects have the same covariance matrix $\mathbf{V}$ (i.e., all $\mathbf{V}_i = \mathbf{V}$), and this is done henceforth.

Confidence intervals and statistical tests concerning θ_f in (2) can be obtained by using the t distribution if it is also assumed that the $\{f_{il}\}$ for the $n = \Sigma_{i=1}^{G} n_i$ subjects have independent normal distributions. The overall mean

$$\bar{f} = \left\{ \sum_{i=1}^{G} \sum_{l=1}^{n_i} f_{il}/n \right\} \qquad (6)$$

is an unbiased estimator of θ_f. An unbiased estimator of its variance is given by $(\hat{v}_f/n)$, where

$$\hat{v}_f = \sum_{i=1}^{G} \sum_{l=1}^{n_i} (f_{il} - \bar{f})^2/(n-1). \qquad (7)$$

Thus, the $100(1-\alpha)\%$ confidence interval for θ_f is given by

$$\bar{f} - t\,[\hat{v}_f/n]^{1/2} \leq \theta_f \leq \bar{f} + t\,[\hat{v}_f/n]^{1/2}, \qquad (8)$$

where $t = t_{1-(\alpha/2)}\{(n-1)\}$ is the $100\{1-(\alpha/2)\}$ percentile of the t distribution with $(n-1)$ degrees of freedom. The hypothesis $H_0: \theta_f = 0$ can be tested by rejecting it if 0 is not included in the interval (8), i.e., if

$$\left| \bar{f}/[\hat{v}_f/n]^{1/2} \right| \geq t. \qquad (9)$$

Analogous one-sided confidence intervals to (8) and hypothesis tests to (9) can be constructed by using $t = t_{1-\alpha}(n-1)$ for the specified direction only.

When the overall sample size n is sufficiently large (e.g., $n > 30$), $\bar{f}$ can have an approximately normal distribution on the basis of central limit theory. For such situations, (8) and (9) are applicable with $t = t_{1-(\alpha/2)}(\infty)$ under the more general assumption that the $\{f_{il}\}$ are independent random variables with a common distribution. Also, the common distribution assumption (and hence the homogeneous variance assumption) can be relaxed to hold within each of the respective groups if $\hat{v}_f$ in (7) is replaced by

$$\tilde{v}_f = \sum_{i=1}^{G} \left(\frac{n_i^2}{n^2} \right) \sum_{l=1}^{n_i} \frac{(f_{il} - \bar{f}_i)^2}{(n_i - 1)n_i}$$

$$= \sum_{i=1}^{G} \frac{n_i}{n^2} \hat{v}_{f,i}, \qquad (10)$$

where $\bar{f}_i = \{\Sigma_{l=1}^{n_i} n_i f_{il}/n_i\}$ is the sample mean of the $\{f_{il}\}$ for the ith group and $\hat{v}_{f,i}$ is the sample variance. The large sample counterparts of (8) and (9) with (10) replacing (7) are of particular interest when the $\{f_{il}\}$ represent contrasts among scored values for ordinally scaled categorical data, e.g., $f_{il} = -1, 0, 1$ for pairwise differences between dichotomous observations $\{y_{ijl} = 0, 1\}$ and $f_{il} = -3, -2, -1, 0, 1, 2, 3$ for pairwise differences between integer scores $\{y_{ijl} = 1, 2, 3, 4\}$ for observations with respect to four ordinal categories. For additional discussion, see Guthrie [31], Koch et al. [50], Stanish et al. [84].

Nonparametric methods provide another way to proceed when the normal distribution assumption for the $\{f_{il}\}$ is not realistic. Symmetry and independence of the distributions of the $\{f_{il}\}$ are sufficient to support usage of the *sign test*; if the $\{f_{il}\}$ can also be ranked, the *Wilcoxon signed rank test* is applicable. Related confidence intervals can be constructed under the further assumption that the $\{f_{il}\}$ have a common, continuous distribution; see Conover [14]. Otherwise, when ratio or percent change comparisons are of interest, the symmetry condition can often be satisfied by replacing the $\{f_{il}\}$ by analogous contrasts for logarithms.

Case 2: Multiple Groups that May Interact With Conditions. The methods discussed for case 1 are appropriate for designs where there is only one group or it can be assumed that the effect of conditions is the same in every group. In most situations, however, it will be of interest to test this assumption; frequently, in fact, the question of whether there is a (group × condition) interaction is of interest in its own right and can be assessed by comparing the $\{f_{il}\}$ among groups.

When the contrast functions $\{f_{il}\}$ are normally distributed, the hypothesis of no group differences in their expected values

$$E\{f_{il}\} = \sum_{j=1}^{d} a_j \theta_j + \sum_{j=1}^{d} a_j (\xi\theta)_{ij}$$
$$= \theta_f + (\xi\theta)_{f,i}$$
$$= \Delta_{f,i}, \qquad (11)$$

can be tested by one-way analysis of variance; relevant background variables, concomitant variables, or functions of them can be taken into account by analysis of covariance methods. For situations where the normal distribution assumption for the $\{f_{il}\}$ is not realistic, other strategies are available. If the sample sizes for each group are sufficiently large for the within-group means $\{\bar{f}_i\}$ to have approximately normal distributions via central-limit theory, then the Wald statistic methods described in Koch et al. [50]. Alternatively, randomization of the subjects to groups or a common family for the distributions of the $\{f_{il}\}$ with location shifts being the only source of across-group variation is sufficient to support usage of nonparametric methods like the Kruskal–Wallis statistic; see Koch [44], Koch et al. [46], Lehmann [54], Puri and Sen [68].

When there is (group × condition) interaction, the contrasts $\{f_{il}\}$ may have different expected values $\{\Delta_{f,i}\}$ for the G groups. As a result, the expected value of their overall mean $\bar{f}$ in (6) is a weighted average of the $\{\Delta_{f,i}\}$, the weights being the proportions $\{(n_i/n)\}$ of the subjects in the respective groups, i.e.,

$$E\{\bar{f}\} = \sum_{i=1}^{G} (n_i/n) \Delta_{f,i}. \qquad (12)$$

Sometimes these weights are reasonable, sometimes other weights $\{w_i\}$ are preferable [e.g., $w_i = (1/G)$ for all i], and sometimes usage of any weights at all is considered undesirable because alternative choices lead to seemingly different conclusions. When the $\{f_{il}\}$ have independent normal distributions with the expected value structure in (11) and the homogeneous variance structure in (4), a confidence interval analogous to (8) can be formed for the weighted average $\Delta_{f,w} =$

$\Sigma_{i=1}^{G} w_i \Delta_{f,i}$. It is

$$f_w - t\{\hat{v}_{f,w}\}^{1/2} \leq \Delta_{f,w} \leq f_w + t\{\hat{v}_{f,w}\}^{1/2}, \quad (13)$$

where

$$f_w = \sum_{i=1}^{G} w_i \bar{f}_i,$$

$$\hat{v}_{f,w} = \bar{v}_f \sum_{i=1}^{G} (w_i^2/n_i)$$

with $\bar{v}_f = \Sigma_{i=1}^{G} \Sigma_{l=1}^{n_i} (f_{il} - \bar{f}_i)^2/(n-G)$, and $t = t_{1-(\alpha/2)}\{(n-G)\}$. Here, the pooled within-group variance $\bar{v}_f$ rather than $\hat{v}_f$ in (7) is used to estimate v_f since it accounts for the potential variation between groups for the expected values $\{\Delta_{f,i}\}$. Thus, the application of (13) to $\bar{f}$ in the context of ((12)) can be viewed as providing a more robust interval for θ_f when (2) is viewed as approximately true rather than strictly true; a limitation of (13) is that it can give a confidence interval undesirably wider than (8) for small samples (e.g., $n - G < 10$). Otherwise, when the assumption of normal distributions is not realistic for the $\{f_{il}\}$, large-sample or nonparametric methods can be formulated in ways that yield results similar in spirit to (13).

Another summary function of interest, which is not a contrast, is the within-subject mean

$$\bar{y}_{i \cdot l} = \left(\sum_{j=1}^{d} y_{ijl}/d \right) = \mathbf{1}'_d \mathbf{y}_{il}/d, \quad (14)$$

where $\mathbf{y}_{il} = (y_{i1l}, \ldots, y_{idl})'$ and $\mathbf{1}_d$ is the $(d \times 1)$ vector of 1's. It allows between group comparisons of the average responses over conditions, i.e., the

$$E\{\bar{y}_{i \cdot l}\} = \left\{ \sum_{j=1}^{d} \mu_{ij}/d \right\} = \bar{\mu}_{i \cdot}, \quad (15)$$

where $\mu_{ij} = E\{y_{ijl}\}$. When all $(\xi\theta)_{ij} = 0$ in the model (1), comparisons between the $\{\bar{\mu}_{i \cdot}\}$ are the same as comparisons between the group effects $\{\xi_i\}$ because

$$\bar{\mu}_{i \cdot} = (\mu + \bar{\theta}) + \xi_i, \quad (16)$$

with $\bar{\theta} = (\Sigma_{j=1}^{d} \theta_j/d)$, in this case. For more general situations with (group × condition) interaction, comparisons among the $\{\bar{\mu}_{i \cdot}\}$ may need to be interpreted cautiously since tendencies for one group to have higher responses than another for some conditions and lower responses for other conditions will be absorbed in the average over conditions. The methods of analysis for the $\{\bar{y}_{i \cdot l}\}$ are similar to those described for contrasts. They are one-way analysis of variance when the $\{\bar{y}_{i \cdot l}\}$ have independent normal distributions with homogeneous variance $v_m = \mathbf{1}'_d \mathbf{V} \mathbf{1}_d/d^2$, and its large-sample or nonparametric counterparts when these assumptions are not realistic. Also, extensions for covariance analysis allow relevant explanatory variables to be taken into account; see Snedecor and Cochran [82] or Neter et al. [67] for the normal distribution framework and Koch et al. [47] and Quade [69] for the nonparametric framework.

In this section, we have discussed estimation and testing of summary functions as if each were dictated by the nature of the conditions and were to be evaluated separately. However, if several summary functions are to be tested and simultaneous inference is an important issue, *multiple comparisons methods* may be used to control the overall significance level. A useful, simple approach is the Bonferroni method (see Miller [60]) in which for ν comparisons, the overall significance level α is obtained by testing each comparison at the nominal significance level (α/ν).

The methods described in this section have been focused on linear functions for complete data situations, but their applicability is broader. The same strategies, particularly the large sample or nonparametric approaches, can be directed at more complicated functions such as ratios or medians. When some subjects have incomplete data (i.e., data for some conditions are missing), functions can be defined to take the observed missing data pattern into account. One approach is to restrict attention to the subset of subjects with the necessary data for determining a particular function; e.g., for a pairwise difference, restriction to those subjects with responses to both conditions. However, the populations represented by this subset of subjects and those for whom the function could not be determined must be equivalent in order for selection bias to be avoided; the justification for such equivalence can involve stratification or a statistical model. Alternatively, attention can be directed at functions that have meaningful definitions for all subjects; e.g., the last observation or the median in a longitudinal study. Since such functions may have complicated distributions, large sample or nonparametric methods may be more appropriate for their analysis; see Koch et al. [46,47].

21.3.3 Multivariate Methods for Within-Subject Functions

Some analysis questions involve the simultaneous consideration of $u \geq 2$ within-subject linear functions $\mathbf{f}_{\mathbf{A},il} = \mathbf{A}'\mathbf{y}_{il}$, where $\mathbf{A}$ is a $(d \times u)$ specification matrix. These can be addressed by the multivariate extensions of the previously discussed univariate methods. For example, when there is only one group or no (group × condition) interaction in the model (1), the overall hypothesis of no differences between condition effects is equivalent to

$$E\{(y_{ijl} - y_{idl})\} = 0$$
$$\text{for } j = 1, 2, \ldots, (d-1). \quad (17)$$

The corresponding matrix formulation is directed at $u = (d-1)$ linear functions $\mathbf{f}_{\mathbf{A},il}$ for which $\mathbf{A}' = [\mathbf{I}_u, -\mathbf{1}_u]$; here $\mathbf{I}_u$ is the uth-order identity matrix and $\mathbf{1}_u$ is the $(u \times 1)$ vector of 1's. When the $\{\mathbf{f}_{\mathbf{A},il}\}$ have independent multivariate normal distributions with homogeneous covariance structure $\mathbf{A}'\mathbf{V}_i\mathbf{A} = \mathbf{V}_{\mathbf{A},i} = \mathbf{V}_{\mathbf{A}}$, the expression of the hypothesis in (17) as $E\{\mathbf{f}_{\mathbf{A},il}\} = \mathbf{0}_u$, where $\mathbf{0}_u$ is a $(u \times 1)$ vector of 0's, allows it to be tested by Hotelling's T^2 statistic. More specifically, for the pooled groups, let

$$\bar{\mathbf{f}}_{\mathbf{A}} = \frac{1}{n}\sum_{i=1}^{G}\sum_{l=1}^{n_i} \mathbf{f}_{\mathbf{A},il}$$
$$= \mathbf{A}'\left\{\frac{1}{n}\sum_{i=1}^{G}\sum_{l=1}^{n_i} \mathbf{y}_{il}\right\} = \mathbf{A}'\bar{\mathbf{y}} \quad (18)$$

denote the mean vector of the $\{f_{\mathbf{A},il}\}$ and let

$$\bar{\mathbf{V}}_{\mathbf{A}} = \frac{1}{(n-G)}$$
$$\times \sum_{i=1}^{G}\sum_{l=1}^{n_i}(\mathbf{f}_{\mathbf{A},il} - \bar{\mathbf{f}}_{\mathbf{A},i})(\mathbf{f}_{\mathbf{A},il} - \bar{\mathbf{f}}_{\mathbf{A},i})'$$
$$= \mathbf{A}'\left\{\frac{1}{(n-G)}\right.$$
$$\left.\times \sum_{i=1}^{G}\sum_{l=1}^{n_i}(\mathbf{y}_{il} - \bar{\mathbf{y}}_i)(\mathbf{y}_{il} - \bar{\mathbf{y}}_i)'\right\}\mathbf{A}$$
$$= \mathbf{A}'\bar{\mathbf{V}}\mathbf{A}, \quad (19)$$

where $\bar{\mathbf{y}}_i = \sum_{l=1}^{n_i}(\mathbf{y}_{il}/n_i)$ and $\bar{\mathbf{f}}_{\mathbf{A},i} = \mathbf{A}'\bar{\mathbf{y}}_i$, denote the pooled within groups unbiased estimator of $\mathbf{V}_{\mathbf{A}}$. The rejection region with significance level α for the hypothesis (17) concerning $u = (d-1)$ linear functions is

$$\frac{(n-G-u+1)}{(n-G)u}T^2 \geq \mathcal{F}, \quad (20)$$

where $\mathcal{F} = \mathcal{F}_{1-\alpha}\{u, (n-G-u+1)\}$ is the $100(1-\alpha)$ percentile of the $\mathcal{F}$ distribution with $\{u, (n-G-u+1)\}$ degrees of freedom, and

$$T^2 = n\bar{\mathbf{f}}_\mathbf{A}' \overline{\mathbf{V}}_\mathbf{A}^{-1} \bar{\mathbf{f}}_\mathbf{A}$$
$$= n\bar{\mathbf{y}}' \mathbf{A} \{\mathbf{A}' \overline{\mathbf{V}}_\mathbf{A} \mathbf{A}\}^{-1} \mathbf{A}' \bar{\mathbf{y}} \quad (21)$$

is the Hotelling T^2 statistic. Here, it is assumed that $\mathbf{V}_\mathbf{A}$ is nonsingular and $n \geq (G+u)$ so that $\overline{\mathbf{V}}_\mathbf{A}$ is almost certainly nonsingular.

An important property of T^2 is that its value stays the same across all nonsingular transformations of the $\{\mathbf{f}_{\mathbf{A},il}\}$. Thus, it is invariant for any specification of the hypothesis (17) or corresponding choice of $\mathbf{A}$ that is a basis of contrast space [i.e., any $(d \times (d-1))$ matrix $\mathbf{A}$ such that rank $(\mathbf{A}) = (d-1)$ and $\mathbf{A}' \mathbf{1}_d = \mathbf{O}_{(d-1)}$], and so it provides a well defined overall test procedure for the hypothesis of no differences among condition effects.

The application of the Roy–Bose [76] method for multiple comparisons to the distribution of T^2 in (21) allows simultaneous confidence intervals to be constructed for all linear contrasts $\theta_f = \sum_{j=1}^d a_j \theta_j$ between condition effects at an overall $100(1-\alpha)\%$ level. Such intervals, which are analogous to those from the *Scheffé method*, have the form

$$\mathbf{a}' \bar{\mathbf{y}} - \mathcal{T}\{\mathbf{a}' \overline{\mathbf{V}}_\mathbf{A} \mathbf{a}/n\}^{1/2}$$
$$\leq \theta_f \leq \mathbf{a}' \bar{\mathbf{y}} + \mathcal{T}\{\mathbf{a}' \overline{\mathbf{V}}_\mathbf{A} \mathbf{a}/n\}^{1/2}, \quad (22)$$

where $\mathbf{a} = (a_1, a_2, \ldots, a_d)'$ and

$$\mathcal{T} = \left\{ \frac{(n-G)u}{(n-G-u+1)} \right.$$
$$\left. \times [\mathcal{F}_{1-\alpha}\{u, (n-G-u+1)\}] \right\}^{1/2} \quad (23)$$

with $u = (d-1)$; such intervals can also be formed with respect to lower-dimensional subspaces with $u < (d-1)$.

A potential limitation of the multivariate procedures (20) and (22) is that their effectiveness for overall inference purposes requires enough subjects to provide a stable estimate of $\mathbf{V}_\mathbf{A}$ (e.g., $n - G - u \geq 15$); a somewhat smaller number is sufficient if the multivariate counterpart of $\hat{v}_f$ in (7) is used to estimate $\mathbf{V}_\mathbf{A}$ instead of $\overline{\mathbf{V}}_\mathbf{A}$ (e.g., $n - u \geq 15$). Also, complete response vectors are required (that is, each subject must be observed under all conditions). An advantage of these methods is their applicability under minimal covariance structure assumptions.

Large-sample and nonparametric counterparts to the multivariate procedures (20) and (22) are similar in spirit to those described for the univariate procedures (8) and (9). When the overall sample size n is sufficiently large (e.g., $n \geq d + G + 30$), $\bar{\mathbf{f}}_\mathbf{A}$ has approximately a multivariate normal distribution. Thus, T^2 has the χ^2 distribution with u degrees of freedom. Also, potential heterogeneity among the covariance structures $\{\mathbf{V}_{\mathbf{A},i}\}$ can be taken into account by refinements analogous to (10). As for univariate methods, such analysis is of particular interest for categorical data. Nonparametric methods for multivariate comparisons among conditions can be based on extensions of the sign test and Wilcoxon signed ranks test; see Koch et al. [46] for discussion and references.

For studies with multiple groups, the hypothesis of no (group × condition) interaction can be tested by multivariate analysis of variance methods when the $\{\mathbf{f}_{\mathbf{A},il}\}$ have independent multivariate normal distributions with homogeneous covariance structure; see Cole and Grizzle [13], Gill [27], Morrison [63], and Timm [86,87]. Strategies for evaluating condition effects in the potential presence of (group × condition) interaction are analogous to (13). When the assumption of multivariate normality is not realistic, large sample or nonparametric procedures are available; see Koch [44], Koch et al. [46,50], and Puri and Sen [68].

Table 1: Repeated measures analysis of variance.[a]

Source of Variation	Degrees of Freedom (df)	Sums of Squares	Mean Square
Groups	$(G-1)$	$SS(G) = \sum_{i=1}^{G} dn_i(\bar{y}_{i\cdot\cdot} - \bar{y}_{\cdot\cdot\cdot})^2$	$MS(G) = SS(G)/(G-1)$
Subjects within groups	$(n-G)$	$SS(S) = \sum_{i=1}^{G} \sum_{l=1}^{n_i} d(\bar{y}_{i\cdot l} - \bar{y}_{i\cdot\cdot})^2$	$MS(S) = SS(S)/(n-G)$
Total among subjects	$(n-1)$	$SS(AS) = \sum_{i=1}^{G} \sum_{l=1}^{n_i} d(\bar{y}_{i\cdot l} - \bar{y}_{\cdot\cdot\cdot})^2$	
Conditions	$(d-1)$	$SS(C) = \sum_{j=1}^{d} n(\bar{y}_{\cdot j\cdot} - \bar{y}_{\cdot\cdot\cdot})^2$	$MS(C) = SS(C)/(d-1)$
Groups × conditions	$(G-1)(d-1)$	$SS(GC) = \sum_{i=1}^{G} \sum_{j=1}^{d} n_i$ $(\bar{y}_{ij\cdot} - \bar{y}_{i\cdot\cdot} - \bar{y}_{\cdot j\cdot} + \bar{y}_{\cdot\cdot\cdot})^2$	$MS(GC) = SS(GC)/\{(G-1)(d-1)\}$
Conditions × subjects within groups (error)	$(n-G)(d-1)$	$SSE = \sum_{i=1}^{G} \sum_{l=1}^{n_i} \sum_{j=1}^{d}$ $(y_{ijl} - \bar{y}_{ij\cdot} - \bar{y}_{i\cdot l} + \bar{y}_{i\cdot\cdot})^2$	$MSE = SSE/\{(n-G)(d-1)\}$
Total within subjects	$n(d-1)$	$SS(WS) = \sum_{i=1}^{G} \sum_{l=1}^{n_i} \sum_{j=1}^{d} (y_{ijl} - \bar{y}_{i\cdot l})^2$	
Total among observational units	$(nd-1)$	$SST = \sum_{i=1}^{G} \sum_{l=1}^{n_i} \sum_{j=1}^{d} (y_{ijl} - \bar{y}_{\cdot\cdot\cdot})^2$	

[a] Here, $n = \sum_{i=1}^{G} n_i$, $\bar{y}_{ij\cdot} = (1/n_i) \sum_{l=1}^{n_i} y_{ijl}$, $\bar{y}_{i\cdot l} = (1/d) \sum_{j=1}^{d} y_{ijl}$, $\bar{y}_{i\cdot\cdot} = (1/d) \sum_{j=1}^{d} \bar{y}_{ij\cdot}$, $\bar{y}_{\cdot j\cdot} = (1/n) \sum_{i=1}^{G} n_i \bar{y}_{ij\cdot}$, and $\bar{y}_{\cdot\cdot\cdot} = (1/n) \sum_{i=1}^{G} n_i \bar{y}_{i\cdot\cdot}$.

21.3.4 Repeated Measures Analysis of Variance

The number of subjects n in split-plot experiments and changeover design studies is often small (e.g., $n \leq 10$) because of cost constraints for the total number of observational units partitioned among them. For such situations, univariate or multivariate analysis of within-subject functions may be unsatisfactory for assessing condition effects because critical values like t in (8) or (13) or $\mathcal{F}$ in (20) may be so large that they excessively weaken the effectiveness of the corresponding confidence intervals and test procedures; also multivariate methods are not applicable if $n \leq u$. One way to resolve this problem is to use methods based on more stringent assumptions that enable the covariance structure of contrasts $\mathbf{f}_{\mathbf{A},il}$ to be estimated with larger degrees of freedom and thereby provide smaller critical values of t or F.

An assumption that can be realistic for many split-plot experiments and changeover studies is that the elements of the covariance matrices of the response vectors satisfy (5) with $v_{*,jj} = v_{*,0}$ for all j and $v_{*,jj'} = 0$ for all $j \neq j'$; i.e., all diagonal elements of $\mathbf{V}_i$ are equal to $(v_{i,*} + v_{*,0})$, and all other elements are equal to $v_{i,*}$. Covariance matrices with this structure are said to have the property of *compound symmetry*. An underlying model for which it holds is

$$y_{ijl} = \mu_{ij} + s_{i*l} + e_{ijl}, \quad (24)$$

where the $\{s_{i*l}\}$ are independent subject effects with $E\{s_{i*l}\} = 0$ and $\text{Var}\{s_{i*l}\} = v_{i,*}$, the $\{e_{ijl}\}$ are independent response errors with $E\{e_{ijl}\} = 0$ and $\text{Var}\{e_{ijl}\} = v_{*,0}$, and the $\{s_{i*l}\}$ and $\{e_{ijl}\}$ are mutually independent.

For split-plot experiments, randomization of conditions within subjects and the interchangeable nature of observational units (e.g., subsamples, littermates) often provide justification for the model (24); for change-over design studies, the response process needs to be sufficiently stable that its conditional distributions within subjects are independent and have homogeneous variance for the respective periods.

When the model (24) applies, a set of contrasts $\mathbf{f}_{\mathbf{A},il} = \mathbf{A}'\mathbf{y}_{il}$ will have variance

$$\text{Var}(\mathbf{f}_{\mathbf{A},il}) = \mathbf{V}_{\mathbf{A}} = (\mathbf{A}'\mathbf{A})v_{*,0}. \quad (25)$$

Moreover, if $\mathbf{A}$ is orthonormal (i.e., its columns are mutually orthogonal and have unit length), then $\mathbf{V}_{\mathbf{A}} = \mathbf{I}_u v_{*,0}$. This property of $\mathbf{V}_{\mathbf{A}}$ is called *sphericity* or *circularity* (see Huynh and Feldt [35] and Rouanet and Lepine [75]); it is implied by compound symmetry, which in turn is implied by the model (24), but it can arise under somewhat more general conditions for the covariance structure of the $\{\mathbf{y}_{il}\}$.

The usual analysis of variance calculations for the model (24) are displayed in Table 1. When the covariance matrix satisfies the sphericity condition with $u = (d-1)$ and the $\{s_{i*l}\}$ and $\{e_{ijl}\}$ are normally distributed, the mean-square error (MSE) is an unbiased estimate of $v_{*,0}$, and the hypothesis of no (group × condition) interaction can be tested with the rejection region

$$\text{MS}(GC)/\text{MSE}$$
$$\geq \mathcal{F}_{1-\alpha}[(d-1)(G-1),$$
$$(d-1)(n-G)]. \quad (26)$$

When the conditions correspond to the cross-classification of two or more factors or have some other relevant structure, there usually is interest in assessing (group × condition) interaction for one or more subsets of contrasts. For a set of u contrasts $\{\mathbf{f}_{\mathbf{A},il}\}$, the numerator of the counterpart test to (26) would be

$$\text{MS}(GC, \mathbf{A})$$
$$= \sum_{i=1}^{G} \frac{n_i(\bar{\mathbf{f}}_{\mathbf{A},i} - \bar{\mathbf{f}}_{\mathbf{A}})'(\bar{\mathbf{f}}_{\mathbf{A},i} - \bar{\mathbf{f}}_{\mathbf{A}})}{[u(G-1)]}, \quad (27)$$

where $\bar{\mathbf{f}}_{\mathbf{A},i} = (\sum_{l=1}^{n_i} \mathbf{f}_{\mathbf{A},il}/n_i)$ and $\bar{\mathbf{f}}_{\mathbf{A}} = \sum_{i=1}^{G}(n_i \bar{\mathbf{f}}_{\mathbf{A},i}/n)$. The denominator would be MSE if sphericity applied to the entire $(d-1)$-dimensional space for contrasts or it would be

$$\mathrm{MS}(E, \mathbf{A}) = \sum_{i=1}^{G}\sum_{l=1}^{n_i} \frac{(\mathbf{f}_{\mathbf{A},il} - \bar{\mathbf{f}}_{\mathbf{A},i})'(\mathbf{f}_{\mathbf{A},il} - \bar{\mathbf{f}}_{\mathbf{A},i})}{[u(n-G)]} \quad (28)$$

if it apppplied only to the u-dimensional subspace $\mathbf{A}$; in the former case, the critical value of $\mathcal{F}$ would be that appropriate to $u(G-1)$ and $(d-1)(n-G)$ degrees of freedom (df), and in the latter, it would be for df $= [u(G-1), u(n-G)]$. It should be noted that the test based on (28) should be used for situations that involve more than two components of variance in models analogous to (24); e.g., split-split-plot experiments or studies that involve several sources of random variation.

When there is no (group × condition) interaction, the overall null hypothesis (17) of no differences between conditions can be tested with the rejection region

$$\mathrm{MS}(C)/\mathrm{MSE} \geq \mathcal{F}_{1-\alpha}[(d-1), (d-1)(n-G)]. \quad (29)$$

For the subset of u contrasts $\{\mathbf{f}_{\mathbf{A},il}\}$ the numerator is replaced by

$$\mathrm{MS}(C, \mathbf{A}) = n\bar{\mathbf{f}}'_{\mathbf{A}}\bar{\mathbf{f}}_{\mathbf{A}}/u,$$

and either MSE or (28) can be the denominator in accordance with the previous discussion for (27). If MSE is used, df $= [u, (d-1)(n-G)]$, while if (28) is used, df $= [u, u(n-G)]$.

When it can be assumed that there is no (group × condition) interaction and that the within-subject means $\{\bar{y}_{i \cdot l}\}$ have homogeneous variances across groups, the hypothesis of no differences between group effects can be tested by the rejection region

$$\mathrm{MS}(G)/\mathrm{MS}(S) \geq \mathcal{F}_{1-\alpha}[(G-1), (n-G)]. \quad (30)$$

Note that the denominator in (30) for the comparisons between groups is different from that used in (26) and (29) for the comparisons between conditions. This is because the common variance of the $\{\bar{y}_{i \cdot l}\}$ usually involves both between-subject and within-subject components of variation. Specifically, relative to the model (24) with all $v_{i,*} = v_s$

$$\mathrm{Var}\{\bar{y}_{i \cdot \cdot}\} = \frac{v_s}{n_i} + \frac{v_{*,0}}{n_i d}, \quad (31)$$

whereas (25) indicates that the variances of contrasts only involve $v_{*,0}$. Typically, more powerful results are provided for comparisons between conditions than comparisons between groups; the underlying considerations are the extent to which pairwise differences between the $\{\bar{y}_{i \cdot \cdot}\}$ have larger variances than pairwise differences between the $\{\bar{y}_{\cdot j \cdot}\}$ and the extent to which the denominator degrees of freedom for (30) is smaller than those for (26) and (29); see Jensen [38] for discussion of related issues.

The previous discussion of repeated measures analysis of variance was based on several assumptions. These included complete response vectors, normal distributions with homogeneous covariance matrices, and the sphericity structure in (25). Also, the question of whether there is (group × condition) interaction is important. When all of these assumptions hold, then usage of these methods and their counterparts for subsets of within-subject contrasts is reasonable. Otherwise, alternative procedures need to be applied. Three general strategies that are of interest for this purpose are summarized as follows:

1. Regression methods for situations with normal distributions and sphericity

2. Nonparametric methods for situations where normal distributions do not apply

3. Methods for situations where sphericity does not apply

Strategy 1: Regression Methods for Situations with Normal Distributions and Sphericity. When response vectors are incomplete or more complex models are required to account for period effects, carryover effects or concomitant variables, within-subject contrasts can be analyzed by least-squares methods. Suppose the lth subject in the ith group is observed for d_{il} conditions; consideration is given to contrasts $\{\mathbf{f}_{il} = \mathbf{A}'_{il}\mathbf{y}_{il}\}$, where the $\{\mathbf{A}'_{il}\}$ are orthonormal $\{(d_{il}-1) \times d_{il}\}$ matrices. Sphericity is assumed and so the respective covariance matrices of the $\{\mathbf{f}_{il}\}$ have the form $\{\mathbf{I}_{(d_{il}-1)}v_{*,0}\}$. Let the variation within and between the $E\{\mathbf{f}_{A,il}\} = \psi_{il}$ be described by the linear model $\psi_{il} = \mathbf{Z}_{il}\boldsymbol{\beta}$, where the $\{\mathbf{Z}_{il}\}$ are known $\{[(d_{il}-1) \times t]\}$ submatrices of the full-rank specification matrix $\mathbf{Z} = [\mathbf{Z}'_{11}, \ldots, \mathbf{Z}'_{1n_1}, \ldots, \mathbf{Z}'_{G,n_G}]'$ for the respective subjects and $\boldsymbol{\beta}$ is the $(t \times 1)$ vector of unknown parameters. Inferences concerning $\boldsymbol{\beta}$ are then based on multiple regression methods; such analysis is illustrated for Example 20.3.2.

An equivalent approach involves the application of least-squares methods to the $\{y_{ijl}\}$ with the subject effects $\{s_{i*l}\}$, but no group effects or background variables for subjects, forced into the model. Linkage of the parameters of such a model to the ψ_{il} and $\boldsymbol{\beta}$ needs to be identified; for related discussion, see Schwertman [80]. This analysis strategy is not appropriate for inferences concerning between-subject sources of variation (e.g., groups) unless (24) holds with all $v_{i,*} = 0$; such questions need to be addressed either by analyses of within-subject means like (14) or in the setting of maximum likelihood or related methods for a general multivariate linear model for the $\{y_{ijl}\}$.

Strategy 2: Nonparametric Methods. For situations where the normal distribution assumption for the $\{y_{ijl}\}$ is not realistic and the response vectors could be incomplete, randomization-based nonparametric methods are appropriate for comparisons between conditions if they have been randomly assigned within subjects or the distributions of the $\{(\mathbf{y}_{il} - \boldsymbol{\mu}_i)\}$ are invariant under within-subject permutations. Examples of such methods are the Cochran [11] statistic for dichotomous data, the Friedman [25] two-way rank analysis of variance statistic for ordinal data, and various extensions; see Darroch [18], Koch et al. [46], Myers et al. [65], White et al. [91]. Exact probability levels can be determined for small n, and chi-square approximations can be used when n is at least moderate.

Strategy 3: Methods Where Sphericity Does Not Apply. The sphericity structure in (25) may not be realistic for several reasons. For complex split-plot experiments with several levels of randomization with respect to a hierarchy of observational units, model (24) may need to be extended to include additional components of variance. It is then possible for sphericity to hold for subsets of orthonormal contrasts; these can then be separately analyzed by methods analogous to (27) and (28).

On the other hand, for longitudinal studies and changeover studies, sphericity is often contradicted by the inherent tendency for closely adjacent observational units to be more highly correlated than remote ones, i.e., for there to be a simplex serial correlation pattern among responses to the conditions. A general strategy here is to use extensions of multivariate models that describe the variation of the $\{\mu_{ij}\}$ across both groups and conditions and specify patterned covariance structures $\{\mathbf{V}_i\}$ with respect to underlying variance components; see Laird and Ware [52], Cook and Ware [16], Ware [90]. This approach can be particularly advantageous for longitudinal studies with an at least moderately large number of subjects whose response vectors are possibly incomplete.

When n is small, d is moderate, and response vectors are complete, corrections can be applied to the test procedures (26) and (29) to adjust for lack of sphericity. The F critical values are applied, but the degrees of freedom of both numerator and denominator are reduced by multiplying by the factor

$$\epsilon = \frac{\{\text{trace}(\mathbf{V_A})\}^2}{[(d-1)\{\text{trace}(\mathbf{V_A^2})\}]} \quad (32)$$

for orthonormal contrasts $\mathbf{A}$ with $\text{rank}(d-1)$; see Box [9] and Greenhouse and Geisser [29]. Since $[1/(d-1)] \leq \epsilon \leq 1$, use of $1/(d-1)$ for ϵ will provide a conservative test; similarly, if attention were restricted to $u = \text{Rank}(\mathbf{A})$ orthonormal contrasts, then $(d-1)$ would be replaced by u in (32) and $(1/u)$ would be the conservative lower bound. For longitudinal studies with correlation ρ^δ for observational units δ units apart, Wallenstein and Fleiss [89] discuss the use of less conservative lower bounds for ϵ. Greenhouse and Geisser [29], Huynh and Feldt [36], and Rogan et al. [74] all suggest estimators of ϵ based on sample estimates of $\mathbf{V_A}$; also, see Huynh [34] and Maxwell and Arvey [57] for further consideration of approximate tests.

Since lack of sphericity can cause the results of repeated measures analysis of variance to be potentially misleading (see Boik [8] and Maxwell [56]), statistical tests concerning it are of some interest. However, methods for this purpose (see Anderson [4], Mauchley [55], and Mendoza [59]) are sensitive to nonnormal distributions for the data (particularly outliers), and their power is relatively weak for small or moderate n (see Boik [8] and Keselman et al. [43]). Thus, usage of repeated measures analysis of variance is only recommended when sphericity can be presumed on the basis of subject matter knowledge and research design structure (e.g., split-plot experiments); otherwise, univariate analysis of within-subject functions is preferable.

Example 21.3.1 A Split-Plot Experiment with Litters of Baby Rats. This experiment was undertaken to compare plasma fluoride concentrations (PFCs) for $G = 6$ groups of litters of baby rats. These groups corresponded to two age strata (6 day-old and 11 day-old) within which three doses for intraperitoneal injection of fluoride (0,10, 0.25, 0.50 µg per gram of body weight) were investigated. For each age stratum, $n_0 = 3$ litters of baby rats were fortuitously assigned to each of the three doses. The $n = 18$ litters are the subjects for this study. Six rats from each litter were assessed; two fortuitously selected rats were sacrificed at each of the three post-injection time conditions (15, 30, 60 minutes) for PFC measurement. The response values for the respective conditions are the averages of the natural logarithms of the PFCs for the corresponding pairs of baby rats; thus, pairs of baby rats are the observational units. The data are displayed in Table 2.

The means for each of the $d = 3$ conditions are shown in Table 3 for each group. They indicate that PFC tends to decrease over time for each (age × dose) group and tends to decrease with lower doses for each (age × time). The 11-day-old rats show a greater decrease over time in response to the higher doses than do the 6-day-old rats. These aspects of the data provide the motivation for considering three orthonormal summary functions of the data from each litter. Function 1 reflects the average response for the three time periods; func-

Table 2: Average logarithms of plasma fluoride concentrations from pairs of baby rats in split-plot study.

Age (days)	Dose (μg)	Litter	Minutes Postinjection		
			15	30	60
6	0.50	1	4.1	3.9	3.3
6	0.50	2	5.1	4.0	3.2
6	0.50	3	5.8	5.8	4.4
6	0.25	4	4.8	3.4	2.3
6	0.25	5	3.9	3.5	2.6
6	0.25	6	5.2	4.8	3.7
6	0.10	7	3.3	2.2	1.6
6	0.10	8	3.4	2.9	1.8
6	0.10	9	3.7	3.8	2.2
11	0.50	1	5.1	3.5	1.9
11	0.50	2	5.6	4.6	3.4
11	0.50	3	5.9	5.0	3.2
11	0.25	4	3.9	2.3	1.6
11	0.25	5	6.5	4.0	2.6
11	0.25	6	5.2	4.6	2.7
11	0.10	7	2.8	2.0	1.8
11	0.10	8	4.3	3.3	1.9
11	0.10	9	3.8	3.6	2.6

Table 3: Means and estimated standard errors for Llgarithms of plasma fluoride concentrations and orthonormal functions for groups of baby rats in split-plot study.

Age (days)	Dose (μg)	Minutes Postinjection			Orthonormal Function		
		15	30	60	Average	Trend	Curvature
6	0.50	5.02	4.53	3.65	7.62	0.97	0.16
6	0.25	4.63	3.89	2.83	6.56	1.27	0.13
6	0.10	3.44	2.97	1.89	4.79	1.09	0.25
11	0.50	5.85	4.36	2.85	7.54	2.12	0.01
11	0.25	4.89	3.65	2.29	6.25	1.84	0.05
11	0.10	3.65	2.99	2.12	5.06	1.08	0.08
Estimated SE for all groups		0.41	0.53	0.36	0.70	0.20	0.20

tion 2 is the linear trend across log time; and function 3 is a measure of any lack of linearity (or curvature) in the trend. The functions are specified by

$$\mathbf{U}' = \begin{bmatrix} (1,1,1)/\sqrt{3} \\ (1,0,-1)/\sqrt{2} \\ (-1,2,-1)/\sqrt{6} \end{bmatrix}$$

$$= \begin{bmatrix} \mathbf{1}_3'/\sqrt{3} \\ \mathbf{A}' \end{bmatrix}; \quad (33)$$

and their means are shown on the right side of Table 4.

Separate analyses of variance can be undertaken for the three functions as long as it is reasonable to assume normality and across-group homogeneity of variances for each of them. Results are shown in the corresponding columns of Table 4. There is a significant effect of dose on the "average response," a significant overall linear trend, and a nearly significant effect of dose on the size of the linear trend. In addition, there is a significant (age × dose) interaction for the linear trend; this corresponds to the tendency for the difference between ages to increase with dose. There is no evidence of curvature in the time trend or of differences in curvature due to age or dose.

The application of repeated measures analysis of variance to these data presumes that the response vectors $\{\mathbf{y}_{il}\}$ for the respective litters have multivariate normal distributions with the same covariance matrix $\mathbf{V}$ and that sphericity holds. Normality and equality of covariance matrices seem reasonable for this example. The pooled within-group estimate for $\mathbf{V}$ is

$$\overline{\mathbf{V}} = \frac{1}{12} \sum_{i=1}^{6} \sum_{l=1}^{3} (\mathbf{y}_{il} - \bar{\mathbf{y}}_i)(\mathbf{y}_{il} - \bar{\mathbf{y}}_i)'$$

$$= \begin{bmatrix} 0.5040 & 0.5318 & 0.3279 \\ 0.5318 & 0.8337 & 0.5116 \\ 0.3279 & 0.5116 & 0.3816 \end{bmatrix}.$$

Its counterpart for the orthonormal functions defined by U' in (33) is

$$\overline{\mathbf{V}}_\mathbf{U} = \mathbf{U}'\overline{\mathbf{V}}\mathbf{U}$$

$$= \begin{bmatrix} 1.4873 & 0.0582 & 0.2756 \\ 0.0582 & 0.1148 & -0.0237 \\ 0.2756 & -0.0237 & 0.1171 \end{bmatrix}.$$

The lower right hand (2 × 2) block of $\overline{\mathbf{V}}_\mathbf{U}$ is the estimated covariance matrix $\overline{\mathbf{V}}_\mathbf{A} = \mathbf{A}'\overline{\mathbf{V}}\mathbf{A}$ for the within-subject contrasts $\{\mathbf{A}'\mathbf{y}_{il}\}$. Its compatibility with the assumption of sphericity can be confirmed in the following way: diagonal structure is supported by the non-significance ($p > 0.100$) of the correlation of -0.20 between the "linear trend" vs. "curvature" (t test with 11 df); then equality of diagonal elements is supported by the nonsignificance ($p > 0.100$) of the F test with (12,12) df for their ratio. An overall test of sphericity is provided by BMDP2V [37]; it was nonsignificant for $\overline{\mathbf{V}}_\mathbf{A}$ with $p = 0.79$. As an additional consideration, the significance ($p = 0.014$) of the correlation of 0.66 between "average response" vs. "curvature" (t test with 11 df) can be interpreted as contradicting compound symmetry for $\mathbf{V}$.

Because the assumption of sphericity is reasonable, the repeated measures analysis of variance results are shown in Table 5. Note that the between-litter part of the table reproduces the information for the "average response" in Table 5; also, the within-litter part of the table could have been obtained by adding the sums of squares from the orthonormal linear and curvature contrasts. The time and (age × time) effects are significant because of the significance of the linear trend function.

Other methods of analysis could be applied to this example. If the assumption of sphericity were unrealistic, time effects and their interactions with other sources of variation could be assessed by multivariate analysis of variance methods. If the PFC data for the litters did not appear to have a multivariate normal distribution, nonpara-

Table 4: Analysis of variance mean squares and significance indicators for orthonormal functions.

Source of Variation	df	Separate Function		
		Average	Trend	Curvature
Overall mean	1	715.349[a]	35.036[a]	0.231
Age	1	0.008	1.460[a]	0.081
Dose	2	10.618[a]	0.431[c]	0.012
Age × dose	2	0.124	0.500[b]	0.003
Within-groups error	12	1.487	0.115	0.117

[a]Significant results with $p < 0.01$.
[b]Results with $0.01 < p < 0.05$.
[c]Suggestive results with $0.05 < p < 0.10$.

Table 5: Analysis of variance mean squares and significance indicators for repeated measures analysis of variance.

Source	df	MS
Age	1	0.008
Dose	2	10.618[a]
Age × dose	2	0.124
Litter within group	12	1.487
Time	2	17.633[a]
Time × age	2	0.770[a]
Time × dose	4	0.222
Time × age × dose	4	0.252[b]
Within litters	24	0.116

[a]Significant results with $p < 0.01$.
[b]Suggestive results with $p = 0.104$.

metric rank methods could be used; see Koch et al. [46].

Example 21.3.2 A Multiperiod Changeover Study. Peak heart rate responses were obtained during four evaluation periods for $G = 2$ sequence groups of subjects. One group with $n_1 = 9$ subjects received drug A during the first treatment period and drug B during the second treatment period, while the other group with $n_2 = 11$ subjects received the opposite regimen; a pretreatment period preceded the first treatment period and a drug-free period occurred between it and the second treatment period. The response values for the drug-free period were based on one visit for each subject while those for the other three periods were based on averages for two visits. The mean vectors $\{\bar{y}_i\}$ and estimated covariance matrices $\{\hat{V}_i\}$ for each sequence group are shown in Table 6. This information is the underlying framework for the subsequent discussion.

A preliminary statistical model of interest here has the structure in Table 7, where μ denotes a common reference parameter for the pretreatment status of both treatment groups; π_1, π_2, π_3 are period effects for the first treatment period, the drug-free period, and the second treatment period, respectively; τ is the direct drug A vs. drug B differential effect, γ_1 is the drug A vs. drug B differential carryover effect for the drug-free period; and γ_2 is the drug A vs. drug B differential carryover effect for the second treatment period. Also π_1 includes the effect of drug B as the reference treatment, π_2 includes its carryover effect to the drug-free period, and π_3 includes its carryover effect to the second treatment period. Four orthogonal within-subject functions that are useful for analyses pertaining to this model have the following specifications with respect to the seven visits of the study:

F1. The average over all seven visits

F2. The difference between drug-free and the pretreatment average

F3. The difference between the second treatment period average and the first treatment period average

F4. The difference between the average over the first treatment period and the second treatment period vs. the average over pretreatment and the drug-free period

These four functions are obtained from the four period summary framework for the data by the linear transformation matrix

$$\mathbf{U}' = \begin{bmatrix} \frac{2}{7} & \frac{2}{7} & \frac{1}{7} & \frac{2}{7} \\ -1 & 0 & 1 & 0 \\ 0 & -1 & 0 & 1 \\ -\frac{2}{3} & \frac{1}{2} & -\frac{1}{3} & \frac{1}{2} \end{bmatrix}. \quad (34)$$

The application of the transformation $\mathbf{U}'$ to the model in Table 7 yields the expected value structure shown in Table 8 for the functions **F1–F4** for each sequence group and the difference between them.

Univariate and Multivariate Analyses of Within-Subject Functions. If differential carryover effects are present, the estimate of the direct differential effect τ for the two treatments can have substantially larger variance than if they are negligible. For this reason, the usage of changeover designs typically presumes that carryover effects are negligible. We assess this assumption for this example through tests of hypotheses about the parameters γ_1, γ_2. If $\gamma_1 = \gamma_2 = 0$, then the difference between the sequence groups should have expected value zero for functions F1, F2, and F4. The two-sample t tests for these functions each have 18 df; their $p$values are 0.18, 0.72, and 0.14, respectively. Since all of these results are nonsignificant, the assumption that $\gamma_1 = \gamma_2 = 0$ is supported.

Table 6: Means and covariance matrices for peak heart rate of subjects in two sequence groups of a changeover study.

Evaluation Period	Sequence A : B ($n_1 = 9$)					Sequence B : A ($n_2 = 11$)				
	Mean	Covariance Matrix				Mean	Covariance Matrix			
Pretreatment	104	142	133	73	137	117	403	160	206	148
1st treatment	108		392	61	134	95		173	191	152
Drug-free	105		Symmetric	131	113	116		Symmetric	362	168
2nd treatment	92				174	115				168

Table 7: Preliminary model for changeover study.

Sequence Group	Pretreatment	1st Treatment	Drug-Free	2nd Treatment
A : B	μ	$\mu + \pi_1 + \tau$	$\mu + \pi_2 + \gamma_1$	$\mu + \pi_3 + \gamma_2$
B : A	μ	$\mu + \pi_1$	$\mu + \pi_2$	$\mu + \pi_3 + \tau$

The hypothesis of no difference between treatments in carryover effects can also be tested in a more comprehensive way with the two-sample Hotelling T^2. The comparison between the two sequence groups of all three functions F1, F2, and F4 is nonsignificant ($p = 0.34$) relative to the F-distribution with (3, 16) d.f. However, F1 incorporates subject effects, and hence can have substantially greater variability than F2 and F4, which are within-subject contrasts. For this reason, the comparison of the two groups for functions F2 and F4 potentially provides a more effective, multivariate assessment of carryover effects; the corresponding two-sample Hotelling T^2 yields $p = 0.35$ relative to the F distribution with (2,17) df. Thus, the overall tests also support the conclusion that the carryover effects are equivalent for the two treatments.

Given that $\gamma_1 = \gamma_2 = 0$, we can test the hypothesis $\tau = 0$ of no difference in effect between drug A and drug B by comparing function F3 for the two sequence groups with a two-sample t test with 18 df. Since the resulting $p < 0.01$, it can be concluded that drug B lowers peak heart rate significantly more than drug A.

Regression Analysis under Sphericity Assumption. Aspects of repeated measures analysis of variance for this example can be illustrated through the orthonormal counterparts of the functions F1–F4 with respect to the seven visits of the study. For these functions, the linear transformation matrix is

$$\begin{bmatrix} \sqrt{7} & 0 & 0 & 0 \\ 0 & \sqrt{\frac{2}{3}} & 0 & 0 \\ 0 & 0 & 1 & 0 \\ 0 & 0 & 0 & \sqrt{\frac{12}{7}} \end{bmatrix} \mathbf{U}'$$

$$= \begin{bmatrix} \sqrt{7}\,\mathbf{u}_1' \\ \mathbf{A}' \end{bmatrix}, \qquad (35)$$

where $\mathbf{U}'$ is defined in (34), $\mathbf{u}_1'$ is its first row, and $\mathbf{A}'$ is the specification matrix for the three orthonormal contrasts that correspond to the last three rows of $\mathbf{U}'$. Under the model (24) for the seven visits of the study, the covariance matrix of the linear functions $\{\mathbf{A}'\mathbf{y}_{il}\}$ has the sphericity structure $\mathbf{V_A} = \mathbf{I}_3 v_{*,0}$. Relative to this background, the pooled within-group estimate

Table 8: Expected value structure of orthogonal functions for changeover study.

Function	Sequence A : B	Sequence B : A	A : B vs. B : A Difference
Overall mean	$\mu + (2\pi_1 + 2\tau + \pi_2 + \gamma_1 + 2\pi_3 + 2\gamma_2)/7$	$\mu + (2\pi_1 + \pi_2 + 2\pi_3 + 2\tau)/7$	$(\gamma_1 + 2\gamma_2)/7$
Drug-free vs. pretreatment	$\pi_2 + \gamma_1$	π_2	γ_1
1st vs. 2nd treatment	$(\pi_3 - \pi_1) - \tau + \gamma_2$	$(\pi_3 - \pi_1) + \tau$	$\gamma_2 - 2\tau$
1st and 2nd treatment vs. drug-free and pretreatment	$(3\pi_1 + 3\tau + 3\pi_3 + 3\gamma_2 - 2\pi_2 - 2\gamma_1)/6$	$(3\pi_1 + 3\tau + 3\pi_3 - 2\pi_2)/6$	$(3\gamma_2 - 2\gamma_1)/6$

for $\mathbf{V_A}$ is

$$\overline{\mathbf{V}}_\mathbf{A} = \mathbf{A}'\overline{\mathbf{V}}\mathbf{A} = \begin{bmatrix} 169 & 12 & -52 \\ 12 & 153 & 65 \\ -52 & 65 & 201 \end{bmatrix}, \quad (36)$$

where $\overline{\mathbf{V}} = (8\hat{\mathbf{V}}_1 + 10\hat{\mathbf{V}}_2)/18$. Its compatibility with the assumption of sphericity can be confirmed in several ways. The diagonal structure for $\mathbf{V_A}$ is supported by the nonsignificance ($p \geq 0.10$) of t tests with 17 df for each of the pairwise correlation estimates from $\overline{\mathbf{V}}_\mathbf{A}$; equality of the diagonal elements of $\mathbf{V_A}$ is supported by the nonsignificance ($p \geq 0.10$) of F tests with (18,18) df for the pairwise ratios of their estimates from $\overline{\mathbf{V}}_\mathbf{A}$. The overall sphericity test from BMDP2V [37] was nonsignificant with $p = 0.51$.

Since sphericity is considered realistic for $\mathbf{V_A}$, the effect parameters $\boldsymbol{\beta} = (\pi_1, \pi_2, \pi_3, \tau, \gamma_1, \gamma_2)'$ of the model in Table 8 can be estimated on a within-subject basis by the application of multiple regression methods to the $\{\mathbf{A}'\mathbf{y}_{il}\}$. The corresponding specification matrix $\mathbf{Z}$ has respective components

$$\mathbf{Z}_{1l} = \begin{bmatrix} (0, 2, 0, 0, 2, 0)/\sqrt{6} \\ (-1, 0, 1, -1, 0, 1) \\ (-3, 2, -3, -3, 2, -3)/\sqrt{21} \end{bmatrix}, \quad (37)$$

for the subjects from the A : B sequence group and respective components

$$\mathbf{Z}_{2l} = \begin{bmatrix} (0, 2, 0, 0, 0, 0)/\sqrt{6} \\ (-1, 0, 1, 1, 0, 0) \\ (-3, 2, -3, -3, 0, 0)/\sqrt{21} \end{bmatrix}, \quad (38)$$

for the subjects from the B : A sequence group. The least-squares estimates for $\boldsymbol{\beta}$ from this framework, their corresponding estimated standard errors, and p values for t tests with 54 df for 0 values are

Parameter	π_1	π_2	π_3
Estimate	-22.3	-0.8	-28.7
SE	4.0	4.9	8.2
p	<0.01	0.87	<0.01
Parameter	τ	γ_1	γ_2
Estimate	26.5	2.6	16.7
SE	5.9	7.3	10.3
p	<0.01	0.72	0.11

(39)

The error mean-square estimate for the within-subject variance component $v_{*,0}$ is $\hat{v}_{*,0} = 174$; it is the average (28) of the diagonal elements of $\overline{\mathbf{V}}_\mathbf{A}$.

The results in (39) indicate that the carryover effect parameters γ_1 and γ_2 can be removed from the model in Table 7 because of their nonsignificance ($p \geq 0.10$), that the difference τ between direct treatment effects is significant ($p < 0.01$), and suggest that period effects are compatible with the constraints $\pi_1 - \pi_3 = 0, \pi_2 = 0$. Thus, peak heart rates during treatment periods with drug A were essentially the same as at pretreatment while those during treatment periods with drug B were significantly lower by about 27 beats per minute; any potential carryover effects of the two treatments were equivalent. These conclusions were also provided by the previously discussed univariate analyses of functions F1–F4. The advantage of the univariate function approach is that its use does not require the assumption of sphericity. However, it lacks comprehensiveness since each function is analyzed separately. For this example, the assumption of sphericity seems reasonable, and so repeated measures model fitting methods are applicable. They have the advantage of providing an effective estimation framework that encompasses the variation of response both across conditions within subjects and across groups of subjects.

Nonparametric Analysis. The application of nonparametric rank methods are of interest because they are based on randomization in the research design rather than on assumptions concerning distributions and covariance structure for the data. It is possible that the covariance matrices for the two sequence groups are not homogeneous; the F test for the comparison of the estimated variances for functions F3 is significant ($p = 0.004$), although such tests for the other functions and for the responses during each period are not. The conclusions from nonparametric analyses agree with those reported here.

Example 21.3.3 A Study Comparing Two Diagnostic Procedures. One thousand subjects were classified according to both a standard version and a modified version of a diagnostic procedure with four ordinally scaled categories: strongly negative, moderately negative, moderately positive, and strongly positive.

Let $\mathbf{n}$ denote the vector of frequencies for the $n = 1000$ subjects and $\mathbf{p} = (\mathbf{n}/n)$ the vector of sample proportions for the 16 possible outcomes in the (4×4) table. If the 1000 subjects in this study can be considered a simple random sample, $\mathbf{p}$ has approximately a multivariate normal distribution. A consistent estimate of the covariance matrix is $\mathbf{V_p} = [\mathbf{D_p} - \mathbf{pp'}]/n$, where $\mathbf{D_p}$ is a diagonal matrix with elements $\mathbf{p}$ on the diagonal.

In the repeated measures context, this study involves one group with two conditions, and so it could be analyzed with a t test for the pairwise difference between conditions if the two responses were normally distributed rather than just ordinal. Thus, attention needs to be given to the analogous comparison of the response category distributions for the two diagnostic procedures. Let $\mathbf{f}$ be the vector of the six marginal proportions for strongly negative, moderately negative, and moderately positive for each diagnostic procedure; proportions for the strongly positive category are not needed because they are linear functions of the others. The vector $\mathbf{f}$ can be obtained from $\mathbf{p}$ by constructing a (6×16) matrix $\mathbf{A}$ such that $\mathbf{f} = \mathbf{Ap}$. The estimates $\mathbf{f}$ and their estimated covariance matrix $\mathbf{V_f} = \mathbf{A V_p A'}$ are shown in Table 9.

Under the hypothesis that the marginal distributions for the two diagnostic procedures are the same (i.e., marginal homogeneity), the Wald statistic

$$Q = \mathbf{f'W'}[\mathbf{W V_f W'}]^{-1}\mathbf{Wf} \quad (40)$$

with $\mathbf{W} = [I_3, -I_3]$ approximately has the chi-square distribution with 3 df; this criterion is analogous to the Hotelling T^2 statistic in (21). Since $Q = 6.68$ approaches significance with $p = 0.083$, the two diagnostic procedures potentially have somewhat different marginal distributions. Repeated measurements of categorical data are discussed in Guthrie [31] and Koch et al. [50]. Other methods for assessing marginal homogeneity and the related hypotheses of symmetry and quasisymmetry are given in Bishop et al. [6] and Gokhale and Kullback [28]; methods for ordinal data are given in Agresti [1] and McCullagh and Nelder [58].

Example 21.3.4 A Study Comparing Two Psychiatric Drugs. A randomized clinical trial was undertaken to compare two drugs for a psychiatric condition. One group of $n_1 = 37$ patients received drug A for 2 months and another group of $n_2 = 37$ patients received drug B for 2 months. Each patient's mental condition was evaluated by three observers at the end of one month and again at the end of the trial. The responses were scored using the ordinal values of (1) unsatisfactory, (2) satisfactory, or (3) good. The mean vectors $\{\overline{\mathbf{y}}_\mathbf{i}\}$ relative to the values 1, 2, 3 and their estimated covariance matrices $\hat{\mathbf{V}}_\mathbf{i}$ for each treatment group are shown in Table 10.

The sampling framework is considered sufficient for the composite mean vec-

Table 9: Estimated marginal proportions and covariance matrix for two diagnostic procedures.

Diagnosis	Response Category	Estimate	Estimated Covariance Matrix $\times 10^6$					
Modified	Strongly negative	0.362	231	−122	−34	144	−39	−31
Modified	Moderately negative	0.339		224	−32	−36	99	5
Modified	Moderately positive	0.093			84	−30	1	28
Standard	Strongly negative	0.394	Symmetric			239	−119	−37
Standard	Moderately negative	0.302					211	−28
Standard	Moderately positive	0.094						85

Table 10: Means and estimated covariance matrices for responses to two treatments for a medical condition.

Treatment Group	Observer	Time	Mean	Estimated Covariance Matrix $\times 10^6$					
A	1	1 month	2.2	5020	2902	566	203	856	1727
A	1	Final	2.4		8183	1727	2191	3497	4048
A	2	1 month	2.8			4498	638	−1219	73
A	2	Final	2.7	Symmetric			5049	3772	1872
A	3	1 month	2.2					9779	5296
A	3	Final	2.5						7284
B	1	1 month	1.9	6081	−39	849	1007	−197	2448
B	1	Final	2.2		7068	1441	928	2468	3001
B	2	1 month	2.4			6633	1599	4284	3415
B	2	Final	2.7	Symmetric			5923	2448	2369
B	3	1 month	1.9					1015	4778
B	3	Final	2.3						9713

tor $\overline{\mathbf{y}} = [\overline{\mathbf{y}}_1', \overline{\mathbf{y}}_2']'$ to have an approximately multivariate normal distribution, and so Wald statistics can be used to test hypotheses concerning groups, observers, time, and their interactions. These statistics are computed via

$$Q = \overline{\mathbf{y}}'\mathbf{W}'[\mathbf{W}\hat{\mathbf{V}}_{\overline{\mathbf{y}}}\mathbf{W}']^{-1}\mathbf{W}\overline{\mathbf{y}}, \qquad (41)$$

where $\hat{\mathbf{V}}_{\overline{\mathbf{y}}}$ is a block diagonal matrix with $\hat{\mathbf{V}}_1$ and $\hat{\mathbf{V}}_2$ as the respective blocks and $\mathbf{W}$ is the specification matrix. Results shown in Table 11 indicate significant variation ($p < 0.01$) between treatment groups, between observers, and between evaluation times; also the (group $\times$ time) interaction is significant ($p = 0.045$). The other sources of variation are interpreted as essentially random since no interaction of observers with group or time was expected on a priori grounds [although the (observer $\times$ time) interaction is recognized to be suggestive with $p = 0.060$].

Table 11: Results of wald statistics for preliminary assessment of group, observer, and time sources of variation.

Source of Variation	W Matrix											Q	df
Group (G)	1	1	1	1	1	-1	-1	-1	-1	-1	-1	9.34^a	1
Observer (O)	1	1	0	0	-1	1	1	0	0	-1	-1	106.90^a	2
	0	0	1	1	-1	0	0	1	1	-1	-1		
Time (T)	1	-1	1	-1	0	1	-1	1	-1	1	-1	40.22^a	1
G × O	1	1	0	0	-1	-1	-1	0	0	1	1	0.83	2
	0	0	1	1	-1	0	0	-1	-1	1	1		
G × T	1	-1	1	-1	0	-1	1	-1	1	1	-1	4.02^b	1
O × T	1	-1	0	0	-1	1	-1	0	0	-1	1	5.62^c	2
	0	0	1	-1	-1	0	0	1	-1	-1	1		
G × O × T	1	-1	0	0	-1	-1	1	0	0	1	-1	2.11	2
	0	0	1	-1	-1	0	0	-1	1	1	-1		

^{a}Significant results with $p < 0.01$.
^{b}Significant results with $0.01 < p < 0.05$.
^{c}Suggestive results with $0.05 < p < 0.10$.

Table 12: Model predicted values (and standard errors) for mean response to treatments A and B for (observer × time) conditions.

	Observer 1		Observer 2		Observer 3	
Group	1 Month	Final	1 Month	Final	1 Month	Final
Treatment A	2.17	2.32	2.68	2.83	2.28	2.44
	(0.06)	(0.06)	(0.05)	(0.05)	(0.06)	(0.07)
Treatment B	1.88	2.18	2.39	2.69	1.99	2.29
	(0.05)	(0.06)	(0.06)	(0.06)	(0.07)	(0.07)

A model that reflects the stated conclusions has the form $E\{\overline{\mathbf{y}}\} = \mathbf{X}\boldsymbol{\beta}$, where

$$\mathbf{X} = \begin{bmatrix} 1 & 0 & 0 & 0 & 0 & 0 \\ 1 & 0 & 0 & 0 & 1 & 0 \\ 1 & 0 & 1 & 0 & 0 & 0 \\ 1 & 0 & 1 & 0 & 1 & 0 \\ 1 & 0 & 0 & 1 & 0 & 0 \\ 1 & 0 & 0 & 1 & 1 & 0 \\ 1 & 1 & 0 & 0 & 0 & 0 \\ 1 & 1 & 0 & 0 & 1 & 1 \\ 1 & 1 & 1 & 0 & 0 & 0 \\ 1 & 1 & 1 & 0 & 1 & 1 \\ 1 & 1 & 0 & 1 & 0 & 0 \\ 1 & 1 & 0 & 1 & 1 & 1 \end{bmatrix} \quad (42)$$

and $\boldsymbol{\beta} = (\beta_1, \beta_2, \beta_3, \beta_4, \beta_5, \beta_6)'$. For this specification, β_1 is a reference value for the expected response score for the classification of observer 1 at one month for treatment A, β_2 is the increment for treatment B, β_3 and β_4 the increments for observers 2 and 3, β_5 the increment for the final evaluation time, and β_6 the increment for the interaction between treatment B and the final evaluation time. Weighted least-squares methods can be used to determine the asymptotically unbiased and asymptotically efficient estimator

$$\mathbf{b} = (\mathbf{X}'\hat{\mathbf{V}}_{\overline{\mathbf{y}}}^{-1}\mathbf{X})^{-1}\mathbf{X}'\hat{\mathbf{V}}_{\overline{\mathbf{y}}}^{-1}\overline{\mathbf{y}} \quad (43)$$

for $\boldsymbol{\beta}$; a consistent estimator for its covariance matrix is $\hat{\mathbf{V}}_\mathbf{b} = (\mathbf{X}'\hat{\mathbf{V}}_{\overline{\mathbf{y}}}^{-1}\mathbf{X})^{-1}$. The goodness of fit of the model specified by (42) is supported by the nonsignificance ($p = 0.16$) of the Wald goodness-of-fit statistic

$$Q = (\overline{\mathbf{y}} - \mathbf{X}\mathbf{b})'\hat{\mathbf{V}}_{\overline{\mathbf{y}}}^{-1}(\overline{\mathbf{y}} - \mathbf{X}\mathbf{b}) = 9.30, \quad (44)$$

relative to the chi-square distribution with 6 df.

Predicted values for the mean response of each group for the $d = 6$ (observer × time) conditions are obtained via $\hat{\mathbf{y}} = \mathbf{X}\mathbf{b}$; these quantities and their estimated standard errors (via square roots of the diagonal elements of $\mathbf{X}\hat{\mathbf{V}}_\mathbf{b}\mathbf{X}'$) are shown in Table 12. They indicate that the response is more favorable for treatment A, and that this tendency is larger at one month. The classifications of observer 2 were higher than those for the other two observers.

Finally, if the covariance matrices for the two groups were not substantially different from each other, the standard multivariate analysis of variance methods discussed in Section 20.3 could be applied here in a large sample context. Additional discussion of the methods illustrated with this example is given in Koch et al. [49,50] and Stanish et al. [84].

Acknowledgments. The authors would like to thank James Bawden for providing the data in Example 20.3.1 and William Shapiro for providing the data in Example 20.3.2. They would also like to express appreciation to Keith Muller for helpful comments with respect to the revision of an earlier version of this entry and to

Ann Thomas for editorial assistance. This research was partially supported by the US Bureau of the Census through Joint Statistical Agreements JSA 83-1, 84-1, and 84-5 and by grant AM 17328 from NI-ADDK.

References

1. Agresti, A. (1984). *Analysis of Ordinal Categorical Data*. Wiley, New York.
2. Allen, D. M. and Cady, F. B. (1982). *Analyzing Experimental Data by Regression*. Lifetime Learning Publications, Belmont, CA.
3. Anderson, R. L. and Bancroft, T. A. (1952). *Statistical Theory in Research*. McGraw-Hill, New York.
4. Anderson, T. W. (1984). *An Introduction to Multivariate Statistical Analysis*, 2nd ed. Wiley, New York.
5. Bennett, C. A. and Franklin, N. L. (1954). *Statistical Analysis in Chemistry and the Chemical Industry*. Wiley, New York.
6. Bishop, Y. M. M., Fienberg, S. E., and Holland, P. W. (1975). *Discrete Multivariate Analysis*. MIT Press, Cambridge, MA.
7. Bock, R. D. (1975). *Multivariate Statistical Methods in Behavioral Research*. McGraw-Hill, New York.
8. Boik, R. J. (1981). *Psychometrika*, **46**, 241–255. (*A priori* tests in repeated measures designs: effects of nonsphericity.)
9. Box, G. E. P. (1954). *Ann. Math. Statist.*, **25**, 484–498.
10. Brown, B. W. (1980). *Biometrics*, **36**, 69–79.
11. Cochran, W. (1950). *Biometrika*, **37**, 256–266.
12. Cochran, W. G. and Cox, G. M. (1957). *Experimental Designs*. Wiley, New York.
13. Cole, J. W. L. and Grizzle, J. E. (1966). *Biometrics*, **22**, 810–828. (Applications of multivariate analysis of variance to repeated measurements experiments.)
14. Conover, W. J. (1971). *Practical Nonparametric Statistics*. Wiley, New York.
15. Constantine, G. and Hedayat, A. (1982). *J. Statist. Plan. Infer.*, **6**, 153–164.
16. Cook, N. and Ware, J. H. (1983). *Ann. Rev. Public Health*, **4**, 1–24.
17. Cox, D. R. (1958). *Planning of Experiments*. Wiley, New York.
18. Darroch, J. N. (1981). *Int. Statist. Rev.*, **49**, 285–307. (The Mantel-Haenszel test and tests of marginal symmetry, fixed-effects, and mixed models for a categorical response.)
19. Dielman, T. E. (1983). *Am. Statist.*, **37**, 111–122. (Pooled cross-sectional and time-series data: a survey of current statistical methodology.)
20. Elashoff, J. D. (1985). *Analysis of Repeated Measures Designs*. BMDP Technical Report 83, BMDP Software, Los Angeles, CA.
21. Federer, W. T. (1955). *Experimental Design*. Macmillan, New York.
22. Federer, W. T. (1975). In *Applied Statistics*, R. P. Gupta, ed., pp. 9–39, North-Holland, Amsterdam.
23. Federer, W. T. (1980, 1981). *Int. Statist. Rev.*, **48**, 357–368; **49**, 95–109, 185–197. (Some recent results in experiment design with a bibliography.)
24. Federer, W. T. and Balaam, L. N. (1972). *Bibliography on Experiment and Treatment Design Pre-1968*. Oliver & Boyd, Edinburgh, Scotland.
25. Friedman, M. (1937). *J. Am. Statist. Assoc.*, **32**, 675–699.
26. Geisser, S. (1980). In *Handbook of Statistics*, Vol. 1, P. R. Krishnaiah, ed., pp. 89–115, North-Holland, Amsterdam.
27. Gill, J. L. (1978). *Design and Analysis of Experiments in the Animal and Medical Sciences*. Iowa State University Press, Ames, IA.
28. Gokhale, D. V. and Kullback, S. (1978). *The Information in Contingency Tables*. Marcel Dekker, New York.

29. Greenhouse, S. W. and Geisser, S. (1959). *Psychometrika*, **24**, 94–112. (On methods in the analysis of profile data.)
30. Grizzle, J. E. and Allen, D. (1969). *Biometrics*, **25**, 357–381.
31. Guthrie, D. (1981). *Psychol. Bull.*, **90**, 189–195.
32. Hedayat, A. and Afsarinejad, K. (1975). In *A Survey of Statistical Design and Linear Models*, J. N. Srivastava, ed., pp. 229–242, North-Holland, Amsterdam.
33. Hedayat, A. and Afsarinejad, K. (1978). *Ann. Statist.*, **6**, 619–628.
34. Huynh, H. (1978). *Biometrika*, **43**, 161–175.
35. Huynh, H. and Feldt, L. S. (1970). *J. Am. Statist. Assoc.*, **65**, 1582–1589. (Conditions under which mean square ratios in repeated-measurement designs have exact F-distributions.)
36. Huynh, H. and Feldt, L. S. (1976). *J. Educ. Statist.*, **1**, 69–82.
37. Jennrich, R., Sampson, P., and Frane, J. (1981). In *BMDP Statistical Software*, W. J. Dixon et al., eds., Chap. 15.2, University of California Press, Los Angeles, CA.
38. Jensen, D. R. (1982). *Biometrics*, **38**, 813–825. (Efficiency and robustness in the use of repeated measurements.)
39. Joiner, B. L. (1981). *Am. Statist.*, **35**, 227–233.
40. Kempthorne, O. (1952). *Design and Analysis of Experiments*. Wiley, New York.
41. Kempthorne, O. (1969). *An Introduction to Genetic Statistics*. Wiley, New York.
42. Kershner, R. P. and Federer, W. T. (1981). *J. Am. Statist. Assoc.*, **76**, 612–619.
43. Keselman, H. J., Rogan, J. C., Mendoza, J. L., and Breen, L. J. (1980). *Psychol. Bull.*, **87**, 479–481.
44. Koch, G. G. (1969). *J. Am. Statist. Assoc.*, **64**, 485–505.
45. Koch, G. G. (1970). *Biometrics*, **26**, 105–128.
46. Koch, G. G., Amara, I. A., Stokes, M. E., and Gillings, D. B. (1980). *Int. Statist. Rev.*, **48**, 249–265. (Some views on parametric and nonparametric analysis for repeated measurements and selected bibliography.)
47. Koch, G. G., Amara, I. A., Davis, G. W., and Gillings, D. B. (1982). *Biometrics*, **38**, 563–595.
48. Koch, G. G., Imrey, P. B., and Reinfurt, D. W. (1972). *Biometrics*, **28**, 663–692.
49. Koch, G. G., Imrey, P. B., Singer, J. M., Atkinson, S. S., and Stokes, M. E. (1985). *Analysis of Categorical Data*. University of Montreal Press, Montreal, Canada.
50. Koch, G. G., Landis, J. R., Freeman, J. L., Freeman, D. H., and Lehnen, R. G. (1977). *Biometrics*, **33**, 133–158. (A general methodology for the analysis of experiments with repeated measurement of categorical data.)
51. Koch, G. G., Gitomer, S. L., Skalland, L., and Stokes, M. E. (1983). *Statist. Med.*, **2**, 397–412.
52. Laird, N. M. and Ware, J. H. (1982). *Biometrics*, **38**, 963–974. (Random-effects models for longitudinal data.)
53. Laycock, P. J. and Seiden, E. (1980). *Ann. Statist.*, **8**, 1284–1292.
54. Lehmann, E. L. (1975). *Nonparametrics: Statistical Methods Based on Ranks*. Holden-Day, San Francisco, CA.
55. Mauchley, J. W. (1940). *Ann. Math. Statist.*, **11**, 204–209.
56. Maxwell, S. E. (1980). *J. Educ. Statist.*, **5**, 269–287.
57. Maxwell, S. E. and Arvey, R. D. (1982). *Psychol. Bull.*, **92**, 778–785.
58. McCullagh, P. and Nelder, J. A. (1983). *Generalized Linear Models*. Chapman & Hall, London, UK.
59. Mendoza, J. L. (1980). *Psychometrika*, **45**, 495–498.
60. Miller, R. (1981). *Simultaneous Statistical Inference*, 2nd ed. McGraw-Hill, New York.

61. Monlezun, C. J., Blouin, D. C., and Malone, L. C. (1984). *Am. Statist.*, **38**, 21–27.
62. Morrison, D. (1970). *Biometrics*, **26**, 281–290.
63. Morrison, D. F. (1976). *Multivariate Statistical Methods*, 2nd ed. McGraw-Hill, New York.
64. Myers, J. L. (1979). *Fundamentals of Experimental Design*, 3rd ed. Allyn & Bacon, Boston, MA.
65. Myers, J. L., DiCecco, J. V., White, J. B., and Borden, V. M. (1982). *Psychol. Bull.*, **92**, 517–525.
66. Nesselroade, J. R. and Baltes, P. B., eds. (1979). *Longitudinal Research in the Study of Behavior and Development*. Academic Press, New York.
67. Neter, J., Wasserman, W., and Kutner, M. H. (1985). *Applied Linear Statistical Models*, 2nd ed. Irwin, Homewood, IL.
68. Puri, M. L. and Sen, P. K. (1971). *Nonparametric Methods in Multivariate Analysis*. Wiley, New York.
69. Quade, D. (1982). *Biometrics*, **38**, 597–611.
70. Rao, C. R. (1965). *Biometrika*, **52**, 447–458. (The theory of least squares when the parameters are stochastic and its application to the analysis of growth curves.)
71. Rao, C. R. (1965). *Linear Statistical Inference and Its Application*. Wiley, New York.
72. Rao, M. N. and Rao, C. R. (1966). *Sankhyā Ser. B*, **28**, 237–258.
73. Reinsel, G. (1982). *J. Am. Statist. Assoc.*, **77**, 190–195.
74. Rogan, J. C., Keselman, H. J., and Mendoza, J. L. (1979). *Br. J. Math. Statist. Psychol.*, **32**, 269–286.
75. Rouanet, H. and Lepine, D. (1970). *Br. J. Math. Statist. Psychol.*, **23**, 147–163.
76. Roy, S. N. and Bose, R. C. (1953). *Ann. Math. Statist.*, **24**, 513–536. (Simultaneous confidence interval estimation.)
77. Roy, S. N., Gnanadesikan, R., and Srivastava, J. N. (1971). *Analysis and Design of Certain Quantitative Multiresponse Experiments*. Pergamon Press, Oxford, UK.
78. Scheffé, H. (1959). *The Analysis of Variance*. Wiley, New York.
79. Schlesselman, J. (1973). *J. Chron. Dis.*, **26**, 561–570.
80. Schwertman, N. C. (1978). *J. Am. Statist. Assoc.*, **73**, 393–396.
81. Searle, S. R. (1971). *Linear Models*. Wiley, New York.
82. Snedecor, G. W. and Cochran, W. G. (1967). *Statistical Methods*, 7th ed. Iowa State University Press, Ames, IA.
83. Snee, R. D., Acuff, S. K., and Gibson, J. R. (1979). *Biometrics*, **35**, 835–848.
84. Stanish, W. M., Gillings, D. B., and Koch, G. G. (1978). *Biometrics*, **34**, 305–317.
85. Thomas, D. R. (1983). *Psychometrika*, **48**, 451–464.
86. Timm, N. H. (1975). *Multivariate Analysis with Applications in Education and Psychology*. Brooks/Cole, Monterey, CA.
87. Timm, N. H. (1980). *Handbook of Statistics*, Vol. 1, P. R. Krishnaiah, ed., pp. 41–87, North-Holland, Amsterdam.
88. Wallenstein, S. and Fisher, A. C. (1977). *Biometrics*, **33**, 261–269.
89. Wallenstein, S. and Fleiss, J. L. (1979). *Psychometrika*, **44**, 229–233.
90. Ware, J. A. (1985). *Am. Statist.*, **39**, 95–101. (Linear models for the analysis of longitudinal studies.)
91. White, A. A., Landis, J. R., and Cooper, M. M. (1982). *Int. Statist. Rev.*, **50**, 27–34.
92. Winer, B. J. (1971). *Statistical Principles in Experimental Design*, 2nd ed. McGraw-Hill, New York.
93. Woolson, R. F., Leeper, J. D., and Clarke, W. R. (1978). *J. Roy. Statist. Soc. Ser. A*, **141**, 242–252.

22. DNA Fingerprinting

Kathryn Roeder

22.1 Introduction

DNA fingerprinting, *DNA typing*, and *DNA profiling* are all terms used to describe techniques that have been employed by forensic scientists to draw inferences from DNA found at crime scenes. The term "fingerprint" is considered a misnomer by some because DNA profiles are not necessarily unique. The culpability of a suspect is based, in part, on whether her or his DNA fingerprint matches the profile obtained at the scene of the crime. The weight of this evidence depends on the probability of obtaining a match by chance.

The recent explosion of molecular technology has transformed many methods of evaluating crime-scene materials and paternity suits (for a review, see Devlin [12]). Just twenty years ago, a blood stain collected at the crime scene could be evaluated for only a handful of protein-based genetic markers, and only when the material was fresh enough to maintain protein activity. Moreover, the markers examined were frequently common in the population. Therefore, the genetic evidence was rarely informative.

The current technology, which allows a forensic scientist to examine DNA markers directly, has altered both the frequency with which genetic markers can be used and their exculpatory/inculpatory value. The former derives from the robustness of DNA, which degrades far more slowly than proteins under almost all environmental conditions. A dramatic demonstration of DNA's robustness came when it was recovered from various fossils that were tens of millions of years old.

The power of the DNA markers springs from the fact that no two individuals are genetically identical, with the exception of identical twins. This does not mean, however, that any location in the human genome would be determinative. In fact, only a small portion of the genome is valuable for forensic inference. Of this portion, forensic scientists have focused predominantly on tandemly repeated segments of DNA consecutive repetitions of a short sequence), because these have extremely high discriminatory power.

A DNA profile consists of a set of measurements of discrete random variables, measured with error. The random variables occur in pairs, one inherited from each parent. Typically three to five different pairs of random variables are measured at different locations (*loci*) on the genome. Forensic scientists declare a *match* if the profiles from two samples are considered to be sufficiently similar. Two samples from the same individual differ only due to measurement error, while samples from different individuals rarely match.

If the samples from the suspect and the crime scene (*evidentiary sample*) do not meet the match criterion, then the case usually does not go to trial. Although there is a possibility of false exclusions, this has not been the focus of attention. Statistical issues arise when the samples do match and some measure of weight of this evidence is presented to the jury. Usually, jurors are presented with an estimate of the probability of obtaining a matching profile from a randomly selected individual from some appropriately selected population, called the *reference population*. This profile probability is usually obtained by multiplying the estimated matching probability for each component of the profile— that is, by assuming independence of the pairs of observations composing the profile.

The estimates of the probability of matching for each component of the profile are based on the distribution of profiles of individuals that have been collected by forensic testing laboratories. Samples are available for each of the major ethnic groups (races). There has been a substantial amount of controversy about these probability calculations (e.g., 22, 7, 28, and 17.)

22.2 DNA Markers

The basic building blocks of DNA are nucleotide bases, adenine, cytosine, guanine, and thyamine, often referenced by their initials, A, C, G, and T. The human genome is composed of about 3 billion base pairs, organized into 46 DNA molecules and associated chromosomal material. Each molecule has many base pairs strung out in a double helix, each strand of the helix having nucleotides linked together like beads on a string. It is impossible using current technology to compare more than a fraction of the DNA of two individuals; however, there is no need to compare more than a small fraction of the genome to differentiate individuals with high probability.

Most DNA profiles produced by forensic scientists are designed to capitalize on regions of the genome that possess substantial interindividual variability. These profiles frequently utilize DNA regions characterized by small repeated sequences of DNA, called *variable number of tandem repeats* (VNTR) loci. The name refers to the fact that each locus consists of segments that are usually identical or extremely similar in sequence and linked consecutively. For almost any VNTR locus, the number of repetitions tends to vary among individuals, so that while one individual might have 6 repeats on one chromosome and 119 at the other, another individual might have 53 and 251 repeats. VNTR markers fall into two categories based on the length of the repeating segment: microsatellites, which have repeating units of only a few base pairs long, and minisatellites, which have repeating units of approximately 9–80 base pairs.

Rather than directly decode the DNA sequence at a VNTR locus, molecular techniques are used to estimate the number of repeats. These techniques are quite accurate for microsatellites, which are also called *short tandem repeats* (STRs). The molecular methodology usually involves the *polymerase chain reaction* (PCR) or a related molecular process to literally produce multiple copies of the locus (*amplify* it) and perhaps a small section of surrounding DNA. The result enables an accurate estimate of the number of repeats in the VNTR.

For minisatellite DNA, the molecular methodology is even more indirect. For these loci, *restriction enzymes*, which can be thought of as a molecular scissors, excise the repeated regions and a small buffer zone. If the repeat region is of length ρ, the fragment has R repeats in tandem, and the buffer zone is of length α, then the length of the fragment is $A = \alpha + R\rho$. Current molecular techniques (agarose gel electrophoresis, followed by autoradiography, chemiluminescence, or some other method to visualize the DNA fragment) cannot determine precisely how many repeats a particular minisatellite VNTR fragment possesses. One observes $X = A + \epsilon$, where ϵ is usually larger than ρ. For instance, a minisatellite VNTR with 249 repeats will rarely be distinguishable from one with 250 repeats. Consequently, rather than classifying *alleles* (distinct types or lengths of DNA) based on repeat number, the sizes of minisatellite alleles are estimated and recorded.

DNA can be obtained from any material that contains nucleated cells: blood, semen, skin, hair roots, and so on. Typically the evidence will consist of a set of "measurements" from several to many specific regions in the DNA. Each locus yields a pair of measurements or alleles, one inherited from each parent. Geneticists call the latter a *single-locus genotype*, and they call the collection of measurements across loci a *multilocus genotype* or simply a *DNA profile*. The allele pairs are determined by the distance the fragments travel on a gel. Visually, the DNA gel measurement of a fingerprint resembles a bar code, where each locus produces one or two dark bands.

Consider the pair of measurements obtained at a given locus. Because of the large number of alleles in the population, most individuals are *heterozygous* (have two distinct alleles at the locus) and produce a double-banded pattern. A single band is sometimes generated at a locus; single-banded patterns may be due to *homozygosity* (two copies of the same allele), to difficulties in distinguishing fragments of similar lengths, or to an allele being too small to be measured [14,34].

22.3 Summarizing the Evidence

To fix ideas, consider the problem of summarizing the genetic evidence assuming the alleles are observable without error. A *multilocus genotype* $\mathcal{G}$ for a particular individual consists of unordered pairs of fragments from each of L loci: $\mathcal{G} = \{(A_1, A_2)_\mathcal{L}, = 1, \ldots, L\}$. At each locus, A_i is a discrete random variable that takes on categorical allele values $\{a(k), k = 1, \ldots, m\}$ with probabilities $\{\gamma(k), k = 1, \ldots, m\}$ that vary across populations. Provided that A_1 is independent of A_2 (*Hardy–Weinberg equilibrium*), the probability of observing the single-locus genotype $\{a(i), a(j)\}$ can be calculated as

$$\Pr(\{a(i), a(j)\}) = \begin{cases} 2\gamma(i)\gamma(j), & i \neq j \text{ (heterozygotes)}, \\ \gamma(i)^2, & i = j \text{ (homozygotes)}. \end{cases} \quad (1)$$

Furthermore, provided the genotypes are independent across loci (*linkage equilibrium*), then the multilocus genotype probability can be obtained by multiplying across loci.

Assume the genetic evidence consists of the multilocus genotype $\mathcal{G}_s$ obtained from the suspect and the multilocus genotype $\mathcal{G}_e$ obtained from the evidentiary sample. The objective is to distinguish between two competing hypotheses:

H_0: The samples were obtained from different individuals.
H_1: The suspect and evidentiary sample were obtained from the same individual.

Notice that neither of these complementary hypotheses contains an evaluation of guilt. In fact, formulations involving guilt and innocence are misleading: like dermal fingerprints, a DNA profile, even if unique, can only place the suspect's DNA at the crime scene.

Assuming that the suspect and evidentiary sample are independently drawn from the reference population, then under H_0, the evidence for a crime can be summarized in the likelihood ratio

$$\mathcal{LR} = \frac{\Pr(\mathcal{G}_s, \mathcal{G}_e | H_1)}{\Pr(\mathcal{G}_s, \mathcal{G}_e | H_0)}$$

$$= \begin{cases} \Pr(\mathcal{G}) \times 1/\Pr(\mathcal{G})^2 \\ = 1/\Pr(\mathcal{G}) \\ \quad \text{if } \mathcal{G}_s = \mathcal{G}_e = \mathcal{G}, \\ 0 \quad \text{if } \mathcal{G}_s \neq \mathcal{G}_e. \end{cases}$$

(2)

Usually the *match probability* $\Pr(\mathcal{G})$, rather than $\mathcal{LR}$, is presented to the jury. This probability is often less than one in a billion. Some consider such minuscule probabilities absurd, given that they are computed from databases consisting of perhaps three thousand profiles. Is there evidence suggesting the probabilities of certain profile matches are truly as small as one in a billion? The answer, emphatically, is yes. Risch and Devlin [31] made all 7.6 million pairwise comparisons of profiles in the FBI database and found no four- or five-locus matches, and only one three-locus match. Based on this analysis, one must conclude that specific profiles are extremely rare and thus match probabilities under H_0 cannot be estimated without making some modeling assumptions.

22.4 Independence and Population Substructure

There are many statistical and population genetic issues involved in computing [Eq. (2)]. For instance, to be strictly correct, the calculations assume independence of alleles constituting a single-locus profile and independence of alleles across loci. Of course, such calculations can yield excellent approximations for profile probabilities even when independence assumptions are not strictly true, as long as they hold approximately.

It is a certainty, in fact, that the independence assumptions are not strictly correct. The most likely violations of independence result from population heterogeneity (also known as *population substructure* [23]). For United States Caucasians, subpopulations might be French, German, Irish, Italian, and so on, as well as individuals of various mixed heritage. These subpopulations certainly differ in their allele frequencies, albeit slightly. Thus, if the database collected by forensic laboratories to serve as a reference population is a reflection of that population, then it follows that independence cannot obtain, at least strictly, in these heterogeneous databases.

Several authors have suggested that population heterogeneity could lead to a serious underestimate of the probability of two DNA profiles matching [19,20,9,10]. Other geneticists and statisticians have countered that, while the argument that heterogeneity causes dependence is theoretically correct, human populations rarely exhibit enough heterogeneity to have a substantial effect on forensic calculations [14,7,8]. Data analysis supports the latter claim (e.g., [36,37]).

To understand the effect that allele distribution variation has on the calculation of genotype probabilities, consider the single-locus case. The model of population substructure assumes independent assortment of alleles within a subpopulation (random mating) and limited matings between subpopulations. This population substructure model is probabilistically equivalent to assuming that the vector of allele probabilities for a given subpopulation $G = (G(1), G(2), \ldots, G(m))$ possibly varies by subpopulation and that, conditional on G, an individual's pair of alleles is sampled independently.

Assuming the population substructure model, the probabilities of observing the genotype $\{a(i), a(j)\}$ in a random draw from a population is correctly calculated as

$$\Pr(\{a(i), a(j)\}) = \begin{cases} 2\gamma(i)\gamma(j) + 2\text{cov}[G(i), G(j)] \\ \quad \text{if } i \neq j, \\ \gamma(i)^2 + \text{var}[G(i)] \\ \quad \text{if } i = j. \end{cases} \quad (3)$$

Assuming that the Hardy–Weinberg equilibrium holds in the entire population is equivalent to assuming that there is no heterogeneity in the population: in other words, discarding the second term in both cases. Clearly, this leads to an underestimate of the probability when $i = j$ and it will usually lead to an overestimate when $i \neq j$.

A genetic model on evolutionary theory, leads to a one-parameter model for the variances and covariances in (3):

$$E[G(i)] = \gamma(i),$$
$$\text{Var}[G(i)] = \theta\gamma(i)[1-\gamma(i)],$$
$$\text{Cov}[G(i), G(j)] = -\theta\gamma(i)\gamma(j) \quad (4)$$

(e.g., see Weir [35]). In genetic parlance θ is analogous to Wright's [40] F_{ST}.

The difference between (1) and (3) depends on the magnitude of the heterogeneity among the subpopulations. If the heterogeneity is small and other sources of dependence can be ignored, then it has little effect on independence and the product of allele probabilities yields an excellent approximation to profile probabilities. In fact, heterogeneity must be quite large, relative to that typical for human genetic data, before it has appreciable effects on profile probabilities when the appropriate reference population is a general population.

22.5 The Affinal Model (Same Subpopulation)

Occasionally a good case can be made for some sort of relatedness between the suspect and the culprit under H_0. In the *affinal model*, the culprit and suspect are assumed to derive from the same subpopulation. Under this model, the appropriate reference population is the subpopulation of the suspect; however, typically, the crime laboratory possesses an insufficient amount of data for this approach. Nevertheless, calculations can be based on the larger reference population [30,2,26,39].

We illustrate these ideas using a one-locus marker. Assume $\mathcal{G}_e = \mathcal{G}_s = \{a(i), a(j)\}$, and calculated $1/LR = \Pr(\mathcal{G}_e|\mathcal{G}_s, H_0)$ by extending the reasoning used to obtain (4), so that

$$\Pr(\mathcal{G}_e|\mathcal{G}_s, H_0) = \begin{cases} 2\frac{[\theta + (1-\theta)\gamma(i)][\theta + (1-\theta)\gamma(j)]}{(1+\theta)(1+2\theta)} \\ \quad \text{if } i \neq j, \\ \frac{[2\theta + (1-\theta)\gamma(i)][3\theta + (1-\theta)\gamma(i)]}{(1+\theta)(1+2\theta)} \\ \quad \text{if } i = j \end{cases} \quad (5)$$

If θ is bigger than 0, then this calculation yields less evidence for H_1 than the independence model. However, most studies indicate that θ is a small number, usually less than 0.005 [33,29,27], and thus the effect on the probability calculation is small.

22.6 The Cognate Model (Same Family)

A more serious concern is based on the defense "the culprit was my brother" [18]. Using standard genetic principles, LR can be calculated when the culprit is assumed to be a relative of the suspect. For example, at a particular locus, identical twins share both alleles *identical by descent* (inherited from the same parent); full sibs (regular brothers) have a 25%, 50%, and

25% chance of sharing both alleles, one allele, and no alleles identical by descent, respectively. Of course, they can also share alleles by chance, which is called *identical by state*. From this we can infer that no matter how polymorphic the genetic markers, there is at least a $(.25)^L$ chance that an individual matches his brother at L loci. Because there is some chance that brothers share alleles identical by state, the probability of a match is slightly greater than .25 at each locus. The exact probability of a match at a single locus is $\{1+2\sum_k \gamma(k)^2 + 2[\sum_k \gamma(k)^2]^2 - \sum_k \gamma(k)^4\}/4$ [38].

In the *cognate model*, the culprit is assumed to be a relative of the suspect with probability c_p of having p alleles identical by descent [11].

If $\mathcal{G}_e = \mathcal{G}_s = \{a(i), a(j)\}$, then

$$\Pr(\mathcal{G}_e|\mathcal{G}_s, H_0) = \begin{cases} c_2 + c_1\gamma(i) + c_0\gamma(i)^2 \\ \quad \text{if } i = j, \\ c_2 + c_1[\gamma(i) + \gamma(j)]/2 + 2c_0\gamma(i)\gamma(j) \\ \quad \text{if } i \neq j. \end{cases}$$

(6)

Clearly an $L\mathcal{R}$ obtained from this calculation is considerably larger than that obtained in (2) when the culprit is allowed to be as closely related as a brother (for full sibs $c_2 = \frac{1}{4}$, $c_1 = \frac{1}{2}$, $c_0 = \frac{1}{4}$). For individuals even so closely related as first cousins, the relatedness effect is already relatively unimportant, as $c_2 = 0$ and $c_1 = \frac{1}{4}$.

Balding and Donnelly [3] have recommended calculating the odds of H_1 by modeling genetic correlations, combining genetic and nongenetic evidence, and allowing for all possible types of relationships between the suspect and the culprit.

22.7 In Practice

How can we allow for the measurement error expected to occur when minisatellite VNTRs are measured? Theoretically, the best way to allow for the measurement error is to construct a continuous version of the likelihood ratio [4,5,16]. The courts found these computations to be overly complex and have resorted to a discrete approximation to the continuous process.

When forensic scientists compare two minisatellite VNTR profiles, they declare a "match" when the fragment lengths are sufficiently similar. The term "match" suggests that the DNA samples are identical. Clearly this is a misnomer. For minisatellites it merely means that the profiles are as similar as they would be expected to be, given measurement error, when the samples are truly from the same individual. It does not mean that the sequences of the complementary alleles are identical, necessarily. A more accurate but more cumbersome terminology would be to say that the two profiles cannot be excluded as coming from the same individual or, simply, that there is a failure to exclude.

For minisatellite VNTRs the fragment lengths are measured with error, with the distribution of errors being dependent on the molecular methodology employed by the testing laboratory. The testing laboratories have characterized these distributions, which appear to be approximately normal. See Budowle et al. [6], Devlin et al. [15], and Berry et al. [5] for details.

For crime-scene materials, environmental factors can cause yet another source of error. Degraded DNA fragments tend to migrate at a somewhat faster or slower rate than fresh DNA. This differential migration is termed *band shifting* [6,24].

Band shifting is an important phenomenon because the DNA from crime-scene material is commonly compared with DNA "fresh" from a suspect. Clearly these comparisons must be undertaken carefully, and the methods differ by laboratory. For instance, in terms of numerical rules, the FBI declares a match if the two fragments differ by less than 5% of the mean of the fragments [6]. The FBI chose this value because it never observed measurement error plus band shifting to create more than a 5% difference between complementary fragment sizes in their published studies. The studies, however, were not exhaustive, so it is not difficult to imagine that larger differences are possible. If such an event occurs, then we obtain a false negative.

Lifecodes Corporation, for a particular molecular method, declares a match if the bands differ by less than 1.8% of the mean. This match criterion, which is equivalent to a 3-standard-deviation (3σ) window, is more stringent than the FBI's because Lifecodes uses different molecular methods and uses monomorphic markers to correct for bandshifting. The FBI does not make such a correction and hence requires a larger window to capture variation due to both measurement error and bandshifting.

To calculate the match probability using the formulas presented earlier, forensic scientists employ a practical approach that involves a process called *binning*. For the *floating-bin* method, an interval (bin) equivalent to the matching window is defined about the evidentiary fragment. Bins are treated as alleles, and the probability of a matching allele is estimated by the proportion of fragments in the database that fall into that bin. Because the bin is determined by the location of the evidentiary fragment, the bin is said to float; hence the name. For the *fixed-bin* method, the range of possible fragment lengths is divided, a priori, into bins.

Many testing laboratories maintain databases collected from the major ethnic groups: Caucasian, African American, Hispanic, and Asian American. The Hispanic database is frequently subdivided by geographic origin, because Hispanics from the western United States have a higher degree of Amerindian heritage, in general, than do Hispanics from the eastern United States. These are samples of convenience, yet they appear to approximate random samples rather closely. Moreover, the different databases are remarkably similar [25,13,1].

If the donor of the evidentiary sample is known to be from a particular ethnic group, say, Caucasian, then the Caucasian database is the natural choice. However, in general, the choice may not be apparent [21]. Consider a murder of an African American that took place in Central Park, New York, with no eyewitnesses. For this case the Caucasian, African American, Hispanic, and Asian American databases seem to be reasonable choices for the reference database. Other databases, such an Amerindian, appear to be irrelevant because Amerindians are a minuscule component of New York's population. Some have argued that the reference database should reflect the ethnicity of the defendant. This conclusion does not follow from any probabilistic argument; however, it does tend to yield conservative estimates of the desired probability (i.e., it favors the accused). Because the defendant's alleles are likely to be most common within his/her own ethnic group, this choice leads to a larger estimate of the probability of a match, on average, than would be obtained if any other database were used. In a given case, it often happens that numerous match probabilities are presented.

Consider a more serious type of error—a sample mixup. It is possible that the wrong DNA samples might be compared? Although forensic scientists operate under strict protocols designed to minimize the chance of such an error, there is little doubt that a sample mixup could occur and that the probability of it is greater than the probability of a chance match. Because the probability of a laboratory error is small, it cannot be accurately estimated from proficiency tests. However, data do suggest that most laboratory errors favor the defendant.

References

1. Balazs, I. (1993). Population genetics of 14 ethnic groups using phenotypic data from VNTR loci. *Proc. 2nd Int. Conf. DNA Fingerprinting*, S. D. J. Pena, R. Chakraborty, J. T. Epplen, and A. J. Jeffries, eds. Birkhauser, Boston.

2. Balding, D. J. and Nichols, R. A. (1994). DNA profile match probability calculation: how to allow for population stratification, relatedness, database selection and single bands. *Forensic Sci. Int.*, **64**, 125–136.

3. Balding, D. J. and Donnelly, P. (1995). Inference in forensic identification. *J. Roy. Statist. Soc. A*, **158**, 21–54.

4. Berry, D. A. (1991). Inferences using DNA profiling in forensic identification and paternity cases. *Statist. Sci.* **6**, 175–205.

5. Berry, D. A., Evett, I. W., and Pinchin, R. (1992). Statistical inference in crime investigations using DNA profiling: Single locus probes. *Appl. Statist.*, **41**, 499–531.

6. Budowle, B., Giusti, A. M., Wayne, J. S., Baechtel, F. S., Fourney, R. M., Adams, D. E., Presley, L. A., Deadman, H. A., and Monson, K. L. (1991). Fixed bin analysis for statistical evaluation of continuous distributions of allelic data from VNTR loci for use in forensic comparisons. *Am. J. Human Genet.*, **48**, 841–855.

7. Chakraborty, R. and Kidd, K. K. (1991). The utility of DNA typing in forensic work. *Science*, **254**, 1735–1739.

8. Chakraborty, R. and Jin, L. (1992). Heterozygote deficiency, population substructure and their implications in DNA fingerprinting. *Hum. Genet.*, **88**, 267–272.

9. Cohen, J. E. (1990). DNA fingerprinting for forensic identification: Potential effects on data interpretation of subpopulation heterogeneity and band number variability. *Am. J. Hum. Genet.*, **46**, 358–368.

10. Cohen, J. E., Lynch, M., and Taylor, C. E. (1991). Forensic DNA tests and Hardy–Weinberg equilibrium. *Science*, **253**, 1037–1038.

11. Cotterman, C. W. (1940). *A Calculus for Statistical Genetics*. Ohio State University, Columbus.

12. Devlin, B. (1993). Forensic inference from genetic markers. *Statist. Methods Med. Res.*, **2**, 241–262.

13. Devlin, B. and Risch, N. (1992). Ethnic differentiation at VNTR loci, with special reference to forensic applications. *Am. J. Hum. Genet.*, **51**, 534–548.

14. Devlin, B., Risch, N., and Roeder, K. (1990). No excess of homozygosity at DNA fingerprint loci. *Science*, **249**, 1416–1420.

15. Devlin, B., Risch, N., and Roeder, K. (1991). Estimation for allele frequencies for VNTR loci. *Am. J. Hum. Genet.*, **48**, 662–676.

16. Devlin, B., Risch, N., and Roeder, K. (1992). Forensic inference from DNA fingerprints. *J. Am. Statist. Assoc.*, **87**, 337–349.

17. Devlin, B., Risch, N., and Roeder, K. (1993). Statistical evaluation of DNA fingerprinting: a critique of the NRC's report. *Science*, **259**, 748–749, 837.

18. Evett, I. W. (1992). Evaluating DNA profiles in the case where the defense is "It was my brother." *J. Forensic Sci. Soc.*, **32**, 5–14.

19. Lander, E. (1989). DNA fingerprinting on trial. *Nature*, **339**, 501–505.

20. Lander, E. S. (1991). Research on DNA typing catching up with courtroom applications. *Am. J. Hum. Genet.*, **48**, 819–823.

21. Lewontin, R. C. (1993). Which population? *Am. J. Hum. Genet.*, **52**, 205.

22. Lewontin, R. C. and Hartl, D. L. (1991). Population genetics in forensic DNA typing. *Science*, **254**, 1745–1750.

23. Li, C. C. (1969). Population subdivision with respect to multiple alleles. *Ann. Hum. Genet.*, **33**, 23–29.

24. McNally, L., et al. (1990). Increased migration rate observed in DNA from evidentiary material precludes the use of sample mixing to resolve forensic cases of identity. *Appl. Theor. Electrophoresis* **5**, 267.

25. Mourant, A. E., Kopec, A. C., and Domainewska-Sobczak, K. (1976). *The Distribution of Human Blood Groups and Other Polymorphisms*. Oxford University Press, London.

26. Morton, N. E. (1992). Genetic structure of reference populations. *Proc. Natl. Acad. Sci. (USA)*, **89**, 2556–2560.

27. Morton, N. E., Collins, A., and Balazs, I. (1993). Bioassay of kinship for hypervariable loci in blacks and caucasians. *Proc. Natl. Acad. Sci. (USA)*, **90**, 1892–1896.

28. National Research Council. (1996). *The Evaluation of Forensic DNA Evidence*. National Academy Press, Washington, DC.

29. Nei, M. and Roychoudhury, A. K. (1982). Genetic relationships and evolution of human races. *Evol. Biol.*, **14**, 1–59.

30. Nichols, R. A. and Balding, D. J. (1991). Effects of population structure on DNA fingerprint analysis in forensic science. *Heredity*, **66**, 297–302.

31. Risch, N. and Devlin, B. (1992). On the probability of matching DNA fingerprints. *Science*, **255**, 717–720.

32. Roeder, K. (1994). DNA fingerprinting: A review of the controversy. *Statist. Sci.*, **9**, 222–278.

33. Roeder, K., Escobar, M., Kadane, J., and Balazs, I. (1993). *Measuring Heterogeneity in Forensic Databases: A Hierarchical Approach to the Question. Technical Report*, Carnegie Mellon University, Pittsburgh, PA.

34. Steinberger, E. M., Thompson, L. D., and Hartmann, J. M. (1993). On the use of excess homozygousity for subpopulation detection. *Am. J. Hum. Genet.*, **52**, 1275–1277.

35. Weir, B. S. (1990). *Genetic Data Analysis*. Sinauer Associates, Sunderland, MA.

36. Weir, B. S. (1992). Independence of VNTR alleles defined by fixed bins. *Genetics*, **130**, 873–887.

37. Weir, B. S. (1992). Independence of VNTR alleles defined as floating bins. *Am. J. Hum. Genet.*, **51**, 992–997.

38. Weir, B. S. (1993). Forensic population genetics and the NRC. *Am. J. Hum. Genet.*, **52**, 437–439.

39. Weir, B. S. (1994). The effects of inbreeding on forensic calculations. *Ann. Rev. Genet.*, **28**, 597–621.

40. Wright, S. (1951). The genetical structure of populations. *Ann. Eugen.*, **15**, 323–354.

23 Epidemics

Klaus Dietz

23.1 Introduction

The early phases of statistics and epidemiology were closely linked: "There is probably no more legitimate use of the instrument of statistics than its application to the study of epidemic diseases" [32]. The term *epidemic* (Greek: "upon the people") means a sudden outbreak of an infectious disease. The number of new cases per week or month, i.e., the *incidence*, rises to a peak and thereafter declines. One talks about a disease as *endemic* if the prevalence of the infection, i.e., the number of infected individuals at a given point in time, stays more or less constant. An infection that is spreading across the borders of countries affecting several continents or even the globe is called *pandemic*.

Epidemiology was originally concerned exclusively with the spread of epidemics. Today it is the study of the distribution and determinants of any disease in a population, the techniques for establishing such knowledge, and application of the results to the control of health problems. The spread of diseases caused by bacteria, viruses, protozoa, fungi, and helminths is investigated by infectious disease epidemiology.

Long before the causative agents of infectious diseases were identified by Pasteur, Koch, and others, mathematical models were used to assess the effect of smallpox on mortality (Daniel Bernoulli) or to describe the shape of epidemic curves in order to predict the future of the epidemic from its initial course (Farr). The first book on the mathematical theory of epidemics was published by Bailey [3] in 1957, and in a revised edition in 1975 [4].

The main uses of mathematical models of infectious diseases are: (1) prediction, (2) understanding, and (3) evaluation of control strategies.

Predictions were and will be in high demand with respect to the pandemic spread of AIDS. This new disease has not only stimulated many medical sciences such as virology, immunology, etc., but also the mathematical theory of infectious diseases and the statistical methodology to deal with censored and truncated data, and with the flexible design of clinical trials. A special section of this chapter is devoted to AIDS (Section 22.6).

Using data about the number of passengers flying between major cities, it was possible to predict the spread of influenza epidemics on a global scale [33].

The most puzzling phenomenon of many infectious diseases is the periodicity of their outbreaks. A striking feature of measles epidemiology is the predominance of 2 year cycles. In some countries, however, one observes yearly cycles; in other countries longer intervals may be observed. Many models have been proposed to explain the periodicity of measles, taking into account either seasonal variations in the contact rate or effects of schooling. Some authors have seen the measles data from New York City as a realization of deterministic chaos instead of a stochastic system (see Schaffer et al. [34], and Grenfell et al. [21]).

Another phenomenon that has been studied statistically as a function of population size is the fadeout of epidemics in relatively isolated populations like Iceland (with a population of less than 300,000) or Australia, particularly before the era of air transport. See Cliff et al. [8] for a comprehensive survey of the epidemiological aspects of measles, including the geographic spread.

The most important applications of epidemic models are devoted to the evaluation of control strategies. The main result is the identification of threshold phenomena according to which the infection cannot spread in the community unless the number of secondary cases caused by one case is greater than one. This means that control interventions do not have to be 100% efficacious and do not need to reach 100% of the population in order to reduce the incidence to zero. Since the mid-1980s there has been a noticeable trend in this theory to attack more and more realistic problems and to include detailed disease-specific aspects into the model assumptions.

The Centers for Disease Control and Prevention of the United States in Atlanta have compiled a list of "vaccine preventable disease issues amenable to mathematical modelling" [13]. With respect to measles, for example, it is proposed to study the effect of a vaccination strategy that applies two doses in urban settings and one dose in rural settings. The evaluation of such strategies is one of the most important motivations for the development of epidemic models, because there is no methodological alternative—the likely outcome of a strategy depends in a highly nonlinear way on the underlying parameters; randomized control trials with whole nations as experimental units are impossible.

23.2 History of Epidemic Models

The historical development started in 1760 with Daniel Bernoulli's calculation of the increase in life expectancy if smallpox could be eradicated by the method of inoculation. In modern terminology, he assumed an endemic situation with a constant force of infection. He correctly took into account that an infection causes immunity in the survivors, so that only susceptible individuals are at risk of infection and subsequent death. He calculated the age-specific proportion of susceptibles, assuming for lack of precise data a constant force of infection and an age-independent case fatality rate. Using differential equations in an ingenious way, he was able to determine the survival curve and the corresponding life expectancy if the competing risk of death from smallpox were eliminated. This approach was (unjustly) criticized (by D'Alembert), but was endorsed and further developed by several mathematicians of his time (Lambert, Tremblay, Duvillard, and Laplace [35]).

The first dynamic model for the spread of epidemics was published in 1889 by P. D. En'ko [11,15] and fitted to the measles epidemics which he recorded at the Alexander Institute of the Smol'ny in St. Petersburg, Russia. In order to determine the size of the next generation of cases, the present number of susceptibles is multiplied by the probability that one susceptible has at least one contact with an infective according to the binomial distribution with a fixed number of contacts and a success probability that equals the prevalence of infectives in the population apart from the one susceptible at risk. A similar approach, which is implicitly based on a Poisson distributed number of contacts, is due to L. J. Reed and W. H. Frost. They used their model in the 1920s in class lectures at The Johns Hopkins University. A description of their model together with other historical material on the mathematical theory of epidemics can be found in Serfling [36].

The threshold concept in the epidemiology of infectious diseases was first announced by Ronald Ross in 1911 in connection with his work on malaria. (He had received the Nobel Prize for Medicine in 1902 for his discovery of the cycle of transmission of malaria by mosquitoes.) Ross argued that eradication of malaria was possible by decreasing the density of mosquitoes below a certain threshold [17].

The first stochastic model that was analyzed analytically was proposed in a seminal paper in 1926 by McKendrick [28]. He derived the distribution of the total number of cases in a household of a given size, assuming a contact rate between the number of susceptibles and the number of infectives and a constant recovery rate into the immune state.

It is remarkable that all the pioneers of mathematical epidemiology mentioned were physicians. A comprehensive history of epidemic theory still has to be written.

23.3 Infectious Disease Data

23.3.1 Surveillance Data

From a global perspective infectious diseases always have been and still are by far the most important cause of death. The earliest routine records of burials often mention the cause of death as smallpox or plague. For specified infectious diseases there exist international, national, and local regulations for the reporting of individual cases. The quality of such data varies greatly from country to country and from time to time. Interpretation and comparability problems are created by reporting delays, underreporting, and diagnostic errors. Farrington et al. [16] describe a statistical algorithm for the automated early detection of outbreaks of infectious disease. Today most notifiable infectious diseases are subject to public health control actions such as vaccination or therapy. Therefore the reported incidence figures cannot be interpreted unless one also has data about the quantity and quality of these interventions.

In August 1994 elimination of polio was declared for the Americas. The World Health Organization has announced a poliomyelitis eradication target to be achieved by the year 2000. The large proportion of subclinical infections leads to the question of how long one has to wait after the last observed case, before one can be sure that silent transmission of polio has ceased and eradication can be certified. This was not such a difficult problem in the case of smallpox, because virtually all infected individuals showed an easily recognizable rash. Wild polio virus could spread unnoticed without causing paralytic cases for a considerable time before new paralytic cases would be detected by surveillance. Stochastic simulation studies have been used by Eichner and Dietz [14] to determine the probability that the wild virus has been extinguished if no cases are observed for a given period of time; one would have to wait for three years in order to be 95% sure that the wild virus is extinguished. Hence the last case of polio infection will never be known if global eradication of polio is in fact successful. Because of the potentially disastrous consequence of premature termination of polio vaccination, one may have to wait even longer before declaring the eradication of polio.

23.3.2 Outbreak Data

Comprehensive surveys about statistical methods appropriate for the analysis of householdspecific outbreak data are given by Bailey [4] and Becker [5]. In order to estimate transmission parameters within and between households one has to define a stochastic model that provides an appropriate setting for statistical inference. Advanced statistical tools such as martingales lead to simple formulas for estimates of contact or removal rates. Some of these procedures assume that the numbers of infectives and susceptibles are known at any time during the outbreak. Often, however, the epidemiological team arrives late at the end of the outbreak and can only determine the total number of cases within each household.

The advantage of outbreak data is the availability of very detailed information on the distribution of the number of cases by age, sex, household composition, and vaccination status. The drawback is that the parameters estimated from a particular outbreak cannot necessarily be generalized to other sections of the population. The fact that an outbreak occurred may be due to special circumstances such as refusal of vaccination because of religious beliefs. Nevertheless, the analysis of individual outbreaks remains an indispensable tool for the field evaluation of vaccination programs.

23.3.3 Seroprevalence Data

For diseases that confer lasting immunity that is detectable by antibodies in the serum, one can estimate for each age group the proportion of the population which has been infected at some time in the past. Models for the interpretation of seroprevalence data usually assume that the infection has achieved an endemic equilibrium such that the risk of acquiring the infection (the force of infection) is constant over time, but may depend on the age of an individual. Following a cohort of newborns, one can then describe the susceptible proportion at a given age as a survivor probability where the hazard rate of dying corresponds to the force of infection. A cross-sectional survey of the population provides for each individual the age at the time of the survey and whether the infection has taken place in the past (left-censored) or has not yet taken place (right-censored). One therefore has to estimate the force of infection from of a sample of data that are all censored.

Nonparametric solutions of this problem are described by Keiding [27]. They involve kernelsmoothing procedures for the age-specific force of infection. Since for such data the precision of the estimates involves $n^{1/3}$ instead of $n^{1/2}$ asymptotics, the sample sizes needed for given accuracy are rather large. From such seroprevalence data one can estimate the age distribution of first infection and derive important parameters like the mean or median age of first infection. The age at first infection is inversely proportional to the transmissibility of the infection. This relationship can be used to estimate parameters that describe the transmission potential of an infection in a completely susceptible population.

23.4 The Basic Reproduction Number

The most important concept in epidemic theory is the average number of secondary cases of an infectious disease that one case could generate in a completely susceptible population. The name and the notation for this concept has varied considerably, but the terminology is converging more and more to the term *basic reproduction number* and the symbol R_0 [9]. A survey is given by Heesterbeek and Dietz [25], and an overview of the associated estimation methods can be found in Dietz [12]. For a homogeneously mixing population the equilibrium proportion of susceptibles equals the reciprocal of the basic reproduction number, because in this situation, on the average, one case produces one secondary case. For infections with lifelong immunity the average age at first infection divided by the life expectancy equals the proportion of susceptibles in the population, because this equals the fraction of life before the infection. Therefore one can estimate the basic reproduction number by the ratio of life expectancy to average age at first infection. The comprehensive reference work of the mathematical approach to the epidemiology and control of human infectious diseases by Anderson and May [2] contains numerous estimates of R_0 using this approach.

One can also estimate R_0 from the final size of an epidemic. If one knows the number of individuals infected during the epidemic and the final proportion of individuals still susceptible, the following formula provides an estimate of R_0 for a homogeneously mixing population

$$R_0 = \frac{\ln u_0 - \ln u_\infty}{u_0 - u_\infty},$$

where u_0 denotes the initial and u_∞ the final proportion of susceptibles.

The practical significance of an estimate of R_0 is that it can be used to determine the lower bound

$$1 - \frac{1}{R_0}$$

for the effective vaccination coverage that would result in elimination of the infection.

The expressions *elimination* and *eradication* are often wrongly used as synonyms to indicate a reduction of infection incidence to a zero level. But epidemic theory shows that a zero incidence may represent either a stable or an unstable state.

A (locally) stable zero equilibrium (elimination) means that the introduction of infections from outside would only lead to a small number of secondary cases and then these small epidemics would soon be extinct, i.e., one would observe a return to the zero level of incidence. In contrast, an unstable zero endemic state (eradication) would lead to major epidemics if new cases were to be introduced into the population. Thus eradication is a global concept.

23.5 Efficacy of Vaccines and Effectiveness of Vaccination Programs

Vaccinations are the most powerful weapon to prevent communicable diseases. They led to the unique success of global smallpox eradication. McLean and Blower [29] introduce three facets of vaccine failure: which they call *take, degree*, and *duration*. The concept of "take" is called "all or nothing" by Halloran et al. [23] and "model 2" by Smith et al. [37]. What is denoted by "degree" in McLean and Blower [29] is called "leaky" by Halloran et al. [23] and "model 1" by Smith et al. [37]. The objective of all these terms is the description of the distribution of remaining susceptibility after vaccination of a susceptible.

The classical notion of *vaccine efficacy* ("model 2" in the sense of Smith et al. [37]) assumes that after vaccination a certain fraction of the vaccinees are successfully vaccinated so that their subsequent susceptibility is reduced to zero. The remaining fraction of the vaccinees have to be considered as complete vaccine failures for whom the susceptibility is still at its original level of one. Most vaccine trials work more or less explicitly with this assumption and either test the null hypothesis that the vaccine has no effect or estimate a confidence interval for the vaccine efficacy in this sense. O'Neill [30] provides formulas and tables for the necessary sample sizes in order to estimate confidence intervals for a vaccine efficacy of specified width. This is very important, because the usual formulas for the determination of sample sizes in cohort studies are based on the hypothesis-testing paradigm.

From a public health point of view it is not sufficient to reject the null hypothesis that the vaccine has no effect. For large-scale use one would prefer an interval estimate for vaccine efficacy that does not contain values below the equilibrium seroprevalence of naturally acquired antibodies. Estimates for vaccine efficacy can be biased if an inappropriate model for the vaccine effect is used to analyze the observations in the vaccinated and the nonvaccinated group. Study designs have to be developed that are appropriate for the estimation of the individual components of this multivariate concept.

After the adoption of a particular vaccine it is necessary to assess the vaccine efficacy in the field, because the results from the vaccine trial may not be applicable to the situation under routine use of the vaccine. There may be operational reasons, such as breakdowns of the cold chain, which can seriously affect the efficacy [31].

For infectious diseases which are transmitted from person to person either directly or indirectly (via vectors such as mosquitoes), vaccination does not only protect the vaccinee against infection, but also the community. There are vaccine-preventable diseases such as tetanus and tickborne encephalitis where the protective effect is restricted to the vaccinee because they are not transmitted from person to person. An excellent review of the concept of herd immunity has been given by Fine [18]. For the contribution of mathematical modeling to vaccination policy, see Fine [19].

Haber et al. [22] introduce measures of effectiveness for a vaccination program in a randomly mixing population. Depending on the vaccination coverage, one has to weight the effects for the vaccinated and nonvaccinated fraction of the population in order to obtain a measure for the population effectiveness. Due to the vaccination program, the force of infection is reduced even for the nonvaccinated individuals. Similarly, for the vaccinated individuals the vaccine effectiveness is determined not only by the vaccine efficacy but also by the reduction in the force of infection for those who are not fully protected. The effectiveness of a vaccination program for the total population is therefore a function of the vaccination coverage and the efficacy of the vaccine at the individual level. If the vaccine efficacy is less than the prevaccination equilibrium seroprevalence of antibodies, then even 100% coverage would not be sufficient to eliminate an infection.

23.6 AIDS

The AIDS pandemic poses unique statistical problems that are due mainly to the fact that the incubation time, i.e., the interval between infection and the onset of full-blown AIDS, has a large variability with a median of about ten years. Even today (1996), nobody knows whether the distribution function will eventually approach 100% or whether there will be a fraction of infected individuals who will never develop AIDS.

23.6.1 AIDS Surveillance

Since the first cases were reported to the U.S. Centers of Disease Control in 1981, the United States and subsequently virtually all countries have established surveillance systems for AIDS cases that involved either voluntary or obligatory reporting of patients who fulfilled a set of diagnostic criteria. From a statistical point of view the interpretation of these data is complicated by the fact that the diagnostic criteria were changed several times by adding further items to the list of syndromes defining full-blown AIDS. Due to the long incubation period, the AIDS incidence does not reflect current infection incidence. A large number of statistical methods were developed to infer HIV incidence from AIDS incidence and to use the estimated HIV incidence for AIDS projections into the future.

Before the AIDS incidence data can be used for this purpose, one has to take into account that the reports are subject to a nonnegligible reporting delay [6]. The distribution of reporting delays does not stay constant over time and is affected by sudden changes in reporting behavior that often result either from administrative deadlines for the distribution of funds or from special publicity campaigns. The greatest problem of the AIDS surveillance data is the unknown proportion of nonreporting. Gail [20] discusses the use of surveillance data for evaluating AIDS therapies.

23.6.2 Estimating HIV Incidence and AIDS Projections

The AIDS incidence can be considered as the convolution of the HIV incidence and the incubation-time distribution. One can estimate HIV incidence if one assumes that the incubation distribution is known and that the reported AIDS incidence is reliable. Deconvolution problems belong to the class of ill-posed problems. The resulting estimates are particularly unreliable for recent HIV incidence, because newly infected individuals hardly contribute to the present AIDS incidence.

A wide variety of methods have been developed to solve this deconvolution problem: parametric methods assuming that the HIV incidence can be described by a simple curve with a few parameters, and nonparametric methods assuming that the HIV incidence is stepwise constant. The analytical methods include the EM algorithm and the use of splines. The estimated HIV incidence is subsequently used to make projections of the AIDS incidence into the future. If one assumes that HIV incidence is reduced to zero after a specified time limit, one obtains a lower bound on the expected AIDS incidence (see Brookmeyer [6]).

23.6.3 Markers and Incubation Times

A large number of studies are devoted to the relation of cofactors and markers with the length of the incubation period: As stated by Brookmeyer and Gail [7]:

> Cofactors are variables that affect the duration of the incubation period and may explain why some infected individuals progress to AIDS faster than other individuals. Markers are variables that track the progression of HIV infection. Markers are consequences of infection, while cofactors are causal agents rather than consequences of disease progression.

According to this definition, age at infection is a cofactor, whereas numbers of CD4 cells are markers. There are many studies that try to assess which markers can be used as surrogate endpoints in clinical trials instead of the appearance of full-blown AIDS. If in a clinical trial one could show that the treatment prevents the number of CD4 cells from declining further, one could terminate the trial earlier and license a potentially effective drug therapy earlier than if one took the appearance of full-blown AIDS as the endpoint of the trial.

23.6.4 Infectivity

The key variable for the understanding of HIV dynamics is the per-contact probability of infecting a susceptible partner, depending on the time since the acquisition of the infection and on the kind of contact. The estimates available for the per-contact probability of transmission are mainly derived from partner studies that try to determine either prospectively or retrospectively the infection status of the partner of an infected index case and the number of contacts that have taken place under risk. One of the many problems associated with such studies is the lack of information about the time of infection of the index case. This problem does not exist in the case of transfusion-induced infections, where the date of infection can be identified.

Taking all studies together, there appears to be agreement that the infection probability per contact is about 3 times higher from male to female than vice versa. Cofactors like the presence of untreated sexually transmitted diseases can increase the infection probability considerably. Estimates for this cofactor yield a relative risk per contact of the order of 10–50 for male-to-female transmission and of 50–300 for female-to-male transmission [24].

A further result that is confirmed by many studies is the high variability of the infection probability per contact. It appears that the infection probability is very high soon after the infection, but the period of high infectivity is short and ends when antibodies appear in the blood. Subsequently the infectivity can be very low for several years and can rise again toward the end of the incubation time [26]. Whether this latter rise has any major epidemiological implication is not known, because the individual may have changed behavior due to counseling in the context of an HIV diagnosis or due to HIV disease.

23.6.5 Transmission Models

Individuals are categorized into various classes according to their contact rate, i.e., their rate of changing partners. Associated with these classes one also has to define a contact matrix, because it is usually not assumed that partners are chosen at random. Numerical studies show that the dynamics of the HIV incidence can vary considerably depending on the assumptions made in these contact matrices. Even multimodal HIV incidence curves are possible when the infection spreads from small contact groups with very high partner change rates to larger groups with smaller partner change rates.

Most models assume that the infection probability per partnership does not depend on the duration of the partnership. This is claimed to be supported by a partner study and a study of transfusion-induced AIDS cases, where there was no correlation between the infection probability of the partner and the number of contacts. This lack of correlation, however, could also be explained by variability of the infection probability between those partnerships. Other models assume that the infectivity probability per partner depends on the partner change rate, and some models explicitly take into account pair formation and duration of partnership [10].

Major concerns, especially in many African countries, concentrate on whether AIDS could turn the growth of a population into a decline if the prevalence of the infection increased above a certain level [1]. Very small changes in the crucial parameter values produce vastly different long-term projections in demographic growth. Locally, a reduction in the growth rate or even a decline in the size of the population after several decades cannot be excluded, but this effect will most likely be compensated for by other populations where a positive growth rate persists.

Modelers have to admit that models cannot be used for long-term predictions, due to the large amount of uncertainty about the individual parameters, but it is often stressed that models help to identify sensitive variables for which further data should be collected. In other words, models may serve for planning future epidemiological studies, which would then help to improve predictions and understanding of the underlying dynamics.

References

1. Anderson, R. M. (1991). Mathematical models of the potential demographic impact of AIDS in Africa. *AIDS*, **5** (Suppl. 1), S37–S44.

2. Anderson, R. M. and May, R. M. (1991). *Infectious Diseases of Humans: Dynamics and Control*. Oxford University Press, Oxford, UK.

3. Bailey, N. T. J. (1957). *The Mathematical Theory of Epidemics*. Griffin, London.

4. Bailey, N. T. J. (1975). *The Mathematical Theory of Infectious Diseases and its Applications*, 2nd ed. Griffin, London.

5. Becker, N. G. (1989). *Analysis of Infectious Disease Data*. Chapman & Hall, London.

6. Brookmeyer, R. (1996). AIDS, epidemics, and statistics. *Biometrics*, **52**, 781–796.

7. Brookmeyer, R. and Gail, M. (1994). *AIDS Epidemiology: A Quantitative Approach*. Oxford University Press, New York.

8. Cliff, A. D., Haggett, P., and Smallman-Raynor, M. R. (1993). *Measles: An Historical Geography of a Major Human Viral Disease. From Global Expansion to Local Retreat, 1840–1990*. Blackwell, Oxford.

9. Diekmann, O., Heesterbeek, J. A. P., and Metz, J. A. J. (1990). On the definition and the computation of the basic reproduction ratio R_0 in models for infectious diseases in heterogeneous populations. *J. Math. Biol.*, **28**, 365–382.

10. Dietz, K. (1988). On the transmission dynamics of HIV. *Math. Biosci.*, **90**, 397–414.

11. Dietz, K. (1988). The first epidemic model: A historical note on P. D. En'ko. *Austral. J. Statist.*, **30A**, 56–65.

12. Dietz, K. (1993). The estimation of the basic reproduction number for infectious diseases. *Statist. Methods Med. Res.*, **2**, 23–41.

13. Dietz, K. (1995). Some problems in the theory of infectious disease transmission and control. In *Epidemic Models: Their Structure and Relation to Data*, D. Mollison, ed., pp. 3–16, Cambridge University Press, Cambridge, UK.

14. Eichner, M. and Dietz, K. (1996). Eradication of poliomyelitis: when can one be sure that polio transmission has been terminated? *Am. J. Epidemiol.*, **143**, 816–822.

15. En'ko, P. D. (1989). On the course of epidemics of some infectious diseases. *Int. J. Epidemiol.*, **18**, 749–755. (Translated from the Russian original of 1889 by K. Dietz.)

16. Farrington, C. P., Andrews, N. J., Beale, A. D., and Catchpole, M. A. (1996). A statistical algorithm for the early detection of outbreaks of infectious disease. *J. Roy. Statist. Soc. A*, **159**, 547–563.

17. Fine, P. E. M. (1975). Ross's a priori pathometry—a perspective. *Proc. Roy. Soc. Med.*, **68**, 547–551.

18. Fine, P. E. M. (1993). Herd immunity: History, theory, practice. *Epidemiol. Rev.*, **15**, 265–302.

19. Fine, P. E. M. (1994). The contribution of modelling to vaccination policy (with discussion by D. J. Nokes). In *Vaccination and World Health*, F. T. Cutts and P. G. Smith, eds., pp. 177–194, Wiley, Chichester, UK.

20. Gail, M. H. (1996). Use of observational data, including surveillance studies, for evaluating AIDS therapies. *Statist. Med.*, **15**, 2273–2288.

21. Grenfell, B. T., Kleczkowski, A. Gilligan, C. A., and Bolker, B. M. (1995). Spatial heterogeneity, nonlinear dynamics and chaos in infectious diseases. *Statist. Methods Med. Res.*, **4**, 160–183.

22. Haber, M., Longini, I. M., Jr., and Halloran, M. E. (1991). Measures of the effects of vaccination in a randomly mixing population. *Int. J. Epidemiol.*, **20**, 300–310.

23. Halloran, M. E., Haber, M., and Longini, I. M., Jr. (1992). Interpretation and estimation of vaccine efficacy under heterogeneity. *Am. J. Epidemiol.*, **136**, 328–343.

24. Hayes, R. J., Schulz, K. F., and Plummer, F. A. (1995). The cofactor effect of genital ulcers on the per-exposure risk of HIV transmission in sub-Saharan Africa. *J. Trop. Med. Hygiene*, **98**, 1–8.

25. Heesterbeek, J. A. P. and Dietz, K. (1996). The concept of R_0 in epidemic theory. *Statist. Neerl.*, **50**, 89–110.

26. Jacquez, J. A., Koopman, J. S., Simon, C. P., and Longini, I. M., Jr. (1994). Role of the primary infection in epidemics of HIV infection in gay cohorts. *J. Acquired Immune Deficiency Syndromes J. AIDS*, **7**, 1169–1184.

27. Keiding, N. (1991). Age-specific incidence and prevalence: a statistical perspective (with discussion). *J. Roy. Statist. Soc. A*, **154**, 371–412.

28. McKendrick, A. G. (1926). Applications of mathematics to medical problems. *Proc. Edinburgh Math. Soc.*, **14**, 98–130.

29. McLean, A. R. and Blower, S. M. (1993). Imperfect vaccines and herd immunity to HIV. *Proc. Roy. Soc. Lond. B*, **253**, 9–13.

30. O'Neill, R. T. (1988). Sample sizes to estimate the protective efficacy of a vaccine. *Statist. Med.*, **7**, 1279–1288.

31. Orenstein, W. A., Bernier, R. H., and Hinman, A. R. (1988). Assessing vaccine efficacy in the field: Further observations. *Epidemiol. Rev.*, **10**, 212–241.

32. Ransome A. (1868). On epidemics, studied by means of statistics of disease. *Br. Med. J.*, **10**, 386–388.

33. Rvachev, L. A. and Longini, I. M., Jr. (1985). A mathematical model for the global spread of influenza. *Math. Biosci.*, **75**, 3–22.

34. Schaffer, W. M., Olsen, L. F., Truty, G. L., Fulmer, S. L., and Graser, D. J. (1988). Periodic and chaotic dynamics in childhood infections. In *From Chemical to Biological Organization*, M. Markus, S. C. Müller, and G. Nicolis, eds., pp. 331–347, Springer, Berlin.

35. Seal, H. L. (1977). Studies in the history of probability and statistics. XXXV Multiple decrements or competing risks. *Biometrika*, **64**, 429–439.

36. Serfling, R. E. (1952). Historical review of epidemic theory. *Hum. Biol.*, **24**, 145–166.

37. Smith, P. G., Rodrigues, L. C., and Fine, P. E. M. (1984). Assessment of the protective efficacy of vaccines against common diseases using case control and cohort studies. *Int. J. Epidemiol.*, **13**, 87–93.

24 Epidemiological Statistics I

M. E. Wise

24.1 Introduction

Epidemiology means, of course, the science of epidemics. But in practice it is no longer about infectious diseases only. The short *Oxford English Dictionary* (1971 edition, corrected 1975) quotes the Sidney Society Lexicon:

> **Epidemic, Adjective.** (1) Of a disease, prevalent among a people or community at a special time and produced by some special causes not generally present in the affected locality. (2) Widely prevalent; universal.

The literal meaning is "upon the population." This includes much of "demography," which is given as "statistics of births, death, diseases, etc." "Epidemiology" as it has grown and matured has reverted to this literal meaning. It is about population studies in medicine in the broadest sense.

In 1849, Snow collected statistics of cholera cases in central London, and related the incidence to where their drinking water came from [28]. Little was then known about infected water. We would now say that he was doing epidemiology. The whole process of how the disease is transmitted in polluted water would now be called the *etiology* of the disease. Once this became known, the pioneer studies had achieved their ends and were forgotten over the years.

Similarly, it was an empirical finding that dairymaids working with cows did not get smallpox. They contracted a mild disease, cowpox, which made them immune. All those vaccinated because of this discovery were inoculated with cowpox.

The empirical stage of our knowledge is certainly not always shortlived. The evidence that "(cigarette) smoking may damage your health" is only epidemiological 30 years after the first results were widely publicized. It is strong evidence because it is made up of many different strands, which separately are breakable. The etiology of bronchitis, emphysema, and lung cancer, as related to cigarette smoking, is still complex and largely unknown.

In other situations, the combined evidence is less strong. "On the basis of epidemiological findings we are now being advised, as ordinary citizens, to avoid tobacco and alcohol; to limit our consumptions of sugar, milk, dairy products, eggs, and fatty meats; to rid ourselves of obesity, and to engage in violent—although some suggest moderate—physical exercise ...." [3]

In the discussion to Snow's paper the seconder of the vote of thanks [26] commented: "There is nothing wrong with this method of formulating hypotheses [i.e., theories of causation of diseases based on epidemiological associations] but too often the authors of such studies succeed in implying that a tenable hypothesis, developed from a careful epidemiological survey, is a proven one. Attribution of this step, by the public, to a statistical argument, is a certain means of discrediting our science."

24.2 Role of the Statistician

Statistics and statisticians have always played a major role, but this has changed. It used to be almost entirely in descriptive as opposed to theoretical statistics, and observational rather than inferential. Now the processes can best be described as descriptive statistics plus modeling. However, "It is *descriptive* statistics and scientific method which have to become fully one" [14].

Typically, the classic book by Bradford Hill [20] consists largely of elementary statistics and, for its time, advanced scientific method. It/he provided very many examples of misuses of statistics comparable with those in a well-known popular exposition [25].

Small-sample statistics and "exact" inference have helped in epidemiology, for example, when the data being interpreted consist of very many small groups. Usually, however, there are disturbing factors that make rigorous inference impossible. Changes in the underlying assumptions, and in the models, make far more difference than that between a moderately good and an optimal method of data analysis or inference.

Experimental design of the kind that was so brilliantly successful in agriculture is generally impracticable for human subjects; even methods used in clinical trials have little place. But there have been exceptional situations where experiments are practically and ethically allowable. For example there were the pioneer studies on coal miners in the Rhondda Fach begun before 1950. This is an isolated valley in the south Wales mining area in Great Britain. It was thought that a disabling disease, pulmonary massive fibrosis, was brought about by a tubercular infection. The coal workers in this population were therefore provided with intensive medical examinations to detect and treat tuberculosis at a very early stage. In a control valley (Aberdare) the corresponding routine medical checks were given. In both populations chest x-rays and various lung physiological variables were measured for many years in the same individuals. This was therefore a *longitudinal* study. Follow-up studies are still being published. Much more information has been obtained than the original plan provided for [4].

A completely contrasting example is interesting. Immediately after World War II, large industrial areas in Europe became very short of food. In an area under British occupation, a population of school children were allocated different kinds of extra rations at random, especially of bread, and the effect on their growth was studied. It appeared that all kinds of bread were equally effective. The question remained, of course, whether results under such abnormal conditions throw any light on normal food values.

Epidemiological methods are used to determine what is generally "normal." In the Rhondda Fach population mentioned above, blood pressures, systolic and diastolic, were measured regularly. These were also compared with those in a representative sample of adults in a rural district not far away, in the Vale of Glamorgan.

This is a good instance where the statistician has more to contribute. The blood pressure of an individual increases with age, and also with his or her weight and arm girth. For much of adult life, evidence from *cross-sectional* studies is that the increase with age is accounted for by the increase in arm girth. These are studies in which the population is measured at many or all ages, but at the same time. In a longitudinal study the same individuals are measured over the years. If they are born within a few years of one another, this population is called a *cohort*. The blood pressure of an individual fluctuates even over a few hours or less. The distribution of blood pressures in people of the same age varies itself with age; the variance increases, and the distribution is of lognormal type. But among old people the distribution is truncated; very high blood pressure reduces the chance of survival.

The example of blood pressures is one where epidemiologists need to know what normal values are. This can be a major problem when lung variables are being investigated, such as forced expiratory volume (FEV) and forced vital capacity (FVC). These are measured and assessed by performance tests; for FEV the subject blows as hard as possible into a bag, after one or two practice blows. Sometimes this improves with practice; sometimes the subject tires. There have been many arguments and analyses on the question: Is one's best performance the best measure? (Or should it be the mean of the best three, or the mean of all except the first two practice blows?) Fortunately it seems to make little difference which is chosen. However, the actual *distribution* of the performances ought to yield extra information. The object is to assess all kinds of external or environmental factors on the lungs, e.g., industrial dusts, atmospheric pollution, cigarette smoking, the weather, the climate, and the season. We also want to detect early signs of chronic bronchitis and asthma. Again we need to know what happens in normal individuals. Age, weight, and height are important, and interesting empirical laws have been found (see, e.g., Fletcher et al. [16]) relating these to the lung variables.

Asthma is also a childhood disease, and polluted air can affect people of all ages. Measurements on growing children lead to additional problems (see Fig. 6). Their FEVs are well correlated with their height but possibly during the pubic spurt that the FEV lags behind. Without going into details, all this should give an idea of what epidemiological statisticians have to be doing now.

24.3 Computing

In 1955 Hollerith punched cards were more indispensable than the hand and the electric calculating machines of this period. Apparatus consisting essentially of many long rods selected cards that had their holes in the required places. The first electronic computers did not bring about much change. Their memories were not large enough to enable moderately complex calculations to be programmed easily, and large amounts of data could not possibly be stored. It was years before programs were available by which existing punched cards could be read into the computer. In the present generation of computers, input, output, memory capacity, and storage capacity are incomparably greater. At the same time the first (or is it the second?) generation of table computers takes care of small-scale work. But it is more necessary than ever to understand what is in a statistical package program. It is so tempting to use an existing program when what is really needed is statistical common sense and instinct. But on balance so much more has become practicable in epidemiological statistics through these new facilities that they are worth all the extra problems they bring with them.

New problems cannot be ignored. Information on individuals that is stored in computers can be misused. The epidemiologist was never in an ivory tower, but is farther away from it now than ever.

24.4 Longitudinal and Cross-Sectional Data and Growth

Data on growth are needed for many problems. They provide a clear example of the difference between cross-sectional and longitudinal data. At the beginning of puberty the growth rate of a child slows down, and after this there is a spurt. This then slowly flattens out. But the time of onset of puberty varies. Its mean is earlier in girls, but for both girls and boys in defined populations it is a distribution. The consequence is that a cross-sectional curve of mean height against age, for populations of boys and girls separately, shows almost no sign of puberty; the structure shown in individual curves is lost [37]. However, the curve for medians may show some structure.

In a related example, it is generally believed that the onset of puberty became earlier in many countries during periods when the general standard of living rose. This would show up in cross-sectional mean heights for different years, for, say, 14-year-olds, although from such data alone, they could simply be becoming taller at all ages. In fact, evidence from countries that have or had compulsory military service indicates that both may be true. The poorer children are shorter as teenagers, they go on growing for longer, but they do not catch up completely.

24.5 Small Risks Undergone by Millions of People: Human Radiation Hazards

The fact that large doses of ionizing radiation could induce serious illnesses was found in those working with radium, and in radiologists using the x-ray apparatus of the time. Epidemiology was scarcely needed for implicating such diseases, in particular skin tumors, cancers of the bone, digestive organs, lungs, and pharynx, and leukemia.

Leukemia is still a rare disease; 30 per million per year was a typical rate round about 1950, but it appeared then to be increasing rapidly, especially in children and in old people.

When the risk to an individual is small, but millions of people are undergoing this risk, this provides good grounds for "retrospective" or backward-looking research. The population under study may be defined to be those with the disease in question who live in a particular region, perhaps a whole country in a particular year. Then the medical histories of everyone in this study population would be obtained, as well as histories of exposure to suspected causes. But this can never prove anything until we know what the unaffected people were typically exposed to. Hence the same questions must be asked of *matched controls*. These should be people comparable in as many ways as possible with the patient, e.g., having about the same age and sex, living in the same district under similar conditions.

Even so, rigorous scientific inference is difficult to achieve with such data. In what is now a classical example, the population under study consisted of young children attacked by a malignant disease in England and Wales in the years 1953, 1954, and 1955. One of such diseases was acute leukemia, which lasted a short time and was usually fatal. Among their mothers, about 10% had undergone diagnostic x-rays while pregnant and about 5% of control mothers (of healthy child controls) had been so x-rayed [34].

Did the x-rays, then, double the risk to the child? This interpretation was generally rejected; it was argued that the mothers were x-rayed because something was possibly going wrong during their pregnancies and the ones with leukemic children were more likely to have some symptoms, and so to be x-rayed, than the mothers of controls.

Broad subdivisions in terms of reasons for x-raying the mothers were therefore made, and cases and controls were compared within each subgroup. In general, the same pattern emerged. The main early papers reporting these findings are inevitably tedious to read, because so much space had to be devoted to all these subcomparisons.

Later, more evidence was obtained by bringing in many more years; in the meantime a number of statisticians have been working on different quantitative interpretations of case–control data. One refinement, which is peculiar to the leukemia data, is worth mentioning, because it is a very instructive example of back-and-forth play among observation, theory, modeling, and medical hypotheses which are then tested on more data. In the firstborn children, the distribution of ages of onset of (acute) leukemia has a definite maximum at 3.5–4 years. (In adult leukemia, also, there is a clear maximum in the fourth year after ionizing radiation that could have caused it.) In cases where the leukemia arises from damage to the fetus that is *not* caused by x-rays, it is well known that the fetus is most vulnerable soon after conception. However, the pregnant mother is most often x-rayed shortly before the child is born—about 8 months later. If, then, the leukemia caused by radiation is initiated 8 months later, is the modal time of its onset about 8 months later than that of nonradiogenic leukemia? Refined analyses that were practicable only when data for many more years were available showed that this was the case. This is indirect evidence, but it is independent of control data and of any possible bias due to the reasons for the x-rays [21,22,32,33].

In later analyses, attempts were made to estimate the doses, or at least the number of times the pregnant mother was x-rayed, to try to quantify the risk factor. It appeared that such doses varied widely between hospitals, and when the first finds were published, these doses tended to be reduced. The original estimate of risk factor of 2:1 was possibly too high, and it was at best an average from very heterogeneous data.

An important negative finding was that in any case the increase in the use of x-rays was not nearly enough to explain the secular increase in childhood leukemia. In particular, the peak in 3- and 4-year-olds was not present at all 20 years earlier. The same group of researchers provided an explanation that depends mainly and convincingly on other epidemiological evidence. The hypothesis was that the new victims of leukemia—at all ages—would have died of pneumonia in the conditions of 20 years earlier. This was supported by evidence from the incidence of deaths from pneumonia at all ages, the secular change in this incidence, and its geographic variation. There is no space to give details of these fascinating studies in which the main contribution of the epidemiological statistics is to biological understanding [21,22,31].

24.6 Human Radiation Hazards and Prospective or Forward-Looking Studies

It can always be rightly said that epidemiology and controlled experiments can never go together. But these objections can be partly met in *prospective* studies. For irradiated pregnant mothers this was in fact tried. The population under study consisted of all such pregnant mothers irradiated in a few large hospitals. Then the medical histories of their children were followed for several years afterward. Their incidence of leukemias was slightly *below* the national average, which was in contradiction to all the findings from the retrospective surveys [7]. Several conflicting explanations were proposed:

1. There were inevitable biases in the retrospective surveys.

2. The numbers in the prospective study were too small.

3. Combined with 2, those particular hospitals had "safer" x-ray apparatus than most others.

Probably the third explanation was right. It was supported by long-term evidence from the retrospective studies mentioned above, in which it was concluded that the risk to the child was decreasing.

A different prospective study yielded very clear results. The population consisted of 13,000 patients who had "poker back" (ankylosing spondylitis; articulation between the joints of the spinal column is gradually lost). Radiotherapy alleviated the symptoms, and there were complete records of doses and the dates they received them. About 40 patients developed leukemia; the expected number from their age and sex distribution was about 4. The calculations were complex because most patients had several treatments, but information was obtained relating doses to increased risks, and on the distribution of time intervals between dose and onset of the disease [6,41].

If data are available over a very long period, many of the problems simplify. For the workers exposed to ionizing radiation at Hanford, Washington, there have been health records available since 1945. More recently, these were analyzed statistically. The findings have caused a stir, because they suggest that most existing models relating dose to effect underestimate the extra risk from various diseases, which, it is suspected, can be caused by small chronic doses (i.e., over long periods) of ionizing radiations, although the risk remains small [24]. This interpretation has, however, led to much argument (see, e.g., Darby [8], Brodsky [1], and Lyon et al. [24]).

For further discussion relevant to some prospective studies.

24.7 Epidemiology and Cigarette Smoking

Before 1950 epidemiologists were asking why the incidence of lung cancer in men in so many countries had been increasing so much. Increases in most other cancers seemed to be attributable to populations aging; age-standardized rates showed much less change. So they looked for some causal factor that had increased during the twentieth century, and cigarette smoking seemed to be the only plausible one. There appeared to be a time lag of decades between the supposed cause and its effect, but this was reasonable. Also, most smokers who began young continued to smoke over many years. Women, too, began to smoke more cigarettes and their lung cancer incidence increased; both occurred much later than with men [36]. Figure 1

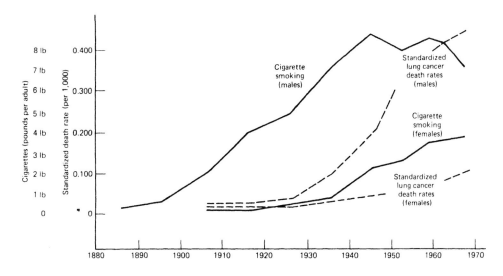

Figure 1: Cigarette smoking and age-standard death rates plotted at the middle of 10- and 5-year periods from 1880 to 1970. (Reproduced by permission from Townsend [40], with thanks to the author and the Royal Statistical Society.)

shows such data for Great Britain for most of this century.

In a pioneer prospective study all British medical doctors were asked to answer a questionnaire; this was kept short and simple and about 70% responded. The study was continued for many years. A clear association was found between incidence of lung cancer and rate of smoking cigarettes. Another feature was that many physicians in this population reduced their smoking, whereas comparable groups in other professions did not. Many years after the study started, incidence among physicians seemed to be becoming increasingly *less* than that of these other groups [12,13].

Another factor that had been increasing over the years was atmospheric pollution. Evidence comparing clean and polluted districts suggested that this does increase the risk of lung cancer. However, miners working underground mostly breathed far more polluted air than city dwellers do. They had a lower risk. As a population they smoked less, being forbidden to smoke underground.

Retrospective studies based on mortality statistics have been done in many countries. All show a high ratio, smokers to nonsmokers, for men and women separately. The ratio increases with the rate they smoke cigarettes. There are large differences between countries. It is also possible that the risk among individuals, even in the same environment within one country, and with the same smoking habits varies greatly.

A majority of medical doctors and medical statisticians are now convinced that cigarette smokers have increased risks of getting many diseases. The majority of the rest of the population, 25 years after the first major findings appeared, have, however, gone on smoking. Many of the rates of increase seem to be leveling off and the effect per cigarette has possibly fallen appreciably.

24.8 Unsolved Problems on the Effects of Cigarette Smoking

Considering the amount and variety of all this evidence, and much more, it is remarkable that so much is still unknown. Many arguments of the critics, the defenders of cigarettes, are instructive to epidemiologists. This writer's view is that they are often right, but that there are still more weaknesses in every alternative to the hypothesis that cigarettes are a causal factor.

Various factors obviously confuse the picture. Any separate groups of people who smoke less are almost certainly different in other ways. One such group studied was a population of Mormons, especially in the state of Utah. They have a low incidence of many diseases that have been related to cigarette smoking. This religious group also drinks no alcohol, tea, or coffee (see discussion in Burch [3]).

Groups responding to questionnaires probably differ from nonrespondents, even if they answer correctly. Estimates of amounts smoked were thought unreliable. There was complementary evidence from different countries relating standardized incidence of lung cancer to consumption of cigarettes per head of population, say, 0, 10, 20, or 30 years earlier. There was a positive correlation, which was much higher with the 20-year time lag than with no time lag; but the individual variations are extremely large.

In the Doll—Hill population of physicians, those saying they inhaled when smoking cigarettes were found to be less at risk than noninhalers smoking at the same rate. This suggests that any effect caused by cigarettes could not be due to direct delivery of an insult to the lung. However, a probable answer to the inhalation paradox, which has not attracted much attention, came from a modest epidemiological study in which it was found that inhalers threw away longer cigarette ends—which were known to contain more carcinogens—than noninhalers. The subjects were asked to supply butt ends but were not told why [17,19].

Other factors are involved, for nonsmokers do get lung cancer. According to Burch [3], one of these factors is inherited, so that those with it are more likely both to smoke heavily and to get various diseases. Most epidemiologists disagree, but the arguments are worth studying.

All the major difficulties relate to quantitative forecasting and modeling. Compared with the hours spent collecting data, much less has been done on playing with the data to locate empirical patterns. There is, in any case, a rapid increase of incidence with age. However, the age of a regular smoker of a constant number of cigarettes per day is perfectly correlated with the total amount he or she has smoked! Unfortunately, there is not much evidence from *ex*-smokers. The overall age-specific death rates of lung cancer, in several countries, and in different years fits

$$y(t) = nrk\theta\tau^{r-1}\exp(-k\tau^r)$$
$$\times \{1 - \exp(-k\tau^r)\}^{n-1}$$

where $\tau = t - \lambda, n = 5, r = 2, \lambda = 2.5$ but different θ and k in the various populations and years; t = age ([2]; see the next section for interpretations of these parameters). These curves have modes that vary around $t = 75$ (Fig. 2). These are not cohort data. They correspond in any case to a positive power law relating incidence to age:

$$y = \theta nrk^n(t - \lambda)^{rn-1}.$$

Such positive power laws are also found when cigarette smokers and nonsmokers are considered separately, and duration of smoking appears to act differently from age, as Figure 3 [39] shows. Some models that have been considered are based on

$$I_x = bNK(x - w)^{K-1},$$

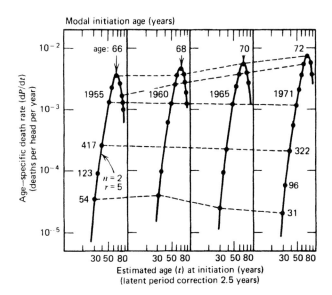

Figure 2: Age-specific recorded death rates (dP/dt) from lung cancer [International classification of Diseases (ICD) 162 and 163] in men in England and Wales against estimated age at initiation for the years 1955, 1960, 1965, and 1971 ($t - \lambda$ in discussion, age t). Log–log plots: some actual numbers of cases are shown for 1955 and 1971. Fits to $y(t)$ as given. (Reproduced by permission from Burch [3] with thanks to the author and the Royal Statistical Society.)

where I_x is the age-specific death rate at age x, N cigarettes are smoked per week, over a period $x - w$. In Figure 3, b is the same for smokers and nonsmokers, and K is about 5.

This equation has been extended by making N a time variable, which is estimated from data on total cigarette consumption in 5-year periods. Townsend [40] made some other refinements, and compared observed and predicted mortality rates such as those in Figure 4. It is interesting that the maxima are predicted, although the agreement is not exact. The interpretation of such maxima is different from that in Figure 2 (see the following section).

24.9 Form of Age Incidence Curves in General

We have been moving from small to large populations. Data on age–incidence curves for many diseases does not need to be obtained from an epidemiological study, but from annual mortality statistics. These are, of course, cross-sectional; we would expect cohort data to be more realistically related to a model. Despite this, these positive power laws, which can be valid over several decades of age, are found everywhere. Burch's collection contains over 200 examples. Mostly these are given in the cumulative form

$$G(t) = \theta[1 - \exp\{-k(t - \lambda)^r\}]^n.$$

Here θ is the proportion of the population that is susceptible, $G(t)$ the proportion that gets the disease at age t or earlier, and λ is assumed to be a constant latent period between the end of an initiation or promotion period and the event actually observed. In this theory n and r are integers, and are related to numbers of clones of cells and numbers of mutations, which finally lead to this event. Often either n or $r = 1$.

Whether Burch's underlying theory is right or not, a great variety of data are described economically in this way; they can be used as input in testing almost anything relevant. In epidemiology it is well-nigh impossible to ignore age distributions in the populations under study.

There are several related models in which observed powers are related to numbers of mutations. It seems certain that a summation of random time intervals is involved. An alternative interpretation for those positive power laws is that they are simply parts of the increasing part of probability curves. Then there is no restriction to integral powers. However, it seems possible to fit the same data to a function like

$$b(t - v)^k,$$

with different values of v, so that k depends on v. On both interpretations the curves have maxima, and the best evidence comes when a decrease at high ages is actually seen, but of course at high ages, say over 60, such a survival population can be very different from that for the age range of, say, 50–60 [5,10,11].

24.10 An Example Where Age Must Be Allowed for: Side Effects from Adding Fluoride to Drinking Water

Obviously, whether fluoridation is effective in preventing dental caries depends on longitudinal population studies of the teeth of children drinking the altered water over several years. There has also been a heated controversy over possible side effects. In the United States, data were available from 1950 to 1970 for 20 large cities, of which 10 had fluoridated water (F$^+$) and 10 did not (F$^-$). The crude death rates from all malignant neoplasms showed a greater increase, starting from near equality, in the

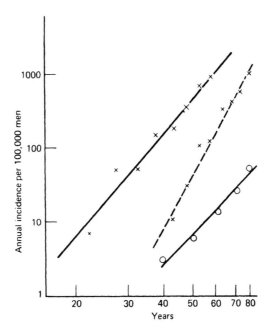

Figure 3: Lung cancer incidence by age and duration of smoking (data from Doll [11]). ×-
-× = cigarette smokers by age; ×—× = cigarette smokers by duration of smoking; ○—
○ = nonsmokers by age. (Reproduced by permission from Townsend [40], with thanks
to the author and the Royal Statistical Society.)

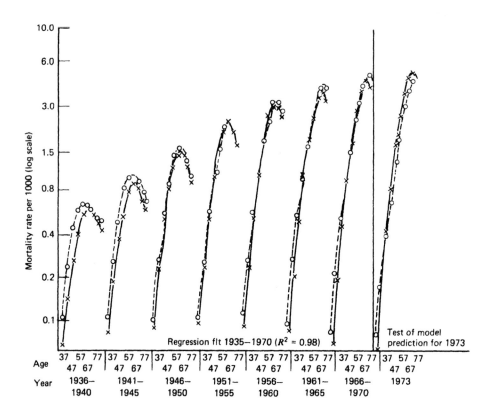

Figure 4: Lung cancer mortality rates related to cohort smoking histories (males). × = actual mortality. (Reproduced by permission from Townsend [40] with thanks to the author and the Royal Statistical Society).

F^+ cities. The two groups of cities, however, had different distributions of sex, age, and race, which *changed* differently during these 20 years. Allowing for these, the excess cancer rate increased during these 20 years by 8.8 per 100,000 in F^+ and by 7.7 in F^-; but in proportional terms the excess cancer rate increased by 1% in F^+ and 4% in F^-. Then it was found that 215 too many cancer deaths had been entered for Boston, F^+, because those for Suffolk county, containing Boston, had been transcribed instead [27].

After this had been put right, the relative increase was 1% less for F^+ than for F^-. On the whole this is in line with other evidence. This example shows, however, how obvious and trivial pitfalls are interspersed with ones that are far from obvious.

This controversy continued [28,30]. It was brought out that data for intervening years should be used if possible, but the way the distributions of sex, age, and race, which were not fully available, changed during these 20 years could be critical. In particular, linear and nonlinear interpolations yield different estimates of expected numbers of deaths.

24.11 Epidemiology and Heart Diseases

Heart diseases are the most common cause of death in all "developed" countries. There is probably more data than for all other topics mentioned so far put together. The data can be both reliable and very dirty. Factors that have been correlated, for example, with death rates in particular (middle or old) age groups include what is eaten, drunk, and smoked and how much, way of life in general, including exercise taken, climate, and population density (see, e.g., Tarpeinin [39] and Townsend and Meade [41]). There are many comparative studies within countries and between countries, which are too numerous to mention. One very experienced group of epidemiologists has based a major study entirely upon comparative statistics for 18 developed countries [35]. Figure 5 shows the kind of data in broad outline. In this study various combinations of variables were tried, principally in multiple regression analyses. One factor that was strongly negatively correlated with the mortality rates was wine consumption per head. No common third variable was found as an alternative explanation for this association. It is particularly strong within Europe, except for Scotland and Finland, but other factors are needed to explain the far higher rates for the United States, Canada, Australia, and New Zealand.

In complete contrast to this approach are prospective longitudinal studies from small areas. Unlike those for leukemia, there are now enough cases. A well-known name is Framingham, a small town in Massachusetts. A population of about 6500 people aged 30–62 in 1948 were studied every 2 years for 20 years; a small number of factors were investigated in relation to the incidence of coronary heart disease: namely, age, weight in relation to height, systolic and diastolic blood pressure, smoking habits, and serum cholesterol level [9,18].

In a more recent study in The Hague, Holland [15], the population was about 1750 of all ages 5–75; an additional factor was whether the women took oral contraceptives. (So far, for those who did, the risk of coronary heart disease seemed about the same, but their blood pressure was higher.)

Population studies have been going on for several years in Vlaardingen (a town in a large industrial/port area near Rotterdam, Holland) and in Vlagdwedde (a wholly rural area in northeast Netherlands). The selected populations are of all ages; in particular, many different mea-

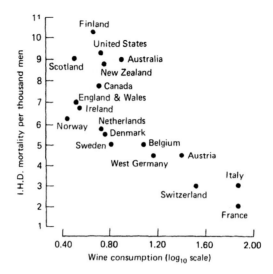

Figure 5: Relationship between ischemic heart disease(IHD) mortality rate in men aged 55–64 and wine consumption. (Reproduced from St. Leger, et al. [36], with thanks to the authors and *The Lancet*.)

surements on lungs, involving blowing and breathing, are made besides the measurements given above. A control group had 117 male conscripts aged 19–20 years. This is part of an extensive field survey on chronic nonspecific lung disease, but much statistical and bioloical information should be valuable apart from this purpose (see, e.g., Sterk et al. [30]).

This kind of data can also tell us more about the range of "normal" physiological variables as functions of age; then criteria for early signs of heart diseases can be detected. Typically, the data are both cross-sectional and longitudinal, for the same individuals are measured at, say, 6-month intervals.

Our final example gives data from growing children at one school, but it illustrates all these points. Figure 6 shows their forced vital capacities, which are related to the cohort data to be more realistically related to a model. Despite this, these positive power laws, which can be valid over several decades of age, are found everywhere. Burch's collection contains over 200 examples. Mostly these are given in the cumulative form growth of their lungs, as a function of height. Each separate line corresponds to one child, measured at 6-month intervals. We want to know what is the normal course of the growth of the lung in an individual. It is not necessarily given by a curve giving the best fit $y(x)$ for all the points.

The problem of describing and modeling such data is typical for epidemiologists today. If the population was of middle-aged people, the x-axis could be, say, blood pressure.

It is also only in the 1980s that the computer plotters became more generally available on smaller machines. It makes it possible to look at far more data quickly before doing any detailed analyses.

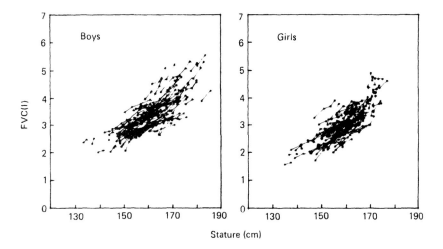

Figure 6: Data on growing children aged 11.5 - 13.5 obtained at half-year intervals at a school in The Hague (St. Aloysius College, Netherlands); FVC = forced vital capacity (lungs). [With acknowledgments to Ch. Schrader (Physiology Laboratory, Leiden).]

24.12 Scope of Epidemiology in the 1980s

On any live problem in this field, active researchers disagree. Such controversies are usually well aired. It is less obvious when a major question is whether the field belongs to epidemiology at all. In relation to typical risk problems, for example on the long-term effects of ionizing radiation or of taking "soft" drugs, one may have to decide whether to study human populations only or also to experiment on animals on a large scale. It is probably the case that in all countries with good welfare services and medical records, studies on human beings are less costly than those on animals. "Record linkage," e.g., for records for an individual obtained at different times and places, is in its early stages, and involves ethical problems of secrecy.

The tendency is now more toward social and "preventive medicine," that is, observing human subjects before they become ill.

Table 1 reminds us that in the age group 15–25, road accidents now provide the largest single cause of death in most of the countries listed. Interpreting accident data seems to belong far more to epidemiology than to, say, traffic engineering.

Epidemiological statisticians studying such data seem less likely to be misled by inappropriate indices. An example is the number of fatal accidents to passengers traveling by air per million miles traveled. But most such accidents take place at or near takeoff or landing; hence the increase in risk for a long flight compared with a short one is very little. On the other hand, such an index is probably reasonable for railway passengers, and possibly so for car drivers. For them we have to distinguish between accidents where another vehicle is or is not involved. Where it is involved, there are more accidents in urban areas where traffic is dense than in rural areas or on motorways. This leads in turn to another misuse of such statistics, in that in towns, motorists have to drive more slowly—namely, to a doubtful conclusion that it is not more risky to drive faster. Naturally, effects of speed have to be com-

Table 1: Correlation coefficients between death-rates and certain variables.[a]

| | Ages 55–64 | | | | | | | | Age 25–34 | |
| | Hypertensive Disease | | Ischemic Heart Disease | | Cerebrovascular Disease | | Bronchitis | | Road Accidents | |
	M	F	M	F	M	F	M	F	M	F
Physicians	−0.38	0.03	−0.51	−0.46	−0.69	−0.36	−0.42	−0.42	0.42	0.15
Nurses	0.10	−0.17	0.65	0.64	0.11	−0.20	0.14	0.25	−0.15	0.06
General Practitioners	−0.62	−0.22	−0.17	−0.26	−0.69	−0.77	−0.67	−0.65	0.10	0.19
Population density	−0.05	0.04	−0.44	−0.45	−0.02	−0.16	0.13	−0.08	0.00	−0.08
Cigarettes	0.26	0.23	0.28	0.44	0.08	0.22	0.44	0.35	0.47	0.34
Alcohol										
Total	−0.09	0.18	−0.70	−0.58	0.15	−0.16	−0.29	−0.28	0.38	0.21
Wine	−0.01	0.23	−0.70	−0.61	0.13	−0.14	−0.32	−0.27	0.13	0.02
Beer	−0.03	−0.09	0.23	0.31	0.14	0.17	0.35	0.22	0.37	0.25
Spirits	−0.38	−0.15	−0.26	−0.32	−0.35	−0.51	−0.57	−0.47	0.30	0.37
Calories	0.43	0.06	0.51	0.61	−0.02	0.36	0.57	0.67	−0.20	−0.03
Total fat	−0.16	−0.40	0.45	0.46	−0.45	−0.16	0.11	0.17	−0.18	0.01
Saturated fat	0.10	−0.17	0.64	0.62	−0.16	0.15	0.30	0.35	−0.08	0.05
Monounsaturated fat	−0.11	−0.42	0.60	0.60	−0.35	−0.05	0.11	0.15	−0.28	−0.17
Polyunsaturated fat	−0.45	−0.28	−0.48	−0.47	−0.51	−0.60	−0.33	−0.30	−0.05	−0.11
Keys' prediction[b]	0.04	−0.19	0.70	0.69	−0.10	0.19	0.24	0.27	0.06	0.14

Source. St. Leger, et al. with thanks to the authors and *The Lancet*.
[a] Obtained from data from the 17 countries shown in Figure 5.
[b] Keys' prediction (Keys et al. [20a]) is a predictive equation for serum-cholesterol.

pared when other things are comparable; data for motorways and streets should not be pooled.

It is even more difficult to set up reliable comparative statistics for cyclists and pedestrians.

This nonmedical example shows up pitfalls that epidemiological statisticians are always being confronted with; consider the population at risk, consider third variables correlated with both of two other variables, ask whether the underlying model is reasonable, with any set of comparative statistics.

In conclusion the main object in this account has been to give instructive examples of what epidemiological statisticians have been and are doing. As such it is far from complete; the geographic and reference coverage is even less so. However, there are very many references in the books and articles in the further reading list, which follows.

24.13 Further Reading

24.13.1 Books

A good introduction with a wealth of data clearly presented is Abraham M. Lilienfeld's *Foundations of Epidemiology* (Oxford University Press, New York, 1976). This can be complemented by Mervyn Susser's *Causal Thinking in the Health Sciences: Concepts and Strategies of Epidemiology* (Oxford University Press, New York, 1973).

Despite its title, at least half the essays in Alvan R. Feinstein's *Clinical Biostatistics* (Mosby, St. Louis, MO, 1977) are relevant to epidemiology. They are selected from feature articles that appear about every 3 months in *Clinical Pharmacology and Therapeutics*. The author's approach is controversial; the articles are very long, but his style is fresh and stimulating and they contain many thorough discussions; those on terminology are particularly recommended.

A more specialized work, but one that contains a great deal on the practical aspects of organizing good population surveys and questionnaires, and on the statistical methods and problems, is *The Natural History of Chronic Bronchitis and Emphysema: An Eight-Year Study of Early Chronic Obstructive Lung Disease in Working Men in London* by Charles Fletcher, Richard Peto, Cecily Tinker, and Frank E. Speizer (Oxford University Press, New York, 1976).

24.13.2 Articles

Two review articles, mainly on statistical methods, are to be recommended also for the very full references (225 and 165 + 26, respectively). They are Charles J. Kowalsky and Kenneth E. Guire's 1974 article "Longitudinal data analysis" (*Growth* **38**, 131–169) and Sonja M. McKinlay's 1975 contribution "The design and analysis of observational studies—a review" (*J. Am. Statist. Assoc.* **70**, 503–520, with comment by Stephen E. Fienberg, pp. 521–523).

For methods currently used, some useful references are as follows:

1. Breslow, N. and W. Powers (1978). *Biometrics*, **34**, 100–105.

2. Holland, W. W. (coordinator), with Armitage, P., Kassel, N., and Premberton, J., eds. (1970). *Data Handling in Epidemiology*. Oxford University Press, New York.

3. Kemper, H. C. G. and van 't Hof, M. A. (1978). *Pediatrics*, **129**, 147–155.

4. McKinlay, S. M. (1975). *J. Am. Statist. Assoc.*, **70**, 859–864.

5. McKinlay, S. M. (1977). *Biometrics*, **33**, 725–735.

6. Prentice, R. (1976). *Biometrics*, **32**, 599–606.

7. van't Hof, M. A., Roede, M. J., and Kowalski, C. J. (1977). *Hum. Biol.*, **49**, 165–179.

Some of these references were provided by P. I. M. Schmitz (Department of Biostatistics, Erasmus University, Rotterdam)—to whom many thanks.

References

1. Brodsky, A. (1979). *Health Phys.*, **36**, 611–628.

2. Burch, P. R. J. (1976). *The Biology of Cancer: A New Approach*. MIP Press, Lancaster, UK.

3. Burch, P. Roy. J. (1978). *J. R. Statist. Soc. A*, **141**, 437–458.

4. Cochrane, A. L., Haley, T. J. L., Moore, F., and Hole, D. (1979). *Br. J. Industr. Med.*, **36**, 15–22.

5. Cook, P. J., Doll, R., and Fellingham, S. A. (1969). *Br. J. Cancer*, **4**, 93–112.

6. Court-Brown, W. M. and Doll, R. (1965). *Br. Med. J.*, **ii**, 1327–1332.

7. Court-Brown, W. M., Doll, R., and Hill, A. B. (1960). *Br. Med. J.*, **ii**, 1539–1545.

8. Darby, S. C. (1979). *Radiat. Prot. Bull.*, **28**, 7–10.

9. Dawber, T. R., Kannel, W. B., and Lyell, L. P. (1963). *Ann. NY Acad. Sci.*, **107**, 539–556.

10. Defares, J. G., Sneddon, I. N., and Wise, M. E. (1973). *An Introduction to the Mathematics of Medicine and Biology*, 2nd ed. North-Holland, Amsterdam/Year Book Medical Publishers, Chicago, pp. 589–601.

11. Doll, R. (1971). *J. Roy. Statist. Soc. A*, **134**, 133–166.

12. Doll, R. and Peto, R. (1976). *Br. Med. J.* **i**, 1525–1536.

13. Doll, R. and Pike, M. C. (1972). *J. Roy. Coll. Phys.* **6**, 216–222.

14. Ehrenberg, A. S. C. (1968). *J. Roy. Statist. Soc. A*, **131**, 201.

15. Erasmus University (1980). *Annual Report of the Institute of Epidemiology*.

16. Fletcher, C., Peto, R., Tinker, C., and Speizer, F. E. (1976). *The Natural History of Chronic Bronchitis and Emphysema: An Eight-Year Study of Early Chronic Obstructive Lung Disease in Working Men in London*. Oxford University Press, New York.

17. Good, I. J. (1962). In *The Scientist Speculates*, I. J. Good, A. J. Mayne, and J. Maynard Smith, eds. Heinemann, London.

18. Gordon, T. and Kannel, W. B. (1970). In *The Community as an Epidemiologic Laboratory: A Case Book of Community Studies*, I. I. Kessler and M. L. Levia, eds., pp. 123–146, Johns Hopkins Press, Baltimore, MD.

19. Higgins, I. T. T. (1964). *Br. J. Industr. Med.*, **21**, 321–323.

20. Hill, A. B. (1971). *Principles of Medical Statistics*, 9th ed. Oxford University Press, London.

21. Keys, A., Anderson, J. T., and Grande, F. (1957). *Lancet*, **ii**, 959.

22. Kneale, G. W. (1971). *Biometrics*, **27**, 563–590.

23. Kneale, G. W. (1971). *Brit. J. Prevent. Soc. Med.*, **25**, 152–159.

24. Lyon, J. L., Klauber, M. R., Gardner, J. W., and Udall, K. S. (1979). *New Engl. J. Med.*, **300**, 397–402.

25. Mancuso, T. F., Stewart, A. H. and Kneale, G. W. (1977). *Health Phys.*, **33**, 369–385.

26. Moroney, M. J. (1965). *Facts from Figures*, 3rd ed. Penguin, Baltimore, MD.

27. Oldham, P. D. (1978). *J. Roy. Statist. Soc. A*, **141**, 460–462.

28. Oldham, P. D. and Newell, D. J. (1977). *Appl. Statist.*, **26**, 125–135.

29. Snow, J. (1936). In *Snow on Cholera*, pp. 1–175, Commonwealth Fund, New York.
30. Sterk, P. J., Quanjer, Ph. H., van der Maas, L. L. J., Wise, M. E., and van der Lende, R. (1980). *Bill. Eur. Physiopath. Resp.*, **16**, 195–213.
31. Stern, G. J. A. (1980). *Appl. Statist.* **29**, 93.
32. Stewart, A. M. (1972). In *Clinics in Haematology*, S. Roath, ed., pp. 3–22, Saunders, London.
33. Stewart, A. M. and Kneale, G. W. (1970). *Lancet*, **ii**, 1185–1188.
34. Stewart, A. M. and Kneale, G. W. (1970). *Lancet*, **ii**, 4–18.
35. Stewart, A. M., Webb, J. W., and Hewitt, D. (1958). *Br. Med. J.*, **i**, 1495–1508.
36. St. Leger, A. S., Cochrane, A. L., and Moore, F. (1979). *Lancet*, **i**, 1017–1020.
37. Stocks, P. (1970). *Br. J. Cancer*, **24**, 215–225.
38. Tanner, J. M., Whitehouse, R. H. and Takaishi, M. (1965). *Arch. Dis. Child.*, **41**, 454–571.
39. Tarpeinen, O. (1979). *Cancer*, **59**, 1–7.
40. Townsend, J. L. (1978). *J. Roy. Statist. Soc. A*, **141**, 95–107.
41. Townsend, J. L. and Meade, T. W. (1979). *J. Epidemiol. Commun. Health*, **33**, 243–247.
42. Wise, M. E. (1962). In *Physicomathematical Aspects of Biology*. (Italian Physical Society), N. Rashevsky, ed. Academic Press, New York.

25 Epidemiological Statistics II

Don McNeil

25.1 Introduction

Epidemiology is the branch of medicine concerned with understanding the factors that cause, reduce, and prevent diseases by studying associations between disease outcomes and their suspected determinants in human populations. It involves taking measurements from groups of subjects and making inferences about relevant characteristics of a wider population typifying the subjects. Since statistics is the science that is primarily concerned with making inferences about population parameters using sampled measurements, statistical methods [38] provide the tools for epidemiological research.

An early example of epidemiological statistics in practice was a study by Louis [21], who investigated the effect of the entrenched medical practice of bloodletting on pneumonia patients in the 1830s, and found evidence that delaying this treatment reduced mortality. Another epidemiological pioneer was Snow [36], who found that the cholera death rate in London in 1854 was 5 times greater among residents drinking water from a particular supplier, thus identifying contaminated water as a risk factor for this disease.

While Louis [20] had already described many of the basic principles underlying experimental research in epidemiology, it was not until the middle of the twentieth century that a major clinical trial was undertaken, when the British Medical Research Council sponsored Hill's investigation [12] of the effect of streptomycin treatment for tuberculosis. In contrast, in the last half-century, as new and more virulent diseases like AIDS have decimated populations, epidemiological research methods have become widely used [3,4,8,11,14,15,19,25,32,35,37–39].

25.2 Data Roles

Each study attempts to answer a research question of interest, involving a target population, using a specified set of research methods. In epidemiological studies, the individual subject is typically the observational unit from which data are collected. Each subject supplies a set of measurements comprising one or more variables. The role of a variable is its definition as an outcome or a possible determinant of that outcome. An *outcome* is a measure of the subject's health status at a particular period of time, whereas a *determinant* is a possible cause of the outcome through its action at an earlier period of time.

Determinants include genetic factors affecting predisposition to disease, such as haemophilia, which may increase the risk of infection through a contaminated blood transfusion, and demographic factors, notably age, gender, occupation, and marital status. They also include environmental and occupational exposures such as contaminated water, asbestos dust, and excess fat and cholesterol in the diet. Behavioral determinants include tobacco and excess alcohol consumption, overexposure to the sun, unsafe sexual practice, and drug addiction. Determinants include preventative measures for health risk reduction, such as a vaccine for combating an infectious disease, screening for breast or prostate cancer, an exercise promotion campaign, or a treatment aimed at curing or alleviating a disease or preventing further deterioration in a person's health. In this general definition, determinants include medications such as aspirin, hypertensive and hormonal drugs, and treatments such as radiotherapy for cancer and surgery for heart disease.

25.3 Concept Maps

It is useful to have a concept map or causal diagram [9; 25, p. 41] to illustrate graphically the roles of variables in a study. Figure 1 shows two examples.

The distinction between a determinant and an outcome is not always clear-cut. Inadequate prenatal care (measured by the number of visits to a clinic) during the first trimester of pregnancy (T1) is usually associated with an increase in the risk of perinatal mortality. However, reduced prenatal care during the third trimester of pregnancy (T3) is usually associated with reduced perinatal mortality. The reason for this apparent anomaly is that any pregnancy complication is likely to give rise to additional prenatal care, so that prenatal care in the first trimester is a determinant, but prenatal care during the third trimester is an outcome. The term *intervening variable* is used to describe a variable on the causal path between a determinant and an outcome. Figure 2 shows a concept map for this example. In this graph, the bold arrows show the true causal relations, while the other arrows show the associations that would be observed in the absence of consideration of the intervening variable.

Typically, a study will have just one outcome variable of interest. If there is more than one outcome variable, these outcomes may be considered separately, and usually the analysis will focus on the one of primary interest. However, often there will be several determinants of interest, even though the research question will focus on the association between a particular determinant and the outcome under consideration.

Consider a study in which the research question is the extent to which calcium deficiency is a risk factor for hip fractures in an elderly population. Osteoporosis is known to increase the likelihood of a hip fracture in this population, and calcium deficiency is also known to cause osteoporosis, as illustrated in Figure 3.

A confounding variable (or confounder) is a determinant that affects the association between a determinant and an outcome. In this case, the association between the determinant (calcium deficiency) and the outcome (hip fracture) is of primary interest, but the intervening variable (osteoporosis) could interfere with the measurement of this association in a study. Dealing with confounding is an important part of epidemiological statistics.

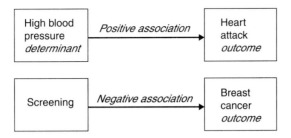

Figure 1: Concept map examples.

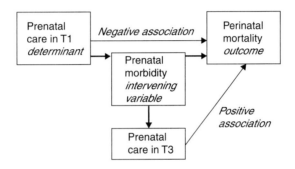

Figure 2: Concept map with an intervening variable.

25.4 Data Types

Variables can also be classified according to their data type. Data types are typically nominal (having two or more distinct categories with no natural order), ordinal (having three or more ordered categories), or interval (measured on a scaled range, also called continuous). Epidemiological outcomes are commonly of binary type, corresponding to the presence or absence of a disease. Binary outcomes are convenient from the medical diagnostic point of view: if a subject is diagnosed as having the disease in question, a specific treatment might be justified, but not otherwise.

Nominal variables with three or more categories often arise in epidemiological studies. Examples of such determinants include a subject's occupation, their country of birth, and marital status. Nominal outcomes include disease diagnosis (type of cancer, such as breast, ovarian, lung, colon, or stomach) and type of lung cancer cell involved (such as small, large, squamous, or adeno-carcinome).

Ordinal variables are useful when classification into just two categories is relatively uninformative. For infants with diarrhea, three categories of disease status (mild, moderate, and severe) are often used, giving four categories in all when disease absence is included. Similarly, a patient's outcome in a cancer trial could be classified as complete response, partial response, no change, or progressive disease. Ordinal determinants commonly measured in epidemiological studies include age group, duration of exposure to an occupational risk factor, and the status of some behavioral variable of interest such as exercise, snoring, smoking, or drug-taking.

Interval outcomes include CD4 count (typically ranging from 600 to 1200 in nor-

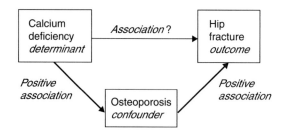

Figure 3: Concept map with a determinant and a confounder.

mal persons, a useful measure of health status for subjects with HIV), and the blood lead concentration for a child exposed to contaminated water or motor vehicle exhaust in a city. The duration of survival and a quality of life index are interval outcomes of major interest in cancer studies. Body mass index is a useful interval outcome in obesity studies. Many determinants are measured in this way: the duration of exposure to an environmental hazard, the dose level of a therapeutic drug, a disabled person's performance status, and the percentage of fat in a diet.

25.5 Disease Measurement

A basic objective in epidemiological research is measuring the level of some disease in a population. If the outcome is simply disease presence or absence, then the *prevalence* is of interest: this is simply the proportion of individuals affected with the disease. For interval outcomes, a statistical summary such as the mean or the median could be used to represent the status of the population. Typical examples include *(1)* the mean birthweight of newborn babies at hospitals in an urban population, and as, e.g., on p. 61 *(2)* the median survival time for patients in an organ transplant program.

A problem with measuring prevalence in a population is that it is likely to reflect what has happened in the past rather than the current situation. In HIV research, the extent to which new cases are occurring is of primary interest, for this measures the effectiveness of preventative programs. Thus the *incidence* of a disease, defined as the proportion of new cases that occur per unit time among persons free of the disease in the population, is another important epidemiological statistic.

25.6 Measuring Associations

Associations between possible determinants and disease outcomes are usually of greater interest than disease levels. For example, knowing whether taking oral contraceptives affects a woman's risk of developing breast cancer is important, and many studies have addressed this question. Where an association has been established, it is often of interest to quantify it more accurately. For example, what is the relative risk that a heavy smoker will develop lung cancer (compared to a nonsmoker)? If both the outcome of interest and the determinant are measured on an interval scale, their strength of association can be expressed as a correlation coefficient. If the outcome is measured on an interval scale and the determinant is categorical, the association can be expressed as a function of the differences in the means of the outcomes associated with the levels of the determinant.

When both variables are categorical, it

is more convenient to express the association as a matrix of conditional probabilities, where a typical element in this matrix is the probability of a specified outcome, given a specified level for the determinant. Equivalently, the conditional probabilities can be expressed as probability ratios (or relative risks). In the simplest situation, in which both the determinant and outcome are binary, this association is expressed as an odds ratio or as a risk difference, as follows.

Denoting the probabilities (or risks) of the ("adverse") outcome for two individuals in two different levels of the determinant by p_1 and p_2, respectively, the *relative risk* is just the ratio p_1/p_2. The *odds* associated with a probability p is defined as $o = p/(1-p)$, so the *odds ratio* ω, say, comparing the two individuals is

$$\omega = \frac{p_1/(1-p_1)}{p_2/(1-p_2)} \qquad (1)$$

The *risk difference*, defined as $p_1 - p_2$, is also widely used, particularly in the public health literature, to measure the risk attributable to the exposure. Similarly, the *attributable risk* proportion is defined as $1 - p_2/p_1$ [32, pp. 203–211].

Where an association is suspected but has not yet been established, the research question is framed in terms of a null hypothesis, which states that there is no association between a specified determinant and a particular disease outcome in the population of interest. On the question of a possible association between oral contraceptive use and breast cancer outcome, the null hypothesis would state that there is no association between these two variables in the target population.

The methods for testing a null hypothesis are essentially statistical. It is important to realize that a study might not provide a conclusive answer to the question, due to the limited size of the sample. Consequently the result of a study is not certain, but probable.

25.7 Types of Studies

Most epidemiological studies require data collection, but this is not essential. A study could be purely deductive, based on logic, starting with known facts or assumptions and arriving at a conclusion. Knowledge of appropriate theory can also reduce the need for data collection.

If a study involves data collection, it is said to be *inductive*, because the data in the sample are used to induce the characteristics of the target population. Inductive studies based on data collection are also called *empirical studies*, and are classified as either quantitative or qualitative.

A quantitative study involves structured data collection. In this case, the same characteristics are measured on all subjects in the study, using a protocol or set of guidelines that are specified in advance. The method of data collection in a quantitative study is often a questionnaire, or an instrument that automatically records the measurements from the sampled subjects. In contrast, a qualitative study is relatively unstructured. It may involve open-ended questioning of subjects, and may be opportunistic in the sense that the answer given to one question determines the next question. A qualitative study may precede a more formal quantitative study, with the aim of getting sufficient information to know what measurements to record in a quantitative study. For example, an investigator wishing to compare a new program for improving the reproductive health of mothers in an aboriginal community in outback Australia would spend some time in the community getting to know the people before embarking on the fully fledged study.

A quantitative study could be purely descriptive, or comparative. Descriptive studies simply aim to measure levels or prevalences, whereas comparative studies measure or test the existence of associations between determinants and outcomes.

Studies are also classified as experimental or observational. An experimental study is one in which the investigator has some control over a determinant. To investigate a possible association between beer consumption and stomach tumor incidence, for example, the investigator could take a sample of laboratory mice and divide them into two groups, force one group to drink beer and deprive the other, and observe each group to compare the incidences of tumors over a period of time.

Observational studies do not involve any control by the investigator over the determinant, and are thus in a sense less rigorous than experimental studies. However, the latter studies require interventions that might be costly, and are not always feasible or ethical, particularly if human subjects are involved.

25.8 Clinical Trials

Experimental studies in epidemiology investigate treatments, such as therapies for cancer or heart disease patients, or interventions, such as screening and health promotion studies. These studies are classified by various factors including the type of subjects and the size and extent of the study. The study is called a *clinical trial* if the subjects are hospital or physicians' patients, whereas the study is a *field trial* if it involves subjects in the community at large. Clinical trials usually involve patients who have some disease or condition, with the objective of investigating and comparing treatments for this condition. In field trials, the subjects usually are disease-free at the time of selection, and the trials aim to compare strategies for prevention.

Clinical trials are classified according to phases of development of new treatments. A phase I trial evaluates the safety of a proposed new treatment, whereas a phase II trial attempts to discover whether a treatment has any benefit for a specific outcome. A phase III trial is used to compare a promising new treatment with a control treatment, which could be no treatment at all. Since patients often react positively to the idea of a treatment (even if it is otherwise ineffective), a placebo, that is, a treatment that looks like a real treatment but contains no active ingredient, is often used instead of no treatment. Phase IV trials are similar to field trials in that they involve monitoring of treatments in the community; conceptually they differ from field trials only in the sense that they are concerned with subjects with some health problem whereas field trials usually focus on prevention.

Experimental studies often involve randomized allocation of subjects to treatment and control groups, an idea proposed by R. A. Fisher in 1923 for comparing treatments at the Rothamsted agricultural research station in Britain. The aim of randomization is to form treatment and control groups that are initially as similar as possible, so that any substantial difference in outcomes observed in these groups cannot be ascribed to factors other than the treatment effects. Randomization ensures that the comparison groups are balanced, not just with respect to known determinants of the outcome, but with respect to all possible risk factors.

An early large field trial was the 1954 study of the Salk vaccine for preventing poliomyelitis [24]. In this study, 400,000 schoolchildren were randomly allocated to receive the vaccine or a placebo, with the result that only 57 cases of polio occurred in the vaccinated group compared with 142 in the control group.

25.9 Cohort Studies

A *cohort study* is similar to a clinical trial, in the sense that the subjects are again selected according to their determinant status, but it is observational rather than experimental (because the investigator has no control over their determinant status). Cohort studies often involve monitoring subjects over an extended period of time, and consequently they are useful for investigating multiple determinants of outcomes. In a classical example, from 1948 onwards residents of Framingham in Massachusetts were continuously monitored with respect to many risk factors and disease outcomes.

If the data collection is *prospective*, as is often the case in cohort studies, a cohort study can be an expensive and time-consuming exercise. Breslow and Day [4] provided a detailed account of the use of cohort studies in cancer research.

25.10 Case–Control Studies

A *case–control study* is similar to a cohort study, but the subjects are selected according to their outcome status rather than the determinant [3,30,34].

Both cohort studies and case–control studies involve differential selection of subjects according to their exposure or disease status. In a cohort study, a group of disease-free subjects exposed to the determinant of interest is first selected, together with a comparable group of subjects not exposed to the determinant, and the subsequent outcome status of the two groups is then compared. In a case–control study, the variables are reversed; first, a group of subjects with the outcome and a comparable outcome-free group are selected, and then the levels of prior exposure in the two groups are compared.

Since the exposure must logically precede the outcome, a cohort study cannot look for outcomes that occur before the exposure, and a case=-control study cannot look for exposures that occur after the outcome. Thus, cohort studies are often said to be *prospective* and case–control studies *retrospective*.

When the disease is rare, a cohort study is inefficient, because a large number of subjects will be needed to obtain sufficiently many adverse outcomes to obtain a conclusive result. In this situation, a case–control study is more efficient because one of the two groups being compared contains only the subjects with the disease, and the control group can be restricted to a comparable number of subjects. By the same token, a case–control design is inefficient when the exposure to a risk factor is rare and the adverse outcome is relatively common.

25.11 Cross-Sectional Studies

In a *cross-sectional study*, there is no differential selection of subjects, by either the determinant or the outcome: one simply selects subjects from the target population, without taking into account the outcome or the determinant. Snow's investigation of risk factors for cholera [36] was a cross-sectional study: here the sample comprised all the residents of a particular area of London where a cholera epidemic had occurred during July and August 1854, and the subjects were classified according to death from cholera (the outcome) and their source of drinking water.

Figure 4 gives a graphical representation summarizing the various types of epidemiological studies.

To summarize, inductive studies involve data collection, in contrast to purely deductive studies. Inductive studies can be qualitative or quantitative, depending on the extent to which similar data are collected from each subjects. Quantitative

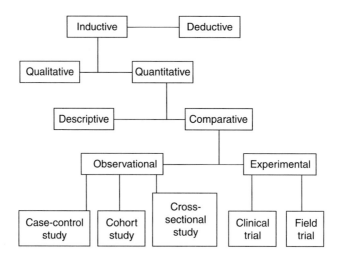

Figure 4: Classification of research studies.

studies can be descriptive or comparative, and comparative studies can be observational or experimental. Observational studies can be of cohort, case-=control, or cross-sectional type, and these are distinguished only by the method of selection of subjects. Since experimental studies involve allocation of subjects to treatment or exposure groups (ideally by randomization), which must occur prior to the occurrence of the outcome, they could be regarded as controlled cohort studies.

25.12   Bias

Two error factors reduce the credibility of a study: *(1)* systematic error, or bias, and *(2)* chance error, or sampling variability.

Bias is a systematic distortion in a measured effect due to a deficiency in the study design. It may arise in one of three ways: *(1)* from poor measurement (information bias), *(2)* because the sample is unrepresentative of the target population (selection bias), or *(3)* from the differential effects of other determinants on the association (confounding).

There are many sources of measurement bias, such as faulty measuring instruments, different standards in different biochemical laboratories, errors made by clinicians in diagnosing diseases, bias by investigators consciously or unconsciously reporting more favorable results for treatments that they believe in, biased reporting of symptoms by patients wishing to please their doctors, memory lapses by subjects in case–control studies when asked to recall past exposure to a risk factor (recall bias), lack of compliance by patients in clinical trials, and poor data quality management.

Some measurement biases can be reduced or eliminated by good study design. For example, blinding of investigators and subjects, so that they don't know which treatment a subject received until after the response has been evaluated, can reduce biased reporting in clinical trials. When the evaluators know the treatment allocation but the subjects do not, the study is said to be *single-blind*. If neither patients nor evaluators know the treatment allocation, the trial is said to be *double-blind*.

Selection bias has two levels. The first occurs when the sample does not represent the target population but still constitutes a representative sample from some restricted population that is a subset of the target population. In this case, the results obtained from the study might not be generalizable to the target population but are valid for a subpopulation. Such studies are said to have internal validity but lack external validity. Because their subjects usually constitute a select group that must satisfy tight eligibility criteria, randomized clinical trials only have internal validity.

The second, more serious, level is *differential selection bias*, which arises when the selection criteria for inclusion in a study vary with respect to a factor related to the outcome. Whenever a cross-sectional study has a low response rate, there is an opportunity for differential selection bias (response bias) because the nonresponders could provide different outcomes to the responders. Such bias often arises in studies in which questionnaires are mailed to sampled subjects.

Sackett [33] classified biases that can arise in case-control studies, which are particularly prone to differential selection bias, because it is difficult to get a control group that is representative of the noncases in the target population.

The third kind of bias in a study is called *confounding bias*, and it arises where the association between a determinant and an outcome is distorted by another determinant. Confounding arises whenever an outcome has two or more determinants that are themselves associated and one is omitted from consideration.

Confounding can make two unrelated variables appear to be related, and can also appear to remove or reverse a valid association.

25.13 Sampling Variability

Sampling variability arises because samples are finite, so even if there is no bias in a study, its conclusion is only probably true. Two statistical measures are associated with the sampling variability—a confidence interval and a P value.

A 95% confidence interval is an interval surrounding an estimate of a population characteristic (such as a mean or a prevalence, or a relative risk or an odds ratio), which contains the population characteristic with probability 0.95. It is thus a measure of the precision with which a population parameter can be determined from a study: the narrower the confidence interval, the greater the precision. The chance that the population parameter will not be located within the given confidence interval is 0.05.

While 0.05 is the conventional statistical false-positive error rate, wider confidence intervals, containing, say, 99% probability, are preferable when several parameters need to be estimated from the same study. Increasing this probability level can ensure that the overall risk of making an incorrect conclusion remains close to 0.05 between these variables in the target population.

A 95% confidence interval for a population parameter may be given approximately and simply as the interval $(T - 1.96 \times \text{SE}, T + 1.96 \times \text{SE})$, where T is the estimate obtained from the study and SE is its standard error, defined as the (estimated) standard deviation of T. This is based on the assumption that the sampling distribution of the estimate T is approximately normal, which can be justified by statistical theory to be so if the sample size is large enough. In the case of the distribution of an odds ratio, the approximation to normality is substantially improved by replacing the odds ratio by its logarithm.

A substantial part of epidemiological statistics is thus concerned with obtaining reliable formulas for the standard errors of the estimates of population parameters that arise in particular situations. If there is no bias, the standard error of an estimate generally decreases in proportion to the inverse square root of the sample size. Symbolically, this result may be expressed as

$$\text{SE} = \frac{c}{\sqrt{n}} \quad (2)$$

where c is a constant and n is the sample size. In particular, this means that you need to quadruple the sample size in order to reduce the width of the confidence interval by a factor of 2, and thus double the precision of the estimate. This result assumes that the observations are independent. In epidemiological studies, data are often clustered, either due to geographic proximity or because repeated measurements are taken on identical or similar subjects. Methods for detecting and adjusting for correlated outcomes have been developed [11,29,39].

A P value is more complex than a confidence interval, in the sense that it involves a null hypothesis, that is, a statement or claim that a target population parameter equals a specified null value.

When there is just one variable of interest, the parameter could be a mean or a prevalence. For a comparative study in which a possible association between two variables of interest may be present, the null hypothesis usually states that there is no association between the variables in the target population.

A P value has the same conceptual basis as a confidence interval, namely, the idea of repeating the study many times under the same conditions, each time using an independent sample based on the same number of subjects. It is the probability, assuming that the null hypothesis is true, that another such study will give an estimate at least as distant from the null value as the one actually observed. The P value is usually calculated as the area in the tails of a normal distribution or as the area in the right tail of a chi-squared distribution.

Although a small P value provides evidence against a null hypothesis, a relatively large P value needs to be supported by a large sample before it provides evidence in favor of a null hypothesis.

25.14 Matching

Both selection bias and confounding, as well as sampling variation, can be reduced by matching, a design technique that involves subdividing the treatment or exposure groups into smaller subgroups, or strata, so that the members of a stratum are homogeneous with respect to specified determinants, such as age or occupational status, which are not themselves of interest in the study. An important special case is the matched pairs design, in which each stratum comprises just two subjects. In some situations it is feasible for each stratum to consist of just one individual, and the corresponding studies are called *crossover studies*, in which the subjects act as their own controls. In a crossover study, the subjects are first divided into two or more treatment groups as in a phase III or IV trial, and after a specified period of time, each subject is given an alternative treatment [1,5,35].

While matching is often used in epidemiological studies to improve precision or reduce bias, it can be unnecessary and even counterproductive. If a covariate is not an independent risk factor for an outcome of interest, matching on it is simply a waste of effort. If a covariate is not an independent risk factor for the outcome but is associated with the exposure factor, matching on it is both wasteful and inefficient, since this will create uninformative matched sets whose members have the same exposure. When a covariate is associated with both the risk factor and the outcome, matching on it can introduce bias. The term *overmatching* is used to describe such situations. Consequently, matching is most effective when the covariate on which the matching is done is a risk factor not associated with the determinant of interest.

In epidemiological studies, it is very common to match with respect to age. The advantage of this strategy is that age is often a strong risk factor, but one that is not of primary interest in its own right. Effective matching will then improve the efficiency of the study by reducing the variation in the outcome due to the matched covariate, and can also reduce bias that might arise from the confounding effect of the covariate.

25.15 Statistical Significance versus Effect Importance

In statistical methodology, the term *significance* is defined by the P value; the smaller the P value, the more significant the result. But a statistically significant result does not necessarily equate to a worthwhile effect. A study could easily fail to detect a worthwhile effect. Alternatively, a result could be statistically significant, but of no practical importance.

Confidence intervals can illustrate and elucidate these apparent anomalies. Figure 5 shows 95% confidence intervals for five different studies involving the estimation of a population parameter. The population parameter has null value 0, and a value of δ or more is considered worthwhile or important. For example, the parameter might be the effect of a new but expensive drug in reducing the risk that an HIV-infected person will die from AIDS within five years. In this case, it might be reasonable to choose δ to be 0.1, on the grounds that any lesser benefit would not outweigh the high cost of the new drug.

In study A, the 95% confidence interval includes the null value, so the result is not statistically significant. However, the effect could be important, because the confidence interval also includes the value δ. In this case, the study is quite inconclusive. A larger study would need to be undertaken.

In study B, the result is statistically significant because the 95% confidence interval does not contain the null value, but the effect might not be important. So the study is still not completely conclusive, and a larger sample is needed to establish the importance of the effect.

In study C, the result is statistically significant and the effect is important. So in this case a conclusive result has been obtained.

In study D, the result is not statistically significant and the effect is not important. Despite the absence of statistical significance, the study is conclusive.

Study E illustrates the final possibility, in which the result is statistically significant but the effect is not important.

Ideally, a study should be large enough to detect a worthwhile effect, but not so large that it can detect an unimportant effect.

25.16 Basic Statistical Methods

For comparative studies, the statistical method is determined by the data types of the determinant X and the outcome Y. If both Y and X are binary, the data comprise a 2×2 contingency table of counts, giving two proportions to be compared, and a single odds ratio measures the strength of the association. If the variables have more than two categories, the data can be summarized using multiple proportions and odds ratios. Logistic regression is appropriate when the outcome is categorical and the determinant is of interval type. For interval outcome data, means rather than proportions are used to summarize the data, and differences of means rather than odds ratios measure the associations of interest. Correlation and regression methods are most convenient for handling data in which both the outcome and

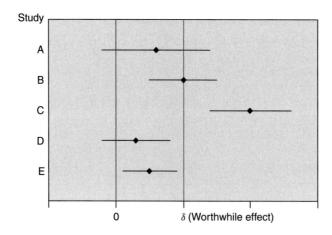

Figure 5: Examples contrasting statistical significance and effect importance.

the determinant are continuous.

As described below, the methods become more complex when a third variable exists.

25.17 Sample Size Determination

An important problem faced by the epidemiological statistician is that of determining the appropriate sample size in a study. This choice depends on the precision needed in estimating the parameter of primary interest in the study. It is also desirable to compute the statistical power, which is the probability of making a correct decision when rejecting the null hypothesis of interest.

Assuming that the sample size is sufficiently large for asymptotic normality properties to apply, the *precision* [22] with which a population parameter can be estimated is defined as the half-width of the confidence interval, that is

$$d = z_{\alpha/2}\text{SE}, \qquad (3)$$

where $z_{\alpha/2}$ is the critical value for the standardized normal distribution corresponding to a two-tailed area α. To determine the sample size required to achieve a specified precision, a formula for the relevant standard error is thus needed.

25.17.1 The Power of a Study

When a major objective of a study is to test a new treatment, rejecting the null hypothesis could lead to a change in health policy. This can happen when a clinical trial comparing a promising new therapy with the standard treatment finds in favor of the new therapy. While there is always a chance of rejecting the null hypothesis when it is true, and this risk (α, the type I error) is conventionally taken to be 5%, the probability of failing to detect a worthwhile benefit (β, the type II error) must also be taken into account.

The power is $(1-\beta)$ of a study is the probability that a worthwhile benefit will be detected. This probability depends on various factors, including the size of the worthwhile benefit, the variability in the data, and the sample size. The power of a study can be expressed in terms of the following parameters.

Suppose that a worthwhile benefit δ exists and that the estimate d of this benefit, based on the data, is approximately normally distributed with mean δ and standard deviation SE, so that $Z = (d - \delta)/\text{SE}$ has a standardized normal distribution. The null hypothesis that δ is 0 is rejected whenever d exceeds $z_{\alpha/2}\text{SE}$ in magnitude. When $\delta = 0$, the null hypothesis is true, and this probability is α, the type I error. But if δ is substantially greater than 0 (greater than SE, say), the null hypothesis is rejected when $|Z + \delta/\text{SE}|$ exceeds $z_{\alpha/2}$, which happens when $Z > z_{\alpha/2} - \delta/\text{SE}$ or $Z < -z_{\alpha/2} - \delta/\text{SE}$. Since $\delta >$ SE, the probability of the second alternative is negligible for reasonable values of α, so the approximate probability of rejecting the null hypothesis is the tail area corresponding to the critical value $z_{\alpha/2} - \delta/\text{SE}$, that is, $1 - \Phi\left(z_{\alpha/2} - \delta/\text{SE}\right)$, where $\Phi(z)$ is the cumulative standardized normal distribution function. This probability is also the power of the study, $1 - \beta$, which equals $\Phi(z_\beta) = 1 - \Phi(-z_\beta)$, so that

$$\delta = \left(z_{\alpha/2} + z_\beta\right)\text{SE}. \qquad (4)$$

This formula assumes that the variability of the estimated benefit d does not depend on δ, and becomes more complicated when this assumption is relaxed. It is quite similar to the precision formula (3), and thus could be used to determine the sample size needed to achieve a specified power. Equation (4) can be inverted to determine the power of a study, given the sample size:

$$1 - \beta = \Phi\left(\delta/\text{SE} - z_{\alpha/2}\right). \qquad (5)$$

25.18 Adjusting for a Stratification Variable

A large body of statistical methods used in epidemiology deals with categorical data involving three factors. The first two factors comprise a determinant X and an outcome Y, and the third is a stratification variable Z. If X, Y, and Z have levels r, c, and s respectively, the data may be represented as an $r \times c \times s$ contingency table.

The associations of X and Y with Z may distort the association between X and Y. When X and Y are both binary, a method due to Cochran [7] is used to test for an association between X and Y after adjusting for Z. This method uses a chi-squared test with one degree of freedom, and assumes that the association is the same for each level of Z. Birch [2] extended the test to general $r \times c \times s$ tables.

When r and s are both equal to 2, the association can be expressed in terms of a single odds ratio. Assuming that this odds ratio is the same in each stratum of Z, Mantel and Haenszel [23] gave a robust estimator of the odds ratio, and Robins et al. [31] derived the variance of its logarithm.

The common odds-ratio assumption can be tested using a chi-squared test with $s-1$ degrees of freedom [3, p. 142]. This test requires that the counts in the individual strata be reasonably large.

Clayton [6] showed that where both X and Y are ordinal, the association could be summarized in terms of a single odds ratio, defined in terms of the aggregated counts for cells with values up to and including each specified value; it is assumed that all these odds ratios are identical in the target population. A chi-squared test comparing the observed counts with expected counts could test this homogeneity assumption.

25.19 Meta-Analysis

If a study is too small, it is unlikely to give a conclusive result. When planning clinical trials, it is conventional to require a power substantially greater than 0.5. But if the research question is important, any properly conducted study gives useful information. The evidence accrued from many small studies can be combined using a method called *meta-analysis*, which first arose in the social science literature [13] and is now widely used in medical research [26,27,37].

Ideally, each contributing study investigates the same treatment or risk factor and the same outcome for similar subjects, and each study will have the same type and quality of design. But selecting the studies to include is difficult because studies, even when they have the same research objective, tend to vary substantially in quality from place to place and at different times. Biases can arise if not all relevant studies are included. In practice, inconclusive or uninteresting studies often remain unpublished, giving rise to publication bias [10]. Undertaking a meta-analysis requires a professional team of scientists.

In general, it is always possible to do a meta-analysis whenever the estimate of the effect and its standard error is available for each contributing study. If y_k is the estimated effect and SE_k its standard error based on the kth study, the combined estimate and its standard error are given by

$$\hat{y} = \frac{\sum w_k y_k}{\sum w_k}, \quad SE(\hat{y}) = \frac{1}{\sqrt{\sum w_k}}, \quad (6)$$

where $w_k = (1/SE_k)^2$.

25.20 Survival Analysis

Survival analysis is a major area of epidemiological statistics concerned with measuring the risk of occurrence of an outcome event as a function of time. It thus focuses on the duration of time elapsed from when a subject enters a study until the event occurs, and uses the survival curve to describe its distribution. Survival analysis is also concerned with the comparison of survival curves for different combinations of risk factors, and uses statistical methods to facilitate this comparison [14,16,18].

In general, survival analysis allows for the proper treatment of incomplete data due to subjects dropping into or out of the study, giving rise to censored (more precisely, right-censored) data. Survival data may be censored because *(1)* the subject withdraws from the study for any reason before experiencing the event ("loss to follow-up"), *(2)* an intervening event occurs prohibiting further observation on the subject, or *(3)* the subject does not experience the event before the study ends (or before an analysis of the results is required). In general, survival analysis allows for the proper treatment of incomplete data due to subjects dropping into or out of the study, giving rise to censored (more precisely, right-censored) data. Survival data may be censored because *(1)* the subject withdraws from the study for any reason before experiencing the event ("loss to follow-up"), *(2)* an intervening event occurs prohibiting further observation on the subject, or *(3)* the subject does not experience the event before the study ends (or before an analysis of the results is required).

When the event of interest occurs, the survival time is conventionally called a *failure time* (even though the event might be a "success," like recovery from some disease).

The *survival curve* is the proportion of subjects surviving beyond a given duration of time t. For a large population in which the survival times range continuously over an interval, this curve will be a smooth function of t that decreases from a maximum value of 1 when t is 0. In practice, the survival curve estimated from a sample of data is a step function that decreases only at the failure times [17].

A useful summary of survival that can be estimated directly from a survival curve is the *median survival time*. This is the survival time exceeded by 50% of the subjects, and is obtained by finding where the survival curve has the value 0.5.

Survival curves compare risks for different groups of subjects by showing the survival curves for the different groups on the same axes. Thus if one curve (for group A, say) is entirely above another (group B), then the subjects in group A have better survival prospects than those in group B. However, if the two curves cross, the situation is more complicated; this means that the relative risk of failure depends on how long a subject survives.

The logrank test [28] provides a P value for testing the null hypothesis that two or more survival functions are identical. This is a special case of Birch's extension of the Mantel–Haenszel–Cochran test, which requires that the population odds ratios in the different strata are the same. In survival analysis, this homogeneity assumption is called the *proportional hazards* assumption.

The proportional hazards model, described in Section 24.23, provides another method for analyzing survival data. It has the advantage that it can handle both continuous and nominal determinants simultaneously. However, in common with other statistical models, it makes additional assumptions about the associations and should be regarded as an accompaniment rather than an alternative to the methods described in this section.

25.21 Logistic Regression

Logistic regression provides a method for modeling the association between a nominal outcome and multiple determinants. It is similar in many ways to linear regression. For m determinants with a binary outcome, it takes the form

$$\ln\left(\frac{p}{1-p}\right) = \alpha + \sum_{j=1}^{m} \beta_j x_j. \quad (7)$$

The only other statistical assumption is that the outcomes are mutually independent. The model is known as *simple logistic regression* [19].

Logistic regression provides a further statistic, the *deviance*, which can assess the statistical significance of a set of determinants in the model. The deviance is defined as $-2\ln L$, where L is the likelihood associated with the data for the fitted parameters. Two logistic regression models are fitted to the data, one containing all the determinants of interest, and the other containing all the determinants except for those being assessed. Asymptotically, the difference between the values of the two deviances has a chi-squared distribution, with the number of degrees of freedom equal to the number of parameters in the determinants being assessed.

The logistic regression model described by Equation (7) can be extended to situations in which the outcome variable is nominal with more than two categories. If these outcome categories are coded as $0, 1, 2, \ldots, c$ and p_k is the probability that an outcome has the value k, the model takes the form, for $0 \leq k \leq c$,

$$p_k = \frac{\exp\left(\alpha_k + \sum_{j=1}^{m} \beta_{jk} x_j\right)}{1 + \sum_{k=1}^{c} \exp\left(\alpha_k + \sum_{j=1}^{m} \beta_{jk} x_j\right)}, \quad (8)$$

and is known as *polytomous logistic regression* [14].

For ordinal outcomes with more than two levels, the logistic model takes a different form. The outcome categories are again coded as $0, 1, 2, \ldots, c$ but p_k is now the probability that an outcome has value *at least* k. Thus, for $0 < k \leq c$, these probabilities are given by

$$\ln\left(\frac{p_k}{1-p_k}\right) = \alpha_k + \sum_{j=1}^{m} \beta_j x_j. \quad (9)$$

This model incorporates an assumption of "proportional odds," meaning that for each determinant, the odds ratios are the same at each cutpoint k.

25.22 Poisson Regression

The Poisson distribution takes nonnegative integer values with probabilities

$$p_k = \frac{\lambda^k}{k!} e^{-\lambda}. \qquad (10)$$

The parameter λ is positive and is the mean of the distribution. This distribution plays an important role in epidemiological research, particularly in studies in which different subjects have different durations of exposure to a risk factor.

Suppose that the probability that a subject experiences an adverse event in a short interval of time δt is constant and equal to $\mu \delta t$. If the outcomes in different intervals are independent, the number of adverse events in a period T containing $N = T/\delta t$ short intervals has a binomial distribution ranging from 0 to N, with probabilities

$$p_k = \frac{N!}{k!(N-k)!} (\mu \delta t)^k (1 - \mu \delta t)^{N-k}.$$

Taking the limit as $\delta t \to 0$, we obtain the Poisson distribution with parameter $\lambda = \mu T$.

The Poisson regression model now arises by expressing the incidence rate μ as the exponential of a linear function of determinants:

$$\lambda = \mu T = \exp\left(\alpha + \sum_{j=1}^{m} \beta_j x_j\right) T. \qquad (11)$$

If the determinants are nominal, Poisson regression is a limiting case of logistic regression, in which the probability of the adverse outcome is infinitesimally small and the number of subjects is correspondingly large. In logistic regression, each subject constitutes a separate experimental unit, whereas in Poisson regression an experimental unit corresponds to a short period of observation on each subject.

In the Poisson model, outcomes are interchangeable in the sense that a group of subjects has the same combination of risk factors, but only the total number of adverse events is important.

25.23 Proportional Hazards Regression

The Poisson regression model given by Equations (10) and (11) specifies the logarithm of the incidence density as a linear function of determinants. This incidence rate μ is the probability that the outcome event of interest occurs in a small interval of time $(t, t + \delta t)$ divided by the length of the interval (δt). Equation (11) can thus be written

$$\mu = \exp\left(\alpha + \sum_{j=1}^{m} \beta_j x_j\right).$$

In the context of survival analysis, this incidence rate is the *hazard function*, and is allowed to depend on the survival time t, simply by allowing the constant parameter α to be a function of t. This model, due to Cox [8], takes the form

$$h(t) = h_0(t) \exp\left(\sum_{j=1}^{m} \beta_j x_j\right), \qquad (12)$$

where $h_0(t)$ is an arbitrary function of the survival time t, called the *baseline hazard function*. It is called the *proportional hazards model*, because the relative risk of an event for two subjects depends only on their determinants, and not on their duration of survival.

Note that if failures were known to occur at the times $t_1, t_2, \ldots, t_q$, $h_0(t)$ could be expressed as a step function with changes at these failure times, and Equation (12) would take the form

$$h(t) = \exp\left(\sum_{i=1}^{q} \alpha_i w_i\right) \exp\left(\sum_{j=1}^{m} \beta_j x_j\right),$$

where each w_i is an indicator variable taking the value 1 for a failure occurring at time t_i and 0 otherwise. It follows that the proportional hazards model is equivalent to the special case of a Poisson regression model in which an additional stratification variable corresponding to the failure time is included.

25.24 Correlated Outcomes

All of the statistical methods described in the preceding sections assume that the outcomes are independent. In practice, data collected in epidemiological studies are often correlated due to clustering. Spatial clustering occurs when subjects are sampled from villages or families sharing particular attributes, or when repeated measurements are taken on the same subjects. Clustering also occurs in time as a result of seasonal effects.

Standard errors of proportions based on correlated data can be corrected using variance inflation factors [29], which specify by how much the sample size of a cluster needs to be increased to compensate for the clustering. In the simplest case, when the correlations between binary outcomes in clusters of size m have equal correlation ρ, the variance inflation factor is $1 + (m-1)\rho$, and the standard error of the log odds ratio is increased by the square root of this factor. This formula shows that even small correlations between outcomes can have a substantial effect in large clusters. For example, if $\rho = 0.1$ and $m = 31$, the standard error is doubled, and the sample size would need to be quadrupled to compensate for the clustering.

Zeger and Liang [39] invented a general and robust method for handling correlated outcomes, which is now widely used in epidemiological statistics. This method, known as generalized estimating equations (GEE), can be applied to all the models described in this section, and has generated a great deal of research both in statistics and epidemiology [11].

References

1. Armitage, P. and Hills, M. (1982). The two-period cross-over trial. *Statistician*, **31**, 119–131.

2. Birch, M. W. (1965). The detection of partial association II: the general case. *J. Roy. Statist. Soc. B*, **27**, 111–124.

3. Breslow, N. E. and Day, N. E. (1980). *Statistical Methods in Cancer Research: Volume I – The Analysis of Case-Control Studies*. IARC, Lyon.

4. Breslow, N. E. and Day, N. E. (1987). *Statistical Methods in Cancer Research: Volume II – The Design and Analysis of Cohort Studies*. IARC, Lyon.

5. Brown, B. W. Jr. (1980). The crossover experiment for clinical trials. *Biometrics*, **36**, 69–70.

6. Clayton, D. G. (1974). Some odds ratio statistics for the analysis of ordered categorical data. *Biometrika*, **61**, 525–531.

7. Cochran, W. G. (1954). Some methods of strengthening the common χ^2 tests. *Biometrics*, **10**, 417–451.

8. Cox, D. R. (1972). Regression models and life tables (with discussion). *J. Roy. Statist. Soc. B*, **34**, 187–220.

9. Cox, D. R. and Wermuth, N. (1996). *Multivariate Dependencies—Models, Analysis and Interpretation*. Chapman & Hall. London.

10. Easterbrook, P. J., Berlin, J. A, Gopalan, R., and Matthews, D. R. (1991). Publication bias in clinical research. *Lancet*, **387**(8746), 867–872.

11. Hardin, J. W. and Hilbe, J. M. (2003). *Generalized Estimating Equations*. Chapman & Hall/CRC Press.

12. Hill, A. B. (1951). The clinical trial. *Br. Med. Bull.*, **7**, 278–287.

13. Hodges, L. V. and Olkin, I. (1885). *Statistical Methods for Meta-analysis*. Academic Press, New York.

14. Hosmer, D. W. and Lemeshow, S. (1989). *Applied Logistic Regression*. Wiley, New York.

15. Hosmer, D. W. and Lemeshow, S. (1999). *Applied Survival Analysis—Regression Modelling of Time to Event Data*. Wiley, New York.

16. Kalbfleisch, J. and Prentice, R. L. (1980). *The Statistical Analysis of Failure Time Data*, Wiley. New York.

17. Kaplin, E. L. and Meier, P. (1958). Nonparametric estimation from incomplete observations. *J. Am. Statist. Assoc.*, **53**, 457–481.

18. Klein, J. P. and Moeschberger, M. L. (1997). *Survival Analysis: Techniques for Censored and Truncated Data*. Springer-Verlag, New York.

19. Kleinbaum, D. G. and Klein, M. (2002). *Logistic Regression: A Self-Learning Text*, 2nd ed. Springer-Verlag, New York.

20. Louis, P. C. A. (1834). *An Essay on Clinical Instruction*, (translated by Peter Martin, S. Highley, London, pp. 26–27).

21. Louis, P. C. A. (1836). *Researches on the Effects of Bloodletting in some Inflammary Diseases, and on the Influence of Tartarized Antimony and Vesication in Pneumonius.* (translated by C. G. Putnum, Hilliard Gray, Boston, MA).

22. Lwanga, S. K. and Lemeshow, S. (1991). *Sample Size Determination in Health Studies. A Practical Manual*. WHO, Geneva.

23. Mantel, N. and Haenszel, W. (1959). Statistical aspects of the analysis of data from retrospective studies of disease. *J. Natl. Cancer Inst.*, **22**, 719–748.

24. Meier, P. (1972). "The biggest public health experiment ever. The 1954 field trial of the Salk poliomyelitis vaccine." In *Statistics. A Guide to the Unknown*, J. M. Tanur, ed., pp. 2–13, Holden Day, San Francisco, CA..

25. Newman, S. C. (2001). *Biostatistical Methods in Epidemiology*. Wiley, New York.

26. Normand, S. L. (1999). Meta-analysis. formulating, evaluating, combining, and reporting. *Statist. Med.*, **18**(3), 321–359.

27. Peto, R. (1987). Why do we need systematic overviews of randomized trials? *Statist. Med.*, **6**, 233–240.

28. Peto, R. and Peto, J. (1972). Asymptotically efficient rank invariance test procedures (with discussion). *J. Roy. Statist. Soc. A*, **135**, 185–206.

29. Rao, J. N. K. and Scott, A. J. (1992). A simple method for the analysis of clustered binary data. *Biometrics*, **48**, 577–585.

30. Robertson, B., Fairley, C. K., Black, J., and Sinclair, M. (2003). Case–control studies. In *Drinking Water and Infectious Disease. Establishing the Links*, P. R. Hunter, M. Waite, and E. Ronchi, eds., pp. 175–182, CRC Press.

31. Robins, J., Breslow, N., and Greenland S. (1986). Estimators of the Mantel-Haenszel variance consistent in both sparse data and large-strata limiting models. *Biometrics*, **42**, 311–323.

32. Sahai, H. and Khurshid. A. (1996). *Statistics in Epidemiology: Methods, Techniques, and Applications*. CRC Press, Boca Raton, FL.

33. Sackett, D. L. (1979). Bias in analytic research. *J. Chron. Dis.*, **32**, 51–63.

34. Schlesselman, J. (1982). *Case-Control Studies*. Oxford University Press, New York.

35. Senn, S. (1990). *Cross-over Trials in Clinical Research*. Wiley, New York.

36. Snow, J. (1855). *On the Mode of Communication of Cholera*. Churchill, London. (Reprinted *in Snow on cholera*: A reprint of two papers. Hafner, New York, 1965).

37. Sutton, A. J., Abrams, K. R., Jones, D. R., Sheldon, T. A., and Song, F. (2000). *Methods for Meta-Analysis in Medical Research*. Wiley, New York.

38. Woodward, M. (1999). *Epidemiology: Study Design and Data Analysis*. Chapman Hall/CRC, Boca Raton, FL.

39. Zeger, S. and Liang, K. (1966). Longitudinal data analysis for discrete and continuous outcomes. *Biometrics*, **42**, 121–130.

26 Event History Analysis

German Rodriguez

26.1 Introduction

Event history analysis is concerned with the study of sequences of life events. A sociologist, for example, may study the process of family formation, following cohorts of individuals as they experience cohabitation, marriage, widowhood, and divorce. A labor economist may be interested in spells of employment and unemployment. An epidemiologist may focus on transitions between health, illness, and death. The special case where one models transitions between two states, usually labeled "alive" and "dead", is the subject of survival analysis. A key aspect of event history data is that the observations are usually right-censored, in the sense that the underlying life histories are still unfolding at the time we would like to analyze the data. Sometimes the process is already underway when observation starts, and the resulting histories are left-truncated. In all cases, however, we require an uninterrupted record of observation between two time points, a condition not always satisfied in panel studies or retrospective surveys.

The term "event history analysis" was introduced in the social science literature in 1976 by Tuma et al. [52]. Sociologists used Markov models for social processes in the 1950s [9] and 1960s [18], and interest in these models in the 1970s was stimulated by the availability of longitudinal data from large-scale social experiments, such as the Seattle and Denver income-maintenance experiment; see the historical remarks in Reference 21. The earliest Markov models assumed constant rates and population homogeneity, and "invariably failed to fit the data" [52]. More realistic models were then developed, allowing transition rates to depend on time and introducing population heterogeneity, first in the context of discrete mixtures—of which the most successful was Blumen et al.'s [9] mover–stayer model of social mobility—and later by explicit modeling of transition rates as functions of observed covariates. These models were generally estimated by maximum likelihood under parametric assumptions for the distribution of transition times (see, for e.g., Ref. 22). Similar developments were also apparent in other disciplines, most notably biostatistics [32] and engineering [36].

The introduction of Cox's [19] proportional hazards model in 1972 sparked the interest of statisticians in models for failure-time data, emphasizing semiparametric models where covariate effects are modeled parametrically but time dependence is treated nonparametrically. Cox's partial likelihood approach was given a solid mathematical footing by the Scandinavian school in the context of point processes and martingale theory [6]. Also in 1972, Nelder and Wedderburn [39] introduced the class of generalized linear models to unify the treatment of regression models for discrete- and continuous-response variables in the exponential family. The close connection between these two developments was soon discovered and had important practical implications that will be noted below. Current research has focused on the introduction of random effects to represent unobserved heterogeneity and/or clustering in so-called frailty models, to be discussed later. Among the neglected areas requiring further research is the estimation of event history models from intermittent records of observation.

In this entry we will adopt the counting process approach to event history analysis initiated by Aalen [1], stressing connections with survival analysis and generalized linear models. For a rigorous account of the theory, see the comprehensive treatise by Andersen et al. [5]. There are several useful journal reviews and books aimed at different audiences, including biostatisticians and epidemiologists [14,16], economists [25,34], sociologists [8,42,51,55], and demographers [49]. A popular elementary introduction is [3].

26.2 Recurrent Events

Consider first a recurrent event that may happen several times to each individual. Let $N_i(t)$ count the number of times that the event has happened to individual i by time t. A realization of $N_i(t)$ is a step function starting at zero and with jumps of size $+1$. We assume that $N_i(t)$ is a counting process with random intensity process $\lambda_i(t)$, such that the probability of the event occurring in a small interval $[t, t+dt)$, conditional on the entire history $\mathcal{F}_{t-}$ of the process just before time t, is $\lambda_i(t)dt$. We further assume that $\lambda_i(t)$ satisfies Aalen's [1] multiplicative intensity model

$$\lambda_i(t) = \alpha_i(t) Y_i(t). \qquad (1)$$

Here $Y_i(t)$ is an observable predictable process taking the value 1 if individual i is observed and at risk at time t and 0 otherwise, and $\alpha_i(t)$ is an unknown hazard function. In a proportional hazards model $\alpha_i(t)$ is further specified as

$$\alpha_i(t) = \alpha_0(t) \exp[\boldsymbol{\beta}' \boldsymbol{x}_i(t)], \qquad (2)$$

where $\alpha_0(t)$ is an unknown baseline hazard and $\exp[\boldsymbol{\beta}' \boldsymbol{x}_i(t)]$ is the relative risk for individual i, depending on a parameter vector $\boldsymbol{\beta}$ and a vector of (possibly time-varying) covariates $\boldsymbol{x}_i(t)$. In the important special case of survival analysis, $N_i(t)$ can jump only once and ((1))–(2) is the counting process formulation of Cox's [19] proportional hazards model; see Reference 6. A key feature of this formulation is that

$$M_i(t) = N_i(t) - \int_0^t \lambda_i(u) du \qquad (3)$$

is a *martingale*, a fact that can be used to establish the properties of tests and estimators using martingale central-limit theory.

26.3 Parametric Estimation

If the baseline hazard $\alpha_0(t)$ depends on a finite-dimensional parameter vector $\boldsymbol{\theta}$, estimation using histories observed in $[0, \tau]$ under fairly general (noninformative) censoring mechanisms proceeds by maximizing the likelihood function

$$L(\boldsymbol{\beta}, \boldsymbol{\theta}) = \prod_{i=1}^{n} [\alpha_i(t_i)]^{\Delta N_i(t)}$$
$$\times \exp\left(-\int_0^\tau \alpha_i(t) Y_i(t) dt\right), \qquad (4)$$

where $\alpha_i(t)$ satisfies (2); [10]; [5, Sec. VII.6].

An important connection with generalized linear models arises if $\alpha_i(t)$ is piecewise constant (see Ref. 33 or 28). Let $0 = \tau_0 < \tau_1 < \cdots < \tau_m = \infty$, and assume that $\alpha_0(t) = \alpha_j$ for $t \in (\tau_{j-1}, \tau_j]$. Suppose further that the vector of covariates x has a fixed value $\boldsymbol{x}_{ij}$ in interval j. Let t_{ij} denote the time at risk for individual i in interval j, and let d_{ij} be the number of times the event happened to individual i in interval j. Then (4) coincides with the likelihood obtained by treating the counts d_{ij} as independent Poisson random variates with means μ_{ij} satisfying the loglinear model

$$\log \mu_{ij} = \log t_{ij} + \log \alpha_j + \boldsymbol{\beta}' \boldsymbol{x}_{ij}. \quad (5)$$

Here $\log t_{ij}$ is an *offset* or known part of the linear predictor. This strategy requires creating a pseudo observation for each interval visited by individual i. Covariates that change at discrete times within an interval can be accommodated by generating a pseudo-observation for each distinct value in each interval. Related developments in discrete time lead to binomial regression models with link logit or complementary log–log; [3,7].

26.4 Semiparametric Estimation

Estimation of the regression parameters $\boldsymbol{\beta}$ without making parametric assumptions about the baseline hazard $\alpha_0(t)$ can be based on the partial likelihood

$$L_p(\beta) = \prod_t \prod_i^n \left(\frac{\exp[\boldsymbol{\beta}' \boldsymbol{x}_i(t)]}{\sum_j \exp[\boldsymbol{\beta}' \boldsymbol{x}_j(t)] Y_j(t)} \right)^{\Delta N_i(t)}, \quad (6)$$

which generalizes Cox's proposal for survival models [20]; the only difference here is that individuals do not leave the risk set when the event occurs. Hypotheses about $\boldsymbol{\beta}$ may be tested by treating (6) as an ordinary likelihood, calculating likelihood ratio tests for nested models, or using Wald tests based on the asymptotic distribution of $\sqrt{n}(\hat{\boldsymbol{\beta}} - \boldsymbol{\beta})$, which can be shown to be multinormal with mean zero and a variance–covariance matrix that can be consistently estimated from the inverse of the observed information matrix.

The integrated baseline hazard $A_0(t) = \int_0^t \alpha_0(u) du$ can be estimated using the Nelson–Aalen estimator

$$\hat{A}_0(t) = \int_0^t \frac{J(u)}{\sum_i \exp[\hat{\boldsymbol{\beta}}' \boldsymbol{x}_i(u)] Y_i(u)} dN(u), \quad (7)$$

where $J(u) = \mathbf{I}\{\sum Y_i(u) > 0\}$ and $N(u) = \sum_i N_i(u)$. To obtain an estimate of the hazard $\alpha_0(t)$ itself one can smooth (7) using, for example, a kernel estimate

$$\hat{\alpha}_0(t) = \frac{1}{b} \int K\left(\frac{t-s}{b}\right) d\hat{A}_0(s) \quad (8)$$

for some bandwidth b. For further details see [5, Chap. VII].

The semiparametric approach is closely related to the Poisson likelihood described above, and can be justified by considering a piecewise exponential model where the intervals have common width w approaching zero (see Refs. 11 and 29). This approach leads to viewing the partial likelihood (6) as a profile likelihood obtained from (4) by maximizing over the nuisance function $\alpha_0(t)$.

26.5 Multistate Models

Consider now a more general situation where an individual can move through a discrete series of states $\mathcal{S} = \{1, \ldots, m\}$.

The counting process framework developed so far can be applied in this setting by focusing on transitions between pairs of states, excluding pairs for which transitions are not possible. Specifically, let $N_{ijk}(t)$ count the number of times that individual i has moved directly from state j to state k by time t. We assume that $N_{ijk}(t)$ is a univariate counting process obeying a multiplicative intensity model with proportional hazards structure, so that

$$\lambda_{ijk}(t) = \alpha_{0jk} \exp[\boldsymbol{\beta}'_{jk}\boldsymbol{x}_i(t)] Y_{ijk}(t). \quad (9)$$

Here $Y_{ijk}(t)$ is a predictable process that takes the value 1 if individual i is observed and at risk of moving from state j to state k at time t, and 0 otherwise. (Usually this means that individual i is observed in state j just before t.) The hazard $\alpha_{0jk}(t)$ is a baseline rate of transition from state j to state k at time t, and in a fully parametric setup would depend on a parameter vector $\boldsymbol{\theta}_{jk}$. Finally, $\exp[\boldsymbol{\beta}'_{jk}\boldsymbol{x}_i(t)]$ is a relative risk of transition from state j to state k for subject i, depending on a parameter vector $\boldsymbol{\beta}_{jk}$ and the usual vector of time-varying covariates $\boldsymbol{x}_i(t)$.

The likelihood function for this process is simply a product of likelihoods, one for each possible transition, and each having the form (4). A similar factorization applies to the partial likelihood (6). The important practical implication of this result is that we can analyze each type of transition separately, using the techniques developed so far. The situation is exactly analogous to what happens in competing risk models, where one can analyze different types of failure separately; [30, Chap. 7].

In the piecewise exponential framework, we need to count events and exposures by type. Let d_{ijkl} count the number of times that individual i moves directly from state j to state k in time interval l, and let t_{ijl} denote the amount of time that individual i is observed at risk in state j during interval l. Note that the time at risk depends on the state of origin but not on the destination. Then the history may be analyzed by treating the d_{ijkl} as independent Poisson observations with means μ_{ijkl} satisfying the loglinear model

$$\mu_{ijkl} = \log t_{ijl} + \log \alpha_{jkl} + \boldsymbol{\beta}'_{jk}\boldsymbol{x}_{il}, \quad (10)$$

where α_{jkl} represents the baseline risk $\alpha_{0jk}(t)$ and $\boldsymbol{x}_{il}$, the vector of covariates $\boldsymbol{x}_i(t)$ for $t \in (\tau_{l-1}, \tau_l]$. In the partial likelihood framework we introduce a separate risk set $\mathcal{R}_j$ for each state of origin and let individuals enter and exit the various risk sets as they move from one state to another.

26.6 Markov and Semi-Markov Models

The multiplicative intensity model (1)–(2) includes as special cases various Markov models, such as a simple Poisson process where events occur with constant rate α, a time-dependent Poisson process where the hazard rate $\alpha(t)$ depends on calendar time, and nonhomogeneous Poisson processes involving fixed covariates x that act multiplicatively on the intensity; [35].

An event history is said to exhibit *occurrence dependence* if the hazard function $\alpha_i(t)$ depends on the number of previous occurrences of the event of interest, or the number of previous visits to a state in more general multistate models. In modeling a woman's reproductive history, for example, we may find that the fertility rate depends on the woman's age t and the number of children she has had by age t.

Occurrence dependence can be modeled parametrically using time-dependent covariates, or nonparametrically using time-dependent strata, as suggested in Reference 43.

A history can also exhibit *duration dependence* if the hazard function $\alpha_i(t)$ depends on the backward recurrence time, or duration since the most recent occurrence of the event. In a woman's reproductive history, for example, the fertility rate may depend not only on the woman's age t but also on the time elapsed since the birth of her youngest child. Models of this type are called *semi-Markov models* and can be accommodated parametrically using time-varying covariates. For example a renewal process where the times between events have independent Weibull distributions can be fitted into the counting process framework by setting $\alpha_0(t) = \alpha$, a constant, and introducing a covariate $x_i(t) = \log(t - T_i)$, where T_i is the time when the event last happened to individual i or 0 if the event has never happened to her. An interesting feature of this particular model is the following. If the history is observed at a random time τ, then the time since the most recent occurrence of the event follows an accelerated life model based on Stacy's generalized gamma distribution [47]. This result has applications in situations where the only information available is the duration since the last occurrence of the event [4].

Duration dependence is also important in multistate models, where analysts often focus on the *sojourn* times or lengths of stays in various states. In studies of labor force dynamics, for example, it is often assumed that the risk of leaving the employed state depends not just on age t but also on how long one has held the present job. Similarly, the rate of transition from the unemployed to the employed state may depend on how long one has been searching for a job. Most social science work to date has, in fact, favored duration in a state over calendar time as the primary time dimension. No special issues arise if duration dependence, or indeed any aspect of earlier transitions, is modeled parametrically (see, e.g., the likelihood construction in Ref. 25). Additional useful sources are References 2 and 31. For an application using piecewise exponential models to analyze birth intervals including occurrence, duration, and lagged duration dependence, in addition to covariate effects [45].

More generally, one can estimate duration effects nonparametrically by assuming that the sojourn times or times between the events are independent, perhaps given the values of time-dependent covariates, and using Cox's partial likelihood [19] with t representing duration of stay in the current state rather than calendar time. This situation is less well understood, however, and the procedure has been justified only in special cases. Voelkel and Crowley [54] have considered the important case where the states can be ordered in such a way that only forward transitions are possible. If an individual can return to an earlier state, however, the martingale argument breaks down, and alternative techniques are needed to establish the necessary asymptotic results [5, Chap. X].

26.7 Frailty Models

So far we have assumed that the observed covariates $\boldsymbol{x}_i(t)$ capture the dependence between the various events experienced by the same individual, but this assumption is rarely realistic. An attractive solution is to introduce a persistent random effect W_i, such that the intensity function for recurrent events *conditional* on $W_i = w_i$ is

$$\lambda_i(t) = \alpha_0(t) \exp[\boldsymbol{\beta}' \boldsymbol{x}_i(t)] w_i Y_i(t). \quad (11)$$

Models incorporating multiplicative random effects on the hazard function are called *frailty models* after Vaupel et al. [53]. They have been used to model unobserved heterogeneity in survival models [24], in Poisson processes [35], and in the analysis of kindred lifetimes [17,23]. For a discussion of issues that arise in modeling unobserved heterogeneity, see References 37 and 44.

Fully parametric estimation requires specification of the baseline hazard $\alpha_0(t)$ and the distribution of the unobservable frailty effect W_i. The gamma distribution with mean 1 and variance ϕ is often chosen because it is flexible and tractable (see, e.g., Ref. 40). Concern that estimated effects might be sensitive to the choice of distribution for the unobservable led Heckman and Singer [24] to propose a nonparametric maximum likelihood estimate for the mixing distribution, while using a (flexible) parametric form for the baseline hazard. Their procedure leads to a finite mixture model where W_i takes values $\xi_1, \ldots, \xi_q$ with probabilities $\pi_1, \ldots, \pi_q$. The estimate often collapses to a mover—stayer model where W_i takes the values $\xi > 0$ and 0 with probabilities π and $1 - \pi$. Of course, estimates can also be sensitive to the choice of baseline hazard, as demonstrated in Reference 50. Frailty models with piecewise constant hazard can be fitted using generalized linear models with random effects, using either approximations to the likelihood [12] or a Bayesian approach based on the Gibbs sampler [56]. In this regard, see also References 15 and 46.

Most statistical work to date has focused on nonparametric estimation of the baseline hazard while retaining the assumption of gamma-distributed frailty. For a fixed value of ϕ, the variance of the unobservable, one can fit the model (11) using an EM algorithm. To start the process, fit the model ignoring unobserved heterogeneity. In the E step, obtain the empirical Bayes estimate of W_i, which is given by

$$\hat{z}_i = \frac{\phi^{-1} + N_i(\tau)}{\phi^{-1} + \int_0^\tau \exp[\boldsymbol{\beta}' \boldsymbol{x}_i(t)] Y_i(t) dA_0(t)}. \quad (12)$$

In the M step, these values are used to estimate β using the partial likelihood (6) and $A_0(t)$ using the Nelson–Aalen estimator (7). The EM algorithm may be modified to estimate ϕ as well, but the likelihood is often not quadratic on ϕ, and it may be preferable to examine the profile likelihood (see Ref. 14). For further work along these lines, see References 41 and 5 (Chap. IX). For a completely different approach to modeling frailty effects using fuzzy sets, see Manton et al. [38, Chap. 5].

26.8 Goodness of Fit

The assessment of goodness of fit is an essential component of model evaluation. Cox [19] suggested including interactions between observed covariates and deterministic functions of time such as $\log t$ or $I\{t \geq t_0\}$ to check for proportionality of hazards. Zucker and Karr [57] estimate time-varying effects using a partial likelihood penalized by a measure of roughness of $\beta(t)$. If the number of observations is large relative to the number of distinct covariate patterns, one can compare the model of interest with a saturated model that estimates a separate parameter for each distinct covariate pattern. An alternative approach is to use stratification, as described in Reference 30.

One can also examine martingale residuals based on the counting process decomposition

$$\hat{M}_i(t) = N_i(t) - \int_0^t \exp[\hat{\boldsymbol{\beta}}' \boldsymbol{x}_i(u)] Y_i(u) d\hat{A}_0(u). \quad (13)$$

These residuals can be used to investigate transformations of the covariates, to identify outliers or points with high leverage, and to assess the proportional hazards assumption; see Therneau et al. [48]. These techniques essentially look separately at each type of transition.

Tuma et al. [52] pioneered the use of diagnostic techniques that look at the entire event history, using the fitted model to predict various aspects of the data. Specifically, in a study of the effect of various levels of income support on marital stability they use the model to predict the probability of being in a given state (e.g., single) at any point in time, the expected number of marriages and dissolutions in any time interval, and the probability of leaving the original marital status in a fixed period.

This strategy was further developed by Heckman and Walker [26,27], who use classical χ^2 tests to compare selected features of the observed and predicted histories. For example, in an analysis of birth intervals using semi-Markov models with and without unobserved heterogeneity, they compare observed and fitted distributions of women by parity at various ages, as well as observed and predicted mean interval lengths. One difficulty with this approach is that in order to predict the number of births that a woman would have by a specified age one needs to evaluate fairly complex multidimensional integrals. The solutions include numerical integration, which appears to be feasible for up to three or four events, and Monte Carlo integration, where each individual's history is simulated repeatedly, say, 100 times, and the results are then averaged to produce a prediction. Both approaches are computationally intensive. A further difficulty, inherent in the use of multiple goodness-of-fit criteria, is that a model that does well in one dimension may perform poorly in another, leaving the analyst with no unambiguous model choice. In an analysis of fertility in Sweden, for example, they find two models that do well at reproducing different aspects of the data: a purely demographic formulation with lagged durations, and a neoclassical model that includes wages and income. Since the models differ only in terms of the regressors, one would be tempted to try an extended model including all predictors. The combined model does better in terms of the maximized likelihood, but is rejected by the χ^2 goodness-of-fit tests.

The problem of reaching conflicting conclusions from different goodness-of-fit criteria is hardly new (see, e.g., the comparison of eight stochastic learning models in Ref. 13). In the present context the use of multiple criteria seems both natural and desirable [26], leading to an informal assessment of competing models based on a hierarchy of features deemed particularly appropriate to the objectives of the analysis.

References

1. Aalen, O. O. (1978). Non-parametric inference for a family of counting processes. *Ann. Statist.*, **6**, 534–545.

2. Aalen, O. O., Borgan, Ø., Keiding, N., and Thormann, J. (1980). Interactions between life history events: Non-parametric analysis of prospective and retrospective data in the presence of censoring. *Scand. J. Statist.*, **7**, 161–171.

3. Allison, P. D. (1984). *Event History Analysis: Regression for Longitudinal Event Data*. Sage Publications, Beverly Hills, CA. (A popular elementary introduction.)

4. Allison, P. D. (1985). Survival analysis of backward recurrence times. *J. Am. Statist. Assoc.*, **80**, 315–322.

5. Andersen, P. K., Borgan, Ø., Gill, R. D., and Keiding, N. (1993). *Statistical Models Based on Counting Processes*. Springer-Verlag, New York. (A comprehensive treatise of event history analysis, including the mathematical foundations and numerous examples and applications.)

6. Andersen, P. K. and Gill, R. D. (1982). Cox's regression model for counting processes: A large sample study. *Ann. Statist.*, **10**, 1100–1120.

7. Arjas, E. and Kangas, P. (1992). A discrete-time method for the analysis of event histories. In *Demographic Applications of Event History Analysis*, J. Trussell, R. Hankinson, and J. Tilton, eds., pp. 253–266, Clarendon Press, Oxford, UK.

8. Blossfeld, H. P., Hamerle, A., and Mayer, K. U. (1989). *Event History Analysis*. Lawrence Erlbaum, Hillsdale, NJ.

9. Blumen, I., Kogan, M., and McCarthy, P. J. (1955). *The Industrial Mobility of Labor as a Probability Process*. Cornell University Press, Ithaca, NY.

10. Borgan, Ø. (1984). Maximum likelihood estimation in parametric counting process models, with applications to censored failure time data. *Scand. J. Statist.*, **11**, 1–16.

11. Breslow, N. E. (1972). Discussion of the paper by D.R. Cox. *J. Roy. Statist. Soc. B*, **34**, 216–217.

12. Breslow, N. E. and Clayton, D. G. (1993). Approximate inference in generalized linear mixed models. *J. Am. Statist. Assoc.*, **88**(421), 9–25.

13. Bush, R. R. and Mosteller, F. (1959). A comparison of eight models. In *Studies in Mathematical Learning Theory*, R. R. Bush and W. K. Estes, eds. Stanford University Press, Stanford, CA.

14. Clayton, D. G. (1988). The analysis of event history data: Review of progress and outstanding problems. *Statist. Med.*, **7**, 819–841. (An excellent review aimed at biostatisticians and epidemiologists. See also Ref. 16 below).

15. Clayton, D. G. (1991). A Monte Carlo method for Bayesian inference in frailty models. *Biometrics*, **47**, 467–485.

16. Clayton, D. G. (1994). Some approaches to the analysis of recurrent event data. *Statist. Methods Med. Res.*, **3**, 244–262.

17. Clayton, D. G. and Cuzick, J. (1985). Multivariate generalizations of the proportional hazards model (with discussion). *J. Roy. Statist. Soc. A*, **148**, 82–117.

18. Coleman, J. S. (1964). *Introduction to Mathematical Sociology*. Free Press, Glencoe, IL.

19. Cox, D. R. (1972). Regression models and life tables (with discussion). *J. Roy. Statist. Soc. B*, **34**, 187–220. (The seminal paper on proportional hazards.)

20. Cox, D. R. (1975). Partial likelihood. *Biometrika*, **62**, 269–276.

21. Fienberg, S. E., Singer, B., and Tanur, J. M. (1985). Large-scale social experimentation in the United States. In *A Celebration of Statistics, the ISI Centenary Volume*, A. C. Atkinson and S. E. Fienberg, eds. Springer-Verlag, New York.

22. Flinn, C. and Heckman, J. (1983). The likelihood function for the multistate—multiepisode model in "models for the analysis of labor force dynamics." In *Advances in Econometrics*, R. Basman and G. Rhodes, eds. JAI Press, Greenwich, CT., 1983.

23. Guo, G. and Rodrıguez, G. (1992). Estimating a multivariate proportional hazards model for clustered data using the EM algorithm, with an application to child survival in Guatemala. *J. Am. Statist. Assoc.*, **87**, 969–976.

24. Heckman, J. J. and Singer, B. (1984). A method for minimizing the impact of distributional assumptions in econometric models for duration data. *Econometrica*, **52**, 271–320. (An influential paper proposing nonparametric estimation of frailty effects.)

25. Heckman, J. J. and Singer, B. (1986). Econometric analysis of longitudinal data. In *Handbook of Econometrics*, Vol. 3, Z. Griliches and M. Intriligator, eds. Elsevier Science, New York. (An excellent review aimed at economists.)

26. Heckman, J. J. and Walker, J. R. (1987). Using goodness of fit and other criteria to choose among competing duration models: A case study of Hutterite data. In *Sociological Methodology 1987*, C. C. Clogg, ed., pp 247–307, Jossey Bass, San Francisco, CA.

27. Heckman, J. J. and Walker, J. R. (1992). Understanding third births in Sweden. In *Demographic Applications of Event History Analysis*, J. Trussell, R. Hankinson, and J. Tilton, eds., pp. 157–208, Clarendon Press, Oxford, UK.

28. Holford, T. R. (1980). The analysis of rates and survivorship using log-linear models. *Biometrics*, **36**, 299–306.

29. Johansen, S. (1983). An extension of Cox's regression model. *Int. Statist. Rev.*, **51**, 258–262.

30. Kalbfleish, J. D. and Prentice, R. L. (1980). *The Statistical Analysis of Failure Time Data*. Wiley, New York.

31. Kay, R. (1984). Multistate survival analysis: An application in breast cancer. *Methods Infer. Med.*, **23**, 157–162.

32. Lagakos, W., Sommer, C. J., and Zelen, M. (1978). Semi-Markov models for partially censored data. *Biometrika*, **65**, 311–318.

33. Laird, N. and Olivier, D. (1981). Covariance analysis of censored survival data using log-linear analysis techniques. *J. Am. Statist. Assoc.*, **76**, 231–240. (Piecewise exponential survival using Poisson models; see also Ref. 28.)

34. Lancaster, T. (1990). *The Econometric Analysis of Duration Data*. Cambridge University Press, Cambridge, UK.

35. Lawless, J. F. (1987). Regression methods for Poisson process data. *J. Am. Statist. Assoc.*, **82**, 808–815.

36. Littlewood, B. (1975). A reliability model for systems with Markov structure. *Ann. Statist.*, **24**, 172–177.

37. Manton, K. G., Singer, B., and Woodbury, M. A. (1992). Some issues in the quantitative characterization of heterogeneous populations. In *Demographic Applications of Event History Analysis*. J. Trussell, R. Hankinson, and J. Tilton, eds., pp. 9–37, Clarendon Press, Oxford, UK.

38. Manton, K. G., Woodbury, M. A., and Tolley, H. D. (1994). *Statistical Applications Using Fuzzy Sets*. Wiley, New York.

39. Nelder, J. A. and Wedderburn, R. W. M. (1972). Generalized linear models. *J. Roy. Statist. Soc. A*, **135**, 370–384. (The original paper introducing the concept.)

40. Newman, J. and McCullogh, C. (1984). A hazard rate approach to the timing of births. *Econometrica*, **52**, 939–961.

41. Nielsen, G. D., Gill, R. D., Andersen, P. K., and Sorensen, T. (1992). A counting process approach to maximum likelihood estimation in frailty models. *Scand. J. Statist.*, **19**, 25–43.

42. Petersen, T. (1995). Analysis of event histories. In *Handbook of Statistical Modeling for the Social and Behavioral Sciences*, G. Arminger, C. C. Clogg, and M. E. Sobel, eds., pp. 453–517, Plenum Press, New York. (An excellent up-to-date review for social scientists.)

43. Prentice, R. L., Williams, B. J., and Peterson, A. V. (1981). On the regression analysis of multivariate failure time data. *Biometrika*, **68**, 373–379.

44. Rodrıguez, G. (1994). Statistical issues in the analysis of reproductive histories using hazard models. In *Human Reproductive Ecology: Interactions of Environment, Fertility, and Behavior*, K. L. Campbell and J. W. Wood, eds., pp. 266–279, New York Academy of Sciences, New York.

45. Rodrıguez, G., Hobcraft, J., McDonald, J., Menken, J., and Trussell, J. (1984). A comparative analysis of the determinants of birth intervals. *Comparative Studies 30*, World Fertility Survey.

46. Sinha, D. (1993). Semiparametric Bayesian analysis of multiple event-time data. *J. Am. Statist. Assoc.*, **88**, 979–983.

47. Stacy, E. W. (1962). A generalization of the gamma distribution. *Ann. Math. Statist.*, **33**, 1187–1192.

48. Therneau, T. M., Grambsch, P. M., and Fleming, T. R. (1990). Martingale-based residuals for survival models. *Biometrika*, **77**, 147–160.

49. Trussell, J., Hankinson, R., and Tilton, J., eds. (1992). *Demographic Applications of Event History Analysis*. Clarendon Press, Oxford. (A collection of demographic papers including three "contests," where pairs of research teams address the same topics using the same data.)

50. Trussell, J. and Richards, T. (1985). Correcting for unmeasured heterogeneity in hazard models using the Heckman–Singer procedure. In *Sociological Methodology 1985*, N. B. Tuma, ed., pp. 242–276, Jossey Bass, San Francisco, CA.

51. Tuma, N. B. and Hannan, M. T. (1984). *Social Dynamics. Models and Methods.* Academic Press, Orlando, FL. (A pioneering book.)

52. Tuma, N. B., Hannan, M. T., and Groeneveld, L. P. (1976). Dynamic analysis of event histories. *Am. J. Socio..*, **84**, 820–854. (Introduced the term "event history.")

53. Vaupel, J. W., Manton, K. G., and Stallard, E. (1979). The impact of heterogeneity in individual frailty on the dynamics of mortality. *Demography*, **16**, 439–454. (The original paper on frailty models.)

54. Voelkel, J. G. and Crowley, J. (1984). Nonparametric inference for a class of semi-Markov models with censored observations. *Ann. Statist.*, **12**, 142–160.

55. Yamaguchi, K. (1991). *Event History Analysis*. Sage Publications, Newbury Park, CA.

56. Zeger, S. L. and Karim, R. M. (1991). Generalized linear models with random effects: a Gibbs sampler approach. *J. Am. Statist. Assoc.*, **86**, 79–86. (Bayesian inference in generalized linear models using Markov chain Monte Carlo methods.)

57. Zucker, D. M. and Karr, A. F. (1990). Nonparametric survival analysis with time-dependent covariate effects: A penalized partial likelihood approach. *Ann. Statist.*, **18**, 329–353.

27 FDA Statistical Programs: An Overview

Charles Anello

At the beginning of the twentieth century [9,13], the Congress of the United States began enacting legislation aimed at protecting the public health against ineffective and/or harmful products. With each new law, the mission of the US Food and Drug Administration (FDA) became better focused. Today, FDA is an important regulatory agency. Many of the FDA's regulatory decisions depend on the quality of the statistical evidence presented to FDA. Consequently, a variety of statistical programs have emerged that enable FDA to carry out its assigned mission. This chapter provides an overview of the FDA's statistical activities in which FDA is engaged. The size of the statistical staff within the various FDA bureaus ranges from one individual in the Bureau of Biologics to over two dozen in the Bureau of Drugs and the Bureau of Foods. Consequently, the level of statistical activity and research varies considerably with the administrative units. For those units where the level of statistical activity is high, examples are presented and a bibliography provided.

In 1906, Congress passed the original Food and Drug Act and a companion bill, the Meat Inspection Act. These laws prohibited the transfer of misbranded and adulterated foods, drinks, and drugs from interstate commerce. This program was administered by Harvey W. Wiley, Chief of the Division of Chemistry, US Department of Agriculture. In 1938, a drug (elixir of sulfanilamide) was associated with the death of 107 persons. This resulted in a new Federal Food, Drug, and Cosmetic Act. The FD&C Act, as it was now called, required the manufacturer of new drugs to show that their products were safe prior to marketing, and required the manufacturer to specify the standards of identity and quality for foods. The FD&C Act allowed FDA to begin the practice of factory inspections and to extend its authority to cover cosmetics and devices. In 1962, Frances O. Kelsey, an FDA medical officer, helped keep the drug Thalidomide off the US market. During the early 1960s, this drug was associated with numerous malformed babies in western Europe. The thalidomide tragedy prompted the US Congress to pass the Kefauver–Harris Drug Amendment, which, for the first time, required drug manufacturers to prove to FDA that their products were effective prior to marketing. During the 1970s FDA's role expanded to include regulation of medical devices, diagnostic and laboratory products, radiation from electronic products, and serum and vaccines. FDA also collaborates with other government agencies [notably the Environmental Protection Agency (EPA)] in the area of toxicologic research.

FDA's statistical programs emphasize the protection of human subjects; the design and analysis of studies of controlled experimental studies; epidemiologic studies, surveys, quality control techniques; and assessment of animal and laboratory studies. Many areas of statistical theory and applications arise within FDA, and the agency statisticians frequently collaborate with scientists from other disciplines. A heavy emphasis is placed on the effectiveness of experimental design (concepts like statistical power are important), the quality of statistical evidence (e.g., the potential for nonsampling errors or bias), and the extent and validity of inference possible. The various FDA statistical programs will be summarized by the organizational unit in which they arise.

The primary responsibility of the statistical staff in the *Bureau of Drugs* is to work with the physicians, chemists, and pharmacologists as a review team member in the evaluation of industry-submitted new drug applications, to design and analyze surveys on the quality of marketed drugs, to participate in postmarketing surveillance programs, and to participate in pharmaceutical research.

The *Bureau of Veterinary Medicine* (BVM) is responsible for the safety and efficacy of animal drugs, feed additives, and animal medical devices. Primary among the Bureau's duties is the responsibility to assure that the public is not exposed to harmful residues in edible animal products as a result of drugs given to food-producing animals. This responsibility is carried out in conjunction with the FDA Bureau of Foods and the US Department of Agriculture. The BVM statistical staff are involved in the review and evaluation of protocols and results of studies dealing with the safety and efficacy of marketed veterinary products. Like the Bureau of Drugs, the BVM statisticians are involved in experimental design, and in the appropriate use of statistical methods to reach valid conclusions. Of special interest is the BVM's work on the statistical aspects of low-level exposure to subacute toxicants, such as carcinogens.

The *Bureau of Medical Devices* (BMD) regulates a variety of manufactured products intended for use in medical diagnosis and therapy, including items such as cardiac pacemakers, intraocular lenses, pregnancy test kits, and blood analyzers. The BMD assures their safety and efficacy through the premarket approval, performance standards, and labeling requirements. Each of these requires studies for the development of scientific evidence. The BMD statistics staff have the general responsibility of assuring that evidence presented to or developed by the Bureau is adequate and statistically valid. Premarket approval applications are reviewed and may incorporate in vitro, animal, and clinical studies. Standards for devices incorporate carefully designed test procedures and sampling plans. Diagnostic devices require statistical models to describe their performance characteristics, such as accuracy, precision, linearity, and detection limit. Appropriate statistical techniques are necessary for determining reference values. The statistics staff also design laboratory experiments and protocols and analyze data for testing device performance.

The statistical activities of the *Bureau of Radiological Health* (BRH) are varied and wide ranging. BRH has as its functions the development of criteria and recommendation of standards for safe limits of radiation exposure, the development of methods and techniques for controlling radiation exposure, planning and conduct of research to determine health effects of radiation exposure, and the conduct of studies to ascertain the needs of educational programs to improve the practice of radiation users. BRH uses data generated by surveys, experimental, and epidemiologic research.

The major statistical work in BRH is in epidemiological and animal research on the effects of radiation on biologic systems. Mathematical modeling is used in the determinations of dose–response functions and risk assessment of long-term effects. The BRH statisticians participate in the determination of use characteristics of medical radiation procedures by initiating national surveys. The statistical problems associated with engineering research and testing are some of the Bureau's more interesting ones. A high level of expertise is needed to deal with issues of underlying probability distributions of dose, development of extreme value distributions, the estimation of the potential health implications, and the evaluation of data from experimental and epidemiologic studies.

Two examples of the statistical research undertaken by the BRH are "A risk/benefit analysis by life table modeling of an annual breast cancer screening program which includes x-ray mammography" by Chiacchierini et al. [2] and "A two-sample test for independence in 2×2 contingency tables with both margins subject to misclassification" by Chiacchierini and Arnold [1]. The first paper presents a life table model to display and numerically evaluate risk and benefits by comparing breast cancer mortality experience in a hypothetical population of 100,000 women age 35 that differ only by age at first screening. The second paper considers a double-sampling scheme for the estimations and testing of 2×2 contingency tables under misclassification in both margins. Maximum-likelihood estimates are given for the error-free and misclassification probabilities.

The *National Center for Toxicologic Research* (NCTR) has as its mission the study of the toxic effects of long-term, low-dose exposure to toxic substances, and the study of the biologic mechanisms of toxicity in order to develop more effective protocols to study safety and to improve prediction of human disease from laboratory animal experiments. These objectives are accomplished by experiments aimed at carcinogenesis, teratogenesis, and/or mutagenesis in a variety of laboratory animal species.

The statistics staff of NCTR participates in the design, analysis, and interpretation of these toxicity studies. They utilize a variety of statistical methods and develop new techniques directed at the analysis of long-term toxicologic studies, the adjustment for competing risks, and the estimation of tumor, teratogenic, or mutagenic incidence rates. They often use special statistical procedures required for ratios of discrete random variables (e.g., number of abnormal fetuses divided by the number of viable fetuses per litter) and techniques for estimating risk at low exposure levels from higher observed experimental dose levels.

Research conducted by the staff of NCTR includes the estimation of time response models such as those presented by Farmer et al. [11] in a paper titled "Dose and time response models for the incidence of bladder and liver neoplasms in mice fed 2-acetylaminofluorene continuously" and presented by Kodell et al. [14] in a paper titled "Estimation of distributions of time to appearance of tumor and time to death from tumor after appearance in mice fed 2-acetylaminofluorene." Kodell et al. [15] developed a nonparametric joint estimation procedure for estimating disease response and survival functions in experiments where animals are sacrificed [3]. Finally, Gaylor and Kodell [12] developed a method for linear interpolation of low-dose risk assessment of toxic substances.

The *Bureau of Biologics* (BOB) is concerned with toxoids, vaccines, antigens, immune sera, allergenic products, and blood and blood products used to treat, prevent, or diagnose disease. The statisticians in the BOB participate in evaluating studies designed to establish the potency and safety of biologic products; they set standards and develop methods of testing. BOB issues licenses of suitable products, and maintains an updated computerized inventory of each product. The staff of BOB review the quality and integrity of information submitted, particularly clinical, laboratory, and production data. BOB develops methodology, conducts tests of safety, efficacy, and potency of vaccines, and surveys possible long-term effects of immunization against viral and rickettsial disease.

The Division of Mathematics in the *Bureau of Foods* is concerned with mathematical and statistical models in the field of sanitation, health, and economics, and in the design and analysis of studies related to food standards, food, and color additives. The *Office of Health Affairs* in BRH is responsible for the review of methodologies used in the collection of FDA's consumer health statistics (obtained through contracts with private firms, universities, or interagency agreements.) The primary focus of this FDA unit is the methodologic review of consumer surveys and the coordination of survey plans with the Office of Management and Budget (OMB). (OMB has the unique role of limiting the response burden on the public from overlapping surveys.) The surveys conducted by OHA/BRH deal with all agency products (including foods, drugs, biologics, etc.) where a rapid turnaround is required. Several references 3–8,10 to specific surveys appear in the bibliography. The survey mechanism is a standing contract which provides for alternative sampling designs to meet varying survey needs. The consumer used in this context is broadly defined and includes the general population and speciality groups (e.g., physicians, dentists, and related health professionals). The surveys are used to provide FDA with reliable information concerning consumers' experiences with and attitudes toward FDA-regulated products.

Finally, the *Office of Scientific Liaison Staff* in the Office of the Commissioner of Food and Drugs has scientists with varying expertise, including biostatistics. The statisticians supporting the Commissioner are involved in developing procedures useful in the assessment of human health risks, especially of carcinogenicity and teratogenicity; statistical models and methods appropriate to these investigations are employed.

Acknowledgments. The following individuals contributed information related to their specific bureaus: V. M. Chinchilli, Scientific Liaison Staff; E. W. Gordon, Bureau of Radiologic Health; H. Lee, Bureau of Medical Devices; J. J. Colaianne, Bureau of Veterinary Medicine; S. C. Rastogi, Bureau of Biologics; R. P. Chiacchierini, Bureau of Radiological Health; D. W. Gaylor, National Center for Toxicologic Research.

References

1. Chiacchierini, R. P., and Arnold, J. C. (1977). *J. Am. Statist. Assoc.*, **72**, 170–174.
2. Chiaccierini, R. P., Lundin, F. E., and Scheidt, P. C. (1980). In *Prevention and Detection of Cancer*, Part II: *Detection*, Vol. 2: *Cancer Detection in Specific Sites*, H. E. Nieburgs, ed., pp. 1741–1762. Marcel Dekker, New York.

3. DHEW, FDA (1965). *Consumer Attitudes toward Over-the-Counter Drug Labels*, Feb.

4. DHEW, FDA (1975). *An Investigation of Consumers' Perceptions of Adverse Reaction to Cosmetic Products*, June.

5. DHEW, FDA (1975). *Survey of Consumers' Perceptions of Patient Package Inserts for Oral Contraceptives*, Sept.

6. DHEW, FDA (1976). *Influenza Consumer Survey*, Feb.

7. DHEW, FDA (1978). *An Evaluation of the Awareness and Reactions to Labeling Changes among Users of In Vitro Diagnostic Products*, Sept.

8. DHEW, FDA (1978). *Comprehension of Selected Labeling Information on Foods, Drugs, and Medical Devices: A Consumer Survey*, Nov.

9. DHEW, FDA (1979). *Milestones in U.S. Food and Drug Law History*, FDA Consumer Memo No. 79-1063.

10. DHEW, FDA and CDC (1978). *A National Survey of the Use of Protein Products in Conjunction with Weight Reduction Diets among American Women*.

11. Farmer, J., Kodell, R. L., and Greenman, D. L. (1979). *J. Environ. Pathol. Toxicol.*, **3**, 55–68.

12. Gaylor, D. W. and Kodell, R. L. (1980). *J. Environ. Pathol. Toxicol.*, **4**, 305–312.

13. Janssen, W. F., FDA historian (1978). From the article "The Food and Drug Law of the United States," *Ind. Santé*, Paris; HEW Publ. (FDA) 79–1054.

14. Kodell, R. L., Farmer, J. H., and Greenman, D. L. (1979). *J. Environ. Path. Toxicol.* **3**, 89–192.

15. Kodell, R. L., Shaw, G. W., and Johnson, A. M. (1982). *Biometrics*, **38**, 43–58.

28 FDA Statistical Programs: Human Drugs

Satya D. Dubey

28.1 Official Statistical Responsibilities

In this chapter statistical activities pertaining to human drugs in the Bureau of Drugs of the Food and Drug Administration (FDA) are described with emphasis on official responsibilities the role of statistics, guidelines, and selected significant problems and research.

The Statistical Program Pertaining to Human Drugs is located in the Division of Biometrics, Office of Biometrics and Epidemiology, Bureau of Drugs, FDA. The Bureau of Drugs develops FDA policy with regard to the safety, effectiveness, and labeling of all drugs for human use. It reviews and evaluates new-drug applications (NDAs) and notices of claimed investigational (exemption for) new drugs (INDs) for human use; develops and implements standards for the safety and effectiveness of all over-the-counter (OTC) drugs; monitors the quality of marketed drugs through product testing, surveillance, and compliance programs; develops and promulgates guidelines on current good manufacturing practices for use by the drug industry; develops and disseminates information and educational material dealing with drugs to the medical community and the public; conducts research and develops scientific standards on the composition, quality, safety, and efficacy of human drugs; collects and evaluates information on the effects and use trend of marketed drugs; monitors prescription drug advertising and promotional labeling to ensure their accuracy and integrity; analyzes data on accidental poisonings and disseminates toxicity and treatment information on household products and medicines; evaluates applications for operation of methadone treatment centers and other activities using methadone or other drugs; and directs the FDA antibiotic and insulin certification program.

These activities routinely produce data from which valid inferences are drawn for regulatory decisions. Data are obtained from clinical trials, laboratory research and experiments, surveys, and epidemiological studies. Consequently, the Division of Biometrics has responsibility for initiating, planning, developing, and implementing the statistical program in the Bureau of Drugs. It provides comprehensive statistical computational and biomathematical consulting services, which include, on a routine basis:

1. The reviews and evaluations of the statistical adequacy of experimental designs for IND or NDA submissions, medical research reports and evidence submitted in support of drug safety and efficacy claims, critiques of statistical analyses conducted in studies related to clinical trials, medical advertising, OTC drugs, drug surveillance, and quality control

2. Collaboration on the statistical design and analysis of morbidity and mortality data associated with the use of drugs

3. Evaluation of statistical methods applied in epidemiological studies; application of statistical methods in support of pharmaceutical research and testing

4. Statistical support in the evaluation of experimental design and bioavailability data

5. Development of sampling plans for field inspection programs and statistical methods for evaluating manufacturing quality control procedures for marketed drugs, development of appropriate statistical methods for conducting, monitoring, and evaluating intramural and extramural research contract projects; evaluation of the quality of data and conclusions presented by interested parties

6. Statistical research on methods and applications

Computational activities include development of scientific computer programs, validation or modification of current statistical methodology programs, and preparation of data for executing existing computer programs on a routine basis.

28.2 Role of Statistics

The 1962 Amendments to the Food, Drug, and Cosmetic Act [6] require that no application to market a new drug shall be approved unless adequate tests show that the drug is safe for use under the conditions presented in the proposed labeling, and substantial evidence from adequate and well-controlled studies demonstrates that the drug is effective for such uses.

Certain essential principles concerning "adequate and well-controlled clinical investigations" appear in the May 8, 1970, *Federal Register Statement* [1, Sec. 314.111(a)(5)(ii)]. These principles are also summarized in the 1977 HEW publication [7]. They include a clear statement of the objective(s) of the study, patient selection criteria, randomization procedure, suitable size of a clinical study, comparability of the patient population studied, concurrent comparison groups, nature of the blinding, methods of observations and quantification, and suitable length of study. Adequate documentation of the design, quality, and accuracy of the clinical data; appropriate use of statistical models; analyses (references of publications used in statistical work); and interpretation of results is required. These principles define the role of statisticians in applying sound statistical theory and methodology to ensure the adequacy of well-controlled clinical investigations. Specifically, statisticians are consulted in the planning, design, execution, and analysis of clinical investigations and clinical pharmacology in order to ensure the validity of estimates of safety and efficacy obtained from these results. They are consulted in drawing valid inferences about drug responses in well-defined target populations.

A statistician in the Bureau of Drugs reviews and evaluates a study protocol to determine whether the proposed study design and statistical procedures:

1. Satisfy the requirements of an adequate and well-controlled clinical investigation

2. Are efficient for the stated objectives without exposing human subjects to unnecessary risks

3. Adequately plan for obtaining valid and accurate data and for monitoring and reporting the study

He or she is a member of the drug review team and is responsible for reviewing and evaluating all submissions that use statistical analyses to make inferences from data. The development of a meaningful review necessitates effective interaction between medical and statistical reviewers and/or other scientists throughout the review process [10,11]. Outside experts are used selectively to perform statistical reviews of IND/NDA submissions, and of published and unpublished medical reports. Prominent biostatisticians are used as voting members or consultants to the various scientific advisory committees to advise on crucial statistical issues. See References. 1, 3, 4, and 7–11.

28.3 Statistical Guidelines

One general statistical guideline has been developed to evaluate the clinical protocols and completed clinical studies done by drug sponsors. Another has been proposed for the use of drug sponsors engaged in the activities of protocol development and NDA submissions. The *IND/NDA Statistical Evaluation Guidelines* became the official document of the Bureau of Drugs on February 23, 1979 [10], for statistical review and evaluation of protocols, investigational new drugs (INDs), new-drug applications (NDAs) including drug efficacy study implementation (DESI), supplemental NDAs, and similar submissions. It is available to outsiders when requested under the Freedom of Information Act. On July 8, 1980, the FDA made available the *General Statistical Documentation Guide for Protocol Development and the NDA Submissions* [8], which sets forth the type of material needed to permit statistical review of protocols and completed clinical studies by the agency.

28.4 Selected Clinical Statistical Problems

Combination drugs is a fruitful area for innovative statistical research. A combination drug consists of two or more chemical ingredients in a fixed quantity that are combined (or mixed) in one dosage form and administered together in a single tablet, etc. For example, for the treatment of muscle spasm and pain caused by strain or sprain or due to a chronic condition, a muscle-relaxing agent may be combined with a pain-reducing agent.

Two or more drugs are combined to achieve

1. An increase in efficacy without increasing adverse effects

2. A reduction in adverse effects without any reduction in efficacy

3. An increase in efficacy and decrease in adverse effects simultaneously

4. A decrease in efficacy and decrease in adverse effects

5. An increase in efficacy and an acceptable increase in adverse effects in order to maximize overall benefits to patients (a muscle relaxant plus an analgesic results in increase in efficacy and increase in adverse effects)

Consider a combination drug with two single ingredients, the first intended for fast onset of drug effect and the second for long duration of action. An example would be a topical anaesthetic intended for dental use where quick onset and lasting anaesthetic effect is desirable.

Sometimes, a third ingredient is developed to provide additionally an increasing percentage of patients experiencing rapid onset of action. Thus combination drug problems offer a challenging area for multivariate statistical research. *Multiple comparisons* and a multiplicity of statistical hypotheses to be tested is not uncommon in clinical research.

In bioavailability studies, there are four basic variables of interest:

1. Area under the plasma-level curve (AUC)

2. Time to achieve first peak

3. First peak plasma level

4. Maximum peak plasma level

Initially, appropriate univariate analyses of variance are carried out on each of these four variables. Since these variables are generally correlated, a correlation matrix for them is computed; if any of them is not well correlated with the remaining three variables while they themselves are found to be highly correlated, a multivariate analysis of variance is performed utilizing only highly correlated variables. Additionally, in order to determine whether a test agent has any statistically significant effect on certain physiological variables of interest, each such variable is analyzed in the multivariate analysis of covariance model. Measurements are taken at the baseline before and after a test agent is administered to selected subjects. Thus the values obtained at n specified time points for each physiological variable are used as the response variables in a $\mathbf{Y}$ matrix; the baseline value is used as a covariate in a multivariate analysis of covariance model, which is formulated as

$$\begin{aligned} \mathbf{Y} = & \; \boldsymbol{\mu} + X\mathbf{1} + \mathbf{O}_i + \mathbf{D}_j + \mathbf{S}_{(ij)k} \\ & + \mathbf{P}_l + \mathbf{T}_m + \mathbf{DT}_{jm} \\ & + \mathbf{PT}_{lm} + \boldsymbol{\epsilon}, \end{aligned}$$

where $\mathbf{Y}$ is an $n \times 1$ vector of response variables (measurements taken at n specified timepoints) for a specified physiological variable of interest (e.g., systolic blood pressure) for a subject assigned to a specific test agent and period,

$\boldsymbol{\mu}$ is an $n \times 1$ vector of n average effects associated with n specified timepoints.

X is a baseline measurement (scalar), for a specified physiological variable of interest, taken prior to the administration of a specific test agent for a specific period providing $X\mathbf{1}$ as an $n \times 1$ vector with $\mathbf{1}$ being a vector of 1s.

$\mathbf{O}_i$ is the effect due to *order* (sequence) in which the test agents are administered at assigned periods.

$\mathbf{D}_j$ is the effect due to the *day* on which the selected subject starts the study.

$\mathbf{S}_{(ij)k}$ is the effect due to *subjects* nested within order (i) and day (j).

$\mathbf{P}_l$ is the effect due to the *period* of the crossover design.

$\mathbf{T}_m$ is the effect due to the *test agent* selected for the study.

$\mathbf{DT}_{jm}$ is the interaction effect between the day and test agent.

$\mathbf{PT}_{lm}$ is the interaction effect due to period and test agent.

$\boldsymbol{\epsilon}$ is an $n \times 1$ error vector associated with the model, defined above.

The number of selected test agents (m) will generally determine the number of periods (l) and the number of sequences or order (i). The number of days (j) and the number of subjects nested (k) within order (i) and day (j) will be generally determined on practical grounds, keeping in mind that each day a sequence of test agents is repeated at least once. Subjects selected for the study are randomized with respect to day and sequence.

Since healthy human beings are used in *bioavailability* or *bioequivalence studies*, sources and magnitudes of observed variability are generally not as much as for severely sick patients used in clinical studies. Therefore, multiresponse multiperiod cross-over design and multivariate covariance analysis often respond well. In many instances, a multiple or sequential range test is applied to make clinically meaningful pairwise comparisons.

In a multiclinic trial different clinics serve different patient subpopulations and use slightly different techniques, despite a common protocol. Analyses based on data combined across investigators should be accompanied by summaries for each investigator and a demonstration of the appropriateness of combining the results.

It is unethical to subject people to risk when a study is too small and cannot be expected to demonstrate the effect desired, or to subject more people to risk than is necessary after the effect has been clearly demonstrated. Clinical investigations are expensive in terms of dollars, time, the commitment of patients and physicians to the study, and physical facilities. During the planning phase the statistician consults with the clinical investigator to determine the size of the difference that is important to detect, the significance level and power of the statistical test, the duration of the study, the acceptable dropout rates, the multiple end-points, and the minimum number of patients necessary for the study.

The interpretation of a study's findings require effective consultation with a clinical investigator. A statistically significant finding may have nothing to do with the drug of interest. It could be due to chance, to nonsampling error, to bias, to an indirect effect not specified, etc. Even if the observed drug effect could be attributed to a particular drug, we need to ask: Do the benefits outweigh the risks? Is the observed difference clinically meaningful?

28.5 Some Nonclinical Statistical Problems

FDA has published several regulations (21 CFR 211) that relate to statistical principles and methodology under the title *Current Good Manufacturing Practices for Finished Pharmaceuticals* [1]. The *United States Pharmacopeia* (USP XX) recommends a method for conducting a content uniformity test [15]. The test is conducted in two stages, the first involving 10 and the second 20 tablets (the second stage may not be necessary). Comer et al. have published papers on the statistical aspects of content uniformity test [2]. They discuss the advantages and disadvantages of attribute measures for setting quality standards. They develop mathematical models for determining the operating characteristics of the USP Content Uniformity Test and suggest how other available information can be used to suggest statistically more powerful procedures.

The pyrogen test (see Ref. 15), sampling and testing of in-process materials and drug products (Sec. 211.110 of Ref. 1), and stability testing (Sec. 211.166 of Ref. 1) involve statistical concepts and procedures. Interest in problems in the drug products quality area has steadily increased, while the demand for statistical services in epidemiological projects, such as postmarketing drug surveillance studies, has grown rapidly.

28.6 Applied Statistical Research

The Division of Biometrics encounters unusual statistical problems that cannot be readily handled by available statistical approaches. These problems provide an opportunity for staff to conduct applied statistical research and publish it in scientific journals. Selected publications by statisticians of the Division of Biometrics are listed in the references.

O'Neill and Chen [13] developed a statistical model that characterizes the onset and termination of treatment response of a subject over a fixed time period jointly. The model was applied to compare the effect of two bronchodilator drugs in subjects with reversible chronic obstructive lung disease.

Fairweather [5] published a statistical procedure for determining the risk of rejecting a satisfactory product when dissolution tests are substantiated for classic blood level studies. One purpose for this kind of investigation is to determine if currently used in vivo methods could be safely replaced by in vitro methods.

O'Neill and Anello [12] proposed a sequential approach to case–control studies. The case–control method has emerged in more recent years as an important epidemiologic tool for studying the potential adverse effects of marketed drugs.

Schuirmann analyzed a collaborative study of the USP Prednisone Dissolution Calibrator in Ref. [14]. The study was undertaken to determine the contributions of several factors to the variability of dissolution values and to determine statistically a range of values within which dissolution results may be expected to fall when obtained under certain conditions.

28.7 Notes

1. This work has been produced by Satya D. Dubey, the author, in the capacity of a US federal government employee, as part of his official duty, is in the public domain, and is not subject to copyright.

2. In 1982, the Bureau of Drugs and the Bureau of Biologics of the FDA were merged, and became the *National Center for Drugs and Biologics*.

References

1. *Code of Federal Regulations, Title 21, Food and Drugs*, Parts 1–499, rev. April 1980. Part 211, pp. 72–89: Sec. 211.110, p. 80; Sec. 211.166, pp. 83–84; and Sec. 314.111 (a)(5)(ii), pp. 106–108 (abbrev. as 21 CFR 314.111). US Government Printing Office, Washington, DC.

2. Comer, J. P., et al. (1970). *J. Pharmacol. Sci.*, **59**, 210–214.

3. Daniel, C., Tutle, E. R., and Kadane, J. A., Federal Statistics (1971). *Statistics and Data Analysis in the Food and Drug Administration. Report of the President's Commission*, Vol. 2, pp. 67–95.

4. Dubey, S. (1981). *Statistics in the Pharmaceutical Industry*, C. Buncher and J. Tsay, eds., Chap. 5, Marcel Dekker, New York.

5. Fairweather, W. R. (1978). *J. Pharmacokinet. Biopharm.*, **5**(4), 405–418.

6. *Federal Food, Drug, and Cosmetic Act as amended* (1962), Sec. 505(d). US Government Printing Office, Washington, DC.

7. *General Considerations for the Clinical Evaluation of Drugs* (1977). DHEW Publ. No. (FDA) 77-3040, pp. 2–4, US Government Printing Office, Washington, DC, Sept.

8. *General Statistical Documentation Guide for Protocol Development and NDA Submissions* (Draft) (1980). Guideline Leader: S. D. Dubey, Chief, Statistical Evaluation Branch, Division of Biometrics. Statistical Evaluation Branch (HFD-232), Bureau of Drugs, Food and Drug Administration, Department of Health and Human Services, April.

9. *Guidelines for the Clinical Evaluation of Anti-inflammatory Drugs (Adults and Children)* (1977). DHEW Publ. No.

(FDA) 78-3054, pp. 24–25, US Government Printing Office, Washington, DC, Sept.

10. *IND/NDA Statistical Evaluation Guidelines* (Staff manual Guide BD 4500.1) (1979). Originator: S. D. Dubey. Food and Drug Administration, Bureau of Drugs, Feb. 23.

11. O'Fallon, J. R., Dubey, S. D., Salsburg, D. S., Edmonson, J. H., Soffer, A., and Colton, T. (1978). *Biometrics*, **34**, 687–695.

12. O'Neill, R. T. and Anello, C. (1978). *Am. Epidemiol.*, **108**(5), 415–424.

13. O'Neill, R. T. and Chen, C. W. (1978). *Biometrics*, **34**, 411–420.

14. Schuirmann, D. J. (1980). *Pharmacopeial Forum*, **6**(1), 75–89.

15. *The United States Pharmacopeia* (20th rev.) (1980), pp. 902–903, Published by the 1979 United States Pharmacopeial Convention, Inc., 12601 Twinbrook Parkway, Rockville, MD 20852; distributed by Mack Publishing Company, Easton, PA 18042.

29

Follow-Up

Lee J. Melton III

29.1 Introduction

Follow-up is the process of locating research subjects and determining whether some outcome of interest has occurred. Follow-up can be directed at a variety of endpoints, can be carried out concurrently or through historical records, can be conducted once or sequentially, and can be of short or long duration. The objective always remains the same, however: to maintain observation of the study subjects and to do so in a manner that avoids introducing bias.

Follow-up is an integral part of the investigation when it is required and should not be undertaken casually. The quality of the follow-up effort may determine the success of trials of therapeutic or prophylactic regimens, epidemiological cohort studies, investigations into the natural history of disease, etc., where subjects must be followed for months, years, or even decades to observe the outcome. Because of its importance, follow-up must be considered when the study is originally designed. The decisions made about the specific follow-up method to be used, the choice of endpoints to measure, and the duration of follow-up desired may have a major impact on the way the rest of the study is conducted.

29.2 Concurrent Follow-Up

Follow-up is most often concurrent; i.e., the subjects are identified in the present and then followed into the future. This method permits the nature of the follow-up to be tailored to the needs of the study, as arrangements are made to contact the subjects at intervals to assess their status. The primary disadvantage is that one may have to wait years for the outcome to be known. Further, concurrent follow-up implies a requirement for continuity of the research team (and funding) so that the necessary contact can be made as scheduled.

The actual follow-up involves two main activities: (1) locating the subject and then (2) collecting the data required for the study. Ideally, these activities are carried out using a rigorously established protocol that is thoroughly and systematically applied to everyone in the study. When this is not the case, systematic errors are easily introduced. These can cause both very severe and extremely subtle problems; see Feinstein [2].

The first step in avoiding such bias is to minimize the loss of subjects by locating each one. This is important because the subjects who are successfully followed until the termination of the study are frequently quite unlike those who are lost to follow-up early on or who cannot be traced at all. Individuals may be lost differentially on the basis of important study variables, including exposure to the factor of interest or risk of developing a particular outcome. For example, patients doing poorly on a specific therapeutic regimen may be overrepresented among study dropouts. The subjects who are not followed may also be more or less likely to have the outcome of interest diagnosed or detected, especially if there are major disparities in medical care available geographically or socioeconomically to the lost subjects. Subjects who are not followed have, in addition, a reduced opportunity to report the occurrence of any particular outcome should it occur. Because the characteristics of the subjects who are easily followed may differ from those lost to follow-up, a low percentage of complete follow-up (less than 80% or 90%) may cast doubt on the validity of conclusions drawn from the investigation.

In terms of locating the subjects, concurrent follow-up studies have two important advantages over historical follow-up. First, all the subjects will have been located at least once for the initial evaluation. Detailed data, such as the identity of next of kin who will always know the subject's whereabouts, can be collected at that time specifically to simplify the task of subsequent follow-up. Second, regular contact with each subject can ensure that changes in name (especially important for women) or address can be identified as they occur. After many years, it becomes extremely difficult to relocate some individuals.

After each subject has been located, one must collect the desired information. The objective of good follow-up is to collect comparable data on each individual. Problems arise when follow-up is not conducted in a uniform manner. Following one subgroup of study patients with direct examination and another indirectly with mailed questionnaires may produce distorted results since the reported frequency of even serious and distinct conditions, such as cancer, is likely to differ between the two groups. Many statistical techniques exist that allow adjustment for different durations of follow-up, but these cannot correct for biases introduced by qualitatively unequal ascertainment of outcome. The opportunity for detection of the outcome should be the same for all subjects.

The nature of the specific endpoint to be measured in determining outcome has a profound effect on the follow-up methods required. The design and implementation of instruments to collect follow-up data, either directly through reexamination of subjects or indirectly through mailed questionnaires, etc., is outside the scope of this entry but some general comments may be useful.

The endpoints to be measured customarily involve mortality (death), morbidity (illness), or a change in some subclinical physiologic or psychological parameter. Mortality is the easiest endpoint to detect through follow-up and is of great interest when survival is the outcome of interest. Mortality can be a very insensitive indicator of other outcomes, however, and should not be used for that purpose. Many cases of a particular disease can occur, for example, for each death recorded. Further, the use of mortality introduces factors associated with survival in addition to others associated with development of the disease in the first place. Because mortality is a relatively infrequent event, its choice as an endpoint generally requires a larger study; because determination of alive or dead status is relatively inexpensive, such a large study may be quite feasible. It may also be possible to take advantage of local, state, and national death registration systems to document the occurrence of death, as discussed below in connection with historical follow-up methods.

Morbidity is the endpoint most often desired in follow-up studies. If the problem of defining and diagnosing the disease can be solved, the difficulty in follow-up is to detect its occurrence. This may require examination of the subject. Reliance on the follow-up letter or questionnaire introduces major questions about the quality of the data collected except in the most specific of disease entities and, even then, steps should be taken to confirm the diagnosis. It is especially hazardous to use surrogates, such as nonspecific physiologic changes or cause of death statements on death certificates, for morbidity determinations.

Change in a physiologic or other parameter may be the most sensitive indicator of the outcome of interest; as such changes may be more frequent than frank clinical disease or death, it may be possible to design a smaller study. However, it is important to realize that an increased sensitivity may involve a concomitantly reduced specificity because not all of the measured changes will reflect the disease process of interest.

The nature of the endpoint to be measured will also dictate the duration of follow-up. If particular outcomes are infrequent or long delayed after the initial event, for example, extended follow-up will be required. The frequency of follow-up is also affected by the choice of endpoints. More frequent follow-up is required if the outcome of interest is transient, and it can improve the completeness of ascertainment and the quality of information concerning dates of events, the details of recent illness, etc. Frequent contact may also help maintain subject interest in the investigation. Regular follow-up is sometimes sacrificed to reduce costs, but this may be false economy if the overall loss to follow-up is greatly increased or the quality of the data collected is compromised.

Problems associated with prospective follow-up of many years, duration have been reported in connection with the Framingham heart study. It was found [4] that losses to follow-up due to death or refusal to be reexamined were greatest among the elderly and, in the case of death, elderly men. Losses due to emigration, on the other hand, were greater among the young. As might be anticipated, loss by death was greatest among those with certain risk factors such as hypertension. The subjects who emigrated seemed somewhat more healthy than those who remained behind. The rate of loss to follow-up was inversely related to the ease with which patients were recruited into the study in the first place. The impact of these losses varied with the endpoint being considered; mortality rates were less affected than observations on the changing natural history of hypertension. It has further been shown [3] that the apparent frequency of various cardiovascular disease endpoints depends on the source of the follow-up data. Compared to clinic data, for example, hospital information overemphasized sudden death and underrepresented angina pectoris.

29.3 Historical Follow-Up

In historical follow-up one identifies the study cohort as of some past date and follows it to the present. This method holds the promise of obtaining follow-up of long duration without having to wait. The main limitation, of course, is finding a "natural experiment" or suitable study cohort selected, described, and identified sufficiently well to support the desired investigation.

Although the activities of historical follow-up, i.e., locating the subjects and collecting data, are the same as for concurrent follow-up, the methods involved may be very different. Ascertainment of outcome is limited to measurements on cohort members who have survived to the present or to data available in existing records.

Thus, mortality is often the most feasible endpoint for this follow-up method.

However, the primary problem in historical follow-up is locating subjects. This generally entails finding a current mailing address or telephone number. Although details depend on the specific study environment, certain procedures are commonly employed. The original study records or medical records, etc., are usually reviewed first to obtain the most recent address, as well as additional leads such as the identity of parents or children or next of kin, referring physician, employer, insurance company, addresses of local relatives, names of witnesses on documents, etc., who should be contacted if the subject cannot be found directly. Correspondence in medical records may also reveal the fact that the subject has died.

A follow-up letter is usually then sent to the last known address. It may request the follow-up data or may simply introduce the research project and alert the subject to a subsequent contact. If no reply is received after a suitable length of time (often 30 days), a second letter may be sent. If the letters are unanswered or returned unclaimed, an attempt should be made to contact the patient directly. This effort can be very expensive in both personnel time and telephone tolls but is often extremely helpful in ensuring a high response rate.

If the foregoing measures have failed to locate the patient (i.e., letters returned by the post office), more imaginative steps are indicated. A search should first be made through local telephone books and city directories. Other possible sources of information are the post office, town and county clerks, unions and professional groups, alumni societies, the motor vehicle bureau, churches, the present resident at the subject's former address, former neighbors identified through a street address directory, etc. If the subject is from a small town, it may be possible to contact the bank, school, police department, or even to call every one in the telephone directory with the same last name. The services of the local credit bureau or a commercial "skip-tracing" firm may also be useful. It may be possible, depending on the nature of the investigation, to enlist the aid of a governmental agency such as the Social Security Administration, Veteran's Administration, Internal Revenue Service, etc. Although these organizations are very concerned with protecting individual privacy, follow-up letters will sometimes be forwarded to the subject without the agency providing an address directly to the research team.

If death is the outcome of interest, obituary columns can be reviewed or local funeral home directors contacted. For subjects who may have died prior to 1979, alphabetical listing of lost individuals can be sent to key state vital statistics bureaus who will usually identify any deaths and provide a death certificate for a fee. For potential deaths in 1979 and after, a similar search can be conducted nationally through the new National Death Index, recently initiated by the Division of Vital Statistics, National Center for Health Statistics.

Many of the methods described above were used in a retrospective cohort study of breast cancer risk after fluoroscopy and have been evaluated in detail by Boice [1]. Vital status was determined most often from city directories or telephone books and somewhat less often from clinical records and searches of state vital statistics data. After addresses were obtained, follow-up information was collected through questionnaires, with an increase in the cumulative response rate with each of three successive mailings and a subsequent telephone call. The characteristics of patients lost to follow-up were substantially different from those of the patients located, and crude mortality was greater for those found early than late. Breast cancer incidence was somewhat greater among those who required a third mailing or telephone call to induce a response than among the subjects who answered the first questionnaire. As Boice points out, however, the direction and magnitude of these biases can be quite different depending on the nature of the cohort being followed as well as on the end point being measured.

References

1. Boice, J. D. (1978). *Am. J. Epidemiol.*, **107**, 127–139. (This is a very detailed discussion of the utility of various follow-up methods to locate subjects for a historical follow-up study.)
2. Feinstein, A. R. (1977). *Clinical Biostatistics.* Mosby, St. Louis. (This text features extensive coverage of the many potential biases in biomedical research and is highly recommended for additional reading. Chapter 6 is perhaps most relevant to the problems associated with follow-up.)
3. Friedman, G. D., Kannel, W. B., Dawber, T. R., and McNamara, P. M. (1967). *Am. J. Public Health*, **57**, 1015–1024.
4. Gordon, T., Moore, F. E., Shurtleff, D., and Dawber, T. R. (1959). *J. Chron. Dis.*, **10**, 186–206. (This and the previous reference cover many of the important features of concurrent follow-up studies.)

30 Frailty Models

David Oakes

The term "frailty," introduced by Vaupel et al. [29], is often used in survival analysis to denote a random unobserved multiplicative factor in the hazard function. Frailty models are one device for introducing heterogeneity ("overdispersion") into survival models, and the term is sometimes used to denote other such devices. This chapter will keep to the narrow definition.

Let W denote a nonnegative random variable with distribution function $K(w)$ and Laplace transform $p(s) = E\exp(-sW) = \int \exp(-sw)dK(w)$. Suppose that conditionally on the value of $W = w$ the nonnegative random variable T has continuous distribution with hazard function

$$
\begin{aligned}
h(t|w) &= \lim_{\Delta \to 0+} \frac{1}{\Delta} \\
&\quad \times \Pr(T \leq t+\Delta | T \geq t, W=w) \\
&= wb(t)
\end{aligned}
$$

for some *baseline* hazard function $b(t)$ corresponding to a survivor function $B(t)$. Then the conditional survivor function of T given W is just $\Pr(T > t|W=w) = B(t)^w$, so that the unconditional survivor function of T is

$$
\begin{aligned}
S(t) &= E\Pr(T > t|W) = E[B(t)]^W \\
&= E\exp\{-W[-\log B(t)]\} = p(u),
\end{aligned} \quad (1)
$$

where $u = -\log B(t)$ is the integrated hazard function for the baseline distribution. For example, if $B(t) = \exp(-\rho t)$, an exponential distribution with constant hazard $b(t) = \rho$, and W has a gamma distribution with unit mean and variance κ, so that $p(s) = (1+\kappa s)^{-1/\kappa}$, then

$$
S(t) = \left(\frac{1}{1+\kappa\rho t}\right)^{1/\kappa},
$$

a Pareto distribution, with decreasing hazard function.

With no further information, the two functions $K(w)$ and $B(t)$ are not determined by the single function $S(t)$. Suppose, however, that in the spirit of Cox's [6] proportional hazards model there is also an observed covariate vector z with coefficient β whose effect on the conditional survivor function of T given $W = w$ is to multiply the hazard by $\psi = \exp(\beta z)$. Then (1) becomes

$$
S_\psi(t) = p(\psi u).
$$

In general it is *not* true that $S_\psi(t) = S_1(t)^\eta$ for some η depending on ψ but not t. The unconditional survivor functions $S_\psi(t)$ will not in general follow a proportional hazards model. This nonproportionality allows identifiability of $K(w)$ and $B(t)$ from the family $S_\psi(t)$, provided W has finite mean (see Elbers and Ridder [8]).

The only exception to nonproportionality is the family of positive stable distributions with Laplace transform $p(s) = \exp(-s^\alpha)(0 < \alpha \leq 1)$ (and infinite mean except in the degenerate case of $\alpha = 1$). For then $S_\psi(t) = S_1(t)^\eta$ with $\eta = \psi^\alpha$, and the unconditional survivor functions also follow a proportional hazards model. The frailty attenuates the effect; however, we have $\beta_\eta = \alpha\beta_\psi$ for the coefficients corresponding to η and ψ. See Hougaard [14–16] and Aalen [1] for descriptions and applications of some frailty models.

The frailty distribution can also be identified, up to a scale factor, from a bivariate frailty model. This is obtained when two random variables are conditionally independent given the value w of an unobserved frailty W and each depends on w through a proportionalhazards model. If $B_j(t)$ are the baseline survivor functions of the two survival times T_j, i.e., their conditional survivor functions given $W = 1$, and $u_j = -\log B_j(t)$ are the corresponding integrated hazards, then the observable joint survivor function is

$$S(t_1, t_2) = \Pr(T_1 > t_1, T_2 > t_2)$$
$$= E \Pr(T_1 > t_1, T_2 > t_2 | W)$$
$$= E \exp\{-W[-\log B_1(t_1)$$
$$- \log B_2(t_2)]\}$$
$$= p(u_1 + u_2).$$

This formula expresses $S(t_1, t_2)$ in terms of $p(s)$ and the functions $B_j(t)$, which are seldom observable. However, it may be reexpressed in terms of the observable marginal survivor functions $S_1(t_1) = S(t_1, 0)$ and $S_2(t_2) = S(0, t_2)$ as

$$S(t_1, t_2) = p[q(S_1(t_1)) + q(S_2(t_2))],$$

where $q(\cdot)$ is the inverse function to the Laplace transform $p(\cdot)$. Thus, bivariate frailty models are a subclass of the *archimedean copula* models of Genest and MacKay [9]. Oakes [26] showed that for all such models the *cross-ratio* function (writing $D_j = \partial/\partial t_j$)

$$\frac{S(t_1, t_2) D_1 D_2 S(t_1, t_2)}{D_1 S(t_1, t_2) D_2 S(t_1, t_2)}$$

depends on (t_1, t_2) only through the survival probability $\nu = S(t_1, t_2)$, so that it may be written as a function $\theta(\nu)$, say. This function has a simple interpretation as the ratio of the conditional hazard rates of the distributions of T_1 given $T_2 = t_2$ and given $T_2 > t_2$, and can be estimated simply even from data that are subject to an arbitrary pattern of right censorship. Clayton [3] introduced this cross-ratio function in the context of the gamma distribution of frailties, for which it is constant. Oakes [26] showed that the function $\theta(\nu)$ characterizes the distribution of W up to a scale factor. A bivariate extreme-value distribution described by Gumbel [12] is obtained for positive stable distribution of frailty. Hougaard [15,16] gives more examples and discussion.

Wild [30] and Huster et al. [17] applied parametric models with gamma frailties to the analysis of matched-pair survival data from medical studies, and also gave some comparisons of asymptotic efficiency. Manatunga and Oakes [20] have undertaken similar work using positive stable frailties. Frailty models have been applied to the study of vaccine efficacy trials [13].

Frailty models can also be useful for repeated events happening to the same individual. Oakes [27] pointed out a connection with classical work of Greenwood and Yule [11] on the equivalence of two models for the occurrence of accidents: (1) the risk of an accident to an individual at time t is $Wb(t)$ irrespective of the number of previous accidents, where W is a gamma-distributed frailty specific to that individual; (2) the risk of an accident at time t to an individual who has already suffered j accidents in $(0, t)$ is $(1 + j\kappa)h(t)$. These two mechanisms lead to the same distributions of accidents for suitable choice of $b(t)$ and $h(t)$.

Other authors (e.g., Marshall and Olkin [21]) have considered models in which the frailties of different subjects may themselves be dependent. Frailties have been used to model dependent censorship in the competing risks problem [31].

Maximum-likelihood estimation with parametric families of frailty distributions, with covariates, is straightforward in principle. Gamma frailties lead to elegant solutions via the EM algorithm [7]. The E step requires only the calculation of the conditional expectations of W and $\log W$ given the observed failure history of each individual or cluster of individuals. These can be obtained explicitly because the relevant conditional distributions of W are also gamma. The M step proceeds as if W and $\log W$ were known. Clayton and Cuzick [5], Klein [19], and Nielsen et al. [25] have discussed semiparametric estimation in a proportional hazards model with gamma frailty [2, Chap. IX], using a combination of the EM algorithm with Johansen's [18] profile likelihood interpretation of Cox's [6] partial likelihood. Murphy [24] has shown consistency of parameter estimates in such settings, but further work on asymptotics is needed. Gibbs sampling and related computational Bayesian methods can be applied, with the W regarded as missing data. (see for e.g., Refs. 4 and 28. McGilchrist and Aisbeitt [23] (see also Ref. 22) discussed a penalized likelihood approach with the individual frailties regarded as nuisance parameters.

Diagnostics for assessing the fit of a frailty model are not yet in widespread use, although Oakes [26] and Genest and Rivest [10] have described two procedures for multivariate survival data without covariates, the former based on estimated values for the cross-ratio function $\theta(\nu)$, the latter based on decompositions of the sample and population versions of Kendall's tau.

References

1. Aalen, O. O. (1988). Heterogeneity in survival analysis. *Statist. Med.*, **7**, 1121–1137.

2. Andersen, P. K., Borgan, O., Gill, R. D., and Keiding, N. (1993). *Statistical Models Based on Counting Processes*. Springer, New York.

3. Clayton, D. G. (1978). A model for association in bivariate survival data and its application in epidemiologic studies of chronic disease incidence. *Biometrika*, **65**, 141–151.

4. Clayton, D. G. (1991). A Monte Carlo method for Bayesian inference in frailty models. *Biometrics*, **47**, 467–485.

5. Clayton, D. G. and Cuzick, J. (1985). Multivariate generalizations of the proportional hazards model. *J. Roy. Statist. Soc. A*, **148**, 82–117.

6. Cox, D. R. (1972). Regression models and lifetables. *J. Roy. Statist. Soc. B*, **34**, 187–220.

7. Dempster, A. P., Laird, N. L., and Rubin, D. B. (1977). Maximum likelihood from incomplete data via the EM algorithm. *J. Roy. Statist. Soc. B*, **39**, 1–38.

8. Elbers, C. and Ridder, G. (1982). True and spurious duration dependence: the identifiability of the proportional hazards model. *Rev. Econ. Stud.*, **XLIX**, 403–409.

9. Genest, C. and MacKay, R. J. (1986). Copules archimèdiennes et familles de lois bidimensioonnelles dont les marges sont donnés. *Can. J. Statist.*, **14**, 145–159.

10. Genest, C. and Rivest, L. P. (1993). Statistical inference procedures for bivariate archimedean copulas. *J. Am. Statist. Assoc.*, **88**, 1034–1043.

11. Greenwood, M. and Yule, K. Y. (1920). An enquiry into the nature of frequency distributions representative of multiple happenings with particular reference to the occurrence of multiple attacks of disease as repeated accidents. *J. Roy. Statist. Soc.*, **83**, 255–279.

12. Gumbel, E. J. (1960). Bivariate exponential distributions. *J. Am. Statist. Assoc.*, **55**, 698–707.

13. Halloran, M. E., Haber, M., and Longini, I. M. (1992). Interpretation and estimation of vaccine efficacy under heterogeneity. *Am. J. Epidemiol.*, **136**, 328–343.

14. Hougaard, P. (1985). Discussion of ref. [5]. *J. Roy. Statist. Soc. A*, **148**, 113.

15. Hougaard, P. (1986). Survival models for heterogeneous populations derived from stable distributions. *Biometrika*, **73**, 387–396.

16. Hougaard, P. (1986). A class of multivariate failure time distributions. *Biometrika*, **73**, 671–678.

17. Huster, J. H., Brookmeyer, R., and Self, S. G. (1989). Modelling paired survival data with covariates. *Biometrics*, **45**, 145–156.

18. Johansen, S. (1983). An extension of Cox's regression model. *Int. Statist. Rev.*, **51**, 258–262.

19. Klein, J. P. (1992). Semiparametric estimation of random effects using the Cox proportional hazards model based on the EM algorithm. *Biometrics*, **48**, 795–806.

20. Manatunga, A. K. and Oakes, D. (1994). Parametric analysis for matched pair survival data.

21. Marshall, A. W. and Olkin, I. (1988). Families of multivariate distributions. *J. Am. Statist. Assoc.*, **83**, 834–841.

22. McGilchrist, C. A. (1993). REML estimation of survival models with frailty. *Biometrics*, **49**, 221–226.

23. McGilchrist, C. A. and Aisbett, C. W. (1991). Regression with frailty in survival analysis. *Biometrics*, **47**, 461–485.

24. Murphy, S. A. (1994). Consistency in a proportional hazards model incorporating a random effect. *Ann. Statist.*, **22**, 712–731.

25. Nielsen, G. G., Gill, R. D., Andersen, P. K., and Sorensen, T. I. A. (1992). A counting process approach to maximum likelihood estimation in frailty models. *Scand. J. Statist.*, **19**, 25–43.

26. Oakes, D. (1989). Bivariate survival models induced by frailties. *J. Am. Statist. Assoc.*, **84**, 487–493.

27. Oakes, D. (1992). Frailty models for multiple event-time data. In *survival Analysis: State of the Art*, J. P. Klein and P. K. Goel, eds., pp. 371–380, Kluwer.

28. Sinha, D. (1993). Semiparametric Bayesian analysis of multiple event-time data. *J. Am. Statist. Assoc.*, **88**, 979–983.

29. Vaupel, J. W., Manton, K. G., and Stallard, E. (1979). The impact of heterogeneity in individual frailty on the dynamics of individual mortality. *Demography*, **16**, 439–454.

30. Wild, C. J. (1983). Failure time models with matched data. *Biometrika*, **70**, 633–641.

31. Zheng, M. and Klein, J. P. (1995). Estimates of marginal survival for dependent competing risks based on an assumed copula. *Biometrika*, **82**, 127–138.

31. Framingham: An Evolving Longitudinal Study

M. A. Woodbury and K. G. Manton

31.1 Introduction

The *Framingham Study* is a long-term study designed both to identify the relation of putative risk factors to circulatory and other disease risks, and to characterize the natural history of chronic circulatory disease processes. It is an interesting example of longitudinal research, wherein limited initial study goals were greatly expanded during the course of data collection. According to Gordon and Kannel [3, p. 124]:

> the Framingham, Massachusetts, study was designed ... to measure ... factors in a large number of "normal" persons ... and to record the time during which these selected factors act and interact before cardiovascular disease results. ... The purpose of the Framingham program ... was the development of case-finding procedures in heart disease. The potential of the Framingham program for epidemiological studies soon became apparent, however, and the program turned increasingly in that direction.

The potential of the Framingham Study for epidemiological studies of a wide variety has been amply demonstrated. The fact that such studies have been extended in the range of etiological factors considered and to include studies of other disease processes, such as malignant neoplasia, attests to the value of longitudinal research (given a high level of imagination and energy of the study directors). Our purpose in this chapter, however, is not to review the range of studies conducted with the Framingham data which have been basic to the identification of serum cholesterol, smoking and elevated blood pressure as important coronary heart disease (CHD) risk factors, but rather briefly to review its initial design and select statistical issues that arise with continuous adaptation of a longitudinal study.

Specifically, we will examine the Framingham Study in terms of the basic data collected as well as the specific analytic and statistical issues that arise in (1) the evaluation of continuous disease processes with periodic and limited data collection and (2) the development of procedures for the analysis of new types of data whose collection was begun after study initiation. We first briefly describe the basic components of the Framingham Study, then describe issues that arise in modeling disease processes with these data, and finally, consider problems inherent in changing and expanding data collection procedures during the course of the study.

31.2 Description of the Framingham Population and Sample

In 1948, when the study began, Framingham was an industrial and trading center of 28,000 persons located 21 miles west of Boston. The town was selected for study because, besides being the location of the first community study of tuberculosis in 1917, there was community interest in the project [3, p. 126]. There was available for Framingham a published town list containing the names of 10,000 persons between the ages of 30 and 59 inclusive. The Bureau of the Census matched the January 1, 1950, town list in the age range 30–59 with the census schedules for April 1, 1950. "Some 89 percent of those on the census schedules were found on the town list" [3, p. 127]. This matching led to the projection of a 10% rate of loss to follow-up. Since it was decided that it was feasible to handle a sample of 6000 persons in the study, this projected low rate dictated that a sample of 6600 persons, or about two-thirds of the eligible population be drawn. Of the 6000 expected to be examined, it was estimated that 5000 of them would be free of cardiovascular disease. It was further expected that 400 would develop cardiovascular disease or die of it by the end of the fifth year of the study, 900 by the end of the tenth year, and 1500 by the end of the twentieth year. It was decided to sample by households and to utilize every person in the chosen age range in each household.

The first major adaption of the study design was forced by a high nonresponse rate. The actual nonresponse rate of 31.2% far exceeded the expected 10%, so the study population was augmented by using 740 of a group of volunteers. "It was at first planned to re-examine only those volunteers who were 'normal' on their initial examination. This plan, however, was not rigorously followed. While 13 people were eliminated for hypertensive or coronary heart disease, these omissions modified the clinical character of the group only trivially" [3, p. 129]. Thus, at the very beginning of the study, a second basic change in the sample design was necessitated, i.e., that, instead of collecting data only on disease-free persons, a sample was drawn with persons manifesting limited cardiovascular conditions. As the study progressed and before any attempt was made to screen out diseased persons, it was decided that persons with cardiovascular disease should be retained in the study since manifest disease could be viewed as a stage in the total disease process. To have removed such persons from the study population, it was concluded in retrospect, would have served to seriously compromise the goal of the study of CHD in a human population.

In addition to the difficulties with sample design, a number of basic measurement issues also had to be resolved during the course of the study. For example, the initial characterization of the sample with respect to the precursors of disease suffered somewhat, since serum cholesterol was not in adequate control until nearly the end of the first measurements examination. As a result, the initial serum cholesterol values were taken from the second exam in the majority of cases. Cigarette-smoking history was not ascertained on the first exam, and the then unknown pressor effect at the first measurement was found to affect the early exam blood pressures. The measurements made on each exam were improved over the life of the study. New measurements were introduced at subsequent examination. For example, diet history and a physical activities study were initiated at exam 4, protein-bound iodine was measured at exams 4 and 5, and a psychological questionnaire was administered on exams 8 and 9. It was concluded that the Framingham Study could have been considered successful even if its only yield was the information derived about measurement phenomena.

The second examination began in earnest in May 1951, more than 2.5 years after the first exam, which began September 29, 1948. Repeated examinations continued at an interval of 2 years. As might be expected, there was a considerable difference in the rate of return to subsequent examinations both within and between the sample and the volunteer groups. According to Gordon and Kannel, it was "surprising to find that where the person remains alive the likelihood of re-examination is about the same in one age–sex group as another" and "that essentially permanent loss to follow-up for reasons other than death has been relatively constant from one examination to another" [3, p. 134].

31.3 Methodological Issues in the Analysis of Disease Processes in Longitudinal Population Studies

The prime value of the Framingham Study is that a large study population was followed for a period of now over 32 years, yielding important time-series information on the change in physiology and risk factors as well as follow-up on disease events. Interestingly, much of the temporal information within the Framingham data on physiological changes has not been fully exploited, despite the explicit desire to assess the natural history of chronic disease processes. One of the reasons for this seems to be the lack of effort to develop appropriate time-series models for human epidemiological data. The primary analytic strategy that has been applied to longitudinal measures of risk has been the logistic multiple regression strategy due to Cornfield [1]. In applying discriminant analysis to CHD risks, Cornfield recognized that the posterior probability was a logistic function of the discriminant score, leading him to accept the logistic response function as the appropriate model of CHD response to risk factors.

One limitation of Cornfield's procedure was the necessity of making the standard discriminant assumption that the risk factors of two groups were normally distributed. Halperin et al. [4] developed conditional maximum likelihood procedures which did not depend on a multivariate normality assumption. These logistic regression strategies have been a primary analytic tool for analyzing the relation of CHD to putative risk factors in the Framingham and other longitudinally followed study populations. Recent attention has been focused on the fact that the logistic function is a mathematically inappropriate functional form to model continuous-time processes such as the development of a chronic disease [8]. The implication of these logical inconsistencies is that the logistic regression coefficients from studies of different length and risk levels cannot be directly compared. One solution that has been proposed is to apply the logistic regression to subintervals of comparable length in the studies to be compared [2,9]. Such strategies are still subject to the difficulty that the logistic coefficients can be confounded with differences in the mean risk of the studies and do not offer a solution for the case where either (1) the follow-ups were conducted for different length intervals, or (2) follow-ups were conducted at irregular intervals. More recent efforts have been directed to the development of methods that are consistent with continuous time changes in risk [5] and one that models physiological changes directly [7].

31.4 Adaptive Nature of Longitudinal Studies

One characteristic of the Framingham Study is that it provided a learning opportunity for the study directors; that is, knowledge gained from the study has been used to enhance the efficiency and utility of the data collection process. Although application of insight acquired during the study is extremely important, it raises a number of difficult statistical issues. First, there are issues that arise because of the modification of data collection procedures. For example, laboratory procedures for assessing serum cholesterol changed over the first few waves of the study. Second, various types of information collection may be either initiated or discontinued. For example, measurements of uric acid ceased after the fourth set of measurements. Third, special studies and measurements are conducted on a one-time basis during the course of the study. Finally, although the biennial measurements of physiological risk variables ceased after 20 years, special survival and other follow-up studies have continued. All these deviations from the initial data collection protocol have introduced interesting but difficult statistical problems.

Three basic statistical models might be employed in the development of analytic means for dealing with these problems. The first is the missing-information principle (EM algorithm), which would permit the utilization of incomplete data [6]. The second principle involves measurement error models and concepts of reliability—concepts well known to psychometricians, but exploited infrequently in biostatistics. A third might be the use of empirical Bayes procedures to integrate data at various levels into a complete model of disease process and to deal with the problem of systematic loss to follow-up and mortality selection.

Applications of such statistical principles and corresponding statistical models are required to make full and efficient use of the rich but sometimes irregular data collected in longitudinal studies of human populations. The central statistical issue seems to be that, in collecting epidemiological data on human health, it is difficult to construct and maintain, in the face of practical exigencies, a study design that permits the application of "simple" statistical procedures. In dealing with human population data, as opposed to data where more stringent experimental controls can be imposed, it appears necessary to apply more general statistical strategies in order to derive inferences from the available data.

References

1. Cornfield, J. (1962). *Proc. Fed. Am. Soc. Exper. Biol.*, **21**, 58–61. (Classical reference for logistic risk.)

2. Cornfield, J. (1978). *Personal communication to Max A. Woodbury.*

3. Gordon, T. and Kannel, W. B. (1970). In *The Community as an Epidemiologic Laboratory*, I. I. Kessler and M. L. Levine, eds. Johns Hopkins University Press, Baltimore, MD. (Good general description of the Framingham Study as originally conceived and as it changed over time.)

4. Halperin, M., Blackwelder, W. C., and Verter, J. I. (1971). *J. Chron. Dis.*, **24**, 125–158. (Alternative method for carrying out logistic regression.)

5. Lellouch, J. and Rakavato, R. (1976). *Int. J. Epidemiol.*, **5**, 349–352. (Loglinear model of mortality risk depending on covariates.)

6. Orchard, T. and Woodbury, M. A. (1971). *Proc. 6th Symp. Mathematical Statistics and Probability*, Vol. 1, pp. 697—715, Berkely, CA., University

of California Press, Berkeley (Methods for maximum-likelihood estimation when data are missing.)

7. Woodbury, M. A., Manton, K. G., and Stallard, E. (1979). *Biometrics*, **35**, 575–585. (Investigation of physiological changes in CHD risks in Framingham.)

8. Woodbury, M. A., Manton, K. G., and Stallard, E. (1981). *Int. J. Epidemiol.*, **10**, 187–197. (Examines issues in risk modeling in longitudinal studies.)

9. Wu, M. and Ware, J. H. (1979). *Biometrics*, **35**, 513–521. (Proposes a strategy for applying logistic regression to longitudinal studies.)

32 Genetic Linkage

Elizabeth Thompson

The genetic material that underlies inherited characteristics consists of linear structures of DNA in the form of a double helix per chromosome. In a diploid organism, chromosomes come in pairs, one deriving from the genetic material of the mother and the other from the father. In the formation of a chromosome that a parent provides to an offspring gamete (sperm or egg cell), several *crossover* events may occur, whereby the transmitted chromosome consists of segments of one parental chromosome together with the complementary parts of the other. Simple genetic characteristics are determined by the specific segments of DNA sequence (*alleles*) at a specific location (*locus*) on a pair of chromosomes. For two distinct locations on a chromosome (two loci), the *recombination parameter* r is the probability that the alleles derive from different parental chromosomes; that is, have different grandparental origins. For this event to occur, there must be an odd number of crossovers between the two loci, in the formation of the offspring gamete. For loci that are very close together on the chromosome, r is close to zero, and alleles at the two loci will show strong dependence in their grandparental origins. The value of r increases with increasing length of chromosome intervening between the two loci, until for loci that are far apart on a chromosome, or are on different pairs of chromosomes, $r = \frac{1}{2}$ and the grandparental origins of alleles at the two loci are independent. Loci for which $r < \frac{1}{2}$ are said to be *linked*. Linkage analysis is the statistical analysis of genetic data, in order to detect whether $r < \frac{1}{2}$, to estimate r, to order a set of genetic loci, and ultimately to place the loci determining genetic traits of interest at correct locations in a genetic map.

In experimental organisms, experiments can be designed such that the grandparental origin of each offspring allele is clear, and very large numbers of offspring can be observed. Estimation of recombination frequencies between pairs of loci is then straightforward. Ordering a set of loci in accordance with the observed values of r is likewise primarily a matter of counting. In 1913 Sturtevant [19] showed that the pattern of recombinations could indeed be well explained by a linear ordering of loci along a chromosome. Fisher [5], in a very early application of likelihood theory and maximum-likelihood estimation [6], also estimated a linear map, on the assumption that no more than one crossover event could occur and hence that recombination frequencies were additive along a chromosome.

Haldane [9] defined genetic map distance as the expected number of crossover events between two loci; this distance measure is additive. *One morgan (unit)* is the length of chromosome in which one crossover event is expected; map distances are normally given in centimorgans (1 cM = 0.01 morgan). For small recombination frequencies there is little difference between recombination frequency r and map distance d. At larger distances, the relationship depends on the pattern of interference – the extent to which one crossover event inhibits others nearby. In the absence of interference, the relationship is $r = \frac{1}{2}(1 - e^{-2d})$, the probability of an odd number of events in distance d morgans when events occur as a Poisson process. Although interference exists, its impact on linkage analyses is slight, and it is often ignored in practice. Many factors influence recombination frequencies. A major one is the sex of the parent; in humans the total female map length of the 22 chromosome pairs (not including the sex chromosomes) is 39 morgans, about 1.5 times the male map length (26.5 morgans). However, this ratio is not constant over the genome. In many linkage analyses, male and female recombination frequencies are estimated jointly, although sometimes they are constrained to be equal, or a fixed relationship may be assumed.

It was not until the 1930s that it became recognized that data on human families could provide useful information for linkage analysis. The problem is one of missing data; in human families individuals may be unavailable for typing, the traits that are observed may have no direct correspondence with the underlying alleles, and even where they do the sharing of common alleles may make grandparental origins of alleles unclear. Thus evaluation of the likelihood involves summing over all the events that could have led to the observed data. Haldane [10] considered use of likelihood to detect linkage, while Fisher [7] addressed the estimation problem more generally, applying ideas of information and efficiency. Smith [18] and Morton [13] introduced the *lod score*, which is $\log_{10}(L(r)/L(\frac{1}{2}))$ the loglikelihood ratio of linkage at a given recombination frequency r relative to that for unlinked loci $r = \frac{1}{2}$. The lod score became the standard criterion for assessing evidence of linkage. Note that the lod score is defined using logarithms to base 10. This also became standard, and is deeply embedded in current methodology in this area, although in multipoint linkage analyses natural logarithms of likelihoods are also used (see *location score* below).

The practical goal of human genetic linkage analysis is to localize the genes contributing to human familial disease traits. To achieve this, linkage of the trait to each of a set of marker loci is tested. These marker loci are of simple known inheritance pattern, and their location in the genome is known through linkage analysis of the marker loci relative to each other. Information of close linkage of a trait gene to a particular marker may then be used to isolate the actual gene, using physical mapping methods, or it may be used to provide genetic counseling. Counseling relies on having typed marker loci closely linked to a disease locus, so that, using family data, risk probabilities may be computed for the disease trait in a fetus, or in advance of symptoms. Often the traits of interest are not simple genetic traits. They may have delayed onset, or there may be incomplete penetrance; individuals carrying the disease allele may never show symptoms. Conversely, there may be individuals who apparently have the disease, but do not carry the gene, there may be several different genetic causes of a disease, or even several genes interacting to produce the observed characteristics. In some cases, the traits of interest may be quantitative.

There are many difficulties in the interpretation of linkage likelihoods resulting from uncertainties in the relationship between the observable trait and the genes underlying it (the trait model), heterogeneity in the genetic causes of the trait, and multiple testing when numerous linked (or unlinked) marker loci are each tested for linkage to the trait of interest. All these complexities have been addressed in the literature. As data on more complex traits are analyzed, the problems of trait model misspecification and trait heterogeneity have become of increasing importance. The excellent text by Ott [16] gives a very thorough review of human genetic linkage, and methods of linkage analysis. One way to avoid the problem of multiple testing is to adopt a Bayesian approach, with a prior probability distribution for the location of a trait gene, converted to a posterior probability in the light of the data. Bayesian methods have a long history in linkage analysis [11], and have more recently attracted more attention. However, the problems of trait model misspecification and heterogeneity are no less with a Bayesian approach, and inference based on the lod score is still the usual approach.

Since 1980, new technology has made available a wealth of new types of genetic markers, based on characteristics of the DNA sequence itself rather than on proteins and enzymes determined by the DNA. These microsatellite markers rely primarily on length variations in DNA, either in terms of numbers of copies of a short repeat sequence, or lengths between occurrences of a given short motif. These genetic markers must first be mapped relative to each other. Then they can be used to localize the genes contributing to traits of interest. There are now thousands of such markers available, and the Human Genome Project goal of a marker map at 1 cM density has been achieved [14]. More recently, even denser markers, single-nucleotide polymorphisms (SNPs) are increasingly used, and with new resequencing technology actual DNA sequence data are becoming available [22].

Statistically, however, this wealth of data complicates analysis, and may not provide additional linkage information. The same markers will not be typed in all studies; even if typed, the same ones will not be informative for linkage. The small recombination frequencies involved means that, unless sample sizes are very large, some intervals between markers may show no recombination. Even on a very large pedigree, there may be no recombinations between the trait locus and several very closely linked markers, so the location of the trait locus among these markers cannot be established. The resolution of linkage methods depends on the number of meioses in the data set, and is typically at best 2–3 cM [2]. Even 1 cM distance is approximately 1 million DNA base pairs — too great a length for current methods of physical mapping.

One approach, that also has a long history in linkage analysis but has gained popularity, is the analysis of associations between a disease allele and a particular marker allele. When a new allele arises by mutation, it does so on some specific chromosome with a specific collection of alleles at nearby markers; that is, there is a specific marker *haplotype* that carries the new disease mutation. Where the loci are very tightly linked, associations of the disease allele with an allele at a linked marker locus may be maintained for many generations before decaying due to recombination events. Despite recent successes in association mapping [20], very large samples of cases and of controls are needed and many difficulties remain, Many such associations exist in populations due to selection, population substructure, and random genetic drift, and few are indicative of linkage, or give a precise estimate of locus order. The power of population data to detect such associations is also highly dependent on the population allele frequencies at the loci considered.

However, in the form of haplotype analysis, the study of association can be very useful in narrowing the region in which a gene is located, once linkage analysis has localized its position as precisely as possible with pedigree data. Provided the disease allele has a single origin in the population considered, the study of haplotypes carrying the allele provides an indication of recombinations occurring or failing to occur over many generations. Far more segregations are implicitly thus observed, than can be explicitly observed in a pedigree study. In experimental organisms, the same effect can be more easily achieved by special study designs involving recombinant inbred lines.

For complex traits, another form of association study is often used to detect linkage. Related affected individuals have an enhanced probability of carrying copies of the same underlying gene causing their affected status, each gene a copy of a single gene in some recent common ancestor. For example, two affected siblings may both carry the same gene received from a parent. The affected relatives will thus have an enhanced probability of carrying the same allele at closely linked marker loci, and such associations can be used to construct a test for linkage. The advantages of such family-based association tests [12] are that they can be done rapidly and easily, and so used as a screening method, and that in the absence of linkage the probability distribution of the test statistics does not depend on the trait model. The disadvantages are that the methods often lack power, that power is highly dependent on the trait locus model, and that the approach provides only a test for linkage detection, not an estimate of the gene location.

Once linkage is detected, a trait locus must be more accurately localized. The most powerful method is to do a *multipoint* linkage analysis, for the hypothesized trait locus against a fixed map of markers. That is, the positions of the marker loci, and their other properties such as marker allele frequencies are assumed known, and a *location score* is computed as a function of the hypothesized position of the trait locus. This location score is again a loglikelihood ratio, although often now, by convention, it is twice the natural logarithm of the likelihood ratio for the trait locus at a given position relative to that with the trait locus unlinked to the local marker framework. Use of multiple marker loci increases the power of the analysis, combining information from markers that are informative in different segregations of the pedigree. Even *interval mapping*, in which a locus is mapped using data on two hypothesized flanking markers, provides much more information than mapping with each marker locus separately.

The usefulness of multipoint linkage analyses is dependent on the accuracy both of the marker map and of the trait model. Although there is now a wealth of markers, their exact positions and the frequencies of their alleles in the study population are often uncertain. Thus there is a need to compute location scores under alternative assumptions about the marker loci. Evaluation of multipoint lod scores is extremely computationally intensive in human pedigrees where there are often missing data and many alternative patterns of gene descent are compatible with the observed data. In recent years much attention has therefore been focused on the computational issues. The efficiency of computer algorithms for exact likelihood evaluation has been much improved, both on large pedigrees [4] and on small [1]. Various approximate methods, including several alternative methods of Monte Carlo estimation of linkage likelihoods have been proposed ([21], and references therein). All involve importance sampling; several involve Markov chain Monte Carlo. Some of the methods are applicable on complex pedigrees; others are directed toward extended but simple pedigrees.

Models for a continuous genome date back to Haldane [9] and to Fisher's *theory of junctions* [8] but become more applicable as a dense marker map becomes ever more available. Rather than considering recombination events between discrete marker loci, it becomes possible to analyze the precise crossover points in a segregation, or the segments of genome shared by relatives, or by individuals having a trait in common. These considerations have led to "genome mismatch scanning" [15], whereby the genomes of affected individuals are compared to find what they do, or do not, have in common. Modern dense SNP marker data permit the detection of genome segments shared due to remote coancestry among individuals not known to be related [3,17]. The power of this new form of association test remains to be investigated, and as in classical linkage analysis will be highly dependent on genetic homogeneity of the trait. As for classical linkage analysis, likelihoods or location scores will be required, and the computational issues will become of even greater significance. Monte Carlo estimation of likelihoods will have a role also in this new situation, in conjunction with other computational approaches to the assessment of the significance of segments of genome shared by known relatives or members of a population who share a given characteristic.

References

1. Abecasis, G. R., Cherny, S. S., Cookson, W. O., and Cardon, L. R. (2002). Merlin–rapid analysis of dense genetic maps using sparse gene flow trees, *Nat. Genet.*, **30**, 97–101.
2. Boehnke, M. (1994). Limits of resolution of genetic linkage studies: Implications for the positional cloning of human disease genes. *Am. J. Hum. Genet.*, **55**, 379–390.
3. Browning, S. R. (2008). Estimation of pairwise identity by descent from dense genetic marker data in a population sample of haplotypes. *Genetics*, **178**, 2123–2132.
4. Cottingham, R. W., Idury, R. M., and Schäffer, A. A. (1993). Faster sequential genetic linkage computations. *Am. J. Hum. Genet.*, **53**, 252–263.
5. Fisher, R.A. (1922). On the systematic location of genes by means of crossover observations. *Am. Natl.*, **56**, 406–411.
6. Fisher, R. A. (1922). On the mathematical foundations of theoretical statistics. *Phil. Trans. Roy. Soc. (Lond.) A*, **222**, 309–368.
7. Fisher, R. A. (1934). The amount of information supplied by records of families as a function of the linkage in the population sampled. *Ann. Eugen.*, **6**, 66–70.
8. Fisher, R. A. (1965). *The Theory of Inbreeding*, 2nd ed., Oliver & Boyd, Edinburgh, UK.
9. Haldane, J. B. S. (1919). The combination of linkage values and the calculation of distances between the loci of linked factors. *J. Genet.*, **8**, 299–309.
10. Haldane, J. B. S. (1934). Methods for the detection of autosomal linkage in man. *Ann. Eugen.*, **6**, 26–65.
11. Haldane, J. B. S. and Smith, C. A. B. (1947). A new estimate of the likage between the genes for colour-blindness and haemophilia in man. *Ann. Eugen.*, **14**, 10–31.
12. Laird, N. M., and Lange, C. (2006). Family-based designs in the age of large-scale gene-association studies. *Nat. Rev. Genet.*, **7**, 385–394.
13. Morton, N. E. (1955). Sequential tests for the detection of linkage. *Am. J. Hum. Genet.*, **7**, 277–318.
14. Murray, J. C., Buetow, K. H., Weber, J. L., et al. (1994). A comprehensive human linkage map with centimorgan density. *Science*, **265**, 2049–2064.

15. Nelson, S. F., McCusker, J. H., Sander, M. A., Kee, Y. Modrish, P., and Brown, P.O. (1993). Genomic mismatch scanning; A new approach to genetic linkage mapping. *Nat. Genet.*, **4**, 11–18.

16. Ott, J. (1999). *Analysis of Human Genetic Linkage,* 3rd. edition. The Johns Hopkins University Press, Baltimore, MD.

17. Purcell, S., Neale, B., Todd-Brown, K., Thomas, L., Ferreira, M. A. R., et al. (2007). PLINK: A tool set for whole-genome association and population-based linkage analyses. *Am. J. Hum. Genet.*, **81**, 559–575.

18. Smith, C. A. B. (1953. Detection of linkage in human genetics. *J. Roy. Statist. Soc. B*, **15**, 153–192.

19. Sturtevant, A. H. (1913). The linear association of six sex-linked factors in *Drosophila*, as shown by their mode of association. *J. Exper. Zool.*, **14**, 43–59.

20. The Wellcome Trust Case Control Consortium (2007). Genome-wide association study of 14,000 cases of seven common diseases and 3,000 shared controls *Nature*, **447**, 661–678.

21. Thompson, E. A. (2000). *Statistical Inferences from Genetic Data on Pedigrees.* NSF-CBMS Regional Conference Series in Probability and Statistics, Vol. 6. Institute of Mathematical Statistics, Beachwood, OH.

22. Wijsman, E. M. (2005). Gene mapping and the transition from STRPs to SNPs. In *Encyclopedia of Genetics, Genomics, Proteomics and Bioinformatics*, L. B. Jorde, P. F. R. Little, M. J. Dunn, and S. Subramaniam eds. Wiley, Hoboken, NJ.

33 Group-Sequential Methods in Biomedical Research

Aiyi Liu, Chunling (Catherine) Liu, and Kai F. Yu

33.1 Introduction

Fixed-size and *sequential* procedures are two popular terms that often appear in statistical and many nonstatistical (biomedical) journals and literatures. For the latter, relevant terms such as *early stopping, interim analyses, interim looks, interim monitoring, stopping boundaries*, and *stopping time*, are also frequently encountered in the literature. A *sequential procedure* in general refers to any statistical method that allows the data collection process to be stopped based on accumulated data. In contrast, termination of the data collection for a fixed-size procedure is independent of the data collected. Thus, technically a fixed size procedure can be viewed as a special case of sequential procedures in the sense that the sample size takes value of a constant integer with probability one. However, the technical treatments of sequential procedures and many of their inherent properties are so different from that of fixed-size procedures, that the two types of procedures are usually viewed as different ones.

Many of the basic concepts and theoretical work on the classical sequential analysis were developed in early 1940s during World War II, in an effort to implement quick yet effective quality control of wartime products that were produced at a massive scale. The seminal paper by Wald [26] concerning the *sequential probability ratio test* (SPRT) procedure laid the foundation and theoretical justification for the application of sequential methods; for more later work related to the SPRT, see, e.g., Wald [27–29], Wald and Wolfowitz [30,31], Girshick [6], Armitage [1,2], and Sobel and Wald [24].

While theory and methods of the classical sequential analysis continued to be developed and their applications widened, it is the past three decades that have witnessed the fast growth of sequential methodology and applications in the biomedical area, most notably in clinical trials and drug development, which, in turn, have provided much of the motivation and stimulus to the development of new sequential methods. Reasons for the use of sequential methods in biomedical research are multiple, and some can be well explained within the context of clinical trials. In a clinical trial it is imperative that inferior drugs, especially those hazardous to patient health, be detected and removed as soon as possible from the trial to avoid further harm, and that effective drugs be approved by regulatory authority so that patients are able to be treated with and benefited from the drugs as early as possible. A sequential procedure monitoring the treatment benefits of the drugs under investigation surely serves the purpose of addressing these ethical concerns. Such a procedure also has the potential to save the study cost by allowing the trial to be terminated earlier than its planned end. The sequential concepts have also been used to reestimate the sample size needed for a study by using interim data to update/modify the assumptions previously made. In genome wide association studies or other biomedical research areas that encounter high-dimensional data, sequential methods are frequently used as a means to deal with the "curse of dimensionality" (large number of variables but small number of experimental units) by eliminating certain insignificant variables based on data from a smaller group of the experimental units. Such elimination helps the investigators to focus only on variables that are statistically significant and considered to be important.

A sequential procedure must have a clear definition of (1) when the accumulated data are analyzed; (2) at each time of doing so, how the data collection (sampling) process will be terminated; (3) and on termination, how the decision will be made regarding the inferential problem, such as rejection of a hypothesis. Termination of the sampling process is usually expressed as a stopping rule that partitions the sample space (of the accumulated data) into two mutually exclusive regions, the stopping region and the continuation region. When the main focus is hypothesis testing concerning a single parameter, as so commonly encountered in clinical trials and other biomedical research areas, the stopping regions are often alternatively expressed in terms of a *triangular boundary* (e.g., Whitehead [37]) or an *error-spending function* [13], and the schedules for interim analyses of the accumulated data are often predetermined and each of such an analysis is often conducted when data from a group of experimental units are collected, resulting in the so-called *group sequential tests*, in contrast to the classical sequential tests where analysis is done after each observation. Research on sequential procedures applicable to biomedical areas has heavily focused on developing theory and methods to define the stopping rule that meet certain requirements. From a decision-making perspective, a sequential procedures should be so defined that the risk (expected loss) of using such a procedure is at most as high as that of using the corresponding fixed-size procedure. In the context of hypothesis testing it usually requires that the sequential procedure maintains the same rate of type I error (false positive) and type II error (false negative) as in a fixed-size procedure. Such a requirement usually comes with the sacrifice that the (maximum) sample size is usually larger than what is needed in a fixed size setting. When two sequential procedures meet the same requirements, the one with smaller expected sample size (in certain parametric space) is considered better than the other. The main purpose of this chapter is thus to provide the readers with a brief overview of many group-sequential procedures developed over the years, focusing on hypothesis testing in particular, for biomedical research in areas such as clinical trials, case–control studies, diagnostic medicine, and genetic association studies.

33.2 The Role of Brownian Motion for Large-Sample Statistics

In sequential testing of statistical hypotheses, controlling error rates in rejecting a null hypothesis requires knowledge of the joint distribution of a series of correlated test statistics. Except for certain special situations, exact calculations for the distribution or the error rates are extremely difficult even when the marginal distributions (used in the setting of fixed-size test) are known or easy to work with. Moreover, many test statistics are based on nonparametric procedures that make it infeasible to perform exact calculations. To this end many sequential calculations are proposed on the basis of large-sample approximation via the use of Brownian motions.

The Brownian motion paradigm, although less familiar to many statisticians, provide a general framework for large sample approximation. Briefly speaking, a Brownian motion $X(t)$ ($t \geq 0$) with drift θ is a stochastic process that satisfies the following properties, $X(0) = 0$; for each t, $X(t)$ is normally distributed with mean θt and variance t, i.e., $X(t) \sim N(\theta t, t)$; and for any $0 < t_1 < t_2 < \cdots < t_k < \infty$, $X(t_1), X(t_2) - X(t_1), \ldots, X(t_k) - X(t_{k-1})$ are mutually independent with $X(t_j) - X(t_i) \sim N(\theta(t_j - t_i), t_j - t_i)$ and $\text{Cov}(X(t_j), X(t_i)) = t_i$, for $i < j$. This latter proposition is usually called *independent increments*. For a specific statistical problem t is usually the Fisher's information and $X(t)$ is usually the test statistic. In general t is an increasing function of the sample size or the number of events in a survival type of analysis. The drift parameters θ is usually formulated according to the problem of interest, for example, as a measure of the difference between two treatment arms in a clinical trial or the odds ratio in a case–control study. Thus the null hypothesis being tested is $H_0 : \theta = 0$, indicating no treatment difference in a clinical trial setting or no disease exposure association in a case–control study. The two-sided hypothesis $\theta \neq 0$ is often considered as the alternative.

Example 1. Consider testing $H_0 : \theta = 0$ concerning a normal population $N(\theta, 1)$ with mean θ and unit variance. Let $X_1, X_2, \ldots$ be the random samples from the population. Then the sample sum $S_n = \sum_{i=1}^{n} X_n$ well resembles a Brownian motion at $t = n$ with drift θ.

Example 2. Consider the two-sample problem of testing the equality of two normal means. Suppose $X_1, X_2, \ldots$ and $Y_1, Y_2, \ldots$ are samples from $N(\mu_X, \sigma^2)$ and $N(\mu_Y, \sigma^2)$, respectively. Define $S_n = \sum_{i=1}^{n}(X_{iA} - X_{iB})/(2\sigma^2)$. Then $S_n \sim N(\theta(n/(2\sigma^2)), n/(2\sigma^2))$, behaves well like a Brownian motion $X(t)$ (at $t = n/(2\sigma^2)$) with drift parameter $\theta = \mu_X - \mu_Y$. Thus the null hypothesis reduces to $H_0 : \theta = 0$.

It has been recognized that almost all statistics encountered in biomedical research, including those based on continuous, discrete, survival, or longitudinal data, missing observations, etc., can be well approximated by a Brownian motion. The Brownian motion framework thus lays the foundation for the development of many sequential methods. Whitehead [36] provided a nice review (in somewhat different notations) of this structure and named it *a unified theory*. He presented details on how the Brownian motion can be constructed for various types of data and test statistics. See also Lan and Zucker [14], Lee et al. [15], and Proschan et al. [20]. For this reason most of the writings below will be confined in the contexts of testing hypothses concerning the drift parameters of a Brownian motion.

33.3 Group-Sequential Testing of a Hypothesis

33.3.1 From Fixed-Size to Group-Sequential

A group-sequential procedure developed for biomedical research is usually a natural extension of the corresponding fixed-size procedure, which is already known to the statisticians. Such extension also provides an easier way to understand the concepts of sequential analysis. The formulation presented below for hypothesis testing provides large-sample approximation for many problems encountered in biomedical research. Consider a sample path in $[0, T]$ of a Brownian motion $X(t)$ with drift parameter θ, where T is a constant. Suppose that the null hypothesis being tested is $H_0 : \theta = 0$ versus a two-sided alternative $\theta \neq 0$. Then with a nominal type I error rate of α, H_0 is rejected if $|X(T)| > \sqrt{T}\Phi^{-1}(1 - \alpha/2)$, where Φ is the standard normal distribution function.

A subsequent group-sequential testing procedure is thus as follows. Suppose that a fixed number of K interim analysis will be conducted and the null hypothesis H_0 will be tested with the first interim analysis being based on the Brownian path in $[0, t_1]$, the second interim analysis being based on the Brownian path in $[0, t_2]$, and so on, with the last analysis on the path in $[0, t_K]$. These interim analyses are conducted sequentially, based on the accumulated paths, and a later analysis is carried out according to the following stopping rule: If at the kth analysis ($k = 1, \ldots, K-1$), $X(t_k) \leq a_k$ or $X(t_k) \geq b_k$, then stop observing the Brownian motion and reject H_0. Otherwise continue to the $(k+1)$th stage by observing an additional path from t_k to t_{k+1}. Data collection will be forced to terminate at the last stage, the Kth analysis, if $a_k < X(t_k) < b_k$ for $1 \leq k \leq K-1$, and the null hypothesis will not be rejected if $a_K < X(t_K) < b_K$. Such a group-sequential test has been widely used in biomedical research, particularly in clinical trials. It should be noted that for these kinds of sequential tests early stopping is only for rejection of the null hypothesis; only at the last stage will the decision be made in terms of acceptance of the hypothesis. If early stopping is also imposed for acceptance of the null hypothesis, then some inner-edge boundaries are needed; see Jennison and Turnbull [11].

It becomes apparent that the group-sequential testing procedure presented above is fully characterized by the following design parameters: K, the total number of interim analysis; $t_k (k = 1, \ldots, K)$, the timepoints at which an interim analysis is performed; and $\{(a_k, b_k) : 1 \leq k \leq K\}$, which define the stopping boundaries and rejection region of the null hypothesis. To determine these $3K + 1$ parameters has been proved to be an extremely difficult task and has become for decades the center of research for group sequential methods. In theory, a total of $3K+1$ constraints (equations) will be needed to uniquely determine these parameters. In practice, however, it may be infeasible to have so many constraints, some of which may not even be of any practical use. Along these lines, many authors have proposed ways of finding these parameters, and eventually resulted in a few excellent books summarizing the research in this area; see Whitehead [35], Jennison and Turnbull [11], and Proschan et al. [20]. In what follows we will give an overview of a number of popular methods that either have been frequently used in biomedical studies or received considerable attention in the literature.

33.3.2 Choosing a Stopping Rule for Group-Sequential Monitoring

In clinical trials or other similar settings a hypothesis-testing procedure must at least meet the error requirements, that is, the type I error is controlled at a desired level, say, α, and the power of the test at certain alternative, say, $\theta = \theta_1$, is at least $1 - \beta$. The fixed size test that rejects H_0 if $|X(T)| > \sqrt{T}\Phi^{-1}(1 - \alpha/2)\sqrt{T}$ thus yields the information needed to achieve the power as

$$T = \left\{ \frac{\Phi^{-1}(1 - \alpha/2) + \Phi^{-1}(1 - \beta)}{\theta_1} \right\}^2.$$

For each specific problem the sample size can be calculated from the functional relationship between T and the sample size. On the other hand, for a group sequential test described above, the type I error and power requirements produce respectively the following two equations:

$$\sum_{k=1}^{K} P_0\{X(t_i) \in (a_i, b_i), i \leq k-1, X(t_k) \notin (a_k, b_k)\} = \alpha,$$

$$\sum_{k=1}^{K} P_{\theta_1}\{X(t_i) \in (a_i, b_i), i \leq k-1, X(t_k) \notin (a_k, b_k)\} = 1 - \beta. \quad (1)$$

Unlike the fixed size case these two equations can not uniquely determine the boundary values a_k, b_k and the group sizes n_k.

One may wonder what would happen if the boundary values were all set to be the common one otherwise used in a fixed size test with significance level α. The answer is quite discouraging: Doing so can severely inflate the type I error rate, and indeed the first term in the type I error equation (1) already equals to α. For example, if setting $b_k = -a_k = \Phi^{-1}(1-\alpha/2)$ for every k, then the type error probabilities for $2 \leq K \leq 10$ are as high as 0.08, 0.11, 0.13, 0.14, 0.15, 0.17, 10.18, 10.19, and 0.20, respectively.

Various methods were proposed to solve the problem of "fewer constraints but many free design parameters." With only type I error and power requirements, many methods try to reasonably reduce the $3K+1$ design parameters to two free parameters, which then can be solved from Equations (1) and (2). Most early developments of group-sequential tests adapt a continuous stopping boundary that is fully determined by a "shape" parameter to a group sequential setting. Pocock's [19] approach uses two symmetric curves $X(t) =$ $\pm$ some constant $\times \sqrt{t}$ and assumes that (1) the total number analyses K is specified, (2) the interim analyses are equally spaced, that is, $t_1 = t_2 - t_1 = \ldots t_K - t_{K-1}$, or $t_k = (k/K)t_K$. Symmetry of the stopping boundaries implies that $b_k = -a_k = c_k$, and that the standardized boundary values are constant, depending only on K and α, that is, $c_k/\sqrt{t_k} = C_P(K, \alpha)$, or $c_k = \sqrt{t} C_P(K, \alpha)$. Under these assumptions the constant $C_P(K, \alpha) = C_P$ can be found by solving

$$\sum_{k=1}^{K} P_0 \left\{ \frac{|X(t_i)|}{\sqrt{t_i}} \leq C_P, i \leq k-1, \right. $$
$$\left. \frac{|X(t_k)|}{\sqrt{t_k}} > C_P\sqrt{t_k} \right\} = \alpha,$$

and after C_P is determined the total information t_K can then be found by solving

$$\sum_{k=1}^{K} P_{\theta_1} \left\{ \frac{|X(t_i)|}{\sqrt{t_i}} \leq C_P, i \leq k-1, \right.$$
$$\left. \frac{|X(t_k)|}{\sqrt{t_k}} > C_P\sqrt{t_k} \right\} = 1 - \beta.$$

The method of O'Brien and Fleming [18] uses Wald's symmetric SPRT boundaries, i.e. two parallel straight lines. Thus in a group-sequential setting the boundary values c_k are constant, also depending only on K and α, that is, $c_k = C_B(K, \alpha)$. This constant, along with the total information t_K, is again determined from the error and power requirements.

While O'Brien and Fleming's [18] method possesses much similarity to the Pocock's [19] method and both maintain the same significance level and power, the former is more conservative, causing the test to be less likely to stop early. Its critical values and the total sample size required are much closer to that for a fixed-size test. It is perhaps for these reasons that the O'Brien and Fleming's [18] method is more preferred in practice.

Wang and Tsiatis [32] extended the group sequential boundaries of Pocock [19] and O'Brien and Fleming [18] by considering $c_k = C_{WT}(K, \alpha, \Delta) t_k^\Delta / t_K^{\Delta - 1/2}$, where Δ defines the shape of the boundary and $C_{WT}(K, \alpha, \Delta)$ is a constant. Thus Pocock's [19] boundary corresponds to $\Delta = \frac{1}{2}$ and O'Brien and Fleming's [18] boundary corresponds to $\Delta = 0$. Therefore the shape parameter Δ allows practitioners to design a group-sequential test that can be more conservative than Pocock's but less conservative than O'Brien and Fleming's.

For various K, α, and Δ the values of the constant C_{WT} are well tabulated in the book by Jennison and Turnbull [11]. Jennison and Turnbull [11] also provide some comparisons among different tests in terms of their maximum sample size and the expected sample sizes at certain alternatives.

33.3.3 More Flexible Methods

The methods discussed above all assume that the interim analysis times are pre-specified and equally spaced. In the conduct of many biomedical studies, this may be quite a limitation. For many practical problems the Fisher information t in the Brownian motion formulation is a function of not only the sample size (number of events in survival analysis) but also certain nuisance parameters. Thus t is estimated at each interim analysis with updated estimates of the nuisance parameters. When the interim analyses are conducted according to real calendar time with equal time intervals, e.g., interim analysis being done after every 12 months, or even when they are conducted with equal number of group sizes, e.g., after every 50 observations (or events), the equally spaced information scales are not guaranteed since the updated estimates of the nuisance parameters vary. As a consequence of the actual unequal information intervals, the type I error can be inflated if these group sequential procedures are used. Furthermore, the actual analysis time, even well planned, may be unpredictable due to unexpected low accrual rate or the unavailability of members of the Data Safety and Monitoring Board (DSMB) to meet on a planned date. These considerations thus call for more flexible group sequential procedures to allow unequal and/or unpredictable information intervals.

Slud and Wei [23] attempted to relax the assumption of equal space via the use of *error spending*. Consider a group-sequential test with a fixed number K of analyses at $t_1, \ldots, t_K$ with symmetric boundaries, i.e. $-a_k = b_k = c_k$; then the type I error at analysis k is

$$\pi_k = P_0\{|X(t_i)| \leq c_i, i \leq k-1, \\ |X(t_k)| > c_k\}, \quad (2)$$

which is better known as *type I error spent at analysis k*. These stagewise error probabilities must sum up to the significance level α so that the type I error rate is controlled. (In contrast, the quantity $\alpha_k = \sum_{i=1}^{k} \pi_i$ is often called the *cumulative* type I error at analysis k.) Once the partition $\{\pi_k, k = 1, \ldots, K\}$ is specified, and information times $t_1, \ldots t_k$ observed, then the boundary values c_k are calculated sequentially according to Equation (3).

The assumptions that K is fixed and the error partition is prespecified were further relaxed by Lan and DeMets [13], which requires only that the total information t_{mx} be fixed. Noting that if $X(t)$ is a Brownian motion with drift θ in $[0, t_{max}]$, then $X(t)/\sqrt{t_{max}}$ is a Brownian motion with drift $\theta^* = \theta\sqrt{t_{max}}$ in $[0, 1]$, Lan and DeMets [13] defined an error spending function in $[0, 1]$ as any function $e(t)$ such that $e(t)$ is increasing in t, $e(0) = 0$, and $e(1) = \alpha$, the given significance level. Using an error spending function $e(t)$, suppose that the first interim analysis is performed with information t_1, not necessarily prefixed, then the boundary value c_1 is obtained from

$$P_0\{|X(t_1)| > c_1\}$$
$$= 2\left\{1 - \Phi\left(c_1/\sqrt{t_1}\right)\right\}$$
$$= e(t_1/t_{\max}),$$

yielding

$$c_1 = \sqrt{t_1}\Phi^{-1}\left\{1 - \frac{1}{2}e(t_1/t_{\max})\right\}.$$

If the test continues to the second analysis and the accumulated information is t_2, then the boundary value c_2 is derived from $P_0\{|X(t_1)| \leq c_1, |X(t_2)| > c_2\} = e(t_2/t_{\max}) - e(t_1/t_{\max})$. Thus the boundary values are determined in a sequential way with c_k obtained from $P_0\{|X(t_i)| \leq c_i, i \leq k-1, |X(t_k)| > c_k\} = e(t_k/t_{\max}) - e(t_{k-1}/t_{\max})$.

This error spending approach offers quite flexibility in conducting a group sequential test since neither the total number K of analysis nor the information time at each analysis need to be prespecified. Lan and DeMets [13] demonstrated that the two special error spending functions, $e(t) = \min\{\alpha \log[1+(e-1)t], \alpha\}$ and $e(t) = \min\{2 - 2\Phi\left(\Phi^{-1}(1-\alpha/2)/\sqrt{t}\right), \alpha\}$, when applied with fixed K and equally spaced interim analyses, approximately give the Pocock's [19] and O'Brien and Fleming's [18] boundary values, respectively. Other spending functions include Hwang et al. [7] with $e(t) = \alpha(1-e^{-\gamma t})/(1-e^{-\gamma})$ if $\gamma \neq 0$ and αt if $\gamma = 0$, and Kim and DeMets [12] (see also Jennison and Turnbull [9,10]) with $e(t) = \min\{\alpha t^\rho, \alpha\}$ for some positive value $\rho > 0$.

There is no universal standard for choosing an error spending function. In general the expected information (sample size) should be small under certain values of θ that are of interest for study objectives. Practical concerns also play a role in choosing an error-spending function. For example, if there are a few secondary endpoints that need to be analyzed then a more conservative error-spending function may be more appropriate since early stopping of the study may limit the power in detecting certain meaningful differences in these secondary endpoints. (The sequential testing only guarantees the power in detecting θ values!)

33.4 Terminal Analysis Following a Sequential Test

Executing a group-sequential test for a biomedical study requires the study be terminated when a stopping boundary is reached, i.e., when the first time $|X(t_k)| > c_k$ for some k. On termination of the study a decision can then be made on whether the null hypothesis $H_0 : \theta = 0$ should be rejected.

The parameter θ is usually the focus of the study aims and thus point estimates and confidence intervals, which measure the magnitude of the parameter, and the P-values, which gives the strength of the study findings against the null hypothesis, should also be provided.

From a frequentist's point of view, the usual maximum likelihood approach in a sequential setting introduces bias because the actual sample size is a random variable that is correlated with the test statistics. As a consequence, suppose the sequential test stops at the kth analysis with information $T = t_k$, then the maximum likelihood estimate of θ, $\hat{\theta}_{ML} = X(T)/T$, is biased and the usual confidence interval, $\hat{\theta}_{ML} \pm \Phi^{-1}(1-\alpha/2)/\sqrt{T}$, no longer warrants coverage probability of $1-\alpha$. To overcome this a number of methods have been developed for proper inference on θ.

33.4.1 Point Estimation

Aiming at reducing the bias of the maximum-likelihood estimate, Whitehead [34] proposed the so-called bias-adjusted estimate, originally in the context of *fully sequential tests* with parallel or triangular stoping boundaries, but quite generally applicable in the group-sequential setting. Write $b(\theta) = E(\hat{\theta}_{ML}) - \theta$, the bias function of the maximum-likelihood estimate. In Whitehead's [34] formulation the bias-adjusted estimate $\hat{\theta}_W$ is to be solved from the equation

$$\hat{\theta}_W + b(\hat{\theta}_W) = \hat{\theta}_{ML}.$$

Whitehead [34] presented considerable numerical results to demonstrate that the bias-adjusted estimate indeed has smaller bias (and mean-squared error as well!) than the maximum-likelihood estimate. Indeed, if the bias function $b(\theta)$ is non-decreasing and can be differentiated at least twice, then the bias-adjusted estimate has uniformly smaller mean-squared error than the maximum-likelihood estimate; see Whitehead [34] and Liu [16].

Another natural approach is to find estimates that do not possess bias. For this purpose Emerson and Fleming [4] used conditional argument to derive an unbiased estimate of θ:

$$\hat{\theta}_{EF} = E\left\{\frac{X(t_1)}{t_1} \bigg| (T, X(T))\right\}.$$

The estimate is identical to the maximum likelihood estimate for $T = t_1$ and is obviously unbiased since $X(t_1)/t_1$ is always so, regardless of the at stage which the test stops. Although $\hat{\theta}_{EF}$ nullifies the bias, numerical results from Emerson and Fleming [4] showed that in general it has a larger mean-squared error than does Whitehead's bias-adjusted estimate, a sacrifice to be made for enforcing unbiasedness. As expected, the two estimates both have smaller bias and mean-squared error than does the maximum-likelihood estimate, as considerable numerical results have shown.

When T is fixed, it is well known that $\hat{\theta}_{ML}$ is the uniformly minimum variance unbiased estimate (UMVUE) of θ, and it is interesting to know whether this remarkable property holds in a group-sequential setting. Liu and Hall [17] showed that there are infinitely many unbiased estimates and none, $\hat{\theta}_{EF}$ particularly included, has uniformly minimum variance. Only when restricted to unbiased estimates that are independent of future stopping rules that $\hat{\theta}_{EF}$ has uniformly minimum variance.

33.4.2 *P* Values

A *P value* is the probability under the null hypothesis of the test statistic taking values more extreme than the current observed one. In a fixed-size test this is usually well defined: Larger values of the test statistic are more extreme than smaller ones. In a group sequential testing being extreme can be interpreted in various ways. Without any doubt, earlier stopping with a larger value of the test statistic surely means more extreme. However, earlier stopping with a relative smaller test statistic can also be viewed as more extreme than later stopping with a relatively larger test statistic.

Several definitions have been proposed in the literature for $(T, X(T))$ to be more extreme than $(T^*, X(T^*))$. This occurs if $X(T) > X(T^*)$ according to Rosner and Tsiatis [21], or if $X(T)/\sqrt{T} > X(T^*)/\sqrt{T^*}$ according to Chang [3], while Emerson and Fleming [4] defined it as $X(T)/T > X(T^*)/T^*$. Siegmund [22], Fairbanks and Madsen [5], and Tsiatis et al. [25] suggested a somewhat different criterion, the so-called *stagewise ordering*. Suppose $T = t_i$ and $T^* = t_{i^*}$. Then $(T, X(T))$ is more extreme than $(T^*, X(T^*))$ if $i < i^*$ and $X(t_i) > b_i$ or $< a_i$ or if $X(t_i) > X(t_{i^*})$ when $i = i^*$.

With each definition, suppose that the test stops at stage k with $X(t_k) = \tilde{x}$, then the upper-tail *P* value is given by

$$P_{\theta=0}\{(T, X(T)) \text{ is more extreme} \\ \text{than } (t_k, \tilde{x})\},$$

and the lower-tail P value is given accordingly. The two-sided P value is twice the smaller of the two tail P values.

The stagewise ordering has some nice properties. It yields P values in agreement with the decision from the hypothesis testing; it is less than the significance level α if and only if the null hypothesis is rejected. Moreover, the stagewise ordering does not depend on the stages beyond the stopping one, making it applicable in some more flexible designs in which future interim analyses are not specified. For these reasons, the stagewise ordering is considered more preferable by many practitioners.

33.4.3 Construction of Confidence Intervals

Lower confidence limit $\theta_L^* = \theta_L^*(T, X(T))$ and upper limit $\theta_U^* = \theta_U^*(T, X(T))$ can be obtained by modifying a fixed size confidence interval. Write $Z(\theta) = (X(T) - T\theta)/\sqrt{T}$, and denote its mean and standard deviation by $\mu(\theta)$ and $s(\theta)$, respectively. When T is fixed, Z is normally distributed with zero mean and unit variance, and serves as a pivot to construct the usual $(1-\alpha)$-level confidence interval $X(T)/T \mp \Phi^{-1}(1-\alpha/2)/\sqrt{T}$. In a group-sequential setting Woodroofe [38] instead proposed using $Z'(\theta) = (Z(\theta) - \mu(\theta))/s(\theta)$ as a pivot, by assuming that it approximately follows a standard normal distribution. This yields an approximate $(1-\alpha)$−level confidence interval for θ as

$$\frac{X(T)}{T} - \frac{\hat{\mu}}{\sqrt{T}} \mp \frac{\hat{s}}{\sqrt{T}}\Phi^{-1}(1-\alpha/2),$$

where $\hat{\mu}$ and $\hat{s}$ are proper estimates of $\mu(\theta)$ and $s(\theta)$, respectively, which are often obtained with θ replaced by its maximum-likelihood estimate.

Woodroofe's [38] method performs well in most cases and the coverage probabilities of the resulting confidence intervals are quite close to the nominal level of $1-\alpha$. However, it does not always guarantee the exact coverage probability of $1-\alpha$.

Confidence intervals with exact coverage can be obtained by utilizing an ordering of the sample space used for calculating the P values. Suppose that the test stops at interim analysis k with information t_k and $X(t_k) = \tilde{x}$. For a given ordering denote by $\lambda(\theta; t_k, \tilde{x})$, the probability that $(T, X(T))$ is at least as extreme as $(t_k, \tilde{x})$. Then the lower confidence limit θ_L^* is the drift value such that $\lambda(\theta_L^*; t_k, \tilde{x}) = \alpha/2$, and the upper confidence limit θ_U^* is the drift value such that $\lambda(\theta_U^*; t_k, \tilde{x}) = 1 - \alpha/2$.

Judging by the criterion that a confidence interval should agree with the hypothesis testing results and contain the maximum-likelihood estimate, the confidence intervals based on the stagewise ordering are more preferable than those based on other orderings; see Jennison and Turnbull [11].

33.4.4 Computing Software

Computation of boundary values, error probabilities, point and interval estimates, and P values involves multiple integration arising from the joint distribution of the Brownian sequence, $X(t_1), \ldots, X(t_k)$. Such computation becomes quite extensive when the number of analysis K is relatively large. Fortunately a number of commercial and free software are available for design and analysis of a sequential trial. Some popular software include EaSt, developed by Cytel Software Corporation (*http://www.cytel.com/Products/East/*); PEST, developed by the Medical and Pharmaceutical Statistics research group led by Whitehead (*http://www.maths.lancs.ac.uk/department/research/statistics/mps/pest*); SAS modules SEQ, SEQSCALE, and SEQSHIFT in SAS version 9.2, developed by the SAS Institute (*http://www.sas.com*);

and an S-PLUS module S+SeqTrial, developed by Insightful Corporation (*http://www.insightful.com/products/ seqtrial*). Two free sources of FORTRAN codes are also available, one from Jennison's homepage on *http://www.bath.ac.uk/ mascj/book/ programs/general*, and the other from *http://www.biostat.wisc.edu/software/ index.htm* developed by a biostatistics group at the University of Wisconsin. Wassmer and Vandemeulebroecke [33] provide a brief overview on software for group-sequential designs.

References

1. Armitage, P. (1947). Some sequential tests of student's hypothesis. *J. Roy. Statist. Soc. Suppl.*, **9**, 250–263.

2. Armitage, P. (1950). Sequential analysis with more than two alternative hypotheses and its relation to discriminant function analysis. *J. Roy. Statist. Soc. B*, **12**, 137–144.

3. Chang, M. N. (1989). Confidence intervals for a normal mean following a group sequential test. *Biometrics*, **45**, 247–254.

4. Emerson, S. S. and Fleming, T. R. (1990). Parameter estimation following sequential hypothesis testing. *Biometrika*, **77**, 875–892.

5. Fairbanks, K. and Madsen, R. (1982). P values for tests using a repeated significance design. *Biometrika*, 69:69–74.

6. Girshick, M. A. (1946a,b). Contributions to the theory of sequential analysis, I, II. *Ann. Math. Statist.*, **17** 123–143, 282–298.

7. Hwang, I. K., Shih, W. J., and DeCani, J. (1990). Group sequential designs using a family of type I error probability spending functions. *Statist. Med.*, **9**, 1439–1445.

8. Jennison, C. (1987). Efficient group sequential tests with unpredictable group sizes. *Biometrika*, **74**, 155–165.

9. Jennison, C. and Turnbull, B. W. (1989). Interim analyses: the repeated confidence interval approach (with discussion). *J. Roy. Statist. Soc. B*, **51**, 305–361.

10. Jennison, C. and Turnbull, B. W. (1990). Statistical approaches to interim monitoring of medical trials: A review and commentary. *Statist. Sci.*, **5**, 299–317.

11. Jennison, C. and Turnbull, B. W. (2000). *Group Sequential Methods With Applications to Clinical Trials*. Chapman & Hall/CRC, New York.

12. Kim, K. and DeMets, D. L. (1987). Design and analysis of group sequential tests based on the type I error spending rate function. *Biometrika*, **74**, 149–154.

13. Lan, K. K. G. and DeMets, D. L. (1983). Discrete sequential boundaries for clinical trials. *Biometrika*, **70**, 659–663.

14. Lan, K. K. G. and Zucker, D. (1993). Sequential monitoring of clinical trials: The role of information and Brownian motion. *Statistics in Medicine*, **12**, 753–765.

15. Lee, S. J., Kim, K., and Tsiatis, A. A. (1996). Repeated significance testing in longitudinal clinical trials. *Biometrika*, **83**, 779–789.

16. Liu, A. (2003). A simple low-bias estimate following a sequential test with linear boundaries. In *Institute of Mathematical Statistics Lecture Notes Monograph Series: Crossing Boundaries: Statistical Essays in Honor of Jack Hall*, J. Kolassa and D. Oakes, eds., Vol. 43, pp. 47–58. IMS, Beachwood, OH.

17. Liu, A. and Hall, W. J. (1999). Unbiased estimation following a group sequential test. *Biometrika*, **86**, 71–78.

18. O'Brien, P. C. and Fleming, T. R. (1979). A multiple testing procedure for clinical trials. *Biometrics*, **35**, 549–556.

19. Pocock, S. J. (1977). Group sequential methods in the design and analysis of clinical trials. *Biometrika*, **64**, 191–199.

20. Proschan, M. A., Lan, K. K. G., and Wittes, J. T. (2006). *Statistical Monitoring of Clinical Trials: A Unified Approach*. Springer-Verlag, New York.

21. Rosner, G. L. and Tsiatis, A. A. (1988). Exact confidence intervals following a group sequential trial: A comparison of methods. *Biometrika*, **75**, 723–729.

22. Siegmund, D. (1978). Estimation following sequential tests. *Biometrika*, **65**, 295–297.

23. Slud, E. V. and Wei, L.-J. (1982). Two-sample repeated significance tests based on the modified Wilcoxon statistics. *J. Amer. Statist. Assoc.*, **77**, 862–868.

24. Sobel, M. and Wald, A. (1949). A sequential decision procedure for choosing one of three hypotheses concerning the unknown mean of a normal distribution. *Ann. Math. Statist.*, **37**, 502–522.

25. Tsiatis, A. A., Rosner, G. L., and Metha, C. R. (1984). Exact confidence intervals following a group sequential test. *Biometrics*, **40**, 797–803.

26. Wald, A. (1943). *Sequential Analysis of Statistical Data: Theory*. Technical Report 75, Statistical Research Group Report b, Columbia University.

27. Wald, A. (1944). On cumulative sums of random variables. *Ann. Math. Statist.*, **15**, 283–296.

28. Wald, A. (1945a). Sequential methods of sampling for deciding between two courses of action. *J. Amer. Statist. Assoc.*, **40**, 277–306.

29. Wald, A. (1945b). Sequential tests of statistical hypotheses. *Ann. Math. Statist.*, **16**, 117–186.

30. Wald, A. and Wolfowitz, J. (1948). Optimum character of the sequential probability ratio test. *Ann. Math. Statist.*, **19**, 326–339.

31. Wald, A. and Wolfowitz, J. (1950). Bayes solutions of sequential decision problems. *Ann. Math. Statist.*, **21**, 82–89.

32. Wang, S. K. and Tsiatis, A. A. (1987). Approximately optimal one-parameter boundaries for group sequential trials. *Biometrics*, **43**, 193–200.

33. Wassmer, G. and Vandemeulebroecke, M. (2006). A brief review on software developments for group sequential and adaptive designs. *Biometr. J.*, **48**, 732–737.

34. Whitehead, J. (1986). On the bias of maximum likelihood estimation following a sequential test. *Biometrika*, **73**, 573–558.

35. Whitehead, J. (1997). *The Design and Analysis of Sequential Clinical Trials*, 2nd ed. Wiley, Chichester, UK.

36. Whitehead, J. (1999). A unified theory for sequential clinical trials. *Statist. Med.*, **18**, 2271–2286.

37. Whitehead, J. (2001). Use of the triangular test in sequential clinical trials,. In *Handbook of Statistics in Clinical Oncology*, J. Crowley, ed. Marcel Dekker, New York.

38. Woodroofe, M. (1992). Estimation after sequential testing: a simple approach for truncated sequential probability ratio test. *Biometrika*, **79**, 347–353.

34 Group-Sequential Tests

B. W. Turnbull

34.1 Introduction

In a group-sequential or multistage procedure, accumulating data are analyzed at periodic intervals after each new *group* of observations has been obtained. There are a maximum of K such analyses, after which the experiment must stop. Here K is a prespecified integer, typically quite small, such as 2, 5, or 10. However, at each of the earlier analyses, termed *interim analyses*, the decision can be made to stop the experiment early if the data accrued at that time so warrant. The procedure then consists of a *stopping rule* and a *terminal decision rule*, the latter specifying what action should be taken on stopping. This action is typically the acceptance or rejection of a statistical hypothesis, but could also be the construction of a P value, or of a point or interval estimate for a parameter of interest. Thus group-sequential procedures offer a compromise between fixed-sample and fully sequential procedures in having the economic and ethical benefits of allowing the opportunity of early stopping without the need for *continuously* monitoring accumulating data, which is often impractical.

Typically, a group-sequential test with as few as 5 or 10 analyses can achieve most of the reduction in *expected* sample size afforded by many fully sequential designs, yet, unlike fully sequential procedures, the *maximum* sample size required is only slightly greater than that of the fixed-sample design. This latter fact can be of importance when planning the experiment. When the interim analyses performed are few and done only after each new group of observations has been obtained, estimation and testing procedures that are based on analytic asymptotic approximations developed for fully sequential designs are often of insufficient accuracy. However, the increasing power of the computer has enabled the use of efficient numerical methods to construct and evaluate group-sequential designs.

Group-sequential methods were first motivated by acceptance sampling schemes in industrial quality control. Dodge and Romig [20] introduced a two-stage acceptance sampling plan for attribute (i.e., binary) data. This plan is defined by six parameters: stage sample sizes n_1, n_2; acceptance numbers c_1, c_2, and rejection numbers d_1, d_2, where $d_1 > c_1 + 1$ and $d_2 = c_2 + 1$. The plan operates as follows. An initial group of n_1 items is taken, and if this contains less than c_1 defectives, the lot is accepted. If more than d_1 defectives are found, the lot is rejected. Otherwise the decision is delayed until a second group of size n_2 is examined, when the accept–reject decision is made according as the total cumulative number of defectives is $\leq c_2$ or $\geq d_2$. This idea is easily generalized to that of a multistage or multiple sampling plan in which up to $K(\geq 2)$ stages are permitted [7]. The multistage plans subsequently developed by the Columbia University Research Group [35] went on to form the basis of the United States military standard MIL-STD 105E [68] for acceptance sampling.

More recently, advances in group-sequential methodology have stemmed from the application to long-term medical trials. Here there is a need to monitor incoming data for ethical reasons, but also for economic and administrative reasons. Whereas continuous monitoring of trial data might be preferable, this is usually impractical, as data safety monitoring boards can only meet at periodic intervals to review the updated results. Although others had earlier proposed group-sequential procedures for medical trials (e.g. Elfring and Schultz [25], McPherson [66], Canner [10]), the major impetus for group-sequential methods came from Pocock [72]. He gave clear guidelines for the application of group-sequential experimental designs and demonstrated the versatility of the approach by showing that the nominal significance levels of repeated significance tests for normal responses can be used reliably for a variety of other responses and situations. In an article appearing shortly after Pocock's paper, O'Brien and Fleming [69] proposed a different class of group-sequential tests. These two papers have formed the basis for designs now in common use in clinical trial applications. Another key contribution was that of Lan and DeMets [58], who facilitated use of group sequential methods by extending their application to situations where the group sizes are unequal and unpredictable. This is particularly important when monitoring survival or longitudinal data, for example. Group-sequential tests can also be used in conjunction with *stochastic curtailment*.

34.2 Two-Sided Tests

Consider first testing hypotheses $H_0 : \theta = 0$ vs. $H_1 : \theta \neq 0$, where θ is a parameter of interest. In a fixed-sample procedure a specified number n of observations are taken and H_0 is rejected in favor of H_1 if and only if some test statistic Z, computable from the observed data, exceeds some critical value c in absolute value. Most two-sided hypothesis-testing situations can be expressed in this form:

Example 34.2.1 In the one-sample normal-data problem, we observe $X_1, \ldots, X_n$, independent normally distributed with unknown mean μ and known variance σ^2. We wish to test $H_0 : \mu = \mu_0$ vs. $H_1 : \mu \neq \mu_0$, where μ_0 is some specified value. Here we take $Z = (\bar{X} - \mu_0)/(\sigma/\sqrt{n})$ and $\theta = (\mu - \mu_0)/\sigma$.

Example 34.2.2 In the two-sample normal-data problem, where we compare two populations, A and B say, we might take observations $X_{A1}, \ldots, X_{An}$ from A and observations $X_{B1}, \ldots, X_{Bn}$ from B. The observations are all independent normally distributed with unknown means μ_A and μ_B in the two populations, respectively, and common known variance σ^2. To test $H_0 : \mu_A = \mu_B$, we set $\theta = (\mu_A - \mu_B)/\sigma\sqrt{2}$ and take $Z = (\bar{X}_A - \bar{X}_B)/(\sigma\sqrt{2}/\sqrt{n})$, where $\bar{X}_A$ denotes the mean of the $\{X_{Ai}\}$ observations and similarly $\bar{X}_B$.

Often, as in the two examples above, Z is normal or approximately normally distributed with mean 0 and variance 1 under H_0. In this case the choice of critical cutoff value $c = z_{\alpha/2}$, ensures that the size or type I error of this fixed sample test is

Table 1: A comparison of two-sided group-sequential tests with four groups.[a]

	Fixed	Pocock [72]	O'Brien and Fleming [69]	Haybittle [40] Peto et al[71].
	Critical Values (Nominal Levels)			
c_1 (α_1)	∞ (0)	2.36 (0.0183)	4.05 (0.0001)	3.00 (0.0027)
c_2 (α_2)	∞ (0)	2.36 (0.0183)	2.86 (0.0042)	3.00 (0.0027)
c_3 (α_3)	∞ (0)	2.36 (0.0183)	2.34 (0.0193)	3.00 (0.0027)
c_4 (α_4)	1.96 (.05)	2.36 (0.0183)	2.02 (0.0434)	1.98 (0.0477)
$N_{\max}/n_f$	1.000	1.183	1.022	1.010
	$ASN(\theta)/n_f$			
$\theta = 0$	1.000	1.156	1.016	1.007
$\theta = \pm 0.5\Delta$	1.000	1.041	0.964	0.978
$\theta = \pm 1.0\Delta$	1.000	0.797	0.767	0.806
$\theta = \pm 1.5\Delta$	1.000	0.437	0.573	0.535
$\theta = \pm 2.0\Delta$	1.000	0.335	0.472	0.372

[a] In this example, $K = 4$, and tests have size $\alpha = 0.05$ at $H_0 : \theta = 0$ and power 0.9 at $\theta = \pm\Delta$ for any given Δ. Entries for the maximum ($N_{\max}$) and expected sample sizes [ASN(θ)] are expressed as the ratio to the sample size n_f that would be required by the fixed-sample test of size α and power $1 - \beta = 0.9$ when $\theta = \pm\Delta$.

Source: Adapted From Reference 50, Table 3.

α. Here $z_{\alpha/2}$ denotes the upper $\alpha/2$ percentage point of the standard normal distribution; for example, if $\alpha = 0.05$, we have $z_{\alpha/2} = 1.96$.

In a group-sequential test with a specified maximum of K stages, we examine the accumulated data at each stage $k (1 \leq k \leq K)$. These comprise k groups of m observations each, say (observation pairs in the case of the two-sample problem). The test statistic Z_k is then computed, based on all the data accrued at that stage. If $|Z_k| \geq c_k$ for critical value c_k, the procedure stops and H_0 is rejected. Otherwise, if $k = K$, the procedure stops with acceptance of H_0, whereas if $k < K$, another group of observations is taken and the procedure continues to stage $k + 1$. The successive critical values (or *discrete boundary values*) $\{c_k, 1 \leq k \leq K\}$ are chosen so that the overall size or type I error probability is equal to the prespecified value α. This condition is

$$\Pr[|Z_1| \geq c_1 \text{ or } \cdots \text{ or } |Z_K| \geq c_K] = \alpha \quad (1)$$

when $\theta = 0$. We define the quantity

$$\pi_k = \Pr[|Z_1| < c_1, \ldots, |Z_{k-1}| < c_{k-1}, |Z_k| \geq c_k]$$

for $1 \leq k \leq K$, which is the probability of stopping at stage k and rejecting H_0. If we compute π_k under $H_0 : \theta = 0$, we have $\pi_1 + \cdots + \pi_K = \alpha$. Then π_k is termed the *exit probability* or *error spent* at stage k. This should not be confused with the *nominal significance level* at the kth stage, which is the marginal probability $\alpha_k = \Pr[|Z_k| \geq c_k | \theta = 0]$ for $1 \leq k \leq K$.

We turn now to the choice of the critical values $\{c_k; 1 \leq k \leq K\}$ for normally distributed statistics $\{Z_k\}$. Clearly, if $K > 1$, we can no longer take each $c_k = z_{\alpha/2}$, for then the type I error of the procedure is inflated. For example, if $K = 5$, and we set $c_1 = \cdots = c_5 = 1.96$, the size of the test, given by the expression (1), is 0.14 and not 0.05. This phenomenon is called the *multiple-looks bias* or the *optional-sampling bias*, and can be subtly present in analysis of experiments if they are really of a sequential nature but this is not recognized. The multiple-looks bias is part of a family of problems of multiplicity in statistics, which include multiple comparisons and variable subset selection. To compensate for repeated testing, Pocock [72] suggested using a common critical value $c_1 = \cdots = c_K = C_P$, say, which is inflated to guarantee (1). This also implies $\alpha_1 = \cdots = \alpha_K = \alpha'$, say, and thus Pocock's test can be considered as a *repeated significance test* as proposed by Armitage [3], since we are repeatedly testing at the same (adjusted) significance level α'. For example, if $K = 5$, then a choice of $C_P = 2.413$, or equivalently $\alpha' = 0.0158$, ensures that the size of the test given by (1) is 0.05. Alternatively, O'Brien and Fleming [69] proposed use of a series of critical values that are decreasing in k, namely, $c_k = \sqrt{K/k} C_{\text{OBF}}$, where the constant C_{OBF} is again chosen to satisfy (1). A convenient source for the constants C_P and C_{OBF} for $\alpha = 0.01, 0.05, 0.10$ and $K = 1, \ldots, 10$ is given in Jennison and Turnbull [48, Table 1].

Once the constants $\{c_k; 1 \leq k \leq K\}$ have been chosen, the power of the test can be computed by evaluating the probability (1) under any given alternative hypothesis, $\theta = \theta_1$. This probability will depend on the group size m and so this can be adjusted to guarantee a specified power, $1 - \beta$ say, at the given alternative θ_1. It may appear that obtaining the size and power of these group-sequential tests from (1) requires evaluation of a multiple integral involving a multivariate normal density, which is difficult numerically, especially when $K > 5$. Fortunately, the device of Armitage et al. [5] allows (1) to be evaluated as a succession of univariate integrals and this greatly facilitates the computation for any value of K. More details are given in Jennison [44]. The same computation yields the exit probabilities $\{\pi_k; 1 \leq k \leq K\}$, and from these the expected sample size (ASN) can be obtained from the expression

$$(1 - \pi_1 - \cdots - \pi_{K-1})mK + \sum_{k=1}^{K-1} mk\pi_k.$$

(This is the expected number of observations per sample in the two-sample problem.)

Table 1 illustrates the differences between the procedures. Here $K = 4$ and $\alpha = 0.05$. For the Pocock and the O'Brien–Fleming procedures, the constants that guarantee (1) are $C_P = 2.36$ and $C_{\text{OBF}} = 2.02$, from which the critical values are calculated. Note that the fixed-sample procedure can be viewed as a four-stage procedure in which $c_1 = c_2 = c_3 = \infty$ and $c_4 = 1.96$. The maximum ($N_{\max}$) and expected sample sizes [ASN(θ)] are expressed as ratios of the sample size, n_f, say, that would be required by the corresponding fixed test with the power 0.9 at the same alternative θ-value. The group size m for either the Pocock or the O'Brien–Fleming procedure is obtained by multiplying the corresponding $N_{\max}$ ratio by n_f and then dividing by $K = 4$. In practice, the resulting number must be rounded up to the next integer.

Although just one example, this table exemplifies the differences between the Pocock and the O'Brien–Fleming tests. The Pocock test has lower ASN when $|\theta|$ is large and the ethical imperative for early stopping is highest. Against this, the Pocock test has a high maximum sample size and a high expected sample size when $|\theta|$ is small. The O'Brien–Fleming test has boundaries (or critical values) that are wide for low k—thus very early stopping is more difficult. However, its maximum sample size is only a little larger than for that of the fixed test. Also, its final nominal significance level α_K is close to α, which means that the confusing situation in which H_0 is accepted but the fixed test would have rejected H_0 is unlikely to arise.

There are clearly many other ways to construct the critical constants $\{c_k; 1 \leq k \leq K\}$ subject to (1). Even stricter requirements for early stopping have been suggested by Haybittle [40] and Peto et al. [71]. These authors recommend using $c_1 = \cdots = c_{K-1} = 3$. Only a very small adjustment to the fixed test's critical value is then needed at the final analysis. The performance characteristics of this test are also included in Table 1. Other families of twosided group sequential tests have been proposed by Fleming et al. [33] and by Wang and Tsiatis [85]. Further comparisons are shown in Jennison and Turnbull [50].

Another possible modification is the introduction of an inner boundary to also allow early termination with acceptance of H_0; group-sequential tests with this feature have been proposed [4,5,27,37,70,89].

34.3 Unequal and Unpredictable Group Sizes

The methodology described so far has assumed that the group sizes m are all equal. In practice they will be unequal or even unpredictable. Pocock [72] suggested that small variations in group sizes might be ignored and the nominal significance levels $\{\alpha_k; 1 \leq k \leq K\}$ appropriate to equal-sized groups be employed at each analysis. Slud and Wei [77] presented an exact solution to this problem in which the total type I error is partitioned between analyses. For a study with K analyses, probabilities $\pi_1, \ldots, \pi_K$, summing to α, are specified and critical values $\{c_k; 1 \leq k \leq K\}$ are found such that the unconditional probability of wrongly rejecting H_0 at analysis k is equal to π_k. These critical values are calculated successively using numerical integration; the kth value depends on the cumulative sample sizes, denoted $n(1), \ldots, n(k)$, say, but not on the as yet unobserved $n(k+1), \ldots, n(K)$.

A similar approach is proposed by Lan and DeMets [58]. Whereas Slud and Wei specify the probabilities $\pi_1, \ldots, \pi_K$ at the outset, Lan and DeMets spend type I error at a prespecified *rate*. Before implementing the Lan–DeMets method, a maximum sample size $N_{\max}$ must be specified. The type I error is then partitioned according to an *error spending* or *use* function, $f(t)$, where f is nondecreasing, $f(0) = 0$ and $f(t) = \alpha$ for $t \geq 1$. The error allocated to analysis k is $\pi_k = f(n(k)/N_{\max}) - f(n(k-1)/N_{\max})$ for $k \geq 1$, and critical values c_k are computed as in Slud and Wei's method. Lan and DeMets [58] and Kim and DeMets [56] propose a variety of functions $f(t)$. One convenient family of functions, namely, $f(t) = \min[\alpha t^\rho, \alpha]$ for $\rho > 0$, provides a good range of Lan–DeMets procedures and includes boundaries that approximate the Pocock and the O'Brien–Fleming tests at $\rho = 0.8$ and $\rho = 3$, respectively, when group sizes are equal. The Lan–DeMets method has flexibility in that the number of analyses, K, need not be fixed in advance, although it is still necessary to specify $N_{\max}$.

34.4 One-Sided Tests

In the one-sided formulation, we test hypotheses that can be put in the form $H_0 : \theta \leq 0$ vs. $H_1 : \theta > 0$. This is often appropriate when it is desired to test if a new "treatment" is better than a standard or control. There are now two sets of critical values $\{c_k, d_k; 1 \leq k \leq K\}$ with $c_k < d_k (1 \leq k \leq K - 1)$ and $c_K = d_K$. At stage k, we stop to accept H_1 if $Z_k \geq d_k$, and stop to accept H_0 if $Z_k \leq c_k$; else the procedure continues to stage $k+1$. The condition $c_K = d_K$ ensures termination at or before stage K. The boundary values $\{c_k, d_k\}$ are said to be *symmetric* if $c_k = -d_k$ for $1 \leq k \leq K$. DeMets and Ware [18,19] proposed asymmetric procedures that were based on modifications of two-sided tests. Whitehead [88] and Whitehead and Stratton [89] describe a one-sided test that has a triangular continuation region when $\sqrt{n(k)}Z_k$ is plotted against the cumulative sample size $n(k)$. The symmetric version has critical values $d_k = -c_k = [a - \Delta n(k)/4]/\sqrt{n(k)}$, where $\Delta > 0$. With normal data, groups of equal size m so $n(k) = km$, and $a = a(m) = -(2/\Delta)\log(2\alpha) - 0.583\sqrt{m}$, this procedure ensures that type I and type II error probabilities are both equal to the specified α at $\theta = -\Delta$ and $\theta = +\Delta$, respectively. (The term $0.583\sqrt{m}$ is needed to correct for overshoot—see Siegmund [76, p. 50].) The required group size m can be found from the condition that $c_K = d_K$, which leads to a quadratic equation to be solved for m. Jennison [43] considered the choice of critical values for a one-sided test that was optimal with respect to a criterion of minimizing the expected sample size. Emerson and Fleming [27] have developed a one-parameter family of symmetric procedures that are almost fully efficient when compared to the optimal tests of Jennison [43]. See also Xiong [91].

34.5 Inference upon Termination

On conclusion of a sequential experiment, a more complete analysis is usually required than a simple accept–reject decision of a hypothesis test. The construction of point and interval estimates of θ is made complicated because the sampling distributions of the maximum-likelihood estimator are typically skewed and multimodal and lack a monotone likelihood ratio, unlike the fixed-sample situation. For instance, in the normal-data Example 33.2.1, the sample mean can have considerable bias when estimating the true mean response. Whitehead [87] has proposed a bias-adjusted point estimate. Emerson and Fleming [28] have described the construction of the uniformly minimum-variance unbiased estimator (UMVUE). Its computation is difficult; a computer program is described by Emerson [26]. There have been several different methods proposed for constructing confidence intervals following a group-sequential test. (For normal data, see Refs. [12, 28, and 83]; for binomial data, see Refs. [13, 21, and 45].) In each case the calculation of the interval estimate is difficult, involving computation of iterated integrals or sums, similar to that used in evaluating (1). A similar computation is needed to compute the P value—see Fairbanks and Madsen [30].

34.6 Repeated Confidence Intervals

Repeated confidence intervals (RCIs) with level $1 - \alpha$ for a parameter θ are defined as sequences of intervals I_k, $k = 1, \ldots, K$, with the property:

$$P_\theta[\theta \in I_k \text{ for all } k \quad (1 \leq k \leq K)]$$
$$= 1 - \alpha \text{ for all } \theta. \quad (2)$$

Here $I_k = (\underline{\theta}_k, = \bar{\theta}_k)$ say, is an interval computed from the information available at analysis, k, $k = 1, \ldots, K$. For the normal-data examples, the $1 - \alpha$ RCI for θ at stage $k (1 \leq k \leq K)$ is simply given by $[(Z_k - c_k)/\sqrt{n(k)}, (Z_k + c_k)/\sqrt{n(k)}]$, where c_k satisfies (1) but with $\alpha/2$ replacing α. The interval I_k provides a statistical summary of the information about the parameter θ at the kth analysis, automatically adjusted to compensate for repeated looks at the accumulating data. Because the coverage probability is guaranteed simultaneously, the probability that I_τ contains θ is also guaranteed to be no less than $1 - \alpha$ for all θ and for any random stopping time τ taking values in $\{1, 2, \ldots, K\}$. The fact that the coverage probability requirement is maintained independent of any stopping rule is very important. In medical trial applications, the stopping boundaries of the group sequential designs discussed in earlier sections are often used only as guidelines, not as strict rules [17]. Other factors, such as side effects, financial cost, and reports of new scientific developments from outside the trial, will often influence the monitoring committee in their decision to stop or to continue the trial. Of course, if the group sequential stopping rule will be strictly followed, then at termination the RCI will be conservative. In that case, we would use the techniques of the previous section to construct the final confidence interval. However, RCIs could still be used to summarize results at intermediate stages—the basic RCI property (2) ensures that these interim results will not be "overinterpreted." *Overinterpretation* refers to the fact that, if the multiple looks bias or optional sampling bias is ignored, results may seem more significant than warranted, which in turn can lead to adverse effects on the conduct of the trial and to pressure to unblind or terminate a study prematurely.

The idea of such "confidence sequences" was originally due to Robbins [73] and was adapted to group sequential procedures by Jennison and Turnbull [46] and by Lai [57]. A thorough treatment of RCIs was given by Jennison and Turnbull [48], who showed how they could be applied to discrete, survival, epidemiological, and other outcome measures.

34.7 Literature

Armitage [1–3] and Bross [8,9] pioneered the use of sequential methods in the medical field, particularly for comparative clinical trials. The shift to group sequential methods did not occur until the 1970s: Elfring and Schultz [25] proposed group sequential designs for comparing two treatments with binary response; McPherson [66] suggested that the repeated significance tests of Armitage et al. [5] might be used to analyze clinical trial data at a small number of interim analyses; Canner [10] used Monte Carlo simulation to find critical values of a test statistic in a survival study with periodic analyses. Two- and three-stage procedures for the case of normal response were developed in References 4, 22, 74, and 75. Hewett and Spurrier [41] have provided a review of specifically two-stage procedures.

In practice, in long-term medical trials there is a tendency of the data safety monitoring committees to delay the termination of the trial to make sure that a "statistically significant trend continues." To formalize this, Canner [10,11] and Falissard and Lellouch [31,32] have described a *succession* procedure, whereby the trial is stopped early only if the results of r successive interim analyses are significant at level α'. Here r is typically a small integer (e.g., 2) and α' is chosen so that the overall size of the test is equal to the specified α. Now, however, the stopping criterion is no longer a function of a sufficient statistic.

The discussion here has concentrated on methods for normal responses. However, since group-sequential tests are often based on sums of observations, by the central-limit theorem such sums will be approximately normal even if the responses are not normally distributed, provided the number of responses is not too small. More generally, a test designed for normal responses may be applied whenever a sequence of test statistics has approximately the same distribution as that of a sequence of sums of independent normal variates. One important example that meets this requirement is the sequence of log rank statistics for testing the equality of two survival distributions with accumulating censored survival data (e.g., see Refs. 39, and 81). Lin and Wei [64] propose an alternative procedure for survival data. Group-sequential tests and repeated confidence intervals for the estimated median of survival distributions have also appeared [47,55]. For survival data with covariates, procedures have been described [38,63,82,84]. Group-sequential procedures for longitudinal data have also been discussed [6,36,59,60,86,90]. The problem of monitoring bivariate or multivariate data has also been treated [15,16,53,79,80]. The problem of comparing three or more treatments has been considered most recently by Hughes [42] and by Follman et al. [34]. Group-sequential tests for equivalence trials have also been proposed that are based on the use of repeated confidence intervals (see Durrleman and Simon [23], and Jennison and Turnbull [54]).

For comparing two treatments with dichotomous responses, it is possible to obtain exact methods, rather than normal approximations (e.g., Lin et al. [65], Coe and Tamhane [14]). For stratified binomial data, Jennison and Turnbull [51] describe the construction of repeated confidence intervals for the Mantel–Haenszel odds-ratio estimator. For the case of normal observations with unknown variance, exact group sequential versions of the t, χ^2, and F tests are described in Jennison and Turnbull [52]. Exact methods for linear rank statistics have been developed by Mehta et al. [67].

The Bayesian approach to inference in sequential designed experiments is more straightforward than the frequentist approach. Bayesian inference procedures based on posterior distributions are unaffected by the sequential nature of the experimental design. Spiegelhalter et al. [78] and Lewis and Berry [61] have discussed frequentist properties of a Bayesian approach. Assigning costs to an incorrect decision and to each observation taken, Eales and Jennison [24] formulated a Bayes sequential decision problem and found that the optimal solution is quite similar to the classical group-sequential designs described earlier. For a detailed comparison of various statistical approaches to interim monitoring of sequential trials, see Jennison and Turnbull [49].

Literature Update

This chapter was written in 1996. Naturally, a number of research publications on the subject have appeared since then; however, the basic ideas and methods remain as outlined here. Additionally there have been several books that give a comprehensive treatment of the theory, methods and applications of group sequential tests, e.g., Jennison and Turnbull [S1] and Proschan, Lan and Wittes [S2].

Group-sequential designs are flexible and adaptive in the sense that [(1)] they allow for unequal and unpredictable group sizes through use of error-spending functions; [(2)] can accommodate nuisance parameters through use of information monitoring; and [(3)]) above all, respond to accumulating positive or negative results by the use of stopping boundaries.

However, the future group sizes and their timing are *not* allowed to depend on the current interim outcome value or else the type I error may be elevated (see Lan and DeMets [S3], Jennison and Turnbull [S4], Proschan and Hunsberger [S5]). To respond to this feature, a new class of so-called *adaptive group sequential tests* has been proposed. These tests do permit midcourse changes in the design based interim outcome data while still preserving the pre-specified type I error. In particular, such adaptive changes in the sample size, sampling frequencies and group sizes are permitted. Decisions typically need to be based on a non-standard test statistic, namely, a weighted combination of Z statistics from the different stages. For details, the reader is referred to Bauer and Köhne [S6], Fisher [S7], Lehmacher and Wassmer [S8], Cui et al. [S9], and Müller and Schäfer [S10]. The procedures can be somewhat inefficient (see Jennison and Turnbull [S11, S12]) and in rare circumstances they can lead to paradoxical results (see Burmann and Sonneson [S13]). However, depending on the circumstance, these two drawbacks may sometimes be considered a worthwhile price to be paid for the gains in flexibility.

Finally, it should be noted that the ideas of group sequential tests can be extended to a much wider variety of situations than that of just deciding between two hypotheses. These applications include multivariate outcomes, regression problems, crossover designs, equivalence trials, experiments with several non-inferiority margins, multiple treatments, dose selection, "enrichment" and trials. For further reading, see Jennison and Turnbull [S1, S14] and Schmidli et al. [S15], for example.

References

1. Armitage, P. (1954). Sequential tests in prophylactic and therapeutic trials. *Quart. J. Med.*, **23**, 255–274.

2. Armitage, P. (1958). Sequential methods in clinical trials. *Amer. J. Public Health*, **48**, 1,395–1,402.

3. Armitage, P. (1975). *Sequential Medical Trials*. Blackwell, Oxford.

4. Armitage, P. and Schneiderman, M. (1958). Statistical problems in a mass screening program. *Ann. NY Acad. Sci.*, **76**, 896–908.

5. Armitage, P., McPherson, C. K., and Rowe, B. C. (1969). Repeated significance tests on accumulating data. *J. Rpy. Statist. Soc. A*, **132**, 235–244.

6. Armitage, P., Stratton, I. M., and Worthington, H. V. (1985). Repeated significance tests for clinical trials with a fixed number of patients and variable follow-up. *Biometrics*, **41**, 353–359.

7. Bartky, W. (1943). Multiple sampling with constant probability. *Ann. Math. Statist.*, **14**, 363–377.

8. Bross, I. (1952). Sequential medical plans. *Biometrics*, **8**, 188–205.

9. Bross, I. (1958). Sequential clinical trials. *J. Chron. Dis.*, **8**, 349–365.

10. Canner, P. L. (1977). Monitoring treatment differences in long-term clinical trials. *Biometrics*, **33**, 603–615.

11. Canner, P. L. (1984). Monitoring long-term clinical trials for beneficial and adverse treatment effects. *Commun. Statist. A*, **13**, 2369–2394.

12. Chang, M. N. (1989). Confidence intervals for a normal mean following a group sequential test. *Biometrics*, **45**, 247–254.

13. Chang, M. N., and O'Brien, P. C. (1986). Confidence intervals following group sequential tests. *Controlled Clin. Trials*, **7**, 18–26.

14. Coe, P. R. and Tamhane, A. C. (1993). Exact repeated confidence intervals for Bernoulli parameters in a group sequential clinical trial. *Controlled Clin. Trials*, **14**, 19–29.

15. Cook, R. J. (1994). Interim monitoring of bivariate responses using repeated confidence intervals. *Controlled Clin. Trials*, **15**, 187–200.

16. Cook, R. J. and Farewell, V. T. (1994). Guidelines for minority efficacy and toxicity responses in clinical trials. *Biometrics*, **50**, 1,146–1,152.

17. DeMets, D. L. (1984). Stopping guidelines vs stopping rules: a practitioner's point of view. *Commun. Statist. Theory Methods A*, **13**(19), 2395–2418.

18. DeMets, D. L. and Ware, J. H. (1980). Group sequential methods for clinical trials with one-sided hypothesis. *Biometrika*, **67**, 651–660.

19. DeMets, D. L. and Ware, J. H. (1982). Asymmetric group sequential boundaries for monitoring clinical trials. *Biometrika*, **69**, 661–663.

20. Dodge, H. F. and Romig, H. G. (1929). A method for sampling inspection. *Bell Syst. Tech. J.*, **8**, 613–631.

21. Duffy, D. E. and Santner, T. J. (1987). Confidence intervals for a binomial parameter based on multistage tests. *Biometrics*, **43**, 81–93.

22. Dunnett, C. W. (1961). The statistical theory of drug screening. In *Quantitative Methods in Pharmacology*, pp. 212–231, North-Holland, Amsterdam.

23. Durrleman, S. and Simon, R. (1990). Planning and monitoring of equivalence studies. *Biometrics*, **46**, 329–336.

24. Eales, J. D. and Jennison, C. (1992). An improved method for deriving optimal one-sided group sequential tests. *Biometrika*, **79**, 13–24.

25. Elfring, G. L. and Schultz, J. R. (1973). Group sequential designs for clinical trials. *Biometrics*, **29**, 471–477.

26. Emerson, S. S. (1993). Computation of the minimum variance unbiased estimator of a normal mean following a group sequential trial. *Comput. Biomed. Res.*, **26**, 68–73.

27. Emerson, S. S. and Fleming, T. R. (1989). Symmetric group sequential designs. *Biometrics*, **45**, 905–923.

28. Emerson, S. S. and Fleming, T. R. (1990). Parameter estimation following group sequential hypothesis testing. *Biometrika*, **77**, 875–892.

29. Etzioni, R. and Pepe, M. (1994). Monitoring of a pilot toxicity study with two adverse outcomes. *Statist. Med.*, **13**, 2,311–2,322.

30. Fairbanks, K. and Madsen, R. (1982). P values for tests using a repeated significance test design. *Biometrika*, **69**, 69–74.

31. Falissard, B. and Lellouch, J. (1992). A new procedure for group sequential analysis. *Biometrics*, **48**, 373–388.

32. Falissard, B. and Lellouch, J. (1993). The succession procedure for interim analysis: extensions for early acceptance of H_0 and for flexible times of analysis. *Statist. Med.*, **12**, 41–67.

33. Fleming, T. R., Harrington, D. P., and O'Brien, P. C. (1984). Designs for group sequential tests. *Controlled Clin. Trials*, **5**, 348–361.

34. Follman, D. A., Proschan, M. A., and Geller, N. L. (1994). Monitoring pairwise comparisons in multi-armed clinical trials. *Biometrics*, **50**, 325–336.

35. Freeman, H. A., Friedman, M., Mosteller, F., and Wallis, W. A. (1948). *Sampling Inspection*. McGraw-Hill, New York.

36. Geary, D. N. (1988). Sequential testing in clinical trials with repeated measurements. *Biometrika*, **75**, 311–318.

37. Gould, A. L. and Pecore, V. J. (1982). Group sequential methods for clinical trials allowing early acceptance of H_0 and incorporating costs. *Biometrika*, **69**, 75–80.

38. Gu, M. and Ying, Z. (1993). Sequential analysis for censored data. *J. Am. Statist. Assoc.*, **88**, 890–898.

39. Harrington, D. P., Fleming, T. R., and Green, S. J. (1982). Procedures for serial testing in censored survival data. In *Survival Analysis*, J. Crowley and R. A. Johnson, eds., pp. 269–286, Institute of Mathematical Statistics, Hayward, CA.

40. Haybittle, J. L. (1971). Repeated assessment of results in clinical trials of cancer treatment. *Brit. J. Radiol.*, **44**, 793–797.

41. Hewett, J. E. and Spurrier, J. D. (1983). A survey of two stage tests of hypotheses: Theory and applications. *Commun. Statist. A*, **12**, 2307–2425.

42. Hughes, M. D. (1993). Stopping guidelines for clinical trials with multiple treatments. *Statist. Med.*, **12**, 901–915.

43. Jennison, C. (1987). Efficient group sequential tests with unpredictable group sizes. *Biometrika*, **74**, 155–165.

44. Jennison, C. (1994). Numerical computations for group sequential tests. In *Computing Science and Statistics*, Vol. 25, M. Tarter and M. D. Lock, eds., pp. 263–272m Interface Foundation of America.

45. Jennison, C. and Turnbull, B. W. (1983). Confidence intervals for a binomial parameter following a multistage test with application to MIL-STD 105D and medical trials. *Technometrics*, **25**, 49–58.

46. Jennison, C. and Turnbull, B. W. (1984). Repeated confidence intervals for group sequential clinical trials. *Controlled Clin. Trials*, **5**, 33–45.

47. Jennison, C. and Turnbull, B. W. (1985). Repeated confidence intervals for the median survival time. *Biometrika*, **72**, 619–625.

48. Jennison, C. and Turnbull, B. W. (1989). Interim analyses: the repeated confidence interval approach (with discussion). *J. Roy. Statist. Soc. B*, **51**, 305–361.

49. Jennison, C. and Turnbull, B. W. (1990). Statistical approaches to interim monitoring of medical trials: A review and commentary. *Statist. Sci.*, **5**, 299–317.

50. Jennison, C. and Turnbull, B. W. (1991). Group sequential tests and repeated confidence intervals. In *Handbook of Sequential Analysis*, B. K. Ghosh and P. K. Sen, eds., Chap. 12, pp. 283–311, Marcel Dekker, New York.

51. Jennison, C. and Turnbull, B. W. (1991). A note on the asymptotic joint distribution of successive Mantel–Haenszel estimates of the odds ratio based on accumulating data. *Sequential Anal.*, **10**, 201–209.

52. Jennison, C. and Turnbull, B. W. (1991). Exact calculations for sequential t, χ^2 and F tests. *Biometrika*, **78**, 133–141.

53. Jennison, C. and Turnbull, B. W. (1993). Group sequential tests for bivariate response: Interim analyses of clinical trials with both efficacy and safety endpoints. *Biometrics*, **49**, 741–752.

54. Jennison, C. and Turnbull, B. W. (1993). Sequential equivalence testing and repeated confidence intervals, with applications to normal and binary response. *Biometrics*, **49**, 31–43.

55. Keaney, K. M. and Wei, L. J. (1994). Interim analyses based on median survival times. *Biometrika*, **81**, 270–286.

56. Kim, K. and DeMets, D. L. (1987). Design and analysis of group sequential tests based on the type I error spending rate function. *Biometrika*, **74**, 149–154.

57. Lai, T. L. (1984). Incorporating scientific, ethical and economic considerations into the design of clinical trials in the pharmaceutical industry: A sequential approach. *Commun. Statist. Theory Methods A*, **13**(19), 2,355–2,368.

58. Lan, K. K. G. and DeMets, D. L. (1983). Discrete sequential boundaries for clinical trials. *Biometrika*, **70**, 659–663.

59. Lee, J. W. and DeMets, D. L. (1991). Sequential comparison of changes with repeated measurements data. *J. Am. Statist. Assoc.*, **86**, 757–762.

60. Lee, J. W. and DeMets, D. L. (1992). Sequential rank tests with repeated measurements in clinical trials. *J. Am. Statist. Assoc.*, **87**, 136–142.

61. Lewis, R. J. and Berry, D. A. (1994). Group sequential clinical trials: A classical evaluation of Bayesian decision-theoretic designs. *J. Am. Statist. Assoc.*, **89**, 1,528–1,534.

62. Lin, D. Y. (1991). Nonparametric sequential testing in clinical trials with incomplete multivariate observations. *Biometrika*, **78**, 120–131.

63. Lin, D. Y. (1992). Sequential log rank tests adjusting for covariates with the accelerated life model. *Biometrika*, **79**, 523–529.

64. Lin, D. Y. and Wei, L. J. (1991). Repeated confidence intervals for a scale change in a sequential survival study. *Biometrics*, **47**, 289–294.

65. Lin, D. Y., Wei, L. J., and DeMets, D. L. (1991). Exact statistical inference for

group sequential trials. *Biometrics*, **47**, 1,399–1,408.

66. McPherson, K. (1974). Statistics: The problem of examining accumulating data more than once. *New Engl. J. Med.*, **290**, 501–502.

67. Mehta, C. R., Patel, N., Senchaudhuri, P., and Tsiatis, A. (1994). Exact permutational tests for group sequential clinical trials. *Biometrics*, **50**, 1,042–1,053.

68. MIL-STD-105E (1989). *Military Standard Sampling Procedures and Tables for Inspection by Attributes*. US Government Printing Office, Washington, DC.

69. O'Brien, P. C. and Fleming, T. R. (1979). A multiple testing procedure for clinical trials. *Biometrics*, **35**, 549–556.

70. Pampallona, S. and Tsiatis, A. A. (1994). Group sequential designs for one-sided and two-sided hypothesis testing with provision for early stopping in favor of the null hypothesis. *J. Statist. Plan. Infer.*, **42**, 19–35.

71. Peto, R., Pike, M. C., Armitage, P., Breslow, N. E., Cox, D. R., Howard, S. V., Mantel, N., McPherson, K., Peto, J., and Smith, P. G. (1976). Design and analysis of randomized clinical trials requiring prolonged observation of each patient. I. Introduction and design. *Br. J. Cancer*, **34**, 585–612.

72. Pocock, S. J. (1977). Group sequential methods in the design and analysis of clinical trials. *Biometrika*, **64**, 191–199.

73. Robbins, H. (1970). Statistical methods related to the law of the iterated logarithm. *Ann. Math. Statist.*, **41**, 1,397–1,409.

74. Roseberry, T. D. and Gehan, E. A. (1964). Operating characteristic curves and accept–reject rules for two and three stage screening procedures. *Biometrics*, **20**, 73–84.

75. Schneiderman, M. (1961). Statistical problems in the screening search for anti-cancer drugs by the National Cancer Institute of the United States. In *Quantitative Methods in Pharmacology*, pp. 232–246, North-Holland, Amsterdam.

76. Siegmund, D. (1985). *Sequential Analysis*. Springer-Verlag, New York.

77. Slud, E. V. and Wei, L.-J. (1982). Two-sample repeated significance tests based on the modified Wilcoxon statistic. *J. Am. Statist. Assoc.*, **77**, 862–868.

78. Spiegelhalter, D. J., Freedman, L. S., and Parmar, M. K. B. (1994). Bayesian approaches to clinical trials (with discussion). *J. Roy. Statist. Soc. A*, **157**, 357–416.

79. Su, J. Q. and Lachin, J. M. (1992). Group sequential distribution-free methods for the analysis of multivariate observations. *Biometrics*, **48**, 1033–1042.

80. Tang, D., Gnecco, C., and Geller, N. L. (1989). Design of group sequential clinical trials with multiple endpoints. *J. Am. Statist. Assoc.*, **84**, 776–779.

81. Tsiatis, A. A. (1981). The asymptotic joint distribution of the efficient scores test for the proportional hazards model calculated over time. *Biometrika*, **68**, 311–315.

82. Tsiatis, A. A., Boucher, H., and Kim, K. (1995). Sequential methods for parametric survival models. *Biometrika*, **82**, 165–173.

83. Tsiatis, A. A., Rosner, G. L., and Mehta, C. R. (1984). Exact confidence intervals following a group sequential test. *Biometrics*, **40**, 797–803.

84. Tsiatis, A. A., Rosner, G. L., and Tritchler, D. L. (1985). Group sequential tests with censored survival data adjusting for covariates. *Biometrika*, **72**, 365–373.

85. Wang, S. K. and Tsiatis, A. A. (1987). Approximately optimal one-parameter boundaries for group sequential trials. *Biometrics*, **43**, 193–200.

86. Wei, L. J., Su, J. Q., and Lachin, M. J. (1990). Interim analyses with repeated measurements in a sequential clinical trial. *Biometrika*, **77**, 359–364.

87. Whitehead, J. (1986). Supplementary analysis at the conclusion of a sequential clinical trial. *Biometrics*, **42**, 461–471.

88. Whitehead, J. (1992). *The Design and Analysis of Sequential Clinical Trials*, 2nd ed. Ellis Horwood, Chichester, UK.

89. Whitehead, J. and Stratton, I. (1983). Group sequential clinical trials with triangular continuation regions. *Biometrics*, **39**, 227–236.

90. Wu, M. C. and Lan, K. K. G. (1992). Sequential monitoring for comparison of changes in a response variable in clinical studies. *Biometrics*, **48**, 765–779.

91. Xiong, X. (1995). A class of sequential conditional probability ratio tests. *J. Am. Statist. Assoc.*, **90**, 1,463–1,473.

Supplemental Bibliography

[S1] Jennison, C. and Turnbull, B. W. (2000). *Group Sequential Methods with Applications to Clinical Trials*. Chapman & Hall/CRC, Boca Raton, FL.

[S2] Proschan, M.A., Lan, K. K. G. and Wittes, J. T. (2007). *Statistical Monitoring of Clinical Trials: A Unified Approach*. Springer-Verlag, New York.

[S3] Lan, K. K. G. and DeMets, D. L. (1989). Changing frequency of interim analysis in sequential monitoring. *Biometrics*, **45**, 1017–1020.

[S4] Jennison, C. and Turnbull, B. W. (1991). Group sequential tests and repeated confidence intervals. In *Handbook of Sequential Analysis* B. K. Ghosh and P. K. Sen, eds., pp. 283–311, Marcel Dekker, New York.

[S5] Proschan, M. A. and Hunsberger, S. A. (1995). Designed extension of studies based on conditional power. *Biometrics*, **51**, 1315–1324.

[S6] Bauer, P. and Köhne, K. (1994). Evaluation of experiments with adaptive interim analyses. *Biometrics*, **50**, 1029–1041. Correction: *Biometrics*, **52**, (1996), 380.

[S7] Fisher, L.D. (1998). Self-designing clinical trials. *Statist. Med.* **17**, 1551–1562.

[S8] Lehmacher, W. and Wassmer, G. (1999). Adaptive sample size calculation in group sequential trials. *Biometrics*, **55**, 1286–1290.

[S9] Cui, L., Hung, H. M. J., and Wang, S.-J. (1999). Modification of sample size in group sequential clinical trials. *Biometrics*, **55**, 853–857.

[S10] Müller, H.-H. and Schäfer, H. (2001). Adaptive group sequential designs for clinical trials: Combining the advantages of adaptive and of classical group sequential procedures. *Biometrics*, **57**, 886–891.

[S11] Jennison, C. and Turnbull, B.W. (2006a). Adaptive and nonadaptive group sequential tests. *Biometrika* **93**, 1–21.

[S12] Jennison, C. and Turnbull, B.W. (2006b). Efficient group sequential designs when there are several effect sizes under consideration. *Statist. Med.* **35**, 917–932.

[S13] Burman, C-F. and Sonesson, C. (2006). Are flexible designs sound? *Biometrics*, **62**, 664–669.

[S14] Jennison, C. and Turnbull, B. W. (2007). "Adaptive seamless designs: Selection and prospective testing of hypotheses." *J. Biopharm. Statist.*. **17**, 1135–1161.

[S15] Schmidli, H., Bretz, F., Racine, A. and Maurer, W. (2006). Confirmatory seamless phase II/III clinical trials with hypotheses selection at interim: Applications and practical considerations. *Biometr. J.* **48**, 635–643.

35 Grouped Data in Survival Analysis

James A. Deddens and Gary G. Koch

35.1 Introduction

There are many types of investigations in which data for lifetimes (or perhaps waiting times until certain events) are expressed as classifications into a set of mutually exclusive intervals rather than as specific values. Some examples that illustrate such grouped survival data are as follows:

E1. An experiment is undertaken to compare treatments for survival of bacteria infected mice. The mice are inspected every 6 hours for the event of death; see Bowdre et al. [2].

E2. Patients for whom a health disorder was recently treated successfully are evaluated for recurrence by diagnostic procedures at a specific set of follow-up times; see Johnson and Koch [14].

E3. Data from a large study of graft survival after kidney transplant operations are summarized for a cross-classification of donor relationship and match grade in life tables with a specific set of intervals; see Laird and Olivier [18].

E4. A large study of female patients with cancer of the cervix uteri has its data summarized in a life table that encompasses deaths due to cancer of the cervix uteri and deaths due to other causes for a specific set of intervals; see Chiang [4] and Example 12.1 in Elandt-Johnson and Johnson [8].

Grouped survival data arise in studies like E1 and E2 because of the periodic monitoring of subjects in their research designs. This method of observation only provides information on whether the event of interest occurred between two follow-up assessments rather than the exact lifetime. However, for many situations, it is the only feasible method with respect to ethical or resource considerations. Some additional examples for which periodic monitoring is used are the detection of health outcomes (e.g., abnormal electrocardiogram or failure of dental restorations), the emergence of certain psychological characteristics (e.g., memory skills), or the inspection of equipment for maintenance purposes.

For studies like E3 and E4, the grouped survival data occur as a consequence of the structure from life tables. Relatedly, when sample sizes are large, a life table can arise from the aggregation for the level of measurement which is used for actual life times (e.g., times of death expressed in months instead of days). In other words, the concept of grouped survival data applies to situations where a continuous monitoring process yields many ties due to roundoff error. Another consideration is that survival times may be determined in different ways for two or more potential causes of failure (e.g., in days for death and in 3-month intervals for recurrence), and so a life table provides a convenient format for summarizing this information in a common way. The discussion here has emphasized how life tables result from primary survival data for subjects, but often they serve as the only available data source for secondary analyses.

As is often the case for studies of lifetimes, grouped survival information can involve censored data. For example, in a clinical trial E2, a subject might be lost to follow-up or withdrawn from risk at its termination date before experiencing recurrence. There also can be vectors of explanatory variables for subjects with grouped survival data; see E3. These might refer to treatment group, demographic characteristics, or medical history status. They can be either continuous or categorical. Continuous explanatory variables are often grouped into a finite number of categories for situations with grouped survival data. Finally, in some studies, there might be several competing risks as potential causes of failure; see E4.

35.2 Data Structure

Since grouped survival data are based on classifications of the follow-up status of subjects with respect to the internal of occurrence of some event, two general formats for their display are contingency tables and life tables. For the contingency table, there are s rows and $(rc+1)$ columns. The s rows correspond to s distinct samples of subjects and the $(rc+1)$ columns encompass the potential outcomes of final observation during one of r time intervals due to one of c causes or the maintenance of survival of all causes through the entire follow-up period. The entries in the contingency table are the frequencies $\{n_{ijk}\}$, where $i = 1, 2, \ldots, s$ indexes the set of samples, $j = 1, 2, \ldots, r$ indexes the successive time intervals $(y_{j-1}, y_j]$ with $y_0 = 0$, and $k = 1, 2, \ldots, c$ indexes the causes for the termination of observation (e.g., death due to cause of interest, death due to some other cause, lost to follow-up, etc.). Also, the combination $j = r, k = 0$ denotes survival of all causes for all time intervals. The size of the ith sample is

$$n_i = \sum_{j=1}^{r} \sum_{k=1}^{c} n_{ijk} + n_{ir0}.$$

Alternatively, for each of the s samples, a life table can be constructed from the corresponding row of the contingency table. It has r rows from the respective time intervals and $(c+1)$ columns for the c causes and the outcome of survival of all causes during the time interval of the row; thus, the entries in its first c columns are the $\{n_{ijk}\}$ and those in its last column are the numbers of subjects

$$n_{ij0} = \sum_{j'=j+1}^{r} \sum_{k=1}^{c} n_{ij'k} + n_{ir0} \quad (1)$$

who survive all causes through the end of the jth interval (i.e., final observation occurs after the jth interval).

Most applications involve only two causes for termination of observation. One ($k = 1$) corresponds to the occurrence of a failure event such as death or recurrence of a disorder, and the other ($k = 2$) corresponds to a withdrawal from risk (i.e., censoring) event such as lost to follow-up or violations of research design specifications (e.g., usage of supplementary treatments). Also, it is usually assumed that the withdrawal events are unrelated to the failure events. Subsequent discussion is primarily concerned with this specific type of situation. Further consideration of the more general framework with c causes is given in Chiang [5], Elandt-Johnson and Johnson [8], Gail [10], Johnson and Koch [14], and Larson [19].

35.3 Statistical Methods

The main strategies for analyzing grouped survival data are nonparametric methods and model fitting methods based on maximum likelihood and weighted least squares. When only minimal assumptions (e.g., random sampling, data integrity) apply, a non-parametric approach provides useful ways for constructing descriptive estimates and hypothesis tests concerning the observed experience of the respective groups. For example, alternative estimates include the following functions of the frequencies $\{n_{ijk}\}$ in life table format for $c = 2$ causes:

(1) The ratio for the number of failures per unit time

$$f_i = \sum_{j=1}^{r} n_{ij1} \bigg/ \sum_{j=1}^{r} N_{ij}. \quad (2)$$

Here $N_{ij} = (n_{ij0} + 0.5 n_{ij1} + 0.5 n_{ij2})(y_j - y_{j-1})$ denotes the total exposure time for which subjects in the ith sample have risk of failure during the jth interval under the approximation that failures and withdrawals occur, on average, at the midpoints of their intervals.

(2) The actuarial survival rates through the end of the jth interval

$$G_{ij} = \prod_{j'=1}^{j} g_{ij'}, \quad (3)$$

where

$$g_{ij'} = \frac{n_{ij'0} + 0.5 n_{ij'2}}{n_{ij'0} + n_{ij'1} + 0.5 n_{ij'2}}.$$

(3) The actuarial hazard function (or failure rate) for the jth interval

$$h_{ij} = \frac{2(1 - g_{ij})}{(y_j - y_{j-1})(1 + g_{ij})}. \quad (4)$$

The significance of the association of survival with groups can be evaluated through censored data rank tests. In this regard, the logrank test is widely used. For the case of $s = 2$ samples, it can be computed as the Mantel–Haenszel statistic for the set of (2×2) tables

	Failure	Not Failure	Total
Sample 1	n_{1j1}	$n_{1j0} + n_{1j2}$	$n_{1,j-1,0}$
Sample 2	n_{2j1}	$n_{2j0} + n_{2j2}$	$n_{2,j-1,0}$
Total	n_{+j1}	$n_{+j0} + n_{+j2}$	$n_{+,j-1,0}$

(5)

corresponding to the r intervals. A more specific expression is

$$Q_C = \frac{\left\{\sum_{j=1}^{r}(n_{1j1} - m_{1j1})\right\}^2}{\left\{\sum_{j=1}^{r} v_j\right\}}, \quad (6)$$

where $m_{1j1} = \{n_{1,j-1,0} n_{+j1}/n_{+,j-1,0}\}$ is the expected value of n_{1j1} and

$$v_j = \frac{m_{1j1} n_{2,j-1,0}(n_{+j0} + n_{+j2})}{n_{+,j-1,0}(n_{+,j-1,0} - 1)} \quad (7)$$

is its variance under the hypothesis of no association between survival and sample. When the sample sizes $\{n_i\}$ are moderately large (e.g., ≥ 20 with ≥ 10 failures), the log-rank statistic Q_C approximately has the chi-square distribution with 1 degree of freedom. The log-rank statistic for the comparison of s samples is the extension of (6) to a quadratic form with $(n_{1j1} - m_{1j1})$ replaced by the vector for a set of $(s-1)$ samples and with v_j replaced by the corresponding covariance matrix; this statistic approximately has the chi-square distribution with $(s-1)$ degrees of freedom. When the samples are cross-classified with strata based on one or more explanatory variables, the computation of Q_C becomes based on the combined set of (2×2) tables in (5) for all intervals within all strata. For more general consideration of censored data rank tests, see Peto and Peto [20], Prentice [21], and Prentice and Marek [23].

For sufficiently large sample sizes, functions of the life table frequencies $\{n_{ijk}\}$ such as (i), (ii), or (iii) approximately have multivariate normal distributions. Also, consistent estimates for the corresponding covariance matrix can be based on linear Taylor series methods. On this basis, weighted least-squares procedures can be used to fit linear models that describe the variation among these functions; the estimated parameters for such models approximately have multivariate normal distributions, and Wald statistics for goodness of fit approximately have chi-square distributions. Applications are described for the actuarial survival rates (ii) in Koch et al. [16], for the actuarial hazard function (iii) in Gehan and Siddiqui [11], and for functions from a competing risks framework in Johnson and Koch [14].

Maximum-likelihood (ML) methods enable the fitting of models with an assumed structure for the underlying distribution of the grouped survival data. A general specification of such models is

$$\pi_{ijk} = \psi_{ijk}(\boldsymbol{\theta}, \mathbf{x}_{ij}), \tag{8}$$

where the $\{\pi_{ijk}\}$ denote the probabilities of the (j, k)th outcome for a randomly obtained observation from the ith population, $\mathbf{x}_{ij}$ is a $(u \times 1)$ vector of explanatory variables for the ith sample during the jth time interval, and $\boldsymbol{\theta}$ is a vector of unknown parameters. The ML estimates $\hat{\boldsymbol{\theta}}$ are obtained by substituting (8) into the product multinomial likelihood

$$\phi(\{n_{ijk}\}|\{\pi_{ijk}\})$$
$$= \prod_{i=1}^{s} n_i! \frac{\pi_{ir0}^{n_{ir0}}}{n_{ir0}!} \left[\prod_{j=1}^{r} \prod_{k=1}^{2} \frac{\pi_{ijk}^{n_{ijk}}}{n_{ijk}!} \right] \tag{9}$$

for the contingency table format of the data and then maximizing the resulting function; this can be done by solving the equations

$$\frac{\partial}{\partial \boldsymbol{\theta}} \{\log_e[\phi(\{n_{ijk}\}|\boldsymbol{\theta})]\} = \mathbf{0} \tag{10}$$

for $\hat{\boldsymbol{\theta}}$ by iterative methods.

A framework for which the determination of ML estimates is relatively straightforward involves the assumption that the time until the failure event has a piecewise exponential distribution; i.e., for each of the s samples, there are independent exponential distributions with hazard parameters $\{\lambda_{ij}\}$ for the respective time intervals. Some additional assumptions which are usually made for this model are

(1) Withdrawal events occur uniformly in the respective intervals and their censoring process is unrelated to failure events in the noninformative sense of Lagakos [17].

(2) The within-interval probabilities of failure $(\pi_{ij1}/\pi_{i,j-1,0})$ are small, where

$$\pi_{i,j-1,0} = \sum_{j'=j}^{r} \sum_{k=1}^{2} \pi_{ij'k} + \pi_{ir0}$$

denotes the probability of surviving all causes through the end of the $(j-1)$th interval.

From the conditions specified here, it follows that maximizing (9) is approximately equivalent to maximizing the piecewise exponential likelihood function

$$\phi_{PE} = \prod_{i=1}^{s} \prod_{j=1}^{r} \lambda_{ij}^{n_{ij1}} \{\exp(-\lambda_{ij} N_{ij})\}. \tag{11}$$

For the likelihood (11), the relationship of the failure event to the explanatory variables is specified through models for the $\{\lambda_{ij}\}$. A useful model for many applications has the log-linear specification

$$\lambda_{ij} = \exp(\mathbf{x}'_{ij}\boldsymbol{\theta}). \tag{12}$$

Table 1: Survival status of mice.

Interval for Death (in hours)	Number of Deaths	
	Carbenicillin	Cefotaxime
0–6	1	1
6–12	3	1
12–18	5	1
18–24	1	0
24–30	1	2
30–36	0	0
36–48	0	2
48–60	1	1
60–72	0	0
72–96	0	1
Alive at 96	0	1
Total	12	10

A convenient feature of (12) is that the ML estimates $\hat{\boldsymbol{\theta}}$ for its parameters can be readily obtained by Poisson regression computing procedures for fitting loglinear models to sets of observed counts or contingency tables; these include both Newton–Raphson procedures and iterative proportional fitting. Additional discussion of statistical methods for piecewise exponential models is given in Aitkin and Clayton [1], Frome [9], Holford [12,13], Laird and Olivier [18], and Whitehead [24].

Simplification of the loglinear model (12) to

$$\lambda_{ij} = \exp(\eta_j + \mathbf{x}'_i \boldsymbol{\beta}) \qquad (13)$$

has *proportional hazards* structure; here $\exp(\eta_j)$ denotes the constant value of the hazard function within the jth interval for a reference population with $\mathbf{x} = \mathbf{0}$, and $\boldsymbol{\beta}$ is the vector of parameters for the relationship of the hazard function for the ith population with its explanatory variables $\mathbf{x}_i$. The more general formulation in Cox [6] of the proportional hazards model is

$$h(y, \mathbf{x}) = h_0(y)\{\exp(\mathbf{x}'\boldsymbol{\beta})\}, \qquad (14)$$

where y denotes continuous time and $h_0(y)$ is the hazard function for the reference population. When there are no ties, the maximizing of the partial likelihood of Cox [7] with respect to $\boldsymbol{\beta}$ is computationally straightforward, and the resulting estimator $\hat{\boldsymbol{\beta}}$ has the usual ML properties. However, for grouped survival data and other situations with many ties, modified strategies are necessary. One approach is to work with a piecewise exponential model like (13) and use Poisson regression computing procedures. Additional discussion of methods for dealing with ties in analyses with the proportional hazards model is given in Breslow [3], Kalbfleisch and Prentice [15], and Prentice and Gloeckler [22].

Example 35.3.1 The data in Table 1 are from a study of Bowdre et al. [2] to compare treatments for mice infected with *Vibrio vulnificus* bacteria. The survival status of each mouse was assessed at 6, 12, 18, 24, 30, 36, 48, 60, 72, and 96 h. The numbers of deaths per hour $\{f_i\}$ from (2) are

$$f_1 = 0.058 \text{ for carbenicillin},$$
$$f_2 = 0.023 \text{ for cefotaxime}.$$

The logrank (Mantel–Haenszel) statistic $Q_C = 4.52$ from (6) indicates that the survival experience of the cefotaxime group is significantly better.

An example for the piecewise exponential model is the graft survival data (E3) that is discussed in Laird and Olivier [18]. Weighted least squares methods are illustrated in References 11, 14, and 16; applications include the fitting of Weibull and other probability distributions to life tables.

Acknowledgments. The authors would like to thank J. H. Bowdre, J. H. Hull, and D. M. Cocchetto for permission to use the data in Table 1. They also would like to express appreciation to Amy Goulson and Ann Thomas for editorial assistance. This research was partially supported by the US Bureau of the Census through Joint Statistical Agreement JSA 84-5.

References

1. Aitkin, M. and Clayton, D. (1980). *J. Roy. Statist. Soc. C*, **29**, 156–163.
2. Bowdre, J. H., Hull, J. H., and Cocchetto, D. M. (1983). *J. Pharm. Exp. Thera.*, **225**, 595–598.
3. Breslow, N. E. (1974). *Biometrics*, **30**, 89–99.
4. Chiang, C. L. (1961). *Biometrics*, **17**, 57–78.
5. Chiang, C. L. (1968). *Introduction to Stochastic Processes in Biostatistics*. Wiley, New York.
6. Cox, D. R. (1972). *J. Roy. Statist. Soc B*, **34**, 187–220.
7. Cox, D. R. (1975). *Biometrika*, **62**, 269–276.
8. Elandt-Johnson, R. C. and Johnson, N. L. (1980). *Survival Models and Data Analysis*. Wiley, New York.
9. Frome, E. L. (1983). *Biometrics*, **39**, 665–674.
10. Gail, M. H. (1975). *Biometrics*, **31**, 209–222.
11. Gehan, E. A. and Siddiqui, M. M. (1973). *J. Am. Statist. Assoc.*, **68**, 848–856.
12. Holford, T. R. (1976). *Biometrics*, **32**, 587–597.
13. Holford, T. R. (1980). *Biometrics*, **36**, 299–306.
14. Johnson, W. D. and Koch, G. G. (1978). *Int. Statist. Rev.*, **46**, 21–51.
15. Kalbfleisch, J. D. and Prentice, R. L. (1980). *Statistical Analysis of Failure Time Data*. Wiley, New York.
16. Koch, G. G., Johnson, W. D., and Tolley, H. D. (1972). *J. Am. Statist. Assoc.*, **67**, 783–796.
17. Lagakos, S. W. (1979). *Biometrics*, **35**, 139–156.
18. Laird, N. and Olivier, D. (1981). *J. Am. Statist. Assoc.*, **76**, 231–240.
19. Larson, M. G. (1984). *Biometrics*, **40**, 459–469.
20. Peto, R. and Peto, J. (1972). *J. Roy. Statist. Soc. A*, **135**, 185–206.
21. Prentice, R. L. (1978). *Biometrika*, **65**, 167–179.
22. Prentice, R. L. and Gloeckler, L. A. (1978). *Biometrics*, **34**, 57–67.
23. Prentice, R. L. and Marek, P. (1979). *Biometrics*, **35**, 861–867.
24. Whitehead, J. (1980). *Appl. Statist.*, **29**, 268–275.

36 Image Processing

Noel Cressie and Jennifer L. Davidson

36.1 Introduction

Humans rely heavily on the sense of sight to collect their own data, which are then transformed into information by the brain; that information forms the basis of human judgments and actions.

Nonetheless, recording and machine processing of images are relatively recent accomplishments. Although photography dates back to the latter half of the nineteenth century, it was not until the early 1920s that one of the first techniques for transmitting a digital image appeared. The Bartlane cable picture transmission system reduced the time it took to send a picture across the Atlantic Ocean from a week to less than three hours. The picture was converted to electronic impulses by special encoding, transmitted across the transatlantic cable, and reconstructed at the receiving end.

Digital image processing as we know it today began in 1964 at the Jet Propulsion Laboratory in Pasadena, California, when, as part of the United States space program, scientists began using digital computers extensively to correct for camera distortions in digital images of the moon. Since then, the scope of applications for digital image processing has broadened enormously. It now includes automatic recovery of properties of objects for the purpose of human interpretation, as well as enhancement of images for visual interpretation by a human viewer. Image processing has played and will continue to play a part in many advanced technological innovations simply because visual information is so rich and multifaceted.

Engineers and computer scientists have developed a wealth of algorithms to restore clean images from noisy ones, to segment images into homogeneous regions, to extract important features from those images, and to reconstruct three-dimensional scenes from two-dimensional slices or projections. The image processing systems that address these tasks have typically used quite simple statistical methods, although there are alternative approaches that employ statistical methods for spatial processes to obtain optimal solutions.

This entry will be concerned exclusively with digital image processing. An *image* X is an $M \times N$ rectangular array

$$X \equiv \{X(m,n) : m = 0, \ldots, M-1; \\ n = 0, \ldots, N-1\}$$

of picture elements, or *pixels*. At a pixel location (m, n) one observes a pixel value $X(m, n)$ that can take any one of K (e.g., $K = 256$) possible intensity or gray values. For example, X might be a *thematic map-*

per satellite image with a 30 × 30*m* pixel resolution whose gray values are the (discretized) registered intensities in a given band of the electromagnetic spectrum. A grid of such scenes, suitably interpreted, can provide an efficient way to estimate crop yields or to monitor the effects of desertification, forest clearing, erosion, and so forth. Medical imaging offers another important example, where noninvasive diagnostic methods such as magnetic resonance imaging (MRI) provide an image X of metabolic activity in selected parts of the body (e.g., Cho et al. [9]).

The image X could be thought of as an MN-dimensional, multivariate statistic. The thousands of remotely sensed scenes recorded every week lead to massive, multivariate data sets whose statistical analyses can be overwhelming; modern society's ability to collect and store data has outstripped its ability to analyze them. Yet, humans are remarkably adept at image processing within their own vision systems, albeit with considerable person-to-person variability. One might argue that digital image processing is an attempt to remove that subjectivity from what is recognized as a very powerful method of spatial data analysis. Interestingly, this has produced many methods of manipulating image data that do not have a known equivalent in human vision.

As a consequence of the computer's great facility to process data, the processing of image data has become widespread [43]. This is not restricted solely to the visible spectrum of electromagnetic wavelengths. Sensors that detect infrared, ultraviolet, x-rays, and radio waves also produce images, as do signals other than electromagnetic, such as low frequency acoustic waves (sonar images) and photon emission [single-photon emission computed tomography, (SPECT)].

This entry is meant to introduce briefly the area of statistical processing of image data within the context of a large and rapidly expanding engineering literature. Most topics receive at least brief mention.

36.2 Hardware of an Imaging System

The hardware for a general imaging system consists of an image acquisition device, an analog-to-digital signal converter, a digital computer on which the image processing is carried out, a storage device for image data, and an output device such as a printer or camera.

Image data result from spatial sampling of a (conceptually) continuous process. The sampled values represent some physical measurement, such as light intensity (as in photographic images) or the magnitude of signal returned from the interaction between sound and tissue (as in ultrasound images). The sensors that collect the sampled data are often arranged in a rectangular array, whose vertical and horizontal spacing each determines a Nyquist frequency above which aliasing occurs [26]. To deal with this, it is assumed that the image is *bandlimited*, i.e., its Fourier transform is zero outside a bounded region of support.

A common way to acquire an image is through a standard videocamera. There are many sensors in a rectangular array (in 1994, as many as $300000/\text{cm}^2$), each one counting the number of photons that impinge upon it. These sensors, or *charge-coupled devices* (CCDs) as they are known, define the pixels of the digital image. The number of photons is then translated into an analog voltage signal (via Ohm's law: voltage = current × resistance), which is then passed out of the array to a device that converts the analog signal to a digital signal. An analog-to-digital converter (ADC) samples the continuously valued analog intensity signal and produces the one discrete value from a finite set of discrete values that is closest to the analog value. Typical sets for these "intensity" or "gray" values are $\{0, \ldots, 2^b - 1\}$, where b is the number of bits; for example, for $b = 8$ bit data, there are $K = 256$ possible gray values. The resulting pixel value from a given CCD represents an averaging of the signal across a small finite area of the scene. Spatial resolution is improved by increasing the number of CCDs per unit area but space between CCDs is needed for electrical isolation and to prevent scattering of light across neighboring CCDs [43].

Eight bits, or one byte, of data per pixel correspond nicely to the usual ways in which data are stored in computers. However, nonlinearities in the ADC, electronic noise from the camera itself, and degradation over time of the electronic circuitry, can all combine to make the two least significant bits of most standard video images useless. This results in 64 useful gray levels from the original 256, although all the 256 gray levels are retained for further processing and display. When the image data (stored digitally in a computer or a computer storage device) is recalled, a digital-to-analog converter produces voltages that are most commonly displayed using a cathode ray tube (CRT). This reconversion process is relatively noise-free. The CRT displays a monochrome image consisting of black (0) to white (256) with varying shades of gray in between. Because the human eye can visually distinguish only 30 or so gray levels [20], 256 levels is more than adequate for those applications where the human eye will judge the quality of an image.

Limitations in the way CCDs can count photons or electronic noise in circuitry, cabling, and the ADC can result in noisy image data that differ from the true pixel values of the original scene (the signal) that would have been produced had there been no distortion. The ratio of the variability in the signal to that in the noise is the *signal-to-noise ratio* (SNR). When SNR is high, the signal is typically easily discernible to the eye; when SNR is low, small features that have few pixels are often invisible to the eye. Removing noise is an important step in image processing. Unfortunately, it is often impractical to collect many images of the same scene and use averaging to remove the noise. Thus noise removal is usually done on an image-by-image basis.

In the sequel we focus on the mathematical and statistical methods used in image processing. Nevertheless, an understanding of the process of image acquisition, as described briefly above, is important for designing efficient image-processing algorithms.

36.3 Processing of Image Data

While the distinction is not universal, the terms *image processing* and *image analysis* often have a different meaning in the engineering literature. *Image processing* usually refers to operations on an image whose result is another image (e.g., noise removal); such operations are sometimes called filters. *Image analysis* usually refers to the extraction of summary statistics from an image (e.g., area proportion of a given phase in a multiphase image).

This section outlines the steps that typically make up the processing of digital image data (see Fig. 1). For any particular image, the actual steps followed depend on the type of data and what goal the user has in mind.

In Figure 1, the processing algorithms performed within the dotted box are sometimes referred to as *computer vision* algorithms rather than image processing algorithms. Here, the information extracted is a description, interpretation, or representation of a scene. This leads to the more general notion of a *computer vision sys-*

tem (e.g., Ballard and Brown [3]), a part of which is the image processing system. Below the top four boxes in Figure 1 are two boxes whose end result is a verbal or symbolic representation of scene content. For example, on analysis of a scene, a computer vision system could output the description: "The field of view contains croplands of corn, soybeans, and wheat in the following percentages ..."

The role of statistics, particularly spatial statistics, in image processing and image analysis has been small, but its importance is beginning to be realized [5,6,22]. Statistics brings to the image-processing literature the notion of *optimality* of the algorithms that restore, segment, reconstruct, or extract features of images in the presence of uncertainty. That uncertainty, or variability, can arise in a number of ways, such as noise generation of the type previously described.

More controversially, even a "noise-free" image may be regarded as a single realization from an ensemble of possible images. In this case, the variability is derived from decisions as diverse as choice of scene, physical conditions of the medium in which the imaging takes place, and the digitization. These influences have led to an image processing approach based on Bayesian inference and statistical modeling. That is, a prior distribution $\pi(\theta)$ is specified for the "true" pixel intensity values θ. This, along with $f(x|\theta)$, the probability distribution of the noisy image X given θ, allows the posterior distribution $p(\theta|x)$ of θ given $X = x$ to be calculated (in principle) via Bayes' theorem:

$$p(\theta|x) = \frac{f(x|\theta)\pi(\theta)}{\sum_\eta f(x|\eta)\pi(\eta)}. \quad (1)$$

The posterior $p(\theta|x)$ represents current uncertainty about the true image in light of having observed $X = x$, a noisy version of it. Should a second, independent observation $Y = y$ be taken of the same image, $p(\theta|x)$ is updated to reflect the now current uncertainty through

$$p(\theta|x,y) = \frac{f(y|\theta)p(\theta|x)}{\sum_\eta f(y|\eta)p(\eta|x)}.$$

As more and more observations are taken, the posterior probability distribution reflects less and less uncertainty about θ. However, in practice there is typically only one noisy image x from which to make inferences on the true image θ.

The Bayesian statistical approach to image analysis is very powerful, because any feature of the image, say, the boundaries $\omega(\theta)$, can be considered in the same manner. For example, uncertainty in the boundaries is characterized through the posterior $p(\omega(\theta)|x)$, which can be obtained by integration of the joint posterior $p(\theta|x)$ over an appropriate subset of all possible values of θ. In practice, the integration may prove difficult to implement; later, in our discussion of segmentation, we show how $p(\omega|x)$ is obtained (via Bayes' theorem) directly from a prior $\pi(\omega)$ on the boundaries.

Even more generally, Grenander and Miller [22] show how pattern theory can be used to construct prior probability measures on a complex space of scenes, made up of a variable number of deformable templates with textured interiors. In this context, Grenander and Miller apply Bayesian statistical methods to process electron micrograph images of mitochondria and membranes in cardiac muscle cells. These examples serve to illustrate that the Bayesian statistical approach is relevant to all the steps of image processing.

Image restoration corresponds to prediction of the true image θ. Suppose that one incurs a "loss" $L(\theta, \delta(x))$ when the true value is θ but the predictor $\delta(x)$ is used. Then the optimal image $\delta^*(x)$ is defined to be the predictor that minimizes the Bayes risk, or $E[L(\theta, \delta(X))]$, where the expecta-

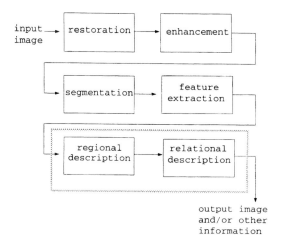

Figure 1: General sequence of steps in image processing and computer vision.

tion is taken over both θ and X. In fact $\delta^*(x)$ minimizes

$$\sum_\theta L(\theta, \delta(X)) p(\theta|x)$$

and hence δ^* depends on $p(\theta|x)$, the posterior distribution of θ given $X = x$. Intuitively speaking, the loss function reflects the relative penalties the image analyst wishes to place on various reconstructions $\delta(x)$ in relation to the true image θ. The choice of loss function may well depend on the goal of the reconstruction (e.g., low misclassification rate, high contrast, differential importance given to subregions of the image).

In general, changing the loss function will lead to a different optimal restoration. If the 0–1 loss function

$$L(\theta, \delta(X)) = \begin{cases} 0 & \text{if } \theta = \delta(x), \\ 1 & \text{if } \theta \neq \delta(x) \end{cases}$$

is used, then $\delta^*(x) = \arg\max_\theta p(\theta|x)$, which is the maximum a posteriori (MAP) estimator. The MAP estimator is simply the mode of the posterior probability density (or mass) function of θ. If the images θ and $\delta(x)$ are regarded as vectors of pixels and the squared-error loss function

$$L(\theta, \delta(X)) = (\theta - \delta(x))^T (\theta - \delta(x))$$

is used, then $\delta^*(x) = E[\theta|X = x]$, the posterior mean of θ.

Probably the most difficult task in Bayesian image processing is the choice of prior $\pi(\theta)$. Because the pixel gray values $\{\theta(m,n) : m = 0, \ldots, M-1; n = 0, \ldots, N-1\}$ are in a spatial array, a prior that quantifies the idea that nearby pixel values tend to be more alike than those far apart is desired. Markov random fields, expressed in their Gibbsian form, offer a very attractive class of prior models [4,16]. Briefly, two pixel locations $\mathbf{u} \neq \mathbf{v}$ are said to be neighbors if $\Pr[X(\mathbf{u}) = x(\mathbf{u})|X(\mathbf{s}) = x(\mathbf{s}) : \mathbf{s} \neq \mathbf{u}]$ depends functionally on $X(\mathbf{v})$. A clique is a set of pixel locations, all of whose elements are neighbors of each other. Then a Markov random field expressed in its Gibbsian form is

$$\pi(\boldsymbol{\theta}) \propto \exp\left[-\sum_{\kappa \in C} V_\kappa(\theta_\kappa)\right],$$

where κ is a clique, $\theta_\kappa \equiv \{\theta(\mathbf{s}) : \mathbf{s} \in \kappa\}$ and the summation is over the set of all cliques C. The potential energy functions

$\{V_\kappa : \kappa \in C\}$ are chosen by the user to generate prior probabilities of various (local) configurations of gray values. For example, unlikely configurations can be down-weighted by setting $V_\kappa(\theta_\kappa)$ to be large for values of θ_κ that give rise to those configurations.

Bayes' theorem using a Markov random field prior $\pi(\theta)$ and a local noise distribution $f(x|\theta)$ yields a posterior $p(\theta|x)$ that is also a Markov random field. Simulating from $p(\theta|x)$ using Gibbs sampling and determining its mode (i.e., the MAP estimator of θ) using simulated annealing is discussed in Geman and Geman [16].

Because the posterior distribution is not generally available in closed form, solutions are only approximate and often computer-intensive. Nevertheless, such methods are gaining acceptance in the image-processing literature; the challenge will be to demonstrate both the power and the practicality of the Bayesian approach in a wide variety of image-processing problems.

In the subsections that follow, the general sequence of steps in image processing (Fig. 1) will be discussed in more detail, with references to the Bayesian statistical approach where appropriate.

36.4 Restoration and Noise Removal

When an image is acquired, it often comes with defects due to the physical limitations of the process used to acquire the data, such as sensor noise, poor illumination, blur due to incorrect camera focus, atmospheric turbulence (for remotely-sensed data), and so on. *Image restoration* refers to the processing of the image data so as to "restore" the observed image back to its true, noise-free state.

Many linear operations have been developed expressly for noise removal in image processing; References. [20, 26, and 36] describe the better known ones. On occasions, the system noise itself is known or can be approximated. For example, in photoelectronic systems, the noise δ in the electron beam current is often modeled [26] as

$$\delta(m,n) = g(m,n)^{1/2}\epsilon_1(m,n) + \epsilon_2(m,n),$$

where (m, n) is the pixel location; g is the signal of the scanning beam current; and ϵ_1 and ϵ_2 are zero-mean, mutually independent, Gaussian white-noise processes. The dependence of the noise δ on the signal g is because the detection and recording processes involve random electron emissions assumed to have a Poisson distribution with mean g. Based on this model, filters can be designed to remove the noise.

Assumptions about the signal usually involve some form of piecewise smoothness over the image domain. For example, discontinuities or change points in intensity is an appropriate assumption when the image is one of distinct objects, within which a smooth function can be used for modeling the objects' texture. This is illustrated by Korostelev and Tsybakov [28], who apply multidimensional nonparametric estimation techniques to obtain minimax estimators of a true, piecewise smooth image intensity. Grenander and Miller [22] take a Bayesian viewpoint, but their models have this same idea of a true image made up of textured objects with sharp boundaries.

On occasion, such as for real-time video sequencing, multiple images of the same scene are available. Averaging of these images can help reduce noise that is random in nature, thus increasing the signal-to-noise ratio.

While linear operations work fairly well for many types of white noise, they do not work well for additive noise that is "spiky" or impulsive in nature (i.e., having large positive or negative values in a spatially compact area). However, the median filter, which is a special case of a rank or order statistic filter [31], does work well on this

Figure 2: (a) Input image with spikey noise; (b) median-filtered image with a 3 × 3 mask.

type of noise. To apply the median filter, choose a local pixel neighborhood, such as a 3 × 3 or a 5 × 5 neighborhood, to act as a window or mask. Then place the mask over the image at each pixel location and compute the median value for the 9 or 25 pixels, respectively, inside the mask. The output image has, at the central location, the median value. In Figure 2b, we show the result of applying the 3 × 3 median filter to the gray-level input image as shown in Figure 2a. The median filter is related to a broad class of nonlinear imaging operations that come from an area known as *mathematical* morphology.

Mathematical morphology unifies the mathematical concepts in image-processing and image analysis algorithms that were being applied in the 1970s to such diverse areas as microbiology, petrography, and metallography [33,45,46]. More recently, a comprehensive *image algebra* for digital images has been developed; it includes discrete mathematical morphology as a subset [40,41].

Mathematical morphology offers a powerful and coherent way of analyzing objects in two or more dimensions [45,47]. Its highly nonlinear operations were first developed for shape analysis of binary images, but the approach extends to noise removal, connectivity analysis, skeletonizing, and edge detection, to name only a few. The mathematical morphological operation of dilation and its dual operation of erosion transform an image (the data) through the use of a structuring element or template.

The template provides a reference shape by which objects in the image can be probed in an organized manner. For the usual input image X, let the template be defined by $\{B(i,j) : (i,j) \in T\}$, where $(i,j) \in T$ implies $(-i,-j) \in T$ and T typically has many fewer elements in it than the MN pixels of X. Then the dilation and erosion of X by B are given, respectively, by

$$\begin{aligned} C(m,n) &= \max\{X(m-i, n-j) \\ &\quad + B(i,j) : (i,j) \in T\}, \\ D(m,n) &= \min\{X(m-i, n-j) \\ &\quad - B(-i,-j) : (i,j) \in T\}. \end{aligned}$$

While these two operations are themselves simple, cascades or combinations of such operations, including set-theoretic ones like set complementation and union (for the binary case), can lead to complex and flexible algorithms for image processing.

The discrete Fourier transform (DFT) and its numerous fast versions are widely used in image processing, particularly for noise reduction [26]. The two-dimensional DFT of an $M \times N$ image X is another

$M \times N$ image, defined by

$$\tilde{X}(i,j) = \sum_{m=0}^{M-1} \sum_{n=0}^{N-1} X(m,n)$$
$$\times \exp\left[-\left(\frac{2\pi i m \sqrt{-1}}{M} + \frac{2\pi j n \sqrt{-1}}{N}\right)\right],$$

where $i = 0, \ldots, M-1$ and $j = 0, \ldots, N-1$. A commonly-used image processing filter is to calculate the DFT of an image and then to select a range of i and j for which $\tilde{X}(i,j)$ is modified to be zero. Filters that pass only the low frequency values, in which $\tilde{X}(i,j)$ is modified to be zero for large values of i and j, are *lowpass filters*, while those that pass only high-frequency values are *high-pass filters*. The DFT is a member of a larger class of operations called *unitary transforms* [36]. Here, the image is represented in terms of basis vectors for the MN-dimensional space; such a representation can be useful in compression and filtering of image data, feature extraction, and other methods of image processing. As the DFT itself is computationally demanding, much effort has been expended to develop fast Fourier transforms for these problems [6].

Reconstruction of image data from projections is required when energy is emitted from a source as a beam and collected at the end of the path by a sensor. The only information given directly by this imaging-acquisition system is the amount of energy being emitted from the source and the sum of the energy as it hits the sensor, after it has gone through the object under study. By rotating the source-sensor assembly around the object, projection views for many different angles can be obtained. Systems that acquire data in this manner are *computerized tomography*, or CT, *systems*. The goal of these image reconstruction problems is to reconstruct the interior of the object from the projection information collected; the principal mathematical tool used is the Radon transformation [17,39,49].

36.5 Image Enhancement

Image enhancement generally refers to the accentuation of edges or gray-level contrast in order to make the data more useful for analysis or display. The distinction between image enhancement and noise reduction is not always clear. For example, the removal of unwanted artifacts with less than a prespecified maximal diameter could be viewed as either.

Consider the local averaging operator, where pixel values in a window or mask are averaged, and the new averaged value replaces the old pixel value at the center of the mask. Figure 3 shows an example of a 3×3 averaging mask, where the weights used in the local averaging are superimposed on the pixel locations in the mask. When the new averaged values instead of the original values are used to calculate the new output values, this gives a moving average. Local averaging is equivalent to a low-pass filter and it tends to blur edges in the image. Highpass filtering is more appropriate for image enhancement; because edges contain mostly high frequency components, it results in a sharpening of edges in the spatial domain. Gonzalez and Woods [20] provide a good description of these techniques.

Manipulation of the contrast in an image is another commonly used enhancement scheme. Here, the gray levels in an image are changed so that contrasts between regions stand out to the eye. Histogram equalization (see, e.g., Jain [26]) provides a way to do this. The histogram of an image is a graph of the relative frequency of occurrence of the gray levels in the image. In histogram equalization, the goal is to transform the original histogram into a uniform one. The original gray values are then mapped to the new gray values according to the transformation. If the gray value $X(m,n)$ is considered to be a nonnegative random variable with proba-

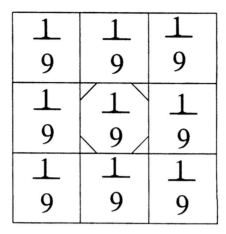

Figure 3: The 3 × 3 local averaging mask. The center pixel is denoted with hash marks.

bility mass function $f(x)$ and cumulative distribution function $F(x)$, then

$$U(m,n) \equiv F(X(m,n)) = \sum_{x=0}^{X(m,n)} f(x)$$

has, by the probability integral transformation, a discrete uniform distribution over (0,1). In practice, the histogram of relative frequencies replaces $f(x)$, and some discretization of $U(m,n)$ is needed to convert it to the new gray values.

When applying histogram equalization to an image, an output image is produced that stretches the low-contrast gray levels in an image. This results in an image with a flat histogram and increased contrast. If this does not produce desired results, then a histogram transform other than the uniform can be tailored by the user. Histogram equalization can be applied at any time in the processing sequence to provide increased contrast to the image data.

36.6 Segmentation

Segmentation is usually performed after restoration and enhancement. *Segmentation* is a broad term describing procedures that break the image up into regions having different properties, as desired by the user. Here we use the terms *segmentation* and *classification* synonymously, although, on occasion, *classification* is used to refer to the regional description following segmentation.

Segmentation of pixels in an image can be viewed from two perspectives: the decomposing of an image into regions based on similarities within each region, such as through topological or gray-value properties; or the separation of an image into regions based on differences between them, such as through edge or boundary detection. Using the first approach, an image can be separated into regions where, for example, the gray values in each region are in a narrow range of values. Approaches such as k-nearest-neighbor classification [14], decision-tree classification [1], or artificial neural networks [23] will achieve this goal.

For the first approach, a Bayesian discriminant analysis is possible. In the simplest form, it classifies each pixel of the image without regard to the intensities of surrounding pixels. Suppose there are H possible classes to which a pixel might belong and let $\{\pi(k) : k = 1, \ldots, H\}$ denote the prior probability of any pixel belonging to that class. At pixel (m, n), the true class is $\theta(m, n)$ and the observed intensity is $X(m, n)$. From the noise distribution, given by density $f(x(m,n)|\theta(m,n))$ and the prior π, a Bayes classification rule declares $\hat{\theta}(m, n) = k^*(x(m, n))$, where $k^*(x(m, n))$ maximizes the posterior distribution

$$p(k|x(m,n)) = \frac{f(x(m,n))|k)\pi(k)}{\sum_{\ell=1}^{H} f(x(m,n))|\ell)\pi(\ell)}$$

with respect to k. Then $\{\hat{\theta}(m, n)\}$ is a segmentation of the image. The segmentation tends to be very patchy, which has led to the development of contextual (spatial) classification techniques that account for neighboring pixel values in the classification criteria [34], Chap. 13].

With binary images (i.e., pixel values taking one of two possible values, namely, zero and one), the topological property of connectivity is a useful tool for segmentation. Two common connectivity notions used in image processing are four-connectivity and eight-connectivity [20]. Let $X(m, n)$ and $X(k, l)$ represent two pixels in a binary image on an $M \times N$ rectangular array. The two pixels are four-neighbors if ((1)) $X(m, n) = X(k, l) = 1$, and (2) $0 < [(m - k)^2 + (n - l)^2] \leq 1$. A four-path from $X(m_1, n_1)$ to $X(m_k, n_k)$ is a sequence of pixels $X(m_1, n_1), X(m_2, n_2), \ldots, X(m_k, n_k)$ such that $X(m_i, n_i) = 1$ for $i = 1, \ldots, k$ and $X(m_i, n_i)$ is a four-neighbor of $X(m_{i+1}, n_{i+1})$ for $i = 1, \ldots, k - 1$. A region is *four-connected* if each pair of pixels in that region has a four-path between them. Similarly, two pixels $X(m, n)$ and $X(k, l)$ are eight-neighbors if they satisfy ((1)) and (2') : $0 < [(m-k)^2+(n-l)^2] \leq 2$. This latter condition simply allows pixels on the diagonals to be considered possible eight-neighbors. Eight-paths and eight-connectivity within a region are defined analogously.

Fast algorithms are available to determine regions of the image that are connected in various ways [2]. An important connectivity scheme used in image processing is eight-connectivity for the white or background pixels (those with gray values of 0), and four-connectivity for the black or object pixels (those with gray values of 1). This avoids paradoxical connectivities that can occur in image processing, such as in a region of the image where sharp corners of two objects almost touch; it should not be possible that both background and object be connected in that region. For a more detailed discussion on digital topology, see Kong and Rosenfeld [27].

The second approach to segmentation is through boundary detection. Here the image is segmented by specifying the boundaries that separate them; for example, the goal may be to segment so that gray values in each region are approximately constant. More generally, regions that have certain texture, shape, or feature properties can also be used to segment an image. One can distinguish between regions in an image through such summary statistics as first- or second-order moments from the normalized histogram, spatial moments of regions, ranges of gray values, frequency content, and connectivity measures for each region [26].

The problem of boundary detection, or edge detection, can be solved using one of two general approaches: difference operators or parametric models. An *edge pixel* is one whose neighboring pixels have a wide variation in gray value from the given pixel's gray value. A difference operator uses local information about the gradient at a pixel location to determine whether that pixel is an edge pixel or not. If the estimated gradient is beyond a specified level, the pixel is declared an edge pixel. Edge detectors include the Sobel, the Robert, the Prewitt, the Kirsch [20], the Marr–Hildreth [32], and the Canny [7] difference operators.

The Marr–Hildreth and Canny operators are examples of multiscale edge detectors. These smooth the image data with a convolution, typically Gaussian, followed by a detection of the zero-crossings of the second derivative of the smoothed image data. Performing this at different values for the variance parameter in the Gaussian convolution, that is, at different "scales," and then combining the results, can result in the detection of boundaries. Although many difference operators are isotropic, with known limitations due to both the isotropy and sensitivity to noise, they nonetheless yield a popular class of edge detectors. See Torre and Poggio [48] for a unified framework for edge detectors based on difference operators.

Another way to detect edges is to use an edge model and compute its degree of match to the image data. While such detectors are usually more computational, they can give better results and more information for further processing. The Hueckel [25], Hough [13], and Nalwa–Binford [35] edge detectors are algorithms based on fitting the image data to a model of the edges, which are fit by a criterion like least squares. The Hueckel edge detector models an edge as a step or ramp function and the Nalwa–Binford edge detector uses a surface described by a piecewise polynomial. The Hough edge detector is actually a more general curve-finding transform that can be used for the analysis of shapes in images. However, there are computational sensitivities associated with the Hough transform that complicate practical implementations [37].

We next present an example of boundary segmentation using a Bayesian statistical model. Non-Bayesian methods of the type described above require additional processing to thin edges and link them, and, as such, they are difficult to implement without much trial and error.

Recall the Bayesian statistical model used in Equation (1) with X denoting the observed (noisy) image and θ the true image. Assume that θ is a Markov random field with realizations that are constant within connected regions of the $M \times N$ rectangular array (regions with constant gray values correspond to objects in the scene). Let ω denote the true boundary image such that $\omega(m,n) = 1$ if a boundary pixel is present at (m, n), and $\omega(m,n) = 0$ otherwise. The goal is to predict ω (optimally); from our loss function formulation, the optimal predictor will be based on the posterior distribution

$$p(\omega|x) \propto f(x|\omega)\pi(\omega),$$

where $\pi(\omega)$ is the prior on the boundary image ω. Boundaries of objects should be closed and thin; this will be guaranteed [24] with a choice of prior taking the form of a Markov random field, where

$$\pi(\omega) \propto \exp\left[-\sum_{\kappa \in C} V_\kappa(\omega_\kappa)\right],$$

C is the set of all cliques, and $V_\kappa(\omega_\kappa) = \infty$ if

1. Boundary pixels are isolated or terminate in any eight-neighborhood contained in κ (this guarantees closed boundaries); or

2. Neighborhood boundary pixels in an eight-neighborhood of a central boundary pixel are vertically, horizontally, or diagonally adjacent to each other, where the eight-neighborhood is contained in κ (this guarantees one-pixel-wide boundaries).

This specification of the prior $\pi(\omega)$ guarantees that its support, and that of the posterior $\pi(\omega|x)$, is contained in Ω_p, the set of all closed and one-pixel-wide boundaries (that is, the set of permissible boundaries).

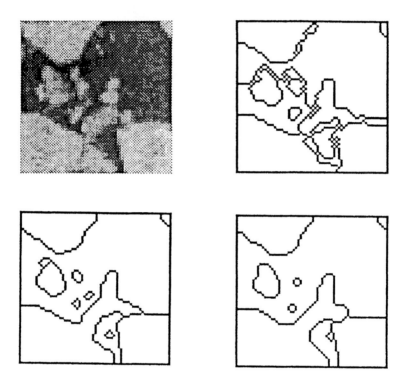

Figure 4: (a) Input image to boundary detection algorithm; (b) boundary image used to start the ICM algorithm; (c) boundary image after one iteration of the ICM algorithm; (d) boundary image output from ICM algorithm [18].

A Gaussian model is assumed for the observed intensities: $X(m,n) = \theta(m,n) + \epsilon(m,n)$, where ϵ is a Gaussian white-noise process with zero mean and variance σ^2, independent of θ. Each $\omega \in \Omega_p$ partitions the $M \times N$ array into disjoint connected regions $\{d_i(\omega) : i = 1, \ldots, K(\omega)\}$, where it is assumed that θ is constant on connected regions.

Define $\mu_i(\omega) \equiv \theta(m,n)$ for $(m,n) \in d_i(\omega)$, and let $l_i(\omega)$ denote the number of distinct sites $(m_{ij}, n_{ij}) \in d_i(\omega)$. Thus

$$f(x|\omega) = \prod_{i=1}^{K(\omega)} \prod_{j=1}^{l_i(\omega)} (2\pi\sigma^2)^{-1/2}$$
$$\times \exp\left[\frac{(x(m_{ij}, n_{ij}) - \mu_i(\omega))^2}{2\sigma^2}\right],$$

which models the probability density function of the observed intensities, assumed to be conditionally independent.

As a consequence of Bayes' Theorem and the Markov random-field form of the prior $\pi(\omega)$, the posterior $p(\omega|x)$ is also a Markov random field with the same neighborhoods and cliques as the prior, but with potential functions modified by the noise distribution $f(x|\omega)$ given above. Recall that the optimal estimator under 0–1 loss is the MAP estimator $\delta^*(x)$ that maximizes $p(\omega|x)$ with respect ω. It is inefficient to search through all the 2^{MN} possible boundary images to find the one in Ω_p that achieves the maximum. A stochastic relaxation method called *simulated annealing* is guaranteed, when implemented properly, to converge to the global maximum [16]. An approximation that is computationally more feasible, called *iterated conditional modes* (ICM), was introduced by Besag [5]. When the ICM algorithm converges, it does so to a local maximum.

Figure 4a shows a 64×64 input image X taken from a larger, synthetic aperture radar (SAR) image of sea ice; the intensity values range from 0 to 255 [24]. Figure 4b is the boundary image used to start the ICM algorithm; notice that it is closed and one pixel wide. Figure 4c shows the result after one iteration of ICM; Figure 4d shows the (locally) optimal boundary image to which ICM converged (after four iterations).

36.7 Feature Extraction

A feature of an image is a descriptor that gives quantitative information about some region within the image. For example, in remote sensing, the gray value $X(m, n)$ might contain information about the reflectivity or transmissivity at pixel location (m, n); in infrared imagery, the amplitude of the signal represents temperatures of the objects in the scene. Other features include those based on a histogram of gray values, on edges, on shapes, on textures, and on transforms.

Once a region in an image has been identified using segmentation techniques, a histogram of the gray values within that region can be calculated. By identifying a random variable G with the normalized histogram, so that

$$\Pr[G = g] \equiv f(g)$$
$$= \frac{\text{number of pixels with gray value } g}{\text{total number of pixels in region}};$$
$$g = 0, \ldots, K-1,$$

one can define histogram features of that region. They include [26]

Moments: $m_i \equiv E[G^i] = \sum_{g=0}^{K-1} g^i f(g)$, $i = 1, 2, \ldots$.

Central moments: $\mu_i \equiv E[(G - m_1)^i] = \sum_{g=0}^{K-1} (g - m_1)^i f(g)$, $i = 1, 2, \ldots$;

Entropy: $H \equiv E[-\log_2 f(G)] = -\sum_{g=0}^{K-1} f(g) \log_2 f(g)$.

The presence or absence of an edge at a given pixel location can be viewed as an edge feature. Also, associated with an edge pixel is a direction of the edge at that pixel location.

Shape features are descriptors that give information on the geometrical structure and shape of an object in the scene. Boundaries, two-dimensional spatial moments, topological properties, skeletons [12], and syntactic methods [26] can all describe shapes. For example, if (a subset of) the edge pixels form a closed-loop boundary, that boundary has a certain geometrical shape.

The boundary, and hence the shape, can be represented in a number of ways. Constructing a chain code involves choosing a starting point on the boundary, tracing the boundary around in sequence until the starting point is reached again, and then encoding the direction vectors between successive boundary pixels into a list [20]. The eight possible directions are given numbers from 1 to 8, and these numbers make up the list; the form of the boundary can be reconstructed exactly from its chain code. Once the boundary is reconstructed, it can then be shifted to its correct position within the image.

Splines can be fitted to boundary points as well. *Fourier descriptors* [26] represent a boundary as a one-dimensional, complex-valued sequence of points and use coefficients of the Fourier series representation to give a unique boundary representation. However, this latter method is sensitive to the starting point, rotation, and scaling of the boundary.

If the interior of a closed-loop boundary is given gray value 1 and the exterior gray value 0, the two-dimensional moments provide shape information [44]. The (p,q)th moment is

$$m_{pq} = \text{avg}\{i^p j^q : X(i,j) = 1\};$$
$$p, q = 0, 1, \ldots,$$

where avg$\{\cdot\}$ denotes the average of the argument set. The central moments are then

$$\mu_{pq} \equiv \text{avg}\{(i - m_{10})^p (j - m_{01})^q : X(i,j) = 1\}; \ p, q = 0, 1, \ldots.$$

For example, m_{10}, m_{01} represent the center of gravity; $\mu_{20}, \mu_{02}, \mu_{11}$ represent moments of inertia; and $\mu_{30}, \mu_{21}, \ldots$ represent asymmetry characteristics of an object.

Related to shape features are topological features, such as the Euler number and the number of holes for binary images [36]. The number of holes in the binary image is counted as the number of regions of white (the background) that are wholly contained within the regions of black (the objects). The Euler number is defined to be the number of connected regions minus the number of holes within the regions.

Texture features include a variety of different measures. The size of texture building blocks (or texels) can be represented by the spatial range of the autocorrelation function (ACF). However, several different textures can have the same ACF, so this is not always a reliable measure [26]. Parameters estimated from spatial statistical models can also be thought of as texture features [11].

Histogram features can give texture information; the gray-level concurrence matrix is a popular, although computationally intensive, method [20]. This matrix gives information regarding the relative positions of pixel values that are spatially close. First, a vector **v** is chosen that describes the relative displacement of pairs of pixel locations in the image. Then, the frequency of occurrence of pairs of pixel gray values are counted for all those pairs of pixel locations that are displaced by vector **v** and are within a given subregion of the image. For a region with K gray values, the concurrence matrix will be a $K \times K$ matrix, where the (i,j)th entry in the matrix counts the number of times gray level i occurs at the tail of vector **v** and gray level j occurs at the head of **v**. The matrix is usually normalized; in this case, the (i,j)th entry gives an estimate of the joint probability that a pair of points at a relative displacement $\mathbf{v}$ will have gray values i and j. The concurrence matrix is essentially a local two-dimensional histogram which, if the number of gray values is more than just a few, can require large amounts of computation time.

Transform features involve representing the region via a standard series expansion that guarantees uniqueness in the coefficients of the transformed signal, and then using information in the transformed image as a feature of the region. The discrete Fourier transform $\tilde{X}$ of an image X is an example of such a feature. After bandpassing certain frequencies and setting the rest of the frequency coefficients equal to zero (e.g., bandpass frequencies within an annular region around the origin), the result can be inverse-transformed back to the spatial domain. Other transformations, such as the Haar wavelet, discrete cosine, and Hotelling (or Karhunen–Loeve) transforms, also provide features [26].

36.8 Other Applications

Inverse problems such as pattern recognition and function approximation occur regularly in the context of image processing. Standard statistical pattern recognition techniques for images are reviewed by Fu [15] and Gonzalez and Thomason [19]; newer developments that use probability and statistics in one form or another include artificial neural networks [8,23,30], fuzzy-logic systems [29], and genetic algorithms [18]. Digital morphing or warping [50], which is a geometric transformation of a digital image used primarily in television advertisements and music videos, is inherently a sampling process. The transformations involve generating a sequence of images, whereby certain strategically sampled gray values remain constant, but their spatial location changes. Interpolation of the data must be performed from frame to frame in the remaining locations, which can involve statistical concepts.

Remote sensing of the earth's surface is now achieved through platforms that allow several different sensors to sample simultaneously the same scene; for example, electromagnetic radiation can be sensed over a number of different frequency bands. Colored images can be considered to be multispectral, corresponding to the intensities of red, green, and blue frequencies. Occasionally, one sees gray-value images converted to color images by assigning colors to various ranges of gray values; these are called pseudocolor *images*. As a consequence of having several pieces of information at one pixel location, the "fusion" (i.e., multivariate analysis) of the vector of image values has developed into a more recent research area. A multivariate statistical approach can be found, for example, in McLachlan [34], and Cressie and Helterbrand [10] summarize the multivariate spatial models that could be used for statistical modeling.

Because image data require huge amounts of memory for processing as well as massive databases for storage, much effort has been expended on efficient compression of data with little loss of information [38]. In addition, efficient ways to encode digital images for transmission over communications channels are being sought. Image compression techniques fall into two main areas: predictive encoding [21] and transform coding [42]. *Predictive encoding* techniques take advantage of redundancy in the image data. *Transform coding* techniques transform the image data so that a large amount of information is packed into a small number of samples or coefficients. Typically there is loss of information when compressing data but, up to a point, this is outweighed by vast improvements in their storage and transmission.

Finally, an area of important activity is computer-vision algorithms. Once the image has been processed at the pixel level, extracting features and other information, regional descriptions can be generated. Higher-level conceptual processing is necessary to identify information in images and these techniques are referred to collectively as *computer vision* techniques [3]. Geometric properties of shapes, such as maximal diameter, can be derived from shape features, texture properties can help define boundaries between regions, and moment values can give properties about regions. Using a priori information whenever possible, one classifies the regions or objects. After classification, relational information between identified objects in the image is output or is used for further processing.

For example, suppose that a region in an image has been identified as containing malignant cells and the region is contained within a certain organ that has also been identified. Then that relational information can be output to the human user, or further processing can be performed to identify the type of cells more precisely. More abstract constructs such as syntactic grammars [15] are also used for extracting relational information from processed images.

References

1. Agin, G. J. and Duda, R. O. (1975). SRI vision research for advaced automation. *Proceedings of the Second USA Japan Computer Conference*, pp. 113–117.

2. Alnuweiri, H. and Prasanna, V. (1992). Parallel architectures and algorithms for image component labeling. *IEEE Trans. Pattern Anal. Machine Intell.*, **14**, 1,014–1,034.

3. Ballard, D. H. and Brown, C. M. (1982). *Computer Vision*. Prentice-Hall, Englewood Cliffs, NJ.

4. Besag, J. (1974). Spatial interaction and the statistical analysis of lattice systems. *J. Roy. Statist. Soc. B*, **36**, 192–225.

5. Besag, J. (1986). On the statistical analysis of dirty pictures. *J. Roy. Statist. Soc. B*, **48**, 259–302.

6. Blahut, R. E. (1985). *Fast Algorithms for Digital Signal Processing*. Addison-Wesley, Reading, MA.

7. Canny, J. (1986). A computational approach to edge detection. *IEEE Trans. Pattern Anal. Machine Intell.*, **8**, 679–697.

8. Cheng, B. and Titterington, D. M. (1994). Neural nets: A review from a statistical perspective. *Statistical Science*, **9**, 2–54.

9. Cho, Z. H., Jones, J. P., and Singh, M. (1993). *Foundations of Medical Imaging*. Wiley, New York.

10. Cressie, N. and Helterbrand, J. D. (1994). Multivariate spatial statistical models. *Geogr. Syst.*, **1**, 179–188.

11. Cross, G. and Jain, A. (1983). Markov random field texture models. *IEEE Trans. Pattern Anal. Machine Intell.*, **5**, 25–39.

12. Davidson, J. L. (1994). Thinning and skeletonizing. In *Digital Image Processing Methods*, E. R. Dougherty, ed., pp. 143–166, Marcel Dekker, New York.

13. Duda, R. and Hart, P. (1972). Use of the Hough transformation to detect lines and curves in pictures. *Commun. ACM*, **15**, 11–15.

14. Duda, R. O. and Hart, P. E. (1973). *Pattern Recognition and Scene Analysis*. Wiley, New York.

15. Fu, K. S. (1982). *Syntactic Pattern Recognition and Applications*. Prentice-Hall, Englewood Cliffs, NJ.

16. Geman, S. and Geman, D. (1984). Stochastic relaxation, Gibbs distributions, and the Bayesian restoration of images. *IEEE Trans. Pattern Anal. Machine Intell.*, **6**, 721–741.

17. Gilbert, P. F. C. (1972). The reconstruction of a three-dimensional structure from projections and its application to electron microscopy: II. Direct methods. *Proc. Roy. Soc. Lond. B*, **200**, 89–102.

18. Goldberg, D. E. (1989). *Genetic Algorithms in Search, Optimization, and Machine Learning*. Addison-Wesley, Reading, MA.

19. Gonzalez, R. and Thomason, M. G. (1978). *Syntactic Pattern Recognition: An Introduction*. Addison-Wesley, Reading, MA.

20. Gonzalez, R. and Woods, R. (1992). *Digital Image Processing*. Addison-Wesley, New York.

21. Graham, R. E. (1958). Predictive quantitizing of television signals. In *IRE Wescon Convention Record*, Vol. 2, Part 2, pp. 147–157.

22. Grenander, U. and Miller, M. (1994). Representations of knowledge in complex systems. *J. Roy. Statist. Soc. B*, **56**, 549–603.

23. Haykin, S. (1994). *Neural Networks: A Comprehensive Foundation*. Macmillan, New York.

24. Helterbrand, J. D., Cressie, N., and Davidson, J. L. (1994). A statistical approach to identifying closed object boundaries in images. *Adv. Appl. Probab.*, **26**, 831–854.

25. Hueckel, M. H. (1971). An operator which locates edges in digital pictures. *J. Assoc. Comput. Mach.*, **18**, 113–125.

26. Jain, A. K. (1989). *Fundamentals of Digital Image Processing*. Prentice-Hall, Englewood Cliffs, NJ.

27. Kong, T. Y. and Rosenfeld, A. (1989). Digital topology: Introduction and survey. *Comp. Vision, Graphics, Image Process.*, **48**, 357–393.

28. Korostelev, A. P. and Tsybakov, A. B. (1993). *Minimax Theory of Image Reconstruction*. Springer Lecture Notes in Statistics, Vol. 82. Springer-Verlag, New York.

29. Kosko, B. (1992). *Neural Networks and Fuzzy Systems*. Prentice-Hall, Englewood Cliffs, NJ.

30. Lippmann, R. P. (1987). An introduction to computing with neural nets. *IEEE Mag. Acoustics, Speech, Signal Process.*, **4**, 4–22.

31. Maragos, P. and Schafer, R. W. (1987). Morphological filters-part II: Their relations to median, order-statistic, and stack filters. *IEEE Trans. Acoustics, Speech, Signal Process.*, **35**, 1170–1184.

32. Marr, D. and Hildreth, E. (1980). Theory of edge detection. *Proc. Roy. Soc. Lond. B*, **207**, 187–217.

33. Matheron, G. (1975). *Random Sets and Integral Geometry*. Wiley, New York.

34. McLachlan, G. J. (1992). *Discriminant Analysis and Statistical Pattern Recognition*. Wiley, New York.

35. Nalwa, V. and Binford, T. (1986). On detecting edges. *IEEE Trans. Pattern Anal. Machine Intell.*, **8**, 699–714.

36. Pratt, W. (1991). *Digital Image Processing*, 2nd ed. Wiley, New York.

37. Princen, J., Illingworth, J., and Kittler, J. (1992). A formal definition of Hough transform: Properties and relationships. *J. Math. Imaging Vision*, **1**, 153–168.

38. Rabbani, M. and Jones, P. W. (1991). *Digital Image Compression Techniques*. SPIE Press, Bellingham, WA.

39. Radon, J. (1917). Ueber die Bestimmung von Funktionen durch ihre Integralwerte Tangs gewisser Mannigfaltigkeiten (On the determination of functions from their integrals along certain manifolds). *Mathematische-Physikalische Kassifaktiones*, **69**, 262–277.

40. Ritter, G. X. (1990). Recent development in image algebra. *Adv. Electron. Electron Phys.*, Vol. 80., pp. 243–308, Academic Press, London.

41. Ritter, G. X., Wilson, J. N., and Davidson, J. L. (1990). Image algebra: An overview. *Comp. Vision Graphics Image Process.*, **49**, 297–331.

42. Roese, J. A., Pratt, W. K., and Robinson, G. S. (1977). Interframe cosine transform image coding. *IEEE Trans. Commun.*, **25**, 1,329–1,339.

43. Russ, J. (1994). *The Image Processing Handbook*. CRC Press, Boca Raton, FL.

44. Schalkoff, R. J. (1989). *Digital Image Processing and Computer Vision*. Wiley, New York.

45. Serra, J. (1982). *Image Analysis and Mathematical Morphology*. Academic Press, London.

46. Serra, J. (1988). *Image Analysis and Mathematical Morphology, Vol. 2: Theoretical Advances*. Academic Press, New York.

47. Sternberg, S. (1986). Grayscale morphology. *Comput. Vision Graphics Image Process.*, **35**, 333–355.

48. Torre, V. and Poggio, T. (1986). On edge detection. *IEEE Trans. Pattern Anal. Machine Intell.*, **8**, 147–163.

49. Vardi, Y., Shepp, L. A., and Kaufman, L. (1985). A statistical model for position emission tomography. *J. Am. Statist. Assoc.*, **80**, 8–20.

50. Wolberg, G. (1994). *Digital Image Warping*. IEEE Computer Society Press, Los Alamitos, Calif.

37. Image Restoration and Reconstruction

Valen E. Johnson

37.1 Introduction

Traditionally, image restoration and reconstruction have been regarded as related tasks having as their common goal the estimation of an image scene. Usually, this goal must be achieved by smoothing images to reduce the effects of noise, and by deconvolving images to account for the imperfect spatial resolution of the sensor.

The tasks of image restoration and reconstruction are differentiated by the type of data processed. Image restoration techniques presume that data are acquired in the image space; that is, the raw data represent a corrupted version of the image scene. In contrast, images are not directly observed in reconstruction problems. Instead, projections of an image are obtained from two or more angles, and the image scene is pieced together from these projections.

X-Ray computed tomography (CT) exemplifies this difference. To generate a CT image, x-ray images, or radiographs, are collected from the same patient at many angles. Each radiograph represents the integrated photon attenuation through the patient at a given angle and, considered individually, can be examined for diagnostic purposes without further processing. However, to obtain a CT image, radiographs collected from many angles around the patient are combined to form an estimate of the attenuation map within the patient's body. Whereas individual radiographs measure total photon attenuation along rays passing through the patient, CT systems reconstruct an estimate of the attenuation coefficients through single pixels (picture elements) or voxels (volume elements) within the patient. Individual radiographs can be *restored*; CT images must be *reconstructed*.

37.2 Image Restoration

Two types of degradation are typically assumed to affect image data. The first source of degradation, commonly referred to as *blurring*, is caused by the imperfect resolution of the imaging system, and occurs when a signal emanating from a single point in the image scene is detected over a finite area on the sensor. If we let x denote a position vector within the image space $\mathcal{I}$, $\mathcal{I} \subset \mathbb{R}^2$, and the image intensity at x by $f(x)$, a general model for blurring may be expressed as

$$b(\boldsymbol{x}) = \int_\mathcal{I} f(\boldsymbol{t})h(\boldsymbol{x},\boldsymbol{t})d\boldsymbol{t}. \quad (1)$$

The function $h(\boldsymbol{x},\boldsymbol{t})$ is the *point spread function*, often assumed to have the form

$$h(\boldsymbol{x},\boldsymbol{t}) = h(\boldsymbol{x}-\boldsymbol{t}),$$

in which case it is termed *spatially invariant* or *shift-invariant*. When the point spread function is shift-invariant, (1) is called a *convolution integral*, and the process by which an image is corrected for this defect is called *deconvolution*.

A second source of image degradation is random fluctuations in signal intensity, and may be referred to as *noise*. If $\Psi(\gamma)$ denotes a parametric distribution of a random variable with parameter γ, then the image intensity at $\boldsymbol{x}$, say $g(\boldsymbol{x})$, that results when the true scene $f(\cdot)$ is subjected to both noise and blurring is often modeled as

$$g(\boldsymbol{x}) = \Psi(b(\boldsymbol{x})). \quad (2)$$

The Gaussian distribution is the most common model for the noise distribution Ψ, in which case $g(\cdot)$ is typically assumed to have mean $b(\boldsymbol{x})$ and constant variance. More recently, however, the Poisson distribution has found increasing application in both astronomy and nuclear medicine imaging.

For digitized images (images represented on a finite $m \times n$ grid), (1) may be reexpressed as

$$\boldsymbol{b} = \boldsymbol{H}\boldsymbol{f}. \quad (3)$$

Here $\boldsymbol{H}$ is the $mn \times mn$ matrix representation of $h(\cdot)$, and $\boldsymbol{b}$ and $\boldsymbol{f}$ are $mn \times 1$ vectors. If, in addition, the noise process (2) is assumed to be additive, the combined model for noise and blur may be written

$$\boldsymbol{g} = \boldsymbol{H}\boldsymbol{f} + \boldsymbol{n}, \quad (4)$$

where $\boldsymbol{n}$ is the noise determined by $\Psi(\gamma)$.

Image restoration seeks to estimate the true scene $\boldsymbol{f}$ from either a blurred image $\boldsymbol{b}$ [(1) or (3)] or a blurred, noise-corrupted image $\boldsymbol{g}$ [(2) or (4)]. The particular restoration technique used depends on the type of degradation assumed [i.e., the form of $h(\cdot)$ and $\Psi(\cdot)$] and on whether the analyst wants to employ prior constraints on the image scene.

Although the number and variety of image modalities preclude a comprehensive discussion of all restoration techniques here, several general approaches should be mentioned. From the standpoint of a statistician, these approaches can be roughly categorized into two groups: those designed primarily for point estimation of the image scene but not necessarily based on an underlying statistical model for image generation, and Bayesian models. The former are referred to here as classical methods. These, though not necessarily motivated by formal models for image generation, can often be regarded as techniques for obtaining high posterior probability points from models implicitly defined within the Bayesian framework.

The most widely used classical methods for image restoration (and reconstruction) are based on Fourier analysis (e.g., Refs. 7, 41). Let script letters denote the Fourier transform of the corresponding italic quantities, and assume that the blurring kernel $h(\cdot)$ is shift-invariant; then (1) represents a convolution operation that may be written in the frequency domain as

$$\mathcal{B}(\nu) = \mathcal{H}(\nu)\mathcal{F}(\nu). \quad (5)$$

In the absence of noise, the true scene $\boldsymbol{f}$ can be estimated by calculating the inverse Fourier transform of $\mathcal{B}(\nu)/\mathcal{H}(\nu)$. However, in the presence of noise, it is necessary to invert the corrupted signal $\boldsymbol{g}$, which is equivalent algebraically to inverting

$$\frac{\mathcal{G}(\nu)}{\mathcal{H}(\nu)} = \mathcal{F}(\nu) + \frac{\mathcal{N}(\nu)}{\mathcal{H}(\nu)}. \quad (6)$$

An obvious difficulty in (6) is that the Fourier transform of the noise generally remains nonnegligible in regions of the frequency domain in which the transform of the point spread function $h(\cdot)$ vanishes. Thus, direct deconvolution of (6) can lead to an estimate of the image scene that is dominated by noise.

To overcome this difficulty, Fourier deconvolution methods often rely on the use of *modified inverse filters*. Rather than deconvolving $\mathcal{G}(\nu)$ by multiplying by $1/\mathcal{H}(\nu)$ and then transforming to the spatial domain, $\mathcal{G}(\nu)$ is instead deconvolved using a nonvanishing filter, say, $\mathcal{M}(\nu)$. The modified inverse filter $\mathcal{M}(\nu)$ is chosen to closely mimic $1/\mathcal{H}(\nu)$ in regions of the frequency domain in which $\mathcal{H}(\nu)$ dominates $\mathcal{N}(\nu)$, and elsewhere is based on properties of the image scene and the noise spectrum. The archetypical example of a modified inverse filter is the Wiener filter or least-squares filter.

Related to the task of deconvolution of noise-corrupted images is the task of filtering, or smoothing, nonblurred images to increase signal-to-noise ratios. As in the deconvolution problem, classical smoothing filters can be linear or nonlinear, recursive or nonrecursive.

The simplest linear filters are equivalent to blurring the image with a shift-invariant point spread function. For example, a simple 3×3 *box filter* replaces the observed value of each pixel in an image matrix with the arithmetic average of it and the values observed at its eight nearest neighbors. A *recursive* box filter iteratively replaces each pixel value with the average of its current value and the eight surrounding values. The (nonrecursive) box filter can be implemented in frequency space by multiplying the Fourier transform of the image by the corresponding transform of the box filter's point spread function. The most common shape of filters for smoothing additive noise is that of a Gaussian density function.

A disadvantage of linear filters is that they are often ineffective at removing salt-and-pepper defects, or shot noise, and they often smooth edges within an image. Nonlinear filters may offer improved performance when these types of degradation are an issue.

A simple example of a nonlinear filter is the median filter. In median filtering, a pixel's value is replaced with the median of its value and values observed at nearby pixels. Like linear filters, nonlinear filters can also be applied recursively. Unlike linear filters, they cannot be implemented in the Fourier domain, a fact that may present computational difficulties if the selected filter has a large window width.

Another class of restoration and deconvolution techniques is based on direct matrix inversion of (4) (e.g., Ref. 2). To overcome the inherent ill-conditioning in this system of equations, methods based on singular-value decompositions are often used. A major obstacle to implementing this class of methods is the computational burden associated with inverting or decomposing the matrix $\boldsymbol{H}$, which in many applications has dimensions in excess of $10^6 \times 10^6$. Fortunately, $\boldsymbol{H}$ often has special structure that facilitates numerical inversion.

37.3 Bayesian Restoration Models

The classical restoration techniques described above have played a dominant role in practical image processing, and will continue to do so. However, considerable effort has been devoted recently to the investigation of formal statistical models for image generation. The advantage of these models is that they provide a framework for investigating quantities like the posterior uncertainty of image-derived quantities, they can incorporate scene-specific prior information, and they enable the use of image data to choose between members of a given class of models.

The most common prior models for Bayesian image analysis are Gibbs distributions (GDs), which are special instances of Markov random fields (e.g., Ref. 3). The interested reader might consult References 4, 5, 6, 13, 14, 16, 17, 20, 21, 24–26, 28, 31–33, and 39, among many others.

An important feature of GDs is that they can be specified locally rather than globally. Thus, the conditional distribution of a random variable defined at a site in an image array, given the values of all other random variables in the array, depends only on the values of the random variables at *neighboring* sites. This a convenient property both from the standpoint of model specification, since the joint prior distribution on all random variables in the array need not be specified explicitly, and for computation, since Markov chain Monte Carlo algorithms (e.g., Refs. 15, 16) can be easily implemented to obtain posterior samples from the image scene.

In specifying a GD as a model for an image scene, several issues must be considered. Foremost is the choice of the neighborhood system and potential functions. The neighborhood system determines the extent of dependence in the conditional distributions of sites given all other neighbors, while the potential functions determine the type and amount of smoothing imposed on the image scene. Indiscriminate smoothing, using, for example, quadratic penalty functions, results in loss of boundary contrast within a region; too little smoothing results in images that can be difficult to interpret due to low signal-to-noise ratios.

Another issue that complicates modeling with GDs is hyperparameter estimation. For most GDs employed as image models, the functional dependence of the partition function on model hyperparameters is not known, and so estimating the posterior distribution of Gibbs hyperparameters is often problematic. Strategies for overcoming this difficulty have been described [4,5,37,38].

Point estimation from Gibbs posteriors on an image scene also can be contentious. Maximum a posteriori (MAP) estimation is generally not computationally feasible (or perhaps desirable), although simulated annealing can, in principle, be used to obtain the MAP estimate [16]. Other choices include the posterior mean, marginal posterior mode, and iterated conditional modes (ICM) [4]. Disadvantages of the posterior mean and marginal posterior mode are that they require extensive simulation from the posterior distribution. On the other hand, ICM (like the related techniques of iterated conditional expectations (ICE) [35] and iterative conditional averages (ICA) [26]) is generally computationally efficient, but may depend critically on the initial value of the system.

Traditionally, the forms of potential functions used for Gibbs image models are invariant with respect to location of the pixels. By making this function location-dependent, scene-specific prior information can be incorporated. For example, in medical imaging anatomical atlases may be used to modify the prior distribution on the correlation pattern in a degraded image (e.g., Refs. 27, 31, 34). However, techniques for modeling differences between the shape of the degraded image and the atlas (or other source of scene-specific information) are not well understood.

Related to GDs that incorporate scene-specific prior information into the restoration of a degraded scene are template models based on deformations of a scene according to an underlying Gaussian process [1,22,23] (see Ref. 36 for a novel template-like approach using a hierarchical specification on nodes indexing closed curves). Unlike GDs, deformable template models are specified through a joint distribution on the image scene. As a result, these models can incorporate a higher level of image understanding than can locally specified GDs. The development of practical algorithms for obtaining point estimates from the posterior distribution of template models and robustifying these models to allow for abnormal scene variation are topics currently under investigation.

37.4 Image Reconstruction

Image reconstruction problems are differentiated from restoration problems in that reconstruction data are not obtained in the image space itself, but instead represent indirect observations of the image, often projected onto a plane perpendicular to the desired image scene. (A notable exception to this is magnetic resonance imaging, in which data are collected in the frequency domain and may be loosely regarded as the Fourier transform of the image.) Here we restrict discussion to reconstruction techniques for projection data. In this restricted setting, reconstruction algorithms can be roughly divided into analytical and statistical methods.

Practical analytic reconstruction methods are based on a well-known relation between the Fourier coefficients of an image and Fourier transforms of the projection data at specified angles. To illustrate this relation, it is useful to reexpress (3) in polar coordinates as follows. Let $b(r, \theta)$ denote the projection of the image scene $f(x, y)$ along a ray at angle θ and distance r from the origin. That is

$$b(r, \theta) = \int_{r,\theta} f(x, y) ds, \quad (7)$$

where s represents path length along the ray indexed by (r, θ). Also, if we let $F(u, v)$ denote the Fourier coefficients of the image scene, defined by

$$\begin{aligned} F(u, v) &= \int_{\mathbb{R}^2} f(x, y) \\ &\quad \times \exp[-2\pi i(ux + vy)] dx\, dy, \end{aligned} \quad (8)$$

then

$$\begin{aligned} F(u, v) &= \int_{\mathbb{R}} b(r, \theta) \exp(-2\pi i k r) dr, \\ k &= \sqrt{u^2 + v^2}, \\ \theta &= \tan^{-1} \frac{v}{u}. \end{aligned} \quad (9)$$

Equation (9) shows that Fourier coefficients of the image scene can be obtained from Fourier transforms of the projection data at specified angles. Thus, an image can be reconstructed by taking the Fourier inverse of the coefficients obtained from projections of the image over all angles.

Several technical points should be noted when implementing this reconstruction prescription. First, (9) is predicated on the assumption that the image is band-limited, i.e., that the image contains no frequencies above some maximum frequency determined from sampling theory (e.g., Ref. 7). Second, the underlying model for the generation of the projection data is assumed to follow (7), which in nearly all applications means that many physical processes that influence data collection must be ignored. For example, in single-photon emission computed tomography (SPECT), photon attenuation through the patient cannot be directly modeled, although various corrections can be applied post hoc (e.g., [9]). Also, (9) does not allow for noise, and reconstructions based on it are generally restored either by prefiltering the projection data or postfiltering the reconstructed image. Finally, the values of the Fourier coefficients obtained from the projection data do not occur on a rectangular grid. Thus, direct Fourier inversion must either be preceded by interpolation to obtain values on a lattice, or be performed indirectly using, for example, backprojection techniques.

Because exact interpolation is computationally expensive, and approximate interpolation methods introduce image artifacts, a technique called *filtered backprojection* (FBP) has become the most commonly used analytic reconstruction method. FBP is a variation of a simple technique known as *backprojection* in which the projection data are "smeared" back across the image plane, i.e., in which the reconstructed image is estimated by

$$\hat{f}(x,y) = \frac{\pi}{m} \sum_{j=1}^{m} b(x\cos\theta_j + y\sin\theta_j, \theta_j), \quad (10)$$

where m denotes the number of (equally spaced) projection angles. Like direct Fourier inversion, FBP is based on a relationship between the backprojected image $\hat{f}(x,y)$ and the Fourier coefficients of the true image $f(x, y)$. From (9) and (10), it follows that the Fourier coefficients of $\hat{f}$, say, $\hat{F}$, satisfy

$$\hat{F}(u,v) = F(u,v)/|k|. \quad (11)$$

Thus, if the Fourier coefficients of the projection data are multiplied by $|k|$, inverted, and backprojected onto the image plane, a reconstruction equivalent to direct Fourier inversion is obtained. Computationally, several methods can be used to accomplish these steps, and convolution backprojection appears to be the most common. A very readable account of these analytic reconstruction methods is provided in Reference 8.

As opposed to analytic reconstruction methods, statistical reconstruction techniques are motivated by the data likelihood function. Since likelihood functions vary between image modalities, positron emission tomography (PET) data are used below to illustrate the general issues involved.

PET reconstruction was popularized in the statistical literature by Vardi et al. [42], who used the EM algorithm to obtain an (almost) maximum-likelihood estimate (MLE) of the image scene. The likelihood function they described arises from the following physical process. A positron-emitting radioisotope is injected into a patient's bloodstream, and concentrates differentially in organs according to each organ's uptake of the chemical to which the isotope is linked. Positrons emitted by the isotope travel a short distance, annihilate with an electron, and generate two photons that travel in nearly opposite directions. A detector ring is placed around some portion of the patient's body, and nearly simultaneous registrations of photons suggest that a positron was emitted near the line, or *tube*, connecting the points at which the photons were registered. If we assume that positrons are emitted independently according to a Poisson distribution with mean intensity $f(x,y)$, constant within pixels, and that the probability that a positron emitted from location (x,y) in the image lattice is detected at tube t is independent of other emissions and is equal to a known constant p_{xy}^t, then the distribution of tube counts, say Y_t, is described by a Poisson distribution

$$Y_t \sim \text{Pois}\left(\sum_{(x,y)} f(x,y) p_{xy}^t\right),$$
$$t = 1, \ldots, T. \quad (12)$$

The statistical reconstruction problem for PET is to perform inference concerning $f(x,y)$ based on the observed data $Y_t, t = 1, \ldots, T$.

A comparison of (7) and (12) reveals the increased flexibility in data modeling that a statistical approach brings to the reconstruction, problem. In (12), the transition probabilities p_{xy}^t are completely arbitrary, implying that the mean of the tube counts need not be assumed equal to the simple ray sum of pixel intensities $f(x,y)$. While this is a significant advantage in PET, it can be even more critical in modalities such as SPECT, where physical effects of scatter and attenuation make the assumptions underlying (7) untenable.

Closer examination of (12) also reveals several disadvantages of statistical reconstruction algorithms. Typically, the number of observations available, in the PET case equal to T, is of the same order of magnitude as the number of pixel intensities that must be estimated. This fact has two important implications. First, maximum-likelihood or maximum a posteriori (MAP) estimates cannot be obtained analytically, but must instead be evaluated numerically. Second, asymptotic properties normally attributed to statistical estimators like the MLE are unlikely to pertain.

Vardi et al. [42] (see also Refs. 30 and 40) overcame the first difficulty by posing the estimation problem within the EM framework (e.g., Ref. 12). By treating the unobserved number of counts originating in each pixel as latent data, they defined a simple EM algorithm that converges to the MLE. Hebert and Leahy [24] and Green [21] extend these ideas to MAP estimation.

The second difficulty is more problematic. In PET, maximum likelihood estimation does not provide useful estimates of the source distribution, and the use of Gibbs priors or other constraints on the image scene has proven necessary. For example, Vardi et al. [42] proposed starting the EM algorithm with a uniform value and stopping iterations after a fixed number of iterations; Coakley [10], Coakley and Llacer [11], and Johnson [29] describe more formal criteria based on cross-validation and the jackknife. Many of the papers cited above on Gibbs priors contain applications of these priors to PET or SPECT data.

37.5 Summary

Expectation–maximization methods have played a dominant role in the development of statistical reconstruction algorithms during the last decade, and classical restoration and reconstruction methods continue to be the most commonly applied image analysis techniques. However, with recent advances in Markov chain Monte Carlo algorithms and faster computers, emphasis in statistical image restoration and reconstruction is gradually shifting away from simple point estimation of an image scene to more complicated image-analytic tasks. For example, posterior samples from image scenes are increasingly being used to assess the uncertainty of image features, and techniques for incorporating scene-specific prior information into image reconstructions, once thought to require excessive computation, are now being considered for routine use. During the next decade, it is likely that these higher-level statistical image analysis tasks will constitute the core of developments in the field.

References

1. Amit, Y., Grenander, U., and Piccioni, M. (1991). Structural image restoration through deformable templates. *J. Am. Statist. Assoc.*, **86**, 376–387.

2. Andrews, H. C. and Hunt, B. R. (1977). *Digital Image Restoration*. Prentice-Hall, Englewood Cliffs, NJ.

3. Besag, J. E. (1974). Spatial interaction and the statistical analysis of lattice systems. *J. Roy. Statist. Soc. B*, **36**, 192–225.

4. Besag, J. E. (1986). On the statistical analysis of dirty pictures. *J. Roy. Statist. Soc. B*, **48**, 259–302.

5. Besag, J. E. (1989). Towards Bayesian image analysis. *J. Appl. Statist.*, **16**, 395–407.

6. Besag, J. E., York, J., and Mollie, A. (1991). Bayesian image restoration with two applications in spatial statistics. *Ann. Inst. Statist. Math.*, **43**, 1–59.

7. Bracewell, R. N. (1978). *The Fourier Transform and Its Applications*, 2nd ed. McGraw-Hill, New York.

8. Brooks, R. A. and Di Chiro, G. (1976). Principles of computer-assisted tomography (CAT) in radiographic and radioisotopic imaging. *Phys. Med. Biol.*, **21**, 689–732.

9. Chang, L. T. (1978). A method for attenuation correction in radionuclide computed tomography. *IEEE Trans. Nucl. Sci.*, **25**, 638–643.

10. Coakley, K. J. (1991). A cross-validation procedure for stopping the EM algorithm and deconvolution of neutron depth profiling spectra. *IEEE Trans. Nucl. Sci.*, **38**, 9–16.

11. Coakley, K. J. and Llacer, J. (1991). The use of cross-validation as a stopping rule and reconstruction of emission tomography images. *Proc. SPIE Medical Imaging V Image Physics Conf.*, pp. 226–233.

12. Dempster, A. P., Laird, N. M., and Rubin, D. B. (1977). Maximum likelihood from incomplete data via the EM algorithm. *J. Roy. Statist. Soc. B*, **39**, 1–38.

13. Derin, H. and Elliot, H. (1987). Modeling and segmentation of noisy and textured images using Gibbs random fields. *IEEE Trans. Pattern Anal. Machine Intell.*, **9**, 39–55.

14. Dubes, R. C. and Jains, A. K. (1989). Random field models in image analysis. *J. Appl. Statist.*, **16**, 131–164.

15. Gelfand, A. E. and Smith, A. F. M. (1990). Sampling-based approaches to calculating marginal densities. *J. Am. Statist. Assoc.*, **85**, 398–409.

16. Geman, S. and Geman, D. (1984). Stochastic relaxation, Gibbs distributions, and the Bayesian restoration of images. *IEEE Trans. Pattern Anal. Machine Intell.*, **6**, 721–741.

17. Geman, S. and McClure, D. E. (1987). Statistical methods for tomographic image reconstruction. *Bull. Int. Statist. Inst.*, **52**, 5–21.

18. Geyer, C. J. (1991). *Reweighting Monte Carlo Mixtures*. Technical Report 568, School of Statistics, University of Minnesota.

19. Geyer, C. J. and Thompson, E. A. (1992). Constrained Monte Carlo for maximum likelihood with dependent data (with discussion). *J. Roy. Statist. Soc. B*, **54**, 657–699.

20. Gindi, G., Lee, M., Rangarajan, A., and Zubal, I. G. (1991). Bayesian reconstruction of functional images using registered anatomical images as priors. In *Lecture Notes in Computer Science 511*, pp. 121–131, Springer-Verlag.

21. Green, P. J. (1990). Bayesian reconstructions from emission tomography data using a modified EM algorithm. *IEEE Trans. Med. Imaging*, **9**, 84–93.

22. Grenander, U. and Manbeck, K. (1993). A stochastic shape and color model for defect detection in potatoes. *J. Comput. Graphical Statist.*, **2**, 131–151.

23. Grenander, U. and Miller, M. (1994). Representations of knowledge in complex systems. *J. Roy. Statist. Soc. B*, **56**, 549–603.

24. Hebert, T. and Leahy, R. (1989). A generalized EM algorithm for 3D Bayesian reconstruction from Poisson data using Gibbs priors. *IEEE Trans. Med. Imaging*, **8**, 194–202.

25. Heikkinen, J. and Högmander, H. (1994). Fully Bayesian approach to image restoration with an application in biogeography. *Appl. Statist.*, **43**, 569–582.

26. Johnson, V. E., Wong, W. H., Hu, X., and Chen, C. T. (1991). Aspects of image using Gibbs priors: boundary modeling, treatment of blurring, and selection of hyperparameters. *IEEE Trans. Pattern Anal. Machine Intell.*, **13**, 412–425.

27. Johnson, V. E. (1993). A framework for incorporating prior information into the

reconstruction of medical images. In *Lecture Notes in Computer Science 687*, pp. 307–321, Springer-Verlag.

28. Johnson, V. E. (1994). A model for segmentation and analysis of noisy images. *J. Am. Statist. Assoc.*, **89**, 230–241.

29. Johnson, V. E. (1994). A note on stopping rules in EM—ML reconstructions of ECT images. *IEEE Trans. Med. Imaging*, **13**, 569–571.

30. Lange, K. and Carson, R. (1984). EM reconstruction algorithms for emission and transmission tomography. *J. Comput. Assist. Tomogr.*, **8**, 306–318.

31. Leahy, R. and Yan, X. (1991). Incorporation of anatomical MR data for improved functional imaging with PET. In *Lecture Notes in Computer Science 511*, pp. 105–120, Springer-Verlag.

32. Levitan, E. and Herman, G. T. (1987). A maximum *a posteriori* probability expectation maximization algorithm for image reconstruction in emission tomography. *IEEE Trans. Med. Imaging*, **6**, 185–192.

33. Molina, R. and Ripley, B. (1989). Using spatial models as priors in astronomical image analysis. *J. Appl. Statist.*, **16**, 193–206.

34. Ouyang, X., Wong, W. H., Johnson, V. E., Hu, X., and Chen, C. T. (1994). Incorporation of correlated structural images in PET image reconstruction. *IEEE Trans. Med. Imaging*, **13**, 627–640.

35. Owen, A. (1986). Comments on "Statistics, images, and pattern recognition" by B. Ripley. *Can. J. Statist.*, **14**, 106–110.

36. Phillips, D. B. and Smith, A. F. M. (1994). Bayesian faces via hierarchical template modeling. *J. Am. Statist. Assoc.*, **89**, 1151–1163.

37. Qian, W. and Titterington, D. M. (1988). Estimation of parameters in hidden Markov models. *Phil. Trans. Roy. Soc. Phys. Sci. Eng. A*, **337**, 407–428.

38. Qian, W. and Titterington, D. M. (1989). On the use of Gibbs Markov chain models in the analysis of images based on second-order pairwise interactive distributions. *J. Appl. Statist.*, **16**, 267–281.

39. Ripley, B. (1986). Statistics, images, and pattern recognition (with discussion). *Can. J. Statist.*, **14**, 83–111.

40. Shepp, L. and Vardi, Y. (1982). Maximum likelihood reconstruction for emission tomography. *IEEE Trans. Med. Imaging*, **1**, 113–122.

41. Walker, J. S. (1988). *Fourier Analysis*. Oxford University Press, New York.

42. Vardi, Y., Shepp, L., and Kaufman. L. (1985). A statistical model for positron emission tomography. *J. Am. Statist. Assoc.*, **80**, 8–20.

38

Imputation and Multiple Imputation

Donald B. Rubin and Nathaniel Schenker[1]

38.1 Introduction

A common technique for handling missing values in a dataset is to impute, that is, fill in, a value for each missing datum. Imputation creates a completed dataset, so that standard methods that have been developed for analyzing complete data can be applied immediately. Thus, imputing for missing values and then using standard complete-data methods of analysis is typically easier than creating specialized methods of analysis for the incomplete data.

Imputation has other advantages in the context of the production of a dataset for use by the public, such as a public-use file from a major survey. The data producer can use specialized knowledge about the reasons for missing data, including confidential information that cannot be released to the public, to help create the imputations. In addition, imputation by the data producer fixes the missing-data problem in the same way for all users, so that consistency of answers across different users employing the same complete-data analysis is ensured. If it were left up to each user to implement some method for handling missing data, the knowledge of the data producer could not be incorporated, and answers for different users with the same complete-data model would not necessarily be consistent, due to the users' varying methods of handling missing data.

Although single imputation, that is, imputing one value for each missing datum, satisfies critical data-processing objectives and can incorporate knowledge from the data producer, it fails to satisfy statistical objectives concerning the validity of the resulting inferences. Specifically, for statistical validity, estimates based on the data completed by imputation should be approximately unbiased for their population estimands, confidence intervals should attain at least their nominal coverages, and tests of null hypotheses should not reject true null hypotheses more frequently than their nominal levels. Before discussing why single imputation cannot generally achieve statistical validity, whereas imputing multiple values (proposed in Rubin [55]) can do so, it is useful to review general considerations for creating imputations.

---

[1] The findings and conclusions in this chapter are those of the authors and do not necessarily represent the views of the National Center for Health Statistics, Centers for Disease Control and Prevention.

38.2 Considerations for Creating Imputations

Little [31] gave a detailed discussion of considerations for creating imputations. One major consideration is that random draws, rather than best predictions, should be used because imputing best predictions will almost always lead to distorted estimates of quantities that are not linear in the data, such as measures of variability and correlation.

A second important consideration is that imputations of missing values should be based on predicting the missing values given all observed values. This consideration is particularly important in the context of a public-use database. Leaving a variable out of the imputation model is equivalent to assuming that the variable is not associated with the variables being imputed, conditionally given all other variables in the imputation model; imputing under this assumption can create biases in subsequent analyses. Because it is not known which analyses will be carried out by subsequent users of a public-use database, ideally it is best not to leave any variables out of the imputation model. It is usually infeasible, of course, to incorporate every available variable, including higher-order terms and interactions, into an imputation model; but it is desirable to condition imputations on as many variables as possible, including variables reflecting the sample design in the case of a complex survey, and to use subject-matter knowledge to help select variables.

Common methods of imputation do not always attend to these two considerations and can lead to seriously invalid answers. For example, a traditional but naive approach replaces each missing value on a variable by the unconditional sample mean of the observed values of that variable, thereby attending to neither of the considerations; it can result in satisfactory point estimates of some quantities, such as unconditional means and totals, but it yields inconsistent estimates of other quantities, such as variances, covariances, quantiles, and measures of shape.

An improvement over unconditional mean imputation is conditional mean imputation. This method typically first classifies cases into cells based on observed variables, and then within each such "adjustment" cell, replaces each missing value on a variable by the sample mean of the observed values of that variable. A generalization of this form of conditional mean imputation is regression imputation, in which the regression (i.e., conditional distribution) of the variable on other observed variables is estimated from the complete cases, and then the resulting prediction equation is used to impute the estimated conditional mean for each missing value.

Although both conditional mean imputation and regression imputation satisfy one major consideration by conditioning on observed variables, neither method satisfies the other major consideration of imputing random draws. An example of a method that does impute random draws is stochastic regression imputation, in which each imputed value is a regression prediction with a random error added, where the random error has variance equal to the estimated residual variance from the regression.

As discussed in Rubin [60] and Rubin and Schenker [66], imputation procedures can be based on explicit models or implicit models, or even combinations. An example of a procedure based on an explicit model is stochastic regression imputation with normally distributed random errors. A common type of procedure based on implicit models is hot-deck imputation, which replaces the missing values for an incomplete case by the values from a matching complete case, where the matching is carried out with respect to variables that are observed for both the incomplete case and complete cases, as with conditional mean imputation. Randomness is often incorporated by drawing the complete case randomly from a pool of matching "donor" cases. Rather than attempting to match cases exactly in hot-deck imputation, it is sometimes useful to define a distance function based on variables that are observed for both complete and incomplete cases and then to impute values for each incomplete case from a complete case that is close. When the distance function is based on the difference between cases on the predicted values of the variables to be imputed, the matching procedure is termed predictive mean matching [58,31,22]. Hot-deck imputation using predictive mean matching based on an explicit prediction model is an example of an imputation procedure that combines aspects of both an implicit model and an explicit model.

The model underlying an imputation procedure, whether explicit or implicit, can be based on the assumption that the reasons for missing data are either ignorable or nonignorable [54]. The distinction between an ignorable and a nonignorable model can be illustrated by the simple example in which there are only two variables, Y and Z, and Z is observed for all cases whereas Y is sometimes missing. Ignorable models assert that a case with Y missing is only randomly different from a complete case having the same value of Z. Nonignorable models assert that there are systematic differences between an incomplete case and a complete case that have identical Z values. An issue with nonignorable models is that, since the missing values cannot be observed, there is no direct evidence in the data to address the assumption of nonignorability. It can be important, therefore, to consider several alternative models and to explore sensitivity of resulting inferences to the choice of model. In practice, almost all imputation models are ignorable. Limited experience suggests that in major surveys with limited amounts of missing data and careful design, ignorable models are satisfactory for most analyses (see, e.g., Rubin et al. [67]). Moreover, including a large number of predictors in the imputation model, as suggested earlier in this section, helps the assumption of ignorability to be tenable because there are many variables in the model to "explain the missingness."

38.3 The Underestimation of Uncertainty with Single Imputation

As mentioned in Section 38.1, although imputation has several advantages, it has a major disadvantage. Because a single imputed value cannot reflect any of the uncertainty about the true underlying value, analyses that treat imputed values just like observed values underestimate uncertainty. Thus, imputing a single value for each missing datum, and then analyzing the completed data using standard techniques designed for complete data, generally will result in standard error estimates that are too small, confidence intervals that undercover, and P values that are too significant; this is true even if the modeling for imputation is conducted carefully and the considerations for imputation discussed previously have been addressed. For example, large-sample results in Rubin and Schenker [64] showed that for simple situations with 30% of the data missing, single imputation under the correct model followed by the standard complete-data analysis results in nominal 90% confidence intervals having actual coverages below 80%. The inaccuracy of nominal levels is even more extreme in multiparameter problems [60, Chap. 4], where nominal 5% tests can easily have rejection rates of 50% or more under the null hypothesis when the fraction of information about the estimand that is missing is 30% or greater.

The fraction of information about an estimand that is missing (commonly called the "fraction of missing information"), a quantity defined in Rubin [60, Chaps. 3 and 4], refers to the loss in information when estimating the estimand due to missing data relative to the amount of information that would exist were the data complete. With ignorable missing data and just one variable, the fraction of missing information is simply the fraction of data values that are missing. When there are several variables, however, the fraction of missing information is often smaller than the fraction of cases that are incomplete because of the ability to predict missing values from observed values.

38.4 Introduction to Multiple Imputation

Multiple imputation [55,60] is an approach that retains the advantages of single imputation while allowing the data analyst to obtain valid assessments of uncertainty. The basic idea is to impute for the missing data 2 or more times using independent draws of the missing values from a distribution that is appropriate under the posited assumptions about the data and the mechanism causing missing data. This process results in two or more datasets, each of which is analyzed using the same standard complete-data method. The analyses are then combined in a simple way that reflects the extra uncertainty due to imputation. Multiple imputations can also be created under several different models to display sensitivity to the choice of missing-data model.

The theoretical motivation for multiple imputation is Bayesian, although the procedure has good properties from a frequentist perspective when applied properly. Consider, for simplicity, the case of an ignorable missing-data mechanism. Let Q be the population quantity of interest, and partition the full data, X, into the observed values X_{obs} and the missing values X_{mis}. If X_{mis} had been observed, inferences for Q would have been based on the complete-data posterior density $p(Q|X_{\text{obs}}, X_{\text{mis}})$. Since X_{mis} is not observed, however, inferences are based on the actual posterior density

$$p(Q|X_{\text{obs}}) = \int p(Q|X_{\text{obs}}, X_{\text{mis}}) \times p(X_{\text{mis}}|X_{\text{obs}}) dX_{\text{mis}}. \quad (1)$$

Equation (1) shows that the posterior density of Q can be obtained by averaging the complete-data posterior density over the posterior predictive distribution of X_{mis}. In principle, multiple imputations are repeated independent draws with density $p(X_{\text{mis}}|X_{\text{obs}})$. Thus, multiple imputation allows the data analyst to approximate equation (1) by analyzing the completed datasets and then combining the analyses.

38.5 Analyzing a Multiply Imputed Dataset

The exact computation of the posterior density (1) by simulation would require that an infinite number of values of X_{mis} be drawn with density $p(X_{\text{mis}}|X_{\text{obs}})$; in addition, $p(Q_0|X_{\text{obs}}, X_{\text{mis}})$ would need to be calculated for every value Q_0. This section gives a simple approximation to (1) for scalar Q [64] that can be used when only a small number of imputations of X_{mis} have been drawn. Procedures for significance testing when Q is multidimensional are given in Rubin [60, Chap. 3], Li et al. [29,30], Meng and Rubin [40], and Little and Rubin [32, Sec. 10.2].

Suppose that if the data were complete, inferences for Q would be based on a point estimate $\hat{Q}$ and a sampling variance estimate $\hat{V}$. When data are missing and there are M sets of imputations for the missing data, the result is M sets of complete-data estimates, say, $\hat{Q}_m$ and $\hat{V}_m$, $m = 1, ..., M$. The point estimate of Q from the multiple imputations is the average of the M complete-data estimates, $\bar{Q} = \sum \hat{Q}_m/M$; and the associated sampling-variance estimate is $T = \bar{V} + (1 + M^{-1})B$, where $\bar{V} = \sum \hat{V}_m/M$ is the average within-imputation variance, and $B = \sum (\hat{Q}_m - \bar{Q})^2/(M-1)$ is the between-imputation variance. These formulas follow from assuming that for the model underlying equation (1), $\hat{Q} \approx E(Q|X_{\text{obs}}, X_{\text{mis}})$ and $\hat{V} \approx V(Q|X_{\text{obs}}, X_{\text{mis}})$, that is, the complete-data point and sampling-variance estimates are close to the complete-data posterior mean and variance of Q, respectively. Under these assumptions, standard relations between conditional and unconditional means and variances imply that the posterior mean and variance given the observed data can be approximated as $E(Q|X_{\text{obs}}) \approx E(\hat{Q}|X_{\text{obs}})$ and $V(Q|X_{\text{obs}}) \approx E(\hat{V}|X_{\text{obs}}) + V(\hat{Q}|X_{\text{obs}})$. The formulas for $\bar{Q}$ and T are simulated versions of these approximations based on the multiple imputations, with the extra factor $(1+M^{-1})$ in the formula for T being an adjustment when the number of imputations, M, is small.

For the case in which a normal reference distribution would be used for complete-data inferences for Q, Rubin and Schenker [64] proposed that multiple-imputation inferences Q be based on a t reference distribution with degrees of freedom $\nu_{RS} = (M-1)\hat{\gamma}^{-2}$, where $\hat{\gamma} = (1+M^{-1})B/T$ is the ratio of the between-imputation component of variance to the total variance. For large M, $\hat{\gamma}$ also estimates the fraction of missing information.

For the case in which the complete-data reference distribution is a t distribution with ν_{com} degrees of freedom, Barnard and Rubin [3] derived the following improved degrees of freedom for multiple-imputation inferences: $\nu_{BR} = [(1/\nu_{RS}) + (1/k)]^{-1}$, where $k = [\nu_{\text{com}}(\nu_{\text{com}} + 1)/(\nu_{\text{com}} + 3)](1 - \hat{\gamma})$.

38.6 Reflecting Appropriate Variability and Choosing M

Ideally, multiple imputations are M independent draws from the posterior predictive distribution of X_{mis} under appropriate modeling assumptions. Such imputations were called "repeated imputations" in Rubin [60, Chap. 3]. In practice, approximations are often used and several important issues arise, such as those discussed in Section 38.2. Two other issues that arise specifically in the context of multiple imputation are how to incorporate appropriate between-imputation variability and how to choose M.

For the case in which a complete-data inference would be valid from the standard respeated-sampling frequentist perspective if there were no missing data, Rubin [60, Chap. 4] specified conditions under which a multiple-imputation procedure produces valid inferences as well and termed such a multiple-imputation procedure "proper" for the complete-data inference. Rubin [63] provided more intuitive statements of the conditions. A key condition is that an imputation procedure must incorporate appropriate variability across the M sets of imputations within a model. Bayesian procedures in general tend to reflect variability fully, and Bayesian reasoning reveals that a two-step paradigm for imputation is useful to ensure that appropriate variability is incorporated. To see this, suppose that a parametric model for the distribution of X_{mis} given X_{obs} has been formulated, with density $p(X_{\text{mis}}|X_{\text{obs}}, \theta)$. Then the distribution to be used for imputation has density

$$p(X_{\text{mis}}|X_{\text{obs}}) = \int p(X_{\text{mis}}|X_{\text{obs}}, \theta) \times p(\theta|X_{\text{obs}})d\theta, \quad (2)$$

where $p(\theta|X_{\text{obs}})$ is the posterior density of θ. It can be seen from Equation (2) that a draw of a value of X_{mis} from its posterior predictive distribution is obtained by first drawing a value of θ from its posterior distribution and then drawing a value of X_{mis} conditional on the drawn value of θ. Fixing θ at a point estimate $\hat{\theta}$ (e.g., the maximum-likelihood estimate) across the M imputations and drawing X_{mis} with density $p(X_{\text{mis}}|X_{\text{obs}}, \hat{\theta})$ generally leads to multiple-imputation inferences that are too sharp, as shown, for example, in Rubin and Schenker [64,65] and Raghunathan and Rubin [47].

The two-step paradigm can be followed in the context of nonparametric methods, such as hot-deck imputation, as well as in the context of parametric models as discussed above. The simple hot-deck procedure that randomly draws imputations for incomplete cases from a donor pool of matching complete cases does not incorporate appropriate variability, because it ignores the sampling variability due to the fact that the population distribution of complete cases is not known but rather is estimated from the complete cases in the sample. Rubin and Schenker [64,66] discussed the use of the bootstrap [8] to make the hot-deck procedure incorporate appropriate variability and called the resulting procedure the "approximate Bayesian bootstrap," because it approximates the Bayesian bootstrap [56]. The two-step procedure first draws a bootstrap sample from the complete cases and then draws imputations randomly from the bootstrap sample. Thus, the bootstrap sampling from the complete cases before drawing imputations is a nonparametric analogue to drawing values of the parameters of the imputation model from their posterior distribution before imputing conditionally on the drawn parameter values.

Other combinations of parametric and non-parametric procedures within the two-step paradigm are possible as well. For example, Heitjan and Little [22] combined an initial bootstrap step with bivariate predictive mean matching imputation in the second step; Dorey et al. [6] used the bootstrap in the first step together with a parametric procedure in the second step; and Schenker and Taylor [73] combined a fully parametric first step with predictive mean matching imputation in the second step.

Another issue to be considered when creating multiple imputations is the choice of M. This choice involves a tradeoff between approximating the posterior distribution (1)—and the corresponding sampling distribution—more precisely, using larger values of M, versus using smaller amounts of computing and storage, as occurs with smaller values of M. The effect of the size of M on precision depends partly on the fraction of missing information. For the moderate fractions of missing information ($<30\%$) that occur with most analyses of data from most large surveys, Rubin [60, Chap. 4] showed that a small number of imputations (say, $M = 3$ or 4) results in nearly fully efficient multiple-imputation estimates. Moreover, empirical and theoretical evaluations have shown that if proper multiple imputations are created, then the resulting inferences generally have close to their nominal coverages or significance levels, even when the number of imputations is moderate. See, for example, Rubin and Schenker [64], Rubin [60, Chap. 4], Rubin and Schenker [65], Clogg et al. [4], Heitjan and Little [22], Ezzati-Rice et al. [10], Schafer and Schenker [70], and Reiter et al. [52].

38.7 Bayesian Iterative Simulation and Multiple Imputation

The two-step paradigm described in the previous section is relatively straightforward to apply when the pattern of missing data is simple, such as when only a single variable has missing values. For complicated patterns of missing data, however, precisely following the paradigm can be infeasible, in large part due to difficulties in drawing a value of the parameter, θ, from its posterior distribution, $p(\theta|X_{\text{obs}})$. In such cases, Bayesian iterative simulation methods [32, Chap. 10], [16, Pt. III]), such as data augmentation [76] and Gibbs sampling [18,15], can facilitate the creation of multiple imputations, and they are especially useful in the context of complicated parametric models. Consider a model for X with a parameter, say θ. The data augmentation (Gibbs sampling) procedure that results in draws from the posterior distribution of θ produces multiple imputations of X_{mis} as well. Specifically, let $\theta^{(t)}$ and $X_{\text{mis}}^{(t)}$ denote the draws of θ and X_{mis} at iteration t in the Gibbs sampler. At iteration $t+1$, a value $\theta^{(t+1)}$ is drawn with density $p(\theta|X_{\text{obs}}, X_{\text{mis}}^{(t)})$, and then a value $X_{\text{mis}}^{(t+1)}$ is drawn with density $p(X_{\text{mis}}|X_{\text{obs}}, \theta^{(t+1)})$. As t approaches infinity, $(\theta^{(t)}, X_{\text{mis}}^{(t)})$ converges to a draw with density $p(\theta, X_{\text{mis}}|X_{\text{obs}})$. Thus, $X_{\text{mis}}^{(t)}$ converges to a draw with density $p(X_{\text{mis}}|X_{\text{obs}})$ and can be used as an imputation of X_{mis} in a multiple-imputation scheme.

In a sense, each iteration of Bayesian iterative simulation follows the two-step paradigm of drawing a value of θ and then drawing a value of X_{mis} conditional on the drawn value of θ. However, with Bayesian iterative simulation, θ is drawn with density $p(\theta|X_{\text{obs}}, X_{\text{mis}}^{(t)})$ rather than $p(\theta|X_{\text{obs}})$. Drawing θ from an approximation to its complete-data posterior distribution is typically more feasible than drawing θ from its observed-data posterior distribution when the pattern of missing data is complicated.

In a multivariate missing-data problem, traditional Bayesian iterative simulation methods begin with the specification of a model for the joint distribution for all of the variables that have missing values. Each iteration of the simulation then cycles through all of the unknown quantities, that is, the missing values of variables and the parameters, drawing a value for each unknown quantity from its distribution conditional on all of the other quantities in the model, with all of the other unknown quantities set at their most recently drawn values.

A model for all of the variables with missing values can be difficult to specify, especially when the variables are of many different types (continuous, categorical, count, etc.), or if there are bounds or other restrictions on the values of the variables. Moreover, the data available for estimation of some parameters of the joint model can be very sparse when there are many variables involved. An extension of traditional Bayesian iterative simulation that has become popular recently avoids these issues by specifying only a model for the conditional distribution of each variable with missing values given all of the other variables, rather than specifying a joint distribution for all of the variables with missing values. The iterative simulation then proceeds in the same fashion as with traditional Bayesian iterative simulation. Such an extension has been termed "multiple imputation by chained equations" [81], "sequential regression multivariate imputation" [45], and "fully conditional specification in multivariate imputation" [80,79].

A theoretical issue with this extension

to Bayesian iterative simulation is that the conditional distributions specified might not be compatible, in the ensemble, with any joint distribution for the variables that have missing values. In such cases, the iterative simulation procedure has the possibility of not converging. However, the methods typically have been found to work well in practice (e.g., see van Buuren et al. [80] and van Buuren [79]).

38.8 More on Conditions for Validity of Inferences with Imputed Data

In order for inferences based on data completed by imputation to be approximately valid, it is first necessary for the inference that would be used if there were no missing data to be appropriate. That is, if the complete-data inference is not approximately valid, imputation generally will not be able to fix the problem.

Even if the complete-data inference is valid, inferences based on data completed by imputation will not be valid unless the imputation procedure is appropriate. This argues for using flexible imputation models and including as many predictors as possible, as discussed earlier. It also argues for checking the imputation model to the extent possible. Abayomi et al. [1] discussed three useful types of diagnostics for imputations: displays of the completed data to check for unusual patterns, comparisons of the distributions of observed and imputed data values, and checks of the fit of the imputation model to the observed data. Because the missing data cannot be observed, however, it is impossible to check completely whether an imputation model is approximately valid. Therefore, it is good practice to display sensitivity to variations in the imputation model, by creating imputations under different assumptions and conducting analyses of the resulting completed data. Indeed, Rubin's [55,60] original discussions of multiple imputation emphasizes not only proper assessment of variability but also imputing under multiple models to display sensitivity of complete-data inferences to the imputation model.

Although use of a valid imputation procedure is generally necessary to ensure validity of inferences based on data completed by imputation, two points are important to note. First, the impact of the imputation procedure on the validity of inferences is dampened by the fact that imputation only affects the missing component of the dataset. Thus, if the imputation procedure is sensible but not perfect, the resulting inferences are often nearly valid as long as the complete-data procedure is approximately valid and the fraction of missing information is not large. Rubin [59] gave the following intuitive, generic example to illustrate this point. Suppose that an analysis problem has 30% missing information, the complete-data analysis procedure is appropriate, and the multiple-imputation procedure is 80% OK (i.e., 20% deficient). Then the analysis of the dataset completed by multiple imputation will be 20% deficient on 30% of the information, or 6% deficient overall, that is, 94% OK.

The second point is that use of multiple imputation can help to protect some inferences from the effects of misspecifying the imputation model. The reason is that the lack of fit of the model typically results in larger estimates, based on the data completed by imputation, of the within- and between-imputation variances than would occur if the imputation model were correctly specified. Thus, when inferences are drawn, the effects of biases in point estimates due to misspecification of the imputation model are offset to some extent by the effects of increased estimates of variance.

The Bayesian justification for multiple imputation, as summarized by equation (1), depends on the complete-data inference ($p(Q|X_{\text{obs}}, X_{\text{mis}})$) being compatible with the imputation model ($p(X_{mis}|X_{\text{obs}})$). Indeed, even if neither the complete-data inference nor the imputation procedure is incorrect, incompatibility of the implied models underlying them can result in biases in the multiple-imputation estimator of sampling variance. Examples of this phenomenon have been given in Rubin and Schenker [65], Fay [12,13], Meng [37,38], Kott [27], Rubin [63], Robins and Wang [53], Nielsen [44], and Kim et al. [25]. Meng [37] provided a general discussion of the situation and coined the term "uncongeniality" for the case in which the imputer's and analyst's models are incompatible.

Little and Rubin [32, Sec. 10.2.4] note that uncongeniality usually leads to conservative inferences. The most common situation discussed in the citations above is when the imputation method uses more information than the complete-data analysis and this information is correct. This tends to result in an analysis of the completed dataset that is, in a sense, more efficient than anticipated, an upward bias in the multiple-imputation estimator of sampling variance, and a conservative inference, such as a confidence interval with true coverage higher than the nominal level. In some instances, however, the multiple-imputation estimator of sampling variance can be biased downward, as demonstrated in examples by Robins and Wang [53] and Nielsen [44]. Such situations typically involve a complete-data analysis procedure that is highly inefficient. In their discussion of Nielsen [44], Meng and Romero [39] noted that a useful requirement to guard against such situations is for complete-data inferences to be "self-efficient," a concept defined in Meng [37]. Intuitively, self-efficiency means that the complete-data inference cannot be made more efficient by excluding a part of the data. Meng and Romero [39] also proposed some diagnostics for self-inefficiency.

Further research in this area and diagnostics for examining the validity of inferences from data completed by imputation will be useful both for theoretical understanding and as an aid to practitioners.

38.9 Some Applications and Different Uses of Multiple Imputation

A search of the literature reveals a very large number of applications of imputation or multiple imputation to a variety of problems. This section briefly describes just a few examples of applying multiple imputation in the context of handling missing data in public-use files, which was a primary motivation for the development of the technique [55,60]. Each example is chosen for its historical significance and/or to illustrate points discussed in the previous sections. After these examples are introduced, some other types of applications of multiple imputation are described.

One of the earliest applications of multiple imputation to public-use data was for the purpose of calibrating industry and occupation codes in 1970 and 1980 census public-use files (see Treiman and Rubin [78], Rubin [57], Weidman [82], Clogg et al. [4], Schenker et al. [74]). Because different coding schemes for narrative responses concerning employment were used for the two censuses, the codes were not comparable across the decades. A relatively small dataset from the 1970 census with codes created under both schemes was used to develop imputation models predicting 1980 codes from 1970 codes and covariates. These models were then used to multiply impute 1980 codes for the 1970 public-use data. In addition to its historical significance, this example illustrates the use of the two-step paradigm discussed in Section 38.6 for fully reflecting variability in imputations when there is a simple pattern of missing data, in the context of imputation models based on logistic regression (see, e.g., Rubin and Schenker [65]). It is also an example in which the imputer's model was incompatible with the analyst's model, in the sense that the "double-coded" dataset used to develop the imputation models was not part of the final data files released for public use. Rubin and Schenker [65] showed that this would tend to result in conservative inferences, a finding consistent with later research described in the previous section.

Another early application of multiple imputation was to missing data in the Survey of Consumer Finances of the Federal Reserve Board (*http://www.federalreserve.gov/pubs/ oss/oss2/scfindex.html*; Kennickell [24]). A significant feature of this application, besides the importance of the data, was its use of Bayesian iterative simulation methods to create the imputations. In fact, the application represents a use of fully conditional specification in multivariate imputation well before the technique was developed and explored more generally. The Federal Reserve Board has continued to use multiple imputation in the Survey of Consumer Finances since the initial application.

Perhaps the first large-scale application of multiple imputation using Bayesian iterative simulation based on full specification of the joint distribution in the imputation model was to missing data on several variables in the Third National Health and Nutrition Examination Survey of the National Center for Health Statistics (*http://www.cdc.gov/nchs/about/major /nhanes/nh3data.htm*; Schafer et al. [69]), a major health survey with both an interview component and a medical examination component. This application illustrates the use of hierarchical models for imputation to account for clustering in the survey.

A final example is to another major health survey, the National Health Interview Survey of the National Center for Health Statistics (*http://www.cdc.gov/nchs/nhis.htm*). There is interest in analyzing relationships between income and health. However, exact responses on income are missing for about 30% of the persons in this survey, and this missingness is related to several characteristics of the persons. Most of the participants who do not provide an exact value of income are willing to specify a coarse income category. The National Center for Health Statistics has created multiple imputations for the income data for every survey year since 1997, using an adaptation of sequential regression multivariate imputation [72]. This application illustrates the incorporation of restrictions into an imputation procedure, such as the bounds implied by coarse categorical responses.

Multiple imputation has been used in many contexts other than handling missing data in public-use files. Examples include estimating latent traits from data with planned missingness, as with matrix sampling in the National Assessment of Educational Progress [41,77], statistical matching of records from different data files [58,48,43], handling various types of censored data [6,11], genetic analysis [14], correcting for misreported data and/or measurement error [19,5,[21], and statistical disclosure control for microdata [61,46,7]. A review of some early applications of multiple imputation is given in Rubin and Schenker [66], and a more recent review of how multiple imputation can be adapted to various types of problems is given in Reiter and Raghunathan [51].

38.10 Other Approaches to Estimating Variability with Imputed Data

Other techniques have been developed for incorporating the extra variability due to missing data into analyses of data completed by imputation, and these are described briefly here. For specific types of imputation procedures and analysis problems, estimation formulas for sampling variances can be derived that incorporate the uncertainty due to imputation; see, for example, Rao and Shao [49], Fay [13], Rao [50], Robins and Wang [53], Schafer and Schenker [70], Lee et al. [28], Schenker [71], and Kim and Fuller [26]. Such techniques, however, are not as broadly applicable as multiple-imputation procedures. In addition, they often need to be customized to the imputation and analysis methods used and/or require the user to have information about the imputation model that is not typically available in public-use datasets.

For the analysis of data from a complex sample survey, the complete-data estimate of sampling variance is sometimes based on the sample variance across ultimate clusters (i.e., the largest sampling units that are treated as independently sampled at random from the population) of a quantity calculated separately for each ultimate cluster. For such cases, Little and Rubin [32, Sec. 5.2] discussed how implementing a single-imputation procedure independently within each ultimate cluster can result in valid estimates of variability from analyses of the completed data. This technique is also somewhat specialized, and the requirement of independence of the imputation procedures across ultimate clusters can lead to sparse data for fitting the imputation models.

A more broadly applicable paradigm for estimating sampling variances when analyzing data completed by imputation is to create multiple resampled versions of the incomplete data via a technique such as balanced repeated replication, the bootstrap, or the jackknife, and then to implement the imputation and point estimation procedures separately for each resampled version, thus creating a set of replicate point estimates. The variability across the replicate point estimates is used to estimate the sampling variance of the point estimate based on the original sample. See Efron [9], Lee et al. [28], and Shao [75] for discussions of techniques based on this paradigm.

Replication techniques are grounded in large-sample theory, and their properties in small-sample problems are questionable, whereas multiple imputation, with its Bayesian underpinnings, can have advantages for small samples. Moreover, the implementation of replication techniques can be made more complicated, and the validity of resulting inferences can be affected, by the fact that the imputation method used for the original sample might not be applicable for every resampled version of the incomplete data. For example, if an imputation procedure for nonresponse involves imputing separately within adjustment cells defined based on completely observed variables, then for a particular resampled version of the incomplete data, an adjustment cell might contain nonrespondents but no respondents. Such situations require either that the imputation method be modified for some resampled datasets or that those resampled datasets be omitted. See Rubin's [62] discussion of Efron [9], and Little and Rubin [32, Example 7.9], for related discussions and examples in the context of applying the bootstrap to incomplete data. Finally, replication techniques typically require a large number of resampled versions of the data to be created, because they rely on the replications to reflect both within- and between-imputation variability. Thus, they have disadvantages compared with multiple imputation, especially in the context of public-use data.

38.11 Summary and Conclusion

Imputation for missing data has major advantages in the production of data sets that are intended for many users, but it is also useful in individual studies since, once accomplished, it enables the investigator to explore immediately the full range of techniques that are available to analyze complete data. In general, single imputation results in underestimation of uncertainty, whereas multiple imputation helps to reflect fully the uncertainty in analyses of the completed data.

This chapter has reviewed general considerations and modeling issues for creating imputations, with an emphasis on multiple imputation, and has outlined how a multiply imputed dataset is analyzed. Conditions have been discussed for the validity of analyses of data completed by imputation. A brief review of applications of multiple imputation has been given. Finally, alternative approaches to estimating the extra variability due to imputation of missing data have been outlined.

Many commercial and non-commercial statistical software packages now have routines for creating and/or analyzing multiply imputed data, and some stand-alone programs have been produced as well. For example,

AMELIA II
(http://gking.harvard.edu/amelia/)
CAT, MIX, NORM, and PAN
(http://www.stat.psu.edu/~jls/misoftwa.html)
IVEware
(http://www.isr.umich.edu/src/smp/ive/)
MICE
(http://www.multiple-imputation.com)
R (http://www.r-project.org)
S+ (http://www.insightful.com)
SAS (http://www.sas.com)
SOLAS (http://www.statsol.ie)
SPSS (http://www.spss.com)
STATA (http://www.stata.com)

These either contain routines or have associated routines supplied by users to create multiple imputations under various models and procedures. Most of the programs just listed also have utilities for analyzing multiply imputed data, as do

SUDAAN
(http://www.rti.org/sudaan/)
WESVAR
(http://www.westat.com/westat/statistical_software/)

This description of software is undoubtedly incomplete and will become more so as new software is developed, which is to be expected for a technique such as multiple imputation, which has been increasing in popularity. Nevertheless, it is hoped that the list of software and Web links will provide readers with useful leads.

Several volumes discuss imputation inside broader discussions of techniques for handling missing data, and they contain information and perspectives that supplement the material in this article. See, for example, Kalton [23], the three books produced by the Panel on Incomplete Data of the Committee on National Statistics [33–35], Schafer [68], Allison [2], Little and Rubin [32], Groves et al. [20], Gelman and Meng [17], McKnight et al. [36], and Molenberghs and Kenward [42].

References

1. Abayomi, K., Gelman, A., and Levy, M. (2008). Diagnostics for multivariate imputations. *J. Roy. Statist. Soc. C (Appl. Statis.)*, **57**, 273–291.

2. Allison, P. D. (2001). *Missing Data*. Sage Publications, Thousand Oaks, CA.

3. Barnard, J. and Rubin, D. B. (1999), Small-sample degrees of freedom with

multiple imputation. *Biometrika*, **86**, 948–955.

4. Clogg, C. C., Rubin, D. B., Schenker, N., Schultz, B., and Weidman, L. (1991). Multiple imputation of industry and occupation codes in census public-use samples using Bayesian logistic regression. *J. Am. Statist. Assoc.*, **86**, 68–78.

5. Cole, S. R., Chu, H., and Greenland, S. (2006). Multiple imputation for measurement-error correction.' *Int. J. Epidemiol.*, **35**, 1074–1081.

6. Dorey, F. J., Little, R. J. A., and Schenker, N. (1993). Multiple imputation for threshold-crossing data with interval censoring. *Statist. Med.*, **12**, 1589–1603.

7. Drechsler, J., Dundler, A., Bender, S., Rässler, S., and Zwick, T. (2008). A new approach for disclosure control in the IAB establishment panel – multiple imputation for a better data access. *Adv. Statist. Anal.*, **92**, 439–458.

8. Efron, B. (1979). Bootstrap methods: Another look at the jackknife. *Ann. Statist.*, **7**, 1–26.

9. Efron, B. (1994). Missing data, imputation, and the bootstrap (with discussion). *J. Am. Statist. Assoc.*, **89**, 463–479.

10. Ezzati-Rice, T., Johnson, W., Khare, M., Little, R., Rubin, D., and Schafer, J. (1995). A simulation study to evaluate the performance of model-based multiple imputation in NCHS health examination surveys. *Proc. 1995 Annual Research Conf*, Bureau of the Census, pp. 257–266.

11. Faucett, C. L., Schenker, N., and Taylor, J. M. G. (2002). Survival analysis using auxiliary variables via multiple imputation, with application to AIDS clinical trial data. *Biometrics*, **58**, 37–47.

12. Fay, R. E. (1992). When are inferences from multiple imputation valid? *American Statistical Association Proceedings of the Section on Survey Research Methods*, pp. 227–232.

13. Fay, R. E. (1996). Alternative paradigms for the analysis of imputed survey data. *J. Am. Statist. Assoc.*, **91**, 490–498. (Discussion: 507–515. Rejoinder: 517–519).

14. Foulkes, A. S., Yucel, R., and Reilly, M. P. (2008. Mixed modeling and multiple imputation for unobservable genotype clusters.' *Statist. Med.*, **27**, 2784–2801.

15. Gelfand, A. E. and Smith, A. F. M. (1990). Sampling-based approaches to calculating marginal densities. *J. Am. Statist. Assoc.*, **85**, 398–409.

16. Gelman, A., Carlin, J. B., Stern, H. S., and Rubin, D. B. (2004), *Bayesian Data Analysis*. Chapman & Hall/CRC, Boca Raton, FL.

17. Gelman, A. and Meng, X.-L. eds. (2004). *Applied Bayesian Modeling and Causal Inference from Incomplete-Data Perspectives*. Wiley, Chichester, UK.

18. Geman, S. and Geman, D. (1984). Stochastic relaxation, Gibbs distributions, and the Bayesian restoration of images. *IEEE Trans. Pattern Anal. Machine Intell.*, **6**, 721–741.

19. Ghosh-Dastidar, B. and Schafer, J. L. (2003). Multiple edit/multiple imputation for multivariate continuous data. *J. Am. Statist. Assoc.*, **98**, 807–817.

20. Groves, R. M., Dillman, D. A., Eltinge, J. L., and Little, R. J. A. eds. (2002). *Survey Nonresponse*. Wiley, New York.

21. He, Y., Yucel, R., and Zaslavsky, A. M. (2008). Here's to your health – misreporting, missing data, and multiple imputation: Improving accuracy of cancer registry databases.' *Chance*, **2**, 55–58.

22. Heitjan, D. F. and Little, R. J. A. (1991). Multiple imputation for the fatal accident reporting system. *J. Roy. Statist. Soc. C (Appl. Statist.)*, **40**, 13–29.

23. Kalton, G. (1982). *Compensating for Missing Survey Data*. ISR Research Report Series, University of Michigan.

24. Kennickell, A. B. (1991). Imputation of the 1989 survey of consumer finances: Stochastic relaxation and multiple imputation. *American Statistical Association Proc. of the Section on Survey Research Methods*, pp. 1–10.

25. Kim, J. K., Brick, J. M., Fuller, W. A., and Kalton, G. (2006). On the bias of the multiple imputation variance estimator in survey sampling. *J. Roy. Statist. Soc. B*, **68**, 509–521.

26. Kim, J. K. and Fuller, W. (2004). Fractional hot deck imputation. *Biometrika*, **91**, 559-578.

27. Kott, P. S. (1995). A paradox of multiple imputation. *American Statistical Association Proc. of the Section on Survey Research Methods*, pp. 380–383.

28. Lee, H., Rancourt, E. and Särndal C. E. (2002). Variance estimation from survey data under single imputation. In *Survey Nonresponse*, R. M. Groves, D. A. Dillman, J. L. Eltinge, and R. J. A. Little eds., pp. 315–328. Wiley, New York.

29. Li, K. H., Meng, X. L., Raghunathan, T. E., and Rubin, D. B. (1991). Significance levels from repeated p-values with multiply imputed data. *Statistica Sinica*, **1**, 65–92.

30. Li, K. H., Raghunathan, T. E., and Rubin, D. B. (1991). Large-sample significance levels from multiply imputed data using moment-based statistics and an F reference distribution. *J. Am. Statist. Assoc.*, **88**, 1014–1022.

31. Little, R. J. A. (1988). Missing data adjustments in large surveys (with discussion). *J. Business Econ Statist.*, **6**, 287–301.

32. Little, R. J. A. and Rubin, D. B. (2002). *Statistical Analysis with Missing Data.* Wiley, Hoboken, NJ.

33. Madow, W. G., Nisselson, H., and Olkin, I. eds. (1983). *Incomplete Data in Sample Surveys*, Vol. 1: *Report and Case Studies.* Academic Press, New York.

34. Madow, W. G. and Olkin, I. eds. (1983). *Incomplete Data in Sample Surveys*, Vol. 3: *Proc. Sympo..* Academic Press, New York.

35. Madow, W. G., Olkin, I., and Rubin, D. B. eds. (1983). *Incomplete Data in Sample Surveys*, Vol. 2: *Theory and Bibliographies.* Academic Press, New York.

36. McKnight, P. E., McKnight, K. M., Sidani, S., and Figureredo, A. J. (2007). *Missing Data: A Gentle Introduction.* Guilford Press, New York.

37. Meng, X.-L. (1994). Multiple-imputation inferences with uncongenial sources of input (with discussion). *Statist. Sci.*, **9**, 538–573.

38. Meng, X.-L. (2002). A congenial overview and investigation of multiple imputation inferences under uncongeniality. In *Survey Nonresponse*, R. M. Groves, D. A. Dillman, J. L. Eltinge, and R. J. A. Little eds., pp. 343-356. Wiley, New York.

39. Meng, X.-L. and Romero, M. (2003). Discussion: Efficiency and self-efficiency with multiple imputation inference," comment on "Proper and improper multiple imputation" by Nielsen. *Int. Statist. Rev.*, **71**, 607–618.

40. Meng, X. L. and Rubin, D. B. (1992). Performing likelihood ratio tests with multiply imputed data sets. *Biometrika*, **79**, 103–111.

41. Mislevy, R. J., Beaton, A., Kaplan, B. A., and Sheehan, K. (1992). Estimating population characteristics from sparse matrix samples of item responses. *J. Ed. Measure.*, **29**, 133–161.

42. Molenberghs, G. and Kenward, M. G. (2007). *Missing Data in Clinical Studies.* Wiley, Chichester, UK.

43. Moriarity, C. and Scheuren, F. (2003). A dote on Rubin's statistical matching using file concatenation with adjusted weights and multiple imputations. *J. Business Econ. Statist.*, **21**, 65–73.

44. Nielsen, S. F. (2003). Proper and improper multiple imputation (with discussion). *Inter. Statist. Rev.*, **71**, 593–627.

45. Raghunathan, T. E., Lepkowski, J. M., van Hoewyk, J., and Solenberger, P. (2001). A multivariate technique for multiply imputing missing values using a sequence of regression models. *Survey Methodol.*, **27**, 85–95.

46. Raghunathan, T. E., Reiter, J. P., and Rubin, D. B. (2003). 'Multiple imputation for statistical disclosure limitation. *J. Official Statis.*, **19**, 1–16.

47. Raghunathan, T. E. and Rubin, D. B. (1997. Roles for Bayesian techniques in survey sampling. *Proc. Survey Methods Section of the Statistical Society of Canada*, pp. 51–55.

48. Rässler, S. (2002). *Statistical Matching: A Frequentist Theory, Practical Applications, and Alternative Bayesian Approaches.* Springer-Verlag, New York.

49. Rao, J. N. K. and Shao, J. (1992). Jackknife variance estimation with survey data under hot deck imputation. *Biometrika*, **79**, 811–822.

50. Rao, J. N. K. (1996). On variance estimation with imputed survey data. *J. Am. Statist. Assoc.*, **91**, 499–506. (Discussion: 507–515. Rejoinder: 519–520.)

51. Reiter, J. P. and Raghunathan, T. E. (2007). The multiple adaptations of multiple imputation. *J. Am. Statist. Assoc.*, **102**, 1462–1471.

52. Reiter, J. P., Raghunathan, T. E., and Kinney, S. K. (2006). The importance of modeling the sampling design in multiple imputation for missing data. *Survey Methodol.*, **32**, 143–149.

53. Robins, J. M. and Wang, N. (2000). Inference for imputation estimators. *Biometrika*, **87**, 113–124.

54. Rubin, D. B. (1976). Inference and missing data (with discussion). *Biometrika*, **63**, 581–592.

55. Rubin, D. B. (1978). Multiple imputations in sample surveys - A phenomenological Bayesian approach to nonresponse (with discussion). *American Statistical Association 1978 Proc. Section on Survey Research Methods*, pp. 20–34.

56. Rubin, D. B. (1981). The Bayesian bootstrap. *Ann. Statist.*, **9**, 130–134.

57. Rubin, D. B. (1983). *Progress Report on Project for Multiple Imputation of 1980 Codes*, manuscript delivered to the Bureau of the Census, the National Science Foundation, and the Social Science Research Council.

58. Rubin, D. B. (1986). Statistical matching and file concatenation with adjusted weights and multiple imputations. *J. Business Econ. Statist.*, **4**, 87–94.

59. Rubin, D. B. (1986). Basic ideas of multiple imputation for nonresponse. *Survey Methodol.*, **12**, 37–47.

60. Rubin, D. B. (1987). *Multiple Imputation for Nonresponse in Surveys.* Wiley, New York.

61. Rubin, D. B. (1993). Discussion: Statistical disclosure limitation. *J. Official Statist.*, **9**, 461–468.

62. Rubin, D. B. (1994). Comment on "Missing data, imputation, and the bootstrap" by Efron. *J. Am. Statist. Assoc.*, **89**, 475–478.

63. Rubin, D. B. (1996). Multiple imputation after 18+ years. *J. Am. Statist. Assoc.*, **91**, 473–489. (Discussion: 507–515. Rejoinder: 515–517.)

64. Rubin, D. B. and Schenker, N. (1986). Multiple imputation for interval estimation from simple random samples with ignorable nonresponse. *J. Am. Statist. Assoc.*, **81**, 366–374.

65. Rubin, D. B. and Schenker, N. (1987). Interval estimation from multiply imputed data: A case study using agriculture industry codes. *J. Official Statist.*, **3**, 375–387.

66. Rubin, D. B. and Schenker, N. (1991). Multiple imputation in health-care databases: An overview and some applications. *Statist. Med.*, **10**, 585–598.

67. Rubin, D. B., Stern, H., and Vehovar, V. (1995). Handling "don't know" survey responses: The case of the Slovenian plebiscite. *J. Am. Statist. Assoc.*, **90**, 822–828.

68. Schafer, J. L. (1997). *Analysis of Incomplete Multivariate Data*, Chapman & Hall, London.

69. Schafer, J. L., Ezzati-Rice, T. M., Johnson, W., Khare, M., Little, R. J. A., and Rubin, D. B. (1996). The NHANES III

multiple imputation project. *American Statistical Association Proc. Section on Survey Research Methods*, pp. 28–37.

70. Schafer, J. L. and Schenker, N. (2000). Inference with imputed conditional means. *J. Am. Statist. Assoc.*, **95**, 144–154.

71. Schenker, N. (2003). Assessing variability due to race bridging: Application to census counts and vital rates for the year 2000. *J. Am. Statist. Assoc.*, **98**, 818–828.

72. Schenker, N., Raghunathan, T. E., Chiu, P.-L., Makuc, D. M., Zhang, G., and Cohen, A. J. (2006). Multiple imputation of missing income data in the national health interview survey. *J. Am. Statist. Assoc.*, **101**, 924–933.

73. Schenker, N. and Taylor, J. M. G. (1996). Partially parametric techniques for multiple imputation. *Comput. Statist. Data Analy.*, **22**, 425–446.

74. Schenker, N., Treiman, D. J., and Weidman, L. (1993). Analyses of public use decennial census data with multiply imputed industry and occupation codes. *J. Roy. Statist. Soc. C (Appl. Statist.)*, **42**, 545–556.

75. Shao, J. (2002). Replication methods for variance estimation in complex surveys with imputed data. In *Survey Nonresponse*, R. M. Groves, D. A. Dillman, J. L. Eltinge, and R. J. A. Little, eds., pp. 303–314. Wiley, New York.

76. Tanner, M. A. and Wong, W. H. (1987). The calculation of posterior distributions by data augmentation. *J. Am. Statist. Assoc.*, **8**, 528–550.

77. Thomas, N. (2000). Assessing model sensitivity of the imputation methods used in the national assessment of educational progress. *J. Educ. Behav. Statist.*, **25**, 351–371.

78. Treiman, D. J. and Rubin, D. B. (1983). *Multiple Imputation of Categorical Data to Achieve Calibrated Public-Use Samples*. Paper proposal to the National Science Foundation.

79. van Buuren, S. (2007). Multiple imputation of discrete and continuous data by fully conditional specification. *Statist. Methods Med. Res.*, **16**, 219–242.

80. van Buuren, S., Brand, J., Groothuis-Oudshoorn, C., and Rubin, D. (2006). Fully conditional specification in multivariate imputation. *J. Statist. Comput. Simul.*, **76**, 1049–1064.

81. van Buuren, S. Oudshoorn C. G. M. (1999), *Flexible Multivariate Imputation by MICE*. TNO Preventie en Gezondheid, TNO/PG 99.054, Leiden.

82. Weidman, L. (1989). *Final Report: Industry and Occupation Imputation*. Report Census/SRD/89/03, Statistical Research Division, Bureau of the Census.

39. Incomplete Data

Roderick J. Little and Donald B. Rubin

39.1 Introduction

Incomplete problems occur frequently in statistics. Indeed, one might view inferential statistics in general as a collection of methods for extending inferences from a sample to a population where the nonsampled values are regarded as missing data.

Although some statistical methods for complete data, such as factor analysis, finite-mixture models, and mixed-model analysis of variance, can be usefully viewed as incomplete data methods [13], we restrict this review to more standard incomplete data problems. For the class of problems reviewed here, we consider "missing data" to be synonymous with "incomplete data." After describing common examples with missing data in the following section, in Section 39.3 we describe techniques for handling these problems. In the last section, we discuss the EM algorithm, an ubiquitous algorithm for finding maximum-likelihood (ML) estimates from incomplete data. Useful reviews of the analysis of incomplete data are given in Afifi and Elashoff [1], Hartley and Hocking [19], Orchard and Woodbury [36], Dempster et al. [13], and Little [29].

39.2 Common Incomplete Data Problems

We first consider problems where missing values are confined to a single outcome variable y, and interest concerns the distribution of y, perhaps conditional on a set of one or more predictor variables x, that are recorded for all units in the sample. Sometimes we have no information about the missing values of y; at other times we may have partial information, for example, that they lie beyond a known censoring point c.

39.2.1 Mechanisms Leading to Missing Values

Any analysis of incomplete data requires certain assumptions about the distribution of the missing values, and in particular how the distributions of the missing and observed values of a variable are related. The work of Rubin [38] distinguishes three cases. If the process leading to missing y values (and, in particular, the probability that a particular value of y is missing) does not depend on the values of x or y, the missing data are called *missing at random* and the observed data are *observed at random*. If the process depends on observed values of x and y but not on missing values of y the missing data are called missing at random, but the observed data are not observed at random. If the process depends on missing values of y, the missing data are not missing at random; in this case, particular care is required in deriving inferences. Rubin [38] formalizes these notions by defining a random variable m that indicates for each unit whether y is observed or missing, and relating these conditions to properties of the conditional distribution of m given x and y. This approach is discussed in Section 39.3.2, on the modeling approach to missing data.

39.2.2 Analysis of Variance

The first incomplete-data problem to receive systematic attention in the statistics literature is that of *missing data in designed experiments*; in the context of agriculture trials, this problem is often called the *missing-plot problem* [3,5]. Designed experiments investigate the dependence of an outcome variable, such as yield of a crop, on a set of factors, such as variety, type of fertilizer, and temperature. Usually, an experimental design is chosen that allows efficient estimation of important effects as well as a simple analysis. The analysis is especially simple when the design matrix is easily inverted, as with complete or fractional replications of factorial designs. The missing-data problem arises when at the conclusion of the experiment, the values of the outcome variable are missing for some of the plots, perhaps because no values were possible, as when particular plots were not amenable to seeding, or because values were recorded and then lost. Standard analyses of the resultant incomplete data assume the missing data to be missing at random, although in practical situations the plausibility of this assumption needs to be checked. The analysis aims to exploit the "near balance" of the resulting dataset to simplify computations. For example, one tactic is to substitute estimates of the missing outcome values and then to carry out the analysis assuming the data to be complete. Questions needing attention then address the choice of appropriate values to substitute and how to modify subsequent analyses to allow for such substitutions. For discussions of this and other approaches, see Healy and Westmacott [21], Wilkinson [50], and Rubin [38,39].

39.2.3 Censored or Truncated Outcome Variable

We have noted that standard analyses for missing plots assume that the missing data are missing at random, that is, the probability that a value is missing can depend on the values of the factors but not on the missing outcome values. This assumption is violated, for example, when the outcome variable measures time to an event (such as death of an experimental animal, failure of a lightbulb), and the times for some units are not recorded because the experiment was terminated before the event had occurred; the resulting data are *censored*. In such cases the analysis must include the information that the units with missing data are censored, since if these units are simply discarded, the resulting estimates can be badly biased.

The analysis of censored samples from the Poisson, binomial, and negative binomial distributions is considered by Hartley [18]. Other distributions, including the normal, lognormal, exponential, gamma, Weibull, extreme-value, and logistic, are covered most extensively in the life-testing literature (for reviews, see Mann et al. [32] and Tsokos and Shimi [49]). Nonparametric estimation of a distribution subject to censoring is carried out by life table methods, formal properties of which are discussed by Kaplan and Meier [25]. Much of this work can be extended to handle covariate information [2,11,16,26]. The EM algorithm, discussed here in the last section, is a useful computational device for such problems.

A variant of censored values occurs when missing values are known to lie within an interval, as when the data are available in grouped form. The analysis of grouped data is discussed by Hartley [18], Kulldorff [26], and Blight [7], among others. Another variant of censored data occurs when the number of censored values is unknown. The resulting data are called truncated, since they can be regarded as a sample from a truncated distribution. A considerable literature exists for this form of data [8,12,13,18].

39.2.4 Sample Survey Data

For the data types discussed in this section, the missing data are not missing at random, but the mechanisms leading to incomplete data are assumed known. For example, the censoring points for censored observations are known. A common and somewhat more intractable problem occurs when the missing data are not missing at random and the mechanism leading to missing data is at best partially known. Incomplete data arising from nonresponse in sample surveys provide an illustration of this kind of problem. For example, nonresponse to a question on household income often depends on the amount of that income, in an unknown way. Restricting the analysis to respondents clearly leads to bias in such situations; given the large samples often available in survey work, this bias is frequently more important than the loss of efficiency of estimation arising from the reduction in sample size.

The effect of survey nonresponse is minimized by (1) designing data collection methods to minimize the level of nonresponse, (2) interviewing a subsample of nonrespondents, and (3) collecting auxiliary information on nonrespondents and employing analytical methods that use this information to reduce nonresponse bias. Models for nonrandomly missing data, as developed by Nelson [34], Heckman [22], and Rubin [40] can also be applied here. Estimates derived from these models, however, are sensitive to aspects of the model that cannot be tested with the available data [17,19,41]. A thorough discussion of survey nonresponse is given in the work of the National Academy of Sciences Panel on Incomplete Data [33].

39.2.5 Multivariate Incomplete Data

The incomplete-data structures discussed so far are *univariate*, in the sense that the missing values are confined to a single outcome variable. We now turn to incomplete-data structures that are essentially *multivariate* in nature.

Many multivariate statistical analyses, including least-squares regression, factor analysis, and discriminant analysis, are based on an initial reduction of the data to the sample mean vector and covariance matrix of the variables. The question of how to estimate these moments with missing values in one or more of the variables is, therefore, an important one. Early literature was concerned with small numbers of variables (two or three) and simple patterns of missing data [1,4]. Subsequently, more extensive datasets with general patterns of missing data have been addressed [6,9,27,34,37,45].

The reduction to first and second moments is generally not appropriate when the variables are categorical. In this case, the data can be expressed in the form of a multiway contingency table. Most of the work on incomplete contingency tables has concerned maximum-likelihood estimation assuming a Poisson or multinomial distribution for the cell counts. Bivariate categorical data form a two-way contingency table; if some observations are available on a single variable only, they can be displayed as a supplemental margin. The analysis of data with supplemental margins is discussed by Hocking and Oxspring [23] and Chen and Fienberg [10]. Extensions to loglinear models for higher-way tables with supplemental margins are discussed in Fuchs [15].

Essentially all the literature on multivariate incomplete data assumes that the missing data are missing at random, and much of it also assumes that the observed data are observed at random. Together these assumptions imply that the process that creates missing data does not depend on any values, missing or observed.

39.2.6 Missing Data in Time Series

Techniques for analyzing time-series data often require observations at regular intervals. In practice, many time series are irregularly spaced, by the nature of the collection process or because otherwise regular series become irregular when some values are missing. The problem of maximum-likelihood estimation for Gaussian stationary processes with missing data has received some attention [23,43] (for other references, see the symposium volume in which Shumway's article [46] appears]. State space representations of the likelihood and the EM algorithm are useful computational tools in this work.

39.3 Methods for Handling Incomplete Data

39.3.1 A Broad Taxonomy of Methods

Methods for handling incomplete data generally belong to one or more of the following categories:

1. Methods that discard units with data missing in some variables and analyze only the units with complete data (e.g., Nie et al. [33]).

2. Imputation-based procedures. The missing values are filled in and the resultant completed data are analyzed by standard methods. For valid inferences to result, modifications to the standard analyses are required to allow for the differing status of the real and the imputed values. Commonly used procedures for imputation include *hot deck* imputation (see Ford [14]), where recorded units in the sample are substituted, *mean* imputation, where means from sets of recorded values are substituted and *regression* imputation, where the missing variables for a unit are estimated by predicted values from regression on the known variables for that unit [9]. A variant of imputation methods produces multiple imputations for each missing value and thereby allows simple adjustments to be made to reflect the differing status of real and imputed values [41,42].

3. Weighting procedures. Randomization inferences from sample survey data without nonresponse are commonly based on *design weights*, which are inversely proportional to the probability of selection. For example, let y_i be the value of the variable y for unit i in the population. Then the population mean is often estimated by

$$\sum \pi_i^{-1} y_i \Big/ \sum \pi_i^{-1}, \qquad (1)$$

where the sums are over sampled units, π_i is the probability of selection for unit i, and π_i^{-1} is the design weight for unit i.

Weighting procedures modify the weights to allow for nonresponse. The estimator (1) is replaced by

$$\sum (\pi_i \hat{p}_i)^{-1} y_i \Big/ \sum (\pi_i \hat{p}_i)^{-1}, \quad (2)$$

where the sums are now over sampled units which respond, and $\hat{p}_i$ is an estimate of the probability of response for unit i, usually the proportion of responding units in a subclass of the sample. Weighting is related to mean imputation; for example, if the design weights are constant in subclasses of the sample, then imputing the subclass mean for missing units in each subclass, or weighting responding units by the proportion responding in each subclass, lead to the same estimates of the population mean, although not the same estimates of sampling variance unless adjustments are made to the data with means imputed. A discussion of weighting with extensions to two-way classifications is provided by Scheuren [45].

4. Model-based procedures. A broad class of procedures is generated by defining a model for the incomplete data and basing inferences on the likelihood under that model, with parameters estimated by procedures such as maximum likelihood. Advantages of this approach are: flexibility; the avoidance of adhocery, in that model assumptions underlying the resulting methods can be displayed and evaluated; and the availability of large-sample estimates of variance based on second derivatives of the log likelihood, which take into account incompleteness in the data. Disadvantages are that computational demands can be large, particularly for complex patterns of missing data, and that little is known about the small-sample properties of many of the large-sample approximations.

39.3.2 The Modeling Approach to Incomplete Data

Any procedure that attempts to handle incomplete data must, either implicitly or explicitly, model the process that creates missing data. We prefer the explicit approach since assumptions are then clearly stated.

The parametric form of the modeling argument can be expressed as follows [38]. Let y_p denote data that are present and y_m data that are missing. Suppose that $y = (y_p, y_m)$ has a distribution $f(y_p, y_m | \theta)$ indexed by an unknown parameter θ. If the missing data are missing at random, then the likelihood of θ given data y_p is proportional to the density of y_p, obtained by integrating $f(y_p, y_m | \theta)$ over y_m:

$$L(\theta | y_p) \propto \int f(y_p, y_m | \theta) \, dy_m. \quad (3)$$

Likelihood inferences are based on $L(\theta | y_p)$. Occasionally in the literature, the missing values y_m are treated as fixed parameters, rather than integrated out of the distribution $f(y_p, y_m | \theta)$, and joint estimates of θ and y_m are obtained by maximizing $f(y_p, y_m | \theta)$ with respect to θ and y_m (e.g., Press and Scott [37] present a procedure that is essentially equivalent to this). This approach is not recommended since it can produce badly biased estimates which are not even consistent unless the fraction of missing data tends to zero as the sample size increases. Also, the model relating the missing and observed values of y is not fully exploited, and if the amount of missing data is substantial, the treatment of y_m as a set of parameters contradicts the general statistical principle of parsimony [29].

An important generalization of (3) is to include in the model the distribution of a vector variable m, which indicates whether each value is observed or missing. The full distribution can be specified as

$$f(m, y_p, y_m | \theta, \phi) = f(y_p, y_m | \theta) f(m | y_p, y_m, \phi), \quad (4)$$

where θ is the parameter of interest and ϕ relates to the mechanism leading to missing data. This extended formulation is necessary for nonrandomly missing data such as arise in censoring problems.

To illustrate (3) and (4), suppose that the hypothetical complete data $y = (y_1, \ldots, y_n)$ are a random sample of size n from the exponential distribution with mean θ. Then

$$f(y_p, y_m | \theta) = \theta^{-n} \exp(-t_n / \theta),$$

where $t_n = \sum_{i=1}^n y_i$ is the total of the n sampled observations. If $r < n$ observations are present and the remaining $n - r$ are missing, then the likelihood ignoring the response mechanism is proportional to the density

$$f(y_p | \theta) = \theta^{-r} \exp(-t_r / \theta), \quad (5)$$

regarded as a function of θ, where t_r is the total of the recorded observations.

Let $m = (m_1, \ldots, m_n)$, where $m_i = 1$ or 0 as y_i is recorded or missing, respectively, $r = \sum m_i$. We consider two models for the distribution of m given y. First, suppose that observations are independently recorded or missing with probability ϕ. Then

$$f(m | y, \phi) = \phi^r (1 - \phi)^{n-r}$$

and

$$f(y_p, m | \theta, \phi) = \phi^r (1 - \phi)^{n-r} \theta^{-r} \exp(-t_r / \theta). \quad (6)$$

The likelihoods based on (5) and (6) differ by a factor $\phi^r (1 - \phi)^{n-r}$, which does not depend on θ, provided that θ and ϕ are distinct; i.e., their joint parameter space factorizes into a θ-space and a ϕ-space. Hence we can base inferences on (5), ignoring the response mechanism.

Suppose instead that the sample is censored, in that only values less than a known censoring point c are observed. Then

$$f(m | y, \phi) = \prod_{i=1}^n f(m_i | y_i),$$

$$f(m_i | y_i) = \begin{cases} 1 & \text{if } m_i = 1 \text{ and } y_i < c \\ & \text{or } m_i = 0 \text{ and } y_i > c, \\ 0 & \text{otherwise.} \end{cases}$$

The full likelihood is then proportional to $f(y_p, m | \theta)$

$$= \prod_{i : m_i = 1} f(y_i | \theta) f(m_i | y_i < c)$$
$$\times \prod_{i : m_i = 0} \Pr(y_i > c | \theta)$$
$$= \theta^{-r} \exp(-t_r / \theta) \exp[-(n - r) c / \theta]. \quad (7)$$

In this case the response mechanism is not ignorable, and the likelihoods based on (5) and (7) differ. In particular, the maximum likelihood estimate of θ based on (5) is t_r / r, the mean of the recorded observations, which is less than the correct maximum likelihood estimate of θ based on (7), namely $[t_r + (n - r) c] / r$. The latter estimate has the simple interpretation as the total time at risk for the uncensored and censored observations divided by the number of failures (r).

39.3.3 Special Data Patterns: Factoring the Likelihood

For certain special patterns of multivariate missing data, maximum likelihood estimation can be simplified by factoring the joint distribution in a way that simplifies the likelihood. Suppose, for example, that the data have the *monotone* or *nested* pattern in Figure 1, where y_j represents a

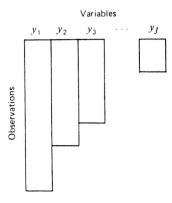

Figure 1: Schematic representation of a monotone (or nested) data pattern.

set of variables observed for the same set of observations and y_j is more observed than $y_{j+1}, j = 1, \ldots, J-1$. The joint distribution of $y_1, \ldots, y_J$ can be factored in the form

$$f(y_1, \ldots, y_J | \theta) \\ = f_1(y_1 | \theta_1) f_2(y_2 | y_1, \theta_2) \cdots \\ f(y_J | y_1, \ldots, y_{J-1}, \theta_J).$$

where f_j denotes the conditional distribution of y_j given $y_1, \ldots, y_{j-1}$, indexed by parameters θ_j. If the parameters $\theta_1, \ldots, \theta_J$ are distinct, then the likelihood of the data factors into distinct complete-data components, leading to simple maximum likelihood estimators for θ [4,37]. Maximum-likelihood estimation with more general patterns of incomplete data can be accomplished by the EM algorithm.

39.4 General Data Patterns: The EM Algorithm

The expectation–maximization (EM) algorithm [13] is an iterative method of maximum-likelihood estimation that applies to any pattern of missing data. Let $l(\theta | y_p, y_m)$ denote the loglikelihood of parameters θ based on the hypothetical complete data (y_p, y_m). Let $\theta^{(i)}$ denote an estimate of θ after iteration i of the algorithm. The $(i+1)$th iteration consists of an E step and an M step. The E step consists of taking the expectation of $l(\theta | y_p, y_m)$ over the conditional distribution of y_m given y_p, evaluated at $\theta = \theta^{(i)}$; that is, the averaged loglikelihood

$$l^*(\theta | y_p, \theta^{(i)}) \\ = \int l(\theta | y_p, y_m) f(y_m | y_p, \theta^{(i)}) dy_m$$

is formed.

The M step consists in finding $\theta^{(i+1)}$, the value of θ that maximizes l^*. This new estimate, $\theta^{(i+1)}$, then replaces $\theta^{(i)}$ at the next iteration. Each step of EM increases the loglikelihood of θ given $y_p, l(\theta | y_p)$. Under quite general conditions, the algorithm converges to a maximum value of the loglikelihood $l(\theta | y_p)$. In particular, if a unique finite maximum-likelihood estimate of θ exists, the algorithm finds it.

An important case occurs when the complete data belong to a regular exponential family. In this case, the E step reduces to estimating the sufficient statistics corresponding to the natural parameters of the distribution. The M step corresponds to maximum likelihood estimation from the hypothetical complete data, with the sufficient statistics replaced by the estimated sufficient statistics from the E step.

The EM algorithm was first introduced for particular problems (e.g., see Hartley [18] for counted data and Blight [7] for grouped or censored data). The regular exponential family case was presented by Sundberg [47]. Orchard and Woodbury [36] discussed the algorithm more generally, using the term "missing-information principle" to describe the link with the complete-data loglikelihood. Dempster et al. [13] introduced the term EM, developed convergence properties, and provided a large body of examples. More recent applications include missing data in discriminant analysis [28] and regression with grouped or censored data [20].

The EM algorithm converges reliably, but it has slow convergence properties if the amount of information in the missing data is relatively large. Also, unlike methods like Newton–Raphson that need to calculate and invert an information matrix, EM does not provide asymptotic standard errors for the maximum likelihood estimates as output from the calculations. Its popularity derives from its link with maximum likelihood for complete data and its consequent usually simple computational form. The M-step often corresponds to a standard method of analysis for complete data and thus can be carried out with existing technology. The E step often corresponds to imputing values for the missing data y_m, or more generally, for the sufficient statistics that are functions of y_m and y_p, and as such relates maximum-likelihood procedures to imputation methods. For example, the EM algorithm for multivariate normal data can be viewed as an iterative version of Buck's [9] method for imputing missing values [6].

Although the EM algorithm is a powerful tool for estimation from incomplete data, many problems remain. For example, nonnormal likelihoods occur more commonly with incomplete data than with complete data, and much remains to be learned about the appropriateness of many incomplete-data methods when applied to real data.

Acknowledgments. This research was sponsored by the U.S. Army under Contract DAAG29-80-C-0041.

References

1. Afifi, A. A. and Elashoff, R. M. (1966). *J. Am. Statist. Assoc.*, **61**, 595–604.
2. Aitkin, M. and Clayton, D. (1980). *Appl. Statist.*, **19**, 156–163.
3. Anderson, R. L. (1946). *Biometrics*, **2**, 41–47.
4. Anderson, T. W. (1957). *J. Am. Statist. Assoc.*, **52**, 200–203.
5. Bartlett, M. S. (1937). *J. Roy. Statist. Soc. B*, **4**, 137–170.
6. Beale, E. M. L. and Little, R. J. A. (1975). *J. Roy. Statist. Soc. B*, **37**, 129–146.
7. Blight, B. J. N. (1970). *Biometrika*, **57**, 389–395.
8. Blumenthal, S., Dahiya, R. C., and Gross, A. S. (1978). *J. Am. Statist. Assoc.*, **73**, 182–187.
9. Buck, S. F. (1960). *J. Roy. Statist. Soc. B*, **22**, 302–306.
10. Chen, T. and Fienberg, S. E. (1974). *Biometrics*, **30**, 629–642.
11. Cox, D. R. (1972). *J. Roy. Statist. Soc. B*, **34**, 187–220.
12. Darroch, J. N. and Ratcliff, D. (1972). *Ann. Math. Statist.*, **43**, 1470–1480.
13. Dempster, A. P., Laird, N. M., and Rubin, D. B. (1977). *J. Roy. Statist. Soc. B*, **39**(1), 1–38.
14. Ford, B. N. (1983). In *Incomplete Data and Sample Surveys, Vol. 2, Theory & Bibliographies*, D. B. Rubin, W. G. Madow, and I. Olkin, eds. Academic Press, New York.
15. Fuchs, C. (1982). *J. Am. Statist. Assoc.*, **77**, 270–278.
16. Glasser, M. (1967). *J. Am. Statist. Assoc.*, **62**, 561–568.
17. Greenlees, W. S., Reece, J. S., and Zieschang, K. D. (1982). *J. Am. Statist. Assoc.*, **77**, 251–261.
18. Hartley, H. O. (1958). *Biometrics*, **14**, 174–194.
19. Hartley, H. O. and Hocking, R. R. (1971). *Biometrics*, **27**, 783–808.
20. Hasselblad, V., Stead, A. G., and Galke, W. (1980). *J. Am. Statist. Assoc.*, **75**, 771–779.
21. Healy, M. and Westmacott, M. (1956). *Appl. Statist.*, **5**, 203–206.

22. Heckman, J. D. (1976). *Ann. Econ. Social Meas.*, **5**, 475–492.

23. Hocking, R. R. and Oxspring, H. H. (1974). *Biometrics*, **30**, 469–483.

24. Jones, R. H. (1980). *Technometrics*, **27**, 389–395.

25. Kaplan, E. L. and Meier, P. (1958). *J. Roy. Statist. Assoc.*, **53**, 457–481.

26. Kulldorff, G. (1961). *Contributions to the Theory of Estimation from Grouped and Partially Grouped Samples*. Almqvist & Wiksell, Stockholm/Wiley, New York.

27. Laird, N. and Olivier, D. (1981). *J. Am. Statist. Assoc.*, **76**, 231–240.

28. Little, R. J. A. (1976). *Biometrika*, **63**, 593–604.

29. Little, R. J. A. (1978). *J. Am. Statist. Assoc.*, **73**, 319–322.

30. Little, R. J. A. (1982). *J. Am. Statist. Assoc.*, **77**, 237–250.

31. Little, R. J. A. and Rubin, D. B. (1983). *Am. Statist.*, **37**,

32. Mann, N. R., Schafer, R. E., and Singpurwalla, N. D. (1974). *Methods for Statistical Analysis of Reliability and Life Data*. Wiley, New York.

33. National Academy of Sciences (1982). *Report of the Panel on Incomplete Data*. National Academy of Sciences, Washington, DC.

34. Nelson, F. D. (1977). *J. Econometr.*, **6**, 581–592.

35. Nie, N. H., Hull, C. H., Jenkins, J. G., Steinbrenner, K., and Bent, D. H. (1975). *SPSS*, 2nd ed. McGraw-Hill, New York.

36. Orchard, T. and Woodbury, M. A. (1972). *Proc. 6th Symp. Mathematical Statistics and Probability*, Vol. 1, pp. 697–715, Berkely, CA. University of California Press, Berkeley.

37. Press, S. J. and Scott, A. J. (1976). *J. Am. Statist. Assoc.*, **71**, 366–369.

38. Rubin, D. B. (1972). *Appl. Statist.*, **21**, 136–141.

39. Rubin, D. B. (1974). *J. Amer. Statist. Assoc.*, **69**, 467–474.

40. Rubin, D. B. (1976). *Biometrika*, **63**, 581–592.

41. Rubin, D. B. (1976). *J. Roy. Statist. Soc. B*, **38**, 270–274.

42. Rubin, D. B. (1977). *J. Am. Statist. Assoc.*, **72**, 538–543.

43. Rubin, D. B. (1978). In *Imputation and Editing of Faulty or Missing Survey Data*, pp. 1–9, US Social Security Administration and Bureau of the Census, Washington, DC.

44. Rubin, D. B. (1980). *Handling Nonresponse in Sample Surveys by Multiple Imputations*. US Department of Commerce, Bureau of the Census Monograph, Washington, DC.

45. Scheuren, F. (1983). In *Incomplete Data and Sample Surveys, Vol. 2: Theory and Bibliographies*, D. B. Rubin, W. G. Madow, and I. Olkin, eds. Academic Press, New York.

46. Shumway, R. H. (1984). In *Proc. Symp. Time Series: Analysis of Irregularly Observed Data*, E. Parzen, ed. Lecture Notes in Statistics, Springer-Verlag, New York.

47. Sundberg, R. (1974). *Scand. J. Statist.*, **1**, 49–58.

48. Trawinski, I. M. and Bargmann, R. E. (1964). *Ann. Math. Statist.*, **35**, 647–657.

49. Tsokos, C. P. and Shimi, I. N. (1977). *The Theory and Applications of Reliability*. Academic Press, New York.

50. Wilkinson, G. N. (1958). *Biometrics*, **14**, 360–384.

40 Interval Censoring

Thomas A. Gerds and Carolina Meier-Hirmer

40.1 Censoring

In survival or time-to-event analysis, interval censoring occurs if the time of an event is the *endpoint* of interest that cannot be observed exactly for each patient but occurred chronologically between two follow-ups. Information of the event status is typically collected by periodically performing clinical examinations or laboratory tests. Events of interest can be the onset of a disease, infection, the recurrence of a tumor, and other changes in the course of a disease. For example, in a *clinical trial/study*, the progression of a tumor can be assessed only at periodic examination times, for example, every 6 months. The true time to progression is hidden somewhere between two adjacent examination times, for example, between 12 and 18 months. Another typical example is the infection time, for example of HIV, when, for the diagnosis, the result of *laboratory tests* is needed. The time to infection is included in an interval defined by the dates of acquisition. Two sources of censoring can be allocated. External censoring is due to the study design that prohibits instantaneous access of the event of interest because the patient is not medically supervised all the time. An example for externally introduced censoring is the arrangement of yearly instead of monthly or daily examinations. On the other hand, the censoring can be due to the natural limits of the experiment. For instance, if the event of interest is the onset of a tumor, or an *adverse drug reaction* becoming visible only after a certain limit of toxication is exceeded.

In this chapter, the response is always the time of an event, the occurrence of which becomes known at examination times. Some special cases shall be distinguished. One speaks of left censoring, if the event of interest occurred before the examination time and of right censoring if the event did not occur until the last examination. The situation with only one examination for each patient is called *case 1* interval censoring, the resulting observations are often called *current status data*.

Left and right censoring can be generalized to "case k" interval censoring for situations where the information from exactly k examinations is available for each patient (k is a positive integer). Since, in the clinical practice, the number of examination times is typically different among patients, most frequently one has to deal with "mixed case" interval censoring. This term refers to situations where some of the observations are exact event times, some are right- or left-censored, and others are really censored to intervals. It is important to emphasize that the term "interval censoring" is often used to generally describe data consisting of such a mixture.

Using artificial data, it is demonstrated in Figure 1 and Table 1 how interval-censored observations can be obtained from the longitudinal data of the examination process. Note that although patients 2 and 4 have the same event time, the observed intervals differ considerably.

A note of caution: complicated censoring schemes arise in the medical practice as well as in other fields. But the connection between the theoretical results and the applications is not yet well developed in all cases or easily available. Moreover, computer programs are not generally available. As a result, ad-hoc methods that may cause biased conclusions are still in use. For instance, a potentially biased analysis would result from using the Kaplan–Meier estimator or the Cox regression model after transforming the interval-censored observations to the right-censoring situation. Replacing intervals with their middle or maximal point, however, approximates the true results only in exceptional cases, for instance, when the observed intervals are generally small and when the accuracy needed in the specific problem is low.

40.2 Classification and Examples

In this section, the types of interval censoring are classified and illustrated more formally.

40.2.1 Case 1 Interval Censoring

The observations are also called *current status data*. Here the information of one examination is available for each patient, and it is known if the event occurred before or after the date of examination. As a result, the observations are either left censored or right censored. *Cross-sectional* studies in clinical or epidemiological projects often result in current status data. Another example are tumor-incidence data in animal experiments where independent individuals are exposed to carcinogens. Typically, a valid histological diagnosis of the onset of a tumor is only possible after death. Thus, the day of death is the only examination time revealing whether a tumor has grown. The limitation of detecting a tumor only if it exceeds a critical size is a technical reason for censoring. For tumor incident experiments, the censoring occurs because of the inability to measure the event of interest exactly (occult tumors). For cross-sectional studies, the censoring is introduced by the study design.

40.2.2 Case k Interval Censoring

For each patient, the results from k examinations are available. Since the same number of examinations are almost never available for all patients, "case k" interval censoring (for k greater than one) occurs rather sparsely in the medical praxis. An example is the supervision of children learning to speak where learned words are collected on a fixed number of questionnaires for each child. However, if single questionnaires are missing for some of the children, and hence the number of available examinations differs, the data have to be considered as "mixed case" interval censored.

40.2.3 Mixed-Case Interval Censoring

For each patient, the event status is known at a differing number of examination times. Therefore, "mixed case" interval-censored data typically consist of

Table 1: Illustration of how to obtain interval-censored observations from hypothetical examination processes.

		Examination Process					True Event Time	Censored Observation
	Time	0	12	30	42	48		
Patient 1	Event	N	N	N	N	N	60	$[48, \infty)$
	Time	0	12	42	—	—		
Patient 2	Event	N	N	Y	—	—	36	$[12, 42]$
	Time	0	—	—	—	—		
Patient 3	Event	Y	—	—	—	—	-10	$(-\infty, 0]$
	Time	0	12	30	54	—		
Patient 4	Event	N	N	N	Y	—	36	$[30, 54]$
	Time	0	30	—	—	—		
Patient 5	Event	N	Y	—	—	—	30	30

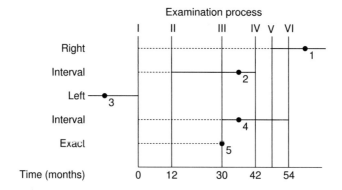

Figure 1: Interval-censored observations (thick lines) corresponding to the artificial data in Table 1. Filled dots in the intervals represent the (unobservable) true event times. The respective type of censoring is marked on the left axis of the diagram. Although observations 2 and 4 have the same true event time, the observed intervals differ considerably.

a mixture of interval-censored and right-censored observations. Sometimes they include left-censored observations and even exact timepoints, if the event happens to occur at the day of examination. An example for mixed-case interval censoring is given by breast cancer studies where the event of interest is the first occurrence of breast retraction. Shortly after therapy, which can be a combination of radiotherapy and chemotherapy, the time interval between two adjacent examinations is typically small but lengthens as the recovery progresses. The exact time of retraction is only known to fall into the interval between two visits or after the last examination time.

40.2.4 Double-Interval Censoring

If the variable of interest is the duration between two events, then both the starting and ending points of the duration can be interval-censored. A well-known example is the incubation time between HIV-infection and AIDS diagnosis, the infection time is interval-censored between the last negative and the first positive antibody test, and the time to AIDS diagnosis is right-censored, when AIDS is not diagnosed within the study time.

40.2.5 Double Censoring

Double censoring refers to situations where the observations are either exact event times, left-censored or right-censored. Double censoring is, therefore, a special case of mixed-case interval censoring.

40.2.6 Interval-Censored Covariates

In the framework of multistate models, the influence of an intermediate event on the main outcome variable is sometimes of interest. For instance, if the endpoint of interest is death, then the recurrence of a tumor can be an influential intermediate event. In this example, the occurrence of the intermediate event is an interval-censored covariate. Situations where the time of the intermediate event is interval censored occur frequently in such frameworks, in particular, for illness–death models.

40.3 Study Design and Statistical Modeling

Considering interval-censored data resulting from periodic follow-up in a clinical trial or longitudinal study, the information is generally the greater the smaller the intervals are between adjacent examinations. The length of the observed intervals evidently influences the power of statistical methods, and it is important for the significance of statistical analysis to gather as much information as possible. However, acquisition on a dense time schedule or even continuous in time can be prohibited for various reasons. Financial costs or stress of patients are typical factors that limit the accuracy of measurement.

The statistical *model* building for censored data proceeds in two steps. In a first step, the survival model is specified, that is, a family of probability distributions that includes the underlying survival function. The second step deals with the dependence of the event time on the censoring scheme. Sometimes it is important to specify a probabilistic model also for the distribution of the censoring process. For convenience, it is often assumed that the censoring is independent of the event time.

Survival models can be parametric, semiparametric, or nonparametric. Besides the differences of the statistical methods used for the three model classes, one must be aware of the well-known tradeoff between bias and variance, which may differ between the three approaches. On one hand, if the parametric model assumptions are violated, the statistical inference can be misleading due to biased estimates. On the other hand, with semiparametric or nonparametric models, the power of the statistical methods is typically low for small or moderate sample sizes, which are often present in clinical studies. In the presence of *covariates*, a suitable survival *regression* model has to be specified. For instance, the proportional hazards model, the proportional odds model, and the accelerated failure time model are frequently used regression models that have extensions to interval-censored data structures.

The task of modeling the censoring mechanism has a similar impact; strong assumptions on the distribution of the censoring mechanism can result in biased inference, whereas allowing general censoring schemes may lead to low statistical power of the statistical procedures. As noted above, the majority of statistical procedures for interval-censored data assumes that the examination times are independent of the event time. This assumption is satisfied for externally scheduled examination times. However, there are situations where the examination process is not independent of the event time. The random censorship assumption is often violated when the data arise from a serial screening and the timing of screening depends on the patient's health status. Or, for time-to-infection data, if the infection can be suspected after certain undertakings or accidents, cool-headed patients would likely urge a laboratory testing. Then, the infection time and the examination time are not independent.

40.4 Statistical Analysis

Several characteristics of survival distributions are of interest in clinical trials: the survival probability function, the difference or ratio of the survival probabilities in two (treatment) groups, and the influence of covariates on the survival probabilities, to name the most important ones. Under the burden of complicated censoring schemes, for each of these cases, a valid estimation method is needed. The methods for interval-censored data are nonstandard and need advanced techniques. Consistent estimators may be available only under restrictive assumptions on the censoring mechanism, and the distributional properties of estimates, confidence intervals, or valid testing procedures are only approximately constructed. To this end, one should note that the development of statistical methods and their distributional properties for interval-censored data are not fully developed and research is an ongoing process. In particular, there are examples with appropriate statistical treatment of dependent interval censoring; see, for example, References 3 and 8. There is also ongoing work in mathematical statistics. The mathematically interested reader is referred to References 11 and 19 and the references given therein. In the remaining section, some of the established statistical methods are named for which it is assumed that the censoring is independent of the event time.

The inference in parametric survival models is relatively straightforward. As a consequence of the independence assumption, likelihood-based methods are applicable, see, for example, Reference 13. In particular, the maximum likelihood estimator has the usual properties, as are, for example, consistency and the usual convergence rate $n^{1/2}$, where n is the sample size. Software for the parametric inference should be available for most of the standard statistic programs.

For the nonparametric estimation of the survival probability function, the so-called nonparametric maximum-likelihood estimator (NPMLE) can be used. The estimator is related to the familiar Kaplan–Meier estimator, which is the nonparametric maximum likelihood estimator for right-censored data. However, the Kaplan–Meier estimator cannot be applied directly and only in exceptional cases to interval-censored data; see the caution at the end of the fist section. NPMLE for interval-censored data is not uniquely defined; any function that jumps the appropriate amount in the so-called equivalence sets represents a NPMLE (see Ref. 15). Briefly, the equivalence sets are found by ordering all the unique left-hand limits and all the unique right-hand limits of all the observed intervals in a dataset, see Reference 14 for details. Outside the equivalence sets the nonparametric estimator defines constant survival probability and the graph is horizontal in these parts. Although the graph of NPMLE is formally undefined in the equivalence sets, some authors visualize NPMLE as if it was a step function, some interpolate between the horizontal parts, and others leave the graph indefinite outside the horizontal parts. Technical differences occur with the computation of NPMLE for the different types of censoring: For "case 1," interval-censored data NPMLE is given by an explicit formula [9]. For the other types of interval censoring, NPMLE has to be computed recursively, see References 15, 20, 4, and 9. For instance, the self-consistency equations developed by Turnbull [20] yield an algorithm which is a special case of the EM algorithm. Turnbull's algorithm is implemented in some of the major statistical software packages (SAS, Splus). The later algorithms achieve improvement concerning stability and computational efficiency (see Refs. 12, 9, and 5).

Unlike the Kaplan–Meier estimator, the NPMLE for interval-censored data converges at a rate slower than $n^{1/2}$, where n is the sample size. In particular, the distribution of this survival function estimator cannot be approximated by a Gaussian distribution. However, at least for case 1 and case 2 interval censoring, the asymptotic distribution of NPMLE has been derived in Reference 9. In some cases, the *bootstrap* provides an alternative method for approximating the distribution of NPMLE [11]. By using such tools, confidence intervals for the survival probability at a fixed time can be constructed. Nonparametric tests for the two group comparison have been proposed in References 18 and 5.

Semiparametric models are prominent for the analysis of regression problems in survival analysis. The frequently used regression models that have extensions for interval-censored data structures are the proportional hazards model [7,10], the proportional odds model [17], and the accelerated failure time model [11]. All these model classes have semiparametric and parametric subclasses. The main difference is that the estimators of the survival curve in the semiparametric models perform as NPMLE, that is, the convergence rate is slow and the distribution is not approximately Gaussian. In case of a parametric survival model, the usual properties retain for the maximum likelihood estimators of the covariate effects. The hypothesis of zero covariate effects can then be tested, for example, under the proportional hazard assumption [7], see also References 6 and 16 for related test statistics.

40.5 Worked Example

In this section, an illustration of the mainstream statistical methods for interval censoring is given. The data are taken from the overview article [14] where the reader also finds comprehensive statistical analysis and a comparison of statistical methods.

The source of the data are clinical studies on the cosmetic effect of different treatments of breast cancer [1,2]. Women with early breast cancer received a breast-conserving excision followed by either radiotherapy alone or a combination of radiotherapy and chemotherapy. The event of interest was the first occurrence of breast retraction. The time interval between two adjacent examinations was in the mean 4–6 months, stretching wider as the recovery progresses. In what follows, treatment 1 corresponds to radiotherapy alone and treatment 2 to a combination of radiotherapy and chemotherapy. It was suggested that additional chemotherapy leads to earlier breast retraction. The data consist of a mixture of interval-censored and right-censored observations. A complete listing taken from Reference 14 is presented in Table 2.

A special diagram has been introduced in Reference 5 for the graphical representation of interval-censored data, see Figure 2. In each treatment group, the interval-censored part of the data is sorted by the length of the observed interval and the right-censored part by the time of the last examination.

Table 3: Test results for the treatment effect in the breast deterioration data.

Test	p Value
Finkelstein [7]	0.004
Dümbgen [5]	0.0028
Sun [18]	0.0043
Parametric	0.0012

Figure 3 compares the parametric estimate of the survival curve in the Weibull survival model to NPMLE. In addition, the Kaplan–Meier estimator was computed by treating the center of the 56 closed intervals, where the right endpoint is not ∞, as if these observations were exact. The graph of the Kaplan–Meier estimator, which is only an ad hoc method in this situation, is also displayed in Figure 3. All estimation methods show late differences between the survival probabilities in the treatment groups. The graphs of the competing methods are quite close for treatment 1, but differ for treatment 2.

To test the hypothesis of no treatment effect, the nonparametric tests proposed in References 5 and 18 are compared to the test under the proportional hazard assumption given in Reference 7, and to the test in a fully parametric survival model (Weibull family). The resulting p values presented in Table 3 are different,

Table 2: Breast cancer retraction data in two treatment arms. The time of first occurrence of breast retraction lies between the left and right endpoints of the intervals; right-censored observations have ∞ as right endpoint.

Treatment 1 ($n = 46$)	[0, 5], [0, 7], [0, 8], [4, 11], [5, 11], [5, 12], [6, 10], [7, 14], [7, 16], [11, 15], [11, 18], [17, 25], [17, 25], [18, 26], [19, 35], [25, 37], [26, 40], [27, 34], [36, 44], [36, 48], [37, 44], [15, ∞), [17, ∞), [18, ∞), [22, ∞), [24, ∞), [24, ∞), [32, ∞), [33, ∞), [34, ∞), [36, ∞), [36, ∞), [37, ∞), [37, ∞), [37, ∞), [38, ∞), [40, ∞), [45, ∞), [46, ∞), [46, ∞), [46, ∞), [46, ∞), [46, ∞), [46, ∞), [46, ∞), [46, ∞)
Treatment 2 ($n = 48$)	[0, 5], [0, 22], [4, 8], [4, 9], [5, 8], [8, 12], [8, 21], [10, 17], [10, 35], [11, 13] [11, 17], [11, 20], [12, 20], [13, 39], [14, 17], [14, 19], [15, 22], [16, 20] [16, 24], [16, 24], [16, 60], [17, 23], [17, 26], [17, 27], [18, 24], [18, 25] [19, 32], [22, 32], [24, 30], [24, 31], [30, 34], [30, 36], [33, 40], [35, 39] [44, 48], [11, ∞), [11, ∞), [13, ∞), [13, ∞), [13, ∞), [21, ∞), [23, ∞) [31, ∞), [32, ∞), [34, ∞), [34, ∞), [35, ∞), [48, ∞)

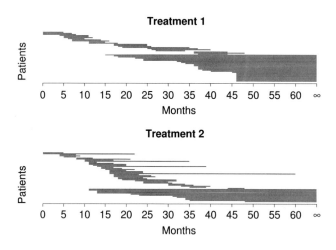

Figure 2: Graphical representation of the breast deterioration data in two treatment groups. Right-censored observations are shown at the end of each treatment group.

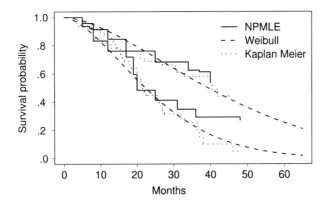

Figure 3: Comparison of estimated survival curves in the treatment groups of the breast deterioration data. The respective upper curves belong to treatment 1 for all methods of estimation.

although, in this example, all methods show significant differences between the treatment arms.

References

1. Baedle, G. F., Come, S., Henderson, C., Silver, B., and Hellman, S. A. H. (1984). The effect of adjuvant chemotherapy on the cosmetic results after primary radiation treatment for early stage breast cancer. *Int. J. Radiat. Oncol., Biol. Phys.*, **10**, 2131–2137.

2. Baedle, G. F., Harris, J. R., Silver, B., Botnick, L., and Hellman, S. A. H. (1984). Cosmetic results following primary radiation therapy for early beast cancer. *Cancer*, **54**, 2911–2918.

3. Betensky, R. A. (2000). On nonidentifiability and noninformative censoring for current status data. *Biometrika*, **87**, 218–221.

4. Dempster, A. P., Laird, N. M., and Rubin, D. B. (1977). Maximum likelihood from incomplete data via the EM algorithm (with discussion). *J. Roy. Stat. Soc. B*, **39**, 1–38.

5. Dümbgen, L., Freitag, S., and Jongbloed, G. (2003). *Estimating a Unimodal Distribution from Interval-Censored Data.* Technical Report, University of Bern.

6. Fay, M. P. (1996). Rank invariant tests for interval censored data under the grouped continous model. *Biometrics*, **52**, 811–822.

7. Finkelstein, D. M. (1986). A proportional hazards model for interval-censored failure time data. *Biometrics*, **42**, 845–854.

8. Finkelstein, D. M., Goggins, W. B., and Schoenfeld, D. A. (2002). Analysis of failure time data with dependent interval censoring. *Biometrics*, **58**(2), 298–304.

9. Groeneboom, P. and Wellner, J. A. (1992). *Information Bounds and Nonparametric Maximum Likelihood Estimation*, DMV-Seminar Vol. 19. Birkhäuser.

10. Huang, J. (1996). Efficient estimation for the proportional hazards model with interval censoring. *Ann. Stat.*, **24**, 540–568.

11. Huang, J. and Wellner, J. A. (1997). Interval censored survival data: A review of recent progress. *Proc. Seattle Symp. 1st in Biostatistics: Survival Analysis.*

12. Jongbloed, G. (1998). The iterative convex minorant algorithm for nonparametric estimation. *J. Comput. Graph. Statist.*, **7**, 310–321.

13. Klein, J. P. and Moeschberger, M. L. (1997). *Survival Analysis—Techniques for Censored and Truncated Data*. Statistics in Biology an Health. Springer.

14. Lindsey, J. C. and Ryan, L. M. (1998). Tutorial in biostatistics—methods for interval-censored data. *Statist. Med.*, **17**, 219–238.

15. Peto, R. (1973). Experimental survival curves for interval-censored data. *Appl. Statist.*, **22**, 86–91.

16. Petroni, G. R. and Wolfe, R. A. (1994). A two-sample test for stochastic ordering with interval-censored data. *Biometrics*, **50**, 77–87.

17. Rossini, A. J. and Tsiatis, A. A. (1996). A semiparametric proportional odds regression model for the analysis of current status data. *J. Am. Statist. Assoc.*, **91**, 713–721.

18. Sun, J. (1996). A nonparametric test for interval-censored failure time data with application to AIDS studies. *Statist. Med.*, **15**, 1387–1395.

19. Sun, J. (2002). *Encyclopedia of Biostatistics: Interval Censoring*, P. Armitage and T. Colton, eds., pp. 2090–2095. Wiley, Hoboken, NJ.

20. Turnbull, B. W. (1976). The empirical distribution function with arbitrarily grouped, censored and truncated data. *J. Roy. Statist. Soc. B*, **38**, 290–295.

41 Interrater Agreement

Pankaj K. Choudhary

41.1 Introduction

Evaluation of agreement between two or more raters measuring the same response is a topic of considerable interest in health sciences research. In practice, a rater may be a measurement method, an instrument, a medical device, a clinical observer, an assay, or even a technique or technology. The response being measured may be continuous, e.g., blood pressure or heart rate, or it may be categorical, e.g., whether a medical condition is present or severity of disease. In either case, the interest is in evaluating if the raters can be used interchangeably, i.e., whether a subject's response measured by one rater can be replaced with a measurement from another other rater without leading to any difference in the clinical interpretation of the measurement. If the raters agree well enough to be used interchangeably, we may prefer the one that is cheaper, less invasive or easier to use.

Thus, the statistical problem here is to quantify the extent of agreement between the raters and assess if it is satisfactory. This chapter describes approaches for agreement evaluation in two raters by focusing on continuous measurements. Reviews of the growing body of literature on this topic can be found in Lin [58], Choudhary and Nagaraja [22], and Barnhart et al. [5]. We do not consider agreement evaluation with categorical measurements. See Banerjee et al. [3], Kraemer et al. [47], and Shoukri [74] for reviews of literature in this area.

Let the random vector (Y_1, Y_2) represent measurements by two raters on a randomly selected subject from the population. We assume that it follows a continuous bivariate distribution with mean (μ_1, μ_2), variance (σ_1^2, σ_2^2), covariance σ_{12} and correlation ρ. The difference $D = Y_1 - Y_2$ follows a continuous distribution with mean $\mu = \mu_1 - \mu_2$ and variance $\tau^2 = \sigma_1^2 + \sigma_2^2 - 2\sigma_{12}$. The raters have *perfect* agreement if $P(Y_1 = Y_2) = 1$ or $\{\mu_1 = \mu_2, \sigma_1^2 = \sigma_2^2, \rho = 1\}$ or equivalently $\{\mu = 0, \tau^2 = 0\}$. In the simplest agreement study, the data consist of a random sample of paired measurements (Y_{i1}, Y_{i2}), $i = 1, \ldots, n$, from the distribution of (Y_1, Y_2). Their differences $D_i = Y_{i1} - Y_{i2}$, $i = 1, \ldots, n$, are a random sample from the distribution of D. In terms of the observed data, perfect agreement means that all the measurement pairs lie on the 45° line through origin or all the differences are zero. The agreement, however, is never perfect in practice. So we evaluate interrater agreement by using the observed data to perform inference on a measure of agreement that is a function of the parameters of the distribution of (Y_1, Y_2). Interrater agreement is often called "reproducibility" (see, e.g., Lin [55]). Providing a comparative description of the various approaches for evaluation of this agreement is the main aim of this article.

Prior to mid-1980s, the correlation ρ between Y_1 and Y_2 was a widely used measure of "agreement" in the medical literature [1,11,12]. A statistically significant test of zero correlation was often taken as evidence of agreement. But this was clearly inadequate since two raters measuring the same response would rarely be uncorrelated. Moreover, the correlation is a measure of *linear relationship*, not of *agreement*. The popular early approaches include a paired-t test of $H_0 : \mu_1 = \mu_2$. But it only assesses whether two raters have the same mean—this is not enough for interchangeable use of two raters. Further, as Lin [55] demonstrates, this test can be misleading for agreement evaluation. It may reject H_0 when the scatter around the 45° line is near zero (indicating good agreement) and may accept H_0 if this scatter is very high (indicating poor agreement). The Pitman–Morgan test for $H_0 : \sigma_1^2 = \sigma_2^2$ and the Bradley–Blackwood test for $H_0 : \{\mu_1 = \mu_2, \sigma_1^2 = \sigma_2^2\}$, described in Krummenauer [48], suffer from the same drawback.

As in Lin [56], the correct hypothesis testing formulation for agreement evaluation must have the following form:

H_0 : Raters do not have satisfactory agreement. versus

H_1 : Raters have satisfactory agreement. (1)

This way the probability of type I error that the test controls is the probability of falsely concluding satisfactory agreement and one requires strong evidence against H_0 to reject it. In applications, this hypothesis can be tested using a confidence interval for the agreement measure of interest. In particular, a lower confidence bound should be used if a large value for the measure implies good agreement. On the other hand, an upper confidence bound should be used if a small value for the measure implies good agreement. Frequently, however, two-sided confidence intervals are used in practice, especially when the interest is in examining the extent of agreement, rather than formally testing a hypothesis.

The remainder of this article is organized as follows. Sections 41.2 and 41.3 describe various agreement measures for unreplicated and replicated measurements, respectively. Measures for evaluating intrarater agreement, i.e., the agreement of a rater with itself, are also discussed in Section 41.3. Section 41.4 briefly describes some popular approaches for modeling data and performing inference on an agreement measure. An illustration is provided in Section 41.5. Section 41.6 concludes with a discussion.

41.2 Agreement Measures for Unreplicated Measurements

Let $(\overline{Y}_1, \overline{Y}_2)$, (S_1^2, S_2^2), and S_{12}, respectively denote the sample counterparts (unbiased estimators) of (μ_1, μ_2), (σ_1^2, σ_2^2), and σ_{12}. Also, let $\overline{D} = \overline{Y}_1 - \overline{Y}_2$ and $S_D^2 = S_1^2 + S_2^2 - 2S_{12}$ be the sample counterparts of (μ, τ^2).

41.2.1 Limits of Agreement (LOA)

This index, due to Bland and Altman [11], is hugely popular in the medical literature. This paper is among the most cited statistical papers of all time (see Ryan and Woodall [72]). Their key idea is to consider two raters to be interchangeable if a large proportion (such as 95%) of differences in their measurements are within a clinically acceptable margin, say $\pm\delta_0$. In particular, they estimate the interval

$(\mu - 1.96\tau, \mu + 1.96\tau)$, which contains 95% of the values of D assuming that it follows a $\mathcal{N}(\mu, \tau^2)$ distribution, using the so-called 95% LOA, $(\overline{D} - 1.96S_D, \overline{D} + 1.96S_D)$. In applications, typically one infers satisfactory agreement when these LOA fall within the thresholds $\pm\delta_0$, where $\delta_0 > 0$. Bland and Altman [11,12] recommend specifying δ_0 in advance and note that its choice is a matter of clinical judgement that will depend on the application. However, in practice, δ_0 is rarely specified in advance. Instead, after computing the LOA, the practitioner judges whether or not the differences within the limits large enough to be clinically important.

An integral part of this analysis is the plot of means $(Y_1+Y_2)/2$ versus differences D, popularly known as the *Bland–Altman plot*. Two horizontal lines corresponding to the 95% LOA are also added to this plot. It is an excellent supplement to the usual scatterplot of Y_1 versus Y_2. Hawkins [41] describes real examples of how it can be used in diagnosing departures from various model assumptions, such as the non-constant mean and variance, presence of outliers, etc., and discusses ways to deal with them.

The simplicity and the intuitive appeal of the LOA approach partly explain why it is so popular among the medical researchers. But the usual practice of comparing $(\overline{D} - 1.96S_D, \overline{D} + 1.96S_D)$ — the *estimate* of the parameter interval $(\mu - 1.96\tau, \mu + 1.96\tau)$ — with the margin of acceptable differences ignores the uncertainty in the estimate. Statistically speaking, it is analogous to deducing whether μ is near zero by comparing the descriptive statistic $\overline{D}$ with zero, without taking its standard error into account. A more appropriate approach would be to use a confidence interval estimate for the interval $(\mu - 1.96\tau, \mu + 1.96\tau)$ so that the uncertainty in the estimation is also accounted for in the inference. This confidence interval will actually be a *tolerance interval* (see Guttman [35]) for the distribution of D — a topic discussed in Section 41.2.2. Bland and Altman [11,12] do provide separate two-sided confidence intervals for the individual limits $\mu \pm 1.96\tau$, but they are rarely used in practice perhaps because of the difficulty in interpreting the two intervals together. Because of this reason, Lin et al. [61] recommend two one-sided confidence bounds, an upper bound for $\mu + 1.96\tau$ and a lower bound for $\mu - 1.96\tau$, and show that using them to evaluate agreement is equivalent to an approximate test of the hypothesis (1) where the alternative H_1 represents $\{-\delta_0 < \mu - 1.96\tau$ and $\mu + 1.96\tau < \delta_0\}$. Liu and Chow [62] provide an exact test for this hypothesis.

The LOA approach is generalized in Bland and Altman [13,14] to accommodate replicate measurements and handle the case when the mean of the difference depends on the magnitude of measurements. Bland and Altman [13] also describe a nonparametric analog of the LOA for use when the normality assumption is suspect.

41.2.2 Total Deviation Index (TDI), Tolerance Interval (TI) and Coverage Probability (CP)

Just like the LOA approach, these approaches are also based on the intuitively appealing idea that two raters can be considered interchangeable if a large proportion of their differences (say, π_0) lies within a clinically acceptable margin (say, $\pm\delta_0$). Assume momentarily that D follows a $\mathcal{N}(\mu, \tau^2)$ distribution and $\pi_0 = 0.95$. Then, essentially the LOA approach asks if the interval $(\mu - 1.96\tau, \mu + 1.96\tau)$ that covers the *middle* 95% portion of D's distribution is within $(-\delta_0, \delta_0)$, or equivalently, if $(-\delta_0, \delta_0)$ covers more than the middle 95%

of D's distribution. The approaches in this section are a slightly liberal variation on the same theme: they ask if the interval $(-\delta_0, \delta_0)$ covers more than *any* 95% portion of D's distribution.

The TDI, introduced by Lin [57], is defined as the π_0th percentile of $|D|$, say TDI(π_0), for a given large proportion π_0 (> 0.5). Generally $0.80 \leq \pi_0 \leq 0.95$ in applications. It is nonnegative, and its smaller values indicate better agreement. Let U be a level $(1-\alpha)$ upper confidence bound for TDI(π_0). Then, the interval $[-U, U]$ is a level $(1-\alpha)$ TI for the distribution of D that has probability content π_0 (see Guttman [35], Choudhary and Nagaraja [26]):

$$P(F\{U\} - F\{-U\} \geq \pi_0) \geq 1 - \alpha.$$

Here F is the cumulative distribution function of D. This TI estimates the range of π_0 proportion of the measurement differences in the population. The practitioner could then infer satisfactory agreement if there are no large clinically meaningful differences in this interval. Thus the advance specification of the margin δ_0 is not really needed. However, when it is available, $U < \delta_0$ implies satisfactory agreement. Moreover, assessing agreement in this way is equivalent to performing a level α test of the hypotheses (1), where

$$H_0 : \text{TDI}(\pi_0) \geq \delta_0 \text{ vs.}$$
$$H_1 : \text{TDI}(\pi_0) < \delta_0. \qquad (2)$$

An alternative to this TDI approach measures agreement using the coverage probability (CP) of the interval $(-\delta_0, \delta_0)$, $P(|D| \leq \delta_0) = \text{CP}(\delta_0)$, say, for a given δ_0. This index ranges between 0 and 1, with larger values implying better agreement. It was introduced by Lin et al. [60]. For a specified (δ_0, π_0), the agreement can be evaluated by testing the hypotheses

$$H_0 : \text{CP}(\delta_0) \leq \pi_0 \text{ vs.}$$
$$H_1 : \text{CP}(\delta_0) > \pi_0, \qquad (3)$$

where satisfactory agreement is inferred on rejection of H_0. Hypotheses (2) and (3) are the same since CP$(\delta_0) \leq \pi_0$ is equivalent to TDI$(\pi_0) \geq \delta_0$. Thus, for given (δ_0, π_0), the TDI and the CP are equivalent approaches to evaluate agreement. Despite this equivalence, however, the TDI approach may be preferable from a practical viewpoint since it can be implemented without having to explicitly provide a δ_0.

The above approaches are not tied to the normality of D. In principle, they can be used with any continuous distribution for D. Nevertheless, when $D \sim \mathcal{N}(\mu, \tau^2)$, TDI and CP have closed-form expressions

$$\text{TDI}(\pi_0) = \tau \left\{ \chi_1^2 \left(\pi_0, \frac{\mu^2}{\tau^2} \right) \right\}^{1/2},$$

$$\text{CP}(\delta_0) = \Phi\left(\frac{(\delta_0 - \mu)}{\tau}\right)$$
$$- \Phi\left(\frac{(-\delta_0 - \mu)}{\tau}\right), \qquad (4)$$

where $\chi_1^2(\pi_0, \Delta)$ denotes the π_0th percentile of a noncentral chisquare distribution with one degree of freedom and noncentrality parameter Δ, and Φ denotes the cumulative distribution function of a $\mathcal{N}(0,1)$ distribution. In this normal setting, it is instructive to compare a TI for $\pi_0 = 0.95$ with the 95% LOA ($\overline{D} - 1.96 S_D, \overline{D} + 1.96 S_D$). The former interval covers 95% values of D with probability $(1-\alpha)$, whereas the latter simply *estimates* $(\mu - 1.96\tau, \mu + 1.96\tau)$ — an interval that covers the middle 95% values of D. Moreover, by virtue of being a confidence interval, the TI takes into account of the uncertainty in estimation, unlike the LOA. Due to this reason, a TI is expected to be wider than the LOA. A TI interpretation is also possible for the 95% LOA provided 1.96 in its definition is replaced by an appropriately chosen constant $k_n > 1.96$ (see Guttman [35]).

So far in the agreement literature, the inference for TDI and CP has only been studied assuming normality for D. Lin [57] argues that testing (2) is hard. So he approximates the TDI as a constant times $\left(E(D^2)\right)^{1/2}$ and suggests a large-sample test for the approximated hypothesis. This approximation is good only when (μ/σ) is small. But making this assumption to justify the use of TDI's approximation may be excessive, especially since it involves parameters that we are trying to learn about. Moreover, for large samples, it may happen that H_0 in (2) is false but Lin's test accepts it and vice versa with probability near one (see Choudhary and Nagaraja [26]). Lin et al. [60] provide a large-sample test for (3), but the simulations in Choudhary and Nagaraja [26] show that it does not work well for small or moderate samples. Moreover, due to the approximations, these tests for (2) and (3) may lead to different conclusions for the same data even though the two hypotheses are equivalent. Wang and Hwang [77] also provide a test of (3), which works well for small samples but not for large samples. The alternatives proposed by Choudhary and Nagaraja [26], including an exact test, overcomes these concerns.

The approach of Choudhary and Nagaraja [26] has been extended in Choudhary and Ng [27] and Choudhary [19] to incorporate a continuous covariate, and in Choudhary [21] to deal with repeated and longitudinal measurements. Choudhary [20] uses TDI when the measurements are subject to left-censoring due to lower detection limits. Lin et al. [54] have extended the approximate TDI/CP approaches to deal with repeated measurements and multiple raters.

41.2.3 Mean Squared Deviation (MSD)

This measure is defined as $E(D^2) = \mu^2 + \sigma^2$. It is nonnegative, with smaller values implying better agreement. Perfect agreement is equivalent to MSD = 0. Despite being a direct and intuitive measure of agreement, an MSD by itself is not popular among practitioners probably because of the difficulty in its interpretation — how small a value for it would indicate an acceptable amount of agreement. Note that if one can come up with acceptable values for $|\mu|$ and τ, the sum of their squares would be an acceptable value for the MSD.

The MSD is a popular starting point among developers of agreement measures. Indeed, the measures such as the concordance correlation [55], the approximations to TDI and CP [57,60,54], and the coefficient of individual agreement [7] are scaled versions of an MSD. Moreover, due to its mathematical tractability, it has proved especially useful for comparing the extent of agreement in pairs of raters, in particular, to find the rater that agrees most with a reference rater (see Hutson et al. [42] and Choudhary and Nagaraja [23,24]). Lin et al. [60], Barnhart et al. [5] and Haber and Barnhart [37] consider extensions of MSD to deal with multiple raters and replicate measurements.

41.2.4 Concordance Correlation Coefficient (CCC)

This measure, due to Lin [55,56], is defined as

$$\begin{aligned} \text{CCC} &= 1 - \frac{\text{MSD}}{\text{MSD}'} \\ &= \frac{2\rho\sigma_1\sigma_2}{(\mu_1 - \mu_2)^2 + \sigma_1^2 + \sigma_2^2}. \end{aligned} \quad (5)$$

where MSD$'$ = MSD assuming Y_1 and Y_2. It ranges between $[-1, 1]$, with high values indicating better agreement. It combines two components of disagreement: the first is correlation ρ, which Lin calls the "precision" component, that measures how close the measurements are to the line of best fit; and the second is the "accuracy" component that lies in $(0, 1)$ and measures how close the best fit line is to the 45° line by assessing differences in means and variances of Y_1 and Y_2.

The following properties of CCC are evident: (a) $0 \leq |\text{CCC}| \leq |\rho| \leq 1$, and CCC and ρ have the same sign. (b) CCC $= \rho$ is equivalent to $\{\mu_1 = \mu_2, \sigma_1 = \sigma_2\}$. (c) CCC $= 0$ and $\rho = 0$ are equivalent, meaning no agreement. (d) CCC $= 1$ is equivalent to MSD $= 0$ implying *perfect* agreement. (e) CCC $= -1$ is equivalent to $P(Y_1 = -Y_2) = 1$, which is called perfect negative agreement. In practice, the CCC is rarely negative because the two raters are trying to measure the same underlying true quantity. It is generally estimated by its sample counterpart

$$\widehat{\text{CCC}} = \frac{2S_{12}}{(\overline{Y}_1 - \overline{Y}_2)^2 + S_1^2 + S_2^2}. \quad (6)$$

It is apparent from its definition in (5) that CCC is a chance-corrected measure — it quantifies agreement between raters in excess of what is expected by chance alone and this "chance agreement" is taken to mean independence of raters. Due to this correction, $|\text{CCC}| \leq |\rho|$. So a low correlation always implies a low agreement. That this may be misleading is illustrated by Haber and Barnhart [36] through the following example of Zegers [80]. Suppose two teachers assign the grades $(8,8), (8,9), (9,8), (9,9)$ to four students on a scale of 0 to 10. Clearly, their agreement is quite good, but $\widehat{\text{CCC}}$ is zero because the sample correlation is zero. See also Zhong and Shao [81] for a similar example.

Haber and Barnhart [36] argue against the idea of taking chance agreement to mean independence. The raters will be independent if, e.g., they use independent random number generators to assign measurements to subjects regardless of their true values. But, in practice, this is not what we mean by chance agreement. The raters will agree to some extent simply because they measure the same subject. A more appropriate definition of chance agreement would be the conditional independence of raters given that they are measuring the same true value. But using it in (5) to correct for chance does not lead to a simple index. So they suggest an alternative called the coefficient of individual agreement, to be presented in Section 41.3.2, that does not involve chance-correction.

Fay [30] illustrates another issue with CCC's chance correction through the following example. Consider two bivariate distributions for (Y_1, Y_2) with $\sigma_1^2 = 1 = \sigma_2^2$ and $\rho = -0.1$. Take $\mu_1 = 0 = \mu_2$ in the first case and $(\mu_1, \mu_2) = (-2, 2)$ in the second case. Clearly, the agreement is worse in the second case, but the CCC does not reflect this since it is higher in the second case (-0.01) than the first case (-0.1). Although $\rho < 0$ is a rare occurrence in practice, Fay notes that in this case increasing difference between marginal distributions of Y_1 and Y_2 may incorrectly lead to higher chance agreement term, i.e., MSD assuming Y_1 and Y_2 are independent, which in turn implies higher CCC. He suggests replacing this term in CCC's definition with $E_{Z_1} E_{Z_2} (Z_1 - Z_2)^2 = \sigma_1^2 + \sigma_2^2 + (\mu_1 - \mu_2)^2/2$, where Z_1 and Z_2 are independent draws from the distribution of the random variable $0.5Y_1 + 0.5Y_2$.

The CCC is notorious for being heavily influenced by the between-subject variation in the data (or the range of measurements). It increases quickly when this variation is increased even if the differences in measurements are held constant. Many authors, including Bland and Altman [12], Müller and Büttner [65], Atkinson and Nevill [2], and Barnhart et al. [7], provide real data examples that have weak interrater agreement but have near-one values for the CCC, just because the between-subject variation in the data is much larger compared to the within-subject variation.

Therefore the CCC warrants a careful interpretation. Moreover, as Lin and Chinchilli [59] note, two data sets with similar CCC values do not have comparable agreement unless their measurement ranges are also similar. Carrasco and Jover [15] recommend adjusting the CCC by removing a portion of the between-subject variation by including subject-specific covariates such as race, gender, age, etc., in the model for measurements (see Section 41.4).

Despite the aforementioned drawbacks, the CCC has attracted much further research. King and Chinchilli [44] presented a generalized CCC by replacing MSD in its definition with a general function of the form $E(g(D))$, and produced robust versions of CCC by taking particular choices of the g function. In King and Chinchilli [43], some additional choices of this function are taken to show that CCC reduces to kappa statistics, which are popular measures of agreement for categorical data (see Shoukri [74]). This way they unify the treatment of agreement evaluation for both continuous and categorical data. They also introduced a stratified version of CCC to take into account of explanatory factors. Barnhart and Williams [4] showed how to adjust CCC for covariates. Liao [52] proposed a modification of CCC that replaces the MSD in its definition with a certain concept of mean squared area. This approach is extended in Liao [53] for the case when the distribution of (Y_1, Y_2) and hence the extent of agreement depend on a continuous covariate. Li and Chow [51] generalized CCC to measure agreement when the measurements are curves. Barnhart et al. [10] developed CCC for measurements that are left-censored because of lower detection limits. Liu et al. [63] and Guo and Manatunga [34] have introduced CCC for time-to-event data that are subject to right censoring. Carrasco et al. [16] compared various approaches for inference on CCC when the data are skewed. Additional extensions of CCC to deal with replicated measurements and more than two raters are mentioned in Sections 41.3 and 41.6, respectively.

41.2.5 Intraclass Correlation Coefficient (ICC)

The term intraclass correlation actually refers to a group of coefficients that measure relationship among measurements of a common class (see, e.g., Shrout and Fleiss [75]). There are numerous versions of ICCs, depending upon the model assumed for the data and how the ICC is going to be used. See McGraw and Wong [64] for a thorough review on this topic. Here we describe only one ICC that is appropriate for measuring agreement.

Assume that the observed measurements Y_{ij}, $i = 1, \ldots, m$, $j = 1, 2$, from two raters can be modeled using the following two-way mixed model

$$Y_{ij} = \mu_j + T_i + \epsilon_{ij}, \qquad (7)$$

where μ_j is the fixed mean of the jth rater, T_i is the random effect of the ith subject, ϵ_{ij} is the random error, and T_i and ϵ_{ij} are mutually independent mean zero and variances σ_B^2 and σ_W^2, respectively. These variances represent the between- and the within-subject variations. Define an ICC associated with this model as:

$$\text{ICC} = \frac{\sigma_B^2}{\sigma_\mu^2 + \sigma_B^2 + \sigma_W^2}, \qquad (8)$$

where $\sigma_\mu^2 = \sum_{j=1}^2 (\mu_j - \overline{\mu})^2$ is the variability in the rater means (see, e.g., McGraw and Wang [64], Carrasco and Jover [15]). This index ranges between 0 and 1, with higher values indicating better agreement.

Several authors, including Lin [55] noted that the estimates of CCC and an ICC tended to be similar. Nickerson [66] pointed out that the estimator of the above ICC given in McGraw and Wang [64] in terms of mean squares reduces to

$$\widehat{\text{ICC}} = \frac{S_{12}}{(\overline{Y}_1 - \overline{Y}_2)^2 + S_1^2 + S_2^2 - S_D^2/n}, \quad (9)$$

which differs from the CCC's estimator in (6) only in the S_D^2/n term in the denominator. She argues that S_D^2/n tends to be small in applications, resulting in similar values for the two estimators. Finally, it was Carrasco and Jover [15], who established that the CCC is actually the above ICC by noting that, under the model (7), $\sigma_B^2 = \sigma_{12}$, $(\sigma_B^2 + \sigma_W^2) = (\sigma_1^2 + \sigma_2^2)/2$, $\sigma_\mu^2 = (\mu_1 - \mu_2)^2/2$, and substituting them in (8) gives the CCC in (5). They also suggest using the estimator (9) for CCC instead of the usual (6) to reduce the bias in estimation.

Barnhart et al. [5] provide a detailed comparison of CCC with several ICCs. St. Laurent [76] develops an ICC for the case when one of the raters is a gold standard. He calls it *gold-standard correlation* and presents it as an alternative to the CCC. Harris et al. [40] discuss a family of estimators for this correlation. Lee et al. [49] develop an ICC assuming that raters are random. Quan and Shih [69] suggest using within-subject coefficient for variation as an alternative to ICC.

Many authors, including Bland and Altman [12], Müller and Büttner [65], and Atkinson and Nevill [2] have noted that the ICCs tend to be overly sensitive to the between-subject variation. So one should avoid concluding satisfactory agreement if the lower confidence bound of an ICC is greater than a predetermined cutoff, say, 0.75, as has been suggested by Lee et al. [49].

41.3 Agreement Measures for Replicated Measurements

In this section, we assume that there are replicate (or repeated) measurements by each rater on every subject. We first consider the case when the true underlying measurement does not change during replication. This assumption is rather strong and does not hold, e.g., when blood pressure measurements are taken on different days. Repeated measurements also allow us to evaluate agreement of a rater with itself, i.e., the *intrarater agreement*. This is especially important since the raters who do not agree well with themselves cannot be expected to agree well with others (see Bland and Altman [13], Hawkins [41]).

Let the random vectors (Y_1, Y_1') and (Y_2, Y_2') respectively denote two replicate measurements by raters 1 and 2 on a randomly selected subject from the population. Also, let Y_{ijk} be the kth measurement from the jth rater on the ith subject, where $k = 1, \ldots, m_{ij} (\geq 2)$, $j = 1, 2$, and $i = 1, \ldots, n$. These data are *unpaired* [18] in that the time order of the replicates is immaterial. This is the case, for example, when a blood sample is divided into four parts and each rater measures the cholesterol level twice using one part each time. These data are assumed to follow a general model of the form,

$$Y_{ijk} = T_{ij} + \epsilon_{ijk}, \quad (10)$$

where T_{ij} is the unobservable true measurement from the jth rater on the ith subject and ϵ_{ijk} is the error term. It is also assumed that (T_{i1}, T_{i2}) is independently distributed as (T_1, T_2), which follows a bivariate distribution with mean (μ_1, μ_2), variance $(\sigma_{B1}^2, \sigma_{B2}^2)$ and correlation ν; ϵ_{ijk} is independently distributed with mean zero and variance σ_{Wj}^2; and the true measurements are mutually independent of the errors. The variances σ_{Bj}^2 and σ_{Wj}^2 represent the between-subject and the within-subject variation, respectively. These assumptions imply that (Y_1, Y_2) and (Y_j, Y_j') have bivariate distributions with following properties:

$$\begin{aligned} E(Y_j) &= E(Y_j') = \mu_j, \\ \operatorname{var}(Y_j) &= \sigma_{Bj}^2 + \sigma_{Wj}^2 \\ &= \operatorname{var}(Y_j') = \sigma_j^2, \\ \operatorname{cov}(Y_1, Y_2) &= \operatorname{cov}(T_1, T_2) \\ &= \nu \sigma_{B1} \sigma_{B2} = \sigma_{12}, \\ \operatorname{cov}(Y_j, Y_j') &= \sigma_{Bj}^2, \; j=1,2. \end{aligned} \quad (11)$$

If, in addition, normality is assumed for T_{ij} and ϵ_{ijk} in (10), then (Y_1, Y_2) and (Y_j, Y_j') will have bivariate normal distributions.

Now, noting that the agreement measures defined in Section 41.2 are functions of parameters of (Y_1, Y_2), it is straightforward to generalize them to deal with repeated measurements: Simply substitute the above expressions for $(\mu_1, \mu_2, \sigma_1^2, \sigma_2^2, \sigma_{12})$ in their respective formulas. The resulting indices are said to measure *total agreement* — a term coined by Barnhart et al. [9] This approach of extending agreement measures has been used by a number of authors, often assuming special cases of model (10). They include Bland and Altman [13,14] for LOA; Carrasco and Jover [15], Barnhart et al. [9], Quiroz [70] and Lin et al. [54] for CCC; Lin et al. [54] for MSD and approximate TDI/CP; and Choudhary [21] for TDI.

The model (10) is not appropriate, e.g., when the repeated measurements are longitudinal or when the true measurement changes over time. In this case, we assume that paired measurements from both raters are available at each observation time. To deal with this kind of data, Chinchilli et al. [18] and King et al. [45,46] develop versions of CCC, and Bland and Altman [13,14] discuss how to compute the LOA. An alternative approach in this case is to explicitly model the data using time as a covariate and let the agreement measure be a function of time. This has been utilized in Choudhary [19,21] for TDI and can used for other measures as well.

The remainder of this section is organized as follows. We first discuss measures for evaluating intrarater agreement assuming model (10) for the data. Then, in Section 41.3.2, we present a measure of interrater agreement that by definition requires replicate measurements.

41.3.1 Measures of Intrarater Agreement

Since the indices that measure agreement between two raters are functions of parameters of (Y_1, Y_2), it is natural to use the same parameter functions of (Y_j, Y_j') to measure agreement of jth rater with itself. Substituting the expressions for means, variances and covariance of (Y_j, Y_j'), given in (11), in the definitions of the measures in Section 41.2 gives us the following indices for measuring intrarater agreement of the jth rater:

$$\begin{aligned} 95\%\,\text{LOA}_j &= \pm 1.96 \left(2\sigma_{Wj}^2\right)^{1/2}, \\ \text{TDI}_j(\pi_0) &= \left(2\sigma_{Wj}^2 \chi_1^2(\pi_0, 0)\right)^{1/2}, \\ \text{CP}_j(\delta_0) &= \Phi\left\{\delta_0 / \left(2\sigma_{Wj}^2\right)^{1/2}\right\} \\ &\quad -\Phi\left\{-\delta_0 / \left(2\sigma_{Wj}^2\right)^{1/2}\right\}, \\ \text{MSD}_j &= 2\sigma_{Wj}^2, \\ \text{CCC}_j &= \frac{\sigma_{Bj}^2}{\sigma_{Bj}^2 + \sigma_{Wj}^2}, \end{aligned}$$

where $j = 1, 2$. It may be noted that $\text{TDI}_j(0.95) = |95\%\,\text{LOA}_j|$. Further, the version of ICC appropriate for this purpose is the *reliability* of rater j (see Chap. 1 of Fleiss [31]). It equals CCC_j and represents the correlation between two replicates from the jth rater on a subject. It can also be interpreted as the proportion of variance of rater j due to between-subject variance. The assumption of normality is also implicit in the first three indices. The first four indices are one-to-one functions of $\text{var}(Y_j - Y_j') = 2\sigma_{Wj}^2$, whereas the last one involves both σ_{Wj}^2 and the between-subject variation σ_{Bj}^2. In either case, a small value for the measurement error variance σ_{Wj}^2 implies good intrarater agreement.

Several authors have studied these measures, often assuming special cases of model (10). In particular, Bland and Altman [13] consider LOA_j, where it is called *repeatability coefficient* for rater j; Lin et al. [54] study all the measures except the LOA_j; Barnhart et al. [9] study CCC_j; and Choudhary [21] considers TDI_j.

41.3.2 Coefficient of Individual Agreement (CIA)

This measure of interrater agreement, proposed by Barnhart et al. [7] incorporates a measure of intrarater agreement in its definition. It is a slight modification of the *coefficient of interobserver variability* due to Haber et al. [38]. It distinguishes between whether or not one rater serves as a reference.

First, consider the case when there is no reference rater and assume that the error variations of both raters are acceptably small. Then, in essence, the CIA considers two raters to be interchangeable if the distribution of (Y_1, Y_2) is similar to those of (Y_1, Y_1') and (Y_2, Y_2'). In other words, the measurements from two raters behave like the replicate measurements from the same raters. It is formally defined as

$$\begin{aligned}
\text{CIA}_N &= [\{E(Y_1 - Y_1')^2 + E(Y_2 - Y_2')^2\}/2] \\
&\quad \times [E(Y_1 - Y_2)^2]^{-1} \\
&= \{\sigma_{W1}^2 + \sigma_{W2}^2\} \\
&\quad \times \{(\mu_1 - \mu_2)^2 + (\sigma_{B1} - \sigma_{B2})^2 \\
&\quad + 2(1 - \nu)\sigma_{B1}\sigma_{B2} + \sigma_{W1}^2 \\
&\quad + \sigma_{W2}^2\}^{-1},
\end{aligned} \quad (12)$$

where the parameters are defined in (11).

Clearly, this index lies between 0 and 1, with higher values indicating better agreement. Note that $\text{CIA}_N = 0$ implies a complete lack of agreement since it corresponds to a limiting situation where

$$\begin{aligned}
E(T_1 - T_2)^2 &= \{(\mu_1 - \mu_2)^2 \\
&\quad + (\sigma_{B1} - \sigma_{B2})^2 \\
&\quad + 2(1 - \nu)\sigma_{B1}\sigma_{B2}\},
\end{aligned}$$

the disagreement between the *true* unobserved measurements, approaches infinity. The index is also zero when $\sigma_{W1}^2 = 0 = \sigma_{W2}^2$, which implies that (Y_1, Y_2) and (T_1, T_2) have the same distribution, and this indicates disagreement whenever $E(T_1 - T_2)^2 > 0$. On the other extreme, $\text{CIA}_N = 1$ is equivalent to $E(T_1 - T_2)^2 = 0$ or $P(T_1 = T_2) = 1$, meaning that the two raters agree perfectly on the *true* unobserved measurements. In general, however, this does not imply good agreement on the *observed* measurements (Y_1, Y_2), due to the possibly different and large error variations σ_{W1}^2 and σ_{W2}^2. This is why Barnhart et al. [5] emphasize first establishing that σ_{W1}^2 and σ_{W2}^2 are acceptable, i.e., the raters agree sufficiently well with themselves. Only then a high value for CIA_N indicates good agreement between the observed measurements from the raters.

Next, consider the case when there is a reference rater, say, rater 2, and assume that its within-subject variation is acceptably small. The CIA regards the raters to be interchangeable if the distributions of (Y_1, Y_2) and (Y_2, Y_2') are similar — i.e., the measurements from two raters are like the replicate measurements from the reference rater. It is defined as follows:

$$\begin{aligned}\text{CIA}_R &= \frac{E(Y_2 - Y_2')^2}{E(Y_1 - Y_2)^2} \\ &= \{2\sigma_{W2}^2\} \\ &\quad \times \{(\mu_1 - \mu_2)^2 + (\sigma_{B1} - \sigma_{B2})^2 \\ &\quad + 2(1-\nu)\sigma_{B1}\sigma_{B2} \\ &\quad + \sigma_{W1}^2 + \sigma_{W2}^2\}^{-1}. \end{aligned} \quad (13)$$

It ranges between 0 and 2. Even though it is a CIA, its interpretation is quite different from that of CIA_N. Firstly, comparing (12) and (13) we see that $\text{CIA}_R \leq 2\,\text{CIA}_N$. Secondly, if $\sigma_{W2}^2 \leq \sigma_{W1}^2$ — as may be the case since the rater 2 serves as a reference, $\text{CIA}_R \leq 1$. Moreover, $\text{CIA}_R = 1$ is equivalent to $\{P(T_1 = T_2) = 1, \sigma_{W1}^2 = \sigma_{W2}^2\}$, which means perfect agreement on the *true* unobserved measurements and the same error variability for the two raters. On the other hand, if $\text{CIA}_R > 1$, rater 1 is clearly better than rater 2 since the former has a smaller error variance than the latter. Thus, assuming that σ_{W2}^2 is acceptably small and $\sigma_{W2}^2 \leq \sigma_{W1}^2$, a value of CIA_R that is slightly less than or equal to one indicates good agreement between the raters.

Barnhart et al. [7] mention CIA's close connection with the FDA [32] reference-scaled criterion for establishing individual bioequivalence of two drugs. This is the reason why this index is called a coefficient of *individual* agreement. They also exploit this connection to suggest a cutoff for CIA that would indicate acceptable agreement. The authors also discuss a relation between CIA and an index of Shao and Zhong [81]. See Haber and Barnhart [36] for a motivation behind this index that, unlike a CCC, does not involve correction of chance agreement. Further, based on an extensive comparison of CIA and CCC, Barnhart et al. [8] conclude that the former is less influenced by the magnitude of between-subject variation than the latter. But just as a large between-subject variation may inflate CCC, a large within-subject variation may also inflate CIA. This is why it is important to first ensure that the within-subject variations of the raters are acceptable before evaluating their agreement using CIA. More recently, Haber and Barnhart [37] generalized the CIA approach to incorporate disagreement functions such as the mean absolute deviation in the definitions (12) and (13) of CIA in place of usual the mean squared deviation.

Haber and Barnhart [37] describe a simple nonparametric approach for estimating CIAs. They also discuss large-sample approaches for constructing confidence intervals for this index. These procedures require replicates from both raters in case of CIA_N and from just the reference rater in case of CIA_R.

41.4 Modeling Measurements and Inference on Agreement Measures

In this section, we briefly summarize some models that are frequently used in the literature. First, consider the case of unreplicated measurements, where Y_{ij}, $j = 1, 2$, $i = 1, \ldots, n$, represent the observed data. A general model for these data is (7), $Y_{ij} = \mu_j + T_i + \epsilon_{ij}$, with the addition that the error variances may be different for the two raters, i.e., $\text{var}(\epsilon_{ij}) = \sigma_{Wj}^2$. This is a *mixed-effects* model since μ_j is a fixed effect and T_i is a random effect. It is also known as a *variance components model*. In this model, the random effect induces dependence in the measurements from two raters. The model cannot have a subject × rater interaction effect since it gets confounded with the error term. The assumptions in (7) imply that (Y_{i1}, Y_{i2}), $i = 1, \ldots, n$, are independently distributed as (Y_1, Y_2), which follows a bivariate distribution with $E(Y_j) = \mu_j$, $\text{var}(Y_j) = \sigma_j^2 = \sigma_B^2 + \sigma_{Wj}^2$ and $\text{cov}(Y_1, Y_2) = \sigma_{12} =$

σ_B^2. Substituting these in the definitions of the interrater agreement measures in Section 41.2 expresses them as functions of model parameters. Although this model allows estimation of σ_{Wj}^2 and hence the evaluation of intrarater agreement, this inference does not tend to be reliable. It is best done with replicated data.

Next, consider the case of replicated measurements and assume that the general model (10), $Y_{ijk} = T_{ij} + \epsilon_{ijk}$, holds for the observed data Y_{ijk}, $k = 1, \ldots, m_{ij} (\geq 2)$, $j = 1, 2$, $i = 1, \ldots, n$. Oftentimes one further assumes that $T_{ij} = \mu_j + T_i + I_{ij}$, where μ_j and T_i are as in the previous model; I_{ij} is a random effect representing the subject × rater interaction following a distribution with mean zero and variance σ_{Ij}^2; and T_i, I_{ij} and the errors are mutually independent. Due to this additional assumption, σ_{Bj}^2 and $cov(Y_1, Y_2)$ defined in (11) become $\sigma_B^2 + \sigma_{Ij}^2$ and σ_B^2, respectively.

In both cases, the standard approach for agreement evaluation consists of two steps. First, estimate the model parameters and plug them in the formula for the agreement measure of interest to get its estimate. Then, use the large-sample theory (see Lehmann [50]) to get a confidence interval (or bound) for the measure, which is then used to evaluate agreement. (An exception is the exact inference in case of TDI/CP — see Choudhary and Nagaraja [26]). The following are three popular approaches for estimating model parameters:

- *Nonparametric approach*: Use the method of moments, i.e., replace population means, variances and covariance with their sample counterparts or use the ANOVA-type methodology. No additional assumptions are made. It has been used, e.g., by King and Chinchilli [44] for CCC and by Barnhart et al. [7] for CIA.

- *Semiparametric approach*: Specify estimating equations for means, variances and covariance and use the GEE approach (see Hardin and Hilbe [39]). No additional assumptions are made regarding the distributions of random effects and errors. It has been used, e.g., by Barnhart and Williamson [4] and Barnhart et al. [9] for CCC and by Lin et al. [54] for several measures.

- *Parametric approach*: Assume normality for random effects and errors. The parameters are estimated using either the maximum likelihood method or the method of moments. It has been used for all the agreement measures.

Although the normality-based parametric approach makes the strongest assumptions among the three, it is also the most flexible. It offers a straightforward approach to include covariates in the model (see, e.g., Carrasco and Jover [15], Choudhary [21]). Moreover, since every major statistical package has capability to fit mixed-effects models, such models are easy to fit. Of course, this approach would be appropriate only when the normality assumption is reasonable. This is especially true for measures such as TDI/CP and LOA whose formulas derived assuming normality is generally not valid for other distributions. Frequently this assumption may be reasonable after a transformation of the data. Section 41.7 briefly describes how to compute confidence bounds for an agreement measure assuming normality for the data.

Often the above models are simplified by making additional assumptions such as no interaction even with replication [70] or equal error variance [15] or equal interaction variance [79]. When the measurements are paired and the agreement measure is based on differences in measurements, e.g., LOA and TDI/CP, sometimes one directly models the differences instead of modeling the whole data (see, e.g., Refs.

13, 60, 27, and 21).

41.5 Illustration

We now use the systolic blood pressure data from Altman and Bland [1] to illustrate agreement evaluation. The data set has 85 subjects and on each subject three replicate measurements (in mmHg) are made in quick succession by each of two experienced observers (J and R) using a sphygmomanometer and by a semi-automatic blood pressure monitor. We will only compare two raters — observer J (rater 1) and the machine (rater 2) — since we have not discussed measuring agreement in multiple raters (see Barnhart et al. [5]). The data will be analyzed using the statistical software R (by R Development Core Team [71]).

Figure 1 provides a trellis plot of the data. The two raters do not appear to agree well as there is not much overlap between their readings. In particular, the machine's readings tend to be higher and have higher within-subject variability than the observer's readings. Moreover, the between-subject variation appears substantially higher than the within-subject variation. The rater × subject interaction is also evident.

We now fit model (10) to these data assuming normality for errors and random effects. The computations were done using the nlme package [67] in R. The diagnostics suggested by Pinheiro and Bates [68] indicate a reasonably good model fit. The maximum likelihood estimates of parameters in (10) are $(\hat{\mu}_1, \hat{\mu}_2) = (127.4, 143.0)$, $(\hat{\sigma}_{W1}^2, \hat{\sigma}_{W2}^2) = (37.4, 83.1)$, $(\hat{\sigma}_{B1}^2, \hat{\sigma}_{B2}^2) = (924.0, 971.3)$ and $\hat{\nu} = 0.83$. Hence $(\hat{\sigma}_1^2, \hat{\sigma}_2^2) = (961.4, 1054.4)$ and $\hat{\rho} = 0.79$. Thus the machine's estimated mean is about 15 mmHg higher than the observer's mean and its error variation is more than twice of the observer's variation. These estimates confirm the impressions from Figure 1.

Next, we compute the estimates and 95% confidence bounds for TDI, CCC and CIA using the methodology described in Section 41.7. The results are presented in Table 1. The upper confidence bound is presented for TDI whereas lower confidence bounds are presented for the other two. We assume $\pi_0 = 0.95$ for the TDI.

Since $\hat{\sigma}_{W2}^2/\hat{\sigma}_{W1}^2 = 2.2$, observer J has substantially better intrarater agreement than the machine. The same conclusion is reached with the indices in Table 1. The magnitudes of these indices can tell us how strong the agreement is. The CCC estimates and bounds are large (≥ 0.89), but they are probably inflated because of the substantially large between-subject variation in these data compared to the within-subject variation. On the other hand, from the TI interpretation of a TDI bound (see Section 41.2.2), the difference between two readings from observer J is estimated to be within ±18.6 for 95% of subjects. This interval widens to ±27.6 for the machine. Since a difference of about 30 mmHg in blood pressure readings will lead to differing clinical interpretations of the measurements [13], the agreement of the machine with itself is clearly not satisfactory. The agreement of observer J with himself is also not as high as one would hope. For this reason and as the mean estimates of raters differ by 15 mmHg, we do not expect satisfactory agreement between them. Indeed, the bound for TDI (0.95) is 55.2, clearly indicating that the two raters cannot be used interchangeably. CIA's bound of 0.14, which is quite small, leads to the same conclusion. In contrast, due to the large between-subject variation, this conclusion is not so clear-cut on the basis of CCC whose bound is 0.62.

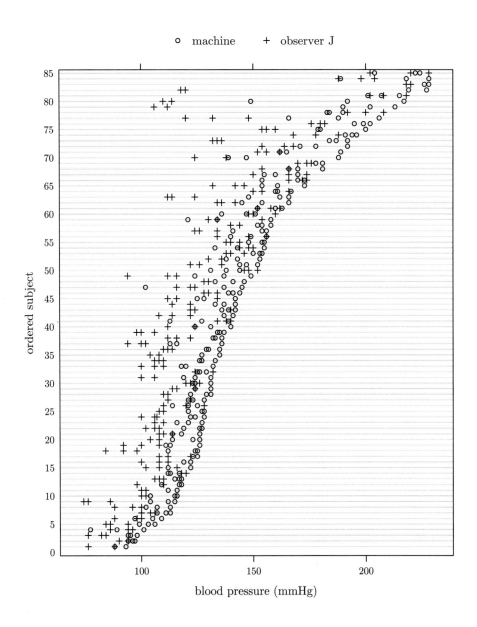

Figure 1: A trellis plot of the blood pressure data. The subjects are ordered according their maximum reading.

Table 1: Estimates and 95% confidence bounds for various measures of agreement. The upper bound is provided for TDI, and lower bounds are provided for the others. The CIA version for intrarater agreement is not available.

	TDI (0.95)	CCC	CIA
Intrarater agreement			
Observer J	17.0 (18.6)	0.96 (0.95)	—
Machine	25.3 (27.6)	0.92 (0.89)	—
Interrater agreement			
Observer J–machine	50.1 (55.2)	0.70 (0.62)	0.18 (0.14)

41.6 Summary and Discussion

In this chapter, we have presented a comparative discussion of various approaches to evaluate agreement between two methods of continuous measurement. We have argued that the popular LOA approach is quite intuitive, but the common practice of simply using the estimates of $\mu \pm 1.96\,\tau$ instead of their confidence bounds to judge interchangeability is not statistically sound. Further, when separate bounds are used for the two limits, their simultaneous interpretation becomes hard. The TDI (or TI) approach provides a better alternative since it is similar in spirit to the LOA but does not have these concerns. The CP approach is also equivalent to the TDI. At present, however, agreement evaluation using these measures require the assumption of normality for the measurements or their differences.

The CCC (or ICC) approach can be implemented without the normality assumption but it may be misleading, due to its hypersensitivity to the between-subject variation or the measurement range. The CCC may be low when the agreement is high and vice versa. The CIA provides an attractive alternative to it when the measurements are replicated. This index, however, may be hard to interpret, especially when its value is not near the extremes. For example, it is unclear how much the methods agree if CIA equals 0.75. Such an issue does not arise with TDI since it clearly indicates how large the differences are in a specified proportion of the population. Nevertheless, one does need to specify this proportion.

The above approaches are "aggregated" in that they combine the various indicators of disagreement, such as a difference in means, a difference in variances and a low correlation, into a single index. Hence when a lack of agreement is inferred, its cause may be unclear without further statistical investigation. To deal with this issue Choudhary and Nagaraja [25] suggest a "disaggregated" approach that involves constructing confidence intervals for individual disagreement indicators and combining the inferences using the intersection-union principle (see Casella and Berger [17, p. 380]). Overall, since the different agreement measures look at different aspects of the distribution of measurements, in our opinion, it is generally a good idea to compare inferences from several approaches.

Frequently in applications it is of interest to evaluate agreement among more than two raters. Lin [55], King and Chinchilli [43], Barnhart et al. [6], and Lin et al. [54] discuss versions of agreement measures that summarize the overall level of agreement among all raters in a single index. Williamson et al. [78] propose a bootstrap procedure for testing equality of two CCCs. St. Laurent [76], Hutson et al. [42], and Choudhary and Nagaraja [23,24] assume that there is a reference method in the comparison and consider the problem of finding the method that agrees most with it. Choudhary and Yin [28] develop methods for multiple comparisons, i.e., simultaneously comparing the extent of agreement between all pairs of methods of concern, and ranking them on the basis of their extent of agreement.

Our focus has only been on the analysis of agreement studies. We have not dealt with the issue of how to plan such studies, in particular, the determination of sample sizes. Many authors have developed sample size formulas for use when the measurements are not to be replicated. They include Lin [56] for CCC, Choudhary and Nagaraja [26] for TDI (or TI/CP), and Lin et al. [61] for LOA. But it is a good idea to replicate the measurements on each subject by every method so that the intrarater agreement can be evaluated and compared with the extent of interrater agreement. Yin et al. [79] discuss a general approach for determining both the number of subjects and the number of replicates for studies involving such simultaneous evaluation.

We conclude by noting that the literature on agreement evaluation is not fully developed for handling many issues of practical interest. They include dealing with functional and multivariate measurements, nonnormal data and data containing outliers, covariates, and the case when the standard model assumptions are not satisfied.

41.7 Appendix

We now briefly describe how to compute a large-sample confidence bound for any scalar measure of agreement, say γ, assuming the mixed-effects model (10) for the data together with the normality of random effects and errors. Let $\hat{\theta}$ be a column vector of parameters $(\mu_1, \mu_2, \sigma_{B1}^2, \sigma_{B2}^2, \nu, \sigma_{W1}^2, \sigma_{W2}^2)$ in this model and $\hat{\theta}$ be its MLE. Also, let $\hat{\gamma}$ be the MLE of γ obtained by plugging-in θ with $\hat{\theta}$ in its expression, and $f(\theta)$ be the likelihood function of θ.

When the number of subjects n is large, it is well-known that $\hat{\theta}$ approximately follows a $\mathcal{N}(\theta, I^{-1})$ distribution, where $I = -(\partial^2 \log f(\theta)/\partial \theta^2)|_{\theta=\hat{\theta}}$ is the observed Fisher information matrix of θ. Further, from an application of the delta method, $\hat{\gamma}$ approximately follows a $\mathcal{N}(\gamma, GI^{-1}G')$ distribution, where $G = (\partial \gamma/\partial \theta)|_{\theta=\hat{\theta}}$ is the gradient of γ with respect to θ and G' is the transpose of G (see Lehmann [50]). Hence a lower confidence bound and an upper confidence bound for θ, each with approximate confidence level $(1-\alpha)$, can be computed as $L = \hat{\gamma} - z_{1-\alpha}(GI^{-1}G')^{1/2}$ and $U = \hat{\gamma} - z_{\alpha}(GI^{-1}G')^{1/2}$, respectively. Here z_α denotes the αth percentile of a $\mathcal{N}(0,1)$ distribution.

Using bootstrap t critical points in place of $z_{1-\alpha}$ and z_α generally lead to more accurate bounds [21,29]. For improved accuracy, it is also recommended to first apply a normalizing transformation to γ, compute the bounds on the transformed scale, and then apply the inverse transformation to get the bounds on the original scale. In particular, the log transformation in case of TDI [57] and the Fisher's z transformation in case of CCC [55] are known to work well.

The above computations can be programmed in the statistical software R (R Development Core Team [71]) using the `nlme` package of Pinheiro et al. [67] for fitting mixed models and the `numDeriv` package of Gilbert [33] for computing numerical derivatives. They can also be programmed in SAS® software of SAS Institute Inc.

References

1. Altman, D. G. and Bland, J. M. (1983). Measurement in medicine: The analysis of method comparison studies. *Statistician*, **32**, 307–317.

2. Atkinson, G. and Nevill, A. (1997). Comment on the use of concordance correlation to assess the agreement between two variables. *Biometrics*, **53**, 775–777.

3. Banerjee, M., Capozzoli, M., McSweeney, L., and Sinha, D. (1999). Beyond Kappa: A review of interrater agreement measures. *Can. J. Statist.*, **27**, 3–23.

4. Barnhart, H. X. and Williamson, J. M. (2001). Modeling concordance correlation via GEE to evaluate reproducibiltiy. *Biometrics*, **57**, 931–940.

5. Barnhart, H. X., Haber, M. J., and Lin, L. I. (2007a). An overview on assessing agreement with continuous measurement. *J. Biopharm. Statist.*, **17**, 529–569.

6. Barnhart, H. X., Haber, M. J., and Song, J. (2002). Overall concordance correlation coefficient for evaluating agreement among multiple observers. *Biometrics*, **58**, 1020–1027.

7. Barnhart, H. X., Kosinski, A. S., and Haber, M. J. (2007). Assessing individual greement. *J. Biopharm. Statist.*, **17**, 697–719.

8. Barnhart, H. X., Lokhnygina, Y., Kosinski, A. S., and Haber, M. J. (2007c). Comparison of concordance correlation coefficient and coefficient of individual agreement in assessing agreement. *J. Biopharm. Statist.*, **17**, 721–738.

9. Barnhart, H. X., Song, J., and Haber, M. J. (2005). Assessing intra, inter and total agreement with replicated readings. *Statist. Med.*, **24**, 1371–1384.

10. Barnhart, H. X., Song, J., and Lyles, R. H. (2005). Assay validation for left-censored data. *Statist. Med.*, **24**, 3347–3360.

11. Bland, J. M. and Altman, D. G. (1986). Statistical methods for assessing agreement between two methods of clinical measurement. *Lancet*, **i**, 307–310.

12. Bland, J. M. and Altman, D. G. (1990). A note on the use of the intraclass correlation coefficient in the evaluation of agreement between two methods of measurement. *Comput. Biol. Med.*, **20**, 337–340.

13. Bland, J. M. and Altman, D. G. (1999). Measuring agreement in method comparison studies. *Statist. Methods Med. Res.*, **8**, 135–160.

14. Bland, J. M. and Altman, D. G. (2007). Agreement between methods of measurement with multiple observations per individual. *J. Biopharm. Statist.*, **17**, 571–582.

15. Carrasco, J. L. and Jover, L. (2003). Estimating the generalized concordance correlation coefficient through variance components. *Biometrics*, **59**, 849–858.

16. Carrasco, J. L., Jover, L., King, T. S., and Chinchilli, V. M. (2007). Comparison of concordance correlation coefficient estimating approaches with skewed data. *J. Biopharma. Statist.*, **17**, 673–684.

17. Casella, G. and Berger, R. (2002). *Statistical Inference*, 2nd ed. Duxbury Press, Pacific Grove, CA.

18. Chinchilli, V. M., Martel, J. K., Kumanyika, S., and Lloyd, T. (1996). A weighted concordance correlation coefficient for repeated measurement designs. *Biometrics*, **52**, 341–353.

19. Choudhary, P. K. (2007a). Semiparametric regression for assessing agreement using tolerance bands. *Comput. Statist. Data Anal.*, **51**, 6229–6241.

20. Choudhary, P. K. (2007b). A tolerance interval approach for assessment of agreement with left censored data. *J. Biopharm. Statist.*, **17**, 583–594.

21. Choudhary, P. K. (2008). A tolerance interval approach for assessment of agreement in method comparison studies with repeated measurements. *J. Statistical Plan. Inf.*, **138**, 1102–1115.

22. Choudhary, P. K. and Nagaraja, H. N. (2004). Measuring agreement in method comparison studies—review. In *Advances in Ranking and Selection, Multiple*

Comparisons, and Reliability N. Balakrishnan, N. Kannan, and H. N. Nagaraja, eds., pp. 215–244. Birkhauser, Boston.

23. Choudhary, P. K. and Nagaraja, H. N. (2005). Selecting the instrument closest to a gold standard. *J Statist. Plan. Inf.*, **129**, 229–237.

24. Choudhary, P. K. and Nagaraja, H. N. (2005). A two-stage procedure for selection and assessment of agreement of the best instrument with a gold standard. *Sequential Analy.*, **24**, 237–257.

25. Choudhary, P. K. and Nagaraja, H. N. (2005c). Assessment of agreement using intersection-union principle. *Biometr. J.*, **47**, 674–681.

26. Choudhary, P. K. and Nagaraja, H. N. (2007). Tests for assessment of agreement using probability criteria. *J. Statist. Plan. Inf.*, **137**, 279–290.

27. Choudhary, P. K. and Ng, H. K. T. (2006). A tolerance interval approach for assessment of agreement using regression models for mean and variance. *Biometrics*, **62**, 288–296.

28. Choudhary, P. K. and Yin, K. (2009). Bayesian and frequentist methodologies for analyzing method comparison studies with multiple methods. Under review for publication.

29. Davison, A. C. and Hinkley, D. V. (1997). *Bootstrap Methods and Their Application.* Cambridge University Press, New York.

30. Fay, M. P. (2005). Random marginal agreement coefficients: Rethinking the adjustment for chance when measuring agreement. *Biostatistics*, **6**, 171–180.

31. Fleiss, J. L. (1986). *The Design and Analysis of Clinical Experiments.* Wiley, New York.

32. Food and Drug Administration (FDA) (2001). *Guidance for Industry: Bioanalytical Method Validation.* http://www.fda.gov/cder/Guidance/.

33. Gilbert, P. (2006). *numDeriv: Accurate Numerical Derivatives.* R package version 2006.4-1.

34. Guo, Y. and Manatunga, A. K. (2007). Nonparametric estimation of the concordance correlation coefficient under univariate censoring. *Biometrics*, **83**, 164–172.

35. Guttman, I. (1988). Statistical tolerance regions. In *Encyclopedia of Statistical Sciences*, Vol. 9, S. Kotz, N. L. Johnson, and C. B. Read, eds., pp. 272–287. Wiley, New York.

36. Haber, M. J. and Barnhart, H. X. (2006). Coefficients of agreement for fixed observers. *Statist. Methods Med. Res.*, **15**, 255–271.

37. Haber, M. J. and Barnhart, H. X. (2008). A general approach to evaluating agreement between two observers or methods of measurement from quantitative data with replicated measurements. *Statist. Methods Med. Res.*, **17**, 151–169.

38. Haber, M. J., Barnhart, H. X., Song, J., and Gruden, J. (2005). Observer variability: A new approach in evaluating interobserver agreement. *J. Data Sci.*, **3**, 69–83.

39. Hardin, J. W. and Hilbe, J. M. (2003). *Generalized Estimating Equations.* Chapman & Hall/CRC, Boca Raton, FL.

40. Harris, I. R., Burch, B. D., and St. Laurent, R. T. (2001). A blended estimator for measure of agreement with a gold standard. *J. Agric. Biol. Environ. Statist.*, **6**, 326–339.

41. Hawkins, D. M. (2002). Diagnostics for conformity of paired quantitative measurements. *Statist. Med.*, **21**, 1913–1935.

42. Hutson, A. D., Wilson, D. C., and Geiser, E. A. (1998). Measuring relative agreement: Echocardiographer versus computer. *J. Agric. Biol. Environ. Statist.*, **3**, 163–174.

43. King, T. S. and Chinchilli, V. M. (2001a). A generalized concordance correlation coefficient for continuous and categorical data. *Statist. Med.*, **20**, 2131–2147.

44. King, T. S. and Chinchilli, V. M. (2001b). Robust estimators of the concordance

correlation coefficient. *J. Biopharm. Statist.*, **11**, 83–105.

45. King, T. S., Chinchilli, V. M., and Carrasco, J. L. (2007a). A repeated measures concordance correlation coefficient. *Statist. Med.*, **26**, 3095–3113.

46. King, T. S., Chinchilli, V. M., Wang, K.-L., and Carrasco, J. L. (2007b). A class of repeated measures concordance correlation coefficients. *J. Biopharm. Statist.*, **17**, 653–672.

47. Kraemer, H. C., Periyakoil, V. S., and Noda, A. (2002). Kappa coefficients in medical research. *Statist. Med.*, **21**, 2109–2129.

48. Krummenauer, F. (1999). Intraindividual scale comparison in clinical diagnostic methods: A review of elementary methods. *Biometr. J.*, **41**, 917–929.

49. Lee, J., Koh, D., and Ong, C. N. (1989). Statistical evaluation of agreement between two methods for measuring a quantitative variable. *Comput. Biol. Med.*, **19**, 61–70.

50. Lehmann, E. L. (1999). *Elements of Large Sample Theory.* Springer-Verlag, New York.

51. Li, R. and Chow, M. (2005). Evaluation of reproducibility for paired functional data. *J. Multivar. Anal.*, **93**, 81–101.

52. Liao, J. J. Z. (2003). An improved concordance correlation coefficient. *Pharm. Statist.*, **2**, 253–261.

53. Liao, J. J. Z. (2005). Agreement for curved data. *J. Biopharm. Statist.*, **15**, 195–203.

54. Lin, L., Hedayat, A. S., and Wu, W. (2007). A unified approach for assessing agreement for continuous and categorical data. *J. Biopharm. Statist.*, **17**, 629–652.

55. Lin, L. I. (1989). A concordance correlation coefficient to evaluate reproducibility. *Biometrics*, **45**, 255–268. Corrections: **56**; 324–325 (2000).

56. Lin, L. I. (1992). Assay validation using the concordance correlation coefficient. *Biometrics*, **48**, 599–604.

57. Lin, L. I. (2000). Total deviation index for measuring individual agreement with applications in laboratory performance and bioequivalence. *Statist. Med.*, **19**, 255–270.

58. Lin, L. I. (2003). Measuring agreement. In *Encyclopedia of Biopharmaceutical Statistics*, 2nd ed., pp. 561–567. Marcel Dekker, New York.

59. Lin, L. I. and Chinchilli, V. (1997). Rejoinder to the letter to the editor from Atkinson and Nevill. *Biometrics*, **53**, 777–778.

60. Lin, L. I., Hedayat, A. S., Sinha, B., and Yang, M. (2002). Statistical methods in assessing agreement: Models, issues, and tools. *J. Am. Statist. Assoc.*, **97**, 257–270.

61. Lin, S. C., Whipple, D. M., and Ho, C. S. (1998). Evaluation of statistical equivalence using limits of agreement and associated sample size calculation. *Commun. Statist. Theory Methods*, **27**, 1419–1432.

62. Liu, J.-P. and Chow, S.-C. (1997). A two one-sided tests procedure for assessment of individual bioequivalence. *J. Biopharm. Statist..*, **7**, 49–61.

63. Liu, X., Du, Y., Teresi, J., and Hasin, D. S. (2005). Concordance correlation in the measurements of time to event. *Statist. Med.*, **24**, 1409–1420.

64. McGraw, K. O. and Wong, S. P. (1996). Forming Inferences about some intraclass correlation coefficients. *Psychol. Methods*, **1**, 30–46, 390.

65. Müller, R. and Büttner, P. (1994). A critical discussion of intraclass correlation coefficients. *Statist. Med.*, **13**, 2465–2476.

66. Nickerson, C. A. (1997). Comment on "A concordance correlation coefficient to evaluate reproducibility". *Biometrics*, **53**, 1503–1507.

67. Pinheiro, J., Bates, D., DebRoy, S., Sarkar, D., and the R Core team (2008). *nlme: Linear and Nonlinear Mixed Effects Models.* R package version 3.1-88.

68. Pinheiro, J. C. and Bates, D. M. (2000). *Mixed-Effects Models in S and S-PLUS.* Springer-Verlag, New York.

69. Quan, H. and Shih, W. J. (1996). Assessing reproducibility by the within-subject coefficient of variation with random effects models. *Biometrics*, **52**, 1195–1203.

70. Quiroz, J. (2005). Assessment of equivalence using a concordance correlation coefficient in a repeated measurements design. *J. Biopharm. Statist.*, **15**, 913–928.

71. R Development Core Team (2008). *R: A Language and Environment for Statistical Computing*. R Foundation for Statistical Computing. Vienna, Austria.

72. Ryan, T. P. and Woodall, W. H. (2005). The most-cited statistical papers. *J. Appl. Statist.*, **32**, 461–474.

73. Shao, J. and Zhong, B. (2004). Assessing the agreement between two quantitative assays with repeated measurements. *J. Biopharma. Statist.*, **14**, 201–212.

74. Shoukri, M. M. (2004). *Measures of Interobserver Agreement*. Chapman & Hall/CRC, Boca Raton, FL.

75. Shrout, P. E. and Fleiss, J. L. (1979). Intraclass correlations: Uses in assessing reliability. *Psychol. Bull.*, **86**, 420–428.

76. St. Laurent, R. T. (1998). Evaluating agreement with a gold standard in method comparison studies. *Biometrics*, **54**, 537–545.

77. Wang, W. and Hwang, J. T. G. (2001). A nearly unbiased test for individual bioequivalence problems using probability criteria. *J. Statist. Plan. Inf.*, **99**, 41–58.

78. Williamson, J. M., Crawford, S. B., and Lin, H.-M. (2007). Resampling dependent concordance correlation coefficients. *J. Biopharm. Statist.*, **17**, 685–696.

79. Yin, K., Choudhary, P. K., Varghese, D., and Goodman, S. R. (2008). A Bayesian approach for sample size determination in method comparison studies. *Statist. Med.*, **27**, 2273–2289.

80. Zegers, F. E. (1991). Coefficients for Interrater Agreement. *Appl. Psychol. Meas.*, **15**, 321–333.

81. Zhong, B. and Shao, J. (2003). Evaluating the agreement of two quantitative assays with repeated measurements. *J. Biopharm. Statist.*, **13**, 75–86.

42 Kaplan–Meier Estimator I

Arthur V. Peterson, Jr.

42.1 Introduction

The *Kaplan–Meier* (KM) estimator, or *product-limit* estimator, of a distribution function or survival function is the censored-data generalization of the empirical distribution function.

42.2 Censored-Data Problem

The censored-data problem arises in many medical, engineering, and other settings, especially follow-up studies, where the outcome of interest is the *time* to some event, such as cancer recurrence, death of the patient, or machine failure. An example of such a follow-up study in a medical setting would be a clinical trial investigating the efficacy of a new treatment for lung cancer patients; an example in an engineering setting might be a life-testing experiment investigating the lifetime distribution of electric motors. In the censored-data problem, the (independent) outcomes $X_i \sim F_i(\cdot)$, $i = 1, 2, \ldots, n$, that are pertinent for inference on the distribution functions $F_i(\cdot)$, $i = 1, 2, \ldots, n$ are, unfortunately, not all fully observed. Some of them are partially observed, or *right-censored*, due to curtailment of the follow-up. The curtailment may be either of a planned or accidental nature; examples of censoring in the medical setting include loss to follow-up, dropout, and termination (or interim analysis) of the study.

Typically observed in these cases are

$$\begin{aligned} T_i &= \min(X_i, C_i) \\ \delta_i &= I[T_i = X_i], \end{aligned} \qquad (1)$$

that is, the smaller of the failure time of interest X_i and a censoring time C_i, and the indicator of whether the observed time T_i is the result of censoring ($T_i = C_i$) or not ($T_i = X_i$). Observations T_i for which $\delta_i = 0$ are called *censored times*, and observations T_i for which $\delta_i = 1$ are called *uncensored times*, or *failures*. The censoring times may be fixed or random. Although the problem is symmetric in X_i and C_i, the aim of the inference is the distribution of the X_i's; the role of the C_i's is that of interfering with full observation of the X_i's.

In censored-data problems the distribution function $F_i(\cdot)$ are often related by a regression model $F(\cdot|z_i)$, or specified as one of k distribution functions . This chapter discusses only the simplest case, the one-sample problem $F_1(\cdot) = \cdots = F_n(\cdot) = F(\cdot)$. The aim of the inference will be the common distribution function $F(\cdot)$ of the X_i's.

42.3 Kaplan–Meier Estimator

In the one-sample problem with censored data (t_i, δ_i), $i = 1, 2, \ldots, n$, the Kaplan–Meier [18] estimator of the (common) *survival function* (or *reliability function* in engineering settings) $S(\cdot) = P(X_i > \cdot)$ is

$$\hat{S}(t) = \begin{cases} \prod_{i:t_{i:n} \leq t} \left(\frac{n-i}{n-i+1}\right)^{\delta_{(i)}} & \text{for } t \leq t_{n:n} \\ \begin{pmatrix} 0 & \text{if } \delta_{(n)} = 1 \\ \text{undefined} & \text{if } \delta_{(n)} = 0 \end{pmatrix} & \text{for } t > t_{n:n}, \end{cases} \quad (2)$$

where the $t_{i:n}$, $i = 1, 2, \ldots, n$, denote the observed times t_i arranged in increasing order of magnitude $t_{1:n} \leq t_{2:n} \leq \cdots \leq t_{n:n}$, and where $\delta_{(i)}$ denotes the censoring indicator for $t_{i:n}$. In the case of ties among the $t_{i:n}$, the usual convention is that failures [$\delta_{(i)} = 1$] precede censorings [$\delta_{(i)} = 0$].

An alternative [but equal to (2)] expression for the KM estimator (where it is defined) is particularly useful in the presence of tied failure times:

$$\hat{S}(t) = \prod_{j:t_{(j)} \leq t} \left(1 - \frac{d_j}{n_j}\right), \quad (3)$$

where the $t_{(j)}$ denote the *ordered, distinct* failures $t_{(1)} \leq t_{(2)} \leq \cdots \leq t_{(k)}$, the d_j denote the number of failures at $t_{(j)}$, and the n_j denote the number of items $\#\{i : t_i \geq t\}$ still alive just before time t.

Table 1 illustrates the computation of the KM estimator, and Figure 1 displays the estimator, for a dataset of remission durations of leukemia patients. Here, the time X_i of interest is the time from remission to relapse, the data are $(t_i, \delta_i), i = 1, 2, \ldots, 21$, and it is desired to estimate the relapse-free survival function $P(X_i > t) = $ probability that an individual is relapse-free (i.e., still in remission) at time t following remission.

The KM estimator (2), like the empirical distribution function estimator, is a step function with jumps at those times t_i that are uncensored. If $\delta_i = 1$ for all i, $i = 1, 2, \ldots, n$ (i.e., no censoring occurs), the KM estimator reduces to a step function with jumps of height d_j/n at each of the $t_{(j)}$, $j = 1, 2, \ldots, k$, which is the usual empirical distribution function.

Some authors adopt the convention of defining the KM estimator to be zero for $t > t_{n:n}$ when $\delta_{(n)} = 0$. Whereas such a convention has advantages of definiteness and simplicity, it is arbitrary; it is usually best in data presentations to specify the undefined character of the KM estimator in this range rather than specify it to be zero. Of course, if we make the reasonable specification that the estimator retain the properties of a survival function in this range, then it must be nonincreasing, nonnegative, and right-continuous.

Under the nonpredictive-censoring assumption discussed below, the KM estimator can be motivated in several useful ways. This estimator is:

1. The generalized maximum-likelihood estimator [18] in the same sense that the empirical distribution function is in the case of uncensored data. (The sense in which the empirical distribution function, and its censored-data generalization the KM estimator, are "maximum-likelihood estimators" among the class of unrestricted distribution functions has been addressed by Kiefer and Wolfowitz [20], Johansen [6], and Scholz [20].)

2. The limit of life-table (data grouped in time intervals) *estimators* [4] as the time intervals increase in number and go to zero in length [18]. In fact, the central idea of the KM estimator as a limit of life-table estimators was present in the early actuarial literature [1].

Table 1: Illustration of computation of the KM estimator for the remission data of Freireich et al. [11] from a clinical trial in acute leukemia.[a]

j	Ordered Distinct Failure Times: $t_{(j)}$	Number of Individuals Alive Just before Time $t_{(j)}$: n_j	Number of Individuals Dying at Time $t_{(j)}$: d_j	Factor Contributed at $t_{(j)}$ to KM Estimator $(1 - \frac{d_j}{n_j})$	KM Estimator for $t \in [t_{(j)}, t_{(j+1)})$ $\hat{S}(t)$	Greenwood Variance Estimator $\widehat{\mathrm{Var}}(\hat{S}(t))$
1	6	21	3	18/21	0.857	0.0058
2	7	17	1	16/17	0.807	0.0076
3	10	15	1	14/15	0.753	0.0093
4	13	12	1	11/12	0.690	0.0114
5	16	11	1	10/11	0.627	0.0130
6	22	7	1	6/7	0.538	0.0164
7	23	6	1	5/6	0.448	0.0181
	35					Undefined after $t = 35$

[a] The (ordered) remission times in weeks on the 21 chemotherapy patients were 6, 6, 6, 6*, 7, 9*, 10, 10*, 11*, 13, 16, 17*, 19*, 20*, 22, 23, 25*, 32*, 32*, 34*, 35* (*denotes a censored observation).

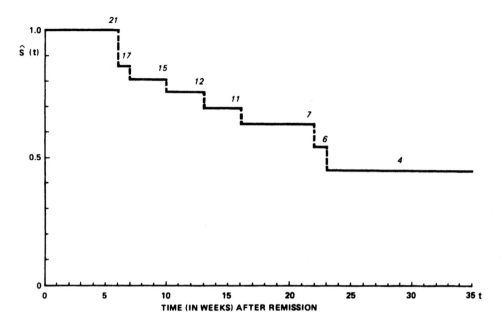

Figure 1: Kaplan–Meier estimator for the acute leukemia remission duration dataset of Table 1. (Indicated also are the numbers of individuals alive at various times t.)

3. (Related to item 2) the estimator obtained from a *product of estimators of conditional probabilities* [18].

4. The *"self-consistent" estimator* [5] $\hat{S}(\cdot)$ defined, by analogy with the empirical survival function in the case without censoring, as

$$\hat{S}(t) = \frac{1}{n}\left(\#\{t_i : t_i > t\} + \sum_{t_i \leq t} a_i(t)\right),$$

where the fractions

$$a_i(t) = \begin{cases} \hat{S}(t)/\hat{S}(t_i) & \text{if } \delta_i = 0 \\ 0 & \text{if } \delta_i = 1 \end{cases}$$

are estimates of $P(X_i > t | T_i = t_i, \delta_i)$.

5. The *redistribute-to-the-right estimator* [5], defined by an algorithm that starts with an empirical distribution that puts mass $1/n$ at each observed time t_i, and then moves the mass of each censored observation by distributing it equally to all observed times to the right of it.

6. A *natural function of two empirical subsurvival functions* [23]; that is, the survivor function $S(\cdot)$ of X can be expressed [under condition (4) below] as a certain function Φ of the subsurvival functions $S_0^*(t) \equiv P(T > t, \delta = 0)$ and $S_1^*(t) \equiv P(T > t, \delta = 1)$:

$$S(t) = \Phi[S_0^*(\cdot), S_1^*(\cdot), t].$$

The KM estimator $\hat{S}(t)$ is just $\Phi[\hat{S}_0^*(\cdot), \hat{S}_1^*(\cdot), t]$, where

$$\hat{S}_0^*(s) \equiv (1/n)\sum_{i=1}^n I[t_i > s, \delta = 0]$$

and

$$\hat{S}_1^*(s) \equiv (1/n)\sum_{i=1}^n I[t_i > s, \delta = 1]$$

are the empirical subsurvival functions for $S_0^*(\cdot)$ and $S_1^*(\cdot)$, respectively.

42.4 Appropriateness

Of crucial importance to the appropriateness of the K–M estimator (and of most other censored data methods as well) is that for each individual the censoring *must not be predictive of future* (*unobserved*) *failure*. Specifically, it must be true for each individual at each time t that

$$\Pr(X \in [t, t+dt) | X \geq t) \\ = \Pr(X \in [t, t+dt) | X \geq t, C \geq t); \quad (4)$$

that is, that the instantaneous probability of failure at time t given survival to t is unchanged by the added condition that censoring has not occurred up to time t (e.g., Kalbfleisch and MacKay [17]). As discussed in Chapter 5 of Kalbfleisch and Prentice [8], this condition is equivalent to specifying that for each individual the instantaneous probability of censoring does not depend on the future failure times of this or other individuals.

Unfortunately, the truth of (4) cannot be tested from the censored data (1) alone (Tsiatis [32] and many others). In practice, a judgment about the truth of (4) should be sought based on the best available understanding of the nature of the censoring. For example, end-of-study censoring might typically be expected to meet (4), whereas censoring that is a dropout due to factors related to imminence of failure (e.g., taking as censoring the time of termination of life testing of a machine that shows signs of overheating) would not be expected to meet (4). A judgment on whether certain loss-to-follow-up circumstances would be expected to satisfy (4) are typically difficult to make, even when the reasons for loss-to-follow-up are known, and thus provide one incentive for strong efforts toward complete follow-up in cohort studies. Inattention to the possibility that the censoring mechanism might be predictive of failure can be disastrous: the KM estimator can be grossly in error in the situation where censoring is predictive of failure [22].

42.5 Properties, Variance Estimators, Confidence Intervals, and Confidence Bands

Under random censorship the process $n^{1/2}[\hat{S}(\cdot) - S(\cdot)]$ has the asymptotic distribution of a Gaussian process (e.g., Breslow and Crowley [3]). Meier [21] discusses corresponding results for the case of fixed censorship. Other aspects of the asymptotic behavior of the KM estimator have been the subject of numerous investigations [1,8–10,25,32,33].

The asymptotic normality of $n^{1/2}(\hat{S}(t) - S(t))$ provides a basis for approximating the finite-sample distribution of the KM estimator $\hat{S}(t)$ by a normal distribution. Alternatively available is a maximum-likelihood estimator of this distribution, termed the *bootstrap* distribution [6].

In particular, estimators of the (finite-sample) variance of the KM estimator $\hat{S}(t)$ at a specified t are readily available. An estimate of the asymptotic variance of $\hat{S}(t)$ provides the well-known Greenwood [14] estimated variance for $\hat{S}(t)$:

$$\widehat{\mathrm{var}}(\hat{S}(t)) = \hat{S}^2(t) \sum_{t_{(j)} \leq t} \frac{d_j}{n_j(n_j - d_j)}.$$

Closely related to the Greenwood estimated variance is Efron's [6] bootstrap estimated variance, which is the variance of the KM estimator's bootstrap distribution. Also, a conservative estimator of the KM variance is discussed by Peto et al. [25].

Using the Greenwood estimated variance (or one of its alternatives), approximate confidence intervals for $S(t)$ can be obtained, based on the asymptotic normality either of $\hat{S}(t)$ itself [24,27,30] or of other functions, such as $\log[-\log \hat{S}(t)]$, that have no range restrictions and / or whose distribution may be more nearly normal.

Simultaneous confidence intervals, or *confidence bands*, for the survival function $S(\cdot)$, based on the asymptotic equivalence of the KM process $n^{1/2}(\hat{S}(\cdot) - S(\cdot))$ to Brownian motion processes, have been developed by Gillespie and Fisher [13] and by Hall and Wellner [5].

42.6 Nonparametric Quantile Estimation Based on the Kaplan–Meier Estimator

The entire estimated survival curve $\hat{S}(\cdot)$, together with standard errors or confidence intervals, is usually a good choice for the presentation of survival data with censoring. Nevertheless, summary statistics such as location estimates are sometimes also useful. With censored data the median, or 0.5 quantile, is a common choice as a location estimator. It is superior to the mean, which is highly sensitive to the right tail of the survival distribution, where estimation tends to be imprecise due to censoring. Other quantiles can be useful in summarizing different aspects of the estimated survival distribution.

In the censored-data problem, the maximum likelihood estimator for the pth quantile

$$F^{-1}(p) = S^{-1}(1-p)$$
$$\equiv \inf\{t : S(t) \leq 1-p\}$$

is conveniently available from the KM estimator $\hat{S}(\cdot)$:

$$\hat{S}^{-1}(1-p) \equiv \inf\{t : \hat{S}(t) \leq 1-p\}.$$

The asymptotic distribution of this quantile estimator has been determined by Sander [29] and Reid [27]. However, the asymptotic variance is a function of the failure distribution *density* at the point $S^{-1}(1-p)$. Because of the difficulty in estimating a density it is difficult to obtain from asymptotic results an approximate estimator for the variance of the quantile estimator. Methods for estimating the finite-sample variance of quantile estimators from censored data include the jackknife [22] and bootstrap [6] methods.

Approximate confidence limits for a pth quantile $S^{-1}(1-p)$ based on the asymptotic normality of $\hat{S}(t)$ for a range of t's have been proposed by Brookmeyer and Crowley [4], Emerson [8], and Simon and Lee [31]. Also, Efron [6] has proposed using percentiles of the bootstrap distribution of $\hat{S}^{-1}(1-p)$ for confidence limits for $S^{-1}(1-p)$.

Acknowledgments. This work was supported by Grants GM-28314 and CA-15704 from the National Institutes of Health.

References

1. Aalen, O. (1978). *Ann. Statist.*, **6**, 534–545.
2. Böhmer, P. E. (1912). *Rapports, Mémoires et Procès-verbaux de Septième Congrès International d'Actuaires*, Vol. 2, pp. 327–343, Amsterdam.
3. Breslow, N. E. and Crowley, J. (1974). *Ann. Statist.*, **2**, 437–453.
4. Brookmeyer, R. and Crowley, J. (1982). *Biometrics*, **38**, 29–41.
5. Cutler, S. J. and Ederer, F. (1958). *J. Chron. Dis.*, **8**, 699–713.
6. Efron, B. (1967). *Proc. 5th Berkeley Symp. Mathematics Statistics and Probability*, Vol. 4, pp. 831–853, Berkely, CA., University of California Press, Berkeley.
7. Efron, B. (1981). *J. Am. Statist. Assoc.*, **76**, 312–319.

8. Emerson, J. (1982). *Biometrics*, **38**, 17–27.

9. Földes, A. and Rejtö, L. (1979). *Asymptotic Properties of the Nonparametric Survival Curve Estimators under Variable Censoring*. Preprint of the Mathematical Institute of the Hungarian Academy of Sciences.

10. Földes, A. and Rejtö, L. (1981). *Ann. Statist.*, **9**, 122–129.

11. Földes, A., Rejtö, L., and Winter, B. B. (1980). *Periodica Math. Hung.*, **11**, 233–250.

12. Freireich, E. O., et al. (1963). *Blood*, **21**, 699–716. (An example of censored data in a medical follow-up setting.)

13. Gillespie, M. J. and Fisher, L. (1979). *Ann. Statist.*, **7**, 920–924.

14. Greenwood, M. (1926). The natural duration of cancer. *Reports on Public Health and Medical Subjects*, Vol. 33, pp. 1–26, H. M. Stationary Office, London. (Of historical interest, this paper presents the Greenwood estimator of the variance of the KM estimator.)

15. Hall, W. J. and Wellner, J. A. (1980). *Biometrika*, **67**, 133–143.

16. Johansen, S. (1978). *Scand. J. Statist.*, **5**, 195–199.

17. Kalbfleisch, J. D. and MacKay, R. J. (1979). *Biometrika*, **66**, 87–90.

18. Kalbfleisch, J. D. and Prentice, R. L. (1980). *The Statistical Analysis of Failure Time Data*. Wiley, New York. (Written for the practicing statistician, this very readable book provides an excellent treatment of the analysis of censored data. Topics include the Kaplan–Meier estimator, the comparison of survival curves, and regression analysis with censored data.)

19. Kaplan, E. L. and Meier, P. (1958). *J. Am. Statist. Assoc.*, **53**, 457–481. (More than 50 years later, this paper is still an informative and motivated description and discussion of the Kaplan–Meier estimator.)

20. Kiefer, J. and Wolfowitz, J. (1956). *Ann. Math. Statist.*, **27**, 887–906.

21. Meier, P. (1975). *Perspectives in Probability and Statistics*, J. Gani, ed. Applied Probability Trust, Sheffield, England, UK.

22. Miller, R. G. (1974). *Jackknifing Censored Data*. Technical Report 14, Department of Statistics, Stanford University, Stanford, CA.

23. Peterson, A. V., Jr. (1976). *Proc. Natl. Acad. Sci. (USA)*, **73**, 11–13.

24. Peterson, A. V., Jr. (1977). *J. Am. Statist. Assoc.*, **72**, 854–858.

25. Peto, R., et al. (1977). *Br. J. Cancer*, **35**, 1–39. (This popular paper includes a technical motivation, description, and illustration of the Kaplan–Meier estimator.)

26. Phadia, E. G. and Van Ryzin, J. (1980). *Ann. Statist.*, **8**, 673–678.

27. Reid, N. (1979). *Ann. Statist.*, **9**, 78–92.

28. Rothman, K. J. (1978). *J. Chronic Dis.*, **31**, 557–560.

29. Sander, J. (1975). *The Weak Convergence of Quantiles of the Product-Limit Estimator*. Technical Report 5, Deptartment of Statistics, Stanford University, Stanford, Calif.

30. Scholz, F. W. (1980). *Can. J. Statist.*, **8**, 193–203.

31. Simon, R. and Lee, Y. K. (1982). *Cancer Treat. Rep.*, **66**, 67–72.

32. Tsiatis, A. (1975). *Proc. Natl. Acad. Sci. (USA)*, **72**, 20–22.

33. Wellner, J. A. (1982). *Ann. Statist.*, **10**, 595–602.

34. Winter, B. B., Földes, A., and Rejtö, L. (1978). *Problems Control Inf. Theory*, **7**, 213–225.

43 Kaplan–Meier Estimator II

Grace L. Yang

43.1 Introduction

Kaplan and Meier [44] proposed a nonparametric estimator for estimating the survival probability $S(t) = P[X > t]$ of a nonnegative random variable X. The estimator is based on a right-censored sample. The term "survival" originated in biostatistics, where one measures the efficacy of a medical treatment by the patient's probability of survival beyond a specified time, e.g., the 5-year survival probability for a certain type of cancer.

More generally, X is a lifetime measurement, or the length of time to the occurrence of an event, or the first-passage time from one state to another, e.g., the lifetime of a human being in life-table constructions, or time to failure of a mechanical system, or durability of a product in reliability studies. In reliability theory, the survival probability is more appropriately called the *reliability function*. The application of the Kaplan–Meier estimator is not limited to lifetime measurements only. For instance, it is used in astronomy, in which X represents the luminosity of a star (see, e.g., Feigelson and Baber [28], Woodroofe [78]).

There are special sampling constraints in collecting lifetime data, which often result in censored measurements of X. Typically there is a data collection period for X. The lifetime X will be censored if the collection period is shorter than X. A simple example of medical follow-up study illustrates the nature of the right-censoring mechanisms. Consider heart transplant patients who enter into the study immediately after the surgery at time ν_j for $j = 1, 2, \ldots$ during the follow-up period $[0, T]$. Since the study terminates at a preset time T, the follow-up time of the jth patient is the minimum of $T - \nu_j$ and his survival time X_j. We say that X_j is *right-censored* if $X_j > T - \nu_j$. A typical dataset consists of some completely observed survival times and some right-censored ones. To estimate the survival probability, one could not simply ignore the censored observations without biasing the estimate.

Kaplan and Meier [44] proposed an estimator of the survival probability that accommodates right-censored observations. It is the estimator that has optimal large-sample properties. The paper stimulated tremendous interest in research on censored data and its applications. For a historical account, see Breslow [10].

Due to its product form and a certain "limit" relationship to the usual life-table estimates, Kaplan and Meier [44] called their estimator the "product limit" (PL) estimator and traced the idea of product limit back to Böhmer [9]. Andersen and Borgan [6] suggested that it dates back to Karup [45]. The product form of dependent terms makes the analysis challenging. Some results in the 1958 Kaplan–Meier paper are obtained by heuristic arguments; major theoretical development of the estimator and its generalizations followed its publication.

This chapter is an update of the ESS entry by Peterson [61], who provided a brief account of the development prior to 1982 and a computational illustration of the Kaplan–Meier estimator with a leukemia dataset.

43.2 The Right-Censoring Model

Let X be a nonnegative random variable representing the lifetime of an individual under investigation. Let $F(t) = P[X \leq t]$ be its distribution function (df) with $F(0) = 0$, and $S(t) = 1 - F(t)$ its *survival function*. Let C be a nonnegative random variable independent of X and with df F_c. Under right censoring one observes only $\delta = I[X \leq C]$, the indicator of the event $[X \leq C]$, and $Z = X \wedge C$, the minimum of X and C. Thus X is completely observable if and only if $X \leq C$, i.e., when $\delta = 1$; otherwise X is known to exceed the censoring variable C. The right-censoring model is the joint distribution of Z and δ given by

$$Q(t,1) = ZP[Z \leq t, \delta = 1]$$
$$= \int_0^t P[C \geq u]\, dF(u), \quad (1)$$
$$Q(t,0) = P[Z \leq t, \delta = 0]$$
$$= \int_0^t P[X \geq u]\, dF_c(u). \quad (2)$$

It follows that X will be completely observed with probability $Q(\infty, 1) = P[\delta = 1]$ and partially observed with probability $Q(\infty, 0) = P[\delta = 0]$. The integrals are the Lebesgue–Stieltjes integrals over $(0, t]$.

43.3 The Kaplan–Meier Estimator

Construction of the Kaplan–Meier (hereafter KM) estimator of S is constructed based on a sample of n independent random vectors (Z_j, δ_j) for $j = 1, \ldots, n$, where Z_j and δ_j have the distribution Q given by (1) and (2). Let $Z_{(1)} \leq Z_{(2)} \leq \cdots \leq Z_{(n)}$ denote the ordered values of the Z_j's, and $\delta_{[j]}$ the concomitant of $Z_{(j)}$, i.e., $\delta_{[k]} = \delta_i$ if $Z_{(k)} = Z_i$. The ties are ordered arbitrarily among themselves. The KM estimator $\hat{S}_n(t)$ is defined by the product

$$\hat{S}_n(t) = 1 - \hat{F}_n(t)$$
$$= \prod_{k=1}^n \left[1 - \frac{\delta_{[k]}}{n-k+1}\right]^{I[Z_{(k)} \leq t]},$$
$$0 \leq t < \infty. \quad (3)$$

This formula is self-adjusted for tied $Z_{(j)}$. The estimator $\hat{S}_n(t)$ is a right-continuous decreasing step function, and is strictly positive on $[Z_{(n)}, \infty)$ if the largest observation is censored ($\delta_{[n]} = 0$). Some authors set $\hat{S}_n(t)$ equal to 0 on $[Z_{(n)}, \infty)$ regardless of whether $\delta_{[n]} = 0$ or 1. Different versions affect the convergence and bias, to be discussed later.

An alternative way of constructing the KM estimator is to use the self-consistency criterion [27,71].

43.4 Model Identifiability

Identifiability addresses the question: If the distribution Q, given in (1) and (2), is completely known, can one determine the survival function S? If S cannot be determined in this extreme situation of completely known Q, one would not expect to obtain a reasonable estimate for S from an estimated Q. For instance, it is known that if S is not identifiable under the model Q, then S cannot be estimated consistently.

The answer is negative if X and C are not stochastically independent, for then there will be multiple solutions of S that satisfy the model Q. For a related discussion, see Tsiatis [70]. If X and C are independent, then $S(t)$ is identifiable under Q for t in the interval $[0, \beta_z]$, where β_z is the smaller of the two upper boundaries β and β_c of X and C, respectively. Because of lack of identifiability in the absence of independence, the assumption of independence cannot be tested from the data; see, e.g., Robertson and Uppuluri [65]. Justification of independence has to come from the physical interpretation of the model.

The study of identifiability as well as of the KM estimator is greatly facilitated by the use of the cumulative hazard function $\Lambda(t)$ of X (Aalen [1]) and the Doléans–Dadé exponential formula [25]. The cumulative hazard function is best understood by first defining it for a discrete random variable X. Let $F(t_-) = \lim_{s \uparrow t} F(s) = P[X < t]$, so that the survival function $S(t_-) = 1 - F(t_-) = P[X \geq t]$. Let $\Delta f(t)$ denote the difference $f(t) - f(t_-)$ of a function $f(t)$. Suppose that X takes on positive values $x_1, x_2, x_3, \ldots$; the hazard rate of X at time t is defined by the conditional probability:

$$P[X = t | X \geq t]$$
$$= \frac{\Delta F(t)}{S(t_-)}$$
$$= \begin{cases} P[X = x_k]/P[X \geq x_k] \\ \quad \text{if } t = x_k, \\ \quad k = 1, 2, \ldots, \\ 0 \quad \text{otherwise.} \end{cases}$$

This is called the *force of mortality* in life tables and the *failure rate* in reliability theory. It is a trivial fact but important to note that the conditioning event is $[X \geq x_k]$ and not $[X > x_k]$; the latter would make the conditional probability zero and useless.

The cumulative hazard function (chf) of X at time t is the sum of the hazard rates

$$\Lambda(t) = \sum_{u \leq t} \frac{\Delta F(u)}{S(u_-)} \quad \text{for} \quad t \geq 0, \quad (4)$$

with the convention that $0/0 = 0$. In terms of the difference notation Δ, the hazard rate is

$$\Delta \Lambda(t) = \frac{\Delta F(t)}{S(t_-)},$$
$$\Lambda(t) = \sum_{u \leq t} \Delta \Lambda(u), \quad t \geq 0. \quad (5)$$

The functions $\Lambda(t)$ and $S(t)$ determine each other uniquely, and $S(t)$ is the product of conditional probabilities

$$S(t) = \prod_{0 \leq u \leq t} [1 - \Delta \Lambda(u)], \quad t \geq 0. \quad (6)$$

The general definition of cumulative hazard function for an arbitrary distribution function F is given by the Lebesgue–Stieltjes integral

$$\Lambda(t) = \int_0^t \frac{1}{S(u_-)} dF(u), \quad t \geq 0, \quad (7)$$

or equivalently

$$S(t) = 1 - \int_0^t S(u_-) d\Lambda(u). \quad (8)$$

43.5 The Inversion Formula

The equivalence of F (or S) and Λ is proved by the following inversion formula deduced from the Doléans–Dadé exponential formula [25]. Given $\Lambda(u)$, the unique solution of (8) is

$$S(t) = \exp[-\Lambda^c(t)] \prod_{u \leq t} [1 - \Delta \Lambda(u)],$$
$$t \geq 0, \quad (9)$$

where $\Lambda^c(t) = \Lambda(t) - \sum_{u \leq t} \Delta\Lambda(u)$ is the continuous part of $\Lambda(t)$. For a proof, see Liptser and Shiryayev [54, Lemma 18.8]. See also Aalen and Johansen [2], Gill [30], Wellner [76], and Shorack and Wellner [67].

If $S(t)$ is a step function, (9) reduces to (6). If $S(t)$ is continuous, $\Lambda(t) = -\log S(t)$, a familiar form. Furthermore, if $S(t)$ is differentiable, the derivative $\lambda(t) = d\Lambda(t)/dt$ exists and is called the *hazard rate*; of course, it is no longer a conditional probability.

43.6 Identifiability

Under independence, the survival function $S_z(t)$ of $Z = \min(X, C)$ is the product

$$S_z(t) = S(t)S_c(t). \quad (10)$$

Thus

$$Q(t,1) = \int_0^t S_c(u_-)dF(u)$$
$$= \int_0^t S_z(u_-)d\Lambda(u). \quad (11)$$

Equating $Q(t, 1)$ to the last integral and solving for Λ yields

$$\Lambda(t) = \int_0^t S_z^{-1}(u_-)dQ(u,1). \quad (12)$$

Thus the model Q determines uniquely the cumulative hazard function of X in the interval $0 \leq t < \beta_z$. Applying the inversion formula (9), one immediately obtains the survival function $S(t)$ for t in the interval support of Z.

43.7 Finite-Sample Properties of the KM Estimator

The properties of the KM estimator $\hat{S}_n(t)$ are known mostly for large samples. Finite-sample results have been difficult to obtain. It is known that $\hat{S}_n(t)$ is the maximum-likelihood estimate; see, e.g., Shorack and Wellner [67, p. 333], Johansen [42]. The estimator is biased. Using the martingale method, Gill [30] (1980) obtained a formula for the bias $b(t) = E\hat{F}_n(t) - F(t)$ [his (3.2.16)], and it is generally negative. Employing a different approach of reversed supermartingales, Stute and Wang [68] showed under very weak conditions that the bias $b(t) \leq 0$ for any finite sample size n and any fixed $t < \beta_z$, the upper boundary of Z, and that $E\hat{F}_n(t)$ converges from below to $F(t)$, provided certain conditions on the boundary β_z are met.

Explicit formulas for the finite-sample moments of the KM estimator are available for the proportional hazards model. They were obtained by Chen et al. [16] for the version of KM estimator $\hat{S}_n(t)$ that assumes a zero value for $t \geq Z_{(n)}$. Wellner [76] showed that the exact moments for $\hat{S}_n(t)$ defined by (3) generally yield smaller biases and variances. The proportional hazards model assumes that in the model Q [defined in (1) and (2)], the censoring distribution $S_c(t) = [S(t)]^\theta$, where $S(t)$ is continuous and θ a positive number.

For arbitrary continuous $S(t)$ and $S_c(t)$, Chang [15] obtained formulas for the second, third, and fourth moments with accuracy $O(n^{-2})$.

43.8 Consistency, Asymptotic Normality, and the Strong Law

The KM estimator is uniformly strongly consistent and asymptotically normal for arbitrary distributions S and S_c. More precisely, as $n \to \infty$,

$$\sup_{0 \leq t < \beta_z} |\hat{S}_n(t) - S(t)| \to 0$$

with probability 1.

Also, for $t \in [0, T]$ with $S_z(T) > 0$, the process $\sqrt{n}[\hat{S}_n(t) - S(t)]$ converges in distribution to a Gaussian process $S(t)W(t)$, where $W(t)$ is a Gaussian process having $EW(t) = 0$ and covariance function

$$\begin{aligned} C(s,t) &= EW^2(s) \\ &= \int_0^s \frac{dQ(u,1)}{S_z^2(u_-)[1 - \Delta\Lambda(u)]}, \\ & \quad 0 \le s \le t \le T. \end{aligned}$$

The limit process $S(t)W(t)$ has fixed discontinuities at jumps of $S(t)$. Thus the asymptotic variance $V(t)$ of the KM estimator is given by $\mathrm{Var}(\sqrt{n}\hat{S}_n(t)) = S^2(t)C(t,t)$. According to (12), it can be written in terms of Λ as

$$V(t) = \begin{cases} S^2(t) \int_0^t \frac{d\Lambda(u)}{S_z(u_-)(1-\Delta\Lambda(u))}, \\ \quad \text{general } S \text{ and } S_c, \\ S^2(t) \int_0^t S_z^{-1}(u) d\Lambda(u), \\ \quad \text{continuous } S \text{ and} \\ \quad \text{general } S_c, \\ S(t)F(t), \\ \quad \text{continuous } S \text{ and} \\ \quad \text{under no censoring} \\ \quad [S_c(u) = 1 \text{ for all finite } u]. \end{cases}$$
(13)

Since $S_z(u) = S(u)S_c(u)$, censoring increases the variance by a factor of $S_c^{-1}(u)$ in the integrand, and the increase is progressively worse as u increases.

Normal approximation can be poor in the case of heavy censoring, particularly near the upper boundary. In that case, a Poisson approximation may work better [76].

There is a large literature on asymptotic results of $\hat{S}_n(t)$ under various conditions. We mention several that use different approaches. Breslow and Crowley [11] used the traditional method of weak convergence of stochastic processes. Aalen [1] reformulated the estimation problem in terms of counting processes and martingales. Lo and Singh [55] obtained a representation of the KM estimator by a sum of bounded i.i.d. random variables with a negligible remainder term. From this asymptotic normality readily follows. Refinements of the Lo–Singh representation are provided by Major and Rejtö [56]. Generalizations using the martingale approach appeared in Gill [30,31]. The proofs given in Shorack and Wellner [67, Chap. 7] rely heavily on special construction (see Csörgö and Révész [19], Komlós et al. [47,48]) and the Doléans–Dadé exponential. The estimate can also be studied by using product integration (see Gill and Johansen [33]).

Using reversed supermartingales, Stute and Wang [68] proved the strong law and convergence in mean of the KM integrals $\int_0^\infty \phi(u)\hat{F}_n(du)$, where ϕ is an integrable function with respect to F. The result is very general and implies the strong consistency of $\hat{F}_n$ or $\hat{S}_n$ and many other functions of $\hat{F}_n$. Results on rates of convergence and strong approximations include Burke et al. [12] and Csörgö and Horváth [21]. Functional laws of the iterated logarithm for the KM estimator have been investigated by Gu and Lai [36] and others.

43.9 Asymptotic Optimality

Among all regular estimating sequences of $F(t)$ for $t \in [0,T]$ with $T < \beta_z$, the asymptotic normal distribution of $\hat{F}_n(t)$ has the smallest variance in the sense of the Hájek—Le Cam convolution theorem [37,51]. That is, the limiting distribution of any regular estimating sequence is a convolution of the limiting distribution of $\sqrt{n}[\hat{F}_n(t) - F(t)]$ with another distribution. This was proved by Wellner [75]. Furthermore, the estimate $\hat{F}_n(t)$ is asymptotic minimax with respect to bowl-shaped loss functions in the sense of the Hájek–Le Cam asymptotic minimax theorem (see Hájek [38], Le Cam [51,52]), as shown by Wellner [75] and Millar [58]. Similar nonparametric results for the uncensored case were obtained by Beran [8] and Millar [57]. Discussions of asymptotic optimality may be found in Le Cam and Yang [53].

43.10 Counting Process Formulation and Martingales

Aalen [1] introduced multivariate counting processes into the study of lifetime data under a variety of censoring mechanisms, of which the right-censoring model is a special case. This extremely fruitful approach will be sketched. Let

$$N(t) = \sum_{j=1}^{n} I[Z_j \leq t, \delta_j = 1], \quad (14)$$

$$Y(t) = \sum_{j=1}^{n} I[Z_j \geq t], \quad t \geq 0. \quad (15)$$

These two counting processes count the number of failures and the number of individuals at risk of failure at time t, respectively. The sample counterpart of the model $Q(t, 1)$ and the survival function $S_z(t)$ are the empirical distribution functions

$$Q_n(t, 1) = N(t)/n,$$
$$S_{n,z}(t_-) = Y(t)/n, \quad (16)$$

respectively. Substituting (16) in (12), we obtain the empirical cumulative hazard function

$$\Lambda_n(t) = \int_0^t \frac{dN(u)}{Y(u)} = \int_0^t \frac{dQ_n(t,1)}{S_{n,z}(u_-)}, \quad (17)$$

with differences $\Delta\Lambda_n(t) = \delta_{[k]}/(n-k+1)$ if $t = Z_{(k)}$, and 0 otherwise. The cumulative hazard $\Lambda_n(t)$ is a step function. Applying the inversion formula (9) yields the corresponding survival function

$$\hat{S}_n(t) = \prod_{0 \leq s \leq t} [1 - \Delta\Lambda_n(s)], \quad t \geq 0, \quad (18)$$

which is precisely the KM estimator (3) in a slightly different form. Aalen showed that

$$M(t) = \frac{1}{\sqrt{n}}\left(N(t) - \int_0^t Y(u) d\Lambda(u)\right)$$

is a square-integrable martingale with respect to an appropriately chosen collection of nondecreasing σfields $\{F_t; t \geq 0\}$ and that the difference between the empirical and the true cumulative hazard function can be written as

$$\sqrt{n}[\Lambda_n(t) - \Lambda(t)]$$
$$= \sqrt{n}\left(\int_0^t \frac{dQ_n(t,1)}{S_{n,z}(u_-)} - \int_0^t \frac{dQ(t,1)}{S_z(u_-)}\right)$$
$$= \int_0^t \frac{dM(u)}{S_{n,z}(u_-)}.$$

Since $S_{n,z}(u_-)$ is predictable, the last integral and hence $\sqrt{n}[\Lambda_n(t) - \Lambda(t)]$ is a martingale with respect to $\{F_t\}$. It converges weakly to a Gaussian process for $t \in [0, T]$, as can be shown by applying the central limit theorem for martingales ([63],[41]).

For an historical account of the introduction of counting process theory into survival analysis, see Aalen [3] and Gill [32].

One can derive the asymptotic normality for $\hat{S}_n(t)$ from that of the estimates $S_n(t) = \exp[-\Lambda_n(t)]$. This $S_n(t)$ is not the KM estimate $\hat{S}_n(t)$. The difference between the two is

$$0 \leq S_n(t) - \hat{S}_n(t)$$
$$= \exp\left(-\sum_{s \leq t} \Delta\Lambda_n(s)\right)$$
$$- \prod_{0 \leq s \leq t} [1 - \Delta\Lambda_n(s)]$$
$$\leq \frac{1}{2}\sum_{j=1}^{B}[\Delta\Lambda_n(Z_{(j)})]^2$$
$$\leq \frac{1}{2}\sum_{j=1}^{B}(n-j+1)^{-2}$$
$$\leq \frac{1}{2n}\frac{\overline{B}}{n^{-1}+1-\overline{B}},$$

where B is the binomial number of Z_j that are $\leq t$, and $\overline{B}$ is its proportion. The difference is therefore asymptotically negligible. An early application of $S_n(t)$ can be found in Altshuler [5], which studies the competing risks of death in experimental animals that were exposed to a variety of carcinogens.

Alternatively, asymptotic normality of $\hat{S}_n(t)$ can be proved directly by using martingale methods.

43.11 Confidence Intervals and Confidence Bands

Asymptotic normality can be used to construct confidence intervals for $S(t)$ for any fixed t in the range of normal convergence.

An estimated variance can be obtained by replacing unknown quantities in $V(t)$ by the respective KM estimate (18) and the empirical distributions $S_{n,z}$ and $Q_n(u,1)$ defined in (16). Since $\hat{S}_n$ and Λ_n are step functions, the first variance formula in (13) should be used. The resulting variance estimate is

$$\hat{V}(t) = [\hat{S}_n(t)]^2 \sum_j \frac{\Delta \Lambda_n(Z_{(j)})}{1 - \Delta \Lambda_n(Z_{(j)})}$$
$$= [\hat{S}_n(t)]^2 \sum_i \frac{d_i}{r_i(r_i - d_i)}. \quad (19)$$

The first sum extends to all $Z_{(j)} \leq t$. The second expression is the classical Greenwood formula [35] where the sum is over those $Z_{(j)} \leq t$ that are uncensored, with d_i denoting the number of failures among the r_i individuals who are at risk at time $Z_{(i)}$. The approximated $(1-\alpha)\%$ confidence intervals have confidence limits

$$\hat{S}_n(t) \pm c_\alpha [\hat{V}(t)]^{1/2},$$

where c_α is the $(\alpha/2)$th percentile of the standard normal distribution. Further discussion on Greenwood's formula and other modifications of the variance estimate may be found in Cox and Oakes [18]. For a variance estimate obtained by bootstrap, see Akritas [4].

The asymptotic Gaussian process can be transformed into a Brownian bridge, as noted by Efron [27]. Using this transformation, confidence bands for $S(t)$ over a fixed interval $[0,T]$ for T smaller than the largest uncensored observation have been constructed; see, e.g., Gillespie and Fisher [34], Hall and Wellner [39], Gill [30], Burke et al. [12], Nair [60], and Csörgő and Horváth [21]. See also Gill [3132] and Ying [81] for discussion of the extension of the bands.

43.12 The Quantile Process of the KM Estimator

A systematic exposition of the quantiles of the KM estimator is given in Csörgő [20, Chap. 8]. Weak convergence and confidence intervals for the median were studied by Sander [66]. See also Reid [64]. Cheng [17] obtained a Bahadur-type representation for quantiles.

43.13 Large Deviations

A Cramér type of large-deviation result for the KM estimator is available. Veraverbeke [72] showed that the relative error of the normal approximation of the upper tail probability of the KM estimator tends to zero as $n \to \infty$ in a manner similar to that of the i.i.d. uncensored case. Another type of large-deviation result has been obtained by Dinwoodie [24], who showed that the tail of the KM estimator decays at an exponential rate as $n \to \infty$.

43.14 Generalizations

Analytical tools that have become available for studying the KM estimator are instrumental for developing statistical methods for analyzing more complex censored data. Parallel to the development of the usual uncensored empirical distribution function, the KM estimator has been used in goodness of fit [46], in biometric functions [79], in reliability theory [26], in constructing optimum minimum distance estimators, [80],[82], in estimating regression coefficients [49], in testing new better than used [50], in cross-sectional sampling [40], and in a variety of censored data [71], including the right-censoring-left-truncation model [78],[74] and the doubly censored model [14], in which the lifetime is subject to either right or left random censoring.

In a different direction, Zhou [83] relaxed the i.i.d. condition by allowing the censoring variables to have different distributions, and proved the strong consistency of the KM estimator.

The two-dimensional extension of the KM estimator has been investigated by several authors, e.g., Campbell [13], Tsai et al. [69], and Dabrowska [23]. The extension refers to nonparametric estimation of a bivariate lifetime distribution in which each component lifetime is subject to a possible right censoring. In the estimate of the bivariate distribution, the KM estimator typically appears as the one-dimensional marginal distributions. Although these bivariate estimates are known to be consistent, Pruitt [62] pointed out a deficiency: Some of the better-known bivariate estimates may assume negative values for any finite sample size and therefore cannot be a proper survival distribution. The problem is complex (see Ref. [62]).

The literature is fast growing; the references provided here are necessarily incomplete. The reader may consult *The Current Index to Statistics* and books either on survival analysis or with extensive discussion on the KM estimator. The latter include Refs. 59, 43, 18, 67, 29, 7 and 32.

References

1. Aalen, O. O. (1978). Non-parametric inference for a family of counting processes. *Ann. Statist.*, **6**, 701–726.

2. Aalen, O. O. and Johansen, S. (1978). An empirical transition matrix for non-homogeneous Markov chains based on censored observations. *Scand. J. Statist.*, **5**, 141–150.

3. Aalen, O. O. (1997). Counting processes and dynamic modeling. In *Festschrift for Lucien Le Cam: Research Papers in Probability and Statistics*, D. Pollard, E. Torgersen, and G. Yang, eds. Springer, New York.

4. Akritas, M. G. (1986). Bootstrapping the Kaplan—Meier estimator. *J. Am. Statist. Assoc.*, **81**, 1032–1038.

5. Altshuler, B. (1970). Theory for the measurement of competing risks in animal experiments. *Math. Biosci.*, **6**, 1–11.

6. Andersen, P. K. and Borgan, O. (1985). Counting process models for life history data: a review. *Scand. J. Statist.*, **12**, 97–158.

7. Andersen, P. K., Borgan, O., Gill, R. D., and Keiding, N. (1993). *Statistical Models Based on Counting Processes*. Springer-Verlag, New York.

8. Beran, R. (1977). Estimating a distribution function. *Ann. Statist.*, **5**, 400–404.

9. Böhmer, P. E. (1912). Theorie der unabhängigen Wahrscheinlichkeiten. *Rapp. Mém. et Procés-Verbaux* 7e *Congr. Int. Act.*, Amsterdam, Vol. 2, pp. 327–343.

10. Breslow, N. E. (1991). Introduction to Kaplan and Meier (1958) nonparametric estimation from incomplete observations. In *Breakthroughs in Statistics*, S. Kotz and N. L. Johnson, eds., vol. II, pp. 311–318, Springer-Verlag, New York.

11. Breslow, N. and Crowley, J. (1974). A large sample study of the life table and product limit estimates under random censorship. *Ann. Statist.*, **2**, 437–453.

12. Burke, M. D., Csörgö, S., and Horvath, L. (1981). Strong approximations of some

biometric estimates under random censorship. *Z. Wahrsch. Verw. Geb.*, **56**, 87–112.

13. Campbell, G. (1981). Nonparametric bivariate estimation with randomly censored data. *Biometrika*, **68**, 417–422.

14. Chang, M. N. and Yang, G. L. (1987). Strong consistency of a nonparametric estimator of the survival function with doubly censored data. *Ann. Statist.*, **15**, 1536–1547.

15. Chang, M. N. (1991). Moments of the Kaplan–Meier estimator. *Sankhya A*, **53**, 27–50.

16. Chen, Y. Y., Hollander, M., and Lansberg, N. A. (1982). Small-sample results for the Kaplan–Meier estimator. *J. Am. Statist. Assoc.*, **77**, 141–144.

17. Cheng, K. F. (1984). On almost sure representation for quantiles of the product limit estimator with applications. *Sankhya A*, **46**, 426–443.

18. Cox, D. R. and Oakes, D. (1984). *Analysis of Survival Data*. Chapman & Hall, New York.

19. Csörgö, M. and Révész, P. (1975). Some notes on the empirical distribution function and the quantile process. In *Limit Theorems of Probability Theory*, Révész, P. ed., Colloq. Math. Soc. J. Bolyai 11, pp. 59–71, North-Holland, Amsterdam.

20. Csörgö, M. (1983). *Quantile Processes with Statistical Applications*. SIAM, Philadelphia.

21. Csörgö, S. and Horvath, L. (1983). The rate of strong uniform consistency for the product-limit estimator. *Z. Wahrsch. Verw. Geb.*, **62**, 411–426.

22. Csörgö, S. and Horváth, L. (1986). Confidence bands from censored samples. *Can. J. Statist.*, **14**, 131–144.

23. Dabrowska, D. M. (1988). Kaplan–Meier estimate on the plane. *Ann. Statist.*, **16**, 1475–1489.

24. Dinwoodie, I. H. (1993). Large deviations for censored data. *Ann. Statist.*, **21**, 1608–1620.

25. Doléans-Dadé, C. (1970). Quelques applications de la formula de changement de variables pour les semimartingales. *Z. Wahrsch. Verw. Geb.*, **16**, 181–194.

26. Doss, H., Freitag, S., and Proschan, F. (1989). Estimating jointly system and component reliabilities using a mutual censorship approach. *Ann. Statist.*, **17**, 764–782.

27. Efron, B. (1967). The two sample problem with censored data. *Proc. 5th Symp. Mathematical Statistics and Probability*, Vol. 4, pp. 831–853, Berkely, CA., L. LeCam and J. Neyman, eds. University of California Press, pp. 831–853.

28. Feigelson, E. D. and Baber, G. J., eds. (1992). *Statistical Challenges in Modern Astronomy*. Springer-Verlag, New York.

29. Fleming, T. R. and Harrington, D. P. (1991). *Counting Processes and Survival Analysis*. Wiley, New York.

30. Gill, R. D. (1980). *Censoring and Stochastic Integrals*. Mathematical Centre Tracts, Vol. 124. Amsterdam.

31. Gill, R. D. (1983). Large sample behavior of the product-limit estimator on the whole line. *Ann. Statist.*, **11**, 49–58.

32. Gill, R. (1994). Lectures on survival analysis. In *École d'Eté de Probabilitiés de Saint Flour XXII-1992*, P. Bernard, ed. *Springer Lecture Notes in Math.* 1581, pp. 115–242.

33. Gill, R. and Johansen, S. (1990). A survey of product-integration with a view towards application in survival analysis. *Ann. Statist.*, **18**, 1501–1555.

34. Gillespie, M. and Fisher, L. (1979). Confidence bands for the Kaplan–Meier survival curve estimate. *Ann. Statist.*, **7**, 920–924.

35. Greenwood, M. (1926). The natural duration of cancer. In *Reports on Public Health and Medical Subjects*, Vol. 33. H. M. Stationary Office, London.

36. Gu, M. G. and Lai, T. L. (1990). Functional laws of the iterated logarithm for the product-limit estimator of a distribution function under random censorship or truncation. *Ann. Probab.*, **18**, 160–189.

37. Hájek, J. (1970). A characterization of limiting distributions of regular estimates. *Z. Wahrsch. Verw. Geb.*, **14**, 323–330.

38. Hájek, J. (1972). Local asymptotic minimax and admissibility in estimation. *Proc. 6 Symp. Mathematical Statistics and Probability*, Vol. 1, pp. 175—194, L. Le Cam and J. Neyman, eds. University of California Press, Berkeley, CA.

39. Hall, W. J. and Wellner, J. A. (1980). Confidence bands for a survival curve from censored data. *Biometrika*, **67**, 133–143.

40. He, S. and Yang, G. L. (1995). Estimating a lifetime distribution under different sampling plans. In *Statistical Decision Theory and Related Topics V*, J. Berger and S. Gupta, eds., pp. 73–85, Springer-Verlag, New York.

41. Helland, I. S. (1982). Central limit theorems for martingales with discrete or continuous time. *Scand. J. Statist.*, **9**, 79–94.

42. Johansen, S. (1978). The product limit estimator as maximum likelihood estimator. *Scand. J. Statist.*, **5**, 195–199.

43. Kalbfleisch, J. D. and Prentice, R. L. (1980). *The Statistical Analysis of Failure Time Data*. Wiley, New York.

44. Kaplan, E. L. and Meier, P. (1958). Nonparametric estimation from incomplete observations. *J. Am. Statist. Assoc.*, **53**, 457–481.

45. Karup, I. (1893). *Die Finanzlage der Gothaischen Staatsdiener-Wittwen-Societät*. Dresden.

46. Kim, J. H. (1993). Chi-square goodness-of-fit tests for randomly censored data. *Ann. Statist.*, **21**, 1621–1639.

47. Komlós, J., Major, P., and Tusnády, G. (1975). An approximation of partial sums of independent rv's and the sample distribution function, I. *Z. Wahrsch. Verw. Geb.*, **32** 111–131.

48. Komlós, J., Major, P., and Tusnády, G. (1976). An approximation of partial sums of independent rv's and the sample distribution function, II. *Z. Wahrsch. Verw. Geb.*, **34**, 33–58.

49. Koul, H., Susarla, V., and Van Ryzin, J. (1981). Regression analysis with randomly right-censored data. *Sequential Anal.*, **5**, 85–92.

50. Kumazawa, Y. (1987). On testing whether new is better than using randomly censored data. *Ann. Statist.*, **15**, 420–426.

51. Le Cam, L. (1972). Limits of experiments. *Proc. 6th Symp. Mathematics Statistics and Probability*, Vol. 1, pp. 245–261, Berkeley, CA., L. Le Cam and J. Neyman, eds. University of California Press, Berkeley.

52. Le Cam, L. (1979). On a theorem of Hájek. In *Contributions to Statistics*, Jaroslav Hájek Memorial Volume, J. Jureckova, ed., pp. 119–135, Reidel, Dordrecht.

53. Le Cam, L. and Yang, G. L. (1990). *Asymptotics in Statistics: Some Basic Concepts*. Springer-Verlag, New York.

54. Lipster, R. S. and A. N. Shiryayev (1978). *Statistics of Random Processes, II: Applications*. Springer-Verlag, Berlin.

55. Lo, S. H. and Singh, K. (1986). The product-limit estimator and the bootstrap: some asymptotic representations. *Probab. Theory Rel. Fields*, **71**, 455–465.

56. Major, P. and Rejtö, L. (1988). Strong embedding of the estimator of the distribution function under random censorship. *Ann. Statist.*, **16**, 1113–1132.

57. Millar, P. W. (1979). Asymptotic minimax theorems for the sample distribution. *Z. Wahrsch. Verw. Geb.*, **48**, 233–252.

58. Millar, P. W. (1983). The minimax principle in asymptotic theory. In *École d'Eté de Probabilitiés de Saint Flour XI*, P. L. Hennequin, ed. *Lecture Notes in Math.* 976, pp. 76–262, Springer-Verlag, New York.

59. Miller, R. G., Jr.,Gong, G., and Munoz, A. (1981). *Survival Analysis*. Wiley, New York.

60. Nair, V. N. (1984). Confidence bands for survival functions with censored data: A comparative study. *Technometrics*, **14**, 265–275.

61. Peterson, A. V., Jr. (1983). Kaplan–Meier estimator. In *Encyclopedia of Statistical Sciences*, Vol. 4, pp. 346–352, Wiley, New York.

62. Pruitt, R. C. (1991). On negative mass assigned by the bivariate Kaplan–Meier estimator. *Ann. Statist.*, **19**, 443–453.

63. Rebolledo, R. (1980). Central limit theorems for local martingales. *Z. Wahrsch. Verw. Geb.*, **51**, 269–286.

64. Reid, N. (1981). Estimating the median survival time. *Biometrika*, **68**, 601–608.

65. Robertson, J. B. and Uppuluri, V. R. R. (1984). A generalized Kaplan–Meier estimator. *Ann. Statist.*, **12**, 366–371.

66. Sander, J. M. (1975). *The Weak Convergence of Quantiles of the Product Limit Estimator*. Tech. Rep., 5, Department of Statistics, Stanford University.

67. Shorack, G. R. and Wellner, J. A. (1986). *Empirical Processes with Applications to Statistics*. Wiley, New York.

68. Stute, W. and Wang, J. -L. (1993). The strong law under random censorship. *Ann. Statist.*, **21**, 1591–1607.

69. Tsai, W. -Y., Leurgans, S., and Crowley, J. (1986). Nonparametric estimation of a bivariate survival function in the presence of censoring. *Ann. Statist.*, **14**, 1351–1365.

70. Tsiatis, A. (1975). A nonidentifiability aspect of the problem of competing risks. *Proc. Natl. Acad. Sci. (USA)*, **72**, 20–22.

71. Turnbull, B. W. (1976). The empirical distribution function with arbitrarily grouped, censored and truncated data. *J. Roy. Statist. Soc. B*, **38**, 290–295.

72. Veraverbeke, N. (1995). *Cramér type large deviations for survival function estimators*. Limburgs Universitair Centrum, Universitaire Campus, B-3590 Diepenbeek, Belgium.

73. Wang, J.-G. (1987). A note on the uniform consistency of the Kaplan–Meier estimator. *Ann. Statist.*, **15**, 1313–1316.

74. Wang, M.-C., Jewell, N. P., and Tsai, W.-Y. (1986). Asymptotic properties of the product limit estimate under random truncation. *Ann. Statist.*, **14**, 1597–1605.

75. Wellner, J. A. (1982). Asymptotic optimality of the product limit estimator. *Ann. Statist.*, **10**, 595–602.

76. Wellner, J. A. (1985). A heavy censoring limit theorem for the product line estimator. *Ann. Statist.*, **13**, 150–162.

77. Wieand, H. S. (1984). Application of nonparametric statistics to cancer data. In *Handbook of Statistics. Vol. 4: Nonparametric Methods*, P. R. Krishnaiah and P. K. Sen, eds., pp. 771–790, North-Holland/Elsevier, Amsterdam, New York.

78. Woodroofe, M. (1985). Estimating a distribution function with truncated data. *Ann. Statist.* **13**, 163–177.

79. Yang, G. L. (1977). Life expectancy under random censorship. *Stoch. Process. Appl.*, **6**, 33–39.

80. Yang, S. (1991). Minimum Hellinger distance estimation of parameter in the random censorship model. *Ann. Statist.*, **19**, 579–602.

81. Ying, Z. (1989). A note on the asymtotic properties of the product-limit estimator on the whole line. *Statist. Probab. Lett.*, **7**, 311–314.

82. Ying, Z. (1992). Minimum Hellinger-type distance estimation for censored data. *Ann. Statist.*, **20**, 1361–1390.

83. Zhou, M. (1991). Some properties of the Kaplan—Meier estimator for independent nonidentically distributed random variables. *Ann. Statist.*, **19**, 2266–2274.

44 Landmark Data

K. V. Mardia

44.1 Introduction

Landmark data naturally arise when studying a shape (or form) where the total shape description is contained in a set of *landmarks* (points of correspondence) on an object. The shape of an object is the geometrical information remaining after location, scale, and rotational effects are removed. If scale information is not removed, then we are led to *size and shape* (or form).

There are different types of landmarks. A *biological landmark* is a point assigned by experts that corresponds between organisms in some biological way, e.g., the junction of two sutures on a skull. *Mathematical landmarks* are points located on an object according to some mathematical or geometric property of the object, e.g., a point of high curvature or an extreme point. The six points on the outline of a mouse vertebra (T2 bone) given in Figure 1 are such points. *Pseudolandmarks* are points constructed on an object either around the outline or in between biological or mathematical landmarks, e.g., the nine equispaced points between a pair of adjacent landmarks along the mouse vertebra boundary in Figure 1. The landmarks are also termed *labeled points, vertices, control points, anchor points*, etc., depending on the context.

Objects are everywhere—natural and human-made. It is therefore not surprising that their study has absorbed anthropologists, biologists, engineers, image analysts, computer vision experts, and so on. From the turn of the century, various studies on skulls have been cited in *Biometrika*, starting from the very first volume ([16]). The data were usually distances between landmarks (lengths, widths) or angles between landmarks, rather than landmarks themselves, following the Frankfurt Concordat of 1882 [2], which standardized anthropometric measurements. For distances, multivariate techniques have been used, but since the late 1990s, there have been many key developments in the area to allow us to work on the landmark data directly. Also, the advances in technology of measuring landmarks have been helpful, e.g., landmarks from digitized objects. Of course, if there were no constraints on the landmarks (e.g. for shape), we could use standard multivariate analysis, but some caveats are necessary even for positive variables. In general, the statistical methodology required for such work is inherently non-Euclidean.

One is led to the standard questions of how to describe shape space, how to define mean shape, how to measure shape variability, how to carry out group

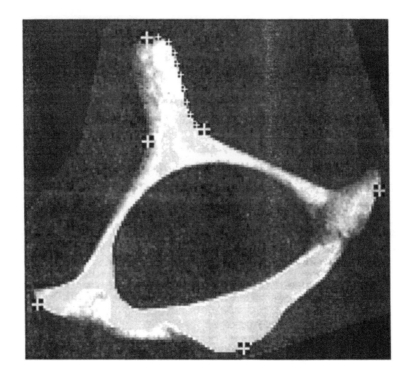

Figure 1: A mouse vertebra bone (T2) with six landmarks and nine pseudolandmarks between each pair of landmarks (only shown for one pair).

comparisons, and so on. We answer these questions mainly for two-dimensional shapes, since complex-variable representations simplify the presentation considerably.

The book by D'Arcy Thompson [55] of 1917 is the classic text on problems of shape analysis in biological frameworks. For landmark-based methods, the real progress has come from (1) a probabilistic and differential geometric point of view originating with Kendall [22], whose work was motivated by ley lines and central place theory, and (2) Bookstein [4] for a statistical viewpoint motivated by biological examples. Their work led to an explicit formulation of shape spaces. Goodall [18] incorporated Procrustes shape ideas; Mardia and Dryden [35,36] opened up shape distribution theory. Kent [24] put forward an explicit framework for tangent shape space. Distance-based methods have been put forward by Lele [29] and Stoyan [53].

44.2 Shape Description

Two objects are said to have the same shape if they differ only by a location, scale, and rotation shift [22]. The simplest example is the shape of a triangle in two dimensions. The commonest way to represent its shape is by using angles, which are invariant under these transformations. However, there are some drawbacks when the triangle is collinear. An alternative method is to take a baseline joining two vertices, translate one of these vertices to the origin, rotate the baseline to be horizontal and then rescale the baseline to unit length. The third vertex after these operations determines the shape coordinates for the triangle. These are Bookstein shape coordinates [4]; Galton [17] used a similar "registration" for face identification.

Consider the most general case of a labeled set of k points in m-dimensional Euclidean space, $\mathbb{R}^m (k \geq m+1)$, represented by a $k \times m$ matrix, X say. For any location vector $c \in \mathbb{R}^m$, a special orthogonal $m \times m$ matrix G satisfying det $G = 1, GG^T = G^T G = I$, and scalar $r > 0$, X has the same shape as $rXG + 1_k c^T$, where 1_k is a vector of ones. The fact that location, scale, and rotation can be removed in different orders means that various intermediate landmarks can be formed en route to obtaining shape coordinates. Location and scale effects are easy to eliminate directly. Let $H, (k-1) \times k$, be the Helmert matrix without the row $k^{-1/2} 1^T$, that is, H has for its jth row, the k elements $(d_j, d_j, \ldots, d_j, -jd_j, 0, \ldots, 0)$, where $d_j = -[j(j+1)]^{-1/2}$ occurs j times, and $j = 1, \ldots, k-1$. Then $Y = HX$ is invariant under location (Y is called the *derived landmarks matrix*), and scale effects are removed by the standardization $Z = HX/||HX||$; Z is called the matrix of *preshape landmarks* or the preshape matrix. The quantity $||HX||^2$ is the sum of the squared elements of HX, and represents the squared *size* of X. The matrix $(I - k^{-1} 11^T)X$ is the *centered landmarks* matrix, and $X/||HX||$ is the *scaled original landmarks*.

The case $m = 2$ is of considerable practical importance; it can be simplified using complex coordinates. Use a vector $u \in \mathbb{C}^k$ in place of the matrix X, $e^{i\theta}$ in place of G, and $z = Hu/||Hu||$, where $\sum z_j = 0, \sum |z_j|^2 = 1$. The preshape landmark vector is $z = (z_1, \ldots, z_{k-1})^T$. Two popular sets of coordinates are used for shape. Kendall's coordinates [22] for k points in $\mathbb{R}^2$ are

$$z_j/z_1, \quad j = 2, \ldots, k-1.$$

In this setting, shape space can be identified with the $(k-2)$-dimensional complex projective space $\mathbb{C}P^{k-2}$, which can be considered as the unit sphere in $\mathbb{C}^{k-1}$, with z equivalent to $ze^{i\theta}$ for all θ. Alternatively, according to Bookstein [4,6], there are also the $k-2$ complex shape coordinates

$$\frac{u_j - u_1}{u_2 - u_1}, \quad j = 3, \ldots, k.$$

Landmarks 1 and 2 form the baseline. For a general coordinate system, see Section 44.8 below.

44.3 Mean Shape

Mean shape can be defined as the Procrustes mean shape [18]: scaling and rotation are used to best match landmarks together after centering them. Let $u_1, \ldots, u_n$ be the observed landmark vectors on $u \in \mathbb{C}^k$, and let $z_1, \ldots, z_n$ be the corresponding preshape landmark vectors. The Procrustes mean shape $g \in C$ can be simplified to $g \in \mathbb{C}P^{k-2}$, where g is the dominant eigenvector of $\sum z_j \bar{z}_j$ [24], and $\bar{z}$ is the transpose of the complex conjugate. For display purposes, it is better to use

$$H^T g, \quad ||g|| = 1,$$

so that there is one point for each original landmark.

Consider the T2 mouse vertebra of Figure 1. A random sample of 23 outlines of such bones was taken from a large group of mice [11]. Figure 2 shows the Procrustes mean shape for the T2 bones (1) using only six landmarks, and (2) using six landmarks and 54 pseudolandmarks ($k = 60$ in total). Note that there is considerable refinement in the mean shape using pseudolandmarks. The estimate can be made robust against outliers [25]. For the triangle case, the Procrustes mean can be obtained explicitly using a suitable transformation [33].

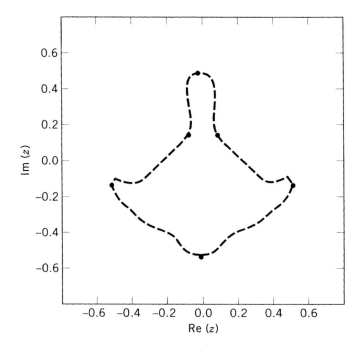

Figure 2: Procrustes mean shape for the T2 bones: (1) 60 landmarks (dash line), (2) 6 landmarks (solid circles).

44.4 Shape Variability

The question of measuring variability around the mean shape is not yet fully resolved, except when the data are concentrated or the underlying model is well specified. When the data are concentrated, we can map them into an appropriate tangent space [24], which will then allow us to use classical multivariate techniques. Let $z = (z_1, \ldots, z_{k-1})^T$ be the preshape landmarks, and let γ be a mean shape. Then we can select a tangent projection such that the configuration is rotated to be as close as possible to γ before projection, i.e.

$$v = (I - \gamma\gamma^*)ze^{i\alpha}, \qquad (1)$$

where α minimizes $(\gamma - e^{i\alpha}z)^*(\gamma - e^{i\alpha}z)$; hence $\alpha = -\theta$, where $\theta = \arg(\gamma^*z)$. Dryden and Mardia [14] extend this to any number of dimensions.

As it is common in multivariate analysis to use principal components (PCs) to get a parsimonious summary of the data, we can apply the technique to landmark data. It can be shown that an isotropic distribution for landmarks about a mean gives rise to an isotropic distribution in the tangent space to the mean, so that standard PC analysis is valid in the tangent space [34]. This result is of great significance, and underlies the development of the tangent space by Kent [24]. However, the result does not hold for the Bookstein coordinate system.

We now indicate how to carry out and display results for the PC analysis in the tangent space for two dimensions. Take γ to be the sample Procrustes mean shape in (1). Let $V(z) = (x_1, y_1, \ldots, x_{k-1}, y_{k-1})^T$ be the vector of real variables from the vector z, and $V(v)$ the corresponding tangent projection vector. Carry out the standard PC analyses on $V(v_i)$, $i = 1, \ldots, n$. Let

$V(\gamma_i)$ be the ith PC with the eigenvalue λ_i, arranged in descending order.

We can now switch back to the original configuration by obtaining preshape $z = \gamma_i^{(P)}$ for $v = \gamma_i$ from the linearized inverse of (1) and plot $H^T\gamma \pm c\lambda_i^{1/2}H^T\gamma_i^{(P)}$ for various values of c, say, $c = -3, -2, -1, 0, 1, 2, 3$. This animated sequence of shapes can be described as the *dynamic representation* of the ith PC, $i = 1, 2, \ldots, k-1$. This idea is due to Cootes et al. [10]. Of course, the first few PCs should suffice. Figure 3 shows a pictorial representation of the shape PC analysis for the T2 bones data of the preceding section. The first PC measures the thickness of the neural spine (the protrusion on the right of the bones). The second PC measures the bend at the end of the neural spine together with the size of the bump on the far left of the bone. The third PC measures the size of the bump on the far left of the bone and the curvature of the bone either side of the neural spine.

There are some other approaches to measuring variability:

1. Bookstein [7] recommends PC analysis with respect to a bending energy metric (see below).

2. Lele [28] has given a distance-based method (see below) for estimating covariance for a factorized covariance $\Sigma = \Sigma_1 \otimes \Sigma_2$ in the analysis of forms.

3. Dryden and Mardia [11] model landmarks as being multinormally distributed in $\mathbb{R}^2$, with a general covariance matrix, and derive the shape distribution. Both the shape and the covariance parameters can then be estimated by maximum likelihood.

44.5 Shape Comparisons and Deformation

Various other multivariate analysis techniques can be applied for concentrated data in a suitable tangent space. For example, we may wish to compare the within-group shape variation with the between-group variation. An appropriate framework to study within-group variation is via the Procrustes tangent space at the mean shape based on the whole data. For some studies, see Kent [25] and Dryden and Mardia [14].

An alternative approach is through partial and relative warps, which relies on fitting a shape [5] and fits in well with the general idea of biological grids or warping.

Warping of one image into another has become a well-known topic. Underlying such work is the deformation of one grid into another. Such grids were employed by D'Arcy Thompson [55], who used them to map one species into another. Underlying these grids are mappings from $\mathbb{R}^2$ to $\mathbb{R}^2$.

Suppose $x_i = (x_i[1], x_i[2])$, $i = 1, \ldots, k$, are the new landmarks in the second image corresponding to the old landmarks $t_i = (t_i[1], t_i[2])$, $i = 1, \ldots, k$, of the first image. A mapping can be constructed by using $(\Phi_1(t), \Phi_2(t))$ such that

$$\Phi_1(t_i) = x_i[1],$$
$$\Phi_2(t_i) = x_i[2], \quad i = 1, \ldots, k.$$

Hence it is instructive to look at a class of interpolators $\Phi_1(\cdot)$ and $\Phi_2(\cdot)$. We could, for example, use the same form for both interpolators, e.g., both thin-plate splines. We describe this approach through *kriging*, which is an extension of the thin-plate spline (see Mardia et al. [42]). Suppose that there is a stochastic process $X(t)$ that follows a stationary (or intrinsic) random field model, with known covariance function (or conditionally positive definite covariance function) $\sigma(h)$ and linear drift (for simplicity). Then the kriging interpolator

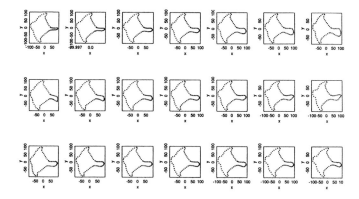

Figure 3: Three rows of series of T2 vertebra shapes along the three principal components (PCs). The ith row shows the shapes at $-3, -2, -1, 0, 1, 2,$ and 3 standard deviations along the ith PC.

with $X(t_i) = x_i[1]$ is given by

$$\hat{X}(t) = a_0 + a_1 t[1] + a_2 t[2] + \sum_{i=1}^{k} b_i \sigma(t - t_i),$$

where $a = (a_0, a_1, a_2)^T$, $b = (b_1, \ldots, b_k)^T$ are the solutions of

$$\begin{bmatrix} \Sigma & D \\ D^T & 0 \end{bmatrix} \begin{bmatrix} b \\ a \end{bmatrix} = \begin{bmatrix} x_{[1]} \\ 0 \end{bmatrix}$$

with $x_{[1]} = (x_{1[1]}, \ldots, x_{k[1]})^T$:

$$\Sigma_{i,j} = \sigma(t_i - t_j), \quad D^T = \begin{pmatrix} 1 & \cdots & 1 \\ t_1 & \cdots & t_k \end{pmatrix}.$$

For $\sigma(h) = |h|^2 \log |h|$, we have the standard thin-plate spline case. Similarly, we can write a predictor with $X(t_i) = x_i[2], i = 1, \ldots, k$, and thus we have completed a mapping. There is no mathematical guarantee that the resulting map will be bijective (with no folding); in practice it will be if so the deformation is not too severe.

Let us write

$$\begin{bmatrix} \Sigma & D \\ D^T & 0 \end{bmatrix}^{-1} = \begin{bmatrix} K^{11} & K^{12} \\ K^{21} & K^{22} \end{bmatrix},$$

the (generalized) *bending energy matrix*. We can study the behavior of the individual predictors through this matrix, which is the same for both predictors $\Phi_1(\cdot)$ and $\Phi_2(\cdot)$. Let γ_j be the jth eigenvector of K^{11}, where the eigenvalues λ_j are arranged in descending order. The γ_j are called the *principal warps*.

The kriging interpolator at $(X(t_1), \ldots, X(t_n))^T = \gamma_j$ becomes

$$\hat{X}(t) = a_0 + a_1 t[1] + a_2 t[2] + \lambda_j \gamma_j^T \sigma(t),$$

where $(\sigma(t))_i = \sigma(t - t_i)$. Hence, on plotting $\gamma_j^T \sigma(t)$ against t, the principal warps in this form are informative. In fact, $P_1(t) = \gamma_1^T \sigma(t)$ has the highest bending energy (stiffest) as measured by λ_1, whereas $P_{k-3}(t) = \gamma_{k-3}^T \sigma(t)$ has the lowest bending energy (is least stiff) as measured by λ_{k-3}. Thus $P_1(t)$ contains the most local warping features whereas $P_{k-3}(t)$ contains the least local warping features, and so on. Thus the magnitudes of the eigenvalues are inversely related to geometric scale. However, $P_{k-3}(t)$ captures features that are not as global as those measured by the affine transformation. The latter has been taken out, i.e., at least three eigenvalues are zero (corresponding to infinite scale). Then

$$\lambda_i X^T \gamma_i P_i(t)$$

is defined as the ith *partial warp*, $i = 1, \ldots, k-3$, viz., the scores corresponding to the ith principal warp.

Bookstein [5] has made effective use of the thin-plate splines in deformation. For other applications, see Sampson et al. [50] and Mardia and Hainsworth [39]. Bookstein and Green [8,9] have incorporated the derivative constraints in a spline setting; they use finite differences instead of derivatives at certain places, because $\sigma(h)$ is not twice continuously differentiable at $h = 0$. An extension to higher derivatives including curvature requires the kriging formulation given by Mardia et al. [42].

44.5.1 Procrustes Analysis and Bending Energy Analysis

Bookstein [7] provides a link between Procrustes analysis and bending energy analysis. Let $z \in \mathbb{C}^{k-1}$ be the preshape landmarks; let H_1 be the tangent Procrustes space and H_2, the thin-plate spline tangent space of complex dimensions $k-1$ and $k-2$, respectively. Then we can select $\nu \perp H_2$ such that $\{\nu\} \oplus H_2 = H_1$.

Let S be the real sample covariance matrix of rank $2k - 4$ in $\mathbb{R}^{2k}$ in H_1. Let B^- be the Moore–Penrose generalized inverse (the inverse of B restricted to H_2). Bookstein [6] argues that the eigenstructure of BS will amplify the significance of small-scale features and the eigenstructure of B^-S will amplify the importance of large-scale features. This type of decomposition is known as *relative warp analysis* and has proved popular in biological applications.

44.6 Shape Distributions

The development of shape distributions has followed the two main approaches used in directional distributions from the multivariate normal distribution [30]:

1. The *marginal approach*, where we integrate out the nondirectional variables as in the derivation of the offset normal distribution.

2. The *conditional approach*, where the nondirectional variables are held constant; e.g., for the von Mises density we fix the length in a suitable bivariate normal distribution.

Both approaches have produced useful distributions. The work really started with Mardia and Dryden's distribution [35,36] following the marginal distribution approach. Kent [24] adopted a conditional approach leading to the complex Bingham distribution. We now describe these approaches.

44.6.1 Marginal Approach

Let $X_1, \ldots, X_k$ be independently distributed points in $\mathbb{R}^2$ with $X_i \sim N(\mu_i, \Sigma)$ for $i = 1, \ldots, k$. For this case there are expressions for the corresponding shape distributions, now known as *Mardia–Dryden distributions* [10,31,35,36]. For the case $k = 3$ [the identification of the shape space with the sphere S^2 with $X_i \sim N(\mu_i, \sigma^2 I)$, $i = 1, 2, 3$], the probability density function of the shape distribution [31] is

$$f(x; \lambda, \kappa) = [1 + \kappa(\lambda^T x + 1)]$$
$$\times \exp[\kappa(\lambda^T x - 1)],$$
$$|\lambda| = 1, \quad \kappa \geq 0,$$

where x lies on S^2 and λ is the "mean shape" [31]. For all κ this density is quite close to a Fisher density with the same mean, and hence Mardia [31] recommended the use of the Fisher distribution for shapes.

Goodall and Mardia [19] have obtained distributions for higher dimensions. In particular, in their Theorems 4 and 5 they derive an important approximation for small variations. Dryden and Mardia [13] have given shape and size distributions; for a further review of the latter, see Reference 12.

44.7 Conditional Approach: The Complex Bingham Distribution

We have noted that the preshape $z = (z_1, z_2, \ldots, z_{k-1})$ lies on a complex sphere $\mathbb{C}S^{k-2}$, i.e., $z^*z = 1$, where z is the complex conjugate transpose of z. One way to construct an appropriate distribution is to condition the complex multivariate normal distribution $\mathbb{C}N_k(0, A)$, where A is Hermitian (i.e., $A = A^*$). Conditioning it on $z^*z = 1$ gives rise to the *Bingham distribution* for complex variables, first introduced by Kent [24]. This distribution on $\mathbb{C}S^{k-2}$ (or specifically on $\mathbb{C}P^{k-2}$) is given by

$$f(z) = C(A)^{-1} \exp(z^* A z), \quad z \in \mathbb{C}S^{k-2},$$

where the matrix A is $(k-1) \times (k-1)$ Hermitian and $C(A)$ is the normalizing constant, for which an explicit expression is available.

The complex Bingham distribution is analogous to the real Bingham distribution and in fact is a particular case of it. It has the property that $f(e^{i\theta} z) = f(z)$ and is thus invariant under rotation of the preshape z. This property therefore makes the distribution suitable for shape analysis (location and scale are already removed, as z is on the preshape sphere). Kent [24] also commented on its limitations for use in shape analysis in that imposing a complex symmetry assumption might not be realistic. However, the distribution does provide an elegant way to get some insight into shape data. In particular, for $k = 2$ it is related to the Fisher distribution [24,33]. Further, the Procrustes shape mean is the maximum likelihood estimator of the dominant eigenvector of A. Also, considering $A = 0$, we are led to a test of randomness of shape [33]. A general Riemannian structure of Euclidean shape space is described by Le and Kendall [27] to provide probabilistic shape measures.

44.8 Shapes in Higher Dimensions

The QR decomposition, often used in multivariate analysis, provides an alternative way of obtaining the shape of a configuration X which provides shape coordinates in any dimension. The similarity transformations are removed as follows.

Again, let $X(k \times m)$ contain the raw landmarks and $Y = HX$ the derived landmarks. To remove orientation consider the QR decomposition of Y, namely

$$Y = T\Gamma,$$
$$\Gamma \in V_{n,m}, V_{n,m} V_{n,m}^T = I_n,$$
$$\|T\| > 0,$$

where $n = \min(k-1, m)$ and $T[(k-1) \times n]$ is lower triangular with nonnegative diagonal elements $T_{ii} \geq 0, i = 1, \ldots, n$, and $\|T\|^2 = \operatorname{tr} T^T T$. The most important case in practice is when $k > m$. In that case, $\Gamma \in O(m)$ and $|\Gamma| = \pm 1$, so that the QR decomposition removes orientation and reflection. To remove orientation only, we have that $\Gamma \in SO(m), |\Gamma| = +1$, and T_{mm} is unrestricted, and then $T, \{T_{ij} : 1 \leq i \leq j \leq m\}$, is called the *size and shape* (or *form*) if $\Gamma \in SO(m)$, and the *reflection size and shape* if $\Gamma \in O(m)$. To remove scale, we divide by the Euclidean norm $\|T\|$ of T, i.e., define W by

$$W = T/\|T\|, \quad \|T\| > 0.$$

We call W the *shape* of our configuration if $\Gamma \in SO(m)$, and the *reflection shape* if $\Gamma \in O(m)$. (If $k \leq m$, then these distinctions are irrelevant.) If $||T|| = 0$, then the landmarks are coincident and the shape is not defined.

In practice, the problem for higher dimensions is not straightforward in its visual or its computational aspect. Some examples for three dimensions are given in Goodall and Mardia [19] and Dryden and Mardia [14]. Another problem arises when the data are matched, e.g., landmarks in growth studies in two consecutive years. Solutions to such problems are developed in Mardia and Walder [46,47] and Prentice and Mardia [49].

44.9 Distance-Based Method

Let $W_{h1,h2} = \sum_{j=1}^{n} d_j^2(h_1, h_2)$, where $d_j^2(h_1, h_2)$ is the Euclidean squared distance between the scaled original landmarks h_1 and h_2 for the jth sample, $j = 1, \ldots, n$. Then, following the classical multidimensional scaling (MDS) method (e.g. Mardia et al. [41]), calculate

$$B = -\frac{1}{2}\left(I_k - \frac{1}{k}1_k 1_k^T\right) \\ \times W \left(I_k - \frac{1}{k}1_k 1_k^T\right).$$

Let a_j be the eigenvalues (in descending order) and f_j be eigenvectors of B, scaled so that $f_j^T f_j = a_j$. Then an estimated average shape can be defined by $f_1 \pm if_2$. In the distance-based method, we cannot distinguish between the configuration $x_j + iy_j$ and its reflection $x_j - iy_j$. But for concentrated data the appropriate choice is not difficult. This method is described in Kent [24], which applies for both form and shape; for form, no prescaling is necessary.

Lele [28], Stoyan [53], and Stoyan and Stoyan [54] also use interpoint distances and MDS for form and shape. Their method basically involves estimating population distances using a method of moments under a Gaussian assumption. The estimate is consistent for form [29] but not robust, since estimation of distances involves fourth moments. However, this modification removes the bias in estimating distances. An advantage of distance-based methods is that these can be extended easily to higher dimensions.

Mardia and Dryden [37] have discussed bias issues for various averages, especially for the triangle case. If one takes the mean based on Bookstein's coordinates, the bias can be removed under the isotropic model. There has been considerable discussion on consistency of the Procrustes mean shape, starting from Lele [29]. Kent and Mardia [26] have shown that under isotropic errors the Procrustes mean shape is consistent, but for nonisotropic errors it need not be consistent, and for large errors it can be arbitrarily inconsistent. Under isotropic normality, the Procrustes form mean is consistent up to a scale factor.

However, a balanced view of the subject has to be taken. Kent [24] has shown that many methods of calculating a mean shape are equivalent for highly concentrated data. For a numerical example in ref. [24], even with moderate variability in data, various mean shapes are almost identical. The main goal of shape analysis is measuring shape changes, and in this case any possible biases will tend to cancel out, so that bias issues are not critical. Besides, for a well-specified model the maximum likelihood estimators are in general not only consistent but also fully efficient.

44.10 Shapes in Image Analysis

Mardia et al. [43,45] have reviewed various statistical models and prior distributions underlying model-based object recognition. Mainly, these papers review how the prior beliefs on shapes are incorporated in object recognition. The two recent edited volumes *Statistics and Images* [32,40] also highlight the current state of the art. Following Bookstein's ideas, warping and averaging have been discussed in Mardia and Hainsworth [39]. In addition, these authors [21,39] attempt to automate landmark location by using outline templates and gray-level templates, respectively. Another key development in prior distributions is in the active shape models of Cootes et al. [10].

Grenander and his workers have produced a wealth of new ideas [20]. Amit et al. [1] proposed a method of object recognition using landmark free methods. Phillips and Smith [48] have introduced hierarchical Bayesian models, e.g., look for the head first, then the face, then the eyes, and so on. The location of the eyes is conditional on the location of the face, which in turn is conditional on the location of the head. For a statistical approach using scale space, see Wilson and Johnson [56]. Mardia and Qian [44] use polar representation of shape for two-dimensional objects; the modified complex autoregressive (CAR) and autoregressive moving average (ARMA) models are also natural [43,51]. A neural net for shape is proposed in Ansari and Li [3].

44.11 Concluding Remarks

There are software packages to carry out shape analysis; for example, one can access the program of Professor Rohlf's group on *http://life.bio.sunysb.edu/morph/software.html* and *ftp://life.bio.sunysb.edu/morphmet*. There is also a morphometric electronic mailing list that provides a focal point for morphometric activities (send the message "subscribe morphmet ⟨ your name ⟩" to *listserv@cunyvm.cuny.edu* to subscribe to morphmet). For projective shape, see Mardia et al. [42].

Various areas still require further investigation, e.g., nonparametric methods for shape, analysis of longitudinal data, scale-space approaches, study of shape landmarks with covariates such as directions, and robust shape analysis (see also Mardia and Gill [38]). The research monograph of Dryden and Mardia [15] provides an up-to-date treatment of these subjects from a statistical viewpoint.

Acknowledgments. I am grateful to Ian Dryden and Alistair Walder for their helpful comments.

References

1. Amit, Y., Grenander, U., and Piccioni, M. (1992). Structural image restoration through deformable templates. *J. Am. Statist. Assoc.*, **86**, 376–387.

2. Anonymous (1882). *Frankfurter Verständigung über ein gemeinsames craniometrisches Verfahren Correspondez-Blatt deutsch-anthrop. Gesellsch.*, **xiv**, 1.

3. Ansari, N. and Li, K. (1993). Landmark-based shape recognition by a modified Hopfield neural network. *Pattern Recogn.*, **26**, 531–542.

4. Bookstein, F. L. (1986). Size and shape spaces for landmark data in two dimensions (with discussions). *Statist. Sci.*, **1**, 181–242. (First mainstream statistical paper to outline the basics of shape analysis with biological and medical applications. It contains a good discussion on the state of the art at that time.)

5. Bookstein, F. L. (1989). Principal warps: Thin plate splines and the decomposition of deformations. *IEEE Trans. Pattern Anal. Machine Intell.*, **11**, 567–585.

6. Bookstein, F. L. (1991). *Morphometric Tools for Landmark Data: Geometry and Biology.* Cambridge University Press, Cambridge, UK. (Text discussing the shape analysis of biological landmark

data, with geometrical insight and pictorial representation.)

7. Bookstein, F. L. (1995). Metrics and symmetries of the morphometric synthesis. In *Current Issues in Statistical Shape Analysis*, K. V. Mardia and C. A. Gill, eds., pp. 139–153, Leeds University Press, Leeds, UK.

8. Bookstein, F. L. and Green, W. D. K. (1993). A feature space for edgels in images with landmarks. *J. Math. Imaging Vision*, **3**, 231–261.

9. Bookstein, F. L. and Green, W. D. K. (1993). A thin-plate spline and the decomposition of deformations. In *Mathematical Methods in Medical Imaging*, Vol. II, D. C. Wilson and J. N. Wilson, eds. *Proc. SPIE*, **2035**, 14–28.

10. Cootes, T. F., Taylor, C. J., Cooper, D. H., and Graham, J. (1992). Training models of shapes from sets of examples. *Proc. British Machine Vision Conf., Leeds*, pp. 9–18, Springer-Verlag, New York.

11. Dryden, I. L. and Mardia, K. V. (1991). General shape distributions in a plane. *Adv. Appl. Probab.*, **23**, 259–276.

12. Dryden, I. L. and Mardia, K. V. (1991). Theoretical and distributional aspects of shape analysis. In *Probability Measures on Groups X*, H. Heyer, ed. Plenum, New York, pp. 95–116.

13. Dryden, I. L. and Mardia, K. V. (1992). Size and shape of landmark data. *Biometrika*, **79**, 57–68.

14. Dryden, I. L. and Mardia, K. V. (1993). Multivariate shape analysis. *Sankhyā A*, **55**(part 3), 460–480.

15. Dryden, I. L. and Mardia, K. V. (1998). *Statistical Shape Analysis*. Wiley, Chichester, UK. (Text emphasizing a blend of statistical theory and applications in shape analysis.)

16. Fawcett, C. D. and Lee, A. (1901–1902). A second study of the variations and correlations of the human skull, with special reference to the Naqada Crania. *Biometrika*, **1**, 408–467.

17. Galton, F. (1907). Classification of portraits. *Nature*, **76**, 617–618.

18. Goodall, C. R. (1991). Procrustes methods in the statistical analysis of shape (with discussion). *J. Roy. Statist. Soc. B*, **53**, 285–339. (The Procrustes least-squares approach to shape analysis, with wide-ranging discussion from various speakers.)

19. Goodall, C. R. and Mardia, K. V. (1993). Multivariate aspects of shape theory. *Ann. Statist.*, **21**, 848–866.

20. Grenander, U. and Miller, M. I. (1994). Representations of knowledge in complex systems (with discussion). *J. Roy. Statist. Soc., B*, **56**(4), 549–603. (One of many papers from Grenander and coworkers on the use of deformable templates for high-level image analysis, with interesting discussion.)

21. Hainsworth, T. J. and Mardia, K. V. (1993). A Markov random field restoration of image sequences. In *Markov Random Fields Theory and Applications*, R. Chellapa and A. K. Jain, eds., pp. 409–445, Academic Press, Boston.

22. Kendall, D. G. (1984). Shape-manifolds, Procrustean matrices and complex projective spaces. *Bull. Lond. Math. Soc.*, **16**, 81–121. (The deep differential geometric foundations of shape theory, with important invariant-measure results.)

23. Kendall, D. G. (1989). A survey of the statistical theory of shape. *Statist. Sci.*, **4**, 87–120. (Discussion of developments since Bookstein's paper [4].)

24. Kent, J. T. (1994). The complex Bingham distribution and shape analysis. *J. Roy. Statist. Soc. B*, **56**, 285–299. (Elegant and simple approach for shape analysis of landmarks in a plane, which fully exploits complex arithmetic.)

25. Kent, J. T. (1997). Data analysis for shapes and images. *J. Statist. Plan. Inference*, **57**, 181–193.

26. Kent, J. T. and Mardia, K. V. (1996). Consistency of Procrustes estimators. *J. Roy. Statist. Soc. B*, **59**, 281–290.

27. Le, H. and Kendall, D. G. (1993). The Riemannian structure of Euclidean shape spaces: A novel environment for statistics. *Ann. Statist.*, **21**, 1225–1271.
28. Lele, S. (1991). Some comments on coordinate free and scale invariant methods in morphometrics. *Am. J. Phys. Anthropol.*, **85**, 407–418.
29. Lele, S. (1993). Euclidean distance matrix analysis (EDMA): Estimation of mean form and mean form difference. *Math. Geol.*, **25**, 573–602.
30. Mardia, K. V. (1972). *Statistics of Directional Data.* Academic Press, London.
31. Mardia, K. V. (1989). Shape analysis of triangles through directional techniques. *J. Roy. Statist. Soc. B*, **51**, 449–458.
32. Mardia, K. V., ed. (1994). *Statistics and Images*, 2. Carfax, Abingdon, Oxford. (Collection of state-of-the-art articles from the interface between statistics and image analysis.)
33. Mardia, K. V. (1995). Directional Statistics and Shape Analysis. *Res. Rep. STAT 95/24*, Department of Statistics, University of Leeds, Leeds, UK.
34. Mardia, K. V. (1995). Shape advances and future perspectives. In *Current Issues in Statistical Shape Analysis*, K. V. Mardia and C. A. Gill, eds., Leeds, pp. 57–75. Leeds University Press.
35. Mardia, K. V. and Dryden, I. L. (1989). Shape distributions for landmark data. *Adv. Appl. Probab.*, **21**, 742–755. (The first of a series of papers on shape distributions for Gaussian points in a plane.)
36. Mardia, K. V. and Dryden, I. L. (1989). Statistical analysis of shape data. *Biometrika*, **76**, 271–281.
37. Mardia, K. V. and Dryden, I. L. (1994). Shape averages and their bias. *Adv. Appl. Probab.*, **26**, 334–340.
38. Mardia, K. V. and Gill, C. A., eds. (1995). *Current Issues in Statistical Shape Analysis*. Leeds University Press. (Proceedings of the 1995 Leeds Annual Statistics Research Workshop: papers from various disciplines, including statistics, morphometry, and machine vision.)
39. Mardia, K. V. and Hainsworth, T. J. (1993). Image warping and Bayesian reconstruction with grey-level templates. In *Statistics and Images*, Vol. 1, pp. 283–294, K. V. Mardia and G. Kanji, eds. Carfax, Abingdon, Oxford.
40. Mardia, K. V. and Kanji, G. K., eds. (1993). *Statistics and Images*, Vol. 1. Carfax, Abingdon, Oxford.
41. Mardia, K. V., Kent, J. T., and Bibby, J. M. (1979). *Multivariate Analysis*. Academic Press, London.
42. Mardia, K. V., Kent, J. T., Goodall, C. R., and Little, J. (1996). Kriging and splines with derivative information. *Biometrika*, **83**, 207–221.
43. Mardia, K. V., Kent, J. T., and Walder, A. N. (1991). Statistical shape models in image analysis. In *Computer Science and Statistics: Proc. 23rd Symp. Interface*, E. M. Keramidas, ed., pp. 550–557, Interface Foundation of North America, Fairfax Station, VA.
44. Mardia, K. V. and Qian, W. (1995). A Bayesian method of compact object recognition from noisy images. *IMA Conf. Proc. Complex Stochastic Systems and Engineering Applications*, M. Titterington, ed., pp. 155–166.
45. Mardia, K. V., Rabe, S., and Kent, J. T. (1995). Statistics, shape and images. *IMA Conf. Proc. Complex Stochastic Systems and Engineering Applications*, M. Titterington, ed., pp. 155–166.
46. Mardia, K. V. and Walder, A. N. (1994). Shape analysis of paired landmark data. *Biometrika*, **81**, 185–196.
47. Mardia, K. V. and Walder, A. N. (1994). Size-and-shape distributions for paired landmark data. *Adv. Appl. Probab.*, **26**, 893–905.
48. Phillips, D. B. and Smith, A. F. M. (1994). Bayesian faces via hierarchical template modeling. *J. Am. Statist. Soc.*, **89**, 1151–1163.
49. Prentice, M. J. and Mardia, K. V. (1996). Shape changes in the plane for landmark data. *Ann. Statist.*, **23**, 1960–1974.

50. Sampson, P. D., Lewis, S., and Guttorp, P. (1991). Computation and interpretation of deformations for landmarks data in morphometrics and environmetrics. In *Computer Science and Statistics: Proc. 23rd Symp. Interface*, E. M. Kermadas, ed., pp. 534–541, Interface Foundation of North America, Fairfax Station, VA.

51. Sekita, I., Kurita, T., and Otsu, N. (1992). Complex autoregressive model for shape recognition. *IEEE Trans. Pattern Anal. Machine Intell.*, **14**, 489–496.

52. Small, C. G. (1996). *The Statistical Theory of Shape*. Springer, New York. (Elegant mathematical and geometric treatment of the subject.)

53. Stoyan, D. (1990). Estimation of distances and variances in Bookstein's landmark model. *Biometr. J.*, **32**, 843–849.

54. Stoyan, D. and Stoyan, H. (1994). *Fractals, Random Shapes and Point Fields*. Wiley, Chichester, UK.

55. Thompson, D. W. (1917). *On Growth and Form*. Cambridge University Press. (Abridged edition, 1961.)

56. Wilson, A. G. and Johnson, V. E. (1994). Priors on scale-space templates. *Mathematical Methods in Medical Imaging III. Proc. Society of Photo-optical Instrumentation Engineers*, Vol. 2299, 161–168.

45 Longitudinal Data Analysis

Burton Singer

45.1 Domain of Longitudinal Data Analysis

Longitudinal data analysis is a subspecialty of statistics in which individual histories—interpreted as sample paths, or realizations, of a stochastic process—are the primary focus of interest. A wide variety of scientific questions can only be addressed by utilizing longitudinal data together with statistical methods which facilitate the detection and characterization of regularities across multiple individual histories. Some examples are:

1. *Persistence.* Do persons who vote according to their political party identification in one presidential election (e.g., 1956) tend to vote this way in a subsequent election [14]? Do persons or firms who are repeatedly victimized by criminals always tend to be victimized in the same way [15,35]?

2. *Structure of individual time paths.* Are there simple functions—e.g., low-order polynomials—that characterize individuals' changes in systolic blood pressure with increasing age in various male cohorts [45]? This question arises in the study of factors associated with the onset of coronary heart disease [13].

3. *Interaction among events.* Are West African villagers infected with one species of malaria parasite [e.g., *Plasmodium falciparum* (Pf)] more resistant to subsequent infection with another species [*Plasmodium malariae* (Pm)] than if Pf was not already present in their peripheral blood [11,32]?

4. *Stability of multivariate relationships.* For neurologists making prognoses about the recovery of patients from nontraumatic coma, are the same neurological indicators useful at admission to a hospital, 24 hours later, 3 days, and 7 days, or does the list of key prognostic indicators change over time—and in what manner [27]?

In this chapter we present some examples of analytical strategies that can assist research workers in answering questions such as these. Our aim is to exhibit the general flavor of longitudinal data analysis, as well as to illustrate how the idiosyncrasies of a scientific problem suggest different methodologies.

45.2 Some History

Statistical methods that are especially suited to the quantitative study of individual histories have their roots in John Graunt's first attempt to construct a life table. However, the development of a diverse set of techniques for measuring the dynamics of vector-valued stochastic processes—as opposed to a single positive random variable (e.g., waiting time until death)—is a phenomenon of the twentieth century, primarily stimulated by the large longitudinal field studies initiated in the 1920s. Among the most influential of these investigations were:

1. L. M. Terman's follow-up study of California schoolchildren who scored in the top 1% of the national IQ distribution. Initiated in 1921, a principal aim of the study was to follow the original sample of 857 boys and 671 girls into adult life to assess whether high IQ was a good predictor of success in later life. For details on the design and early analyses of these data, see Terman et al. [42] and Terman and Oden [41].

2. E. Sydenstricker's Hagerstown morbidity study, initiated in December 1921, with follow-up on almost 2000 households through March 1924, had a central aim of assessing sickness incidence over a sufficiently long period of time to distinguish it from sickness prevalence. In this connection, see Sydenstricker [39]. The Hagerstown sample was subsequently followed up in 1941, and some straightforward descriptive analyses suggested the important relation that in this population chronic illness led to poverty rather than the reverse implication.

The natural conceptual framework in which to consider alternative analyses of these and other longitudinal data is the theory of stochastic processes. However, in the 1920s this subject was very much in its infancy, and the first attempts to utilize process models in longitudinal analyses did not occur until the late 1940s and early 1950s. In the social sciences, Tinbergen [43] presented a graphical caricature of a finite-state process with intricate causal relations among the states and across time which triggered P. Lazarsfeld's 1954 utilization of this framework to study voting behavior and opinion change in election campaigns. This work can be viewed as a precursor to Blumen et al.'s [9] classic study of interindustry job mobility utilizing Markov and simple mixtures of Markov models to describe the individual dynamics. For a superb and up-to-date review of subsequent utilization of stochastic process models for the study of intraindividual dynamics in the social sciences, the reader should consult Bartholomew [5]. A useful and comprehensive presentation of longitudinal analysis methods and associated substantive problems in economics are the papers in *Annals de l'INSEE* [4]. Despite this extensive development most longitudinal analyses in the social sciences from the 1920s through the present involve the estimation and interpretation of correlation coefficients in linear models relating a multiplicity of variables to each other and to change over time. For critiques of this technology, see Rogosa [36] and Karlin et al. [22]. For an imaginative application, see Kohn and Schooler [25]. A balanced appraisal of path analysis by its originator and insights appear in Wright [50].

In medicine and public health, the Hagerstown morbidity study anticipated the more recent major longitudinal data collections such as the Framingham study of atherosclerotic disease [13], the University Group Diabetes Program evaluations of oral hypoglycaemic agents [18], and the World Health Organization field surveys of malaria in Nigeria [31]. Analyses exploiting the availability of individual histories, or portions of them, in these and other studies have primarily been of two types:

1. Estimation of age-dependent incidence rates; i.e., the expected number of occurrences of a given event per unit time per individual at risk of the event (see, e.g., Bekessey et al. [6].

2. Survival analysis, where the waiting time until occurrence of an event is the primary dependent variable of interest. For details on the analysis of survival data, which can be viewed as representing the duration of one episode in a single state in what may be a multistate stochastic process. Insightful examples appear in Crowley and Hu [12] and Menken et al. [30].

Analyses where the modeling of continuous functions of time are of interest, or where the goal is to characterize transition rates between discrete states and assess the influence of a variety of covariates on them, are of relatively recent vintage. Particularly important in facilitating such analyses is the literature on growth curves—e.g., Rao [34] and the references in Ware and Wu [45], Foulkes and Davis [17], and McMahon [29]—and the recently developed nonparametric methods for the analysis of counting processes (see, e.g., Aalen et al. [2]).

45.3 Designs and Their Implications for Analysis

In empirical applications, testing whether specific classes of processes describe the occurrence of events or the evolution of a continuous variable is best facilitated by observing, in full, many realizations of the underlying process $X(t)$ for all t in an interval $[T_1, T_2]$. Examples of such data are the work histories in the Seattle and Denver Income Maintenance Experiments [44], the fertility histories in the Taichung IUD experiment [28], and the job vacancy histories for ministers in Episcopalian churches in New England [46]. In most substantive contexts, however, ascertaining the exact timing of each occurrence of an event for each individual is either impossible, economically infeasible, or both. Observations usually contain gaps and censoring relative to a continuously evolving process. Some examples of this situation are:

1. In the Framingham study of atherosclerotic disease, individuals were examined once every 2 years, at which times symptoms of illness, hospitalizations, or other events occurring between examinations were recorded (retrospective information). In addition, a physical examination, some blood studies, and other laboratory work (current information) were completed. One topic of considerable interest is the intraindividual dynamics of systolic blood pressure. This is a continuous-time and continuous-state process that can be modeled only using the biennial samples, i.e., measurements made at the examinations. Such data represent fragmentary information about the underlying process. Lazarsfeld and Fiske [26] introduced the terminology, "panel" study, to refer to this kind of data collection. An associated body of statistical techniques is frequently referred to as *methods of panel analysis*.

2. The WHO field surveys of malaria in Nigeria—the Garki Project [31]—involved the collection of a thick blood film every 10 weeks for 1.5 years from individuals in eight village clusters. The blood films were examined for presence or absence of any one or more of three species of malaria parasite together with an estimate of the density of *P. falciparum* parasites if they were present. Data on the mosquito vectors, including person-biting rates

measured via human bait, were collected in some of the villages every 5 weeks in the dry season and every 2 weeks in the wet season. Thus the dynamics of intraindividual infection and of parasite transmission between humans and mosquitos can be modeled only from partial information about a continuously evolving process.

3. In Taeuber et al.'s [40] residence history study, observations are taken retrospectively on current residence, first and second prior residence, and birthplace of individuals in particular age cohorts. Analyses in which duration of residence is a dependent variable of interest must accommodate censoring on the right for current residence. Furthermore, characterization of the pattern of adult residence histories is complicated by the fact that initial conditions are unknown for persons who have occupied more than three residences beyond, for example, age 18.

A feature of modeling with such fragmentary data is that algebraic characterizations of the datasets that can possibly be generated by given continuous-time models are frequently very difficult to obtain. On the other hand, these characterizations are, of necessity, the basis of tests for compatibility of the data with proposed models. In addition, estimation of quantities such as rates of occurrence of events per individual at risk of the event at a given time is made complicated by the fact that some of the occurrences are unobserved. This necessitates estimation of rates that have meaning within stochastic process models that are found to be compatible with the observed data.

45.4 Analytical Strategies—Examples

It is our view that the flavor of longitudinal data analysis is best conveyed by a variety of examples. It is to be understood, however, that the issues raised in each example are applicable to a much wider range of studies.

Nonparametric Estimation of Integrated Incidence Rates and Assessment of Possible Relationships between Events. Aalen et al. [2] utilize retrospective data on 85 female patients at the Finsen Institute in Copenhagen to assess whether hormonal changes in connection with menopause or similar artifically induced changes in ovarian function might affect the development of a chronic skin disease, *pustulosis palmoplantaris*. They propose a stochastic compartment model of the possible disease dynamics and mortality which can be summarized by the directed graph in Figure 1.

Since the etiology of *pustulosis palmoplantaris* is unknown, there is no defensible basis for proposing a very restrictive parametric family of stochastic process models as candidates to describe the movement of persons among nodes on the graph in Figure 1. However, the question of possible influence of natural and induced menopause on the outbreak of disease only requires comparisons of the age-dependent rates of transition per person at risk, $r_{O,D}(t), r_{M,MD}(t), r_{I,ID}(t)$, for the transitions $O \to D, M \to MD$, and $I \to ID$, respectively. To this end we first define the integrated rate of transition from a state labeled i to a state labeled j in a general stochastic compartment model as $A_{ij}(s,t) = \int_s^t r_{ij}(u)du$. Then we bring in

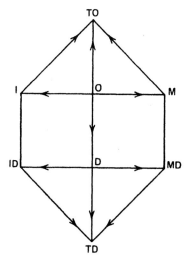

Figure 1: Caricature of compartment model of skin disease, menopause, and mortality. O, no event has occurred; M, natural menopause has occurred; D, disease has been detected; I, induced menopause has occurred; TO, dead without disease; TD, dead with disease.

the nonparametric estimator

$$A_{ij}(s,t) = \sum_{k: s \leq t_k^{(i,j)} < t} [Y_i(t_k^{(i,j)})]^{-1}, \quad (1)$$

where

$$\begin{aligned} Y_i(t) &= \text{number of individuals in} \\ & \quad i \text{ state at age } t, \\ t_k^{(i,j)} &= \text{age of } k\text{th transition from} \\ & \quad \text{state } i \text{ to state } j \text{ with} \\ & \quad 0 < t_1^{(i,j)} < t_2^{(i,j)} < \cdots. \end{aligned}$$

As verified in a remarkable paper by Aalen [1], (1) is an unbiased, consistent, asymptotically normal estimator of $A_{ij}(s,t)$ for quite general discrete-state, continuous-time stochastic processes. In particular, no strong assumptions about dependencies across time—such as the

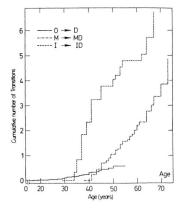

Figure 2: Estimated integrated "conditional" intensities for outbreak of pustulosis palmoplantaris before and after natural and induced menopause. (From Aalen et al. [2].)

Markov property—are necessary to validate the desirable statistical properties of (1). A good estimator of var $\hat{A}_{ij}(s,t)$ is given by

$$\sum_{k: s \leq t_k^{(i,j)} < t} [Y_i(t_k^{(i,j)})]^{-2}. \quad (2)$$

Applying (1) to the Finsen Institute data yields the estimates of the integrated rates $A_{O,D}(0,t), A_{M,MD}(0,t)$ and $A_{I,ID}(0,t)$ shown in Figure 2.

The graph suggests, and more formal tests (see Aalen et al. [2] for details) support, the conclusion that induced and natural menopause increase the chance of appearance of pustulosis palmoplantaris. Additional analyses, based on $A_{D,MD}(0,t)$ and $A_{O,M}(0,t)$ utilizing data from the Finsen Institute and a Norwegian study allowing for estimation of the age distribution for natural menopause, indicated that the outbreak of the disease does *not* influence the occurrence of the natural menopause.

The implications of this example for longitudinal data analysis generally are as follows:

1. If the individual level dynamics are interpretable as realizations of a discrete-state, continuous-time stochastic process, and there is no well-developed substantive theory to guide the modeling, then estimation of quantities such as integrated transition rates is best carried out using nonparametric methods such as (1) and (2). This at least provides unbiased estimates of basic descriptive quantities—i.e., the integrated rates—which must constrain any subsequent modeling based on proposed subject-matter theory.

2. Graphical displays of integrated rates such as Figure 2 are a particularly useful guide to parametric specification of the transition rates $r_{ij}(t)$. Figure 2 suggests that curves of the form ct^α with $c > 0$ and $\alpha > -1$ would be appropriate for the Finsen Institute data.

3. If the underlying process is actually Markovian—see Anderson and Goodman [3] for formal tests—then $\{r_{ij}(t)\}$ may be interpreted as entries in a one-parameter family of intensity matrices, $R(t)$, governing the Kolmogorov forward and backward differential equations for the transition probabilities of continuous-time inhomogeneous Markov chains:

$$\frac{\partial P(s,t)}{\partial t} = P(s,t)R(t)$$

$$\frac{\partial P(s,t)}{\partial s} = -R(s)P(s,t),$$

Here $P(t,t) = I$ and $P(s,t)$ has entries interpreted as $p_{ij}(s,t) = \Pr(X(t) = j | X(s) = i)$, and $(X(t), t \geq 0)$ is a realization of the Markov chain and i and j index states.

4. Nonparametric procedures such as (1) and (2) require observations on a process over an interval of time, i.e., knowledge of $(x(t,l), 0 \leq t \leq T_l)$ = (observed realization of the underlying process for individuals labeled $l = 1, 2, \ldots$, etc.). The interval T_l may, in many studies, vary across individuals. Some mild grouping of events—e.g., in the Aalen et al. [2] study, we only know that menopause occurred in some small age range—still allows for reasonable estimates of integrated rates. However, appropriate analogues of these procedures for estimation in panel designs where there are multiple unobserved transitions in a discrete-state, continuous-time process remain to be developed.

Malaria Parasite Interaction in a Common Human Host: An Example of Panel Analysis to Assess Specific Theoretical Proposals. As part of a field study of malaria in Nigeria (see Molineaux and Gramiccia [31]), blood samples from persons in eight clusters of villages were collected on eight occasions, with each pair of surveys separated in time by approximately 10 weeks. For each person at each survey, a blood sample is examined to assess whether that individual is infected with either one or both of two species of malaria parasite called *Plasmodium falciparum* (Pf) and *Plasmodium malariae* (Pm), respectively. We define a four-state stochastic process for an individual's infection status where the states are,

State	Pf	Pm
1	−	−
2	−	+
3	+	−
4	+	+

and (−) means absence of parasites, (+) means presence of parasites.

A question of considerable importance is whether the presence of Pf in an individual's blood makes him (her) more or less resistant to infection by Pm than if there is no Pf. In principle, this question should be answerable using the methods from Aalen's theory of counting processes described in the previous example. Especially, we should first assess whether the possibly time-dependent transition rates per individual at risk of a transition, $r_{12}(t)$ and $r_{34}(t)$, satisfy $r_{12}(t) \equiv r_{34}(t)$. If this null hypothesis is rejected and our estimates suggest that $r_{12}(t) > r_{34}(t)$, then the evidence would favor an interpretation of parasite interaction in which the presence of Pf inhibits the acquisition of Pm. Similarly, a competitive effect would be suggested if $r_{43}(t) > r_{21}(t)$—i.e., the presence of Pf promotes the loss of Pm.

Because data on infection status is collected only every 10 weeks, there can be multiple unobserved transitions among the four states in the continuous-time infection status process. In particular, the times of occurrence of transitions, which are necessary for utilization of (1) and (2), cannot be determined. One response to this problem of missing data relative to a continuously evolving process is to estimate the rates $r_{ij}(t)$ within a class of models that are at least compatible with the observed data. To this end we introduce a time series of time-homogeneous Markov chains, each of which is a candidate for describing the unobserved dynamics between a pair of successive surveys. These models are constrained by the requirement that the transition probabilities of the chain are—to within-sampling variability—coincident with the conditional probabilities $\Pr(X(k\Delta) = i_k | X((k-1)\Delta) = i_{k-1})$ for $k = 0, 1, 2, \ldots, 7$. Here $X(t), 0 \leq t \leq 7\Delta$, is the four-state infection status process; $0, \Delta, 2\Delta, \ldots, 7\Delta$ are the survey dates, $\Delta = 10$ weeks, and i_k may be any one of the states (1,2,3,4) in which an individual is observed at time $k\Delta$.

We illustrate this strategy and its implications in detail in a 4×4 table of transition counts from the WHO survey of malaria in Nigeria mentioned above (see Table 1). The entries in the table are denoted by $N_{ij}(3\Delta, 4\Delta) =$ (number of persons in state i at time 3Δ and in state j at time 4Δ)—e.g., $N_{31}(3\Delta, 4\Delta) = 77$. If these observations are generated by a continuous-time homogeneous Markov chain, there must be a 4×4 stochastic matrix $\mathbf{P}(3\Delta, 4\Delta)$ representable as $\mathbf{P}(3\Delta, 4\Delta) = e^{\Delta \mathbf{R}}$ and such that $N_{ij}(3\Delta, 4\Delta) \approx N_{i+}(3\Delta, 4\Delta)(e^{\Delta \mathbf{R}})_{ij}$. Here $N_{i+}(3\Delta, 4\Delta) = \sum_{j=1}^{4} N_{ij}(3\Delta, 4\Delta) =$ (number of persons in state i at time 3Δ) and $\mathbf{R}$ is a matrix whose entries satisfy $r_{ij} \geq 0$ for $i \neq j$ and $\sum_{j=1}^{4} r_{ij} = 0$. The off-diagonal entries, r_{ij}, are the transition rates per person at risk, and they are constrained by the model to be constant during the 10-week interval Δ. We also introduce the more restricted class of models where $r_{14} = r_{23} = r_{32} = r_{41} = 0$. These zero elements on the minor diagonal of $\mathbf{R}$ exclude the possibility that both Pf and Pm would either be gained or be lost simultaneously.

Introducing the goodness-of-fit measure

$$G^2 = -2 \sum (\text{observed frequency}) \times \log \left(\frac{\text{frequency predicted by the model}}{\text{observed frequency}} \right)$$

$$= -2 \sum_{i,j} N_{ij} \log \left(\frac{N_{i+}(e^{\Delta \mathbf{R}})_{ij}}{N_{ij}} \right),$$

we calculate $\mathbf{R}$, which minimizes this quantity subject to the constraint $r_{14} = r_{23} = r_{32} = r_{41} = 0$. For the data in Table 1, we find that the constrained Markovian model fits the data well and that

$$\hat{\mathbf{R}} = \begin{bmatrix} -0.751 & 0.116 & 0.635 & 0 \\ 3.351 & -3.351 & 0 & 0 \\ 0.764 & 0 & -0.970 & 0.206 \\ 0 & 0.621 & 1.946 & -2.567 \end{bmatrix}.$$

The surprising feature of this matrix is that

$$0.116 = \hat{r}_{12} < \hat{r}_{34} = 0.206$$
$$1.946 = \hat{r}_{43} < \hat{r}_{21} = 3.351.$$

Table 1: Transitions in infection status from survey 4 to survey 5 (i.e., $k = 3$ and 4) for all individuals Aged 19–28 years present at both surveys.

State at Survey 4	State at Survey 5			
	1	2	3	4
1	340	14	171	7
2	21	2	9	0
3	77	3	103	13
4	16	2	20	4

Source: Cohen and Singer [11].

This suggests that, contrary to expectations based on previous literature [10], the presence of Pf *promotes* the acquisition and *reduces* the loss of Pm. Thus there is a cooperative rather than a competitive effect. The same calculations applied to many other 4×4 tables indicated that in each instance a constrained, time-homogeneous Markov model fits the data; and the empirical regularity $\hat{r}_{12} < \hat{r}_{34}$ holds in this population regardless of season or age of the individuals. On the other hand, $\hat{r}_{43} < \hat{r}_{21}$ tends to hold for younger persons, but for individuals over age 44 we typically find $\hat{r}_{43} > \hat{r}_{21}$. In this modeling strategy time variation in the transition rates is measured only by their variation across different pairs of successive surveys. It is important to emphasize, however, that this does not necessarily imply that the infection histories across all eight surveys are representable as a time-inhomogeneous Markov chain. In fact, tests of the Markov property on these data reveal that there is dependence in the infection statuses across several surveys.

A next step in the study of parasite interactions should be the estimation of the transition rates $r_{ij}(t)$ within a model that is compatible with the frequencies

$$n_{i_0, i_1, \ldots, i_7}$$
$$= \text{(number of individuals with infection status } i_0 \text{ at time 0,}$$
$$i_1 \text{ at time } \Delta, i_2 \text{ at time}$$
$$2\Delta, \ldots, \text{etc.}\text{)}.$$

Furthermore, there should be an assessment of whether the important qualitative conclusions about cooperative, as opposed to competitive, effects still hold up in a model which accounts for non-Markovian dependence. It is at this stage that we need procedures analogous to (1) and (2) tailored to data where there are gaps in the observation relative to a continuously evolving process. This is an important but currently unresolved research problem.

The strategy employed in this example is based on a philosophy about modeling longitudinal microdata that is in sharp contrast to the methodology utilized in the example in the preceding section. In particular, we have here adopted the view that one should:

1. Begin with very simple, somewhat plausible classes of models as candidates to describe some portion of the observed data and within which the unobservable dynamics are well defined—e.g., the time series of time-homogeneous Markov chains where each separate model only describes unobserved dynamics between a pair of successive surveys and fits the observed transitions.

2. Estimate and interpret the parameters of interest—e.g., the transition rates r_{ij}—within the simplified models, and then assess whether these models can, in fact, account for finer-grained detail such as the sequence frequencies $n_{i_0, i_1, \ldots, i_7}$.

3. Typically, the original proposed models—they are usually first-order

Markovian across a wide range of subject matter contexts—which may adequately represent data based on pairs of consecutive surveys will not account for higher-order dependences. Such dependences tend to be the rule rather than the exception in longitudinal microdata. We then look for structured residuals from the simple models to guide the selection of more realistic and interpretable specifications (see, e.g., Singer and Spilerman [38] for a discussion of this kind of strategy in a variety of sociology and economics investigations).

The repeated fitting of models and then utilizing structured residuals to guide successively more realistic model selection is a strategy that, on the surface, seems to be very reasonable. However, the process frequently stagnates after only one or two stages because the possible explanations for given structured residuals are usually too extensive to be helpful by themselves. One really needs, in addition, specific subject-matter theories translated into mathematics to guide the model selection process. Unfortunately, in most fields where analysis of longitudinal microdata is of interest, the development of substantive theory is quite weak.

The potential danger of the foregoing strategy, even in the use of transition rate estimates, such as in the malaria example, is that parameter estimates may be biased simply as a result of model misspecification. The biases, in turn, can lead to incorrect conclusions about relationships between events. However, if a process is observed continuously over an interval—producing what is frequently referred to as *event history data*—then this possibility can be avoided by utilizing methods such as (1) and (2), which are not based on strong, substantively indefensible assumptions about dependences in the underlying process. In this connection, an important research problem is to provide guidance about the class of situations for which rates from a time series of Markovian models are good approximations to the corresponding rates estimated within quite general counting processes as in Aalen's [1] theory.

Growth Curves, Polynomial Models, and Tracking. There is frequently a sharp distinction between what one can learn from repeated cross-sectional surveys as opposed to prospective longitudinal designs. An instance of this arises if one asks whether the time trend of population means from repeated cross sections in any way reflects the structure of individual time paths. That the answer is often negative is illustrated by a comparison of the time trend in systolic blood pressure (SBP) for two male cohorts in the Framingham study with the pattern of individual serial measurements. As mentioned previously, data in the Framingham study were collected biennially on the same individuals over a 12-year period. Averaging across persons' SBP measurements at each examination yields the linear trend in Figure 3.

This pattern could also be obtained if independent random samples—cross-sectional data collection—had in fact been utilized. However, an analysis of individual change in SBP over time indicates that these serial measurements are best represented as cubic functions of time. Thus the points in the linear pattern in Figure 3 may be interpreted as averages across cubic functions sampled at discrete times. The cubic relationships, characterizing individual SBP dynamics, are not retrievable from repeated cross sections.

A systematic strategy for ascertaining the foregoing relationships proceeds according to the following steps:

1. If $t_1 < t_2 < \cdots < t_K$ are the examination dates, then introduce the $K \times K$ matrix Φ whose rows ϕ'_l, $l = 0, 1, \ldots,$

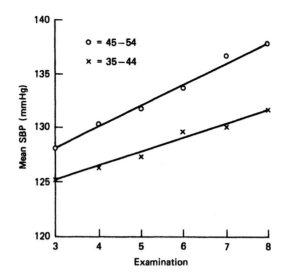

Figure 3: Mean systolic blood pressure (SBP) at examinations 3–8 in Framingham Heart Study: men aged 35–44 and 45–54 at examination 2. (From Ware and Wu [45].)

$K-1$ are the orthogonal polynomials of degree l on $(t_1, \ldots, t_K)$. Let

$$\mathbf{Y}_i = \begin{bmatrix} y_i(t_1) \\ \vdots \\ y_i(t_K) \end{bmatrix}$$

denote the vector of observations for the ith individual and consider the family of models for $\mathbf{Y}_i$ where

$$y_i(t_j) = \sum_{l=0}^{L} \beta_{li}\phi_l(t_j) + \epsilon_{ij},$$
$$L = 0, 1, \ldots, K-1,$$

and ϵ_{ij} is a residual to be interpreted as a value of a normally distributed random variable with mean 0 and variance-covariance structure satisfying the assumptions

$$\mathbf{Y}_i|\boldsymbol{\beta}_i \sim N(\boldsymbol{\Phi}\boldsymbol{\beta}_i, \sigma^2 \mathbf{I}) \quad (3)$$
$$\boldsymbol{\beta}_i \sim N(\boldsymbol{\beta}, \boldsymbol{\Lambda}). \quad (4)$$

Here $N(\boldsymbol{\mu}, \boldsymbol{\Sigma})$ denotes the multivariate normal distribution with mean vector $\boldsymbol{\mu}$ and covariance matrix $\boldsymbol{\Sigma}$. We define $b_{li} = \phi'_l \mathbf{Y}_i$ and determine the degree L of the polynomial describing the time trend in the population as the largest l for which $Eb_{li} \neq 0$. (See Ware and Wu [45, p. 429] for formal tests.) We then estimate $E\mathbf{Y}$ using $\boldsymbol{\Phi}'_L \bar{b}$, where $\bar{b}$ is the vector of sample means of the first L orthogonal polynomial coefficients. This produced the linear relationship shown in Figure 3, based on the Framingham data.

2. The specification (3) and (4) implies that $\mathbf{Y}_i \sim N(\mathbf{X}\boldsymbol{\beta}, \mathbf{X}\boldsymbol{\Lambda}\mathbf{X}^T + \sigma^2\mathbf{I})$, where $\mathbf{X}$ is a design matrix whose columns are orthonormal polynomials on $(t_1, \ldots, t_K)$. This suggests that in order to estimate polynomial representations for individuals we should test the data for compatibility with the covariance structure $\mathbf{X}\boldsymbol{\Lambda}\mathbf{X}^T + \sigma^2\mathbf{I}$ against the alternative of an arbitrary covariance matrix, $\boldsymbol{\Sigma}$. Here $\boldsymbol{\Lambda}$ and σ^2 are estimated by maximum likelihood. By finding the polynomial of

lowest degree for which a likelihood ratio test fails to reject $\mathbf{X\Lambda X}^T + \sigma^2 \mathbf{I}$ as an adequate model, we choose a polynomial model for the individual curves. For details on this kind of procedure, see Rao [34]. When applied to the male cohorts—ages 35–44 and 45–54, respectively, at examination 2—we obtain cubic functions of age characterizing SBP measurements.

An important application of polynomial growth curve models arises in the problem of "tracking." The papers by Ware and Wu [45], Foulkes and Davis [17], and McMahon [29] should each be consulted for alternative views about the concept of tracking. Here one is interested in ascertaining whether an initial ordering of individual observations persists over a prescribed interval of time. For example, in order to carry out early identification of persons at risk of cardiovascular disease it is necessary to know whether children with high blood pressure also tend to have high blood pressure as they grow older. Alternatively, in developmental psychology there is a substantial literature dealing with individual differences and their possible stability with increasing age for persons in the same birth cohort. L. Terman's classic follow-up study of gifted children raises the question of stability of individual differences in terms of performance on tests. This is precisely the question of tracking as set forth in the biometry literature and that has received in-depth consideration in the context of longitudinal data analysis only in the past few years.

For an assessment of tracking in a time interval $[T_1, T_2]$ we introduce the index

$$\begin{aligned}\gamma(T_1, T_2) &= \Pr(f(t, \boldsymbol{\beta}_j) \geq f(t, \boldsymbol{\beta}_i) \\ &\quad \text{for all } t \text{ in } [T_1 T_2] \\ &\quad \text{or } f(t, \boldsymbol{\beta}_j) < f(t, \boldsymbol{\beta}_i) \\ &\quad \text{for all } t \text{ in } [T_1, T_2]),\end{aligned}$$

where $f(t, \boldsymbol{\beta}_i)$ and $f(t, \boldsymbol{\beta}_j)$ are time paths of two randomly chosen individuals. With polynomial specifications, $f(t, \boldsymbol{\beta}_i)$ may be written in the form

$$f(t, \boldsymbol{\beta}_i) = \beta_{i0} + \beta_{i1} t + \cdots + \beta_{iL} t^L.$$

Foulkes and Davis [17] propose the rule that no tracking will be said to occur if $\gamma(T_1, T_2) < \frac{1}{2}$. Then for estimated values in the interval $(\frac{1}{2}, 1)\gamma$ may be interpreted as a measure of the extent of tracking.

In an interesting application of this idea, Foulkes and Davis [17] utilize data assembled by Grizzle and Allen [20] to assess the quantity of coronary uric potassium (in milliequivalents per liter) following a coronary occlusion in three groups of dogs. The assessment is based on measurements taken at 2-minute intervals during the first 13 minutes following coronary occlusion. The three populations consist of (1) 9 control dogs, (2) 10 dogs with extrinsic cardiac denervation three weeks prior to coronary occlusion, and (3) 8 dogs subjected to extrinsic cardiac denervation immediately prior to coronary occlusion. They find that for individuals in each population, the amount of potassium is representable as a cubic function of time. Furthermore, the tracking index $\gamma(1, 13)$ is estimated to be 0.444, 0.711, and 0.500 in groups 1–3, respectively, with corresponding standard errors given by 0.059, 0.060, and 0.094. This suggests that only the group (2) dogs track—i.e., those with extrinsic cardiac denervation three weeks prior to coronary occlusion.

Repeat Victimization: Detecting Regularities in Turnover Tables. One of the original aims of the National Crime Survey [33] was the measurement of annual change in crime incidents for a limited set of major crime categories. However, longitudinal analysis is facilitated by the fact that the basic sample is divided into six rotation groups of approximately 10,000 housing units each. The occupants of each housing unit are interviewed every 6 months over a 3-year period. For individuals victimized at least once in a given 6-month interval, a detailed record of their victimization history in that period is collected. This retrospective information forms the basis of individual victimization histories. For a detailed critique of the NCS design and of the measurement of criminal victimization generally, see Penick and Owen [33] and Fienberg [15].

A question of considerable interest and importance is whether persons, or households, that are victimized 2 or more times within a 3-year period tend to be victimized in the same or a similar manner. A useful first cut at this kind of question can be developed by preparing a turnover table of the frequency with which a particular succession of crimes is committed on the same individual or household. To this end, Table 2 lists the number of successive victimizations for eight major crime categories in households with two or more victimizations from July 1, 1972, to December 31, 1975. Although there is no natural order relation among these categories, similar types of crime are listed, to the extent possible, in adjacent rows and columns. In particular, crimes of personal violence are in rows 1–3, while those involving theft without personal contact are in rows 5–7.

Informal examination of the table suggests that in repeatedly victimized households there is a strong propensity for persons to be victimized in the same way two or more times. Formal support for this claim, together with some refinements to crimes of similar character, can be obtained by the following strategy.

1. First test whether the transition probabilities p_{ij} = Pr (second victimization is of type j|first victimization is of type i) are such that the row proportions in Table 2 are homogeneous. This hypothesis is, as you would expect, clearly rejected.

2. Prepare a new table of counts in which the diagonal entries in Table 2 are deleted and again test the hypothesis of homogeneous row proportions. This constrained specification is much closer to satisfying the baseline model of homogeneous row proportions than the original data.

3. Delete cells from the original table in diagonal blocks in each that list crimes of similar type. Then test the hypothesis of homogeneous row proportions on the reduced table. A pattern of deletions that yields a table consistent with this hypothesis is shown in Figure 4. See Fienberg [15] for more details on this strategy as applied to victimization data.

What this analysis suggests is that relative to a table with homogeneous row proportions, there is elevated repeat victimization involving crimes of similar type. This example is the prototype of strategies that use residuals from simple baseline models—quasi-independence or quasi-homogeneity—to detect special regularities in turnover tables.

45.5 Brief Guide to Other Literature

The remarks and examples in the two preceding sections can, at best, convey a rudimentary impression of the issues involved in analyses of longitudinal data. Some important topics—e.g., strategies for incorporating measurement error into process models and the introduction of continuous-time, continuous-state processes as covariates in survival analyses—were hardly mentioned at all. Thus it seems prudent to

Table 2: Repeat victimization data for eight major crime categories[a]

First Victimization in Pair	Second Victimization in Pair								
	Rape	Assault	Robbery	Purse Snatching/ Pocket Picking	Personal Larceny	Burglary	Household Larceny	Motor Vehicle Theft	Totals
Rape	26	50	11	6	82	39	48	11	273
Assault	65	2,997	238	85	2,553	1,083	1,349	216	8,586
Robbery	12	279	197	36	459	197	221	47	1,448
Purse snatching/ pocket picking	3	102	40	61	243	115	101	38	703
Personal larceny	75	2,628	413	229	12,137	2,658	3,689	687	22,516
Burglary	52	1,117	191	102	2,649	3,210	1,973	301	9,595
Household larceny	42	1,251	206	117	3,757	1,962	4,646	391	12,372
Motor vehicle theft	3	221	51	24	678	301	367	269	1,914
Total	278	8,645	1,347	660	22,558	9,565	12,394	1,960	57,407

Source: Reiss [35].

[a] Reported crimes by households with two or more victimizations while in survey July 1, 1972, to December 31, 1975.

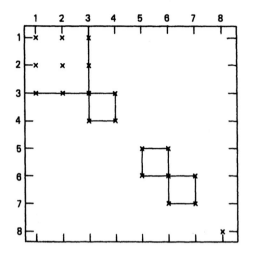

Figure 4: Deletion pattern applied to Table 2 that yields a table with transition probabilities such that row proportions are homogeneous. ×, cell deleted.

mention some of the literature that an interested reader could follow up to become acquainted with other aspects of longitudinal data analysis.

1. For an enlightening discussion of biases due to measurement error in panel surveys, see Williams and Mallows [48] and Williams [47]. A nice discussion of measurement models and unobservables with particular emplasis on the psychology literature appears in Bentler [7].

2. Tradeoffs between data collection design, particularly prospective vs. retrospective studies, are important to understand, and very nicely discussed in Schlesselman [37]. The impact of different designs on analytical strategies is lucidly treated in Hoem and Funck-Jensen [21] with particular emphasis on demography. See also Goldstein [19]. Experimental studies in biology, medicine, and psychology called *repeated measurement de-*

signs represent an important area discussed in the review papers of Koch et al. [23],24].

3. The introduction of time-varying covariates in survival models, particularly as stochastic process models, is an important and quite recent development. For an insightful discussion emphasizing problems in labor economics, see Flinn and Heckman [16]. In the context of medical epidemiology, the analysis by Woodbury et al. [49] is quite illuminating.

References

1. Aalen, O. (1978). *Ann. Statist.*, **6**, 701–726.

2. Aalen, O., Borgan, Ø., Keiding, N., and Thormann, J. (1980). *Scand. J. Statist.*, **7**(4), 161–171.

3. Anderson, T. W. and Goodman, L. (1957). *Ann. Math. Statist.*, **28**, 89–110.

4. *Annals de l'INSEE* (1978). *The Econometrics of Panel Data*, pp. 30–31, INSEE, Paris.

5. Bartholomew, D. J. (1982). *Stochastic Models for Social Processes*, 3rd ed. Wiley, New York.

6. Bekessy, A., Molineaux, L., and Storey, J. (1976). *Bull. WHO*, **54**, 685–693.

7. Bentler, P. (1980). *Annu. Rev. Psychol.*, **31**, 419–456.

8. Berelson, B., Lazarsfeld, P., and McPhee, W. (1954). *Voting*. University of Chicago Press, Chicago.

9. Blumen, I., Kogan, M., and McCarthy, P. J. (1955). *The Industrial Mobility of Labor as a Probability Process*. Cornell Stud. Ind. Labor Relations, Vol. 6. Cornell University Press, Ithaca, NY.

10. Cohen, J. E. (1973). *Quart. Rev. Biol.*, **48**, 467–489.

11. Cohen, J. E. and Singer, B. (1979). In *Lectures on Mathematics in the Life Sciences*, Vol. 12, S. Levin, ed., pp. 69–133,

American Mathematical Society, Providence, RI.

12. Crowley, J. and Hu, M. (1977). *J. Am. Statist. Assoc.*, **72**, 27–36.

13. Dawber, T. (1980). *The Framingham Study*. Harvard University Press, Cambridge, MA.

14. Duncan, O. D. (1981). In *Sociological Methodology*, S. Leinhardt, ed. Jossey-Bass, San Franscisco, CA.

15. Fienberg S. (1980). *Statistician*, **29**, 313–350.

16. Flinn, C. J. and Heckman, J. J. (1982). *Adv. Econometr.*, **1**, 35–95.

17. Foulkes, M. A. and Davis, C. E. (1981). *Biometrics*, **37**, 439–446.

18. Gilbert, J. P., Meier, P., Rümke, C. L., Saracci, R., Zelen, M., and White, C. (1975). *J. Am. Med. Assoc.*, **231**, 583–608.

19. Goldstein, H. (1979). *The Design and Analysis of Longitudinal Studies*. Academic Press, New York.

20. Grizzle, J. and Allen, D. (1969). *Biometrics*, **25**, 357–381.

21. Hoem, J. and Funck-Jensen (1982). In *Multidimensional Mathematical Demography*, K. Land and A. Rogers, eds. Academic Press, New York.

22. Karlin, S., Cameron, E., and Chakraborty, R. (1983). *Am. J. Hum. Genet.*, **35**, 695–732.

23. Koch, G., Amara, I., Stokes, M., and Gillings, D. (1980). *Int. Statist. Rev.*, **48**, 249–265.

24. Koch, G., Gillings, D., and Stokes, M. (1980). *Annu. Rev. Public Health*, **1**, 163–225.

25. Kohn, M. and Schooler, C. (1978). *Am. J. Sociol.*, **84**, 24–52.

26. Lazarsfeld, P. F. and Fiske, M. (1938). *Public Opinion Quart.*, **2**, 596–612.

27. Levy, D. E., Bates, D., Caronna, J., Cartlidge, N., Knill-Jones, R., Lapinski, R., Singer, B., Shaw, D., and Plum, F. (1981). *Ann. Intern. Med.*, **94**, 293–301.

28. Littman, G. and Mode, C. J. (1977). *Math. Biosci.*, **34**, 279–302.

29. McMahon, C. A. (1981). *Biometrics*, **37**, 447–455.

30. Menken, J., Trussell, T. J., Stempel, D., and Balakol, O. (1981). *Demography*, **18**, 181–200.

31. Molineaux, L. and Gramiccia, G. (1980). *The Garki Project: Research on the Epidemiology and Control of Malaria in the Sudan Savanna of West Africa*. WHO, Geneva.

32. Molineaux, L., Storey, J., Cohen, J. E., and Thomas, A. (1980). *Am. J. Trop. Med. Hygiene.*, **29**, 725–737.

33. Penick, B. K. and Owens, M. E. B., eds. (1976). *Surveying Crime* (Report of Panel for the Evaluation of Crime Surveys.) National Academy of Sciences, Washington, DC.

34. Rao, C. R. (1965). *Biometrika*, **52**, 447–458.

35. Reiss, A. J. (1980). In *Indicators of Crime and Criminal Justice: Quantitative Studies*, S. Fienberg and A. Reiss, eds. U.S. Government Printing Office, Washington, DC.

36. Rogosa, D. (1980). *Psychol. Bull.*, **88**, 245–258.

37. Schlesselman, J. J. (1982). *Case Control Studies*. Oxford University Press, Oxford, UK.

38. Singer, B. and Spilerman, S. (1976). *Ann. Econ. Soc. Meas.*, **5**, 447–474.

39. Sydenstricker, E. (1927). A Study of Illness in a General Population Group. *Hagerstown Morbidity Stud. No. 3, Public Health Rep.*, p. 32.

40. Taeuber, K. E., Chiazze, L., Jr., and Haenszel, W. (1968). *Migration in the United States*. US Government Printing Office, Washington, DC.

41. Terman, L. M. and Oden, M. H. (1959). *The Gifted Group at Mid-Life. Genetic Studies of Genius*, Vol. 5. Stanford University Press, Stanford, CA.

42. Terman, L. M., Burks, B. S., and Jensen, D. W. (1930). *The Promise of Youth*. Genetic Studies of Genius, Vol. 3. Stanford University Press, Stanford, CA.

43. Tinbergen, J. (1940). *Rev. Econ. Stud.*, 73–90.

44. Tuma, N., Hannan, M., and Groeneveld, L. (1979). *Am. J. Sociol.*, **84**, 820–854.

45. Ware, J. H. and Wu, M. C. (1981). *Biometrics*, **37**, 427–437.

46. White, H. C. (1970). *Chains of Opportunity*. Harvard University Press, Cambridge, MA.

47. Williams, W. H. (1978). In *Contributions to Survey Analysis and Applied Statistics*, H. A. David, ed., pp. 89–112, Academic Press, New York.

48. Williams, W. H. and Mallows, C. L. (1970). *J. Am. Statist. Assoc.*, **65**, 1338–1349.

49. Woodbury, M. A., Manton, K. G., and Stallard, E. (1979). *Biometrics*, **35**, 575–585.

50. Wright, S. (1983). *Am. J. Hum. Genet.*, **35**, 757–768.

46 Meta-Analysis

Larry V. Hedges

Meta-analysis is the use of statistical methods in reviews of related research studies. For example, meta-analyses often involve the use of statistical methods to combine estimates of treatment effects from different research studies that investigate the effects of the same or related treatments. Meta-analysis is distinguished from primary analysis (the original analysis of a dataset) and secondary analysis (reanalysis of another's data) by the fact that meta-analyses do not usually require access to raw data, but only to summary statistics. Thus the data points for meta-analyses are summary statistics, and a sample of studies in meta-analysis is analogous to a sample of subjects or respondents in primary analysis.

Meta-analysis in the social and medical sciences began as an attempt to better utilize the evidence from increasingly large numbers of independent experiments. Important scientific and policy questions often stimulate dozens of research studies designed to answer essentially the same question. Meta-analysis provides a procedure for extracting much of the information in these studies to provide a broader base for conclusions than is possible in any single study and to increase statistical power.

Meta-analyses are much like original research projects and involve the same general stages of research procedures such as problem formulation, data collection, data evaluation, data analysis and interpretation, and presentation of results [4]. Subjective decisions about procedure are necessary at each stage, and the use of an overall plan or protocol is often useful to avoid biases by constraining procedural variations. Such a protocol describes the details of procedures in each state of the meta-analysis.

Problem formulation is the process of stating precise questions or hypotheses in operational terms that make it possible to decide whether a given study examines the hypothesis of interest. Even studies that are superficially similar usually differ in the details of treatment, controls, procedures, and outcome measures, and consequently problem formulation involves the specification of the range of acceptable variation in these details. For example, consider studies of the effects of a drug. The treatment consists of administration of the drug but what dosage level, schedule and modality of administration, etc., should count as instances of the "same" treatment? Should control conditions involve placebos, alternative treatments, or the conventional treatment (and if so, are there variations in the conventional treatment)? Outcomes like death rates seem unambiguous, but calculation of death rates may not be identical across studies. For example, should deaths from all causes be included or should only deaths from the disease under treatment be considered (and if so what about deaths due to side effects of treatment)? The breadth of the question or hypothesis of the meta-analysis has implications for the range of acceptable variation in constructs of treatment, control, outcome, and so on.

Data collection in meta-analysis consists of assembling the studies that may provide relevant data and extracting an estimate of effect magnitude from each study. Sometimes (as in studies of the efficacy of a new drug) all of the studies that were conducted are immediately available, but more frequently studies must be obtained from the published or unpublished literature in a substantive research area. The method used to obtain the sample of studies requires careful attention because some procedures used to search for studies may introduce biases. For example, published journal articles may be a selected sample of the set of all research studies actually conducted, overrepresenting studies that yield statistically significant results [7, Chap. 14; 11]. Statistical corrections for selection are sometimes possible [7], but exhaustive enumeration (and then perhaps sampling) of studies is usually advisable.

Data collection also involves the selection of an estimate of effect magnitude from each study. Typical indices of effect magnitude compare a treatment group with a control group via indices such as a raw mean differences, raw or transformed differences in incidence proportions, risk ratios, odds ratios, or correlation coefficients. The standardized mean difference $\delta = (\mu_1 - \mu_2)/\sigma$, where σ is a within-group standard deviation, is frequently used in the social and behavioral sciences as a scale-free index of treatment effect or as an index of overlap between distributions (i.e., a one-dimensional Mahalanobis distance).

Data evaluation is the process of deciding which possibly relevant studies should be included in the data analysis. Sometimes the criteria for study inclusion will be straightforward (such as all randomized clinical trials) but frequently inclusion criteria that are more subjective (such as well-controlled and possibly nonrandomized trials). Because serious biases can arise following decisions about study inclusion, protocols for decisions about inclusion of studies may be useful in just the same way as they are necessary in decisions for inclusion of patients in clinical trials. Such protocols may also enhance the reliability of decisions about study quality, which may otherwise be quite low [4,9]. Some investigators also suggest the use of "blind" ratings of study quality in which raters are not aware of the *findings* of the studies that they rate. Empirical methods for detecting outliers or influential observations may also have a role in data evaluation. Such methods sometimes reveal problem studies that were initially overlooked [7].

Data analysis is the process of combining estimates of effect magnitude across studies. The analyses vary depending on the conceptualization of between-study variability used in the data analysis model. Different models lead to different interpretations of results and can lead to very different estimates of the precision of results. The models differ primarily in whether they treat between-study differences as fixed or random and consequently whether between-study variability should be incorporated into the uncertainty of the combined result.

Let $\delta_1, \ldots, \delta_k$ be effect magnitude parameters from k studies, and let $d_1, \ldots, d_k$ and $S_1, \ldots, S_k$, respectively, be the corresponding sample estimates and their approximate standard errors. The simplest and most common procedures treat

$\delta_1, \ldots, \delta_k$ as if $\delta_1 = \cdots = \delta_k = \delta$ and estimate the common effect δ. The procedures most frequently used to estimate δ involve a weighted mean $\bar{d}$ in which the ith weight is proportional to $1/S_i^2$. The weighted sum of squares $\Sigma(d_i - \bar{d})^2/S_i^2$ about the weighted mean $\bar{d}$ is often used as a statistic to test the consistency of the δ_i across studies. (see Hedges and Olkin [7] or Cochran [2]). A test of homogeneity of effects may reveal that the assumption of a common δ is unrealistic. In this situation one of three approaches to the situation is usually used.

The fixed-effects approach treats between-study variation in effect magnitudes as if it were a function of known explanatory variables that arise as characteristics of studies (e.g., aspects of particular treatments, controls, procedures, or outcome measures). Fixed-effects analyses usually estimate the vector $\boldsymbol{\beta}$ of unknown coefficients in a linear model of the form

$$\delta_i = \mathbf{x}_i'\boldsymbol{\beta}, \qquad i = 1, \ldots, k,$$

where $\mathbf{x}_i$ is a vector of study characteristics [7]. The estimates of $\boldsymbol{\beta}$ provide insight into the relationship between study characteristics and effect magnitude. Tests of the goodness of fit of the model (analogous to the test of homogeneity given above) can help to determine the degree to which the data are consistent with the model. For example, Becker and Hedges [1] used a linear model to estimate the historical trend in gender differences in cognitive abilities by using the year that a study was conducted to explain between-study variations in the magnitude of sex differences in cognitive performance.

An alternative to fixed-effects models are simple random-effects models in which the effect magnitudes $\delta_1, \ldots, \delta_k$ are modeled via $\delta_i = \delta + \eta_i$, where $\eta_1, \ldots, \eta_k$ are treated as independent random variables with a mean of zero and a variance of σ_η^2. This conceptualization is particularly appealing when the treatments used in studies exhibit substantial uncontrolled variability that may plausibly be treated as "random." Random-effects analyses of this sort usually concentrate on estimation of δ and the variance component σ_η^2 (see Hedges and Olkin [7] or Cochran and Cox [3]). For example, Schmidt and Hunter [10] used a random-effects model to study variation in (population) correlation coefficients that were used as indicators of the validity of psychological tests used in personnel selection.

More complex random-effects or mixed models involve the use of both explanatory variables and random effects. The effect magnitudes might be modeled via

$$\delta_i = \mathbf{x}_i'\boldsymbol{\beta} + \eta_i, \qquad i = 1, \ldots, k,$$

where $\mathbf{x}_i$ is a vector of known study characteristics, $\boldsymbol{\beta}$ is a vector of coefficients, and $\eta_1, \ldots, \eta_k$ are independent, identically distributed random variables with zero mean and variance σ_η^2. Such models are sometimes more realistic when effect magnitudes are more variable than would be expected from fixed-effects models with no random contribution.

The presentation of the results of a meta-analysis should include a description of the formal protocol as well as any other steps of problem formulation, data collection, data evaluation, and data analysis that are likely to affect results. It is usually helpful to provide a brief tabular summary that presents relevant characteristics of each of the individual studies along with the computed index of effect magnitude for each study.

References

1. Becker, B. J. and Hedges, L. V. (1984). *J. Educ. Psychol.*, **76**, 583–587.

2. Cochran, W. G. (1954). *Biometrics*, **10**, 101–129.

3. Cochran, W. G. and Cox, G. M. (1957). *Experimental Design*, 2nd ed. Wiley, New York. (Chapter 14 is a discussion of the analysis of series of experiments.)

4. Cooper, H. M. (1984). *The Integrative Literature Review*. Sage, Beverly Hills, CA. (Includes a discussion of problem formulation, data collection, and data evaluation.)

5. Fleiss, J. L. (1973). *Statistical Methods for Rates and Proportions*. Wiley, New York. (Chapter 10 is a discussion of combining analyses of contingency tables.)

6. Glass, G. V., McGaw, B., and Smith, M. L. (1981). *Meta-Analysis in Social Research*. Sage, Beverly Hills, CA. (A good introduction to the perspective of meta-analysis although its treatment of statistics is somewhat dated.)

7. Hedges, L. V. and Olkin, I. (1985). *Statistical Methods for Meta-Analysis*. Academic, New York. (A comprehensive review of statistical methods for meta-analysis using standardized mean differences and correlation coefficients.)

8. Light, R. J. and Pillemer, D. (1984). *Summing Up: The Science of Reviewing Research*. Harvard University Press, Cambridge, MA. (A good treatment of conceptual issues in combining research results.)

9. Rosenthal, R. (1984). *Meta-Analytic Procedures for Social Research*. Sage, Beverly Hills, CA. (A very clear introduction to selected statistical procedures for meta-analysis.)

10. Schmidt, F. L. and Hunter, J. (1977). *J. Appl. Psychol.*, **62**, 529–540.

11. Sterling, T. C. (1959). *J. Amer. Statist. Ass.*, **54**, 30–34.

47. Missing Data: Sensitivity Analysis

Geert Molenberghs

47.1 Introduction

Early work on missing values was largely concerned with algorithmic and computational solutions to the induced lack of balance or deviations from the intended study design [1,10]. General algorithms such as expectation–maximization (EM) [5], and data imputation and augmentation procedures [38] combined with powerful computing resources have largely provided a solution to this aspect of the problem. There remains the very difficult and important question of assessing the impact of missing data on subsequent statistical inference.

When referring to the missing-value or nonresponse process, we will use terminology of Rubin [37] and Little and Rubin [26] (Chapter 6). A nonresponse process is said to be *missing completely at random* (MCAR) if the missingness is independent of both unobserved and observed data and *missing at random* (MAR) if, conditional on the observed data, the missingness is independent of the unobserved measurements. A process that is neither MCAR nor MAR is termed *missing not at random* (MNAR). In the context of likelihood inference, and when the parameters describing the measurement process are functionally independent of the parameters describing the missingness process, MCAR and MAR are *ignorable*, while a nonrandom process is nonignorable.

Many methods are formulated as selection models [26] as opposed to pattern–mixture modeling (PMM; [22,23]). A selection model factors the joint distribution of the measurement and response mechanisms into the marginal measurement distribution and the response distribution, conditional on the measurements. This is intuitively appealing since the marginal measurement distribution would be of interest also with complete data. Little and Rubin's taxonomy is most easily developed in the selection setting. Parameterizing and making inference about the effect of treatment and its evolution over time is straightforward in the selection model context.

In the specific case of a clinical trial setting, standard methodology used to analyze longitudinal data subject to nonresponse is mostly based on such methods as *last observation carried forward* (LOCF), *complete case analysis* (CCA), or simple forms of imputation. This is often done without questioning the possible influence of these assumptions on the final results, even though several authors have written about this topic. A relatively early account is given in Heyting, et al. [12]. Mallinckrodt et al. [27,28] and Lavori, et al. [21] propose direct-likelihood and multiple-imputation methods respectively, to deal with incomplete longitudinal data. Siddiqui and Ali [42] compare direct-likelihood and LOCF methods.

In realistic settings, the reasons for dropout are varied and it is therefore difficult to fully justify on a priori grounds the assumption of MAR. At first sight, this calls for a further shift toward MNAR models. However, some careful considerations have to be made, the most important one of which is that no modeling approach, whether MAR or MNAR, can recover the lack of information that occurs because of the incompleteness of the data. First, under MAR, a standard analysis would follow if one would be entirely sure of the MAR nature of the mechanism. However, it is only rarely that such an assumption is known to hold [33]. Nevertheless, ignorable analyses may provide reasonably stable results, even when the assumption of MAR is violated, in the sense that such analyses constrain the behavior of the unseen data to be similar to that of the observed data. A discussion of this phenomenon in the survey context has been given in Rubin, et al. [40]. These authors argue that in well-conducted experiments (some surveys and many confirmatory clinical trials), the assumption of MAR is often to be regarded as a realistic one. Second, and very important for confirmatory trials, an MAR analysis can be specified a priori without additional work relative to a situation with complete data. Third, while MNAR models are more general and explicitly incorporate the dropout mechanism, the inferences they produce are typically highly dependent on the untestable and often implicit assumptions built in regarding the distribution of the unobserved measurements,

given the observed ones. The quality of the fit to the observed data need not reflect at all the appropriateness of the implied structure governing the unobserved data. This point is irrespective of the MNAR route taken, whether a parametric model of the type of Diggle and Kenward [6,43] or a semiparametric approach such as in Robins, et al. [36] is chosen. Hence, in any incomplete-data setting, there cannot be anything that could be termed a *definitive analysis*. On the basis of these considerations, we recommend to always explore the impact of deviations from the MAR assumption on the conclusions using sensitivity analysis tools.

We could define a sensitivity analysis as one in which several statistical models are considered simultaneously and/or where a statistical model is further scrutinized using specialized tools (such as diagnostic measures). This rather loose and very general definition encompasses a wide variety of useful approaches. The simplest procedure is to fit a selected number of (nonrandom) models that are all deemed plausible or one in which a preferred (primary) analysis is supplemented with a number of variations. The extent to which conclusions (inferences) are stable across such ranges provides an indication about the belief that can be put into them. Variations to a basic model can be constructed in different ways. The most obvious strategy is to consider various dependencies of the missing data process on the outcomes and/or on covariates. Alternatively, the distributional assumptions of the models can be changed.

47.2 Sensitivity Analysis for Contingency Tables

At the technical level, it is not difficult to formulate models for the MNAR setting, that is, models in which the probability of an outcome being missing depends on unobserved values. The observed data likelihood is then obtained by integrating over the distribution of the missing data. Little [25] provides a review of such approaches. However, there is a fundamental interpretational problem. Molenberghs, et al. [30] provided examples, in the contingency table setting, where different MNAR models that produce the same fit to the observed data are different in their prediction of the unobserved counts. This implies that such models cannot be examined using data alone. Indeed, even if two models fit the observed data equally well, one still needs to reflect on the plausibility of the assumptions made. A number of issues are listed in Reference 30. Similar problems manifest themselves in the continuous setting.

Such problems with MNAR models do not imply, however, that they are of no value. In the first place, many of these issues apply equally well to MAR models, which have no a priori justification: an MAR model can usually be formulated as a special member of a general family of MNAR models, although it may be easier to fit. It might be argued then that one role of MNAR models is to supplement information obtained from the MAR model. The concept of fitting a single model is then replaced by that of *sensitivity analysis*, where several plausible MNAR models are contrasted. This route has been advocated by Vach and Blettner [46].

Thus, a natural way to proceed is to acknowledge the inherent ambiguity and explore the range of inferences that is consistent with the gap in our knowledge. Kenward, et al. [16] have attempted to formalize this idea. See also Reference 35. Indeed, while there is a formal mathematical statistical framework for imprecision (variance, standard errors, sampling distributions, confidence intervals, hypothesis tests, and so on), most implementations of sensitivity analysis have remained ad hoc. These authors have developed a simple framework to formalize sensitivity concepts. To this end, a language is needed to describe *ignorance* (due to incompleteness of the data) and the familiar *imprecision* (due to finite sampling) and to combine both into *uncertainty*.

While the conduct of an informal sensitivity analysis is enlightening, it does not remove all concerns. Indeed, there is no guarantee that, by considering a number of models, the resulting family of intervals will provide a good coverage of all (nonignorable) models within a class of plausible models. A formal sensitivity analysis strategy that addresses this issue is discussed next.

It is useful to distinguish between two types of *statistical uncertainty*. The first one, *statistical imprecision*, is due to finite sampling. However, even if all would have been included, there would have been residual uncertainty because some fail to report at least one answer. This second source of uncertainty, due to incompleteness, will be called *statistical ignorance*.

Statistical imprecision is classically quantified by means of estimators (standard error and variance, confidence region, etc.) and properties of estimators (consistency, asymptotic distribution, efficiency, etc.). In order to quantify statistical ignorance, it is useful to distinguish between complete and observed data. Let us focus on two binary questions. There are 9 observed cell counts, whereas the hypothetical complete data would have 16 counts.

A sample produces empirical proportions representing the probabilities with error. This imprecision disappears as the sample size tends to infinity. What remains is ignorance regarding the redistribution of all but the first four probabilities over the missing outcomes value. This leaves ignorance regarding any probability in which at least one of the first or second indices is equal to 0, and hence regarding any derived parameter of scientific interest.

For such a parameter, θ, a region of possible values that is consistent with the observed data is called a *region of ignorance*. Analogously, an observed incomplete table leaves ignorance regarding the would-be observed complete table, which in turn leaves imprecision regarding the true complete probabilities. The region of estimators for θ consistent with the observed data provides an estimated region of ignorance. The $(1-\alpha)100\%$ *region of uncertainty* is a larger region in the spirit of a confidence region, designed to capture the combined effects of imprecision and ignorance. Various ways for constructing regions of ignorance and regions of uncertainty are conceivable [47].

47.3 Selection Models and Local Influence

Particularly within the selection modeling framework, there has been an increasing literature on nonrandom missing data. At the same time, concern has been growing precisely about the fact that models often rest on strong assumptions and relatively little evidence from the data themselves. This point was already raised by Glynn, et al. [9], who indicate that this is typical for the so-called selection models, whereas it is much less so for a pattern-mixture model. Much of the debate on selection models is rooted in the econometrics literature, in particular, Heckman's selection model [11]. Draper [7] and Copas and Li [4] provide useful insight in model uncertainty and nonrandomly selected samples. Vach and Blettner [46] study the case of incompletely observed covariates in logistic regression.

Because the model of Diggle and Kenward [6] fits within the class of selection models, it is fair to say that it raised, at first, too high expectations. This was made clear by many discussants of the paper. It implies that, for example, formal tests for the null hypothesis of random missingness, although technically possible, should be approached with caution.

In response to these concerns, there is growing awareness of the need for methods that investigate the sensitivity of the results with respect to the model assumptions in selection models. See, for example, References 8, 18, 20, 24, 30, 34, and 39. Still, only few actual proposals have been made. Moreover, many of these are to be considered useful but ad hoc approaches. Whereas such informal sensitivity analyses are an indispensable step in the analysis of incomplete longitudinal data, it is desirable to conduct more formal sensitivity analyses.

In any case, fitting a nonrandom dropout model should be subject to careful scrutiny. The modeler needs to pay attention, not only to the assumed distributional form of his or her model [15,23], but also to the impact one or a few influential subjects may have on the dropout and/or measurement model parameters. Because fitting a nonrandom dropout model is feasible by virtue of strong assumptions, such models are likely to pick up a wide variety of influences in the parameters describing the nonrandom part of the dropout mechanism. Hence, a good level of caution is in place.

Verbeke et al. [50] and Molenberghs et al. [32] adapted the model of Diggle and Kenward [6] to a form useful for sensitivity analysis. Such a sensitivity analysis method, based on local influence [3,45], is relatively easy to conduct, provides interpretable quantities, and opens up an number of roots, many of them still unexplored, to focus on particular quantities. Verbeke and Molenberghs [49] made a comparison with a more conventional global influence analysis [2]. Van Steen et al. [48] and Jansen et al. [14] proposed similar approaches for categorical data settings.

47.4 Pattern-Mixture Modeling Approach

Fitting pattern-mixture models can be approached in several ways [44]. It is important to decide whether pattern-mixture and selection modeling are to be contrasted with one another or rather the pattern-mixture modeling is the central focus [29].

In the latter case, it is natural to conduct an analysis, and preferably a sensitivity analysis, *within* the pattern-mixture family. The key area where sensitivity analysis should be focused is on the unidentified components of the model and the way(s) in which this is handled. Before doing so, it is relevant to reflect on the somewhat "paradoxical" nature of this underidentification. Assume we have two measurements where Y_1 is always observed and Y_2 is either observed ($t=2$) or missing ($t=1$). Let us further simplify the notation by suppressing dependence on parameters and denote $g(t|y_1, y_2) = f(t|y_1, y_2)$, $p(t) = f(t)$, and $f_t(y_1, y_2) = f(y_1, y_2|t)$. Equating the selection model and pattern-mixture model factorizations yields

$$f(y_1,y_2)g(d=2|y_1,y_2) = f_2(y_1,y_2)p(t=2),$$
$$f(y_1,y_2)g(d=1|y_1,y_2) = f_1(y_1,y_2)p(t=1).$$

Since we have only two patterns, this obviously simplifies further to

$$f(y_1,y_2)g(y_1,y_2) = f_2(y_1,y_2)p,$$
$$f(y_1,y_2)[1-g(y_1,y_2)] = f_1(y_1,y_2)[1-p],$$

of which the ratio yields

$$f_1(y_1,y_2) = \frac{1-g(y_1,y_2)}{g(y_1,y_2)}\frac{p}{1-p}f_2(y_1,y_2).$$

All selection model factors are identified, as are the pattern-mixture quantities on the right-hand side. However, the left-hand side is not entirely identifiable. We can further separate the identifiable from the nonidentifiable quantities:

$$f_1(y_2|y_1) = f_2(y_2|y_1)\frac{1-g(y_1,y_2)}{g(y_1,y_2)} \times \frac{p}{1-p}\frac{f_2(y_1)}{f_1(y_1)}. \quad (1)$$

In other words, the conditional distribution of the second measurement given the first one, *in the incomplete first pattern*, about which there is no information in the data, is identified by equating it to its counterpart from the complete pattern modulated via the ratio of the "prior" and "posterior" odds for dropout $[p/(1-p)$ and $g(y_1,y_2)/(1-g(y_1,y_2))$ respectively] and via the ratio of the densities for the first measurement.

Thus, although an identified selection model is seemingly less arbitrary than a pattern-mixture model, it incorporates *implicit* restrictions. Indeed, precisely these are used in Equation (1) to identify the component for which there is no information. This clearly illustrates the need for sensitivity analysis. It also shows that Equation (1) can be used as a tool to reflect on the meaningfulness of the chosen parameterization in one family, by considering its implications in the other family.

Within the pattern-mixture family, we can consider three strategies to deal with under identification.

Strategy 1. Little [22–24] advocated the use of identifying restrictions and presented a number of examples. We will outline a general framework for identifying restrictions. This strategy allows construction of MAR and certain MNAR counterparts within the PMM family.

Strategy 2. As opposed to identifying restrictions, model simplification can be done in order to identify the parameters. The advantage is that the number of parameters decreases, which is desirable since the length of the parameter vector is a general issue with pattern-mixture models. Indeed, Hogan and Laird [13] noted that in order to estimate the large number of parameters in general pattern-mixture models, one has to make the awkward requirement that each dropout pattern occurs sufficiently often. For example, trends can be restricted to functional forms supported by the information available within a pattern (e.g., a linear or quadratic time trend is easily extrapolated beyond the last obtained measurement). Alternatively, one can let the parameters vary across patterns in a controlled parametric way. Thus, rather than estimating a separate time trend within each pattern, one could, for example, assume that the time evolution within a pattern is unstructured, but parallel across patterns. This is effectuated by treating pattern as a covariate.

While the second strategy is computationally simple, there is a price to pay. Indeed, simplified models, qualified as "assumption rich" by Sheiner, et al. [41], are also making untestable assumptions, just as in the selection model case. In the identifying-restrictions setting, on the other hand, the assumptions are clear from the start, in agreement with the aforementioned paradox.

A final observation, applying to both strategies, is that pattern-mixture models do not always automatically provide estimates and standard errors of marginal quantities of interest, such as overall treatment effect or overall time trend. Thus, selection model quantities need to be determined by appropriately averaging over the pattern-mixture model quantities and their standard errors can be derived using delta method ideas.

In line with the results obtained by Molenberghs et al. [31], we restrict attention to monotone patterns. In general, let us assume we have $t = 1, \ldots, T$ dropout patterns where the dropout indicator, introduced earlier, is $d = t + 1$. For pattern t, the complete data density is given by

$$\begin{aligned} & f_t(y_1, \ldots, y_T) \\ & = f_t(y_1, \ldots, y_t) f_t(y_{t+1}, \ldots, \\ & \quad y_T | y_1, \ldots, y_t). \end{aligned} \quad (2)$$

The first factor is clearly identified from the observed data, while the second factor is not. It is assumed that the first factor is known or, more realistically, modeled using the observed data. Then, identifying restrictions are applied in order to identify the second component.

While, in principle, completely arbitrary restrictions can be used by means of any valid density function over the appropriate support, strategies that relate back to the observed data deserve privileged interest. One can base identification on all patterns for which a given component, y_s, is identified, producing

$$\begin{aligned} & f_t(y_1, \ldots, y_T) \\ & = f_t(y_1, \ldots, y_t) \prod_{s=0}^{T-t-1} \\ & \quad \times \left[\sum_{j=T-s}^{T} \omega_{T-s,j} \right. \\ & \quad \left. \times f_j(y_{T-s} | y_1, \ldots, y_{T-s-1}) \right]. \end{aligned} \quad (3)$$

Let us consider three special but important cases. Little [22] proposes complete-case missing value (CCMV) restrictions, which uses the following identification:

$$\begin{aligned} & f_t(y_s | y_1, \ldots y_{s-1}) \\ & = f_T(y_s | y_1, \ldots y_{s-1}), \\ & s = t + 1, \ldots, T. \end{aligned} \quad (4)$$

In other words, information that is unavailable is always borrowed from the completers. This strategy can be defended in cases where the bulk of the subjects are complete and only small proportions are assigned to the various dropout patterns. Also, extension of this approach to non-monotone patterns is particularly easy. Alternatively, the nearest identified pattern can be used:

$$\begin{aligned} f_t(y_s|y_1,\ldots y_{s-1}) \\ = f_s(y_s|y_1,\ldots y_{s-1}), \\ s = t+1,\ldots,T. \end{aligned} \quad (5)$$

We will refer to these restrictions as *neighboring-case missing values (NCMV)*.

The third special case will be (ACMV) available-case missing-value restrictions. Thus, ACMV is reserved for the counterpart of MAR in the PMM context. The corresponding $\boldsymbol{\omega}_s$ vectors can be shown to have components

$$\omega_{sj} = \frac{\alpha_j f_j(y_1,\ldots,y_{s-1})}{\sum_{\ell=s}^{T} \alpha_\ell f_\ell(y_1,\ldots,y_{s-1})}, \quad (6)$$

where α_j is the fraction of observations in pattern j.

Indeed, Molenberghs et al. [31] have shown that ACMV (in the pattern-mixture setting) and MAR (in the selection model setting) are equivalent in the case of dropout. This equivalence is important in that it enables us to make a clear connection between both frameworks. By implication, CCMV and NCMV are of MNAR type. However, the MNAR family is huge and the need may exist to construct further subsets, which are broader than MAR. The entire class of such models will be termed *missing non-future-dependent* (MNFD). While they are natural and easy to consider in a selection model context, there exist important examples of mechanisms that do not satisfy MNFD, such as shared-parameter models [25,51].

Kenward, et al. [19] have shown there is a counterpart to MNFD in the pattern-mixture context. Precisely, the MNFD selection models obviously satisfy

$$\begin{aligned} f(r=t|y_1,\ldots,y_T) \\ = f(r=t|y_1,\ldots,y_{t+1}). \end{aligned} \quad (7)$$

Note that MAR is a special case of MNFD, which in turn is a subclass of MNAR.

Within the PMM framework, we define *non-future dependent missing-value* (NFMV) restrictions as follows:

$$\begin{aligned} \forall t \geq 2, \forall j < t-1: \\ f(y_t|y_1,\ldots,y_{t-1}, r=j) \\ = f(y_t|y_1,\ldots,y_{t-1}, r \geq t-1). \end{aligned} \quad (8)$$

NFMV is not a single set of restrictions, but rather leaves one conditional distribution per incomplete pattern unidentified:

$$f(y_{t+1}|y_1,\ldots,y_t, r=t). \quad (9)$$

In other words, the distribution of the "current" unobserved measurement, given the previous ones, is unconstrained. Kenward, et al. [19] have shown that, for longitudinal data with dropouts, MNFD and NFMV are equivalent.

For pattern t, the complete data density is given by

$$\begin{aligned} f_t(y_1,\ldots,y_T) \\ = f_t(y_1,\ldots,y_t) f_t(y_{t+1}|y_1,\ldots,y_t) \\ f_t(y_{t+2},\ldots,y_T|y_1,\ldots,y_{t+1}). \end{aligned} \quad (10)$$

It is assumed that the first factor is known or, more realistically, modeled using the observed data. Then, identifying restrictions are applied in order to identify the second and third components. From the data, estimate $f_t(y_1,\ldots,y_t)$. The user has full freedom to choose

$$f_t(y_{t+1}|y_1,\ldots,y_t). \quad (11)$$

Substantive considerations can be used to identify this density. Or a family of densities can be considered by way of sensitivity analysis. Using Equation (8), the densities $f_t(y_j|y_1,\ldots,y_{j-1})$, $(j \geq t+2)$ are identified. This identification involves not only the patterns for which y_j is observed, but also the pattern for which y_j is the current, the first unobserved measurement.

Thus, it follows that $f_t(y_1, \ldots, y_{t+1})$ is identified from modeling and choice. Next, NFMV states that

$$f_t(y_s|y_1, \ldots, y_{s-1}) = f_{(\geq s-1)}(y_s|y_1, \ldots, y_{s-1}), \quad (12)$$

for $s = t+2, \ldots, T$. A general expression can be shown to be

$$f_t(y_s|y_1, \ldots y_{s-1}) = \sum_{j=s-1}^{T} \omega_{sj} f_j(y_s|y_1, \ldots y_{s-1}), \\ s = t+2, \ldots, T, \quad (13)$$

with

$$\omega_{sj} = \frac{\alpha_j f_j(y_1, \ldots, y_{s-1})}{\sum_{\ell=s-1}^{T} \alpha_\ell f_\ell(y_1, \ldots, y_{s-1})}. \quad (14)$$

Choosing ω_{sj} that differ from the ones specified above yields missing data mechanisms that do depend on future observations. In a sensitivity analysis, it can be envisaged that the impact of such departures on substantive conclusions might be explored. Indeed, in the general MNAR case, the conditional distribution of the unobserved measurements given the observed ones needs to be determined by means of assumptions. Under NFMV, only the conditional distribution of the *first* ("current") unobserved outcome given the observed ones needs to be identified by assumption. Thus, when MNFD is deemed plausible, one combines the flexibility of a broad class of models with a sensitivity space that is reasonably easy to manage. In the special case of MAR, the conditional distributions of the unobserved outcomes are completely identified by means of ACMV and there is no further room for sensitivity analysis.

Two obvious mechanisms, within the MNFD family but outside MAR, are FD1 [i.e., choose Eq. (11) according to CCMV] and FD2 [i.e., choose Eq. (11) according to NCMV]. Since the other densities are to be identified using Equation (13), FD1 and FD2 are strictly different from CCMV and NCMV.

Acknowledgments. We gratefully acknowledge support from Belgian IUAP/PAI network "Statistical Techniques and Modeling for Complex Substantive Questions with Complex Data."

References

1. Afifi, A. and Elashoff, R. (1966). Missing observations in multivariate statistics I: review of the literature. *J. Am. Statist. Assoc.*, **61**, 595–604.

2. Chatterjee, S. and Hadi, A. S. (1988). *Sensitivity Analysis in Linear Regression*. Wiley, New York.

3. Cook, R. D. (1986). Assessment of local influence. *J. Roy. Statist. Soc. B*, **48**, 133–169.

4. Copas, J. B. and Li, H. G. (1997). Inference from non-random samples (with discussion). *J. Roy. Statist. Soc. B*, **59**, 55–96.

5. Dempster, A. P., Laird, N. M. and Rubin, D. B. (1977). Maximum likelihood from incomplete data via the EM algorithm (with discussion). *J. Roy. Statist. Soc. B*, **39**, 1–38.

6. Diggle, P. J. and Kenward, M. G. (1994). Informative drop-out in longitudinal data analysis (with discussion). *Appl. Statist.*, **43**, 49–93.

7. Draper, D. (1995). Assessment and propagation of model uncertainty (with discussion). *J. Roy. Statist. Soc. B*, **57**, 45–97.

8. Fitzmaurice, G. M., Molenberghs, G., and Lipsitz, S. R. (1995). Regression models for longitudinal binary responses with informative dropouts. *J. Roy. Statist. Soc. B*, **57**, 691–704.

9. Glynn, R. J., Laird, N. M., and Rubin, D. B. (1986). Selection modelling versus mixture modelling with non-ignorable nonresponse. In *Drawing Inferences from Self Selected Samples*, H. Wainer, ed., pp. 115–142, Springer-Verlag, New York.

10. Hartley, H. O. and Hocking, R. (1971). The analysis of incomplete data. *Biometrics*, **27**, 783–808.

11. Heckman, J. J. (1976). The common structure of statistical models of truncation, sample selection and limited dependent variables and a simple estimator for such models. *Ann. Econ. Soc. Meas.*, **5**, 475–492.

12. Heyting, A., Tolboom, J., and Essers, J. (1992). Statistical handling of dropouts in longitudinal clinical trials. *Statist. Med.*, **11**, 2043–2061.

13. Hogan, J. W. and Laird, N. M. (1997). Mixture models for the joint distribution of repeated measures and event times. *Statist. Med.*, **16**, 239–258.

14. Jansen, I., Molenberghs, G., Aerts, M., Thijs, H., and Van Steen, K. (2003). A local influence approach applied to binary data from a psychiatric study. *Biometrics*, **59**, 410–419.

15. Kenward, M. G. (1998). Selection models for repeated measurements with nonrandom dropout: an illustration of sensitivity. *Statist. Med.*, **17**, 2723–2732.

16. Kenward, M. G., Goetghebeur, E. J. T., and Molenberghs, G. (2001). Sensitivity analysis of incomplete categorical data. *Statist. Model.*, **1**, 31–48.

17. Kenward, M. G. and Molenberghs, G. (1998). Likelihood based frequentist inference when data are missing at random. *Statist. Sci.*, **12**, 236–247.

18. Kenward, M. G. and Molenberghs, G. (1999). Parametric models for incomplete continuous and categorical longitudinal studies data. *Statist. Methods Med. Res.*, **8**, 51–83.

19. Kenward, M. G., Molenberghs, G., and Thijs, H. (2003). Pattern-mixture models with proper time dependence. *Biometrika*, **90**, 53–71.

20. Laird, N. M. (1994). Discussion to Diggle, P.J. and Kenward, M.G.: informative dropout in longitudinal data analysis. *Appl. Statist.*, **43**, 84.

21. Lavori, P. W., Dawson, R., and Shera, D. (1995). A multiple imputation strategy for clinical trials with truncation of patient data. *Statist. Med.*, **14**, 1913–1925.

22. Little, R. J. A. (1993). Pattern-mixture models for multivariate incomplete data. *J. Am. Statist. Assoc.*, **88**, 125–134.

23. Little, R. J. A. (1994). A class of pattern-mixture models for normal incomplete data. *Biometrika*, **81**, 471–483.

24. Little, R. J. A. (1994). A class of pattern-mixture models for normal incomplete data. *Biometrika*, **81**, 471–483.

25. Little, R. J. A. (1995). Modeling the drop-out mechanism in repeated measures studies. *J. Am. Statist. Assoc.*, **90**, 1112–1121.

26. Little, R. J. A. and Rubin, D. B. (1987). *Statistical Analysis with Missing Data*. Wiley, New York. '

27. Mallinckrodt, C. H., Clark, W. S., Carroll, R. J., and Molenberghs, G. (2003). Assessing response profiles from incomplete longitudinal clinical trial data under regulatory considerations. *J. Biopharm. Statist.*, **13**, 179–190.

28. Mallinckrodt, C. H., Sanger, T. M., Dube, S., Debrota, D. J., Molenberghs, G., Carroll, R. J., Zeigler Potter, W. M., and Tollefson, G. D. (2003). Assessing and interpreting treatment effects in longitudinal clinical trials with missing data. *Biol. Psychiatry*, **53**, 754–760.

29. Michiels, B., Molenberghs, G., Bijnens, L., Vangeneugden, T., and Thijs, H. (2002). Selection models and pattern-mixture models to analyze longitudinal quality of life data subject to dropout. *Statist. Med.*, **21**, 1023–1041.

30. Molenberghs, G., Goetghebeur, E. J. T., Lipsitz, S. R., and Kenward, M. G.

(1999). Non-random missingness in categorical data: strengths and limitations. *Am. Statist.*, **53**, 110–118.

31. Molenberghs, G., Michiels, B., Kenward, M. G., and Diggle, P. J. (1998). Missing data mechanisms and pattern-mixture models. *Statist. Neerl.*, **52**, 153–161.

32. Molenberghs, G., Verbeke, G., Thijs, H., Lesaffre, E., and Kenward, M. G. (2001). Mastitis in dairy cattle: influence analysis to assess sensitivity of the dropout process. *Comput. Statist. Data Anal.*, **37**, 93–113.

33. Murray, G. D. and Findlay, J. G. (1988). Correcting for the bias caused by dropouts in hypertension trials. *Statist. Med.*, **7**, 941–946.

34. Nordheim, E. V. (1984). Inference from nonrandomly missing categorical data: an example from a genetic study on Turner's syndrome. *J. Am. Statist. Assoc.*, **79**, 772–780.

35. Raab, G. M. and Donnelly, C. A. (1999). Information on sexual behaviour when some data are missing. *Appl. Statist.*, **48**, 117–133.

36. Robins, J. M., Rotnitzky, A., and Scharfstein, D. O. (1998). Semiparametric regression for repeated outcomes with nonignorable non-response. *J. Am. Statist. Assoc.*, **93**, 1321–1339.

37. Rubin, D. B. (1976). Inference and missing data. *Biometrika*, **63**, 581–592.

38. Rubin, D. B. (1987). *Multiple Imputation for Nonresponse in Surveys*. Wiley, New York.

39. Rubin, D. B. (1994). Discussion to Diggle, P.J. and Kenward, M.G.: informative dropout in longitudinal data analysis. *Appl. Statist.*, **43**, 80–82.

40. Rubin, D. B., Stern, H. S., and Vehovar, V. (1995). Handling "don't know" survey responses: the case of the Slovenian plebiscite. *J. Am. Statist. Assoc.*, **90**, 822–828.

41. Sheiner, L. B., Beal, S. L., and Dunne, A. (1997). Analysis of nonrandomly censored ordered categorical longitudinal data from analgesic trials. *J. Am. Statist. Assoc.*, **92**, 1235–1244.

42. Siddiqui, O. and Ali, M. W. (1998). A comparison of the random-effects pattern mixture model with last observation carried forward (LOCF) analysis in longitudinal clinical trials with dropouts. *J. Biopharm. Statist.*, **8**, 545–563.

43. Smith, D. M., Robertson, B., and Diggle, P. J. (1996). *Object-oriented Software for the Analysis of Longitudinal Data in S*. Technical Report MA 96/192, Department of Mathematics and Statistics, University of Lancaster, Lancaster, UK LA1 4YF.

44. Thijs, H., Molenberghs, G., Michiels, B., Verbeke, G., and Curran, D. (2002). Strategies to fit pattern-mixture models. *Biostatistics*, **3**, 245–265.

45. Thijs, H., Molenberghs, G., and Verbeke, G. (2000). The milk protein trial: influence analysis of the dropout process. *Biometrical J.*, **42**, 617–646.

46. Vach, W. and Blettner, M. (1995). Logistic regresion with incompletely observed categorical covariates–investigating the sensitivity against violation of the missing at random assumption. *Statist. Med.*, **12**, 1315–1330.

47. Vansteelandt, S., Goetghebeur, E., Kenward, M. G., and Molenberghs, G. (2003). Ignorance and uncertainty regions as inferential tools in a sensitivity analysis; submitted for publication.

48. Van Steen, K., Molenberghs, G., Verbeke, G., and Thijs, H. (2001). A local influence approach to sensitivity analysis of incomplete longitudinal ordinal data. *Statist. Model. Int. J.*, **1**, 125–142.

49. Verbeke, G. and Molenberghs, G. (2000). *Linear Mixed Models for Longitudinal Data*. Springer-Verlag, New York.

50. Verbeke, G., Molenberghs, G., Thijs, H., Lesaffre, E., and Kenward, M. G. (2001). Sensitivity analysis for nonrandom dropout: A local influence approach. *Biometrics*, **57**, 7–14.

51. Wu, M. C. and Bailey, K. R. (1989). Estimation and comparison of changes in

the presence of informative right censoring: Conditional linear model. *Biometrics*, **45**, 939–955.

48. Multiple Testing in Clinical Trials

Alexei Dmitrienko and Jason C. Hsu

48.1 Introduction

Multiplicity problems caused by multiple analyses performed on the same dataset arise frequently in a clinical-trial setting. The following are examples of multiple analyses encountered in clinical trials:

- *Multiple comparisons.* Multiple testing is often performed in clinical trials involving several treatment groups. For example, most phase II trials are designed to assess the efficacy and safety of several doses of an experimental drug compared to a control.

- *Multiple primary endpoints.* Multiplicity can be caused by multiple criteria for assessing the efficacy profile of an experimental drug. Multiple criteria are required to accurately characterize various aspects of the expected therapeutic benefits. In some cases, the experimental drug is declared efficacious if it meets at least one of the criteria. In other cases, drugs need to produce significant improvement with respect to all of the endpoints; for example, new therapies for the treatment of Alzheimer's disease are required to demonstrate their effects on both cognition and global clinical improvement.

It is commonly recognized that failure to account for multiplicity issues can inflate the probability of an incorrect decision and could lead to regulatory approval of inefficacious drugs and increased patient risks. For this reason, regulatory agencies mandate a strict control of the false-positive (type I error) rate in clinical trials and require that drug developers perform multiple analyses with a proper adjustment for multiplicity. To stress the importance of multiplicity adjustments, the draft guidance document entitled *Points to Consider on Multiplicity Issues in Clinical Trials* released by the European Committee for Proprietary Medicinal Products on September 19, 2002 states that

> a clinical study that requires no adjustment of the Type I error is one that consists of two treatment groups, that uses a single primary variable, and has a confirmatory statistical strategy that prespecifies just one single null hypothesis relating to the primary variable. All other situations require attention to the potential effects

of multiplicity.

Because of these regulatory concerns, multiplicity adjustment strategies have received much attention in the clinical trial literature. This chapter provides a brief review of popular approaches to performing multiple analyses of clinical trial data. It outlines main principles underlying multiple-testing procedures and introduces single-step and stepwise multiple tests widely used in clinical applications. See References 8, 21, and 11 for a comprehensive review of multiple-decision theory with clinical trial applications.

Throughout the article, $H_{01}, \ldots, H_{0k}$ will denote the k null hypotheses and $H_{A1}, \ldots, H_{Ak}$ denote the alternative hypotheses tested in a clinical study. The associated test statistics and p values will be denoted by $T_1, \ldots, T_k$ and $p_1, \ldots, p_k$, respectively.

48.2 Concepts of Error Rates

In order to choose an appropriate multiple-testing method, it is critical to select the definition of correct and incorrect decisions that reflect the objective of the study.

48.2.1 Comparisonwise Error Rate

In the simple case when each hypothesis is tested independently, the comparisonwise error rate is controlled at a significance level α (e.g., 0.05 level) if each H_{0i} is tested so that the probability of erroneously rejecting H_{0i} is no more than α. Utilizing the law of large numbers, it can be shown that in the long run the proportion of erroneously rejected null hypotheses does not exceed α. However, if the k null hypotheses are true, the probability of rejecting at least one true null hypothesis will be considerably greater than the significance level chosen for each individual hypothesis. Thus, if a correct decision depends on correct inference from all k tests, the probability of an incorrect decision will exceed α.

48.2.2 Experimentwise Error Rate

An early attempt to alleviate this problem and achieve a better control of the probability of an incorrect decision was to consider each experiment as a unit and define the experimentwise error rate. The experimentwise error rate is said to be controlled at α if the probability of rejecting at least one true null hypothesis does not exceed α when the k null hypotheses are simultaneously true. Control of the experimentwise error rate is sometimes referred to as the *weak control of the familywise error rate*. Note, however, that, in terms of the probability of making an incorrect decision, $H_{01}, \ldots, H_{0k}$ all being true is not always the worst-case scenario. Suppose, for example, that $H_{01}, \ldots, H_{0(k-1)}$ are true but H_{0k} is false. Then, any multiple-testing method for which the probability of incorrectly rejecting at least one null hypothesis when all the null hypotheses are true is no more than α, but the probability of rejecting at least one of $H_{01}, \ldots, H_{0(k-1)}$, given that they are true and H_{0k} is false, is greater than α, protects the experimentwise error rate. It is obvious from this example that preserving the experimentwise error rate does not necessarily guarantee that the probability of an incorrect decision is no greater than α.

48.2.3 Family-Wise Error Rate

Due to the described limitation of the experimentwise error rate, clinical researchers rely on a more stringent method for controlling the probability of an incorrect decision known as the strong control of the familywise error rate (FWER). The FWER is defined as the probability of erroneously rejecting any true null hypothesis in a family regardless of which and how many other null hypotheses are true. This definition is essentially based on the maximum experimentwise error rate for any subset of the k null hypotheses and, for this reason, FWER-controlling tests are sometimes said to preserve the maximum type I error rate.

48.2.4 False Discovery Rate

Another popular approach to assessing the performance of multiple tests, known as the *false discovery rate* (FDR), is based on the ratio of the number of erroneously rejected null hypotheses to the total number of rejected null hypotheses [2]. To be more precise, the FDR is said to be controlled at α if the expected proportion of incorrectly rejected (true) null hypotheses is no more than α:

$$E\left(\frac{\text{Number of true } H_{0i} \text{ rejected}}{\text{Total number of } H_{0i} \text{ rejected}}\right) \leq \alpha.$$

FDR-controlling tests are useful in multiplicity problems involving a large number of null hypotheses (e.g., multiplicity problems arising in genetics) and are becoming increasingly popular in preclinical research. It is important to point out that the FDR is uniformly larger than the FWER, and thus controlling the FDR may not control the probability of an incorrect decision. In fact, in confirmatory studies, it is often possible to manipulate the design of the clinical trial so that any conclusion desired can be almost surely inferred without inflating the FDR [5].

48.3 Union–Intersection Testing

Most commonly, multiple-testing problems are formulated as union–intersection (UI) problems [16], meaning that one is interested in testing the global hypothesis, denoted by H_{0I}, which is the intersection of k null hypotheses versus the union of the corresponding alternative hypotheses, denoted by H_{AU}. As an illustration, consider a dose-finding study designed to test a low and high doses of an experimental drug (labeled L and H) to placebo (P). The primary endpoint is a continuous variable with larger values indicating improvement. Let μ_P, μ_L, and μ_H denote the mean improvement in the placebo, low dose, and high dose groups, respectively. The individual null hypotheses tested in the trial are $H_L : \mu_L \leq \mu_P$ and $H_H : \mu_H \leq \mu_P$. In this setting, an UI approach would test $H_{0I} : \mu_L \leq \mu_P$ and $\mu_H \leq \mu_P$ versus $H_{AU} : \mu_L > \mu_P$ or $\mu_H > \mu_P$. According to the UI-testing principle, the global hypothesis H_{0I} is tested by examining each of its components individually, rejecting H_{0I} if at least one of the components is rejected. Tests of homogeneity, which one learns in elementary statistics courses, such as the F test, tend to be UI tests. The following is a brief overview of popular methods for constructing UI tests.

48.3.1 Single-Step Tests Based on Univariate p Values

These tests (e.g., the Bonferroni and Šidák tests) are intuitive, easy to explain to nonstatisticians, and, for this reason, are frequently used in clinical applications. The Bonferroni adjustment for testing H_{0i} amounts to computing an adjusted p value given by kp_i. Similarly, the Šidák-adjusted p value for H_{0i} is equal to $1 - (1 - p_i)^k$. The adjusted p values are then compared to α and the global hypothesis H_{0I} is rejected if at least one adjusted p value is no greater than α. Another example of a test based on univariate p values is the Simes test [18]. The adjusted Simes p value for the global hypothesis H_{0I} is $k \min(p_{[1]}, p_{[2]}/2, \ldots, p_{[k]}/k)$, where $p_{[1]}, \ldots, p_{[k]}$ are ordered p values, that is,

$p_{[1]} \leq \cdots \leq p_{[k]}$. It is easy to see from this definition that the Simes test is uniformly more powerful than the Bonferroni test in the sense that the former rejects H_{0I} every time the latter does. Although the Simes test has a power advantage over the Bonferroni test, one needs to remember than the Simes test does not always preserve the overall type I error rate. It is known that the size of this test does not exceed α when p_1, ..., p_k are independent or positively dependent [17]. It is important to keep in mind that tests based on univariate p values ignore the underlying correlation structure and become very conservative when the test statistics are highly correlated or the number of null hypotheses is large, for example, in clinical trials with multiple-outcome variables.

48.3.2 Parametric Single-Step Tests

The power of simple tests based on univariate p values can be improved considerably when one can model the joint distribution of the test statistics T_1, ..., T_k. Consider, for example, the problem of comparing k doses of an experimental drug to a control in a one-sided manner. Assuming that T_1, ..., T_k follow a multivariate normal distribution and larger treatment differences are better, Dunnett [4] derived a multiple test that rejects H_{0i} if $T_i \geq d$, where d is the $100(1-\alpha)\%$ percentile of $\max(T_1, \ldots, T_k)$. Dunnett's method also yields a set of simultaneous one-sided confidence intervals for the true mean treatment differences δ_1, ..., δ_k:

$$\delta_i > \widehat{\delta_i} - ds\sqrt{2/n}, \quad i = 1, \ldots, k,$$

where s is the pooled sample standard deviation and n is the common sample size per treatment group.

48.3.3 Resampling-Based Single-Step Tests

A general method for improving the performance of tests based on univariate p values was proposed by Westfall and Young [21]. Note first that the adjusted p value for H_{0i} is given by $P\{\min(P_1, \ldots, P_k) \leq p_i\}$. In this equation, $P_1, \ldots, P_k$ denote random variables that follow the same distribution as p_1, ..., p_k under the assumption that the global hypothesis H_{0I} is true. The joint distribution of the p values is unknown and can be estimated using either permutation or bootstrap resampling. The advantage of using resampling-based testing procedures is that they account for the empirical correlation structure of the individual p-values and thus are more powerful than the Bonferroni and similar tests. Furthermore, unlike the Dunnett test, the resampling-based approach does not rely on distributional assumptions. When carrying out resampling-based tests, it is important to ensure that the *subset pivotality condition* is met. This condition guarantees that the resampling-based approach preserves the FWER at the nominal level. The subset pivotality condition is met in most multiple-testing problems for which pivotal quantities exist; however, it may not be satisfied in the case of binary variables, for example, see Reference 21 for more details.

48.4 Closed Testing

A cornerstone of multiple hypotheses testing has been the closed-testing principle of Marcus et al.[14]. The principle has provided a foundation for a variety of multiple-testing methods and has found a large number of applications in multiple-testing problems arising in clinical trials. Examples of such applications include procedures for multiple treatment comparisons and multiple outcome variables [1,13], testing a dose–response relationship in dose ranging trials [15], and gatekeeping strategies for addressing multiplicity issues arising in clinical trials with multiple primary and secondary endpoints [3,20].

The closed-testing principle is based on a hierarchical representation of the multiplicity problem in question. To illustrate, consider the null hypotheses H_L and H_H from the dose-finding trial example. In order to derive a closed test for this multiple-testing problem, construct the *closed family of null hypotheses* by forming all possible intersections of the null hypotheses. The closed family contains H_L, H_H, and $H_L \cap H_H$. The next step is to establish *implication relationships* in the closed family. A hypothesis that contains another hypothesis is said to imply it, for example, $H_L \cap H_H$ implies both H_L and H_H. The closed-testing principle states that an FWER-controlling testing procedure can be constructed by testing each hypothesis in the closed family using a suitable level-α test. A hypothesis in the closed family is rejected if its associated test and all tests associated with hypotheses implying it are significant. For example, applying the closed-testing principle to the dose-finding trial example, statistical inference proceeds as follows:

- If $H_L \cap H_H$ is accepted, the closed test has to accept H_L and H_H because $H_L \cap H_H$ implies H_L and H_H.

- If $H_L \cap H_H$ is rejected but not H_L or H_H, the inference is at least one of the two alternative hypotheses is true, but we cannot specify which one.

- If $H_L \cap H_H$ and H_H are rejected but H_L is accepted, one concludes that H_H is false, that is, $\mu_H > \mu_P$. Similarly, if $H_L \cap H_H$ and H_L are rejected but H_H is accepted, the null hypothesis H_L is declared to be false, that is, $\mu_L > \mu_P$.

- Last, if $H_L \cap H_H$, H_L, and H_H are rejected, the inference is that $\mu_L > \mu_P$ and $\mu_H > \mu_P$.

Now, in order to construct a multiple-testing procedure, one needs to choose a level α significance test for the individual hypotheses in the closed family. Suppose, for example, that the individual hypotheses are tested using the Bonferroni test. The resulting closed-testing procedure is equivalent to the stepwise testing procedure proposed by Holm [9]. The Holm procedure relies on a sequentially rejective algorithm for testing the ordered null hypotheses $H_{[01]}, \ldots, H_{[0k]}$ corresponding to the ordered p values $p_{[1]} \leq \cdots \leq p_{[k]}$. The procedure first examines the null hypothesis associated with the most significant p value, that is, $H_{[01]}$. This hypothesis is rejected if $p_{[1]} \leq \alpha/k$. Further, $H_{[0i]}$ is rejected if $p_{[j]} \leq \alpha/(k-j+1)$ for all $j = 1, \ldots, i$. Otherwise, the remaining null hypotheses $H_{[0i]}, \ldots, H_{[0k]}$ are accepted and testing ceases. Note that $H_{[01]}$ is tested at the α/k level and the other null hypotheses are tested at successively higher significance levels. As a result, the Holm procedure rejects at least as many (and possibly more) null hypotheses as the Bonferroni test it was derived from. This example shows that by applying the closed-testing principle to a single-step test one can construct a more powerful stepwise test that maintains the FWER at the same level.

The same approach can be adopted to construct stepwise testing procedures based on other single-step tests. For example, the popular Hochberg and Hommel testing procedures can be thought of as closed-testing versions of the Simes test [7,10]. It is worth noting that the Hommel procedure is uniformly more powerful than the Hochberg procedure and both of the two procedures preserve the FWER at the nominal level only when the Simes test does, that is, under the assumption of independence or positive dependence.

In the parametric case, an application of the closed-testing principle to the Dunnett test results in the stepwise Dunnett test defined as follows. Consider again the comparison of k doses of an experimental drug to a control in a one-sided setting. Let $T_{[1]}, \ldots, T_{[k]}$ denote the ordered test statistics ($T_{[1]} \leq \cdots \leq T_{[k]}$) and d_i be the $100(1-\alpha)\%$ percentile of $\max(T_1, \ldots, T_i), i = 1, \ldots, k$. The stepwise Dunnett test begins with the most significant statistic and compares it to d_k. If $T_{[k]} \geq d_k$, the null hypothesis corresponding to $T_{[k]}$ is rejected and the second most significant statistic is examined. Otherwise, the stepwise algorithm terminates and the remaining null hypotheses are accepted. It is easy to show that the derived stepwise test is uniformly more powerful than the single-step Dunnett test. An important limitation of the closed-testing principle is that it does not generally provide the statistician with a tool for constructing simultaneous confidence intervals for parameters of interest. For instance, it is not clear how to set up simultaneous confidence bounds for the mean differences between the k dose groups and control group within the closed-testing framework.

The closed-testing principle can also be used in the context of resampling-based multiple tests to set up stepwise testing procedures that account for the underlying correlation structure.

48.5 Partition Testing

The partitioning principle introduced in references 6 and 19 can be viewed as a natural extension of the principle of closed testing. The advantage of using the partitioning principle is twofold. Partitioning procedures are sometimes more powerful than procedures derived within the closed-testing framework and, unlike closed-testing procedures, they are easy to invert in order to set up simultaneous confidence sets for parameters of interest. To introduce the partitioning principle, consider k null hypotheses tested in a clinical trial and assume that H_{0i} states that $\theta \in \Theta_i$, where θ is a multidimensional parameter and Θ_i is a subset of the parameter space. Partition the union of $\Theta_1, \ldots, \Theta_k$ into disjoint subsets Θ_J, $J \subset \{1, \ldots, k\}$, which can be interpreted as the part of the parameter space in which exactly $H_{0i}, i \in J$, are true and the remaining null hypotheses are false. Now define null hypotheses corresponding to the constructed subsets, that is, $H_J^* : \theta \in \Theta_J^*$, and test them at level α. Since these null hypotheses are mutually exclusive, at most one of them is true. Therefore, even though no multiplicity adjustment is made, the resulting multiple test controls the FWER at the α level.

To illustrate the process of carrying out partitioning tests, consider the null hypotheses $H_L : \mu_L \leq \mu_P$ and $H_H : \mu_H \leq \mu_P$ from the dose-finding trial example. The union of H_L and H_H is partitioned into three hypotheses:

$$H_1^* : \mu_L \leq \mu_P \text{ and } \mu_H \leq \mu_P,$$
$$H_2^* : \mu_L \leq \mu_P \text{ and } \mu_H > \mu_P,$$
$$H_3^* : \mu_L > \mu_P \text{ and } \mu_H \leq \mu_P.$$

Testing each of the three hypotheses with a level-α significance test results in the following decision rule:

- If H_1^* is accepted, neither H_L nor H_H can be rejected; otherwise, infer that $\mu_L > \mu_P$ or $\mu_H > \mu_P$.

- If H_1^* and H_2^* are rejected, one concludes that $\mu_L > \mu_P$. Likewise, rejecting H_1^* and H_3^* implies that $\mu_H > \mu_P$.

- Finally, if H_1^*, H_2^*, and H_3^* are rejected, the inference is that $\mu_L > \mu_P$ and $\mu_H > \mu_P$.

Although this decision rules appears to be similar to the closed-testing rule, it is important to point out that the partitioning principle does not deal with the hypotheses in the closed family (i.e., H_L, H_H, and $H_L \cap H_H$) but rather with hypotheses H_1^*, H_2^*, and H_3^* defined above.

Due to the choice of mutually exclusive null hypotheses, partitioning tests can be inverted to derive a confidence region for the unknown parameter θ. Recall that the most general method for constructing a confidence set from a significance test is defined as follows. For each parameter point θ_0, test $H_0 : \theta = \theta_0$ using an level α test and then consider the set of all parameter points θ_0 for which $H_0 : \theta = \theta_0$ is accepted. The set obtained is a $100(1 - \alpha)\%$ confidence set for the true value of θ. This procedure corresponds to partitioning the parameter space into subsets consisting of a single parameter point and can be used for constructing simultaneous confidence limits associated with various stepwise tests. Consider, for example, confidence limits for the mean treatment differences between k dose groups and a control group [19]. If the largest mean difference is not significant, $(T_{[k]} < d_k)$, the one-sided limits for the true mean differences $\delta_1, \ldots, \delta_k$ are given by

$$\delta_i > \hat{\delta}_i - d_k s \sqrt{2/n}, \quad i = 1, \ldots, k,$$

and testing stops. Otherwise, one infers that $\delta_{[k]} > 0$ and examines the second largest difference. At the jth step of the stepwise test, the one-sided limits for $\delta_{[1]}, \ldots, \delta_{[k-j+1]}$ are

$$\delta_{[i]} > \hat{\delta}_{[i]} - d_k s \sqrt{2/n},$$
$$i = 1, \ldots, k-j+1,$$

if the corresponding test statistic is not significant $(T_{[k-j+1]} < d_{k-j+1})$ and $\delta_{[k-j+1]} > 0$ otherwise. Comparing the resulting testing procedure to the stepwise Dunnett test derived in Section 47.4 using the closed-testing principle, it is easy to see that the partitioning principle extends the closed-testing framework by enabling clinical researchers to set up confidence limits for treatment–control differences. The partitioning principle can also be used for constructing confidence sets in a much more general context, for example, confidence intervals for fixed-sequence testing methods arising in dose-finding studies and other clinical applications [12].

References

1. Bauer, P. (1991). Multiple testings in clinical trials. *Statist. Med.*, **10**, 871–890.

2. Benjamini, Y., and Hochberg, Y. (1995). Controlling the false discovery rate—a practical and powerful approach to multiple testing. *J. Roy. Statist. Soc. B*, **57**, 289–300.

3. Dmitrienko, A., Offen, W., and Westfall, P. H. (2003). Gatekeeping strategies for clinical trials that do not require all primary effects to be significant. *Statist. Med.*, **22**, 2387–2400.

4. Dunnett, C. W. (1955). A multiple comparison procedure for comparing several treatments with a control. *J. Am. Statist. Assoc.*, **50**, 1096–1121.

5. Finner, H., and Roter, M. (2001). On the false discovery rate and expected Type I errors. *Biomed. J.*, **43**, 985–1005.

6. Finner, H., and Strassburger, K. (2002). The partitioning principle: a powerful tool in multiple decision theory. *Ann. Statist.*, **30**, 1194–1213.

7. Hochberg, Y. (1988). A sharper Bonferroni procedure for multiple significance testing. *Biometrika*, **75**, 800–802.

8. Hochberg, Y. and Tamhane, A. C. (1987). *Multiple Comparison Procedures*. Wiley, New York.

9. Holm, S. (1979). A simple sequentially rejective multiple test procedure. *Scand. J. Statist.*, **6**, 65–70.

10. Hommel, G. (1988). A stagewise rejective multiple test procedure based on a modified Bonferroni test. *Biometrika*, **75**, 383–386.

11. Hsu, J. C. (1996). *Multiple Comparisons: Theory and Methods*. Chapman & Hall, London.

12. Hsu, J. C., Berger, R. L. (1999). Stepwise confidence intervals without multiplicity adjustment for dose-response and toxicity studies. *J. Am. Statist. Assoc.*, **94**, 468–482.

13. Lehmacher, W., Wassmer, G., and Reitmeir, P. (1991). Procedures for two-sample comparisons with multiple endpoints controlling the experiment-wise error rate. *Biometrics*, **47**, 511–521.

14. Marcus, R., Peritz, E., and Gabriel, K. R. (1976). On closed testing procedure with special reference to ordered analysis of variance. *Biometrika*, **63**, 655–660.

15. Rom, D. M., Costello, R. J., and Connell, L. T. (1994). On closed test procedures for dose-response analysis. *Statist. Med.*, **13**, 1583–1596.

16. Roy, S. N. (1953). On a heuristic method for test construction and its use in multivariate analysis. *Ann. Statist.*, **24**, 220–238.

17. Sarkar, S., and Chang, C. K. (1997). Simes' method for multiple hypothesis testing with positively dependent test statistics. *J. Am. Statist. Assoc.*, **92**, 1601–1608.

18. Simes, R. J. (1986). An improved Bonferroni procedure for multiple tests of significance. *Biometrika*, **63**, 655–660.

19. Stefansson, G., Kim, W.-C., and Hsu, J. C. (1988). On confidence sets in multiple comparisons. In *Statistical Decision Theory and Related Topics IV*, S. S. Gupta and J. O. Berger, eds., pp. 89104, Academic Press, New York.

20. Westfall, P. H., and Krishen, A. (2001). Optimally weighted, fixed sequence, and gatekeeping multiple testing procedures. *J. Stat. Plan. Inf.*, **99**, 25–40.

21. Westfall, P. H. and Young, S. S. (1993). *Resampling-based Multiple Testing: Examples and Methods for P-value Adjustment*. Wiley, New York.

49. Mutation Processes

W. J. Ewens

A *mutation* is a heritable change in the genetic material. The consequences of mutation have been studied in great detail from the mathematical point of view in the subject of population genetics, and the theory of mutation processes has found its greatest application in genetic and evolutionary areas. Nevertheless, mutation processes can be described in abstract terms, and this allows an application of the theory beyond genetics and biological evolution.

The essential elements in the structure of a mutation process are a population of individuals (in genetics, genes), each individual being of one or another of a set of types (in genetics, alleles), a well-defined model describing the formation of one generation of individuals from the parental generation, and a mutation structure describing the probability that a mutant offspring of a parent of given type should be of any other type. The population is normally assumed to be of large and fixed size and the number of possible types is in some models a fixed finite number and in other models, infinite. In biological evolutionary theory, one also allows the possibility of selection, that is, of differential reproduction rates of different types, but here we do not consider this generalization. Attention is paid to time-dependent and also to stationary properties of the process, and also to properties of samples of individuals taken from the population, in particular at stationarity.

Properties of mutation processes may be studied either retrospectively, by considering properties of the ancestor sequence of any sample of individuals in the current population (see, in particular, Kingman [5]) or prospectively, by considering lines of descent from any such sample (see, in particular, Griffiths [1,2]). The two approaches can be unified largely through the concept of *time reversibility* (see Tavaré [7] for a review of these and associated matters).

When there exists a finite number m of types, with symmetric mutation structure, the stationary distribution of the frequencies $x_1, \ldots, x_m$ of the m types is (in large populations) in the Dirichlet form

$$f(x_1, \ldots, x_m) = \text{const} \prod_i x_i^{\theta/(m-1)-1},$$
$$\Sigma x_i = 1, \quad x_i \geq 0.$$

Here $\theta = cNu$, where c is a constant depending on the model assumed for the formation of each new generation (often $c = 1, 2,$ or 4); N is the population size, and u the mutation rate for each individual. There exists no nontrivial limit for the distribution of each frequency as

$m \to \infty$. Nevertheless, a limiting concept does exist if we focus on the order statistics $x_{(1)}, x_{(2)}, \ldots$; for any fixed j, there exists a nondegenerate limiting distribution for the first j order statistics. This is the marginal distribution of the first j components of the so-called Poisson-Dirichlet distribution with parameter θ introduced by Kingman [4]. From this distribution, one may find properties of the *infinite-type process* and thus the m-type and the infinite-type models may be related through standard convergence arguments as $m \to \infty$.

It is also possible, and often simpler, to proceed directly to the infinite-type process. Here all mutants are regarded as being of an entirely novel type, and the concept of stationarity refers to patterns of type frequencies rather than the frequency of any specific type. Stationarity properties may be found by a retrospective analysis using [5] the concept of the *N-coalescent*. This is a Markov chain of equivalence relations in which, for any $i, j = 1, \ldots, N (i \neq j)$, we have $i \sim j$ at step s of the chain if individuals i and j at time 0 have a common ancestor at time $-s$. A backward Kolmogorov argument, watching the equivalence classes formed by a sample of n individuals during each ancestor generation until a common ancestor for all n individuals is reached, shows that the probability that in the sample there exist k different types ($k = 1, 2, \ldots, n$), in such a way that β_i types are represented by exactly i individuals ($\Sigma \beta_i = k, \Sigma i \beta_i = n$), is

$$n! \theta^k \Big/ \left\{ \theta^{[n]} \prod (\beta_i! i^{\beta_i}) \right\},$$

where
$$\theta^{[n]} = \theta(\theta+1) \ldots (\theta + n - 1).$$

This is the Ewens–Karlin–McGregor–Kingman–Watterson sampling formula, which has many applications in population genetics theory. Note that, given a sample of n individuals yielding an observed value for the vector $(\beta_1, \beta_2, \ldots, \beta_n)$, the statistic k is sufficient for the parameter θ, and that the conditional distribution of $(\beta_1, \beta_2, \ldots, \beta_n)$, given k, is of the form

$$\text{const} \times \left\{ \prod \beta_i! i^{\beta_i} \right\}^{-1}.$$

From this it follows that, although the mutation process is symmetric and the selection of individuals to be parents is random, the most likely observed configurations of $(\beta_1, \ldots, \beta_n)$, given k, are those where one type predominates, together with a small number of types at low frequency. This unexpected conclusion may be explained by considering the times at which the various types in the sample first arose in the population.

The retrospective analysis also yields results on the "ages" of types represented in a sample. For example, if from the sample of n individuals, we take a subsample of m, the probability that the oldest type present in the sample is represented in the subsample is $m(n+\theta)/\{n(m+\theta)\}$. The probability distribution of the frequency of the oldest type in the sample is [3]

$$\Pr(j \text{ individuals of oldest type})$$
$$= \theta \binom{n-1}{j-1} \Big/ \left\{ n \binom{n+\theta-1}{j} \right\},$$
$$j = 1, 2, \ldots, n.$$

The probability that there exist m types in the population older than the oldest type in the sample is [6]

$$\Pr(m \text{ older types})$$
$$= \frac{n}{n+\theta} \left(\frac{\theta}{n+\theta} \right)^m,$$
$$m = 0, 1, 2, \ldots.$$

A large variety of similar results are described in the references.

The prospective, as opposed to the retrospective, properties of the infinite-type process have been found by Griffiths [1,2]. These include time-dependent analogs of the sampling formula (1) as well as properties of samples taken from the population t generations apart. The latter include the distribution of the number and frequencies of types common to both samples. An important adaptation of these results concerns properties of samples taken from two different populations that split from a common stock $\frac{1}{2}t$ generations in the past. A sufficient statistic for t is the set of type frequencies in common in the two populations. By time reversibility arguments, the properties of the age of the oldest type are identical to corresponding properties of the time that the current types present survive. Given j types in the population at any time, the distribution of the time until the first type is lost may be found. This time has exponential distribution with mean $2[j(j + \theta - 1)]^{-1}$.

A final question in the infinite-type model is to find the probability that the most frequent type in the population is also the oldest. This is neatly answered by a time reversal, since the probability in question is identical to the probability that, of the current types, the most frequent will survive the longest and is thus the mean frequency of the most frequent type. This may be found from the Poisson–Dirichlet distribution of $x_{(1)}$. Details are given by Watterson and Guess [8], who also provide further similar examples.

References

1. Griffiths, R. C. (1979). *Adv. Appl. Probab.*, **11**, 310–325.
2. Griffiths, R. C. (1980). *Theor. Popul. Biol.*, **17**, 37–50.
3. Kelly, F. (1977). *J. Appl. Probab.*, **13**, 127–131.
4. Kingman, J. F. C. (1975). *J. Roy. Statist. Soc. B*, **37**, 1–22.
5. Kingman, J. F. C. (1982). *J. Appl. Probab.*, **19A**, 27–43.
6. Saunders, I., Tavaré, S., and Watterson, G. A. (1984). *Adv. Appl. Probab.*, **16**, 471–491.
7. Tavaré, S. (1984). *Theoret. Popul. Biol.*, **26**, 119–164.
8. Watterson, G. A. and Guess, H. A. (1977). *Theor. Popul. Biol.*, **11**, 141–160.

50 Nested Case-Control Sampling

Ørnulf Borgan

50.1 Introduction

Cox's regression model is one of the cornerstones in modern survival analysis, and is the method of choice when one wants to assess the influence of risk factors and other covariates on mortality or morbidity. Estimation in Cox's model is based on a partial likelihood, which, at each observed death or disease occurrence ("failure"), compares the covariate values of the failing individual to those of all individuals at risk at the time of the failure. In large epidemiological cohort studies of a rare disease, Cox regression requires collection of covariate information on all individuals in the cohort even though only a small fraction of these actually get diseased. This may be very costly, or even logistically impossible.

Cohort sampling techniques where covariate information is collected for all failing individuals ("cases"), but only for a sample of the nonfailing individuals ("controls") then offer useful alternatives that may drastically reduce the resources that need to be allocated to a study. Further, since most of the statistical information is contained in the cases, such studies may still be sufficient to give reliable answers to the questions of interest.

The most common cohort sampling design is nested case–control sampling. Here one compares each case with a small number of controls selected at random from those at risk at the case's failure time, and a new sample of controls is selected for each case. A different type of cohort sampling design is *case–cohort sampling*. For this design one selects at the outset of the study a random sample of control individuals (the subcohort), who are used as controls throughout the study (provided they are still at risk). In this entry we focus on the nested case–control design.

We first indicate the relation between this form of case–control sampling and more classical case–control designs and sketch the development of the subject. To fix ideas, we then describe in more detail a particular nested case–control study. We review the Cox model, describe how the nested case–control data are collected, and present methods for statistical inference. We conclude with a note on efficiency, some remarks on extensions of the nested case–control design, and a brief comparison between nested case–control sampling and case–cohort sampling.

50.2 Nested Case–Control Studies and Other Case–Control Designs

The theory for case–control studies for a binary response variable (diseased, not diseased) dates back to the work of Cornfield [9] in the early 1950s, and proceeds via the landmark 1959 paper by Mantel and Haenszel [14] to the implementation of the logistic regression model and the development of conditional logistic regression for matched case–control data in the 1970s. The monograph by Breslow and Day [7] gives an extensive exposition of this "classical" case-control theory; Breslow [6] provides a nice historical account.

Age or other timescales play no role in the statistical models on which the classical case–control theory is based, so this important aspect of a study has to be accounted for by stratification or time matching. This is different from a nested case–control study, where Cox's regression model is used to model the occurrence of failures, and where the controls are sampled from the risk sets. Further, in a "classical" case–control study the population from which the controls are sampled is often not well defined, whereas a nested case–control study is performed within a well-defined cohort. This places the nested case–control design between a classical case–control study and a full cohort analysis.

The nested case–control design was suggested in 1977 by Thomas [18] as a tool to reduce error checking and the computational burden for the analysis of large cohorts. He proposed to base inference on a modification of Cox's partial likelihood, given in (6) below. This suggestion was supported by Prentice and Breslow [17], who derived the same expression as a conditional likelihood for time-matched case–control sampling from an infinite population. A more decisive, but heuristic, argument was provided by Oakes [15], who showed that (6) is a partial likelihood when the sampling of controls is performed within the actual finite cohort. After more than 10 years, Goldstein and Langholz [10] proved rigorously that the estimator of the regression coefficients based on Oakes' partial likelihood enjoys large-sample properties similar to those of ordinary maximum-likelihood estimators. Later Borgan et al. [3] gave a more direct proof along the lines of Andersen and Gill [2], using a marked point process formulation. We indicate below how this approach also solves the problem of estimating the baseline hazard rate function from nested case–control data.

50.3 An Example

The nested case–control design has been used in many studies to avoid the collection of covariate information for the full cohort, or to reduce error checking and the computational burden in the analysis of large cohorts. It is now recognized that most time-matched case–control studies, ubiquitous in epidemiological research, are indeed nested case–control studies where the cohort is given as the (sometimes not well defined) population within a given geographic area. In order to fix ideas in the subsequent discussion, we take a closer look at one such study.

The International Agency for Research on Cancer in Lyon, France, maintains a register of 21,183 workers from 11 countries (Australia, Austria, Canada, Denmark, Finland, Germany, Italy, the Netherlands, New Zealand, Sweden, and the United Kingdom) exposed to phenoxy herbicides, chlorophenols, and dioxins. In a cohort analysis, an increased mortality of soft-tissue sarcomas was found among exposed subjects. In order to examine the effect of exposure to various chemicals more fully, a nested case–control study was undertaken, where, for each of the 11 cases of soft-tissue sarcoma (all males), five controls were sampled at random from those from the same country and of the same age as the case [11]. The degrees of exposure of the 11 cases and the 55 controls to a number of chemicals were reconstructed through the use of individual job records and of detailed company exposure questionnaires and company reports.

In principle, this information could have been collected for the complete cohort of 21,183 individuals, but this would have implied an enormous amount of work. Further, as the main limitation of the data is the small number of cases, such an effort would mostly have been in vain. In fact, the nested case–control study based on the 11 + 55 cases and controls provides 83% efficiency relative to the full cohort data for testing associations between single exposures and the disease (see below).

50.4 Model and Data

Consider a cohort of n individuals, and denote by $\lambda_i(t) = \lambda(t; \mathbf{x}_i(t))$ the hazard rate function at time t for an individual i with vector of covariates $\mathbf{x}_i(t) = (x_{i1}(t), \ldots, x_{ip}(t))'$. Here the time variable t may be (as in the example), times since employment, or some other time relevant to the problem at hand. The covariates may be time-fixed (such as gender) or time-dependent (such as cumulative exposure to a chemical), and they may be indicators for categorical covariates (such as the exposure groups "nonexposed," "low," "medium," and "high") or numerical (as when the actual amount of exposure is recorded).

Cox's regression model relates the covariates of individual i to its hazard rate function by

$$\lambda_i(t) = \lambda_0(t) e^{\boldsymbol{\beta}' \mathbf{x}_i(t)}. \qquad (1)$$

Here $\boldsymbol{\beta} = (\beta_1, \ldots, \beta_p)'$ is a vector of regression coefficients, while the baseline hazard rate function $\lambda_0(t)$, corresponding to an individual with all covariates identically equal to zero, is left unspecified.

Sometimes one adopts a stratified version of (1), where the baseline hazard rate function may differ between strata, while the regression coefficients are assumed the same across strata. Thus, in the example, it may be reasonable to stratify according to country of residence, since both exposure and occurrence or recognition of disease may differ by country. In order to simplify the presentation, we will concentrate on the model (1) with no stratification, and only comment on the modifications for the situation with stratification when relevant.

The individuals in the cohort may be followed over different periods of time, i.e., our observations may be subject to left truncation and/or right censoring. The risk set $\mathcal{R}(t)$ is the collection of all individuals who are under observation just before time t, and $n(t) = \#R(t)$ is the number at risk at that time. We let $t_1 < t_2 < \cdots$ be the times when failures are observed and, assuming that there are no tied failures, denote by i_j the index of the individual who fails at t_j (a few ties may be broken at random).

A nested case–control sample is then obtained as follows. At each failure time t_j, one selects by simple random sampling without replacement $m - 1$ individuals (controls) from the $n(t_j) - 1$ nonfailing individuals in $\mathcal{R}(t_j)$. The sampled risk $\tilde{\mathcal{R}}(t_j)$ then consists of the case i_j and these $m - 1$ controls. Covariate information is collected for all individuals in the sampled risk sets, but is not needed for the remaining individuals in the cohort. The sampling is done independently at the different failure times, so an individual may be a member of more than one sampled risk set.

A basic assumption is that truncation and censoring, as well as the sampling of controls, are independent in the sense that the additional knowledge of which individuals have entered the study, have been censored, or have been selected as controls before any time t does not carry information on the risks of failure at t (see Ref. 1, Sects. III.2–3 and Ref. 3 for a general discussion). For a small time interval $[t, t+dt)$, this assumption and (1) imply that

$$\Pr(i \text{ fails in } [t, t+dt)|\mathcal{F}_{t-})$$
$$= e^{\beta' \mathbf{x}_i(t)} \lambda_0(t) dt \qquad (2)$$

if individual i is at risk just before time t. Here the *history* $\mathcal{F}_{t-}$ contains information about observed failures, entries, exits, and changes in covariate values in the cohort, as well as information on the sampling of controls, up to but not including time t. (Not all of this information will actually be available to the researcher in a nested cas–control study.)

In the example, controls were not selected from all individuals at risk, but only from those at risk in the same country as the case. (Since age is used as the timescale when forming the risk sets, the controls will also have the same age as the case.) This way of sampling the controls is related to the stratified Cox model discussed earlier. When the stratified model applies, the sampling of controls should be restricted to those at risk in the same stratum as the case. We say that the controls are *matched* by the stratification variable ("country" in the example).

50.5 Estimation

Estimation of the regression coefficients in (1) is based on a partial likelihood that may be derived heuristically as follows. Denote by $\tilde{\mathcal{R}}(t)$ the sampled risk set were a failure to occur at t, and let $\mathcal{P}(t)$ be the collection of all possible sampled risk sets at that time. Then $\mathcal{P}(t)$ is the set of all $\binom{n(t)}{m}$ subsets of $\mathcal{R}(t)$ of size m. Consider a set $\mathbf{r} \in P(t)$ and an individual $i \in \mathbf{r}$. Then by (2), since the $m-1$ controls are sampled by simple random sampling without replacement from the $n(t)-1$ nonfailing individuals in $\mathcal{R}(t)$, we obtain

$$\Pr(i \text{ fails in } [t, t+dt), \tilde{R}(t) = \mathbf{r}|\mathcal{F}_{t-})$$
$$= \Pr(\tilde{R}(t) = \mathbf{r}|i \text{ fails at } t, \mathcal{F}_{t-})$$
$$\quad \times \Pr(i \text{ fails in } [t, t+dt)|\mathcal{F}_{t-})$$
$$= \binom{n(t)-1}{m-1}^{-1} e^{\beta' \mathbf{x}_i(t)} \lambda_0(t) dt. \quad (3)$$

Now the sampled risk set equals $\mathbf{r}$ if one of the m individuals in $\mathbf{r}$ fails, and the remaining $m-1$ individuals are selected as controls. Therefore $\Pr$(one individual in $\mathbf{r}$ fails in $[t, t+dt)$,

$$\tilde{R}(t) = \mathbf{r}|\mathcal{F}_{t-})$$
$$= \binom{n(t)-1}{m-1}^{-1}$$
$$\quad \times \sum_{l \in \mathbf{r}} e^{\beta' \mathbf{x}_l(t)} \lambda_0(t) dt. \quad (4)$$

Dividing (3) by (4), it follows that

$\Pr[i$ fails at $t|$ one individual in $\mathbf{r}$ fails at t,

$$\tilde{R}(t) = \mathbf{r}, \mathcal{F}_{t-}] = \frac{e^{\beta' \mathbf{x}_i(t)}}{\sum_{l \in \mathbf{r}} e^{\beta' \mathbf{x}_l(t)}}. \quad (5)$$

Multiplying together conditional probabilities of the form (5) for all failure times t_j, cases i_j, and sampled risk sets $\tilde{\mathcal{R}}(t_j)$, we arrive at the partial likelihood

$$L(\beta) = \prod_{t_j} \frac{e^{\beta' \mathbf{x}_{i_j}(t_j)}}{\sum_{l \in \tilde{R}(t_j)} e^{\beta' \mathbf{x}_l(t_j)}}. \quad (6)$$

This is similar to the full cohort partial likelihood, except that the sum in the denominator is taken over the sampled risk set $\tilde{\mathcal{R}}(t_j)$ in place of the full cohort risk set $\mathcal{R}(t_j)$. Inference concerning β, using the usual large-sample likelihood methods, can be based on the partial likelihood (6). The above heuristic derivation of (6) is essentially the one given by Oakes [15]. Borgan et al. [3] made this argument rigorous using a marked point processes formulation.

The partial likelihood (6) also applies to the stratified Cox model when the controls are matched by the stratification variable, i.e., sampled from the same stratum as the case. Further, (6) is of the same form as the conditional likelihood for logistic regression with $m-1$ matched controls per case [7, Chap. 7]. Thus standard software for conditional logistic regression can be used for data analysis.

The cumulative baseline hazard rate function $\Lambda_0(t) = \int_0^t \lambda_0(u)du$ can be estimated [3,4] by

$$\hat{\Lambda}_0(t) = \sum_{t_j \leq t} \left(\sum_{l \in \tilde{R}(t_j)} e^{\hat{\boldsymbol{\beta}}' \mathbf{x}_l(t_j)} \frac{n(t_j)}{m} \right)^{-1}, \quad (7)$$

where $\hat{\boldsymbol{\beta}}$ is the maximum partial likelihood estimator maximizing (6). The estimator (7) is of the same form as that used for cohort data. However, since nested case–control data only use information from a sample of those at risk, the contribution for each subject in the sampled risk set, including the case, is weighted by the reciprocal of the proportion sampled. The estimator (7) is almost unbiased when averaged over all possible failure and sampled risk set occurrences.

When we have only a small number of strata, the stratum-specific cumulative baseline hazard rate functions for the stratified Cox model may be estimated by a slight modification of (7). All that is required is that the sum be restricted to those failure times t_j when a failure in the actual stratum occurs, and that the $n(t_j)$ be taken to be the numbers at risk in this stratum. When there are many strata, however, there may be too little information in each stratum to make estimation of the stratum-specific cumulative baseline hazard rate functions meaningful.

50.6 Relative Efficiency

Goldstein and Langholz [10] were the first to carry out a rigorous study of the asymptotic properties of the maximum partial likelihood estimator $\hat{\boldsymbol{\beta}}$ maximizing (6). Based on the asymptotic distributional results, they also presented a study of the asymptotic efficiency of the maximum partial likelihood estimator for nested case–control data relative to the estimator based on the full cohort partial likelihood. When $\boldsymbol{\beta} = 0$, the asymptotic covariance matrix of the nested case–control estimator equals $m/(m-1)$ times the asymptotic covariance matrix of the full cohort estimator, independent of censoring and covariate distributions. Thus the efficiency of the nested case–control design relative to the full cohort is $(m-1)/m$ for testing associations between single exposures and disease—a result that has been known for some time for binary covariates based on the time-matched case–control study paradigm [8]. In the example with 5 controls per case, this yields the relative efficiency of $\frac{5}{6} = 83\%$ mentioned earlier.

When $\boldsymbol{\beta}$ departs from zero, and when more than one regression coefficient has to be estimated, the efficiency of the nested case–control design may be much lower than given by the $(m-1)/m$ efficiency rule [8,10]. For example, with one binary covariate for exposure with relative risk $e^{\beta} = 4$, the relative efficiency of the nested case–control design with one control per case is about $\frac{1}{4}$ when 10% of the cohort is exposed, rather than $\frac{1}{2}$ as the rule suggests.

Properly normalized, the estimator (7) for the cumulative baseline hazard rate function converges weakly to a Gaussian process [3]. This asymptotic distributional result makes it possible to study the asymptotic efficiency of (7) relative to the full cohort estimator, but such an efficiency study has yet to be performed. Preliminary studies by the author of this entry indicate, however, that the relative efficiency of (7) is much higher than that of the maximum partial likelihood estimator $\hat{\beta}$.

50.7 Extensions

Estimation of the cumulative hazard rate function for an individual with given time-fixed covariate values has been studied [4] and extended to time-varying covariate histories [13]. The latter work also described how the results presented earlier extend to regression models where (1) is replaced by $\lambda_i(t) = \lambda_0(t) r(\boldsymbol{\beta}, \mathbf{x}_i(t))$ for some relative risk function $r(\boldsymbol{\beta}, \mathbf{x}_i(t))$, and discussed estimation of absolute risk without and in the presence of competing risks. Estimation of excess risk from nested case–control data using Aalen's nonparametric linear regression model was discussed in Reference 5.

In a nested case–control study, the controls are selected by simple random sampling. An alternative is to select the controls by stratified random sampling. This design, termed *countermatched sampling*, may reduce the estimation uncertainty in situations of practical interest [12]. Using a marked point process formulation, a general framework for the sampling of controls, incorporating the nested case–control design and countermatched sampling as special cases, was introduced and studied in Reference 3.

Related to nested case–control studies is the case-cohort design [16]. Here a subcohort C is selected by simple random sampling from the entire cohort at the outset of the study. Covariate information is collected for the members of C as well as for cases occurring outside this subcohort. Estimation of β is based on a pseudolikelihood having the same form as (6), but with $\tilde{\mathcal{R}}(t_j)$ replaced by $[C \cap R(t_j)] \cup \{i_j\}$ for the sums in the denominator. Since this estimation is not based on a partial likelihood, the usual large-sample likelihood methods do not apply; hence the analysis of data from a case–cohort study is more cumbersome than the analysis of nested case–control data. For most studies involving a single disease, the case–cohort and the nested case–control design seem to have about the same efficiency. The main potential of the case–cohort design therefore lies in situations where multiple disease endpoints are to be evaluated, because disease-free members of the subcohort may serve as controls for the disease cases of each type.

References

1. Andersen, P. K., Borgan, Ø., Gill, R. D., and Keiding, N. (1993). *Statistical Models Based on Counting Processes*. Springer-Verlag, New York.

2. Andersen, P. K. and Gill, R. D. (1982). Cox's regression model for counting processes: A large sample study. *Ann. Statist.*, **10**, 1100–1120.

3. Borgan, Ø., Goldstein, L., and Langholz, B. (1995). Methods for the analysis of sampled cohort data in the Cox proportional hazards model. *Ann. Statist.*, **23**, 1749–1778.

4. Borgan, Ø. and Langholz, B. (1993). Nonparametric estimation of relative mortality from nested case-control studies. *Biometrics*, **49**, 593–602.

5. Borgan, Ø. and Langholz, B. (1997). Estimation of excess risk from case-control data using Aalen's linear regression model. *Biometrics*, **53**, 10–17.

6. Breslow, N. E. (1996). Statistics in epidemiology: the case-control study. *J. Am. Statist. Assoc.*, **91**, 14–28.

7. Breslow, N. E. and Day, N. E. (1980). *Statistical Methods in Cancer Research. Vol. 1—The Analysis of Case-Control Studies.* IARC Scientific Publications 32. International Agency for Research on Cancer, Lyon.

8. Breslow, N. E., Lubin, J. H., Marek, P., and Langholz, B. (1983). Multiplicative models and cohort analysis. *J. Am. Statist. Assoc.*, **78**, 1–12.

9. Cornfield, J. (1951). A method of estimating comparative rates from clinical data. Applications to cancer of the lung, breast and cervix. *J. Natl. Cancer Inst.*, **11**, 1269–1275.

10. Goldstein, L. and Langholz, B. (1992). Asymptotic theory for nested case-control sampling in the Cox regression model. *Ann. Statist.*, **20**, 1903–1928.

11. Kogevinas, M., Kauppinen, T., Winkelmann, R., Becher, H., Bertazzi, P. A., Bueno-de-Mesquita, H. B., Coggon, D., Green, L., Johnson, E., Littorin, M., Lynge, E., Marlow, D. A., Mathews, J. D., Neuberger, M., Benn, T., Pannett, B., Pearce, N., and Saracci, R. (1995). Soft tissue sarcoma and non-Hodgkin's lymphoma in workers exposed to phenoxy herbicides, chlorophenols, and dioxins: Two nested case-control studies. *Epidemiology*, **6**, 396–402.

12. Langholz, B. and Borgan, Ø. (1995). Countermatching: A stratified nested case-control sampling method. *Biometrika*, **82**, 69–79.

13. Langholz, B. and Borgan, Ø. (1997). Estimation of absolute risk from nested case-control data. *Biometrics*, **53**, 767–774.

14. Mantel, N. and Haenszel, W. (1959). Statistical aspects of the analysis of data from retrospective studies of disease. *J. Natl. Cancer Inst.*, **22**, 719–748.

15. Oakes, D. (1981). Survival times: aspects of partial likelihood (with discussion). *Int. Statist. Rev.*, **49**, 235–264.

16. Prentice, R. L. (1986). A case-cohort design for epidemiologic cohort studies and disease prevention trials. *Biometrika*, **73**, 1–11.

17. Prentice, R. L. and Breslow, N. E. (1978). Retrospective studies and failure time models. *Biometrika*, **65**, 153–158.

18. Thomas, D. C. (1977). Addendum to: Methods of cohort analysis: appraisal by application to asbestos mining. By F. D. K. Liddell, J. C. McDonald, and D. C. Thomas. *J. Roy. Statist. Soc. A*, **140**, 469–491.

51 Observational Studies

S. M. McKinlay

51.1 Introduction

The term *observational* is employed to denote a type of investigation that can be describedsomewhat negatively as *not an experiment*. A general definition of such a study was presented by Wold, who, after setting down three criteria for a controlled experiment, defined observational data as those in which at least one of the three following conditions is violated:

1. The replications of the experiment are made under similar conditions (so as to yield an internal measure of uncontrolled variation).

2. The replications are mutually independent.

3. The uncontrolled variation in the replication is subjected to randomization in the sense of Fisher [16, p. 30].

Observational studies involving explanation rather than just description were defined as the collection of data in which the third criterion, randomization, is not possible [16, p. 37]:

> ... it is the absence of randomization that is crucial, for since this device is not available there is no clear-cut distinction between factors which are explicitly accounted for in the hypothetical model and disturbing factors which are summed up in the residual variation.

Wold included in this class of studies all those concerned with the investigation of cause and effect and/or the development of predictive models.

In a seminal discussion of this type of investigation, Cochran [1] suggested two main distinguishing characteristics:

1. The objective is the investigation of possible cause–effect relationships.

2. This is implemented by the comparison of groups subject to different "treatments" that were preassigned in a nonrandom manner.

Most broadly, the term *observational* can apply to any investigation that is not an experiment, including descriptive surveys of populations, the essential characteristic being that the subjects or material under investigation are not manipulated in any way. Following the lead of Wold and Cochran, among others, however, the term has become most closely associated with that subset of studies investigating a hypothesized cause–effect relationship. This more narrow definition is assumed in the discussion presented here.

The two major defining characteristics of an observational study proposed by Cochran and quoted earlier are in a real sense in conflict. A cause–effect relationship can be established only under appropriately controlled experimental conditions that include randomization. Without this control of variability, the inference that A causes B cannot be made and one is restricted to the weaker inference that A and B are associated. If this association includes a temporal sequence (e.g., A is always observed to occur before B), then this is suggestive of a cause-effect relationship. It is important to understand that data from an observational study cannot be used to demonstrate a cause–effect relationship—only to suggest one (for possible testing in an experiment). For further discussion of the unique inferential power of experiments, the reader is referred to articles by Kempthorne and McKinlay [5,9] as well as to the appropriate entries in this volume.

Observational studies are variously designed and titled, but fall into either of two categories—prospective and retrospective. These categories and aspects of design associated with them are discussed in the following section.

51.2 Prospective

This type of study is closest in design to an experiment. Groups are formed in a nonrandom (usually self-selected) manner according to categories or levels of a hypothesized cause (factor F) and subsequently observed with respect to the outcome of interest Y. The causal factor F can be referred to as the *design* or *independent* variable. The outcome Y can also be termed the *dependent* variable. A well-known example is the prospective study of the association between smoking habit and lung cancer [4].

Groups of individuals who did and did not smoke (F) were identified and followed for a fixed period. All cases of lung cancer (Y) were recorded for each group in that time. The temporal sequence of F and Y suggested (but did not demonstrate) a cause–effect relationship. The equivalent experimental study would have required the identification of subjects who had never smoked to be randomly assigned to one of two groups, either to maintain their nonsmoking status or to start and continue to smoke over the observation period.

The major inferential weakness of this observational design is that subjects self-select themselves into groups. It is possible, for example, that the predisposition toward smoking in some individuals is also a causal factor for lung cancer and that the observed association between smoking and lung cancer is merely reflective of such a common cause. Other terms that have been used to describe prospective studies of this type are *cohort* and *quasi-experiment*.

The difference between the type of prospective, observations study described here and descriptive longitudinal or *panel* studies should be noted. Panel studies (of opinions, health, etc.) describe changes in a population over time without predetermined groups for comparison.

The use of the term *prospective* to describe a type of study design must also not be confused with the type of data collection. A prospective observational study may be designed entirely from pre-existing records. For example, if sufficient information is available in a set of records (e.g., medical or occupational) to identify smokers and nonsmokers at the start of recordkeeping and to identify diagnosed lung cancer within a specified period, then a prospective study of the association between lung cancer and smoking can be designed from such records. Smokers and nonsmokers can be defined and their lung cancer experience observed prospectively over the period covered by the records. Investigations of associations of various prenatal factors with fetal outcome (e.g., see McKinlay and McKinlay [8] and Neutra et al. [11]) illustrate this design.

51.3 Retrospective

This type of design is particularly applicable in the epidemiology of chronic diseases or conditions, for which the period of observation between introduction of the hypothesized cause and appearance of the effect can be many years or even decades, and/or for which the effect is rare (as in many cancers).

The distinguishing characteristic of this design is that the groups for comparison are defined on an observed outcome and differences between the groups are sought on a hypothesized cause (or causes). For example, the initial investigations reporting an association between smoking and lung cancer in Britain [3] were comparing groups with and without lung cancer (the outcome or effect) with respect to potential causes. The roles of dependent and independent (design) variables are reversed. The dependent variable Y in a retrospective study is a hypothesized cause, while the independent or design variable F is the outcome or effect being investigated.

Although the term *retrospective* refers primarily to the design, it also frequently applies to the method of data collection, particularly when the cause(s) may have preceded the outcome chronologically, sometimes by many years. Heavy reliance on memory recall is a recurring disadvantage of this design, which must be weighed against the disadvantage of an alternative long and costly prospective study. A topical example that illustrates this design problem is the investigation of the role of diet in early adulthood on the subsequent development of selected cancers, requiring diet recall of 20 years or more.

The retrospective design does have a considerable advantage over equivalent prospective studies in terms of efficiency, particularly when the rate of occurence of the effect (outcome) is relatively low (e.g., endometrial cancer, cancer of the cervix, with rates under 5/1000). Numbers per group required for a prospective study may rapidly exceed 20 times the number required for the equivalent retrospective study. An excellent, detailed discussion of the advantages and disadvantages of the two designs is provided by Schlesselman [12].

Other terms for a retrospective observational study include *case–control* and *ex post facto*. The latter also has been used to denote a matched-pair design [14].

51.4 Control of Variation

A recurring issue in the design and analysis of observational studies (both retrospective and prospective) is the selective control of variation due to covariables ("intervening" or "confounding" variables).

Techniques that, in experimental design and analysis, are used to increase efficiency in the comparison, have the supplementary role of controlling for potential bias in observational comparisons—a role assumed by randomization in experiments. The relative effectiveness of such techniques as pre- or poststratification, pair matching, and covariance adjustments has been investigated notably since the early 1970s. Two comprehensive reviews provide full discussions of this issue [2,6]. Results of this research show that pair matching is seldom more effective than analytic adjustments on independent samples. Moreover, pair matching may be very costly in terms of locating viable matches and/or discarding unmatchables [7], although it remains a popular technique in many fields.

51.5 Role of Observational Studies

This type of study has two major purposes: to establish the need for a subsequent experiment; and to provide optimal information in situations not amenable to the conduct of an experiment (for practical or ethical reasons). This design is not an alternative to a controlled, randomized experiment as it does not evaluate cause–effect relationships directly.

Retrospective studies, in particular, can be cost-effective preliminary investigations that provide the rationale for more costly experiments—especially when the observed associations are equivocal or contradictory. More recent retrospective and prospective studies of the association between cardiovascular disease and use of oral contraceptives [13,15], for example, have stimulated the instigation of experiments to study the effect of this hormone combination on such cardiovascular disease risk factors as blood pressure and blood lipids [10].

The second important role for observational studies is as a less powerful substitute for experiments when the latter are not feasible. For example, it is not ethical to randomly assign subjects to smoking or nonsmoking groups in order to observe the relative lung cancer incidence. To investigate the impact of water composition on a community's health, it may not be feasible to randomly change water constituents in participating towns because of such constraints as variable water sources and the need for town council approval. In such a situation, the observational study may provide information quickly and cheaply concerning a potential cause–effect relationship. In some instances, this type of design—albeit inconclusive—may provide the only information on a cause–effect relationship.

References

1. Cochran, W. G. (1965). *J. Roy. Statist. Soc. A*, **128**, 234–266.
2. Cochran, W. G. and Rubin, D. B. (1973). *Sankhyā A*, **35**, 417–446. (Excellent review on controlling bias in observational studies.)
3. Doll, R. and Hill, A. B. (1952). *Br. Med. J.*, **2**, 1271–1286.
4. Doll, R. and Hill, A. B. (1964). *Br. Med. J.*, **1**, 1399–1410, 1460–1467.
5. Kempthorne, O. (1977). *J. Statist. Plan. Infer.*, **1**, 1–25.
6. McKinlay, S. M. (1975). *J. Am. Statist. Assoc.*, **70**, 503–520. (Comprehensive review on design and analysis of observational studies.)
7. McKinlay, S. M. (1977). *Biometrics*, **33**, 725–735.
8. McKinlay, J. B. and McKinlay, S. M. (1979). *Epidemiol. Community Health*, **33**, 84–90.
9. McKinlay, S. M. (1981). *Milbank Memorial Fund Quart. Health Soc.*, **59**, 308–323.
10. *National Institutes of Health, National Institute of Child Health and Human Development*. RFP No. NICHD-CE-82-4, Feb. 1, 1982.
11. Neutra, R. R., Fienberg, S. E., Greenland, S., and Friedman, E. A., (1978). *New Engl. J. Med.*, **299**, 324–326.
12. Schlesselman, J. J. (1982). *Case-Control Studies: Design, Conduct, Analysis*. Oxford University Press, New York. (A comprehensive book on the retrospective design that provides an excellent discussion of its advantages relative to the prospective design.)
13. Slone, D., Shapiro, S., Kaufman, D. W., Rosenberg, L., Miettinen, O. S., and Stolley, P. D. (1981). *New. Engl. J. Med.*, **305**, 420–424.
14. Thistlethwaite, D. C. and Campbell, D. T. (1960). *J. Educ. Psychol.*, **51**, 309–317.

15. Vessey, M. P. and Doll, R. (1968). *Br. Med. J.*, **2**, 199–205.
16. Wold, H. (1956). *J. Roy. Statist. Soc. A*, **119**, 28–60.

52. One- and Two-Armed Bandit Problems

Donald A. Berry

52.1 Introduction

Suppose there are two available treatments for a certain disease. Patients arrive one at a time, and one of the treatments must be used on each. Information as to the effectiveness of the treatments accrues as they are used. The overall objective is to treat as many of the patients as effectively as possible. This seemingly innocent but important problem is surprisingly difficult, even when the responses are dichotomous: success–failure. It is a version of the two-armed bandit.

A bandit problem in statistical decision theory involves sequential selections from k stochastic processes (or "arms," machines, treatments, etc.). Time may be discrete or continuous, and the processes themselves may be discrete or continuous. The processes are characterized by parameters that are typically unknown. The process selected for observation at any time depends on the previous selections and results. A decision procedure (or strategy) specifies which process to select at any time for every history of previous selections and observations. A utility is defined on the space of all histories. This provides a definition for the utility of a strategy in the usual way by averaging over all possible histories resulting from that strategy.

Most of the literature, and most of this chapter, deals with discrete time. In this setting, each of the k arms generates an infinite sequence of random variables. Making an observation on a particular sequence is called a *pull* of the corresponding arm. The classical objective in bandit problems is to maximize the expected payoff value $\sum_1^\infty \alpha_i Z_i$, where Z_i is the variable observed at stage i and the α_i are known nonnegative numbers (usually $\alpha_i \geq \alpha_{i+1}$ is assumed) with $\sum_1^\infty \alpha_i < \infty$; $(\alpha_1, \alpha_2, \ldots)$ is called a *discount sequence*. A strategy is *optimal* if it yields the maximal expected payoff. An arm is optimal if it is the first pull of some optimal strategy.

52.2 Finite-Horizon Bernoulli Bandits

Historically, the most important discount sequence has been "finite-horizon uniform": $\alpha_1 = \cdots = \alpha_n = 1, \alpha_{n+1} = \cdots = 0$, and most of the literature deals with Bernoulli processes. The objective is then to maximize the expected number of successes in the first n trials.

The finite-horizon two-armed Bernoulli

bandit was first posed by Thompson [28]. It received almost no attention until it was studied by Robbins [22], and from a different point of view by Bradt et al. [10]. Robbins [22] suggested a selection strategy that depends on the history only through the last selection and the result of that selection (i.e., pull the same arm after a success and switch after a failure), and compared its effectiveness with that of random selections. This originated an approach called "finite memory" [12,17,26]. The decision maker's choice at any stage can depend only on the selections and results in the previous r stages.

Bradt et al. [10] considered information given as a joint probability distribution of the Bernoulli parameters p_1 and p_2 (a so-called Bayesian approach) and characterized optimal strategies for the case in which one parameter is known a priori. With such an approach, a strategy requires (and "remembers") only the sufficient statistics: the numbers of successes and failures on the two arms. Most of the recent bandit literature (and the remainder of the article) takes the Bayesian approach. It is not that most researchers in bandit problems are "Bayesians"; rather, Bayes' theorem provides a convenient mathematical formalism that allows for adaptive learning, and so is an ideal tool in sequential decision problems.

52.2.1 Myopic Strategies

Feldman [14] solved the Bernoulli two-armed bandit in the finite-horizon uniform setting for a deceptively simple initial distribution on the two parameters: both probabilities of success are known, but not which goes with which arm. Feldman showed that *myopic* strategies are optimal: At every stage, pull the arm with greater expected immediate gain (the unconditional probability of success with arm j is the prior mean of p_j). Feldman's result was extended in different directions by Fabius and van Zwet [13], Berry [2], Kelley [18], and Rodman [24].

It is important to recognize that myopic strategies are not optimal—or even good—in general. As a simple example, suppose p_1 is known to be $\frac{1}{2}$ (pulling arm 1 is like tossing a fair coin) and p_2 is either 1 or 0 (the other coin is either two-headed or two-tailed); let r be the initial probability that $p_2 = 1$. The fact that a single pull of arm 2 reveals complete information makes the analysis of this problem rather easy. If $r < \frac{1}{2}$, then a myopic strategy indicates pulls of arm 1 indefinitely and has utility $n/2$. On the other hand, pulling arm 2 initially, and indefinitely if it is successful and never again if it is not, results in n successes with probability r and an average of $(n-1)/2$ successes with probability $1 - r$. The advantage of this strategy over the myopic is

$$rn + \tfrac{1}{2}(1-r)(n-1) - \tfrac{1}{2}n$$
$$= \tfrac{1}{2}\{r(n+1) - 1\},$$

which is positive for $r > 1/(n+1)$. In this and other bandit problems, it may be wise to sacrifice some potential early payoff for the prospect of gaining information that will allow for more informed choices later. But the "information vs. immediate payoff" question is seldom as clear as it is in this example.

52.2.2 "Stay with Winner" Rule

Partial characterizations of optimal strategies in the two-armed bandit with finite-horizon uniform discounting were given by Fabius and van Zwet [13]. Berry [2] gave additional characterizations when p_1 and p_2 are independent a priori. One such is the "stay with a winner rule"; if the arm pulled at any stage is optimal and yields a success, then it is optimal at the next stage as well. (Bradt et al. [10] give a counterexample to this result when the arms are dependent.) Nothing can be said in general about the arm to pull following a failure; the simplest example of staying with a loser involves an arm whose success rate is known, for then a failure contains the same information as a success.

The stay-with-a-winner rule, which characterizes optimal strategies, is to be contrasted with play-the-winner rules [20,27]. These are complete strategies and are based on "finite memory" as described earlier, where the memory length is one. That is, the same arm is pulled following a success and the other following a failure. Such a strategy is optimal only in very special circumstances; for example, $n = 2$ and the distribution of (p_1, p_2) is exchangeable.

52.2.3 Solution by Dynamic Programming

Since the only uncertainty arises after a failure, one might say that the problem is half solved. But the stay-with-a-winner rule, while picturesque, is little help in finding optimal strategies. One must weigh the utility of all possible histories when deciding which arm to pull. The standard method of solution for such problems is dynamic programming or backward induction. To determine optimal strategies, one first finds the maximal conditional expected payoff (together with the arm or arms that yield it) at the very last stage, for every possible $(n-1)$-history (sequence of pulls and results), optimal and otherwise. Here, "conditional" refers to the particular history. Proceeding to the penultimate stage, one maximizes the conditional expected payoff from the last two observations for every possible $(n-2)$-history. Continuing backward—remembering the optimal arms at each partial history—gives all optimal strategies. The problem is four-dimensional since that is the dimension of a minimal sufficient statistic. But a computer program requiring on the order of $n^3/6$ storage locations is possible.

52.3 Geometric Discounting

Much recent literature (notably Refs. 15, 16, 19, 29, and 30) has dealt with geometric discounting: $\alpha_m = \beta^{m-1}$, where $\beta \geq 0$ and, usually, $\beta < 1$. In economic applications, for example, one may wish to assume that the rate of inflation does not change over time. Or, in a medical setting, a new and obviously better treatment may be discovered at each stage (or the disease could spontaneously disappear) with constant probability $1 - \beta$. Although it is quite special, geometric discounting is important, in part because, except for a multiplicative factor, it is invariant under a time shift. In addition, when the discount sequence is unknown, there are many instances in which the decisionmaker should act as though the discount sequence is geometric [4].

In k-armed bandits the attractiveness of an arm depends in general on the other arms available. But when the discount sequence is geometric (and the arms are independent), Gittins and Jones [16] showed that the arms can be evaluated separately by assigning a real number, a "dynamic allocation index" or "Gittins index," to each arm—any arm with the largest index is optimal. In view of this result, the desirability of a particular arm can be calculated by comparing the arm and various known arms; the Gittins index can be determined as the expectation of that known arm for which both arms are optimal in a two-armed bandit problem (this is quite different from the expectation of the unknown arm; see Berry and Fristedt [5]). In practice, it may not be possible to evaluate a Gittins index exactly, but it can be approximated by truncating and using backward induction.

Robinson [23] compares various strategies obtained for geometric discounting when the actual discount sequence is the finitehorizon uniform for large n. He finds that, for his criterion, a rule proposed by Bather [1] performs quite well.

52.4 One-Armed Bandits

When $k = 1$, the decision problem is trivial — the only available arm is pulled forever! The so-called one-armed bandit is formulated differently. There is a single process with unknown characteristics, and the decisionmaker has the option of stopping observation on the process at any time. Having this option makes the problem the same as a two-armed bandit, in which the second process is known with expectation 0 (or some number greater than 0 if there is a cost of observation). Stopping observation at stage i in the original formulation is analogous to choosing the known process at stages subsequent to i in the two-armed bandit. For this reason, two-armed bandits with one arm known are sometimes called "one-armed bandits." Whether this is appropriate depends on the discount sequence.

52.4.1 Regular Discounting

When the characterisitics of one arm are known, Berry and Fristedt [5] showed that an optimal strategy is to pull the known arm indefinitely once it is pulled for the first time, provided the discount sequence is nonincreasing and *regular*

$$\sum_{j}^{\infty} \alpha_i \sum_{j+2}^{\infty} \alpha_i \le \left(\sum_{j+1}^{\infty} \alpha_i \right)^2$$

for $j = 1, 2, \ldots$. Moreover, if the discount sequence is not regular there is a distribution for the unknown arm such that all optimal strategies call for switches from the known arm [5]. Therefore, the term "one-armed bandit" is appropriate for this problem if and only if the discount sequence is regular.

Examples of regular discount sequences are $(1, 1, \ldots, 1, 0, 0, \ldots)$-finite-horizon, $(1, \beta, \beta^2, \ldots)$-geometric, and $(2, 1, 1, 0, 0, \ldots)$. Nonregular sequences include $(10, \beta, \beta^2, \ldots)$ for $0 < \beta < 1$, $(3, 1, 0, 0, \ldots)$, and $(2, 1, 1, 1, 0, 0, \ldots)$.

52.4.2 Chernoff's Conjecture

A conjecture of Chernoff [11] attempts to relate one- and two-armed bandits. For a given finite-horizon two-armed bandit, consider a modified problem in which arm 1 must be pulled forever if it is ever pulled once. If arm 1 is optimal in the modified problem, then Chernoff conjectured that it is also optimal in the unmodified problem. Since the modified problem is the equivalent of a one-armed bandit, this would imply that the solution of a two-armed bandit problem is partially determined by the solution of a corresponding one-armed bandit. The validity of the conjecture is easy to resolve for many regular discount sequences, but its truth remains an open question in the finite-horizon setting.

52.5 Continuous Time

The preceding problems are set in discrete time. Chernoff [11] considered a continuous version in which the arms are continuous-time processes (in particular, independent Wiener processes with unknown means and known variances). An arm is observed, payoff accumulates equal to the value of the process, and information about the process accumulates continuously until a switch is made and the other process observed. Observation continues until some fixed time has elapsed. A less than helpful characteristic of every optimal strategy in this problem is that almost every switch is accompanied by an uncountable number of switches within every time interval of positive duration that includes the switch.

52.6 Other Objectives

The preceding discussion dealt with the objective of maximizing $\sum \alpha_i Z_i$. At least two different objectives have been considered in Bernoulli bandit problems.

52.6.1 Maximizing Successes First Failure

Berry and Viscusi [9] consider k-armed Bernoulli bandits in which observation ceases with the first failure. The discount sequence may not be monotonic. This version is analytically simpler than classical bandits. Generalizing Berry and Viscusi [9], Berry and Fristedt [8] show that there is always an optimal strategy using a single arm indefinitely ("no switching") whenever the arms are independent and the discount sequence is *superregular*:

$$\alpha_{i+1}/\alpha_i \geq \alpha_{i+j+1}/\alpha_{i+j},$$

for all positive integers i and j. Superregular sequences are also regular; the geometric and finite-horizon uniform are both superregular.

When the discount sequence is superregular, the class of strategies to be considered is so restricted (to k in number) that an optimal strategy is easy to find. For example, suppose $\alpha_m > 0$ and $\alpha_i = 0$ for $i \neq m$. Then an arm is optimal if it has largest mth moment (average of p_j^m with respect to the prior distribution of the success rate p_j of arm j).

52.6.2 Reaching A Goal

Berry and Fristedt [6,7] consider the problem of maximizing the probability of achieving n successes minus failures before m failures minus successes in a two-armed bandit (with no discounting). The problem is a sequential allocation version of "gambler's ruin" and is very difficult, with solutions having various nonintuitive characterisitics. Speaking somewhat loosely, when one arm has a known success rate, there are circumstances in which it is optimal to use the known arm until "ruin" is approached, at which time the other, riskier arm is used [6].

52.6.3 Fields of Application

In addition to being classified in statistical decision theory, bandit problems are sometimes classified in the areas of learning processes, sequential stochastic design, control theory, online experimentation, and dynamic programming. Since the decision problem typically involves weighing immediate gain against the possibility of obtaining information to increase the potential for future gain, they have attracted much interest in the fields of psychology [21], economics [25], biomedical trials [3], and engineering [31] among others.

As an example, consider the medical trial setting in the initial paragraph of this article. For the sake of simplicity, assume the length of the trial is known to be n and the responses are dichotomous with independent uniform priors on the two success rates. Table 1 gives the expected proportion of successes for an optimal strategy for selected values of n. Optimal strategies are difficult to find and specify. However, Berry [3] gives an easy-to-use strategy whose expected success proportion approximates that of an optimal strategy.

References

1. Bather, J. A. (1981). *J. Roy. Statist. Soc. B*, **43**, 265–292.

Table 1: The expected proportion of successes for an optimal strategy for selected values of n.

n	1	5	10	50	100	∞
Maximum success proportion	0.500	0.578	0.602	0.640	0.649	0.667

2. Berry, D. A. (1972). *Ann. Math. Statist.*, **43**, 871–897.

3. Berry, D. A. (1978). *J. Am. Statist. Assoc.*, **73**, 339–345.

4. Berry, D. A. (1983). In *Mathematical Learning Models—Theory and Algorithms*, U. Herkenrath et al., eds., pp. 12–25, Springer-Verlag, New York.

5. Berry, D. A. and Fristedt, B. E. (1979). *Ann. Statist.*, **7**, 1086–1105.

6. Berry, D. A. and Fristedt, B. E. (1980). *Adv. Appl. Probab.*, **12**, 775–798.

7. Berry, D. A. and Fristedt, B. E. (1980). *Adv. Appl. Probab.*, **12**, 958–971.

8. Berry, D. A. and Fristedt, B. E. (1983). *Stoch. Process. Appl.*, **15**, 317–325.

9. Berry, D. A. and Viscusi, W. K. (1981). *Stoch. Process. Appl.*, **11**, 35–45.

10. Bradt, R. N., Johnson, S. M., and Karlin, S. (1956). *Ann. Math. Statist.*, **27**, 1060–1070.

11. Chernoff, H. (1968). *Sankhyā A*, **30**, 221–252.

12. Cover, T. M. and Hellman, M. E. (1970). *IEEE Trans. Inform. Theory*, **16**, 185–195.

13. Fabius, J. and van Zwet, W. R. (1970). *Ann. Math. Statist.*, **41**, 1906–1916.

14. Feldman, D. (1962). *Ann. Math. Statist.*, **33**, 847–856.

15. Gittins, J. C. (1979). *J. Roy. Statist. Soc. B*, **41**, 148–177.

16. Gittins, J. C. and Jones, D. M. (1974). In *Progress in Statistics*, J. Gani et al., eds., pp. 241–266, North-Holland, Amsterdam.

17. Isbell, J. R. (1959). *Ann. Math. Statist.*, **30**, 606–610.

18. Kelley, T. A. (1974). *Ann. Statist.*, **2**, 1056–1062.

19. Kumar, P. R. and Seidman, T. I. (1981). *IEEE Trans. Automat. Control*, **26**, 1176–1184.

20. Nordbrock, E. (1976). *J. Am. Statist. Assoc.*, **71**, 137–139.

21. Rapoport, A. and Wallsten, T. S. (1972). *Ann. Rev. Psychol.*, **23**, 131–176.

22. Robbins, H. (1952). *Bull. Am. Math. Soc.*, **58**, 527–536.

23. Robinson, D. R. (1983). *Biometrika*, **70**, 492–495.

24. Rodman, L. (1978). *Ann. Probab.*, **6**, 491–498.

25. Rothschild, M. (1974). *J. Econ. Theory*, **9**, 185–202.

26. Smith, C. V. and Pyke, R. (1965). *Ann. Math. Statist.*, **36**, 1375–1386.

27. Sobel, M. and Weiss, G. H. (1970). *Biometrika*, **57**, 357–365.

28. Thompson, W. R. (1933). *Biometrika*, **25**, 275–294.

29. Whittle, P. (1980). *J. Roy. Statist. Soc. B*, **42**, 143–149.

30. Whittle, P. (1982). *Optimization over Time, Dynamic Programming and Stochastic Control*, Vol. I, Wiley, Chap. 14, New York.

31. Witten, I. H. (1976). *J. Franklin Inst.*, **301**, 161–189.

53 Ophthalmology

Bernard Rosner

53.1 Introduction

Ophthalmology is the branch of medical science concerned with ocular phenomena. The types of research studies in ophthalmology are similar to those encountered in general biostatistical work, including epidemiological studies, clinical trials, small laboratory studies based on either humans or animals, and genetic studies. For the biostatistician, the one important distinction between ophthalmologic data and data obtained in most other medical specialties is that information is usually collected on an *eye-specific* rather than a *person-specific* basis. This method of data collection has both advantages and disadvantages from a statistical point of view. We will discuss this in the context of (1) studies involving a comparison of two or more treatments such as in a clinical trial and (2) observational studies.

53.2 Clinical Trials

Usually one wishes to compare two or more treatment groups where one (or more) of the treatment groups may serve as a control group(s). This is generally done in other medical specialties by either (1) randomly assigning treatments to different groups of individuals or (2) assigning both treatments to the same individual at different points in time using a crossover design and making intraindividual comparisons. To compare treatment groups in method 1, one either (1) relies on the randomization to minimize differences between treatment groups on other covariates, (2) performs additional multivariate analyses such as regression analyses to adjust for differences in treatment groups that may emerge despite the randomization, or (3) matches individuals in different treatment groups on at least a limited number of independent variables at the design stage so as to minimize extraneous treatment group differences. One always has the problem that none of these methods may adequately control for confounding, since one may not anticipate the appropriate potential confounding variables at the data collection phase. Design (2), if feasible, seemingly overcomes many of these problems since intra-individual rather than interindividual comparisons are being made. However, in a crossover study, one must assume that the effect of one treatment is not carried over to the time of administration of the next treatment, an assumption that is not always easy to justify. In addition, a crossover study requires more commitment from the patient, and the problem of loss to follow-up becomes more important. In ophthalmologic work, two treatments can be assigned randomly to the left and right eyes, thus preserving the best features of the crossover design, that is, enabling intraindividual rather than interindividual comparisons, while eliminating the undesirable feature of the carryover effect, since both treatments typically are administered simultaneously. The administration of the two treatments simultaneously also improves patient compliance over that of a typical crossover study.

53.3 Observational Studies

In contrast to the clinical trial situation, the collection of data on an eye-specific basis has both advantages and disadvantages in terms of observational studies. An important advantage is that if responses on two eyes of an individual are treated as replicates, then one has an easily obtained independent estimate of intraindividual variability. Such estimates are more difficult to obtain when data are obtained on a person-specific basis. However, an important disadvantage is that collection of data on an eye-specific basis complicates even the most elementary of analyses such as the assessment of standard errors of means or proportions, and, in particular, poses challenging problems for conducting multivariate analyses. For this reason, many ophthalmologists often disregard this problem and present their results in terms of *distributions over eyes* rather than *distributions over individuals*. The observations in these eye-specific distributions are treated as independent random variables and standard statistical methods are used thereby. The problem with this formulation is that observations on two eyes of an individual are highly but not perfectly correlated random variables. Thus the assumption of independence results in standard errors that are generally underestimated and significance levels that are too extreme, rendering some apparently statistically significant results actually not significant. A conservative approach that is sometimes used to avoid this problem is to select only one of the two eyes of each individual for analysis and then proceed with standard statistical methods. The "analysis" eye may be a randomly selected eye or may be an eye selected as having the better (or worse) visual function of the two eyes. This is a valid use of such data, but is possibly inefficient since two eyes contribute somewhat more information than one eye but not as much information as two independent observations. This problem has been discussed by Ederer [2]. One possible solution for normally distributed outcome variables is offered by the following intraclass correlation model [3]

$$\begin{aligned} y_{ijk} &= \mu + \alpha_i + \beta_{ij} + e_{ijk}, \\ i &= 1,\ldots,g, \quad j=1,\ldots,P_i, \\ k &= 1,\ldots,N_{ij}, \end{aligned} \quad (1)$$

where i denotes group, j denotes individual within group, k denotes eye within individual, α_i is a fixed effect, $\beta_{ij} \sim N(0, \sigma_\beta^2)$, $e_{ijk} \sim N(0, \sigma^2)$, and one permits a variable number of individuals per group (P_i) and a variable number of eyes available for analysis for each individual (N_{ij}). The data layout for the model in (1) is given in Table 1, where $\bar{y}_{ij.} = \sum_k y_{ijk}/N_{ij}$, $\bar{y}_{i..} = \sum_j \bar{y}_{ij.}/P_i$, $\bar{y}_{\cdots} = \sum_i P_i \bar{y}_{i..}/\sum_i P_i$, $P = \sum_i P_i$, $N = \sum_i \sum_j N_{ij}$. Typically, one would wish to test the hypothesis H_0: all $\alpha_i = 0$ vs. H_1: at least one $\alpha_i \neq 0$. An exact solution to this problem is difficult because of the unbalanced design whereby individuals can contribute either one or two eyes to the analysis. However, the method of unweighted means is a reasonable approximate method in this case since $\max N_{ij}^{-1/2}/\min N_{ij}^{-1/2} \leq 2$ for ophthalmologic work [5, p. 367]. Thus, one obtains the expected mean squares given in Table 1, where each individual is assumed

Table 1: Data layout under the intraclass correlation model.

Source of Variation	Mean Square	df	Expected Mean Square
Between groups	$\dfrac{\sum P_i(\bar{y}_{i..} - \bar{y}_{...})^2}{g-1} \equiv \mathrm{MSG}^a$	$g-1$	$\dfrac{\sum_i P_i(\alpha_i - \bar{\alpha})^2}{g-1} + \sigma_\beta^2 + (\sigma^2/\bar{N})$
Between persons within groups	$\dfrac{\sum\sum(\bar{y}_{ij.} - \bar{y}_{i..})^2}{P-g} \equiv \mathrm{MSP}^b$	$P-g$	$\sigma_\beta^2 + (\sigma^2/\bar{N})$
Between eyes within persons	$\dfrac{\sum\sum\sum(y_{ijk} - \bar{y}_{ij.})^2}{N-P} \equiv \mathrm{MSE}^c$	$N-P$	σ^2

^{a}Mean-square groups.
^{b}Mean-squared persons.
^{c}Mean-squared error.
Source: Reproduced from Reference 3 with permission.

to contribute $\bar{N} = P[\sum_i \sum_j (1/N_{ij})]^{-1}$ eyes to the analysis. It follows that an appropriate test statistic for the preceding hypotheses is given by $\lambda = \mathrm{MSG}/\mathrm{MSP} \sim F_{g-1, P-g}$ under H_0, where one would reject H_0 if $\lambda > F_{g-1, P-g, 1-\alpha}$. Specific groups ($i_1$ and i_2) can be compared by computing the test statistic

$$u_{i_1, i_2} = \frac{(\bar{y}_{i_1..} - \bar{y}_{i_2..})}{\left[\mathrm{MSP}(P_{i_1}^{-1} + P_{i_2}^{-1})\right]^{1/2}}$$

and rejecting the null hypothesis that the group means are equal if $|u_{i_1, i_2}| > t_{P-g, 1-\alpha/2}$, $i_1, i_2 = 1, \ldots, g$, $i_1 \neq i_2$.

These methods are illustrated in the following dataset obtained from an outpatient population of 218 persons aged 20–39 with retinitis pigmentosa (RP) who were seen at the Massachusetts Eye and Ear Infirmary from 1970 to 1979 [1]. Patients were classified according to a detailed genetic pedigree into the genetic types of autosomal dominant RP(DOM), autosomal recessive RP(AR), sex-linked RP(SL), and isolate RP(ISO), the purpose being to compare patients in the four genetic types on the basis of selected ocular characteristics. For the sake of simplicity, only one person from the age group 20–39 was selected from each family; if more than one affected person in this age range was available in a given family, then a randomly selected affected person was chosen, thus yielding 218 individuals from 218 distinct families. In Table 2, we present an analysis comparing the four genetic types on spherical refractive error. The analysis is based on the subgroup of 212 individuals who had information on spherical refractive error in at least one eye, of whom 210 had information on both eyes and 2 had information on one eye. All refractive errors were determined with retinoscopy after cycloplegia.

There are overall significant differences between the four groups ($\lambda \equiv \mathrm{MSG}/\mathrm{MSP} = 3.68 \sim F_{3, 208}$, $p = 0.013$). From the t statistics in Table 2(b), we see that this overall difference is completely attributable to the lower spherical refractive error in the SL group as compared with the other three groups. The estimated intraclass correlation between eyes over all groups was 0.969!

For comparative purposes, a one-way ANOVA was also performed on the same data where the data from two eyes of an in-

Table 2: Comparison of spherical refractive error of RP patients by genetic type using the intraclass correlation model in Equation (1).

(a) Overall ANOVA Table

Source of Variation	Sum of Squares	df	Mean Square	F Statistic
Between groups	133.59	3	44.53	3.68 ($p = 0.013$)
Between persons within groups	2518.45	208	12.11	
Within persons	80.49	210	0.383	

(b) Descriptive Statistics: t Statistics and p Values for Comparisons of Specific Groups

Group (i)	Mean ($\bar{y}_{i..}$)	Estimated Standard Error $(\mathrm{MSP}/P_i)^{1/2}$	Number of Persons (P_i)
DOM	[a]0.127	0.658	28
AR	−0.831	0.778	20
SL	−3.299	0.820	18
ISO	−0.842	0.288	146

	Comparison Group		
Group (i)	AR	SL	ISO
DOM	0.941+ (NS)	3.259 ($p = 0.001$)	1.350 (NS)
AR	—	2.183 ($p = 0.030$)	0.013 (NS)
SL	—	—	−2.826 ($p = 0.005$)

[a]t Statistic.
Source: Reproduced from Reference 3 with permission.

Table 3: Comparison of spherical refractive error of RP patients by genetic type assuming independence between eyes for an individual.

(a) Overall ANOVA Table

Source of Variation	Sum of Squares	df	Mean Square	F Statistic
Between groups	267.94	3	99.21	8.45 ($p = 0.00026$)
Within groups	4906.83	418	11.73	

(b) Descriptive Statistics: t Statistics, and p Values for Comparisons of Specific Groups

Group (1)	Mean ($\bar{y}_{i..}$)	Estimated Standard Error	Number of Eyes (N_i)
DOM	+0.386	0.466	54
AR	−0.831	0.542	40
SL	−3.299	0.571	36
ISO	−0.842	0.201	292

Group (1)	Comparison Group		
	AR	SL	ISO
DOM	1.704 (NS)	4.999 ($p < 0.001$)	2.421 ($p = 0.016$)
AR	—	3.135 ($p = 0.002$)	0.018 (NS)
SL	—	—	−4.059 ($p < 0.001$)

Source: Reproduced from Reference 3 with permission.

dividual were treated as independent random variables, thus yielding a sample of 422 eyes. The results from this analysis are presented in Table 3.

All significant differences found in Table 2 (both overall and between specific groups) are much more significant in Table 3. The p value for the overall comparison between groups is 50 times larger in Table 2 than it is in Table 3 ($p = 0.013$ vs. $p = 0.00026$). In addition, the spherical refractive errors of the DOM and ISO group that were not significant in Table 2 become significant in Table 3 ($p = 0.016$). Clearly, the assumption of independence between eyes is inappropriate for this dataset and has a major impact on the analysis.

A model similar to the one given in Equation (1) has been developed for the case of a binomially distributed outcome variable [3]. These methods have been extended so that one can perform multiple linear regression and multiple logistic regression analyses in the context of ophthalmologic data [4].

References

1. Berson, E. L., Rosner, B., and Simonoff, E., (1980). *Am. J. Ophthalmol.*, **89**, 763–775. (An outpatient population of retinitis pigmentosa and their normal relatives. Risk factors for genetic typing and detection derived from their ocular examinations.)

2. Ederer, F. (1973). *Arch. Ophthalmol.*, **89**, 1–2. (Shall we count numbers of eyes or numbers of subjects?)

3. Rosner, B. (1982). *Biometrics*, **38**, 105–114. (Statistical methods in ophthalmology with an adjustment for the intraclass correlation between eyes.)

4. Rosner, B. (1985). *Biometrics*, **41** (in press).

5. Searle, S. R. (1971). *Linear Methods*. Wiley, New York.

54 Panel Count Data

Jianguo Sun

54.1 Introduction

This chapter discusses statistical analysis of panel count data arising from recurrent event or event history studies that concern the occurrence of some recurrent events. The data from these studies are often referred to as *event history data* and they can occur in many areas, including reliability studies and social sciences [25,41]. The event history data can be generally classified into two types. One is from the studies that monitor study subjects continuously and thus they provide information on the times of all occurrences of recurrent events. These data are usually referred to as recurrent event data [7]. The other type is the so-called panel count data discussed here, which arise when study subjects are examined or observed only at discrete timepoints and thus give only the numbers of occurrences of the events between observation times.

Examples of recurrent event data include data on occurrences of the hospitalizations of intravenous drug users [43], occurrences of the same infection such as recurrent pyogenic infections among inherited disorder patients [23], repeated occurrences of certain tumors, and warranty claims for a particular automobile [19]. These examples become examples of panel count data if the continuous observation scheme is changed to a discrete observation scheme. The panel count data could occur for different reasons. For example, they may arise because it is too expensive, impossible, or not realistic to conduct continuous follow-ups of study subjects.

To give a specific example of panel count data, consider the data discussed in Thall and Lachin [40] and given in dataset IV of Appendix A in Sun [30], among others. They arose from a clinical trial on the use of the natural bile acid chenodeoxycholic acid for the dissolution of cholesterol gallstones. The data include the observed information on the incidence of nausea from the first-year follow-up on 111 patients with floating gallstones in high-dose and placebo groups. The original study has 10-year follow-ups and consists of three groups, placebo, low dose, and high dose. Nausea is an unpleasant sensation vaguely referred to the epigastrium and abdomen, often culminating in vomiting, and it is very commonly associated with gallstone disease. During the study, the patients were scheduled to return for clinic observations at 1, 2, 3, 6, 9, and 12 months during the first year follow-up. At each visit, they were asked to report the total number of the incidences of nausea among other symptoms that had occurred between successive visits. That is, the observed data include actual visit times and the numbers of the incidences or occurrences of nausea between the visits. As expected, actual visit or observation times differ from patient to patient. Another example of panel count data, which has been discussed by many authors, can be found in dataset V of Appendix A in Sun [30]. It arose from a bladder cancer study consisting of the patients with superficial bladder tumors, and is discussed further below.

Many authors have discussed statistical analysis of recurrent event data [6,21,27,42]. In particular, Andersen et al. [1] published an excellent book that provides comprehensive coverage of counting process approaches for the analysis of recurrent event data. More recently, Cook and Lawless [7] also provided a relatively complete and comprehensive review of the whole recurrent event data literature.

There exists relatively sparse literature on the analysis of panel count data. Among the authors who considered this, Kalbfleisch and Lawless [18] discussed the fitting of Markov model to panel count data, and Sun and Kalbfleisch [33] and Wellner and Zhang [45] investigated estimation of the mean function of the underlying counting process that yields observed panel count data. For treatment comparison based on panel count data, Thall and Lachin [40] presented a procedure that transforms the problem to a multivariate comparison problem, while Sun and Fang [31] proposed a nonparametric approach. Sun and Wei [36] and Zhang [47] investigated regression analysis of panel count data. Note that in addition to the amount of relevant information available, another difference between recurrent event data and panel count data is the observation process. For the former, the observation process means the length of the whole follow-up, while for the latter, it also includes a sequence of consecutive observation times. Most of the methods mentioned above or given in the literature assume that the observation process is independent of the underlying counting process generating recurrent event or panel count data.

To analyze recurrent event data, it is common and convenient to characterize the occurrences of recurrent events by counting processes and to model the intensity process of the counting process [1]. On the other hand, for the analysis of panel count data, it is usually more convenient to work directly on the mean function of the counting processes conditional on covariate processes, due to the incomplete nature of observed information. In this case, a natural and simple approach is to fit the panel count data to parametric Poisson processes or mixed parametric Poisson processes. For example, Hinde [13] and Breslow [2] discussed regression analysis of Poisson count data, and Thall [39] gave some regression approaches for mixed Poisson processes. An alternative parametric approach is to treat the data as longitudinal count data and to use the generalized estimating equation approach [8]. In this chapter, we focus on nonparametric and semiparametric approaches that regard observed data as realizations of some underlying counting processes.

In the following, we will first consider in Section 54.2 nonparametric estimation of the mean function of counting processes giving rise to panel count data assuming that study subjects come from a homogeneous population. For the problem, several procedures will be discussed. In Section 54.3 we will study the treatment comparison based on panel count data. We will formulate the problem by comparing the mean functions of different counting processes and discuss some nonparametric procedures. In Section 54.4, we will consider regression analysis of panel count data with the focus on the proportional mean model in modeling the underlying counting processes that will be defined later.

Throughout Sections 54.2–54.4, we will assume that the observation process is independent of the underlying counting process of interest completely or conditional on the covariate process. Of course, this may not hold in practice. In Section 54.5, we will discuss situations where the two processes may be related with each other with the focus on regression analysis. Note that in practice, one could regard panel count data as a special type of longitudinal data and apply the methodology developed for general longitudinal data. However, a major drawback for this is that one would miss the special structure of panel count data. Also there often exist some questions of interest about panel count data that cannot be answered from the longitudinal data perspective.

54.2 Nonparametric Estimation

In this section, we consider an event history study that involves n independent subjects from a homogeneous population and in which each subject gives rise to a counting process $N_i(t)$, representing the total number of the occurrences of the recurrent event of interest from subject i up to time t, $i = 1, ..., n$. Define $\mu(t) = E\{N_i(t)\}$, the mean function of the processes N_i's, and assume that the goal is to estimate $\mu(t)$. For the ith subject, suppose that one observes $N_i(t)$ only at $0 < t_{i,1} < \cdots < t_{i,m_i}$ with $n_{i,j} = N_i(t_{i,j})$, the observed value of $N_i(t)$ at time $t_{i,j}$, $j = 1, \ldots, m_i$, $i = 1, \ldots, n$. That is, only panel count data are available, and the observed data are

$$\{(t_{i,j}, n_{i,j}); j = 1, \ldots, m_i, i = 1, \ldots, n\}.$$

Note that here different subjects can have different numbers of observations and also different observation times.

In the following, we will first describe a general and simple estimator of $\mu(t)$, the isotonic regression estimator. A generalization of the isotonic regression estimator and some other estimates are then discussed.

54.2.1 Isotonic Regression Estimator (IRE)

To describe the IRE, we first consider a simple situation in which $t_{i,j} = s_j$ for all $i = 1, ..., n$ with $m_i = m$. That is, all subjects have the same observation timepoints and the same numbers of observations. This can be the case in a follow-up study with prespecified observation timepoints and in which all subjects follow the prespecified observation schedule. It is obvious that for the situation, a natural estimate of $\mu(t)$ at the observation time s_l is given by the sample mean

$$\frac{\sum_{i=1}^{n} N_i(s_l)}{\sum_{i=1}^{n} I(s_l \leq t_{i,m_i})}$$

$$= \sum_{j=1}^{l} \left\{ \sum_{i=1}^{n} I(s_j \leq t_{i,m_i})[N_i(s_j) - N_i(s_{j-1})] \right\}$$

$$\times \left\{ \sum_{i=1}^{n} I(s_j \leq t_{i,m_i}) \right\}^{-1}.$$

Note that it is clear that one can estimate $\mu(t)$ only at the observation times s_l's and for any timepoint between s_{l-1} and s_l, the mean function $\mu(t)$ can take any value between the estimated values of $\mu(t)$ at s_{l-1} and s_l, where $s_0 = 0$. Assume that $\mu(t)$ is a step function with jumps only at the s_l's. Then the sample mean estimate given above can be rewritten as

$$\int_0^t \frac{\sum_{i=1}^{n} dN_i(s)}{\sum_{i=1}^{n} I(s \leq t_{i,m_i})},$$

which is the Nelson–Aalen estimator of $\mu(t)$ for recurrent event data [1].

Now we consider the general situation where subjects may have different observation times as well as different numbers of observations. In this case, it is clear that one can still estimate $\mu(t)$ by using the sample mean based on the observed values at each individual observation time. However, the resulting estimate may not be valid as it may not have the nondecreasing property that $\mu(t)$ has. To fix this, Sun and Kalbfleisch [33] proposed the following IRE. Let $s_1 < \cdots < s_m$ denote the ordered and different time points of all observation timepoints $\{t_{i,j}\}$ and w_l and $\bar{n}_l$ the number and mean value, respectively, of the observations made at s_l, $l = 1, \ldots, m$. The IRE, denoted by $\hat{\mu}_I = (\hat{\mu}_{I,1}, \ldots, \hat{\mu}_{I,m})$, of $\mu(t)$ at the s_l's is defined as $\mu_I = (\mu_1, \ldots, \mu_m) = (\mu(s_1), \ldots, \mu_m(s_m))$ that minimizes the weighted sum of squares

$$\sum_{l=1}^{m} w_l (\bar{n}_l - \mu_l)^2 \tag{1}$$

subject to the order restriction $\mu_1 \leq \cdots \leq \mu_m$. It can be easily seen that if $\bar{n}_1 \leq \ldots \leq \bar{n}_m$, then we have $\hat{\mu}_{I,l} = \bar{n}_l$, $l = 1, \ldots, m$. Also, for the simple situation discussed above, the IRE reduces to the sample mean estimator given before.

By definition, the IRE defined above is the isotonic regression of $\{\bar{n}_1, \ldots, \bar{n}_m\}$ with weights $\{w_1, \ldots, w_m\}$ [28]. Thus $\hat{\mu}_{I,l}$ has a closed form given by

$$\hat{\mu}_{I,l} = \max_{r \leq l} \min_{s \geq l} \frac{\sum_{v=r}^{s} w_v \bar{n}_v}{\sum_{v=r}^{s} w_v}$$
$$= \min_{s \geq l} \max_{r \leq l} \frac{\sum_{v=r}^{s} w_v \bar{n}_v}{\sum_{v=r}^{s} w_v}$$

using the max–min formula for the isotonic regression [28]. In practice, for the determination of $\hat{\mu}_I$, a number of algorithms are available such as the pool adjacent violators and the up-and-down algorithms. The use of these algorithms will yield a sequence of the so-called blocks S_r's that are the increasing and adjacent partition of $\{1, \ldots, m\}$ such that

$$\hat{\mu}_{I,l} = \frac{\sum_{v \in S_r} w_v \bar{n}_v}{\sum_{v \in S_r} w_v}$$

for $l \in S_r$ and that the $\hat{\mu}_{I,l}$'s are a nondecreasing sequence. In the following, we will define the IRE $\hat{\mu}_I(t)$ as the nondecreasing step function with possible jumps only at the s_l's and $\hat{\mu}_I(s_l) = \hat{\mu}_{I,l}$, $l = 1, \ldots, m$.

To illustrate the IRE described above, consider the set of panel count data given in Table 1 from a reliability study on the loss of feedwater flow over 30 nuclear plants. Among others, Gaver and O'Muircheartaigh [10] and Sun and Kalbfleisch [33] discussed the data. For each plant, the available information consists of the observation time (one point per plant) and the number of losses of feedwater flow observed up to the observation time. To estimate the population average of the cumulative number of losses of feedwater flow, note that there are a total of 10 different observation time points, that is, $m = 10$. For the determination of the IRE, we first calculated the number of observations (w_l) and the sample mean of the numbers of the losses of feedwater flow ($\bar{n}_l$) at each observation timepoint. The application of the pool adjacent violators algorithm yielded the IRE given in Figure 1. To give an idea of how the algorithm works, Figure 1 also includes the sample means of the numbers of observed losses ($\bar{n}_l$ vs s_l). It can be clearly seen that the IRE is obtained by pooling the $\bar{n}_l$'s according to the order restriction.

Note that for the nuclear plant data discussed above, there is only one observation for each subject; that is, $m_i = 1$. Such data are commonly referred to as *current status data* and often occur in demographical studies and reliability experiments among others [9,20,32].

The IRE defined above can also be derived as a nonparametric maximum pseudolikelihood estimator [45]. To see this, as-

Table 1: Observed numbers of loss of feedwater flow from 30 nuclear plants.

			Observation Time t_i (in years) and Observed Number n_i								
Plant	t_i	n_i	Plant	t_i	n_i	Plant	t_i	n_i	Plant	t_i	n_i
1	15	4	9	4	13	17	2	11	25	1	1
2	12	40	10	3	4	18	2	1	26	3	10
3	8	0	11	4	27	19	2	0	27	2	5
4	8	10	12	4	14	20	1	3	28	4	16
5	6	14	13	4	10	21	1	5	29	3	14
6	5	31	14	2	7	22	1	6	30	11	58
7	5	2	15	3	4	23	5	35			
8	4	4	16	3	3	24	3	12			

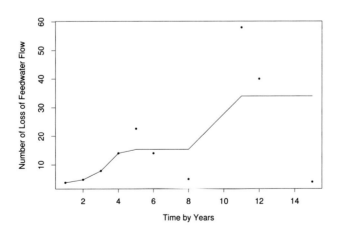

Figure 1: IRE of the average number of losses of feedwater flow.

sume that the $N_i(t)$'s are nonhomogeneous Poisson processes. If we ignore the dependence of $\{N_i(t_{i,j}), j = 1,...,m_i\}$ for each i, one can construct a pseudologlikelihood function as

$$\sum_{i=1}^{n}\sum_{j=1}^{m_i}[n_{i,j}\log\mu(t_{i,j}) - \mu(t_{i,j})]$$

$$= \sum_{l=1}^{m} w_l(\bar{n}_l \log \mu_l - \mu_l)$$

for the mean function $\mu(t)$. It can be shown that the maximization of the above pseudologlikelihood function is equivalent to the minimization of (1) [28,45].

54.2.2 Generalization of the IRE and Other Estimates

Several other estimates of the mean function $\mu(t)$ have been proposed in the literature. One is the class of estimates given in Hu and Lagakos [14], who suggested minimizing the generalized least-squares function

$$\sum_{i=1}^{n}\sum_{j=1}^{m_i}\sum_{l=1}^{m_i} w(t_{i,j}, t_{i,l})$$
$$\times \{N_i(t_{i,j}) - \mu(t_{i,j})\}\{N_i(t_{i,l}) - \mu(t_{i,l})\}$$

subject to the nondecreasing property of $\mu(t)$, where $(w(t_{i,j}, t_{i,l}))$ is a known $m_i \times m_i$ symmetric weight matrix. By using the identity weight matrix, one will obtain the IRE discussed in the previous subsection. Hu and Lagakos [14] discussed some other choices for the weight matrix.

As an alternative to the least squares type of approaches discussed above, one can also apply the full likelihood approach. For this, assume that the $N_i(t)$'s are nonhomogeneous Poisson processes. Then the log full likelihood function is proportional to

$$\sum_{i=1}^{n}\sum_{j=1}^{m_i}(n_{i,j} - n_{i,j-1})$$
$$\times \log[\mu(t_{i,j}) - \mu(t_{i,j-1})]$$
$$-\sum_{i=1}^{n}\mu(t_{i,m_i}),$$

where $t_{i,0} = 0$ and $n_{i,0} = 0$, and one can naturally estimate $\mu(t)$ by maximizing this log likelihood function. As before, let $s_1 < \cdots < s_m$ denote the ordered distinct observation times in the set $\{t_{i,j}; j = 1,...,m_i, i = 1,...,n\}$. Also let $b_l = \sum_{i=1}^{n} I(t_{i,m_i} = s_l)$ for $l = 1,...,m$ and

$$\tilde{n}_{l,l'} = \sum_{i=1}^{n}\sum_{j=1}^{m_i}(n_{i,j} - n_{i,j-1})$$
$$\times I(t_{i,j} = s_l, t_{i,j-1} = s_{l'}),$$

for $0 \leq l' < l \leq m$. Then the loglikelihood function given above can be rewritten as

$$\sum_{l'=0}^{m-1}\sum_{l=l'+1}^{m}\tilde{n}_{l,l'}\log[\mu(s_l) - \mu(s_{l'})]$$
$$-\sum_{l=1}^{m} b_l \mu(s_l). \quad (2)$$

As pointed out before, only the values of $\mu(t)$ at the s_l's can be estimated and one can define the nonparametric maximum likelihood estimator of $\mu(t)$ as the nondecreasing step function with possible jumps only at the s_l's that maximizes (2). Thus the maximization of the loglikelihood function (2) over functions $\mu(t)$ becomes maximizing (2) over m-dimensional parameter vectors $\mu = (\mu_1,...,\mu_m)$ with $\mu_1 \leq ... \leq \mu_m$, $l = 1,...,m$. It can be easily seen that there is no closed solution for the maximum-likelihood estimate defined here. For the determination of the defined estimate, Wellner and Zhang [45] gave an iterative convex minorant algorithm.

In comparing the IRE and the nonparametric maximum-likelihood estimator of $\mu(t)$, it is easy to see that the latter could be more efficient than the former. Wellner and Zhang [45] studied this by simulation and suggested that this is true for both nonhomogeneous Poisson processes and some other counting processes. A disadvantage of the latter is that its determination is much more involved in terms of programming and requires much more computing time than that of the IRE. In general, the IRE provides a general idea about the shape of the mean function $\mu(t)$, especially for the case where the number of observations for each subject is small. The maximum likelihood estimator should be used, for example, if the nonhomogeneous Poisson assumption seems reasonable.

Note that both the IRE and the nonparametric maximum-likelihood estimator of $\mu(t)$ discussed above are nonnonparametric in the sense that they make no assumption about the dependence of $\{n_{i,j}, j = 1, ..., m_i\}$. Sometimes it may be reasonable to impose some structures on the dependence for estimation of $\mu(t)$. Zhang and Jamshidian [49] considered this and assumed that given a latent variable b_i, $E[N_i(t)|b_i] = b_i \mu(t)$ and $\{n_{i,j}, j = 1, ..., m_i\}$ are independent, $i = 1, ..., n$. Furthermore, by assuming that the b_i's follow a gamma distribution, they developed an EM algorithm for estimation of $\mu(t)$. As the nonparametric maximum-likelihood estimator, this estimate could be more efficient than the IRE but its determination is also much more involved than that of the IRE. Furthermore, it may be difficult to verify the assumed latent variable model in practice.

In addition to the authors mentioned above, Lu et al. [24] also considered the estimation problem discussed here and they studied both pseudolikelihood and likelihood approaches when the mean function $\mu(t)$ can be approximated by the monotone cubic I splines. Hu and Lagakos [14] investigated the same problem for a general response process that includes the counting process as a special case. Note that all approaches described so far directly estimate the mean function $\mu(t)$, for which one has to take into account the monotonic property of $\mu(t)$. In contrast, Thall and Lachin [40] proposed an alternative method that estimates the rate function $d\mu(t)$ first by the empirical estimate and then to estimate $\mu(t)$ by the integral of the obtained rate estimate.

54.3 Nonparametric Comparison

Treatment comparison is one of the common objectives in data analysis and in this section we will consider this problem in the context of panel count data with respect to mean functions. Consider an event history study giving only panel count data and let the $N_i(t)$'s, $t_{i,j}$'s, $n_{i,j}$'s be defined as in the previous section. Also let $\mu_i(t) = E\{N_i(t)\}$, $i = 1, ..., n$, and suppose that the goal is to test the hypothesis $H_0 : \mu_1(t) = ... = \mu_n(t)$. That is, all study subjects have identical mean functions. In the following, we will first discuss the two-sample situation. That is, all study subjects come from two different treatment groups. Then we will consider the situation where there exist k different treatment groups.

54.3.1 Two-Sample Comparison

For subject i, define z_i to be the binary (e.g., 0 or 1) group indicator, $i = 1, ..., n$. Let $\hat{\mu}_I(t)$ denote the IRE of the $\mu_i(t)$'s under the hypothesis H_0. To test H_0, motivated by the t-test statistics, we can use the test statistic

$$U = \sum_{i=1}^{n} z_i \sum_{j=1}^{m_i} \{n_{i,j} - \hat{\mu}_I(t_{i,j})\},$$

the sum of the differences between the observed numbers of the event of interest and the estimated numbers of the event over the group with $z_i = 1$. It is easy to see that if all subjects have only one observation at the same timepoint t_0 ($m_i = 1$, $t_{i,1} = t_0$), we have

$$U = \sum_{i:z_i=1} N_i(t_0) - \frac{\sum_{i=1}^n z_i}{n} \times \sum_{i=1}^n N_i(t_0).$$

Let $\hat{\mu}_I^{(u)}(t)$, $\{s_l^{(u)}\}$ and $\{w_l^{(u)}\}$ be defined as $\hat{\mu}_I(t)$, $\{s_l\}$ and $\{w_l\}$ in the definition of the IRE but based only on subjects with $z_i = u$, $u = 0, 1$. Then the above test statistic U can be rewritten as

$$U = \int w^{(1)}(t) [\hat{\mu}_I^{(1)}(t) - \hat{\mu}_I(t)] \, d\bar{N}^{(1)}(t), \quad (3)$$

where $w^{(1)}(t)$ is a step function with jumps only at the $s_l^{(1)}$'s and $w^{(1)}(s_l^{(1)}) = w_l^{(1)}$ and $\bar{N}^{(1)}(t) = \sum_l I(t \geq s_l^{(1)})$. That is, U is the integrated weighted difference between an individual group estimator $\hat{\mu}_I^{(1)}(t)$ and the overall estimator $\hat{\mu}_I(t)$.

Sun and Kalbfleisch [32] and Sun and Fang [31] investigated the statistic U. They showed that under some regularity conditions and H_0, one can approximate the distribution of $n^{-1/2}U$ by the normal distribution with mean zero and the following variance

$$\hat{\sigma}^2 = \frac{1}{n} \sum_{i=1}^{n} \left[(z_i - \bar{z}) \times \sum_{j=1}^{m_i} \{n_{i,j} - \hat{\mu}_I(t_{i,j})\} \right]^2,$$

where $\bar{z} = \sum_{i=1}^{n} z_i/n$. Hence one can test the hypothesis H_0 using the statistic $U^* = U/(n^{1/2}\hat{\sigma})$ based on the standard normal distribution.

To illustrate the test procedure described above, consider the panel count data discussed in Section 54.1 from a clinical trial on the use of the natural bile acid chenodeoxycholic acid for the dissolution of cholesterol gallstones. Define $z_i = 1$ for patients in the placebo group and $z_i = 0$ otherwise. Let $N_i(t)$ denote the cumulative number of occurrences of nausea up to weeks t for the ith patient, $i = 1, ..., 111$. The application of the test procedure gave $U^* = 0.7328$ based on the data observed up to 58 weeks. This corresponds to a p value of 0.4637 for testing H_0 according to the standard normal distribution. It suggests that the average incidences of nausea did not differ significantly overall between the patients in the placebo and high-dose groups. To compare the two groups graphically, we calculated the separate IRE of the average cumulative numbers of the nausea and present them in Figure 2. The figure indicates that the patients in the placebo group seem to have higher incidences of nausea during the first 10 weeks than those in the high-dose group but the difference disappeared by the end of the first year.

Note that in the test procedure discussed above, the IRE was used for estimation of the mean function $\mu(t)$. As alternatives, one could develop the similar procedures that use the other estimates discussed in Section 54.2.2 instead of the IRE. Furthermore, instead of (3), one could consider the statistic

$$\int w^{(1)}(t) \{\hat{\mu}_I^{(1)}(t) - \hat{\mu}_I(t)\}^2 \, d\bar{N}^{(1)}(t)$$

or

$$\int w(t) \, | \, \hat{\mu}_I^{(1)}(t) - \hat{\mu}_I^{(0)}(t) \, | \, d\bar{N}(t) \quad (4)$$

for testing H_0, where $w(t)$ is a weight function and $\bar{N}(t) = \sum_j I(t \geq s_j)$. Of course,

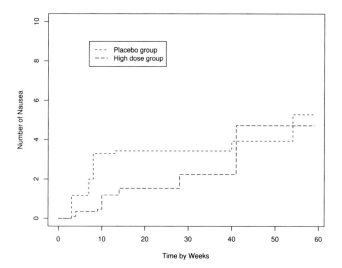

Figure 2: Estimates of the average cumulative counts of episodes of nausea.

one needs to derive the asymptotic null distributions for them.

54.3.2 k-Sample Comparison

In this subsection, we assume that the n study subjects come from k different treatment groups with $\mu^{(l)}(t)$ denoting the common mean function for the subjects in the lth group, $l = 1, ..., k$. Then the hypothesis H_0 can be rewritten as $\mu^{(1)}(t) = ... = \mu^{(k)}(t)$. Let $\hat{\mu}^{(l)}(t)$ denote the IRE of $\mu^{(l)}(t)$ based on the subjects only in the lth group. To test H_0, motivated by the statistic (4), one can compare each $\mu^{(l)}(t)$ ($l > 1$) to $\mu^{(1)}(t)$ and consider the vector of test statistics

$$U_n = (U_{1,2}, U_{1,3}, ..., U_{1,k})^T,$$

where

$$U_{l_1,l_2} = \left(\frac{n_{l_1} n_{l_2}}{n}\right)^{1/2} \times \int w(t) \left\{\hat{\mu}^{(l_1)}(t) - \hat{\mu}^{(l_2)}(t)\right\} d\bar{N}_n(t)$$

with $\bar{N}_n(t) = n^{-1} \sum_{i=1}^{n} \sum_{j=1}^{m_i} I(t_{i,j} \leq t)$ and n_l denoting the number of subjects in group l.

Zhang [48] studied the statistic U_n and showed that under some regularity conditions and the hypothesis H_0, one can approximate the distribution of U_n by the multivariate normal distribution with mean zero and the covariance matrix

$$\hat{\sigma}_n^2 \begin{bmatrix} \frac{n_1+n_2}{n} & \frac{\sqrt{n_2 n_3}}{n} & \cdots & \frac{\sqrt{n_2 n_k}}{n} \\ \frac{\sqrt{n_2 n_3}}{n} & \frac{n_1+n_3}{n} & \frac{\sqrt{n_3 n_4}}{n} & \cdots \\ \vdots & \vdots & \vdots & \vdots \\ \frac{\sqrt{n_2 n_k}}{n} & \frac{\sqrt{n_3 n_k}}{n} & \cdots & \frac{n_1+n_k}{n} \end{bmatrix},$$

where

$$\hat{\sigma}_n^2 = \frac{1}{n} \sum_{i=1}^{n} \left[\sum_{j=1}^{m_i} w(t_{i,j}) \times \{n_{i,j} - \hat{\mu}_I(t_{i,j})\}\right]^2.$$

For the selection of the weight function $w(t)$, Zhang [48] suggested that one can use $w_n(t) = n^{-1} \sum_{i=1}^{n} I(t \leq t_{i,m_i})$ among others.

In addition to the procedures given above, to test H_0, other available procedures include the one given in Thall and Lachin [40], who first partitioned the entire study period into several fixed, consecutive intervals and transformed the observed numbers of events on each subject over each interval into a vector of variables. The comparison was then conducted by using the procedure given in Wei and Lachin [44] for multivariate nonnegative-valued random vectors. A shortcoming of this procedure is that the test result could depend on the selection of the intervals. Another approach for testing H_0 is to apply regression techniques discussed in the next section.

54.4 Regression Analysis

In this section, we will consider regression analysis of panel count data with the focus on the marginal modeling approach. As in the previous sections, we assume that there are n independent subjects and each gives rise to a counting process. Also we assume that for each subject, there exists a p-dimensional vector of covariates denoted by Z_i, assumed to be time-independent. In the following, we will first discuss the Poisson-based approaches that assume that the counting processes of interest are nonhomogeneous Poisson processes. Some estimating equation-based approaches are then described that do not rely on the Poisson process assumption.

54.4.1 The Poisson-Based Procedures

Let the $N_i(t)$'s, $t_{i,j}$'s, $n_{i,j}$'s, and s_l's be defined as before. Define $\mu(t; Z_i) = E\{N_i(t)|Z_i\}$, the mean function given covariates Z_i. In this subsection, we will assume that the $N_i(t)$'s are nonhomogeneous Poisson processes and the mean function $\mu(t; Z)$ has the form

$$\mu(t; Z) = \mu_0(t) \exp(Z'\beta). \qquad (5)$$

Here $\mu_0(t)$ is an unknown baseline mean function, the mean function for subjects with $Z = 0$, and β is a p-dimensional vector of regression parameters. As in Section 54.2.1, by ignoring the dependence of $\{N_i(t_i,), j = 1, ..., m_i\}$ for each i, one can easily derive a pseudo log likelihood function given by

$$l_p(\mu_0, \beta)$$
$$= \sum_{i=1}^{n} \sum_{j=1}^{m_i} \{ n_{i,j} \log \mu_0(t_{i,j}) + n_{i,j} Z_i' \beta$$
$$- \mu_0(t_{i,j}) \exp(Z_i' \beta) \}.$$

To estimate $\mu_0(t)$ and β, it is natural to maximize the pseudo log likelihood function $l_p(\mu_0, \beta)$. For this, let the w_l's and $\bar{n}_l$'s be defined as in Section 54.2 and define

$$\bar{a}_l(\beta)$$
$$= \frac{1}{w_l} \sum_{i=1}^{n} \sum_{j=1}^{m_i} \exp(Z_i' \beta) \, I(t_{i,j} = s_l)$$

and

$$\bar{b}_l(\beta)$$
$$= \frac{1}{w_l} \sum_{i=1}^{n} \sum_{j=1}^{m_i} n_{i,j} Z_i' \beta \, I(t_{i,j} = s_l)$$

for given β, $l = 1, ..., m$. Then the pseudologlikelihood function $l_p(\mu_0, \beta)$ can be rewritten as

$$l_p(\mu_0, \beta)$$
$$= \sum_{l=1}^{m} w_l \{ \bar{n}_l \log \mu_0(s_l) - \bar{a}_l(\beta) \mu_0(s_l)$$
$$+ \bar{b}_l(\beta) \}.$$

As with the nonparametric situation, only the values of $\mu_0(t)$ at the s_l's can be estimated. Let $\hat{\mu}_0(t)$ and $\hat{\beta}$ denote the estimators of $\mu_0(t)$ and β that maximize l_p with $\hat{\mu}_0(t)$ being a nondecreasing step function with possible jumps only at the s_l's. Then the determination of $\hat{\mu}_0(t)$ and $\hat{\beta}$ is equivalent to maximizing $l_p(\mu_0, \beta) = l_p(\mu, \beta)$ over the $(m+p)$ unknown parameters $\mu = (\mu_1, ..., \mu_m)$ and β with $\mu_1 \leq \cdots \leq \mu_m$, where $\mu_l = \mu_0(s_l)$, $l = 1, ..., m$. To maximize $l_p(\mu, \beta)$, Zhang [47] proposed a two-step iterative algorithm that maximizes l_p over μ and β alternatively. Note that for fixed β, the maximization of l_p over μ is equivalent to maximizing

$$\sum_{l=1}^m w_l \bar{a}_l(\beta) \left(\frac{\bar{n}_l}{\bar{a}_l(\beta)} \log \mu_l - \mu_l \right).$$

That is, the $\hat{\mu}_0(s_l)$'s are the isotonic regression estimator of $\{\bar{n}_1/\bar{a}_1(\beta), ..., \bar{n}_m/\bar{a}_m(\beta)\}$ with weights $\{w_1 \bar{a}_1(\beta), ..., w_m \bar{a}_m(\beta)\}$. Thus for given β, we have

$$\hat{\mu}_\beta(s_l) = \max_{r \leq l} \min_{s \geq l} \frac{\sum_{v=r}^s w_v \bar{n}_v}{\sum_{v=r}^s w_v \bar{a}_v(\beta)}$$
$$= \min_{s \geq l} \max_{r \leq l} \frac{\sum_{v=r}^s w_v \bar{n}_v}{\sum_{v=r}^s w_v \bar{a}_v(\beta)}$$

by the max–min formula of the isotonic regression estimate.

For given $\mu_0(t)$ or μ, one can simply use the Newton–Raphson algorithm for estimation of β. It can be easily shown that the pseudologlikelihood function l_p is a concave function of β for given $\mu_0(t)$, and its value increases after each iteration [47]. For the convergence criterion for the two-step algorithm given above, one can compare the relative absolute change of either the log likelihood function l_p between two successive estimators of $\mu_0(t)$ and β or the difference between the two successive estimators. For the variance estimation of the resulting estimates, one could apply the simple bootstrap procedure.

To illustrate the methodology described above, we will apply it to the bladder cancer panel count data mentioned before. The data arose from a bladder cancer study conducted by the Veterans Administration Cooperative Urological Research Group [3,36]. The patients had superficial bladder tumors when they entered the study and these tumors were removed transurethrally. Many patients had multiple recurrences of tumors during the study and these recurrent tumors were also removed transurethrally at the patient's clinic visits. The observed information includes clinical visit times and the numbers of recurrent tumors that occurred between the visits for 85 patients in two treatment groups, placebo (47) and thiotepa (38). In addition, for each patient, two potentially important baseline covariates were recorded, and they are the number of initial tumors and the size of the largest initial tumor. One of the study objectives was to make inference about the effects of the treatments and baseline covariates on tumor recurrence rates.

For the analysis, with respect to patient i, define $N_i(t)$ to be the total number of bladder tumors that had occurred up to month t. Also define Z_{i1} to be the number of initial tumors observed at the beginning of the study, Z_{i2} the size of the largest initial tumor, and $Z_{i3} = 1$ if the patient was in the thiotepa group and 0 otherwise, $i = 1, ..., 85$. Then the application of the method given above yielded $\hat{\beta}_1 = 0.2831$, $\hat{\beta}_2 = -0.0515$, $\hat{\beta}_3 = -1.3574$ with the estimated standard errors being 0.0832, 0.1007, and 0.3695, respectively, based on the simple bootstrap procedure. These results suggest that the thiotepa treatment significantly reduced the recurrence rate of the bladder tumors and the rate was significantly related to the number of initial tumors but not to the size of the largest initial tumor.

For estimation of $\mu_0(t)$ and β, instead of using the pseudologlikelihood function l_p, one may consider to maximize the following full loglikelihood function

$$l(\mu_0,\beta) = \sum_{l'=0}^{m-1}\sum_{l=l'+1}^{m} \tilde{n}_{l,l'}\,\log[\mu_0(s_l) - \mu_0(s_{l'})]$$
$$- \sum_{l=1}^{m} b_l(\beta)\,\mu_0(s_l)$$
$$+ \sum_{i=1}^{n} n_{i,m_i}\,Z_i'\,\beta\,,$$

where $b_l(\beta) = \sum_{i=1}^{n} I(t_{i,m_i} = s_l)\exp(Z_i'\beta)$ and

$$\tilde{n}_{l,l'} = \sum_{i=1}^{n}\sum_{j=1}^{m_i} (n_{i,j} - n_{i,j-1}) \times I(t_{i,j} = s_l, t_{i,j-1} = s_{l'})$$

for $0 \le l' < l \le m$. It is obvious that $l(\mu_0,\beta)$ could yield more efficient estimates than $l_p(\mu_0,\beta)$ [46]. However, the estimates resulting from the former are much complicated and hard to study.

In the above, it was assumed that all study subjects come from a single population and sometimes this may not be true. For example, the subjects may arise from a mixture of G different populations characterized by $N_i(t) = \sum_{g=1}^{G} z_{i,g}\,C_{i,g}(t)$, where $z_{i,g}$ indicates if subject i belongs to the subpopulation or cluster g and $C_{i,g}(t)$ is a nonhomogeneous Poisson process. Among others, Nielsen and Dean [26] investigated this type of situations and provided an example of such panel count data arising from an experiment to test the difference of pheromones in disrupting the mating pattern of the cherry bark tortrix moth.

54.4.2 The Estimating Equation-Based Procedures

In the previous subsection, the underlying counting processes giving rise to panel count data were assumed to be nonhomogeneous Poisson processes. Clearly this may not be true in reality. In this subsection, we present some inference procedures that do not require the Poisson assumption.

Let the $N_i(t)$'s, $t_{i,j}$'s, Z_i's and $\mu(t, Z_i)$ be defined as before and suppose that the mean function $\mu(t, Z)$ satisfies the model (5). For subject i, suppose that there exists a random variable C_i representing the follow-up time on the subject. In some cases, we may have $C_i = t_{i,m_i}$ and sometimes this may not be the case. Define $O_i(t) = \sum_{j=1}^{m_i} I(t_{i,j} \le t)$, a counting process representing the total number of observations on subject i up to time t, $i = 1,...,n$. In the following, we will assume that $N_i(t)$, $O_i(t)$ and C_i are independent of each other given Z_i. Also we will assume that the goal is to make inference about regression parameters β.

To estimate β, for each subject, define a new process

$$\tilde{N}_i(t) = \int_0^t N_i(s)\,dO_i(s)\,,$$

which has possible jumps only at the observation time points $t_{i,j}$'s with respective jump sizes $N_i(t_{i,j})$. It can be easily seen that as for the $O_i(t)$'s, we have recurrent event or complete data for the $\tilde{N}_i(t)$'s rather than panel count data. Also it can be easily shown that

$$E\{\,d\tilde{N}_i(t)|O_i(s), Z_i\,\} = \mu_0(t)\,\exp(Z_i'\beta)\,dO_i(t)\,.$$

Let $o_i(t) = O_i(t) - O_i(t^-)$ and define

$$S^{(j)}(\beta;t) = \frac{\sum_{i=1}^{n} I(c_i \ge t)\,Z_i^{\otimes j}\,\exp(Z_i'\beta)\,o_i(t)}{\sum_{i=1}^{n} o_i(t)}$$

for t with $\sum_{i=1}^{n} o_i(t) > 0$ and $j = 0,1,2$. For estimation of β, motivated by the partial score function with respect to the Cox type of models, Hu et al. [15] proposed to use the following estimating function

$$U_n(\beta; \tilde{N}, w) = \sum_{i=1}^{n} \int_0^\tau w(t)\, I(C_i \geq t) \times \{Z_i - \bar{Z}(t;\beta)\}\, d\tilde{N}_i(t),$$

where $w(\cdot)$ is a weight function as before and $\bar{Z}(t;\beta) = S^{(1)}(\beta;t)/S^{(0)}(\beta;t)$. If all subjects have only one observation at time, say, t_0, then the estimating function given above reduces to

$$U_n(\beta; \tilde{N}, 1) = \sum_{i=1}^{n} Z_i N_i(t_0)$$
$$- \left\{ \sum_{i=1}^{n} \int_0^{t_0} \frac{1}{\sum_{j=1}^{n} \exp(Z_i'\beta)}\, dN_i(t) \right\}$$
$$\times \left\{ \sum_{i=1}^{n} Z_i \exp(Z_i'\beta) \right\}$$

with $w(t) = 1$. Now let $\hat{\beta}_n$ denote the solution to $U_n(\beta; \tilde{N}, 1) = 0$. Hu et al. [15] showed that $\hat{\beta}_n$ is a consistent estimate of β and one can approximate the distribution $\sqrt{n}(\hat{\beta}_n - \beta_0)$ by the multivariate normal distribution with mean zero and the covariance matrix $\hat{\Sigma}_n = \hat{A}(\beta)^{-1}\hat{B}(\beta)\hat{A}(\beta)^{-1}$ with β replaced by $\hat{\beta}_n$, where β_0 denotes the true value of β

$$\hat{A}(\beta) = -\frac{1}{n}\sum_{i=1}^{n}\int_0^\tau I(C_i \geq t)$$
$$\times \left[\frac{S^{(2)}(\beta;t)}{S^{(0)}(\beta;t)} - \bar{Z}(t;\beta)^{\otimes 2}\right] d\tilde{N}_i(t)$$

and

$$\hat{B}(\beta) = \frac{1}{n}\left[\sum_{i=1}^{n}\int_0^\tau \{Z_i - \bar{Z}(t;\beta)\}\, d\hat{M}_i(t;\beta)\right]^{\otimes 2}$$

with

$$\hat{M}_i(t;\beta) = \int_0^t I(C_i \geq s)$$
$$\times \{N_i(s) - \hat{\Lambda}_0(s;\beta)$$
$$\times \exp(Z_i'\beta)\}\, dO_i(s)$$

and

$$\hat{\Lambda}_0(t;\beta) = \frac{\sum_{i=1}^{n} I(C_i \geq t) N_i(t) o_i(t)}{\sum_{i=1}^{n} I(C_i \geq t) \exp(Z_i'\beta) o_i(t)}.$$

For the illustration, we again consider the bladder cancer panel count data analyzed in the previous subsection. Let $N_i(t)$ and Z_i be defined as there, $i = 1,\ldots,85$. By applying the estimation procedure given above, we obtained $\hat{\beta}_{n1} = 0.28$, $\hat{\beta}_{n2} = -0.07$ and $\hat{\beta}_{n3} = -1.36$ with the estimated standard errors of 0.09, 0.12, and 0.45, respectively. These results are similar to those given by the Poisson-based pseudo likelihood estimation procedure discussed above and again indicate that the bladder cancer patients in the thiotepa treatment group had significantly lower recurrence rate of bladder tumors than those in the placebo group. The initial number of tumors could be a useful diagnostic factor but not the size of the largest initial tumor.

In the procedure given above, it was assumed that the covariates Z_i have no effect on the observation process $O_i(t)$. In practice, this may not be true and for this, as with model (5) for $N_i(t)$, one natural way is to assume that

$$E\{O_i(t) \mid Z_i\} = \tilde{\mu}_0(t) \exp(Z_i'\gamma) \qquad (6)$$

for the effect of Z_i on $O_i(t)$, where $\tilde{\mu}_0(t)$ is an unknown baseline mean function and γ denotes the vector of regression parameters as $\mu_0(t)$ and β, respectively. Among others, Hu et al. [15] considered the model (6) together with model (5) and generalized the estimation procedure described above. As the procedure given above, the generalized estimation procedure also does not involve the baseline mean functions $\mu_0(t)$ as well as $\tilde{\mu}_0(t)$. Sun and Wei [36] investigated a more general situation where $N_i(t)$, $O_i(t)$ and C_i may depend on each other, but are independent given Z_i. In particular, they used models (5) and (6) and the proportional hazards model for the effect of covariates on the follow-up time C_i.

In comparison of the Poisson-based and estimating equation-based estimation procedures, it is obvious that the former could be more efficient than the latter if the Poisson process assumption is valid. Of course, in practice, it may be difficult to check or verify the assumption without prior information. On the other hand, the former could be much more complicated than the latter partly because of the involvement or the need of estimation of the baseline mean functions. Another advantage of the estimating equation-based procedures is that they give close formed variance estimation.

Other authors who considered regression analysis of panel count data include Cheng and Wei [5], Ishwaran and James [17], Lawless and Zhan [22], Staniswalls et al. [29], and Sun and Matthews [34]. In particular, Cheng and Wei [5] developed an inference approach similar to the estimation procedure based on $U_n(\beta; \tilde{N}, w)$. Lawless and Zhan [22] and Staniswalls et al. [29] gave some approaches that base the inference on the modeling of the rate function of the underlying counting process instead of the mean function.

54.5 Regression Analysis with Dependent Observation Process

For the methods discussed above, it was assumed that observation times $t_{i,j}$'s or the counting process $O_i(t)$ that characterizes them is independent of the underlying counting process $N_i(t)$ governing the observed panel count data either completely or conditional on covariates. In practice, however, this may not be true. For example, the observation times could be hospitalization times and this could be the case in an observational study concerning the occurrence rate of certain symptoms related a disease under study which can be observed or known only when the patients are in the hospital because of the disease. It is apparent that the patients may come in the hospital simply because of these symptoms, and thus the observation times are related to the occurrence process. A more specific example is given by the bladder cancer panel count data discussed before and for the data, several authors noticed that some patients in the study had more visits than others, suggesting that the occurrence of bladder tumors and the visit may be related [36]. This example will be discussed further below.

In this section, we will assume that $N_i(t)$ and $O_i(t)$ may be correlated and the relationship can be described by some latent variables. First we will describe an estimating equation-based procedure for estimation of covariate effects, and several other available methods will then be briefly discussed.

For each study subject, we assume that there exists a scale latent variable x_i and given x_i and covariates Z_i, the mean function of $N_i(t)$ satisfies

$$\mu_i(t; Z_i) = x_i^\alpha \mu_0(t) \exp(Z_i'\beta) \qquad (7)$$

instead of model (5). Here $\mu_0(t)$ and β are defined as before and α is an unknown parameter. With respect to the observation process, we assume that $O_i(t)$ is a nonhomogeneous Poisson process with the intensity function given by

$$\lambda_i(t; Z_i) = x_i \lambda_0(t) \exp(Z_i'\gamma) \qquad (8)$$

given x_i and Z_i. In the above, $\lambda_0(t)$ is a completely unknown continuous baseline intensity function and γ denotes the vector of regression parameters. Define $\Lambda_0(t) = \int_0^t \lambda_0(s)ds$, the baseline cumulative intensity function. It will be assumed that $\Lambda_0(\tau) = 1$ to avoid the identifiability issue and the x_i's are realizations of a latent variable X satisfying $E(X|Z) = E(X)$. Also it will be assumed that $N_i(t)$ and $O_i(t)$ are independent of each other given x_i and Z_i and both are independent of C_i.

To estimate regression parameters β as well as γ, Sun et al. [38] proposed the following two-step procedure. Let the s_j's be defined as before, d_j the number of the observation times equal to s_j, and n_j the number of the observation times satisfying $T_{il} \leq s_j \leq C_i$, $i = 1, ..., n$. Also define $Z_{1i}' = (Z_i', 1)$, $\gamma_1' = (\gamma', \log E(X))$. Then one can first estimate γ_1 by the estimating equation

$$\sum_{i=1}^{n} w_{1i} Z_{1i} \{ K_i^* \widehat{\Lambda}_0^{-1}(C_i)$$
$$- \exp(Z_{1i}'\gamma_1) \} = 0 \,,$$

where $K_i^* = \tilde{N}_i(t_{m_i})$, the w_{1i}'s are some weights and

$$\widehat{\Lambda}_0(t) = \prod_{s_l > t} \left(1 - \frac{d_l}{n_l} \right).$$

Let $\hat{\gamma}_1$ denote the solution to the estimating equation above. For estimation of β, one can apply the estimating equation

$$\sum_{i=1}^{n} w_{2i} \hat{Z}_{2i} \{ \bar{N}_i - \exp(Z_i'\hat{\gamma})$$
$$\times \exp(\hat{Z}_{2i}'\beta_1) \} = 0, \qquad (9)$$

where the w_{2i}'s are again some weights, $\bar{N}_i = \sum_{l=1}^{m_i} N_i(t_{i,l})$, $\hat{Z}_{2i}' = (Z_i', \log \hat{x}_i, 1)$ with

$$\hat{x}_i = \frac{K_i^*}{\widehat{\Lambda}_0(C_i) \, e^{Z_i'\hat{\gamma}}},$$

and $\beta_1' = (\beta', 1 + \alpha, \log(\theta))$ with $\theta = \int_0^{\tau} \lambda_0(t) P(C_i \geq t) \mu_0(t) \, dt$.

Let $\hat{\beta}^*$ denote the estimator of β in model (7) given by the solution to the equation (9). Sun et al. [38] proved that it is consistent and its distribution can be approximated by a normal distribution. Furthermore, they provided a sandwich estimate of its covariance matrix (not given here).

Now we apply the estimation procedure to the bladder cancer panel count data discussed before with the focus on the 47 patients in the placebo group. In this case, we have two covariates Z_{i1}, the the number of initial tumors, and Z_{i2}, the size of the largest initial tumor. As before, define $N_i(t)$ to be the cumulative number of bladder tumors observed up to time t on patient i, $i = 1, ..., 47$. Assume that the occurrence process of the bladder tumors and the visit process can be described by models (7) and (8), respectively. The application of the procedure gave $\hat{\beta}_1^* = 0.1213$ and $\hat{\beta}_2^* = -0.0044$ with the estimated standard errors being 0.068 and 0.088. As before, the results suggest that the number of initial tumors seems to be positively and significantly related with the tumor recurrence rate but the size of the largest initial tumor does not. The procedure also yielded $\hat{\alpha} = -0.5067$ with the estimated standard error of 0.085, the estimate of the parameter α in model (7) representing the correlation between the tumor occurrence process and the patient visit process. It indicates that the two processes are negatively correlated, meaning that the patients who visited more often had the smaller tumor recurrence rate given other factors. One possible reason for this is that the more visits a patient had, the less time the patient had for new tumor development.

In the estimation procedure described above, it was assumed that $N_i(t)$ and $O_i(t)$ are independent of C_i. The procedure can be easily generalized to situations where $N_i(t)$ and $O_i(t)$ may depend on C_i but be independent of C_i given covariates. In particular, Sun et al. [38] considered the situation and proposed to use the proportional hazards model for C_i. Huang et al. [16] also investigated the same problem and studied the situation where $N_i(t)$ is a non-homogeneous Poisson process with the intensity function having the same form as model (7) and $O_i(t)$ can depend on the latent variable and covariates arbitrarily.

Sometimes $N_i(t)$ and $O_i(t)$ and C_i may still be related even conditional on covariates. Among others, He et al. [11] developed an inference procedure for regression analysis of such panel count data. In the procedure, they used two latent variables u_i and v_i to characterize the relationship and assumed that given u_i, v_i and Z_i, the mean function of $N_i(t)$ and the intensity function of $O_i(t)$ have the forms

$$\mu_i(t; Z_i, u_i, v_i)$$
$$= \mu_0(t) \exp(Z_i' \beta_1 + u_i \beta_2 + v_i \beta_3)$$
$$\lambda_i(t; Z_i, u_i, v_i)$$
$$= \lambda_0(t) \exp(Z_i' \alpha_1 + u_i),$$

respectively. Furthermore, they assumed that the hazard function of C_i is given by

$$\lambda_{0c}(t) \exp(Z_i' \gamma_1 + u_i \gamma_2 + v_i)$$

given u_i, v_i, and Z_i.

54.6 Discussion and Other Topics

The analysis of panel count data is still a relatively new and undeveloped field, and there remain many open problems. In the preceding sections, we discussed three basic aspects of the analysis of panel count data: nonparametric estimation, treatment comparison, and regression analysis. Furthermore, all discussion has been on univariate situations. In practice, of course, there may exist several related recurrent events that are of interest and for which only panel count data are available. Among others, Chen et al. [4] discussed the analysis of multivariate panel count data using a marginal mixed Poisson process approach by assuming that the baseline intensity function is piecewise constant. He et al. [12] studied the same problem and proposed some estimating equation-based approaches.

All methods discussed above are for the mean function of underlying counting processes generating panel count data. As mentioned before, given the structure of panel count data and the amount of observed information, it is much more convenient to deal with the mean function rather than the intensity process or rate function. On the other hand, sometimes one may want to directly model the intensity process or rate function [17,22,29]. For this purpose, however, one usually has to make certain assumptions about the shape of the intensity process or rate function in order to perform nonparametric or semiparametric analysis [35,34].

Another feature of the methods presented above is that they were developed mainly for panel count data in which observation and censoring times differ from subject to subject. For the situation where observation times or intervals are the same for all subjects, the data can be regarded as multivariate data, and any method that accommodates multivariate positive integer-valued response variables can be used for the analysis. This holds even though subjects may miss some intermediate observations and/or drop out of the study early. In this case, the resulting data can be seen as multivariate data with missing values. Also as mentioned before, one can treat panel count data as a special case of longitudinal data and apply the methods developed for longitudinal data. However, these methods may not be able to take into account the special structure of panel count data.

For a recurrent event, instead of the occurrence rate of the event, sometimes one may be interested in the gap time of the event, the time between successive occurrences of the event [37,51]. For the analysis of gap times, although some approaches have been proposed for the case of recurrent event data, there exists little research for panel count data situations. Also there does not seem to exist much research investigating regression analysis of panel count data with time-dependent covariates.

References

1. Andersen, P. K., Borgan, O., Gill, R. D., and Keiding, N. (1993). *Statistical Models Based on Counting Processes*. Springer-Verlag: New York.

2. Breslow, N. E. (1984). Extra-Poisson variation in log-linear models. *Appl. Statist.*, **33**, 38–44.

3. Byar, D. P., Blackard, C., and The Veterans Administration Cooperative Urological Research Group (1977). Comparisons of placebo, pyridoxine, and topical thiotepa in preventing recurrence of stage I bladder cancer. *Urology*, **10**, 556–561.

4. Chen, B. E., Cook, R. J., Lawless, J. F., and Zhan, M. (2005). Statistical methods for multivariate interval-censored recurrent events. *Statist. Med.*, **24**, 671–691.

5. Cheng, S. C. and Wei, L. J. (2000). Inferences for a semiparametric model with panel data. *Biometrika*, **87**, 89–97.

6. Cook, R. J. and Lawless, J. F. (1996). Interim monitoring of longitudinal comparative studies with recurrent event responses. *Biometrics*, **52**, 1311–1323.

7. Cook, R. J. and Lawless, J. F. (2007). *The Analysis of Recurrent Event Data*. Springer-Verlag, New York.

8. Diggle, P. J., Liang, K. Y., and Zeger, S. L. (1994). *The Analysis of Longitudinal Data*. Oxford University Press, New York.

9. Diamond, I. D. and McDonald, J. W. (1991). The analysis of current status data. In *Demographic Applications of Event History Analysis*, J. Trussel, R. Hankinson, and J. Tilton, eds. Oxford University Press, Oxford, UK.

10. Gaver, D. P. and O'Muircheartaigh, I. G. (1987). Robust empirical Bayes analyses of event rates. *Technometrics*, **29**, 1–15.

11. He, X., Tong, X., and Sun, J. (2009). Semiparametric analysis of panel count data with correlated observation and follow-up times. *Lifetime Data Anal.*, **15**, 177–196.

12. He, X., Tong, X., Sun, J., and Cook, R. J. (2008). Regression analysis of multivariate panel count data. *Biostatistics*, **9**, 234–248.

13. Hinde, J. (1982). Compound Poisson regression models. In *GLIM 82: Proc. Int. Conf. Generalized Linear Models*, R. Gilchrist ed., pp. 109–121. Springer-Verlag, Berlin.

14. Hu, X. J. and Lagakos, S. W. (2007). Nonparametric estimation of the mean function of a stochastic process with missing observations. *Lifetime Data Anal.*, **13**, 51–73.

15. Hu, X. J., Sun, J., and Wei, L. J. (2003). Regression parameter estimation from panel counts. *Scandinavian Journal of Statistics*, **30**, 25–43.

16. Huang, C-Y., Wang, M.-C., and Zhang, Y. (2006). Analysing panel count data with informative observation times. *Biometrika*, **93**, 763–775.

17. Ishwaran, H. and James, L. F. (2004). Computational methods for multiplicative intensity models using weighted gamma processes: Proportional hazards, marked point processes, and panel count data. *J. Am. Statist. Assoc.*, **99**, 175–190.

18. Kalbfleisch, J. D. and Lawless, J. F. (1985). The analysis of panel data under

a Markov assumption. *J. Am. Statist. Assoc.*, **80**, 863–871.

19. Kalbfleisch, J. D., Lawless, J. F., and Robinson, J. A. (1991). Methods for the analysis and prediction of warranty claims. *Technometrics*, **33**, 273–285.

20. Keiding, N. (1991). Age-specific incidence and prevalence: A statistical perspective (with discussion). *J. Roy. Statist. Soc. A*, **154**, 371–412.

21. Lawless, J. F. and Nadeau, J. C. (1995). Some simple robust methods for the analysis of recurrent events. *Technometrics*, **37**, 158–168.

22. Lawless, J. F. and Zhan, M. (1998). Analysis of interval-grouped recurrent-event data using piecewise constant rate functions. *Can. J. Statist.*, **26**, 549–565.

23. Lin, D. Y., Wei, L. J., Yang, I., and Ying, Z. (2000). Semiparametric regression for the mean and rate functions of recurrent events. *J. Roy. Statist. Soc. B*, **62**, 711–730.

24. Lu, M., Zhang, Y., and Huang, J. (2007). Estimation of the mean function with panel count data using monotone polynomial splines. *Biometrika*, **94**, 705–718.

25. Nelson, W. B. (2003). *Recurrent Events Data Analysis for Product Repairs, Disease Recurrences, and Other Applications*. ASA-SIAM Series on Statistics and Applied Probability, Vol. 10.

26. Nielsen, J. D. and Dean, C. B. (2008). Clustered mixed nonhomogeneous Poisson process spline models for the analysis of recurrent event panel data. *Biometrics*, **64**, 751–761.

27. Pepe, M. S. and Cai, J. (1993). Some graphical displays and marginal regression analyses for recurrent failure times and time dependent covariates. *J. Am. Statist. Assoc.*, **88**, 811–820.

28. Robertson, T., Wright, F. T., and Dykstra, R. (1988). *Order Restricted Statistical Inference*. Wiley, New York.

29. Staniswalls, J. G., Thall, P. F., and Salch, J. (1997). Semiparametric regression analysis for recurrent event interval counts. *Biometrics*, **53**, 1334–1353.

30. Sun, J. (2006). *The Statistical Analysis of Interval-Censored Failure Time Data*. Springer-Verlag, New York.

31. Sun, J. and Fang, H. B. (2003). A nonparametric test for panel count data. *Biometrika*, **90**, 199–208.

32. Sun, J. and Kalbfleisch, J. D. (1993). The analysis of current status data on point processes. *J. Am. Statist. Assoc.*, **88**, 1449–1454.

33. Sun, J. and Kalbfleisch, J. D. (1995). Estimation of the mean function of point processes based on panel count data. *Statistica Sinica*, **5**, 279–290.

34. Sun, J. and Matthews, D. E. (1997). A random-effect regression model for medical follow-up studies. *Can. J. Statist.*, **25**, 101–111.

35. Sun, J. and Rai, S. N. (2001). Nonparametric tests for the comparison of point processes based on incomplete data. *Scand. J. Statist.*, **28**, 725–732.

36. Sun, J. and Wei, L. J. (2000). Regression analysis of panel count data with covariate-dependent observation and censoring times. *J. Roy. Statist. Soc. B*, **62**, 293–302.

37. Sun, L., Park, D., and Sun, J. (2006). The additive hazards model for recurrent gap times. *Statistica Sinica*, **16**, 919–932.

38. Sun, J., Tong, X., and He, X. (2007). Regression analysis of panel count data with dependent observation times. *Biometrics*, **63**, 1053–1059.

39. Thall, P. F. (1988). Mixed Poisson likelihood regression models for longitudinal interval count data. *Biometrics*, **44**, 197–209.

40. Thall, P. F. and Lachin, J. M. (1988). Analysis of recurrent events: nonparametric methods for random-interval count data. *J. Am. Statist. Assoc.*, **83**, 339–347.

41. Vermunt, J. K. (1997). *Log-Linear Models for Event Histories*. Sage Publications, Newbury Park, CA.

42. Wang, M-C. and Chen Y. Q. (2000). Nonparametric and semiparametric trend

analysis of stratified recurrence time data. *Biometrics*, **56**, 789–794.

43. Wang, Y. and Taylor, M. G. (2001). Jointly modeling longitudinal and event time data with application to Acquired Immunodeficiency Syndrome. *J. Am. Statist. Assoc.*, **96**, 895–905.

44. Wei, L. J. , and Lachin, J. M. (1984). Two-sample asymptotically distribution-free tests for incomplete multivariate observations. *J. Am. Statist. Assoc.*, **79**, 653–661.

45. Wellner, J. A. and Zhang, Y. (2000). Two estimators of the mean of a counting process with panel count data. *Ann. Statist.*, **28**, 779–814.

46. Wellner, J. A., Zhang, Y., and Liu, H. (2004). A semiparametric regression model for panel count data: When do pseudo-likelihood estimators become badly inefficient? *Proc. 2nd Seattle Symp. Biostatistics*, pp. 143–174. Springer, New York.

47. Zhang, Y. (2002). A semiparametric pseudolikelihood estimation method for panel count data. *Biometrika*, **89**, 39–48.

48. Zhang, Y. (2006). Nonparametric k-sample tests with panel count data. *Biometrika*, **93**, 777–790.

49. Zhang, Y. and Jamshidian, M. (2003). The gamma-frailty Poisson model for the nonparametric estimation of panel count data. *Biometrics*, **59**, 1099–1106.

50. Zhang, Y., Liu, W., and Wu, H. (2003). A simple nonparametric two-sample test for the distribution function of event time with interval censored data. *J. Nonparametric Statist.*, **16**, 643–652.

51. Zhao, Q. and Sun, J. (2006). Semiparametric and nonparametric analysis of recurrent events with observation gaps. *Comput. Statist. Data Anal.*, **51**, 1924–1933.

55. Planning and Analysis of Group-Randomized Trials

David M. Murray

55.1 Introduction

Planning a group-randomized trial (GRT) is a complex process. Readers interested in a more detailed discussion might wish to consult Murray's text on the design and analysis of GRTs [6], from which much of this article was abstracted. Donner and Klar's text is another good source of information [1].

55.2 The Research Question

The driving force behind any GRT must be the research question. The question will be based on the problem of interest and will identify the target population, the setting, the endpoints, and the intervention. In turn, those factors will shape the design and analytic plan.

Given the importance of the research question, the investigators must take care to articulate it clearly. Unfortunately, that doesn't always happen. Investigators may have ideas about the theoretical or conceptual basis for the intervention, and often even clearer ideas about the conceptual basis for the endpoints. They may even have ideas about intermediate processes. However, without very clear thinking about each of these issues, the investigators may find themselves at the end of the trial unable to answer the question of interest.

To put themselves in a position to articulate their research question clearly, the investigators should first document thoroughly the nature and extent of the underlying problem and the strategies and results of previous efforts to remedy that problem. A literature review and correspondence with others working in the field are ingredients essential to that process, as the investigators should know as much as possible about the problem before they plan their trial.

Having become experts in the field, the investigators should choose the single question that will drive their GRT. The primary criteria for choosing that question should be (1) whether it is important enough to do and (2) whether this is the right time to do it. Reviewers will ask both questions, and the investigators must be able to provide well-documented answers.

Most GRTs seek to prevent a health problem, so that the importance of the question is linked to the seriousness of that problem. The investigators should document the extent of the problem and the potential benefit from a reduction in that problem.

The question of timing is also important. The investigators should document that the question has not been answered already and that the intervention has a good chance to improve the primary endpoint in the target population. That is most easily done when the investigators are thoroughly familiar with previous research in the area, when the etiology of the problem is well known, when there is a theoretical basis for the proposed intervention, when there is preliminary evidence on the feasibility and efficacy of the intervention, when the measures for the dependent and mediating variables are well developed, when the sources of variation and correlation as well as the trends in the endpoints are well understood, and when the investigators have created the research team to carry out the study. If that is not the state of affairs, then the investigators must either invest the time and energy to reach that state or choose another question.

Once the question is selected, it is very important to put it down on paper. The research question is easily lost in the day-to-day details of the planning and execution of the study, and because much time can be wasted in pursuit of issues that are not really central to the research question, the investigators should take care to keep that question in mind.

55.3 The Research Team

Having defined the question, the investigators should determine whether they have expertise sufficient to deal with all the challenges that are likely to arise as they plan and execute the trial. They should identify the skills that they do not have and expand the research team to ensure that those skills are available. All GRTs will need expertise in research design, data collection, data processing and analysis, intervention development, intervention implementation, and project administration.

Because the team usually will need to convince a funding agency that they are appropriate for the trial, it is important to include experienced and senior investigators in key roles. There is simply no substitute for experience with similar interventions, in similar populations and settings, using similar measures, and similar methods of data collection and analysis. Because those skills are rarely found in a single investigator, most trials will require a team, with responsibilities shared among its members.

Most teams will remember the familiar academic issues (e.g., statistics, data management, intervention theory), but some may forget the very important practical side of trials involving identifiable groups. However, to forget the practical side is a sure way to get into trouble. For example, a school-based trial that doesn't include on its team someone who is very familiar with school operations is almost certain to get into trouble with the schools. A hospital-based trial that doesn't include on its team someone who is very familiar with hospital operations is almost certain to get into trouble with the hospitals; and the same can be said for every other type of identifiable group, population, or setting that might be used.

55.4 The Research Design

The fundamentals of research design apply to GRTs as well as to other comparative designs. Because they are discussed in many familiar textbooks [3–5,7,8], they will be reviewed only briefly here. Additional information may be found in two textbooks on the design and analysis of GRTs [1,6].

The goal in the design of any comparative trial is to provide the basis for valid inference that the intervention as implemented caused the result(s) as observed. To meet that goal, three elements are required. There must be (1) control observations, (2) a minimum of bias in the estimate of the intervention effect, and (3) sufficient precision for that estimate.

The nature of the control observations and the way in which the groups are allocated to treatment conditions will determine in large measure the level of bias in the estimate of the intervention effect. Bias exists whenever the estimate of the intervention effect is different from its true value. If that bias is substantial, the investigators will be misled about the effect of their intervention, as will the other scientists and policy makers who use their work.

Even if adequate control observations are available so that the estimate of the intervention effect is unbiased, the investigator should know whether the effect is greater than would be expected by chance, given the level of variation in the data. Statistical tests can provide such evidence, but their power to do so will depend heavily on the precision of the intervention effect estimate. As the precision improves, it will be easier to distinguish true effects from the underlying variation in the data.

55.5 Potential Design Problems and Methods for Avoiding Them

For GRTs, the four sources of bias that are particularly problematic and should be considered during the planning phase are selection, differential history, differential maturation, and contamination. *Selection bias* refers to baseline differences among the study conditions that might explain the results of the trial. *Bias due to differential history* refers to some external influence that operates differentially among the conditions. *Bias due to differential maturation* reflects uneven secular trends among the groups in the trial favoring one condition or another. These first three sources of bias can either mask or mimic an intervention effect, and all three are more likely given either nonrandom assignment of groups or random assignment of a limited number of groups to each condition.

The first three sources of bias are best avoided by randomization of a sufficient number of groups to each study condition. This will increase the likelihood that potential sources of bias are distributed evenly among the conditions. Careful matching or stratification can increase the effectiveness of randomization, especially when the number of groups is small. As a result, all GRTs planned with fewer than 20 groups per condition would be well served to include careful matching or stratification prior to randomization.

The fourth source of bias is due to contamination that occurs when intervention-like activities find their way into the comparison groups; it can bias the estimate of the intervention effect toward the null hypothesis. Randomization will not protect against contamination; while investigators can control access to their intervention materials, there often is little that they can do to prevent the outside world from introducing similar activities into their control groups. As a result, monitoring exposure to activities that could affect the trial's endpoints in both the intervention and comparison groups is especially important in GRTs. This will allow the investigators to detect and respond to contamination if it occurs.

Objective measures and evaluation personnel who have no connection to the intervention are also important strategies to limit bias. Finally, analytic strategies, such as regression adjustment for confounders, can be very helpful in dealing with any observed bias.

55.6 Potential Analytic Problems and Methods for Avoiding Them

The two major threats to the validity of the analysis of a GRT, which should be considered during the planning phase, are misspecification of the analytic model and low power. Misspecification of the analytic model will occur if the investigator ignores or misrepresents a measurable source of random variation, or misrepresents the pattern of any overtime correlation in the data. To avoid model misspecification, the investigator should plan the analysis concurrent with the design, plan the analysis around the primary endpoints, anticipate all sources of random variation, anticipate the error distribution for the primary endpoint, anticipate patterns of overtime correlation, consider alternate structures for the covariance matrix, consider alternate models for time, and assess potential confounding and effect modification.

Low power will occur if the investigator employs a weak intervention, has insufficient replication, has high variance or intraclass correlation in the endpoints, or has poor reliability of intervention implementation. To avoid low power, investigators should plan a large enough study to ensure sufficient replication, choose endpoints with low variance and intraclass correlation, employ matching or stratification prior to randomization, employ more and smaller groups instead of a few large groups, employ more and smaller surveys or continuous surveillance instead of a few large surveys, employ repeat observations on the same groups or on the same groups and members, employ strong interventions with good reach, and maintain the reliability of intervention implementation. In the analysis, investigators should employ regression adjustment for covariates, model time if possible, and consider post hoc stratification.

55.7 Variables of Interest and Their Measures

The research question will identify the primary and secondary endpoints of the trial. The question may also identify potential effect modifiers. It will then be up to the investigators to anticipate potential confounders and nuisance variables. All these variables must be measured if they are to be used in the analysis of the trial.

In a clinical trial, the primary endpoint is a clinical event, chosen because it is easy to measure with limited error and is clinically relevant [5]. In a GRT, the primary endpoint need not be a clinical event, but it should be easy to measure with limited error and be relevant to public health. In both clinical and GRTs, the primary endpoint, together with its method of measurement, must be defined in writing before the start of the trial. The endpoint and its method of measurement cannot be changed after the start of the trial without risking the validity of the trial and the credibility of the research team. Secondary endpoints should have similar characteristics and also should be identified prior to the start of the trial.

In a GRT, an effect modifier is a variable whose level influences the effect of the intervention. For example, if the effect of a school-based drug-use prevention program depends on the baseline risk level of the student, then baseline risk is an effect modifier. Effect modification can be seen intuitively by looking at separate intervention effect estimates for the levels of the effect modifier. If they differ to a meaningful degree, then the investigator has evidence of possible effect modification. A more formal assessment is provided by a statistical test for effect modification. That is accomplished by including an interaction term between the effect modifier and condition in the analysis and testing the statistical significance of that term. If the interaction is significant, then the investigator should present the results separately for the levels of the effect modifier. If not, the interaction term is deleted and the investigator can continue with the analysis. Proper identification of potential effect modifiers comes through a careful review of the literature and from an examination of the theory of the intervention. Potential effect modifiers must be measured as part of the data collection process so that their role can later be assessed.

A confounder is related to the endpoint, not on the causal pathway, and unevenly distributed among the conditions; it serves to bias the estimate of the intervention effect. There is no statistical test for confounding; instead, it is assessed by comparing the unadjusted estimate of the intervention effect to the adjusted estimate of that effect. If, in the investigator's opinion, there is a meaningful difference between the adjusted and unadjusted estimates, then the investigator has an obligation to report the adjusted value. It may also be appropriate to report the unadjusted value to allow the reader to assess the degree of confounding. The adjusted analysis will not be possible unless the potential confounders are measured. Proper identification of potential confounders also comes through a careful review of the literature and from an understanding of the endpoints and the study population.

The investigators must take care in the selection of potential confounders to select only confounders and not mediating variables. A confounder is related to the endpoint and unevenly distributed in the conditions, but is not on the causal pathway between the intervention and the outcome. A mediating variable has all the characteristics of a confounder, but is on the causal pathway. Adjustment for a mediating variable, in the false belief that it is a confounder, will bias the estimate of the intervention effect toward the null hypothesis.

Similarly, the investigator must take care to avoid selecting as potential confounders variables that may be affected by the intervention even if they are not on the causal pathway linking the intervention and the outcome. Such variables will be proxies for the intervention itself, and adjustment for them will also bias the estimate of the intervention effect toward the null hypothesis.

An effective strategy to avoid these problems is to restrict confounders to variables measured at baseline. Such factors cannot be on the causal pathway, nor can their values be influenced by an intervention that hasn't been delivered. Investigators may also want to include variables measured after the intervention has begun, but will need to take care to avoid the problems described above.

Nuisance variables are related to the endpoint, not on the causal pathway, but evenly distributed among the conditions. They cannot bias the estimate of the intervention effect, but they can be used to improve precision in the analysis. A common method is to make regression adjustment for these factors during the analysis so as to reduce the standard error of the estimate of the intervention effect, thereby improving the precision of the analysis. Such adjustment will not be possible unless the nuisance variables are measured. Proper identification of potential nuisance variables also comes from a careful review of the literature and from an understanding of the endpoint. The cautions described above for the selection of potential confounding variables apply equally well to the selection of potential nuisance variables.

55.8 The Intervention

No matter how well designed and evaluated a GRT may be, strengths in design and analysis cannot overcome a weak intervention. While the designs and analyses employed in GRTs were fair targets for criticism during the 1970s and 1980s, the designs and analyses employed have improved, with many examples of very well-designed and carefully analyzed trials. Where intervention effects are modest or shortlived even in the presence of good design and analytic strategies, investigators must take a hard look at the intervention and question whether it was strong enough.

One of the first suggestions for developing the research question was that the investigators become experts on the problem that they seek to remedy. If the primary endpoint is cigarette smoking among ninth-graders, then the team should seek to learn as much as possible about the etiology of smoking among young adolescents. If the primary endpoint is obesity among Native American children, then the team should seek to learn as much as possible about the etiology of obesity among those young children. If the primary endpoint is delay time in reporting heart attack symptoms, then the team should seek to learn as much as possible about the factors that influence delay time. The same can be said for any other endpoint.

One of the goals of developing expertise in the etiology of the problem is to identify points in that etiology that are amenable to intervention. There may be critical developmental stages, or critical events or influences that trigger the next step in the progression, or it may be possible to identify critical players in the form of parents, friends, coworkers, or others who can influence the development of that problem. Without careful study of the etiology, the team will largely be guessing and hoping that their intervention is designed properly. Unfortunately, guessing and hoping rarely lead to effective interventions.

Powerful interventions are guided by good theory on the process for change, combined with a good understanding of the etiology of the problem of interest. Poor theory will produce poor interventions and poor results. This was one of the primary messages from the community-based heart disease prevention studies, where the intervention effects were modest, generally of limited duration, and often within chance levels. Fortmann et al. noted that one of the major lessons learned was how much was not known about how to intervene in whole communities [2].

Theory that describes the process of change in individuals may not apply to the process of change in identifiable groups. If it does, it may not apply in exactly the same way. Good intervention for a GRT will likely need to combine theory about individual change with theory about group processes and group change.

A good theoretical exposition will also help identify channels for the intervention program. For example, there is strong evidence that recent immigrants often look to long-term immigrants of the same cultural group for information on health issues. This has led investigators to try to use those long-term immigrants as change agents for the latter immigrants.

A good theoretical exposition will often indicate that the phenomenon is the product of multiple influences and so suggest that the intervention operate at several different levels. For example, obesity among schoolchildren appears to be influenced most proximally by their physical activity levels and by their dietary intake. In turn, their dietary intake is influenced by what is served at home and at school and their physical activity is influenced by the nature of their physical activity and recess programs at school and at home. The models provided by teachers and parents are important both for diet and physical activity. This multilevel etiology suggests that interventions be directed at the school food-service, physical education, and recess programs; at parents; and possibly at the larger community.

GRTs would benefit by following the example of clinical trials, where some evidence of feasibility and efficacy of the intervention is usually required prior to launching the trial. When a study takes several years to complete and costs hundreds of thousands of dollars or more, that seems a very fair expectation. Even shorter and less expensive GRTs would do well to follow that advice.

What defines preliminary evidence of feasibility? It is not reasonable to ask that the investigators prove that all intervention and evaluation protocols can be implemented in the population and setting of interest in advance of their trial. However, it is reasonable to ask that they demonstrate that the major components of the proposed intervention can be implemented in the target population; that can be done in a pilot study. It is also reasonable to ask that the major components of the proposed evaluation be feasible and acceptable in the setting and population proposed for the trial; that, too, can be done in a pilot study.

What defines preliminary evidence of efficacy? It is not fair to ask that the investigators prove that their intervention will work in the population and setting of interest in advance of their trial. However, it is fair to ask that they provide evidence that the theory supporting the intervention has been supported in other situations. It is also fair to ask that the investigators demonstrate that similar interventions applied to other problems have been effective. Finally, it is reasonable to ask that the investigators demonstrate that the proposed intervention generates short-term effects for intermediate outcomes related to the primary and secondary endpoints and postulated by the theoretical model guiding the intervention. Such evidence provides reassurance that the intervention will be effective if it is properly implemented.

55.9 Power

A detailed exposition on power for GRTs is beyond the scope of this chapter. Excellent treatments exist, and the interested reader is referred to those sources for additional information. Reference 6, Chapter 9, provides perhaps the most comprehensive treatment of detectable difference, sample size, and power for GRTs [6].

Even so, a few points bear repeating here. First, the increase in between-group variance due to the intraclass variation (ICC) in the simplest analysis is calculated as $1+(m-1)$ICC, where m is the number of members per group; as such, ignoring even a small ICC can underestimate standard errors if m is large. Second, while the magnitude of the ICC is inversely related to the level of aggregation, it is independent of the number of group members who provide data. For both of these reasons, more power is available given more groups per condition with fewer members measured per group than given just a few groups per condition with many members measured per group, no matter the size of the ICC. Third, the two factors that largely determine power in any GRT are the ICC and the number of groups per condition. For these reasons, there is no substitute for a good estimate of the ICC for the primary endpoint, the target population, and the primary analysis planned for the trial, and it is unusual for a GRT to have adequate power with fewer than 8–10 groups per condition. Finally, the formula for the standard error for the intervention effect depends on the primary analysis planned for the trial, and investigators should take care to calculate that standard error, and power, based on that analysis.

55.10 Summary

GRTs are often complex studies, with greater challenges in design, analysis, and intervention than what is seen in other studies. As a result, much care and effort is required for good planning. Future trials will be stronger and more likely to report satisfactory results if they (1) address an important research question, (2) employ an intervention that has a strong theoretical base and preliminary evidence of feasibility and efficacy, (3) randomize a sufficient number of assignment units to each study condition so as to have good power, (4) are designed in recognition of the major threats to the validity of the design and analysis of group-randomized trials, (5) employ good quality-control measures to monitor fidelity of implementation of intervention and measurement protocols, (6) are well executed, (7) employ good process-evaluation measures to assess effects on intermediate endpoints, (8) employ reliable and valid endpoint measures, (9) are analyzed using methods appropriate to the design of the study and the nature of the primary endpoints, and (10) are interpreted in light of the strengths and weaknesses of the study.

References

1. Donner, A. and Klar, N. (2000). *Design and Analysis of Cluster Randomization Trials in Health Research*. Arnold, London.

2. Fortmann, S. P., Flora, J. A., Winkleby, M. A., Schooler, C., Taylor, C. B., and Farquhar, J. W. (1995). Community intervention trials: reflections on the stanford five-city project experience. *Am. J. Epidemiol*; **142**(6), 576–586.

3. Kirk, R. E. (1982). *Experimental Design: Procedures for the Behavioral Sciences*, 2nd ed. Brooks/Cole, Belmont, CA.

4. Kish, L. (1987). *Statistical Design for Research*. Wiley, New York.

5. Meinert, C. L. (1986). *Clinical Trials*. Oxford University Press, New York.

6. Murray, D. M. (1998). *Design and Analysis of Group-Randomized Trials*. Oxford University Press, New York.

7. Shadish, W. R., Cook, T. D., and Campbell, D. T. (2002). *Experimental and Quasi-experimental Designs for Generalized Causal Inference*. Houghton-Mifflin, Boston.

8. Winer, B. J., Brown, D. R., and Michels, K. (1991). *Statistical Principles in Experimental Design*. McGraw-Hill, New York.

56. Predicting Preclinical Disease Using the Mixed-Effects Regression Model

Shan L. Sheng and Larry J. Brant

56.1 Introduction

Human aging is a process that involves behavioral and physiological changes that occur in the individual during the course of time following the attainment of maturity. This process of change affects all members of the population and is a highly individual phenomenon that occurs differently from person to person.

The Baltimore longitudinal study of aging (BLSA), begun in 1958 and administered by the Intramural Research Program of the National Institute on Aging, measures age changes through serial measurements taken on individual members of a homogeneous group of generally well-educated and dedicated community-residing adults [1]. The BLSA recruits men and women aged 17–96 years old to participate in repeated assessments of health, and physical and psychological performance with 2.5-day visits to the research center in Baltimore approximately every 2 years. All participants who have entered the BLSA are volunteers and are given a careful health screening to ensure that they are of excellent health with no known diseases upon entry into the study. Although the participants in the BLSA are not randomly taken from the US population, the study population consists of participants chosen from a large group of waiting-list volunteers so that the age distribution of the BLSA sample is comparable to the age distribution in the US population. Results from the BLSA have shown that some individuals maintain a particular level of function to an extremely old age, others demonstrate a gradual change in the same function, while some show a dramatic decline in function. This pattern of change, of course, need not be the same for different functions in the same individual. Thus, a given individual may show a greater than average increase in systolic blood pressure with age, while maintaining a more gradual and less severe decline in pulmonary function. Nathan Shock, the founder of the BLSA and considered by many to be the "father of gerontology," best summarized this by stating there is no general pattern of aging that applies to all performances, all organ systems, or all individuals [2].

While the incidence of disease increases with age, disease and aging are not synonymous. In contrast to aging, which occurs in everyone with passage of time, disease is a selective event that occurs in only some individuals in the population. Therefore, the successful prediction of preclinical disease can play a very important role in promoting healthy human life during aging. To predict preclinical disease, any method should consider cross-sectional differences between the average values of variables with regard to the different age groups of individuals, and individual longitudinal changes of the individuals within each age group for the classifying variables [3].

An example of the natural heterogeneity that exists in humans is illustrated by examining the distribution of systolic blood pressure (SBP) for 40 and 80-year old male participants of the BLSA (Fig. 1). The figure shows the heterogeneity that exists in SBP measurements for two different age groups of persons measured at approximately 40 and 80 years of age. As expected, the average blood pressure measurements increase with age from 120 mmHg at age 40 to 141 mmHg at age 80. In addition, the variance or heterogeneity increases with age from a standard deviation of 16.0 to 18.5 mmHg. It should also be noted that no individuals have a blood pressure measurement $\geq$ 160 mmHg at age 40, while 21% have a value $\geq$ 160 mmHg at age 80. Figure 1, however, does not show the large amount of within- and between-individual variability that exists in rates of change in systolic blood pressure. In this regard, Figure 2 shows the variability in systolic blood pressure that exists in longitudinal measurements observed on a total of four randomly selected men from the BLSA at each of the ages of 40 and 80 over a 10–20 year period.

As Figure 2 suggests, long-term longitudinal studies are uniquely suited for the study of individual change over time including the effects of aging and other factors that affect change. Because of the importance of the BLSA of longitudinal data, it is necessary to employ satisfactory methods of analyzing such data, which is often characterized by unequally spaced points of observation, missing data, attrition, and time-varying variables.

When individuals have been observed at the same timepoints and when there are few missing observations in the dataset, a multivariate general linear model with simple covariance structure is often appropriate [4,5]. However, when individuals have been observed at different times or when there are many missing values (i.e., the dataset is highly unbalanced with regard to time), the resulting covariance structure is more complicated and better handled by mixed-effects regression or autoregressive models. Autoregressive models [6,7] were originally developed for a single, longtime series where successive measurements are correlated with one another. Asthma attacks or episodes of air pollution are often studied using autoregressive models, since a recent attack or episode increases the chances of an occurrence on subsequent days. For autoregressive models, the correlation between repeated measurements declines as the time between measurements increases. On the other hand, mixed-effects and related models with random effects [8–10] assume that the correlation among repeated measurements of an individual is due to some latent characteristic of an individual that gives him or her higher or lower than average initial level or rate of change.

Brant and colleagues have used mixed models to examine longitudinal patterns of change in various biological and behavioral variables to predict preclinical disease, while controlling for other important contributing risk factors such as age and time of observation [11,12]. Figure 3 illustrates this approach by showing the longitudinal trend in a specific variable or risk

Preclinical Disease Using the Mixed-Effects Regression Model 615

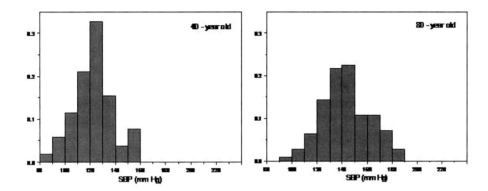

Figure 1: Distributions of systolic blood pressure showing the increasing variability for 40- and 80-year-old BLSA male participants.

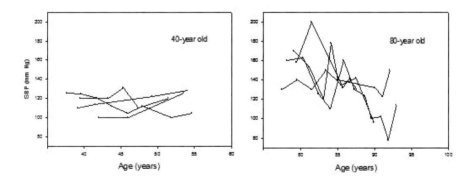

Figure 2: Observed longitudinal systolic blood pressure measurements showing the within- and between-individual variability for 8 BLSA male participants from 40- and 80-year old age groups.

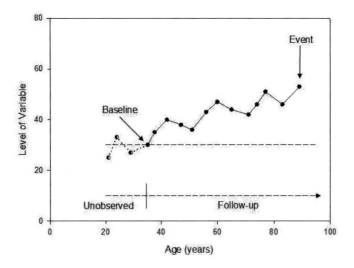

Figure 3: Hypothetical longitudinal data for an individual preceding the development of a failure or disease event.

factor and its relationship, along with the relationship between the initial or baseline measurement of a risk factor, to the occurrence of a future preclinical event, and also suggests the need to describe patterns of change in risk factor over time in different groups with regard to a future event, since different groups will have different levels and patterns of change for the variables in question. By utilizing restricted maximum likelihood (REML) and Bayesian statistical methods and the computation of estimates that include expectation–maximization (EM) and Newton-Raphson procedures, mixed-effects regression models have contributed greatly to the study of individual patterns of change by providing flexibility, allowing for unbalanced designs and incomplete data, and modeling the correlation between repeated observations for an individual.

In health clinics, medical practitioners often collect biological measurements sequentially or one examination at a time for use in the detection of specific diseases. This chapter uses collected longitudinal measurements of biological variables or risk factors to screen for the early progression of disease, specifically Alzheimer's disease, prostate cancer, and coronary heart disease. A classification method is given based on a linear mixed-effects regression model that classifies individuals with repeated biological measurements into nondiseased (normal) or diseased diagnostic groups. In the process of screening for preclinical disease using the mixed-effects regression model, the predicted starting levels and predicted rates of change of biological variables over the follow-up period are examined for individuals in each of several diagnostic states. Using posterior probabilities, each individual can be examined sequentially and classified into one of several diagnostic states for preclinical disease. The preclinical screening method is first described using a single prediction variable and is later extended to a multivariate prediction model based on a model with two correlated biological prediction variables.

56.2 Single-Variable Prediction Model

The linear mixed-effects (LME) regression model for the longitudinal data is employed in the classification or prediction procedure to classify the participants by calculating posterior probabilities of membership in the different diagnostic groups of a particular disease. Assuming that there are n diagnostic groups for N individuals in the population, the vector of longitudinal measurements follows the LME as

$$y_i = X_i\beta + Z_i b_i + \varepsilon_i \quad i = 1, \ldots, N \quad (1)$$

where y_i denotes the n_i-dimensional response vector of measurements of a biological variable on the ith individual, X_i and Z_i are the corresponding design matrices for the fixed and random effects, respectively. The design matrix X_i usually contains columns corresponding to controlling variables such as age at first examination, terms related to the follow-up time since the first measurement, the index variables that indicate the diagnostic group for individual i, and some interaction terms if necessary. The terms involving age at first examination give a measure of cross-sectional differences, while those involving time provide a measure of longitudinal differences [3]. The matrix Z_i includes the intercept and terms related to the follow-up time. The p-dimensional vector β and the q-dimensional vector b_i are vectors of the fixed and random effects parameters. The vector of random effects b_i has a multivariate normal distribution with mean 0 and positive definite covariance matrix D. The vector of measurement errors ε_i is assumed to be independent and normally distributed with mean 0 and variance-covariance matrix $\sigma^2 I_i$ where I_i is an identity matrix of dimension $n_i \times n_i$. Furthermore, it is assumed that ε_i and b_j $(i, j = 1, 2, \ldots, N)$ are independent.

The procedure for classifying the kth individual is to first fit the model (1) to the data in each diagnostic group g, omitting participant k, which yields estimates $\hat{\beta}_{(k)}$, $\hat{D}_{(k)}$, and $\hat{\sigma}^2_{(k)}$, and second to compute the marginal distribution for each diagnostic group for this individual k.

This marginal distribution is then given by

$$y_{k|group\,g} \sim N\left(X_k \hat{\beta}_{g(k)}, Z_k \hat{D}_{(k)} Z_k^T + \hat{\sigma}^2_{(k)} I_k\right)$$

and the corresponding density is denoted by $f_{g(k)}(y_k)$, where $\hat{\beta}_{g(k)}$ denotes a subset of $\hat{\beta}_{(k)}$ for diagnostic group g. Given prior probabilities p_g, $g = 1, \ldots, n$, for the n diagnostic groups, and applying Bayes theorem, the posterior probability that individual k belongs to group g is given by

$$p_{kg} = \frac{p_g f_{g(k)}(y_k)}{\sum_{j=1}^n p_j f_{j(k)}(y_k)}, \quad (2)$$

in which the prior probabilities p_g are usually replaced by the observed proportions of individuals in each diagnostic group. Note that although the LME (1) specifies the response vector conditionally on the vector b_i of random effects, the classification is based on the marginal distribution obtained from integrating over the random effects.

The classification process proceeds for individual k by first calculating the posterior probabilities in Equation (2) using the first two response measurements, and then sequentially repeating the process by adding one measurement at a time until the classification stopping rule shown in Figure 4 is met or all the measurements have been used for individual k.

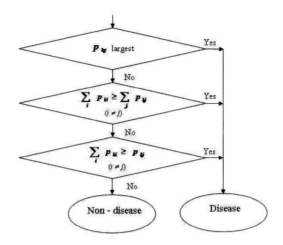

Figure 4: Classification stopping rule for the kth individual; g is diagnostic group ($g = 1, \ldots, n$), i belongs to the disease the group i, and j belongs to the non-disease group $j (i \neq j)$.

56.3 Predicting Alzheimer's Disease

Alzheimer's disease (AD) is a common cause of dementia, affecting approximately 4 million Americans, and is a leading cause of death in the United States [13]. The Benton visual retention test (BVRT) is one test of cognition that has been used by investigators to evaluate potential AD patients [14,15].

The BVRT is a test of cognition that requires the reproduction of geometric figures [16]. Because the BVRT requires spatial conception, immediate recall, and visuomotor reproduction, it is often used in conjunction with other neuropsychologic tests as an indicator of dementia. Zonderman et al. [17] demonstrated that immediate visual memory performance can have a prognostic long-term importance for as many as 22 years prior to the development of disease. In practice, screening for dementia usually begins with evaluation of the patient according to a set of cognitive test scores. Those selected as cognitively impaired are further examined by a clinician who makes the final diagnosis with regard to dementia.

Data for this Alzheimer's study are from participants of the BLSA. All participants who entered the BLSA were given a careful health screening to ensure that they were in excellent health with no known diseases, and had a series of biomedical and psychological examinations including BVRT measurements and neurologic evaluation. All participants were followed to obtain information regarding their health status, especially the information related to neurologic and other disease events. This monitoring continued over time regardless of the collection of BVRT measurements, and diagnoses of AD were made without considering the BVRT scores.

This study includes only BLSA participants who were determined to be free of AD before 1988, and who were actively enrolled in the study at that time [12]. After 1988, 20 female and 57 male subjects in this study population were clinically diagnosed with AD. Individuals with only one BVRT measurement or who had evidence of AD before the first BVRT measurement were excluded from the analysis because these participants did not have adequate time to develop a BVRT trend before diagnosis. The remaining AD cases were matched by age of initial examination, number of repeated BVRT measurements, and the length of follow-up to 32 female and 88 male controls who showed no evidence of any type of dementia throughout the study period. Separate disease classifications were performed for female and male participants since males tended to have slightly higher levels of education than females.

To classify a subject, the longitudinal data from the remaining subjects is used to fit the LME model. The general form of the model for the prediction of AD disease using BVRT measurements is

$$\begin{aligned} \text{BVRT}_{ijg} &= \beta_0 \text{Fage}_i + (\beta_{1g} + b_{1i}) \\ &+ (\beta_{2g} + b_{2i})t_{ij} \\ &+ (\beta_{3g} + b_{3i})t_{ij}^2 + \varepsilon_{ijg}, \\ i &= 1, \ldots, N, \\ j &= 1, \ldots, n_i, \\ g &= 1, 2, \end{aligned}$$

where BVRT_{ijg} denotes the jth measurement of BVRT in the diagnostic group g ($g = 1$ for normal, 2 for AD) on the ith individual. Fage_i is the age at first measurement for individual i. The variable t_{ij} represents the time (in years) since the first BVRT measurement and t_{ij}^2 denotes time^2. The corresponding measurement error term is ε_{ijg}. In this model the normal and AD groups have a different t_{ij}^2 coefficient, as well as different t_{ij} coefficients and different intercepts or initial values. The random effect b_{1i} allows each individual to have his or her own initial BVRT value, and the random effects b_{2i} and b_{3i} allow each individual to have his or her own coefficient of the t_{ij} and t_{ij}^2 terms, respectively. Figure 5 shows the average trends of BVRT for 60-year-old individuals 14 years prior to a clinical diagnosis. While the t_{ij}^2 term was statistically significant in the model, the predicted lines show little curvature for this age range.

If $\boldsymbol{y}_k$ represents the vector of all BVRT measurements on the kth individual and $f_{g(k)}(\boldsymbol{y}_k)$ represents the LME marginal distribution of $\boldsymbol{y}_{k|\text{group } g}$ for diagnostic group g (normal or AD) for individual k, the LME marginal distribution $f_{g(k)}(\boldsymbol{y}_k)$ is determined from the longitudinal data by omitting the data of individual k. The procedure for classifying individual k is to first fit the LME model in each diagnostic group g, and to use the resulting parameter and variance-covariance estimates to compute the marginal distribution for the normal and AD diagnostic groups for this individual k. Next, Bayes' formula is used to calculate the posterior probability that individual k belongs to group g and is given by

$$p_{kg} = \frac{p_g f_{g(k)}(\boldsymbol{y}_k)}{(p_1 f_{1(k)}(\boldsymbol{y}_k) + p_2 f_{2(k)}(\boldsymbol{y}_k))},$$

where the prior probabilities p_g are estimated by the observed proportions of the individuals in each diagnostic group.

The posterior probabilities from the LME model are used to classify all subjects using different diagnostic criteria or cutoff probability values for an AD classification. As the cutoff value decreases, the proportion of correctly classified AD cases (sensitivity) increases. Also, the proportion of normal subjects incorrectly classified as AD (1-specificity) increases. Figure 6 gives the receiver operating characteristic (ROC) curves corresponding to the classification results for men and women. The area under the ROC curve (AUC) is 0.73 for females and 0.65 for males. In addition, the diagnostic criteria using the point on the ROC curve located nearest to the top

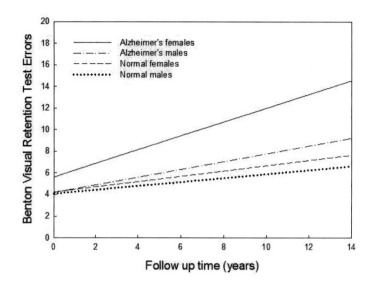

Figure 5: Predicted average 14-year trends in BVRT scores for 60-year-old BLSA females and males in the Alzheimer's and normal groups based on the LME model.

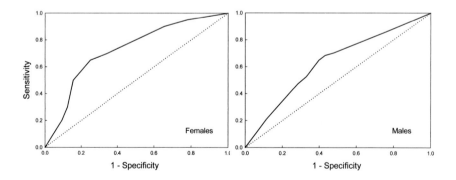

Figure 6: ROC curves based on the single-variable prediction model of BRVT scores for BLSA females and males in the Alzheimer's study.

left corner of the plot is chosen as a cut-off value that jointly optimizes the errors in classification represented by sensitivity and 1-specificity. For both female and male subjects, this occurs for a cutoff probability value of 0.42.

The classification results corresponding to the optimal classification error rates show that the method overall correctly classifies 37 of 52 (71.2%) of the female subjects and 90 of 145 (62.1%) of the male subjects. In terms of sensitivity, 13 of 20 (65.0%) of the female cases and 37 of 57 (64.9%) of the male cases were correctly classified, while for specificity, of the normal subjects 24 of 32 (75.0%) for the female and 53 of 88 (60.2%) for the male subjects were correctly classified. Using Mantel–Haenszel test for matched samples, the classification rates for the LME prediction method are better than those expected by chance alone for both female ($P = 0.011$) and male ($P = 0.007$) subjects. For the AD cases that were correctly classified, a mean lead time between the classification and the actual diagnosis was 2.7 years (95% confidence interval 1.25–4.15) for female cases and 6.7 years (95% confidence interval 4.33–9.07) for male cases.

56.4 Predicting Prostate Cancer

Prostate cancer is one of the common causes of cancer deaths in men. After lung and stomach cancer, it accounts for the largest yearly number of new non-skin cancer cases reported worldwide [18]. In the United States, for example, prostate cancer is the most common clinically diagnosed non-skin cancer with about 1 in 10 American men eventually getting a positive diagnosis. Prostate-specific antigen (PSA) is a glycoprotein that is produced by prostatic epithelium and can be measured in serum samples by immunoassay [19]. The PSA levels correlate with the volume of cancer in the prostate and it has been found to be useful in the management of men with prostate cancer. Pearson et al. [20] described patterns of change in PSA in the normal, benign prostatic hyperplasia (BPH), local cancer, and metastatic cancer groups using LME for ln(1+PSA).

Longitudinally collected PSA levels have been found to be a useful predictor of prostate cancer in the BLSA [11]. All participants in the PSA study were continually monitored to obtain health status information related to prostate disease and other disease events. In the case of death, information regarding the individual was obtained from a personal physician and autopsy information was recorded when available. The data in the study consist of 342 BLSA male participants with no evidence of cancer at the start of the study and with at least 10 years of follow-up. The 342 participants used in the study include 275 with no evidence of prostate disease, 26 who develop local or regional prostate cancer, 8 who develop metastatic cancer, and 33 who develop BPH. None of the 342 patients showed evidence of prostate disease at the beginning of the follow-up period, and all PSA measurements were taken prior to a clinical diagnosis of prostate disease. The clinical diagnoses of prostate disease or cancer occurred sometime during the follow-up period and after which no PSA measurements were used in the study. Figure 7 displays the observed repeated PSA measurements from 5 systematically selected normal or nondiseased males (right graph) and 5 systematically selected males who subsequently develop local cancer (left graph), based on the ordered means of all PSA measurements for each individual at the minimum, first quartile, median, third quartile and maximum in each group. The graph illustrates the difference in trends over time for the normal (non-diseased) and local cancer (diseased) groups. The figure suggests that

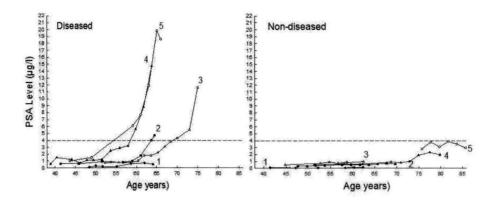

Figure 7: Observed repeated PSA measurements from 10 systematically selected individuals with prostate cancer (diseased) and without prostate cancer (nondiseased) at minimum (1), first quartile (2), median (3), third quartile (4), and maximum (5) values.

PSA measurements for approximately 25% of the local cancer cases give little evidence for the detection of prostate cancer. However, for the remaining local cases and the normal cases, a prediction approach using level and rate of change in PSA appears to be a viable method for assisting clinicians in the diagnosis of prostate cancer.

For the prediction of prostate cancer using PSA, the LME model in the general form is

$$\ln(1 + \text{PSA}_{ijg})$$
$$= \beta_0 \text{Fage}_i + (\beta_{1g} + b_{1i})$$
$$+ (\beta_{2g} + b_{2i})t_{ij}$$
$$+ (\beta_{3g} + b_{3i})t_{ij}^2 + \varepsilon_{ijg},$$
$$i = 1, \ldots, N, \; j = 1, \ldots, n_i,$$
$$g = 1, \ldots, 4$$

where PSA_{ijg} denotes the jth measurement of PSA in the diagnostic group g (g = 1 for normal, 2 for BPH, 3 for local or regional prostate cancer, 4 for metastatic prostate cancer) on the ith male. Fage_i is the age at first examination for individual i. The variable t_{ij} represents the time (in years) since the first PSA measurement and t_{ij}^2 denotes time^2. The corresponding measurement error term is ε_{ijg}. For some diagnostic groups, the term of t_{ij}^2 was not necessary in the model since the terms β_{3g} and b_{3i} were found not to have a statistical level of significance. This model was obtained through a backward elimination procedure where the intercept, t_{ij} and t_{ij}^2 coefficients are allowed to vary between the four diagnostic groups.

The classification process proceeds for individual k by first calculating the posterior probabilities in (2) using the first two PSA measurements, and then sequentially repeating the process by adding one measurement at a time until the classification stopping rule (Fig. 4) is met or all the measurements have been used. If p_{k1}, p_{k2}, p_{k3}, and p_{k4} represent the posterior probabilities for the normal, BPH, local/regional, and metastatic diagnostic groups, since the aim of the classification procedure is to determine when an individual has developed preclinical cancer, the classification rule seeks to give the greatest chance of detecting the disease. In particular, if the posterior probability of either of the cancer groups is the largest, then the patient is classified as a cancer, i.e., if the posterior probabilities for either the local/regional group (p_{k3}) or the metastatic group (p_{k4}) are the largest of the four probabilities. Next, if the combined probabilities of the cancer groups are larger than the combined probabilities of the noncancer groups, or larger than either the normal or BPH probabilities, i.e., if $p_{k3} + p_{k4} \geq p_{k1} + p_{k2}$, $p_{k3} + p_{k4} \geq p_{k1}$, or $p_{k3} + p_{k4} \geq p_{k2}$, then individual k is classified as a cancer case. Otherwise, the individual is considered as a noncancer case. If the participant has not been classified as a cancer case by his final PSA measurement, the individual is classified as either normal or BPH based on his posterior probabilities at the last measurement.

Figure 8 shows the predicted average trends of PSA levels for men at the average starting age of approximately 50 years with respect to years of follow-up time for the four diagnostic groups. Both the normal and BPH (noncancerous) groups show only a small linear (constant) increase in PSA with time. However, the local and metastatic cancer groups show evidence of an exponential trend over time, where this exponential increase is larger in the metastatic group than in the local cancer group.

The classification results for prostate cancer from the longitudinal PSA measurements of the 342 men show that overall 277 of 342 (81.0%) males are correctly classified. If the BPH and normal groups are combined together to obtain noncancer and cancer groups, then 297 of 342 (86.8%) males are correctly classified, while if no distinction is made between local and metastatic cancer, then the classification rate increases to 88.3% (302/342). In this final grouping, 21 of 34 (61.8%) of the cancer cases (sensitivity) and 281 of 308 (91.2%) of the noncancer cases (specificity) were correctly classified in the prediction procedure. Finally, an examination of the time between the individual's age at classification and the actual age at the clinical diagnosis of cancer (lead time) in those cancers correctly classified gives estimated mean lead times of 6.1 ± 2.9 years and 7.7 ± 3.9 years for the local and metastatic cancer groups respectively. Figure 9 shows the ROC curve for the prostate example where the AUC is 0.92.

56.5 Multivariate Prediction Model

The procedure to classify the participants by calculating posterior probabilities of membership in the different diagnostic groups of a particular disease based on the LME prediction model for a single variable is extended to the multivariate mixed-effects regression model (MLME). The MLME model allows the prediction of preclinical disease based on the two or more predictor responses that might be correlated with each other [21,22].

Assume that there are n diagnostic groups with m predictor responses, and each of the m responses for the ith individual can be expressed as the n_i-dimensional vector $\boldsymbol{y}_{is}$, $s = 1, \ldots, m$. The MLME model for the vector of longitudinal measurements of all m response predictors for individual i is

$$\boldsymbol{y}_i^* = \boldsymbol{X}_i^* \boldsymbol{\beta}^* + \boldsymbol{Z}_i^* \boldsymbol{b}_i^* + \boldsymbol{\varepsilon}_i^*, \quad i = 1, \ldots, N, \quad (3)$$

where the $mn_i \times 1$ vector $\boldsymbol{y}_i^* = \left[\boldsymbol{y}_{i1}^T, \boldsymbol{y}_{i2}^T, \ldots, \boldsymbol{y}_{im}^T\right]^T$ denotes the longitudinal measurements of the m predictor responses for the ith individual. Comparing formula (3) with formula (1) for the single variable prediction model, $\boldsymbol{X}_i^*$ and $\boldsymbol{Z}_i^*$ in (3) are diagonal matrices with m blocks where block s ($s = 1, \ldots, m$) in $\boldsymbol{X}_i^*$ or $\boldsymbol{Z}_i^*$ is similar to $\boldsymbol{X}_i$ or $\boldsymbol{Z}_i$ in (1). The matrix $\boldsymbol{X}_i^*$ usually contains controlling variables such as age at the first examination, terms related to the follow-up time since

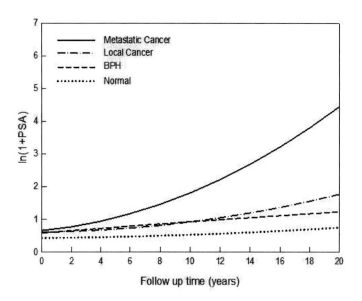

Figure 8: Predicted average 20-year trends for BLSA males at the average starting age representing the four diagnostic groups from the PSA data based on the LME model.

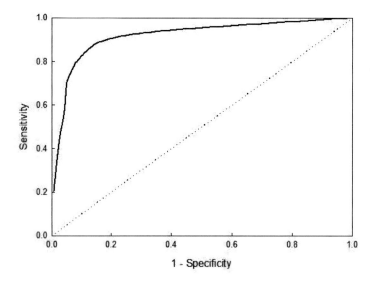

Figure 9: ROC curve based on the single variable prediction model of PSA for BLSA males in the prostate cancer study.

the first measurement, the index variables that denote the possible diagnostic group and the predictor response, and some interaction terms if necessary, where the matrix Z_i^* may contain the intercept and terms related to the follow-up time for individual i. The p^*-dimensional vector β^* and the q^*-dimensional vector b_i^* are the vectors of the fixed and random effects for all of m predictor responses. The vector of random effects b_i^* has a multivariate normal distribution with mean $\mathbf{0}$ and $q^* \times q^*$ positive definite covariance matrix D^*, which allows the inherent covariance structure for the multivariate responses with a variance within each of the m responses and a covariance between each pair of responses for the random effects in model 2. The vector of measurement errors ε_i^* is a $mn_i \times 1$ vector equal to $[\varepsilon_{i1}^T, \varepsilon_{i2}^T, \ldots, \varepsilon_{im}^T]^T$ for individual i with ε_{is} being a $n_i \times 1$ error vector for response s, and is normally distributed with mean $\mathbf{0}$ and variance–covariance matrix R_i ($mni \times mn_i$). The matrix R_i is $\Sigma \otimes I_i$, where I_i is an identity matrix of dimension $n_i \times n_i$, Σ is the $m \times m$ matrix, and $\otimes$ stands for the Krönecker product. The general form of Σ is an unstructured covariance matrix assuming that there is correlation between ε_{is} and ε_{it} ($s \neq t$, $s, t = 1, \ldots, m$), and the simplified form is diagonal with an assumption of independence among $\varepsilon_{i1}, \varepsilon_{i2}, \ldots, \varepsilon_{im}$. In addition, it is assumed that there is no correlation between ε_i^* and ε_j^*, b_i^* and b_j^*, or ε_i^* and b_j^*, where $i \neq j$, $i, j = 1, \ldots, N$. Finally, the classification procedure for individual k based on the MLME model approach is similar to the procedure based on the LME model stated in previous section, except that the estimates used in the computation are derived from the MLME model.

56.6 Predicting Coronary Heart Disease

Coronary heart disease (CHD) is a leading cause of death and disability in many populations. For example, studies suggest that CHD will occur in almost 50% of all middle-aged American men during their lifetime [23]. Experience has shown success in prevention through changes in factors known to be related to causes of CHD. Thus, identifying individuals at risk for CHD can lead to an intervention strategy that is appropriate for a given individual depending on his level of risk [24,25].

Longitudinal data from BLSA volunteers, who had no evidence of CHD at the beginning of the study, were examined to measure changes in CHD risk factors and to check for the occurrence of CHD during the follow-up period. Data obtained from individuals after a diagnosis of CHD were excluded from the study. The study participants were divided into those who eventually developed CHD and those who remained free of any evident sign of CHD during the study period. The development of CHD during the course of the study is defined as a myocardial infarction by history or by pathologic Q waves on resting electrocardiogram or sudden cardiac death. It is well known that individuals with high levels of blood pressure and/or cholesterol have a high incidence of CHD [26]. Thus, the risk factors of mean arterial pressure (MAP) measured in mmHg and approximated using the formula MAP = (SBP + 2DPB)/3 and 12-hour fasting total serum cholesterol (mg/dL) are considered with regard to the example given here for CHD classification or prediction.

The BLSA sample used for this example includes 980 male participants with the average age at the first measurement of 56.9 years in the range (30.7, 73.9 years). Among them, 191 individuals developed CHD with the average diagnostic age of

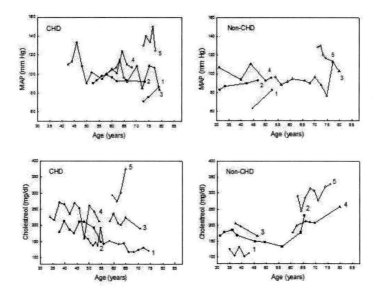

Figure 10: Observed repeated mean arterial pressure (MAP) and cholesterol measurements for systematically selected individuals who develop CHD and those who do not develop CHD at minimum (1), first quartile (2), median (3), third quartile (4), and maximum (5) values.

68.7 years in the range of (35.4, 86.8 years). The mean follow-up time was 8.6 years (with the maximum time as 33.3 years) for the CHD group and 11.1 years (with the maximum time as 35.6 years) for the non-CHD group. The interval between repeated measurements for both groups was approximately 2 years, and the correlation between total cholesterol and MAP at the first measurement was 0.17 ($P < 0.0001$). Figure 10 shows the observed MAP (top panels) and total cholesterol (bottom panels) measurements for those individuals whose overall average values were approximately at the minimum, first quartile, median, third quartile and maximum for CHD cases (left panels) and non-CHD cases (right panels). The four panels demonstrate the relatively high degree of variability in the measurements for both within and between the subjects used in this example. In general, the CHD cases appear to be overall higher than the non-CHD cases for both variables.

The general MLME prediction model for CHD using total cholesterol and MAP as the predictor response variables is

$$\begin{aligned} y_{ijgs} &= \beta_{10gs}\text{Fage}_i + \beta_{11gs} + \beta_{12gs}t_{ij} \\ &+ \beta_{13gs}t_{ij}^2 + \beta_{14gs}\text{Fage}_i t_{ij} \\ &+ \beta_{15gs}\text{Fage}_i t_{ij}^2 + b_{1si} \\ &+ b_{2si}t_{ij} + b_{3si}t_{ij}^2 + \varepsilon_{ijgs} \\ i &= 1,\ldots,N,\ j = 1,\ldots,n_i, \\ g &= 1,2,\ s = 1,2, \end{aligned}$$

where y_{ijgs} denotes the jth measurement of predictor response s for individual i ($s = 1$ for MAP, 2 for total cholesterol) in the diagnostic group g ($g = 1$ for non-CHD, 2 for CHD). Fage$_i$ represents the age at the first examination for individual i. The variable t_{ij} stands for the time (in years) and t_{ij}^2 for time^2. The corresponding measurement error term is ε_{ijgs}. For some subgroups, the statistically insignificant estimates of the fixed and random effects were omitted in the modeling procedure. Also, current cigarette smoking was considered in the model, but since only about 5% of the 980 males were smokers, smoking was not found to be significant, and thus was not included in the final model. Table 1 gives the fixed effects estimates for the MLME prediction model shown above. Using the estimates in Table 1, the average starting values for a non-CHD 30-year-old and 50-year-old participant are 89.5 and 94.9 mmHg for MAP, and 191.3 and 200.9 mg/dL for total cholesterol, respectively. Also, there are significant differences in the rate of changes over time for MAP between CHD cases and noncases ($P = 0.0039$), while for total cholesterol significant differences between CHD and non-CHD individuals exist with regard to both age at the first visit ($P < 0.0001$) and time in study ($P = 0.0001$). In addition, Table 2 shows all variance and covariance estimates of the within and between random effects for MAP and cholesterol used in the above model (left matrix), and the variance and covariance estimates for the measurement errors corresponding to MAP ($\hat{\sigma}_{\varepsilon 1}^2 = 77.59$) and total cholesterol ($\hat{\sigma}_{\varepsilon 2}^2 = 451.90$) (right matrix).

Figure 11 shows the predicted longitudinal trends corresponding to MAP (left panel) and total cholesterol (right panel) from the MLME model for individuals developing and not developing CHD and entering the study at approximately age of 50 years with a 12-year follow-up period. As seen in Table 1 (CHD estimates), MAP is 3.4 mmHg higher in individuals with CHD compared to those with non-CHD and increases more rapidly for those developing CHD. In the case of total cholesterol, the effects or terms used for this variable involve higher order terms with regard to CHD, age at the first examination (Fage) and time (see Table 1), and so CHD cases for those individuals starting the study at age 50 are on average 23.9 mg/dL higher than those individuals not developing CHD

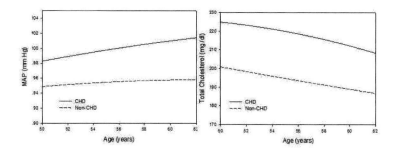

Figure 11: Predicted average 12-year trends for 50-year-old BLSA males developing CHD and those not developing CHD from the bivariate mixed-effects model with total cholesterol and mean arterial pressure (MAP).

(non-cases) during the 12-year follow-up period. Proceeding as in the univariate examples, the optimal value of sensitivity and 1-specificity from the corresponding ROC curve (see Fig. 12, AUC = 0.71) gives a sensitivity of 70% and a specificity of 63% for the CHD bivariate prediction method using MAP and total cholesterol. For the CHD group, the estimated mean lead time for those diagnosed with CHD between the age at classification and actual age at the clinical diagnosis is 8.0 ±8.0 years. While this chapter only gives a bivariate example for the MLME model, the MLME model can easily be extended to a situation with more than two predictor variables. In doing so, more terms need to be included and estimated in the model and thus the necessary computational time increases as well.

56.7 Discussion

In this chapter mixed-effects models, along with posterior probabilities of disease outcomes calculated from Bayes theorem, are used to classify or to make predictions for individuals to be into known or identified disease states. All models include age at first examination (Fage) and time since first examination (time) as independent or predictor variables, as well as indicator variables for the possible disease states. When necessary, interaction terms involving Fage and time with the disease state variables are also included in the model. Those terms involving age at first examination give a measure of cross-sectional differences, while those involving time provide a measure of longitudinal differences. In addition, other variables related to the disease outcome (e.g., cigarette smoking and alcohol use), along with diet and family history may also be included in a model. When such information is sufficiently available and is measured accurately, the inclusion of any important predictors can reduce the random-effects variability and possibly improve classification results.

In addition to obtaining measures of sensitivity and specificity, the mean lead time between being in a preclinical disease state and the clinical diagnostic state for individuals with the disease outcome should be reported when evaluating the results from a prediction classification method. Also, when evaluating prediction or classification models, one should examine both the discrimination and calibration aspects of the model [27,28]. Usually, the ROC curve and its area (AUC) or c-statistic are used to measure how well a procedure or model discriminates or classifies individuals into di-

Table 1: Fixed effects estimates for the bivariate prediction model with cholesterol and mean arterial pressure (MAP) for BLSA males in the CHD study.

Group	Effect	MAP (mmHg)			Cholesterol (mg/dL)		
		Estimates	SE	P Value	Estimates	SE	P Value
Non-CHD	Intercept	81.38	1.22	< 0.0001	176.87	4.30	< 0.0001
	Fage	0.27	0.025	< 0.0001	0.48	0.088	< 0.0001
	Time	1.49	0.20	< 0.0001	0.53	0.31	0.085
	Time^2	-0.031	0.0076	< 0.0001	0.0075	0.0061	0.22
	Fage×Time	-0.027	0.0041	< 0.0001	-0.036	0.0055	< 0.0001
	Fage×Time^2	0.00048	0.00017	0.0044			
CHD	Intercept	84.78	1.53	< 0.0001	245.81	11.81	< 0.0001
	Fage	0.27	0.025	< 0.0001	-0.43	0.20	0.037
	Time	1.67	0.21	< 0.0001	1.091	0.47	0.020
	Time^2	-0.031	0.0076	< 0.0001	-0.056	0.015	0.0003
	Fage×Time	-0.027	0.0041	< 0.0001	-0.036	0.0055	< 0.0001
	Fage×Time^2	0.00048	0.00017	0.0044			

Table 2: The covariance matrix (left) of random effects for MAP intercept, MAP time, MAP time^2, cholesterol intercept, cholesterol time, and cholesterol time^2, as well as the covariance matrix (right) for the measurement errors of MAP and cholesterol, for the bivariate prediction model with cholesterol and MAP for BLSA males in the CHD study.

$$\begin{bmatrix} 67.04 & -2.33 & 0.033 & 3.65 & 1.35 & -0.087 \\ -2.33 & 0.58 & -0.014 & 7.058 & -0.28 & 0.0058 \\ 0.033 & -0.014 & 0.00043 & -0.20 & 0.013 & -0.00030 \\ 3.65 & 7.058 & -0.20 & 931.76 & -9.78 & -0.41 \\ 1.35 & -0.28 & 0.013 & -9.78 & 3.93 & -0.11 \\ -0.087 & 0.0058 & -0.00030 & -0.41 & -0.11 & 0.0046 \end{bmatrix} \quad \begin{bmatrix} 77.59 & 8.80 \\ 8.80 & 451.90 \end{bmatrix}$$

agnostic groups. In regard to calibration, measures of how well observed proportions agree with predicted probabilities can also be reported when evaluating the accuracy of a model (e.g., see Lemeshow and Hosmer [29]). Finally, Pepe et al. [30] cautions that present statistical methods for evaluating risk predictors are not satisfactory. Thus, the researcher should carefully consider all the different tools available when evaluating any prediction or classification procedure.

In the case of Alzheimer's disease, the AUC or c-statistic was 0.73 and 0.65 for females and males, respectively. Using the optimal diagnostic criteria from the ROC curves, the overall percent correctly classified was 71.2% for females and 62.1% for males. Both of these values were better than percentages expected by chance alone. The corresponding sensitivities were approximately the same (65%) for both females and males, while females had a higher specificity (75%) than males (60%). On the other hand, males had a greater lead time of 6.7 years (95% confidence interval 4.3–9.1) than the 2.7 years (95% confidence interval 1.2–4.1) for females.

For the prostate disease example in the BLSA male population, the percentage of males correctly classified using the LME approach varied from 81% to 88%, depending on how individuals in the different diagnostic groups of normal, BPH, local and metastatic cancer are combined together. Also, it should be noted that the ability of the prediction or classification model to correctly classify individuals depends on how accurate the disease groups can be clinically diagnosed and how well the predictor variables provide accurate and precise measurements of the biological functions they represent. For example, in the prostate example, PSA gives a relatively high specificity (91.2%) but a lower sensitivity (61.8%). One reason for this is that PSA remains relatively low in about 33% of all prostate cancer cases [31]. The LME predictions gave an ROC curve with a high AUC of 0.92. Finally, for the local and metastatic cancer groups, the estimated mean lead times were 6.1 ± 2.9 years and 7.7 ± 3.9 years, respectively.

Results for the bivariate mixed model prediction method using MAP and cholesterol to predict CHD gave a sensitivity of 70% and a specificity of 63% at the optimal value from the ROC curve given in Figure 12. Also, for those with CHD that were classified as CHD, the estimated mean lead time was 8.0 ± 8.0 years. The AUC for the bivariate CHD example using the mixed model was 0.71. This AUC value can no doubt be made larger by extending the bi-

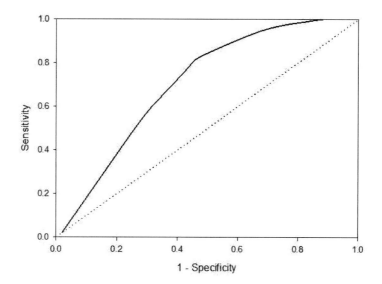

Figure 12: ROC curves based on the multivariate prediction model using total cholesterol and mean arterial pressure (MAP) for BLSA males in the CHD study.

variate MLME model to a higher dimension, or perhaps by obtaining a better bivariate model for the data by adding additional regression variables that are related to MAP and total cholesterol. For example, in the case of the bivariate model summarized in Table 1, MAP was modeled with interaction terms involving age at the first examination (Fage) by time and time^2 and CHD by time; while the model for cholesterol involved interaction terms for Fage by time and CHD with Fage, time and time^2. In constructing LME or MLME models, it is important to maintain a well-formulated model in relation to the higher- with the corresponding lower-order terms [32].

Finally, both the univariate and multivariate mixed-model prediction methods simultaneously estimate the longitudinal trajectories of all the biological variable(s) or marker(s), which are considered in the classification process, while accounting for the correlation between and within the measurements for the different variables. As in the case of the single or univariate classification method, the resulting predicted multivariate probability density values are used to calculate posterior probabilities of each individual being in the different diagnostic groups. These probabilities are calculated at each measurement time for each individual in the dataset, and a decision is made on the basis of a preassigned classification rule to group the individual into a disease or disease-free group. In conclusion, the prediction of preclinical disease using the mixed-effects regression model provides a valuable statistical tool for application in the life and health sciences.

Acknowledgment. This research was supported entirely by the Intramural Research Program of NIH, National Institute on Aging.

References

1. Shock, N. W., Greulich, R. C., Andres, R., Arenberg, D., Costa, P. T., Lakatta, E. G., and Tobin, J. D. (1984). *Normal Human Aging: The Baltimore Longitudinal Study of Aging*, NIH Publication No. 84-2450.

2. Shock, N. W. (1985). The physiological basis of aging. In *Frontiers in Medicine: Implications for the Future*, pp. 300–312, Human Sciences Press, New York.

3. Morrell, C. H., Brant, L. J., and Ferrucci, L. (2009). Model choice can obscure results in longitudinal studies. *J. Gerontol. Biol. Sci. Med. Sci.*.

4. Johnson, R. A. and Wichern, D. W. (1988). *Applied Multivariate Statistical Analysis*, Prentice Hall, Englewood Cliffs, NJ.

5. Anderson, T. W. (1984). *An Introduction to Multivariate Statistical Analysis*, Wiley, New York.

6. Kowalski, C. J. and Guire, K. E. (1974). Longitudinal data analysis. *Growth*, **38**, 131.

7. Rosner, B., and Munoz, A. (1988). Autoregressive modeling for the analysis of longitudinal data with unequally spaced examinations. *Statist. Med.*, **7**, 59.

8. Rao, C. R. (1965). The theory of least squares when the parameters are stochastic and its applications to the analysis of growth curves. *Biometrika*, **52**, 447.

9. Laird, N. M. and Ware, J. H. (1982). Random effects models for longitudinal data. *Biometrics*, **38**, 963.

10. Lindstrom, M. J. and Bates, D. M. (1988). Newton–Raphson and EM algorithms for linear mixed-effects models for repeated-measures data. *J. Am. Statist. Assoc.* **83**, 1014.

11. Brant, L. J., Sheng, S. L., Morrell, C. H., Verbeke, G. N., Lesaffre, E., and Carter, H. B. (2003). Screening for prostate cancer by using random-effects models. *J. Roy. Statist. Soc. A* **166**, 51.

12. Brant, L. J., Sheng, S. L., Morrell, C. H., and Zonderman, A. B. (2005). Data from a longitudinal study provided measurements of cognition to screen for Alzheimer's disease. *J. Clin. Epidemiol.*, **58**, 701–707.

13. Progress Report on Alzheimer's Disease (1999). *National Institute on Aging*. Bethesda, MD: National Institutes of Health.

14. Fabrigoule, C., Rouch, I., Taberly, A., Letenneur, L., Commenges, D., Mazaux, J. M., Orgogozo, J. M., and Dartigues, J. F. (1998). Cognitive process in preclinical phase of dementia. *Brain*, **121**, 135–141.

15. Kawas, C. H., Corrada, M. M., Brookmeyer, R., Morrison, A., Resnick, S. M., Zonderman, A. B., and Arenberg, D. (2003). Visual memory predicts Alzheimer's disease more than a decade before diagnosis. *Neurology*, **60**, 1089–1093.

16. Benton, A. L. (1963). *The Revised Visual Retention Test: Clinical and Experimental Applications*. Psychological Corp, New York.

17. Zonderman, A. B., Giambra, L. M., Arenberg, D., Resnick, S. M., and Costa P. T. (1995). Changes in immediate visual memory predict cognitive impairment. *Arch. Clin. Neuropsychol.*, **10**, 111–123.

18. Pisani, P., Parkin, D. M., Bray, F., and Ferlay, J. (1999). Estimates of the worldwide mortality from 25 cancers in 1990. *Int. J. Cancer*, **83**, 18–29.

19. Wang, M. C., Valenzuela, L. A., Murphy, G. P., and Chu, T. M. (1979). Purification of a human prostate specific antigen. *Invest. in Urol.*, **17**, 159–163.

20. Pearson, J. D., Morrell, C. H., Landis, P. K., Carter, H. B., and Brant, L. J. (1994). Mixed-effects regression models for studying the natural history of prostate disease. *Statist. Med.*, **13**, 587–601.

21. Shah, A., Laird, N., and Schoenfeld, D. (1997). A random-effects model for multiple characteristics with possibly missing

data. *J. Am. Statist. Assoc.*, **92**, 775–779.

22. Morrell, C. H., Brant, L. J., Sheng, L. S., and Najjar, S. (2003). Using multivariate mixed-effects models to predict hypertension. *Proc. Am. Statist. Assoc.*, 1–5.

23. Goldberg, R. J. (2009). To the Framingham data, turn, turn, turn. *Circulation*, **119**, 1189–1191.

24. Greenland, P., Smith, S. C., and Grundy, S. M. (2001). Improving coronary heart disease risk assessment in asymptomatic people—role of traditional risk factors and noninvasive cardiovascular tests. *Circulation*, **104**, 1863–1867.

25. Grundy, S. M. (1999). Primary prevention of coronary heart disease integrating risk assessment with intervention. *Circulation*, **100**, 988–998.

26. Lewington, S. and Clarke, R. (2005). Combined effects of systolic blood pressure and total cholesterol on cardiovascular disease risk. *Circulation*, **112**, 3373–3374.

27. Cook, N. R. (2007). Use and misuse of the receiver operating characteristic curve in risk prediction. *Circulation*, **115**, 928–935.

28. Cook, N. R. (2008). Statistical evaluation of prognostic versus diagnostic models: Beyond the ROC curve. *Clin. Chemistry*, **54**, 17–23.

29. Lemeshow, S. and Hosmer, D. W. (1982). A review of goodness. *Am. J. Epidemiol.*, **115**, 92 106 (.)

30. Pepe, M. S., James, H., and Gu, J. W. (2007). Letter regarding article, "Use and misuse of receiver operating characteristic curve in risk prediction." *Circulation*, **116**, 132.

31. Carter, H. B. and Pearson, J. D. (1994). Evaluation of changes in PSA in the management of men with prostate cancer. *Semin. Oncol.*, **21**, 554–559.

32. Morrell, C. H., Pearson, J. D., and Brant, L. J. (1997). Linear transformations of linear mixed-effects models. *Am. Statist.*, **51**, 338–343.

57. Predicting Random Effects in Group-Randomized Trials

Edward J. Stanek III

57.1 Introduction

The main objective in a group-randomized trial is a comparison of the expected response between treatments, where the expected response for a given treatment is defined as the average expected response over all groups in the population. Often, there may be interest in the expected response for a particular group included in the trial. Since a random sample of groups is included in the trial, this is usually represented as a random effect. In this context, predicting the expected response for an individual group requires predicting a random effect.

For example, in a study of the impact of teaching paradigms on substance use of high school students in New Haven, Connecticut, high schools were randomly assigned to an intervention or control condition. The main evaluation of the intervention was a comparison of student response between intervention and control averaged over all high schools. There was also interest in student response at particular high schools. Since high schools were randomly assigned to conditions, the difference in response for a particular high school from the population average response is a random effect.

57.2 Background

Since the early work on analysis of group-randomized trials [22,31,47], models for response have included both fixed and random effects. Such models are called *mixed models*. Fixed effects appear directly as parameters in the model, while group effects are included as random variables. The random effects have mean zero and a nonzero variance. An advantage of the mixed model is the simultaneous inclusion of population parameters for treatments while accounting for the random assignment of groups in the design.

Historically, since the main focus of a trial is comparison of treatments over all groups, and groups are assigned at random to a treatment, the effect of an assigned group was not considered to be of interest. This perspective has resulted in many authors [7,23,26,32,35] limiting discussion to estimation of fixed effects. Support for this position stems from the fact that a group cannot be guaranteed to be included in the study. This fact, plus the result that the average of the group effects is zero, has been sufficient for many to limit discussion of group effects to estimating the group variance, not particular group effects.

Nevertheless, when conducting a group-randomized trial, it is natural to want to estimate response for a particular group. Owing to the limitations of random effects in a mixed model, some authors [8,34] suggest that such estimates can be made, but they should be based on a different model. The model is conditional on the group assignment, and represents groups as fixed effects. With such a representation, response for individual groups can be directly estimated. However, the fixed effect model would not be suitable for estimating treatment effects, since the evaluation would not be based on the random assignment of groups to treatments.

The apparent necessity of using different models to answer two questions in a group-randomized trial has prompted much study [30]. A basic question is whether a mixed model can be used to predict the mean of a group that has been randomly assigned to a treatment. If such prediction were possible, the mixed model would provide a unified framework for addressing a variety of questions in a group-randomized trial, including prediction of combinations of fixed and random effects. A number of workers [12,16,24,29] argue that such prediction is possible. The predictor is called the *best linear unbiased predictor* (BLUP) [9]. Moreover, since the BLUP minimizes the expected mean squared error (EMSE), it is optimal.

In light of these results, it may appear that methods for prediction of random effects are on firm ground. There are, however, some problems. The BLUP of a realized group is closer to the overall sample treatment mean than the sample group mean, a feature called *shrinkage*. This shrinkage results in a smaller average mean squared error (MSE). The reduction in EMSE is often attributed to "borrowing strength" from the other sample observations.

However, the best linear *unbiased* predictor of a realized group is biased, while the sample group mean is unbiased. The apparent contradiction in terms is due to two different definitions of bias. The BLUP are *unbiased* in the sense that there is zero average bias over all possible samples (i.e., unconditionally). The sample group mean is unbiased conditional on the realized group. A similar paradox may occur when considering the EMSE, where the MSE for the *best* linear unbiased predictor may be larger than the MSE of the group mean for a realized group. This can occur since for the BLUP, the EMSE is *best* in the sense that the average MSE is minimum over all possible samples (i.e., unconditionally), while for the sample group mean, the MSE is evaluated for a realized group [41]. These differences lead to settings, such as when the distribution of group means is bimodal, where the BLUP seems inappropriate for certain groups [44].

For such reasons, predictors of random effects are somewhat controversial. We do not resolve such controversies here. Instead, we attempt to clearly define the problem, and then describe some competing models and solutions. Statistical research in this area is active, and new insights and controversies may still emerge.

57.3 Methods

We provide a simple framework that may be helpful for understanding these issues in the context of a group-randomized trial. We first define a study population, along with population/group parameters. This provides a finite population context for the group-randomized trial. Next, we explicitly represent random variables that arise from random assignment of groups, and sampling subjects within a group. This context provides a framework for discussion of various mixed models and assumptions. We then proceed to outline development of predictors of random effects. We limit this development to the simplest setting. Finally, we conclude with a broader discussion of other issues.

57.4 Modeling Response for a Subject

Suppose a group-randomized trial is to be conducted to evaluate the impact of a substance abuse prevention program in high schools. We assume that for each student, a measure of the perception of peer substance abuse can be obtained from a set of items of a questionnaire administered to the student as a continuous response. We denote the kth measure of response (possibly repeated) of student t in high school s by

$$Y_{stk} = y_{st} + W_{stk}, \quad (1)$$

indexing students by $t = 1, \ldots, M$ in each of N high schools, indexed by $s = 1, \ldots, N$. Measurement error (corresponding to test, retest variability) is represented by the random variable W_{stk}, and distinguishes y_{st} (a fixed constant representing the expected response of student t) from Y_{stk}. The subscript k indicates an administration of the questionnaire, where potentially $k > 1$. The average expected response over students in school s is defined as $\mu_s = (1/M) \sum_{t=1}^{M} y_{st}$, while the average over all schools is defined as $\mu = (1/N) \sum_{s=1}^{N} \mu_s$. We will refer to μ_s as the latent value of school s. We limit discussion to the simplest setting where each school has the same number of students, an equal number of schools are assigned to each intervention, an equal number of students are selected from each assigned school and a single measure of response is made on each selected student.

These definitions provide the context for defining the impact of a substance abuse prevention program. To do so, we imagine that each student could be measured both with and without the intervention. If no substance abuse program is implemented, let the expected response of a student be y_{st}; if an intervention program is in place, let the expected response be y_{st}^*. The difference, $y_{st}^* - y_{st} = \delta_{st}$ represents the effect of the intervention on student t in high school s. The average of these effects over students in school s is defined as $\delta_s = (1/M) \sum_{t=1}^{M} \delta_{st}$, while the average over all schools is defined as $\delta = \frac{1}{N} \sum_{s=1}^{N} \delta_s$. The parameter δ is the main parameter of interest in a group-randomized trial.

To emphasize the effect of the school and of the student in the school, we define $\beta_s = (\mu_s - \mu)$ as the deviation of the latent value of school s from the population mean and $\varepsilon_{st} = (y_{st} - \mu_s)$ as the deviation of the expected response of student t (in school s) from the school's latent value. Using these definitions, we represent the expected response of student t in school s as

$$y_{st} = \mu + \beta_s + \varepsilon_{st}. \quad (2)$$

This model is called a *derived model* [19]. The effect of the intervention on a student can be expressed in a similar manner as $\delta_{st} = \delta + \delta_s^* + \delta_{st}^*$ where $\delta_s^* = \delta_s - \delta$ and $\delta_{st}^* = \delta_{st} - \delta_s$. Combining these terms, the expected response of student t in school s receiving the intervention is $y_{st}^* = \mu + \delta + \beta_s^* + \varepsilon_{st}^*$ where $\beta_s^* = \beta_s + \delta_s$ and $\varepsilon_{st}^* = \varepsilon_{st} + \delta_{st}^*$. When the effect of the intervention is equal for all students (i.e., $\delta_{st} = \delta$), we represent the expected response for student t (in school s) by

$$y_{st} = \mu + x_s \delta + \beta_s + \varepsilon_{st},$$

where x_s is an indicator of the intervention, taking a value of one if the student receives the intervention, and zero otherwise. We assume the effect of the intervention is equal for all students to simplify subsequent discussions. The latent value for school s under a given condition is given by $\mu + x_s \delta + \beta_s$.

57.5 Random Assignment of Treatment and Sampling

The first step in a group-randomized trial is random assignment of groups to treatments. We assume that there are two treatments (intervention and control), with n groups (i.e., schools) assigned to each treatment. A simple way to conduct the assignment is to randomly permute the list of schools, assigning the first n schools to the control, and the next n schools to intervention. We index the school's position in the permutation by $i = 1, \ldots, N$, referring to control schools by $i = 1, \ldots, n$, and to the intervention schools by $i = n+1, \ldots, 2n$. A sample of students in a school can be represented in a similar manner by randomly permuting the students in a school, and then including the first $j = 1, \ldots, m$ students in the sample. We represent the expected response (over measurement error) of a student in a school as Y_{ij}, using the indices for the positions in the permutations of students and schools. Note that Y_{ij} is a random variable since the actual school and student corresponding to the positions will differ for different permutations. Once a permutation (say, of schools) is selected, the school that occupies each position in the permutation is known. Given a selected permutation, the school is a "fixed" effect; models that condition on the schools in the sample portion of a permutation of schools will represent schools as fixed effects.

Schools are represented as random effects in a model that does not condition on the permutation of schools. The resulting model is an unconditional model. We represent the expected response for a school assigned to a position explicitly as a random variable, and account for the uncertainty in the assignment by a set of indicator random variables, $U_{is}, s = 1, \ldots, N$ where U_{is} takes the value of one when school s is assigned to position i, and the value zero otherwise. Using these random variables, $\sum_{s=1}^{N} U_{is}\mu_s$ represents a random variable corresponding to the latent value of the school assigned to position i. Using $\beta_s = (\mu_s - \mu)$ and noting that for any permutation, $\sum_{s=1}^{N} U_{is} = 1$, $\sum_{s=1}^{N} U_{is}\mu_s = \mu + B_i$, where $B_i = \sum_{s=1}^{N} U_{is}\beta_s$ represents the random effect of the school assigned to position i in the permutation of schools.

We use the random variables U_{is} to represent permutations of schools, and a similar set of indicator random variables, $U_{jt}^{(s)}$ that take on a value of one when the jth position in a permutation of students in school s is occupied by student t, and zero otherwise to relate y_{st} to Y_{ij}. For ease of exposition, we refer to the school that will occupy position i in the permutation of schools as the *primary sampling unit* (PSU) i, and to the student that will occupy position j in the permutation of students in a school as *secondary sampling unit* (SSU) j. PSUs and SSUs are indexed by positions (i and j), whereas schools and students are indexed by labels (s and t) in the finite population. As a consequence, the random variable corresponding to PSU i and SSU j is given by

$$Y_{ij} = \sum_{s=1}^{N} \sum_{t=1}^{M} U_{is} U_{jt}^{(s)} y_{st}.$$

Using the representation of expected response for an SSU in a PSU, we obtain

$$Y_{ij} = \mu + x_i\delta + B_i + E_{ij},$$

noting that $\sum_{t=1}^{M} U_{jt}^{(s)} = 1$, $\delta \sum_{s=1}^{N} U_{is} x_s = x_i \delta$ since the treatment assigned to a position depends only on the position, and $E_{ij} = \sum_{s=1}^{N} \sum_{t=1}^{M} U_{is} U_{jt}^{(s)} \varepsilon_{st}$. Adding measurement error, the model is given by

$$Y_{ijk} = \mu + x_i\delta + B_i + E_{ij} + W_{ijk}^*,$$

where $W_{ijk}^* = \sum_{s=1}^{N} \sum_{t=1}^{M} U_{is} U_{jt}^{(s)} W_{stk}$. This model includes fixed effects (i.e., μ

57.6 Intuitive Predictors of the Latent Value of a Realized Group

Suppose that there is interest in predicting the latent value of a realized group (i.e., school), say, the first selected group, $\mu + B_i = \sum_{s=1}^{N} U_{is}\mu_s$, where $i = 1$. Before randomly assigning schools, since we do not know which school will be first, the expected value of the first PSU is a random variable represented by the sum of a fixed (i.e., μ) and a random effect (i.e., B_i where $i = 1$). Once the school corresponding to the first PSU has been randomly assigned, the random variables U_{is} for $i = 1$ and $s = 1, \ldots, N$ will be realized. If school s^* is assigned to the first position, then the realized values, that is, $U_{1s} = u_{1s}, s = 1, \ldots, N$, are $u_{1s} = 0$ when $s \neq s^*$ and $u_{1s} = 1$ when $s = s^*$; the parameter for the realized random effect will correspond to β_{s^*}, the deviation of the latent value of school s^* from the population mean.

We discuss methods for predicting the latent value of a realized PSU assigned to the control condition (i.e., $i \leq n$) in the simplest setting when there is no measurement error. The model for SSU j in PSU i is the simple random-effects model:

$$Y_{ij} = \mu + B_i + E_{ij}. \qquad (3)$$

The latent value of PSU i is represented by the random variable $\mu + B_i$. The parameter for the latent value of school s is $\mu_s = (1/M)\sum_{t=1}^{M} y_{st}$. We wish to predict the latent value for the school corresponding to PSU i, which we refer to as the latent value of the realized PSU.

It is valuable to develop some intuitive ideas about the properties of predictors. For school s, we can represent the latent value as the sum of two random variables, that is, $\mu_s = (1/M)(\sum_{j=1}^{m} Y_{sj} + \sum_{j=m+1}^{M} Y_{sj})$, where $Y_{sj} = \sum_{t=1}^{M} U_{jt}^{(s)} y_{st}$ represents response for SSU j in school s. Let $\bar{Y}_{sI} = (1/m)\sum_{j=1}^{m} Y_{sj}$ and $\bar{Y}_{sII} = (1/M-m)\sum_{j=m+1}^{M} Y_{sj}$ represent random variables corresponding to the average responses of SSUs in the sample and remainder, respectively. Then $\mu_s = f\bar{Y}_{sI} + (1-f)\bar{Y}_{sII}$, where the fraction of students selected in a school is given by $f = (m/M)$. If school s is a control school (i.e., one of the first n PSUs), then the average response for students selected from the school, $\bar{Y}_{sI}$, will be realized after sampling and the only unknown quantity in the expression for μ_s will be $\bar{Y}_{sII}$, the average response of students not included in the sample. Framed in this manner, the essential problem in predicting the latent value of a realized PSU is predicting the average response of the SSUs not included in the sample.

The predictor of the latent value of a realized PSU will be close to the sample average for the PSU when the second-stage sampling fraction is large. For example, representing a predictor of $\bar{Y}_{sII}$ by $\hat{\bar{Y}}_{sII}$, and assuming that $f = 0.95$, the predictor of the latent value of school s is $\hat{\mu}_s = 0.95\bar{Y}_{sI} + 0.05\hat{\bar{Y}}_{sII}$. Even poor predictors of $\bar{Y}_{sII}$ will only modestly affect the predictor of the latent value. This observation provides some guidance for assessing different predictors of the latent value of a realized random effect. As the second-stage sampling fraction is allowed to increase, a predictor should be closer and closer to the average response of the sample SSUs for the realized PSU.

57.7 Models and Approaches

We provide a brief discussion of four approaches that have been used to predict the latent value of a realized group, limiting ourselves to the simplest setting given by (3). The four approaches correspond to Henderson's approach [16,18], a Bayesian approach [35], a superpopulation model approach [33,37], and a random permutation model approach [37]. An influential paper by Robinson [29] and its discussion brought prediction of realized random effects into the statistical limelight. This paper identified different applications (such as selecting sires for dairy cow breeding, Kalman filtering in time-series analysis, Kriging in geographic statistics) where the same basic problem of prediction of realized random effects arose, and directed attention to the BLUP as a common solution. Other authors have discussed issues in predicting random effects under a Bayesian setting [3], random effects in logistic models [14,42,43], and in applications to blood pressure [27], cholesterol, diet, and physical activity [41]. Textbook presentations of predictors are given in References 10, 25, 28, and 35. A general review in the context of longitudinal data is given in Reference 36.

57.8 Henderson's Mixed Model Equations

Henderson developed a set of mixed-model equations to predict fixed and random effects [16]. The sample response for model (3) is organized by SSUs in PSUs, resulting in the model

$$\mathbf{Y} = \mathbf{X}\alpha + \mathbf{Z}\mathbf{B} + \mathbf{E}, \qquad (4)$$

where $\mathbf{Y}$ is an $nm \times 1$ response vector, $\mathbf{X} = \mathbf{1}_{nm}$ is the design matrix for fixed effects, and $\mathbf{Z} = \mathbf{I}_n \otimes \mathbf{1}_m$ is the design matrix for random effects, where $\mathbf{1}_{nm}$ denotes an $nm \times 1$ column vector of ones, $\mathbf{I}_n$ is an $n \times n$ identity matrix, and $\mathbf{A}_1 \otimes \mathbf{A}_2$ represents the Krönecker product formed by multiplying each element in the matrix $\mathbf{A}_1$ by $\mathbf{A}_2$ [11]. The parameter $\alpha = E(Y_{ij})$ is a fixed effect corresponding to the expected response, while the random effects are contained in $\mathbf{B} = (B_1 B_2 \cdots B_n)'$. We assume that $E(B_i) = 0$, $\text{Var}(\mathbf{B}) = \mathbf{G}$, $E(E_{ij}) = 0$, and $\text{Var}(\mathbf{E}) = \mathbf{R}$ so that $\mathbf{\Sigma} = \text{Var}(\mathbf{ZB} + \mathbf{E}) = \mathbf{ZGZ}' + \mathbf{R}$. Henderson proposed estimating α (or a linear function of the fixed effects) by a linear combination of the data, $\mathbf{a}'Y$. Requiring the estimator to be unbiased and have minimum variance lead to the generalized least squares estimator $\hat{\alpha} = (\mathbf{X}'\mathbf{\Sigma}^{-1}\mathbf{X}')^{-1}\mathbf{X}'\mathbf{\Sigma}^{-1}\mathbf{Y}$. Difficulties in inverting $\mathbf{\Sigma}$ (in more complex settings) lead Henderson to express the estimating equations as the solution to two simultaneous equations known as *Henderson's mixed-model equations* [18]:

$$\mathbf{X}'\mathbf{R}^{-1}\mathbf{X}\hat{\alpha} + \mathbf{X}'\mathbf{R}^{-1}\mathbf{Z}\hat{\mathbf{B}} = \mathbf{X}'\mathbf{R}^{-1}\mathbf{Y}$$
$$\mathbf{Z}'\mathbf{R}^{-1}\mathbf{X}\hat{\alpha} + (\mathbf{Z}'\mathbf{R}^{-1}\mathbf{Z} + \mathbf{G}^{-1})\hat{\mathbf{B}} = \mathbf{Z}'\mathbf{R}^{-1}\mathbf{Y}.$$

These equations are easier to solve than the generalized least-squares equations since the inverse of $\mathbf{R}$ and $\mathbf{G}$ are often easier to compute than the inverse of $\mathbf{\Sigma}$. The mixed-model equations were motivated by computational needs and arose from a matrix identity [17]. The vector $\hat{\mathbf{B}}$ was a byproduct of this identity, and had no clear interpretation.

If $\hat{\mathbf{B}}$ could be interpreted as the predictor of a realized random effect, then the mixed-model equations would be more than a computational device. Henderson [15] provided the motivation for such an interpretation by developing a linear unbiased predictor of $\alpha + B_i$ that had minimum prediction squared error in the context of the joint distribution of $\mathbf{Y}$ and $\mathbf{B}$. Using

$$E\begin{pmatrix}\mathbf{Y}\\\mathbf{B}\end{pmatrix} = \begin{pmatrix}\mathbf{X}\alpha\\\mathbf{0}_n\end{pmatrix} \text{ and}$$

$$\text{var}\begin{pmatrix}\mathbf{Y}\\\mathbf{B}\end{pmatrix} = \begin{pmatrix}\mathbf{\Sigma} & \mathbf{ZG}\\\mathbf{GZ}' & \mathbf{G}\end{pmatrix}. \quad (5)$$

Henderson showed that the BLUP of $\alpha + B_i$ is $\hat{\alpha} + \hat{B}_i$ where $\hat{B}_i$ is the ith element of $\hat{\mathbf{B}} = \mathbf{GZ}'\mathbf{\Sigma}^{-1}(\mathbf{Y} - \mathbf{X}\hat{\alpha})$. Using a variation of Henderson's matrix identity, one may show that this is identical to the predictor obtained by solving the second of Henderson's mixed model equations for $\hat{\mathbf{B}}$. With the additional assumptions that $\mathbf{G} = \sigma^2\mathbf{I}_n$ and $\mathbf{R} = \sigma_e^2\mathbf{I}_{nm}$, where $\text{Var}(B_i) = \sigma^2$ and that $\text{Var}(E_{ij}) = \sigma_e^2$, the expression for $\hat{B}_i$ simplifies to $\hat{B}_i = k(\bar{Y}_i - \bar{\bar{Y}})$, with $k = \frac{\sigma^2}{\sigma^2 + \sigma_e^2/m}$, $\bar{Y}_i = \frac{1}{m}\sum_{j=1}^{m} Y_{ij}$, and $\bar{\bar{Y}} = \frac{1}{n}\sum_{i=1}^{n}\bar{Y}_i$. The coefficient k is always less than one, and "shrinks" the size of the deviation of sample mean for the ith PSU from the sample PSU average. The predictor of $\alpha + \mathrm{B}_i$ is given by

$$\hat{\alpha} + \hat{B}_i = \bar{\bar{Y}} + k(\bar{Y}_i - \bar{\bar{Y}}).$$

Henderson's mixed-model equations arise from specifying the joint distribution of $\mathbf{Y}$ and $\mathbf{B}$. Only first- and second-moment assumptions are needed to develop the predictors. As discussed in Reference 35, Henderson's starting point was not the sample likelihood. However, if normality assumptions are added to Equation (5), then the conditional mean, $E(\alpha + \mathrm{B}_i|\bar{Y}_i = \bar{y}_i) = \alpha + k(\bar{y}_i - \alpha)$. Replacing α by the sample average, $\bar{\bar{y}}$, the predictor of $\alpha + B_i$ conditional on $\mathbf{Y}$ is $\bar{\bar{y}} + k(\bar{y}_i - \bar{\bar{y}})$.

57.9 Bayesian Estimation

The same predictor can be obtained using a hierarchical model with Bayesian estimation [35]. Beginning with Equation (4), we classify the terms in the model as observable (including $\mathbf{Y}, \mathbf{X}$, and $\mathbf{Z}$), and unobservable (including $\alpha, \mathbf{B}$, and $\mathbf{E}$). We consider the unobservable terms as random variables, and hence use a different notation to distinguish α from the corresponding random variable A. The model is given by

$$\mathbf{Y} = \mathbf{X}A + \mathbf{ZB} + \mathbf{E}$$

and a hierarchical interpretation arises from considering the model in stages [10,28]. At the first stage, assume that the clusters corresponding to the selected PSUs are known, so that $A = \alpha_0$ and $\mathbf{B} = b_0$. At this stage, the random variables $\mathbf{E}$ arise from selection of the SSUs. At the second stage, assume A and $\mathbf{B}$ are random variables, and have some joint distribution. The simplest case (which we consider here) assumes that the unobservable terms are independent and that $A \sim N(\alpha, \tau^2), \mathbf{B} \sim N(\mathbf{0}_n, \mathbf{G})$, and $\mathbf{E} \sim N(\mathbf{0}_{nm}, \mathbf{R})$, where $\mathbf{G} = \sigma^2\mathbf{I}_n$ and $\mathbf{R} = \sigma_e^2\mathbf{I}_{nm}$. Finally, we specify the prior distributions for $\alpha, \tau^2, \sigma^2, \sigma_e^2$. We condition on $\mathbf{Y}$ in the joint distribution of the random variables to obtain the posterior distribution. The expected value of the posterior distribution is commonly used to estimate parameters.

We simplify the problem considerably by assuming that σ^2, and σ_e^2 are constant. We represent the lack of prior knowledge of the distribution of A by setting $\tau^2 = \infty$. With these assumptions, the Bayesian estimate of $\mathbf{B}$ is the expected value of the posterior distribution, that is, $\hat{\mathbf{B}} = \mathbf{GZ}'\mathbf{\Sigma}^{-1}(\mathbf{Y} - \mathbf{X}\hat{\alpha})$. This predictor is identical to the predictor defined in Henderson's mixed-model equation.

57.10 Superpopulation Model Predictors

Predictors of the latent values of a realized group have also been developed in a finite population survey sampling context [33]. To begin, suppose we consider a finite population of M students in each of N schools as the realization of a potentially very large group of students, called a *superpopulation*. We do not identify students or schools explicitly in the superpopulation, and refer to them instead as PSUs and SSUs. Nevertheless, there is a correspondence between a random variable, Y_{ij}, and the corresponding realized value for PSU i and SSU j, y_{ij}, which would identify a student in a school. Predictors are developed in the context of a probability model specified for the superpopulation. This general strategy is referred to in survey sampling as model-based inference.

The random variable corresponding to the latent value for PSU i (where $i \leqslant n$) given by $\bar{Y}_i = (1/M)\sum_{j=1}^{M} Y_{ij}$, can be divided into two parts, $\bar{Y}_{i,\mathrm{I}}$ and $\bar{Y}_{i,\mathrm{II}}$ such that $\bar{Y}_i = f\bar{Y}_{i,\mathrm{I}} + (1-f)\bar{Y}_{i,\mathrm{II}}$, where $\bar{Y}_{i,\mathrm{I}} = (1/m)\sum_{j=1}^{m} Y_{ij}$ corresponds to the average response of SSUs that will potentially be observed in the sample, $\bar{Y}_{i,\mathrm{II}} = (1/M-m)\sum_{j=m+1}^{M} Y_{ij}$ is the average response of the remaining random variables, and $f = (m/M)$. Predicting the latent value for a realized school simplifies to predicting an average of random variables not realized in the sample. For schools that are selected in the sample, only response for the students not sampled need be predicted. For schools not included in the sample, a predictor is needed for all students' response.

Scott and Smith assume a nested probability model for the superpopulation, representing the variance between PSUs as σ^2, and the variance between SSUs as σ_e^2. They derive a predictor for the average response of a PSU that is a linear function of the sample, unbiased, and has minimum MSE. For selected PSUs, the predictor simplifies to a weighted sum of the sample average response, and the average predicted response for SSUs not included in the sample, $(m/M)\bar{y}_i + (\frac{M-m}{M})\hat{\bar{Y}}_{i,\mathrm{II}}$, where $\hat{\bar{Y}}_{i,\mathrm{II}} = \bar{\bar{y}} + k(\bar{y}_i - \bar{\bar{y}})$ and $k = \frac{\sigma^2}{\sigma^2 + \sigma_e^2/m}$. This same result was derived by Scott and Smith under a Bayesian framework. For a sample school, the predictor of the average response for students not included in the sample is identical to Henderson's predictor and the predictor resulting from Bayesian estimation. For PSUs not included in the sample, Scott and Smith's predictor reduces to the simple sample mean, $\bar{\bar{y}}$.

There is a substantial conceptual difference between Scott and Smith's predictors and the mixed-model predictors. The difference is due to the direct weighting of the predictors by the proportion of SSUs that need to be predicted for a realized PSU. If this proportion is small, the resulting predictor will be close to the PSU sample mean. On the other hand, if only a small portion of the SSUs are observed in a PSU, Scott and Smith's predictor will be very close to the mixed model predictors. There is a substantial conceptual difference between Scott and Smith's predictors and the mixed-model predictors. The difference is due to the direct weighting of the predictors by the proportion of SSUs that need to be predicted for a realized PSU. If this proportion is small, the resulting predictor will be close to the PSU sample mean. On the other hand, if only a small portion of the SSUs are observed in a PSU, Scott and Smith's predictor will be very close to the mixed model predictors.

57.11 Random Permutation Model Predictors

An approach closely related to Scott and Smith's superpopulation model approach can be developed from a probability model that arises from sampling a finite population. Such an approach is based on the two-stage sampling design and is design-based [38]. Since selection of a two-stage sample can be represented by randomly selecting a two-stage permutation of a population, we refer to models under such an approach as random permutation models. This approach has the advantage of defining random variables directly from sampling a finite population. Predictors are developed that have minimum expected mean square error under repeated sampling in a similar manner as those developed by Scott and Smith. In a situation comparable to that previously described for Scott and Smith, predictors of the realized latent value are nearly identical to those derived by Scott and Smith. The only difference is the use of a slightly different shrinkage constant. The predictor is given by

$$\frac{m}{M}\bar{y}_i + \left(\frac{M-m}{M}\right)\hat{\bar{Y}}^*_{i,\mathrm{II}},$$

where

$$\begin{aligned}
\hat{\bar{Y}}^*_{i,\mathrm{II}} &= \bar{\bar{y}} + k^*(\bar{y}_i - \bar{\bar{y}}), \\
k^* &= \frac{\sigma^{*2}}{\sigma^{*2} + \sigma_e^2/m}, \\
\sigma^{*2} &= \sigma^2 - \frac{\sigma_e^2}{M}.
\end{aligned}$$

57.12 Practice and Extensions

All of the predictors of the latent value of a realized group include shrinkage constants in the expressions for the predictor. An immediate practical problem in evaluating the predictors is estimating this constant. In a balanced setting, simple method of moment estimates of variance parameters can be substituted for variance parameters in the shrinkage constant. Maximum likelihood, or restricted maximum likelihood estimates for the variance are also commonly substituted for variance parameters in the prediction equation. In the context of a Bayesian approach, the resulting estimates are called *empirical Bayes estimates*. Replacing the variance parameters by estimates of the variance will inflate the variance of the predictor using any of the approaches. Several methods have been developed that account for the larger variance [4,13,20,21,46].

In practice, groups in the population are rarely of the same size, or have identical within group variances. The first three approaches can be readily adapted to account for such unbalance when predicting random effects. The principal difference in the predictor is replacement of σ_e^2 by σ_{ie}^2, the SSU variance within SSU i, when evaluating the shrinkage constant. The simplicity in which the methods can account for such complications is an appeal of the approaches. Predictors of realized random effects can also be developed using a random permutation model that represents the two-stage sampling. When SSU sampling fractions are equal, the predictors have a similar form with the shrinkage constant constructed from variance component estimates similar to components used in two-stage cluster sampling variances [2]. Different strategies are required when second-stage sampling fractions are unequal [39].

In many settings, a simple response error model will apply for a student. Such response error can be included in a mixed model such as Equation (4). When multiple measures are made on some students in the sample, the response error can be separated from the SSU variance component. Bayesian methods can be generalized to account for such additional variability by adding another level to the model hierarchy. Superpopulation models [1] and random permutation models [38] can also be extended to account for response error.

Practical applications often involve much more complicated populations and research questions. Additional hierarchy may be present in the population (i.e., school districts). Variables may be available for control corresponding to districts, schools, and students. Measures may be made over time on the same sample of students, or on different samples. Response variables of primary interest may be continuous, categorical, or ordinal. Some general prediction strategies have been proposed and implemented [6,10,48] using mixed models and generalized mixed models [25], often following a Bayesian paradigm. Naturally, the hypotheses and approaches in such settings are more complex. There is active research in these areas, which should lead to clearer guidance in the future.

57.13 Discussion and Conclusions

The latent value of a group is a natural parameter of interest in a group-randomized trial. While such a parameter may be readily understood, development of an inferential framework for predicting such a parameter is not easy. Many workers struggled with ideas underlying interpretation of random effects in the midtwentieth century. Predictors have emerged largely on the basis of computing strategies and Bayesian models since the 1980s. Such strategies have the appeal of providing answers to questions that have long puzzled statisticians. Computing software (based on mixed model and Bayesian approaches) is widely available and flexible, allowing multilevel models to be fitted with covariates at different levels for mixed models. Although flexible software is not yet available for superpopulation model or random permutation model approaches, there is some evidence that when sampling fractions are small (< 0.5), predictors and their MSE are very similar [38].

Whether the different approaches predict the latent value of a realized group is a basic question that can still be asked. All of the predictors have the property of shrinking the realized group mean toward the overall sample mean. While the predictors are unbiased and have minimum expected MSE, these properties hold over all possible samples, not conditionally on a realized sample. This use of the term "unbiased" differs from the popular understanding. For example, for a realized group, an unbiased estimate of the group mean is the sample group mean, while the BLUP is a biased estimate of the realized group mean. The rationale for preferring the biased estimate is that in an average sense (over all possible random effects), the MSE is smaller. Since this property refers to an average over all possible random effects, it does not imply smaller MSE for the realized group [44]. In an effort to mitigate this effect, Raudenbush and Bryk [28] suggest including covariates that model realized group parameters to reduce the potential biasing effect. Alternative strategies, such as conditional modeling frameworks have been proposed [5,40,45], but increase in complexity with the complexity of the problem. While there is increasing popularity of models that result in BLUP for realized latent groups, the basic questions that plagued researchers about interpretation of the predictors in the late twentieth century remain for the future.

References

1. Bolfarine, H. and Zacks, S. (1992). *Prediction Theory for Finite Populations*. Springer-Verlag, New York.

2. Cochran, W. (1977). *Survey Sampling*, Wiley, New York.

3. Cox, D. R. (2000). The five faces of Bayesian statistics. *Calcutta Statist. Assoc. Bull.*, **50**, 199–200.

4. Das, K., Jiang, J., and Rao, J. N. K. (2001). *Mean Squared Error of Empirical Predictor*. Technical Report, School of Mathematics and Statistics, Carleton University, Ottawa, Canada.

5. Dawid, A. P. (1979). Conditional independence in statistical theory. *J. Roy. Statist. Soc.*, **41**, 1–31.

6. de Leeuw, J. and Kreft, I. (2001). Software for multilevel analysis, In *Multilevel*

Modelling of Health Statistics, A. Leyland and H. Goldstein, eds. Wiley, Chichester.

7. Donner, A. and Klar, N. (2000). *Design and Analysis of Cluster Randomized Trials in Health Research*. Arnold, London.

8. Eisenhart, C. (1947). The assumptions underlying the analysis of variance. *Biometrics*, **3**, 1–21.

9. Goldberger, A. S. (1962). Best linear unbiased prediction in the generalized linear regression model. *Am. Statist. Assoc. J.*, **57**, 369–375.

10. Goldstein, H. (2003). *Multilevel Statistical Modeling*, 3rd ed. Kendall's Library of Statistics 3, Arnold, London.

11. Graybill, F. A. (1983). *Matrices with Applications in Statistics*. Wadsworth International, Belmont, CA.

12. Harville, D. A. (1976). Extension of the Gauss-Markov theorem to include the estimation of random effects. *Ann. Statist.*, **4**, 384–395.

13. Harville, D. A. and Jeske, D. R. (1992). Mean squared error of estimation or prediction under general linear model. *J. Am. Statist. Assoc.*, **87**, 724–731.

14. Heagerty, P. J. and Zeger, S. L. (2000). Marginalized multilevel models and likelihood inference. *Statist. Sci.*, **15**, 1–26.

15. Henderson, C. R. (1963). Selection index and expected genetic advance. In *Statistical Genetics and Plant Breeding*. National Academy of Sciences—National Research Council.

16. Henderson, C. R. (1984). *Applications of Linear Models in Animal Breeding*. University of Guelph, Guelph, Canada.

17. Henderson, H. V. and Searle, S. R. (1981). On deriving the inverse of a sum of matrices. *SIAM Rev.*, **23**, 53–60.

18. Henderson, C. R., Kempthorne, O., Searle, S. R., and von Krosigk, C. M. (1959). The estimation of environmental and genetic trends from records subject to culling. *Biometrics*, 192–218.

19. Hinkelmann, K. and Kempthorne, O. (1994). *Design and Analysis of Experiments, Vol. 1, Introduction to Experimental Design*. Wiley, New York.

20. Kackar, R. N. and Harville, D. A. (1984). Approximations for standard errors of estimators of fixed and random effects in mixed linear models. *J. Am. Statist. Assoc.*, **79**, 853–862.

21. Kass, R. E. and Steffey, D. (1989). Approximate Bayesian inference in conditionally independent hierarchical models (Parametric empirical Bayes models). *J. Am. Statist. Assoc.*, **84**, 717–726.

22. Kempthorne, O. (1975). Fixed and mixed models in the analysis of variance. *Biometrics*, **31**, 437–486.

23. Kirk, R. E. (1995). Experimental design. In *Procedures for the Behavioral Sciences*, 3rd ed. Brooks/Cole, New York.

24. Littell, R. C., Milliken, G. A., Stroup, W. W., and Wolfinger, R. D. (1996). *SAS System for Mixed Models*. SAS Institute, Cary, NC.

25. McCulloch, C. E. and Searle, S. R. (2001). *Generalized, Linear, and Mixed Models*. Wiley, New York.

26. Murray, D. M. (1998). *Design and Analysis of Group-randomized Trials*. Oxford University Press, New York.

27. Rabinowitz, D. and Shea, S. (1997). Random effects analysis of children's blood pressure data. *Statist. Sci.*, **12**, 185–194.

28. Raudenbush, S. R. and Bryk, A. S. (2002). *Hierarchical Linear Models: Applications and Data Analysis Methods*, 2nd ed. Sage Publications, London.

29. Robinson, G. K. (1991). That BLUP is a good thing: The estimation of random effects. *Statist. Sci.*, **6**, 15–51.

30. Samuels, M. L., Casella, G., and McCabe, G. P. (1991). Interpreting blocks and random factors. *J. Am. Statist. Assoc.*, **86**, 798–821.

31. Scheffe, H. (1956). A "mixed model" for the analysis of variance. *Ann. Math. Statist.*, **27**, 23–36.

32. Scheffé, H. (1959). *Analysis of Variance*. Wiley, New York.

33. Scott, A. and Smith, T. M. F. (1969). Estimation in multi-stage surveys. *J. Am. Statist. Assoc.*, **64**, 830–840.

34. Searle, S. (1971). *Linear Models*. Wiley, New York.

35. Searle, S. R., Casella, G., and McCulloch, C. E. (1992). Prediction of random variables. In *Variance Components*, Chap. 7. Wiley, New York.

36. Singer, J. M. and Andrade, D. F. (2000). Analysis of longitudinal data. In *Handbook of Statistics*, E. P. K. Sen and C. R. Rao, eds. Elsevier Science BV, New York.

37. Stanek, E. J. I. and Singer, J. M. (2003). *Estimating Cluster Means in Finite Population Two Stage Clustered Designs*. International Biometric Society, Eastern North American Region.

38. Stanek, E. J. I. and Singer, J. M. (2003). *Predicting Random Effects from Finite Population Clustered Samples with Response Error*. Department of Biostatistics/Epidemiology, University of Massachusetts, Amherst, MA.

39. Stanek, E. J. I. and Singer, J. M. (2003). *Predicting Realized Cluster Parameters from Two Stage Samples of an Unequal Size Clustered Population*. Department of Biostatistics/Epidemiology, University of Massachusetts, Amherst, M. A.

40. Stanek, E. J. I. and O'Hearn, J. R. (1998). Estimating realized random effects. *Commun. Statist.- Theory Methods*, **27**, 1021–1048.

41. Stanek, E. J. I., Well, A., and Ockene, I. (1999). Why not routinely use best linear unbiased predictors (BLUPs) as estimates of cholesterol, per cent fat from Kcal and physical activity?. *Statist. Med.*, **18**, 2943–2959.

42. Ten Have, T. R. and Localio, A. R. (1999). Empirical Bayes estimation of random effects parameters in mixed effects logistic regression models. *Biometrics*, **55**, 1022–1029.

43. Ten Have, T. R., Landis, J. R., and Weaver, S. L. (1995). Association models for periodontal disease progression: A comparison of methods for clustered binary data. *Statist. Med.*, **14**, 413–429.

44. Verbeke, G. and Molenberghs, G. (2000). *Linear Mixed Models for Longitudinal Data*. Springer, New York.

45. Verbeke, G., Spiessens, B., and Lesaffre, E. (2001). Conditional linear mixed models. *Am. Stat.*, **55**, 25–34.

46. Wang, J. and Fuller, W. A. (2003). The mean squared error of small area predictors constructed with estimated area variances. *J. Am. Statist. Assoc.*, **98**, 716–723.

47. Wilk, M. B. and Kempthorne, O. (1955). Fixed, mixed, and random models. *Am. Statist. Assoc. J.*, **50**, 1144–1167.

48. Zhou, X. -H., Perkins, A. J., and Hui, S. L. (1999). Comparisons of software packages for generalized linear multilevel models. *Am. Statist.*, **53**, 282–290.

58 Probabilistic and Statistical Models for Conception

Bruno Scarpa

58.1 Introduction

The interest in statistical analysis of reproduction dates back to Galton [35]. He studied aristocratic English families, trying to understand the reasons for the subfertility they were suffering.

However, the probabilistic nature of conception was first introduced and studied by Corrado Gini in 1923, when, at the Ordinary Assembly of the Reale Istituto Veneto di Scienze Lettere ed Arti, in Venice (Italy), he gave a communication on his first researches about fecundability in women [37]. His first definition states: "I call *fecundability of the woman* the probability that the ovule of the woman, once he became mature, is fertilized, in a married couple." In a second talk in 1924 he was more precise stating: "I call fecundability of the woman the probability that a married woman conceives during a menstrual cycle, in absence of any contraception practice and of any abstention from acts of sexual intercourse in order to limit procreation."

Following Gini's intuition, several scientists, from different fields, proposed and discussed statistical models to analyze fecundability and find the determinants of probabilities of conceptions. Many different interests from different scientific disciplines combine together in analyzing the dynamics of human reproduction. Demographers proposed different models estimating the distribution of fecundability in a population, supplying tools to compare it between populations. Epidemiologists are more interested in models to identify preventable causes of reproductive dysfunction, such as environmental factors or toxic exposures, while reproductive physiologists concentrate on understanding causes for important variation in human fertility and in characterizing factors that can modify it through intended interventions, such as contraceptive methods or clinical treatment for infertility. Statistical modeling for fecundability supports all these substantive researches by attempting to link the individual and aggregate levels of analysis, and helping to understand the influence of both behavioral and biological aspects.

The need for nontrivial statistical models has always been clear to the researcher interested in modeling the different aspects of human fertility. Gini itself [38] observed that "all married women capable of becoming pregnant does not have the same fecundability. This condition that cannot be disregarded does not interfere with reaching a measurement of fecundability; however, the way to take is a bit longer." The attempt to model heterogeneity, across women and within different cycles of the same woman, has been one of the main goals faced by statisticians. However, a considerable degree of residual heterogeneity remains unexplained, and modelization of fecundability data is still a challenging task.

In the following we will give a sketch of the main paths that have been followed in proposing statistical and probabilistic model for studying conception and human fertility. We will start by presenting the models for aggregated populations, mainly developed by demographers who were interested in biological and behavioral sources of variability in fertility and in comparisons of fertility levels among populations. Most of these models require only aggregated data and do not need for individual information. Often more detailed data are available, sometimes only for subsamples of populations or for selected groups of people. In order to take advantage of the more specific information available, models on waiting time to conception have been developed. Thus, we discuss how hazard models have been used largely for studying changing fecundability with concurrent variables, taking into account heterogeneity, in particular analyzing the *time to pregnancy*, that is, the time required for noncontracepting, sexually active couples to achieve conception, and supplying also tools to identify the effects of infertility treatments or potential toxic exposure. Finally we will discuss models that relate the cycle-specific coital histories of couples to the probability of conception. These models need to be based on very specific and detailed data about the life of the involved couples, including daily information on cycle characteristics such as menses and ovulation and on behavioral data such as unprotected intercourse incidents.

A short discussion of some applications and usage of statistical models for conception is also provided. The first section is a short introduction to some biological background in order to better understand the need for statistical models.

58.2 Biological Background

A mature, fecund woman, of reproductive age, releases a single ovum from the ovary in each menstrual cycle, in a process called *ovulation*. Fertilization of an ovum, its union with a sperm, can occur only within a few hours immediately after its release, so there must be intercourse near the time of ovulation. If the ovum is not fertilized, a menstrual discharge of blood and uterine tissue follows about 2 weeks later.

The duration of the cycle and its preovulatory phase varies from one cycle to another and from onewoman to another. It also varies with other characteristics of the woman, such as age or environmental exposures.

In addition to the male sperm being capable of fertilizing the ovum, several biological factors in the female must also be favorable. Sperm may survive in the female reproductive tract for few days; therefore fertilization may take place only if coitus occurs during a short interval of about six days around the ovulation [77,19]. The ovum must be viable, and about a week after fertilization, it must be transported to the uterus, where it becomes implanted in the uterine wall. The uterine endometrium must be adequately prepared hormonally to support implantation, and the implanted conceptus must be able to survive; in fact, in some cases, a fertilized ovum is expelled rapidly without any interruption of menses. Schwartz et al. [63] call "cycle viability" the fact that all these factors are favorable to gestation.

Once a woman conceives, ovulation is suppressed throughout her pregnancy. If pregnancy is followed by lactation, ovulation is usually suppressed for at least part of the time that the woman is lactating, otherwise ovulation resumes about 6 weeks after the pregnancy is ended. It is possible that age, order of pregnancies, or other internal or external factors are associated with variation of the duration of nonovulating period after pregnancy.

Statistical analyses of this process have been developed by considering mainly the following relevant variables: (1) the duration of the process, the time, measured from a starting point such as marriage or from the onset of sexual relations, during which the reproductive process can continue; (2) the fecundability, the probability of conception in a menstrual cycle; (3) the time that an ovulating woman, living in a sexual union, waits before she conceives (this is a function of the woman fecundability); and (4) the various outcomes of pregnancy and the probability of each outcome. In addition to live births, traditionally pregnancies that end in fetal death or stillbirths or early pregnancy losses are of interest, but among outcomes it is possible to include also the sex of the baby and number of twins. A fifth variable is the duration of the nonfertile intervals associated with conceptions.

Statistical models relate these outcome variables with endogenous and exogenous factors that may affect conception. In particular, they may be grouped as follows:

1. *Biological factors*. Heterogeneity may be a physiological disorder or illness affecting fertility.

2. *Environmental factors*. Heterogeneity may be observed as a consequence of environmental factors or toxic exposure.

3. *Behavioral factors*. Frequency and time of intercourse on particular days relative to ovulation have a major impact on fecundability.

4. *Purely random* heterogeneity. Part of the differences observed between cycles and women are attributable only to random variability.

Note that some of these factors, such as age or exposure to a potential toxicant, can be directly observed, but others are unobservable and manifest themselves in observed variability in the outcome variables between cycles with the same intercourse pattern and dependence among outcomes within multiple cycles from the same couple.

Often, besides the principal effects, potential interactions between biological, environmental, and behavioral factors are considered in models; for instance, the length of the woman's fertility window and the length of sperm life, which are biological factors, modify the effect of intercourse pattern on the probability of conception.

58.3 Population Models

58.3.1 Fertility

Demographers have long been interested in measuring and modeling fertility in populations and in studying biological sources of its variability. Most of their analyses start from definitions of fertility indices, such as *age-specific fertility rate*, the ratio between the observed number of births to women of a specific age and the observed number of woman-years at the same age. These quantities are a powerful tool in analyzing the effect of age in fertility and in comparisons between populations.

An important measure of total reproductive output of a population can be computed as the sum of all the age-specific fertility rates multiplied by the length of the age interval. This quantity is called *total fertility rate* (TFR) and is an estimate of the expected number of offspring ever born to a random selected woman who survives to the end of the reproductive span, given that she has, at each age, the age-specific fertility rates observed at that time. The TFR has several interesting properties that make it a reference index in demography. Besides its simplicity, the TFR is a pure fertility measure; it is not influenced by other variables such as age or sex composition of the population or by mortality. Moreover, it is an aggregate measure that can be interpreted in terms of an individual woman's expected lifetime reproductive experience; this characteristic is especially important in modeling the relationship between physiology and behavior at the individual level and population dynamics at the aggregate level. Many other indices have been defined and are used to study nonparametrically the characteristics of fertility and its relations with relevant factors (see, e.g., Campbell [9]).

In terms of practical usage, when these indices are computed from finite samples, they represent estimates with some degree of sampling error. Except under very restrictive conditions (see, Wood [81, p. 25]), we don't have explicit results about the sampling properties of fertility indices; nevertheless, it is possible to obtain the standard errors of the estimates and confidence intervals by using a resampling approach such as bootstrap or jackknife (e.g., see Efron and Tibshirani [34]; Davison and Hinkley [16]).

58.3.2 Heterogeneity Fertility

Even if it is often calculated for subpopulations, TFR as a measure of individual fertility does not account for the differences among women within each population. Moreover, a certain amount of variability of individual fertility can be attributable to simply random variability between women (or couples). Different probabilistic models have been used in order to describe this irreducible source of variability.

By assuming the simplified hypotheses (1) that all women survive to menopause, (2) that the probability of reproducing at each age is the same for all of them, and (3) that successive births of each woman are independently distributed through time, the total number of live births n for a woman is distributed as a Poisson random variable, with probability function $p(n) = (\lambda^n/n!)e^{-\lambda}$, where λ is the mean number (and the variance) of live births in the population.

Such a model is clearly oversimplified, implying, for instance, that there is no systematic difference in fertility between women in the population. We can relax this hypothesis by supposing that the mean of livebirths λ vary among subpopulations; for instance, if $\lambda \sim \mathcal{G}(\theta, \gamma)$, where $\mathcal{G}(\theta, \gamma)$ denotes a gamma distribution with shape parameter $\theta > 0$ and scale $\gamma > 0$, the distribution of n is a negative binomial with probability function

$$p(n) = \frac{(\theta + n - 1)!}{(\theta - 1)!n!} \left(\frac{\gamma}{1+\gamma}\right)^n \times \left(1 - \frac{\gamma}{1+\gamma}\right)^\theta$$

with mean $E(n) = \theta\gamma$ and variance $\text{Var}(n) = \theta\gamma(1 + \gamma)$. Since γ is strictly positive, the variance of this distribution is greater than the mean, in contrast with the Poisson case.

By fitting these very simple distribution models to data on human fertility, scientists have found that they are appropriate for a large number of populations (e.g., Brass [6]).

58.3.3 Effects of Fertility Control and Other Variables

Among all causes of variability in level of fertility within and among human populations, the age pattern plays a consistent role. In fact, if observed in natural populations, the pattern of the age specific fertility rate from real population shows a monotone decline from a peak in the early twenties, going to near zero by the late forties following a convex shape [81]. However, this curve results very different if couples in a population deliberately control their fertility. In fact, in controlled fertility populations a more rapid drop of fertility rate with age and a concave shape of the curve are observed. These differences seem to reflect the fact that most couples using fertility control, once they attain the desired family size, often early in marriage, terminate their reproduction through contraception.

Louis Henry [43] suggested a standardized age pattern of marital fertility in populations in which there is little or no voluntary control of births; he called "natural fertility" the fertility observed in absence of control. In 1974, Coale and Trussel [10] proposed a semiparametric model where a single parameter describes the effect of deliberate fertility control. They observed that marital fertility either follows natural fertility (if deliberate birth control is not practiced), or departs from natural fertility in a way that increases with age according to a typical pattern. In a population in which fertility is voluntarily controlled, the marital fertility at each age, $r(a)$, is related to the natural fertility, $n(a)$, following the relation

$$r(a) = n(a)Me^{mv(a)},$$

where the natural age specific fertility rate $n(a)$ multiplied by M, a scale parameter setting the level of natural fertility at some arbitrarily chosen age, is discounted by an exponential term containing a scale parameter m determining the degree of fertility control and the function $v(a)$, which express the standard effect of control at each age a, that is, the tendency for older women in a population practicing contraception to effect particularly large reduction in fertility below the natural level.

Coale and Trussel [10] treated $n(a)$ and $v(a)$ as given, and in that case the estimate of M and m follow directly by applying usual statistical techniques, for example taking logs of the equation of the model we obtain a simple linear regression equation

$$\log\left[\frac{r(a)}{n(a)}\right] = \log M + mv(a)$$

that can be fit by standard least squares. Broström [8] has derived maximum likelihood estimators for M and m by assuming that marital fertility varies, at each age, according to a Poisson random variable. The Coale–Trussel model has been discussed and many authors do not consider it to be a correct model for the complex measure of the effect of fertility control on the level of fertility (e.g., Wilson et al. [80]).

Also, from a purely statistical point of view, there is positive confounding of the estimates of M and m, at least under the maximum likelihood approach, since the covariance of the estimates is positive and is quite large with respect to the sampling variances of the estimates [8]. Therefore, the interpretation of the estimates should be taken with caution. Another drawback of the model is the choice of the Poisson distribution for the maximum-likelihood approach. In fact, as we have seen, it is hard to believe that a heterogeneous population may satisfy this assumption and the hypothesis that characterize it. Nevertheless, the Poisson is the only distribution used for birth process with only one parameter; all other theoretical distributions involve fitting at least two parameters, which requires more data than just age-specific marital fertility rates. Note that in this context models are developed to fit on aggregated data; this is why they need to be parsimonious. Similar models could be also fitted to more detailed data, now largely available at least for specific populations, considering more accurate models, with more appropriate distribution models, not only for marital fertility. Little has been done yet in this direction, and further analyses are needed. In conclusion, for all the aforementioned reasons, demographers suggest to use the Coale–Trussel model only for exploratory analysis.

Following the Henry intuition, Page [52] observed that control of fertility depends primarily on previous fertility experience, and in a population in which childbearing occurs predominantly within marriage, duration of marriage may be considered as a proxy variable for fertility control. Thus he proposed a model similar to that of Coale and Trussel [10] by including this variable among the inputs. Let us call $L(t)$ the general level of marital fertility at time t, averaged over all ages and marriage durations;, $v(a,t)$, a factor characteristic of age a at time t representing the age pattern shared by all marriage cohorts at time t; and $u(d,t)$, the duration pattern shared by all age groups, a factor characteristic of duration d at time t, the Page model for marital fertility rate for women at age a, and marriage duration d at time t is $r(a,d,t) = L(t)v(a,t)u(d,t)$.

By analyzing real data and considering the difficulties in estimating such a model for all t, Page proposed a modified model in which the effect of age does not depend on t and the effect on marriage duration has a simple parametric form $r(a,d,t) = L(t)v(a)e^{bd}$, where b is a parameter characterizing the latent level of fertility control for couples married at time $t-d$. Estimates and inference for the parameters can be performed as for the Coale–Trussel model, by taking the logarithms, or hypothesizing a Poisson distribution for the marital fertility rate, by maximum likelihood.

Note that for populations in which marriage does not precede childbearing, and if other proxies of fertility control are available, it is possible to fit a similar model. For example, sometimes data are available on a number of previous pregnancies, or time from cohabitation.

Other models have been developed for relating fertility with age and other variables. For example, following a more empirical approach, Brass [6] proposed to fit a third degree polynomial or a growth curve, such as the Gompertz curve, for the cumulative fertility rates [71].

58.3.4 Proximate Determinants of Fertility

A different approach to modeling aggregate fertility phenomenon has been developed while trying to identify the relative effect on the fertility of some intermediate variable which is considered determinant. Davis and Blake [15] in a pioneering paper and later Bongaarts [4,5] and others (viz., Hobcraft and Little [44]; Palloni [53]) proposed models that account for variability in fertility evaluating all the single factors that potentially affect reproduction, including social, economic, and cultural conditions besides behavioral and physiological characteristics.

This approach identifies a short number of factors that directly affect fertility while all other characteristics and conditions have only an indirect effect through the "direct determinants"; and they propose a model for analyzing the relationships between intermediate fertility variables and the level of fertility.

Following Bongaarts, the eight direct de-

Table 1: Bongaarts proximate determinants of fertility

I. Exposure factors

1. Proportion married: this variable measures the proportion of women of reproductive age that engages in sexual intercourse regularly. For convenience, the term "marriage" is used by Bongaarts.

II. Deliberate marital fertility control factors

2. Contraception: any deliberate parity-dependent practice, including abstention and sterilization, undertaken to reduce the risk of conception. Thus defined the absence of contraception and induced abortion implies the existence of natural fertility.
3. Induced abortion: any practice that deliberately interrupts the normal course of gestation.

III. Natural marital fertility factors

4. Lactational infecundity: the duration of the period of infecundability after a pregnancy depends on the duration and intensity of lactation.
5. Frequency of intercourse: this variable measures normal variations in the rate of intercourse, including temporary separation or illness, and excluding voluntary abstinence to avoid pregnancy.
6. Sterility: a couple may become sterile before the woman reaches menopause for reasons other than contraceptive sterilization.
7. Spontaneous intrauterine mortality: a proportion of all conceptions does not result in a live birth.
8. Duration of the fertile period: a woman is able to conceive for only a short period in the middle of the menstrual cycle when ovulation takes place.

terminants are grouped into three classes (see Table 1).

Other authors [15,81] propose a different list of direct determinants, each of them specifying a particular interest or approach to the physiological or cultural views of fertility.

To obtain the total fertility rate (TFR), intermediate variables are considered sequentially in influencing the total fecundity rate (TF), the theoretical maximum fertility that could be achieved in the absence of all limiting factors:

$$\text{TFR} = C_m\, C_c\, C_a\, C_i\, \text{TF},$$

where C_m is the proportion of married couples and may be estimated as the weighted age specific marital fertility rate $C_m = (\sum_a m(a)g(a))/\sum_a g(a)$, where $g(a)$ is the proportion of women already married at age a and $m(a)$ is the age specific marital fertility, the ratio between the observed number of births to women of a specific age and the observed number of married woman-years at the same age. C_c is a coefficient of deliberate fertility control among married couples, and is obtained as function of the average proportion of married women currently using contraception and of the average contraceptive effectiveness (the proportion by which the monthly probability of conception is reduced as the result of contraceptive practice). C_a measures the influence of the induced abortion on TFR and is obtained as the ratio of the observed TFR to the estimated TFR without induced abortion $C_a = \text{TFR}/(\text{TFR} + A)$ where A is the average number of births prevented per woman. Finally, C_i is a measure of lactational infecundity and depends on i the average duration of infecundity from birth to the first postpartum ovulation. A good approximation is $C_i = 20/(18.5 + i)$, where 20 months is an estimate of the average birth interval without breastfeeding and $18.5 + i$ months the same quantity with breastfeeding. According to Bongaart's results, very little variability in total fecundity remains to be explained by the other intermediate variables, and since the interest is in interpopulation variation in fertility, he concluded that they do not have an appreciable effect on it.

The Bongaarts model has been applied by many analysts (e.g., Kalule-Sabiti [44], Singh et al. [67], Ross et al. [57], and many others) because of its simplicity, in particular in applications to aggregate data. The main results of its application is related to the estimate of the relations and of the relative importance of the intermediate variables in affecting the level of fertility. Nevertheless, it has several limitations. It is not a probabilistic model, it involves a large number of simplifying assumptions and does not allow for estimating the effect of secondary determinants of fertility; there is no direct connection between aggregated and individual data, making it difficult to measure the differences within a population; and finally, there are difficulties in estimating interactions between the determinants and its effect on fertility.

The diffusion of individual data and the availability of the results of high-quality studies on fertility and reproductive health with detailed observations, have moved most of the more recent research in more sophisticated models that can provide more precise estimates of fertility parameters and can permit comparative studies of infertility treatments or potentially toxic exposures. In the following sections we will discuss some of these models.

58.4 Fecundability and Time to Pregnancy

58.4.1 Homogeneous Population

Gini [37] proposed a simple model relating fecundability with the distribution of conception waits in the homogeneous case, in which fecundability is constant among couples and over time. Under this simplified hypothesis, by denoting as p the fecundability, the probability that a randomly selected couple will conceive for the first time in the tth cycle of exposure to intercourse, while in a fecund state, is a geometric random variable with density function

$$f(t) = (1-p)^{t-1}p,$$

where time is considered discrete, which is a reasonable assumption by considering the cycle as time units; nevertheless, many models have been proposed considering continuous time (see, e.g., Sheps and Menken [66]). The expectation of this random variable is $E(t) = 1/p$ and the variance $\text{Var}(t) = /(1-p)/p^2$. Note that, even with the hypothesis of homogeneity and despite the fact that most couples conceive within a few months of initiating exposure, if we suppose for example a realistic $p = 0.2$, this model allows for a fairly large fraction of couples (about 5%) who must wait more than one year before conceiving; also, it is clear that the upper tail of the distribution reaches zero only asymptotically (there is always some small chance that a couple will never conceive). Therefore, even in the homogeneous case, a considerable variability in conception waits is expected.

However, the relationship between fecundability and the distribution of conception waiting times depends on the characteristics of the couples involved. To consider the effect of random variation in fecundability between cycles of the same couple, we can assume that the fecundability p is a random variable with density $g(p)$ by supposing that $g(p)$ is the same across couples. Now, with each month, the expected value of p for any randomly selected couple is $E(p) = \int_0^1 pg(p)dp$ and $f(t) = [E(1-p)]^{t-1} E(p)$, which is still a geometric random variable; thus, if we suppose a random selection for p, for a single woman, requiring that fecundability values be independent from its values in the previous months, the time to conception has the same distribution of time to conception for a woman with constant fecundability equal to the mean of $g(p)$.

Considering the problem of estimate p, Sheps [64] noted that if K is the maximum number of cycles of follow-up for the data available, then the likelihood of the data is completely specified by the first K moments of p. The maximum-likelihood estimates for these moments are simply functions of the number of couples conceiving at each cycle.

Weinberg and Gladen [74] propose an alternate way to write the likelihood in this case, by decomposing each term into a series of cycle-specific conditional rate. They call

$$\mu_j = E(p|t > j-1) = 1 - \frac{E[(1-p)^j]}{E[(1-p)^{j-1}]}$$

the conditional mean of the fecundability p after $j-1$ not conceiving cycles, and they show that the contribution to the likelihood of each couple can be written as $f(\mu_j; t) = \mu_t \prod_{j=1}^{t-1}(1-\mu_j)$, and can be seen as the result of a sequence of Bernoulli trials with parameters $\mu_j, j = 1, 2, \ldots$. The expression of likelihood as a function of the conditional mean has some advantages in defining models depending on covariates, as will be clear later.

58.4.2 Heterogeneous Population

Sheps [64] assumes that although the fecundability of each couple is constant over time, or varies only randomly, the fecundability of different couples in the population is not identical, but varies systematically between couples according to the probability density function $h(p)$ with mean μ_p and variance σ_p^2. In this case she assumes that, for each of n observed couples, a specific value $p_j, i = 1 \ldots, n$ from this distribution is characteristic of the jth couple from the beginning of observations until she conceives (while in the previous case p was constant among couples). In general Sheps [64] showed that the distribution of conception waits under systematic among-couple heterogeneity follows the *compound geometric distribution*, with density function

$$f(t) = \int_0^1 p(1-p)^{t-1} h(p) dp,$$

mean $E(t) = 1/H_p$ and variance $\text{Var}(t) = E(t)[1 - E(t)] + 2\text{Var}(1/p)$ where H_p is the harmonic mean of p values among couples $H_p = n[\sum_{j=1}^n (1/p_j)]^{-1}$.

Note that, with respect with the homogeneous case, this model has a variance greater by $2\text{Var}(1/p)$, which measures how much variation in average waiting times to conception exists between couples. Also in general H_p, the harmonic mean, is less or equal to $E(p)$, and the two have the same value only when $\sigma_p^2 = 0$. Thus, in presence of variability between couples ($\sigma_p^2 > 0$), the mean conception waiting time is longer than what it would otherwise be expected. Thus, heterogeneity has the effect of inflating both mean and variance of the waiting times to conception.

By comparing the probability functions of waiting times in homogeneous and heterogeneous cases, it can be shown (e.g., see Wood [81]) that, whatever distribution is supposed for the random effect of fecundability among couples, both waiting time distributions have a single mode at the first month, but after that the curve for the heterogeneous case drops off more rapidly, the difference between the two curves depending on the variance of p in the heterogeneous case. Therefore the probability to observe short conception waits is higher in the homogeneous case. However, at some point, the two probability curves intersect, and the probability of observing long conception waiting times is greater in the heterogeneous case. Intuitively this result can be explained as a selection bias. In fact couples with higher fecundability are expected to conceive sooner than couples with low fecundability, but while time passes, couples with high values of p conceives and are no longer at risk of a first conception, while the remaining couples are increasingly selected for low fecundability; thus the distribution of p among couples, who have not yet conceived at a given time, is different from its distribution in the original population.

In order to fit these models to suitable datasets a specific form of $h(p)$, the density function of fecundability, needs to be postulated. Different choices have been proposed, such as gamma or beta [43] or tree points discrete [69] or triangular. The beta distribution seems the most widely used [81]: choosing $p \sim \mathcal{B}(a,b)$, where $\mathcal{B}(a,b)$ denotes the beta distribution with shape parameters a and b so that the mean is $a/(a+b)$ and the variance is $ab/((a+b)^2(a+b+1))$, the distribution for the waiting times to conception is easily obtained as

$$f(t) = \frac{1}{B(a,b)} \int_0^1 p^a (1-p)^{b+t-2} dp$$
$$= \begin{cases} \frac{a}{a+b} & y=1 \\ \frac{ab(b+1)\cdots(b+t-2)}{(a+b)(a+b+1)\cdots(a+b+t-1)} & y \geq 2, \end{cases}$$

where $B(\alpha,\beta) = \int_0^1 t^{\alpha-1}(1-t)^{\beta-1} dt$ is the beta integral. Many drawbacks have been shown in estimating fecundability by using such models (see, e.g., Sheps and Menken [66]; Heckman and Singer [40]; Wood [81]), in particular related to the difficulty to distinguish between situations in which the fecundability is constant for each woman and when it decreases, at least for an appreciable proportion of women. Heckman and Walker [41] and Heckman et al. [42] presented an extended review of these models concerning mixtures of geometric distributions, by proposing the imposition of impose additional criteria in order to solve this nonidentification problem.

58.4.3 Time-to-Pregnancy Models with Covariates

Weinberg and Gladen [74] extended the beta binomial model to include couple specific covariates. By considering the likelihood as a sequence of Bernoulli trials, they use a generalized linear model framework [50], with an inverse link function in time and binomial distribution; by identifying subpopulations by a common level l of the covariate considered, they propose, for each couple i, that the mean fecundability of each subpopulation l follows a beta distribution $\mu_{li} \sim \mathcal{B}(a_l, b_l)$. They also considers the case in which a subpopulation has zero probability of conception (a sterile subpopulation).

The beta geometric distribution seems to be very flexible and describes situations where the risk ratio crosses in time. For example, Crouchley and Dassios [11] applied such a model showing that the previous contraceptive pill increases fecundability for less fertile women but decreases it for the more fertile ones. However, this model presents some drawbacks, related to the characteristics of the beta distribution. In fact, for this random variable, an increase in variance modifies the shape of the density, and overdispersion creates artificially unlikely U-shaped fecundability distributions among couples [22], which hardly seems to make biological sense.

A semiparametric model of the marginal hazards, based on a discrete-time version of the Cox model, has been proposed by Weinberg et al. [73], but in this model, variability among couples other than the one attributable to covariates is not allowed. Other extensions of this model have been proposed by Ridout and Morgan [56] that robustify the model against digit preference among women (typically observed in retrospective studies) and by Crouchley and Dassios [11] that developed a quantile ratio that has an individual-level interpretation. Boldsen and Schaumburg [3] proposed a similar model treating time to pregnancy as a continuous variable.

All these models do not allow for time-dependent covariates that are often relevant in measuring fecundability and related quantities; age or accruing environmental exposure are typical examples. Some authors proposed regression models where both fixed and random effects are interpreted at the couples level (see Scheike and Jensen [62]; Dunson and Zhou [22]; Ecochard and Clayton [31]; Dunson and Neelon [27]). They propose discrete-time survival models with random effects, by considering the sequence of Bernoulli trials and covariates as time-dependent. If random-effect and time-dependent covariates act additively on the same scale, the model for the conditional mean of the fecundability of couple i at time j is $\mu_{ij} = h\left(\mathbf{u}_{ij}^T \alpha + a_i\right)$, where $h(\cdot)$ is a defined link function, $\mathbf{u}_{ij}$ is the vector of the covariates for couple i, possibly time-dependent, α is the vector of parameter describing the fixed effects, and a_i is the random effect associated with couple i.

Ecochard and Clayton [31] and Scheike and Jensen [62] use as link $h(\cdot)$ the complementary log–log function because of its specific place in time-to-pregnancy models as special cases of time-to-event data on a discrete scale, and for the computational advantages for integration over the distribution of the random effect that has, in this case, an explicit solution [47]. Note that if, as observed in most studies, fecundability is low (usually it is estimated between 10% and 30%), relative risk and odds ratios are very similar, so that the choice of $h(\cdot)$ does not modify the interpretation of the parameters in the time to pregnancy type models. With the complementary log–log link function, the model can be written

$$-\log(1-\mu_{ij}) = \xi_i \exp(\mathbf{u}_{ij}^T \alpha),$$

with $\xi_i = \exp(a_i)$, showing that ξ_i is a random multiplier, often termed *frailty*, for the ith couple. The ξ_i are assumed to be independent and identically distributed from an unknown distribution. A typical hypothesis for this distribution is $\xi_i \sim \mathcal{G}(1,\gamma)$, where $\mathcal{G}(a,b)$ denotes the gamma distribution with mean ab and variance ab^2.

Ecochard and Clayton [32] extended this model including more than one random effect. It sometimes happens that a simple hierarchy in two levels is not sufficient to describe data, for example, they [32] discuss the artificial insemination by donor where the sperm of each donor is used to inseminate several women, and each woman generally receives sperm from different donors in successive cycles within each attempt at pregnancy. In general, M crossed hierarchies may be considered by writing the model

$$\mu_{ij} = h\left(\mathbf{u}_{ij}^T \alpha + \sum_{m=1}^M a_{m(i)}\right),$$

where $\exp(a_{m(i)}) \sim \mathcal{G}(1,\gamma_m)$. Another extension has been proposed to consider the random effect of some covariates; for example, age may have a strong negative effect on fecundability for some women and no effect for others. Such an extension leads to a generalized linear mixed-model framework

$$\mu_{ij} = h\left(\mathbf{u}_{ij}^T \alpha + \mathbf{v}_{ij}^T a_i\right),$$

where $a_i \sim \mathcal{N}(0, \Sigma)$ and v_{ij} is a vector of covariates having a random effect and α and a_i are the vectors of parameters describing the fixed and random effects.

58.4.4 Pearl Index

In studies of conceptions, and more often in comparative analysis of contraceptive methods, it is useful to have some summary index to collect most of relevant aspects of the *quality*, in terms of effectiveness, of a method. The most used of such indexes is the *Pearl index* or *Pearl rate* [54], also known as *pregnancy rate*. This index is the ratio between the number of conceptions of each single woman in a sample and her *exposure time*, which is defined as the length of time during which she is exposed to the risk of conceiving. If n couples are followed from marriage to the first conception and n_t is the number of couples conceiving at month t, the Pearl index may be defined as $P = (\sum_{t=1}^{T} n_t)/(\sum_{t=1}^{T} t n_t)$, where T is the time when the last couple conceived. If this time is not observed and the study is censored at time c ($c < T$), the Pearl index becomes

$$P = \frac{\sum_{t=1}^{c} n_t}{\sum_{t=1}^{c} t n_t + c\left(n - \sum_{t=1}^{c} n_t\right)}.$$

When fecundability p is constant between cycles of a same woman and across women, the Pearl index is the maximum-likelihood estimator of p. Otherwise, when fecundability is heterogeneous, the expected value of the index moves from the arithmetic mean of p, when the length of observation for all the women is one, and tend to the harmonic mean of p when all the observation lengths go to infinity.

Like all the single-number indices, the Pearl rate is often considered too rough to be used in comparisons, and in many studies it is presented in junction with nonparametric estimates of survival curves such as life-table or Kaplan–Meier estimators that do not assume a constant hazard rate (e.g., see Sheps [65]; Potter [55]).

58.5 Day-Specific Probabilities of Conception

58.5.1 The Homogeneity Models

Detailed prospective studies finalized in estimating fecundability are now available for research purposes. In these studies data are recorded not only on a couple-specific level (e.g., age and demographic variables, medical and reproductive data, occupational exposure, past use of drugs, or past use of contraception) and on a cycle-specific level (e.g., occurrence of conception, reproductive intentions), but also on a day-specific level. At this level couples record daily markers of ovulation (such as urine, serum or saliva hormones, basal body temperature, mucus observation); menstrual or other vaginal bleeding; use of alcohol, tobacco, drugs or herbs, biological exposure assessments; and sexual intercourse or genital contact, with possible use of barriers, withdrawal, or spermicide. The multilevel nature of this data structure requires specific statistical methods to be developed, but makes it possible to assess the effects of various exposures, demographic and behavioral factors, and their interactions by a measure of human fecundity not depending on intercourse behavior, unlike the other measures that we previously discussed.

If for every cycle only one intercourse act is observed, we can obtain direct estimates of the day-specific probability of conception as the ratio of (1) instances in which intercourse acts in one day resulted in conception to (2) the total number of acts of intercourse of the same day [78]. Since day-specific probabilities provide a direct measure of biologic fecundity, the knowledge of biological aspects of menstrual cycle can help in developing models for it. In particular, ovulation is the key event in the menstrual cycle, and it is biologically well known that chances of conception are higher if intercourse occurs in the days around ovulation. Therefore, it is preferable to refer the direct estimates of the probability of conception to the day of ovulation, or to a proxy of it, rather than to the first day of the cycle [13]. In this framework it would be straightforward to obtain models that relate these probabilities to covariates by using a logistic regression, maybe with a couple-specific random effect included to account for within couple dependency.

However, often in these studies, multiple acts of intercourse occur in many cycles, and it is not possible to attribute conception to a single act. The first model for accounting for this problem was suggested by Peter Armitage to Barrett and Marshall [1]. They assume that batches of sperm introduced into the reproductive tract on different days in a single cycle are statistically independent in attempting to fertilize the egg. If p_k is the probability of conception in a cycle with intercourse only on day k (the so called *day-specific probability of conception*) relative to the available estimate of ovulation o, then the probability of conception for couple i in cycle j is

$$P_{ij} = 1 - \prod_{k \in \mathcal{K}(o)} (1 - p_k)^{X_{ijk}}$$

where $\mathcal{K}(o)$ is the set of days around ovulation when intercourse may result in pregnancy and X_{ijk} is the indicator of the intercourse, so that $X_{ijk} = 1$ if intercourse occurred on day k in menstrual cycle j and $X_{ijk} = 0$ otherwise. The basic idea of this model is that the probability of not conceiving in the entire cycle, $1 - P_{ij}$, is the product of the (assumed independent) probabilities of not conceiving in each day when intercourse occurs, $(1 - p_k)^{X_{ijk}}$. The estimate of the parameters p_k of such a model can be obtained by maximum likelihood by assuming P_{ij} as the expected value of a Bernoulli distribution for the conception in each cycle.

Some years later Schwartz et al. [63] observed that other biological factors, addition to the timing of intercourse also impact conception. They propose modifying the Barrett–Marshall model by multiplying it by a factor A that includes all the other effects; they call it "cycle viability," but it is clear that it also reflects male factors (e.g., the presence of motile sperm) and interaction effects (e.g., ability of sperm to fertilize ovum, survival of embryo to detection). Therefore A can be seen as the fraction of cycles that are even susceptible to possible conception. Thus, the model proposed by Schwartz et al. [63] is

$$P_{ij} = A \left[1 - \prod_{k \in \mathcal{K}(o)} (1 - p_k)^{X_{ijk}} \right].$$

Under this model, p_k is interpreted as the probability of conception in a viable cycle with intercourse only on day k. Maximum-likelihood estimates can be obtained for p_k and A by assuming P_{ij} as the expectation of a Bernoulli random variable describing the conception. Royston [58], Wilcox et al. [77], Colombo and Masarotto [13]), and Colombo et al. [14], using different data, have found an estimate of A of less than 50%.

Covariate effects on A were considered first by Royston [58] by expressing A as a linear function of a single covariate and later by Weinberg et al. [74], who developed an approach to allow the modeling of A to be in the class of the generalized linear models allowing for couple-specific and cycle-specific covariates, $A_{ij} = h(\mathbf{u}_{ij}^T \alpha)$, where $\mathbf{u}_{ij}$ is a vector of woman- and cycle-specific covariates and α is the vector of the corresponding parameters. They used an EM algorithm [17] by defining two hypothetical unobservable indicator variables Z_j indicating the cycle viability and S_j, an indicator for whether pregnancy would have occurred if the cycle were viable.

Some covariates ahve no or only minimal effect on the cycle viability A but may interact with time to ovulation, having a different effect on each p_k; for example, daily mucus quality, local contraception, or the use of alcohol or caffeine on the day of intercourse may affect that day-specific probability of conception. Zhou and Weinberg [82] proposed a latent marginal probability model that adds day specific covariates

$$P_{ij} = A_{ij}\left[1 - \prod_{k \in \mathcal{K}(o)} \left(1 - g\left(\mathbf{w}_{ijk}^T \beta\right)\right)^{X_{ijk}}\right].$$

where $g(\cdot)$ is a link function and $\mathbf{w}_{ijk}$ is a vector of covariates supposed to modify the effect of intercourse on day k with respect to ovulation. In order to estimate such a model, they [82] introduced an EM algorithm that maximized a complete-data pseudolikelihood.

Royston [58] and Weinberg and Wilcox [75] proposed extensions of the Schwartz et al. approach by considering survival times of sperm and egg, modeling them by a parametric function that became elements for the fecundability model.

58.5.2 Models with Heterogeneity

All the models described in this section consider day-specific probability of conception p_k and cycle viability A (and all the related parameters expressing dependence on covariates) as fixed, while we already know that unobserved heterogeneity is present and should be considered in the models.

Zhou et al. [83] introduced a model assuming that cycle viability varies as a couple specific random effect. They assume that all the cycles from the same couple share a common latent random cycle viability ω_i and, conditional on ω, outcomes are considered independent within a couple. They assume that $\omega_i \sim \mathcal{B}(a,b)$, with the a and b chosen, requiring that the expectation $E(\omega_{ij}) = h(\mathbf{u_i}^T \alpha)$, where $h(\cdot)$ is a link function. Conditional on ω_i the conception probability for cycle j of couple i is assumed to be

$$P_{ij} = \omega_i(\mathbf{u_i})$$
$$\times \left[1 - \prod_{k \in \mathcal{K}(o)} \left(1 - g\left(\mathbf{w}_{ijk}^T \beta\right)^{X_{ijk}}\right)\right].$$

All the parameters can be estimated via maximum likelihood by modifying the EM algorithm for the fixed-effect case.

The choice of the beta distribution in this model presents the same drawbacks we have seen for the time-to-pregnancy analysis, related to the U-shaped estimated cycle viability for large variances. Dunson and Zhou [22] observed, in addition to this limitation, that these models do not take into account the presence, if observed, of a sterile subpopulation. They propose using a latent structure for modeling fecundability from a mixture of subpopulations, by using a threshold quantal response model [51] to describe the effects of endogenous and exogenous factors on the cycle viability probability. If π is the proportion of couples that are sterile, the conditional probability of conception, with sterile sub-population incorporated, in the jth cycle of couple i, is

$$P_{ij} = (1-\pi)\Phi(\alpha_i + \mathbf{u}_{ij}^T \beta)$$
$$\times \left[1 - \prod_{k \in \mathcal{K}(o)} (1-p_k)^{X_{ijk}}\right].$$

Inference for such a model is carried out following a Bayesian approach by using a

Gibbs sampler algorithm with appropriate prior distributions.

Often the ovulation day is not directly observed but is estimated by using some proxies such as basal body temperature shift (e.g., see Barrett and Marshall [1]) or mucus peak [14]. In this case the models for the per-cycle probability of conception are affected by the error in estimating the day of ovulation; Dunson et al. [24] extend the Dunson–Zhou model to account for error in estimating ovulation by mixing across the measurement error distribution, and fit it by maximum likelihood via EM algorithm. Moreover, in studies on human behavior, such as the one performed in analyzing fecundability, the presence of a portion of missing data characterize the data. In this context Dunson and Weinberg [20] propose a Bayesian version of the Schwartz et al. model that allows for missing data on intercourse days.

58.5.3 Identifiability

As we have seen, in the Schwartz et al. model, and in its extensions, all the parameters, the cycle viability A and the day-specific probability p_k, include female and male factors and interaction effects, and it is not clear which factors relate directly to each [23]. From a statistical point of view, this means that it is difficult to separately estimate A and all the p_k because of collinearity in these parameters. By observing that the highest p_k results close to one in most of the fitted Schwartz et al. models to actual data, Dunson (2001) proposes a simple modification of that model, setting the highest p_k equal to one. Such a solution should solve the problem of estimability and identifiability of the joint estimates. The model also incorporates the constraints that the p_k increase in time to the maximum and then decrease. Specifically, the probability of conception in a menstrual cycle is

$$P_{ij} = A\left\{X_{ijM} + (1 - X_{ijM}) \times \left[1 - \prod_{k \in \mathcal{K}(o)} (1-p_k)^{X_{ij,k+M}}\right]\right\},$$

where M is the most fertile day relative to ovulation. In this case A is interpreted as the probability of conception given intercourse on the most fertile day and p_k is the ratio of the probability of conception given intercourse on only day k to the probability of conception given intercourse on the most fertile day. An extension of the model incorporates covariates as fixed and random effects on both A and the p_k and has been applied in analyzing the effect of the follicular phase length and of the age on fecundity [25,68]. Inference is done in a Bayesian framework through a data augmentation Markov chain Monte Carlo (MCMC) algorithm involving both Gibbs sampling [36] and Metropolis–Hastings [39] steps.

Considering the problem of the weak identifiability and computational instability of the Schwartz et al. model, Dunson and Stanford [28] proposed a different approach, based on the Barrett–Marshall [1] model, incorporating fixed and random effects in the same linear predictor at day-specific parameters. Following this hierarchical model, the probabilities of conception is

$$P_{ij} = 1 - \prod_{k \in \mathcal{K}(o)} (1 - p_{ijk})^{X_{ijk}},$$
$$p_{ijk} = 1 - \exp\left\{-\xi_i \exp(\mathbf{u}_{ijk}^T \beta)\right\},$$
$$\xi_i \sim \mathcal{G}(\phi, 1/\phi),$$

where p_{ijk} is the day-specific probability of conception in cycle j from couple i given intercourse only on day k, ξ_i is a fecundability multiplier for couple i, $\mathcal{G}(a,b)$ denotes the gamma distribution with mean ab and variance ab^2, and β is a vector of regression coefficients. The choice of setting the two parameters of the gamma distribution as inversely related implies that $E(\xi_i) = 1$, which avoids nonidentifiability between $E(\xi_i)$ and the day-specific parameters. The parameter ϕ is the inverse of the variance of ξ_i, $\phi = 1/\text{Var}(\xi_i)$, and it is interpretable as a measure of the level of the variability between couples. This model can be simply modified to allow for a sterile subpopulation, by modifying the distribution of ξ_i, which could incorporate a point mass at 0. A particularly appealing characteristic of this model is that the marginal probability of conception, integrating out the couple-specific frailty ξ_i, has a simple closed form:

$$P(\text{conception at cycle } j \text{ of a generic woman})$$
$$= 1 - \left(\frac{\phi}{\phi + \sum_{k \in \mathcal{K}(o)} X_{ijk} \exp(\mathbf{u}_{ijk}^T \beta)} \right)^{\phi}.$$

Thus the marginal day-specific conception probability in a cycle with intercourse only on day k and predictors $\mathbf{u}$ is

$$p_k = 1 - \left(\frac{\phi}{\phi + \exp(\mathbf{u}^T \beta)} \right)^{\phi}.$$

In a Bayesian framework, estimates of the parameters, selection of variables, and order-restricted inference are straightforward when using an efficient MCMC algorithm with auxiliary variables, by mixing Gibbs sampling steps with a Metropolis step for updating ϕ. This model has been widely used in clinical and epidemiological applications; for example, Bigelow et al. [2] and Scarpa et al. [59] apply it to demonstrate the advantage of using mucus observation as predictor of conception, and Scarpa and Dunson [60] propose a Bayesian decision-theoretic approach based on this model for searching for optimal rules for timing intercourse to achieve pregnancy.

Many other refinements of the Barrett–Marshall [1] model have been proposed to tackle practical and methodological problems involved in fertility modeling, by considering specific aspects, such as error measurement in identifying the ovulation day (e.g., Royston [58]; Dunson and Weinberg [21]; Dunson et al. [24]), evolution of the day-specific probability of conception before and after the most fertile day [58,23,75], length of the fertility window [24,25], joint analysis of cycles with both complete and incomplete intercourse information [20,26], and specific analysis for in vitro fertilization studies [84,18]. Some review papers have been published (e.g., see Zhou [85]; Weinberg and Dunson [76]) discussing the characteristics of many of these models.

58.6 Data

The organization of retrospective and prospective studies to collect data for analysis is a crucial aspect in this context. For all the models that we have described, some reliable data are needed. Tingen et al. [70] review methodological and statistical approaches in study design, focusing on the data required and the practical advantages and disadvantages of each design.

Some detailed prospective studies have been organized in the past and have compiled datasets of high quality for conception analysis. They describe two different situations. A large part of them involve normal fertile couples, recruited from the general population [77,49] or among couples who use of natural family planning methods [12–14]. A second group of studies collected data from infertile couples treated by intrauterine artificial insemination with the husband's spermatozoa [48] or artificial insemination by donor [30]. Ecochard [33] gives a detailed description of some of these datasets.

58.7 Models for Markers

In this chapter we concentrated mainly on models for fecundability even though, as we have seen, the estimate of probabilities of conception depends strongly on many other aspects of the biological, behavioral and social aspects of human life. However, a good model for conception requires models for many of these other aspects. For example, we have seen how crucial is the identification of the day of ovulation, which is often identified by observable markers such as basal body temperature or cervical mucus. Statistical models for such markers are therefore needed in order to improve the efficiency of conception estimates. Many papers have been published while considering such aspects; for example, Dunson and Colombo [29] considered cervical mucus and developed a Bayesian hierarchical model for describe its structure observed across the menstrual cycle, and Scarpa and Dunson [61] analyzed basal body temperature curves across the menstrual cycle and propose a Bayesian nonparametric hierarchical model that includes in the functional prior the biological information available.

58.8 Discussion

The author has reviewed statistical and probabilistic models and strategies for analyzing data for conception. A great progress is evident from the population models described in Section 57.3, based on aggregated data, to the direct models based on data obtained from detailed prospective studies. This advance has been possible mainly because of the availability of specific data and the capability of storage of modern computers. Moreover, the growth of computation capabilities has made possible the actual evaluation of complex statistical models biologically based. Nevertheless, going back to the seminal papers on fecundability by Gini [37,38] and reading them, it is amazing to that they discuss all the main topics that have been studied in the most recent decades, applying theories from most of the results that have been actually obtained from data during these years.

It is also interesting to outline the importance of the collection of "good" data in this field. Since the information to be collected is very sensible, quality of data is not easily achieved. The availability of some quality dataset is due mainly to the will, the patience and the honesty of some scientists, such as John Marshall, Bernardo Colombo, Allen Wilcox, and others, who for many years were responsible for the collection and the quality of data.

Much work still remains to be done. Nonlinear and nonparametric modeling of fertility parameters, methods for handling measurements errors, and more biological plausible models, are some methodologically line of research. Also, computational aspects need to be addressed; in Bayesian framework with complex models and a relatively large amount of data, as is typical in this context, MCMC algorithms are quite time-consuming and research is needed to make inference procedures faster. Also, new biological hypotheses need to be verified. For instance, Wilcox et al. [79] found evidence that intercourse and ovulation may not be independent, and other authors suppose that time to pregnancy and behavioral variables are related to the sex of the baby. These and many other hypothesis need to be verified; models need to be proposed and specific data need to be collected.

References

1. Barrett, J. C. and Marshall, J. (1969). The risk of conception on different days of the menstrual cycle. *Pop. Stud.*, **23**, 455–461.

2. Bigelow, J. L., Dunson, D. B., Stanford, J. B., Ecochard, R., Gnoth, C., and Colombo, B. (2004). Mucus observations in the fertile window: a better predictor of conception than timing of intercourse. *Hum. Reprod.*, **19**, 889–892.

3. Boldsen, J. L. and Schaumburg, I. (1990). Time to pregnancy—a model and its application. *J Biosoc Sci*, **22**, 255–262.

4. Bongaarts, J. (1978). A framework for analyzing the proximate determinants of fertility. *Pop. Devel. Rev.*, **4**, 105–132.

5. Bongaarts, J. (1982). The fertility-inhibiting effects of the intermediate fertility variables. *Stud. Family Plan.*, **13**, 179–189.

6. Brass, W. (1958). The distribution of births in human populations. *Popul. Stud.*, **12**, 51–72.

7. Brass, W. (1960). The graduation of fertility distributions by polynomial functions. *Popul. Stud.*, **14**, 148–162.

8. Broström, G. (1985). Practical aspects on the estimation of the parameters in Coale's model for marital fertility. *Demography*, **22**, 625–631.

9. Campbell, A. A. (1983.) *Manual of Fertility Analysis.* Churchill Livingstone, Edinburgh.

10. Coale, A. J. and Trussel, J. (1974). Model fertility schedules: Variations in the age structure of childbearing in human populations. *Popul. Index*, **40**, 185–258.

11. Crouchley, R. and Dassios, A. (1998) Interpreting the beta geometric in comparative fecundability studies. *Biometrics*, **54**, 161–167.

12. Colombo, B., Miolo, L., and Marshall, J. A. (1993) A data base for biometric research on changes in basal body temperature in the menstrual cycle. *Statistica*, **53**, 563–572.

13. Colombo, B. and Masarotto G. (2000). Daily fecundability: first results from a new data base. *Demogr. Res.*, **3**, 5.

14. Colombo, B., Mion, A., Passarin, K., and Scarpa, B. (2006). Cervical mucus symptom and daily fecundability: First results from a new data base. *Statist. Meth. Med. Res.*, **15**, 161–180.

15. Davis, K. and Blake, J. (1956). Social structure and fertility: An analytic framework. *Econ. Devel. Cultural Change*, **4**, 211–235.

16. Davison, A. C. and Hinkley, D. V. (1997). *Bootstrap Methods and their Application.* Cambridge University Press.

17. Dempster, A. P., Laird, N. M., and Rubin, D. B. (1977). Maximum likelihood from incomplete data via the EM algorithm. *J. Roy. Statist. Soc. B*, **39**, 1–38.

18. Dukic, V. and Hogan, J. (2002). A hierarchical Bayesian approach to modeling embryo implantation following in vitro fertilization. *Biostatistics*, **3**, 361–77.

19. Dunson, D. B., Weinberg, C. R., Perreault, S. D., and Chapin, R. E. (1999). Summarizing the motion of self-propelled cells: Applications to sperm motility. *Biometrics*, **55**.

20. Dunson, D. B. and Weinberg, C. R. (2000a). Accounting for unreported and missing intercourse in human fertility studies. *Statis. Med.*, **19**, 665–679.

21. Dunson, D. B. and Weinberg, C. R. (2000b). Modeling of human fertility in presence of measurement error. *Biometrics*, **56**, 288–292.

22. Dunson, D. B. and Zhou, H. A. (2000). Bayesian model for fecundability and sterility. *J. Am. Statist. Assoc.*, **95**, 1054–1062.

23. Dunson, D. B. (2001). Bayesian modeling of the level and duration of fertility in the menstrual cycle. *Biometrics*, **57**, 1067–1073.

24. Dunson, D. B., Weinberg, C. R., Baird, D. D., Kesner, J. S., and Wilcox, A. J. (2001). Assessing human fertility using several markers of ovulation, *Statist. Med.*, **20**, 965–978.

25. Dunson, D. B., Colombo, B., and Baird, D. D. (2002). Changes with age in the level and duration of fertility in the menstrual cycle. *Hum. Reprod.*, **17**, 1399–1403.

26. Dunson, D. B. (2003). Incorporating heterogeneous intercourse records into time to pregnancy models. *Math. Popul. Stud.*, **10**, 127–43.

27. Dunson, D. B. and Neelon, B. (2003). Bayesian inference on order-constrained parametersin generalized linear models. *Biometrics*, **59**, 286–295.

28. Dunson, D. B. and Stanford, J. B. (2005). Bayesian inferences on predictors of conception probabilities. *Biometrics*, **61**, 126–133.

29. Dunson, D. B. and Colombo, B. (2003). Bayesian modeling of markers of day-specific fertility. *J. Am. Statist. Assoc.*, **98**, 28–37.

30. Ecochard, R. and Clayton, D. G. (1998). Multi-level modelling of conception in artificial insemination by donor. *Statist. Med.*, **17**, 1137–1156.

31. Ecochard, R. and Clayton, D. G. (2000). Multivariate parametric random effect regression models for fecundability studies. *Biometrics*, **56**, 1023–1029.

32. Ecochard, R. and Clayton, D. G. (2001). Fitting complex random effect models with standard software using data augmentation: application to a study of male and female fecundability. *Statist. Model.*, **1**, 319–331.

33. Ecochard, R. (2006). Heterogeneity in fecundability studies: issues and modelling. *Statist. Methods Med. Res.*, **15**, 141–160.

34. Efron, B. and Tibshirani, R. (1994). *An Introduction to the Bootstrap.* Chapman & Hall/CRC.

35. Galton, F. (1869). *Hereditary Genius: An Inquiry Into Its Laws and Consequences.* MacMillan, London.

36. Gelfand, A. E. and Smith, A. F. M. (1990). Sampling-based approaches to calculating marginal densities. *J. Am. Statist. Assoc.*, **85**, 398–409.

37. Gini, C. (1923). Prime ricerche sulla "fecondabilità" della donna, *Atti del Reale Istituto Veneto di Scienze, Lettere ed Arti*, **LXXXIII**, 315–344.

38. Gini, C. (1924). Nuove ricerche sualla "fecondabilità" della donna. *Atti del Reale Istituto Veneto di Scienze, Lettere ed Arti*, **LXXXIV**, 269–308.

39. Hastings, W. K. (1970). Monte Carlo sampling methods using Markov chains and their applications. *Biometrika*, **57**, 97-109.

40. Heckman, J. J. and Singer, B. (1984). A method for minimizing the impact of distributional assumptions in econometric models for duration data. *Econometrica*, **52**, 271 320.

41. Heckman, J. J. and Walker, J. R. (1990). Estimating fecundability from data on waiting times to first conception. *J. Am. Statist. Assoc.*, **85**, 283–294.

42. Heckman, J. J., Robb, R. and Walker, J. R. (1990). Testing the mixture of exponentials hypothesis and estimating the mixing distribution by the method of moments. *J. Am. Statist. Assoc.*, **85**, 582–589.

43. Henry, L. (1961). Some data on natural fertility. *Eugen. Quart.*, **8**, 81–91.

44. Hobcraft, J. and Little J. A. (1984). Fertility exposure analysis: A new method for assessing the contribution of proximate determinants to fertility differentials. *Popul. Stud.*, **38**, 21–45.

45. Hougaard, P. (1986). Survival models for heterogeneous populations derived from stable distributions. *Biometrika*, **73**, 387–396.

46. Kalule-Sabiti, I. (1984). Bongaarts' proximate determinants of fertility applied to group data from the Kenya fertility survey 1977/78. *J. Biosoc. Sci.*, **16**, 205–218.

47. Keiding, N., Andersen, P. K., and Klein, J. P. (1997). The role of frailty models and accelerated failure time models in describing heterogeneity due to omitted covariates. *Statist. Med.*, **5**, 215–224.

48. Mathieu, C., Ecochard, R., Bied, V., Lornage, J., and Czyba, J. C. (1995). Cumulative conception rate following intrauterine artificial insemination with husband's

spermatozoa: influence of husband's age. *Hum. Reprod.*, **10**, 1090–1097.

49. Maruani, P. and Schwatz, D. (1983). Sterility and Fecundability estimation. *J. Theor. Biol.*, **105**, 211–219.

50. McCullough, P. and Nelder, J. A. (1989). *Generalized Linear Models, 2nd ed.* Chapman & Hall, London.

51. Morgan, B. J. T. (1992). *Analysis of Quantal Response Data*. Chapman & Hall, London.

52. Page, H. J. (1977). Patterns underlying fertility schedules: A decomposition by both age and marriage duration. *Popul. Stud.*, **31**, 85–106.

53. Palloni, A. (1984). Assessing the effects of intermediate variables on birth interval-specific measures of fertility. *Popul. Stud.*, **31**, 85–106.

54. Pearl, R. (1933). Factors in human fertility and their statistical evaluation. *Lancet*, **222**, 607–611.

55. Potter, R. G. (1966). Application of life table techniques to measurement of contraceptive effectiveness. *Demography*, **3**, 297–304.

56. Ridout, M. and Morgan, B. (1991). Modeling digit preference in fecundability studies. *Biometrics*, **47**, 1423–1433.

57. Ross, J. L., Blangero J., Goldstein, M. C., and Schuler, S. (1986). Proximate determinants of fertility in Kathmandu Valley, Nepal: An anthropological case study. *J. Biosoc. Sci.*, **18**, 179–196.

58. Royston, J. P. (1982). Basal body temperature, ovulation and the risk of conception, with special reference to the lifetimes of sperm and egg. *Biometrics*, **38**, 397–406.

59. Scarpa, B., Dunson, D. B., and Colombo B. (2006). Cervical mucus secretions on the day of intercourse: An accurate marker of highly fertile days. *Eur. J. Obstet. Gynecol. Reprod. Biol.*, **125**, 72–78.

60. Scarpa, B. and Dunson, D. B. (2007). Bayesian methods for searching for optimal rules for timing intercourse to achieve pregnancy, *Statist. Med.*, **26**, 1920–1936.

61. Scarpa, B. and Dunson, D. B. (2009). Bayesian hierarchical functional data analysis via contaminated informative priors. *Biometrics*, in press.

62. Scheike, T. H. and Jensen, T. K. (1997). A discrete survival model with random effects: an application to time to pregnancy. *Biometrics*, **53**, 318–329.

63. Schwartz, D., MacDonald, P. D. M., and Heuchel, V. (1980). Fecundability, coital frequency, and the viability of ova. *Popul. Stud.*, **34**, 397–400.

64. Sheps, M. C. (1964). On the time required for conception. *Popul. Stud.*, **18**, 85–97.

65. Sheps, M. C. (1966). Characteristics of a ratio used to estimate failure rates: Occurrences per person year of exposure. *Biometrics*, **22**, 310–321.

66. Sheps, M. C. and Menken, J. A. (1973). *Mathematical Models of Conception and Birth*. University of Chicago Press, Chicago and London.

67. Singh, S., Casterline, J. B., and Cleland, J. G. (1985). The proximate determinants of fertility: Sub-national variations. *Popul. Stud.*, **39**, 113–135.

68. Stanford, J. B., White, G. L., and Hatasaka, H. (2002). Timing intercourse to achieve pregnancy: current evidence. *Obstet. Gynecol.*, **100**, 1333–1341.

69. Tietze, C. (1959). Differential fecundity and effectiveness of contraception. *Eugen. Rev.*, **50**, 231–234.

70. Tingen, C., Stanford, J. B., and Dunson, D. B. (2004). Methodologic and statistical approaches to studying human fertility and environmental exposure. *Environ. Health Perspect.*, **112**, 87–93.

71. United Nations (1983). *Manual X: Indirect Techniques for Demographic Estimation*. United Nations publication, Sales No. E.83.XIII.2.

72. Weinberg, C. R. and Gladen, B. C. (1986). The beta-geometric distribution applied to comparative fecundability studies. *Biometrics*, **42**, 547–560.

73. Weinberg, C. R., Baird, D. D., and Wilcox, A. J. (1994). Sources of bias in studies of time to pregnancy. *Statist. Med.*, **13**. 671–81.

74. Weinberg, C. R., Gladen, B. C., and Wilcox, A. J. (1994b). Models relating the timing of intercourse to the probability of conception and the sex of the baby. *Biometrics*, **50**, 358–367.

75. Weinberg, C. R. and Wilcox, A. J. (1995). A model for estimating the potency and survival of human gametes in-vivo. *Biometrics*, **51**, 405–412.

76. Weinberg, C. R. and Dunson, D. B. (2000). Some issues in assessing human fertility. *J. Am. Statist. Assoc.*, **95**, 300–303

77. Wilcox, A. J., Weinberg, C. R., and Baird D. D. (1995). Timing of intercourse in relation to ovulation: probability of conception, survival of pregnancy, and sex of the baby. *New Engl. J. Med.*, **333**, 1517–1521.

78. Wilcox, A. J., Dunson, D. B., Weinberg, C. R., Trussell, J., and Baird, D. D. (2001). Likelihood of conception with a single act of intercourse: Providing benchmark rates for assessment of postcoital contraceptives. *Contraception*, **63**, 211–215.

79. Wilcox, A. J., Baird, D. D., Dunson, D. B., Weinberg, C. R. (2004). On the frequency of intercourse around ovulation: evidence for biological influences. *Hum. Reprod.*, **19**, 1539–1543.

80. Wilson, C., Oeppen, J., and Pardoe, M. (1988). What is natural fertility? The modeling of a concept. *Popul. Index*, **54**, 4–20.

81. Wood, J. W. (1994). *Dynamics of Human Reproduction. Biology, Biometry, Demography.* Aldine De Gruyter, New York.

82. Zhou, H. and Weinberg, C. R. (1996). Modeling conception as an aggregated Bernoulli outcome with latent variables via the EM algorithm. *Biometrics*, **52**, 945–954.

83. Zhou, H., Weinberg, C. R., Wilcox, A. J., and Baird, D. D. (1996). Random effects model for cycle viability in fertility studies. *J. Am. Statist. Assoc.*, **91**, 1413–1422.

84. Zhou, H. and Weinberg, C. R. (1998). Evaluating uterine receptivity and embryo viability for couples undergoing in vitro fertilization. *Statist. Med.*, **17**, 1601–12.

85. Zhou, H. (2006). Statistical models for human fecundability. *Statist. Methods Med. Res.*, **15**, 181–194.

59 Probit Analysis

J. R. Ashford

59.1 Introduction

Quantal response analysis is concerned with the relationship between one or more dependent or response variables on one hand and a set of independent or explanatory variables on the other, in the situation in which the response is assessed in terms of a discrete, qualitative, or categorical variable. The simplest form of quantal variable involves a dichotomy, with two response categories that may be assigned the conventional labels 0 and 1, respectively. More generally, response may be recorded in terms of a polychotomy with several classes or categories, which may or may not reflect some implied underlying ordering.

59.2 Applications

In the past, the main applications of quantal response analysis have been in biometry and medical statistics, including pharmacology, toxicology, occupational medicine, and follow-up studies of groups of patients with particular characteristics, conditions, or treatment regimes. Quantal models have commonly been concerned with the death of a human being or other biological organism or the appearance of a particular symptom, in response to a stimulus such as a drug, a poison, a period of exposure to a hazardous environment, or merely the passage of time. Many biometric applications involve just one response variable and just one stimulus, although some consideration has been given both to multiple quantal responses and mixtures of several stimuli administered jointly.

More recently, there have been an increasing number and range of economic applications, in a variety of contexts that include laborforce participation, choice of occupation, migration, transportation, housing, education, legislation, and criminology. Many of the variables of interest in this field either are naturally expressed in a discrete or categorical form or are recorded in this way. Because of the complexities of the situation, large numbers of explanatory variables are commonly used; in addition, the growing availability and sophistication of survey data has led to a sharp increase in the raw material for such studies. Both developments reflect the growing use and power of modern computing facilities.

There is an extensive literature on the applications of quantal response analysis in biometry; (e.g., see Refs. 6, 8, and 13). Economic analysis is less well served in this context, although Daganzo [7] dealt with one specific application—transport model choice—and Amemiya [1] produced an excellent survey of the use of qualitative response models in both econometrics and biometry.

59.3 Framework for Quantal Response Models

A general framework for the formulation of quantal response models in biometry was provided by Ashford [3], who proposed that a biological organism should be regarded as a system comprising several subsystems, $S_1, S_2, \ldots, S_k$. When a stimulus consisting of d components, $D_1, D_2, \ldots, D_d$, is applied at level $\mathbf{z} = (z_1, z_2, \ldots, z_d)$, the organism manifests a reaction $\mathbf{y} = (y_1, y_2, \ldots, y_k)$, where y_i, $i = 1, 2, \ldots, k$, refers to the underlying response of subsystem S_i; y_i may in principle be expressed in quantitative terms, but is not necessarily observable in this form. For example, human subjects are assessed in terms of several distinct respiratory symptoms, each of which corresponds to the quantal response of a particular subsystem. The same framework may also be employed in economic analysis, although the interpretation of the components reflects the field of application.

The underlying stimulus–response relation may then be represented in terms of an additive model of the form

$$\mathbf{y} = \mathbf{x}(\mathbf{z}) + \boldsymbol{\epsilon}. \tag{1}$$

In this expression, $\mathbf{x}(\mathbf{z})$ is the vector *effects function* of the stimulus $\mathbf{z}$. The random vector $\boldsymbol{\epsilon}$ represents the "tolerance" or "susceptibility" of the individual subject and is distributed with joint cumulative distribution function (cdf) $T(\boldsymbol{\epsilon})$ and joint probability density function (pdf) $t(\boldsymbol{\epsilon})$ in the population of subjects. In general, the tolerance components $(\epsilon_i, \epsilon_j), (i, j = 1, 2, \ldots, k; i \neq j)$, will be positively correlated, since they reflect different characteristics of the same subject.

59.4 Relation between Quantitative and Quantal Responses

The value of the underlying reaction $\mathbf{y}$ will determine the particular quantal or semiquantal category of the observed system response. Suppose that the k-variate reaction space is divided into mutually exclusive and exhaustive regions corresponding to the various categories. In the simplest situation, just one subsystem S is involved; for such a subsystem quantal response, the two regions correspond to the subdivision of the $y$axis at a point that, without loss of generality, may be taken as zero. The probability of subsystem quantal response may then be expressed as

$$\begin{aligned} P(\mathbf{z}) &= P\{y \geq 0\} \\ &= P\{x(\mathbf{z}) + \epsilon \geq 0\} \\ &= 1 - T(-x(\mathbf{z})), \end{aligned} \tag{2}$$

where $T(\epsilon)$ is the (marginal) cdf of ϵ.

If we make the plausible assumption that $T(\epsilon)$ is symmetrical, (2) may be expressed in the form

$$P(\mathbf{z}) = T(x(\mathbf{z})). \tag{3}$$

The form of the right-hand side of (3) emphasizes that, on the basis of a quantal response alone [i.e., using only the information contained in $P(\mathbf{z})$], it is not possible to determine the forms of both the tolerance distribution and the effects function without ambiguity, since for any given $P(\mathbf{z})$ and specified marginal cdf $T(\epsilon)$, the form of $x(\mathbf{z})$ is uniquely determined.

A semiquantal response involving s ordered categories $C_1, C_2, \ldots, C_s$ may also be represented in terms of the response of a single subsystem S. The boundaries of the various categories may be defined as the points $\eta_1, \eta_2, \ldots, \eta_{s-1}$ on the $y$axis, where $\eta_1 < \eta_2 < \cdots < \eta_{s-1}$, the rth response category C_r corresponding to all points y in the interval (η_{r-1}, η_r), where η_0 and η_s are taken as $-\infty$ and $+\infty$, respectively. Without loss of generality, any one of the η_i may be set as equal to zero. The probability that the response falls within category C_r is then

$$\begin{aligned} P_r(\mathbf{z}) &= P\{\eta_{r-1} \leq y_1 < \eta_r\} \\ &= P\{\eta_{r-1} - x(\mathbf{z}) \leq \epsilon \\ &\qquad \leq \eta_r - x(\mathbf{z})\} \\ &= T\{\eta_r - x(\mathbf{z})\} \\ &\quad - T\{\eta_{r-1} - x(\mathbf{z})\}. \end{aligned} \quad (4)$$

Quantal and semiquantal responses defined in relation to a complete system may also be represented in terms of the underlying reaction of more than one subsystem. Thus a "weakest link" type of system might involve the quantal response of any one of k constituent subsystems. The corresponding probability is

$$\begin{aligned} P(\mathbf{z}) &= P\{\text{any } y_i > 0\} \\ &= 1 - P\{\text{all } y_i \leq 0\} \\ &= 1 - T(-x_1(\mathbf{z}), -x_2(\mathbf{z}), \ldots, \\ &\qquad -x_k(\mathbf{z})). \end{aligned} \quad (5)$$

In general, if the quantal response corresponds to a region R in the underlying reaction space, the corresponding probability is

$$p(\mathbf{z}) = \int\int \cdots \int_R t(\mathbf{x}(\mathbf{z}) - \boldsymbol{\epsilon}) d\epsilon_1 d\epsilon_2 \cdots d\epsilon_k. \quad (6)$$

Semiquantal responses may similarly be defined in terms of a mutually exclusive and exhaustive set of regions. Such categories will not in general reflect any intrinsic ordering of the various response variables, although, for example, regions defined by a set of parallel hyperplanes would in some sense imply an underlying gradation.

59.5 The Tolerance Distribution: Probits and Logits

Various algebraic forms have been proposed for the distribution of the tolerances ϵ in (1). Turning first to the marginal distribution of a particular component, say ϵ, the following types of distribution have been employed in the analysis of quantal data.

59.5.1 Linear Probability Model

$$T(\epsilon) = \begin{cases} 0, & \epsilon \leq 0, \\ \epsilon, & 0 < \epsilon \leq 1, \\ 1, & \epsilon > 1. \end{cases} \quad (7)$$

This model restricts the range of variation of the tolerance component, which may well be unrealistic in many applications. This deficiency must be balanced against the advantage of computational simplicity. The model has been found useful in the preliminary analysis of economic data.

59.5.2 Probit Model

The scale used for the measurement of the underlying response is arbitrary. Given that the mean of the tolerance distribution has already been set equal to zero, the variance may be equated to unity and the tolerance may be assumed to be a standardized variable. The most widely used tolerance distribution of this type is the standardized normal

$$T(\epsilon) = \Phi(\epsilon) = \int_{-\infty}^{\epsilon} (1/\sqrt{2\pi}) e^{-(1/2)t^2} dt, \quad (8)$$

which forms the basis of probit analysis. The *probit* of the proportion p is defined in terms of the inverse of Φ as

$$p^* = \Phi^{-1}(p) + 5. \quad (9)$$

The factor 5 was originally included in (9) for convenience of manual computation, to avoid the occurrence of negative values of the probit. In the more modern usage of the term "probit," this factor is excluded from the definition.

59.5.3 Logistic Model

An alternative "smooth" standardized distribution to the normal is the logistic distribution, employed in the form

$$T(\epsilon) = L(\lambda, \epsilon) = \epsilon^{-\lambda \epsilon}/(1 + \epsilon^{-\lambda \epsilon}). \quad (10)$$

Taking $\lambda = \pi/\sqrt{3}$, the variance of the tolerance distribution may be set equal to unity. However, some authors consider that closer overall agreement with the standardized normal distribution may be obtained by setting $\lambda = 1.6$. The inverse of the relation $L(1, \epsilon)$ defines the *logit* transformation

$$l = \ln[p/(1-p)]. \quad (11)$$

Cox [6, p. 33] has suggested a modified form of (11) that avoids infinite values of the logit corresponding to $p = 0$ and $p = 1$.

The logit and probit transformations are similar for values of p that lie toward the center of the range and differ substantially only for values close to zero or unity. In practical applications, extremely large numbers of observations are required to differentiate between the two transformations [5]. The logit transformation has the advantage of greater ease of computation, which must be set against the fact that the probit, by virtue of its derivation from the normal distribution, is closer to the mainstream of statistical analysis.

A two-parameter class of tolerance distributions has been proposed [14] that includes both the normal and logistic distributions as special cases

$$t(m_1, m_2, \epsilon)$$
$$= \frac{\exp(m_1 \epsilon)\{1 + \exp \epsilon\}^{-(m_1 + m_2)}}{\beta(m_1, m_2)}, \quad (12)$$

where β represents the beta function. The logistic model is given by $m_1 = m_2 = 1$, while (12) converges to a normal distribution as $m_1, m_2 \to \infty$. Other limiting special cases are the extreme minimal value ($m_1 = 1, m_2 \to \infty$) and the extreme maximum value ($m_1 \to \infty, m_2 = 1$).

59.6 Multivariate Tolerance Distributions

The assumption of a standardized multivariate normal distribution for the joint distribution of several tolerance components corresponds to multivariate probit analysis [2]. This has the advantage of standardized normal univariate marginal distributions, as used in probit analysis, coupled with a representation of the association between pairs of tolerance components in terms of the correlation coefficient. Representations based on a multivariate logistic distribution have also been proposed. Morimune [11] compared the Ashford–Sowden bivariate probit model with the Nerlove–Press [12] multivariate logit model in an application concerned with home ownership and number of rooms in house, and concluded that the probit was to be preferred.

59.7 Ordered Response Variables: Rankits

In circumstances such as sensory tests for degrees of preference, the response is measured by a variable that can be placed in rank order, but cannot be assigned a quantitative value. The (subjective) ranks may be replaced by the corresponding *rankits* [10]. The rankit of the rth in rank order from a sample of n observations is defined as the expected value of the rth-order statistic of a random sample of size n from a standardized normal distribution. This transformation is in a sense a generalization of the probit transformation and produces a response variable that may be regarded as quantitative in form.

59.8 Mixed Discrete–Continuous Response Variables: Tobits

Response may be recorded in terms of a variable that is capable of being measured continuously over only part of the real line. For example, household expenditures on luxury goods are zero at low income levels. Individuals available for employment may be working full-time or for a certain number of hours part-time or may be unemployed during a particular week. The general "two-limit" Tobit model envisages an underlying quantitative response which is only measurable in categorical form at the lower and upper limits of the range of variation. If one or other of the extreme categories is not present, the Tobit model is called "right-truncated" or "left-truncated," as appropriate. Tobit models invariably involve a standardized normal distribution of tolerances and the probabilities of obtaining one or other of the categorical responses may be calculated on the basis of (2) and (8).

59.9 The Effects Function

The effects function $\mathbf{x}(\mathbf{z})$ in (1) represents the average contribution of the stimulus $\mathbf{z}$ to the response $\mathbf{y}$ of the subject. In general, this function and its individual components will include parameters $\boldsymbol{\theta} = (\theta_1, \theta_2, \ldots, \theta_s)$, which must be estimated. The form of $\mathbf{x}(\mathbf{z})$ and its components $x_i(\mathbf{z})$ reflects the particular application. In most but not all applications it is reasonable to suppose that any given component of the effects function will be a monotonically increasing function of the level of any particular component of the stimulus, for fixed levels of any other components. Usually, in the absence of any stimulus all components of the effects function will be zero. Linear functions of the form $\mathbf{z}'\boldsymbol{\theta}$ of the components z_i of the stimulus $\mathbf{z}$ have been commonly used, conforming to the classical general linear model. Nonlinear functional forms may, however, be appropriate in certain circumstances.

The simplest situation, commonly used in the biological assay of a single drug, involves a single subsystem S for which the effects function is expressed in terms of the logarithm of the dosage of a single drug. The underlying dosage–response relation then takes the form

$$y = \theta_1 + \theta_2 \ln z_1 + \epsilon, \qquad (13)$$

where ϵ is a standardized variable and the corresponding probability of a quantal response is, from eq. (3),

$$P(z_1) = T(\theta_1 + \theta_2 \ln z_1). \qquad (14)$$

Both the response y and the stimulus z may be subject to errors of observation. Suppose that the observed value of y (which forms the basis of the observed quantal response) is denoted $y' = y + \phi_y$ and that the observed stimulus $(\ln z_1)' = \ln z_1 + \phi_z$. The underlying dosage–response relation (13) may then be expressed in the form

$$y' = \theta_1 + \theta_2 \ln z_1 + \epsilon', \qquad (15)$$

where $\epsilon' = \epsilon_1 + \phi_y - \theta_2 \phi_z$. Given that ϕ_y and ϕ_z follow independent normal distributions with zero mean and variances σ_y^2 and σ_z^2, respectively, ϵ' is itself normally distributed with zero mean and variance $(1 + \sigma_y^2 + \theta_2^2 \sigma_z^2)$. This means that for a standardized normal distribution of tolerances ϵ the corresponding observed dosage–response relation is

$$y'' = \frac{y'}{\sqrt{(1+\sigma_y^2+\theta q_z^2\sigma_z^2)}}$$

$$= \frac{(\theta_1+\theta_2\ln z)}{\sqrt{(1+\sigma_y^2+\theta_z^2\sigma_z^2)}} + \epsilon. \quad (16)$$

Comparison of (13) and (16) with (14) shows that the effect of errors of observation is to reduce the slope of the quantal response relationship between the inverse of the cdf of the tolerance distribution and the log dose, but the position corresponding to a 50% response, the LD50, is unchanged.

In bioassay more complex forms of the effects function have been proposed to represent stimuli with several components, corresponding to mixtures of drugs or poisons, which are becoming increasingly important in both pharmacology and toxicology (see, e.g., Ref. 9). An approach based on a simplified representation of the underlying mode of biological action of the stimulus has been put forward by Ashford (see Ref. 3), presenting models based on drug-receptor theory that permit both monotonic and nonmonotonic dosage–response relationships for a single-component stimulus and a simple representation of the phenomena of synergism and antagonism.

The majority of examples in economics arise from the representation of events whose outcome is determined by the decisions of a customer, a product, or some other economic unit. There is an implicit assumption that rational decisions are made which optimize some inherent utility [15]. An effects function $\mathbf{x}'\boldsymbol{\theta}$ is commonly thought to be appropriate.

59.10 Estimation and Hypothesis Testing

The main objective of the statistical analysis of quantal response data is the assessment of the parameters $\boldsymbol{\theta}$ contained within the effects function, although in certain applications further parameters relating to the tolerance distribution are also of interest, including the correlation coefficients associated with a multivariate normal distribution of tolerances. As far as the raw data are concerned, a distinction is often made between structures in which only one or a few observations correspond to any given level of stimulus and situations in which many observations are involved. If possible, the number of observations at any stimulus level should be sufficient for large-sample theory to apply in order to minimize the effects of sampling variations on the observed proportions of quantal or semiquantal responses.

The most commonly used estimation procedure is maximum likelihood (ML), which may be applied to both data structures. Suppose that for any particular value z_i of the stimulus or independent variable, there are n_i observations, of which r_i manifest the quantal response of some subsystem S. (If there is only one observation at any particular stimulus level, r_i will be zero or unity.) Assuming that the observations are independent and applying the binomial theorem on the basis of (3), the loglikelihood may be written (apart from a constant term) in the form

$$L = \sum_i \{r_i \ln T(x(\boldsymbol{\theta}, \mathbf{z}_i)) + (n_i - r_i) \\ \times \ln(1 - T(x(\boldsymbol{\theta}, \mathbf{z}_i)))\}. \quad (17)$$

The ML estimator of $\boldsymbol{\theta}$ is given by the solution of the simultaneous equations

$$\frac{\partial L}{\partial \theta_k}$$

$$= \sum_i \frac{\{r_i - n_i T(x(\boldsymbol{\theta}, \mathbf{z}_i))\}}{T(x(\boldsymbol{\theta}, \mathbf{z}_i))\{1 - T(x(\boldsymbol{\theta}, \mathbf{z}_i))\}} \\ \times t(x(\boldsymbol{\theta}, \mathbf{z}_i))\frac{\partial x(\boldsymbol{\theta}, \mathbf{z}_i)}{\partial \theta_k} = 0, \quad (18)$$

where $k = 1, 2, \ldots, s$.

In general, explicit algebraic solutions of (18) do not exist and the ML estimators must be obtained by an iterative method. For the probit or logit models, L is globally concave, which implies that the ML estimator is unique if it is bounded; hence any iterative procedure that is guaranteed to converge to a stationary point must also converge to the global maximum. Various methods of iteration have been used. Powell's method has the advantage of involving the calculation of the function L, but not of its derivatives, in contrast to the Newton–Raphson method and the method of scoring, which both involve the first derivatives.

For a semiquantal response, the loglikelihood involves the application of the multinomial distribution in conjunction with expressions of the form (4) in a generalization of (17). For the Tobit model, the loglikelihood involves the sum of components corresponding to the discrete and continuous elements. The loglikelihoods corresponding to multiple subsystem and system quantal responses (which reflect the state of several subsystems) may also be defined in terms of the binomial or multinomial distributions.

Under certain general conditions the ML estimator is consistent and asymptotically normal. The large-sample variance–covariance matrix of the ML estimators may be calculated as the inverse of the matrix $[E\{-\partial^2 L/\partial \theta_k \partial \theta_l\}]$, where $k, l = 1, 2, \ldots, s$. Tests of hypotheses about the parameters $\boldsymbol{\theta}$ may be based on the ML estimators using Wald's test and the likelihood ratio test.

Other forms of estimation have been proposed, largely in an attempt to reduce the computational difficulties associated with the ML method. The most widely used is the modified minimum logit χ^2 method [4].

References

1. Amemiya, T. (1981). *J. Econ. Lit.*, **19**, 1483–1536.
2. Ashford, J. R. (1970). *Biometrics*, **26**, 535–546.
3. Ashford, J. R. (1981). *Biometrics*, **37**, 457–474.
4. Berkson, J. (1955). *J. Am. Statist. Assoc.*, **50**, 130–162.
5. Chambers, E. A. and Cox, D. R. (1967). *Biometrika*, **54**, 573–578.
6. Cox, D. R. (1970). *The Analysis of Binary Data*. Methuen, London.
7. Daganzo, C. (1979). *Multinomial Probit*. Academic Press, New York.
8. Finney, D. J. (1971). *Probit Analysis*, 3rd ed. Cambridge University Press, Cambridge, UK.
9. Hewlett, P. S. and Plackett, R. L. (1979). *The Interpretation of Quantal Response in Biology*. Edward Arnold, London.
10. Ipsen, J. and Jerne, N. K. (1944). *Acta Pathol.*, **21**, 343–361.
11. Morimune, K. (1979). *Econometrica*, **47**, 957–976.
12. Nerlove, M. and Press, S. J. (1973). *Univariate and Multivariate Log-Linear and Logistic Models*. Mimeograph No. R-1306-EDA/NIH, Rand Corporation, Santa Monica, CA.
13. Plackett, R. L. (1974). *The Analysis of Categorical Data*. Charles Griffin, London.
14. Prentice, R. L. (1976). *Biometrics*, **32**, 761–768.
15. Thurstone, L. (1927). *Psychol. Rev.*, **34**, 273–286.

60 Prospective Studies

Raymond S. Greenberg

60.1 Introduction

In epidemiology, the term "prospective study" usually designates a nonexperimental research design in which all the phenomena under observation occur after the onset of the investigation. This approach is known by a variety of other names, such as *follow-up, incidence, longitudinal,* and *cohort studies*. Regardless of the title preferred, the prospective approach may be contrasted with other nonexperimental studies that are based on historical information.

The usual plan of prospective research is illustrated in Figure 1. After initiation of the study, subjects are enrolled according to the level of exposure to the main independent variable. Typically, one group is defined as "exposed" and the comparison group consists of "unexposed" subjects. The group assignment in a prospective study is determined by observations on naturally occurring exposures. The reliance on natural exposures may be contrasted with experimental studies, in which the exposure status is assigned by randomization.

After the study groups are defined in a prospective investigation, the subjects are followed forward in time for the development of the outcome variable. Then the frequency of the outcome among exposed subjects is compared against the frequency of the outcome among unexposed subjects.

Example. One of the most noteworthy examples of the prospective approach was the investigation of atherosclerotic heart disease in Framingham, Massachusetts [1]. This study was started in 1948 by the US Public Health Service for the purpose of evaluating possible determinants of heart disease. Framingham was chosen as the site of investigation because it had a stable, cooperative population, with good access to local and referral medical facilities. A sample of 6500 persons between the ages of 30 and 60 years was chosen from a population of about 10,000 persons in that age group. Over 5000 persons without clinical evidence of atherosclerotic heart disease agreed to participate in this research. Each subject was examined at the outset and was reexamined at intervals of 2 years. The primary outcome was the development of atherosclerotic heart disease. Over the 20 years of follow-up several risk factors were identified, including hypertension, elevated serum cholesterol, and cigarette smoking [6].

678 *Prospective Studies*

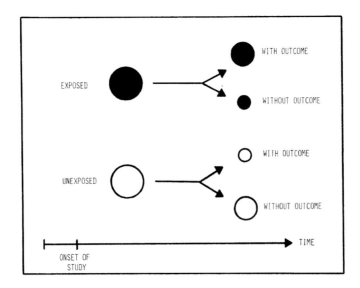

Figure 1: Schematic diagram of a prospective study (shaded areas represent subjects exposed to the antecedent factor; unshaded areas correspond to unexposed subjects).

60.2 The Selection of Subjects

A variety of factors must be considered in the choice of the study population for prospective research. In particular, the subjects must be accessible, willing to cooperate, and traceable over time. The following sources of subjects are employed commonly in prospective research.

1. *General population.* A prospective study can be performed in an entire community, as illustrated by the Framingham Study. The choice of a community for investigation may be governed by the stability and demographic features of the population, public support for the study, and the availability of resources over a sufficient period of time to evaluate the outcome.

2. *Special population.* A prospective study can be conducted among persons who are affiliated with a particular group. For example, Doll and Hill [2] studied the relationship between cigarette smoking and lung cancer among physicians on the British Medical Register who were living in the United Kingdom. The choice of physicians as a study population offered the advantages of a group that was interested in health research and could be traced through professional organizations.

3. *Clinical population.* A prospective study can be performed among patients at specific clinical facilities. As an illustration, the Collaborative Perinatal Project was a prospective study of more than 54,000 pregnant women who received prenatal care and delivered at one of 12 cooperating hospitals between 1959 and 1966 [5]. This population was used to evaluate risk factors for perinatal and infant mortality, as well as for infant morbidity.

Table 1: Data layout for calculation of the incidence density ratio (IDR).

	Exposed Subjects	Unexposed Subjects	Total
Number of specific outcome(s)	a	b	$m = a + b$
Population time of follow-up	L_1	L_0	$T = L_1 + L_0$

Table 2: Data layout for calculation of the risk ratio (RR).

	Exposed Subjects	Unexposed Subjects	Total
Number with outcome	a	b	$m_1 = a + b$
Number without outcome	c	d	$m_0 = c + d$
Total	$n_1 = a + c$	$n_0 = b + d$	$N = a + b + c + d$

60.3 The Measurement of Effect

The analysis of a prospective study is based on a comparison of the frequencies of outcomes among exposed and unexposed subjects. The magnitude of effect usually is quantified with one of the following measures.

1. Incidence density ratio (IDR). The IDR is defined as the incidence rate among the exposed divided by the incidence rate among the unexposed. The standard data layout for this type of analysis is portrayed in Table 1. For this analysis, the symbols a and b represent the number of specific outcomes among exposed and unexposed subjects, respectively. The length of observation on each individual, summed over all subjects in the study group are represented by L_1 and L_0, respectively. Thus, the incidence density rate of the outcome among exposed persons is estimated by a/L_1. Similarly, the incidence density of the outcome among unexposed persons is estimated by b/L_0. The IDR is estimated from

$$\widehat{\mathrm{IDR}} = \frac{a/L_1}{b/L_0}.$$

When there is no association between the exposure and outcome of interest, the IDR equals unity. An IDR less than unity implies a reduced incidence of the outcome among exposed subjects, compared with the unexposed. Conversely, an IDR greater than unity indicates that exposed subjects have an elevated incidence of the outcome. A large-sample test of difference from the null value can be constructed from the normal approximation to the binomial distribution [3]:

$$z = \frac{a - m(L_1/T)}{[m(L_1/T)(L_0/T)]^{1/2}}.$$

2. Risk ratio (RR). The RR is defined as the risk of the outcome among exposed divided by the risk among unexposed. The standard data layout for this type of analysis is depicted in Table 2. The symbols a and b have the previously defined interpretations. The symbols c and d represent the number of subjects who do not develop the outcome among exposed and unexposed persons, respectively. The risk of the outcome among the exposed is estimated by a/n_1. Similarly, the risk of the outcome among the unexposed is estimated by b/n_0. Thus, the RR is estimated by

$$\widehat{\mathrm{RR}} = \frac{a/n_1}{b/n_0}.$$

Interpretation of the value of the RR is identical to that for the IDR. A large-sample test for this measure, based on the hypergeometric distribution, uses the Mantel–Haenszel chi-square statistic [4]:

$$\chi^2_{(1)} = \frac{(N-1)(ad-bc)^2}{m_1 m_0 n_1 n_0}.$$

Table 3: Guidelines for interpretation of incidence density and risk ratios.

Value	Effect of Exposure
0–0.3	Strong benefit
0.4–0.5	Moderate benefit
0.6–0.8	Weak benefit
0.9–1.1	No effect
1.2–1.6	Weak hazard
1.7–2.5	Moderate hazard
≥ 2.6	Strong hazard

As indicated, the IDR and RR have the same scale of potential values. Although there are no absolute rules for the interpretation of these measures, Table 3 may serve as a rough guideline. For both measures, the absolute lower limit is zero, which results when there are no outcomes among exposed subjects. The upper limit is positive infinity, which results when there are no outcomes among unexposed subjects.

60.4 Relative Merits of Prospective Studies

When compared with other methods of nonexperimental research, the prospective design offers several advantages: (1) incidence rates of the outcome can be measured, (2) there is a clear temporal relationship between the purported cause and effect in prospective research, (3) exposures that are uncommon can be evaluated by selective sampling of exposed persons, (4) the amount and quality of information collected are not limited by the availability of historical records or recall of subjects, and (5) multiple effects of a given exposure can be assessed.

Nevertheless, prospective studies also have important limitations. Large numbers of subjects may be needed to evaluate rare outcomes, and prospective study may involve years of observation to evaluate chronic processes. A prolonged investigation may suffer from logistical problems, such as turnover of the research staff and variations in the methods used to measure the outcome. Moreover, it is difficult to follow subjects over protracted time periods, and these individuals may change their exposure levels over time. Finally, prospective research is relatively expensive when compared with alternative nonexperimental methods. Ultimately, the decision to conduct a prospective study is governed by the purpose of the investigation, the nature of the phenomena under observation, and the resources available for the study.

Acknowledgments. The preparation of this entry was supported in part by a contract from the National Cancer Institute (N01-CN-61027) to the author, and the manuscript was typed by Ms. Vickie Thomas.

References

1. Dawber, T. R., Meadors, G. F., and Moore, F. E. (1951). *Am. J. Public Health*, **41**, 279–286.

2. Doll, R. and Hill, A. B. (1964). *Br. Med. J.*, **1**, 1399–1410.

3. Kleinbaum, D. G., Kupper, L. L., and Morgenstern, H. (1982). *Epidemiologic Research*, p. 286–, Lifetime Learning Publications, Belmont, CA.

4. Mantel, N. and Haenszel, W. (1959). *J. Natl. Cancer Inst.*, **22**, 719–748.

5. Niswander, K. and Gordon, M. (1972). *The Women and Their Pregnancies*. National Institutes of Health Publication 73–379. US Government Printing Office, Washington, DC.

6. Truett, J., Cornfield, J., and Kannel, W. (1967). *J. Chron. Dis.*, **20**, 511–524.

61 Quality Assessment for Clinical Trials

David Moher

Commercial airline pilots have enormous responsibilities. They are accountable for expensive equipment, undergo extensive training and lifelong learning, and must abide by international standards regardless of where they fly to and from. Part of this standardization includes going through a checklist of tasks prior to any flight departure. For the flight to become airborne, the checklist must be completed. A completed checklist provides the flight crew with important information regarding the status of the aircraft and results in the safety and well-being of the passengers.

Clinical trialists share several similarities with airline pilots. Many of them are investigators (and principal investigators) of large multicenter randomized controlled trials (RCTs). It is not uncommon for these studies to cost millions of dollars to complete, and this is typically the end stage of a long development process that can easily cost in excess of $100 million. Clinical trialists would have often completed advanced training in epidemiology and very definitely have the safety and welfare of trial participants uppermost in their minds. What seems to be missing from the trialist's arsenal is standardization of the process of conducting such studies. There has been some consensus on how RCTs should be reported [26], although only a very small percent (2.5%) of the approximately 20,000 health care journals have brought in such reporting standards.

Without an agreed-on checklist of standard operating procedures, clinical trialists are left in the precarious position of operating in a partial and loose vacuum. Although eager to do the best possible job, it becomes increasingly difficult to know whether the investment of monies results in a quality product, namely, a report of an RCT whereby the estimates of the intervention's effectiveness are free of bias (i.e., systematic error).

Quality is a ubiquitous term, even in the context of RCTs. Importantly, it is a construct, and not something that we can actually put our hands on. It is a gauge for how well the RCT was conducted and/or reported. In the context of this chapter, the discussions around quality are related to internal validity, and it is defined as "the confidence that the trial design, conduct, analysis, and presentation has minimized or avoided biases in its intervention comparisons" [24].

It is possible to conduct an RCT with

excellent standards and report it badly, thereby giving the reader an impression of a low-quality study. Alternatively, an RCT may be conducted rather poorly but may be very well reported, providing readers with the impression of "high" quality. Typically, the only way we can ascertain the quality of an RCT is to examine its report. It is the only tangible evidence as to how such studies are conducted. In order to be able to reasonably rely on the quality of a trial's report, it must be a close facsimile of how the trial was conducted.

Early investigations pointed in this direction [13,23]. Liberati and colleagues assessed the quality of reports of 63 published RCTs using a 100-item quality assessment instrument developed by Thomas C. Chalmers [7]. To evaluate whether the quality of reports closely resembled their conduct, the investigators interviewed 62 (of 63) corresponding authors. The average quality ratings went from 50% (reports) to 57% (conduct), suggesting a similar relationship between the quality of reports and their conduct.

However, latest studies suggest that this relationship may not be as simple as previously thought [14,36]. Soares and colleagues assessed the quality of 58 published reports of terminated phase III RCTs completed by the Radiation Therapy Oncology Group, comparing them to the quality of the protocols for which they had complete access. These investigators reported that in six of the seven quality items examined there was a substantially higher reporting in the protocol compared to what was published. For example, only 9% of published reports provided information on the sample size calculation, yet this information was readily available in 44% of the Group's protocols.

This study is important for several reasons, not the least of which is that the authors had access to all the protocols from which the published reports emanated. If we are to gain a more thorough understanding of the relationship between trial reporting and conduct, it is important that journals and their editors support the publication of RCT protocols ([12], Douglas G Altman, persona communication). Such information provides readers with a full and transparent account of what was planned and what actually happened. Although these results are impressive and might suggest more of a chasm between reporting and conduct, caution is advised. This is a more recent publication, and time is required to permit replication and extension by other groups.

Historians may well view the first 50 years of reporting of RCTs with some amazement. They will encounter what might be described as a cognitive dissonance: a disconnection between the increased sophistication of the design of these studies and the apparent lack of care—disastrous in some cases—with which they were reported. Unfortunately, despite much evidence that trials are not all performed to high methodological standards, many surveys have found that trial reporting is often so inadequate that readers cannot know exactly how the trials were done [31–35].

There are three approaches to assess the quality of RCT reports: components, checklists, and scales. *Component* assessment focuses on how well a specific item is reported. Such an approach has the advantage of being able to complete the assessment quickly and is not encumbered with issues surrounding other methods of quality assessment. Allocation concealment is perhaps the best-known example of component quality assessment [29]. *Checklists* combine a series of components together, hopefully items that are conceptually linked. There is no attempt to combine the items together and come up with an overall summary score. The CONSORT Group developed a 22-item checklist to help investigators improve the quality of reporting their trials [26]. Scales are similar to checklists with one major difference: Scales are developed to sum their individual items together and come up with an overall summary score. Many scales have been developed [19,24]. Jaded and colleagues developed a three-item scale to assess the quality of reports of RCTs in pain [18]. Assessing the quality of trial reports using scales is considered contentious by some, presently [4,21].

Using 25 different scales, Jüni and colleagues assessed the quality of 17 trial reports included in a systematic review examining the thromoprophylaxis of low-molecular-weight heparin (LMWH) compared to standard heparin. They reported little consistency across the scales and found that different scales yielded different assessments of trial quality. When these quality scores were incorporated into the meta-analytic analysis, it resulted in different estimates of the intervention's effectiveness (i.e., LMWH was "apparently" effective using some scales and ineffective using others). Psychometric theory would predict this, and practice shows it. This observation is important — if different scales are applied to the same trial, inconsistent quality assessments and estimates of an intervention's effectiveness, often in the opposite direction, can result. Given the unfortunate process by which quality scales have been developed, this finding is only altogether unexpected. Indeed, previous research pointed in this direction [24]. In the four years since this publication, there has been no published replication or extension of this work. Until this happens, the results should be cautiously interpreted.

Scales can provide holistic information that may not be forthcoming with the use of a component approach. It is ironic that the use of scales in the context of systematic reviews is considered so problematic. The science of systematic reviews is predicated, somewhat, on the axiom that the sum is better than the individual parts. Yet, summary measures of quality assessment are considered inappropriate by some.

Pocock and colleagues reported that a statement about sample size was mentioned in only 5 (11.1%) of 45 reports, and that only 6 (13.3%) made use of confidence intervals [28]. These investigators also noted that the statistical analysis tended to exaggerate intervention efficacy because authors reported a higher proportion of statistically significant results in their abstracts compared to the body of the papers. A review of 122 published RCTs that evaluated the effectiveness of selective serotonin reuptake inhibitors (SSRIs) as first-line management strategy for depression found that only one (0.8%) paper described randomization adequately [15]. A review of 2000 reports of trials in schizophrenia indicated that only 1% achieved a maximum possible quality score and there was little improvement over time [37]. Such results are the rule rather than the exception. And until quite lately, such results would have had little impact on clinical trialists and others.

A landmark investigation from 1995 found empirical evidence that results may be biased when trials use inferior methods, or are reported without adequately describing the methods. Notably, the failure to conceal the allocation process is associated with an exaggeration of the effectiveness of an intervention by 30% or more [30]. Schulz and colleagues assessed the quality of randomization reporting for 250 controlled trials extracted from 33 meta-analyses of topics in pregnancy and childbirth, and then analyzed the associations between those assessments and estimated intervention effects. Trials in which the allocation sequence had been inadequately or unclearly concealed yielded larger estimates of treatment effects (odds ratios exaggerated, on average, by 30–40%) than those in which authors reported adequate allocation concealment. These results provide strong empirical evidence that inadequate allocation concealment contributes to bias in estimating treatment effects.

Several studies subsequently examined the relationship between quality and estimated intervention effects [2,20,22,27]. In an important demonstration of the contributions methodologists can make in detecting and understanding the extent by which bias can influence the results of clinical trials, Egger and colleagues have set a standard: completing a series of systematic reviews around methodological questions, such as quality [10]. These authors report that, after pooling 838 RCT reports, the effect of low quality on the estimates of an intervention's effectiveness may be large, in the order of 30%, although there may be some differences in some methodological evaluations. The cause for concern is obvious: treatments may be introduced that are perhaps less effective than was thought, or even ineffective (Fig. 1).

In the mid-1990s, two independent initiatives to improve the quality of reports of RCTs led to the publication of the "original" CONSORT statement that was developed by an international group of clinical trialists, statisticians, epidemiologists, and biomedical editors [3]. The CONSORT statement consists of a checklist and flow diagram. The statement was revised in 2001, fine-tuning its focus on the "simple" two-group parallel design [26] along with an accompanying explanation and elaboration document [1]. The intent of the latter document, known as the E & E, is to help clarify the meaning and rationale of each checklist item (Table 1, Fig. 2).

Two aspects separate CONSORT from previous efforts. First, authors are asked to report particulars about the conduct of their studies because failure to do so clearly is associated with producing biased treatment results. This approach is in keeping with the emergence of evidence-based health care. Second, CONSORT was inclusionary, whereby a wide variety of experts, including clinical trialists, methodologists, epidemiologists, statisticians, and editors, participated in the whole process.

Continual review and updating of CONSORT is essential. To maintain the activities of the CONSORT Group requires considerable effort, and a mechanism has been developed to monitor the evolving literature and help keep the CONSORT statement evidence-based. Some of the items on the CONSORT checklist are already justified by solid evidence that they affect the validity of the trial being reported. Methodological research validating other items is reported in a diverse set of journals, books, and proceedings. In order to bring this body of evidence together, several CONSORT members have formed an "ESCORT" working party. They are starting to track down, appraise, and annotate reports that provide "Evidence Supporting (or refuting) the CONSORT standards On Reporting Trials" (ESCORT). The ESCORT group would appreciate receiving citations of reports our readers consider relevant to any items on our checklist (via the CONSORT Website).

CONSORT has been supported by a growing number of medical and health care journals (e.g., *Canadian Medical Association Journal, Journal of the American Medical Association,* and *British Medical Journal*) and editorial groups, including the International Committee of Medical Journal Editors (ICMJE, The Vancouver Group), the Council of Science Editors (CSE), and the World Association of Medical Editors (WAME). CONSORT is also published in multiple languages. It can be accessed together with other information about the CONSORT group at *www.consort-statement.org.*

There have been some initial indica-

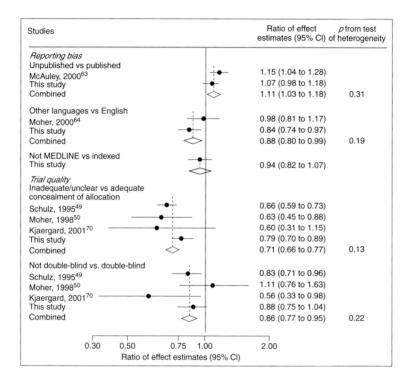

Figure 1: Meta-analysis of empirical studies of reporting bias and trial quality. All studies compared estimates of treatment effects within a large number of meta-analyses and calculated ratios of effect estimates for this purpose. A ratio of estimates below 1.0 indicates that trials with the characteristic (e.g., published in a language other than English) showed a more beneficial treatment effect. (Adapted from Egger et al. by permission of Oxford University Press.)

Table 1: Checklist of items to include when reporting a randomized trials.

Paper Section and Topic	Item Number	Descriptor
Title and abstract	1	How participants were allocated to interventions (e.g., "random allocation," "randomized," or "randomly assigned")
Introduction		
Background	2	Scientific background and explanation of rationale
Methods		
Participants	3	Eligibility criteria for participants and the settings and locations where the data were collected
Interventions	4	Precise details of the interventions intended for each group and how and when they were actually administered
Objectives	5	Specific objectives and hypotheses
Outcomes	6	Clearly defined primary and secondary outcome measures and, when applicable, any methods used to enhance the quality of measurements (e.g., multiple observations, training of assessors)
Sample size	7	How sample size was determined and, when applicable, explanation of any interim analyses and stopping rules
Randomization		
Sequence generation	8	Method used to generate the random allocation sequence, including details of any restriction (e.g., blocking, stratification)
Allocation concealment	9	Method used to implement the random allocation sequence (e.g., numbered containers or central telephone), clarifying whether the sequence was concealed until interventions were assigned.
Implementation	10	Who generated the allocation sequence, who enrolled participants, and who assigned participants to their groups
Binding (masking)	11	Whether participants, those administering the interventions, and those assessing the outcomes were blinded to group assignment; if done, how the success of blinding was evaluated
Statistical methods	12	Statistical methods used to compare groups for primary outcome(s); methods for additional analyses, such as subgroup analyses and adjusted analyses
Results		
Participant flow	13	Flow of participants through each stage (a diagram is strongly recommended); specifically, for each group report the numbers of participants randomly assigned, receiving intended treatment, completing the study protocol, and anlayzed for the primary outcome; describe protocol deviations from study as planned, together with reasons

Table 1: *Continued*

Recruitment	14	Dates defining the periods of recruitment and follow-up
Baseline data	15	Baseline demographic and clinical characteristics of each group
Numbers analyzed	16	Number of participants (denominator) in each group included in each analysis and whether the analysis was by "intention to treat"; state the results in absolute numbers when feasible (e.g., 10:20, not 50%).
Outcomes and estimation	17	For each primary or secondary outcome, a summary of results for each group and the estimated effect size and its precision (e.g., 95% confidence interval)
Ancillary analyses	18	Address multiplicity by reporting any other analyses performed, including subgroup analyses and adjusted analyses, indicating those prespecified and those exploratory
Adverse events	19	All important adverse events or side effects in each intervention group
Discusssion		
Interpretation	20	Interpretation of the results, taking into account study hypotheses, sources of potential bias or imprecision, and the dangers associated with multiplicity of analyses and outcomes.
Generalizability	21	Generalizability (external validity) of the trial findings.
Overall evidence	22	General interpretation of the results in the context of current evidence

tions that the use of CONSORT does improve the quality of reporting of RCTs. Moher and colleagues examined 71 published RCTs, in three journals in 1994 and found that allocation concealment was not clearly reported in 61% ($n = 43$) of the RCTs [25]. Four years later, after these three journals required authors reporting an RCT to use CONSORT, the percentage of papers in which allocation concealment was not clearly reported had dropped to 39% (30 of 77, mean difference = −22%; 95% confidence interval of the difference: −8%, −6%). Devereaux and colleagues reported similar encouraging results in an evaluation of 105 RCT reports from 29 journals [8]. CONSORT "promoter" journals reported a statistically higher number of factors (6.0 of 11) compared to nonpromoter journals (5.1). Egger and colleagues examined the usefulness of the flow diagram by reviewing 187 RCT reports published during 1998 in four CONSORT "adopting" journals, comparing them to 83 reports from a non-adopting journal [11]. They observed that the use of flow diagrams led to better reporting, in general.

Although the simple two-group design is perhaps the most common design reported, a quick examination of any journal issue would indicate that other designs are used and reported. While most elements of the CONSORT statement apply equally to these other designs, certain elements need to be adapted, and in some cases, additional elements need to be added, to adequately report these other designs. The CONSORT Group is now developing CONSORT "extension papers" to fill

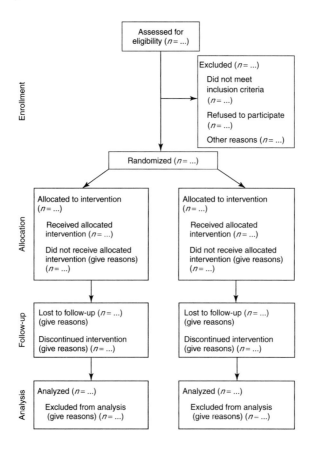

Figure 2: Flow diagram of the progress through the phases of a randomized trial (enrollment, intervention allocation, follow-up, and data analysis).

in the gaps.

A CONSORT extension for reporting randomized cluster (group) designs was published recently [6]. Other trial extension papers in development consider equivalence, noninferiority, multiarmed parallel, factorial (a special case of multiarm), and concurrent within individual trials. The six articles will have a standard structure, mirroring features of previous publications. The CONSORT Group will introduce and explain the key methodological features of that design, consider empirical evidence about how commonly these trials have been used (and misused), and review any published evidence relating to the quality of reporting of such trials. After these brief literature reviews, the Group will provide a design-specific CONSORT checklist (and flow diagram, if applicable), and provide examples of good reporting. Our goal is to publish these extensions in different journals, in the hope of increasing their dissemination throughout the disciplines of clinical medicine.

The poor quality of reporting of harm (safety, side effects) in RCTs has received considerable attention. Among 60 RCTs on antiretroviral treatment with at least 100 patients, only a minority of reports provided reasons and numbers per arm of withdrawals resulting from toxicity, and of participants with severe or life-threatening, clinically adverse events [16]. These observations have been validated in a substantially larger study of 192 trials covering antiretroviral therapy and six other medical fields [17]. Overall, the space allocated to safety in the "Results" section of RCTs was slightly less than that allocated to the names of the authors and their affiliations. The median space was only 0.3 page across the seven medical fields.

To help address these problems, the Group developed a paper similar in format to the other extension papers. Ten recommendations that clarify harm-related issues are each accompanied by an explanation and examples to highlight specific aspects of proper reporting. For example, fever in vaccine trials may be defined with different cutoffs, measured at various body sites, and at different times after immunization [5]. The results of such assessments are obviously problematic. The fourth recommendation asks authors to report whether the measuring instruments used to assess adverse events were standardized and validated. This document is currently under peer review (personal communication, John Ioannidis).

There is an increasing need to standardize many aspects of RCT conduct and reporting. Until such time as this happens, these studies run the ever-increasing risk of misadventure and inappropriate interpretation. For example, Devereaux and colleagues reported on "attending" internal medicine physicians interpretation of various aspects of blinding in the context of trials [9]. The 91 respondents (92% response rate) provided 10, 17, and 15 unique interpretations of single, double, and triple blinding, respectively.

More than 41,000 RCTs are now actively recruiting participants (CenterWatch, *http://www.centerwatch.com/*, accessed Feb. 25, 2004). As their numbers suggest, such studies form an important and central role in the development and maintenance in the delivery of evidence-based health care. If RCTs are to be conducted and reported with the highest possible standards, considerable energies must be spent on improving their conduct and reporting. Only through a well-funded and continuing program of research, evaluation, and dissemination will standard-making groups be able to provide up-to-date knowledge and guidance as to the importance of specific conduct and reporting recommendations.

References

1. Altman, D. G., Schulz, K. F., Moher, D., Egger, M., Davidoff, F., Elbourne, D., Gøtzsche, P. C., and Lang, T., The CONSORT Group (2001). The revised CONSORT statement for reporting randomized trials: Explanation and elaboration. *Ann. Intern. Med.*, **134**, 663–694.

2. Balk, E. M., Bonis, P. A., Moskowitz, H., Schmid, C. H., Ioannidis, J. P., Wang, C., and Lau, J. (2002). Correlation of quality measures with estimates of treatment effect in meta-analyses of randomized controlled trials. *JAMA*, **287**, 2973–2982.

3. Begg, C., Cho, M., Eastwood, S., Horton, R., Moher, D., Olkin, I., Pitkin, R., Rennie, D., Schulz, K., Simel, D., and Stroup, D. (1996). Improving the quality of reporting of randomized controlled trials: The CONSORT statement. *JAMA*, **276**, 637–639.

4. Berlin, J. A. and Rennie, D. (1999). Measuring the quality of trials: The quality of quality scales. *JAMA*, **282**, 1083–1085.

5. Bonhoeffer, J., Kohl, K., Chen, R., et al. (2001). The Brighton Collaboration: Addressing the need for standardized case definitions of adverse events following immunization (AEFI). *Vaccine*, **21**, 298–302.

6. Campbell, M. K., Elbourne, D. R., and Altman, D. G., The CONSORT Group.

The CONSORT statement: extension to cluster randomised trials. *Br. Med. J.*, in press.

7. Chalmers, T. C., Smith, H., Blackburn, B., Silverman, B., Schroeder, B., Reitman, D., and Ambroz, A. (1981). A method for assessing the quality of a randomized control trial. *Controlled Clin. Trials*, **2**, 31–49.

8. Devereaux, P. J., Manns, B. J., Ghali, W. A., Quan, H., and Guyatt, G. H. (2002). The reporting of methodological factors in randomized controlled trials and the association with a journal policy to promote adherence to the Consolidated Standards of Reporting Trials (CONSORT) checklist. *Controlled Clin. Trials*, **23**, 380–388.

9. Devereaux, P. J., Manns, B. J., Ghali, W. A., Quan, H., Lacchetti, C., Montori, V. M., Bhandari, M., and Guyatt, G. H. (2001). Physician interpretations and textbook definitions of blinding terminology in randomized controlled trials. *JAMA*, **285**, 2000–2003.

10. Egger, M., Juni, P., Bartlett, C., Holenstein, F., and Sterne, J. (2003). How important are comprehensive literature searches and the assessment of trial quality in systematic reviews? Empirical study. *Health Technol. Assess.*, **7**, 1–76.

11. Egger, M., Juni, P., and Bartlett, C., The CONSORT Group (2001). Value of flow diagrams in reports of randomized controlled trials. *JAMA*, **285**, 1996–1999.

12. Godlee F. (2001). Publishing study protocols: Making them visible will improve registration, reporting and recruitment. *BMC News Views*, **2**(4), 1.

13. Hadhazy, V., Ezzo, J., and Berman, B. (1999). *How Valuable Is Effort to Contact Authors to Obtain Missing Data in Systematic Reviews.* Presented at The VII Cochrane Colloquium, Rome, Italy, Oct. 5–9, 1999.

14. Hill, C. L., LaValley, M. P., and Felson, D. T. (2002). Discrepancy between published report and actual conduct of randomized clinical trials. *J. Clin. Epidemiol.*, **55**, 783–786.

15. Hotopf, M., Lewis, G., and Normand, C. (1997). Putting trials on trial—the costs and consequences of small trials in depression: A systematic review of methodology. *J. Epidemiol. Community Health*, **51**, 354–358.

16. Ioannidis, J. P. A. and Contopoulos-Ioannidis, D. G. (1998). Reporting of safety data from randomized trials. *Lancet*, **352**, 1752–1753.

17. Ioannidis, J. P. A. and Lau, J. (2001). Completeness of safety reporting in randomized trials: an evaluation of 7 medical areas. *JAMA*, **285**, 437–443.

18. Jadad, A. R., Moore, R. A., Carroll, D., Jenkinson, C., Reynolds, D. J., Gavaghan, D. J., and McQuay, H. J. (1996). Assessing the quality of reports of randomized clinical trials: Is blinding necessary? *Control. Clin. Trials*, **17**, 1–12.

19. Jüni, P., Altman, D. G., and Egger, M. (2001). Systematic reviews in health care: assessing the quality of controlled clinical trials. *Br. Med. J.*, **323**, 42–46.

20. Jüni, P., Tallon, D., and Egger, M. (2000). "Garbage in—garbage out?" Assessment of the quality of controlled trials in meta-analyses published in leading journals. *Proc. 3rd Symp. Systematic Reviews: Beyond the Basics* pp. 19–, St Catherine's College, Oxford. Centre for Statistics in Medicine, Oxford, UK.

21. Jüni, P., Witschi, A., Bloch, R., and Egger, M. (1999). The hazards of scoring the quality of clinical trials for meta-analysis. *JAMA*, **282**, 1054–1060.

22. Kjaergard, L. L., Villumsen, J., and Gluud, C. (2001). Reported methodologic quality and discrepancies between large and small randomized trials in meta-analyses. *Ann. Intern. Med.*, **135**, 982–989.

23. Liberati, A., Himel, H. N., and Chalmers, T. C. (1986). A quality assessment of randomized control trials of primary treatment of breast cancer. *J. Clin. Oncol.*, **4**, 942–951.

24. Moher, D., Jadad, A. R., and Tugwell, P. (1996). Assessing the quality of randomized controlled trials: Current issues

and future directions. *Int. J. Technol. Assess. Health Care*, **12**, 195–208.

25. Moher, D., Jones, A., and Lepage, L., The CONSORT Group (2001). Use of the CONSORT statement and quality of reports of randomized trials: A comparative before and after evaluation? *JAMA* **285**, 1992–1995.

26. Moher, D., Schulz, K. F., and Altman, D. G., The CONSORT Group (2001). The CONSORT statement: Revised recommendations for improving the quality of reports of parallel group randomized trials. *Ann. Intern. Med.*, **134**, 657–662.

27. Moher, D., Pham, B., Jones, A., Cook, D. J., Jadad, A. R., Moher, M., Tugwell, P., and Klassen, T. P. (1998). Does quality of reports of randomised trials affect estimates of intervention efficacy reported in meta-analyses? *Lancet*, **352**, 609–613.

28. Pocock, S. J., Hughes, M. D., and Lee, R. J. (1987). Statistical problems in the reporting of clinical trials. *New Engl. J. Med.*, **317**, 426–432.

29. Schulz, K. F., Chalmers, I., Grimes, D. A., and Altman, D. G. (1994). Assessing the quality of randomization from reports of controlled trials published in obstetrics and gynecology journals. *JAMA*, **272**, 125–128.

30. Schulz, K. F., Chalmers, I., Hayes, R. J., and Altman, D. G. (1995). Empirical evidence of bias: dimensions of methodological quality associated with estimates of treatment effects in controlled trials. *JAMA*, **273**, 408–412.

31. Schulz, K. F and Grimes, D. A. (2002). Sample size slippages in randomised trials: exclusions and the lost and wayward. Lancet, **359**, 781–785.

32. Schulz, K. F. and Grimes, D. A. (2002). Unequal group sizes in randomised trials: guarding against guessing. *Lancet*, **359**, 966–970.

33. Schulz, K. F. and Grimes, D. A. (2002). Blinding in randomised trials: hiding who got what. *Lancet*, **359**, 696–700.

34. Schulz, K. F. and Grimes, D. A. (2002). Allocation concealment in randomised trials: Defending against deciphering. *Lancet*, **359**, 614–618.

35. Schulz, K. F. and Grimes, D. A. (2002). Generation of allocation sequences in randomised trials: Chance, not choice. *Lancet*, **359**, 515–519.

36. Soares, H. P., Daniels, S., Kumar, A., Clarke, M., Scott, C., Swann, S., and Djulbegovic, B. (2004). Bad reporting does not mean bad methods for randomised trials: Observational study of randomised controlled trials performed by the Radiation Therapy Oncology Group. *Br. Med. J.*, **328**, 22–24.

37. Thornley, B. and Adams, C. E. (1998). Content and quality of 2,000 controlled trials in schizophrenia over 50 years. *Br. Med. J.*, **317**, 1181–1184.

62 Repeated Measurements

Geert Verbeke and Emmanuel Lesaffre

62.1 Introduction and Case Study

In medical science, studies are often designed to investigate changes in a specific parameter that is measured repeatedly over time in the participating subjects. Such studies are called *longitudinal studies*, in contrast to *cross-sectional studies*, where the response of interest is measured only once for each individual. As pointed out by Diggle et al. [6], the main advantage of longitudinal studies is that they can distinguish changes over time within individuals (longitudinal effects) from differences among people in their baseline values (cross-sectional effects).

In randomized clinical trials, where the aim is usually to compare the effect of two (or more) treatments at a specific timepoint, the need for and the advantage of taking repeated measures is, at first sight, not very obvious. Indeed, a simple comparison of the treatment groups at the end of the follow-up period is often sufficient to establish the treatment effect(s) (if any) by virtue of the randomization. However, in some instances, it is important to know how the patients have reached their endpoint, that is, it is important to compare the average profiles (over time) between the treatment groups. Further, longitudinal studies can be more powerful than those evaluating the treatments at any one single timepoint. Finally, follow-up studies often suffer from dropout, that is, some patients leave the study prematurely, for known or unknown reasons. In such cases, a full repeated measures analysis will help in drawing inferences at the end, on account of the fact that such analyses implicitly impute the missing values. As a typical example, we consider data from a randomized, double-blind, parallel group, multicenter study for the comparison of two oral treatments (in the sequel coded as A and B) for toenail dermatophyte onychomycosis (TDO). See Reference 5 for more details about this study. TDO is a common toenail infection, difficult to treat, affecting more than 2% of the population. Antifungal compounds classically used for treatment of TDO need to be taken until the whole nail has grown out healthy. However, new compounds have reduced the treatment duration to three months. The aim of the present study was to compare the efficacy and safety of two such new compounds, labelled A and B, and administered during 12 weeks.

In total, 2×189 patients were randomized, and distributed over 36 centers. Subjects were monitored during 12 weeks (3 months) of treatment and further up to a

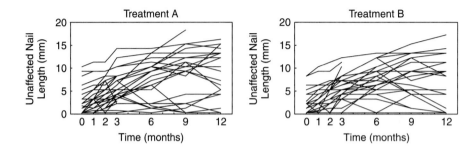

Figure 1: Toenail data: Individual profiles of 30 randomly selected subjects in each treatment arm.

Table 1: Toenail data: number and percentage of patients with severe toenail infection, for each treatment arm separately.

	Group A			Group B		
	# Severe	# Patients	Percentage	# Severe	# Patients	Percentage
Baseline	54	146	37.0%	55	148	37.2%
1 month	49	141	34.7%	48	147	32.6%
2 months	44	138	31.9%	40	145	27.6%
3 months	29	132	22.0%	29	140	20.7%
6 months	14	130	10.8%	8	133	6.0%
9 months	10	117	8.5%	8	127	6.3%
12 months	14	133	10.5%	6	131	4.6%

total of 48 weeks (12 months). Measurements were taken at baseline, every month during treatment, and every 3 months afterward resulting in a maximum of 7 measurements per subject. As a first response parameter, we consider the unaffected nail length in millimeters (one of the secondary endpoints in the study), measured from the nailbed to the infected part of the nail, which is always at the free end of the nail. Obviously, this response will be related to the toe size. Therefore, we will include here only those patients for whom the target nail was one of the two big toenails. This reduces our sample under consideration to 146 and 148 subjects, respectively. Individual profiles for 30 randomly selected subjects in each treatment group are shown in Figure 1. Our second parameter will be severity of the infection, coded as 0 (not severe) or 1 (severe). The question of interest was whether or not the percentage of severe infections decreased over time, and whether that evolution was different for the two treatment groups. A summary of the number of patients in the study at each timepoint, and the number of patients with severe infections, is given in Table 1.

A key issue in the analysis of longitudinal data is that outcome values measured repeatedly within the same subjects tend to be correlated, and this correlation structure needs to be taken into account in the statistical analysis. This is easily seen with paired observations obtained from, for example, a pretest/posttest experiment. An obvious choice for the analysis is the paired *t*test, based on the subject-specific difference between the two measurements. While an unbiased estimate for the treatment effect can also be obtained from a two-sample *t*test, standard errors and hence also *p*values and confidence intervals obtained from not accounting for the correlation within pairs will not reflect the correct sampling variability, and hence will lead to wrong inferences.

In general, classical statistical procedures assuming independent observations cannot be used in the context of repeated measurements. In this chapter, we will give an overview of the most important models useful for the analysis of clinical trial data, and widely available through commercial statistical software packages. In Section 62.2, we will first focus on linear models for Gaussian data. In Section 62.3, we will discuss models for the analysis of discrete outcomes. Section 62.4 deals with some design issues, and we end Section 62.5 with some concluding remarks.

62.2 Linear Models for Gaussian Data

With repeated Gaussian data, a general, and very flexible, class of parametric models is obtained from a random-effects approach. Suppose that an outcome Y is observed repeatedly over time for a set of persons, and also that the individual trajectories are of the type shown in Figure 2. Obviously, a regression model with intercept and linear time effect seems plausible to describe the data of each person separately. However, the trajectories for different persons tend to have different intercepts and different slopes. One can therefore assume that the jth outcome Y_{ij} of subject i ($i = 1, \ldots, N, j = 1, \ldots, n_i$), measured at time t_{ij} satisfies $Y_{ij} = \tilde{b}_{i0} + \tilde{b}_{i1} t_{ij} + \varepsilon_{ij}$. Assuming the vector $\boldsymbol{\tilde{b}_i} = (\tilde{b}_{i0}, \tilde{b}_{i1})'$ of person-specific parameters to be bivariate normal with mean $(\beta_0, \beta_1)'$ and 2×2 covariance matrix D and assuming the errors ε_{ij} to be normal as well, this leads to a so-called linear mixed model. In practice, one will often formulate the model as

$$Y_{ij} = (\beta_0 + b_{i0}) + (\beta_1 + b_{i1})t_{ij} + \varepsilon_{ij},$$

with $\tilde{b}_{i0} = \beta_0 + b_{i0}$ and $\tilde{b}_{i1} = \beta_1 + b_{i1}$, and the new random effects $\boldsymbol{b_i} = (b_{i0}, b_{i1})'$ are now assumed to have mean zero.

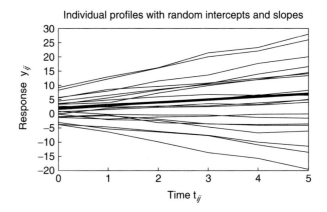

Figure 2: Hypothetical example of continuous longitudinal data that can be accurately described by a linear mixed model with random intercepts and random slopes. The thin lines represent the observed subject-specific evolutions. The bold line represents the population-averaged evolution. Measurements are taken at six timepoints: 0, 1, 2, 3, 4, 5.

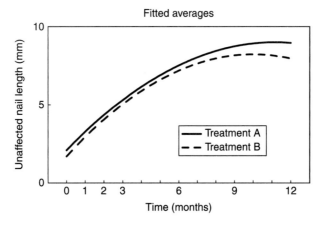

Figure 3: Toenail data: fitted average profiles based on model (3).

The above model can be viewed as a special case of the general linear mixed model, which assumes that the outcome vector $\boldsymbol{Y_i}$ of all n_i outcomes for subject i satisfies

$$\boldsymbol{Y_i} = X_i\boldsymbol{\beta} + Z_i\boldsymbol{b_i} + \boldsymbol{\varepsilon_i}, \quad (1)$$

in which $\boldsymbol{\beta}$ is a vector of population-average regression coefficients called *fixed effects*, and $\boldsymbol{b_i}$ is a vector of subject-specific regression coefficients. The $\boldsymbol{b_i}$ are assumed normal with mean vector 0 and covariance D, and they describe how the evolution of the ith subject deviates from the average evolution in the population. The matrices X_i and Z_i are $(n_i \times p)$ and $(n_i \times q)$ matrices of known covariates. Note that p and q are the numbers of fixed and subject-specific regression parameters in the model, respectively. The residual components $\boldsymbol{\varepsilon_i}$ are assumed to be independent $N(\boldsymbol{0}, \Sigma_i)$, where Σ_i depends on i only through its dimension n_i. Model (1) naturally follows from a so-called two-stage model formulation. First, a linear regression model is specified for every subject separately, modeling the outcome variable as a function of time. Afterward, in the second stage, multivariate linear models are used to relate the subject-specific regression parameters from the first-stage model to subject characteristics such as age, gender, and treatment. Estimation of the parameters in (1) is usually based on maximum likelihood (ML) or restricted maximum likelihood (REML) estimation for the marginal distribution of $\boldsymbol{Y_i}$ which can easily be seen to be

$$\boldsymbol{Y_i} \sim N(X_i\boldsymbol{\beta}, Z_i D Z_i' + \Sigma_i). \quad (2)$$

Note that model (1) implies a model with very specific mean and covariance structure, which may or may not be valid, and hence needs to be checked for every specific dataset at hand. Note also that, when $\Sigma_i = \sigma^2 I_{n_i}$, with I_{n_i} equal to the identity matrix of dimension n_i, the observations of subject i are independent conditionally on the random effect $\boldsymbol{b_i}$. The model is therefore called the conditional independence model. Even in this simple case, the assumed random-effects structure still imposes a marginal correlation structure for the outcomes Y_{ij}. Indeed, even if all Σ_i equal $\sigma^2 I_{n_i}$, the covariance matrix in (2) is not the identity matrix, illustrating that, marginally, the repeated measurements Y_{ij} of subject i are not assumed to be uncorrelated. Another special case arises when the random effects are omitted from the model. In that case, the covariance matrix of $\boldsymbol{Y_i}$ is modeled through the residual covariance matrix Σ_i. In the case of completely balanced data, that is, when n_i is the same for all subjects, and when the measurements are all taken at fixed timepoints, one can assume all Σ_i to be equal to a general unstructured covariance matrix Σ, which results in the classical multivariate regression model.

Inference in the marginal model can be done using classical techniques including approximate Wald tests, $t$tests, Ftests, or likelihood ratio tests. Finally, Bayesian methods can be used to obtain "empirical Bayes estimates" for the subject-specific parameters $\boldsymbol{b_i}$ in (1). See References 18, 12, 13, 14, 19, 39, and 40 for more details about estimation and inference in linear mixed models.

As an illustration, we analyze the unaffected nail length response in the toenail example. The model proposed by Verbeke, et al. [41] assumes a quadratic evolution for each subject, with subject-specific intercepts, and with correlated errors within subjects. More formally, they assume that Y_{ij} satisfies

$$Y_{ij}(t)$$
$$= \begin{cases} (\beta_{A0} + b_i) + \beta_{A1}t + \beta_{A2}t^2 + \varepsilon_i(t), \\ \quad \text{in group A} \\ (\beta_{B0} + b_i) + \beta_{B1}t + \beta_{B2}t^2 + \varepsilon_i(t), \\ \quad \text{in group B,} \end{cases}$$
$$(3)$$

where $t = 0, 1, 2, 3, 6, 9, 12$ is the time in

the study, expressed in months. The error components $\varepsilon_i(t)$ are assumed to have common variance σ^2, with correlation of the form $\text{corr}(\varepsilon_i(t), \varepsilon_i(t-u)) = \exp(-\varphi u^2)$ for some unknown parameter φ. Hence, the correlation between within-subject errors is a decreasing function of the time span between the corresponding measurements. Fitted average profiles are shown in Figure 3. An approximate F test shows that, on average, there is no evidence for a treatment effect ($p = 0.2029$).

Note that, even when interest was only in comparing the treatment groups after 12 months, this could still be done based on the above fitted model. The average difference between group A and group B, after 12 months, is given by $(\beta_{A0} - \beta_{B0}) - 12(\beta_{A1} - \beta_{B1}) + 12^2(\beta_{A2} - \beta_{B2})$. The estimate for this difference equals 0.80 mm ($p = 0.0662$). Alternatively, a two-sample t test could be performed based on those subjects which have completed the study. This yields an estimated treatment effect of 0.77 mm ($p = 0.2584$) illustrating that modeling the whole longitudinal sequence also provides more efficient inferences at specific timepoints.

62.3 Models for Discrete Outcomes

Whenever discrete data are to be analyzed, the normality assumption in the models in the previous section is no longer valid, and alternatives need to be considered. An extensive discussion on various models for discrete repeated measurements is given by Molenberghs and Verbeke [31]. The classical route, in analogy to the linear model, is to specify the full joint distribution for the set of measurements $Y_{ij}, \ldots, Y_{in_i}$ per individual. Clearly, this implies the need to specify all moments up to order n_i. Examples of marginal models can be found in References 1, 2, 7, 8, 20, 30, and 31.

For longer sequences, especially, and/or in cases where observations are not taken at fixed timepoints for all subjects, specifying a full likelihood and making inferences about its parameters, traditionally done using maximum likelihood principles, can become very cumbersome. Therefore, inference is often based on a likelihood obtained from a random-effects approach. Associations and all higher-order moments are then implicitly modeled through a random-effects structure. This will be discussed in Section 62.3.1. A disadvantage is that the assumptions about all moments are made implicitly, and are very hard to check. As a consequence, alternative methods have been suggested, which require the specification of only a small number of moments, leaving the others completely unspecified. In a large number of cases, one is primarily interested in the mean structure, whence only the first moments need to be specified. Sometimes, there is also interest in the association structure, quantified for example using odds ratios or correlations. Estimation is then based on so-called generalized estimating equations, and inference no longer directly follows from maximum likelihood theory. This will be explained in Section 62.3.2. In Section 62.3.3, both approaches will be illustrated in the context of the toenail data. A comparison of both techniques will be presented in Section 62.3.4.

62.3.1 Generalized Linear Mixed Models (GLMM)

As discussed in Section 62.2, random effects can be used to generate an association structure between repeated measurements. This can be exploited to specify a full joint likelihood in the context of discrete outcomes. More specifically, conditionally on a vector $\boldsymbol{b_i}$ of subject-specific regression coefficients, it is assumed that all responses Y_{ij} for a single subject i are independent,

Table 2: Toenail data: parameter estimates (standard errors) for a generalized linear mixed model (GLMM) and a marginal model (GEE).

Parameter	GLMM Estimate (SE)	GEE Estimate (SE)
Intercept group $A(\beta_{A0})$	−1.63 (0.44)	−0.72 (0.17)
Intercept group $B(\beta_{B0})$	−1.75 (0.45)	−0.65 (0.17)
Slope group $A(\beta_{A1})$	−0.40 (0.05)	−0.14 (0.03)
Slope group $B(\beta_{B1})$	−0.57 (0.06)	−0.25 (0.04)
Random intercepts SD (σ)	4.02 (0.38)	

satisfying a generalized linear model with mean $\mu_{ij} = h(\boldsymbol{x}_{ij}'\boldsymbol{\beta} + \boldsymbol{z}_{ij}'\boldsymbol{b}_i)$ for a prespecified link function h, and for two vectors $\boldsymbol{x}_{ij}$ and $\boldsymbol{z}_{ij}$ of known covariates belonging to subject i at the jth timepoint. Let $f_{ij}(y_{ij}|\boldsymbol{b}_i)$ denote the corresponding density function of Y_{ij}, given $\boldsymbol{b}_i$. As for the linear mixed model, the random effects $\boldsymbol{b}_i$ are assumed to be sampled from a normal distribution with mean vector $\boldsymbol{0}$ and covariance D. The marginal distribution of $\boldsymbol{Y}_i$ is then given by

$$f(\boldsymbol{y}_i) = \int \prod_{j=1}^{n_i} f_{ij}(y_{ij}|\boldsymbol{b}_i) f(\boldsymbol{b}_i) \boldsymbol{b}_i, \quad (4)$$

in which dependence on the parameters $\boldsymbol{\beta}$ and D is suppressed from the notation. Assuming independence across subjects, the likelihood can easily be obtained, and maximum-likelihood estimation becomes available.

In the linear model, the integral in (4) could be worked out analytically, leading to the normal marginal model (2). In general, however, this is no longer possible, and numerical approximations are needed. Broadly speaking, we can distinguish between approximations to the integrand in (4), and methods based on numerical integration. In the first approach, Taylor series expansions to the integrand are used, simplifying the calculation of the integral. Depending on the order of expansion and the point around which one expands, slightly different procedures are followed. See References 3, 21, 22, and 43 for an overview of estimation methods. In general, such approximations will be accurate whenever the responses Y_{ij} are "sufficiently continuous" and/or if all n_i are sufficiently large. This explains why the approximation methods perform poorly in cases with binary repeated measurements with a relatively small number of measurements available for all subjects [43]. In such examples, numerical integration proves especially useful. Of course, a wide kit of numerical integration tools, available from the optimization literature, can be applied. A general class of quadrature rules selects a set of abscissas and constructs a weighted sum of function evaluations over those. See References 15, 16, and 34 for more details on numerical integration methods in the context of random-effects models.

62.3.2 Generalized Estimating Equations (GEE)

Liang and Zeger [23] proposed the so-called *generalized estimating equations* (GEEs), which require only the correct specification of the univariate marginal distributions, provided, one is willing to adopt "working" assumptions about the association structure. More specifically, a generalized linear model [28] is assumed for each response Y_{ij}, modeling the mean μ_{ij} as $h(\boldsymbol{x}_{ij}'\boldsymbol{\beta})$ for a prespecified link function h, and a vector x_{ij} of known covariates. In case of independent repeated measurements, the classical score equations for the estimation of β are well known to be

$$S(\beta) = \sum_i \frac{\partial \mu_i'}{\partial \beta} V_i^{-1} (Y_i - \mu_i) = 0, \quad (5)$$

where $\mu_i = E(Y_i)$ and V_i is a diagonal matrix with $v_{ij} = \text{Var}(Y_{ij})$ on the main diagonal. Note that, in general, the mean–variance relation in generalized linear models implies that the elements v_{ij} also depend on the regression coefficients β. Generalized estimating equations are now obtained from allowing nondiagonal "covariance" matrices V_i in (5). In practice, this comes down to the specification of a 'working correlation matrix' which, together with the variances v_{ij}, results in a hypothesized covariance matrix V_i for Y_i.

Solving $S(\beta) = 0$ is done iteratively, constantly updating the working correlation matrix using moment-based estimators. Note that, in general, no maximum-likelihood estimates are obtained, since the equations are not first-order derivatives of some loglikelihood function for the data under some statistical model. Still, very similar properties can be derived. More specifically, Liang and Zeger [23] showed that $\hat\beta$ is asymptotically normally distributed, with mean β and with a covariance matrix that can easily be estimated in practice. Hence, classical Wald-type inferences become available. This result holds provided that the mean was correctly specified, whatever working assumptions were made about the association structure. This implies that, strictly speaking, one can fit generalized linear models to repeated measurements, ignoring the correlation structure, as long as inferences are based on the standard errors that follow from the general GEE theory. However, efficiency can be gained from using a more appropriate working correlation model [27].

The original GEE approach focuses on inferences for the first-order moments, considering the association present in the data as nuisance. Later on, extensions have been proposed that also allow inferences about higher-order moments. See References 24, 25, and 35 for more details on this.

62.3.3 Application to the Toenail Data

As an illustration of GEE and GLMM, we analyze the severity of infection binary outcome in the toenail example. We will first apply GEE, based on the marginal logistic regression model

$$\log\left[\frac{P(Y_i(t) = 1)}{1 - P(Y_i(t) = 1)}\right]$$
$$= \begin{cases} \beta_{A0} + \beta_{A1}t, & \text{in group } A \\ \beta_{B0} + \beta_{B1}t, & \text{in group } B \end{cases}. \quad (6)$$

Furthermore, we use an unstructured 7×7 working correlation matrix. The results are reported in Table 2, and the fitted average profiles are shown in the top graph of Figure 4. Based on a Wald-type test, we obtain a significant difference in the average slope between the two treatment groups ($p = 0.0158$).

Alternatively, we consider a generalized linear mixed model, modeling the association through the inclusion of subject-specific (random) intercepts. More specifically, we will now assume that

$$\log\left[\frac{P(Y_i(t) = 1 | b_i)}{1 - P(Y_i(t) = 1 | b_i)}\right]$$
$$= \begin{cases} \beta_{A0} + b_i + \beta_{A1}t, & \text{in group } A \\ \beta_{B0} + b_i + \beta_{B1}t, & \text{in group } B \end{cases}. \quad (7)$$

with b_i normally distributed with mean 0 and variance σ^2. The results, obtained using numerical integration methods, are also

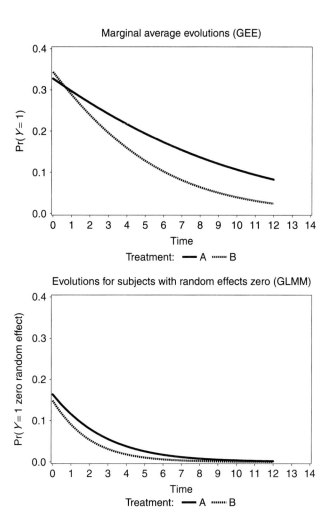

Figure 4: Toenail data. Treatment-specific evolutions: (a) marginal evolutions as obtained from the marginal model (6) fitted using GEE; (b) evolutions for subjects with random effects in model (7) equal to zero.

reported in Table 2. As before, we obtain a significant difference between β_{A1} and β_{B1} ($p = 0.0255$).

62.3.4 Marginal Versus Hierarchical Parameter Interpretation

Comparing the GEE results and the GLMM results in Table 2, we observe large differences between the parameter estimates. This suggests that the parameters in both models need to be interpreted differently. Indeed, the GEE approach yields parameters with a population-averaged interpretation. Each regression parameter expresses the average effect of a covariate on the probability of having a severe infection. Results from the generalized linear mixed model, however, require an interpretation conditionally on the random effect, i.e., conditionally on the subject. In the context of our toenail example, consider model (7) for treatment group A only. The model assumes that the probability of severe infection satisfies a logistic regression model, with the same slope for all subjects, but with subject-specific intercepts. The population-averaged probability of severe infection is obtained from averaging these subject-specific profiles over all subjects. This is graphically illustrated in Figure 5. Clearly, the slope of the average trend is different from the subject-specific slopes, and this effect will be more severe as the subject-specific profiles differ more, i.e., as the random-intercepts variance σ^2 is larger. Formally, the average trend for group A is obtained as

$$\begin{aligned}
P(Y_i(t) &= 1) \\
&= E[P(Y_i(t) = 1|b_i)] \\
&= E\left[\frac{\exp(\beta_{A0} + b_i + \beta_{A1}t)}{1 + \exp(\beta_{A0} + b_i + \beta_{A1}t)}\right] \\
&\neq E\left[\frac{\exp(\beta_{A0} + \beta_{A1}t)}{1 + \exp(\beta_{A0} + \beta_{A1}t)}\right].
\end{aligned}$$

Hence, the population-averaged evolution is not the evolution for an 'average' subject, i.e., a subject with random effect equal to zero. The bottom graph in Figure 4 shows the fitted profiles for an average subject in each treatment group, and these profiles are indeed very different from the population-averaged profiles shown in the top graph of Figure 4 and discussed before. In general, the population-averaged evolution implied by the GLMM is not of a logistic form any more, and the parameter estimates obtained from the GLMM are typically larger in absolute value than their marginal counterparts [33]. However, one should not refer to this phenomenon as bias since the two sets of parameters target at different scientific questions. Note that this difference in parameter interpretation between marginal and random-effects models immediately follows from the nonlinear nature, and therefore is absent in the linear mixed model, discussed in Section 62.2. Indeed, the regression parameter vector $\boldsymbol{\beta}$ in the linear mixed model (1) is the same as the regression parameter vector modeling the expectation in the marginal model (2).

62.4 Design Considerations

So far, we have focused on the analysis of longitudinal data. In the context of a clinical trial, however, one is usually first confronted with design questions. This involves the number of patients to be included in the study, the number of repeated measurements to be taken for each patient, as well as the time-points at which measurements will be scheduled. Which design will be "optimal" depends on many characteristics of the problem. In a cross-sectional analysis, such as the comparison of endpoints between several treatment groups, the power typically depends on the alternative to be detected, and the variance in the different treatment groups. In a longitudinal context, however, power

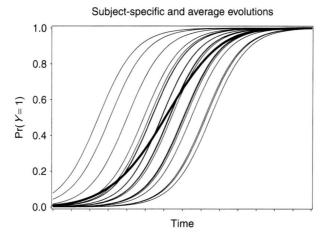

Figure 5: Graphical illustration of a random-intercepts logistic model. The thin lines represent the subject-specific logistic regression models. The bold line represents the population-averaged evolution.

will depend on the complete multivariate model that will be assumed for the vector of repeated measurements per subject. This typically includes a parametric model for the average evolution in the different treatment groups, a parametric model for how the variability changes over time, as well as a parametric model for the association structure. Not only is it difficult in practice to select such models prior to the data collection; power calculations also tend to highly depend on the actual parameter values imputed in these models. Moreover, unless in the context of linear mixed models (see Refs. 17 and 38), no analytic power calculations are possible, and simulation-based techniques need to be used instead. Therefore, power analyses are often performed for the cross-sectional comparison of endpoints, while the longitudinal analyses are considered additional, secondary analyses.

62.5 Concluding Remarks

No doubt repeated measurements occur very frequently in a variety of contexts. This leads to data structures with correlated observations, hence no longer allowing standard statistical modeling assuming independent observations. Here, we gave a general overview of the main issues in the analysis of repeated measurements, with focus on a few general classes of approaches often used in practice, and available in many commercially available statistical software packages. A much more complete overview can be found in Reference 6. Many linear models proposed in the statistical literature for the analysis of continuous data are special cases of linear mixed models discussed in Section 62.2. See References 39 and 40 for more details. We did not discuss nonlinear models for continuous data, but the nonlinearity implies important numerical and interpretational issues similar to those discussed in Section 62.3 for discrete-data models, and these are discussed in full detail in References 4 and 42. An overview of many models for discrete data can be found in Reference 8. One major approach to the analysis of correlated data is based on random-effects models, both for continuous as well as discrete outcomes. These models are presented in full detail in Reference 34.

A variety of models is nowadays available for the analysis of longitudinal data. A recent overview is given by Fitzmaurice et al. [9]. Most models pose very specific, often restrictive, assumptions about the data at hand. In many other contexts, procedures for model checking or for testing goodness of fit have been developed. For longitudinal data analysis, relatively few techniques are available, and it is not always clear to what extent inferences rely on the underlying parametric assumptions. See References 37 and 40 for a selection of available methods for model checking, and for some robustness results, in the context of linear mixed models. Since model checking is far from straightforward, attempts have been made to relax some of the distributional assumptions (see, e.g., Refs. 9, 11, and 36).

Finally, it should be noted that many applications involving repeated measures will suffer from missing data, i.e., measurements scheduled to be taken are not available, for a variety of known or (often) unknown reasons. Technically speaking, the methods that have been discussed here can handle such unbalanced data structures but, depending on the chosen analysis, biased results can be obtained if the reason for missingness is related to the outcome of interest. See References 26, 29, 35, and 40 for missing-data issues.

Nowadays, generalized estimating equations and mixed models can be fitted using a variety of (commercially available) software packages, including MIXOR, MLwiN, and Splus. However, in the context of clinical trials, the SAS procedures GENMOD (for GEE analyses), MIXED (for linear mixed models), and NLMIXED (for generalized linear and non-linear mixed models) are probably the most flexible and best documented procedures, and are therefore the most widely used ones.

Acknowledgments. The authors gratefully acknowledge support from Fonds Wetenschappelijk Onderzoek-Vlaanderen Research Project G.0002.98 Sensitivity Analysis for Incomplete and Coarse Data and from Belgian IUAP/PAI network Statistical Techniques and Modeling for Complex Substantive Questions with Complex Data.

References

1. Althan, P. M. E. (1978). Two generalizations of the binomial distribution. *Appl. Statist.*, **27**, 162–167.

2. Bahadur, R. R. (1961). A representation of the joint distribution of responses of p dichotomous items. In *Studies in Item Analysis and Prediction*, H. Solomon, ed. Stanford University Press, Stanford, CA.

3. Breslow, N. E. and Clayton, D. G. (1993). Approximate inference in generalized linear mixed models. *J. Am. Statist. Assoc.*, **88**, 9–25.

4. Davidian, M. and Giltinan, D. M. (1995). *Nonlinear Models for Repeated Measurement Data*. Chapman & Hall.

5. De Backer, M., De Keyser, P., De Vroey, C., and Lesaffre, E. (1996). A 12-week treatment for dermatophyte toe onychomycosis:terbinafine 250 mg/day vs. itraconazole 200 mg/day—a double-blind comparative trial. *Br. J. Dermatol.*, **134**, 16–17.

6. Diggle, P. J., Liang, K. Y., and Zeger, S. L. (1994). *Analysis of Longitudinal Data*. Clarendon Press, Oxford.

7. Efron, B. (1986). Double exponential families and their use in generalized linear regression. *J. Am. Statist. Assoc.*, **81**, 709–721.

8. Fahrmeir, L. and Tutz, G. (1994). *Multivariate Statistical Modelling Based on*

Generalized Linear Models. Springer Series in Statistics. Springer-Verlag, New York.

9. Fitzmaurice, G., Davidian, M., Verbeke, G., and Molenberghs, G. eds. (2009). *Longitudinal Data Analysis.* Chapman and Hall/CRC.

10. Ghidey, W., Lesaffre, E., and Eilers, P. (2004). Smooth random effects distribution in a linear mixed model. *Biometrics*, **60**, 945–953.

11. Ghidey, W., Lesaffre, E., and Verbeke, G. (2009). A comparison of methods for determining the random effects distribution of a linear mixed models. *Statist. Meth. Med. Res.*, in press.

12. Harville, D. A. (1974). Bayesian inference for variance components using only error contrasts. *Biometrika*, **61**, 383–385.

13. Harville, D. A. (1976). Extension of the Gauss-Markov theorem to include the estimation of random effects. *Ann. Statist.*, **4**, 384–395.

14. Harville, D. A. (1977). Maximum likelihood approaches to variance component estimation and to related problems. *J. Am. Statist. Assoc.*, **72**, 320–340.

15. Hedeker, D. and Gibbons, R. D. (1994). A random-effects ordinal regression model for multilevel analysis. *Biometrics*, **50**, 933–944.

16. Hedeker, D. and Gibbons, R. D. (1996). MIXOR, a computer program for mixed-effects ordinal regression analysis. *Comput. Methods Programs Biomed.*, **49**, 157–176.

17. Helms, R. W. (1992). Intentionally incomplete longitudinal designs: methodology and comparison of some full span designs. *Statist. Med.*, **11**, 1889–1913.

18. Henderson, C. R., Kempthorne, O., Searle, S. R., and VonKrosig, C. N. (1959). Estimation of environmental and genetic trends from records subject to culling. *Biometrics*, **15**, 192–218.

19. Laird, N. M. and Ware, J. H. (1982). Random-effects models for longitudinal data. *Biometrics*, **38**, 963–974.

20. Lang, J. B. and Agresti, A. (1994). Simultaneously modeling joint and marginal distributions of multivariate categorical responses. *J. Am. Statist. Assoc.*, **89**, 625–632.

21. Lavergne, C. and Trottier, C.. Sur l'estimation dans les modèles linéaires généralisés à effects aléatoires (2000). *Rev. Statist. Appl.*, **48**, 49–67.

22. Lesaffre, E. and Spiessens, B. (2001). On the effect of the number of quadrature points in a logistic random-effects model: An example. *Appl. Statist.*, **50**, 325–335.

23. Liang, K. Y. and Zeger, S. L. (1986). Longitudinal data analysis using generalized linear models. *Biometrika*, **73**, 13–22.

24. Liang, K. Y., Zeger, S. L., and Qaqish, B. (1992). Multivariate regression analyses for categorical data. *J. Roy. Statist. Soc. B*, **54**, 3–40.

25. Lipsitz, S. R., Laird, N. M., and Harrington, D. P. (1991). Generalized estimating equations for correlated binary data: Using the odds ratio as a measure of association. *Biometrika*, **78**, 153–160.

26. Little, R. J. A. and Rubin, D. B. (1987). *Statistical Analysis with Missing Data.* Wiley, New York.

27. Mancl, L. A. and Leroux, B. G. (1996). Efficiency of regression estimates for clustered data. *Biometrics*, **52**, 500–511.

28. McCullagh, P. and Nelder, J. A. (1989). *Generalized Linear Models.* second edition, Chapman & Hall, London.

29. Molenberghs, G. and Kenward, M. G. (2007). *Missing Data in Clinical Studies.* Wiley, Chichester, UK.

30. Molenberghs, G. and Lesaffre, E. (1994). Marginal modelling of correlated ordinal data using a multivariate plackett distribution. *J. Am. Statist. Assoc.*, **89**, 633–644.

31. Molenberghs, G. and Lesaffre, E. (1999). Marginal modelling of multivariate categorical data. *Statist. Med.*, **18**, 2237–2255.

32. Molenbergs, G. and Verbeke, G. (2005). *Models for Discrete Longitudinal Data.* Springer-Verlag, New York.

33. Neuhaus, J. M., Kalbfleisch, J. D., and Hauck, W. W. (1991). A comparison of cluster-specific and population-averaged approaches for analyzing correlated binary data. *Int. Statist. Rev.*, **59**, 25–30.

34. Pinheiro, J. C. and Bates, D. M. (2000). *Mixed Effects Models in S and S-plus.* Springer-Verlag, New York.

35. Prentice, R. L. (1988). Correlated binary regression with covariates specific to each binary observation. *Biometrics*, **44**, 1033–1048.

36. Verbeke, G. and Lesaffre, E. (1996). A linear mixed-effects model with heterogeneity in the random-effects population. *J. Am. Statist. Assoc.*, **91**, 217–221.

37. Verbeke, G. and Lesaffre, E. (1997). The effect of misspecifying the random effects distribution in linear mixed models for longitudinal data. *Comput. Statist. Data Anal.*, **23**, 541–556.

38. Verbeke, G. and Lesaffre, E. (1999). The effect of drop-out on the efficiency of longitudinal experiments. *Appl. Statist.*, **48**, 363–375.

39. Verbeke, G. and Molenberghs, G. (1997). *Linear Mixed Models in Practice: A SAS-oriented Approach.* No. 126. Lecture Notes in Statistics. Springer-Verlag, New York.

40. Verbeke, G. and Molenberghs, G. (2000). *Linear Mixed Models for Longitudinal Data.* Springer Series in Statistics. Springer-Verlag, New York.

41. Verbeke, G., Lesaffre, E., and Spiessens, B. (2001). The practical use of different strategies to handle dropout in longitudinal studies. *Drug Inform. J.*, **35**, 419–434.

42. Vonesh, E. F. and Chinchilli, V. M. (1997). *Linear and Nonlinear Models for the Analysis of Repeated Measurements.* Marcel Dekker, New York.

43. Wolfinger, R. D. (1998). *Towards practical application of generalized linear mixed models.* In B. Marx and H. Friedl, eds., *Proc. 13th Int. Workshop on Statistical Modeling*, pp. 388–395, New Orleans, LA, USA, July 27–31.

44. Wolfinger, R. D. and O'Connell M. (1993). Generalized linear mixed models: A pseudo-likelihood approach. *J. Statist. Comput. Simul.*, **48**, 233–243.

63 Reproduction Rates

P. R. Cox

A population is reckoned to reproduce itself if its numbers are maintained from generation to generation. If the number of births between time $t-x$ and time $t-x+\delta t$ is written as $^{t-x}B\delta t$, then the number of births to which they give rise in their turn x years later, between age x and age $x+\delta x$, may be written as

$$^{t-x}B\delta t \left[{}^t_x p_0 \cdot {}^t\phi_x \delta x \right],$$

where $^t_x p_0$ expresses the chance of survival from birth of age x and $^t\phi_x$ is the fertility rate at age x. The total number of births between times t and $t+\delta t$ is thus $\int_{x=0}^{\infty} {}^{t-x}B\delta t \cdot {}^t_x p_0 \cdot {}^t\phi_x dx$, but this also equals $^t B\delta t$. If now $_x p_0$ and ϕ_x are constant in time, and the annual geometric rate of growth in the number of births is written as r, then

$$^t B\delta t = \int_0^\infty {}^{t-x}B\delta t \cdot {}_x p_0 \cdot \phi_x \, dx$$

$$= {}^t B\delta t \int_0^\infty e^{-rx} \cdot {}_x p_0 \cdot \phi_x \, dx,$$

and so

$$1 = \int_0^\infty e^{-rx} \cdot {}_x p_0 \cdot \phi_x \, dx.$$

This equation applies to one sex or the other. It has exactly one real solution, besides a number of complex roots in conjugate pairs. It can be shown that, as t increases, the behavior of $^t B$ is dominated by the real root, and the proportion of the sex in each age group tends toward a constant amount. Thus, for large t, the age distribution is constant, and it follows that so also is the rate of growth r. Such a population has been called a *stable* population.

The equation shown immediately above expresses the rate of population growth r in terms of mortality and fertility rates. It can be solved approximately by expanding the exponential

$$e^{-rx} = 1 - rx + r^2 x^2/2! - \cdots$$

and thus arriving at

$$1 = \int_0^\infty {}_x p_0 \cdot \phi_x dx$$

$$- \int_0^\infty r x \cdot {}_x p_0 \cdot \phi_x dx + \cdots.$$

Because r is small, its second and higher powers may be ignored for an approximate answer. Writing the first two terms on the right-hand side of the equation for convenience as m_1 and m_2, we have

$$m_1 - r m_2 \doteq 1$$

or

$$m_1 \{1 - (m_2/m_1)r\} \doteq 1.$$

Taking logarithms, this leads to $\log_e m_1 + \log_e \{1 - (m_2/m_1)r\} \doteq 0$. Expanding the logarithmic series and once again disregarding powers of r above the first, $\log_e m_1 - (m_2/m_1)r \doteqdot 0$. Hence

$$m_1 \doteqdot e^{rm_2/m_1}.$$

So m_1 is a near measure of the population increase in time m_2/m_1. Now m_2/m_1 represents the average age at bearing a child, that is, the average length of a generation. Thus m_1 can be regarded as a measure of the rate of reproduction, i.e., $\int_0^\infty {}_xp_0 \cdot \phi_x \, dx$. Its value exceeds, equals, or falls short of unity as population is growing, steady, or falling, respectively. It represents the average number of offspring born, in a generation, of the same sex as the parent.

Both ${}_xp_0$ and ϕ_x differ in value between the sexes, and so reproduction rates are not the same for men and women. Attempts to find a mathematical solution common to both have not proved wholly satisfactory. Men marry, on the whole, later in life than women do, and so they are older when the children are born. Thus the male length of a generation is longer than the female. There may also be sex differences in the proportion marrying. Male reproduction rates exceed female rates, but it is not often possible to calculate them, as few national statistical systems classify births by age of the father. Reproduction rates for women are, however, often available, and they are normally quoted in one of two forms "Net" rates are those assessed according to the formula $\int_x p_0 \cdot \phi_x dx$; "gross" reproduction rates disregard the element of mortality and thus represent $\int \phi_x dx$. They enable comparisons in time or between countries to be made, free from the complications of differences in survivorship. Table 1 shows some specimen figures and Table 2 shows how gross reproduction rates for women have varied in certain years. There is a considerable contrast in these data, not only in the size of the figures, but also in their trends. In some countries there has been a sharp downward movement; in others there has been only a slight fall, while in one the rates have risen and then fallen. The high rates are those experienced in the developing areas and are akin to those for European countries in past centuries. As high fertility is often a concomitant of high mortality, the net reproduction rates corresponding to the above would show a narrower range of variation.

Differences in reproduction rates can, by a simple algebraic process, be analyzed into their component parts where the data are available. However, demographers do not now make very much use of such rates. They represent little more than a convenient index of current fertility and are not sufficiently stable or illustrative of underlying trends to have predictive value. Considerable efforts were made at one time to improve them by processes of standardization, for instance, in respect to marriage and legitimacy, or by the use of fertility rates analyzed by marriage duration. The variety of answers obtained demonstrated that nothing steady or fundamental was being produced to give a reliable guide to the future.

From a historical perspective, however, it is of value to measure, wherever data are available for women born in each period of time, the number of female births to which they gave rise in due course at each adult age; in other words, to make a study of replacement in cohort or generation form. The following are some estimated generation *replacement rates* for England and Wales:

Year of Birth	Replacement Rate
1848–1853	1.36
1868–1873	1.09
1888–1893	0.81
1908–1913	0.70
1928–1933	0.96
1948–1953	1.04

These rates allow for the effects of mortality, falls in which, with the passage of time,

Table 1: Specimen figures.

Country and Period	Female Reproduction Rate	
	Net	Gross
Ghana, 1960	2.3	3.3
Ireland, 1968	1.83	1.91
France, 1931–1935	0.90	1.06

Table 2: Gross reproduction rates for women.

	1950	1955	1960	1965	1970
Brazil		2.76		2.61	2.42
Finland	1.55	1.42	1.31	1.18	0.89
Luxemburg	0.90	1.01	1.13	1.19	0.97
Netherlands	1.50	1.48	1.51	1.47	1.26
Singapore	3.06	3.17	2.77	2.25	1.51
Tunisia			3.34	3.34	3.12

assisted in the rise in the figures from their low point of 0.70.

References

1. Henry, L. (1976). *Population, Analysis and Models*. Edward Arnold, London. (Translated from the French, it illustrates the calculation of reproduction rates.)

2. Keyfitz, N. and Smith, D. (1977). *Mathematical Demography*. Springer-Verlag, Berlin, Germany. (Contains reprints of original papers on stable population theory and on the relationships between male and female reproduction rates.)

3. Pollard, J. H. (1973). *Mathematical Models for the Growth of Human Populations*. Cambridge University Press, London. (Includes a rigorous analysis of the deterministic population models of Lotka.)

4. Royal Commission on Population (1950). *Reports and Selected Papers of the Statistics Committee*. H. Ma. Stationery Office, London. (Discusses the measurement of reproductivity and attempts to improve reproduction rates.)

64. Retrospective Studies

Raymond S. Greenberg

64.1 Introduction

In nonexperimental research, historical information can be used to evaluate the relationship between two or more study factors. When all of the phenomena under investigation occur prior to the onset of a study, the research may be described as retrospective in design. That is to say, the study findings are based upon a look backward in time. The retrospective approach has proved useful for research in the social, biological, and physical sciences. Regardless of the field of inquiry, there are certain contexts in which a retrospective study is appropriate. The most obvious reason to conduct a retrospective investigation is to evaluate the effects of an event that occurred in the past and is not expected to recur in the future. Similarly, it may be necessary to resort to historical information for the study of events that occur infrequently, or are difficult to predict in advance. For example, much of the information on the health effects of ionizing radiation was obtained through retrospective studies of persons exposed to the bombings of Hiroshima and Nagasaki [3]. Retrospective studies are often performed to evaluate deleterious exposures, such as ionizing radiation, when an experiment with human subjects cannot be morally justified.

A retrospective approach also offers logistical advantages for the study of factors that are separated by a prolonged period of time. With the use of historical information, the investigator does not have to wait for the occurrence of the outcome variable. As an illustration, consider research on a suspected cause of cancer in humans. For many carcinogens, the median timespan between first exposure and the subsequent clinical detection of cancer is more than a decade [1]. A study that begins at the time of exposure to a suspected carcinogen would require years to evaluate the subsequent risk of cancer. In contrast, a retrospective investigation could be undertaken after the occurrence of the clinical outcomes, thereby saving time and resources.

64.2 Types of Retrospective Studies

Retrospective research may be subclassified according to the sequence of observations on the factors of interest. In a retrospective cohort study (also known as *historical cohort study*), the subjects are entered by level of past exposure to the antecedent variable. In the usual situation, two groups of subjects are defined: exposed and unexposed subjects (Fig. 1). Then,

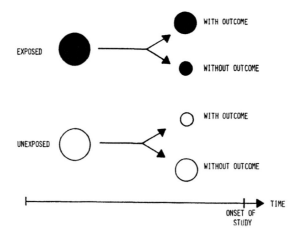

Figure 1: Schematic diagram of a retrospective cohort study. Shaded areas represent subjects exposed to the antecedent factor; unshaded areas correspond to unexposed subjects.

the subsequent frequencies of the outcome variable in these groups are compared. Although the subjects are traced forward in time, the study is retrospective because all of the phenomena under consideration occurred prior to the onset of investigation.

Example of a Retrospective Cohort Study. The evaluation of the long-term health effects of occupational exposures often is performed with the retrospective cohort method. As a typical example, consider a study conducted by McMichael and colleagues [8] on mortality in workers in the rubber industry. These authors used employment records to identify a group of active and retired employees of a tire factory who were alive on January 1, 1964. Through various record sources, the subsequent mortality experience of these 6678 workers was determined for a 9-year period of time. The age and sex-specific mortality rates of the US population were used for comparison. The authors found that the rubber workers had excessive numbers of deaths from cancers of the stomach, prostate, and hematopoietic tissues [8].

A second type of retrospective study is referred to as the *case–control method*, (also known as the *case–compeer*, or *case–referent study*). In a case–control investigation, the subjects are entered according to the presence or absence of the outcome variable (Fig. 2). By convention, persons who possess the outcome are labeled as "cases" and persons who do not possess the outcome are termed "controls." The case and control groups then are compared with respect to previous exposure to the antecedent factor. Thus, in a case–control study, the natural sequence of events is reversed and subjects are traced from outcome to a preceding exposure.

Example of a Case–Control Study. Often, the evaluation of a suspected cause of disease is performed with the case–control method. As a typical example, consider a study conducted by Clarke and colleagues [4] to evaluate the purported association between cigarette smoking and cancer of the cervix. A total of 181 women with cervical cancer (i.e., cases) and 905 women without cervical cancer (i.e., controls) were interviewed to determine previous smoking history. The authors found that a significantly greater proportion of

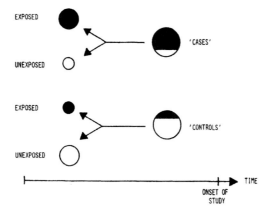

Figure 2: Schematic diagram of a case–control study. Shaded areas represent subjects exposed to the antecedent factor; unshaded areas correspond to unexposed subjects.

cases had smoked cigarettes, when compared with controls [4]. Of course, this finding does not prove that cigarette smoking causes cervical cancer. One cannot exclude the possibility that this association occurred because cigarette smoking is related to a correlate of cervical cancer, such as sexual practices.

64.3 The Frequency of Study Factors and Retrospective Research

In nonexperimental research, the choice of a study design often is dictated by the characteristics of the factors under investigation. As indicated previously, a retrospective approach is useful when the study factors are separated by a prolonged period of time. Also, consideration must be given to the relative frequencies of the study factors in the source population. When exposure to the putative causal factor is rare in the general population, a study sample that is selected randomly will have few exposed subjects. For this situation, disproportionate sampling rates of exposed and unexposed subjects are desirable. In a retrospective cohort study, the investigator can fix the prevalence of exposure, because the subjects are sampled contingent on the presence or absence of the antecedent factor.

Example of a Retrospective Cohort Study of a Rare Exposure. The poly(brominated biphenyls) (PBBs) are chemicals that are used commercially as fire retardants. Although heavy exposure to these compounds is uncommon, concern has been raised about the health effects of such exposures. Bahn and coworkers [2] used historical information to identify a group of 86 men that were employed at a firm that manufactured PBBs. A separate group of 89 unexposed persons was chosen from two other industries and the community. Evidence of primary thyroid dysfunction was found in four exposed subjects, as compared with none of the unexposed persons [2].

When the effect of interest is rare in the general population, a randomly selected sample will have few subjects with the outcome under study. For this situation, disproportionate sampling rates for the outcomes are desirable. In a case–control study, the investigator can fix the prevalence of the effect of interest, because the subjects are sampled contingent upon the presence or absence of the outcome.

Example of a Case–Control Study of a Rare Outcome. The toxic shock syndrome (TSS) is a life-threatening illness that is related to certain types of staphylococcal infections. Among menstruating women, the population at greatest risk of TSS, this disease is still uncommon, with an estimated annual incidence of 144 cases per million women [6]. To evaluate the relationship between use of tampons and subsequent risk of TSS, Kehrberg and colleagues [6] conducted a case-control study. These authors found that the prior use of tampons was significantly more frequent among the 29 women with TSS, as compared to the 91 women without TSS [6].

When both the supposed cause and effect of interest are rare in the general population, the standard retrospective methods often lack sufficient statistical power to evaluate the association of these factors. In this situation, the investigator may choose a hybrid design, which combines the cohort and case–control sampling procedures. Although a variety of hybrid studies might be envisioned, consider one particular approach. First, historical information is used to identify a group with a moderate baseline rate of exposure to the suspected causal factor. Then, subjects within this cohort are sampled contingent upon the presence or absence of the outcome of interest. This hybrid approach has been described as a case–control study nested within a cohort [7].

Example of a Case–Control Study within a Cohort. To evaluate the relationship between heavy exposure to certain heavy-metal oxides and subsequent risk of prostate cancer, Goldsmith and colleagues [5] first identified an occupational group with a moderate baseline rate of exposure to these compounds. Then, within this cohort, 88 men with prostate cancer (i.e., cases) and 258 men without prostate cancer (i.e., controls) were sampled. The authors subsequently found that the level of exposure to heavy-metal oxides and organic accelerators was significantly more frequent among cases, as compared with controls [5].

64.4 Other Considerations in Retrospective Research

Aside from the temporal pattern and frequencies of the study factors, the choice of a retrospective study design may be influenced by other considerations. For instance, case–control studies are relatively inexpensive and can be conducted in a short period of time. Also, the case–control approach allows the simultaneous evaluation of multiple suspected causal factors. However, these advantages must be weighed against the following limitations of case–control studies:

1. Only one outcome can be evaluated.
2. The rates of the outcomes within exposure groups cannot be estimated.
3. The method is especially susceptible to certain types of systematic errors.
4. The manner in which cases and controls are selected can introduce bias into the study results.
5. A causal relationship cannot be established by a single study.

With the retrospective cohort method, more than one outcome can be evaluated. Also, the rates of the various outcomes within exposure groups can be estimated with this design. However, the retrospective cohort approach has the following limitations:

1. Only one suspected causal factor can be evaluated.

2. The task of determining outcomes within cohorts can be tedious and time-consuming.

3. The reliance on recall or historical records may limit the type of information available and even lead to erroneous conclusions.

4. A causal relationship cannot be established by a single study.

Ultimately, the choice of a research strategy is affected by the goals of the study and the nature of the factors under investigation. For many scientific questions, a retrospective design provides an expedient and appropriate method of evaluation.

Acknowledgments. The preparation of this entry was supported in part by National Cancer Institute Contract N01-CN-61027. The manuscript was typed by Ms. Vickie Thomas.

References

1. Armenian, H. K. and Lilienfeld, A. M. (1974). *Am. J. Epidemiol.*, **99**, 92–100.

2. Bahn, A. K., Mills, J. L., Snyder, P. J., et al. (1980). *New Engl. J. Med.*, **302**, 31–33.

3. Beebe, G. W., Kato, H., and Land, C. E. (1978). *Radiat. Res.*, **75**, 138–201.

4. Clarke, E. A., Morgan, R. W., and Newman, A. M. (1982). *Am. J. Epidemiol.*, **115**, 59–66.

5. Goldsmith, D. F., Smith, A. H., and McMichael, A. J. (1980). *J. Occup. Med.*, **22**, 533–541.

6. Kehrberg, M. W., Latham, R. H., Haslam, B. T., et al. (1981). *Am. J. Epidemiol.*, **114**, 873–879.

7. Kupper, L. L., McMichael, A. J., and Spirtas, R. (1975). *J. Am. Statist. Assoc.*, **70**, 524–528.

8. McMichael, A. J., Spirtas, R., and Kupper, L. L. (1974). *J. Occup. Med.*, **16**, 458–464.

65. Sample Size Determination for Clinical Trials

Man-Lai Tang

65.1 Introduction

Determining sample size has long become a very important issue in clinical trials because unnecessarily large samples may result in increased research costs, longer study duration, and more subjects exposed to the risk of an experiment, while mistakenly small samples may lead to scientifically useless results (see, for example, Altman [2]). For instance, many studies in randomized controlled trials have not considered sample size or statistical power as important issues in study design. Moher et al. [12] reported that most trials with negative results did not have large enough sample sizes to detect a 25% or a 50% relative difference. As a result, sample size calculation prior to performing an experiment is particularly important and practically required in clinical trials.

Generally, there are two types of sample size formulas, namely sample size formulas that achieve a prespecified power level and sample size formulas that control a prespecified confidence width. For sample size based on power achievement, practitioner is going to determine the sample size so that a prespecified power level can be achieved by a test or statistic for the hypotheses of interest at a given nominal/significant level. For sample size based on width control, practitioner is going to determine the sample size so that a prespecific expected width can be controlled by a confidence interval estimator for the parameter of interest at a given coverage level. Unfortunately, some sample size formulas in the existing literature were presented without explicitly stating the statistics/tests or confidence interval estimators being based, nor the hypotheses or parameter being investigated. Therefore, practitioners may conduct a test or construct a confidence interval based on a sample size derived from another test or confidence interval estimator. As a result, sample size calculation performed in the earlier design stage would become useless and even misleading.

In this chapter, we will present some useful sample size formulas based on achieving a prespecified power level or controlling a prespecified confidence width (wherever available) for some popular applications in general clinical trials. The sample size formulas will be discussed for two general data types, namely the continuous and binary data. Sample size formulas for (1) the one-sample problem for mean and proportion, (2) the two-sample problem for comparing means and proportions, (3) the multiple-sample problem for comparing means and proportions, (4) multiple-arm dose–response trials, (5) multiple regression analysis, and (6) multiple logistic regression analysis will be discussed from Sections 65.2–65.7. Conclusions will be drawn in Section 65.8.

65.2 Sample Size for One-Sample Problem

65.2.1 Continuous Clinical Endpoint

Let $Y_1, Y_2, \ldots, Y_n$ be a random sample from an approximately normal distribution with mean μ and variance σ^2 [denoted as $N(\mu, \sigma^2)$].

Confidence Interval Estimation. The $(1-\alpha)100\%$ confidence interval for the population mean μ is given by

$$[\bar{Y} - z_{\alpha/2}\sigma/\sqrt{n},\ \bar{Y} + z_{\alpha/2}\sigma/\sqrt{n}]$$

if σ is known or sample size $n \geq 30$, where $\bar{Y} = \sum_{i=1}^{n} Y_i/n$ and $z_{\alpha/2}$ is the upper $(\alpha/2)$ critical point from the standard normal distribution $N(0,1)$, or

$$[\bar{Y} - t_{n-1,\alpha/2}s/\sqrt{n},\ \bar{Y} + t_{n-1,\alpha/2}s/\sqrt{n}]$$

if σ is unknown and sample size $n < 30$, where $t_{n-1,\alpha/2}$ is the upper-$(\alpha/2)$ critical point from the Student-t distribution with $(n-1)$ degrees of freedom and $s^2 = \sum_{i=1}^{n}(Y_i - \bar{Y})^2/(n-1)$ is the sample variance estimate of σ^2.

Recall that the margin of error represents the maximum difference between the observed sample mean $\bar{Y}$ and the true value of the population mean μ. It can also be regarded as the half of the confidence width. The crude sample size necessary to produce results accurate to a specified confidence level (say, $1-\alpha$) and margin of error (say, E) is given by

$$n = \left[\frac{z_{\alpha/2}\sigma}{E}\right]^2.$$

If the resultant sample size obtained from this formula is less than 30, a more accurate method for estimating the sample size is to iteratively evaluate the following formula with respect to n:

$$n = \left[\frac{t_{n-1,\alpha/2}\sigma}{E}\right]^2.$$

That is, start with an initial guess for n, plug in the above formula, and iteratively solve for n.

Significance Testing. For testing the following hypotheses:

$$H_0: \mu = \mu_0 \text{ vs. } H_A: \mu \neq \mu_0,$$

one can use the test statistic

$$z = \frac{\bar{Y} - \mu_0}{\sigma/\sqrt{n}} \quad \text{if } \sigma \text{ is known or } n \geq 30,$$

or

$$t = \frac{\bar{Y} - \mu_0}{s/\sqrt{n}} \quad \text{if } \sigma \text{ is unknown and } n < 30.$$

One will reject the null hypothesis H_0 if $|z| > z_{\alpha/2}$ or $|t| > t_{n-1,\alpha/2}$. Given a fixed nominal/significance level (i.e., type I error probability) α and the population standard deviation σ, the sample size n required for obtaining a test of $H_0: \mu = \mu_0$ vs. $H_A: \mu = \mu_A$ ($\neq \mu_0$) with a type II error probability no larger than β (or, with power at least $1 - \beta$) is given by

$$n = \left[\frac{(z_{\alpha/2} + z_\beta)\sigma}{\mu_0 - \mu_A}\right]^2.$$

If the alternative is one-sided [e.g., $H_A: \mu = \mu_A$ ($> \mu_0$)], then the desired sample size n is simply

$$n = \left[\frac{(z_\alpha + z_\beta)\sigma}{\mu_0 - \mu_A}\right]^2.$$

Again, if the resultant sample sizes obtained from the above formulae are less than 30, a more accurate method for estimating the sample sizes is to iteratively evaluate the following formula with respect to n

$$n = \left[\frac{(t_{n-1,\alpha/2} + t_{n-1,\beta})\sigma}{\mu_0 - \mu_A}\right]^2$$

or

$$n = \left[\frac{(t_{n-1,\alpha} + t_{n-1,\beta})\sigma}{\mu_0 - \mu_A}\right]^2.$$

65.2.2 Binary Clinical Endpoint

Suppose Y follows a binomial distribution with parameters n and π [denoted as Bin(n, π)].

Confidence Interval Estimation. Agresti and Coull [1] considered an improved $(1 - 0.05)100\%$ (i.e., 95%) confidence interval for a single binomial proportion π by adding two pseudo-successes and two pseudofailures. Brown et al. [4] generalized Agresti–Coull's idea to the general $(1 - \alpha)100\%$ confidence interval by constructing the modified, recentered estimator $\tilde{\pi} = (Y + 0.5z_{\alpha/2}^2)/\tilde{n}$ of π, where $\tilde{n} = n + z_{\alpha/2}^2$. The generalized Agresti–Coull $(1 - \alpha)100\%$ confidence interval is based on $\tilde{\pi}$ given by

$$[\tilde{\pi} - z_{\alpha/2}\sqrt{\tilde{\pi}(1-\tilde{\pi})/\tilde{n}},$$
$$\tilde{\pi} + z_{\alpha/2}\sqrt{\tilde{\pi}(1-\tilde{\pi})/\tilde{n}}].$$

This interval is highly recommended due to its simplicity and its stable coverage characteristic [4].

Given a prespecified $(1 - \alpha)$ coverage level and an anticipated value of π, say, π_0, Piegorsch [13] showed that the sample size n that will lead to an interval width of some desired value ω_0 is given by

$$n = \frac{z_{\alpha/2}^2 \pi_0(1 - \pi_0)}{(\omega_0/2)^2} - z_{\alpha/2}^2.$$

It is noteworthy that the first term in this formula is the most popular sample size formula suggested in most elementary statistics textbooks based on the unstable Wald confidence interval. Moreover, n always guarantees a reduction in the popular sample size requirement (based on the Wald confidence interval) by $z_{\alpha/2}^2$ subjects with significantly improved coverage performance.

Significance Testing. To test the hypotheses

$$H_0: \pi = \pi_0 \text{ vs. } H_A: \pi \neq \pi_0,$$

one can use the test statistic

$$z = \frac{Y/n - \pi_0}{\sqrt{\pi_0(1-\pi_0)/n}}.$$

One will reject the null hypothesis H_0 if $|z| > z_{\alpha/2}$. Given a fixed nominal/significance level α, the sample size n required for obtaining a test of $H_0: \pi = \pi_0$ vs. $H_A: \pi = \pi_A \ (\neq \pi_0)$ with power at least $1 - \beta$ is given by

$$n = \left[\frac{z_{\alpha/2}\sqrt{\pi_0(1-\pi_0)} + z_\beta\sqrt{\pi_A(1-\pi_A)}}{\pi_A - \pi_0}\right]^2.$$

If the alternative is one-sided [e.g., $H_A: \pi = \pi_A \ (> \pi_0)$], then the desired sample size n is simply

$$n = \left[\frac{z_\alpha\sqrt{\pi_0(1-\pi_0)} + z_\beta\sqrt{\pi_A(1-\pi_A)}}{\pi_A - \pi_0}\right]^2.$$

65.3 Sample Size for Two-Sample Problem

65.3.1 Continuous Clinical Endpoint

Let $Y_{1,1}, Y_{1,2}, \ldots, Y_{1,n_1}$ be a random sample from an approximately N(μ_1, σ_1^2) and

$Y_{2,1}, Y_{2,2}, \ldots, Y_{2,n_2}$ be a random sample from an approximately $N(\mu_2, \sigma_2^2)$. Suppose that $n_1/n_2 = k$ and $n_2 = n$.

Confidence Interval Estimation. The $(1-\alpha)100\%$ confidence interval for the population mean difference $\mu_1 - \mu_2$ is given by

$$[(\bar{Y}_1 - \bar{Y}_2) - z_{\alpha/2}\sqrt{\sigma_1^2/(kn) + \sigma_2^2/n},$$
$$(\bar{Y}_1 - \bar{Y}_2) + z_{\alpha/2}\sqrt{\sigma_1^2/(kn) + \sigma_2^2/n}],$$

if σ_1 and σ_2 are known, where $\bar{Y}_1 = \sum_{i=1}^{n_1} Y_{1,i}/(kn)$ and $\bar{Y}_2 = \sum_{i=1}^{n_2} Y_{2,i}/n$, then

$$[(\bar{Y}_1 - \bar{Y}_2) - z_{\alpha/2}\sqrt{s_1^2/(kn) + s_2^2/n},$$
$$(\bar{Y}_1 - \bar{Y}_2) + z_{\alpha/2}\sqrt{s_1^2/(kn) + s_2^2/n}],$$

if σ_1 and σ_2 are unknown, where $s_j^2 = \sum_{i=1}^{n_j}(Y_{j,i} - \bar{Y}_j)^2/(n_j - 1)$, $j = 1, 2$, or

$$[(\bar{Y}_1 - \bar{Y}_2) - z_{\alpha/2}s_p\sqrt{(k+1)/(kn)},$$
$$(\bar{Y}_1 - \bar{Y}_2) + z_{\alpha/2}s_p\sqrt{(k+1)/(kn)}],$$

if $\sigma_1 = \sigma_2$ are unknown, and $n_1 < 30$ and $n_2 < 30$, where $s_p^2 = [(n_1 - 1)s_1^2 + (n_2 - 1)s_2^2]/(n_1 + n_2 - 2)$.

Given that the population standard deviations for two populations are respectively σ_1 and σ_2, to obtain a $(1-\alpha)100\%$ confidence interval for the population mean difference $\mu_1 - \mu_2$ with a margin of error no greater than E, the desired sample size for n_2 is given by

$$n_2 = (\sigma_1^2/k + \sigma_2^2)\left[\frac{z_{\alpha/2}\sigma}{E}\right]^2,$$

with $n_1 = kn_2$.

Significance Testing. To test the hypotheses

$$H_0: \mu_1 - \mu_2 = 0 \text{ vs. } H_A: \mu_1 - \mu_2 \neq 0,$$

one can use the test statistic

$$z = \frac{\bar{Y}_1 - \bar{Y}_2}{\sqrt{\sigma_1^2/(kn) + \sigma_2^2/n}}$$

if σ_1 and σ_2 are known, then

$$z = \frac{\bar{Y}_1 - \bar{Y}_2}{\sqrt{s_1^2/(kn) + s_2^2/n}}$$

if σ_1 and σ_2 are unknown, or

$$z = \frac{\bar{Y}_1 - \bar{Y}_2}{s_p\sqrt{1/(kn) + 1/n}}$$

if $\sigma_1 = \sigma_2$ are unknown, and $n_1 < 30$ and $n_2 < 30$.

One will reject the null hypothesis H_0 if $|z| > z_{\alpha/2}$. Given a fixed nominal/significance level α and the population standard deviations σ_1 and σ_2, the sample size n_2 required for obtaining a test of H_0: $\mu_1 - \mu_2 = 0$ vs. H_A: $\mu_1 - \mu_2 = \Delta \ (\neq 0)$ with power at least $1 - \beta$ is given by

$$n_2 = [\sigma_1^2/k + \sigma_2^2]\left[\frac{z_{\alpha/2} + z_\beta}{\Delta}\right]^2,$$

and $n_1 = kn_1$. If the alternative is one-sided [e.g., H_A: $\mu_1 - \mu_2 = \Delta (>0)$], then the desired sample size n is simply

$$n_2 = [\sigma_1^2/k + \sigma_2^2]\left[\frac{z_\alpha + z_\beta}{\Delta}\right]^2.$$

If the resultant sample sizes obtained from the above formulas are less than 30 and the standard deviations from both populations are identical, a more accurate method for estimating the sample sizes is to iteratively evaluate the following formula with respect to n_2:

$$n_2 = [\sigma_1^2/k + \sigma_2^2]$$
$$\times \left[\frac{t_{(k+1)n_2-2,\alpha/2} + z_{(k+1)n_2-2,\beta}}{\Delta}\right]^2,$$

or

$$n_2 = [\sigma_1^2/k + \sigma_2^2]$$
$$\times \left[\frac{t_{(k+1)n_2-2,\alpha} + z_{(k+1)n_2-2,\beta}}{\Delta}\right]^2.$$

65.3.2 Binary Clinical Endpoint

Suppose that Y_1 follows $\text{Bin}(n_1, \pi_1)$ and Y_2 follows $\text{Bin}(n_2, \pi_2)$ with $n_1/n_2 = k$ and $n_2 = n$. Let $\hat{\pi}_1 = Y_1/n_1$ and $\hat{\pi}_2 = Y_2/n_2$.

Confidence Interval Estimation. Similar to the generalized Agresti–Coull confidence interval for a single proportion, the generalized Agresti–Coull $(1-\alpha)100\%$ confidence interval for $\pi_1 - \pi_2$ can be computed by

$$[\tilde{\pi}_1 - \tilde{\pi}_2 - z_{\alpha/2}$$
$$\times \sqrt{\tilde{\pi}_1(1-\tilde{\pi}_1)/\tilde{n}_1 + \tilde{\pi}_2(1-\tilde{\pi}_2)/\tilde{n}_2},$$
$$\tilde{\pi}_1 - \tilde{\pi}_2 + z_{\alpha/2}$$
$$\times \sqrt{\tilde{\pi}_1(1-\tilde{\pi}_1)/\tilde{n}_1 + \tilde{\pi}_2(1-\tilde{\pi}_2)/\tilde{n}_2}],$$

where $\tilde{\pi}_i = (Y_i + 0.5 z_{\alpha/2}^2)/\tilde{n}_i$ and $\tilde{n}_i = n_i + z_{\alpha/2}^2$, $i = 1, 2$.

Given a prespecified $(1 - \alpha)$ coverage level and anticipated values of π_1 and π_2, say, π_{10} and π_{20}, the desired sample size n for population 2 that will lead to an interval width of some desired value ω_0 can be obtained by solving the following equation with respect to n:

$$\frac{\omega_0^2}{4z_{\alpha/2}^2} = \frac{\pi_{10}(1-\pi_{10})}{kn + z_{\alpha/2}^2} + \frac{\pi_{20}(1-\pi_{20})}{n + z_{\alpha/2}^2}.$$

For balanced design (i.e., $k = 1$), the desired sample sizes are

$$n_1 = n_2 = n$$
$$= \frac{4z_{\alpha/2}^2}{\omega_0^2}[\pi_{10}(1-\pi_{10}) + \pi_{20}(1-\pi_{20})]$$
$$- z_{\alpha/2}^2.$$

For imbalanced design (i.e., $k \neq 1$), an upper bound for the desired sample sizes can be shown to be

$$n_2 = n = \frac{4z_{\alpha/2}^2}{\omega_0^2}\left[\frac{\pi_{10}(1-\pi_{10})}{k} + \pi_{20}(1-\pi_{20})\right],$$

with $n_1 = kn$.

Significance Testing. To test the hypotheses

$$H_0 : \pi_1 - \pi_2 = 0 \text{ vs. } H_A : \pi_1 - \pi_2 \neq 0,$$

Fleiss [7] proposed the following test statistic

$$z = \frac{|\hat{\pi}_1 - \hat{\pi}_2| - (k+1)/(2kn)}{\sqrt{\bar{\pi}(1-\bar{\pi})(k+1)/(kn)}},$$

where $\bar{\pi} = (\hat{\pi}_2 + k\hat{\pi}_1)/(k+1)$. One will reject the null hypothesis H_0 if $|z| > z_{\alpha/2}$. Given a fixed nominal/significance level α, the sample size n required for obtaining a test of H_0: $\pi_1 - \pi_2 = 0$ vs. H_A: $\pi_1 - \pi_2 \neq 0$ with power at least $1 - \beta$ is given by

$$n = \frac{n'}{4}\left[1 + \sqrt{1 + \frac{2(k+1)}{n'k|\pi_1 - \pi_2|}}\right]^2,$$

where

$$n' = \{[z_{\alpha/2}\sqrt{(k+1)\bar{\pi}(1-\bar{\pi})}$$
$$+ z_\beta\sqrt{k\pi_2(1-\pi_2) + \pi_1(1-\pi_1)}]^2\}$$
$$\times \{k(\pi_1 - \pi_2)^2\}^{-1},$$

and $\bar{\pi} = (\pi_2 + k\pi_1)/(k+1)$. If the alternative is one-sided (e.g., H_A: $\pi_1 - \pi_2 > 0$), then the desired sample size is n with n' being replaced by

$$n' = \{[z_\alpha\sqrt{(k+1)\bar{\pi}(1-\bar{\pi})}$$
$$+ z_\beta\sqrt{k\pi_2(1-\pi_2) + \pi_1(1-\pi_1)}]^2\}$$
$$\times \{k(\pi_1 - \pi_2)^2\}^{-1}.$$

65.4 Sample Size for Multiple-Sample Problem

65.4.1 Continuous Clinical Endpoint

Let X_{ij} be the outcome of the jth subject from the ith treatment group, $i = 1, \ldots, k$, $j = 1, \ldots, n_i$. We assume that the data follow the following one-way analysis of variance (ANOVA) model

$$x_{ij} = \mu_i + \epsilon_{ij},$$

where μ_i represents the fixed effect of the ith treatment and ϵ_{ij} is a random error in observing X_{ij}. It is assumed that ϵ_{ij}'s are i.i.d. normal random variables with mean 0 and variance σ^2. Denote $n_T = n_1 + \cdots + n_k$.

Pairwise Comparison. In practice, it is often of interest to compare means among treatments under study. Thus, we consider the following null and alternative hypotheses:

$$H_{0ii'} : \mu_i = \mu_{i'} \text{ vs. } H_{Aii'} : \mu_i \neq \mu_{i'}$$
$$(i, i' = 1, \ldots, k; i \neq i').$$

Let $\Delta_{ii'}$ denote the difference between μ_i and $\mu_{i'}$ that is worth detecting. To test the null hypothesis $H_{ii'}$, we consider the following statistic:

$$t_{ii'} = \frac{\bar{X}_i - \bar{X}_{i'}}{s\sqrt{1/n_i + 1/n_{i'}}}.$$

One will reject the null hypothesis $H_{0ii'}$ if $|t_{ii'}| > t_{n_i+n_{i'}-2, \alpha/(2L)}$ with $L = k(k+1)/2$. Suppose that we consider a balanced design. For controlling the overall type I error rate at the desired significance level α, the sample size n required to achieve power $1 - \beta$ for detecting a clinical meaningful difference between μ_i and $\mu_{i'}$ (i.e., $\Delta_{ii'}$) is given by

$$n = \max\{n_{ii'}, \text{ for } i, i' = 1, \ldots, k; i \neq i'\}$$

with

$$n_{ii'} = 2\left[\frac{(z_{\alpha/(2L)} + z_\beta)\sigma}{\Delta_{ii'}}\right]^2.$$

It is noteworthy that the above well-known multiple t-test is appropriate for examining the $H_{0ii'}$ if the control of the comparisonwise type I error rate (i.e., the expected rate of erroneously rejecting true elementary hypotheses) is required. On the other hand, the Tukey's T procedure is recommended if the experimentwise type I error rate (i.e., the probability of erroneously rejecting at least one true elementary hypothesis) is to be controlled. In the latter case, the same test statistic (i.e., $t_{ii'}$) and sample size formula (i.e., $n_{ii'}$) can be adopted except that the critical points $t_{n_i+n_{i'}-2,\alpha/(2L)}$ and $z_{\alpha/(2L)}$ should be replaced by $q_{k,n_T-k,\alpha}/\sqrt{2}$ and $q_{k,\infty,\alpha}/\sqrt{2}$, respectively. Here, $q_{d_1,d_2,\alpha}$ is the upper α critical point from the Studentized range distribution with parameter d_1 and d_2 number of degrees of freedom. In practice, there are only τ ($< L$) comparisons of interest. In this case, the above test procedures and sample size formulas are still valid, with L replaced by τ.

Simultaneous Comparison. To test the following ANOVA hypotheses

$$H_0 : \mu_1 = \mu_2 = \cdots = \mu_k, \text{ vs.}$$
$$H_A : \mu_i \neq \mu_{i'} \text{ for some } 1 \leq i < i' \leq k,$$

we consider the following statistic

$$F = \frac{\text{MSTr}}{\text{MSE}}$$

where $\text{MSTr} = \text{SSTr}/(k-1)$ with $\text{SSTr} = \sum_{i=1}^{k} n_i \bar{X}_i - n_T \bar{X}$ and $\bar{X} = \sum_{i=1}^{k} \sum_{j=1}^{n_j} X_{ij}/n_T$. The null hypothesis H_0 is rejected at the α level of significance if $F > F_{k-1, n_T-k, \alpha}$, which is the upper α quantile of the F distribution with degrees of freedom $k-1$ and $n-k$. Again, we consider only the balanced design with $n_i = n$ for $i = 1, \ldots, k$. Given a fixed nominal/significance level α, the sample size n required for obtaining a test of the above ANOVA hypotheses with power at least $1 - \beta$ can be obtained by solving the following equation

$$\chi^2_{k-1}(\chi_{k-1,\alpha}|\lambda = n\Delta) = \beta,$$

where $\chi^2_{n-1,\alpha}$ is the upper α quantile of the χ^2 distribution with $k-1$ degrees of freedom, $\Delta = \sum_{i=1}^{k}(\mu_i - \bar{\mu})^2/\sigma^2$, $\bar{\mu} =$

$\sum_{i=1}^{k} \mu_i/k$, and $\chi^2_{k-1}(.|\lambda)$ is the cumulative distribution function of the non-central χ^2 distribution with degrees of freedom $(k-1)$ and non-centrality parameter λ.

65.4.2 Binary Clinical Endpoint

Let X_{ij} be the binary outcome of the j-th subject from the ith treatment group, $i = 1, \ldots, k$, $j = 1, \ldots, n_i$. Assume that $\Pr(X_{ij} = 1) = \pi_i$. Define $\hat{\pi}_i = \sum_{j=1}^{n_i} X_{ij}/n_i$ and $n_T = n_1 + \cdots + n_k$. In practice, it is often of interest to compare proportions among treatments under study. Thus, we consider the following null and alternative hypotheses:

$$H_{0ii'} : \pi_i = \pi_{i'} \text{ vs. } H_{Aii'} : \pi_i \neq \pi_{i'}$$
$$(i, i' = 1, \ldots, k; i \neq i').$$

Let $\Delta_{ii'}$ denote the difference between π_i and $\pi_{i'}$ that is worth detecting. To test the null hypothesis $H_{0ii'}$, we consider the following statistic:

$$v_{ii'} = 2 \frac{\arcsin \sqrt{\hat{\pi}_i} - \arcsin \sqrt{\hat{\pi}_{i'}}}{\sqrt{1/n_i + 1/n_{i'}}}.$$

One will reject the null hypothesis $H_{0ii'}$ if $|v_{ii'}| > z_{\alpha/(2L)}$ for the control of the comparisonwise type I error rate; or $> q_{k,n_T-k,\alpha}/\sqrt{2}$ for the control of the experimentwise type I error rate. Here, $L = k(k+1)/2$. Suppose that we consider a balanced design. For controlling the overall type I error rate at the desired significance level α, the sample size n required to achieve power $1 - \beta$ for detecting a clinical meaningful difference between π_i and $\pi_{i'}$ (i.e., $\Delta_{ii'}$) is given by

$$n = \max\{n_{ii'}, \text{ for } i, i' = 1, \ldots, k; i \neq i'\}$$

with

$$n_{ii'} = \left[\frac{(z_{\alpha/(2L)} + z_\beta)}{\sqrt{2}(\arcsin(\Delta_{ii'}))} \right]^2.$$

Again, for controlling the experimentwise type I error rate the same sample size formula (i.e., $n_{ii'}$) can be adopted except that the critical point $z_{\alpha/(2L)}$ should be replaced by $q_{k,\infty,\alpha}/\sqrt{2}$. If only τ ($<L$) comparisons are of interest, the above test procedures and sample size formulas are still valid with L being replaced by τ.

For a thorough discussion on the sample size formulas for pairwise and many-one multiple comparison in the parametric, nonparametric and binomial cases, one can consult Horn and Vollandt [8].

65.5 Sample Size for Multiple-Arm Dose–Response Trial Problem

Unlike the ANOVA problem considered in the previous section, the ith treatment group, $i = 1, \ldots, k$, considered in this section corresponds to the ith dose level of a chemical/drug in dose–response studies. Usually, we include $i = 0$ to represent the control group. In general, prior information about the dose–response curve such as linear, monotonic or U-shaped is available. In what follows, we will consider sample size calculation in multiple-arm dose–response trials with normal and binary responses.

65.5.1 Normal Response

Let X_{ij} be the outcome of the jth subject exposed to the ith dose level of a chemical or drug, $i = 0, \ldots, k$, $j = 1, \ldots, n_i$. We assume that the data $\{X_{i1}, X_{i2}, \ldots, X_{i,n_i}\}$ follow the normal distribution with mean μ_i and variance σ^2 for $i = 0, \ldots, k$. Hence, μ_i represents the population mean response for dose level i. The null hypothesis of no dose effect can be written as follows

$$H_0 : \mu_0 = \mu_1 = \cdots = \mu_k$$

or
$$H_0' : \sum_{i=0}^{k} c_i \mu_i = 0$$

where contrasts $\{c_0, \ldots, c_k\}$ satisfy the condition $\sum_{i=0}^{k} c_i = 0$. It should be noted that if the null hypothesis H_0' is rejected for some $\{c_0, \ldots, c_k\}$ satisfying $\sum_{i=0}^{k} c_i = 0$, then H_0 is also rejected. In particular, we are interested in the following alternative hypothesis:

$$H_A : \sum_{i=0}^{k} c_i \mu_i = \epsilon$$

with $\epsilon \neq 0$. To test H_0', we consider the following statistic

$$z = \frac{\sum_{i=0}^{k} c_i \bar{X}_i}{\sqrt{(s^2/n_T) \sum_{i=0}^{k} (c_i^2/f_i)}},$$

where

$$s^2 = \frac{(n_0 - 1) S_0^2 + \cdots + (n_k - 1) S_k^2}{n_T - k - 1},$$

with $n_T = n_0 + \cdots + n_k$, $\bar{X}_i = \sum_{j=1}^{n_i}(X_{ij}/n_i)$, $S_i^2 = \sum_{j=1}^{n_i}(X_{ij} - \bar{X}_i)^2/(n_i - 1)$, and $f_i = n_i/n_T$ for $i = 0, \ldots, k$. Here, f_i is the sample size fraction for the ith dose level. One will reject the null hypothesis H_0' if $z > z_\alpha$. Given a fixed significance level α and the common population standard deviation σ, the total sample size n_T required for obtaining a test of $H_0' : \sum_{i=0}^{k} c_i \mu_i = 0$ vs. $H_A : \sum_{i=0}^{k} c_i \mu_i = \epsilon$ with power at least $1 - \beta$ is given by

$$n_T = \left[\frac{(z_\alpha + z_\beta)\sigma}{\epsilon}\right]^2 \sum_{i=0}^{k} \frac{c_i^2}{f_i}.$$

65.5.2 Binary Response

Let π_i be the proportion of response in the ith dose level, $i = 0, \ldots, k$, $j = 1, \ldots, n_i$. Let n_i be the number of subjects exposed to dose level i and X_i the corresponding number of subjects who have the response of interest. We are interested in testing the following hypotheses

$$H_0 : \pi_0 = \pi_1 = \cdots = \pi_k$$

against

$$H_A : \sum_{i=0}^{k} c_i \pi_i = \epsilon$$

where contrasts $\{c_0, \ldots, c_k\}$ satisfy the condition $\sum_{i=0}^{k} c_i = 0$ and $\epsilon \neq 0$. To test H_0, we consider the following statistic

$$z = \frac{\sum_{i=0}^{k} (c_i/n_i) X_i}{\sqrt{\hat{\pi}(1 - \hat{\pi}) \sum_{i=0}^{k} (c_i^2/n_i)}},$$

where $\hat{\pi} = \sum_{i=0}^{k} X_i/n_T$ and $n_T = n_0 + \cdots + n_k$. One will reject the null hypothesis H_0 if $z > z_\alpha$. Again, let $f_i = n_i/n_T$ for $i = 0, \cdots, k$. Given a fixed significance level α, the total sample size n_T required for obtaining a test of $H_0 : \pi_0 = \pi_1 = \cdots = \pi_k$ vs. $H_A : \sum_{i=0}^{k} c_i \pi_i = \epsilon$ with power at least $1 - \beta$ is given by

$$n_T = \left[\left\{z_\alpha \sqrt{\sum_{i=0}^{k} \frac{c_i^2}{f_i} \bar{\pi}(1 - \bar{\pi})} \right.\right.$$
$$\left.\left. + z_\beta \sqrt{\sum_{i=0}^{k} \frac{c_i^2}{f_i} \pi_i(1 - \pi_i)} \right\}\right.$$
$$\left. \times \{\epsilon\}^{-1}\right]^2,$$

where $\bar{\pi} = \sum_{i=0}^{k} f_i \pi_i$.

Suppose that the score for the ith dose level is known to be d_i for $i = 0, \ldots, k$. Assume that the probability of response follows a linear trend in logistic scale, i.e.,

$$\pi_i = \frac{e^{\gamma + \lambda d_i}}{1 + e^{\gamma + \lambda d_i}},$$

$i = 0, \ldots, k$. Here, γ is the intercept and λ is the slope for the well-known logistic dose–response model. The hypothesis test problem for monotonic increasing response with respect to dose can be stated as

$$H_0 : \lambda = 0 \quad \text{vs.} \quad H_A : \lambda = \epsilon > 0.$$

Similarly, $\epsilon < 0$ corresponds to monotonic decreasing response with respect to dose. Let $\bar{d} = \sum_{i=0}^{k} d_i n_i / n_T$ and $U = \sum_{i=0}^{k} (d_i - \bar{d}) X_i$. To test H_0, one can consider the following trend test statistic

$$z = \frac{U - \Delta/2}{\sqrt{\hat{\pi}(1-\hat{\pi}) \sum_{i=0}^{k} (d_i - \bar{d})^2 n_i}},$$

where $\Delta/2 = (d_i - d_{i-1})/2$ is the continuity correction for equal-spaced doses. It is noteworthy that the famous Cochran–Armitage test statistic is identical to the square of z. One will reject the null hypothesis H_0 if $z > z_\alpha$. Let $r_i = n_i/n_0$. Given a fixed significance level α, the sample size n_0 for the control group required for obtaining a test of $H_0 : \lambda = 0$ vs. $H_A : \lambda > 0$ with power at least $1 - \beta$ is given by

$$n_0 = \frac{n_0^*}{4} \left[1 + \sqrt{1 + 2\frac{\Delta}{A n_0^*}} \right]^2,$$

with

$$n_0^* = \frac{1}{A} \left[z_\alpha \sqrt{\bar{\pi}(1-\bar{\pi}) \sum_{i=0}^{k} r_i (d_i - \bar{d})^2} \right.$$
$$\left. + z_\beta \sqrt{\sum_{i=0}^{k} \pi_i (1-\pi_i) r_i (d_i - \bar{d})^2} \right]^2,$$

where $A = \sum_{i=0}^{k} \pi_i r_i (d_i - \bar{d})$. It is noted that the sample size formula of the trend test depends on the specified alternative.

65.6 Sample Size for Multiple Regression Analysis

Consider a random sample of n observations $(x_{i1}, x_{i2}, \ldots, x_{ik}, y_i)$, $i = 1, \ldots, n$. The $p = k + 1$ variables are assumed to satisfy the following linear model

$$\begin{aligned} y_i &= \beta_0 + \beta_1 x_{i1} + \beta_2 x_{i2} + \cdots \\ &\quad + \beta_k x_{ik} + \epsilon_i, \ i = 1, \ldots, n, \end{aligned}$$

where y is the dependent variable, $(x_1, x_2, \ldots, x_k)$ are k independent variables, ϵ is the random error term, and $(\beta_0, \beta_1, \ldots, \beta_k)$ are the $k + 1$ unknown regression coefficient parameters. Here, we assume that $E(\epsilon_i) = 0$ and $\text{Var}(\epsilon_i) = \sigma^2$, and $\text{Cov}(\epsilon_i, \epsilon_j) = 0$ for $i \neq j$.

Confidence Interval Estimation. A $(1 - \alpha)100\%$ confidence interval for a single population standardized regression coefficient, say β_j, can be computed as

$$\left[\hat{\beta}_j - t_{n-k-1, \alpha/2} \right.$$
$$\times \sqrt{\frac{1 - \hat{R}^2}{(n-k-1)(1-\hat{R}^2_{XX_j})}},$$
$$\hat{\beta}_j + t_{n-k-1, \alpha/2}$$
$$\left. \times \sqrt{\frac{1 - \hat{R}^2}{(n-k-1)(1-\hat{R}^2_{XX_j})}} \right],$$

where $\hat{\beta}_j$ is the standardized least-squares regression coefficient estimate, j represents a specific regressor ($j = 1, \ldots, k$), $\hat{R}^2$ is the observed multiple correlation coefficient of the model, and $\hat{R}^2_{XX_j}$ represents the observed multiple correlation coefficient predicting the jth regressor (X_j) from the remaining $k - 1$ regressors (see Cohen and Cohen [5]).

Kelly and Maxwell [10] suggested that the desired sample size n, such that the confidence interval around a particular population regression coefficient, β_j, will have an expected confidence width ω_0, is given by

$$n = \left(\frac{z_{\alpha/2}}{\omega_0/2}\right)^2 \left(\frac{1-R^2}{1-R^2_{XX_j}}\right) + k + 1,$$

where R^2 represents the population multiple correlation coefficient predicting the dependent variable Y from the k independent variables and $R^2_{XX_j}$ represents the population multiple correlation coefficient predicting the jth independent variable from the remaining $k-1$ independent variables. Noticing that there is approximately only a 50% chance that the above interval will be no larger than the specified width, the following sample size formula is suggested

$$n = \left(\frac{z_{\alpha/2}}{\omega_0/2}\right)^2 \left(\frac{1-R^2}{1-R^2_{XX_j}}\right) \times \left(\frac{\chi^2_{n-1,\gamma}}{n-k-1}\right) + k + 1,$$

where γ represents the desired degree of uncertainty of the computed confidence interval being the specified width. In other words, one will be approximately $100(1-\gamma)\%$ confident that the β_j of interest will have a corresponding confidence interval width that is no larger than specified.

Significance Testing. For testing the following hypotheses involving one of the population standardized regression coefficient, say, β_j

$$H_0: \beta_j = 0 \text{ versus } H_A: \beta_j \neq 0,$$

One can consider the following test statistic

$$F = t^2 = \frac{\hat{\Delta}r_j^2/1}{(1-\hat{R}^2)/(n-k-1)},$$

where $\hat{\Delta}r_j^2$ represents the explained variance attributed to the jth independent variable when entered last in the regression equation. One can reject the null hypothesis H_0 at the α level of significance if $F > F_{1,n-k-1,\alpha}$. Given a fixed nominal/significance level α, Milton [15] showed that the sample size n required for the above F test to detect $H_0: \beta_j = 0$ is given by

$$n = \frac{t^2_{n-k-1,\alpha/2}(1-R^2)}{\Delta r_j^2} + k + 1,$$

where Δr_j^2 is the minimum addition to R^2 when the jth variable is entered last, which, if attained, will ensure a statistically significant regression coefficient.

65.7 Sample Size for Multiple Logistic Regression Analysis

Consider a random sample of n observations $(x_{i1}, x_{i2}, \ldots, x_{ik}, y_i)$, $i = 1, \ldots, n$. Here, y_i is the binary outcome of the ith subject. For instance, $y_i = 1$ if the ith subject has a certain disease; $= 0$ otherwise. The $p = k$ variables [i.e., $\mathbf{x} = (\mathbf{x_1}, \ldots, \mathbf{x_k})$] may include exposure and confounder. Here, the probability of the binary outcome variable y_i is modeled via a multivariate logistic regression model as the following conditional probability

$$Pr(y=1|\mathbf{x}) = \frac{e^{\beta_0+\beta_1\mathbf{x_1}+\cdots+\beta_k\mathbf{x_k}}}{1+e^{\beta_0+\beta_1\mathbf{x_1}+\cdots+\beta_k\mathbf{x_k}}}$$

where $(\beta_0, \beta_1, \ldots, \beta_k)$ are the $k+1$ unknown logistic regression coefficient parameters. Suppose that we would like to test the following hypotheses of β_j:

$$H_0: \beta_j = 0 \text{ vs. } H_A: \beta_j \neq 0.$$

Let $\hat{\beta}_j$ be the maximum-likelihood estimate of β_j, V the inverse of the $(k+1)\times(k+1)$ Fisher information matrix for $(\beta_0, \beta_1,$

..., β_k), V_j the corresponding (j, j)th diagonal element of matrix V evaluated at the alternative and $\hat{V}_j$ the variance (i.e., V_j) evaluated at $\hat{\beta}_j$. To test H_0, Demidenko [6] considered the following Wald test statistic:

$$z = \frac{\sqrt{n}\hat{\beta}_j}{\sqrt{\hat{V}_j}}.$$

One will reject the null hypothesis H_0 at the α level of significance if $|z| > z_{\alpha/2}$. Given a fixed significance level α, the sample size n required for obtaining a test of $H_0 : \beta_j = 0$ vs. $H_A : \beta_j = \beta_{jA} (\neq 0)$ with power at least $1 - \beta$ is given by

$$n = \frac{(z_\alpha + z_\beta)^2 V_j}{\beta_{jA}^2}.$$

In particular, let consider the following simplest logistic regression model with only one binary exposure x (i.e., $k = 1$)

$$Pr(y = 1|x) = \frac{e^{\beta_0 + \beta_1 x}}{1 + e^{\beta_0 + \beta_1 x}},$$

and x_i takes value 1 with probability π_x and 0 with probability $1 - \pi_x$. Here, e^{β_1} represents the odds ratio (OR) and the null hypothesis means that there is no association between the outcome variable (i.e., y) and exposure variable (i.e., x). Therefore, given a fixed significance level α, the sample size n required for obtaining a test of $H_0 : \beta_1 = 0$ vs. $H_A : \beta_1 = \beta_{1A}(\neq 0)$ with power at least $1 - \beta$ is given by

$$n = \frac{(z_\alpha + z_\beta)^2 V_1}{\beta_{1A}^2}$$

with

$$V_1 = \frac{\pi_x(1 + A)^2 B + (1 - \pi_x)(1 + AB)^2}{\pi_x(1 - \pi_x) AB},$$

where $A = e^{\beta_0}$ and $B = e^{\beta_1}$ as the alternative OR.

65.8 Discussion

In this chapter, we discuss some commonly used sample size formulas for clinical trials. These formulas can be classified into two main categories: sample size formulas that achieve a prespecified power level and that control a prespecified confidence width. For sample size formulas that achieve a prespecified power level, the sample size determination usually involves some unknown population parameters. For instance, in the two-sample problem with continuous clinical endpoint discussed in Section 65.3, they are σ^2 and $\mu_1 - \mu_2$, respectively. In this case, one can obtain an estimate of the unknown population variance (i.e., σ^2), for example, the sample variance, from a pilot study in the population of interest. However, this estimate could be unreliable and the substitution of this estimate in the proposed sample size formula may lead to the actual power of the planned study being less than the planned power (see, e.g., Browne [3]) if the pilot study are not large enough. To overcome this problem, Browne [3] suggested calculating an upper one-sided confidence limit for the population variance from the pilot data in accordance with the planned power and then adopting this upper limit in the sample size determination for the planned clinical trial. Similarly, it is unwise to base a sample size determination on the observed difference in a pilot study as a substitution of $\mu_1 - \mu_2$. In this case, the clinical trial should be powered to detect a minimal clinically important difference as suggested by Jaeschke et al. [9]. Another approach is to consult experts in the field to decide what difference would require to be demonstrated for them to adopt a new treatment or drug in view of its costs and risks (see, e.g., Spiegelhalter and Freedman [14]). Similar issues also exist for sample size formulas that control a prespecified confidence width. In this case, one can improve the sample size calculation by accounting for the stochastic nature of the estimates that are used to substitute the unknown population parameters. For this purpose, one may want to ensure that with a certain probability the width of the confidence interval is no more than a prespecified value and this can be accomplished by including a tolerance probability in the calculation (see, e.g., Kupper and Hafner [11]).

It is very common that different sample size formulas are proposed for the same particular problem in existing literature. However, readers are reminded that each sample size formula should correspond to a specific statistic (for formula achieves a prespecified power level) or confidence interval estimator (for formula controls a prespecific confidence width). Most importantly, different statistics or confidence interval estimators are usually derived under different assumptions. As a result, planned power or confidence width may not be guaranteed if different statistics or confidence interval estimators are used. We therefore urge clinical planners to stick with the statistic or confidence interval estimator from which the sample size formula is derived.

Finally, it is noteworthy that the larger the sample size, the larger the power, the greater the precision and hence the higher the possibility of establishing an effect (if it truly exists). In other words, it is possible to identify very small effects (e.g., proportion differences between treatments) if sample size is sufficiently large. However, it can be questioned whether it is correct, from an ethical perspective, to perform such a large-scale study to identify an effect that may be small and not clinically relevant. This may, in fact, be both incorrect and dangerous, and one must always remember that any new treatment may be potentially harmful. As a result, sample size determination should always be conducted in the methodologic design of any clinical trial.

References

1. Agresti, A. and Coull, B. A. (1998). Approximate is better than "exact" for interval estimation of binomial proportions. *Am. Statist.*, **54**, 119–126.

2. Altman, D. G. (1980). Statistics and ethics in medical research, III: How large a sample? *Br. Med. J.*, **281**, 1336–1338.

3. Browne, R. H. (1995). On the use of a pilot sample for sample size determination. *Statist. Med.*, **14**, 1933–1940.

4. Brown, L. D., Cai, T. T., and DasGupta, A. (2001). Interval estimation for a binomial proportion. *Statist. Sci.*, **16**, 101–133.

5. Cohen, J. and Cohen, P. (2001). *Applied Multiple Regression/Correlation Analysis for the Behavioral Sciences, 2nd ed.* Erlbaum, Hillsdale, NJ.

6. Demidenko, E. (2007). Samples size determination for logistic regression revisited. *Statist. Med.*, **26**, 3385–3397.

7. Fleiss, J. L. (1981). *Statistical Methods for Rates and Proportions, 2nd ed.* Wiley, New York.

8. Horn, M. and Vollandt, R. (2000). A survey of sample size formulas for pairwise and many-one multiple comparisons in parametric, nonparametric and binomial case. *Biometr. J.*, **42**, 27–44.

9. Jaeschke, R., Singer, J., and Guyatt, G. H. (1989). Measurement of health issues. Ascertaining the minimal clinically important difference. *Controlled Clin. Trials*, **10**, 407–415.

10. Kelly, K. and Maxwell, S. E. (2001). Sample size for multiple regression: Obtaining regression coefficients that are accurate, not simply significant. *Psychol. Methods*, **8**, 305–321.

11. Kupper, L. L. and Hafner, K. B. (1989). How appropriate are popular sample size formulas? *Am. Statist.*, **43**, 101–105.

12. Moher, D., Dulberg, C. S., and Wells, G. A. (1989). Statistical power, sample size, and their reporting in randomized controlled trials. *J. Am. Med. Assoc.*, **272**, 122–124.

13. Piegorsch, W. W. (2004). Samples sizes for improved binomial confidence intervals. *Comput. Statist.. Data Anal.*, **46**, 309–316.

14. Spiegelhalter, D. J. and Freeman, L. S. (1986). A predictive approach to select the size of a clinical trial, based on subjective clinical opinion. *Statist. Med.*, **5**, 1–13.

15. Milton, S. (1986). A sample size formula for multiple regression studies. *Pub. Opin. Quart.*, **50**, 112–118.

66. Scan Statistics

Joseph Glaz and Joseph I. Naus

66.1 Introduction

A public health official investigates unusual cancer clusters; a molecular biologist focuses on neighboring translocation breaks on a DNA strand, or scans a DNA sequence for clusters of palindromes; a communicable disease agency bases a bioterrorism surveillance plan on disease syndrome clusters. Professionals in a variety of biological and health science disciplines scan times (locations) of occurrence of events and focus on multiple events happening close in time (space).

Several statistics arise from scanning a time interval $(0, T)$ and observing that there exists a subinterval of length t that contains n events. Among these scan statistics are $n(t)$, the largest number of points within any subinterval of length t; $p(n)$, the length of the smallest interval containing n points; and $W(n, t)$, the waiting time until one first observes n events all occurring within a subinterval of length t. A related set of scan statistics arises from scanning the outcomes of a series of trials. We illustrate the statistics for these two cases.

Events Occurring in Time. A health official reviewing a community's records for the year is startled to find five cases of cancer occurring in the 10-day period from April 21 through April 30. In the past the community has averaged about 36 cases per year, which is similar to other communities. Here $n(10) = 5$, $p(5) = 10$, and $W(5, 10) = 120$.

Events as Outcomes of Trials. A blood production facility processes units of whole-blood components. Under normal conditions about 2% of the processed units do not comply with standards. On a particular day, of 500 units produced, 11 failed inspection. Reviewing the records, the quality control manager noted that six of the failed units were all among 100 consecutive units processed. Here $n(100) = 6$, $p(6) = 100$, and if the last of the six noncomplying units occurred on the 280th processed unit, $W(6, 100) = 280$. Both the public health official and the quality control manager are interested in whether the observed clusters are unlikely to have happened by chance.

Scan statistics have been used retrospectively to test for unusual clustering of HIV, Down's syndrome, livestock diseases, suicides, sudden infant death syndrome, specific cancers (breast, brain, and others). The threat of a bioterrorist attack, as in the anthrax-laden mail in the aftermath of September 11, 2001, has led to setting up prospective ongoing surveillance monitoring. Scan-type statistics are especially useful for early warnings in cases where there would be multiple attacks or individuals affected within a short time, and where it will be obvious within a fixed period (e.g., the time for lab tests to come back) that what was observed was not normal chance variation in syndromes.

The distributions of the scan statistics $n(t)$, $p(n)$, and $W(n,t)$ are related:

$$\begin{aligned} \Pr(n(t) \geq n) &= \Pr(p(n) \leq t) \\ &= \Pr(W(t) \leq T). \end{aligned}$$

These probabilities depend on the underlying process distribution chosen. The distributions of the scan statistics have been derived for the Poisson, Bernoulli, and other processes. The Poisson and Bernoulli processes provide simple models of chance variation for events occurring in time and as outcomes of trials, respectively. Actually each of the processes leads to several chance models appropriate for different situations.

The Poisson Process. Given events occurring in time according to a Poisson process with mean λ, the number of events (points) k occurring in any interval $(0, T)$ has the Poisson distribution. Conditional on there being N points from the Poisson process in $(0, T)$, the N points are independently and identically distributed (i.i.d.) according to the uniform distribution on $(0, T)$. The scan statistic probability $\Pr(n(t) \geq n|$ given N points in $(0, T))$, is different from the unconditional probability $\Pr(n(t) \geq n)$. Denote and abbreviate these probabilities $P(n; N, t/T) = P$ and $P^*(n; \lambda T, t/T) = P^*$, respectively. Each of these probabilities has different testing and other applications. In a case where no comparable average is known, one would use the conditional level of significance P; in a case where the average is known and is different from the total, or in sequential monitoring over a year, which may be stopped as soon as $n(t) \geq n$, so that N is not known, one would use P^*.

The probabilities P and P^* are in general complex to compute, and a variety of formulas, approximations, bounds, and tables for different cases and generalizations have been developed.

66.2 History and Literature

In this discussion we refer extensively to a more recent cluster of four books dealing with scan statistics. The book by Glaz, Naus, and Wallenstein (GNW) [32] is a comprehensive survey giving extensive applications, details on history, formula, methods of proof, and over 600 references. The first six chapters of this book is designed for the applied practitioner. The book edited by Glaz and Balakrishnan (GB) [29] presents results and applications on the scan statistic written by leading researchers. The book by Balakrishnan and Koutras (BK) [4] focuses on results for scan statistics in sequences of trials. The book by Glaz, Pozdnyakov, and Wallenstein (GPW) [33] gives recent results and applications of scan statistics to areas of science and technology, including the health sciences.

The conditional probability P was studied early in photographic science, where it was hypothesized that a silver speck on film would develop if k or more photons are absorbed within the decay time t of the nucleus. $P = P(n; N, t/T)$ depends on n, N, and the ratio $r = t/T$. The special cases $P(2; N, r)$, the cumulative distribution function (cdf) of the smallest gap between any two of the points, and $P(N; N, r)$, the cdf of the sample range for uniformly distributed variables were known for a long time. The probability P could be written as an integral that is very complicated to evaluate except in simple cases; for $n = 2$ and $n = N$ the probability is a polynomial in r. For other n, the probability is piecewise polynomial in r, with different polynomials for r in different ranges. Naus [60] uses a combinatorial approach to derive P that gives a simple formula for $n > N/2$, one polynomial in r for $r \leq 0.5$, and another polynomial for $r > 0.5$. The approach is generalized in Naus [61] for $r = 1/L$ and in Huntington and Naus [41] for all n, N, r. However, for large N and small r, the general results involve the sum of a large number of determinants of large matrices. Wallenstein and Naus [94] use the general results to develop formula and give tables for the case $N/3 < n \leq N/2$. Neff and Naus [66] review these results, develop computationally feasible approaches, and give extensive tables of the probabilities and polynomials for $N \leq 19$, allowing evaluation for a full range of n and r. Huffer and Lin [38] develop an approach based on spacings to extend the range of computation of exact probabilities and the piecewise polynomials. For many practical applications where tables are not available, there was still a need for simple approximations that do not involve extensive computing. The simplest approximation is by Wallenstein and Neff [96], and more precise approximations and bounds are developed in References 24–26. A large deviation approximation is given in Reference 53, and asymptotic results in Reference 17. There are various asymptotic and approximate results for multiple scan statistics and for scan statistics on the circle. The exact and approximate approaches and results are detailed in Chapters 2, 8, 9, 10 and 17 of GNW [32], and Chapters 6, 7, and 9 in GB [29].

The probability P^* arose in certain studies of visual perception, investigations of the rate of n-fold accidental coincidences in counters, probability of overflow in queueing, and other applications. Prior to the 1960s, exact formulas for P^* were known for only a few special cases. Newell [70] derives asymptotic results for $\Pr(n(t) \geq n)$ for the more general process where the waiting times between points are independently and identically distributed; the special case where the interpoint distance distribution is exponential is the Poisson case. The asymptotic results provided only rough approximations for P^*, and a better simple approximation is provided by Alm and Naus [1,2]. Naus [63, Theorem I] gives exact expressions for P^* for $r = \frac{1}{2}, \frac{1}{3}$, and a highly accurate approximation based on these results for all r. Janson [43] provides sharp bounds for P^*, and shows that the approximation in Naus [63] falls within these bounds. Samuel-Cahn [83] gives a series of approximations to the expectation of the waiting time $W(n, t)$ until the scan interval contains n points for general point processes with independent and identically distributed interarrival times of points. See Chapters 3 and 11 in GNW [32].

For the Bernoulli trials model, Roberts [82] develops various quality control tests and Troxell [90] gives various acceptance sampling plans based on scan statistics. Other authors [52,26,27,42] derive exact results. Naus [63] gives a highly accurate approximation, and Naus [31] gives tight bounds. Asymptotic results appear in the literature under the title of Erdos–Renyi, or Erdos–Renyi–Shepp laws of large numbers. Loader [53] develops a large deviation approximation. See Chapters 4, 12, and 13 of GNW [32], Chapter 2 in GB [29], and the approaches and variations in BK [4]. An important application of discrete scan statistics arises in matching in DNA or protein sequences viewed as a sequence of letters from an alphabet. The scientists look for similar parts of different sequences to suggest commonality of functions, or genetic material that is preserved. Given two sequences aligned by a global criterion, scientists scan for closely matching subsequences. Given two aligned sequences

$(X_1, X_2, ..., X_N)$ and $(Y_1, Y_2, ..., Y_N)$, the sequences match in position j if $X_j = Y_j$. Let $Z_j = 1$ if $X_j = Y_j$, and be 0 otherwise. If in the Bernoulli sequence $(Z_1, Z_2, ..., Z_N)$ there are n 1s within some t consecutive trials, this implies that there is an aligned subsequence of length t that matches on n of the letters, and the scan statistic $n(t) \geq n$. Approximations for the discrete scan statistic for a variety of chance models have been developed motivated by this application, see, for example, References 3 and 46, Chapter 13 in GB [29], and Chapter 6 in GNW [32].

Another important application of discrete scan statistics arises in reliability of linear systems of N components, where the system fails if anywhere within the system there are t consecutive components with at least n failures. The components can have different probabilities of failure and have Markov dependence. A variety of computational algorithms have been developed to compute discrete scan statistics probabilities. See References 4, 22, and 35 for a review of many of these and related results.

66.3 Higher-Dimensional Scans

Researchers in many health science fields scan for unusual clusters in two or more dimensions. An epidemiologist is drawn to a spatial cluster of a human or animal disease. A radiologist uses medical imaging to focus on unusual clusters in three dimensions. In two and higher dimensions results have been derived for rectangular, circular, triangular, and other shaped windows scanning various-shaped regions. Scan statistic distributions have been derived for Poisson, uniform, and other distributions of points over continuous space, and for binomial, hypergeometric distributions on a two-dimensional grid of points.

Given N points uniformly distributed over the unit square, let $S(u, v)$ denote the two-dimensional scan statistic, the maximum number of points in any subrectangle with sides of length u and height v parallel to the sides of the unit square. Let $P(n; N, u, v)$ denote $\Pr(S(u, v) \geq n)$. Loader [53] uses large deviation theory to derive a reasonable approximation for u, v, and $P(n; N, u, v)$ small, and Tu [91] develops an alternative approximation. Naiman and Priebe [59] and Priebe et al. [81] develop efficient Monte Carlo methods using importance sampling to estimate $P(n; N, u, v)$.

The corresponding unconditional on N version of the two-dimensional problem, lets the number of points in the unit square follow a Poisson process with mean λ, and denotes $\Pr(S(u, v) \geq n)$ by $P^*(n; \lambda, u, v)$. Alm [1,2] gives a simple approximation for $P^*(n; \lambda, u, v)$. In certain applications the events can occur only at a discrete set of points in space, or the method of observation limits the observed events to a grid of positions. The simplest case deals with a rectangular lattice of independently and identically distributed Bernoulli random variables. A discrete two-dimensional scan statistic looks at the maximum number of 1s in any $k \times m$ subrectangle with sides parallel to the sides of the lattice. Darling and Waterman [15] derive limit laws, and Chen and Glaz [9,10] derive approximations for this scan statistic. For a detailed discussion of the above two-dimensional cases, see Chapters 5 and 16 in GNW [32] and Chapters 2, 5, and 10 in GB [29].

66.4 Scan Statistics with Nonuniform Background

Most of the exact, and many of the approximate, distributional results for the classical fixed-window scan statistic were developed under the assumption of a constant null distribution background. For the case where there are incidence trends over time, or varying population densities over space, the p-values must be adjusted. Weinstock [97] stretches the timeframe (area) to deal with nonuniform background rates. Naus and Wallenstein [65] give a P-scan procedure to determine p-values for retrospective or prospective, continuous, discrete, or grouped temporal data that adjusts for nonuniform background. Naus and Wallenstein [65,95] illustrate the use of the P-scan model to temporal surveillance for bioterrorism. Kulldorff developed Web-based programs for application of the algorithms and simulation of the scan statistic distribution (*http://dcp.nci.nih.gov/BB/SaTScan.html*). The free, versatile, and well documented SaTScan simulation program (*www.satscan.org*) described in the next section handles temporal and spatial data for the non-uniform background case by allowing input of background population.

66.5 Scan Statistics with Variable-Size Windows

The previous sections deal with the classical scan statistic with a fixed-size window. In some cases there is a natural choice based on the application or balancing out periodicities (for example, using $w = 7$ days to balance out day-of-week effects). In other cases, the researcher may want to use a scan statistic that measures the most unusual cluster for any of several scanning window sizes. For the constant-background case, Loader [53] developed a large deviation approximation for p-values for variable-size-window scan statistics. Naus and Wallenstein [65] develops an accurate approximation for the p-values for the continuous temporal variable-size-window scan statistic (or alternatively for a range of cluster sizes) for the constant-background case. For a given set of pairs of scanning window sizes and cluster sizes $\{p_n, n\}, n = a, \ldots, b$, the experiment-wide error rate is given by $P\left(\bigcup_{n=a}^{b} E_n\right)$, where E_n denotes the event $(p(n) \leq p_n)$. The Hunter upper bound

$$P\left(\bigcup_{n=a}^{b} E_n\right)$$
$$\leq \sum_{n=a}^{b} P(E_n)$$
$$- \sum_{n=a}^{b-1} P(E_n | E_{n+1}) P(E_{n+1}),$$

employed in Naus and Wallenstein [64], yields a useful approximation.

For the constant background case, GNW [32] and Naus and Wallenstein [65] state simple approximations for $P(E_n)$ for the continuous, binary trial, and ratchet scans for the constant background retrospective and prospective cases. Approximations for $P(E_n | E_{n+1})$ can also be found for these cases. Naus and Wallenstein [64] give an approximation for the constant background continuous case. Glaz and Zhang [34] applied this approach for the discrete scan statistic for i.i.d. $0-1$ Bernoulli trials, and generalized it to the two-dimensional discrete scan statistic. Pozdnyakov et al. [78] applied a martingale approach to find p-values for variable window discrete scan statistics for i.i.d. nonnegative integer-valued observations. For the variable-size-window, non-constant (null)-background case, simulation of a generalized likelihood ratio temporal and spatial scan statistic has been typically used; Glaz and colleagues [31,32] and Huffer and Lin [38] develop algorithms for handling variable-size scanning windows in two dimensions. Kulldorff and Williams [50] established an excellent, versatile, well documented, and free SaTScan software (www.satscan.org). This software has been used widely in epidemiology, biosurveillance, and the health sciences. There is currently research into making the simulations more efficient, particularly for the spatial, and spatial-temporal cases. See for example, Reference 68.

66.6 The Conditional Probability P

The probability P is piecewise polynomial in r, with different polynomials in different ranges of r. For example, for $n > N/2$, there is one polynomial for $r \leq \frac{1}{2}$, and a different polynomial for $r > \frac{1}{2}$. The polynomial for $n > N/2$, $p \leq \frac{1}{2}$ is particularly simple and can be written in terms of binomials and sums of binomials:

$$P(n, N; r) = 2 \sum_{i=n}^{N} \binom{N}{n} r^i (1-r)^{N-i}$$
$$+ [(n/r) - N - 1] \binom{N}{n}$$
$$\times r^n (1-r)^{N-n}.$$

Wallenstein and Neff [96] points out that this exact formula can be used as an approximation for $P(n, N; r)$ even when $n \leq N/2$; the approximation is accurate for $P(n; N, w) < 0.05$, and for even larger values.

General formulas are available for all n, N, r, but these involve sophisticated programming. Simplified formulas for special cases and tables of the probabilities and polynomials and approximations are available (see Section 66.2).

66.7 The Unconditional Probability P^*

The probability P^* is important in studies of visual perception where it is assumed that photons arrive at the eye receptors according to a Poisson process. It is conjectured that several photons all arriving within a time t causes the neurons to discharge. What is of particular interest is the distribution of $W(n, t)$, the waiting time until discharge.

Except for simple cases, exact formulas for P^* involve complex computer evaluation, and various asymptotic results and bounds are available. The following approximation [63] is highly accurate in many situations. Let $Q^* = 1 - P^*$, $L = T/t$, $Q_2^* = Q^*(n, 2\lambda, 1/2)$ and $Q_3^* = Q^*(n, 3\lambda, \frac{1}{3})$; then $Q^*(n, L\lambda, 1/L) \approx Q_2^*(Q_3^*/Q_2^*)^{L-2}$. The values of Q_2^* and Q_3^* can be read from tables [66] or computed from simple formulas [63]. For example, to approximate $P^*(5, 36; \frac{10}{365})$, note that $n = 5$, $L\lambda = 36$, $1/L = 10/365$, so $\lambda = 0.9863$. Then $Q_2^* = 0.98434962$, $Q_3^* = 0.9727576$, and $P^*(5, 36; \frac{10}{365})$ is approximately 0.35. The health official's observed cluster of five cancer cases within some consecutive 10-day period of the year could easily have arisen by chance.

An accurate approximation [63] to the expectation of $W(n, t)$ is as follows: without loss of generality define the units to make $t = 1$ and let λ denote the expected number of points per unit. Then $E(W(n, 1)) = 2 + Q_2^*/\ln(Q_2^*/Q_3^*)$, where Q_2^* and Q_3^* are found as above. (In the above example, $\lambda = 0.9863$ per unit of 10 days, Q_2^* and Q_3^* are as given, and the expected waiting time until getting a cluster of five cases within 10 days is 85 units of 10 days, or 850 days.)

66.8 Sequence of Trials: Bernoulli Model

Given a sequence of T Bernoulli trials, each trial can result in a success with probability p and in a failure with probability $1 - p$, where $0 \leq p \leq 1$. The occurrence of at least n successes within some t consecutive trials is called an $n : t$ quota. Let $n(t)$ be the maximum number of successes within any scanning sequence consisting of t consecutive trials. $\Pr(n(t) \geq n)$ is the probability of observing an $n : t$ quota in a sequence of T Bernoulli trials. The special case of a $t:t$ quota is a success run of length t and $\Pr(n(t) \geq t)$ is the probability of at least one success run of length t. For iid Bernoulli trials $\Pr(n(t) \geq n)$ can be evaluated using a Markov chain embedding method developed by Fu [19] and Fu and Koutras [21] and discussed in great detail in BK [4]. Until recently these computations were feasible for a restricted range of the parameters n and t. For example, $\Pr(n(t) \geq n)$ could not be evaluated for $t = 50$ and $n = 25$ since the dimension of the state space of the embedded Markov chain could be as high as $2^{t-1}+1$ [4, Chap. 9] or $\binom{t}{n-1}+1$ [8]. In Fu [20], the construction of the state space of the embedded Markov chain is modified so that the dimension of the state space is at most $2t$. Therefore, $Pr(n(t) \geq n)$ can be evaluated for any value of n, t and T. Approximations and inequalities for $\Pr(n(t) \geq n)$ are discussed in Chapter 13 in GNW [32] and in Chapter 9 in BK [4]. Fu's method [20] can be employed to evaluate $\Pr(n(t) \geq n)$ for Bernoulli trials modeled by a homogeneous Markov chain. Results for this model are also presented in Glaz [23] and in Section 9.3 in BK [4].

In some applications it is known that a sequence of T i.i.d. Bernoulli trials results in N successes and $T - N$ failures. In this case the $\binom{T}{N}$ arrangements of the successes and failures are equally likely. The probability of observing an $n:t$ quota, conditional on the event that N successes have occurred in T trials, is referred to as the "generalized birthday probability" or the "probability of a conditional discrete scan statistic." Generalized birthday probabilities can be evaluated exactly for a restricted range of the parameters [62]. Accurate inequalities for the generalized birthday probabilities are given in Chen et al. [11]. These results and additional approximations are discussed in Chapters 4 and 12 in GNW [32] and Section 9.4 in BK

[4]. For t and T both large relative to n and N, the conditional probability to observe a quota in the i.i.d. Bernoulli case can be approximated by P. For the quality control example, the probability of a 6 : 100 quota given 11 defective items in 500 trials can be approximated by $P(6, 11; \frac{100}{500})$, which by the simple formula ($n > N/2, r < 0.5$) is equal to 0.198. The exact generalized birthday probability is 0.185 and the lower and upper bounds (based on Ref. 11) are both equal to 0.185.

66.9 Integer-Valued Random Variables

Let $X_1, \ldots, X_T$ be a sequence of i.i.d. nonnegative integer-valued random variables. Let $n(t)$ denote the maximum of $X_i + \cdots + X_{i+t-1}$, for $1 \leq i \leq T-t+1$. $n(t)$ is referred to as a *discrete scan statistic*. Approximations and inequalities for $P'(n, T; t) = \Pr(n(t) \geq n)$ are presented in Chapter 2 in GB [29] and in Chapter 13 in GNW [32]. Numerical results have been derived for the charge problem ($X_i = \pm 1, 0, 1 \leq i \leq T$) and the binomial and Poisson models. Discrete scan statistics have applications in many areas of science, including bioinformatics, biosurveillance, epidemiology, medical sciences, meteorology, molecular biology, reliability theory, and quality control (for references, see GB [29], GNW [32], and GPW [33]).

66.10 The Circle

Given N points distributed at random over the circle of unit circumference, let $P_c = P_c(n, N; t)$ denote the probability that a scanning arc of length t somewhere contains at least n points. Exact results for P_c are available for a restricted range of the parameters (Sec. 8.12 in GNW [32]). Approximations and inequalities have been discussed in References 63, 96, and 11. Note that on the circle, if the arc of length

t nowhere contains as many as n points, then the complementary arc of length $1-t$ must contain at least $N-n+1$ points, and knowing the distribution of the maximum number of points in a scanning arc gives the distribution of the minimum. The minimum number of points in a scanning arc is also a scan statistic and has been studied under coverage problems (Sec. 8.12.2 in GNW [32]).

Consider a sequence of T i.i.d. nonnegative integer valued random variables arranged in a circular fashion. Let $n(t)$ be the maximum sum of any consecutive subsequence of length t. Let $P_c^{'}(n,T;t) = \Pr(n(t) \geq n)$. Approximations for $P_c(n,T;t)$ have been investigated [10] for the binomial and Poisson models. For the Bernoulli model, approximations and inequalities have been derived for both conditional [10] and the unconditional [11] cases.

66.11 Two-Dimensional Scan Statistics

Let $[0, R_1] \times [0, R_2]$ be a rectangular region. Let $h_i = R_i/T_i > 0$, where T_i are positive integers, $i = 1, 2$. In many applications the exact locations of the observed events in the rectangular region are unknown. What is usually available are the counts in small rectangular subregions. For $1 \leq i \leq T_1$ and $1 \leq j \leq T_2$, let X_{ij} be the number of events that have been observed in the rectangular subregion $[(i-1)h_1, ih_1] \times [(j-1)h_2, jh_2]$. We are interested in detecting unusual clustering of these events under the null hypothesis that X_{ij} are independent and identically distributed nonnegative integer-valued random variables from a specified distribution. For $1 \leq i_1 \leq T_1 - t_1 + 1$ and $1 \leq i_2 \leq T_2 - t_2 + 1$, define $Y_{i_1,i_2} = \sum_{j=i_2}^{i_2+t_2-1} \sum_{j=i_1}^{i_1+t_1-1} X_{ij}$ to be the number of events in a rectangular region comprising of t_1 by t_2 adjacent rectangular subregions with area h_1h_2 and the southwest corner located at the point $((i_1-1)h_1, (i_2-1)h_2)$. If Y_{i_1,i_2} exceeds a preassigned value of n, we will say that n events are clustered within the inspected region. We define a *two dimensional discrete scan statistic* as the largest number of events in any t_1 by t_2 adjacent rectangular subregions with area h_1h_2 and the southwest corner located at the point $((i_1-1)h_1, (i_2-1)h_2)$:

$$S_{t_1,t_2}^{'} = \max\{Y_{i_1,i_2}; 1 \leq i_1 \leq T_1 - t_1 + 1, \\ 1 \leq i_2 \leq T_2 - t_2 + 1\}.$$

A especially interesting case is when X_{ij} are independent and identically distributed Bernoulli random variables and $n = t_1t_2$. In this case we are evaluating the probability that a $t_1 \times t_2$ rectangular grid has all 1s. This extends the notion of a run of 1s to two dimensions. Approximations and inequalities for $P(S_{t_1,t_2}^{'} = t_1t_2)$ are discussed in Chapter 16 in GNW [32].

$S_{t_1,t_2}^{'}$ is used for testing the null hypothesis of randomness that assumes the X_{ij}'s are i.i.d. binomial random variables with parameters n_0 and $0 < p_0 < 1$ or i.i.d. Poisson random variables with mean $\theta_0 > 0$, respectively. Approximations for $P(S_{t_1,t_2}^{'} \geq n)$ are discussed in Chapter 16 in GNW [32]. Applications for the two-dimensional discrete scan statistics are discussed in Chapters 5 and 16 in GNW [32].

In some applications one is interested in approximations for the distribution of the two dimensional discrete scan statistic, conditioned on the number of events that have occurred in the rectangular region $[0, R_1] \times [0, R_2]$. In Section 16.2 in GNW [32] approximations and inequalities are derived for the distribution of the conditional scan statistic. These approximations and inequalities yield approximations and inequalities for the expected size and standard deviation of the conditional scan statistic. These approximations and inequalities are based on simulation algorithms.

Let $\{X(s,t); 0 \leq s,t < \infty\}$ be a two-dimensional homogeneous Poisson process with intensity λ. For $0 \leq s,t < \infty$ and $0 \leq u,v < \infty$, let $Y_{s,t}(u,v)$ be the number of points in the rectangle $[s, s+u) \times [t, t+v)$, of dimension u by v with a southwest location at s,t. For $0 < u < T_1 < \infty$ and $0 < v < T_2 < \infty$, the two-dimensional *scan statistic*, $S(u,v) = \max\{Y_{s,t}(u,v), 0 \leq s \leq T_1 - u, 0 \leq t \leq T_2 - v\}$, denotes the largest number of points in any rectangle of dimension u by v within $[0, T_1) \times [0, T_2)$. Approximations for the distribution of $S(u,v)$ have been derived in the articles by Alm [1,2].

Let $X_{1,1}, ..., X_{1,N}$ and $X_{2,1}, ..., X_{2,N}$ be independent and identically distributed observations from a uniform distribution on the interval $[0,1)$. For $1 \leq i \leq N$, let $\mathbf{X}_i = (X_{1,i}, X_{2,i})$. Then $\mathbf{X}_1, ..., \mathbf{X}_N$ can be viewed as N random points in the unit square $[0,1) \times [0,1)$. For $0 < u,v < 1$, $0 \leq s < 1 - u$ and $0 \leq t < 1 - v$, let $Y_{s,t}(u,v)$ be the number of points in the rectangle $[s, s+u) \times [t, t+v)$, of dimension u by v with a southwest location at s,t. The conditional two-dimensional scan statistic is the largest number of points in any rectangle of dimension u by v within the unit square $[0,1) \times [0,1)$. Approximations, inequalities, and exact results for the distribution of this conditional two dimensional scan statistics are presented in Chapters 5 and 16 in GNW [32].

66.12 Multiple-Scan Statistics

Let $X_{(1)} \leq \cdots \leq X_{(N)}$ be the ordered observations of N points uniformly distributed on the unit interval. Define

$$\xi = \sum_{j=1}^{N-n+1} I_j,$$

where

$$I_j = \begin{cases} 1, & X_{(j+n+1)} - X_{(j)} \leq t \\ 0, & X_{(j+n+1)} - X_{(j)} > t. \end{cases}$$

The random variable ξ counts the number of intervals of length t or less that contain n points. It has been referred to in the statistical literature as a multiple scan statistic [30]. Note that $P(n; N, t) = P(\xi \geq 1)$. Approximations for the distribution of ξ and its extension to the case when N has a Poisson distribution are discussed in Chapter 17 in GNW [32].

A compound Poisson approximation for the distribution of a multiple scan statistic has been investigated for a sequence of i.i.d. binomial and Poisson random variables [10]. In BK [4], for the case of i.i.d. 0-1 Bernoulli trials, a Markov chain embedding method has been employed to derive the distribution of multiple-scan statistics. A two-dimensional discrete multiple-scan statistic and an approximation for its distribution is discussed in Section 16.1.8 in GNW [22]. Poisson-type approximations for multiple scan statistics for observations uniformly distributed over a region in a d-dimensional space are discussed by Mansson [54–56]. Cucula [14] has proposed a new approach based on m-order spacings in a temporal point process, for approximating the distribution of a variable window multiple scan statistic. The empirical distribution of this statistic is evaluated via a Monte Carlo procedure.

66.13 Spatial Scan Statistics

Modeling and inference of spatial data is one the most active research areas in probability and statistics. It has many applications in science and technology, including animal science, biosurveillance, ecology, environmental science, epidemiology, food sciences, image analysis, medicine, and urban and regional planning. The use of spatial scan statistics in two- or higher-dimensional regions have been discussed in the literature [6,32,47–49,74,79–81,85,93].

66.14 Bioinformatics and Genetic Epidemiology

Bioinformatics and genetic epidemiology are two of the major areas in the health sciences where scan statistics have been successfully employed. Hoh and Ott [36,37] have developed a variable-size-window scan statistic based on a sequence of combined information of ordered markers on chromosomes. They [36] applied this scan statistic to a genome screen with autism families, yielding a new susceptibility region. A scan statistic based on an ordered sequence of values of a linkage statistic, for a marker with those of surrounding markers, has been employed by DeWan and Ott [18]. They have obtained improved linkage results for a previously analyzed schizophrenic dataset.

Sun et al. [87,88], have developed scan statistics for detection of significant patterns of single-nucleotide polymorphism (SNP) associations in their chromosomal context. The identification of significant patterns of SNP's are of great importance to the assessment of genetic influence on risks of various diseases.

Levin et al. [52] have developed a model based scan statistic for detecting extreme chromosomal regions of gene expression data, using a compound Poisson model.

Chan and Zhang [7] introduced a scan statistic for weighted observations. Accurate approximations for this statistic have been derived and applied to analyses of DNA sequences.

66.15 Process Control and Monitoring

Quality control and monitoring of events in health sciences have received an increased level of attention. Methods based on scan statistics, in both one and two dimensions have been investigated by researchers in health sciences. Lachenbruch et al. [51] discussed a scan statistic for a Bernoulli model in the area of blood product manufacturing. In this area scan statistics based on binomial and Poisson models might prove to be of importance as well. Kulldorff et al. [49] have developed a novel approach using a tree-based scan statistic for database disease surveillance.

Cumulative sum (CUSUM) type statistics, as well as scan statistics, have been employed extensively in the area of industrial quality control [28]. In References 42, 45, and 86, a CUSUM-type framework has been adapted, based on scan statistics, to monitor events in the area of health sciences, such as occurrences of certain diseases or outbreak of rare infections in both one- and multidimensional datasets.

Perez et al. [75] have employed spatial scan statistics to monitor and identify diseases in veterinary sciences.

66.16 Bayesian Scan Statistics

Bayesian spatial scan statistics have been introduced in References 98 and in 69. Bayesian scan statistics, for variable-size rectangular windows, have been investigated [98] for observed events modeled by two-stage hierarchical models with nonconjugates. These scan statistics are based on a sequence of Bayes factors and their p-values, obtained via simulation. In Reference 69, univariate Bayesian scan statistics, via an approach of generalized likelihood ratio type statistics, have been developed for conjugate priors. In Reference 67, univariate Bayesian scan statistics have been generalized to multivariate Bayesian scan statistics. Moreover, an agent based Bayesian scan statistic, related to an explicit Bayesian network, has been investigated in References 16 and 44. They developed effective algorithms for evaluating these Bayesian spatial scan statistics and applied to several interesting data sets in the areas of spatial detection and disease surveillance.

66.17 Theoretical Methods

The derivation of exact results, approximations, and inequalities for the distribution of scan statistics discussed in this chapter involve sophisticated and intricate theoretical methods and techniques. Among these are included Bonferroni-type inequalities in Chapters 9, 12, and 13 in GNW [32], Poisson and compound Poisson approximations in Chapters 10–13, 16, and 17 in GNW [32], finite Markov chain embedding in BK [4], combinatorial and geometric probability methods in Chapter 9 in GNW [32], false discovery rate (FDR) [76,77], large deviation [7,53], linear programming [38], Markov chain modeling [38], Monte Carlo testing [48], order statistics and spacings in GNW [32] and Cucula [14], saddle point approximations [6], and symbolic computing [38].

References

1. Alm, S. E. (1997). On the distribution of scan statistics of a two-dimensional Poisson processes. *Adv. Appl. Probab.*, **29**, 1–18.

2. Alm, S. E. (1998), Approximation and simulation of the distribution of scan statistics for Poisson processes in higher dimensions. *Extremes*, **1**, 111–126.

3. Arratia, R., Gordon, L. and Waterman, M. S. (1990). The Erdos Renyi law in distribution for coin tossing and sequence matching. *Ann. Statist.*, **18**, 539–570.

4. Balakrishnan, N. and Koutras, M. V. (2001). *Runs and Scans with Applications*, Vol. I. Wiley, New York.

5. Chan, H. P. and Lai, T. L. (2002). Boundary crossing probabilities for scan statistics and their applications to change-point detection, *Methodol. Comput. Appl. Probab.*, **4**, 317–336.

6. Chan, H. P. and Lai, T. L. (2003). Saddlepoint approximations and nonlinear boundary crossing probabilities of Markov random walks, *Ann. Appl. Probab.*, **13**, 395–429.

7. Chan, H. P. and Zhang, N. R. (2007). Scan statistics with weighted observations. *J. Am. Statist. Assoc.*, **102**, 595–602.

8. Chang, J. C., Chen, R. J., and Hwang, F. K. (2001). A minimal-automation-based algorithm for the reliability of Con(d,k,n) system. *Methodol. Comput. Appl. Probab.*, **3**, 379–386.

9. Chen, J. and Glaz, J. (1996). Two dimensional discrete scan statistics. *Statist. Probab. Lett.*, **31**, 59–68.

10. Chen, J. and Glaz, J. (1999). Approximations for discrete scan statistics on the circle. *Statist. Probab. Lett.*, **44**, 167–176.

11. Chen, J., Glaz, J., Naus, J., and Wallenstein, S. (2001). Bonferroni-type inequalities for conditional scan statistics. *Statist. Probab. Lett.*, **53**, 67–77.

12. Christiansen, L. E., Andersen, J. S., Wegener, H. C., and Madsen, H. (2006). Spatial scan statistics using elliptic windows. *J. Agric. Biol. Environ. Sci.*, **11**, 411–424.

13. Coulston, J. W. and Riitters, K. H. (2003). Geographic analysis of forest health indicators using spatial scan statistics. *Environ. Manage.*, **31**, 764–773.

14. Cucula, L. (2008). A hypothesis-free multiple scan statistic with variable window. *Biomet. J.*, **50**, 299–310.

15. Darling, R. W. R. and Waterman, M. S. (1986). Extreme value distribution for the largest cube in random lattice. *SIAM J. Appl. Math.*, **46**, 118–132.

16. Das, K., Schneider, J., and Neill, D. B. (2008). Detecting anomalous groups in categorical data sets, submitted for publication, Carnegie Mellon University, School of Computer Science.

17. Dembo, A. and Karlin, S. (1992). Poisson approximations for r-scan processes. *Ann. Appl. Probab.*, **2**, 329–357.

18. DeWan, A. and Ott, J. (2004). Reanalysis of a genome scan for schizophrenia loci using multigenic methods. *Hum. Hered.*, **57**, 191–194.

19. Fu, J. C. (1986). Reliability of consecutive-k-out-of-n-F system with (k-1) step Markov dependence. *IEEE Tran. Reliab.*, **35**, 602–606.

20. Fu, J. C. (2001). Distribution of the scan statistic for a sequence of bi-state trials. *J. Appl. Probab.*, **38**, 908–916.

21. Fu, J. C. and Koutras, M. V. (1994). Distribution theory of runs: A Markov chain approach. *J. Am. Statist. Assoc.*, **89**, 1050–1058.

22. Fu, J. C. and Lou, W .Y. (2003). *Distribution Theory of Runs and Patterns and Its Applications*. World Scientific. River Edge, NJ.

23. Glaz, J. (1983). Moving window detection for discrete data. *IEEE Trans. Inform. Theory*, **IT-29**(3), 457–462.

24. Glaz, J. (1989) Approximations and bounds for the distribution of the scan statistic, *J. Am. Statist. Assoc.*, **84**, 560–569.

25. Glaz, J. (1992). Approximations for tail probabilities and moments of the scan statistic. *Comput. Statist. Data Anal.*, **14**, 213–227.

26. Glaz, J. (1993). Approximations for tail probabilities and moments of the scan statistic. *Statist. Med..*, **12**, 1845–1852.

27. Glaz, J. (1996). Discrete scan statistics with applications to minefield detection. *Proc. Conf. SPIE*, Orlando FL, SPIE, Vol. 2765, pp. 420-4-29.

28. Glaz, J. (2007). Scan statistics, in *Encyclopedia of Statistics in Quality and Reliability*, F. Ruggeri, R. Kenett, and F. W. Faltin, eds., pp. 1761–1766 Wiley, Chichester, UK.

29. Glaz, J. and Balakrishnan, N., eds. (1999). *Scan Statistics and Applications*. Birkhauser, Boston.

30. Glaz, J. and Naus, J. (1983). Multiple clusters on the line. *Commun. Statist. Theory Methods*, **12**, 1961–1986.

31. Glaz J. and Naus J. (1991) Tight bounds and approximations for scan statistic probabilities for discrete data. *Ann. Appl. Probab.*, **1**, 306–318.

32. Glaz, J., Naus, J., and Wallenstein, S. (2001). *Scan Statistics*. Springer-Verlag, New York.

33. Glaz, J., Pozdnyakov, V., and Wallenstein, S. (2009). *Scan Statistics: Methods and Applications*. Birkhauser Publishers, Boston.

34. Glaz, J. and Zhang, Z. (2004). Multiple window discrete scan statistics. *J. Appl. Statist.*, **31**, 967–980.

35. Godbole, A. and Papastavridis, S. G., eds. (1994). *Runs and Patterns in Probability*. Kluwer Academic Publishers, Netherlands.

36. Hoh, J. and Ott, J. (2000). Scan statistics to scan markers for susceptibility genes. *Proc. Nat. Acad. Sci. (USA)*, **97**, 9615–9617.

37. Hoh, J. and Ott, J. (2003). Mathematical multi-locus approaches to localizing complex human trait genes., *Nat. Rev. Genet.*, **4**, 701–709.

38. Huffer, F. W. and Lin, C. T. (1997). Computing the exact distribution of the extremes of sums of consecutive spacings. *Comput. Statist. Data Analy.*, **26**, 117–132.

39. Huntington, R. J. (1974). Distributions for clusters in continuous and discrete cases. *Northeast Sci. Rev.*, **4** (W. H. Long and M. Chatterjii, eds.), 153–161.

40. Huntington, R. J. (1976). Mean recurrence times for k successes within m trials. *J. Appl. Probab.*, **13**, 604–607.

41. Huntington, R. and Naus, J. I. (1975) A simpler expression for Kth nearest neighbor Coincidence probabilities. *Ann. Probab.*, **3**, 894–896.

42. Ismail, N. A., Pettitt, A. N., and Webster, R. A. (2003). "Online" monitoring

and retrospective analysis of hospital outcomes based on scan statistics. *Statist. Med.*, **22**, 2861–2876.

43. Janson, S. (1984). Bounds on the distributions of extremal values of a scanning process. *Stoch. Process. Appl.*, **18**, 313–328.

44. Jiang, X., Neill, D. B., and Cooper, G. F. (2008). A Bayesian Network Model for Spatial Event Surveillance, submitted for publication, University of Pittsburgh, Department of Biomedical Informatics.

45. Joner Jr, M. D., Woodall, W. H., and Reynolds Jr, M. R. (2007). Detecting a rate increase using a Bernoulli scan statistic. *Statist. Med.*, **27**, 2555–2575.

46. Karlin, S. and Ost, F. (1987). Counts of long aligned word matches among random letter sequences. *Adv. Appl. Probab.*, **19**, 293–251.

47. Kulldorff, M. (1997). A spatial scan statistic. *Commun. Statist. Theory Methods*, **26**, 1481–1496.

48. Kulldorff, M. (1999). Spatial scan statistics: Models, calculations and applications. In *Scan Statistics and Applications*, J. Glaz and N. Balakrishnan, eds., pp. 303–322. Birkhauser, Boston.

49. Kulldorff, M., Fang, Z., and Walsh, S. J. (2002). A tree-based scan statistic for data base disease surveillance. *Biometrics*, **59**, 323-331.

50. Kulldorff, M. and Williams, G. (1997). *SatScan v. 1.0. Software for the Space and Space-Time Scan Statistics.* National Cancer Institute, Bethesda, MD.

51. Lachenbruch, P. A., Foulkes, M. A., Williams, A. E. and Epstein, J. S. (2005). Potential use of the scan statistic for quality control in blood product manufacturing. *J. Biopharm. Statist.*, **15**, 353–366.

52. Levin, A. M., Ghosh, D., Cho, K. R., and Kardia. S. L. R. (2005). A model-based scan statistic for identifying extreme chromosomal regions of gene expression in human tumors. *Bioinformatics*, **21**, 2867–2874.

53. Loader, C. (1991). Large deviation approximations to the distribution of scan statistics. *Adv. Appl. Probab.*, **23**, 751–771.

54. Mansson, M. (1999a). Poisson approximation in connection with clustering of random points. *Ann. App. Probab.*, **9**, 465–492.

55. Mansson, M. (1999b). On Poisson approximation for continuous multiple scan statistics in two dimensions. In *Scan Statistics and Applications*, J. Glaz and N. Balakrishnan, eds. Birkhauser, Boston.

56. Mansson, M. (2000). On compound Poisson approximation for sequence matching. *Combin. Probab. Comput.*, **9**, 529–548.

57. Modarres, R. and Patil, G. P. (2007). Hotspot detection with bivariate data. *J. Statist. Plan. Infer.*, **137**, 3643–3654.

58. Nagarwalla, N. (1996). A scan statistic with a variable window. *Statist. Med.e*, **15**, 845–850.

59. Naiman, D. Q. and Priebe, C. (2001) Computing scan statistic p-values using importance sampling, with applications to genetics and medical image analysis. *J. Comp. Graph. Statist.*, **10**, 296–328.

60. Naus, J. (1965). The distribution of the size of the maximum cluster of points on a line. *J. Am. Statist. Assoc.*, **60**, 532–538.

61. Naus, J. (1966). Some probabilities, expectations, and variances for the size of the largest clusters and smallest intervals. *J. Am. Statist. Assoc.*, **61**, 1191–1199.

62. Naus, J. I. (1974) Probabilities for a generalized birthday problem. *J. Am. Statist. Assoc.*, **69**, 810–815.

63. Naus, J. I. (1982). Approximations for distributions of scan statistics. *J. Am. Statist. Assoc.*, **77**, 177–183.

64. Naus, J. and Wallenstein, S. (2004). Multiple window and cluster size scan procedures. *Methodol. Comput. Appl. Probab.*, **6**, 389–400.

65. Naus, J. and Wallenstein, S. (2005). Temporal surveillance using scan statistics, *Statist. Med.,* **25**, 311–324.
66. Neff, N. and Naus, J. (1980). *Selected Tables in Mathematical Statistics*, Vol. 6: *The Distribution of the Size of the Maximum Cluster of Points on a Line.* American Mathematical Society, Providence, RI.
67. Neill, D. B., and Cooper, G. F. (2008). A multivariate Bayesian scan statistic for early event detection and characterization, *Machine Learning*, in press.
68. Neill, D. B. and Moore, A. W. (2004). A fast multiresolution method for detection of significant spatial disease clusters. *Adv. Neural Inform. Process. Syst.,* **10**, 651–658.
69. Neill, D. B., Moore, A. W., and Cooper, G. F. (2006). A Bayesian spatial scan statistic, *Adv. Neural Inform. Process. Syst.,* **18**, 1003–1010.
70. Newell, G. F. (1963). Distribution for the smallest distance between any pair of Kth nearest-neighbor random points on a line. *Time series analysis, Proc. Time Series Analysis Conf.*, Brown University, M. Rosenblatt, ed., pp. 89–103. Wiley, New York.
71. Nødtvedt, A., Guitian, J., Egenvall, A., Emanuelson, U., and Pfeiffer, D. U. (2007). The spatial distribution of atopic dermatitis cases in a population of insured Swedish dogs. *Prevent. Vet. Med.,* **78**, 210–222.
72. Odoi, A., Martin, S. W., Michel, P., Middleton, D., Holt, J., and Wilson, J. (2004). Investigation of clusters of giardiasis using GIS and a spatial scan statistic. *Int. J. Health Geogr.,* **3**, 1–11.
73. Ozdenerol, E., Williams, B. L., Kang, S. U., and Magsumbol, M. S. (2005). Comparison of spatial scan statistic and spatial filtering in estimating low birth weight clusters. *Int. J. Health Geogr.,* **4**, 1–11.
74. Patil, G. P., Bishop, J., Myers, W. L., Vraney, R. and Wardrop, D. (2002). Detection and delineation of critical areas using echelons and spatial scan statistics with synoptic data. *Environ. Ecol. Statist.,* **11**, 139–164.
75. Perez, A. M., Ward, M. P., Torres, P., and Ritacco, V. (2002). Use of spatial statistics and monitoring data to identify clustering of bovine tuberculosis in Argentina. *Prevent. Vet. Med.,* **56**, 63–74.
76. Perone-Pacifico, M., Genovese, C., Verdinelli, I., and Wasserman, L. (2004). False discovery control for random fields. *J. Am. Statist. Assoc.,* **99**, 1002–1014.
77. Perone-Pacifico, M., Genovese, C., Verdinelli, I., and Wasserman, L. (2007). Scan clustering: A false discovery approach. *J. Multivar. Anal.,* **98**, 1441–1469.
78. Pozdnyakov, V., Glaz, J., Kulldorff, M., and Steele, J. M. (2005). A martingale approach to scan statistics. *Ann. Instit. Statist. Math.,* **57**, 21–37.
79. Priebe, C. E., Olson, T., and Healy, Jr., D. M. (1997). A spatial scan statistic for stochastic scan partitions. *J. Am. Statist. Assoc.,* **92**, 1476–1484.
80. Priebe, C. E. and Chen, D. (2001). Spatial scan density estimates. *Technometrics,* **43**, 73.
81. Priebe, C. E., Naiman, D. Q., and Cope, L. M. (2001) Importance sampling for spatial scan analysis: Computing scan statistic p-values for marked point processes. *Comput. Statist. Data Anal.,* **35**, 475–485.
82. Roberts, S. W. (1958). Properties of Control chart zone tests. *Bell Sys. Tech. J.,* **37**, 83–114.
83. Samuel-Cahn, E. (1983). Simple approximations to the Expected waiting time for a cluster of any given size, for point processes. *Adv. App. Probab.,* **15**, 21–38.
84. Saperstein, B. (1972). The generalized birthday problem. *J. Am. Statist. Assoc.,* **67**, 425–428.
85. Siegmund, D. and Yakir, B. (2000). Tail probabilities for the null distribution of scanning statistics, *Bernoulli,* **6**, 191–213.

86. Sonesson, C. (2007). A CUSUM framework for detection of space-time disease clusters using scan statistics. *Statist. Med.*, **26**, 4770–4789.

87. Sun, Y. V., Jacobsen, D. M., and Kardia, S. L. R. (2006). ChromoScan: a scan statistic application for identifying chromosomal regions in genomic studies. *Bioinformatics*, **22**, 2945–2947.

88. Sun, Y. V., Levin, A. M., Boerwinkle, E., Robertson, H., and Kardia, S. L. R. (2006). *Genet. Epidemiol.*, **30**, 627–635.

89. Tango, T. and Takahashi, K. (2005). A flexibly shaped spatial scan statistic for detecting clusters. *Int. J. Health Geogr.* **3**, 1-11.

90. Troxell, J. R. (1980). Suspension systems for small sample inspections. *Technometrics*, **22**, 517–533.

91. Tu, I. P. (1997). *Theory and Applications of Scan Statistics*. Ph.D. thesis, Stanford University.

92. Viel, J.-F., Floret, N., and Mauny, F. (2005). Spatial and space-time scan statistics to detect low rate clusters of sex ratio. *Environ. Ecol. Statist.*, **12**, 289–299.

93. Wallenstein, S., Gould, M. S., and Kleinman, M. (1989). Use of the scan statistic to detect time-space clustering. *Am. J. Epidemiol.*, **130**, 1057–1064.

94. Wallenstein, S. and Naus, J. (1974). Probabilities for the size of largest clusters and smallest intervals. *J. Am. Statist. Assoc.*, **69**, 690–697.

95. Wallenstein, S. and Naus, J. (2004). Statistics for temporal surveillance of bioterrorism. In *Syndromic Surveillance: Reports from a National Con;. MMWR*, **53**(Suppl.), 74–78.

96. Wallenstein, S. and Neff, N. (1987). An approximation for the distribution of the scan statistic. *Statist. Med.*, **6**, 197–207.

97. Weinstock, M. (1981). A generalized scan statistic for the detection of clusters. *Int. J. of Epidemiol.*, **10**, 289–293.

98. Zhang, Z. and Glaz, J. (2008). Bayesian variable window scan statistics. *J. Statist. Plan. Inf.*, **138**, 3561–3567.

67
Semiparametric Analysis of Competing-Risk Data

R. Sundaram and N. Balakrishnan

67.1 Introduction and Notation

Competing-risk data are typically encountered in medical studies and in reliability studies. In clinical trials, issues of great interest include assessing the effect of treatment and risk factors on the long-term outcomes, such as time to death and/or disease progression. During the follow-up, the patients may die from causes other than the disease under investigation. Such data where the types of failure are mutually exclusive from each other are commonly referred to as competing-risk data. For example, in cancer clinical trials, patients being followed up for relapse may die before relapse occurs; hence, relapse and death without antecedent relapse constitute the competing failure types. The goals of such study include estimating the probabilities and evaluating the temporal patterns of failing from particular causes over time and identifying potential risk factors for each failure type separately. An example of such data can be found in a clinical study involving 205 patients operated on for malignant melanoma; 71 patients died, of whom 57 were recorded as having died of the disease and 14 from causes unrelated to the disease during the follow-up [2]. As more than a quarter of the patients were 65 years or older, they were often more likely to die from other possible causes. As different causes of death may relate to distinct prognostic factors, it is desirable to perform a regression analysis of survival time for each cause. Another example from reproductive health can be encountered in studying progression of human labor. In such studies dealing with time to progression of spontaneous labor, where labor due to medical intervention (e.g., delivery by cesarean section) or membrane rupture leading to labor are treated as other causes. Here, it is of importance to analyze the effect of some known risk factors such as the woman's age, body mass index (BMI), previous history of hypertension, or diabetes on each type of labor.

Typically in these situations, the response can be classified in terms of failure from disease of interest and/or non-disease-related causes. So in the competing-risk framework, each individual is exposed to K distinct types of risks and the eventual failure can be attributed to precisely one of the risks. Suppose that each subject has an underlying continuous failure time $\tilde{T}$ that may be subject to censoring. The cause of failure $\epsilon \in \{1,\ldots,K\}$ is observed along with covariate information. The cause-specific hazard function for a subject with a covariate vector z is defined by

$$\lambda_k(t|z) = \lim_{\Delta t \to 0} \frac{P\left(t \leq \tilde{T} < t+\Delta t, \epsilon=k | \tilde{T} \geq t, z\right)}{\Delta t}$$

for $k = 1,\ldots,K$. The cause-specific hazard function $\lambda_k(t|z)$ is the instantaneous rate of occurrence of the kth failure cause in the presence of all causes of failure given z. Note that the instantaneous rate of occurrence of failure (irrespective of cause) is given by $\lambda(t|z) = \sum_{k=1}^{K} \lambda_k(t|z)$. Hence, the overall survival function can be represented as usual by

$$S(t|z) = \exp\left(-\sum_{k=1}^{K} \int_0^t \lambda_k(u|z) du\right).$$

To predict the survival probability of the kth type of failure time with a particular set of covariates z, a natural approach may be to use $S_k(t|z) = \exp\{-\int_0^t \lambda_k(u|z)\,du\}$. However, this quantity has no simple probability interpretation when the competing risks are dependent [18, Chap. 7]. An appealing alternative to S_k is the cumulative incidence function, defined by

$$F_k(t|z) = P(\tilde{T} \leq t, \epsilon = k | z).$$

The cumulative incidence function $F_k(t|z)$ is the probability of the subject failing from cause k in the presence of all the competing risks given z.

Note that $F_k(t|z)$ can be expressed in terms of $\lambda_l(t|z)$, where $l = 1,\ldots,K$. It is easy to show that

$$F_k(t|z) = \int_0^t S(u|z) \lambda_k(u|z)\,du \quad (1)$$

with the overall survival function $S(t|z) = \exp(-\int_0^t \sum_{l=1}^{K} \lambda_l(u|z)\,du)$. The quantities of considerable interest are the cause-specific hazards and the cause-specific cumulative incidence function. Thus, it is natural to estimate the cumulative incidence function through the cause-specific hazard function. The cumulative incidence function is also referred to as *absolute risk*. This is a quantity of considerable interest to clinicians dealing with cancer or other diseases of interest. For example, absolute cancer risk is the probability that an individual with given risk factors and a given age will develop cancer over a defined period of time. Substantial effort has been exerted in developing statistical models to predict the absolute risk of developing specific cancers among average-risk individuals on the basis of their risk and protective factors for cancer. Examples of these factors include race, age, sex, BMI, family history of cancer, history of tobacco use, use of aspirin and nonsteroidal anti-inflammatory drugs, physical activity, use of hormone replacement therapy, reproductive factors, history of cancer screening, and dietary factors.

A popular breast cancer risk assessment model was developed in National Cancer Institute, National Institutes of Health called the "Gail model." This is a popular model used by clinicians in addressing questions for the patients. Developing statistical models that estimate the probability of developing cancer over a defined period of time will help clinicians identify individuals at higher risk of specific cancers, allowing for earlier or more frequent screening and counseling of behavioral changes to decrease risk. These types of models also are useful for designing future chemoprevention and screening intervention trials in individuals at high risk for specific cancers in the general population. Much effort has been put forth in modeling the cause-specific hazards, especially those based on semiparametric models, to assess the effects of covariates on the hazard. Then, the estimated values are used to predict the cumulative incidence functions for specific causes. We will discuss various models available in the literature using this approach. Next, we will discuss some issues where the cause associated with an event time is missing; in other words, one cannot identify the specific reason of death (outcome) leading to so-called masked cause competing-risk data. Finally, we will also discuss the limitation in modeling the cumulative incidence function via cause-specific hazards.

67.2 Semiparametric Models

We first begin by introducing some notation that will be used throughout this chapter. Suppose that there are K distinct failure types. Let T_{ki} be the kth latent failure time for individual i, where $k = 1, \ldots, K$ and $i = 1, \ldots, n$. However, one can only observe $(T_i, \epsilon_i, \delta_i)$, where $T_i = \min(\tilde{T}_i, C_i)$, $\tilde{T}_i = \min_k\{T_{ki}, k = 1, \ldots, K\}$. Here, ϵ_i is the failure type indicator ($\epsilon_i = k$ if $\tilde{T}_i = T_{ki}$), δ_i is the censoring indicator ($\delta_i = 1$ if $\tilde{T}_i$ is observed and 0 otherwise), and C_i is the independent censoring variable. One usual assumption, which is important for the identifiability issue, is that T_{ki} does not equal T_{li} for $k \neq l$, $k, l = 1, \ldots, K$. In other words, a subject cannot encounter multiple failures at the same instant. Note that this assumption does not require the competing causes to be independent.

67.2.1 Modeling Cause-Specific Hazard Function

We first begin by discussing various approaches that have been used in literature. Typically, one models the effect of covariates using the proportional hazards model (see Cox [6]; Anderson and Gill [3]). Cheng et al. [5] have studied the estimation of the cumulative incidence function based on Cox's regression model in a competing risks model. In this approach, the cause-specific hazard function for T_{ki}, given covariates x and z, is modeled by

$$\lambda_{ki}(t|z) = \lambda_{0k}(t)\exp(z^T\beta_k), \quad (2)$$

for $k \in \{1, 2, \ldots, K\}$. The estimators $\hat{\beta}_k$ of the regression coefficients β_k can be estimated by the maximizing the partial likelihood, and the baseline cause-specific cumulative hazard $\Lambda_{0k}(t)$ can be estimated by the Breslow and Crowley [4] estimator $\hat{\Lambda}_{0k}(t)$. Then, the cause-specific cumulative incidence function $F_k(t|z_0)$ can be estimated consistently by

$$\hat{F}_k(t|z_0)$$
$$= \int_0^t \hat{S}(u|z_0) d\hat{\Lambda}_k(u)$$
$$= \sum_{i=1}^n \delta_i I\{\epsilon_i = k\}$$
$$\times \frac{\hat{S}(\tilde{T}_i|z_0) I\{\tilde{T}_i \leq t\} \exp(\hat{\beta}_k' z_0)}{\sum_{j=1}^n I\{\tilde{T}_i \leq \tilde{T}_j\} \exp(\hat{\beta}_k' Z_j)},$$

where $\hat{S}(t|z_0) = \exp(-\sum_{k=1}^K \hat{\Lambda}_k(t|z_0))$ and $\hat{\Lambda}_k(t|z_0) = \hat{\Lambda}_{0k}(t)\exp(\hat{\beta}_k' z_0)$,

$$\hat{\Lambda}_{0k}(t)$$
$$= \sum_{i=1}^n \delta_i I\{\epsilon_i = k, \tilde{T}_i \leq t\}$$
$$\times \left\{\sum_{j=1}^n I\{\tilde{T}_i \leq \tilde{T}_j\} \exp(\hat{\beta}_k' Z_j)\right\}^{-1}.$$

The large-sample properties including convergence to Gaussian distribution in the limit have been established for the estimated cumulative incidence function. This, in turn, is used to construct confidence intervals/bands for the cumulative incidence function.

In practice, the proportionality assumption under the proportional hazards model may be too restrictive. This motivates alternative approaches for modeling the cause-specific hazard. An important alternative is the additive risk model. Shen and Cheng [30] presented an approach to estimating the cumulative incidence function under the additive risk model [20]

$$\lambda_k(t|z) = \lambda_{0k}(t) + z^T \beta_k, \quad (3)$$

where $\lambda_{0k}(\cdot)$ is an unspecified baseline hazard function for the kth failure type and β_k is an unknown parameter vector. Observe that the cause-specific cumulative hazard function for $z = z_0$ can be expressed as follows:

$$\Lambda_k(t|z_0) = \Lambda_{0k}(t) + t z^T \beta_k.$$

Denote by $Y_i(t) = I\{\tilde{T}_i \geq t\}$ the at-risk process for the ith patient. Let $\delta_{ki} = \delta_i I\{\epsilon_i = k\}$ and $N_{ki}(t) = \delta_{ki} I\{\tilde{T}_i \leq t\}$ indicate whether an event of type k has been observed for subject i by time t. Using the Lin–Ying [20] estimation approach, the regression coefficient β_k can be estimated by $\hat{\beta}_k$

$$= \left[\sum_{i=1}^{n} \int_0^\infty Y_i(t) \left\{ Z_i - \frac{\sum_{i=1}^{n} Z_i Y_i(t)}{\sum_{i=1}^{n} Y_i(t)} \right\}^{\otimes 2} \right]^{-1}$$

$$\times \left[\sum_{i=1}^{n} \int_0^\infty \left\{ Z_i - \frac{\sum_{i=1}^{n} Z_i Y_i(t)}{\sum_{i=1}^{n} Y_i(t)} \right\} dN_{ki}(t) \right],$$

and the baseline cause-specific hazard can be estimated by

$$\hat{\Lambda}_{0k}(t) = \sum_{i=1}^{n} \int_0^t \frac{1}{\sum_{i=1}^{n} Y_i(t)} dN_{ki}(u)$$

$$- \int_0^t \hat{\beta}_k' \frac{\sum_{i=1}^{n} Z_i Y_i(u)}{\sum_{i=1}^{n} Y_i(u)} du.$$

Consequently, the cause-specific cumulative incidence function $F_k(t|z_0)$ can be consistently estimated by

$$\hat{F}_k(t|z_0) = \int_0^t \hat{S}(u|z_0) d\hat{\Lambda}_k(u|z_0)$$

$$= \int_0^t \exp\left(-\sum_{k=1}^{K} \Lambda_k(u|z_0)\right) d\hat{\Lambda}_k(u|z_0),$$

where $\hat{\Lambda}_k(t|z_0) = \hat{\Lambda}_{0k}(t) + t z^T \hat{\beta}_k$. The large sample properties of the above estimators can be established using standard martingale techniques.

Often, the above model provides only a rough summary of the effect of covariates, because in this model the influence of the covariates has been restricted to be constant over time; i.e., time-varying effects of covariates cannot be captured by this model. One alternate way to deal with this is the flexible extension of the additive risk model, which allows some covariates to be modeled parametrically and others to be modeled nonparametrically. The suggested approach is to study the semiparametric additive model based on McKeague and Sasieni [25] in the competing-risk setup, which may be more appropriate in some applications. See Hyun et al. [16] for details. To be specific, under the semiparametric additive risk model the cause-specific hazard function, given covariates x and z, takes the form

$$\lambda_k(t|x, z) = x^T \alpha_k(t) + z^T \beta_k, \quad (4)$$

where the covariates are partitioned into x—a p-dimensional covariate vector with time-varying effects, and z—a q-dimensional covariate vector with time-constant effects. Here $\alpha_k(t)$ is a p-dimensional locally integrable function, and β_k is a q-dimensional regression vector. The first component of x may be set to 1 to allow for a general baseline hazard. When some covariates are known or anticipated to yield time-varying effects, they can be investigated nonparametrically. The proposed estimators as well their large sample properties are discussed in complete detail in the next section.

Alternatively, for modeling all the effects nonparametrically, Aalen et al. [1] have proposed a model for the cause-specific hazard, where the effects of covariates are allowed to be modeled nonparametrically:

$$\lambda_k(t|x,z) = x^T(t)\alpha_k(t), \quad (5)$$

Note by taking the first component of x to be identity, we can get the baseline hazard function. Recently, Scheike and Zhang [28] have also considered a flexible model to reflect the time-varying effects in which the time-varying effect is additive and the fixed effect is multiplicative. In other words, their model is as follows:

$$\lambda_k(t|x,z) = \exp(z^T\beta_k) * (x^T\alpha_k(t)), \quad (6)$$

Here, the baseline hazard can be identified by incorporating 1 as the first component of the covariate z associated with the time-varying effects.

Note that the above approaches have all proposed methods to assess the effect of covariates on cumulative incidence function through modeling cause-specific hazards. However, one's interest in practical situations is often to study the effect of covariates on the cause-specific cumulative incidence function $F_k(\cdot|z)$ defined in (1). This is achieved by estimating the respective parameters in models (2)–(6); consequently, the kth cause-specific hazards $\lambda_k(t|z)$ can be estimated by $\hat{\lambda}_k(t|z)$. Consequently, the overall survival function $S(t|z)$ can be estimated by

$$\hat{S}(t|z) = \exp\left(-\sum_{k=1}^{K}\hat{\Lambda}_k(t|z)\right),$$

where $\hat{\Lambda}_k(t|z)$ denotes to the kth cause-specific cumulative hazard function. Using these estimates, one can estimate the kth cause-specific cumulative incidence function by

$$\hat{F}_k(t|z) = \int_0^t \hat{S}(t|z) d\hat{\Lambda}_k(t|z).$$

In order to assess the effect of change in covariate values from $z = z_1$ to $z = z_2$, one has to estimate $\hat{F}_k(t|z_1)$ and $\hat{F}_k(t|z_2)$ and construct confidence intervals and bands for assessing the differences and/or their ratios. We now discuss this in more detail.

67.2.2 Direct Modeling of Cause-Specific Cumulative Incidence Functions

In the previous section, we discussed the standard analysis approach of analyzing competing-risk data. However, sometimes it is desirable to have modeling approach that will allow for the direct interpretation of the regression coefficients on the cumulative incidence function of a specific cause. This is motivated by the practical applications where the effect of a covariate on a cause-specific hazard of failure of a specific type may be very different from the effect on the corresponding cumulative incidence function; see Gray [12] and Pepe [26]. In fact, a covariate may be highly significant for the cause-specific hazard, but might have no significant effect on the cause-specific cumulative incidence function. Fine and Gray [8] have proposed a proportional hazards model for the cumulative incidence function. Their model was originally proposed for the kth type subdistribution hazard function [12] as follows:

$$\lambda_k^{CI}(t|z) = \lambda_{0k}(t)\exp(\beta' z),$$

where

$$\lambda_k^{CI}(t|z) = \lim_{\Delta t \to 0} \{ P(t \leq \tilde{T} < t + \Delta t,$$
$$\epsilon = k | \tilde{T} \geq t \cup (\tilde{T} \leq t \cap K \neq k), z) \}$$
$$\times \{\Delta t\}^{-1}.$$

Note that λ_k^{CI} is the hazard for the improper random variable $T^* = I\{K = k\}\tilde{T} + (1 - I\{K = k\}) \times \infty$. Note that this model can be equivalently expressed in terms of the cumulative incidence function for cause k given by

$$F_k(t|z) = 1 - \exp\left(-\exp(z^T \beta_k)\right.$$
$$\left.\times \log\left(\int_0^k (u) du\right)\right).$$

Fine and Gray [8] proposed a partial likelihood approach for estimating the regression coefficients β_k independently of $\lambda_{0k}(t)$ in which the risk set for type k failure is constructed so that subjects who have already encountered events that are different from type k are always included in the future risk sets of type k events. This is different from the traditional analysis through cause-specific hazards where individuals with event types different from type k are precluded from all future risk sets. Note that the proposed model only incorporates time-independent covariates. The literature for direct modeling approaches is very much limited. There are some nonparametric methods available in the literature, as well as parametric approaches (e.g., Jeong and Fine [17]; Fine [9]; Gray [12]).

67.2.3 Modeling in Presence of Masked Cause of Failure

In some competing-risk setup, it may so happen that the cause of death may not be known for a certain part of the sample. In such circumstances, patients are known to die, but the cause of death information is not available for some individuals; for example, whether death is attributable to the cause of interest or other causes may require documentation with information that is not collected or lost, or the cause may be difficult for investigators to determine for some patients [2]. In such cases, excluding the missing observations from the analysis or treating them as censored may yield biased estimates and erroneous inferences.

For this part of the discussion, we restrict our attention to two types of causes, i.e., $K = 2$. This can be achieved by treating it as type I failure, the failure of interest, and grouping other causes of failure as type II. So, the observation for subject i is given by $(\tilde{T}_i, C_i, \epsilon_i, \delta_i, \overline{Z_i(\tilde{T}_i)})$. Let $U_i = 1$ if the failure type is known for individual i, 0 otherwise. If $\delta_i = 0$, i.e., the subject is censored, which is known, hence $U_i = 1$. Note that if $U_i = 1$, this implies $\epsilon_i = 1$ or 2.

Goetghebeur and Ryan [11] proposed the following model under the assumption that the probability of missing cause of death may depend on time but not on covariates:

$$\lambda_1(t|z,x) = \lambda_0(t) \exp(\xi + \beta' x),$$
$$\lambda_2(t|z,x) = \lambda_0(t) \exp(\gamma' z).$$

Here (z, x) denote the observed covariates that may influence the underlying cause-specific hazards. Furthermore, given an individual with noncensored failure time $\tilde{T}_i$, the missingness indicator U_i is assumed independent of the failure indicator ϵ_i and is independent of the covariates (z_i, x_i):

$$P(U_i = 1 | \tilde{T}_i = t, \epsilon_i, \bar{Z}_i, \bar{X}_i, \delta_i = 1)$$
$$= \pi(\psi, t, Z_i, X_i).$$

This assumption is akin to the missingness at random [21, p. 90] in the sense that the probability of missingness is allowed to depend on only all or part of the observed data.

The estimation procedure that they proposed utilizes two types of partial likelihood, L and L^*, which are defined below. We introduce some relevant notation first. Denote by $Y_i(t) = I\{\tilde{T}_i \geq t\}$, the risk indicator by time t and by $N_{ki}(t) = I\{\tilde{T}_i \leq t, \epsilon_i = k, \delta_i = 1\}$, the counting process counting the number of failures of type k that occur by time t. Further, denote by $N_{ui}(t) = I\{\tilde{T}_i \leq t, U_i = 1\}$ the counting process, counting the number of failures of unknown cause. Denote $A_i = (Z_i, X_i), \theta = (\xi, \beta, \gamma), r_1(\theta, t, A_i) = \exp(\xi + \beta' x_i), r_2(\theta, t, A_i) = \exp(\gamma' x)$.

Then the partial likelihood, based on the conditional probabilities of a specific event given that one event of that type occurs from the risk set at that time is given by

$$L = \prod_{t \geq 0} \prod_{i=1}^{n}$$
$$\times \left[\prod_{k=1}^{2}\{\pi(\theta, t, A_i) r_k(\theta, t, A_i)\}^{dN_{ki}(t)}\right.$$
$$\times \left\{\sum_{j=1}^{n} Y_j(t) \pi(\psi, t, A_j)\right.$$
$$\times \left. r_k(\theta, t, A_j)\right\}^{-dN_{ki}(t)}\Bigg]$$
$$\times \left\{(1 - \pi(\theta, t, A_i)) \sum_{k=1}^{2}\right.$$
$$\times \left. r_k(\theta, t, A_i)\right\}^{dN_{ui}(t)}$$
$$\times \left\{\sum_{j=1}^{n} Y_j(t)(1 - \pi(\psi, t, A_j))\right.$$
$$\times \left.\sum_{k=1}^{2} r_k(\theta, t, A_j)\right\}^{-dN_{ui}(t)}.$$

The estimates are the solution to the score equations obtained above.

Alternatively, the partial likelihood L^* is based on a more informative partial likelihood. This is based on the conditional probabilities of an event of specified type, given that one event occurs, but without conditioning on the type of event, is given by

$$L^* = \prod_{t \geq 0} \prod_{i=1}^{n}$$
$$\times \left[\prod_{k=1}^{2}\left\{\frac{\pi(\theta, t, A_i) r_k(\theta, t, A_i)}{\sum_{j=1}^{n} Y_j(t) r_k(\theta, t, A_j)}\right\}^{dN_{ki}(t)}\right]$$
$$\times \left\{\frac{(1-\pi(\theta,t,A_i))\sum_{k=1}^{2} r_k(\theta,t,A_i)}{\sum_{j=1}^{n} Y_j(t)\sum_{k=1}^{2} r_k(\theta,t,A_j)}\right\}^{dN_{ui}(t)}.$$

Note that L^* can be expressed as the product of two terms $L^* = L^*(\psi)L^*(\theta)$, where

$$L^*(\psi)$$
$$= \prod_{t \geq 0} \prod_{i=1}^{n} \pi(\psi, t, A_i)^{dN_{1i}(t)+dN_{2i}(t)}$$
$$\times (1 - \pi(\psi, t, A_i))^{dN_{ui}(t)}$$

and

$$L^*(\theta) = \prod_{t \geq 0} \prod_{i=1}^{n}$$
$$\times \frac{\left\{\prod_{k=1}^{2} r_k(\theta, t, A_i)^{dN_{ki}(t)}\right\} r.(\theta, t, A_i)^{dN_{ui}(t)}}{\sum_{j=1}^{n} Y_j(t) r.(\theta, t, A_j)^{dN_{1i}(t)+dN_{2i}(t)+dN_{ui}(t)}}.$$

Therefore, if the parameters modeling the missingness mechanism ψ and those for the competing risks model θ are separate, one can estimate θ based on $L^*(\theta)$ only.

They showed that the resulting estimator is robust against misspecification of the proportional baseline hazards assumption and retains high efficiency with respect to the estimator based on the Dewanji partial likelihood L^* only. However, when the missingness probability also depends on covariates, the Goetghebeur–Ryan method needs to model the missingness mechanism explicitly (e.g., assuming a logistic model) and estimate the associated parameters along with other model parameters. Therefore, their estimator might be biased under the misspecification of the missingness mechanism. The partial likelihood L^* has been studied by many investigators (see Holt [15]; Kalbfleisch and Prentice [18]; Dewanji [7]). Later, Lu and Tsiatis [23] showed that the estimator based on L^* is not only consistent and asymptotically normal but also semiparametric efficient and does not require modeling of the missingness mechanism even when missingness depends on covariate information.

Other approaches including the multiple-imputation approach to deal with the missing cause of failure, have been studied by Lu and Tsiatis [22], and the linear transformation model has been studied by Gao and Tsiatis [10].

67.3 Statistical Inference under the Additive Risks Model

Here, we illustrate the asymptotic theory behind the class of semiparametric models discussed in the previous section. We focus in particular on the McKeague–Sasieni [25] additive model (4). The results presented here are discussed in further detail in Hyun et al. [16]. Recall, that (4) assumes that the cause-specific hazard function for T_{ki}, given covariates x and z is modeled by

$$\lambda_{ki}(t|x,z) = x^T \alpha_k(t) + z^T \beta_k, \quad (7)$$

for $k \in \{1, 2, \ldots, K\}$. Let $Y_i(t) = I(T_i \geq t)$ indicate the at-risk process of the ith individual, where $I(\cdot)$ is the indicator function. We organize the covariates into design matrices

$$X(t) = (Y_1(t)x_1, \ldots, Y_n(t)x_n)^T$$

and

$$Z(t) = (Y_1(t)z_1, \ldots, Y_n(t)z_n)^T.$$

Let $\delta_{ki} = \delta_i I(\epsilon_i = k)$. Then $N_{ki}(t) = \delta_{ki} I(T_i \leq t)$ indicates whether an event of type k has been observed by time t for individual i. Define

$$N_k(t) = (N_{k1}(t), \ldots, N_{kn}(t))^T,$$
$$\lambda_k(t) = (\lambda_{k1}(t), \ldots, \lambda_{kn}(t))^T,$$

the n-dimensional counting process and its intensity. Using the estimating procedures proposed by McKeague and Sasieni [25], each β_k and the cumulative time-varying effect $A_k(t) = \int_0^t \alpha_k(u)\,du$, $k = 1, \ldots, K$ can be consistently estimated by $\widehat{\beta}_k$ and $\widehat{A}_k(t)$, while treating all the failure time T_i with $\epsilon_i \neq k$ as censored observation. To be specific,

$$\widehat{\beta}_k = \left[\int_0^\infty Z^T(t) H(t) Z(t)\,dt\right]^{-1}$$
$$\times \int_0^\infty Z^T(t) H(t)\,dN_k(t),$$

$$\widehat{A}_k(t) = \int_0^t X^-(u)$$
$$\times \left(dN_k(u) - Z(u)\widehat{\beta}_k\,du\right),$$

where $H(t) = I_n - X(t)X^-(t)$. Here I_n is the $n \times n$ identity matrix and $X^-(t)$ is the generalized inverse of $X(t)$, defined as $X^-(t) = \left(X^T(t) X(t)\right)^{-1} X^T(t)$.

By combining these we can consistently estimate the cumulative incidence function for the cause 1, for example, with a particular set of covariates x_0 and z_0 by

$$\widehat{F}_1(t|x_0, z_0)$$
$$= \int_0^t \widehat{S}(u|x_0, z_0)\,d\widehat{\Lambda}_1(u|x_0, z_0),$$

where

$$\widehat{S}(t|x_0, z_0) = \exp\left(-\sum_{k=1}^K \widehat{\Lambda}_k(t|x_0, z_0)\right)$$

is an estimator of the overall survival probability $S(t|x_0, z_0)$ and $\widehat{\Lambda}_k$ is the estimator of the cumulative cause-specific hazard function,

$$\widehat{\Lambda}_k(t|x_0, z_0) = \int_0^t x_0^T\,d\widehat{A}_k(u) + t z_0^T \widehat{\beta}_k.$$

We can show that
$$\sqrt{n}\left(\widehat{F}_1(t|x_0,z_0) - F_1(t|x_0,z_0)\right)$$
is asymptotically equivalent to a sum of square integrable martingales $U_1(t|x_0,z_0)$ given by

$$\begin{aligned}
U_1&(t|x_0,z_0) \\
&= \sqrt{n}\sum_{i=1}^{n}\epsilon_{1i}(t|x_0,z_0), \\
&= \sqrt{n}\Bigg\{\sum_{i=1}^{n}\int_0^t S(u|x_0,z_0)x_0^T \\
&\quad \times \left(X(u)^T X(u)\right)^{-1} x_i\, dM_{1i}(u) \\
&\quad - \int_0^t S(u|x_0,z_0)x_0^T X^-(u) \\
&\quad \times Z(u)\, du\, C^{-1} D_{1i} \\
&\quad - \int_0^t S(u|x_0,z_0)\, du\, z_0^T C^{-1} D_{1i} \\
&\quad - \sum_{k=1}^{K}\Bigg(\int_0^t H_1(t,u)x_0^T \\
&\quad \times \left(X(u)^T X(u)\right)^{-1} x_i\, dM_{ki}(u) \\
&\quad - \int_0^t H_1(t,u)x_0^T X^-(u) \\
&\quad \times Z(u)\, du\, C^{-1} D_{ki} \\
&\quad - \int_0^t H_1(t,u)\, du\, z_0^T C^{-1} D_{ki}\Bigg)\Bigg\},
\end{aligned}$$
(8)

where
$$\begin{aligned}
M_{ki}(t) &= N_{ki}(t) - \int_0^t Y_i(u) \\
&\quad \times \{x_i^T \alpha_k(u) + z_i^T \beta_k\}\, du, \\
H_k(t,u) &= F_k(t|x_0,z_0) - F_k(u|x_0,z_0), \\
C &= \int_0^\infty Z^T(t) H(t) Z(t)\, dt, \\
D_{ki} &= \int_0^\infty \{z_i - Z^T(t)X(t) \\
&\quad \times \left(X^T(t)X(t)\right)^{-1} x_i\} \\
&\quad dM_{ki}(t)
\end{aligned}$$

for $k \in \{1,2,\ldots,K\}$ and $i=1,2,\ldots,n$. Furthermore, using the martingale central limit theorem, this converges in distribution to a Gaussian process. The martingale representation of $U_1(t|x_0,z_0)$ can also be used to construct a consistent estimator of the asymptotic variance function. The variance function at time t can be consistently estimated by

$$\widehat{\sigma}_1^2(t|x_0,z_0) = n\sum_{i=1}^{n}\left(\widehat{\epsilon}_{1i}(t|x_0,z_0)\right)^2,$$

obtained by replacing all the terms with their empirical versions and the parameters β_k and A_k by their consistent estimates $\widehat{\beta}_k$ and $\widehat{A}_k$.

Next, combining the asymptotic normality of $\widehat{F}_1(t|x_0,z_0)$ and the consistent estimator $\widehat{\sigma}_1^2(t|x_0,z_0)$ for the asymptotic variance, we can construct $(1-\alpha)\times 100\%$ confidence interval for $F_1(t;x_0,z_0)$ as

$$\begin{aligned}
g^{-1}(g(\widehat{F}_1(t|x_0,z_0)) &\pm n^{-1/2} \\
\times g'(\widehat{F}_1(t|x_0,z_0))&\widehat{\sigma}_1(t|x_0,z_0)z_{\alpha/2}),
\end{aligned}$$

where g is a smooth function chosen such that it retains the range of the distribution F_1, g' is its continuous derivative, and g^{-1} denotes the inverse function of g. Typically, the transformations g are chosen for constructing confidence interval for distributions so as to retain their range as well as to improve the coverage probability. Observe that using the functional delta method, the asymptotic normality of $g(\widehat{F}_1(t|x_0,z_0))$ follows from that of $\widehat{F}_1(t|x_0,z_0)$, but with asymptotic variance given by $g'^2(F_1(t|x_0,z_0))\sigma_1^2(t|x_0,z_0)$.

Another quantity of considerable interest is the confidence band for the cumulative incidence function. However, in order to construct the $(1-\alpha)\times 100\%$ confidence band for $F_1(t|x_0,z_0)$ we need to investigate the distribution of the supremum of the process U_1. This is analytically challenging, but not intractable, as U_1 does not have an independent increment structure. Alternatively, one can adapt the general procedure suggested by Lin et al. [19] to get an approximation of the distribution of the process U_1. We approximate the distribution of $U_1(t|x_0, z_0)$ by a zero-mean Gaussian process, denoted by $\widehat{U}_1(t|x_0, z_0)$, whose distribution can be easily generated through simulations. We replace $\{M_{ki}(t)\}$ in (8) with $\{N_{ki}(t)G_i\}$, and the other unknown quantities in (8) with their respective sample estimators to yield $\widehat{U}_1(t|x_0, z_0)$, where $\{G_i : i = 1, \ldots, n\}$ are independent standard normal variables. The asymptotic equivalence of $U_1(t|x_0, z_0)$ and $\widehat{U}_1(t|x_0, z_0)$ is established by the fact that

$$E\{M_{ki}(t)\} = 0,$$
$$\mathrm{Var}\{M_{ki}(t)\} = E\{N_{ki}(t)\}.$$

To approximate the distribution of $U_1(t|x_0, z_0)$, we simply obtain a large number, say, M, of realizations from $\widehat{U}_1(t|x_0, z_0)$ by repeatedly generating random samples $\{G_i\}$, while fixing the data $\{(T_i, \delta_{ki}, x_i, z_i) : k = 1, \ldots, K, i = 1, \ldots, n\}$ at their observed values.

To construct a $(1-\alpha) \times 100\%$ confidence band for $F_1(t|x_0, z_0)$ on $[t_1, t_2]$, we need to estimate the cutoff value $c_{\alpha/2}$ such that

$$P\left[\sup_{t_1 \leq t \leq t_2} |\widehat{U}_1(t|x_0, z_0)/\widehat{\sigma}_1(t|x_0, z_0)| > c_{\alpha/2}\right] = \alpha$$

through replicates of $\widehat{U}_1(t|x_0, z_0)$. Here t_1 and t_2 ($t_1 < t_2$) can be any time points between $(0, t_0)$, where $t_0 = \inf\{t : EY_1(t) = 0\}$.

67.4 A Comprehensive Analysis of Malignant Melanoma Data

We now illustrate the methodology discussed in the previous section with an analysis of data on 205 malignant melanoma patients observed during the years 1962–1977 [2]. Among these 205 patients, 57 patients died from malignant melanoma, 14 patients died from causes other than malignant melanoma, and the remaining 134 patients were alive at the end of follow-up. The covariates considered here were tumor thickness (mean: 2.92 mm; standard deviation: 2.96 mm), ulceration status (90 present and 115 not present), age (mean—52 years; standard deviation—17 years) and gender (79 male and 126 female). In this analysis, the covariates of tumor thickness and age were standardized. The main interest in this melanoma study is to predict the patient-specific cumulative incidence probability for the melanoma death. The analysis presented here is based on the findings of Hyun et al. [16].

We first begin with fitting the proportional hazards model to the competing risks malignant melanoma data. Cheng et al. [5] have provided estimation procedures for the cumulative incidence function by using the usual Cox regression model to relate the cause-specific hazard function to the covariates, and the malignant melanoma data have been analyzed using the Cox model for the cause-specific hazard function. We conducted goodness-of-fit tests of the Cox model with the data to test for the proportionality assumption with respect to covariates: age, sex, tumor thickness and ulceration status (see Table 1).

The output is based on the Lin et al. [19] score process test for proportionality. See also Figure 1. Note that the proportionality assumption for the covariates—tumor thickness and ulceration status—are not appropriate for the melanoma data. Alternatively, one can use the approach laid out on pages 96–97 of Kalbfleisch and Prentice [18]. Here, they propose the derived covariates approach to identify the time-varying effects. This approach requires knowledge of the shape of the time-varying effect for

Table 1: Test for proportionality.

	sup $|\hat{U}(t)|$	p value H_0
Prop(age)	1.81	0.554
Prop(sex)	2.10	0.380
Prop(tumor thickness)	3.18	0.032
Prop(ulceration)	3.66	0.008

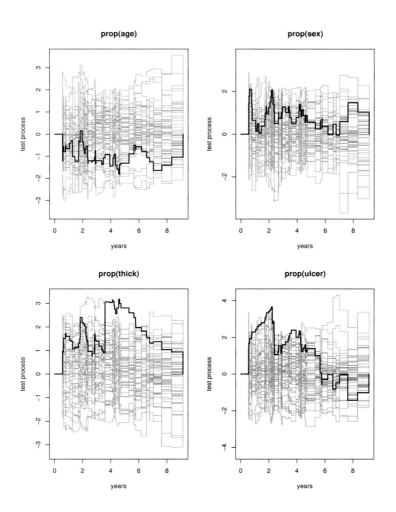

Figure 1: Score processes with 50 simulated processes under the model.

instance, linearly changing with time e.g., $a + bt$ and also of the change point (if any), i.e., $a + bt$ for $t \leq t_0$ and $a + ct$ for $t > t_0$, where a, b, c are the regression coefficients. These may not be the appropriate shapes, and one has to test for the shape being linear or must know the change point t_0 (i.e., when the shape changes).

First, we consider the popular alternative to the Cox model—the additive model (as proposed by Lin and Ying [20] and Shen and Cheng [30]) approach to estimating the cumulative incidence function. However, these models assume that covariates have time-constant effect. So, we first conducted goodness-of-fit tests of the additive model with the melanoma data using the score test procedures described below. See Martinussen and Scheike [24] for details. We investigated the time-varying effect or constant effect of the pth covariate by conducting the following hypothesis test

$$H_0 : \beta_p(t) = b, \ 0 \leq t \leq \tau$$

based on the semiparametric additive model

$$\lambda(t) = x(t)\beta(t) + z(t)\gamma.$$

Defining the cumulative regression coefficient as $B_p(t) = \int_0^t \beta(u)du$, the abovementioned test can be equivalently reformulated as

$$H_0 : B_p(t) = bt, \ 0 \leq t \leq \tau.$$

The following two tests were computed:

Test 1: $\sup_{0 \leq t \leq \tau} |\hat{B}_p(t) - \hat{B}_p(\tau)\frac{t}{\tau}|$

Test 2: $\int_0^\tau \left(\hat{B}_p(t) - \hat{B}_p(\tau)\frac{t}{\tau}\right)^2 dt.$

The tests showed that tumor thickness does not have a time-constant effect over time. The score test processes shown in Figure 3 seems to indicate that ulceration status also has slightly time-varying effect. See Figures 2 and 3. (See also Table 2.)

So, under the additive models for risk, age, and sex seem to have time-constant effects and tumor thickness and ulceration seem to have time-varying effects. The additive model that McKeague and Sasieni [25] and Hyun et al. [16] have studied consider an additive model with effects of some covariates constant across time and effects of some other covariates time-varying. This model does not assume anything about the shape of the time-varying effect. It is, in fact, modeled nonparametrically by $\alpha_k(t), 0 \leq t \leq \tau$, see Equation (4). As we have been already shown, Cox's model is not an appropriate choice for the melanoma data, and without making any further assumption on the shape of the time-varying effect, this additive model allows some covariates to have time-varying effects and some to have time-constant effects. So, based on this preliminary analysis, the final model considers the effect of gender and age as time constant effects, and tumor thickness and ulceration status as time-varying nonparametric effects. Based on this final model, the cumulative effects of the baseline estimate, tumor thickness, and ulceration, with 95% pointwise confidence intervals and bands, are given in Figure 4. The cumulative regression function for tumor thickness shows a positive effect in the first 4 years, which then seems to flatten out. Ulceration shows a positive effect within the first 4 years and a slower effect after that. The 95% confidence interval of the constant coefficients for gender and age are $(-0.1021, 0.1450)$ and $(-0.0535, 0.0691)$, respectively.

Next, we estimated the cumulative incidence function of malignant melanoma for a 52-year-old female patient with ulceration and tumor thickness of 6.76 mm (90th percentile of tumor thickness), and display its 95% confidence intervals and bands under the semiparametric additive model in

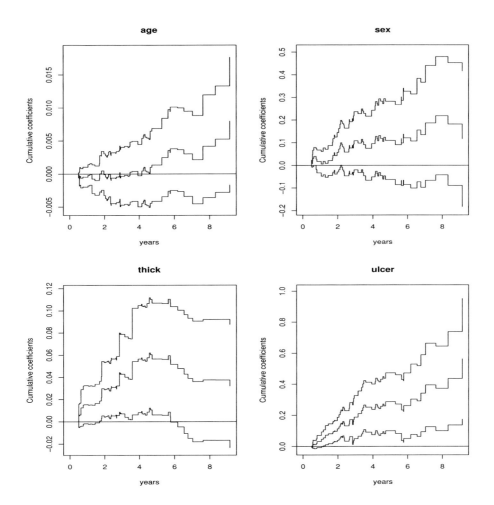

Figure 2: Estimated cumulative regression functions.

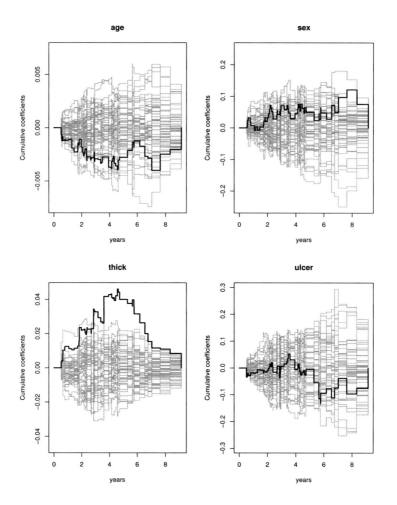

Figure 3: Test processes for testing constant effects with 50 simulated processes.

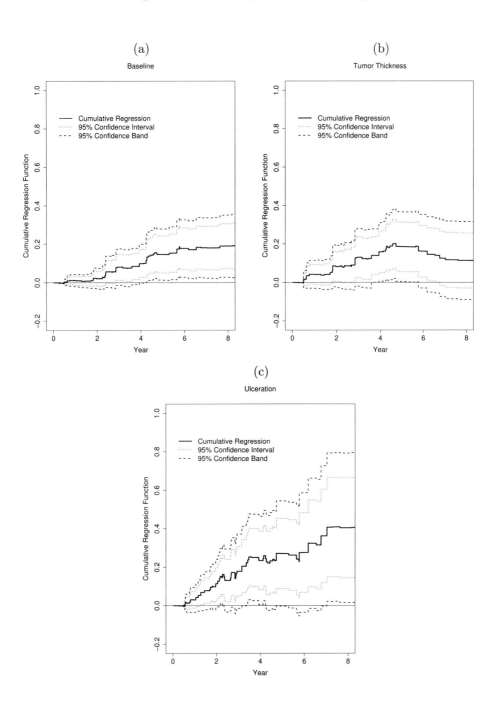

Figure 4: Cumulative time-varying effects for malignant melanoma death based on the semiparametric model with baseline, tumor thickness, and ulceration as nonparametric part.

Table 2: Test for time-invariant effects.

	$\sup |B(t) - (t/\tau)B(\tau)|$	p value $H_0 : B(t) = bt$
(Intercept)	0.15100	0.437
Age	0.00398	0.319
Sex	0.12100	0.350
Tumor Thickness	0.04600	0.004
Ulceration	0.13100	0.493
	$\int (B(t) - (t/\tau)B(\tau))^2 dt$	p value $H_0 : B(t) = bt$
(Intercept)	$2.90e-02$	0.614
Age	$5.18e-05$	0.185
Sex	$2.81e-02$	0.359
Tumor Thickness	$5.62e-03$	0.001
Ulceration	$2.69e-02$	0.575

Figure 5. We used a transformation of $g(\cdot) = \log(-\log(\cdot))$.

67.4.1 Software

Regression on cause-specific hazards can be performed in any package that includes the Cox proportional hazards model. An option to fit stratified Cox models needs to be included if we want to fit or test for equality of covariate effects for different transitions. Cumulative incidence curves in a competing risks setting can be estimated in S-Plus/R (*cmprsk* library) and Stata (*stcompet.ado* module). The Stata Website also provides further explanations on fitting competing risks models (see http://www.stata.com/support/faqs/stat/stmfail.html). Rosthøj et al. [27] have written a set of SAS macros that allows to translate results from a Cox model on cause-specific hazards into cumulative incidence curves for some choice of covariate values (see http://www.pubhealth.ku.dk/∼pka). It also calculates standard errors. A Cox

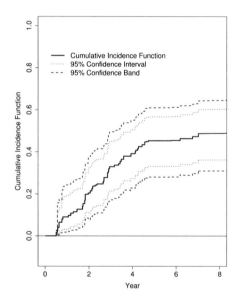

Figure 5: Predicted cumulative incidence function of malignant melanoma death for a 52-year-old female with ulceration and tumor thickness of 6.76 mm with 95% confidence intervals and bands.

null model without covariates can be used to obtain a single cumulative incidence curve for the whole group. The R package *mstate*, can also be used for competing risks. The R package *timereg* fits various nonproportional hazards models in the competing-risk setup and is very useful in incorporating time-varying covariates (see *http://staff.pubhealth.ku.dk/~ts/time-reg.html*).

67.5 Discussion

Epidemiological and clinical studies are becoming more complex, with the primary endpoint of interest being not just one event, but a collection of events. Consequently, competing-risk type data are encountered very frequently. We have discussed various approaches for inference-based on right-censored competing-risk data. We have a rich collection of semiparametric models for analyzing the competing risks data via modeling cause-specific hazards. However, one can clearly see that the literature gets very limited when one wants to model the effect of covariates directly on the cause-specific cumulative incidence functions; only proportional hazards models and transformation models are available. Even in dealing with proportional hazards models, only time-constant covariates can be incorporated into the models. Thus there is a need to investigate further models and develop software to be able to analyze data using alternative models. In addition, one may encounter different types of censoring such as current status, where a subject is observed only once and one knows whether a failure of a certain type has occurred only by that time. This situation has been the focus of recent study where the asymptotic properties of the maximum-likelihood estimates of the underlying cumulative incidence functions have been investigated [13,14]. Furthermore, other situations arise where the underlying event history of interest consists of both recurrent events processed with terminal endpoint of interest, i.e., failure has competing causes. For example, in colon cancer patients, one may be interested in repeated occurrence of polyps in colons with the patient dying due to either primary event of interest (colon cancer) or due to competing causes (ex. cardiovascular diseases). In order to properly analyze such data, one would have to jointly model the recurrent event process with the terminal event that is observed in competing-risk setup. This is becoming more of an issue as better treatments are available for treating cancer patients (e.g., prostate cancer, breast cancer) increasing their survival rate and consequently putting the patient at a higher risk of dying due to other competing causes of death. Last, but not the least even in the classical type of right censored competing risk data, one may not be able to precisely measure the time to failure, i.e., it is susceptible to measurement errors. For example, in the competing-risk example of time to spontaneous labor, the actual time since conception is never truly identified as there are no precise biomarkers to identify the start of pregnancy. Consequently, this results in measurement error in the survival time of interest, which can potentially have a significant impact on distinguishing truly preterm deliveries of infants from full term deliveries.

In conclusion, even though considerable literature is available on modeling competing-risk data in the classical sense, much work needs to be done to be able to analyze more complex data structures that incorporate various types of censoring or include other longitudinal processes that are observed alongside the competing-risk data.

Acknowledgment. The first author's research is supported in part by the intramural research program of *Eunice Kennedy*

Shriver National Institute of Child Health and Human Development, NIH, DHHS.

References

1. Aalen, O. O., Borgan, O., and Fekjaer, H. (2001). Covariate adjustment of event histories estimated from Markov chains: The additive approach. *Biometrics*, **57**, 993–1001.

2. Andersen, P. K., Borgan, Ø., Gill, R. D., and Keiding, N. (1993). *Statistical Models Based on Counting Processes*. Springer-Verlag, New York.

3. Andersen, P. K. and Gill, R. D. (1982). Cox's regression model for counting processes: A large sample study. *Ann. Statist.*, **10**, 1100–1120.

4. Breslow, N. and Crowley, J. (1974). A large sample study of the life table and product limit estimates under random censorship. Collection of articles dedicated to Jerzy Neyman on his 80th birthday. *Ann. Statist.*, **2**, 437–453.

5. Cheng, S. C., Fine, J. P., and Wei, L. J. (1998). Prediction of cumulative incidence funciton under the proportional hazards model. *Biometrics*, **54**, 219–228.

6. Cox, D. R. (1972). Regression models and life tables (with discussion). *J. Roy. Statist. Soc. B*, **34**, 187–220.

7. Diwanji, A. (1992). A note on a test for competing risks with missing failure type. *Biometrika*, **79**, 855–857.

8. Fine, J. P. and Gray, R. J. (1999). A proportional hazards model for the subdistribution of a competing risk. *J. Am. Statist. Assoc.*, **94**, 496–509.

9. Fine, J. P. (2001). Regression modelling of competing crude failure probabilities. *Biostatistics*, **2**, 85–97.

10. Gao, G. and Tsiatis, A. A.(2005). Semiparametric estimators for the regression coefficients in the linear transformation competing risks model with missing cause of failure. *Biometrika*, **92**, 875-891.

11. Goetghebeur, E. and Ryan, L. (1990). A modified log rank test for competing risks with missing failure type. *Biometrika*, **77**, 207–211.

12. Gray, R. J. (1988). A class of K-sample tests for comparing the cumlative incidence function of a competing risk. *Ann. Statist.*, **16**, 1141–1154.

13. Groeneboom, P., Maathuis, M. H., and Wellner, J. A. (2008). Current status data with competing risks: consistency and rates of convergence of the MLE. *Ann. Statist.*, **36**, 1031–1063.

14. Groeneboom, P., Maathuis, M. H., and Wellner, J. A. (2008). Current status data with competing risks: limiting distribution of the MLE. *Ann. Statist.*, **36**, 1064–1089.

15. Holt, J.D. (1978). Competing risk analyses with special reference to matched pair experiments. *Biometrika*, **65**, 159-165.

16. Hyun, S., Sun, Y., and Sundaram, R. (2009). Assessing cumulative incidence functions under the semiparametric additive risk model. Available at *http://www.math.uncc.edu/preprint/2009/2009_03c.pdf*.

17. Jeong, J-H. and Fine, J. P. (2006). Direct parametric inference for the cumulative incidence function. *App. Statist.*, **55**, 187–200.

18. Kalbfleisch, J. D. and Prentice, R.L. (1980). *The Statistical Analysis of Failure Time Data*. Wiley, New York.

19. Lin, D. Y., Wei, L. J., and Ying, Z. (1993). Checking the Cox model with cumulative sums of martingale-based residuals. *Biometrika*, **80**, 557–572.

20. Lin, D. Y. and Ying, Z. (1994). Semiparametric analysis of the additive risk model. *Biometrika*, **81**, 61–71.

21. Little, R. J. A. and Rubin, D. B. (1987). *Statistical Analysis with Missing Data*. Wiley, New York.

22. Lu, K. and Tsiatis, A. A. (2001). Multiple imputation methods for estimating regression coefficients in the competing risks model with missing cause of failure. *Biometrics*, **57**, 1191-1197.

23. Lu, K. and Tsiatis, A. A. (2005). Comparison between two partial likelihood approaches for the competing risks model with missing cause of failure. *Lifetime Data Anal.*, **11**, 29–40.

24. Martinussen, T. and Scheike, T. H. (2006). *Dynamic Regression Models for Survival Data*. Springer-Verlag, New York.

25. McKeague, I. W. and Sasieni, P. D. (1994). A partly parametric additive risk model. *Biometrika*, **81**, 501–514.

26. Pepe, M. S. (1991). Inference for events with dependent risks in multiple endpoint studies. *J. Am. Statist. Assoc.*, **86**, 770–778.

27. Rosthøj, S., Andersen, P. K., and Abildstrm S. Z. (2004). SAS macros for estimation of the cumulative incidence functions based on a Cox regression model for competing risks survival data. *Comput. Methods Programs Biomed.*, **74**, 69-75.

28. Scheike, T. H., Zhang, M. J., and Gerds, T. (2008). Predicting cumulative incidence probability by direct binomial regression. *Biometrika*, **95**, 1–16, DOI: *10.1093/biomet/asm096*.

29. Scheike, T. H. and Zhang, M. J. (2003). Extensions and Applications of the Cox–Aalen Survival Model. *Biometrics*, **59**, 1036–1045.

30. Shen, Y. and Cheng, S. C. (1999). Confidence bands for cumulative incidence curves under the additive risk model. *Biometrics*, **55**, 1093–1100.

68 Size and Shape Analysis

James E. Mosimann

68.1 Introduction

The terms "size" and "shape" have been used in a variety of contexts (often biological) in statistical and quantitative analyses of data; an indication of some of this variety will be given later. Here the approach will be of broad applicability in the probability modelling and statistical analysis of positive random variables.

The thrust of size and shape analysis is to consider models and methods for the analysis of random proportions or ratios. If proportions are regarded as fixed values, the concepts considered here are less relevant. However, when the proportions or ratios are random quantities, the concepts that follow assume importance. The results are applicable to any vector of positive random variables.

In many scientific studies, the data are random proportions or ratios that represent dimensionless quantities in a physical sense. When these data represent the same physical dimensions (say lengths, counts) and are expressed in the same units (say, millimeters, dozens), their ratios are dimensionless numbers, which can be viewed as expressing "shape" in some general sense.

Thus given any two similar plane triangles of different size, any ratio of the sides of one triangle will be the same as the corresponding ratio in the other, reflecting their similarity. They have the same "shape." Now let (A,B,C) be random variables representing the respective lengths of corresponding sides of random triangles. Then random ratios like A/C and B/C represent the "shape" of the triangle, while the length of one of the sides, say, A, might be chosen to represent "size."

Alternatively, suppose that one has measurements of the length, width, and height of the shell, or carapace, of a number of turtles of the same species. Let the random vector (L,W,H) represent these measurements. For some species of turtles, recently hatched individuals are nearly round when viewed from above so that the ratio of length to width L/W is nearly 1. Subsequently length grows faster than width, so that the ratio L/W is greater in larger turtles (with L, W, and H all large) [36]. Thus there is a positive association of the random "shape" variable L/W with each of the random "size" variables L, W, and H, respectively. In biology the study of such associations of dimensionless shape variables, like L/W, with variables containing scale or size information is a part of the field of allometry. (For references see Refs. 20–22, 24, 28, 33, 36, 38, and 49.)

As a further illustration, let $(T_1, \ldots, T_k)$

be the intervals of time between the occurrence of successive events in a Poisson process. Then $T_1, \ldots, T_k$ are mutually independent exponential variables with a common scale parameter, say, λ. Let $S_k = T_1 + \cdots + T_k$. Then S_k is statistically independent of the shape vector of proportions $(T_1/S_k, \ldots, T_k/S_k)$, reflected in the fact that S_k is a sufficient statistic for the scale parameter λ, which itself contains all size information. For examples of studies of statistics sufficient for scale or size under various models, see References 17, 25, and 59. The basic concepts that follow are developed in References 38, 40, 41, and 45.

68.2 Shape Vectors and Size Variables

A *positive* vector in k-dimensional Euclidean space is one whose coordinates are all positive. The set of all such positive vectors is denoted by R_+^k; for example, R_+^1 is the set of positive real numbers. Forc each positive vector, a pair of associated variables is defined: a scalar "size variable," which contains all size or scale information, and a "shape vector," which contains all shape or ratio information. A *size variable* is a scalar-valued function G from R_+^k to R_+^1 with the homogeneity property that $G(\alpha \mathbf{x}) = \alpha G(\mathbf{x})$ for all $\mathbf{x} \in R_+^k$ and all $\alpha > 0$. (Thus, if each coordinate of the vector $\mathbf{x}$ is expressed in millimeters, then the size variable G is expressed in millimeters.) A *shape vector* is a vector-valued function $\mathbf{z}$ from R_+^k to R_+^k, defined using a given size variable G by $\mathbf{z}(\mathbf{x}) = \mathbf{x}/G(\mathbf{x})$ for all $\mathbf{x} \in R_+^k$.

Examples of size variables (in notation used throughout) are

The ordinary sum

$$S_k = \sum_{i=1}^{k} x_i;$$

The geometric mean

$$M_k = \left(\prod_{i=1}^{k} x_i\right)^{1/k};$$

The distance

$$D_k = \left(\sum_{i=1}^{k} x_i^2\right)^{1/2};$$

The maximum coordinate

$$\max_k = \max(x_1, \ldots, x_k);$$

The last coordinate

$$C_k = x_k.$$

Corresponding to each size variable is a shape vector. Thus $\mathbf{x}/S_k$ is the usual vector of proportions, while $\mathbf{x}/D_k$ is the vector of direction cosines. Other shape vectors, such as $\mathbf{x}/M_k$, have no particular name, but the name *ratio vector* will be adopted for a shape vector where $\mathbf{x}$ is divided by a single coordinate, e.g. $\mathbf{x}/x_k$.

68.3 Size Variables Represent A Broad Class

The image of any size variable is R_+^1, the positive real numbers. The image of a shape vector is never all of R_+^k, but is that subset of positive vectors constrained so that the defining size is 1, because by homogeneity, $G(\mathbf{z}) = G(\mathbf{x}/G(\mathbf{x})) = G(\mathbf{x})/G(\mathbf{x}) = 1$. This subset of vectors of size 1 can be regarded of as the unit sphere of the size variable G, e.g., the surface of the positive simplex if $G = S_k$. A size variable can be defined by arbitrarily determining a unit sphere, that is, by selecting a point on each positively directed ray from the origin, and assigning size 1 to that point. By homogeneity then, the size variable is defined everywhere in R_+^k.

While any size variable must be continuous along a ray because of homogeneity, size variables need not be continuous functions on R_+^k. The class of size variables is a very broad one that includes nonmeasurable functions unsuitable for defining random variables. Excluding these, the class remains very broad. Thus there is a one–one correspondence of values of a given shape vector with the positive rays from the origin. Simply put, a shape vector names the rays, while a size variable reveals where a given vector terminates on its particular ray.

68.4 Regular Sequences of Size Variables

Functions can be defined that relate shape and size in a lower-dimensional space with shape and size, respectively, in a higher-dimensional space. To define a "related pair" of size variables consider a k-dimensional positive vector $\mathbf{x}$, and append to it a new measurement x_{k+1}. Let G_k be a size variable defined using $\mathbf{x}$ alone (i.e., from R_+^k); let G_{k+1} be defined using $\mathbf{x}'_{k+1} = (\mathbf{x}'; x_{k+1})$, so that G_{k+1} is defined from R_+^{k+1}. If one knows only G_k and G_{k+1}, one also can know the pair G_k, x_{k+1}, and conversely, knowing G_k and x_{k+1}, one can also know G_k and G_{k+1}, then G_k and G_{k+1} are said to be *related*. The idea is that all scale information needed to capture G_{k+1} is encapsulated in G_k and x_{k+1}, and correspondingly, one can recapture x_{k+1} from G_k and G_{k+1} alone. For example, S_k and S_{k+1} are related, but $\max_k$ and $\max_{k+1}$ are not (x_{k+1} cannot be recaptured from $\max_k$ and $\max_{k+1}$). Also, $C_k, C_{k+1}, M_k, M_{k+1}$, and D_k, D_{k+1} are all related pairs, but the pair M_k, S_{k+1} is not.

One consequence of having a related pair, G_k, G_{k+1}, is that knowledge of k-dimensional shape and of the ratio G_{k+1}/G_k ensures knowledge of $(k+1)$-dimensional shape without any scale information.

A sequence of size variables $G_1, \ldots, G_{k+1}$ is *regular* if each contiguous pair G_j, G_{j+1} is a related pair. (Here G_1 is defined from R_+^1 based on the first coordinate of $\mathbf{x}$; G_2 from R_+^2 based on the first two coordinates of $\mathbf{x}$; and so forth.)

68.5 Size Ratios of Regular Sequence Represent Shape

When $(G_1, \ldots, G_{k+1})$ is a regular sequence of size variables, there is an inverse function by which the vector $\mathbf{x}$ may be known from $(G_1, \ldots, G_{k+1})$. The vector of size ratios $(G_2/G_1, \ldots, G_{k+1}/G_k)$ is also in one–one correspondence with the $(k+1)$-dimensional rays. Therefore the size ratios of a regular sequence represent shape as well as any shape vector [40].

Any data vector $(x_1, \ldots, x_{k+1})$ can be represented by the size ratios with size appended $(G_2/G_1, \ldots, G_{k+1}/G_k, G_{k+1})$. An advantage of this is that when the latter is partitioned as two vectors, the first $(G_2/G_1, \ldots, G_r/G_{r-1})$ represents r-dimensional shape while involving none of the coordinates from $r+1$ to k. In contrast, the second vector $(G_{r+1}/G_r, \ldots, G_{k+1}/G_k, G_{k+1})$ represents equally well either the higher-dimensional size variables $(G_r, \ldots, G_{k+1})$ or r-dimensional size and the higher coordinates $(G_r, x_{r+1}, \ldots, x_{k+1})$. These representations hold when the size sequence is regular, but need not be valid when it is not.

68.6 Independence of Shape and Size; Isometry and Neutrality

Let $\mathbf{X}_k$ be a k-dimensional positive random vector. It follows from the preceding discussion that if some random shape vector, say, $Z = X_k/G(X_k)$, is statistically independent of a random variable, say, H, then every random shape vector [including, say, $U = X_k/D_k(X_k)$] is independent of H. Thus it is possible to speak unambiguously of the independence of shape and any random H. (This implies that the probability distribution of the rays does not change given different values of H.) When H is a size variable, so that shape is independent of size, then one has *isometry with respect to H*. Because the regular size ratios of X_k may be used to represent the rays as well as any shape vector, isometry with respect to H occurs if and only if the size ratio vector for every regular size sequence is independent of H. Also, because the lower size ratios represent lower-dimensional shape (e.g., G_2/G_1, two-dimensional shape), then H independent of k-dimensional shape implies H independent of $(k-1)$-dimensional shape, and hence H must thereby be independent of the shape defined by any subset of the coordinates of X_k, without regard to their order within X_k.

Consider X_{k+1} of $k+1$ coordinates and denote its first r coordinates by X_r. Then the size ratios $(G_2/G_1, \ldots, G_r/G_{r-1})$ represent the r-dimensional shape of X_r. The condition that r-dimensional shape be independent of the vector of higher-dimensional size variables $(G_r, \ldots, G_{k+1})$ is *isometry with respect to* $G_r, \ldots, G_{k+1}$. The weaker condition that r-dimensional shape be independent of the higher-dimensional size ratios $(G_{r+1}/G_r, \ldots, G_{k+1}/G_k)$ is *neutrality with respect to* $G_r, \ldots, G_{k+1}$. Isometry with respect to $G_r, \ldots, G_{k+1}$ implies neutrality with respect to $G_r, \ldots, G_{k+1}$.

Complete isometry with respect to $G_2, \ldots, G_{k+1}$ occurs when r-dimensional shape is independent of the vector of higher size variables $(G_r, \ldots, G_{k+1})$, for $r = 2, \ldots, k+1$. Complete isometry is equivalent to the mutual independence of the k size ratios and size: $G_2/G_1, \ldots, G_{k+1}/G_k, G_{k+1}$. Correspondingly, *complete neutrality* is equivalent to the mutual independence of the size ratios $G_2/G_1, \ldots, G_{k+1}/G_k$. The concept of neutrality is scale-free, and applicable to shape vectors alone. Nonetheless it is a concept applicable to any positive random vector, including those containing scale information. Thus the size ratios are the same, whether determined directly from X_{k+1} or from any of its shape vectors.

A pivotal result, (given in Ref. 38, with proof completed in Ref. 45) can be simply stated: The shape of X_k can be statistically independent of at most one size variable $G(X_k)$. Thus, if some shape vector $Z(X_k)$ is independent of a size variable $G(X_k)$, then shape cannot be independent of any other size variable $H(X_k)$. The only exception to this sweeping statement is virtually no exception at all; shape is also independent of any other size variable that, with probability 1, is a scalar multiple of $G(X_k)$.

A comparable result holds for related size ratios. As before, let X_k denote the initial k coordinates of X_{k+1}, and suppose that G_k, G_{k+1} and H_k, H_{k+1} are related pairs. If k-dimensional shape is independent of the ratio $G_{k+1}(X_{k+1})/G_k(X_k)$, then it cannot be independent of the ratio $H_{k+1}(X_{k+1})/H_k(X_k)$ [40]. Again the only exception is when G_k is a scalar multiple of H_k with probability 1.

68.7 Isometry and Neutrality Characterize Distributions

Specification of a size variable or of a regular sequence of size variables is thus necessarily important in questions of isometry or neutrality, in striking contrast to the representation of shape, which, for questions of independence, may be specified by any shape vector or vector of "regular size ratios," that is, size ratios from a regular sequence.

Because the independence of shape and size is such a strong condition, it is not surprising that the additional condition of mutual independence of the coordinate variables themselves results in a specific distributional form. Consider the characterization of the gamma distribution given by Lukacs [34]: Suppose (1) that X_1 and X_2 are independent positive (nondegenerate) random variables and also (2) that X_1/X_2 is independent of $X_1 + X_2$; then X_1 and X_2 each have a gamma distribution with the same scale parameter λ. Condition 2 above is simply the independence of two-dimensional shape [say, $(X_1/X_2, 1)$] and size $S_2 = X_1 + X_2$.

The result proved by Lukacs for two dimensions holds also in k dimensions. Thus (1) the mutual independence of positive and nondegenerate $X_1, \ldots, X_k$ along with (2) the independence of k-dimensional shape and S_k implies that each X_i has a gamma distribution with the same scale parameter [37]. If one has mutually independent, not necessarily identically distributed, positive variables, and shape independent of M_k, then each coordinate variable must follow a lognormal distribution. If size is D_k, then each coordinate must have a generalized gamma distribution. For these and other results see Mosimann [38] and James [25]; see also Ferguson [17]. James identified the class of size variables for which both (1) shape may be independent of size, and simultaneously (2) the original coordinates may be mutually independent.

Neutrality properties can also be used to characterize distributions. Thus, suppose that $\mathbf{X}_{k+1}$ is constrained so that $S_{k+1}(\mathbf{X}_{k+1}) = 1$ with probability 1, and that $\mathbf{X}_{k+1}$ is completely neutral with respect to $S_2, \ldots, S_{k+1}$. Then consider a vector whose coordinate variables are simply those of $\mathbf{X}_{k+1}$ in reverse order. If the latter is completely neutral with respect to $S_2, \ldots, S_{k+1}$, then $\mathbf{X}_{k+1}$ must follow a Dirichlet distribution, the characterizations of which by Darroch and Ratcliffe [14], Fabius [15], and James and Mosimann [27] all concern neutrality properties with respect to the additive sequence $S_1, \ldots, S_{k+1}$ for $\mathbf{X}_{k+1}$ along with at least one of its permutations. A subfamily of the multivariate lognormal family is similarly characterized by neutrality properties with respect to the multiplicative sequence $M_1, \ldots, M_{k+1}$ [27,41].

A concept closely related to additive neutrality is that of F independence; see Darroch and James [12]. A more recent discussion of additive neutrality is given in Ref. 26.

Briefly, then, whether the positive measurements are independent or not, there are no probability distributions on R_+^k for which two truly distinct size variables are both independent of shape. The specification of a size variable is crucial; even when working with scale-free shape vectors alone, and forming concepts of independence based on size ratios alone, the choice of a size sequence is crucial. Independence of the rays in the lower-dimensional space, and a size ratio that determines the rays in the next higher-dimensional space, can occur only with respect to at most one related pair of size variables.

68.8 Diversity of Hypotheses for Isometry (Neutrality)

Let $\mathbf{X}_k$ be a k-dimensional positive random vector and $\mathbf{Y}_k$, the random vector whose coordinates are, respectively, logarithms of the coordinates of $\mathbf{X}_k$. Suppose that $\mathbf{Y}_k$ follows a multivariate normal distribution with covariance matrix $\boldsymbol{\Sigma}$ and mean vector $\boldsymbol{\mu}$. Then $\mathbf{X}_k$ follows a multivariate lognormal distribution with parameters $\boldsymbol{\Sigma}$ and

$\boldsymbol{\mu}$, and hypotheses of isometry can be discussed through examination of $\boldsymbol{\Sigma}$ and its submatrices. Let $\boldsymbol{\Sigma}_r$ denote the covariance matrix of the initial r coordinates of $\mathbf{Y}_k$, so that $\boldsymbol{\Sigma}_r$ constitues the upper left $r \times r$ elements of $\boldsymbol{\Sigma}$. The following statements are all equivalent:

M1. r-dimensional shape, say, $(X_1/X_r, \ldots, X_{r-1}/X_r, 1)$, is independent of the r-dimensional geometric mean $M_r = (X_1 \cdots X_r)^{1/r}$.

M2. $(Y_1 - Y_r, \ldots, Y_{r-1} - Y_r)$ is independent of $T_r = Y_1 + \cdots + Y_r$.

M3. $\operatorname{Cov}(Y_i - Y_r, T_r) = 0, i = 1$ to $r - 1$.

M4. $\operatorname{Cov}(Y_i, T_r) = \operatorname{Cov}(Y_r, T_r), i = 1$ to $r - 1$.

M5. The column totals of $\boldsymbol{\Sigma}_r$ are the same.

M6. $(1, \ldots, 1)$ is an eigenvector of $\boldsymbol{\Sigma}_r$.

Statement M5 shows that the independence of r shape and the geometric mean M_r is equivalent to the columns of $\boldsymbol{\Sigma}_r$ having the same column totals. Since the column totals of $\boldsymbol{\Sigma}_r$ may be the same without those of $\boldsymbol{\Sigma}_{r+1}$ being the same, r-dimensional shape may be independent of M_r without affecting the independence or lack thereof of $(r + 1)$-dimensional shape and M_{r+1}. Thus complete isometry with respect to $M_2, \ldots, M_k$ occurs if the column totals of $\boldsymbol{\Sigma}_r$ are the same for $r = 2, \ldots, k$.

The hypotheses above reflect the specific order represented by $\mathbf{X}_k$. If the coordinates of $\mathbf{X}_k$ are permuted, then isometry with respect to $M_2, \ldots, M_k$ represents different hypotheses. The situation is complex, indeed.

If, instead of multiplicative isometry with respect to $M_2, \ldots, M_k$, we consider coordinate isometry (with respect to $C_2, \ldots, C_k$), then the following statements are equivalent:

C1. r-dimensional shape, say, $(X_1/X_r, \ldots, X_{r-1}/X_r, 1)$, is independent of the rth coordinate $C_r = X_r$.

C2. $(Y_1 - Y_r, \ldots, Y_{r-1} - Y_r)$ is independent of Y_r.

C3. $\operatorname{Cov}(Y_i - Y_r, Y_r) = 0, i = 1, \ldots, r - 1$.

C4. $\operatorname{Cov}(Y_i, Y_r) = \operatorname{Var}(Y_r), i = 1$ to $r - 1$.

As a consequence, if for any pair (X_i, X_j), $\operatorname{cov}(Y_i, Y_j) = \operatorname{var}(Y_i)$, then two-dimensional shape $(X_j/X_i, 1)$ is independent of X_i. [If both M1 and C1 are true, then the linear combination

$$Y_r - \{1/(r-1)\}(Y_1 + \cdots + Y_{r-1})$$

has variance zero, so that X_r/M_r is degenerate as required.]

It is thus clear that isometry and neutrality hypotheses for loglinear size variables are readily framed within the multivariate lognormal context. For example the appropriate multiple correlation coefficient can be used to test such hypotheses, as well as to measure the degree of association of size with shape [40,43,46].

Unfortunately, neither additive isometry nor neutrality (with respect to $S_2, \ldots, S_k$) can occur within the multivariate lognormal family, so that such hypotheses are not readily studied in this framework [41]. A related result is that for every nonsingular member of the multivariate lognormal family, there exists a unique loglinear size variable that is independent of shape [52].

Aitchison [2] has studied the (closely related) additional concept of "subcompositional independence" by assumption of multivariate lognormality for a variety of log ratios, which can be expressed as log ratios of additive size variables. The term *additive logistic distribution* is the distribution of proportions $\mathbf{X}_k/S_k$ when the ratios $\mathbf{X}_k/X_k$ follow a multivariate lognormal distribution, while the *multiplicative logistic distribution* is the distribution of proportions when logs of functions of additive size ratios of the permuted vector $(X_k, \ldots, X_2, X_1)$ follow a multivariate normal distribution.

68.9 Practical Considerations

In any statistical distribution, k-dimensional shape can be independent of at most one size variable or related size ratio, but in the sample, shape may appear to be independent of more than one size variable. Thus suppose that all positive probability occurs only on a single ray from the origin in R_+^k. Then shape is constant, and thereby independent of all size variables. (In fact, any size variable must be a scalar multiple of any other with probability 1.) If we now perturb the situation by letting the positive probability occur in a long very narrow ellipse, say, with the ray as its major axis, then shape will be "nearly" independent of a variety of size variables, and there would be little prospect of detection of the nonindependence in moderate-sized samples. In the sample, shape could appear to be independent of more than one size variable.

Tests of independence in practice are often tests of zero covariance. Thus, while shape cannot be independent of S_k in the multivariate lognormal family, if the logs $Y_1, \ldots, Y_k$ are independently and identically distributed, each $P_i = X_i/S_k$ is uncorrelated with S_k.

Such a distinction is unnecessary when a multivariate lognormal assumption is made for various size ratios of $\mathbf{X}$ or for $\mathbf{X}$ itself. The most successful analyses involving size and shape are those in which the range of methods based on an underlying multivariate or univariate normal distribution can be applied; see Aitchison [1,2] and Aitchison and Shen [4].

Prior to 1970 there was no emphasis on the need to specify a particular size variable for study. Jolicoeur [28] proposed that the major axis of the ellipse of concentration of the log measurements be used to represent the allometric relationship (see Gould [20,21], Huxley [24], Reeve and Huxley [49], Griffiths and Sandland [22], and Jolicoeur [29]). He defines isometry to be the condition that this axis be the equiangular line. From statement M6, this is completely consistent with the results here, but it does implicitly select M_k, the geometric mean of the measurements, as size. Hopkins [23] and Sprent [57,58] present models in which the observed covariance matrix is the sum of a "model" or structural matrix plus an error matrix. Again, the definitions here are applicable, but to the structural portion of the model. The presence of an error matrix often destroys the ability to test certain isometric hypotheses, but Mosimann et al. [46] were able to study some isometric hypotheses in the presence of error. (See also Jolicoeur and Heusner [30].)

68.10 The Choice of a Size Variable

The choice of a size variable should be specific for a given study, and will ideally be based on underlying processes thought to generate the random quantities under analysis. In studies of random proportions or ratios it seems particularly important to think about the processes that generate the measurements being studied. Methods for studying such measurements will likely differ, based on the underlying model, which may dictate which size variables are appropriate. The examples that follow are illustrative of the possibilities, and are by no means definitive.

Let the random vector (L,W,H) represent the length, width, and height, respectively, of the carapace, or shell, of turtles of a given species. The three turtle measurements are perpendicular to each other and thus $M_3 = (LWH)^{(1/3)}$ may be a reasonable measurement of "volume" (how reasonable, for three distinct species, was studied empirically by Mosimann [36]). The hypothesis that three-dimensional shape is independent of volume M_3 is distinct from the hypothesis that two-dimensional shape $(L/W, 1)$ is independent of "transverse sectional area" $M_2 = (LW)^{(1/2)}$. Rearranging the order of the measurements changes the two-dimensional space represented by the first two coordinates. Thus, for (W, H, L), the size variable M_2 becomes $(WH)^{(1/2)}$, or "cross-sectional area." It makes a difference. In the map turtle, two-dimensional shape $(W/H, 1)$ is isometric with respect to cross-sectional area, $(WH)^{(1/2)}$, but, since length grows faster than height or width, two-dimensional shape $(W/L, 1)$ is not isometric with respect to $(LW)^{(1/2)}$ [36; 43, p. 447].

On the other hand, in the study of proportions of various midline measurements of scales (or scutes) of turtles [35; 11, p. 204], all measurements are taken along a single physical dimension, and a natural size variable is the sum S_k.

In another context, let $\mathbf{T}' = (T_1, \ldots, T_k)$ represent the time intervals between successive events in a Poisson process. Here there is a single time dimension, and S_k is a natural size variable. (The T_i's are independent exponential random variables and exhibit complete isometry with respect to $S_2, \ldots, S_k$ for every permutation of $\mathbf{T}$.)

In a study of geographic variation in blackbirds, Mosimann and James [43] used standard measurements available from museum "study-skins" to examine geographic variation in bill and wing–tail proportions and size. Various size variables were studied, and a coordinate size variable, wing length, exhibited the highest between/within ratio of variance components. One advantage of considering a separate definition of shape vector was that it was possible to relate shape to a measurement of size (weight) apart from the linear measurements themselves. Jungers [32], in studying limb proportions of primates, gives reasons for choosing weight as size. In the same volume, Ford and Corruccini [18] discuss practical difficulties that arise with this choice. Baron and Jolicoeur [5] use body weight as a size variable against which the relative volumes of various portions of the brain are studied in comparing species of bats and insectivorous mammals.

Veitch [60] thoughtfully uses a multivariate linear model to study changes in the proportions of claws and appendages of two species of fiddler crabs. The size variable chosen is the geometric mean of carapace length and breadth. Reyment [50] compared relations of three-dimensional shape with the geometric mean of height, breadth, and diameter, as well as diameter alone, in three species of fossil ammonites. Fordham and Bell [19] consider the behaviour of the geometric mean size variable with certain principal components in an artificial dataset. Sampson and Siegel [52] define a "residual" size variable, with an interesting interpretation of the allometry* of antlers in the Irish elk.

Seebeck [53,54] has used linear models in the analysis of size and shape. His work has focused on growth and body composition in cattle and other livestock. In Ref. 53, a variety of covariance analyses of log shape show that, when size is taken into account, faster growing animals have a lower yield of carcass. Seebeck's studies exploit, in an interesting way, the use of loglinear size variables embedded in experimental designs permitting the application of the general linear model.

Darroch and Mosimann [13] present canonical discriminant (and principal-component) analyses for log shape that are invariant when applied to any log shape vector formed from a loglinear size variable. The component analyses are the same as those presented for composition data by Aitchison [3], based on a different argument.

The choice of a suitable size variable for chemical composition data seems important. For such data, both Connor and Mosimann [11] and Aitchison [2] implicitly chose the sum S_k. Whether such a choice is a suitable one may vary, depending on the processes generating the data. A point worth noting is that unequal changes of the scale of measurement affect additive isometry and neutrality. Thus, if chemical compositions by weight are additively neutral (or isometric), then the comparable compositions by volume cannot be so [45].

Unequal changes of scale are important in other analyses involving proportions. Differential production of pollen by different kinds of plants may affect interpretation of fossil pollen diagrams [16,39,42]. In such studies a weighted sum (with positive weights adding to 1) may be as appropriate for size as the sum S_k.

Finally, investigators in different scientific fields may habitually normalize their data in different ways; for example, as proportions $\mathbf{X}/S_k$ or ratios $\mathbf{X}/X_k$. It would be unwise to assume without reflection that the particular size used for the normalization is relevant. The underlying processes may be such that any of a variety of size variables, and their related size ratios, could be the random quantities pertinent to the scientific investigation.

68.11 Relationships to Other Concepts of Size and Shape

The terms *size* and *shape* are often used in a distinct, but closely related, sense in the statistical literature. If in discriminant or principal-component analysis one linear combination of the coordinate variables is identified as a size variable, then other variables orthogonal to size may be designated as shape variables. By construction, the shape variables are uncorrelated with size, and cannot be used to study the association of size with shape (see Rao [48], Jolicoeur and Mosimann [31], many examples in Reyment et al. [51], and Burnaby [10]). In a typical situation the investigator wishes to classify organisms into one of several populations or species, and to classify small and large individuals from the same species correctly. Hence the shape variables are chosen for the purpose of classification, and should not depend on size. The need for such variables stems from the lack of isometry in the first place; although useful for classification, they generally contain a mixture of scale and scale-free information.

Penrose [47] and Spielman [56] have partitioned the squared distances represented by Pearson's coefficient of racial likeness and Mahalanobis' D^2 into size and shape components, respectively. Mosimann and Malley [44] show that Penrose's partition is based on the size variable S_k, while that of Spielman is based on D_k. In both cases the shape component may reflect scale information [56,44].

Reyment et al. [51] present a particularly informative discussion of a host of applications of multivariate methods in morphometry, including uses of the concept of size and shape variables. The work of Blum [6], Siegel and Benson [55], and much of the work of Bookstein and collaborators [7–9] is relevant to matching points and choosing measurements that are most suitable for description of shape as an outline. Important points relevant to pattern recognition are revealed in such studies, particularly with respect to meaningful measurements to describe the deformation of one outline to another. Whatever measurements best define such deformation, if the choice results in a vector of positive random variables, whose coordinates are potentially expressible in the same scale, then the concepts of size and shape presented here will be relevant. In this chapter, the vector of measurements is taken as given, and may or may not be associated with outlines in the physical world. The methods are applicable to discrete and continuous measurements.

D. G. Kendall describes models for generating random polygonal figures in the plane. Consider triangles; after translation, scaling, and rotation, Kendall defines a space in which each triangle shape corresponds to a single point. He treats the cases of labeled and unlabeled vertices. *Labeled vertices* are identified externally to the analysis so that permutations of the labels are not allowed. *Unlabeled vertices* are not so identified, and, in effect, all permutations of "labels" are permitted. In a typical application with triangles in morphometrics, the vertices and the distances between them are labeled externally to the analysis and therefore may not lose their identities under permutations. For example, suppose three points of an animal are observed from the lateral aspect (tip of snout, posterior of eye, rear of jaw) with the corresponding distances (snout–jaw, jaw–eye, eye–snout). The distances (3,4,5) and (5,3,4) have distinct appearances and distinct shape vectors, but would correspond to the same point in a shape space for unlabeled vertices.

Models may be constructed with labeled vertices whose placement is restricted so that permutations are not relevant. (In such cases shape vectors correspond to shapes of both labeled and unlabeled triangles.) An example of unlabeled vertices is that of triangles formed by the positions of three similar stones in a field. Kendall's methods represent a major development for understanding randomly generated polygonal shapes.

References

1. Aitchison, J. (1981). In *Statistical Distributions in Scientific Work*, Vol. 4, C. Taillie, G. P. Patil, and B. A. Baldessari, eds., pp. 147–156, Reidel, Dordrecht, The Netherlands.

2. Aitchison, J. (1982). *J. Roy. Statist. Soc. B*, **44**, 139–177.

3. Aitchison, J. (1983). *Biometrika*, **70**, 57–65.

4. Aitchison, J. and Shen, S. M. (1980). *Biometrika*, **67**, 261–272.

5. Baron, G. and Jolicoeur, P. (1980). *Evolution*, **34**, 386–393.

6. Blum, H. (1973). *J. Theor. Biol.*, **38**, 205–287.

7. Bookstein, F. L. (1978). *The Measurement of Biological Shape and Shape Change*. Lecture Notes in Biomathematics, **24**. Springer-Verlag, Berlin.

8. Bookstein, F. L. (1984). *J. Theor. Biol.*, **107**, 475–520.

9. Bookstein, F., Chernoff, B., Elder, R., Humphries, J., Smith, G., and Strauss, R. (1985). *Morphometrics in Evolutionary Biology*, Special Publication 15. The Academy of Natural Sciences of Philadelphia, Philadelphia, PA.

10. Burnaby, T. P. (1966). *Biometrics*, **22**, 96–110.

11. Connor, R. J. and Mosimann, J. E. (1969). *J. Am. Statist. Assoc.*, **64**, 194–206.

12. Darroch, J. N. and James, I. R. (1974). *J. Roy. Statist. Soc. Ser. B*, **36**, 467–483.

13. Darroch, J. N. and Mosimann, J. E. (1985). *Biometrika*, **72**, 241–252.

14. Darroch, J. N. and Ratcliff, D. (1971). *J. Am. Statist. Assoc.*, **66**, 641–643.

15. Fabius, J. (1973). *Ann. Statist.*, **1**, 583–587; Correction: **9**, 234 (1981).

16. Fagerlind, F. (1952). *Botaniska Notiser*, **2**, 185–224.

17. Ferguson, T. S. (1962). *Ann. Statist.*, **33**, 986–1001.

18. Ford, S. M. and Corruccini, R. S. (1985). In *Size and Scaling in Primate Biology*, W. L. Jungers, ed., pp. 401–435.
19. Fordham, B. G. and Bell, G. D. (1978). *Math. Geol.*, **10**, 111–139.
20. Gould, S. J. (1966). *Biol. Rev.*, **41**, 587–640.
21. Gould, S. J. (1977). *Ontogeny and Phylogeny*. The Belknap Press of Harvard University Press, Cambridge, MA.
22. Griffiths, D. A. and Sandland, R. L. (1982). *Growth*, **46**, 1–11.
23. Hopkins, J. W. (1966). *Biometrics*, **22**, 747–760.
24. Huxley, J. S. (1932). *Problems of Relative Growth*. The Dial Press, New York.
25. James, I. R. (1979). *Ann. Statist.*, **7**, 869–881.
26. James, I. R. (1981). In *Statistical Distributions in Scientific Work*, Vol. 4, C. Taillie, G. P. Patil, and B. A. Baldessari, eds., pp. 125–136, Reidel, Dordrecht, The Netherlands.
27. James, I. R. and Mosimann, J. E. (1980). *Ann. Statist.*, **8**, 183–189.
28. Jolicoeur, P. (1963). *Biometrics*, **19**, 197–499.
29. Jolicoeur, P. (1984). *Biometrics*, **40**, 685–690.
30. Jolicoeur, P. and Heusner, A. A. (1971). *Biometrics*, **27**, 841–855.
31. Jolicoeur, P. and Mosimann, J. E. (1960). *Growth*, **24**, 339–354.
32. Jungers, W. L. (1985). In *Size and Scaling in Primate Biology*. W. L. Jungers, ed. pp. 401–435.
33. Kapur, B. D. and Patil, G. P. (1979). *J. Indian Statist. Assoc.*, **17**, 69–82.
34. Lukacs, E. (1955). *Ann. Math. Statist.*, **26**, 319–324.
35. Mosimann, J. E. (1956). *Miscellaneous Publications of the Museum of Zoology*, No. 97, pp. 1–43, University of Michigan, Ann Arbor.
36. Mosimann, J. E. (1958). *Rev. Can. Biol.*, **17**, 137–228.
37. Mosimann, J. E. (1962). *Biometrika*, **49**, 66–82.
38. Mosimann, J. E. (1970). *J. Am. Statist. Assoc.*, **65**, 930–945.
39. Mosimann, J. E. (1970). In *Random Counts in Scientific Work*, Vol. 3, pp. 1–30, G. P. Patil, ed. The Pennsylvania State University Press, University Park.
40. Mosimann, J. E. (1975). In *Statistical Distributions in Scientific Work*, Vol. 2, G. P. Patil, S. Kotz, and J. K. Ord, eds., pp. 187–217, Reidel, Dordrecht, The Netherlands.
41. Mosimann, J. E. (1975). In *Statistical Distributions in Scientific Work*, Vol. 2, G. P. Patil, S. Kotz, and J. K. Ord, eds., pp. 219–239, Reidel, Dordrecht, The Netherlands.
42. Mosimann, J. E. and Greenstreet, R. L. (1971). In *Statistical Ecology*, Vol. 1, G. P. Patil, E. C. Pielou, and W. E. Waters, eds., pp. 23–58, The Pennsylvania State University Press, University Park.
43. Mosimann, J. E. and James, F. C. (1979). *Evolution*, **33**, 444–459.
44. Mosimann, J. E. and Malley, J. D. (1979). In *Statistical Ecology*, Vol. 7, L. Orloci, C. R. Rao, and W. M. Stiteler, eds., pp. 175–189, International Co-operative Publishing House, Fairland, MD.
45. Mosimann, J. E. and Malley, J. D. (1981). In *Statistical Distributions in Scientific Work*, Vol. 4, C. Taillie, G. P. Patil, and B. A. Baldessari, eds., pp. 137–145, Reidel, Dordrecht, The Netherlands.
46. Mosimann, J. E., Malley, J. D., Cheever, A. W., and Clark, C. B. (1978). *Biometrics*, **34**, 341–356.
47. Penrose, L. S. (1954). *Ann. Eugen.*, **18**, 337–343.
48. Rao, C. R. (1964). *Sanhkyā A*, **26**, 329–358.
49. Reeve, E. C. R. and Huxley, J. S. (1945). In *Essays on Growth and Form*, W. E. Le Gros Clark and P. B. Medawar, eds., pp. 121–156, Oxford University Press, Oxford, UK.

50. Reyment, R. A. (1982). *Stockholm Contrib. Geol.*, **37**, 201–214.
51. Reyment, R. A., Blackith, R. E., and Campbell, N. A. (1984). *Multivariate Morphometrics*, 2nd ed. Academic Press, New York.
52. Sampson, P. D. and Siegel, A. F. (1985). *J. Am. Statist. Assoc.*, **80**, 910–914.
53. Seebeck, R. M. (1983). *Animal Product.*, **37**, 53–66.
54. Seebeck, R. M. (1983). *Animal Product.*, **37**, 321–328.
55. Siegel, A. F. and Benson, R. H. (1982). *Biometrics*, **38**, 341–350.
56. Spielman, R. S. (1973). *Am. Naturalist*, **107**, 694–708.
57. Sprent, P. (1968). *Biometrics*, **24**, 639–656.
58. Sprent, P. (1972). *Biometrics*, **28**, 23–27.
59. Tate, R. F. (1959). *Ann. Math. Statist.*, **30**, 341–366.
60. Veitch, L. G. (1978). *Math. Sci.*, **3**, 35–45.

69 Stability Study Designs

Tsae-Yun Daphne Lin and Chi Wan Chen

69.1 Introduction

The *stability* of a drug substance or drug product is the capacity of the drug substance or drug product to remain within the established specifications to ensure its identity, strength, quality, and purity during a specified period of time. The US Food and Drug Administration (FDA) requires that the shelf life (also referred to as *expiration dating period*) must be indicated on the immediate container label for every human drug and biologic on the market.

A good stability study design is the key to a successful stability program. From statistical perspectives, there are several elements that need to be considered in planning a stability study design. First, the stability study should be well designed so that the shelf life of the drug product can be estimated with a high degree of accuracy and precision. Second, the stability design should be chosen so that it can reduce bias and identify and control any expected or unexpected source of variations. Third, the statistical method used for analyzing the data collected should reflect the nature of the design and provide a valid statistical inference for the established shelf life.

In many cases, the drug product may have several different strengths packaged in different container sizes due to medical needs. To test every batch under all factor combinations on a real-time, long-term testing schedule can be expensive and time-consuming. As an alternative, a reduced design (e.g., bracketing, matrixing) may be considered as an efficient method for reducing the amount of testing needed while still obtaining the necessary stability information. A number of experimental designs that can capture the essential information to allow the determination of a shelf life while requiring less sampling and testing are available. Some of these designs require relatively few assumptions about the underlying physical and chemical characteristics, and others require a great many.

Stability requirements for the worldwide registration of pharmaceutical products have changed dramatically. A series of guidelines on the design, conduct, and data analysis of stability studies of pharmaceuticals have been published by the ICH (International Conference on Harmonization). ICH Q1A(R2) [11] defines the core stability data package that is sufficient to support the registration of a new drug application in the tripartite regions of the European Union, Japan, and the United States. ICH Q1D [12] provides guidance on reduced designs for stability studies. It outlines the circumstances under which a bracketing or matrixing design can be used. ICH Q1E [7] describes the principles of stability data evaluation and various approaches to statistical analysis of stability data in establishing a retest period for the drug substance or a shelf life for the drug product, which would satisfy regulatory requirements.

In this chapter, the discussion will focus on the statistical aspects of stability study designs in relation to the ICH guidelines. The statistical analysis of stability data will not be covered here. In Section 69.2, different types of experimental designs, such as complete factorial design, fractional factorial design, will be exemplified. In Section 69.3, several commonly used criteria for design comparison will be presented. In Section 69.4, the role of stability study protocol will be emphasized. In Section 69.5, the statistical and regulatory considerations on the selection of stability study design, in particular, a full design versus bracketing or matrixing design will be discussed. Finally, a conclusion will be drawn in Section 69.6.

69.2 Stability Study Designs

The design of a stability study, based on testing a limited number of batches of a drug substance or product, is intended to establish a retest period or shelf life applicable to all future batches of the drug substance or product manufactured under similar circumstances. Tested batches should, therefore, be representative in all respects, such as formulation, container and closure, manufacturing process, manufacturing site, and source of drug substance, for the population of all production batches, and conform to the quality specification of the drug product. The stability study should be well designed so that the shelf life of the product can be estimated with a high degree of accuracy and precision.

A typical stability study consists of several design factors (e.g., batch, strength, container size) and several response variables (e.g., assay, dissolution rate) measured at different timepoints under a variety of environmental conditions. One can categorize the stability designs as full, bracketing, or matrixing designs by the number of timepoints or levels of design factors as shown in Table 1. The following sections describe the characteristics, advantages, and disadvantages of various designs.

69.2.1 Full Design

Several different types of experimental design can be applied to the stability study. A *full design* (also referred to as a *complete factorial design*) is one in which samples for every combination of all design factors are tested at all time points as recommended in ICH Q1A(R2), for example, a minimum of 0, 3, 6, 9, 12, 18, 24 months, and yearly thereafter for long-term testing and 0, 3, and 6 months for accelerated testing.

As mentioned above, ICH has issued a series of guidelines on the design and conduct of stability studies. ICH Q1A(R2) recommends that the drug substance and product be stored at the long-term condition [e.g., 25°C ±2°C/60% RH ±5% RH or 30°C ±2°C/65% RH ±5% RH (where RH = relative humidity)] that is reflective of the storage condition intended for the container label. Unless the drug substance or product is destined for freezer storage, it should also be stored at the accelerated condition for 6 months, and tested minimally at 0, 3, and 6 months. The guideline also states that if long-term studies are conducted at 25°C ±2°C/60% RH ±5% RH and "significant change" occurs at any time during 6 months' testing at the accelerated storage condition, additional testing at the intermediate storage condition

Table 1: Types of stability designs.

	All Levels of Design Factors	Partial Levels of Design Factors
All timepoints	Full design	Bracketing design or matrixing design on factors
Partial timepoints	Matrixing design on time points	Matrixing design on timepoints and factors

Table 2: An example of a stability protocol using a full design according to the ICH Q1A(R2) guideline.

Number of batches	3
Number of strengths	3
Number of container sizes	3
Number of container closure systems	4
Long-term[a], intermediate, and accelerated	Long-term, 25°C ±2 °C/60% RH ±5% RH
	Intermediate, 30°C ±2°C/65% RH ±5% RH
	Accelerated, 40°C ±2°C/75% RH ±5% RH
Timepoints (months)	Long-term, 0, 3, 6, 9, 12, 18, 24, 36
	Intermediate, 0, 6, 9, 12
	Accelerated, 0, 3, 6
Total number of samples tested	1188
Total number of samples tested[b]	1620

[a]The long-term condition could be 25°C ±2°C/60% RH ±5% RH or 30°C ±2°C/65% RH ±5% RH.

[b]If long-term studies are conducted at 25°C ±2°C/60% RH ±5% RH and "significant change" occurs at any time during 6 months' testing at the accelerated storage condition, then the intermediate testing is needed.

should be conducted and evaluated.

A product may be available in several strengths and each of the strengths may be packaged in more than one container closure system and several container sizes. In this case, the resources needed for stability testing are considerable. Table 2 illustrates an example of a stability protocol using a full design for a drug product manufactured in three strengths and packaged in three container sizes and four container closure systems. This example shows that it will require 1188 test samples for long-term and accelerated stability testing. In addition, if long-term studies are conducted at 25°C ±2°C/60% RH ±5% RH and "significant change" occurs at any time during 6 months' testing at the accelerated storage condition, intermediate testing is needed and the number of test samples will be increased to 1620.

Table 3 shows an example of a simple full design. This example describes a protocol for long-term stability testing (25°C/60% RH) of a drug product manufactured in three strengths (25, 50, and 100 mg) and packaged in three container sizes (10, 50, and 100 mL). Samples for every combination of all design factors, that is, strength, batch, and container size, are tested at all time points. Hence, this example is a complete factorial design and the total number of samples tested is $N = 3 \times 3 \times 3 \times 8 = 216$.

As shown in the above examples, a full design involves complete testing of all factor combinations at all time-points, which can be costly. The pharmaceutical industry would like to apply a reduced design so that not every factor combination will be tested at every time-point. ICH Q1D listed some principles for situations in which a reduced design can be applied. However, before a reduced design is considered, certain assumptions should be assessed and justified. The potential risk of establishing a shorter retest period or shelf life than could be derived from a full design due to the reduced amount of data collected should be considered.

69.2.2 Reduced Design

As defined in the ICH Q1D guidelines, a *reduced design* (also referred to as a *fractional factorial design*) is one in which samples for every factor combination are not all tested at all timepoints. Any subset of a full design is considered a reduced design. Bracketing and matrixing designs are two most commonly used reduced designs.

In 1989, Wright [26] proposed using the factorial designs in stability studies. Nakagaki [22] used the term *matrix and bracket* in his 1991 presentation. In 1992, Nordbrock [23], Helboe [13] and Carstensen [2] published articles discussing several methods for reducing the number of samples tested from the chemistry and economic aspects of stability studies. Nordbrock investigated various types of fractional factorial designs and compared these designs on the basis of the power of detecting a significant difference between slopes. He concluded that the design that gives acceptable performance and has the smallest sample size can be chosen on the basis of power. Lin [17] investigated the applicability of matrixing and bracketing approaches to stability studies. She concluded that the complete factorial design is the best design when the precision of shelf life estimation is the major concern, but indicated that matrixing designs could be useful for drug products with less variability among different strengths and container sizes. With regard to the statistical analysis of data from complex stability studies, Fairweather [9], Ahn [1], Chen [5], and Yoshioka [28] and their colleagues have proposed different procedures for testing and classifying stability data with multiple design factors.

Table 3: Example of a full stability study design.

Granulation Batch	Strength	Container Size		
		10 mL	50 mL	100 mL
A	25	T	T	T
	50	T	T	T
	100	T	T	T
B	25	T	T	T
	50	T	T	T
	100	T	T	T
C	25	T	T	T
	50	T	T	T
	100	T	T	T

T = Sample tested at 0, 3, 6, 9, 12, 18, 24, 36 months.

69.2.3 Bracketing Design versus Matrixing Design

ICH Q1D states that a reduced design can be a suitable alternative to a full design when multiple design factors are involved in the product being evaluated. However, the application of a reduced design has to be carefully assessed, taking into consideration any risk to the ability of estimating an accurate and precise shelf life or the consequence of accepting a shorter-than-desired shelf life. Bracketing and matrixing designs are two most commonly used reduced designs. The reduced stability testing in a bracketing or matrixing design should be capable of achieving an acceptable degree of precision in shelf-life estimation without losing much information.

The terms of bracketing and matrixing are defined in the ICH Q1D as follows:

- *Bracketing:* The design of a stability schedule such that only samples on the extremes of certain design factors (e.g., strength, container size) are tested at all timepoints as in a full design. The design assumes that the stability of any intermediate levels is represented by the stability of the extremes tested.

- *Matrixing:* The design of a stability schedule such that a selected subset of the total number of possible samples for all factor combinations is tested at a specified timepoint. At a subsequent timepoint, another subset of samples for all factor combinations is tested. The design assumes that the stability of each subset of samples tested represents the stability of all samples at a given timepoint.

As defined above, bracketing and matrixing are two different approaches to designing a stability study. Either design has its own assumptions, advantages, and disadvantages. The applicability of either design to stability study generally depends on the manufacturing process, stage of development, assessment of supportive stability data, and other factors as described in ICH Q1D. The following additional points regarding the use of a bracketing or matrixing design in a stability study can be considered.

In a bracketing design, samples of a given batch for a selected extreme of a factor are analyzed at all time points. Therefore, it is easier to assess the stability pattern in a bracketing study than in a matrixing study where samples of a given batch are often tested at fewer timepoints. If all selected strengths or container sizes tested show the same trend, it can be concluded with a high degree of certainty that the stability of the remaining strengths or container sizes is represented, or bracketed, by the selected extremes.

In a matrixing design, the samples to be tested are selected across all factor combinations. This procedure may be less sensitive to assessing the stability pattern than bracketing, due to the reduced timepoints. Therefore, a matrixing design is more appropriate for confirming a prediction or available stability information, and is more in the following situations: (1) in the later stages of drug development when sufficient supporting data are available, (2) for stability testing of production batches, and (3) for annual stability batches. One of the advantages of matrixing over bracketing is that all strengths and container sizes are included in the stability testing.

The general applicability of bracketing and matrixing has been discussed in the literature (e.g., Lin [17–20], Fairweather [8], Chen [4], Chambers [3], Helboe [14], and Yoshioka [27]). From statistical and regulatory perspectives, the applicability of bracketing and matrixing depends on the type of drug product, type of submission, type of factor, data variability, and product stability. The factors that may be bracketed or matrixed in a stability study are outlined in ICH Q1D. This ICH guideline also briefly discusses several conditions that need to be considered when applying these types of design. A bracketing or matrixing design may be preferred to a full design for the purpose of reducing the number of samples tested, and consequently the cost. However, the ability to adequately predict the product shelf life by these types of reduced designs should be carefully considered.

In general, a matrixing design is applicable if the supporting data indicate predictable product stability. Matrixing is appropriate when the supporting data exhibit only small variability. However, where the supporting data exhibit moderate variability, a matrixing design should be statistically justified. If the supportive data show large variability, a matrixing design should not be applied.

A statistical justification could be based on an evaluation of the proposed matrixing design with respect to its power to detect differences among factors in the degradation rates or its precision in shelf life estimation.

If a matrixing design is considered applicable, the degree of reduction that can be made from a full design depends on the number of factor combinations being evaluated. The more factors associated with a product and the more levels in each factor, the larger is the degree of reduction that can be considered. However, any reduced design should have the ability to adequately predict the product shelf life.

An example of a bracketing design is given in Table 4. Similar to the full design in Table 3, this example is based on a product available in three strengths and three container sizes. In this example, it should be demonstrated that the 10- and 100-mL containers truly represent the extremes of the container sizes. In addition, the 25 and 100 mg strengths should also represent the extremes of the strengths. The batches for each selected combination should be tested at each time point as in a full design. The total number of samples tested will be $N = 2 \times 2 \times 3 \times 8 = 96$.

An example of a matrixing-on-timepoints design is given in Table 5. The description of the drug product is similar to that in the previous examples. Three time codes are used in two different

Table 4: Example of a bracketing design.

Batch	Strength	Container Size		
		10 mL	50 mL	100 mL
A	25	T[a]	—	T
	50	—	—	—
	100	T	—	T
B	25	T	—	T
	50	—	—	—
	100	T	—	T
C	25	T	—	T
	50	—	—	—
	100	T	—	T

[a] Sample tested at 0, 3, 6, 9, 12, 18, 24, 36 months.

Table 5: Example of a matrixing-on-time-points design.

Batch	Strength	Container Size		
		10 mL	50 mL	100 mL
A	25	T1	T2	T3
	50	T3	T1	T2
	100	T2	T3	T1
B	25	T2	T3	T1
	50	T1	T2	T3
	100	T3	T1	T2
C	25	T3	T1	T2
	50	T2	T3	T1
	100	T1	T2	T3

Complete One-Third Design							
Time Code	Timepoint (Months)						
T1	0	3	—	—	12	—	—	36
T2	0	—	6	—	—	18	—	36
T3	0	—	—	9	—	—	24	36

Complete Two-Thirds Design							
Time Code	Timepoint (Months)						
T1	0	3	—	9	12	—	24	36
T2	0	3	6	—	12	18	—	36
T3	0	—	6	9	—	18	24	36

designs. As an example of a complete one-third design, the time points for batch B in strength 100 mg and container size 10 mL are 0, 9, 24, and 36 months. The total number of samples tested for this complete one-third design will be $N = 3 \times 3 \times 3 \times 4 = 108$.

Table 6 illustrates an example of matrixing on both timepoints and factors. Similar to the previous example, three time codes are used. As an example, batch B in strength 100 mg and container 10 mL is tested at 0, 6, 9, 18, 24, and 36 months. The total number of samples tested for this example will be $N = 3 \times 3 \times 2 \times 6 = 108$.

69.2.4 Other Fractional Factorial Designs

Fractional factorial designs other than those discussed above may be applied. Since bracketing and matrixing designs are based on different principles, the use of bracketing and matrixing in one design should be carefully considered. If this type of design is applied, scientific justifications should be provided.

69.3 Criteria for Design Comparison

Several statistical criteria have been proposed for comparing designs and choosing the appropriate design for a particular stability study. Most of the criteria are similar in principles but different in procedures. This section will discuss several commonly used criteria for design comparison. Since the statistical method used for analyzing the data collected should reflect the nature of the design, the availability of a statistical method that provides a valid statistical inference for the established shelf life should be considered when planning a study design.

The optimality criteria have been applied by several authors (e.g., Nordbrock [23], Ju and Chow [15]) to the selection of stability design. These optimality criteria are statistical efficiency criteria and have been developed and widely used (e.g., by Kiefer [16] and Fedorov [10]) in other scientific areas for many years. Hundreds of journal articles on these optimality criteria have been published.

Let Y denote the result of a test attribute, say, assay. The following model can be used to describe Y:

$$Y = X\beta + \varepsilon;$$

X is the design matrix, β is the coefficient vector, and ε is the residuals vector. The least-squares solution of this matrix equation with respect to the model coefficient vector is $\beta = (X'X)^{-1}X'Y$, and $X'X$ is called *the information matrix* because its determinant is a measure of the information content in the design.

Several different optimality criteria are used for design comparison, and among them the D-optimality and A-optimality are the most commonly used. The basic principle behind these criteria is to choose a design that is optimal with respect to the precision of the parameter estimators. For example, say X_1 and X_2 are the design matrices for two different fractional factorial designs. A D-optimality criterion is that if $\text{Det}(X_1'X_1) > \text{Det}(X_2'X_2)$, then design X_1 is said to be preferred to design X_2. An A-optimality criterion is that if $\text{Tr}(X_1'X_1) > \text{Tr}(X_2'X_2)$, then design X_1 is said to be preferred to design X_2. The D-optimal design is used most often in experimental designs.

If we apply the D-optimality concept directly to stability studies, then the selection of observations (i.e., timepoints) that give the minimum variance for the slope is to place one-half at the beginning of the study and one-half at the end. However, in reality, the linearity of the test attribute–time may not hold for many chemical and physical characteristics of the drug product. Hence, the statistically designed sta-

Table 6: Example of a matrixing on timepoints and factors.

Batch	Strength	Container Size		
		10 mL	50 mL	100 mL
A	25	T1	T2	T3
	50	—	—	—
	100	T2	T3	T1
B	25	T2	T3	T1
	50	—	—	—
	100	T3	T1	T2
C	25	T3	T1	T2
	50	—	—	—
	100	T1	T2	T3

Time Code	Timepoint (Months)							
T1	0	3	—	9	12	—	24	36
T2	0	3	6	—	12	18	—	36
T3	0	—	6	9	—	18	24	36

bility studies should not only apply these basic optimality principles but also include several intermediate time points (e.g., 3 and 6 months) as a check for linearity. However, depending on the drug product, manufacturing process, marketing strategy, and other factors, different designs may be chosen and no single design is optimal in all cases. Because analyses are typically done at several different times (e.g., at the time a registration application for the new drug product is filed, or yearly for a marketed product), the choice of design should also take into account that the analysis will be done after additional data are collected.

Nordbrock [23] proposed to choose a design on the basis of the power for detecting a significant difference among slopes. The method he proposed consisted of three steps. The first step is to list slopes that must be compared, that is, to list factors and factor interactions (for slopes) that may affect stability. The second step is to list some alternative designs with reduced sample sizes. Some general experimental design considerations are used to derive these alternative designs. The third step is to compare statistical properties of the designs on the basis of the power of contracts of interest. He then suggested that "among designs with the same sample size, the design with the highest power is the best design. One way to select a design for a particular study is to choose the desired power, and then, from designs having at least the desired power, the design with the smallest sample size is the best design."

The primary objective of a stability study is to establish a shelf life for the drug product, and not to examine the effect of factors used in the experiment. Thus, the design chosen on the basis of the power of contracts of interest as proposed by Nordbrock may not be the best choice. As an alternative, Ju and Chow [15] proposed a design criterion on the basis of the precision of the shelf life estimation. Mathematically, this criterion of choosing a design is based on the following comparison:

Design X_A is considered to be better

than design X_B if

$$x(t)'(X_A'X_A)^{-1}x(t) < x(t)'(X_B'X_B)^{-1}x(t) \text{ for } t \geq t_\varepsilon,$$

where $x(t)$ is the design matrix X at time t, and t_ε is chosen on the basis of the true shelf life. This paper utilized the shelf lifem estimation procedure developed by Shao and Chow [25] that a random effect model was used. This criterion also takes into account the batch variability; hence, it could be applied to the balanced cases and could be extended to unbalanced cases.

On the basis of the above criterion and comparison results, Ju and Chow then proposed that, "For a fixed sample size, the design with the best precision for shelf life estimation is the best design. For a fixed desired precision of shelf life estimation, the design with the smallest sample size is the best design."

Murphy [21] introduced the uniform matrix designs for drug stability studies and compared it to standard matrix designs. The strategy of the uniform matrix design is to delete certain times (e.g., the 3, 6, 9, and 18 months); therefore, testing is done only at 12, 24, and 36 months. This design has the advantages of simplifying the data entry of the study design and eliminating timepoints that add little to reducing the variability of the slope of the regression line. The disadvantage is that, if there are major problems with the stability, there is no early warning because early testing is not done. Further, it may not be possible to determine if the linear model is appropriate.

Murphy used five efficacy measures (design moment, D-efficiency, uncertainty, G-efficiency, and statistical power) to compare uniform matrix designs to other matrix designs. Three of the criteria (moment, D-efficiency, and power) are well known and widely used. Uncertainty is a measure of the reliability of the shelf life estimates, which is dependant on the reliability of the estimated slopes. Another measure of uncertainty is the width of a confidence interval on the fitted slope, which, for a fixed amount of residual variation, is dependent only on the number and spacing of the data points. G-efficiency is a relatively new concept, which is defined as below:

$$G - \text{efficiency} = 100[(p/n)^{1/2}]/\sigma_{\max},$$

where

n = number of points in the design
p = number of parameters
$\sigma_{\max}$ = maximum standard error of prediction over the design

On the basis of comparisons of five different criteria, Murphy stated that, "the uniform matrix designs provide superior statistical properties with the same or fewer design points than standard matrix design."

Pong and Raghavarao [24] compared the power for detecting a significant difference between slopes and the mean-square error to evaluate the precision of the estimated shelf life of bracketing and matrixing designs. They found the power of both designs to be similar. On the basis of the conditions under study, they concluded that bracketing appears to be a better design than matrixing in terms of the precision of the estimated shelf life.

Evidently, the abovementioned criteria are based on different procedures. Thus, the final choice of design might be different by using different criteria. In general, in addition to chemical and physical characteristics of the drug product, regulatory and statistical aspects need to be considered. Ideally, one would like to choose a design that is optimal with respect to the precision of the shelf life estimation and the power of detecting meaningful effects.

69.4 Stability Protocol

A good stability study design is the key to a successful stability program. The program should start with a stability protocol that specifies clearly the study objective, the study design, batch and packaging information, specifications, timepoints, storage conditions, sampling plan, statistical analysis method, and other relevant information. The protocol should be well designed and followed rigorously, and data collection should be complete and in accordance with the protocol. The planned statistical analysis should be described in the protocol to avoid the appearance of choosing an approach to produce the most desirable outcome at the time of data analysis. Any departure from the design makes it difficult to interpret the resulting data. Any changes made to the design or analysis plan without modification to the protocol or after examination of the data collected should be clearly identified.

69.5 Basic Design Considerations

69.5.1 Impact of Design Factors on Shelf Life Estimation

A drug product may be available in different strengths and different container sizes. In such a case, stability designs for long-term stability studies will involve the following design factors: strength, container size, and batch. As mentioned by Nordbrock [23] and Chow [6], it is of interest to examine several hypotheses, such as the following:

1. Degradation rates among different container sizes are consistent across strengths.
2. Degradation rates are the same for all container sizes.
3. Degradation rates are the same for all strengths.
4. Degradation rates are the same for all batches.

Hence, there is a need to investigate the impact of main effects (e.g., strength, container size, batch) and interaction effects on the stability of the drug product under long-term testing. In constructing a stability design, one needs to consider to what extent it is acceptable to pool data from different design factors. For example, if there is a statistically significant interaction effect between container size and strength, then a separate shelf life should be estimated for each combination of container size and strength. On the other hand, if there is no statistically significant interaction effect, that is, if the degradation rates for all container sizes, strengths, and batches are the same, it will be acceptable to pool all results and calculate one single shelf life for all container sizes, strengths and batches produced under the same circumstances. A design should be adequately planned such that it is capable of detecting the possible significant interaction effects and main effects.

A full design can provide not only valid statistical tests for the main effects of design factors under study but also better precision in the estimates for interactions. Hence, the precision of the estimated drug shelf life for a full design is better than a reduced design. A reduced design is preferred to a full design for the purpose of reducing the test samples and, consequently, the cost. However, it has the following disadvantages. First, one may not be able to evaluate some interaction effects for certain designs. For example, for a 2^{4-1} fractional factorial design, two-factor effects are confounded with each other; hence, one may not be able to determine whether the data should be pooled. Second, if there are interactions between two factors, such as the strength by container size, the data cannot be pooled to establish a single shelf life. It is recommended that a separate shelf life for each combination of strength and container size be established. However, no shelf-life estimation for the missing factor combinations can be assessed. Third, if there are many missing factor combinations, there may not be sufficient precision for the estimated shelf-life.

It is generally impossible to test the assumption that the higher-order terms are negligible. Hence, if the design does not permit the estimation of interactions or main effects, it should be used only when it is reasonable to assume that these interactions are very small. This assumption must be made on the basis of theoretical considerations of the formulation, manufacturing process, chemical and physical characteristics, and supporting data from other studies.

Thus, to achieve a better precision of the estimated shelf life, a design should be so chosen as to avoid possible confounding and/or interaction effects. Once the design is chosen, statistical analysis should reflect the nature of the design selected.

69.5.2 Sample Size and Sampling Considerations

The total number of samples tested in a stability study should be sufficient to establish the stability characteristics of the drug product and to estimate the shelflife of the drug product with an acceptable degree of precision.

The total number of test samples needed in a stability study generally depends on the objective and design of the study. For example, for a drug product available in a single strength and a single container size, the choice of design is limited; that is, the design should have three batches of the product and samples should be tested every 3 months over the first year, every six months over the second year, and then annually. For drug products involving several design factors, several different types of design, such as full, bracketing, matrixing designs, can be chosen. In general, the number of design factors planned in the study, the number of batches, data variability within or across design factors, and the expected shelf life for the product all need to be considered when choosing a design. In addition, available stability information, such as the variability in the manufacturing process and the analytical procedures, also need to be evaluated.

The estimation of shelf life of a drug product is based on testing a limited number of batches of a drug product. Therefore, tested batches should be representative in all respects as discussed previously.

69.5.3 Other Issues

The purpose of selecting an appropriate stability design is to improve the accuracy and precision of the established shelf life of a drug product. Background information, such as regulatory requirements, manufacturing process, proposed specification, and developmental study results, are helpful in the design of a stability study.

The choice of stability study design should reflect on the formulation and manufacturing process. For example, the study design and associated statistical analysis for a product available in three strengths made from a common granulation will be different from those made from different granulations with different formulations.

The stability study design should be capable of avoiding bias and achieving minimum variability. Therefore, the design should take into consideration variations from different sources. The sources of variations may include individual dosage units, containers within a batch, batches, analytical procedures, analysts, laboratories, and manufacturing sites. In addition, missing values should be avoided and the reason for the missing values should be documented.

When choosing a matrixing design, one cannot rely on the assumption that the shelf life of the drug product is the same for all design factors. If the statistical results show that there is a significant difference among batches and container sizes, one cannot rely on certain combinations of batch and container size or on the statistical model to provide reliable information on the missing combinations. The shelf-life must be calculated for the observed combinations of batch and container size. The shortest observed shelf life is then assigned to all container sizes

69.6 Conclusions

A good stability study design is the key to a successful stability program. The number of design factors planned in the study, number of batches, data variability within or across design factors, and the expected shelf life for the product all need to be considered when choosing a design. The ICH guidelines, such as Q1A(R2), Q1D, and Q1E, should be perused and the recommendations therein should be followed.

A reduced design, such as bracketing and matrixing, can be a suitable alternative to a full design when multiple design factors are involved in the product being evaluated. However, the application of a reduced design has to be carefully assessed, taking into consideration any risk to the ability of estimating an accurate and precise shelf life or the consequence of accepting a shorter-than-desired shelf life. When the appropriate study design is chosen, the total number of samples tested in a stability study should be sufficient to establish the stability characteristics of the drug product and to estimate the shelf life of the drug product with an acceptable degree of precision. To achieve a better precision of the estimated shelf life, a design should be chosen to avoid possible confounding and/or interaction effects. Once the design is chosen, the statistical method used for analysis of the data collected should reflect the nature of the design and provide a valid statistical inference for the established shelf life.

Acknowledgments. The authors would like to thank the FDA CDER Office of Biostatistics Stability Working Group for support and discussion on the development of this manuscript.

References

1. Ahn, H. Chen, J., and Lin, T. D. (1997). A two-way analysis of covariance model for classification of stability data. *Biomet. J.*, **39**(5), 559–576.

2. Carstensen, J. T. Franchini, M. and Ertel, K. (1992). Statistical approaches to stability protocol design. *J. Pharm. Sci.*, **81**(3), 303–308.

3. Chambers, D. (1996), Matrixing/bracketing. US industry views. *Proc. EFPIA Symp. Advanced Topics in Pharmaceutical Stability Testing Building on the ICH Stability Guideline.* EFPIA, Brussels.

4. Chen, C. (1996), US FDA's perspective of matrixing and bracketing, *Proc. EFPIA Symp. Advanced Topics in Pharmaceutical Stability Testing Building on the ICH Stability Guideline.* EFPIA, Brussels.

5. Chen, J. Ahn, H. and Tsong, Y. (1997). Shelf life estimation for multi-factor stability studies. *Drug Inform. J.*, **31**(2), 573–587.

6. Chow, S. C. and Liu, J. P., eds. (1995) *Statistical Designs and Analysis in Pharmaceutical Science*, Marcel Dekker, New York.

7. *Draft ICH Consensus Guideline Q1E Stability Data Evaluation.* Food and Drug Administration; Center for Drug Evaluation and Research and Center for

Biologics Evaluation and Research, Jan. 2003.

8. Fairweather, W. R. and Lin, T. D. (1999) Statistical and regulatory aspects of drug stability studies: An FDA perspective. *In International Stability Testing*, David Mazzo., ed., pp. 107–132 Interpharm Press.

9. Fairweather, W. Lin, T. D., and Kelly, R. (1995). Regulatory, design, and analysis aspects of complex stability studies. *J. Pharm. Sci.*, **84**(11), 1322–1326.

10. Fedorov, V. V. (1972). *Theory of Optimal Experiments*, translated and edited by W. J. Studden and E. M. Klimko, eds. Academic Press, New York.

11. *Guidance for Industry: ICH Q1A(R2), Stability Testing of New Drug Substances and Products.* Food and Drug Administration, Center for Drug Evaluation and Research and Center for Biologics Evaluation and Research, Jan. 2003.

12. *Guidance for Industry: ICH Q1D Bracketing and Matrixing, Designs for Stability Testing of New Drug Substances and Products.* Food and Drug Administration, Center for Drug Evaluation and Research and Center for Biologics Evaluation and Research, Jan. 2003.

13. Helboe, P. (1992). New designs for stability testing programs: Matrix or factorial designs. Authorities viewpoint on the predictive value of such studies. *Drug Inform. J.*, **26**, 629–634.

14. Helboe, P. (1999) Matrixing and bracketing designs for stability studies: An overview from the European perspective. *In International Stability Testing*, David Mazzo, ed., pp. 135–160, Interpharm Press.

15. Ju, H. L. and Chow, S. C. (1995). On stability designs in drug shelf life estimation. *J. Biopharm. Statist.*, **5**(2), 201–214.

16. Kiefer, J. (1959) Optimal experimental designs. *J. Roy. Statist. Soc. B*, **21**, 272–319.

17. Lin, T. D. (1994). Applicability of matrix and bracket approach to stability study design. *Proc. of the American Statistical Association, Biopharmaceutical Section*, pp. 142–147.

18. Lin, T. D. (1999). Statistical considerations in bracketing and matrixing. *Proc. IBC Bracketing and Matrixing Conf.*, London.

19. Lin, T. D. (1999). *Study Design, Matrixing and Bracketing*, AAPS Workshop on Stability Practices in the Pharmaceutical Industry—Current Issues, Arlington, VA

20. Lin, T. D. and Fairweather, W. R. (1997) Statistical design (bracketing and matrixing) and analysis of stability data for the US market. *Proc. IBC Stability Testing Conf.*, London.

21. Murphy, J. R. (1996). Uniform matrix stability study designs. *J. Biopharm. Statist.*, **6**(4), 477–494.

22. Nakagaki, P. (1990). AAPS Annual Meeting.

23. Nordbrock, E. (1992). Statistical comparison of stability study designs. *J. Biopharm. Statist.*, **2**, 91–113.

24. Pong, A. and Raghavarao, D. (2000). Comparison of bracketing and matrixing designs for a two-year stability study. *J. Pharm. Statist.*, **10**(2), 217–228.

25. Shao, J. and Chow, S. C. (1994). Statistical inference in stability analysis. *Biometrics*, **50**(3), 753–763.

26. Wright, J. (1989) Use of factorial designs in stability testing. *Proc. Stability Guidelines for Testing Pharmaceutical Products: Issues and Alternatives*, AAPS Meeting.

27. Yoshioka S. (1999) Current application in Japan of the ICH stability guidelines: Does Japanese registration require more than others do?. *In International Stability Testing*, David Mazzo., ed., pp. 255–264, Interpharm Press.

28. Yoshioka, S. Aso, Y., and Kojima, S. (1997). Assessment of shelf-life equivalence of pharmaceutical products. *Chem. Pharm. Bull.*, **45**, 1482–1484.

70 Statistical Analysis of DNA Microarray Data

Susmita Datta

70.1 Introduction

The expression of a gene is measured by the abundance of its mRNA level. It is possible to measure the expression level of thousands of genes at the same time with microarray technology. The DNA microarray usually consists of a microscopic slide on which the DNA molecules are chemically bonded. Labeled nucleic acids are made from the biological sample mRNA and then hybridized to the DNA on the array. The abundance of the labeled nucleic acids is measured after hybridization. In this chapter, we discuss very briefly the major statistical issues and methods in analyzing microarray data sets. An expanded overview of selected statistical techniques in microarray data analysis can be found in Reference 7.

70.2 cDNA versus Oligonucleotide Chips

At the moment, there are primarily two types of microarray chips. One is the cDNA array, and the other one is the oligonucleotide (also known as the *Affymetrix*) array.

Complementary DNA (cDNA) microarrays are constructed by printing target cDNAs on a small glass surface using computer-controlled robotics. In a single-slide microarray experiment, one can compare the transcript abundance of genes in two mRNA samples. Red (Cy5)- and green (Cy3)-labeled mRNA samples are hybridized to the same slide. Fluorescent intensities are measured to indicate the expression levels. There could be multiple slides to compare the transcript abundance of two or more types of mRNA samples hybridized to different slides. Typically, microarray data are expressed as a ratio of expression levels in an active cell to that in a vegetative cell. Thus, a ratio larger than one represents a gene that was expressed, and a ratio smaller than one represents a gene that was repressed, at a given timepoint, during a biological process [10,22].

Oligonucleotide (Affymetrix) chips usually come with predetermined probes depending on the type of experiment. First, the set of oligonucleotide probes to be synthesized is defined, on the basis of the ability to hybridize to the target loci or genes of interest. Probe arrays are manufactured by Affymetrix proprietary, light-directed chemical synthesis processes. The amount of hybridization by a target gene is measured by the intensity of fluorescent dyes. Probes that most clearly match the target generally produce signals stronger than those that have mismatches. A statistic called the "average difference" is reported for each target. A large-average-difference signifies that the gene is expressed, and a small or negative average difference indicates that the gene is not expressed. This measure has been criticized by statisticians, and a modified form is currently used by Affymetrix.

70.3 Preprocessing of Data

It has been observed by Yang et al. [28,29], among others, that there is often a systematic bias in the intensities of the red and green dyes. Possible reasons for this type of effect have been listed as physical properties of the dyes, efficiency of dye incorporation procedures, scanner settings, and so on. Likewise, in the context of replicate experiments, the gene expression ratio levels may have different spreads (variances), due to the differences in experimental conditions. It has been argued that some type of normalization (scaling) to the expression ratios (which corresponds to a bias correction in the log scale) be made before other statistical techniques are used. A good account of normalization and the relevant software for the cDNA data can be found in Reference 12.

The issue of preprocessing and calculation of a "true" measure of gene expressions for the Affymetrix chips has received quite a bit of attention. Microarray Suite software MAS 5.0 by Affymetrix estimates the expression as a weighted average of log probe pair differences. The idea of averaging different probe intensities for the same gene has been questioned by various statisticians. One of the reasons is that individual probes in a probe set have very different hybridization kinetics. A good measure of expression should incorporate information about probe characteristics, on the basis of the affinity of each probe across chips and the amount of gene present in a sample. These principles are the basis of the multichip models. In Reference 16, a number of such measures are compared and a robust multiarray average (RMA) of background adjusted, normalized, and log-transformed perfect match values have been advocated. Apparently, inspired by this work, Affymetrix is planning a 2004 release of its own version of the multichip model.

70.4 Statistical Clustering

A typical microarray reading consists of expression levels of thousands of genes often observed over various timepoints during the course of a biological data process. These, in turn, have been used in statistical clustering of genes into groups on the basis of the similarity of their expression patterns. It has been shown in various situations [5,10,23] that the resulting groups contain genes with similar biological functions. Another closely related statistical technique, namely, classification or discriminant analysis, has also been used for similar purposes. Tissue samples are being clustered on the basis of their expression matrix as well.

Hierarchical clustering with correlation similarity [15] has been the most commonly used clustering technique with microarray data. With time, however, several other algorithms such as K-means, Diana, model-based clustering, self-organizing maps, and so on are finding their way into the microarray literature. Issues such as how to choose one amongst several clustering methods? and how to validate a clustering method? are being addressed [4,8,18,28].

A number of supervised learning techniques (e.g., linear discriminant analysis, quadratic discriminant analysis, support vector machines, partial least-squares discriminant analysis, neural networks, k-nearest neighbors, recursive partitioning) have been applied to various microarray datasets as well. Once again, comparative studies regarding the accuracy and robustness of various classification techniques prove to be useful [3,11].

70.5 Detection of Differentially Expressed Genes

One of the major goals of microarray experiments is to identify the set of genes that are differentially expressed in two or more tissue samples (e.g., normal vs. cancer cells). Statistical tests of significance such as the two sample t tests [13] and ANOVA-based tests [19] are often used. The issue of multiple comparison and the control of overall error rate is an important problem in these analyses since the number of genes involved could run into thousands. Stepdown P-value corrections using permutation or resampling methods have been advocated for this purpose [13] following the general resampling technique of Westfall and Young [7]. Bayesian and empirical Bayes methods have also been proposed [1,14,20]. Datta et al. [10] introduced a combination of empirical Bayes and the stepdown procedures to control the overall familywise error rate under the complete null while increasing the sensitivity of the procedure. Another popular choice of error rate control is that of the false discovery rate (FDR) [4,25].

70.6 Regression Techniques

Clustering and classification techniques are primarily data exploratory tools. However, modeling and statistical inference for the interrelationship between expression levels among various genes is still largely unknown. Kato et al. [8] used regression techniques to infer about gene regulation networks. Principal-component regression or partial least-squares regression appears to be attractive for application to microarray datasets where the simultaneous expression levels of many genes are collected each at a few timepoints (or individuals). Microarray data on sporulation of budding yeast (*Saccharomyces cerevisiae*) have been analyzed by partial least-squares regression in Datta [6]. Moreover, the Cox proportional hazards model and partial least squares have been used to predict patient survival using gene expression data [21]. Bar-Joseph et al. [2] had the interesting idea of modeling each gene's expression profile by a spline curve in which the gene-specific spline coefficients are taken to be random effects within a cluster of "similar" genes (say, those grouped together by a clustering algorithm).

70.7 Bootstrap

Bootstrap and permutation techniques are being used in the computation of sampling distribution and P-value adjustments plus validation of clustering algorithms. Normal-distribution-based parametric bootstrap [24] and ANOVA-based residual bootstrap [18] have been proposed in the microarray data context.

References

1. Baldi, P. and Long, A. D. (2001). A Bayesian framework for the analysis of microarray expression data: regularized t-test and statistical inferences of gene changes. *Bioinformatics*, **17**, 509–519.

2. Bar-Joseph, J., Gerber, G., Gifford, D. K., Jakkola, T. S., and Simon, I. (2002). A new approach to analyzing gene expression time series data. Proc. 6th Annual Int. *Conf. Research in Computational Molecular Biology* (RECOMB), pp. 39–48. ACM Press, New York.

3. Ben-Dor, A., Bruhn, L., Friedman, N., Nachman, I., Schummer, M., and Yakhini, Z. (2000). Tissue classification with gene expression profiles. *J. Comput. Biol.*, **7**, 559–583.

4. Benjamini, Y. and Hochberg, Y. (1995). Controlling the false discovery rate: A practical and powerful approach to multiple testing. *J. Roy. Statist. Soc. B*, **57**, 289–300.

5. Chen, G., Jaradat, S. A., Banerjee, N., Tanaka, T. S., Ko, M. S. H., and Zhang, M. Q. (2002). Evaluation and comparison of clustering algorithms in analyzing ES cell gene expression data. *Statist. Sin.*, **12**, 241–262.

6. Chu, S., DeRisi, J., Eisen, M., Mulholland, J., Botstein, D., Brown, P. O., and Herskowitz, I. (1998). The transcriptional program of sporulation in budding yeast. *Science*, **282**, 699–705.

7. Datta, S. (2001). Exploring relationships in gene expressions: a partial least squares approach. *Gene Express.*, **9**, 249–255.

8. Datta, S. (2003). Statistical techniques for microarray data: A partial overview. *Commun. Statist. Theory Methods*, **32**, 263–280.

9. Datta, S. and Datta, S. (2003). Comparisons and validation of statistical clustering techniques for microarray gene expression data. *Bioinformatics*, **19**, 459–466.

10. Datta, S., Satten, G. A., Benos, D. J., Xia, J., Heslin, M., and Datta, S. (2004). An empirical Bayes adjustment to increase the sensitivity of detecting differentially expressed genes in microarray experiments. *Bioinformatics*, **20**, 235–242.

11. DeRisi, J. L., Vishwanath, R. I., and Brown, P. O. (1997). Exploring the metabolic and genetic control of gene expression on a genomic scale. *Science*, **278**, 680–686.

12. Dudoit, S., Fridlyand, J., and Speed, T. P. (2002). Comparison of discrimination methods for the classification of tumors using gene expression data. *J. Am. Statist. Assoc.*, **97**, 77–87.

13. Dudoit, S. and Yang, Y. H. (2003). R packages for the analysis of cDNA microarray data. In *The Analysis of Gene Expression Data: Methods and Software*, G. Parmigiani, E. S. Garrett, R. A. Irizarry, and S. L. Zeger, eds. pp. 78–101, Springer, New York.

14. Dudoit, S., Yang, Y. H., Speed, T. P. and Callow, M. J. (2002b). Statistical methods for identifying differentially expressed genes in replicated cDNA microarray experiments. *Statist. Sin.*, **12**, 111–139.

15. Efron, B., Tibshirani, R., Storey, J. D., and Tusher, V. (2001). Empirical Bayes analysis of a microarray experiment. *J. Am. Statist. Assoc.*, **96**, 1151–1160.

16. Eisen, M., Spellman, P. T., Botstein, D., and Brown, P. O. (1998). Cluster analysis and display of genome-wide expression patterns. *Proc. Natl. Acad. Sci. (USA)*, **95**, 14863–14867.

17. Irizarry, R. A., Hobbs, B., Collin, F., Beazer-Barclay, Y. D., Antonellis, K. J., Scherf, U., and Speed, T. P. (2003). Exploration, normalization and summaries of high density oligonucleotide array probe level data. *Biostatistics*, **4**, 249–264.

18. Kato, M., Tsunoda, T., and Takagi, T. (2000). Inferring genetic networks from DNA microarray data by multiple regression analysis. *Genome Inform. Ser Workshop Genome Inform.*, **11**, 118–128.

19. Kerr, M. K. and Churchill, G. (2001). Bootstrapping cluster analysis: assessing the reliability of conclusions from microarray experiments. *Proc. Natl. Acad. Sci. (USA)*, **98**, 8961–8965.

20. Kerr, M. K., Martin, M., and Churchill, G. A. (2000). Analysis of variance for gene expression microarray data. *J. Comput. Biol.*, **7**, 819–837.

21. Newton, M. A., Kendziorski, C. M., Richmond, C. S., Blattner, F. R., and Tsui, K. W. (2001). On differential variability of expression ratios: Improving statistical inference about gene expression changes from microarray data. *J. Comput. Biol.*, **8**, 37–52.

22. Nguyen, D. V. and Rocke, D. M. (2002). Partial least squares proportional hazard regression for application to DNA microarray survival data. *Bioinformatics*, **18**, 1625–1632.

23. Schena, M., Shalon, D., Heller, R., Chai, A., Brown, P. O., and Davis, R. W. (1996). Parallel human genome analysis: Microarray-based expression monitoring of 1000 genes. *Proc. Natl. Acad. Sci. (USA)*, **93**, 10614–10619.

24. Spellman, P. T., Sherlock, G., Zhang, M. Q., Iyer, V. R., Anders, K., Eisen, M. B., Brown, P. O., Botstein, D., and Futcher, B. (1998). Comprehensive identification of cell cycle-regulated genes of the yeast *Saccharomyces cerevisiae* by microarray hybridization. *Molec. Biol. Cell.*, **12**, 3273–3297.

25. Storey, J. D. (2002) A direct approach to false discovery rates. *J. Roy. Statist. Soc. B.*, **64**, 479–498.

26. van der Laan, M. J. and Bryan, J. (2001). Gene expression analysis with the parametric bootstrap. *Biostatistics*, **2**, 445–461

27. Westfall, P. H. and Young, S. S. (1993). *Resampling Based Multiple Testing: Examples and Methods for p-value Adjustment*. Wiley, New York.

28. Yang, Y. H., Dudoit, S., Luu, P., Lin, D. M., Peng, V., Ngai, J., and Speed, T. P. (2002). Normalization for cDNA microarray data: A robust composite method addressing single and multiple slide systematic variation. *Nucleic Acids Res.*, **30**(4), e15.

29. Yang, Y. H., Dudoit, S., Luu, P., and Speed, T. P. (2001). Normalization for cDNA Microarray Data. In *Microarrays: Optical Technologies and Informatics, Proceedings of SPIE* 4266, M. L. Bittner, Y. Chen, A. N. Drosel, and E. R. Dougherty, eds., pp. 141–152. SPIE, - Bellingham, WA.

30. Yeung, K., Haynor, D. R., and Ruzzo, W. L. (2001). Validating clustering for gene expression data. *Bioinformatics*, **17**, 309–318.

71 Statistical Genetics

Susmita Datta

Genetics is an increasingly important subdiscipline of biology, where laws of statistics and probability play an essential role. *Heredity* and *variation* are two major components of *genetics*. The term *heredity* is a phenomenon by which characters are transmitted from one generation to the next by the reproductive process. Even though the parents contribute the same hereditary material to their offspring, it is not necessary that all of the offspring produced be identical. The variability that is produced by heredity is called *genetic variation*. Genetic differences among individuals can be studied at two levels: individual and population. The science of genetics that is studied at an *individual* level is often called *Mendelian genetics*, named after Gregor J. Mendel. It is called *population genetics* when it is done at a population level. If quantitative characters are involved, then it is termed *quantitative* or *biometrical genetics*. Population genetics includes the field of quantitative genetics. Genetic attributes at an individual or population level can be studied only in terms of statistical parameters to summarize population characteristics, and hence the science of population genetics is mostly statistical in nature. The study of theoretical population genetics is predominantly statistical and mathematical in nature and is called *statistical genetics*. R. A. Fisher, J. B. S. Haldane, and S. Wright laid the foundation for statistical genetics. J. F. Crow, Motoo Kimura, Masatoshi Nei, and others have advanced the field further.

We will first introduce the genetic model introduced by Mendel. Certain observable traits or *phenotypes* are determined by *genes* inherited from parents. Each gene can take a number of different forms called *alleles*, say, $A_1, A_2, A_3, \ldots$. Usually, in higher organisms, an individual has two genes for each trait, and these genes are defined as his/her *genotypes*. When two genes are identical, for example, $A_1 A_1$ or $A_2 A_2$, then the genotype is called *homozygous*, otherwise it is called *heterozygous*, for example, $A_1 A_2$. If the phenotype of $A_1 A_2$ is the same as that of $A_1 A_1$, then A_1 is called the *dominant allele* and A_2 is the *recessive* one. If the phenotype associated with $A_1 A_2$ is different from that of either $A_1 A_1$ or $A_2 A_2$, then A_1, A_2 are called *codominant*. During reproduction, an individual receives one of the two genes from the mother with equal probability and similarly from the father. This is known as *Mendel's first law (or law of segregation)*. Mendel also proposed that segregation of the genes for one trait is not affected by segregation of genes of other traits (*laws of independent assortment*). As geneticists have later discovered, these laws do not always hold.

Departures from the first law are studied under segregation analysis. Suppose that a disease is caused by a mutant, disease-causing allele D at an autosomal locus and the normal (wild-type) allele is denoted by d. Autosomal dominant diseases are usually rare, and we will assume that frequency of allele D is low. Most affected individuals are therefore expected to have genotype Dd rather than DD. Thus, most matings between an affected and an unaffected individual will be of type $Dd \times dd$. According to Mendel's first law, the offspring of this mating type have a probability $\frac{1}{2}$ of having the disease, which should therefore be the probability of disease among the offspring of affected individuals. Now suppose that in a random sample of mating of affected and unaffected individuals it is seen that among n offspring r of them are affected by the disease. We wish to test the hypothesis of autosomal dominant disease, that is, checking whether the segregation ratio is $\frac{1}{2}$. According to the binomial distribution, suppose that there are n trials and p is the success probability of being affected and X is the random variable representing the number of affected individuals. Then

$$P(X = x) = \binom{n}{x} p^x (1-p)^{n-x}.$$

The problem therefore reduces to testing whether the success probability p equals $\frac{1}{2}$. A large sample test for this can be constructed using a normal approximation to binomial with

$$\mu = np,$$
$$\sigma^2 = np(1-p).$$

The test statistic is

$$Z = \frac{X - np}{(np(1-p))^{1/2}}$$

or in terms of a chi-square distribution

$$Z^2 = \frac{(X - np)^2}{np(1-p)},$$

which is of the form

$$Z^2 = \sum \frac{(O - E)^2}{E},$$

where O, E are the observed and the expected counts, respectively. One can also form a likelihood ratio test where the likelihood is

$$L(p) = \binom{n}{r} p^r (1-p)^{n-r}.$$

The test statistic is $2(\ln L_1 - \ln L_0)$, where $\ln L_0$ is the log-likelihood under the null hypothesis of $p = \frac{1}{2}$. This test statistic has a chi-square distribution.

Estimation of the segregation ratio p based on a random sample reduces to that of estimation of a binomial success probability. Segregation ratios in family studies can be estimated by maximum likelihood estimation under truncated binomial distribution with complete ascertainment [6]. Numerical methods like the EM (expectation–maximization) algorithm and Newton–Raphson and Fisher scoring methods, can be used for finding the maximum-likelihood estimate for the segregation ratio. Under incomplete ascertainment, a modified maximum-likelihood method can be used.

Segregation ratios in simple specific mating types can be explained by Mendelian segregation. However, if a population consists of offspring of mixtures of different mating types, then mating type frequencies arise from random mating and the ratios of the different genotypes follow a mathematical rule by the English mathematician Hardy [8] and German physician Weinberg [12]. It says that the relative frequencies of the genotypes will remain in equilibrium following subsequent generations under the random mating. For example, let there be alleles $A_1, A_2, \ldots, A_m$ with frequencies $p_1, p_2, \ldots, p_m$. Under the Hardy–Weinberg equilibrium, the frequency of a homozygous genotype $A_i A_i$ in the next generation is p_i^2, and that of the heterozygous genotype such as $A_i A_j (i \neq j)$ is $2 p_i p_j$.

There are several reasons for deviations from the Hardy–Weinberg equilibrium. One of the reasons could be misspecified genetic basis of the trait. The other reasons may be nonrandom mating, for example, only opposite sex-pairs can produce offspring, and force of selection, for example, some of the genotypes are preferentially selected in the sampling process.

Another area of genetics where statistical methodologies are widely used is estimation and testing of genetic linkage values. Laws of independent assortment by Mendel works only for loci of traits allocated on separate chromosomes. On the other hand, alleles on the same chromosome can exhibit a property called *linkage*. For example, when two loci A, B are considered, a gamete may be of the same parental type $A_f B_f$ or $A_m B_m$, where f,m stand for father and mother, respectively. There could also be gametes $A_f B_m$ or $A_m B_f$. Such gametes are called nonparental recombinants. A biological process called *crossover* in meiosis is responsible for recombination. The recombination fraction between the two loci is defined as the probability that the gamete is a recombinant and it is denoted by θ. For the two loci being on different chromosomes, it is conceivable that the recombinant and nonrecombinant gametes are equally likely to occur and $\theta = \frac{1}{2}$. For the loci on the same chromosomes, the closer the two loci, the smaller the chance that there will be crossover between them during meiosis, and there will be lesser chance of recombination than nonrecombination and hence $\theta < \frac{1}{2}$. Two loci with $\theta < \frac{1}{2}$ are said to be "in linkage". Two loci are more tightly linked for smaller values of θ. The genetic map distance between the two loci is defined as the expected number of crossovers between them on a single chromatid during meiosis. The mathematical relationship between a genetic map distance m and recombination fraction θ is called the *map function*.

In linkage analysis, the parameter of primary interest is the recombination fraction or a set of recombination fractions for multilocus analysis. Other parameters such as allele frequencies, penetrance, and so on are treated as nuisance parameters. These parameters are not jointly estimated with the recombination fractions; they have to be specified on the basis of prior knowledge. Misspecification of these parameters may affect the robustness of statistical tests for detecting linkage. Likelihood-based linkage analysis for pedigree data can be handled by the Elston–Stewart algorithm [4]. The Lander–Green algorithm [9] can be used for large numbers of loci. The nonparametric approach is more appropriate for linkage analysis for complex diseases like heart disease, depression, and diabetes. For nonparametric methods of linkage analysis, allele sharing is an important concept. There are two different types of allele-sharing: identity-by-state (IBS) or identity-by descent (IBD). Two alleles of the same form (same DNA sequence) are called IBS; additionally, if they have descended from the same ancestral allele then they are said to be IBD. A class of nonparametric tests for linkage can be constructed on the basis of IBD configuration and IBD sharing functions of the affected members in pedigrees for a complex disease [14]. In parametric linkage analysis, it is customary to summarize the results in terms of *lod* scores instead of p values. The *lod* score at θ is defined as the logarithm (base 10) of the ratio of the likelihood value at θ to the likelihood value at $\theta = \frac{1}{2}$. A commonly used measure of linkage of a pedigree data is the expected *lod* score at the true value

θ. The enormous success of linkage analysis can be attributed to powerful statistical tools.

Let us illustrate the concept of linkage in a random sample of three-generation pedigrees. Suppose that there are N fully informative gametes and R recombinants. The maximum likelihood estimate of the recombination fraction is

$$\widehat{\theta} = \min\left(\frac{R}{N}, \frac{1}{2}\right).$$

Under the null hypothesis of no linkage, the expected number of recombinants and nonrecombinants are both $N/2$. So a chi-squared test statistic is formed via

$$T = \frac{(R - N/2)^2}{N/2} + \frac{(N - R - N/2)^2}{N/2}$$
$$= \frac{(N - 2R)^2}{N}.$$

This provides a one-tailed test for linkage, with one degree of freedom. There could be an alternative likelihood ratio test for the linkage. The multilocus situation for linkage analysis also can be done. The purpose of multilocus linkage analysis is map building of a chromosomal region consisting of multiple marker loci. Another purpose is the detection of a disease locus using established maps of marker loci.

One more important area of statistical genetics is the study of allelic associations. Association analysis between diseases and different polymorphisms has existed for a long time, and with the advent of the human genome project, it will play an important role in medical sciences. Consider two loci A, B with alleles $A_1, A_2, \ldots, A_m$ and $B_1, B_2, \ldots, B_n$ with population frequencies $p_1, p_2, \ldots, p_m$ and $q_1, q_2, \ldots, q_n$, respectively. The joint probability distribution of these loci is characterized by the mn possible haplotypes, for example, $A_1 B_1, A_1 B_2, \ldots, A_m B_n$ and their relative frequencies $h_{11}, h_{12}, \ldots, h_{mn}$, say. Fix a pair of indices i and j; $1 \leq i \leq m, 1 \leq j \leq n$. If there is no association between the two loci, then the cell frequency of haplotype $A_i B_j$ can be obtained from the marginal frequencies

$$h_{ij} = p_i q_j.$$

If this equality does not hold, then the two loci are not independent of each other and the difference $h_{ij} - p_i q_j$ is a measure of the allelic association, commonly called a *linkage disequilibrium*. Other names for this are *gametic phase disequilibrium* or *allelic disequilibrium*. If one locus is nuclear and the other is mitochondrial, then it is also called a *cytonuclear disequilibrium*. Linkage disequilibrium or cytonuclear disequilibrium can be generated and maintained by population forces like random genetic drift, mutation, selection, population admixture, and so on. Deterministic mathematical models are difficult to explain the dynamics of these disequilibria. Stochastic processes (e.g., Markov chain) have been used to model these; see References 1–3 and 13 . Association analysis can be used as a tool of fine mapping of disease loci as a follow-up of linkage analysis [2].

Linkage analysis of loci that are responsible for quantitative traits (QTL) is a harder statistical problem. Detection of a true QTL based on statistical association is often difficult. Often positive results from association studies might be due to linkage disequilibrium.

Complex segregation analysis is yet another area of statistical genetics that uses sophisticated statistical tests to detect major locus effects of complex traits. These tests involve a population genetic model of the trait involving allelic frequencies, penetrance, and other parameters. Usually, multigenerational pedigree data are needed for these studies and ascertainment corrections are needed. In spite of its complexity, complex segregation analysis seldom has good power for the detection of dichotomous traits. However, it has better power for detecting quantitative and polychotomous traits.

Statistical methods have been used in the construction of phylogenetic trees. Nei [10] provides a detailed description of constructing phylogenies. These trees depict the evolutionary pathways of related species. A mother node of multiple branches corresponding to different species represents a common ancestor. Often a genetic distance between two species is calculated using the observed number of molecular bases in their genetic sequences. A standard hierarchical clustering (e.g., UPGMA) is then run to cluster species and construct a dendrogram (phylogenetic tree) using this distance. Modification of this scheme includes the algorithm by Fitch and Margoliash [7] that incorporates the concept of operational taxonomic units. Likelihood methods of tree construction also exist [5].

Other types of statistical genetic data analysis fall under the related area of statistical genomics. Examples include sequence analysis, microarray data analysis, and proteomics.

References

1. Basten, C. J. and Asmussen, M. A. (1997). The exact test for cytonuclear disequilibria. *Genetics*, **146**, 1165–1171.
2. Bodmer, W. F. (1986). Human genetics: The molecular challenge. *Cold Spring Harb. Symp. Quant. Biol.*, **L1**, 1–13.
3. Datta, S., Fu, Y. X., and Arnold, J. (1996). Dynamics and equilibrium behavior of cytonuclear disequilibria under genetic drift, mutation, and migration. *Theor. Popul. Biol.*, **50**, 298–324.
4. Elston, R. C. and Stewart, J. (1971). A general model for the analysis of pedigree data. *Hum. Hered.*, **21**, 523–542.
5. Felsenstein, J. (1981). Evolutionary trees from DNA sequences: A maximum likelihood approach. *J. Molec. Evol.*, **17**, 368–376.
6. Fisher, R. A. (1934). The effect of methods of ascertainment upon the estimation of frequencies. *Ann. Eugen.*, **6**, 13–25.
7. Fitch, W. M. and Margoliash, M. (1967). Construction of phylogenic trees. *Science*, **155**, 279–284.
8. Hardy, G. H. (1908). Mendelian proportions in a mixed population. *Science*, **28**, 49–50.
9. Lander, E. S. and Green, P. (1987). Construction of multilocus genetic maps in humans. *Proc. Natl. Acad. Sci. (USA)*, **24**, 2363–2367.
10. Nei, M. (1987). *Molecular Evolutionary Genetics*. Columbia University Press, New York.
11. Sham, P. (1998). *Statistics in Human Genetics*. Arnold, London.
12. Weinberg, W. (1908). Uber den nachweis der vererbung beim menschen. *Jahreshefte Vereins vaterlandische Naturkunde Wuttemberg*, **64**, 368–382.
13. Weir, B. S. (1990). *Genetic Data Analysis*. Sinauer, Sunderland, UK.
14. Whittemore, A. S. and Halpern, J. (1994). A class of tests for linkage using affected pedigree members. *Biometrics*, **50**, 109–117.

72 Statistical Methods in Bioassay

Robert K. Tsutakawa

72.1 Introduction

Bioassay refers to the process of evaluating the potency of a stimulus by analyzing the responses it produces in biological organisms. Examples of a stimulus are a drug, a hormone, radiation, an environmental effect, and various forms of toxicants. Examples of biological organisms are experimental animals, human volunteers, living tissues, and bacteria.

When a new drug is introduced, we are often interested in how it compares with a standard drug. One means of approaching this problem is to use the two drugs on living organisms from a common population. On the other hand, when an insecticide is being considered for use on an unfamiliar insect population, we may be more interested in the tolerance of the insect population to the insecticide than in the relative strength of the insecticide compared to another. These are examples of two basic problems in bioassay. One is the evaluation of the relative potency of an unknown drug to a standard. The other is the estimation of the stimulus response.

The response to a stimulus may often be classified as quantitative or quantal. If the response can be measured, as would be the case if we were studying weight changes following the use of a vitamin, it is quantitative. On the other hand, if the response is all or nothing, as would be the case of death or survival, it is quantal.

The development of statistical methods in bioassay has paralleled the development of statistics in general. Not only are the familiar techniques of regression analysis and maximum likelihood extensively used in bioassay, but a number of statistical problems have originated in bioassay and have stimulated the general development of statistics. The companion books by Finney [14,16], which are now in their third edition, have contributed greatly toward unifying the field and introducing many biomedical researchers, as well as seasoned statisticians, to current practices in bioassay. Reference 14 deals primarily with the probit model for quantal responses, and Reference 16, with bioassay more generally, including more recent developments in the field.

One of the earliest applications of the normal distribution and normal deviate to quantal responses was Fechner's [12] psychophysical experiment on human sensitivity to various physical stimuli. Probit analysis, which is based on normal deviates, was widely used for estimating quantal response curves in the late 1920s and early 1930s and put into its present form by R. A. Fisher [17]. During this period the need for standardizing various drugs, hormones, and toxicants brought about statistical methods for estimating relative potencies [22]. The multitude of competing methods for estimating parameters of quantal response curves were unified under the theory of RBAN estimation by Neyman [25]. A number of sequential designs for quantal responses were proposed in the 1950s and 1960s [9] and a limited number of Bayesian methods were introduced in the 1970s [18]. The extensive use of animals for measuring concentrations of hormones and enzymes have been recently replaced by radioimmunoassay and related procedures known for their high precision [27].

72.2 Estimation of Relative Potency from Quantitative Responses

Consider a test drug T being compared to a standard drug S. T and S are said to be similar if there is a constant $\rho > 0$ such that the distribution of the response to dose z_2 of the test drug is equivalent to that of dose $z_1 = \rho z_2$ of the standard drug. When the drugs are similar, ρ is called the *relative potency* of T to S.

In many dose–response experiments the quantitative response Y is linearly related to the log dose, $x = \log z$, and we can assume the model

$$E[Y|x] = \alpha + \beta x, \qquad (1)$$

where α and β are unknown real-valued parameters. If S and T are similar, then the responses Y_1 and Y_2 to S and T, respectively, are related by

$$E[Y_1|x] = \alpha + \beta x$$
$$E[Y_2|x] = \alpha + \beta \log \rho + \beta x \qquad (2)$$

for some α, β, and ρ. This follows from the assumption that dose z of T is equivalent to dose ρz of S. The estimation of ρ based on this model is called a *parallel-line assay*, in contrast to a *slope-ratio assay* [16], which is based on a model where the regression is linear with respect to $x = z^\lambda$, for some $\lambda \neq 0$.

Given the intercepts $\alpha_1 = \alpha$ and $\alpha_2 = \alpha + \beta \log \rho$ and the common slope β, the horizontal distance between the regression lines (2) is $\log \rho = (\alpha_2 - \alpha_1)/\beta$. Thus if one has estimates a_1, a_2, and b of α_1, α_2, and β, he or she may estimate $\log \rho$ by $M = (a_2 - a_1)/b$ and ρ by $\log^{-1}(M)$.

A typical experiment for a parallel-line assay consists of a series of n_1 observations $(Y_{11}, \ldots, Y_{1n_1})$ at log doses $(x_{11}, \ldots, x_{1n_1})$ of the standard and n_2 observations $(Y_{21}, \ldots, Y_{2n_2})$ at log doses $(x_{21}, \ldots, x_{2n_2})$ of the unknown. In addition to the linearity assumption, suppose that the observations are independent and normally distributed with common variance σ^2. Then α_1, α_2, β and σ^2 may be estimated by the standard least-squares method, and a γ-level confidence interval for $\log \rho$ is given by

$$(\bar{x}_1 - \bar{x}_2) + (1-g)^{-1}$$
$$\times \left\{ M - \bar{x}_1 + \bar{x}_2 \pm \frac{st}{b} \right.$$
$$\times \left[(1-g)\left(\frac{1}{n_1} + \frac{1}{n_2}\right) \right.$$
$$\left.\left. + \frac{(M - \bar{x}_1 + \bar{x}_2)^2}{\sum_{i=1}^{2} \sum_{j=1}^{n_i}(x_{ij} - \bar{x}_i)^2} \right]^{1/2} \right\}, \quad (3)$$

where

$$\bar{x}_i = \sum_{j=1}^{n_i} x_{ij}/n_i,$$

$$s^2 = \sum_{i=1}^{2} \sum_{j=1}^{n_i} \frac{(Y_{ij} - a_i - b_i x_{ij})^2}{(n_1 + n_2 - 3)},$$

$$g = \frac{t^2 s^2}{b^2 \sum_{i=1}^{2} \sum_{j=1}^{n_j}(x_{ij} - \bar{x}_i)^2},$$

and t is the $(1+\gamma)/2$ quantile of the t-distribution with $n_1 + n_2 - 3$ df.

The expression (3) follows from the celebrated Fieller's theorem [13], which has been widely used in deriving confidence intervals for ratios of two parameters whose estimators have a bivariate normal distribution. We note that small values of g indicate a significant departure of β from 0. However, large values of g, (i.e., $g > 1$) indicate that β is not significantly different from 0 and the confidence region is the complement of the interval defined by (3) and is of little practical value.

As an example, consider an estrogen hormone assay illustrated by Brownlee [7] using data derived from Emmons [11]. A total of 33 rats were assigned to three levels of the standard and two levels of the unknown. The response variable Y is some linear function of the logarithm of the weights of the rats' uteri and the independent variable x the logarithm of 10 times the dose. (See Table 1.)

In this example the estimated variance about the regression lines is $s^2 = 153.877$ and the estimated regression coefficients are $a_1 = 67.7144$, $a_2 = 32.8817$, and $b = 50.3199$. Thus the estimated $\log \rho$ is $M = -0.6722$. Moreover, $g = 0.307316$ and the 95% confidence interval for $\log \rho$ based on (3) is $(-0.8756, -0.4372)$. Thus the estimate of ρ is 0.213 and its 95% confidence interval is $(0.133, 0.365)$. For computational details and tests for the validity of assumptions, see Brownlee [7, pp. 352–358].

72.3 Quantal Response Models

When the response to a drug (or other stimulus) is quantal, it is often reasonable and convenient to assume that each member of the population has a tolerance level such that the member responds to any dose greater than this level and does not to any lesser dose. This gives rise to the concept of a continuous tolerance distribution, F, which has the characteristics of a cdf. In particular, the probability of a response from a randomly chosen member at dose x (usually measured in log units) is $F(x)$ and the probability of nonresponse is $(1-F(x))$. In practice, this model is sometimes generalized to include an unknown proportion C of the population consisting of those who will respond in the absence of any stimulus and an unknown proportion D consisting of those that are immune to the stimulus and will not respond to any finite dose. When the responses are deaths and survivals, the dose that is lethal to 50% of the population is called LD50. More generally ED05, ED50, and ED90 denote doses that affect 5%, 50% and 90% of the population.

Relative to a suitable transformation of the dose, such as the logarithm, the tolerance distribution is often approximated by a normal or logistic distribution, i.e.,

$$F(x) = \Phi(\alpha + \beta x) \qquad (4)$$

or

$$F(X) = \psi(\alpha + \beta x), \qquad (5)$$

where

$$\Phi(t) = \int_{-\infty}^{t} \frac{1}{\sqrt{2\pi}} \exp(-w^2/2)\, dw$$
$$\psi(t) = [1 + \exp(-t)]^{-1},$$

$-\infty < t < \infty$, and α and β are unknown parameters $-\infty < \alpha < \infty$, $0 < \beta < \infty$. Given the probability of response P, the probit is defined for the normal (probit model) by the value of y such that $P = \Phi(y-5)$ and the logit for the logistic (logit model) by $y = \ln[P/(1-P)]$. Berkson [4] promoted the use of the logit model because of its similarity to the normal and numerical tractability. In view of the more recent development of computational facilities, the numerical convenience is usually negligible and the choice between the two

Table 1: Uterine weights in coded units.

Standard Dose (mg)			Unknown Dose (mg)	
0.2	0.3	0.4	1.0	2.5
73	77	118	79	101
69	93	85	87	86
71	116	105	71	105
91	78	76	78	111
80	87	101	92	102
110	86		92	107
	101			102
	104			112

Source: Brownlee [7].

is generally not crucial. Prentice [26] has introduced other distributions, including those that are skewed and may be better suited for modeling extreme percent points such as the ED99.

72.4 Estimation of Quantal Response Curves

A typical experiment for quantal responses consists of taking n_i independent observations at log dose x_i and observing the frequency of responses $r_i, i = 1, \ldots, k$, where the number k of dose levels is generally at least two, and more than two when the model is being tested. If we assume that the tolerance distribution $F(x|\theta)$ belongs to a family of distributions with parameter θ, the loglikelihood function is given by

$$\sum_{i=1}^{k} \{r_i \log F(x_i|\theta) + (n_i - r_i) \times \log[1 - F(x_i|\theta)]\}.$$

For most of the commonly used models, including the probit and logit, the maximum-likelihood estimators of θ do not have explicit expressions and estimates must be found numerically by iterative schemes such as the Newton–Raphson. The probit method originally proposed by Fisher [17] is still a widely used method for finding such estimates and has been described in detail by Finney [14].

There are a number of other estimators, such as the minimum chi-square and weighted least squares, which belong to the RBAN family [25] and hence have the same asymptotic efficiency as the maximum likelihood estimator. Their relative merits for small samples have not been studied extensively and the choice among them is usually a matter of convenience.

For the probit model, with x the log dose, log ED50 is estimated by $\hat{\mu} = a/b$, where (a,b) is the maximum likelihood estimate of (α, β). The γ-level confidence interval for log ED50, valid for large samples, is given by

$$\hat{\mu} + \frac{g}{1-g}(\hat{\mu} - \bar{x}) \pm \frac{K}{b(1-g)}$$

$$\times \left[\frac{1-g}{\sum_{i=1}^{k} n_i w_i} + \frac{(\hat{\mu} - \bar{x})^2}{\sum n_i w_i (x_i - \bar{x})^2} \right]^{1/2},$$

where

$$\bar{x} = \sum_{i=1}^{k} n_i w_i x_i \bigg/ \sum_{i=1}^{k} n_i w_i,$$

$$w_i = z_i^2 / P_i Q_i,$$

$$z_i = (2\pi)^{1/2} \exp[-\tfrac{1}{2}(a+bx_i)^2],$$

$$g = \frac{K^2}{b^2 \sum_{i=1} n_i w_i (x_i - \bar{x})^2},$$

and K is the $(1+\gamma)/2$ quantile of the standard normal distribution. As in the case of relative potency, the interval should be used only when $g < 1$.

Among the several nonparametric (distribution-free) estimators of the ED50 that have been used, the Spearman–Kärber estimator [23,28] is not only quite simple to use but is also considered quite efficient [5]. If the levels are ordered such that $x_1 < \cdots < x_k$, this estimator is defined by

$$\tilde{\mu} = \sum_{i=1}^{k-1}(p_{i+1} - p_i)(x_i + x_{i+1})/2,$$

provided that $p_1 = 0$ and $p_k = 1$, where $p_i = r_i/n_i, i = 1,\ldots,k$. If $p_1 > 0$, then an extra level is added below x_1, where no responses are assumed to occur. Similarly, if $p_k < 1$, an extra level is added above x_k, where responses are only assumed to occur. When the levels are equally spaced with interval d, the variance of $\tilde{\mu}$ is estimated by

$$\text{var}(\tilde{\mu}) = d^2 \sum p_i(1-p_i)/(n_i - 1),$$

provided that $n_i \geq 2, i = 1, \ldots, k$.

The maximum likelihood and Spearman–Kärber methods may be illustrated using data from Finney [14] on the toxicity of quinidine to frogs, given in Table 2. The dose was measured in units of 10^{-2} mL per gram of body weight. If the probit model with respect to log dose is assumed, maximum likelihood estimates may be computed using a SAS program [3]. The resulting maximum-likelihood estimates are $a = -1.313$, $b = 4.318$, and log LD50 $= 1.462$. The 95% confidence interval for log LD50 is $(1.398, 1.532)$. On the other hand, the Spearman–Kärber estimates are $\tilde{\mu} = 1.449$ and $\text{var}(\tilde{\mu}) = 9.334 \times 10^{-4}$. (The estimate here uses an additional level at log dose $= 1.879$, where $p_i = 1$ is assumed.)

Table 2: Toxicity of quinidine to frogs.

Log Dose	n	r	p
1.0000	26	0	0.0000
1.1761	24	2	0.0833
1.3522	23	9	0.3913
1.5276	24	17	0.7083
1.7033	26	20	0.7692

Source: Finney [14].

There are a few nonparametric procedures for estimating the entire tolerance distribution F. One approach is to use the isotone regression method for estimating ordered binomial parameters [2]. Under this method the estimates of F at the points $x_1 < \cdots < x_k$ are given by

$$\hat{F}(x_i) = \min_{i \leq v \leq k} \max_{l \leq u \leq i} \left(\sum_{\nu=u}^{v} r_\nu \bigg/ \sum_{\nu=u}^{v} n_\nu \right),$$

$i = 1, \ldots, k$. Between these points and outside the interval $[x_1, x_k]$ the estimate $\hat{F}$ may be defined arbitrarily, subject to the constraint that $\hat{F}$ be nondecreasing.

Another parametric approach, which is Bayesian, assumes that F is a random distribution function, whose distribution is defined by a Dirichlet process, and uses the Bayes estimate of F with respect to a suitable loss function. Antoniak [1] gives a theoretical discussion of the Dirichlet prior and resulting posterior, as well as references to related works.

72.5 Estimation of Relative Potency from Quantal Responses

When quantal responses satisfy the probit or logit model with respect to log dose and the drugs are similar, a parallel-line assay may be performed to estimate the relative potency of an unknown to a standard. Under the probit model, for example, the condition of similarity requires that the probabilities of response at log dose x must satisfy

$$F_1(x) = \Phi(\alpha_1 + \beta x),$$
$$F_2(x) = \Phi(\alpha_2 + \beta x),$$

for the standard and unknown, respectively, for some parameters α_1, α_2, and $\beta > 0$. In this case, $\alpha_2 = \alpha_1 + \beta \log \rho$ or $\log \rho = (\alpha_2 - \alpha_1)/\beta$.

If we have two independent series of independent quantal response observations, one for the standard and the other for the unknown, the joint likelihood function may be formed and the maximum-likelihood estimates a_1, a_2, and b of α_1, α_2, and β may be computed by the Newton–Raphson or similar method. These estimates may be used to estimate $\log \rho$ by $M = (a_2 - a_1)/b$. Use of Fieller's theorem gives an approximate confidence interval for $\log \rho$ and ρ. Computational details and numerical illustrations are given in Finney [16].

72.6 Design of the Experiment

The design of the experiment for bioassay involves the selection of dose levels and the allocation of the living organisms to these levels in order to obtain experimental results that can be used to answer some predetermined question.

In estimating the relative potency, the objective may be to select the design that minimizes the width of the confidence interval for a given confidence level and total sample size. If a parallel-line assay is being considered, it is important to have a rough estimate of ρ so that doses chosen for the unknown will give results comparable to those of the standard. One recommended approach is to use the same number of dose levels, usually between 3 and 6, for both the unknown and standard and the same number of organisms at each level. For each level of the standard, the level of the unknown should be chosen so that the predicted outcome will be like that of the standard. Although there are certain advantages to using dose levels that cover a wide range for qualitative responses, this does not hold for quantal responses since responses at extreme levels are quite predictable (being all responses or all nonresponses) and little information is gained. (See Finney [16] for more detailed instructions.)

In estimating the ED50 from quantal responses, again the objective may be to minimize the width of the confidence interval. For both the probit and Spearman–Kärber methods, it is generally recommended that an equal number of observations be taken at three to six equally spaced levels that are between 0.5 and 2 standard deviations (of F) apart. The specific recommendations vary according to the number of levels to be used, total sample size available, and previous information. (See Finney [14] for the probit method and Brown [6] for the Spearman–Kärber.)

A number of sequential designs have been proposed as a means of allocating N experimental units to different levels as information becomes available. The Robbins–Monro process [9] and up-and-down method [10], together with their variations, are the most intensively studied sequential procedures for estimating the ED50. For the up-and-down method, one observation at a time is taken, starting at some initial level x_0 and successively at levels $x_{i+1} = x_i \pm d$, where d is the step size chosen to be close to the standard deviation of F, a $+$ used if the ith observation is a nonresponse and a $-$ if it is a response. Under this scheme, the experimental dose levels tend to fluctuate about the ED50 and the average, $\Sigma_{i=2}^{N+1} x_i/N$, is used to estimate the ED50. For the probit model this estimator approximates the maximum-likelihood estimator and has better efficiency than do the fixed sample designs [8,10]. The up-and-down method, when modified by using several observations at a time, not only reduces the number of trials but has been shown to provide even greater efficiency [29].

The selection of dose levels for estimating ED50 and extreme percent points, such as the ED90, of F may also be based on Bayesian principles. Freeman [18] has proposed a sequential method for estimating the ED50 of a one-parameter logit model, where the slope β is assumed known, and Tsutakawa [30] has proposed nonsequential designs for logit models, with unknown slope. These methods depend on the explicit and formal use of a prior distribution of the parameters and are aimed at minimizing the posterior variance of the percent point of interest.

72.7 Related Areas

Some of the techniques that have been described above have been modified and extended to more complex problems. We will briefly describe some of these where further work can be expected.

The analysis of quantal responses to combinations or mixtures of two or more drugs introduces many additional problems, since we must consider not only the relative proportion of the drugs but also the interaction of the drugs. When there are two drugs and one is effectively a dilution of the other, or they act independently, it is not difficult to extend the models for single drugs to those for a mixture. However, when the drugs interact, either antagonistically or synergistically, the model building becomes considerably more complex. Hewlett and Plackett [21] have discussed different modes in which the effect of the mixture of drugs may depend on the amount of drugs reaching the site of the action.

In many quantal response studies the time of response is an important variable. For example, in a carcinogenic experiment animals are often exposed to different doses of a carcinogen and kept under observation until they develop a tumor, they die, or the experiment is terminated. For a discussion of such experiments and related references, see Hartley and Sielken [20].

Statistical techniques for bioassay have been used in radioimmunoassay and related techniques for measuring minute concentrations of hormones and enzymes. Radioactivity counts, resulting from antigen–antibody reactions, are observed at different doses of a ligand in order to estimate a dose–response curve. Under appropriate transformations, the counts are often related to the dose by a logistic model with unknown asymptotes (generally different from 0 and 1). For transformations, weighted least-squares methods, and references, see Rodbard and Hutt [27] and Finney [15].

The probit and logit models have also been applied to mental testing and latent trait analysis, where human subjects with different mental abilities respond to questions of different degrees of difficulty [24]. When the responses are classified correct or incorrect, the probability of a correct answer to a particular question usually depends on the ability of the subject and can often be approximated by one of these models. In such cases, the ability of each subject can be estimated by using the joint response to several questions. See Hambleton and Cook [19] for a review of this area and related references.

References

1. Antoniak, C. E. (1974). *Ann. Statist.*, **2**, 1152–1174.
2. Ayer, M., Brunk, H. D., Ewing, G. M., Reid, W. T., and Silverman, E. (1955). *Ann. Math. Statist.*, **26**, 641–647.
3. Barr, A. J., Goodnight, J. H., Sall, J. P., and Helwig, J. T. (1976). *A User's Guide to SAS76*. SAS Institute, Raleigh, NC.
4. Berkson, J. (1944). *J. Am. Statist. Assoc.*, **39**, 357–365.
5. Brown, B. W., Jr. (1961). *Biometrika*, **48**, 293–302.
6. Brown, B. W., Jr. (1966). *Biometrics*, **22**, 322–329.
7. Brownlee, K. A. (1965). *Statistical Theory and Methodology in Science and Engineering*, 2nd ed. Wiley, New York.
8. Brownlee, K. A., Hodges, J. L., Jr., and Rosenblatt, M. (1953). *J. Am. Statist. Assoc.*, **48**, 262–277.
9. Cochran, W. G. and Davis, M. (1965). *J. Roy. Statist. Soc. B*, **27**, 28–44.
10. Dixon, W. J. and Mood, A. M. (1948). *J Am. Statist. Assoc.*, **43**, 109–126.
11. Emmons, C. W. (1948). *Principles of Biological Assay*. Chapman & Hall, London.
12. Fechner, G. T. (1860). *Elemente der Psychophsysik*, Breitkopf und Hartel, Leipzig. (Translated into English in 1966 by H. E. Adler, Holt, Rinehart and Winston, New York.)
13. Fieller, E. C. (1940). *J. Roy. Statist. Soc. Suppl.*, **7**, 1–64.
14. Finney, D. J. (1971). *Probit Analysis*, 3rd ed. Cambridge University Press, Cambridge, UK.
15. Finney, D. J. (1976). *Biometrics*, **32**, 721–740.
16. Finney, D. J. (1978). *Statistical Methods in Biological Assay*, 3rd ed. Macmillan, New York.
17. Fisher, R. A. (1935). *Ann. Appl. Biol.*, **22**, 134–167.
18. Freeman, P. R. (1970). *Biometrika*, **57**, 79–89.
19. Hambleton, R. K. and Cook, L. L. (1977). *J. Educ. Meas.*, **14**, 75–96.
20. Hartley, H. O. and Sielken, R. L., Jr. (1977). *Biometrics*, **33**, 1–30.
21. Hewlett, P. S. and Plackett, R. L. (1964). *Biometrics*, **20**, 566–575.
22. Irwin, J. O. (1937). *J. Roy. Statist. Soc. Suppl.*, **4**, 1–48.
23. Kärber, G. (1931). *Arch. Exper. Pathol. Pharmakol.*, **162**, 480–487.
24. Lord, F. M. and Novick, M. R. (1968). *Statistical Theories of Mental Test Scores*. Addison-Wesley, Reading, MA.
25. Neyman, J. (1949). *Proc. Berkeley Symp. Mathematical Statistics and Probability*, pp. 239–273, University of California Press, Berkeley.
26. Prentice, R. L. (1976). *Biometrics*, **32**, 761–768.
27. Rodbard, D. and Hutt, D. M. (1974). *Radioimmunoassay and Related Procedures in Medicine*, Vol. 1, pp. 165–192, International Atomic Energy Agency, Vienna.
28. Spearman, C. (1908). *Br. J. Psychol.*, **2**, 227–242.
29. Tsutakawa, R. K. (1967). *J. Am. Statist. Assoc.*, **62**, 842–856.
30. Tsutakawa, R. K. (1980). *Appl. Statist.*, **29**, 25–33.

73 Statistical Modeling of Human Fecundity

Rajeshwari Sundaram, Germaine M. Buck Louis, and Sungduk Kim

73.1 Introduction

Human *fecundity* describes the capability of producing offspring. The quantitative form of fecundity is referred to as fecundability and was first introduced by Gini [20]. *Fecundability*, the probability of conception per menstrual cycle for a sexually active couple not using birth control, is of great interest both to reproductive scientists and couples trying to conceive or avoid pregnancy. Demographers are interested in estimating the fecundability distribution in a population and in comparing across populations. Reproductive epidemiologists, in contrast, are interested in models for human fecundity and fertility to track differences across time and place and to identify preventable causes of reproductive dysfunction. Moreover, interest also lies in the relationship of fecundity to fertility (as measured by live births, plurality of births of secondary sex ratios). Also, the interest stems from suspicion in the scientific community that environmental factors are negatively affecting human fertility. In fact, recently Buck Louis et al. [7] have found a negative association between antiestrogenic polychlorinated biphenyls found in preconception serum collected from women and their corresponding time to pregnancy. Statistical modeling for fecundability has been evolving as the scientific research in fertility studies advances. In this chapter, we give an overview of the historic development of modeling and give some current research directions.

We first introduce some biological background related to conception. In women of reproductive age, a single ovum is normally released from the ovary in each menstrual cycle, in a process called *ovulation*. For pregnancy to occur, there must be intercourse near the time of ovulation. In addition to the male sperm being capable of fertilizing the ovum, several biological factors in the female must also be favorable. The ovum must be viable and must be transported to the uterus at the right time. The uterine endometrium must be adequately prepared under hormonal influences to support implantation, and the implanted conceptus must be able to survive to the point of detection. The constellation of such factors being favorable to gestation has been termed "cycle viability." Clearly, the timing and frequency of intercourse will have a major impact on the probability of conception. There are many other factors, both endogenous and exogenous that may affect conception. Some of these factors are unobservable, but manifest themselves in variability in the conception rate between cycles with the same intercourse pattern and dependence among outcomes within multiple cycles from the same couple. Other factors, such as age and exposure to a potential toxicant, can be observed directly.

In a typical fertility study, a cohort of couples will be followed up for a fixed time period after their enrollment in the study, usually after the discontinuation of contraception. The pregnancy status for each menstrual cycle is observed. Most such detailed studies include not only information on the occurrence of menstrual periods but also daily records indicating any intercourse, together with a means of identifying the day when, in each menstrual cycle, ovulation is most likely to have occurred. For example, the Early Pregnancy Study (EPS) enrolled couples with no known fertility problems who were ready to begin a pregnancy. The couples enrolled in the study after they stopped contraception. Women collected a first urine specimen each morning and kept a daily diary recording whether they had had unprotected intercourse in the preceding 24 hours and whether they had had any menstrual bleeding during that time. Participation continued through 8 weeks into a pregnancy or for 6 months if no clinical pregnancy was identified. Most women were in their midtwenties to early thirties. Urine specimens were assayed for levels of metabolites of ovarian hormones, and the results were used to identify the day of ovulation. Multiple other prospective studies have been conducted in United States and in European countries trying to ascertain human fecundity. One such study is the New York State Angler Prospective Pregnancy Study (principal investigator: Dr Germaine Buck Louis, NICHD, NIH), where the couples were followed for longer duration than the EPS study, in fact up to 12 at-risk menstrual cycles. This study was in fact a feasibility study for a much larger study being conducted by NICHD, NIH on assessing the effects of chemical exposures on fecundity: longitudinal investigation of the fertility and environment. More details about this study can be found at *http://www.lifestudy.us*.

The outcome variable in these fertility studies is naturally discrete; that is, a success or failure is observed for each menstrual cycle depending on whether a pregnancy is achieved during that cycle. Since different intercourse patterns exist across couples and also across different menstrual cycles within a couple, the resulting pregnancy cannot be directly linked to a particular days intercourse because sperm from several intercourse acts around the time of ovulation could potentially contribute to the success of a pregnancy. For conception cycles in which there are intercourse acts on multiple days, the day that led to conception is inherently unidentifiable; sperm are known to survive at least several days and still be capable of fertilizing a human ovum. For a non-pregnancy cycle, however, it is not clear whether the failure of conception is due to the couple's specific pattern of intercourse, that is, the timing and frequency of intercourse in that cycle, or due to the failure of one of the biological factors from either male or female. Clearly, in a menstrual cycle where pregnancy is observed, both the "timing of intercourse" and "cycle viability" factors have to be right. Consequently, faced with the complex nature of fertility data, greater importance has been placed on the statistical modeling for fecundability that can simultaneously take into account the intercourse patterns in relation to ovulation and the biological factors included to estimate the probabilities of conception associated with specific days. In this chapter, we present an overview of the development of fecundability models. We next present an analysis of a prospective pregnancy study. In Section 73.4, we discuss in detail the biological challenges encountered in addressing fecundity. We conclude with a discussion of future research.

73.2 Statistical Models

The statistical models proposed in the literature can be grouped mainly into two different classes of models: (1) time-to-pregnancy models; and (2) the Barrett–Marshall–Schwartz (BMS) class of models and some extensions thereof.

We first begin by reviewing the first class of models. This class of model focuses on fecundability, ascertained through time to pregnancy. The number of menstrual cycles required to achieve a clinical pregnancy is called *time to pregnancy* (TTP). There is some debate as to the definition of TTP. Some studies include all of the menstrual cycles in this quantity, while others include only "at risk" menstrual cycles (those in which unprotected intercourse occurred during the fertile window). A less common variation is to count the number of months rather than the menstrual cycles because the length of a menstrual cycle can vary greatly even within one woman. For the remainder of this chapter, we will define TTP as the number of menstrual cycles required to achieve a clinical pregnancy, regardless of whether they were at risk. It is well known that there is underlying heterogeneity among couples with respect to TTP. That is, two couples trying to conceive may have very different endpoints, varying anywhere from one menstrual cycle to an indefinite number of cycles in the case of an infertile couple. Some of the risk factors thought to be associated with TTP are age, parity (number of previous live births), and cigarette, drug, and alcohol use. These observed covariates explain some of the heterogeneity of TTP among couples. However, some unexplained heterogeneity remains, due to the limitation of biomarkers for identifying time of ovulation and other unexplained biological factors. This necessitates building better statistical models to ascertain the unmeasured heterogeneity in TTP among couples.

Suppose that there are n couples in a study. For each couple i ($i = 1, \ldots, n$) and each cycle j ($j = 1, \ldots, c$), the outcome of conception Y_{ij} can take one of the two possible values: 1 if conception occurs at cycle j, 0 otherwise. We denote the indicator of intercourse by X_{ijk}, which takes value 1 if intercourse occurs on day k of cycle j by couple i or is 0 otherwise. Also, the time to conception, i.e., the number of menstrual cycles to conceive by couple i, will be denoted by T_i.

73.2.1 Population Averaged Models

An earliest model for fecundability is given by

$$Y_{ij} \sim \text{Bernoulli}(\mu_i),$$
$$\mu_i = \mu, \forall i = 1, \ldots, n.$$

With this model, under the assumption of constant fecundability across cycles during the study period, the conception delays can be modeled by a geometric distribution [20]. Instead of having identical values of μ_i (homogeneous population), the class of models proposed by demographers allowed μ_i to vary within a couple, i.e., the population is heterogenous population. This was achieved by introducing random-effect models, which accounts for both the effect of heterogenous predictors of fecundability and heterogeneity of sexual behavior among couples.

Henry [24] proposed to use a conjugate beta distribution to model the heterogeneity in fecundability, i.e., account for differing values of fecundability among couples. This led to the beta geometric model:

$$\mu_i \sim \beta(\mu, \theta)$$

here μ and θ are respectively the mean and the shape parameter, the latter is the function of variance that accounts for the variability of the unobserved heterogeneity among couples. Various nonparametric choices of the random effects have been proposed starting with the work of Tietze [43], who assumed a three-point distribution of fecundability

$$\mu_i \sim \text{dpd}(m)$$

where dpd() represents the discrete probability distribution and $m = 3$ represents three mass points. Potter [31] subsequently extended this approach to include an arbitrary triangular density of fecundability

$$\mu_i \sim \text{triangular}(\mu, \theta).$$

Sheps [37] and Sheps and Menken [38] extended these investigations generalizing the results to distributions of practically any shape. Heckman and Walker [23] and Heckman et al. [22] presented an extensive review of mixture of geometric distributions in the context of fecundability modeling. Weinberg and Gladen [47] introduced covariates into the beta geometric models, which we will discuss detail below.

73.2.2 Discrete Survival Models

An alternate class of models based on viewing TTP as a discrete survival time have been proposed. Scheike and Jensen [34] and proposed model based on the discrete hazard rate associated with the discrete survival time, TTP. Conditional on their p-dimensional covariate x_{it} that may depend on time, the hazard rate $\lambda_i(t)$ of a couple's waiting time to conceive T_i, is defined as

$$\lambda_i(t) = P(T_i = t | T_i \geq t).$$

Scheike and Jensen [34] have proposed the following model:

$$\lambda_i(t) = 1 - \exp(-\exp(x'_{it}\beta)).$$

Note that if the covariates do not depend on time, the time index can be omitted, i.e., $x_i = x_{it}$. Observe that, in particular if the coefficient $\beta = (\gamma_1, \cdots, \gamma_m, \alpha_1, \cdots, \alpha_m)$, where $m, q > 0$, $x'_{it}\beta = \gamma_t + x'_i \alpha$. Note that this model is the grouped version of the usual continuous-time Cox regression model, where γ_t is the grouped baseline hazard rate. To see this, let U denote a continuous survival time and consider the following m intervals $[a_0, a_1(, \ldots, [a_{m-1}, \infty($ partitioning $[0, \infty($, and define the grouped version of U by $U_d = k+1$ if U is observed in $[a_k, a_{k+1}($. Assuming that U is modelled by a Cox regression hazard rate

$$\lambda(t|x) = \lambda_0(t) \exp(x'\alpha),$$

it then follows that the grouped version U_d has discrete hazard rate

$$\begin{aligned} P(T = t | T \geq t) \\ = 1 - \exp(-\exp(\gamma_t + x'\alpha)), \end{aligned}$$

where $\gamma_t = \log(\exp(\int_0^{a_t} \lambda_0(u)du))$.

Further observe that the model implies

$$\begin{aligned} P(T_i = t) \\ = \lambda_i(t) \prod_{i=1}^{t-1}(1 - \lambda_i(j)) \\ = \lambda_i(t) \exp\left(-\sum_{j=1}^{t-1} \exp(x'_{ij}\beta)\right). \end{aligned}$$

Thus,

$$P(T_i \geq t) = \exp\left(-\sum_{j=1}^{t-1} \exp(x'_{ij}\beta)\right).$$

Fahrmeir and Tutz [17] present a discussion of this model as well as other alternatives. A logistic model has been proposed by Thompson [42]:

$$\text{logit}(\lambda_i(t)) = x'_{it}\beta.$$

This model is similar to that of Cox [11]. The popular beta-geometric model for TTP as proposed by Weinberg and Gladen [47] can be viewed as

$$\log(\lambda_i(t)) = x'_{it}\beta.$$

However, one disadvantage of this model is that when β varies across $(-\infty, \infty)$, $\lambda_i(t|x)$ is not restricted to the range $[0,1]$ of a probability.

The traditional beta-geometric model assumes that each subject (here couple) given an unobserved random rate p_i has a constant hazard rate over time:

$$\lambda_i(t|p_i) = p_i.$$

This implies that the waiting time given p_i follows a geometric distribution:

$$P(T_i = t|p_i)$$
$$= p_i \prod_{j=1}^{t-1}(1-p_i) = p_i(1-p_i)^{t-1}.$$

When the subject rate p_i is sampled from a beta distribution, it follows that the unconditional observed hazard rate is

$$\lambda(t) = 1/(\alpha + \mu(t-1)),$$

where α and μ can be expressed in terms of the parameters of the underlying beta distribution that describes heterogeneity between couples. The beta-geometric distribution can be modified to include covariates by

$$1/\lambda_i(t) = \alpha + \mu(t-1) + x'_i\beta.$$

This extension makes the interpretation of the model valid only for subpopulations based on the covariates that do not change over time. Note that continuous covariates will partition the data so finely that the number of subpopulations may be equal to the number of couples. The above models can all be fitted using maximum-likelihood estimation by the use of standard software available for generalized linear models because the above model implies that a waiting time has the same likelihood as a sequence of Bernoulli distributed outcomes with probability of success $\lambda_i(t)$ (see, e.g., Fahrmeir and Tutz, [17, p. 322]). The only differences between the models are the choice of link function and the assumption on the time dependence. All the above models are population-based models and that parameters should be interpreted accordingly.

In modeling fecundity, one has to be cognizant of the fact that their exists considerable heterogeneity among couples, due partly to various unobserved biological factors and lack of complete understanding of various factors (biological and behavioral) that may potentially influence fecundability. In order to account for some of these issues, Scheike and Jensen [34] included a couple-specific random effect R_i resulting in the following model

$$\lambda_i(t) = 1 - \exp(-\exp(R_i + x'_{it}\beta)),$$

under the assumption R_i independent of the covariates. Here, the random effect R_i denotes residual variation when the covariates have been accounted for, and may be thought of as accounting for unobserved covariates and further biological variation. In the above random effects model, the parameters relating the covariates to a discrete hazard rate can be interpreted as the effect of the covariate on the chance of conceiving for a subject (couple), whereas the parameters in the population models presented in the previous section relate the covariates to effects on the population.

Further, observe that the distribution function of the observed TTP under the assumption of gamma distribution with mean 1 and variance ν for the $\exp(R_i)$ distribution is given by

$$P(T_i = t)$$

$$= \left(\frac{1}{\nu \sum_{j=1}^{t-1} \exp(x'_{ij}\beta) + 1} \right)^{1/\nu}$$

$$- \left(\frac{1}{\nu \sum_{j=1}^{t} \exp(x'_{ij}\beta) + 1} \right)^{1/\nu},$$

and the discrete hazard rate is given by

$$\lambda_i(t) = 1 - \left(\frac{\nu \sum_{j=1}^{t-1} \exp(x'_{ij}\beta) + 1}{\nu \sum_{j=1}^{t} \exp(x'_{ij}\beta) + 1} \right)^{1/\nu}.$$

The estimation of the discrete random effects survival model can be fitted by maximum-likelihood estimation through Fisher scoring, which can be implemented by iteratively reweighted least squares. In fact, the likelihood of the TTP based on the above model is same as that of a sequence of Bernoulli distributed outcomes with probability of success $\lambda_i(t)$.

73.2.3 Models for Repeat Fecundability

An issue of considerable interest to reproductive epidemiologists is whether the fecundability experience repeats over multiple pregnancies. In other words, one is interested in joint modeling of TTPs to answer questions of such as

1. What is the probability that longer TTP during first pregnancy is associated with longer TTP in the following consecutive pregnancy?

2. What is the probability that shorter TTP during the first pregnancy is associated with shorter TTP in the following consecutive pregnancy?

3. What is the probability of crossover experience, i.e., shorter TTP in the first and longer TTP in the second or longer TTP in the first and shorter TTP in the second?

The random-effects model proposed by Scheike and Jensen [34] lends itself to easy extension to deal with the multivariate scenario. This is possible under the assumptions that the dependence between multiple TTPs for a couple is captured by the random effect. Conditional on subject i's random effect R_i, the observed (two) TTPs T_1 and T_2 for couple i are assumed independent and described by

$$P(T_i^1 = t_1, T_i^2 = t_2 | R_i)$$

$$= \prod_{p=1}^{2} \left\{ \exp\left(-\exp(R_i) \sum_{j=1}^{t_p - 1} \exp(x'_{ipj}\beta)\right) \right.$$

$$\left. - \exp\left(-\exp(R_i) \sum_{j=1}^{t_p} \exp(x'_{ipj}\beta)\right) \right\},$$

where x_{ipj} are covariates associated with the pth TTP. Again, the estimation of model coefficients can be fitted by maximum-likelihood estimation through Fisher scoring, which can be implemented by iteratively reweighted least squares. In fact, the likelihood of the multiple TTPs can be viewed as the likelihood of a sequence of Bernoulli distributed outcomes with probability of success $\lambda_i^1(t)$ and $\lambda_i^2(t|T_i^1)$.

In many epidemiological studies where the main interest is in assessing fecundability, the first TTP is retrospectively ascertained and the couples are followed prospectively for the second TTP. In addition, there is a period in between, where the couple is not attempting to achieve another pregnancy (between-pregnancy attempts). At the end of the between-pregnancy attempts period, a new TTP time is initiated. Scheike et al. [35] have studied joint modeling in such situations.

The above models all treat time to pregnancy as a discrete right-censored survival time. However, they do not account for more biological and behavioral information that is typically available in the prospective pregnancy studies. For instance, these studies capture information on a daily level within a menstrual cycle concerning various hormonal levels such as the lutenizing hormone, which is useful in identifying the days that a woman is most fertile so that intercourse could potentially result in conception. In addition, some studies also capture information on cervical mucus, which has been proved to be conducive for conception by assisting in the transportation of sperm through the hostile environment of the vagina. Also, on the behavioral side, daily level information on whether a couple had intercourse on a specific day is tracked. This aspect is what distinguishes TTP from a typical discrete right-censored data because if a couple does not have intercourse during the window of opportunity in a menstrual cycle, i.e., during the most fertile days of a menstrual cycle, the couple cannot conceive (the risk of conception is in fact zero for that cycle for the couple). In order to address some of the limitations, significant literature exists where the authors have focused on modeling fecundability through modeling the probability of conception in a specific menstrual cycle by a couple.

73.2.4 Modeling of Day-Specific Conception Probabilities

We first begin by introducing some notation. Suppose that there are n couples in a study. For a given couple in the study, let X_{ijk} be an indicator variable, where $X_{ijk} = 1$ denotes the occurrence of intercourse on day k in menstrual cycle j by couple i. Here, the index k is aligned to the estimated day of ovulation. The earliest statistical model for addressing the timing of intercourse effect is due to Barrett and Marshall. At the suggestion of Peter Armitage, Barrett and Marshall [3] assumed that the batches of sperm introduced on different days in a single cycle are statistically independent. Let p_k denote the probability of conception on day k; then the probability of conception for any menstrual cycle can be simply expressed as

$$\begin{aligned} \Pr(\text{conception in cycle } j | \mathbb{X}_{ij}) \\ = \ & P(Y_{ij} = 1 | \mathbb{X}_{ij}) \\ = \ & 1 - \prod_{k=1}^{K}(1 - p_k)^{X_{ijk}}. \end{aligned}$$

where $\mathbb{X}_{ij} = \{X_{ij1}, \ldots, X_{ijK}\}$, the intercourse pattern during the "fertile window," the window of opportunity around the day of ovulation when a couple can possibly conceive if they have unprotected intercourse. They used this model to study a cohort of British couples who had had at least one child and were using the rhythm method to avoid further pregnancy. The estimation of parameter $\{p_k\}$ was carried out by maximizing the likelihood function, that is, a product of probabilities based on the binary outcomes defined above across all menstrual cycles.

Noting that the conception also depends on certain biological factors in addition to the timing of intercourse, Schwartz et al. [36] extended the original Barrett and Marshall model such that the probability of conception for a given menstrual cycle j is

$$\begin{aligned}\Pr(\text{conception in cycle } j|\mathbb{X}_{ij}) &= P(Y_{ij}=1|X_{ijk}) \\ &= A\left[1-\prod_{k=1}^{K}(1-p_k)^{X_{ijk}}\right],\end{aligned}$$

where A is the probability that the cycle is viable; p_k under this model is interpreted as the conditional probability that conception would occur if there were sexual intercourse only on day k, given that the cycle is viable. Cycle nonviability can be regarded as a kind of immunity, where A is the fraction of cycles that are even susceptible to possible conception. Note that the viability in the Barrett–Marshall model is assumed as 1 and reduced fecundity is included in p_ks.

Wilcox et al. [52] analyzed the Early Pregnancy Study data using the above model and found that the estimated day-specific probabilities increased fairly monotonically up to a maximum on the estimated day of ovulation and then fell to 0 on the next day. The probabilities were statistically significantly different from 0 only in a six-day interval ending on the estimated day of ovulation, thus suggesting that the sperm are capable of maintaining their viable life for up to 5 days in the female reproductive tract, whereas the survival of the ovum may be much shorter. The estimate for A was 0.37, indicating that many ovulatory cycles are nonviable.

Royston [33] and Dunson [12] have shown that there is an identifiability problem in the BMS model and that the maximum of p_k is statistically indistinguishable from 1. Consequently, Dunson proposed constraining the value of p_k at the most fertile day to 1, clarifying the interpretation of A and p_ks, respectively. Under this constraint, A, typically defined as the cycle viability parameter, can be interpreted as the probability of conception given that the intercourse occurred on the most fertile day of the cycle; hence it can be referred to as the "susceptibility" parameter and p_k becomes the ratios of the daily conception probabilities for other days of the cycle relative to the peak probability.

Various extensions of the BMS model have been proposed in the literature. Royston [33] allowed A to vary linearly with one covariate

$$A_i = u'_i\alpha,$$

where u is limited to one covariate and α represents the corresponding covariate. Zhou and Weinberg [53] proposed a generalized linear model to relate A to the covariates that are couple- and/or cycle-specific:

$$A_{ij} = h_1(u'_{ij}\alpha).$$

In addition, the day-specific effects of covariate on p_k have been added giving an important flexibility to the BMS-type model:

$$\begin{aligned}\Pr(\text{conception on cycle } j|\mathbb{X}_{ij}) &= h_1(u'_{ij}\alpha) \\ &\times \left[1-\prod_{k=1}^{K}(1-h_2(w'_{ijk}\beta))^{X_{ijk}}\right],\end{aligned}$$

where w_{ijk} denotes a vector of covariates, which may modify the effect of intercourse on day k of the cycle. Even with these modifications, considerable residual heterogeneity remains predominant in analyzing fecundability. This is attributable mainly to the following: (1) know factors of heterogeneity are not included in the model for parsimony or (2) lack of information and some known factors are measured with measurement error.

Dunson and Stanford [14] have proposed a further extension of the BMS model, introducing fixed and random effects in modeling p_ks and absorbing the "susceptibility" factor A_{ij}, which is absorbed in the model for p_k by allowing it to vary by couple as well as by cycle for a specific day.

$$\Pr(\text{conception on cycle } j | \mathbb{X}_{ij}, b_i)$$
$$= 1 - \prod_k (1 - h_2(w'_{ijk}\beta + b_i))^{X_{ijk}}, \quad (1)$$

where b_i denotes the couple-specific random effect. Dunson proposed a log–log link function h_2 and the random effect $\exp(b_i)$ as a gamma distribution with mean 1 and variance γ. This results in a closed form for the marginal probability of conception given the intercourse pattern:

$$\Pr(\text{conception on cycle } j | \mathbb{X}_{ij})$$
$$= 1 - \left(\frac{\gamma}{\gamma + \sum_{k=1}^{K} X_{ijk} \exp(w'_{ijk}\beta)} \right)^\gamma.$$

In all the BMS models discussed above, an important assumption is that one can clearly identify the "fertile window," that is, the window of opportunity when a couple can potentially conceive if they have unprotected intercourse. As mentioned previously, Wilcox et al. [52] have shown the window to be of length 6 days when the probability of conception is statistically significantly nonzero. In particular, one assumes the fertile window to range within $\{-5, -4, -3, -2, -1, 0, 1\}$, where 0 refers to the day of ovulation. Based on this knowledge, in all the BMS models, the product term is restricted to these days (the other days are given zero chance for conception). However, tremendous uncertainty exists in this concept of the fertile window, which we will revisit later in the chapter.

Another issue with the abovementioned extensions of the BMS models are that they are conditional on the observed intercourse pattern. This implies that the interpretable quantities are the effects of day-specific covariates on the day-specific probabilities, as well as the day-specific probabilities of conception. However, in order to present the probability of conception in a cycle, one can discuss only with respect to a specific intercourse pattern. This is not interpretable for a couple seeking advice on their chance of conception in a specific cycle. Similarly, it may be difficult for the clinicians to convey to the couple their chances of conception. Furthermore, it is also well known that the intercourse behavior is very heterogenous among couples and may as well vary from cycle to cycle for a couple. This necessitates looking at marginal probability of conception ("unconditional on intercourse behavior"). Dunson [13] has studied the joint modeling of intercourse and fecundability under a variant of the Schwartz model, which does not require the intercourse pattern to be fixed. However, these authors assume that the couple's acts of intercourse on consecutive days are independent. This may be restrictive in the application. Recently, Kim et al. [27] have proposed a joint model of intercourse and fecundability, where modeling (1) the probability of conception without conditioning on the intercourse behavior and (2) the heterogeneity of the intercourse behavior in the fertility window by accounting for dependence of intercourse acts on consecutive days within a couple.

Taking expectation of Equation (1) throughout with respect to the distribution of $\mathbb{X}_{ij}$, they propose the following model for the probability of conception in a cycle j:

$$\Pr(Y_{ij} = 1 | \rho, \lambda^*)$$
$$= 1 - \prod_{k=1}^{K} \left(1 - \rho^1_{ijk} \lambda^*_{ijk}\right), \quad (2)$$

where $\rho_{ijk} = (\rho^1_{ijk}, \rho^0_{ijk})$ and $\rho^1_{ijk} = P(X_{ijk} = 1)$. They proposed the following model for the day-specific conception probabilities through the following reparameterization

$$\lambda^*_{ijk} = \frac{\exp(\lambda_{ijk})}{1+\exp(\lambda_{ijk})},$$

using a generalized linear mixed effects model. Observe that with the above reparameterization, $-\infty < \lambda_{ijk} < \infty$. Let $\lambda = (\lambda_{ijk}, k = 1,\ldots,K, j = 1,\ldots,n_i, i = 1,\ldots,I)'$ and $= (b_1,\ldots,b_I)'$. Incorporating this, their proposed conception model is given by

$$\begin{aligned}\Pr(Y_{ij} &= 1|\rho,,\lambda,\mathbb{Z}_{ij})\\ &= 1 - \prod_{k=1}^{K} \frac{1+\rho^0_{ijk}\exp(\lambda_{ijk})}{1+\exp(\lambda_{ijk})}\\ \lambda_{ijk} &= b_i + Z'_{ijk}\beta + \epsilon_{ijk}, \quad (3)\end{aligned}$$

where the subject-specific effect $b_i \sim F_b$ and the overall error $\epsilon_{ijk} \sim F$, b_i and ϵ_{ijk} are assumed independent, and F_b and F are the cumulative distribution function. This proposed model has several nice properties. Observe that the underlying latent variable has a random-effects model structure and that the reparametrized λ_{ijk} has the support of R. This representation facilitates an easy implementation of the Gibbs sampling algorithm. Furthermore, (3) defines a rich class of conception models. For example, F could belong to a class of scale mixtures normal distribution (see Chen and Dey [8]). In the next section, we illustrate the analysis of NYSAC Prospective Pregnancy Study using the methods and models discussed by Kim et al..

73.3 Analysis of a Prospective Pregnancy Study: New York Angler Cohort Study

In this section, we discuss the analysis first presented by Kim et al. [27] to illustrate the methods and models discussed in this chapter. We analyzed the New York State Angler Prospective Pregnancy Cohort Study [7]. The study used a prospective cohort design to recruit women aged 20–34 years who participated in the overall New York State Angler Cohort Study (NYSACS). This was a population-based cohort comprising licensed anglers aged 18–40 who were randomly selected from 16 contiguous counties along Lakes Erie and Ontario [45]. The study cohort comprised 113 women who reported planning pregnancies within 6 months; however, 14 women were excluded since they were found to be pregnant at baseline. The final sample comprised 99 women, 83 of whom returned daily diary information on menstruation, sexual intercourse, home pregnancy test results, and covariates believed to impact female fecundity. In addition, a baseline interview was conducted. Given the absence of a biomarker of ovulation for the study, the Ogino–Knaus method was used for estimating the likely date of ovulation by counting back 14 days from the end of the cycle [25,30]. Then, the fertile window was defined as starting 8 days before the day of ovulation and 3 days postovulation. This analysis focuses on intercourse patterns during fertile window, conception and covariates: age of mother at her enrollment, parity–previous births (yes/no), and daily smoking and alcohol usage patterns of the woman. Table 1 lists the characteristics of the cohort at enrollment by pregnancy status. In order to help the numerical stability in the implementation of the Markov chain Monte Carlo (MCMC) sampling algorithm, all the covariates were standardized.

73.3.1 Intercourse Pattern Analysis

In this subsection we analyze data corresponding to the intercourse pattern of couples in the NYSACPPS. A generalized linear model with structural equation model

Table 1: Description of cohort at enrollment by pregnancy status ($n = 83$).

Characteristic	Pregnant ($n = 62$) n (%)	Not Pregnant ($n = 21$) n (%)
Age (in years)		
20–29	25 (40)	8 (38)
30–34	37 (60)	13 (62)
Mean (SD)	30.1 (2.37)	30.0 (2.73)
Parity		
Nulliparous	15 (24)	8 (38)
Multiparous	47 (76)	13 (62)
Average number of intercourse acts in a FW[a] while trying to conceive		
	3.36	3.84
Smoking while trying to conceive		
	14 (29)	6 (29)
Alcoholic beverage consumption while trying to conceive		
	49 (79)	19 (90)

[a] Fertile window.

was fitted to the intercourse data:

$$X_{ijk} = \begin{cases} 1 & \text{if } w_{ijk} > 0, \\ 0 & \text{if } w_{ijk} \leq 0, \end{cases}$$
$$w_{ijk} = \mu_k + \alpha'_k \omega_k \eta_{ij} + \phi' U_{ijk} + \delta_{ijk}, \qquad (4)$$

where $\delta_{ijk} \sim N(0, \sigma_k^2)$, μ_k is the overall mean effect due to response k (intercourse on day k), α_k is a p_k-dimensional column vector of coefficients loading on η_{ij}, an r-dimensional vector of the latent variable, ω_k is the $p_k \times r$ fixed loading matrix that controls the dependence of response k (intercourse on day k) on the set of latent variables, and ϕ is the q-dimensional vector of regression coefficients corresponding to U_{ij}, a q-dimensional vector of covariates. Note that the probability of intercourse is modeled by

$$P(X_{ijk} = 1 | \eta_{ij}, U_{ijk}) = \Phi(\mu_k + \alpha'_k \omega_k \eta_{ij} + \phi' U_{ijk}),$$

where the day-specific covariates of interest are U_{ijk} = (smoke (yes/no), consume alcohol (yes/no), parity (yes/no), Age)'. For the New York State Angler Prospective Pregnancy Study, loading of the 12 manifest variables (fertile window length here is 12 days) in the measurement model, as well as the structural model of the SEM, is illustrated in Figure 1. The fertile window is split into three parts: FW1—the first part comprises days −8 through −5 prior to ovulation; FW2—the second part comprises days −4 through 1 (days around ovulation) and FW3: the third part comprises days 2 and 3 postovulation. In other words, the fertile window is split into an early phase, the window around ovulation and postovulation. This is motivated by the findings of Wilcox et al. [52] that an increase in frequency of intercourse occurs around ovulation reflecting biological changes in libido.

The path connections considered in the structural equation part are all forward in time as it is not reasonable to assume that the dependence structure would have a backward-in-time influence. Here, the dimension p_k of α_k is one and γ is a $r \times r$ matrix with $r = 3$. The parameters of interest in the model (4) are $\theta = (\mu, \alpha, \phi, \gamma, \sigma_\alpha^2)'$, where $\gamma = (\gamma_1, \gamma_2, \gamma_3)'$ is

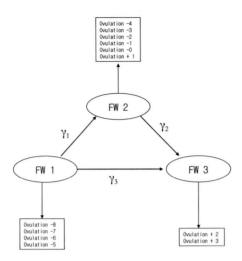

Figure 1: Path diagram for the New York State Angler Prospective Pregnancy Cohort.

a *three*-dimensional vector of the parameters in (4). Note that γ_1 gives a measure of influence of FW1 on FW2 and γ_2 gives the influence on FW3; similarly, γ_2 gives the influence of FW2 on FW3. The parameter μ_k denotes the overall main effect for intercourse on day k, and the parameters ϕ indicate the influence of the covariate U_{ijk} associated with them. The way to interpret the parameters ϕ is that positive values indicate an positive association of intercourse while the corresponding covariate and negative values indicate negative association of intercourse with the corresponding covariates.

Table 2 reports the posterior means, the posterior standard deviations, and the 95% highest posterior density (HPD) intervals of the model parameters under intercourse model. Observe that the parameters associated with the covariates: smoke and consume alcohol with the corresponding 95% HPD intervals of (0.0426, 0.1435) and (0.0213, 0.1032), respectively, indicating an association between increased intercourse activity and both smoking and alcohol consumption. However, no association of occurrence of intercourse activity with respect to parity and age is observed, which may be due to the fact that every couple in the cohort under investigation has intentions of becoming pregnant. Furthermore, to investigate the SEM model, note that 95% HPD for γ_1 and γ_2 indicate that they are significantly different from 0 whereas γ_3 is not significantly different from 0. This indicates that the intercourse pattern on days -8 through -5 (FW1) do influence the intercourse acts on days -4 through $+1$ (FW2) but do not significantly influence the intercourse activity on days $+2, +3$ (FW3); similarly, intercourse acts on days -4 through $+1$ (FW2) influence intercourse activity on days $+2, +3$ (FW3).

Table 3 presents the posterior means, the posterior standard deviations, and the 95%

Table 2: Posterior estimates under proposed intercourse models for New York State Angler Pregnancy data.

Variable	Posterior Estimates			Variable	Posterior Estimates		
	Estimate	SE	95% HPD Interval		Estimate	SE	95% HPD Interval
μ_{-8}	-1.0109	0.1050	(-1.2258, -0.8165)	$\alpha_{-8,1}$	0.6815	0.1469	(0.4098, 0.9776)
μ_{-7}	-0.8288	0.0760	(-0.9711, -0.6767)	$\alpha_{-7,1}$	0.3554	0.1120	(0.1460, 0.5814)
μ_{-6}	-0.8054	0.0765	(-0.9556, -0.6574)	$\alpha_{-6,1}$	0.3863	0.1086	(0.1825, 0.6053)
μ_{-5}	-0.7375	0.0726	(-0.8775, -0.5942)	$\alpha_{-5,1}$	0.3462	0.1082	(0.1328, 0.5552)
μ_{-4}	-0.5468	0.0705	(-0.6837, -0.4063)	$\alpha_{-4,2}$	0.4372	0.1077	(0.2214, 0.6449)
μ_{-3}	-0.5511	0.0692	(-0.6892, -0.4186)	$\alpha_{-3,2}$	0.3561	0.1070	(0.1473, 0.5636)
μ_{-2}	-0.5728	0.0668	(-0.7035, -0.4434)	$\alpha_{-2,2}$	0.2448	0.0948	(0.0580, 0.4307)
μ_{-1}	-0.5566	0.0721	(-0.6989, -0.4183)	$\alpha_{-1,2}$	0.4652	0.1118	(0.2511, 0.6948)
μ_0	-0.5933	0.0730	(-0.7342, -0.4479)	$\alpha_{-0,2}$	0.4606	0.1095	(0.2557, 0.6842)
μ_1	-0.5063	0.0649	(-0.6300, -0.3762)	$\alpha_{1,2}$	0.2310	0.0946	(0.0551, 0.4265)
μ_2	-0.6200	0.0715	(-0.7566, -0.4770)	$\alpha_{2,3}$	0.3837	0.1087	(0.1763, 0.5973)
μ_3	-0.6741	0.0794	(-0.8326, -0.5217)	$\alpha_{3,3}$	0.5163	0.1251	(0.2846, 0.7658)
ϕ_1 : Smoke	0.0938	0.0257	(0.0426, 0.1435)	γ_1	3.0224	1.2167	(1.0704, 5.4342)
ϕ_2 : Alcohol	0.0628	0.0209	(0.0213, 0.1032)	γ_2	2.3392	1.3717	(0.0229, 5.0733)
ϕ_3 : Parity	-0.0028	0.0283	(-0.0578, 0.0536)	γ_3	0.7406	2.0028	(-3.4541, 4.5386)
ϕ_4 : Age	-0.0559	0.0275	(-0.1070, -0.0006)	σ_α^2	0.1920	0.0923	(0.0604, 0.3710)

Table 3: Day-specific overall intercourse probabilities.

Day	Estimate	SE	95% HPD Interval
−8	0.20475	0.01900	(0.16874, 0.24280)
−7	0.22067	0.01958	(0.18383, 0.25963)
−6	0.22926	0.01973	(0.19176, 0.26832)
−5	0.24596	0.02026	(0.20680, 0.28629)
−4	0.31032	0.02162	(0.26927, 0.35336)
−3	0.30420	0.02171	(0.26266, 0.34682)
−2	0.29147	0.02159	(0.24906, 0.33345)
−1	0.30911	0.02165	(0.26694, 0.35150)
0	0.29731	0.02163	(0.25753, 0.34191)
1	0.31316	0.02187	(0.27143, 0.35691)
2	0.28393	0.02141	(0.24214, 0.32497)
3	0.27718	0.02138	(0.23617, 0.31931)

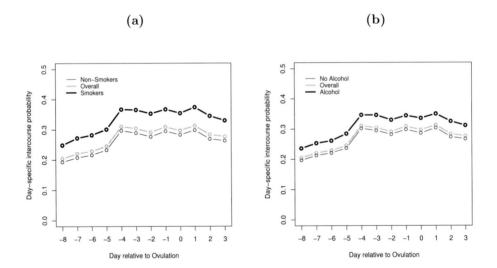

Figure 2: Day-specific intercourse probability plots.

HPD intervals for the day-specific intercourse probabilities. Figure 2(a) represents the plot of posterior mean day-specific probability of intercourse comparing the effect of smoking and Figure 2(b) represents the plot of posterior mean day-specific probability of intercourse comparing the effect of alcohol consumption (yes/no). Observe that the day-specific probability of intercourse increases as the days progress from FW1 to FW2 and then decreases from FW2 to FW3, supporting the findings of Wilcox et al. [51] that there is increase in intercourse in fertile window (their definition of fertile window corresponds to FW2).

73.3.2 Conception Model

The model (3) jointly with the model for intercourse (4), based on the covariates cigarette smoking, alcohol consumption, parity, and age, was fitted to the fecundability data. The covariates considered here are $Z_{ijk} = [1(k=-8), 1(k=-7), \ldots, 1(k=3)$, cigarette smoking, alcohol consumption, parity, age]$'$ for the conception model. The coefficients associated with the first 12 covariates correspond to baseline day-specific conception probabilities. Furthermore, the the correspondence of β to $k=-8,\ldots,-5$ and $k=2,3$ is constrained to be smaller than the $\beta_k, k=-4,\ldots,1$, but otherwise unconstrained within FW1, FW2, and FW3. This is a reasonable assumption as the day-specific conception probabilities should reduce the further one gets from the day of ovulation; in fact, it should be zero if the ovulation marker is precise. Table 4 presents the posterior means, the posterior standard deviations, and the 95% HPD intervals of the model parameters under the conception model. Note that smoking has a significantly negative effect on day-specific conception probabilities and parity has a positive effect on the day-specific conception probabilities. However, alcohol consumption and age did not turn out to be significant. Table 5 presents the posterior means, the posterior standard deviations, and the 95% HPD interval of the day-specific conception probabilities. Figure 3(a) presents the plot of posterior mean day-specific conception probability comparing the effect of smoking, and Figure 3(b) presents the plot of posterior mean day-specific conception probability comparing the effect of parity.

73.4 Biological Issues around Fecundity

In this section, we review biological issues associated with human fecundity. It is important to understand the measurement error issues and heterogeneity around various quantities that play an important role in building a model for fecundity. We will discuss in more detail the biology of the menstrual cycle, identify and discuss and issues around ovulation, and, define the fertile window. These three factors represent of the major biological components in conception.

We first begin with the definition of menstruation. The age at menarche is a developmental milestone representing the lower bound of reproductive age in women. However, menarche does not necessarily denote female fecundity, in part given a period of subfecundity until ovulation is initiated. There is variability in menstruation both within and between women. A prospective study with longitudinal measurement suggests that 95% of women have cycles between 22 and 36 days [18]. Cycles outside this range are typically considered irregular.

The menstrual cycle is divided into three phases when pertaining to changes in the endometrium: bleeding, proliferative phase, and secretory phase. The ovarian cycle is divided into two phases reflecting

Table 4: Posterior estimates under proposed conception models for New York State Angler Pregnancy data.

Variable	Estimate	SE	95% HPD Interval
Smoking	-0.5376	0.2352	(-1.0081, -0.0885)
Alcohol consumption	-0.2982	0.4324	(-1.1619, 0.4512)
Parity	0.8094	0.2894	(0.2691, 1.4004)
Age	0.3599	0.2234	(-0.0907, 0.7912)
β_{-8}	-7.8514	2.1547	(-12.1945, -3.3510)
β_{-7}	-6.5851	1.5663	(-9.4338, -3.6214)
β_{-6}	-5.6370	1.2758	(-8.2139, -3.3881)
β_{-5}	-4.9760	1.0962	(-7.1841, -2.9482)
β_{-4}	-3.3593	1.1588	(-5.6670, -1.1267)
β_{-3}	-3.1821	1.0600	(-5.1985, -1.0779)
β_{-2}	-2.9635	1.1254	(-5.1081, -0.5505)
β_{-1}	-3.2792	1.1043	(-5.4620, -1.1141)
β_0	-3.3667	1.0776	(-5.4288, -1.3445)
β_1	-3.2139	1.0141	(-5.2967, -1.2940)
β_2	-5.4665	1.3961	(-8.4758, -3.0529)
β_3	-7.0515	2.0007	(-10.9714, -3.3612)
σ_b^2	0.0115	0.0786	(-0.1022, 0.1120)
σ_{b*}^2	0.0094	0.2178	(0.0001, 0.0158)

Table 5: Day-specific overall conception probabilities.

Day	Estimate	SE	95% HPD Interval
-8	0.00729	0.01326	(0.00000, 0.03427)
-7	0.01394	0.01874	(0.00006, 0.05277)
-6	0.02386	0.02290	(0.00052, 0.06989)
-5	0.03636	0.02660	(0.00164, 0.08807)
-4	0.11451	0.07737	(0.00543, 0.27863)
-3	0.12535	0.07573	(0.00942, 0.28233)
-2	0.14441	0.09289	(0.00592, 0.33133)
-1	0.11977	0.07812	(0.00658, 0.27798)
0	0.11270	0.07040	(0.00693, 0.24911)
1	0.12051	0.07074	(0.00901, 0.25935)
2	0.02812	0.02397	(0.00014, 0.07617)
3	0.01229	0.01715	(0.00001, 0.04944)

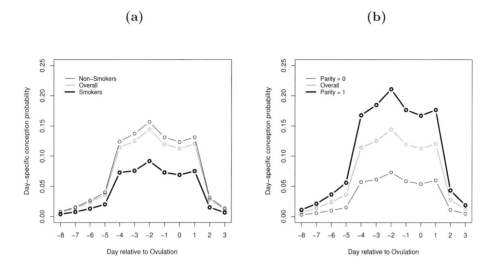

Figure 3: Day-specific conception probability plots.

what is occurring within the ovary: follicular and luteal phases. *Ovulation* is the event that separates these two phases and the proliferative and secretory phases of the menstrual cycle.

Day 1 of the menstrual cycle is defined as the first day of bleeding (although different women may interpret "bleeding" differently). Estradiol concentrations rise in the proliferative phase, and the inherent negative-feedback mechanism underlying the hypothalamus–pituitary–gonadal (HPG) axis reduces follicle stimulating hormone (FSH) during the later stages of this phase. Rising estradiol concentrations during the proliferative phase has a positive feedback on the production of luteinizing hormone (LH) and is believed to be responsible for triggering the LH surge. Following ovulation, the corpus luteum (CL) produces progesterone, estradiol, and 17-hydroxyprogesterone.

Classic longitudinal studies on menstrual cycles found considerable variability at the two ends of the woman's reproductive lifespan—initially after the onset of menarche and 2–3 years prior to menopause. So, aging affects menstrual cycle lengths in an inverse fashion [44]. Also, greater variability in menstrual cycle lengths have been reported after the first few cycles following childbirth and breastfeeding [54], after discontinuing hormonal contraception [21], during times of stress [4,19], and in relation to select lifestyle factors such as excess exercise, obesity, smoking and diet [32,39].

Ovulation. Release of the ovum from the follicle is defined as ovulation. Originally thought to be a gentle release, it is now considered a potentially traumatic event for the ovarian surface. For some time, the clinical critical window was said to be days 10–17, but more recent data fail to support this window largely because of dates (i.e., last menstrual period or next expected period).

Transvaginal daily ultrasonography is considered the gold standard for detecting ovulation in that there is sonographic verification of the collapse of the follicle indicative of ovulation. Since the 1950s, a series of proxy measures have been used, including (1) calendar-based methods, (2) basal body temperature (BBT), (3) cervical mucus changes, and (4) hormonal profiles. Possibly more accurate methods became available in the 1980s with the advent of rapid immunoassay test sticks for the detection of human chorionic gonadotropin (hCG) and lutenizing hormone (LH). In the 1970s, the World Health Organization attempted to determine the relation between hormone markers and ovulation, and reported the surge in LH to be the best marker of impending ovulation, and the rise in estradiol the beginning of the fertile period [49,50]. Subsequently, estrone-3-glucuronide (E3G) was reported to be a predictive urinary estrogen metabolite for the start of fertile window [1,6]. Other approaches tracking changes in hormones to estimate ovulation include (1) use of the ratio of urinary estrogen to progesterone metabolites (E/P algorithm) [2]; (2) peak serum LH [41], and (3) use of midcycle peak urinary FSH [28]. The accuracy of using these three hormonal approaches was evaluated by Li and colleagues [28] using daily ultrasonography for identifying follicular collapse (gold standard for determining ovulation). The authors reported that all approaches had errors, but the distributions of errors differed by choice of algorithm. More recently, Chen et al. [8] estimated urinary progesterone metabolites for rapid rise during the early luteal phase [pregnanediol-3-glucuronide (PdG)-rise algorithm].

It is important to note that differences in findings across studies may reflect the choice of hormones measured (parent or metabolites), media (saliva, urine or serum) used for analysis, study population, or duration and frequency of prospective measurement.

Fertile Window. The term "fertile window" implies that there is a specific time during the menstrual/ovarian cycles when fertilization can occur. A cycle at risk for pregnancy requires vaginal–penile intercourse during this window. The absence of a biomarker of ovulation (LH surge is only a proxy that ovulation is likely) makes it difficult to establish this window with accuracy.

Presumably, the length of the fertile window is determined by the lifespan of the gametes within the female reproductive tract and the time of ovulation. Survival of spermatozoa is affected by the quality of the woman's cervical mucus. Preparation of endometrium for implantation is critical and is under the control of LH and progesterone. Currently, it is estimated that the oocyte has a short period of viability after ovulation, possibly as short as a few hours.

While some authors have reported the fertile window to begin 5 days before ovulation and to end on the day of ovulation [52], others have reported much more variability in the length of the window ranging from < 1 to > 5 days [26]. Possible explanations for these differences may reflect choice of study sample such as restricting the study to fertile women or the inclusion of couples seeking medical treatment for impaired fecundity, respectively, or the method for estimating ovulation. Wilcox et al. [52] relied on the ratio of estrogen to progesterone, while Keulers and colleagues [26] utilized serial vaginal ultrasonography in combination with postcoital tests and sperm–mucus penetration tests. If correct, this interpretation may reflect less variation for healthy fertile women in comparison to women experiencing conception delays or other fecundity-related problems. Lynch and colleagues [29] summarized studies attempting to estimate day-specific probabilities of conception. Fertile windows ranged from 7 days before conception to 4 days following ovulation across seven authors. Specifically, authors relying on basal body temperature to estimate ovulation reported a range in the fertile window from 4 to 7 days before ovulation to 1 to 4 days following ovulation [3,36,33,10]. Using peak day of cervical mucus secretion, the fertile window ranged from 5 to 7 days before ovulation to 3 to 4 days following ovulation [10]. Using the ratio of estrogen to progesterone with or without the LH surge, respectively, the fertile window was 5 days before ovulation and the day of ovulation only [52,16]. At a minimum, the findings are supportive of additional research aimed at quantifying the within- and between-subject variation. In addition, to the best of the authors' knowledge, no studies have assessed the fertile window from a couple-based perspective with the inclusion of male factors in the model.

Another reason for considering a variable fertile window is the longstanding recognition of the inherent heterogeneity in menstrual cycle length. For example, the follicular phase is reported to vary from 11 to 27 days and the luteal phase from 7 to 15 days when BBT is used for estimating ovulation [46]. In fact, most variation in follicular phase is believed to be associated with ovulatory disorders, including defects in the corpus luteum or inadequate production of estrogen or progesterone [40]. Variability in the luteal phase is reported to be considerably less than in the follicular phase, but it remains unclear how these two patterns relate to each other.

More recently, various studies have used a fertility monitor to assess the fertile window. Behre et al. [5] conducted a prospective study of 53 (150 cycles) women aged 19–39 years to test the fertility monitor's accuracy against the gold standard—transvaginal ultrasonography with serum hormone measurements. For 91% of cycles, agreement between the LH surge and ultrasonography (U/S) confirmed ovulation; 91% of cycles had ovulation during the 2 peak monitor days and 97% during the 2 peak days plus the 1 day following high fertility; 76% of cycles with ovulation were confirmed by U/S on the second day of peak. Serum LH surge was detected in 139 of 149 cycles; in 51% cycles, ovulation occurred 1 day after the serum LH surge and 43% within 2 days after the LH surge. Ovulation occurred in 96% of cycles during the 3-day interval from start of the detection of LH surge. It is important to note that LH is not a marker of ovulation but a marker that ovulation is likely (and pending).

Furthermore, differences in the length and variability of follicular and luteal phases across studies may reflect differences in the proxy markers for estimating ovulation as summarized by Fehring and colleagues [18]. Specifically, Fehring et al. [18] assessed inter- and intra-subject variability in menstrual cycle and how the phases of the menstrual cycle related to each other. The mean cycle length was 28.9 ± 3.3 days with 95% of cycles between 22-36 days [95% CI 21.8–36.2]. The mean follicular length counting first day of menses as day 1 of the cycle to (including) the day of ovulation as measured with a peak monitor reading was 16.5 ± 3.4 (with 95 cycles having follicular length 10–22 days). The mean luteal phase was 12.4 ± 2 days with 95% cycles within 9–16 days. The mean length of menses was 5.8 ± 2.9 days with 95% of cycles having lengths of 3–8 days.

In summary, the menstrual cycle and the ovarian cycle and the phases within them all have considerable variability. They also depend on the approach to measure them. This in turn effects the detection of ovulation (in absence of a true biomarker) and the fertile window lengths. These quantities are in turn subject to both inter- and intra-subject variation.

73.5 Further Extensions

In Section 73.2 we reviewed the most commonly used approaches for modeling human fecundability. They can be broadly classified into two groups: (1) the discrete survival approach and (2) the Barrett–Marshall–Schwartz approach. The first group of models are easy to implement but do not take into account the biology and behavior aspects. The second group of models takes into account the biology as well as behavior aspect of the data structure but makes some strong assumptions concerning the "fertile window" and ovulation.

In the literature, various authors have dealt with the issue of incorporating behavior, especially intercourse, into the discrete survival model by including a covariate that sums the number of intercourse acts in that cycle. This does account for behavior to some extent, but is far from addressing the biological aspect, as in the absence of any intercourse activity during the fertile window of the menstrual cycle by the couple, the hazard for conception in that cycle should be zero. This is obviously not the case by simply including the number of intercourse acts into the model as a covariate. One has to go back to the basics and redefine the counting processes that form the building block in estimating right-censored data. In particular, one needs to perhaps modify the definition of the *risk set* at a specific cycle by incorporating findings on whether the couple had any unprotected intercourse during the fertile window. This will lead to a nonmonotonic risk set. As is well known, the asymptotic theory of the estimators based on right censored data makes extensive use of this property. So, further attention needs to be paid to the estimates of survival and hazard function of TTP. Furthermore, other quantity of interest for reproductive epidemiologists and infertility treatment clinicians is success probability of conception on specific days. It would be interesting and worthwhile to see how this information can be extracted from the hazard models.

In the case of Barrett–Marshall–Schwartz-type modeling of human fecundity, the definition of fertile window as well as identification of the day of ovulation plays an important role. Dunson et al. [15,16] have studied the impact of measurement error in identifying the day of ovulation and marker of ovulation. As pointed out in the discussion in Section 73.4, many of the quantities measured to identify menstrual cycle, ovulation detection, and fertile window are all fraught with measurement error and with significant within-subject and between-subject variability, thus necessitating need for more sophisticated statistical modeling of human conception that address these biological issues with inherent measurement error and variability.

Acknowledgment. The authors' research was supported by the intramural research program of National Institutes of Health, *Eunice Kennedy Shriver* National Institute of Child Health and Human Development.

References

1. Adlercreutz, H., Brown, J., Collins, W., et al., (1982). The measurement of urinary steroid glucuronides as indices of the fertile period on women. *J. Steroid Biochem.*, **17**, 695–702.

2. Baird, D. D., Weinberg, C. R., Wilcox, A. J., McConnaughey, D. R., Musey, P. I., and Collins, D. C. (1991). Hormonal profiles of natural conception cycles ending in early, unrecognized pregnancy loss. *J. Clin. Endocrinol. Metab.*, **72**, 793–800.

3. Barrett, J. C. and Marshall, J. (1969). The risk of conception on different days of the menstrual cycle. *Popul. Stud.*, **23**, 455-461.

4. Barsom, S. H., Mansfield, P. K., Koch, P. B., Gierach, G., West, S. G. (2004).

Association between psychological stress and menstrual cycle characteristics in perimenopausal women. *Women's Health Issues*, **14**, 235—241.

5. Behre, H. M., Kuhlage, J., Gabner, C., Sonntag, B., Schem, C., Schneider, H. P. G., and Nieschlag, E. (2000). Prediction of ovulation by urinary hormone measurements with the home use Clear-Plan Fertility Monitor: Comparison with transvaginal ultrasound scans and serum hormone measurements. *Hum. Reprod.*, **15**, 2478–2482.

6. Branch, C. M., Collins, P. O., and Collins, W. P. (1982). Ovulation prediction: changes in the concentrations of urinary estrone-3-glucuronide, estradiol-17 beta-glucuronide and estriol-16 alpha-glucuronide during conceptional cycles. *J. Steroid Biochem.*, **16**, 345–347.

7. Buck Louis, G., Dmochowski, J., Lynch, C. D., Kostyniak, P. J., McGuinness, B. M., and Vena, J. E. (2008). Polychlorinated biphenyl concentrations, lifestyle and time-to-pregnancy. *Hum. Reprod.* (advance access published Oct. 21, 2008; *doi:10.1093/humrep/den373*).

8. Chen, M.-H. and Dey, D. K. (1998). Bayesian modeling of correlated binary responses via scale mixture of multivariate normal link functions. *Sankhya, A*, **60**, 322–343.

9. Chen, C., Wang, X., Wang, L., Yang, F., Tang, G., Xing, H., Ryan, L., Lasley, B., Overstreet, J. W., Stanford, J.B., and Xu, X. (2005). Effect of environmental tobacco smoke on levels of urinary hormone markers. *Environ. Health Perspec.*, **113**(4), 412–417.

10. Colombo, B. and Masarotto, G. (2000). Daily fecundability: First results from a new data base. *Demogr. Res.*, **3/5**, 39 (*www.demographic-research.org/volumes/vol3/5*).

11. Cox, D. R. (1972). Regression models and life tables. *J. Roy. Statist. Soci. B*, **34**, 406–424.

12. Dunson, D. B. and Weinberg, C. R. (2000). Modelling human fertility in the presence of measurement error. *Biometrics*, **56**, 288–292.

13. Dunson, D. B. (2001). Bayesian modeling of the level and duration of fertility in the menstrual cycle. *Biometrics*, **57**, 1067–1073.

14. Dunson, D. B., Weinberg, C. R., Baird, D. D., Kesner, J. S., and Wilcox, A. J. (2001). Assessing human fertility using several markers of ovulation. *Statist. Med.*, **20**, 965–978.

15. Dunson, D. B. (2003). Incorporating heterogeneous intercourse records into time to pregnancy models. *Math. Popul. Stud.*, **10**, 127–143.

16. Dunson, D. B. and Stanford, J. B. (2005). Bayesian inference on predictors of conception probabilities. *Biometrics*, **61**, 126–133.

17. Fahrmeir, L. and Tutz, G. (1994). *Multivariate Statistical Modelling Based on Generalised Linear Models*. Springer-Verlag, New York.

18. Fehring, R.J., Schneider, M., and Raviele, K. (2006). Variability in the phases of the menstrual cycle. *J. Obstet. Gynecol. Neonatal Nur.*, **35**, 376–384.

19. Fenstser, L., Waller, K., Chen, J., Hubbard, A. E., Windham, G., Elkin, E., et al. (1999). Psychological stress in the workplace and menstrual function. *Am. J. Epidemiol.*, **149**, 127–134.

20. Gini, C. (1926). Decline in the birth rate and fecundability of women. *Eugen. Rev.*, **17**, 258–274.

21. Gnoth, C., Frank-Hermann, P., Schmoll, A., Godehardt, E., and Freundl, G. (2002). Cycle characteristics after discontinuation of oral contraceptives. *J. Gynecol. Endocrinol.*, **16**, 307–317.

22. Heckman, J. J., Robb, R., and Walker, J. R. (1990). Testing the mixture of exponentials hypothesis and estimating the mixing distribution by the method of moments. *Journal of the American Statistical Association*, **85**, 582–589.

23. Heckman, J. J. and Walker, J. R. (1990). Estimating fecundability from data on

waiting times to first conception. *J. Am. Statist. Assoc.*, **85**, 283–294.
24. Henry, L. (1953). Fondements theóriques des mesueres de la fécondabilité naturelle. *Int. Statist. Rev.*, **21**, 135–153.
25. Knaus, H. (1929). Eline neue methods zur bestimmung des ovulationstermines. *Zentralbl F. Gynak*, **53**, 219.
26. Keulers, M. J., Hamilton, C. J. C. M., Franx, A., Evers, J. L. H., and Bots, R. S. G. M. (2007). The length of the fertile window is associated with the chance of spontaneously conceiving an ongoing pregnancy in subfertile couples.*Human Reproduction*, **22**, 1652-1656.
27. Kim, S. D., Sundaram, R., and Louis, G. B. (200x). Joint modeling of intercourse behavior and human fecundability using Structural equation models. Under review, available upon request at *sundaramr2@mail.nih.gov*.
28. Li, H., Chen, J., Overstreet, J. W., Nakajima, S. T., Lasley, B. L. (2002). Urinary follicle-stimulating hormone peak as a biomarker for estimating the day of ovulation. *Fertil. Steril.*, **77**, 961–966.
29. Lynch, C. D., Jackson, L. W., Buck Louis, G. M. (2006). Estimation of the day-specific probabilities of conception: current state of the knowledge and the relevance for epidemiological research. *Paediatr. Perinatal Epidemiol.*, **20**(Suppl.1), 3–12.
30. Ogino, K. (1930). Ovulationstermin and konzeptionstermin. *Zentralbl F. Gynak*, **54**, 464–479.
31. Potter, R. G. (1960). Length of the observation period as a factor affecting the contraceptive rate. *Milbank Memorial Fund Quart.*, **38**, 140–152.
32. Rowland, A. S., Baird, D. D., Long, S., Wegeinka, G., Harlow, S. D., Alavanja, M., et al. (2002). Influence of medical conditions and lifestyle factors on menstrual cycle. *Epidemiology*, **13**, 668–674.
33. Royston, J. P.(1982). Basal body temperature, ovulation and the risk of conception, with special reference to the lifetimes of sperm and egg. *Biometrics*, **38**, 397–406.
34. Scheike, T. H. and Jensen, T. K. A discrete survival model with random effects: an application to time to pregnancy. *Biometrics*, **53**, 318–329.
35. Scheike, T. H., Petersenm, J. H., and Martinussen, T. (1999). Retrospective ascertainment of recurrent events: An application to Time to Pregnancy. *J. Am. Statist. Assoc.*, **94**, 713–725.
36. Schwartz, D., MacDonald, P. D. M., and Heuchel, V. Fecundability, coital frequency and viability of ova. *Popul. Stud.*, **34**, 397-400.
37. Sheps, M. C.(1964). On the time required for conception. *Popul. Stud.*, **18**, 85–97.
38. Sheps, M. C. and Menken, J. A. (1973). *Mathematical Models of Conception and Birth*. University of Chicago Press, Chicago.
39. Solomon, C. G., Hu, F. B., Dunaif, A., Rich-Edwards, J., Willet, W. C., Hunter, D. J. et al. (2001). Long or highly irregular menstrual cycles as a marker for risk of type 2 diabetes mellitus. *J. Am. Med. Assoc.*, **286**, 2421–2426.
40. Speroff, L. and Fritz, M. A. (2005). *Clinical Gynecology Endocrinology and Infertility*. Lippincott-Williams & Wilkins, Philadelphia.
41. Stewart, D.R., Overstreet, J.W., Nakajima, S.T., and Lasley, B.L. (1993). Enhanced ovarian steroid secretion before implantation in early human pregnancy. *J. Clin. Endocrinol. Metab.*, **76**, 1470–1476.
42. Thompson, W.(1977). On the treatment of grouped observations in life studies. *Biometrics*, **33**, 463–470.
43. Tietze, C. (1959). Differential fecundity and effectiveness of contaception. *Eugen. Rev.*, **50**, 231–234.
44. Treloar, A. E., Boynton, R., Behn, B., and Brown, B. (1967). Variation of the human menstrual cycle through repoprductive life. *Int. J. Fertil.*, **12**, 77–126.
45. Vena, J. E., Buck, G. M., Kostyniak, P., Mendola, P., Fitzgerald, E., Sever,

L., Freudenheim, J., Greizerstein, H., Zielezny, M., Mcreynolds, J. and Olson, J. (1996). The New York Angler Cohort Study: Exposure characterization and reproductive and developmental health. *Toxicol. Indust. Health*, **12**, 1–8.

46. Vollman, R. (1977). *The Menstrual Cycle*. Saunders, Philadelphia.

47. Weinberg, C. R. and Gladen, B. C. (1986). The beta-geometric distribution applied to comparative fecundability studies. *Biometrics*, **42**, 547–560.

48. Weinberg, C. R., Baird, D. D., and Wilcox, A. J. (1994). Source of bias in studies at time to pregnancy. *Statist. Med.*, **13**, 671–681. 547-560.

49. WHO Task Force on Methods for Determination of the Fertile Period (1980). Temporal relationships between ovulation and defined changes in the concentration of plasma estradiol-17?, luteinising hormone, follicle stimulating hormone and progesterone. I. Probit analysis. *Am. J. Obstet. Gynecol.*, **138**, 383–390.

50. WHO Task Force on Methods for Determination of the Fertile Period (1980). Temporal relationships between ovulation and defined changes in the concentration of plasma estradiol-17?, luteinising hormone, follicle stimulating hormone and progesterone. II. Histologic dating. *Am. J. Obstet. Gynecol.*, **139**, 886–895.

51. Wilcox, A. J., Weinberg, C. R., and Baird, D .D. (1995). Timing of intercourse in relation to ovulation:probability of conception, survival of pregnancy, and sex of the baby. *New Engl. J. Med.*, **333**, 1517–1521.

52. Wilcox, A. J., Baird, D. D., Dunson, D. B., Mcconnaughey, D. R., Kesner, J. S., and Weinberg, C. R. (2004). On the frequency of intercourse around ovulation: evidence for biological influences. *Hum. Reprod.*, **19**, 1539-1543.

53. Zhou, H. and Weinberg, C. R. (1996). Modeling conception as an aggregated Bernoulli outcome with latent variables via the EM algorithm. *Biometrics*, **52**, 94–954.

54. Zinaman, M. and Stevenson, W. (1991). Efficacy of the symptothermal method of natural family planning in lactating women after the return of menses. *Am. J. Obstet. Gynecol.*, **162**(Suppl. 2), 2037–2039.

74
Statistical Quality of Life

Mounir Mesbah

74.1 Introduction

The World Health Organization ([48] defines *quality of life* as: "an individual's perception of his/her position in life in the context of the culture and value systems in which he/she lives, and in relation to his/her goals, expectations, standards and concerns. It is a broad-ranging concept, incorporating in a complex way the person's physical health, psychological state, level of independence, social relationships, and their relationship to salient features of their environment."

In epidemiological surveys and/or clinical trials, the *health related quality of life* (HrQoL) is often considered as a global subjective health indicator. In either setting, epidemiology, or clinical trial, its measurement and statistical analysis has remained an important issue, since the important paper of Cox et al. [11]. In clinical trials, the primary endpoint is generally the survival (duration of life) or another biological/physiological efficiency variable, while the HrQoL is an important secondary endpoint. It is considered as the most important outcome among all other patient-reported outcomes (PRO).

Linda Bren [5] from the Food and Drug Administration (FDA) defines PRO as a

- A patient-reported outcome (PRO) is a measurement of any aspect of a patient's health status that comes directly from the patient, without the interpretation of the patient's responses by a physician or anyone else.

She also indicates its relationship with HrQoL:

- PROs and Quality of Life PRO measurements are sometimes confused with quality of life measurements. Quality of life is a broad concept referring to all aspects of a person's well-being. PRO instruments are used to measure quality of life, but they also can focus much more narrowly-for example, on a single symptom, such as pain.

Measurement of HrQoL is usually assessed through a patient questionnaire, where item (or question, or variable) responses are often categorical. In this chapter, we present mathematical methods used in the statistical validation and analysis of the HrQoL. These methods are based on the statistical validation of some essential properties induced by measurement models linking the observed responses and unobserved latent HrQoL variables.

In Section 74.2, some important measurement model used in HrQoL research are introduced. First, a classical model, the parallel model, describing the linear unidimensionality of a set of quantitative variables, is presented. This model is probably one of the most popular classical measurement model.

The parallel model is a parametric model usually justified when item responses are quantitative. It can be interpreted as specific factorial analysis model with only one latent component. Such factorial analysis models are often used in the HrQoL research, as quality of life is always considered as a multidimensional concept.

The extension of these models to categorical response is straight forward through graphical models. Modern measurement models and graphical modeling are presented in the same section. These models can be either parametric or nonparametric models. They are generally nonlinear. Within that section, we show how some important inequalities involving the Kullback–Leibler measure of association among conditionally independent variables can be very helpful in the process of validation. In the same section, we introduce the Rasch model. The Rasch model can be considered as a natural extension of the parallel model to the nonlinear case and the standard for unidimensional measurement models. Thus, it must be used as a "docking" target in building unidimensional scores.

Based on the results presented in sections 74.2, statistical validation of health related quality of life measurement models is considered in depth in Section 74.3. First, we define reliability of a measurement and give its expression and the expression of reliability of the sum of item responses under a parallel model, which is estimated by the Cronbach alpha coefficient. Then the backward reliability curve is presented, its correlation with the notion of unidimensionality is explained, and consequently, how it can be used to empirically determine the unidimensionality of a set of variables is discussed.

The Cronbach alpha coefficient is well known as a reliability or internal consistency coefficient, but is of little help in the process of validation of questionnaires. On the other hand, the backward reliability curve can be very helpful in the assessment of unidimensionality, which is a crucial measurement property. We explain why, when such a curve does not increase, a lack of unidimensionality in a set of questions is strongly suspected.

Differential instrument functioning or invariance of measurement across groups is an important property, which is also addressed in Section 74.3. The problem of multiplicity when a great number of statistical tests are performed during the validation process of a HrQoL questionnaire is also discussed.

Analysis of health-related quality-of-life variation between groups is tackled in Section 74.4. Direct statistical analysis of latent scores through a global latent regression model is briefly discussed, then longitudinal analysis of HrQoL and finally joint analysis of HrQoL and survival.

Section 74.5 is devoted to a few important issues in HrQoL field that appeared recently as a new challenge to statistical research.

74.2 Measurement Models of Health-Related Quality of Life

74.2.1 Classical Unidimensional Models for Measurement

Latent variable models involve a set of observable variables $A = \{\mathbf{X}_1, \mathbf{X}_2, \ldots, \mathbf{X}_k\}$ and a latent (unobservable) variable θ, which may be either unidimensional (i.e., scalar) or vector-valued of dimension $d \leq k$. In such models, the dimensionality of A

is defined by the number d of components of θ. When $d = 1$, the set A is unidimensional.

In a HrQoL study, measurements are taken with an instrument: the questionnaire. It is made up of questions (or items). The random response of a subject i to a question j is noted $\mathbf{X}_{ij}$. The random variable generating responses to a question j is noted, without confusion $\mathbf{X}_j$.

The parallel model is a classical latent variable model describing the unidimensionality of a set $A = \{\mathbf{X}_1, \mathbf{X}_2, \ldots, \mathbf{X}_k\}$ of quantitative observable variables. Define $\mathbf{X}_{ij}$ as the measurement of subject i, $i = 1, \ldots, n$, given by a variable $\mathbf{X}_j$, where $j = 1, \ldots, k$, then:

$$\mathbf{X}_{ij} = \tau_{ij} + \varepsilon_{ij}, \qquad (1)$$

where τ_{ij} is the true measurement corresponding to the observed measurement $\mathbf{X}_{ij}$ and ε_{ij} a measurement error. Specification of τ_{ij} as

$$\tau_{ij} = \beta_j + \theta_i,$$

defines the parallel model. In this setting, β_j is an unknown fixed parameter (nonrandom) effect of variable j and θ_i is an unknown random parameter effect of subject i.

Each variable is generally assumed with zero mean and unknown standard error σ_θ. The zero-mean assumption is an arbitrary identifiability constraint with consequence in interpretation of the parameter; its value must be interpreted comparatively to the mean population value. *In our setting, θ_i is the true latent health-related quality of life that the clinician or health scientist wants to measure and analyze.* It is a zero mean individual random part of all observed subject responses X_{ij}, the same whatever is the variable X_j (in practice, a question j of a HrQoL questionnaire). ε_{ij} are independent random effects with zero mean and standard error σ corresponding to the additional measurement error. Moreover, the true measure and the error are assumed uncorrelated: $\text{cov}(\theta_i, \varepsilon_{ij}) = 0$. This model is known as the *parallel model*, because the regression lines relating any observed item $X_j, j = 1, \ldots, k$ and the true unique latent measure θ_i are parallels.

Another way to specify model (1) is through conditional moments of the observed responses. So, the conditional mean of a subject response is specified as follows:

$$E[X_{ij}/\theta_i; \beta_j] = \beta_j + \theta_i. \qquad (2)$$

Again, β_j, $j = 1, \ldots, k$ are fixed effects, and θ_i, $i = 1, \ldots, n$ are independent random effects with zero mean and standard error σ_θ. The conditional variance of a subject response is specified as

$$\text{Var}[X_{ij}/\theta_i; \beta_j] = \text{Var}(\varepsilon_{ij}) = \sigma^2. \qquad (3)$$

These assumptions are classical in experimental design. This model defines relationships between different kinds of variables: the observed score X_{ij}, the true score τ_{ij}, and the error ε_{ij}. It is interesting to make some remarks about assumptions underlying this model. The random part of the true measure given by the response of individual i to a question j is the same regardless of the variable j. θ_i does not depend on j. The model is unidimensional. One can assume that in their random part all observed variables (questions X_j) are generated by a common unobserved (θ_i). More precisely, let $X_{ij}^* = X_{ij} - \beta_j$ the calibrated version of the response to item j of person i. Models (2) and (3) can be rewritten

$$E[X_{ij}^*/\theta_i; \beta_j] = \theta_i; \; \forall j, \qquad (4)$$

with the same assumptions on β and θ and with the same conditional variance model.

Another important consequence of the previous assumptions, when the distribution is normal, is a conditional independence property: whatever j and j', two observed items X_j and $X_{j'}$ are independent conditional to the latent θ_i. So, even when normality cannot be assumed, it is essential to specify this property.

74.2.2 Multidimensional Models and Factorial Analysis

Factorial analysis models generalize the previous simple parallel model

$$X_j = \beta_j + \theta + \varepsilon_j$$

(where the subject subscript i is omitted without risk of confusion) from one true component θ to p true components θ_l, with $1 < l < p$.

First, we remark that

$$\begin{aligned} X_j &= \beta_j + \theta + \varepsilon_j \Leftrightarrow X_j - \beta_j \\ &= \theta + \varepsilon_j \Leftrightarrow X_j^* = \theta + \varepsilon_j. \end{aligned} \quad (5)$$

In a general factorial analysis model, the observed item is a linear function of p latent variables:

$$X_j = a_{11}\theta_1 + a_{12}\theta_2 + \cdots + a_{1p}\theta_p + E_j. \quad (6)$$

In the factor analysis setting, this is usually written as:

$$X = AU + E, \quad (7)$$

where A is the factor loading matrix and U and E are independent.

Principal-component analysis (PCA) can be considered as a particular factorial analysis model with $p = k$, and without error terms (E is not in the model). In PCA, components (θ_l) are chosen orthogonal ($\theta_l \perp \theta_m$) and with decreasing variance (amount of information). In practice, a varimax rotation is often performed after a PCA to allow a better interpretation of the latent variable in terms of the original variables. It allows a clear clustering of the original variables in subsets (unidimensional). In Section 74.3.2 we will show how this can be *checked* using a graphical tool, the backward reliability curve.

Parallel as well factor analysis models are members of *classical* measurement models. They deal mainly with quantitative continuous responses, even if some direct adaptations of these models to more general responses are available today. In the next section, we present the *modern* approach, which includes the classical one as a special case. Within this approach, qualitative and quantitative responses can be treated indifferently. Some useful general properties that are not well known but are very important for the validation process of questionnaires are also presented. We introduce the Rasch model, and show how it can be interpreted as a nonlinear parallel model, which is more appropriate when responses are categorical.

74.2.3 ModernMeasurement Models and Graphical Modeling

Modern ideas about measurement models are more general. Instead of arbitrarily defining the relationship between the observed and the true latent (variables) as an additive function (of the true latent and the error), they focus only on the joint distribution of the observed and the true variables $f(X, \theta)$. We do not need to specify any kind of distance between X and θ. The error E and its relation to the observed X and the latent θ could be anything!

This leads us logically to graphical modeling. Graphical modeling aims to represent the multidimensional joint distribution of a set of variables by a graph. We will focus on conditional independence graphs. The interpretation of an independence graph is easy. Each multivariate distribution is represented by a graphic, which is composed of nodes and edges between nodes. Nodes represent one-dimensional random variables (observed or latent, i.e., nonobserved) while a missing edge between two variables means that those two variables are independent conditionally on the rest (all other variables in the multidimensional distribution). Since the pioneering work of Lauritzen and Wermuth [29], many monographs on graphical modeling have become available [17].

One way to define *latent unidimensionality* in the context of the graphical model is straightforward; a set of variables X are unidimensional, if there exist one and only one *scalar* latent variable θ such that each variable X is related to θ and only to θ. In Figure 1a, the set of variables $X_1, X_2, \ldots, X_9$ is unidimensional. In Figure 1b, the set of variables $X_1, X_2, \ldots, X_9$ is bidimensional. The unidimensionality is a consequence of the dimension of θ. The word *latent* means more than the fact that θ is not observed (or hidden). It means that θ is causal. The observed items X_j are caused by the true unobserved θ and no other variable! This causal property is induced by the conditional independence property. If X_j is independent of $X_{j'}$ conditionally on θ, then knowledge of θ is enough. Such directed graphical models are also known are *causal graphics* or *Bayesian networks*.

Measure of Association and Graphical Models. Let $K(f, g)$ be the Kullback–Leibler information between two distributions with respective density function f and g:

$$K(f, g) = \int f(x) \log(\frac{f(x)}{g(x)}) dx \quad (8)$$

The *Kullback–Leibler measure of association* (KI) between two random variables X and Y with respective marginal distribution f_x and f_y and with joint distribution f_{xy} is given by

$$KI(X, Y) = K(f_{xy}, f_y f_x). \quad (9)$$

In the same way, the measure of association between two variables X and Y conditionally on a third one Z is the *Kullback–Leibler measure of conditional association* $(KI((X, Y)/Z)$, which using similar straightforward notations, is given by

$$\begin{aligned} KI((X,Y)/Z) &= K(f_{xyz}, f_{y/z} f_{x/z} f_z) \\ &= K(f_{xyz}, f_{yz} f_{xz}/f_z). \end{aligned} \quad (10)$$

Theorem 74.2.1 *Let X, Y, and Z be three random variables such that X is independent of Y conditionally on Z. Then, under mild general regularity conditions, we have*

1. $KI((X, Y)/Z) = 0$

2. $KI((Y, Z)/X) = KI((Y, Z) - KI((X, Y); \quad KI((X, Z)/Y) = KI((X, Z) - KI((X, Y)$

3. $KI((X, Y) \leq KI((X, Z)$ and $KI((X, Y) \leq KI((Y, Z)$

PROOF: Conditions 1 and 2 can be easily derived. Condition 3 is a direct consequence of 1, 2, and the Cauchy–Schwartz inequality, $[K(X, Y)$ is always positive]. The interpretation of 3 is as follows. If we use KI as measure of association, we can easily show that the marginal association between two variables related by an edge in the graph G is stronger than the marginal association between two unrelated variables.

Remarks:

1. If (X, Y) is normally distributed, then $KI(X, Y)$ is a monotonic function of $\rho^2(X, Y)$, the square of the correlation coefficient. So $KI(X, Y)$ can be considered as a generalization of $\rho^2(X, Y)$.

2. If (X, Y, Z) is normally distributed, then $KI(X, Y/Z)$ is a monotonic function of $\rho^2(X, Y/Z)$, the square of the partial correlation coefficient. So, $KI(X, Y/Z)$ can be considered as a generalization of $\rho^2(X, Y/Z)$.

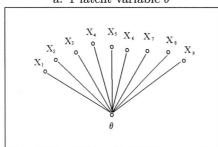 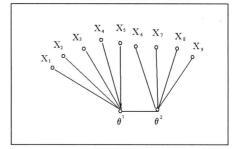

Figure 1: Graphical unidimensional and bidimensional models.

Using result 3 of Theorem 74.2.1, and the collapsibility property of a graphical model (see Frydenbergh [21]; Mesbah et al. [36]), one can derive the following useful consequences:

1. In Figure 1a, the marginal association between any observed item X and the latent variable θ is stronger than the association between two observed items.

2. In Figure 1b, the marginal association between any observed item X and its own latent variable θ is stronger than the association between that item X and another latent variable (other dimensions).

These two relationships between marginal measures of association are useful characterizations of the conditional independence property, which is a core property of latent variable models.

Remarks: Under the parallel model presented in Section 74.2, we have $\text{Corr}(X_j, X_{j'}) = \rho$ and $\text{Corr}(X_j, \theta) = \sqrt{\rho}$; then

$$\text{Corr}^2(X_j, \theta) = \rho \geq \text{Corr}^2(X_j, X_{j'}) = \rho^2.$$

This is a direct consequence of the fact that, under normality and parallel model assumption, items are independent conditionally to the latent variable. Consequences 1 and 2 (above) are very helpful in the process of questionnaire validation.

The graphical model framework is helpful in explaining relationships between variables when some of them are observed and others are not. Historically, the Rasch model, which we will introduce in the next section, was established earlier (in the 1960s), mainly as a measurement model more appropriate for categorical responses, which occurs frequently in HrQoL questionnaires. Nevertheless, its correlation with graphical models through the conditional independence properties that it includes, is more recent.

The Rasch Model. The parallel model presented in Section 74.2 is a linear mixed model. When item responses are binary, ordinal, or categorical, the parallel model is inappropriate. For instance, when the item response is a Bernoulli variable X_{ij} taking values x_{ij} [coded, e.g., 0 (failure, false, or "no") or 1 (success, correct, or "yes")], theories of exponential family and generalized linear models [31] suggest an adapted generalized linear model alternative to model (2). Instead of the linear model

$$E[X_{ij}/\theta_i; \beta_j] = \beta_j + \theta_i, \quad (11)$$

we define the generalized linear model, using a canonical link associated to Bernoulli distribution,

$$\text{Logit}(E[X_{ij}/\theta_i; \beta_j]) = \beta_j + \theta_i, \quad (12)$$

where as previously, β_j is a fixed effect and θ_i are independent random effects with zero mean and standard error σ_θ. This model is known as the *mixed Rasch model*. Its classical version, with θ_i assumed as a fixed parameter, was introduced and popularized by the Danish mathematician George Rasch [44] with the expression below. It is probably the most popular modern measurement model in the psychometric context, where it is used mainly as a *measurement model*. Under the Rasch model framework, the probability of the response given by a subject i to a question j is

$$P(X_{ij} = x_{ij}/\theta_i; \beta_j) = \frac{\exp(x_{ij}(\theta_i - \beta_j))}{1 + \exp(\theta_i - \beta_j)}. \quad (13)$$

The variable θ_i is the person parameter: it measures the ability of an individual n, on the latent trait. It is the true latent variable in a continuous scale. It is the true score that we want to obtain, using the *instrument* (questionnaire) including k items (questions) allowing us to estimate the true measurement (HrQoL) θ_i of person i. β_j is the item parameter. It characterizes the level of difficulty of the question. The Rasch model is member of the *item response theory models* [20]. The *partial credit model* [3,32] is another member of same the family: it is the equivalent to the Rasch model for ordinal categorical responses. Let $P_{ijx} = P(X_{ij} = x)$; then

$$P_{ijx} = \frac{\exp\left(x\theta_i - \sum_{l=1}^{x} \beta_{jl}\right)}{\sum_{h=0}^{m_j} \exp\left(h\theta_i - \sum_{l=1}^{h} \beta_{jl}\right)}, \quad (14)$$

for $x = 1, 2, \ldots, m_j$ (where m_j is the umber of levels of item j); $i = 1, \ldots, N$ (number of subjects); $j = 1, \ldots, k$ (number of items).

Some Rasch model properties are

1. *Monotonicity* of the response probability function

2. *Local sufficiency*—sufficiency of the total individual score for the latent parameter (considered as fixed parameter)

3. *Local independence* (items are independent conditional to the latent)

4. *Nondifferential item functioning* (conditional to the latent, items are independent from external variables)

The first property is an essential property for latent models. It is included in the Rasch model through the logistic link. The Mokken model [39] does not assume the logistic link, but assumes a nonparametric monotone link function: this is appealing for HrQoL field, but relaxing the logistic link, we loose the sufficiency property 2 of the total individual score, which is the most interesting characteristic property of Rasch model in the HrQoL field. This property justifies use of the simple scores as surrogates for the latent score. Kreiner and Christensen [27] focus on this sufficiency property and define a new class of nonparametric models: the graphical Rasch model. The last properties (3 and 4) are not included nor specific in the Rasch model, but added general latent models properties.

Considering the latent parameter as a fixed parameter leads to the joint maximum likelihood method which, in this context, can be inconsistent [20]. Conditional maximum likelihood method based on the sufficiency property gives consistent and asymptotically normal estimates for item parameters [2].

When the latent parameter is clearly assumed as random, estimation of (β, σ^2) can be obtained by the marginal maximum-likelihood method. In HrQoL practice, the distribution of the latent parameter is generally assumed as Gaussian with zero population mean and unknown variance σ^2. The likelihood function can be easily derived after marginalizing over the unobserved random parameter, the joint distribution of item responses, and the latent variable and using local independence property:

$$L(\beta, \sigma^2) = \frac{1}{(\sqrt{2\pi\sigma^2})^K}$$
$$\times \prod_{i=1}^{K} \left\{ \int_{-\infty}^{+\infty} \prod_{j=1}^{J} \frac{\exp((\theta - \beta_j) x_{ij})}{1 + \exp(\theta - \beta_j)} \right.$$
$$\left. \times \exp\left(\frac{-\theta^2}{2\sigma^2}\right) d\theta \right\}. \quad (15)$$

Estimation of β parameters can be obtained using Newton–Raphson and numerical integration techniques or the EM algorithm followed by Gauss–Hermite quadrature [24,20].

Dorange et al. [14] or Hardouin and Mesbah [25] indicate solutions using a recent SAS procedure devoted to estimation of nonlinear mixed model, PROC NLMIXED.

An alternative and faster way to estimate β parameters is to use the *generalized estimating equation* (GEE) approach (GEE) approach [18,19]. GEE methods [30,43,6], also known as *partial quasi likelihood methods*, can be considered as a *semi parametric* method because it consists of estimation without full specification of the joint distribution; only the marginal distribution at each point or the two first marginal moments are specified. The correlation is treated as a nuisance. A working covariance matrix for the repeated observations (in the Rasch setting: item responses) is introduced. The GEE approaches are based on first-order expansion methods (Taylor series expansion) around $\theta_i = \hat{\theta}_i$ (or Laplace approximation), or around $\theta_i = 0$.

Under the Rasch model, the likelihood function is fully specified. So, there is no need to use a method such as GEE devoted to partially specified models. It is, however, well known that a full likelihood analysis for such mixed models is hampered by the need for numerical integrations. To overcome such integration problems, following useful approximations of the likelihood and after derivation of its first marginal moments, Feddag et al. [18] developed a GEE algorithm for estimation of the β and σ^2 parameters. Feddag and Mesbah [19] extended the method to estimation of the covariance matrix of a multidimensional latent trait. These results are useful in the longitudinal or multidimensional latent case.

Estimation of Latent Parameters. Estimation of item parameters is generally the main interest in psychometrical area. Calibration of the HrQoL is the preliminary goal. When item parameters are known (or assumed as fixed and known), estimation of the latent parameter is straightforward. One easy method is simply to maximize the classical joint likelihood method assuming the latent parameter fixed. As item parameter is supposedly known, there is no inconsistency. Another popular estimator of latent parameter is the Bayes estimator, given by the posterior mean of the latent distribution [1]. Other estimators can be obtained. Mislevy [38] proposes a nonparametric Bayesian estimator for the latent distribution in the Rasch model. Martynov and Mesbah [33] give a nonparametric estimator of the latent distribution in a mixed Rasch model.

The posterior distribution of the latent

parameter is defined as follows:

$$P\left(\theta_i/x_i,\beta\right)$$
$$= \frac{P(X_i = x_i/\theta_i,\beta)g(\theta_i)}{\int P(X_i = x_i/\theta_i,\beta)g(\theta_i)d\theta_i}. \quad (16)$$

The Bayesian modal estimator is $\widehat{\theta}_i$, the value of θ_i that maximizes the posterior distribution, while the Bayes estimator is given by

$$\widehat{\theta}_i = \int \theta_i P\left(\theta_i/x_i,\beta\right)g(\theta_i)d\theta_i. \quad (17)$$

In the next section, we show, how, using results shown in Sections 74.2 and 74.3, validation of questionnaires and construction of scales can be performed.

74.3 Validation of Health-Related Quality-of-Life Measurement Models

74.3.1 Reliability of an Instrument: Cronbach Alpha Coefficient

A measurement instrument gives us values that we call *observed measure*. The reliability ρ of an instrument is defined as the ratio of the true over the observed measure. Under the parallel model, one can show that the reliability of any variable X_j (as an instrument to measure the true value) is given by

$$\rho = \frac{\sigma_\theta^2}{\sigma_\theta^2 + \sigma^2}, \quad (18)$$

which is also the constant correlation between any two variables. This coefficient is also known as the intraclass coefficient. The reliability coefficient ρ can be easily interpreted as a correlation coefficient between the true and the observed measure.

When the parallel model is assumed, the reliability of the sum of k variables equals

$$\tilde{\rho}_k = \frac{k\rho}{k\rho + (1-\rho)}. \quad (19)$$

This formula is known as the *Spearman–Brown* formula. Its maximum-likelihood estimator, under the parallel model and normal distribution assumption, is known as *Cronbach's alpha coefficient* (CAC)(see Cronbach [12] and Kristof [28]). It is expressed as

$$\alpha = \frac{k}{k-1}\left(1 - \frac{\sum_{j=1}^{k} S_j^2}{S_{\text{tot}}^2}\right), \quad (20)$$

where

$$S_j^2 = \frac{1}{n-1}\sum_{i=1}^{n}(X_{ij} - \overline{X_j})^2$$

and

$$S_{\text{tot}}^2 = \frac{1}{nk-1}\sum_{i=1}^{n}\sum_{j=1}^{k}(X_{ij} - \overline{X})^2.$$

Under the parallel model, the *joint* covariance matrix of the observed items X_j and the latent trait θ is

$$V_{X,\theta} = \begin{pmatrix} \sigma_\theta^2 + \sigma^2 & \cdots & \cdots \sigma_\theta^2 & \sigma_\theta^2 \\ \vdots & \vdots & \vdots & \vdots \\ \sigma_\theta^2 & \sigma_\theta^2 & \sigma_\theta^2 + \sigma^2 & \sigma_\theta^2 \\ \sigma_\theta^2 & \cdots & \sigma_\theta^2 & \sigma_\theta^2 \end{pmatrix},$$

and the *joint* correlation matrix in of the observed items X_j and the latent trait θ is

$$R_{X,\theta} = \begin{pmatrix} 1 & \rho & \cdots & \cdots \rho & \sqrt{\rho} \\ \rho & 1 & \rho & \cdots \rho & \sqrt{\rho} \\ \vdots & \vdots & \vdots & \vdots & \vdots \\ \rho & \cdots & \rho & 1 & \sqrt{\rho} \\ \sqrt{\rho} & \cdots & \cdots & \sqrt{\rho} & 1 \end{pmatrix}.$$

The *marginal* covariance V_X and correlation matrix R_X of the k observed variables X_j, under the parallel model, are

$$V_X = \begin{pmatrix} \sigma_\theta^2 + \sigma^2 & \sigma_\theta^2 & \cdots & \cdots \sigma_\theta^2 \\ \sigma_\theta^2 & \sigma_\theta^2 + \sigma^2 & \sigma_\theta^2 & \cdots \sigma_\theta^2 \\ \vdots & \vdots & \vdots & \vdots \\ \sigma_\theta^2 & \cdots & \sigma_\theta^2 & \sigma_\theta^2 + \sigma^2 \end{pmatrix}$$

and

$$R_X = \begin{pmatrix} 1 & \rho & \cdots & \cdots \rho \\ \rho & 1 & \rho & \cdots \rho \\ \vdots & \vdots & \vdots & \vdots \\ \rho & \cdots & \rho & 1 \end{pmatrix}.$$

This tyep of structure is known as *compound symmetry*. It is easy to show that the reliability of the sum of k items given in (19) can be expressed as:

$$\tilde{\rho}_k = \frac{k}{k-1}\left[1 - \frac{\text{Tr}(V_X)}{J'V_X J}\right]. \quad (21)$$

with J a vector with all component equal 1, and

$$\alpha = \frac{k}{k-1}\left[1 - \frac{\text{Tr}(S_X)}{J'S_X J}\right], \quad (22)$$

where S_X is the observed variance, empirical estimation of S_X. Even in the more recent literature there has been understandable confusion between the Cronbach alpha as a population parameter (theoretical reliability of the sum of items) and its sample estimate. Exact distribution of α under Gaussian parallel model and its asymptotic approximation are well known [46]. This is explained more precisely as follows.

Exact Distribution of Cronbach Alpha. Assuming the parallel model as a Gaussian distribution of the latent and error component, we have

$$\frac{1}{1-\tilde{\rho}_k}(1-\alpha) \sim F_n^{n(k-1)}, \quad (23)$$

where $F_n^{n(k-1)}$ is the Fisher distribution with n and $k-1$ degree of freedom. A direct consequence is that, under the same assumption, the exact population mean and the variance of α are as follows:

$$\begin{aligned} E(\alpha) &= \frac{n\tilde{\rho}_k - 2}{n-2}; \\ \text{Var}(\alpha) &= \frac{2(1-\tilde{\rho}_k)^2 n(nk-2)}{(k-1)(n-2)^2(n-4)}\tilde{\rho}_k. \end{aligned}$$
$$(24)$$

Asymptotical Distribution of Cronbach Alpha. When the Gaussian distribution cannot be assumed, but the parallel form remains, the following results are obtained:

$$\begin{aligned} E(\alpha) &\to \tilde{\rho}_k; \\ n\text{Var}(\alpha) &\to \frac{2(1-\tilde{\rho}_k)^2 k}{(k-1)}\tilde{\rho}_k; \\ \alpha &\to \tilde{\rho}_k; \end{aligned} \quad (25)$$

$$\frac{\sqrt{n}}{2}\ln(1-\alpha) \sim N\left(\frac{1}{2}\ln(1-\tilde{\rho}_k); \frac{k}{2(k-1)}\right). \quad (26)$$

Principal-Component Analysis and Cronbach Alpha. It is easy to show a direct connection between CAC and the percentage of variance of the first component in principal-component analysis (PCA), which is often used to assess unidimensionality [40]. The PCA is based mainly on analysis of the latent roots of V_X or R_X (or, in practice, their sample estimate). The matrix R_X has only two different latent roots; the greater root is $\lambda_1 = (k-1)\rho + 1$, and the other multiple roots are $\lambda_2 = \lambda_3 = \lambda_4 = \cdots = 1 - \rho = (k - \lambda_1)/(k-1)$. So, using the Spearman–Brown formula, we can express the reliability of the sum of the k variables as

$$\tilde{\rho}_k = \frac{k}{k-1}\left(1 - \frac{1}{\lambda_1}\right).$$

This clearly indicates a monotonic relationship, which is estimated by α (CAC) and the first latent root λ_x, which in practice is estimated by the corresponding value of the observed correlation matrix and thus the percentage of variance of the first principal component in a PCA. Thus, CAC is also considered as a measure of unidimensionality.

Nevertheless, this measure is not very useful, because, as is easily shown, using the Spearman–Brown formula (6), under the parallel model assumption, the reliability of the total score is an increasing function of the number of variables. Moreover, this coefficient lies between 0 and 1. Zero value indicates a totally unreliable scale, while unit value means that the scale is perfectly reliable. Of course, in practice, these two scenarios never occur! In the next section, we show how to build and use a more operational and more valid criterion: the backward reliability curve.

74.3.2 Unidimensionality of an Instrument: Backward Reliability Curve

Statistical validation of unidimensionality can be performed through a goodness-of-fit test of the parallel model or Rasch model. There is a vast literature on the subject, using classical or modern methods. These goodness-of-fit tests are generally very powerless because their null hypotheses do not focus on unidimensionality; they include indirectly other additional assumptions (e.g., normality for parallel models, local independence for Rasch models), so the deviation from these null hypothesis is not specifically a unidimensionality departure.

In the following, we present a graphical tool, helpful in the step of checking the unidimensionality of a set of variables. It consist on a curve to be drawn in a stepwise manner, using estimates of reliability of sub scores (total of a sub set included in the starting set).

The first step uses all variables and compute their CAC. Then, at every successive step, one variable is removed from the score. The removed variable is that one which leaves the score (remaining set of variables) with a maximum CAC value among all other CAC of remaining sets checked at this step. This procedure is repeated until only two variables remains. If the parallel model is true, increasing the number of variables increases the reliability of the total score which is estimated by Cronbach's alpha. Thus, a decrease of such a curve after adding a variable would cause us to suspect strongly that the added variable did not constitute a unidimensional set with variables already in the curve. This algorithm was successfully used in various previous medical applications (see Moret et al. [40]; Curt et al. [13]; Nordman et al. [41]). We present the algorithm in detail below.

Let $A = A_k$ be a set of k random variables:

$$A_k = \{X_1, X_2, \ldots, X_k\}.$$

By definition k, the number of variables in set A is by definition, card(A). First, let us give few definitions

Definition 74.3.1 Cronbach alpha coefficient of the set A. Let X_{ij} be the measurement of subject i ($i = 1, \ldots, n$), given by a variable X_j, where $j = 1, \ldots, k$. Define the CAC of the set A_k based on responses of subjects i, $i = 1, \ldots, N$, of a population P, as

$$\alpha_{A_k} = \alpha(X_{i1}, X_{i2}, \ldots, X_{ik}; \\ i = 1, \ldots, N)$$

$$= \frac{k}{k-1}\left(1 - \frac{\sum_{j=1}^{k} S_j^2}{S_{tot}^2}\right), \quad (27)$$

where

$$S_j^2 = \frac{1}{N-1}\sum_{i=1}^{N}(X_{ij} - \overline{X_j})^2$$

$$S_{tot}^2 = \frac{1}{Nk-1}\sum_{i=1}^{N}\sum_{j=1}^{k}(X_{ij} - \overline{X})^2.$$

Definition 74.3.2 *The maximum backward one step Cronbach alpha coefficient* of the set A, is MBOSC$(A) = \alpha_{k,\max}$, where

$$\alpha_{k,\max} = \max_{j/j=1,\ldots,k}\{\alpha_{A_k - \{X_j\}}\}. \quad (28)$$

Definition 74.3.3 *The worst backward one step variable of the set A is the variable* WBOS$(A) = X_{j_{\text{worst}}}$ *such that* $\alpha_{A_k - \{X_{j_{\text{worst}}}\}} = \alpha_{k,\max}$:

$$X_{j_{\text{worst}}} | \alpha_{A_k - \{X_{j_{\text{worst}}}\}} \\ = \alpha_{k,\max} = \max_{j/j=1,\ldots,k}\{\alpha_{A_k - \{X_j\}}\}. \quad (29)$$

Definition 74.3.4 *Backward reliability function* (BRF) *of a set A of random variables f_{BCA}^A is the unique empirical real function following the properties a, b, and c below.*

Property a. Let x be a real number, $f_{BCA}^A : x \mapsto f_{BCA}^A(x)$ is not defined for neither $x < 2$ nor $x > \text{card}(A)$.

Property b. For any integer x such that $1 < x < \text{card}(A) + 1$, f_{BCA}^A can be computed using the algorithm below.

Property c. For any real y and any integer x such that $x < y < x+1$, $f_{BCA}^A(y)$ is obtained by linear interpolation between x and $x+1$:

$$f_{BCA}^A(y) = f_{BCA}^A(x)[2-y] \\ + f_{BCA}^A(x+1)[y-1].$$

Algorithm for Evaluating the *BRF* **for Integer Values.**

1. For $x = \text{card}(A)$, $f_{BCA}^A(x) = \alpha_A$.

2. For $x = \text{card}(A) - 1$, $f_{BCA}^A(x) = $ MBOSC(A).

3. If card$(A) > 2$, then $A \leftarrow [A - \text{WBOS}(A)]$ (replace set A by $[A - \text{WBOS}(A)]$) and go to 1.

4. If card$(A) < 3$, then stop.

Theorem 74.3.1 *If the BRF is not an increasing function, then the set A of items do not follow a parallel model* in population P.

PROOF: The proof of this theorem is a direct consequence of the Spearman–Brown formula (19). As indicated in Section 74.3.1, $\tilde{\rho}_k$, the reliability of the total score (sum of k variables), is an increasing monotone function of k, the number of variables in the model.

So, if the *population* reliability (or theoretical reliability) of the sum of any number of variables is not an increasing function, this means that the set of variables do not follow the model. The proof is completed, when one remark that the Cronbach alpha coefficient α_{A_k} is the maximum-likelihood estimate of $\tilde{\rho}_k$, through his expression as a direct function of the *consistent* maximum-likelihood estimate of ρ and of the number of items.

Remark: Application of Theorem 74.3.1 to real data is an essential tool in the validation process of a HrQoL questionnaire. When one develops a HrQoL questionnaire, generally the main goal is to measure some unidimensional latent subjective traits (e.g., as sociability, mobility). Use of the BRF in empirical data is very helpful in detecting nonunidimensional subsets of items. When the BRF is not an increasing function, one can remove one or more items to get an increasing function. So, if the reduced set gives an increasing function, it is in some sense, *more valid in term of unidimensionality* than the previous one.

Figure 2 is an example of application to a real dataset. This data set comes from ENTRED, a large national survey ($N = 3198$) about Quality of Life of Diabetics in France, using a random sample of diabetic patients contacted by mail. The HrQoL measurement instrument used was the "diabetic health profile" [9]. It consists on a set of 27 ordinal questions (4 levels), split into three dimensions: "barriers to activity" (13 items), "Psychological distress" (14 items) and "desinhibited eating" (5 items). The backward reliability curve of the "psychological distress" dimension is shown in Figure 2. Removing item 22 leads to a perfect increasing curve. The content of this item is "Do you look forward to the future?"

74.3.3 Construction of HrQoL Scales

From Reliability to Unidimensionality. Measuring individual quality of life is frequently done by computing one or various scores. This approach assumes that the set of items being considered represent a single dimension (one score) or multiple dimensions (multiple scores). These scores can be considered as statistics, a function of individual measurements (for instance, item responses). They must have good statistical properties.

The Cronbach alpha as an indicator of reliability of an instrument is probably one of the most widely used tools in the HrQoL field or more generally in applied psychology. The major problem with Cronbach α as a reliability coefficient is the lack of a clear scientific rule to decide whether a score (based on a set of items) is reliable. We need to establish a threshold to decide whether that the score is reliable. Following Nunnally [42], a scale is satisfactory when it has a minimal Cronbach's alpha value around 0.7. The "Nunnally rule" is an empirical rule without any clear scientific justification. *So, reliability is not a direct operational indicator*. The Spearman–Brown formula (6) is a direct consequence of parallel model assumptions. It implies that, when adding an item, or more generally increasing the number of items, the reliability of the sum of item responses must increase. This property is, of course, a *population property* characterizing the parallel model. Its sampling version is probably less regular. The Cronbach α coefficient is the sampling estimate of reliability of the sum of item responses. So, use of the backward reliability curve as an empirical rule to validate graphically the parallel model and thus, *unidimensionality* of a set of items, is straightforward.

Use of the backward reliability curve to *find a unidimensional set of items* must be done in an *exploratory* way. It is a fast way to find *suspect* items, i.e., those items that must be removed to ensure an increasing curve and thus a parallel model. It can also be used in a *confirmatory* way to a given set that is supposedly unidimensional. When a given set of items have a nice backward reliability curve (i.e., smoothly increasing in in much the same way to the one theoretical Spearman–Brown curve), one can perform some statistical goodness-of-fit tests to check specific underlying properties. This consists mainly in validating the *compound symme-*

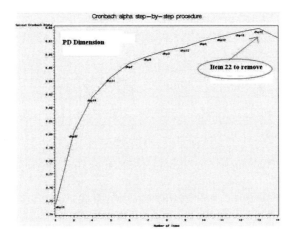

Figure 2: Backward reliability curve of psychological distress dimension.

try structure of the covariance matrix of the items, including assumption of equality of item variances and item-latent variances. When the item responses are binary or ordinal, one can test some underlying properties of the Rasch model [23]. In practice, this is rarely done, because of the lack of implementation of such tests in most of general statistical software. Under the Rasch model a reliability coefficient close to the Cronbach alpha can be derived [24]. It can be interpreted in the same way as in parallel models. A backward reliability curve can be used as a first step followed by a goodness-of-fit test of the Rasch model.

Hardouin and Mesbah [25] used a multidimensional Rasch model and Akaike information, in a stepwise procedure, to obtain, in an exploratory manner, unidimensional clusters of binary variables.

Usually, in actual HrQoL research, simpler validation techniques are performed. More details are given in the next section.

Specificity and Separability of Scores. Measurement models considered here are very simple models based on the unidimensionality principle. They can be defined as Rasch-type models: a parallel model for quantitative items and Rasch or partial credit model for ordinal items. Each *unidimensional* set of items is related to one and only one latent variable. There is no confusion between "concepts," so an item cannot be related directly to two latent variables. An item can be related to another latent variable only through its own latent variable. It is, of course, a strong property, difficult to obtain in practice. HrQoL questionnaires are built using questions drawn with words, and often health concepts (psychological, sociological, or even physical concepts) are not clearly separated. Nevertheless, measurement is generally considered as the beginning of science, and science is hard to achieve.

So, correlations between each item and all unidimensional scores must be computed. This can be considered as part of the internal validation in a multidimensional setting, to ensure the *separability* of the subsets. We must check the following for any item: (1) *specificity*—there is a strong correlation between that item and its own score; and (2) *separability*—the item correlation between that item and its own score is higher than the correlation between the same item and scores built on other dimensions.

This is a direct consequence of Section 74.2.3. The first property is another view of internal consistency condition of the *subscale*. Under the parallel model, that correlation is the same regardless of the item, and it is also known as the *intraclass coefficient*. The Cronbach alpha is a monotone function of that value. It must be evaluated for each subscale. Item correlations between each item and all subsores must be tabulated.

Graphical Latent Variable Models for HrQoL Questionnaires. *Graphical latent variable models* for scales can be easily defined as graphical models [29] built on multivariate distribution of variables with three kind of nodes:

- Those corresponding to observed or manifest variables corresponding to items or questions
- Those corresponding to unobserved or hidden variables corresponding to latent variables
- Those corresponding to other external variables

Figure 3 shows two examples, one (Figure 3a), with 13 items related to three latent variables and without external variables, and the second with 9 items related to two latent variables and two external variables Y and Z. The part of the graphic relating items and their corresponding latent variable is a graph where, as previously, items are not, 2×2, related by an edge. They are related only to the latent variable. One must have also the following properties:

1. *Monotonicity.* The marginal distribution of an item conditional to its latent variable must be a monotonous function of the latent variable.

2. *Nondifferential item functioning.* This is a graphical property. There are no direct edges between nodes corresponding to any item and another latent variable or between any item and any external variable.

74.4 Analysis of Quality of Life Variation between Groups

74.4.1 Construction of HrQoL Scores or Global Analysis

Development and validation of a HrQoL questionnaire is generally hard work, requiring more than one survey and many real sets of data. When the structure of the questionnaire is stabilized, i.e., when the clustering of the items in a subset of unidimensional items is clearly defined, one needs simple rules for analyzing data of studies, including the HrQoL questionnaire simultaneously with other external variables. So, a HrQoL questionnaire, like any instrument, must include *guidelines* for the statistical analysis step. Most of the time, for ease to use, only simple rules based on computing simple scores are included:

1. *Sum of item responses.* This score is a sufficient statistics for the latent parameter under the Rasch model. Under the parallel model, its reliability is estimated by the Cronbach alpha coefficient. It is the simplest and easiest score to derive.

2. *Weighted sum of item responses.* This is more complicated than the previous score. The weights are generally fixed and obtained with a principal component analysis previously performed in a "large representative population."

3. *Percentage of item responses.* This score is similar to the first, with a

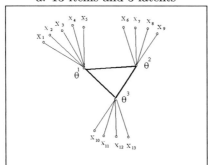 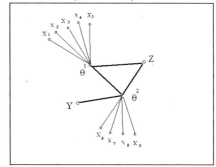

a. 13 items and 3 latents   b. 9 items, 2 latents, 2 Covariates

Figure 3: Graphical latent variable model.

different range of values, from 0% to 100%. When a dimension includes k ordinal items with responses coded $0, 1, \ldots, m$ (all items with the same maximum level m), this score is obtained by dividing the first score by km.

Unfortunately, estimation of the latent parameter is rarely suggested in a "guidelines book" of an HrQoL questionnaire, because it requires the use of specific software, including the latent variable estimation section. Weighted-sum scores (item 2 above) require the knowledge of the "good weights" given by the instrument developer, which is generally a marketing device to oblige any user of the questionnaire (e.g., scientists, clinical investigators, pharmaceutical companies) to pay royalties. In practice, these weights are generally obtained in a specific population and are not valid for another one. Use of a score such as type 1 or 3 is, in our opinion, the best way to go, particularly when we do not have easy access to specific software for estimation of Rasch-type models.

74.4.2 Latent Regression of HrQoL

It is usual to analyze HrQoL data with classical linear or generalized linear models where the response are scores of HrQoL built at a first step (measurement step). So, item responses are forgotten and replaced by summary surrogate scores. The analysis is, of course, easier and can be done using classical general software. Generally, one assumes that the distribution of scores is Gaussian, which is facilitated by the fact that most measurement models (parallel, Rasch, etc.) specify a Gaussian distribution for the latent variable. For instance, when the built score is a percentage, one can analyze its relation to other external variables by the mean of a logistic regression model, which allows interesting interpretations in term of odds ratios. Nevertheless, analyzing surrogate scores as "observations" instead of the actual observation, i.e., item responses, can give unsatisfactory results (see Mesbah [34]), mainly in terms of lack of efficiency. So, when analyzing the relationships between the latent HrQoL and any other external variables (survival time, treatment, sex, age, etc.), it could be more efficient to consider a global model, even if one does not need to build new scores or to validate once more the measurement model. In fact, under some additional simple conditions, which in most of real situations can be, easily assumed, improved statistical efficiency may be possible when considering such a global model. Building a global model taking into account the latent trait parameter, *without separation between measurement and analysis steps*, is a promising latent regression approach (see Christensen et al. [8]; Sébille and Mesbah [45]), enabled nowadays by the increasing performance of computers. Nevertheless, this approach needs to be handled with care. The theoretical aspects of each practical case must be well analyzed, with an in-depth investigation of which specific identifiability constraints must be chosen. We have to ensure that this choice does not upset the interpretation of the final results.

74.4.3 Joint Analysis of Quality and Quantity of Life

Joint Analysis of a Longitudinal QoL Variable and An Event Time. The motivation for the following models is a HRQoL clinical trial involving analysis of a longitudinal HRQoL variable and an event time. In such a clinical trial, the longitudinal HRQoL variable is often unobserved at dropout time. The model proposed by Dupuy and Mesbah [16] (DM model) works when the longitudinal HRQoL is directly observed at each time visit, except, of course at dropout time. We propose extending the DM model to the latent context case, i.e., when the HRQoL variable is obtained through a questionnaire.

Let T be a random time to some event of interest, and let Z be the HRQoL longitudinally measured. Let C be a random right-censoring time. Let $X = T \wedge C$ and $\Delta = 1_{\{T \leq C\}}$. Suppose that T and C are independent conditionally on Z

Following the Cox model, the hazard function of T has the form

$$\lambda(t|Z) = \lambda(t) \exp(\beta^T Z(t)) \quad (30)$$

The observations are as follows: $[X_i, \Delta_i, Z_i(u), 0 \leq u \leq X_i]_{1 \leq i \leq n}$.

The unknown parameters are β and $\Lambda(t) = \int_0^t \lambda(u)\,du$. Let us assume that C is noninformative for β and λ. Dupuy and Mesbah [16] suggest a method that assumes a nonignorable missing process, takes into account the unobserved value of the longitudinal HRQoL variable at dropout time, and uses a joint modeling approach of event-time and longitudinal variable.

Dupuy and Mesbah's model assumes that

$$\lambda(t|Z) = \lambda(t) \exp(W(t)\beta_0 Z_{a_d} + \beta_1 Z_d), \quad (31)$$

where

- Z has a density satisfying a Markov property:

$$f_Z(z_j|z_{j-1}, \ldots, z_0; \alpha)$$
$$= f_Z(z_j|z_{j-1}; \alpha), \; \alpha \in \mathbb{R}^p.$$

- C is noninformative for α and does not depend on $Z(t)$.

Let $W(t) = (Z_{a_d}, Z_d)^T$ and $\beta^T = (\beta_0, \beta_1)$ (see also Fig. 4).

The observations are

$$Y_i = (X_i, \Delta_i, Z_{i,0}, \ldots, Z_{i,a_d})_{1 \leq i \leq n}.$$

The unknown parameters of the model are $\tau = (\alpha, \beta, \Lambda)$. There are hidden variables in the model, the missing values of Z at dropout time, Z_{i,a_d}. The objective is to estimate τ from n independent vectors of observations Y_i.

The likelihood for one observation y_i ($1 \leq i \leq n$) is obtained as follows:

$$L^{(i)}(\tau)$$
$$= \int \lambda(x_i)^{\delta_i} \exp\left[\delta_i \beta^T w_i(x_i)\right.$$
$$\left. - \int_0^{x_i} \lambda(u) e^{\beta^T w_i(u)}\,du\right]$$
$$\times f\left(z_{i_0}, \ldots, z_{i_{a_d}}, z_d; \alpha\right) dz_d$$
$$= \int l(y_i, z_d, \tau)\,dz_d,$$

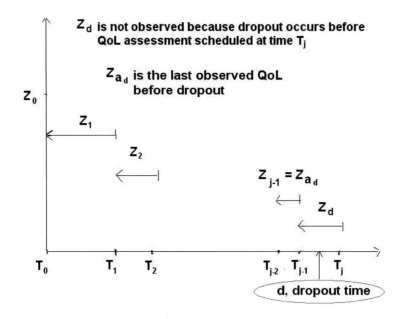

Figure 4: QoL assessments: $t_0 = 0 < \cdots < t_{j-1} < t_j < \cdots < \infty$. Z: takes value $Z(t)$ at time t and constant values Z_j in the intervals $(t_{j-1}, t_j]$. Z_j is not observed until t_j.

The parameter τ is identifiable. First, suppose that the functional parameter τ is a step function $\Lambda_n(t)$ with jumps at event times X_i. Then, taking unknown values $\Lambda_n(X_i) = \Lambda_{n,i}$, we rewrite the likelihood and estimate α, β, and $\Lambda_{n,i}$. The contribution of y_i to the likelihood obtained is now taken to be

$$L^{(i)}(\tau) = \int \Delta\Lambda_{n,i}^{\delta_i} \exp\left[\delta_i \beta^T w_i(x_i) - \sum_{k=1}^{p(n)} \Delta\Lambda_{n,k} e^{\beta^T w_i(x_k)} 1_{\{x_k \leq x_i\}}\right] \times f(z_{i_0}, \ldots, z_{i_{a_d}}, z_d; \alpha) \, dz_d,$$

where $\Delta\Lambda_{n,k} = \Delta\Lambda_n(X_k) = \Lambda_{n,k} - \Lambda_{n,k-1}$, $\Delta\Lambda_{n,1} = \Lambda_{n,1}$ and $X_1 < \ldots < X_{p(n)}$ ($p(n) \leq n$) are the increasingly ordered event times. The maximizer $\hat{\tau}_n$ of $\sum_{i=1}^n \log L^{(i)}(\tau)$ over $\tau \in \Theta_n$ satisfies

$$\sum_{i=1}^n \frac{\partial}{\partial \tau}\left[L_{\hat{\tau}_n}^{(i)}(\tau)\right]_{|\tau = \hat{\tau}_n} = 0.$$

where $L_{\hat{\tau}_n}^{(i)}(\tau) = E_{\hat{\tau}_n}[\log l(Y, Z; \tau)|y_i]$. Let us refer to $\sum_{i=1}^n L_{\hat{\tau}_n}^{(i)}(\tau)$ as the *EM loglikelihood*.

An EM algorithm used to solve the maximization problem is described by Dupuy and Mesbah [16]. A maximizer $\hat{\tau}_n = (\hat{\alpha}_n, \hat{\beta}_n, \hat{\Lambda}_n)$ of $\sum_{i=1}^n \log L^{(i)}(\tau)$ over $\tau \in \Theta_n$ exists and under some additional conditions,

$$(\sqrt{n}(\hat{\alpha}_n - \alpha_t), \sqrt{n}(\hat{\beta}_n - \beta_t), \sqrt{n}(\hat{\Lambda}_n - \Lambda_t)) \sim G,$$

where G is a tight Gaussian process in $l^\infty(H)$ with zero mean and a covariance process $\text{cov}[G(g), G(g^*)]$ (see Dupuy et al. [15]).

From this, we deduce for, instance:

1. $\sqrt{n}(\hat{\beta}_n - \beta_t)$ converges in distribution to a bivariate normal distribution with mean 0 and variance–covariance matrix $\Sigma_{\tau_t}^{-1}$.

2. A consistent estimate of Σ_{τ_t} is obtained.

Similar results are obtained for $\hat{\alpha}_n$ and $\hat{\Lambda}_n$.

Example. Quality of life was assessed among subjects involved in a cancer clinical trial [4]. Quantitative scores were obtained via a HRQoL instrument by autoevaluation. There was two treatment groups and a nonignorable dropout analysis were performed. Results are indicated in Table 1 [35].

In this example HRQoL, except for its value at dropout time, was just considered as an observed continuous score, Z. But in fact, HRQoL is not directly observed. It is an unobserved latent variable. In practice, HRQoL data always consist in a multi-dimensional binary or categorical observed variable termed quality-of-life scale used to measure the true unobserved latent variable HRQoL. From the quality-of-life scale, we can derive HRQoL scores, i.e., statistics. These scores are surrogates of the true unobserved latent variable HRQoL. In the next section, we will extend the previous DM model to the latent case.

Joint Analysis of a Latent Longitudinal HRQoL Variable and an Event Time. When the HRQoL variable z was observed (except for the last unobserved dropout value z_d), the likelihood for one observation $y_i (1 \leq i \leq n)$ was

$$
\begin{aligned}
L^{(i)}(\tau) &= \int \lambda(x_i)^{\delta_i} \exp\left[\delta_i \beta^T w_i(x_i)\right.\\
&\qquad \left. - \int_0^{x_i} \lambda(u) e^{\beta^T w_i(u)} du\right]\\
&\quad \times f\left(z_{i_0}, \ldots, z_{i_{a_d}}, z_d; \alpha\right) dz_d\\
&= \int l(y_i, z_{i_d}, \tau) dz_{i_d},
\end{aligned}
$$

where

$$
\begin{aligned}
y_i &= (x_i, \delta_i, z_{i_0}, \ldots, z_{i_{a_d}})\\
&= (x_i, \delta_i, z_{i_{obs}})
\end{aligned}
$$

and, all the previous statistical inferences, based on the likelihood

$$L^{(i)}(\tau) = \int l(x_i, \delta_i, z_{i_{obs}}), z_d, \tau) dz_d \quad (32)$$

are highly validated by theoretical asymptotic results and efficiently working computer algorithms. In the latent variable context, $z_{i_{obs}}$ is in fact not directly observed. The k-item responses Q_{ij} of a subject i (response or raw vector Q_i) are observed and must be used to recover the latent HRQoL values z_i through a measurement model. The obvious choice in our context is the *Rasch model*, which is, for binary responses:

$$
\begin{aligned}
P(Q_{ij} &= q_{ij} \mid z_i, \zeta_j)\\
&= f(q_{ij}, z_i, \zeta_j) = \frac{e^{(z_i - \zeta_j) q_{ij}}}{1 + e^{z_i - \zeta_j}}.
\end{aligned}
\quad (33)
$$

So, currently, observations are $Y_i = (X_i, \Delta_i, Q_{i_0}, \ldots, Q_{i_{a_D}})_{1 \leq i \leq n}$; with $Q_i = (Q_{i1}, \ldots, Q_{ip})$ for an unidimensional scale of p items. Unknown parameters of the model are $\tau = (\alpha, \beta, \Lambda)$ and nuisance parameters, ζ. The objective is now to estimate τ from n independent vectors of observations Y_i. Let us suppose that the following two assumptions hold:

Table 1: HRQoL analysis in a cancer clinical trial.

	Arm A		Arm B		Test Statistics	
	Random	NI	Random	NI	Random	NI
$\hat{\beta}_0$	-0.16	0.13	-0.17	0.09	0.033	0.35
SE($\hat{\beta}_0$)	0.08	0.08	0.08	0.08	—	—
$\hat{\beta}_1$		-0.36		-0.32	—	-0.37
SE($\hat{\beta}_1$)		0.09		0.09	—	—
$\hat{\alpha}$	0.96	0.95	0.96	0.95	0.35	0.32
SE($\hat{\alpha}$)	0.01	0.01	0.01	0.01	—	—
$\hat{\sigma}_e^2$	0.70	0.71	0.571	0.576	2.19565	2.18126
SE($\hat{\sigma}_e^2$)	0.046	0.047	0.04	0.04	—	—
Loglikelihood	-963.9	-896.4	-927.2	-857.2	-	-

1. The DM analysis model holds for the true unobserved QoL Z and dropout D or survival T.

2. The Rasch measurement model relates the observed response items Q to QoL Z.

First, we have two main issues:

- Specification of a model for the data and the true latent QoL
- Choice of a method of estimation

Similar to the Rasch model, for categorical ordinal responses (with the number of levels m_j different per item), the partial credit model is

$$p_c = P(Q_{ij} = c \mid z_i, \zeta_j)$$
$$= \frac{e^{\left(cz_i - \sum_{l=1}^{c} \zeta_{jl}\right)}}{\sum_{c=0}^{m_j} e^{\left(cz_i - \sum_{l=1}^{c} \zeta_{jl}\right)}} \quad (34)$$

The joint distribution of Q (items), Z (latent), D (time to death or dropout), and T (treatment) can be derived, using only the conditional independence property:

$$f(Q, Z, D, T/Z)$$
$$= \frac{f(Q, Z, D, T)}{f(Z)}$$
$$= \frac{f(Q, Z)}{f(Z)} \times \frac{f(Z, D, T)}{f(Z)}, \quad (35)$$

we have

$$f(Q, Z, D, T/Z) = f(Q/Z) \times f(D, T/Z). \quad (36)$$

Then, without any other assumption, we can specify two models:

- First model:

$$f(Q, Z, D, T)$$
$$= f(Q/Z) \times f(D/Z, T)$$
$$\times f(Z/T) \times f(T) \quad (37)$$

- Second model:

$$f(Q, Z, D, T)$$
$$= f(Q/Z) \times f(Z/D, T)$$
$$\times f(D/T) \times f(T). \quad (38)$$

The right likelihood must be based on the probability function of the observations, i.e., currently, $Y_i = (X_i, \Delta_i, Q_{i_0}, \ldots, Q_{i_{a_D}})_{1 \leq i \leq n}$. The parameters of the model are $\tau = (\alpha, \beta, \Lambda)$ and the nuisance difficulty parameters of the HRQoL questionnaire, ζ. There are nonobserved (hidden) variables in the model (latent Z, missing Q): $(Z_{i_0}, \ldots, Z_{i_{a_D}}, Z_{i_d}, Q_d)_{1 \leq i \leq n}$.

Directly following from the graph of the DMq model, factorization rules of the joint distribution function of the observations

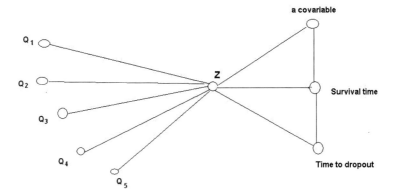

Figure 5: Joint graphical model for HrQoL and survival.

(Y_i), the latent HRQoL (Z) and the missing questionnaire Q_d can now be specified, and then, integrating through the hidden variables, we obtain the likelihood:

$$\begin{aligned}
L^{(i)}(\tau) &= \int \prod_{j=1}^{p} \left[\frac{e^{\left(cz_{i_0} - \sum_{l=1}^{c} \zeta_{jl}\right)}}{\sum_{h=0}^{m_j} e^{\left(cz_{i_0} - \sum_{l=1}^{c} \zeta_{jl}\right)}} \right.\\
&\quad \times \cdots \times \\
&\quad \times \frac{e^{\left(cz_{i_{a_d}} - \sum_{l=1}^{c} \zeta_{jl}\right)}}{\sum_{h=0}^{m_j} e^{\left(cz_{i_{a_d}} - \sum_{l=1}^{c} \zeta_{jl}\right)}} \\
&\quad \times \sum_{c=1}^{p} \left.\frac{e^{\left(cz_{i_d} - \sum_{l=1}^{c} \zeta_{jl}\right)}}{\sum_{h=0}^{m_j} e^{\left(cz_{i_d} - \sum_{l=1}^{c} \zeta_{jl}\right)}} \right] \\
&\quad \times \Delta \Lambda_{n,i}^{\delta_i} \exp\left[\delta_i \beta^T w_i(x_i)\right.\\
&\quad \left. - \sum_{k=1}^{p(n)} \Delta \Lambda_{n,k} e^{\beta^T w_i(x_k)} \mathbf{1}_{\{x_k \leq x_i\}} \right] \\
&\quad \times f(z_{i_0}, \ldots, z_{i_{a_d}}, z_d; \alpha) \\
&\quad dz_{i_0}, \ldots, z_{i_{a_d}}, z_{i_d}.
\end{aligned}$$

The marginalization over the latent variables is similar to the marginalization over the dropout missing value. Computer programs are easily extended. Nevertheless, when the number of latent components is large, computing time can be very long. So, in health applications, a two step approach is generally preferred. (see also Fig. 5.)

74.5 Conclusion

The definition (or construction) of variables and indicators, and the analysis of the evolution of their joint distribution between various populations, times and areas are generally two different, well separated steps of the work for a statistician in the field of health related quality of life.

The first step generally deals with calibration and metrology of questionnaires. Keywords are measurement or scoring, depending on the area of application. Most of the time, the statistical methods used are exploratory. The kinds of models specified are generally structural: classical factorial analysis models or modern item response theory models, focusing on the concept of latent variable and as a direct consequence unidimensionality.

The backward reliability curve can be used as a tool to quickly find a unidimensional set of items, followed by a goodness-of-fit test of unidimensional models to confirm more rigorously the unidimensionality of the identified set of items. When more than one dimension is available, computation of scores and correlations between items and scores is useful to check the separability of dimensions.

The second step is certainly more widely known by most statisticians. Linear, generalized linear, time-series, and survival models are very useful models in this step, where the variables constructed in the first step are incorporated and their joint distribution with the other analysis variables (treatment group, time, duration of life, etc.) is investigated. HrQoL scores, validated during the first step, are then analyzed, with a complete omission of the real observations, i.e., item responses. The latent nature of the HrQol is generally hidden.

Mesbah [34] compared the simple strategy of separating the two steps with the global strategy of defining and analyzing a global model including both the measurement and the analysis step. If, with a real dataset, one find a significant association between a built (from items) score and an external covariate, then the true association, i.e., the one between the external covariate and the true latent trait, is probably larger. So, if the scientific goal is to show an association between the true and the covariate, one does not need to use a global model: just the model with the surrogate built score instead of the true latent. Conclusions about the built score also include for the true. But, if no significant association is found between the built score and the covariate, then the true association could be anything (and perhaps larger). So, one have to consider a global model, even if one does not need to build new scores or to validate the measurement model. Building a global model taking into account the latent trait parameter in a one-step procedure, i.e., without separation between measurement and analysis, is a promising latent regression approach [8,45] allowed by the increasing performance of computers. In the HRQoL field, most of papers are devoted to a two-steps approach, where the HRQoL scores are used instead of the original item response data. Moreover, scientific results are published in different kinds of scientific journals: those devoted to validation of measurements and instruments and numerous others, specializing in analysis of previously validated measurements.

Joint analysis of a longitudinal variable and an event time is nowadays a very active field. Vonesh et al. [47], Cowling [10], Chi and Ibrahim [7] are a few of recent the more papers indicating that "Joint modeling of longitudinal and survival data is becoming increasingly essential in most cancer and AIDS clinical trials." Mainly because of the complexity of the computing programs, there are unfortunately no papers considering a joint model between a longitudinal latent trait and an event time.

Another very popular method used in the nineteenth century was the Q-TWIST (quality-adjusted time without symptoms of toxicity) approach [22], where duration of life was divided into different categories corresponding to various states of health with given utilities. So, it was a weighted (by utility weights or quality-of-life weights) survival analysis. It was a two-step approach, but was criticized mainly because it used utility values, which had, in practice, very poor measurement properties.

Our approach, philosophically introduced and discussed in Mesbah and Singpurwalla [37], can be considered as in the framework for mixed models with a clear interpretation of the random factor by a latent trait previously validated in a measurement step. Items are repeated in measurements of such true latent traits. Computer programs are now available with general softwares [26] that allows one to build and estimate models with nonlinear random-effects models.

Finally, let us conclude by mentioning some important current issues in quality-of-life research, that were not considered in depth in this chapter:

- Economic measures of quality of life
- Equating quality-of-life scores
- Missing data in quality-of-life studies
- Sample size issue in quality-of-life studies
- Test multiplicity in quality-of-life studies.

References

1. Admane, O. and Mesbah, M. (2006). Estimation des paramètres du modèle de Rasch dichotomique. *Annales de L'ISUP*.
2. Andersen, E. B. (1970). Asymptotic properties of conditional maximum likelihood estimators. *J. Roy. Statist. Soc. B*, **32**, 283–301.
3. Andrich, D. (1978). A rating formulation for ordered response categories. *Psychometrika*, **43**, 357–374.
4. Awad, L., Zuber, E., and Mesbah, M. (2002). Applying survival data methodology to analyze longitudinal quality of life data. In *Statistical Methods for Quality of Life Studies, Design, Measurement and Analysis*, M. Mesbah, B. F. Cole, and M. L. T. Lee, eds., pp 231–243. Kluwer Academic Publishing, Boston.
5. Bren, L. (2006). The importance of patient-reported outcomes ...it's all about the patients. *FDA Consumer Magazine*. (Nov.–Dec. 2006); available at *http://www.fda.gov/fdac/features/2006/606_patients.html*.
6. Breslow, N. E. and Clayton, N. E. (1993). Approximate inference in generalized linear mixed models. *J. Am. Statist. Assoc.*, **88**, 9–25.
7. Chi, Y.-Y. and Ibrahim, J. G. (2006). Joint models for multivariate longitudinal and multivariate survival data. *Biometrics*, **62**, 432–445.
8. Christensen, K. B., Bjorner, J. B., Kreiner, S., and Petersen, J. H. (2004). Latent regression in loglinear Rasch models. *Commun. Statist. Theory Methods*, **33**, 1295–1313.
9. Chwalow, J., Meadows, K. , Mesbah, M., and the FASAVED Group (2007). Empirical validation of a quality of life instrument: Empirical internal validation and analysis of a quality of life instrument in French diabetic patients during an educational intervention. In *Mathematical Methods in Survival Analysis, Reliability and Quality of Life*, C. Huber, N. Limnios, M. Mesbah, and M. Nikulin, eds. Wiley, Chichester, UK.
10. Cowling, B. J., Hutton, J. L., and Shaw, J. E. H. (2006). Joint modelling of event counts and survival times. *Appl. Statist.*, **55**, 31–39.
11. Cox, D. R., Fitzpatrick, R., Fletcher, A. E., Gore, S. M., Spiegelhalter, D. J., and Jones, D. R. (1992). Quality-of-life assessment: Can we keep it simple? *J. Roy. Statist. Soc. A*, **155**, 353–393.
12. Cronbach, L. J. (1951). Coefficient alpha and the internal structure of tests. *Psychometrika*, **16**, 297–334.
13. Curt, F., Mesbah, M., Lellouch, J., and Dellatolas, G. (1997). Handedness scale: how many and which items? *Laterality*, **2**, 137–154.
14. Dorange, C., Chwalow, J., and Mesbah, M. (2003). Analyzing quality of life data with the ordinal Rasch model and NLMixed SAS procedure. In *Proc. Int. Conf. Advances in Statistical Inferential Methods (ASIM2003)*, pp. 41–73. Kimep, Almaty.
15. Dupuy, J. F., Grama, I., and Mesbah, M. (2006). Asymptotic theory for the Cox model with missing time dependent covariate. *Ann. Statist.*, **34**.
16. Dupuy, J. F. and Mesbah, M. (2002). Joint modeling of event time and nonignorable missing longitudinal data. *Lifetime Data Anal.*, **8**, 99–115.
17. Edwards, D. (2000). *Introduction to Graphical Modeling*, 2nd ed. Springer-Verlag, New York.

18. Feddag, M. L., Grama, I., and Mesbah, M. (2003). GEE to the logistic mixed models. *Commun. Statist. Theory Methods*, **32**, 851–874.

19. Feddag, M. L. and Mesbah, M. (2005). Generalized estimating equations for longitudinal mixed Rasch model. *J. Statist. Plan. Infer.*, **129**, 159–179.

20. Fisher, G. H. and Molenaar, I. W. (1995). *Rasch Models, Foundations, Recent Developments and Applications*. Springer-Verlag, New York.

21. Frydenberg, M. (1990). Marginalization and collapsibility in graphical interaction models. *Ann. Statist.*, **18**, 790–805.

22. Gelber, R. D., Goldhirsch, A., Cole, B. F., Wieand, H. S., Schroeder, G., and Krook, G. E. (1996). A quality-adjusted time without symptoms or toxicity (Q-TWIST) analysis of adjuvant radiation therapy and chemotherapy for resectable rectal cancer. *J. Natl. Cancer Inst.*, **88**, 1039–1045.

23. Hamon, A., Dupuy, J. F., and Mesbah, M. (2002). Validation of model assumptions in quality of life measurements. In *Goodness of Fit Tests and Model Validity*, C. Huber, N. Nikulin, N. Balakrishnan, and M. Mesbah, eds., pp. 371–386. Birkhauser, Boston.

24. Hamon, A. and Mesbah, M. (2002). Questionnaire reliability under the Rasch model. In *Statistical Methods for Quality of Life Studies. Design, Measurement and Analysis*. M. Mesbah, B. F. Cole, and M. L. T. Lee, eds., pp. 155–168. Kluwer Academic Publishing, Boston.

25. Hardouin, J. B. and Mesbah, M. (2004). Clustering binary variables in subscales using an extended Rasch model and Akaike information criterion. *Commun. Statist. Theory Methods*, **33**, 1277–1294.

26. Hardouin, J. B. and Mesbah, M. (2007). The SAS macro-program %ANAQOL to estimate the parameters of item responses theory models. *Commun. Statist. Simul. Compu.*, **36**, 437–453.

27. Kreiner, S. and Christensen, K. B. (2002). Graphical Rasch models. In *Statistical Methods for Quality of Life Studies. Design, Measurement and Analysis*, M. Mesbah, B. F. Cole, and M. L. T. Lee, eds., pp. 155–168. Kluwer Academic Publishing, Boston.

28. Kristof, W. (1963). The statistical theory of stepped-up reliability coefficients when a test has been divided into several equivalent parts. *Psychometrika*, **28**, 221–238.

29. Lauritzen, S. L. and Wermuth, N. (1989). Graphical models for association between variables, some of which are qualitative and some quantitative. *Ann. Statist.*, **17**, 31–57.

30. Liang, K. Y. and Zeger, S. L. (1986). Longitudinal data analysis using feneralised linear models. *Biometrika*, **73**, 121–130.

31. Mac Cullagh, P. and Nelder, J. (1989). *Generalized Linear Models*. Chapman & Hall, London.

32. Masters, G. N. (1982). A Rasch model for partial credit scoring. *Psychometrika*, **47**, 149–174.

33. Martynov, G. and Mesbah, M. (2006). Goodness of fit test and latent distribution estimation in the mixed Rasch model. *Commun. Statist. Theory Methods*, **35**, 921–935.

34. Mesbah, M. (2004). Measurement and analysis of health related quality of life and environmental data. *Environmetrics*, **15**, 471–481.

35. Mesbah, M., Dupuy, J. F., Heutte, N., and Awad, L. (2004). Joint analysis of longitudinal quality of life and survival processes. In *Handbook of Statistics, Vol. 22, Advances in Survival Analysis*, N. Balakrishnan and C. R. Rao, eds. North Holland, Amsterdam.

36. Mesbah, M., Lellouch, J., and Huber, C. (1991). The choice of loglinear models in contingency tables when the variables of interest are not jointly observed. *Biometrics*, **48**, 259–266.

37. Mesbah, M. and Singpurwalla, N. (2008). A Bayesian ponders "the quality of life." In *Goodness of Fit Tests and Model Validity*. F. Vonta, M. Nikulin, N. Limnios,

and C. Huber-Carol, eds., pp. 369–381. Birkhauser, Boston.

38. Mislevy, R. J. (1984). Estimating latent distribution. *Psychometrika*, **49**, 359–381.

39. Molenaar, I. W. and Sijstma, K. (1988). Mokken's approach to reliability estimation extended to multicategory items. *Kwantitatieve Methoden*, **9**, 115–126.

40. Moret, L., Mesbah, M., Chwalow, J., and Lellouch, J. (1993). Validation interne d'une échelle de mesure: Relation entre analyse en composantes principales, coefficient alpha de Cronbach et coefficient de corrélation intra-classe. *la Revue d'Epidémiologie et de Santé Publique*, **41**, 179–186.

41. Nordman, J. F., Mesbah M., and Berdeaux G. (2005). Scoring of visual field measured through Humphrey perimetry: Principal component, varimax rotation followed by validated cluster analysis. *Invest. Ophtalmol. Visual Sci.*, **48**, 3168–3176.

42. Nunnally, J. (1978). *Psychometric Theory*, 2nd ed. McGraw-Hill, New York.

43. Prentice, R. L. and Zhao, L. P. (1991). Estimating equation for parameters in means and covariances of multivariate discrete and continuous responses. *Biometrics*, **47**, 825–839.

44. Rasch, G. (1960). *Probabilistic Models for Some Intelligence and Attainment Tests*, Danmarks Paedagogiske Institut, Copenhagen.

45. Sébille, V. and Mesbah, M. (2005). Sequential analysis of quality of life Rasch measurements. In *Probability Statistics and Modeling in Public Health In Honor of Marvin Zelen*, M. Nikouline, D. Commenges, and C. Huber, eds., pp. 421–439. Wiley, New York.

46. van Zyl, J. M., Neudecker, H., and Nel, D.G. (2000). On the distribution of the maximum likelihood estimator of Cronbach's alpha. *Psychometrika*, **65**, 271–280.

47. Vonesh, E. F., Greene, T., and Schluchter, M. D. (2006). Shared parameter models for the joint analysis of longitudinal data and event times. *Statist. Med.*, **25**, 143–163.

48. The WHOQoL Group (1994). The development of the World Health Organization Quality of Life Assessment Instrument (the WHOQoL). In *Quality of Life Assessment: International Perspectives*, J. Orley and W. Kuyken, eds.. Springer-Verlag, Heidleberg.

75. Statistics at CDC

William K. Sieber Jr., Donna F. Stroup, and G. David Williamson

75.1 Introduction

The Centers for Disease Control and Prevention (CDC) is recognized as the lead US federal agency for protecting the health and safety of people. It accomplishes this mission by promoting health and quality of life with programs preventing and controlling disease, injury, and disability. In this mission CDC is joined by the Agency for Toxic Substances and Disease Registry (ATSDR), which is charged with evaluating the human health effects of exposure to hazardous substances. Both CDC and ATSDR are components of the US Department of Health and Human Services (HHS). In 2004, CDC undertook an ambitious strategic planning process called the "Futures Initiative" to better meet the challenges to public health in the twenty-first century. Modernizing CDC will enhance its impact on emerging health problems, support its capacity to respond to public health emergencies, and more effectively serve the American public. In this transition, CDC decided to align its priorities and investments under two overarching health protection goals:

- *Health promotion and prevention of disease, injury, and disability.* All people, especially those at higher risk due to health disparities, are to achieve their optimal lifespan with the best possible quality of health in every stage of life.

- *Preparedness.* People in all communities are to be protected from infectious and occupational, environmental, and terrorist threats.

To accomplish these goals, the newly reorganized CDC comprises of four Coordinating Centers (Coordinating Center for Infectious Diseases, Coordinating Center for Health Promotion, Coordinating Center for Health Information and Service, and the Coordinating Center for Environmental Health and Injury Prevention), three Offices (Coordinating Office for Global Health, Coordinating Office for Terrorism Preparedness and Emergency Response, and Office of the Director), and the National Institute for Occupational Safety and Health with responsibilities for birth defects and developmental disabilities, chronic disease prevention and health promotion, environmental health, health statistics, infectious diseases, human immunodeficiency virus (HIV) infection and other sexually transmitted diseases, tuberculosis prevention, injury prevention and control, vaccine-preventable disease, occupational safety and health, public health surveillance, community practice of public health, and terrorism preparedness and emergency response. CDC/ATSDR employs more than 14,000 people in locations throughout the United States and over 50 foreign countries, with headquarters in Atlanta, Georgia. CDC/ATSDR employs approximately 330 mathematical and health statisticians.

The CDC began in 1946 as the Communicable Disease Center and successor to the Office of Malaria Control in War Areas, part of the US Public Health Service. In 1951 it established the Epidemic Intelligence Service (EIS), which is a training program in applied epidemiology. EIS officers investigate all types of epidemics, including chronic disease and injuries. Publication of the Morbidity and Mortality Weekly Report (MMWR), providing weekly data on deaths and morbidity in each state, was transferred to CDC in 1961. CDC has been renamed several times to reflect its growth and dynamic mission, becoming the current Centers for Disease Control and Prevention in 1992. ATSDR was created in 1980 by congressional mandate to implement the health-related sections of laws protecting the public from hazardous wastes and environmental spills of hazardous substances. Thus, ATSDR concentrates its efforts at Superfund sites and works closely with the Environmental Protection Agency as well as with CDC. CDC performs many of the administrative functions for ATSDR, and the CDC Director also serves as the Administrator of ATSDR. In 1987, the National Center for Health Statistics (NCHS) became a part of CDC. NCHS is a federal statistical agency with broad responsibilities to monitor the health of the nation. In addition to conducting a data collection program that encompasses vital statistics, interview surveys, examination surveys, and provider surveys, NCHS prepares the annual report, *Health, United States* [13] submitted by the DHHS Secretary to the President and Congress to present a comprehensive profile of health in America and track key indicators and trends. NCHS is also responsible for advancing the field of health statistics through research into statistical and analytical methods.

75.2 Public Health Data and Statistics at CDC/ATSDR

Since the inception of the agency, an important function of CDC has been compilation, analysis, and interpretation of statistical information to guide actions and policies to improve health. Sources of data include vital statistics records, medical records, personal interviews, telephone and mail surveys, physical examinations, and laboratory testing. Surveillance data have been used to characterize the magnitude and distribution of illness and injury, to track health trends, and for development of standard curves such as growth charts. Beyond the development of appropriate program study designs and analytic methodologies, statisticians have played roles in the development of public health data collection systems and software to analyze collected data.

75.3 Statistics and Research at CDC/ATSDR

The integration of statistics and analytic techniques into public health research is a critical asset to the agency, and has resulted in important applications in various disciplines such as epidemiology, economics, and the behavioral and social sciences. Examples include the use of statistical methods to demonstrate the relationship between removing lead from gasoline and decreased blood lead levels in children and adults [1], economic determinations contributing to folic acid supplementation of foods to decrease birth defects [9], behavioral science methods leading to the development of strategies for the prevention of HIV/AIDS [8], quantitative epidemiological analyses leading to an understanding of the relationships between radon and lung cancer in coal miners [6], and evaluations of the effectiveness of using back belts in reducing back injury claims and back pain [26]. Other areas of continuing statistical contribution include survey planning and analytic methodology, data collection systems, detection algorithms and scan statistics to document health trends and identify emerging health issues, and model development to project disease incidence and injury or numbers of cases prevented through treatment and public health measures during an outbreak. For example, new methodologies have been developed to make it possible to compare population characteristics across data collection programs and over time when there is a change in data collection methods [19] and to quantify disparities in health and health care [10]. Methodological work has also been done on how to deal with high levels of nonresponse on central variables such as income [20]. The National Laboratory for Collaborative Research in Cognition and Survey Measurement applies cognitive methods to questionnaire design research and the testing of data collection instruments to improve data quality [14]. The reliance on statistics for policy and programmatic use and the growing number and diversity of users has required ongoing research and innovative approaches to protect the confidentiality of respondents and data sources while offering the widest possible access to data [3].

Recent CDC activities presenting new analytic challenges include counterterrorism and emerging infectious diseases. CDC statistical programs have contributed to development of methodologies for syndromic surveillance, techniques useful during anthrax investigations and remediation, and approaches to the civilian smallpox vaccination effort. Such techniques have also been useful for emerging infectious diseases such as severe acute respiratory syndrome (SARS) or influenza and in extending analytic capabilities for chronic diseases and in developing approaches to national health report cards (i.e., health assessments or health profiles).

The anthrax investigations of September–December 2001 spurred development of multiple analytic techniques. Maps linking analytic sampling activity with analytic results were developed to better understand the spread and deposition of spore-containing particles. Numbers of cases reported were used to estimate parameters of mathematical epidemiological models, which, in turn, simulated the potential size and duration of the outbreak and the effect of preventative measures [2]. Analyses of environmental sampling information have included comparison of analytic sampling techniques, tests of sensitivity and specificity of various presumptive and confirmatory tests for *Bacillus anthracis*, and use of statistical methods such as survival analysis and dose–response modeling to quantify critical exposure levels [22].

Stochastic simulation has been used to optimize patient flowthrough in clinics dispensing oral antibiotics following a bioterrorism attack [25]. Aberration detection in public health data represents another area of statistical contribution [7]. For example, the Division of Emergency Preparedness and Response at CDC receives vaccination and adverse-event data from several sources. These sources employ both active and passive data collection and provide registry, contraindication, and adverse events information. These disparate data are joined into analyzable elements needed for decisionmaking.

75.4 The CDC/ATSDR Statistical Advisory Group

Coordination of statistical activities across CDC and ATSDR resulted in formal recognition of the statistical community in 1989. In that year, at the request of the CDC Office of the Director, a coordinating group for analytic methods known as the CDC/ATSDR Statistical Advisory Group (SAG) was created. The SAG was established to act in an advisory capacity to the Office of the Director to facilitate and address statistical issues, problems, and opportunities that influenced the quality and integrity of science at CDC, and to increase communication across organizational components. It is one of eight science-related work groups at CDC/ATSDR.

Some activities of the SAG further illustrate the breadth of statistical activity across CDC/ATSDR. Biennial symposia have been held since 1989 on topics of interest to the public health community (clustering methods [24], surveillance [17], evaluation of intervention and prevention strategies [18], use of multi-source data [27], small-area statistics [23], public health decisionmaking [4], emerging statistical issues in public health [21], data structures [11], study design [12], public health policy [15], and health inequities). Each year the SAG recognizes outstanding statistical papers published during the previous year with the CDC/ATSDR Statistical Science Awards. Recent winners included manuscripts on linking geographic data [5] and combining information from surveys [16]. SAG is responsible for advanced statistical/epidemiological training and maintains a Listserv and Internet site (*http://www.cdc.gov/sag*).

Other statistical activities include participation in statistical/protocol review and leadership in the development, procurement, and installation of statistical software available for use by the CDC/ATSDR community. The SAG has provided review and advice on complex statistical and broad scientific issues such as validation of the statistical design of the Vietnam Experience Study of the health of Vietnam veterans, and code-veloped an evaluation of recruitment and retention policies at CDC/ATSDR, for example. Other special requests, such as for development of training materials or requests for interagency collaboration/representation, are also often handled through the SAG. Since 1990 the SAG has sponsored an exhibit booth highlighting statistical activities at CDC/ATSDR that has been displayed at the Joint Statistical Meetings and other conferences for informational and recruiting purposes.

75.5 Future Directions for Statistics at CDC/ATSDR

The critical role of statistics toward accomplishing the mission of CDC/ATSDR will become even more apparent as the agency begins to align its activities around its overarching goals of health promotion and preparedness. The assessment of burden, effectiveness of interventions, cost considerations, and evaluation frameworks will all require rigorous attention to methods of data collection, study design, and analytic technique. The role of statisticians to ensure that quantitative sciences will be most effectively utilized in research and analysis, and in meeting new challenges in CDC/ATSDR's evolving public health mission, will require a reexamination of statistical skills and contributions. A multidisciplinary approach to investigation of public health problems such as counterterrorism or obesity is already being realized. Efficient use of new technologies such as in informatics, Web-based query systems, geographic information systems, and survey data collection methodologies will be key to continued valuable statistical input. Advances in the field of relational databases, for example, and its coupling with Web-based technology, have facilitated improvements in the efficiency of data collection and increases in size and completeness of data available for analysis. The BioSense program currently under development at CDC uses information from both traditional and nontraditional data sources (such as emergency calls and over-the-counter drug sales) for syndromic surveillance. Such use of multisource data to extract maximal information from existing data sources will also require addressing privacy and confidentiality concerns, as well as appropriate methods of communication of important public health findings to the nation.

Acknowledgment. The authors wish to express their appreciation to Dr. Jennifer Madans for her help in preparing this description of statistical activities across CDC and ATSDR. across CDC and ATSDR.

References

1. Annest, J. L., Pirkle, J. L., Makuc, D., Neese, J. W., Bayse, D. D., and Kovar, M. G. (1983). Chronological trend in blood lead levels between 1976 and 1980. *New Engl. J. Med.*, **308**, 1373–1377.

2. Brookmeyer, R. D. and Blades, N. (2003). Statistical models and bioterrorism: Application to the U.S. anthrax outbreak. *J. Am. Statist. Assoc.*, **98**(464), 781–788.

3. Cox, L. H. (2002). Disclosure risk for tabular economic data. In *Confidentiality, Disclosure and Data Access: Theory and Practical Applications for Statistical Agencies*, P. Doyle, J. Lane, J. Theeuwes, and L. Zayatz, eds., pp. 167–183 Elsevier, New York.

4. Falter, K. H., Betts, D. R., Rolka, D. B., Rolka, H. R., and Sieber, W. K., eds. (1999). Symposium on statistical bases for public health decision-making: From exploration to modelling. *Statist. Med.* **18**(23), 3159–3376.

5. Gotway, C. A. and Young, L. J. (2007). A geostatistical approach to linking geographically aggregated data from different sources. *J. Comput. Graph. Statist.*, **16**(1), 115–135.

6. Hornung, R. W. and Meinhardt, T. J. (1987). Quantitative risk assessment of lung cancer in U.S. Uranium miners. *Health Phys.*, **52**(4), 417–430.

7. Hutwagner, L. H., Thompson W. W., Seeman, G. M., and Treadwell, T. (2005). A simulation model for assessing aberration detection methods used in publish health surveillance for systems with limited baselines. *Statist. Med.*, **24**(4), 543–550.

8. Kamb, M. L., Fishbein, M., Douglas, J. M., Rhodes, F., Rogers, J., Bolan, G., Zenilman, J., Hoxworth, A., Malotte, K., Iatesta, M., Kent, C., Lentz, A., Graziano, S., Byers, R., and Peterman, T. (1998). Efficacy of risk-reduction counseling to prevent human immunodeficiency virus and sexually transmitted diseases. *J. Am. Med. Assoc. (JAMA)*, **280**, 1161–1167.

9. Kelly, A. E., Haddix, A. C., Scanlon, K. S., Helmick, C. G., and Mulinare, J. (1996). Cost-effectiveness of strategies to prevent neural tube defects. In *Cost-Effectiveness in Health and Medicine*, M. R. Gold, J. E. Siegel, L. B. Russell, and M. C. Weinstein, eds., pp. 313–348 Oxford University Press, New York.

10. Keppel, K. G., Pearcy, J. N., and Klein, R. J. (2004). *Measuring Progress in Health People 2010*. Healthy People 2010 Statistical Notes No. 25, National Center for Health Statistics, Hyattsville, MD.

11. Lin, L. S., Conn, J. M., Green, T. A., Johnson, C. H., Odencrantz, J. R., and Sieber, W. K., eds. (2003). 8th Biennial CDC and ATSDR symposium on statistical methods: Issues associated with

complicated designs and data structures. *Statist. Med.*, **22**(9), 1359–1626.

12. Lipman, H. (2005). Symposium on statistical methods: Study design and decision making in public health. *Statist. Med.*, **24**(4), 491–669.

13. National Center for Health Statistics (2007). *Health, United States, 2007, with Chartbook on Trends in the Health of Americans (PHS) 2007-1232*. GPO 017-022-01606-4. Hyattsville, MD.

14. National Center for Health Statistics. *Cognitive Methods*. Working Paper Series, www.cdc.gov/nchs/products/pubs.

15. Pallos, L. L., Andrew, M. E., Banerjee, S. N., Lee, T. D., Pals, S. L., and Sieber, W. K. (2007). Symposium on statistical methods: Statistics in public health policy. *Statist. Med.*, **26**(8), 1653-1895.

16. Raghunathan, T. E., Xie, D., Schenker, N., Parsons, V. L., Davis, W. W., Dodd, K. W., and Feue, E. J. (2007). Combining information from two surveys to estimate county-level prevalence rates of cancer risk factors and screening. *J. Am. Statist. Assoc.*, **102**, 474–486.

17. Reynolds, G. H., McGee, D. L., and Stroup, D. F., eds. (1989). Symposium on statistics in surveillance. *Statist. Med.*, **8**(3), 251–400.

18. Reynolds, G. H., ed. (1993). Symposium on statistical methods for evaluation of intervention and prevention strategies, December 1990. *Statist. Med.* **12**(3/4), 191–414.

19. Schenker, N., and Parker, J. D. (2003). From single-race reporting to multiple-race reporting: using imputation methods to bridge the transition. *Statist. Med.*, **22**, 1571–1587.

20. Schenker, N., Raghunathan, T. G., Chu, P., Makuc, D. M., Zhang, G., and Cohen, A. J. (2004). *Multiple Imputation of Family Income and Personal Earnings in the National Health Interview Survey*. National Center for Health Statistics, Hyattsville, MD.

21. Sieber, W. K., Green, T. A., Haugh, G. S., Kresnow, M., Luman, E. T., and Wilson, H. G., eds. (2001). Symposium on emerging statistical issues in public health for the 21st century (2001). *Statist. Med.*, **20**(9/10), 1307–1362.

22. Sieber, W. K., Bennett, J. S., Gillen, M., Shulman, S., Pulsipher, B., Sego, L., and Wilson, J. (2008). *Development of Probabilistic Sampling Options to Supplement Judgemental Approaches Used during Initial Response Sampling Following a Terrorism Incident*. Paper presented at CDC Conf. Emergency Response and Prepardness. Oct. 4–5, 2008, Atlanta, GA.

23. Smith, S. J., ed. (1996). Symposium on small area statistics in public health: Design, analysis, graphic and spatial methods. *Statist. Med.*, **15**(17/18), 1827–1986.

24. Steinberg, K., ed. (1990). National conference on clustering of health events, *Am. J. Epidemiol.* **132**(1), S1–S202.

25. Washington, M. L. (2003). *Optimizing Patient Flow-through in Large-scale Clinics for Dispensing Oral Antibiotics Following a Bioterrorism Attack Using Discrete-Event Computer Simulation*, Paper Presented at 9th. Biennial CDC/ATSDR Symp. on Statistical Methods: Study Design and Decision Making. Jan. 27–29, 2003, Atlanta, GA.

26. Wassell, J. T., Gardner, L. I., Landsittel, D. P., Johnston, J. J., and Johnston, J. M. (2000). A prospective study of back belts for prevention of back pain and injury. *JAMA*, **284**(21), 2727–2732.

27. Williamson, G. D., Massey, J. T., Shulman, H. B., Sieber, W. K., and Smith, S. J., eds. (1995). Symposium on quantitative methods for utilization of multi-source data in public health. *Statist. Med.*, **14**(5/6/7), 447–718.

76 Statistics in Dentistry

Shu L. Cheuk and Roger Weinberg

In order to give their patients the best available treatment, dentists need to be able to interpret the statistics that are used in evaluating new and possibly improved treatments appearing in dental journals. About one-fifth of such articles in the 1981 issues of the *Journal of the American Dental Association*, which has the widest circulation of any dental journal, used inferential statistics to test hypotheses about their assertions. The *Journal of Dental Research* has a smaller circulation, but its research orientation may make it even more important to a dentist searching for improved treatments; approximately half of the articles published in 1981 used inferential statistics to substantiate their analysis.

Most of the statistical methods used by these and other journals can be included in a fairly short list: paired and unpaired t tests, one-way analysis of variance, correlation and regression, and chi-square. Other statistical methods, such as multiple range tests, two-way analysis of variance, and nonparametric analyses (other than chi-square), appear occasionally in the literature. We have listed references to articles containing such infrequently used statistics at the beginning of our bibliography.

Beertsen and Everts used the paired t test [1]. It revealed that freezing the periodontal ligament of one central incisor decreased that ligament's fibroblastic nuclei compared to the ligament of the paired incisor on the other side of the jaw. Although before–after comparisons make use of the paired t test in dentistry as well as in other disciplines, pairing opposite sides of the mandible makes the paired t test particularly suitable to many dental experiments.

Dentists, like other biological scientists, use the unpaired t test to compare a difference between the means of two independent groups. For example, Ranney et al. used that test to show that subjects with severe periodontal disease had a significantly different mean level of IgG immunoglobin in their saliva than did normal subjects [4].

Greenberg et al. used one-way analysis of variance to compare the mean upper-body strengths of groups with different mandibular positions [3]. The F statistic obtained from that experiment failed to show that the groups differed significantly. Although such a negative result is not conclusive (perhaps a larger number of subjects could reveal some significant difference among groups), their article did show that their experiment did not support any relationships between mandibular position and upper-body strength.

Linear regression predicts a dependent

variable's value from the values of an independent variable. Correlation then estimates the strength of association between those two variables. These two methods are commonly applied to predict caries activity from a variable suspected of influencing the incidence (e.g., fluoride concentration in water and lacobacillus count on tooth surfaces) and to predict behavior from an attitudinal survey. An example of the former is Crossner's finding that regression can predict caries activity from salivary lactobacillus counts, and that the correlation is significant [2].

The *chi-square* method is commonly used by dentists when they compare the effects of different kinds of injections, commonly classified as satisfactory or unsatisfactory, or into less than four categories of satisfaction. They then count the number in each category for each kind of injection. A typical example is Walton and Abbott's demonstration that back pressure significantly increased the proportion of satisfactory anesthesia in periodontal ligament injections [5].

Before 1980, dentists needed to understand a fairly restricted list of statistical analyses in order to evaluate evidence for better treatments. That limited list does not now suffice for all articles. The Duncan multiple range test to determine pairwise differences in means for more than two groups, replacing the t test as it should, appears in almost every recent issue of the *Journal of Dental Research*. Nonparametric statistics like the Mann–Whitney U, Wilcoxon's signed rank, and Kendall's correlation coefficient are beginning to appear more often in dental journals.

As dental researchers continue to expand their ability to analyze data and refine their ability to choose the correct tests, the clinicians will have to improve their ability to understand the increasingly complex statistics that appear in those articles. Only by so doing can they hope to keep abreast of current knowledge in the field.

References

1. Beertsen, W. and Everts, V. (1981). *J. Periodont. Res.*, **16**, 524–541.

2. Crossner, C.-G. (1981). *Community Dent. Oral Epidemiol.*, **9**, 182–189.

3. Greenberg, M. S., Cohen, S. G., Springer, P., Kotwick, J. E., and Vegso, J. J. (1981). *J. Am. Dent. Assoc.*, **103**, 576–579.

4. Ranney, R. R., Ruddy, S., Tew, J. G., Welshimer, H. J., Palcanis, K. G., and Segreti, A. (1981). *J. Periodont. Res.*, **16**, 390–402.

5. Walton, R. E. and Abbott, B. J. (1981). *J. Am. Dent. Assoc.*, **103**, 571–575.

77 Statistics in Evolutionary Genetics

K. Simonsen and W. J. Ewens

Statistics and population genetics are cousin subjects, which grew together during the twentieth century in a relationship in which each has decisively influenced the other. Many genetic problems led to the creation of new statistical procedures for their solution, while many central statistical concepts, in particular the concept of the analysis of variance, originated in the genetic realm.

The applications of statistics in genetics fall essentially into four main areas: biometrical analysis, evolutionary theory, human genetics, and plant and animal breeding. We focus here on evolutionary genetics, although many concepts (such as the additive genetic variance) are central to all four areas.

In the decade following the rediscovery of the Mendelian hereditary system in 1900, there was considerable and often acrimonious debate between the group of biometricians around Karl Pearson, who saw little evolutionary significance in the Mendelian system, and those biologists, around Bateson, who saw evolution occurring in a sequence of non-Darwinian quantal jumps through genetic mutation. The biometricians, who as part of their analysis of a gradualist Darwinian process had introduced the statistical concepts of correlation and regression, had noted various regular forms for the correlation of metrical traits such as height between various grades of relatives. Their evolutionary theories centered around these measurements (the *regression to the mean* being an evolutionary concept), and they felt that the correlations they observed could not be reconciled with Mendelian inheritance. It was an early triumph of statistical genetics to show that this reconciliation is possible and natural, and in carrying out the relevant calculations, Fisher [3] laid the groundwork for one of his later major statistical achievements, namely, introduction of the concept of the analysis of variance. The relevant calculations, in the simplest possible case, are as follows.

We consider one gene locus A, and assume that two possible alleles (i.e., gene types) A_1 and A_2 occur at this locus. Any individual then has one or another of the three genotypes A_1A_1, A_1A_2, and A_2A_2. We assume random mating in the population and no complications such as a geographic distribution of the individuals in the population, mutation, etc.

Suppose that we are interested in some measurable character of the individuals in the population, and that all individuals of genotypes $A_1 A_1$ have measurement m_{11} for this character, all individuals of genotype $A_1 A_2$ have measurement m_{12}, and all individuals of genotype $A_2 A_2$ have measurement m_{22}. Then the mean $\overline{m}$ and the variance σ^2 of the measurement are

$$\begin{aligned} \overline{m} &= m_{11}x^2 + 2m_{12}x(1-x) \\ &\quad + m_{22}(1-x)^2, \\ \sigma^2 &= m_{11}^2 x^2 + 2m_{12}^2 x(1-x) \\ &\quad + m_{22}^2(1-x)^2 - \overline{m}^2, \end{aligned}$$

where x is the population frequency of A_1. Further, since there is a one-to-one correspondence between measurement value and genotype, one can write down all possible combinations for this measurement between any pair of relatives (mother–daughter, brother–brother, etc.), together with their probabilities, using as a key part of the argument the Mendelian rules for the transmission of genes from parent to offspring. From this one can find the covariance, and hence the correlation, between the two relatives for the measurement in question. When this is done, a remarkable series of formulae arises [3]. All such correlations are of the form

$$\text{Corr} = \frac{\alpha \sigma_A^2 + \beta \sigma_D^2}{\sigma^2},$$

where α and β are simple constants such as $0, \frac{1}{2}$, and $\frac{1}{4}$ [and more generally $(\frac{1}{2})^k$ for some integer k], and the *additive genetic variance* σ_A^2 and *dominance variance* σ_D^2 are defined by

$$\begin{aligned} \sigma_A^2 &= 2x(1-x)[xm_{11} + (1-2x)m_{12} \\ &\quad - (1-x)m_{22}]^2, \\ \sigma_D^2 &= x^2(1-x)^2 \\ &\quad \times (m_{11} - 2m_{12} + m_{22})^2. \end{aligned}$$

Then $\sigma^2 = \sigma_A^2 + \sigma_D^2$, so that the total variance in the measurement subdivides into two components that enter differently into the correlation between relatives of different degrees.

These calculations refer to a greatly simplified case, but the conclusion just noted carries over to more realistic and complicated cases. Ignoring variance caused by environmental agencies, the total genetic variance in any character may be subdivided into meaningful components, each of which enters into the correlation formula in different proportions, depending on the nature of the relationship of the two persons involved in the correlation. The concept of the subdivision of a total variance into several components, each component having its own interpretation and significance, begins here. The genetical origin of the statistical concept of the analysis of variance is thus clear.

This subdivision of variance is also crucial in evolutionary population genetics, and is more easily explained in that context. The essence of the Darwinian theory is the improvement of a population over successive generations by natural selection, this improvement requiring the existence of variation within the population so that the more fit types can be selected for. If we take the fitness of individuals of any genotype as the measure discussed above, the mean measure $\overline{m}$ is the *mean fitness* of the population. An increase in the mean fitness is taken to be a Darwinian improvement in the population.

It is possible to use evolutionary genetics theory to show that the increase in mean fitness from one generation to another is not the total variance in fitness, but is rather the component σ_A^2 of the total variance in fitness. This conclusion neatly encapsulates in Mendelian terms the Darwinian idea of improvement deriving from variation. However, there is a new feature — only part of the variance in fitness, namely σ_A^2, has an evolutionary significance. If σ_A^2 is zero, there is no evolutionary change in the population. It is possible to show that σ_A^2 is the component of the total variance in fitness caused by *genes within genotypes*. This explains the evolutionary importance of the variance component σ_A^2 — since a parent passes on only one randomly chosen gene to any of his/her offspring, and not his/her complete two-gene genotype, it is only the gene component of the total variation in fitness, namely, σ_A^2, that has an evolutionary significance. Here the concept of the analysis of variance is seen again, leading to a genetic evolutionary paradigm more subtle than that originally envisioned by Darwin.

Recognition that the gene transmitted from parent to offspring is determined randomly, together with recognition of the many chance factors arising in the biological world, led Fisher [4] and [9] to the development of stochastic models of genetic evolution. The model at the center of their similar (but independent) work is now called the *Wright – Fisher model*. In the simplest case, suppose, as above, that the two alleles at a gene locus A are A_1 and A_2. The random variable considered is the number of A_1 genes in any generation, and the Wright – Fisher model is a Markov chain describing the stochastic evolution of this number. Specifically, if the population size is N, and if the number of A_1 genes in a parental generation is i, then in the simplest case (no selection, mutation, geographic population structure, etc.) the number of A_1 genes in the daughter generation has a binomial distribution with parameter $i/(2N)$ and index $2N$. Many results for this and similar models are now known [2].

More generally, there can be an arbitrary number of possible alleles at the locus under consideration. In a significant advance, Malécot [8] developed the theory in which an infinite number of allelic types is allowed. Following the recognition of the gene as a sequence of DNA nucleotides for which an effective infinity of types (i.e., DNA sequences) is, indeed, possible, many authors extended Malécot's work, and there developed a field of molecular evolution, pioneered by Kimura [6], which in effect placed population genetics theory in a molecular framework. These results derive from stochastic process theory and have little direct contact with statistical inference.

With the accumulation of large amounts of DNA data, however, the direction of evolutionary theory took a different, and more statistical, turn. It was no longer felt necessary to confirm the Darwinian theory by investigating the evolution forward in time of a genetic population. This task has been completed. Instead, the DNA data were increasingly used to test hypotheses about the past evolution of a currently observed population and to estimate parameters of that evolution. In this endeavor the coalescent theory of Kingman [7] has played a central role.

Genetic sequences sampled from a population at the same locus will usually be similar but not identical. The similarity is the result of common ancestry, while the differences are attributed to mutations that have arisen since divergence from a common ancestor. These provide a "footprint" of the history of the sample, and can give information about the evolutionary history of the population.

A sample of sequences is the result of two processes with stochastic components. First, the evolutionary processes of mutation, selection, drift, population structure and size fluctuation, mating systems, and other historical events all influence the genetic makeup of a population at the time of sampling. Second, a set of genes sampled from the population has some ancestral relationship, or genealogy. The coalescent framework allows these two processes to be considered simultaneously in making inferences, and provides the sampling distribution for the genetic data.

The coalescent model is most easily described for a simple evolutionary scenario known as the *neutral model*, whose key assumptions are that the population remains at a constant size N, ($2N$ genes), offspring are produced by the random union of gametes, and no allele has a selective advantage over any other. Modifications to this model can be made to encompass many other evolutionary processes, but their analysis is complicated, and typically involves computer simulation.

We say that the lineages of two genes *coalesce* in generation t if, going back in time, their most recent common ancestor (MRCA) existed t generations ago. In the simple model considered above, t has a geometric distribution with mean $2N$. Suppose more generally that we trace back the ancestry of a sample of n genes. At some time two of these ancestries coalesce, the number of generations back to the MRCA of the two genes involved having a geometric distribution with mean $4N/[n(n-1)]$. At the time of that common ancestor we can consider a new sample of size $n-1$, consisting of that common ancestor and the direct ancestors of the other $n-2$ members of the sample. At some time further in the past, two of these ancestries coalesce, the coalescent time having a geometric distribution with mean $4N/[(n-1)(n-2)]$. This process may be repeated through a series of coalescent events on samples of decreasing size, until only one gene remains; this is the MRCA of the entire sample. A sample genealogy is thereby constructed that has two attributes: a topology determined by the order in which pairs coalesce, and branch lengths determined by the coalescence times.

If mutation did not occur, then all the sampled genes would be identical in type to one another and to their common ancestor. Mutation is often modeled as a Poisson process that occurs independently along the branches of the generalogy. Mutations on a branch are inherited by all offspring descended from that branch. Thus the branch lengths affect the number of mutations, and the topology determines their pattern and frequency in the sample.

Various forms of selection and population size fluctuation may be incorporated into the coalescent model by modifying the per-generation probability of coalescence. Population structure and nonrandom mating can be modeled by preferentially coalescing within subgroups. Recombination between genetic loci yields a much more complicated process, wherein both recombination and coalescent events affect the structure of the genealogy. The result is a set of possibly different but correlated genealogies at adjacent loci.

The coalescent theory has yielded many useful theoretical results for the neutral model and for other simple models. It is also a straightforward computational tool for stimulating data under various models; in computer simulation complex models can be considered. The ability to simulate samples of sequences rather than entire populations is quite valuable when a large number of samples must be considered, as when determining empirical distributions for test statistics or performing power studies.

Genetic applications of coalescence theory are given by Hudson [5]; probabilistic and statistical aspects are described by Donnelly and Tavaré [1].

References

1. Donnelly, P. J. and Tavaré, S. (1995). Coalescents and the genealogical structure under neutrality. *Ann. Rev. Genet.*, **29**, 542–551.

2. Ewens, W. J. (1979). *Mathematical Population Genetics*. Springer, Berlin.

3. Fisher, R. A. (1918). The correlation between relatives on the supposition of Mendelian inheritance. *Trans. Roy. Soc. Edinburgh*, **52**, 399–433.

4. Fisher, R. A. (1930). *The Genetical Theory of Natural Selection*. Clarendon, Oxford, UK.

5. Hudson, R. R. (1991). Gene genealogies and the coalescent process. In *Oxford Survey in Evolutionary Theory*, 7, D. Futuyma and J. Antonovics, eds. Oxford University Press, Oxford, UK.

6. Kimura, M. (1957). Theoretical foundations of population genetics at the molecular level. *Theor. Popul. Biol.*, **2**, 174–208.

7. Kingman, J. F. C. (1982). The coalescent. *Stoch. Process. Appl.*, **13**, 235–248.

8. Malécot, G. (1948). *The Mathematics of Heredity Pr(in French)*. Masson, Paris.

9. Wright, S. (1931). Evolution in Mendelian populations. *Genetics*, **16**, 97–159.

78. Statistics in Forensic Science

Colin G. G. Aitken

78.1 Introduction

Statistics in forensic science, *forensic statistics*, requires consideration of scientific evidence under two propositions, that of the prosecution and that of the defense, and is often concerned with the evaluation of so-called transfer evidence or trace evidence for identification purposes.

The process of addressing the issue of whether a particular item came from a particular source is most properly termed *individualization*. "Criminalistics is the science of individualization," as defined by Kirk [21], but established forensic and judicial practices have led to it being termed *identification*. An *identification*, however, is more correctly defined as *the determination of some set to which an object belongs or the determination as to whether an object belongs to a given set* [19,20]. Further discussion is given in Reference 22.

What follows will be described in the context of a criminal case but can be applied to a civil case in which there would be a plaintiff and a defendant. *Transfer evidence* is evidence that is transferred from the criminal to the scene of the crime or vice versa. Such evidence is often in the form of traces of some material, such as DNA, glass or fibers. The consideration of such evidence is best done through the Bayesian paradigm. Interest in the role of statistics in forensic science was fueled considerably by the advent of DNA profiling. However, there has been interest in the Bayesian assessment of transfer evidence since it was explored by Cullison [9] and Finkelstein and Fairley [14,15] and developed by Lindley [24]. Overviews are given by Aitken and Taroni [3] and Robertson and Vignaux [26]. The interpretation of DNA evidence is described by Evett and Weir [12]. Those not enamored of the Bayesian approach will find support for their views in References 29 and [18].

The use of techniques such as significance probabilities and analysis of variance, which may be used in the analysis of scientific results from experiments in forensic science, is not discussed here. Neither are situations involving the formal objective presentation of statistical evidence.

78.2 History

The following conclusion was presented in the appeal in the Dreyfus case by experts Darboux, Appell, and Poincaré [27]:

> Since it is absolutely impossible for us to know the 'a priori' prob-

ability, we cannot say: this coincidence proves that the ratio of the forgery's probability to the inverse probability is a real value. We can only say that, following the observation of this coincidence, this ratio becomes X times greater than before the observation.

This statement expresses the value of the evidence as a factor that converts the prior odds of the guilt of the suspect into posterior odds. It is interesting that this was written over 100 years ago and yet the approach taken is still a matter of keen debate within the forensic scientific community. Three approaches to identification evidence are discussed briefly before the Bayesian approach, which forms the main part of this chapter, is described.

78.2.1 Unsubstantiated Probabilistic Evidence

The case of *People vs. Collins* (a crime committed in 1963 and discussed at length in Fairley and Mosteller [13]) is an example of the use of frequency probabilities for the assessment of evidence. No justification was given for the characteristics chosen for consideration, or for the values of the frequency probabilities of these characteristics or for the independence assumptions used in their multiplication. The errors of evidence evaluation committed in this case set back the cause of robust evidence evaluation for many years.

Discriminating Power. The Bayesian approach to evidence evaluation, described Section 78.3, is applicable to particular cases. A technique known as *discriminating power* provides a general measure of how good a method is at distinguishing two samples of materials from different sources.

As an example, consider a population with a DNA marker system S in which there are k genotypes and in which the jth genotype has relative frequency p_j, such that $p_1 + \cdots + p_k = 1$. Two people are selected at random from this population such that their genotypes may be assumed independent. The probability Q of a match of genotypes between the two people in S is given by

$$Q = p_1^2 + p_2^2 + \cdots + p_k^2.$$

The discriminating power, or probability of discrimination, is $1 - Q$. Discriminating power takes values between 0 (no discrimination, all members of the system belong to one category) and $1 - 1/k$ (maximum discrimination, all k categories are equally likely). A system for which $1 - Q$ is low would not be a good discriminator among individuals. A system for which $1 - Q$ is close to 1, which may happen if k is large and the k categories are all equally likely, would provide a good discriminator amongst individuals. The method does not, however, provide a value for the evidence in a particular case. When the $\{p_1, \ldots, p_k\}$ are not available directly, empirical estimates of discriminating power may be obtained by considering pairs of items from different sources and determining the proportion of members of pairs that are indistinguishable from each other. This proportion provides an estimate of Q; a low value implies discrimination.

78.2.2 Significance Probabilities

These are inappropriate in the evaluation of evidence. A *significance probability* is a measure of the extremeness of an observation, assuming a proposition (null hypothesis) to be true, in relation to an alternative proposition. However, what is relevant in the evaluation of evidence is the likelihood only of what is observed, and not of anything more extreme.

78.3 Evaluation of Evidence

When considering the evaluation of evidence, three principles need to be borne in mind [12]:

1. To evaluate the uncertainty of any given proposition, it is necessary to consider at least one alternative proposition.

2. Scientific interpretation is based on questions such as "What is the probability of the evidence, given the proposition?"

3. Scientific interpretation is conditioned not only by the competing propositions but also by the framework of circumstances within which they are to be evaluated.

It is useful to distinguish between the value of evidence (defined below) and the interpretation, which is how the meaning of the value is conveyed to the court.

Evidence is evaluated using the odds form of Bayes' theorem with due regard to the three principles listed above. The following notation will be used to refer to various concepts:

- P: the proposition put forward by the prosecution (e.g., guilty, contact with crime scene, true father of child)

- D: the proposition put forward by the defense (e.g., not guilty, no contact with crime scene, not father of child)

- E: the evidence to be evaluated

- I: the background information

The odds form of Bayes' theorem relates these three concepts as follows:

$$\frac{\Pr(P \mid E, I)}{\Pr(D \mid E, I)} = \frac{\Pr(E \mid P, I)}{\Pr(E \mid D, I)} \times \frac{\Pr(P \mid I)}{\Pr(D \mid I)}.$$

The three fractions are

- The posterior odds in favor of the prosecution proposition

- The likelihood ratio

- The prior odds in favor of the prosecution proposition

This equation is fundamental to the evaluation of evidence. Note the following:

- The likelihood ratio is the factor that converts the prior odds in favor of the prosecution proposition into the posterior odds in favor of the prosecution proposition; a value for the likelihood ratio of greater than one can be thought to support the prosecution proposition in the sense that, in such a circumstance, the posterior odds in favor of the prosecution proposition are greater than the prior odds. Conversely, a value for the likelihood ratio that is less than one can be thought to support the defence proposition.

- This multiplicative expression can be converted into an additive expression by taking logarithms. The logarithm of the likelihood ratio has been termed by I. J. Good [16] the *weight of evidence*.

Denote the ratio

$$\frac{\Pr(E \mid P, I)}{\Pr(E \mid D, I)}$$

by V, and call this the *value* of the evidence.

Consider a case in which the evidence is the DNA genotyping results G_c and G_s (with $G_c = G_s$) for the crime stain profile and suspect's profile, respectively. Then, when I is omitted throughout for ease of notation, we obtain

$$V = \frac{\Pr(E \mid P)}{\Pr(E \mid D)}$$
$$= \frac{\Pr(G_c, G_s \mid P)}{\Pr(G_c, G_s \mid D)}$$
$$= \frac{\Pr(G_c \mid G_s, P)\Pr(G_s \mid P)}{\Pr(G_c \mid G_s, D)\Pr(G_s \mid D)}$$
$$= \frac{\Pr(G_c \mid G_s, P)}{\Pr(G_c \mid G_s, D)},$$

since the DNA genotype for the suspect is independent of the two propositions. Assuming no errors in the typing and that the prosecution proposition is true, the numerator is 1. For the moment, also assume that the information I is such that, when D is true, knowledge of G_s does not affect the probability of G_c, and let the frequency of G_c in the relevant population be P. Then

$$V = 1/P.$$

This is applicable only in very straightforward cases. More complicated examples are given below.

78.3.1 Combining Evidence

The representation of the value of evidence as a likelihood ratio enables successive pieces of evidence to be evaluated sequentially in a way that makes explicit the dependence amongst different pieces of evidence. The posterior odds from one piece of evidence, (E_1, say) become the prior odds for the following piece of evidence (E_2 say). Thus, again omitting I, we obtain

$$\frac{\Pr(P \mid E_1, E_2)}{\Pr(D \mid E_1, E_2)}$$
$$= \frac{\Pr(E_2 \mid P, E_1)}{\Pr(E_2 \mid P, E_1)} \frac{\Pr(P \mid E_1)}{\Pr(P \mid E_1)}$$
$$= \frac{\Pr(E_2 \mid P, E_1)}{\Pr(E_2 \mid D, E_1)} \frac{\Pr(E_1 \mid P)}{\Pr(E_1 \mid D)}$$
$$\times \frac{\Pr(P)}{\Pr(D)}.$$

This expression generalizes in an obvious way to allow for more pieces of evidence. If the two pieces of evidence are independent, as may be the case with DNA typing systems, this leads to likelihood ratios combining by simple multiplication.

Thus, if V_{12} is the value for the combination of evidence (E_1, E_2) and if V_1 and V_2 are the values for E_1 and E_2 respectively, then

$$V_{12} = V_1 \times V_2.$$

If the weight of evidence is used, different pieces of evidence may be combined by addition. This is a procedure that has an intuitive analogy with the scales of justice.

78.3.2 Fallacies

There are two well-known fallacies associated with evidence interpretation [28]. The prosecutor's fallacy associates a low probability for finding the evidence on an innocent person with a high probability of guilt. The defense fallacy associates a high value for the number of expected people with the evidence being assessed amongst the population from which the criminal is thought to come as a sign that the evidence is not relevant for the assessment of the guilt of the suspect.

78.4 Continuous Measurements

Consider evidence of a form in which measurements may be taken and for which the data are continuous, for example, the refractive index of glass. The value V of the evidence is formally

$$V = \frac{\Pr(E \mid P, I)}{\Pr(E \mid D, I)}.$$

The quantitative part of the evidence is represented by the measurements of the characteristic of interest. Let x denote the measurements on the (known) source of the evidence (e.g., a window), and let y denote the measurements on the receptor object (e.g., clothing of a suspect). For example, if a window is broken during the commission of a crime, the measurements on the refractive indices of m fragments of glass found at the crime scene will be denoted $x_1, \ldots, x_m$ (denoted $\mathbf{x}$). The refractive indices of n fragments of glass found on a suspect will be denoted $y_1, \ldots, y_n$ (denoted $\mathbf{y}$). The quantitative part of the evidence concerning the glass fragments in this case can be denoted by

$$E = (\mathbf{x}, \mathbf{y}).$$

Continuous measurements are being considered and the probabilities Prare therefore replaced by probability density functions f, and again I is omitted for ease of notation.

$$V = \frac{f(\mathbf{x}, \mathbf{y} \mid P)}{f(\mathbf{x}, \mathbf{y} \mid D)}. \quad (1)$$

The value V of the evidence in Equation (1) may be rewritten via

$$\begin{aligned} V &= \frac{f(\mathbf{x}, \mathbf{y} \mid P)}{f(\mathbf{x}, \mathbf{y} \mid D)} \\ &= \frac{f(\mathbf{y} \mid \mathbf{x}, P)}{f(\mathbf{y} \mid \mathbf{x}, D)} \times \frac{f(\mathbf{x} \mid P)}{f(\mathbf{x} \mid D)}. \end{aligned}$$

The measurements $\mathbf{x}$ are those on the source object. Their distribution is independent of whether P or D is true. Thus

$$f(\mathbf{x} \mid P) = f(\mathbf{x} \mid D)$$

and

$$V = \frac{f(\mathbf{y} \mid \mathbf{x}, P)}{f(\mathbf{y} \mid \mathbf{x}, D)}.$$

If D is true, then the measurements $(\mathbf{y})$ on the receptor object and the measurements $(\mathbf{x})$ on the source object are independent. Thus

$$f(\mathbf{y} \mid \mathbf{x}, D) = f(\mathbf{y} \mid D),$$

and

$$V = \frac{f(\mathbf{y} \mid \mathbf{x}, P)}{f(\mathbf{y} \mid D)}. \quad (2)$$

Note that this assumption of independence, $\mathbf{y}$ and $\mathbf{x}$, in the denominator of Equation (1) may not always be true. See Reference 12 for an example concerning DNA profiles and relatives.

An extension of this expression to take account of ideas of transfer (t) and persistence (p) is

$$V = t_0 + \frac{p_0 t_1 f(\mathbf{y} \mid \mathbf{x}, P)}{p_1 f(\mathbf{y} \mid D)}. \quad (3)$$

Here, t_0 is the probability of no transfer, p_0 is the probability of no innocent transfer of evidence, t_1 is the probability of transfer between suspect and crime scene, and p_1 is the probability of an innocent transfer of the evidential material to the suspect but not from the crime scene. These are subjective probabilities and form part of the judgment of the scientist. Extensions of these ideas of transfer and persistence in the context of fiber examination are given in Reference 6.

Often there are two levels of variation to consider: within-source variation and between-source variation. Parameterize the distribution of $\mathbf{x}, \mathbf{y}$ within a source by θ (possibly vector-valued) and the distribution of θ represents the distribution between sources. In this context, an alternative expression for Equation (2) is

$$V = \frac{\int f(y \mid \theta) f(x \mid \theta) f(\theta) d\theta}{\int f(x \mid \theta) f(\theta) d\theta \int f(y \mid \theta) f(\theta) d\theta}.$$

An expression for V in which the density functions for within-source data $f(\mathbf{y} \mid \theta)$ and $f((\mathbf{x}) \mid \theta)$ are taken to be normally distributed and the density functions for between-source data $f(\theta)$ are estimated using kernel density estimation procedures is given in reference 3.

Use of the above formulas requires consideration of the population on which the probability models are based. This is the population from which the criminal is thought to have come. Background information I and the defense proposition D help in its definition.

78.4.1 Consideration of Prior and Posterior Odds

Many important aspects of evidence evaluation are illustrated by what is known as *the island problem*. A crime has been committed on an island. Consider a single stain left at a crime scene. The island has a population of $(N+1)$ individuals. In connection with this crime, a suspect has been identified by other evidence. The genotype G_S of the suspect and the genotype G_C of the crime stain are the same. The probability that a person selected at random from the island population has this genotype is ϕ.

The two hypotheses are

- P: the suspect has left the crime stain.

- D: some other person left the crime stain.

There is considerable debate as to how prior odds may be determined. This is related to a debate about the meaning of the phrase "innocent until proven guilty." Note, from the basic equation, that a value of zero for the prior odds means that the posterior odds will also be zero, regardless of the value of the likelihood ratio. Thus, if "innocent until proven guilty" is taken to mean the prior probability $\Pr(\text{guilty}) = 0$, then the posterior probability $\Pr(\text{guilty} \mid \text{evidence})$ will equal zero, regardless of the value of the likelihood ratio.

A more realistic assessment of the prior odds is to say, before any evidence is presented, that the suspect is as likely to be guilty as anyone else in the relevant population. In the context of this example, this implies that $\Pr(C \mid I) = 1/(N+1)$ [26]. The prior odds are then $1/N$. With some simplifying assumptions, such as that $P(G_C \mid G_S, C, I) = 1$, it can be shown that the posterior odds are then $1/(NP)$. Values for the posterior odds provide valuable information for the debate as to what is meant by "proof beyond reasonable doubt."

An extension to this analysis is to allow for varying match probabilities amongst the inhabitants of the island. Let π_0 be the prior probability that the suspect left the crime stain and identify inhabitant number $(N+1)$ with the suspect. The inhabitants $i = 1, \ldots, N$ are innocent. Let π_i denote the probability that inhabitant i left the crime stain $(i = 1, \ldots, n)$ such that $\sum_{i=1}^{n} \pi_i + \pi_0 = 1$. Assume, as before, that $P(G_C \mid G_S, C, I) = 1$. Now, generalize the match probability ϕ to be, possibly, different for each individual, such that $\Pr(G_C \mid G_S, \bar{C}, I) = \phi_i$. Then

$$\Pr(C \mid G_C, G_S, I) = \frac{\pi_0}{\pi_0 + \sum_{i=1}^{N}(\pi_i \phi_i)}.$$

From this expression, it is possible to allow for different match probabilities and different prior probabilities for different individuals. It is possible to write the expression in terms of $w_i = \pi_i/\pi_0$, where each w_i can be regarded as a weighting function for how much more or less probable than the suspect the ith person is to have left the crime stain, on the basis of the other evidence as recorded in I [4,12].

78.4.2 Paternity

Illustrations of various issues associated with the evaluation of evidence are given in paternity testing. The hypotheses to be compared are

- P: the alleged father is the true father.

- D: the alleged father is not the true father.

The probability that the alleged father would pass the child's nonmaternal genes is compared to the probability that the genes would be passed randomly. Thus, the value

V of the evidence is

$$V = \frac{\text{Probability that the alleged father would pass the genes}}{\text{Probability that the genes would be passed by a random male}}.$$

This may be expressed verbally as "the alleged father is V times more likely than a randomly selected man to pass this set of genes." The evidence can be in several parts $E_1, \ldots, E_n$, where E_i relates to the phenotypes of child, mother, and alleged father under the ith genetic marker system ($i = 1, \ldots, n$). The posterior probability of P, given all the evidence, is then

$$\Pr(P \mid E_1, \ldots, E_n) = \left\{ 1 + \frac{\Pr(D)}{\Pr(P)} \prod_{i=1}^{n} \frac{\Pr(E_i \mid D)}{\Pr(E_i \mid P)} \right\}^{-1}.$$

78.4.3 Verbal Scales

The posterior probability of paternity has been called the *plausibility of paternity* [5]. Evett [11] has provided a verbal scale for the likelihood ratio and Aitken and Taroni [2] have provided a scale based on logarithms for the probability for the prosecution's proposition.

78.5 Future

The determination of the competing propositions is a very difficult and important aspect in the procedure of evaluating and interpreting evidence. These propositions provide the framework within which the evidence is evaluated. Work in which propositions are considered before examination of a potentially recovered piece of evidence in a so-called *preassessment* model is described in Reference 7. Also, a *hierarchy* of propositions of source, activity, and crime is described in Reference 8. The source level concerns analyses and results on measurements on samples only. For example, suppose that a person A allegedly kicked B on the head and blood has been transferred from B to one of A's shoes. The analyses and measurements at the source level would be concerned with propositions of the form, for example, "B is the source of the stain," and "B is not the source of the stain." The activity level is exemplified by "A kicked B" on the head, and "A did not kick B" on the head. The crime level is exemplified by the availability of eye-witness evidence to suggest a crime may have been committed.

Bayesian networks and graphical models are becoming increasingly important in forensic statistics. They allow for practitioners to contribute to the construction of complicated models and to understand easily the propagation of evidence through the network. These graphical approaches to evidence evaluation follow from pioneering work of Wigmore [30] in the legal context and from the work of Lauritzen and Spiegelhalter [23] in the theoretical context. Applications are illustrated in References 1, 10, and 17.

References

1. Aitken, C. G. G. and Gammerman, A. (1989). Probabilistic reasoning in evidential assessment. *J. Forensic Sci. Soc.*, **29**, 303–316.

2. Aitken, C. G. G and Taroni, F. (1998). A verbal scale for the interpretation of evidence. *Sci. Justice*, **38**, 279–281.

3. Aitken, C. G. G. and Taroni, F. (2004). *Statistics and the Evaluation of Evidence for Forensic Scientists* 2nd ed. Wiley, Chichester, UK.

4. Balding, D. J. and Nichols, R. A. (1995). A method for characterizing differentiation between populations at multiallelic loci and its implications for establishing identity and paternity. *Genetics*, **96**, 3–12.

5. Berry, D. A. and Geisser, S. (1996). Inference in cases of disputed paternity. In *Statistics and the Law*, M. H. De Groot,

S. E. Fienberg, and J. B. Kadane, eds., pp. 353–382 Wiley, New York.

6. Champod, C. and Taroni, F. (1999). Interpretation of fibres vidence. In *Forensic Examination of Fibres*, 2nd ed., J. Robertson and M. Grieve, eds., pp. 343–363 Taylor & Francis, London.

7. Cook, R., Evett, I. W., Jackson, G., Jones, P. J., and Lambert, J. A.(1998). A model for case assessment and interpretation. *Sci. Justice*, **38**, 151–156.

8. Cook, R., Evett, I. W., Jackson, G., Jones, P. J., and Lambert, J. A. (1998). A hierarchy of propositions: Deciding which level to address in casework. *Sci. Justice*, **38**, 231–239.

9. Cullison, A. D. (1969). Probability analysis of judicial fact-finding: A preliminary outline of the subjective approach. *Univ. Toledo Law Rev.*, 538–598.

10. Dawid, A. P. and Evett, I. W. (1997). Using a graphical method to assist the evaluation of complicated patterns of evidence. *J. Forensic Sci.*, **42**, 226–231.

11. Evett, I. W. (1987). Bayesian inference and forensic science: Problems and perspectives. *Statistician*, **36**, 99–105.

12. Evett, I. W. and Weir, B. S. (1998). *Interpreting DNA Evidence*. Sinauer, Sunderland, MA.

13. Fairley, W. B. and Mosteller, W. (1977). *Statistics and Public Policy.*, pp. 355–379 Addison-Wesley, London.

14. Finkelstein, M. O. and Fairley, W. B. (1970). A Bayesian approach to identification evidence. *Harvard Law Rev.*, **83**, 489–517.

15. Finkelstein, M. O. and Fairley, W. B. (1971). A comment on "Trial by mathematics." *Harvard Law Rev.*, **84**, 1801–1809.

16. Good, I. J. (1950). *Probability and the Weighing of Evidence*. Charles Griffin & Company Ltd., London.

17. Kadane, J. B. and Schum, D. A. (1996). *A Probabilistic Analysis of the Sacco and Vanzetti Evidence*. Wiley, New York.

18. Kind, S. S. (1994). Crime investigation and the criminal trial: A three chapter paradigm of evidence. *J. Forensic Sci. Soc.*, **34**, 155–164.

19. Kingston, C. R. (1965). Applications of probability theory in criminalistics. *J. Am. Statist. Assoc.*, **60**, 70–80.

20. Kingston, C. R. (1965). Applications of probability theory in criminalistics — II. *J. Am. Statist. Assoc.*, **60**, 1028–1034.

21. Kirk, P. L. (1963). The ontogeny of criminalistics, *J. Criminal Law, Criminol. Police Sci.*, **54**, 235–239.

22. Kwan Q. Y. (1977). *Inference of Identity of Source*. Doctor of Criminology thesis, University of California, Berkeley.

23. Lauritzen, S. and Spiegelhalter, D. (1988). Local computations with probabilities on graphical structures and their application to expert systems. *J. Roy Statist. Soc. B*, **50**, 157–224.

24. Lindley, D. V. (1977). A problem in forensic science. *Biometrika*, **64**, 207–213.

25. Robertson, J. and Grieve, M., eds. (1999). *Forensic Examination of Fibres*, 2nd ed. Taylor & Francis, London.

26. Robertson, B. and Vignaux, G. A. (1995). *Interpreting Evidence: Evaluating Forensic Science in the Courtroom*. Wiley, London.

27. Taroni, F., Champod, C., and Margot, P. A. (1998). Forerunners of Bayesianism in early forensic science. *Jurimetrics J.*, **38**, 183–200.

28. Thompson, W. C. and Schumann, E. L. (1987). Interpretation of statistical evidence in criminal trials. The prosecutor's fallacy and the defence attorney's fallacy. *Law Hum. Behav.*, **11**, 167–187.

29. Tribe, L. (1971). Trial by mathematics: precision and ritual in the legal process. *Harvard Law Rev.*, **84**, 1329–1393.

30. Wigmore, J. (1937). *The Science of Proof: as Given by Logic, Psychology and General Experience and Illustrated in Judicial Trials*, 3rd ed. Little, Brown, Boston, Mass.

79 Statistics in Human Genetics I

Cedric A. B. Smith

79.1 Introduction

For more than 100 years the sciences of human genetics and statistics have been closely related. Sir Francis Galton was very interested in human genetics, partly viewed as a means to eugenics, i.e., the improvement of humanity by selective breeding. He invented the regression and correlation coefficients as tools of statistical investigation in genetics (see Galton [11,12]). These methods were further developed by his friend Karl Pearson. In 1900, Mendel's work was rediscovered, and the classical tools for mathematical and statistical investigation of Mendelian genetics were forged by Sir Ronald Fisher, Sewall Wright, and J. B. S. Haldane in the period 1920–1950.

79.2 Applications

Present-day applications of statistical theory to human genetics could conveniently be classified under four headings.

79.2.1 Theory, Verification, Parameters and Practice

Theory. Theory refers to the precise specification of the probabilistic mechanism of inheritance. Many discrete characters have *simple Mendelian* inheritance. The ABO blood groups are an example. In a slightly simplified treatment they can be considered as determined by three genes, A, B, O, with the rule that any one individual carries exactly two of these, not necessarily different. Thus individuals have six possible pairs of genes, or *genotypes, AA, AO, BB, BO, AB, OO*. Each child gets at random one gene from its mother, and one from its father, so a child of an $AB \times OO$ mating can be AO or BO, each with probability $\frac{1}{2}$.

Continuously (or almost continuously) varying characters such as height or fingerprint ridge count are believed to be due to the combined action of many genes of small effect (*multifactorial* or *polygenic* characters) together with possible environmental effects. Some discrete characters, such as diseases, are regarded as *threshold* characters [8,9]; there is a hypothetical multifactorial continuous character x such that when $x > $ some threshold x_T, the individual is affected by the disease or disability. Other more complicated types of inheritance are conceivable, together with infection, environmental and social inheritance, and combinations of these.

Verification. Verification consists in checking whether a supposed mode of inheritance is compatible with observed family data. Mendelian inheritance can often be checked by goodness of fit*χ^2, e.g., the testing of the $1AO : 1\ BO$ ratio from an $AB \times OO$ mating. Multifactorial inheritance can be tested by the observed correlations between relatives, e.g., mother–daughter and brother–sister correlations. In the simplest situation, when all genes act perfectly additively and there is negligible environmental effect, this correlation is equal to the expected proportion of genes the relatives have in common. Thus, since a daughter inherits just one of each pair of genes carried by her mother, both daughter and mother have half their genes in common, and hence correlation $\frac{1}{2}$. Only a slightly more complicated argument shows that brother and sister have correlation $\frac{1}{2}$, grandparent–grandchild $\frac{1}{4}$, etc. Fingerprint ridge count shows correlations approximating to these [16]. The theoretical correlations for more complicated situations were first found by Fisher [10] in a classical paper, whose conclusions have not been greatly modified by subsequent work.

Estimation of Genetic Parameters. Many genetic parameters are proportions. Examples are the proportion of AO individuals from an $AB \times OO$ mating; the proportion of AB individuals in a population; the proportion of A genes in a population; the mutation rate, or frequency with which a gene changes into a different gene in the next generation; the viability of a genotype, or probability of surviving from birth to reproductive maturity; the proportion of marriages between cousins; and the recombination fraction θ in genetic linkage expressing the frequency with which certain "crossover" events occur, which result in the joining of a part of one chromosome (a body carrying the genes) with a part of another chromosome.

In the simplest cases the observation of x occurrences of an event in n trials gives the obvious estimate x/n of the probability of occurrence, with binomial SE = $\sqrt{x(n-x)/n^3}$. In practice the situation is complicated by various factors, of which the most important is recessivity. Thus, in the ABO blood groups we have anti-A and anti-B sera, which will show the respective presence or absence of the A and B genes in an individual by reacting or not reacting with a drop of her or his blood. But there is no reliable direct test for the presence or absence of O, so that an individual whose blood reacts with anti-A but not with anti-B (so-called blood group A) can be either AA or AO. The gene O which has no direct test for its presence is *recessive*. Thus a mating between individuals of groups A and O can be genetically either $AA \times OO$ or $AO \times OO$, producing a mixed distribution of off-spring, not directly amenable to testing by χ^2. A problem also occurs in the testing of segregation ratios in the case of a rare recessive condition like phenylketonuria. Two apparently normal parents, genetically $Ph.ph \times Ph.ph$, produce on the average 25% phenylketonuric offspring $ph.ph$ (who will be mentally and physically stunted unless specially treated). But the sampling will be biased because families with $Ph.ph \times Ph.ph$ parents who happen not to have any phenylketonuric offspring will not be noticed. Many methods exist for coping with biases due to recessivity, including the fully efficient "counting method" [2,28], and simple nearly efficient alternatives [13,19].

As far as estimation of the "recombination fraction" θ in the genetic linkage investigations is concerned, it is generally most convenient to calculate the loglikelihood function numerically for specified values of θ, such as $0, 0.05, 0.1, 0.15, \ldots, 0.5$ [21,22]. Inferences are thus drawn from the likelihood curve.

Quantitative characters give rise to two distinct estimation problems. The data will consist of measurements on individuals in families of varying sizes. To find the correlations, we must first find the covariances between relatives and their variances. This is equivalent to estimating variance and covariance components, a problem on which there is a multitude of papers (e.g., Sahai [27]; see also Smith [29]). Then these correlations or variance and covariance components have, in turn, to be interpreted as showing the magnitudes of environmental and genetic effects and their interactions. This can be done either by ANOVA techniques [20] or by path coefficients [26]. For example, Rao and Morton claim to have shown that in American children the "hereditability" of IQ is about 70%, which can roughly be interpreted as the proportional contribution of genetic factors, but in adults it is only 35%, although the precision of these estimates is questioned by Goldberger [14].

Practice. Galton envisaged the chief practical use of human genetics as *eugenics*, i.e., the creation of better individuals through the encouragement of suitably desirable parents. Nowadays the practical value is more modestly limited to genetic counseling, i.e., to calculating the risks of inherited disease and abnormality for the benefit of anxious potential parents. This is in principle no more than the calculation of a conditional probability of abnormality given the known data on the family. For Mendelian characters, this uses only the simplest laws of probability [25], although the calculations can be quite complicated in detail. For quantitative and threshold characters it is usual to assume for simplicity that the joint distributions in relatives are multivariate normal. Even so, methods at present available involve formidable problems of numerical integration.

79.2.2 Population Size

In addition to the four classifications given above, we can also classify genetic questions as relating either (1) to individuals and individual families or (2) to populations. Examples of class 1 are Mendelian probabilities, e.g., the fact that children of an $AB \times AB$ mating have probabilities $\frac{1}{4}AA + \frac{1}{2}AB + \frac{1}{4}BB$. The class 2 issues concern the immense field of population genetics, with issues involving the proportions of genes and genotypes in a population, and the effects of inbreeding, migration, selection, random fluctuations, or "drift" in small populations, mutation (an alteration in type of genes passed from parent to child), the effects of population intermarriage, etc.

79.3 Genetic Mechanisms

The fundamental law of human population genetics is the Hardy–Weinberg law of population proportions in a large population mating at random [15,32].

In its original form it stated that if some character was controlled by two genes, G, g, which occur with respective frequencies P and $Q = 1 - P$ in the population, then the frequencies of genotypes are

$$P^2.GG + 2PQ.Gg + Q^2.gg.$$

This is easily extended to more complicated situations; e.g., if the frequency of the three blood group genes are $p.A + q.B + r.O$ ($= 0.26.A + 0.05.B + 0.69.O$ in Britain), then the frequency of genotype AA is p^2 ($= 0.07$), that of AB is $2pq$ ($= 0.03$), and so on. In theory, the Hardy–Weinberg law holds only in quite restricted situations, but in practice it is almost always found to apply to a quite satisfactory approximation.

The most important exception to Hardy–Weinberg relates to the effect of inbreeding on the frequency of rare recessive characters. An individual (I, say) is *inbred* if his or her mother and father share a common ancestor (C) a few generations back. If so, there is the possibility that (regarding any one particular character, such as phenylketonuria) the two genes that I gets from his or her mother and father were both descended from one single gene in C; they are said to be *identical by descent*. The probability F of this happening is I's *coefficient of inbreeding*. Thus if I's parents are first cousins, $F = \frac{1}{16}$.

If the original gene in the common ancestor C happened to be ph, then I would have genotype $ph.ph$, and therefore be phenylketonuric. The probability of this happening is accordingly Fq, where q (approximately 0.006 in Britain) is the frequency of the ph gene. Thus Fq ($\simeq 400 \times 10^{-6}$) is much larger than q^2 ($\simeq 36 \times 10^{-6}$), the probability that a child of unrelated parents is phenylketonuric [24]; that is, inbred children have a raised probability of suffering from recessive abnormalities (in this case the probability is greater by a factor of about $\frac{400}{36} = 11$). Since in the general British population at the time of Munro's investigation roughly one marriage in 100 was between cousins, we would expect that among parents of phenylketonurics about 1 in $\frac{100}{11}$, i.e., 1 in 9 would be pairs of cousins. In fact, Munro found 10 cousin pairs out of 104, in good agreement (especially considering the roughness of the approximations involved).

A study of consanguinity can assist in the analysis of more complicated situations. Thus Chung et al. [4] noted that if in Stevenson and Cheeseman's [31] investigation of congenital deafness in Northern Ireland we consider those families in which both parents are normal, then when there is only one ("isolated") congenitally deaf child, the mean inbreeding coefficient for the children is about 2.8×10^{-3}, whereas when more than one child is affected, it is 8.5×10^{-3}. Since in the general population they estimated the mean inbreeding coefficient to be 0.4×10^{-3}, and since where more than one child in a family is affected it is very plausible that a recessive gene is involved, we can estimate that a proportion $(2.8-0.4)/(8.5-0.4) = 0.3$ of isolated cases are due to recessives.

Data on the survival of inbred individuals also gives information about the frequency of deleterious recessive genes in the population. Morton et al. [23] note that in American families studied by G. L. B. Arner in 1908, a proportion 0.168 of children died before the age of 20 when the parents were first cousins (inbreeding coefficient $F = 0.0625$), but only 0.116 when the parents were unrelated ($F = 0$). The difference $0.168 - 0.116 = 0.052$ is evidently due to deleterious recessives and may be taken as a crude estimate of $0.0625 \sum q$, where $\sum q$ is the total frequency of these all recessive genes, together, and is therefore estimated as $0.052/0.0625 = 0.8$. This calculation assumes that all these recessives are invariably lethal when in homozygous form gg. If they kill in only a proportion of cases, their total frequency must be correspondingly increased, to produce the same number of juvenile deaths. Thus we conclude that although any one particular type of deleterious recessive (such as phenylketonuria) is rare, altogether they are sufficiently numerous to have a high total frequency. (The presentation here of inbreeding calculations is kept deliberately simple; for more careful treatment, see the original papers.)

If a character is damaging to an organism, so that it does not reproduce, the corresponding genes are lost from the population. It is usually plausible that the supply of such genes is maintained by mutation, whereby normal genes are changed into abnormal (usually deleterious) ones. If the new genes show their effect immediately, we can count them and thus "directly" determine the rate of mutation. Otherwise, we have to rely on the principle that in the long run the number of new mutants entering the population by mutation must equal the number lost by premature death. So, with suitable reservations, the mutation rate can be "indirectly" estimated from the number of deaths. In either case it generally turns out that most genes have a mutation rate of the order of once in each 10^5 generations.

79.4 Literature

Further information on these and other topics in human genetical statistics can be obtained from many textbooks, although these tend to concentrate on theoretical population genetics. The books by Cavalli-Sforza and Bodmer [1], Charlesworth [3], Crow and Kimura [5], Elandt-Johnson [6], Ewens [7], Kempthorne [17], Li [18], and Wright [33] seem particularly worth consulting. A good general guide to human genetics is provided by Stern [30].

References

1. Cavalli-Sforza, L. L. and Bodmer, W. F. (1971). *The Genetics of Human Populations*. Freeman, San Francisco. (A very readable and informative textbook.)
2. Ceppellini, R., Siniscalco, M., and Smith, C. A. B. (1955). *Ann. Hum. Genet.*, **20**, 97–115.
3. Charlesworth, B. (1980). *Evolution in Age-Structured Populations*. Cambridge University Press, Cambridge, UK. (The first book to deal with the effects of continuous reproduction.)
4. Chung, C. S., Robison, O. W. and Morton, N. E. (1959). *Ann. Hum. Genet.*, **23**, 357–366.
5. Crow, J. F. and Kimura, M. (1970). *An Introduction to Populations Genetics Theory*. Harper & Row, New York.
6. Elandt-Johnson, R. C. (1971). *Probability Models and Statistical Methods in Genetics*. Wiley, New York. (An elementary introduction to the field, lucidly written.)
7. Ewens, W. J. (1979). *Mathematical Population Genetics*. Springer-Verlag, Berlin.
8. Falconer, D. S. (1960). *Introduction to Quantitative Genetics*. Oliver & Boyd, Edinburgh.
9. Falconer, D. S. (1965). *Ann. Hum. Genet.*, **29**, 51–76. (A classic paper, showing how the methods of quantitative genetics can be applied to discontinuous characters, especially inherited diseases.)
10. Fisher, R. A. (1918). *Trans. Roy Soc. Edinburgh*, **52**, 399–433. (A classic paper.)
11. Galton, F. (1877). *Proc. Roy Inst.*, **8**, 282–301. (Introduces regression.)
12. Galton, F. (1885). *Proc. Roy Soc.*, **45**, 135–145. (Introduces the correlation coefficient from a genetical point of view.)
13. Gart, J. J. (1968). *Am. Hum. Genet.*, **31**, 283–292. (Simple estimation of parameters in a truncated binomial.)
14. Goldberger, A. S. (1978). In *Genetic Epidemiology*, N. E. Morton and C. S. Chung, eds., pp. 195–222. Academic Press, New York.
15. Hardy, G. H. (1908). *Science*, **28**, 49–50. (A classical paper, introducing what is now called the Hardy–Weinberg law of population genetics.)
16. Holt, S. B. (1968). *The Genetics of Dermal Ridges*. Charles C. Thomas, Springfield, IL.
17. Kempthorne, O. (1957). *An Introduction to Genetic Statistics*. Iowa State University Press, Ames. (An excellent introduction to the subject.)

18. Li, C. C. (1976). *First Course in Population Genetics.* Boxwood Press, Pacific Grove, CA (A clear, detailed introduction.)

19. Li, C. C. and Mantel, N. (1968). *Am. J. Hum. Genet.,* **31**, 283–292. (Simple estimation of parameters in a truncated binomial.)

20. Mather, K. and Jinks, J. L. (1971). *Biometrical Genetics,* 2nd ed.

21. Morton, N. E. (1955). *Am. J. Hum. Genet.,* **7**, 277–318.

22. Morton, N. E. (1957). *Am. J. Hum. Genet.,* **9**, 55–75. (These last two papers show how likelihoods for linkage can be quickly evaluated.)

23. Morton, N. E., Crow, J. F., and Muller, H. J. (1956). *Proc. Natl. Acad. Sci. (USA),* **42**, 855–863. (A classical paper, showing how to estimate the average number of deleterious recessives carried by man.)

24. Munro, T. A. (1939). *Proc. 7th Int. Congress of Genetics.* Cambridge University Press, Cambridge, UK. (Investigates the genetics of phenylketonuria.)

25. Murphy, E. A. and Chase, G. A. (1975). *Principles of Genetic Counseling.* Year Book Medical Publishers, Chicago.

26. Rao, D. C. and Morton, N. E. (1978). In *Genetic Epidemiology,* N. E. Morton and C. S. Chung, eds., pp. 145–193 Academic Press, New York.

27. Sahai, H. (1979). *Int. Statist. Rev.,* **47**, 177–222. (A list of papers on variance components.)

28. Smith, C. A. B. (1957). *Ann. Hum. Genet.,* **21**, 254–276. (A simple method for estimating most genetic parameters.)

29. Smith, C. A. B. (1980). *Ann. Hum. Genet.,* **44**, 95–105. (A simple method for estimating correlations between relatives.)

30. Stern, C. (1960). *Principles of Human Genetics,* 2nd ed. Freeman, San Francisco.

31. Stevenson, A. C. and Cheeseman, E. A. (1956). *Ann. Hum. Genet.,* **20**, 177–207.

32. Weinberg, W. (1908). *Jahresh. Ver. vaterl. Naturkd. Württemberg,* **64**, 368–382.

33. Wright, S. (1968, 1969, 1977, 1978). *Evolution and the Genetics of Populations,* Vols. 1–4. University of Chicago Press, Chicago. (A summary of his life's work by America's leading population geneticist.)

80 Statistics in Human Genetics II

Christopher Amos and Rudy Guerra

80.1 Introduction

Molecular genetics changed drastically over the period 1978–1997. Key developments include the discovery of restriction enzymes, allowing scientists to cut and paste DNA fragments; the ability to determine DNA sequences; and the ability to amplify a DNA sequence. These techniques of so-called *recombinant DNA* have fundamentally changed the way in which geneticists conduct research. In particular, a path to uncovering the genetic architecture of human traits—especially diseases—has been paved with the realization that the human genome could eventually be sequenced.

The set of 23 pairs of human chromosomes is the *human genome*. The human haploid genome comprises about three billion DNA base pairs, encoding for between 100,000 and 200,000 genes. Many of these genes are involved in determining traits such as height, eye color, plasma cholesterol concentration, and risk for diseases such as breast cancer. The observable traits, or *phenotypes*, can be classified as either following classical Mendelian inheritance (defined as a simple correspondence between genotype at a single locus and phenotype), complex (non-Mendelian), or quantitative.

Modern application of statistics to human genetics largely centers on (1) genomics, the molecular characterization of the human genome (Human Genome Project) [47,88], and (2) using the resulting (marker) data from genomic research to uncover the genetic basis of human traits. The latter enterprise is generally known as *genetic epidemiology*, the statistical and epistemological aspects of which are reviewed by Thompson [117] and Morton [83], respectively. The following discussion emphasizes the recent high level of activity in genetic epidemiology, especially as it pertains to finding loci underlying complex traits. Searching for trait loci was not practical until the early 1980s.

80.2 Genomics

Genomics is concerned with the mapping and sequencing of DNA fragments to chromosomes. These fragments may be genes or molecular markers, which are loci of neutral DNA variation, or polymorphism not associated with phenotypic variation in traits. For convenience, both types of DNA fragments will be referred to as *markers*. Variant forms of markers are called *alleles*.

The mapping is conducted at different levels of resolution. First, the marker is mapped to a specific chromosome. Second, the position of the marker relative to previously mapped markers is determined. At this stage, the relative positions are determined by recombination; the process is called genetic mapping. Third, a physical map of the marker is constructed, the purpose of which is to allow a set of sequenced DNA fragments to be treated as though it were a contiguous piece of chromosome.

At each level of genomic mapping major statistical problems have had to be addressed, and several still exist. A very important application of genomic mapping is the identification of a breast cancer gene (BRCA1) to a locus spanning 80,000 base pairs [82], of which only 5589 base pairs actually encode the BRCA1 protein.

A standard method of determining the genome location of a marker is linkage analysis. The basic idea is to assess the probability r that alleles from two distinct loci recombine (are not coinherited). Loci that are closely situated on the same chromosome tend to be coinherited, while those far apart on the same chromosome or on different chromosomes should have an equal probability to be coinherited or not coinherited. The recombination fraction r is 0 when alleles at the two loci are completely linked and hence always coinherited. A recombination fraction of 0.5 indicates 50% chance of coinheritance and an absence of genetic linkage.

Significance testing and estimation of r are used to map markers. In experimental organisms the problem is straightforward. However, because of missing-data issues, linkage analysis in humans is more challenging, the main problem being evaluation of a likelihood that must be summed over a large number of events potentially leading to the observed data. More recent approaches to estimation of likelihoods, especially those based on complex pedigrees, use Markov chain Monte Carlo methods to take account of the missing-data aspect of the problem [112,118]. Alternatively, radiation hybrid and physical mapping can be used to order the markers [70,79].

The standard paradigm for finding disease genes depends on finding associations between disease status and marker genotypes within families, using genetic mapping as a first step. It is important to understand that genetic mapping is defined by linkage and hence allows one to identify a *local region* along a chromosome containing the true disease gene(s). Under the current paradigm the next step is to map the disease gene(s) physically by cloning consecutive, overlapping DNA fragments containing the localized region linked to the trait. Once a physical map exists, all genes within the localized region can theoretically be identified. The last step is to determine which gene(s) is associated with the disease. This typically is done by looking for functional alleles distinguishing normal from affected individuals. Physical mapping is inherently a statistical problem [68,86]. Stewart et al. [115] present a current physical map of the human genome.

80.3 Genetic Analysis of Traits

Biometrical methods have been traditionally used to assess heritability and evaluate potential models of genetic transmission. For traits showing a heritable component, with the more recent generation of large numbers of mapped genetic markers, linkage analyses are routinely conducted and genome scans for trait loci are becoming more common.

80.3.1 Biometrical Methods

The foundation of the biometrical method to determine the relative contribution of genetic factors (heritability) to quantitative traits dates back to Fisher [38]. He proposed that the variation in such traits was due to the additive effects of many genes, each contributing a small effect (polygenes), and possibly to environmental effects. Biometrical methods are still routinely used to assess familial aggregation as evidence for heritability of a quantitative trait. For a collection of nuclear families, the estimated slope of a simple regression of mean off-spring on mean parent trait values provides an unbiased estimate of heritability for a continuous trait that is not influenced by shared environmental factors [33]. Because most traits that are fairly well understood are not completely determined by polygenes, more complete and realistic approaches to evaluating correlations among family members have been developed. More recent advances in this area apply statistically robust methods such as quasi-likelihood modeling in familial aggregation analysis [44,75,121,136] or further develop methods for disease phenotypes with variable age of onset [16,40,80].

One standard multifactorial approach, segregation analysis, typically partitions phenotypic variation into variance components associated with an unmeasured major gene, residual polygenes, shared environment, and random environmental effects. This variance components approach also allows for covariates and analysis of *pedigrees* (family trees showing inheritance patterns of specific phenotypes) with more than two generations. In most applications it is assumed that the putative major gene has two alleles under a certain genetic model—for example, a rare and recessive gene. Additional usual assumptions include Hardy–Weinberg equilibrium, no genotype-by-environment interaction, and random mating with respect to trait-causing genes. Elston and Stewart [30] introduced an efficient likelihood-based method to evaluate such genetic models.

Extensions of this model provided a basis for the evaluation of complex pedigrees [71] and linked loci [90]. For extended families, evaluation of any likelihood including both major gene and polygenic components becomes impractical because of the need for large numbers of numerical integrations. Two approaches for evaluating data from extended families are the application of an approximation that avoids numerical integration [50], and the division of pedigrees into nuclear families [62].

Segregation analysis can also be used for assessing disease etiology. In this context segregation analytic models estimate the *penetrance*, or probability that an individual with a susceptible genotype will become affected by a disease; the *sporadic risk*, or probability that an individual with the nonsusceptible genotype nevertheless becomes affected by the disease; and the frequency of susceptibility alleles in the general population. For the majority of diseases, however, the occurrence of genetic susceptibility is sufficiently uncommon that selection without reference to disease status would result in low power to detect genetic effects. However, most patterns of selection (ascertainment) introduce biases into parameter estimates of the genetic analyses. When the ascertainment events are well characterized, these biases can often be accommodated by appropriate mathematical conditioning [14,29,54]. Ascertainment corrections that require limited assumptions about the sampling procedure have been developed [31].

An important alternative to traditional segregation analysis of quantitative or disease traits was proposed by Bonney [10,11], whose so-called *regressive models* have a Markovian basis in that they sequentially condition pedigree data to allow for familial correlations. This is accomplished by specifying a regression of a person's phenotype on the phenotypes siblings of ancestors and/or older siblings, (unmeasured) major gene genotype, and other explanatory variables. Regressive models can specify a variety of sources of nongenetic variation and covariation [23].

Model assessment for segregation analysis is usually accomplished by asymptotic likelihood-ratio tests or by invoking Akaike's information criterion. However, approaches that use either Monte Carlo [110] or the Metropolis algorithm [78] may be more reliable, since they may better characterize the likelihood surface. Adaptations of quasi–likelihood and other robust modeling schemes have been applied [73,127] for segregation analytic procedures also. Although these procedures are robust to distributional assumptions, separately identifying the gene frequency and penetrance parameters can be difficult [135].

80.3.2 Single-Marker Linkage Analysis

When a trait is found to have a substantial genetic component, especially a major gene component, classical linkage studies [94] are conducted to identify coinheritance of a trait with a single genetic marker. For all but the most complex of pedigrees, two-point linkage analysis (which includes a single marker locus and the trait locus for the two points) is trivially performed using modern high-speed computers. A statistically significant result provides evidence of a putative trait locus cosegregating with a marker. Provided a model for the joint effects from two disease susceptibility loci is available, perhaps from population studies or by analogy with animal models, classical genetic linkage analysis assuming a two-locus model is currently feasible [109].

Classical linkage analysis requires the investigator to specify a genetic model for the mode of inheritance of the trait. The genetic models are typically inferred by segregation analysis. For complex diseases or traits, identifying the correct genetic model through segregation analysis is often difficult or impossible, because segregation analyses cannot precisely model either effects from several loci jointly or gene–environment interactions. Many investigators posit a genetic model and then test for genetic linkage, without performing any preliminary segregation analysis. Under rather general assumptions this approach does not lead to excess false positives, provided the model guessing is done prior to any analysis [77,131]. However, incorrectly guessing a model and then performing linkage analysis can lead to a dramatic loss in power to detect a true genetic linkage [17].

Two alternative approaches are sometimes applied. Joint segregation and linkage analysis [12] provides the most efficient method for identifying genetic effects [5,40,41] but requires ascertainment corrections. The MOD score approach maximizes the ratio of the likelihood under linkage (with r less than 0.5) to the likelihood with no linkage ($r = 0.5$). This approach does not require ascertainment correction and provides unbiased estimates of the penetrance and other parameters [76].

80.3.3 Multipoint Linkage Analysis

As a result of the human genome project, hundreds of markers have been mapped on the human genome. Of course, markers have to be properly located on the genome and this requires either multipoint, radiation hybrid, or physical mapping methods [94]. For three hypothetical markers A, B, and C, this means determining the true order of the markers (A–B–C, A–C–B, or $B - A - C$) and their relative genetic or physical distances. Most human pedigrees include some deceased or unavailable individuals. In addition, markers are rarely completely informative, that is, not all individuals are heterozygous for all markers. As a result, in all studies of human pedigrees, for linked markers, determining which alleles are collocated on the same chromosome to form haplotypes is typically impractical. Therefore, evaluation of multiple markers requires the enumeration of probabilities for all possible haplotypes consistent with the observed data.

The likelihood computations eventually become prohibitive, depending on the complexity of the pedigree structures, the informativity of the markers, the number of markers to be considered, and the number of missing individuals. Computations based upon the Elston–Stewart algorithm [30] are linear in the number of individuals and exponential in the number of markers, and can therefore efficiently incorporate data from extended families [19,87]. Alternatively, hidden Markov models—within an EM algorithm—have been developed to evaluate data from multiple markers. These procedures are linear in the number of markers and exponential in the number of individuals and hence can efficiently be used for mapping large numbers of markers in small pedigrees [60,65]. Whereas Ott [91,92] foresaw the potential of the EM algorithm in calculating likelihoods, it was Lander and Green [65] who successfully proposed and implemented it for practical use in multipoint mapping.

Researchers are thus no longer limited to linkage studies involving a trait and a limited number of markers, as they were until about 1980 [13], when it was realized that restriction-fragment-length polymorphisms could yield hundreds of new genetic markers. Instead, investigators can now conduct linkage studies involving hundreds of markers spanning the entire human genome. These types of studies come under various names: linkage mapping, genome or genome-wide scan, genome or genome-wide screen, trait mapping, gene or genetic mapping, and global search.

In genomic work the character to be mapped is typically a new genetic marker, while in genetic epidemiology it is a Mendelian, quantitative, or complex trait. In either case, it was demonstrated in the mid-1980s that simultaneously using multiple, previously mapped markers is more efficient in detecting linkage than the traditional approach of separately testing for linkage at each marker [63,72].

80.3.4 Model-Free Methods

As mentioned, the classical linkage approach is model-based and has limitations. Therefore, model-free or genetically robust methods of analysis, also called *allele-sharing methods*, are increasingly used. The idea is based on simple genetic principles and is practically independent of specifying an underlying genetic model, making the model-free methods particularly appealing for the genetic analysis of complex traits. These methods are based on assessing whether related individuals, say siblings, share more genetic material at a specific locus than would be otherwise expected under no linkage between the marker and a putative trait locus.

For any single locus a pair of full siblings will share no, one, or two alleles from a common ancestor *[identical by descent (IBD)]*) with probabilities .25, .5, and .25, respectively. If a marker locus is closely linked to a trait-affecting locus, then the sib pairs that share two alleles IBD for the marker locus are expected to have the same genotypes at the trait locus, while sib pairs that share no alleles IBD are not expected to have the same trait genotypes. This fact about sib pairs forms the basis of several test statistics for linkage.

For quantitative traits the *Haseman-Elston method* is the most notable [49]. Evidence for linkage is provided by a statistically significant negative slope of a simple regression of squared sibling trait difference on the number of marker alleles shared IBD This method has been used [18,45] to identify two loci associated with high-density lipoprotein cholesterol, a risk factor for heart disease. A similar procedure extending to general pedigrees partitions variance among relatives into components attributable to effects from a locus linked to a genetic marker, to other genetic factors, and to nongenetic factors [2,134]. Model-free and model-based methods of linkage detection for quantitative traits are extensively reviewed by Wijsman and Amos [130].

For quantitative traits in which the major locus has a rather low contribution to the interindividual variation (less than 30% of variability), sampling of extreme individuals improves the power for linkage detection [15,25,104,105]. Analytical studies have shown that extremely discordant sibling pairs provide the greatest improvement in power (relative to other sampling schemes) to detect linkage to a marker under many plausible genetic models [104,105].

For disease traits, affected-sibling and other affected-relative-pair methods are widely used in model-free methods. Evidence for linkage is based on assessing observed marker allele sharing between affected siblings (or other relatives) with respect to that expected under no linkage. Blackwelder and Elston [6] provided power comparisons among three score tests for affected-sib-pair methods. Risch [101–103] extensively developed likelihood-based methods for detecting linkage using affected sib pairs. Holmans [53] provided a test with improved power for linkage detection by using restricted maximum-likelihood procedures. Although the majority of statistical developments are for specific methods for specific relative pairs, procedures combining different types of relative pairs have been developed for diseases [21] and for quantitative traits [3,89]. Affected-sib-pair and pedigree-member methods have been critical in identifying genetic loci involved in complex diseases such as type 2 diabetes [48] and Alzheimer's disease [95].

Genome-wide searches for trait loci can be and are conducted by separately applying model-free linkage tests at each of several hundred markers. An example using sib pair methods is a genome search for loci underlying asthma [20]. When testing each marker separately, optimal designs can be determined so that the cost of a linkage study can be halved in many situations [27,28]. The basic design is based on two stages. First, m markers are typed for each pair of affected relatives. Second, for each marker found to be significant at the first stage, additional markers are typed about their location and further assessed for significance.

For quantitative traits, extensions of the Haseman–Elston linkage method include using two markers [39], and several markers jointly [61] to estimate more precisely the IBD sharing at points along a chromosome. For disease traits, extensions of affected-relative-pair methods based on pointwise testing have also been developed [58,60,61]. Hauser et al. [51] extended Risch's method [101–103] to a likelihood-based interval mapping procedure, not only to detect linkage, but also to exclude regions yielding little evidence of disease loci. Simulation studies indicate that all of these extensions generally provide better power than their single-marker counterparts. Alternatively, methods that evaluate IBD sharing of chromosomal *regions* have been proposed [42,46].

80.3.5 Assessing Significance in Genome Scans

A major issue in conducting genome-wide scans is multiple testing [67]. Specifically, a nominal significance level of 0.05 at each locus tested will yield a genome-wide significance level greater than 0.05. Several investigators have used analytical methods to obtain appropriate critical values to control genome-wide false-positive rates.

Feingold et al. [35] were the first to propose a formal statistical method for human genome scans that took account and advantage of using mapped markers simultaneously. Their assumptions included a single susceptibility locus and having a dense map of identity-by-descent indicators for n affected pairs of relatives. On each chromosome the number X_t of relative pairs having identity by descent at a locus t is a Markov chain on the states $0, 1, \ldots, n$. Using a normalized version of X_t, they proposed to evaluate linkage as a change point (at the trait locus) of a Gaussian process approximating the Markov chain. Their approach controls the genome-wide false-positive rate and allows for power, sample size, and confidence region calculations.

Working with experimental organisms, Lander and Botstein [64] had earlier applied Brownian motion and large deviation theory to obtain a closed-form expression for genome-wide critical values; Feingold et al. [35] noted similarities in the two approaches. The results of Lander and Botstein were later applied to human complex traits [61,66,67]. In particular, rank-based critical values were developed for genome-wide scans for quantitative trait loci [61]. Dupuis et al. [24] extend the work in Reference 35 to multilocus models. Although much has been written about appropriate critical values for genome-wide scans, there is still considerable debate over current results [59,107,120].

Statistical significance for classical linkage analysis and genome scans may appropriately be assessed by Monte Carlo procedures that simulate genetic (probability) mechanisms. The idea is to simulate the inheritance of trait or marker genotypes consistent with observed pedigree structures under a given hypothesis. In human linkage analysis, simulation methods were introduced as a practical way to estimate the power of a proposed linkage study for fully penetrant Mendelian traits [8] and complex traits [96].

To investigate issues other than statistical power, Ott [93] proposed and implemented a more general simulation method for classical linkage analysis. More recently, statistical significance in candidate gene [123] and genome scan investigations [20,67,124] has been assessed by simulation methods.

80.3.6 Localization and Identification of Trait Genes

Although genome scanning is now becoming more common, limitations confine such studies to relatively few large laboratories across the world. When an entire genome scan is not practical for a laboratory, *candidate* regions may be studied. These regions are typically chosen because they contain genes or gene clusters thought likely to influence the trait or risk for disease being studied, or because animal studies suggest that a region may be important in modulating trait expression or disease risk. As genotyping becomes cheaper and more efficient, and as computer programs become available for data manipulation and rapid linkage analysis, more laboratories will be able to conduct genome scans.

Identifying the critical region containing a trait affecting locus for subsequent positional cloning requires the study of a large number of meiotic events [9]. Alternatively, linkage disequilibrium mapping can be used [55,132]. This approach requires that affected individuals be ancestrally related and therefore carry a common mutation. Linkage disequilibrium methods for the localization of quantitative traits have been developed [133].

80.3.7 Association Studies

Association studies evaluate differences in mean values or proportions of affected individuals among classes of individuals defined by genotypes. Individuals need not be related for association studies. By contrast, linkage analysis evaluates covariances among relatives, or association of genotypes with diseases within families.

Association in populations between a marker genotype and a phenotype can occur for several reasons. First, the genotype may have a functional relationship with the disease risk or trait levels. Second, the marker may have no direct effect upon the trait or disease risk, but marker and trait loci may be linked and alleles at each locus may co-occur dependently. Alleles at the two loci may have dependent frequencies as a result of several processes, most commonly the admixture of two distinct populations or the occurrence of a mutation to a trait allele that subsequently became common because of random drift or through preferential selection of the new allele [126]. For linked loci, the disequilibrium at the nth generation is $(1-r)^n$ times the original disequilibrium. If marker and disease (or trait) alleles are not independent (and hence are in disequilibrium), then association will exist between disease risk or trait variability and marker alleles or genotypes.

For unrelated individuals, association studies are analyzed using standard statistical techniques such as analysis of variance or logistic regression. However, genetic association studies are often performed using sampling through families, for a number of reasons. First, when disease causation is influenced by relatively uncommon genetic factors, sampling through families that contain multiple affected individuals can substantially improve the power to identify these genetic effects [105].

Second, if the sample that is obtained contains unobserved population stratification, spurious associations (not related either to an effect of the alleles on the trait or to linkage) may result [32]. However, conditioning on the parental genotypes eliminates this source of possible *noncausal* association (i.e., the association that does not result from effects of a linked genetic locus) [34].

Third, when searching for associations caused by linkage disequilibrium, the study of extended haplotypes may yield more information than evaluation of a single marker. Although haplotypes can be inferred from population-based data [122], the study of family data improves the precision of haplotype assignment and hence of association studies. Whittemore and Halpern [129] discuss optimal designs for association, segregation, and linkage studies.

A variety of methods are available for the statistical evaluation of associations for clustered data [97]. However, many of the methods developed for analyses of clustered data are not readily adapted for analysis of human data, in which the families or clusters vary in size and there may also be missing individuals. Regressive models permit evaluation of the complete data with inclusion of covariates such as effects from marker alleles [10,11]. Alternatively, estimating equations methods can be applied [99]. Both approaches provide an analytical strategy that incorporates data from the entire family, including both affected and unaffected individuals. These methods tacitly assume that all of the families have been sampled from a single ethnically similar population.

Other methods that condition upon the parental genotypes and hence remove any effects from population stratification have been developed [108]. The *transmission disequilibrium test* (TDT) has become a popular method for evaluating evidence for association or linkage disequilibrium between a genetic marker and disease risk [114]. The TDT approach provides a method for testing the hypothesis of no linkage or disequilibrium, against an alternative hypothesis of linkage and disequilibrium. For this test, the alleles transmitted to an affected child are compared with parental alleles not transmitted to that child. The alleles not transmitted to the affected child form a matched control, and methods for analysis of matched data are then applied—for example, McNemar's test. Extensions of this method are available to permit analysis of multiallelic markers [56] and multiple loci [84].

80.3.8 Analysis of Multivariate Traits

Relatively few methods are available for genetic linkage analysis of multivariate traits in families. For family studies of quantitative traits, biometrical and segregation analyses commonly use principal components to evaluate multivariate phenotypes [7].

Alternatively, path-analytic procedures have been extensively developed for multivariate genetic studies using extended families [98,100]. Combined multivariate segregation and linkage analysis is technically feasible [12] and has been implemented using Markov chain Monte Carlo methods [52]. Both the Haseman–Elston [4] and variance components procedures have been extended for application to multivariate traits [1,22].

Extensive methods and software for the biometrical analysis of twin studies have been developed, using structural equation approaches [85], which provide a framework for variance components analysis. The power of structural modeling for multivariate phenotypes in dizygotic or sib pairs has been characterized through simulation studies [81]. Multivariate methods for disease phenotypes have not been as well developed. Whittemore et al. [128] provided a stochastic modeling approach, which permits transitions among disease states.

80.4 Additional Topics

Several other problems in human genetics make use of statistical theory and methods. One that has received renewed attention because of advances in molecular biology is estimating the history of human evolution. A standard method of reconstructing this history is estimation of *evolutionary trees*, typically showing divergence of a set of species from an assumed common ancestor. Using DNA–DNA hybridization, specific genes, or more generally sequences of DNA or RNA, studies of various human populations are used to construct phylogenetic trees. A popular method of estimating both topology and divergence times is maximum likelihood [36]. Among the many statistical issues that arise in inferring phylogenies, a more recent aspect is that of *confidence* for clades (related parts) of an observed tree, for which Felsenstein [37] proposed bootstrap methodology. Efron et al. [26] review criticisms and propose extensions of Felsenstein's bootstrap approach to confidence levels and testing. Additional statistical aspects of inferring evolutionary trees from molecular data are reviewed by Li [74].

There are also forensic applications in human genetics requiring statistical analysis [125]. These are primarily concerned with identifying individuals. Common applications include paternity cases, in which the identity of a putative father is compared with DNA sequences in the child (and mother if available); identification in criminal cases, in which specimens (blood or semen) obtained from the crime are compared with samples from the putative perpetrator; and other parentage issues, for instance when the parents of an orphaned child are not known and a putative relative has been identified [106].

With the substantive advances in genetic technology for genotyping individuals, probability assignments currently yield extremely precise estimates in identification, but numerous technical issues invoke statistical and societal considerations. For instance, forensic applications generally require specification of gene frequencies from a population of potential perpetrators, and there are substantial ethnic variations.

Genetic counseling often requires application of Bayes' theorem. Generally, the Elston–Stewart algorithm provides an efficient tool for incorporating pedigree information and performing the requisite Bayes calculations [30]. In the case of diseases that are strongly selected against and for which new mutations represent a substantial proportion of affected individuals, specialized applications of Bayes' theorem have been developed [119]. The book by Weir provides a comprehensive overview of related issues in population genetics [125].

80.5 Literature

Additional applications of statistics in human genetics are discussed in the ESSU entries "Linkage, Genetic", by Thompson, and "Statistics in evolutionary genetics", by Simonsen and Ewens. In addition to those cited in the original entry "Statistics in human genetics" by Smith, there are more recent general and specialized books, useful to those interested in learning about statistical genetics in more detail: Singer [111] and Griffiths et al. [43] discuss genetics; Lander and Waterman [69], Lange [70], and Speed and Waterman [113] cover many topics from a mathematical or statistical viewpoint; Khoury et al. [57] introduce fundamental concepts and techniques of genetic epidemiology; Thompson [116] discusses pedigree analysis. An overview of biostatistical and epidemiological applications for genetic analysis is provided by Weiss [126]. Biological aspects of human genetics are explained by Thompson et al. [119].

References

1. Almasy, L., Dyer, T. D., and Blangero, J. (1997). Bivariate quantitative trait linkage analysis: pleiotropy versus coincident linkages. *Genet. Epidemiol.*, **14**, 953–958.

2. Amos, C. I. (1994). Robust detection of genetic linkage by variance components analysis. *Am. J. Hum. Genet.*, **54**, 535–543.

3. Amos, C. I. and Elston, R. C. (1989). Robust methods for the detection of genetic linkage for quantitative data from pedigrees. *Genet. Epidemiol.*, **6**, 349–361.

4. Amos, C. I., Elston, R. C., Bonney, G. E., Keats, B. J. B., and Berenson, G. S. (1990). A multivariate method for detecting genetic linkage with application to the study of a pedigree with an adverse lipoprotein phenotype. *Am. J. Hum. Genet.*, **47**, 247–254.

5. Amos, C. I. and Rubin, L. A. (1995). Major gene analysis for diseases and disorders of complex etiology. *Exp. Clin. Immunogenet.*, **12**, 141–155.

6. Blackwelder, W. C. and Elston, R. C. (1985). A comparison of sib-pair linkage tests for disease susceptibility loci. *Genet. Epidemiol.*, **2**, 85–97.

7. Blangero, J. and Konigsberg, L. W. (1991). Multivariate segregation analysis using the mixed model. *Genet. Epidemiol.*, **8**, 299–316.

8. Boehnke, M. (1986). Estimating the power of a proposed linkage study: a practical computer simulation approach. *Am. J. Hum. Genet.*, **39**, 513–527.

9. Boehnke, M. (1994). Limits of resolution of genetic linkage studies: implications

for the positional cloning of human disease genes. *Am. J. Hum. Genet.*, **55**, 379–390.

10. Bonney, G. E. (1984). On the statistical determination of major gene mechanisms in continuous traits: regressive models. *Am. J. Med. Genet.*, **18**, 731–749.

11. Bonney, G. E. (1986). Regressive logistic models for familial disease and other binary traits. *Biometrics*, **42**, 611–625.

12. Bonney, G. E., Lathrop, G. M., and Lalouel, J. M. (1988). Combined linkage and segregation analysis using regressive models. *Am. J. Hum. Genet.*, **43**, 29–37.

13. Botstein, D., White, R. L., Skolnick, M. H., and Davis, R. W. (1980). Construction of a genetic linkage map in man using restriction fragment length polymorphisms. *Am. J. Hum. Genet.*, **32**, 314–331.

14. Cannings, C. and Thompson, E. (1977). Ascertainment in the sequential sampling of pedigrees. *Clin. Genet.*, **12**, 208–212.

15. Carey, G. and Williamson, J. A. (1991). Linkage analysis of quantitative traits: increased power by using selected samples. *Am. J. Hum. Genet.*, **49**, 786–796.

16. Clayton, D. G. (1994). Some approaches to the analysis of recurrent event data. *Statist. Methods Med. Res.*, **3**, 244–262.

17. Clerget-Darpoux, F. M., Bonaiti-Pellie, C., and Hochez, J. (1986). Effects of misspecifying genetic parameters in lod score analysis. *Biometrics,* **42**, 393–399.

18. Cohen, J. C., Wang, Z., Grundy, S. M., Stoesz, M., and Guerra, R. (1994). Variation at the hepatic lipase and apolipoprotein AI/CIII/AIV loci is a major cause of genetically determined variation in plasma HDL cholesterol levels. *J. Clin. Invest.*, **94**, 2,377–2,384.

19. Cottingham, R. W., Idury, R. M., and Schaffer, A. A. (1993). Faster sequential genetic linkage computations. *Am. J. Hum. Genet.*, **53**, 252–263.

20. Daniels, S. E., Bhattacharrya, J. A., et al. (1996). A genome-wide search for quantitative trait loci underlying asthma. *Nature,* **383**, 247–250.

21. Davis, S., Schroeder, M., Goldin, L. R., and Weeks, D. E. (1996). Nonparametric simulation-based statistics for detecting linkage in general pedigrees. *Am. J. Hum. Genet.*, **58**, 867–880.

22. de Andrade, M., Thiel, T. J., Yu, L., and Amos, C. I. (1997). Asssessing linkage in chromosome 5 using components of variance approach: univariate versus multivariate. *Genet. Epidemiol.*, **14**, 773–778.

23. Demenais F. M. (1991). Regressive logistic models for familial diseases: a formulation assuming an underlying liability model. *Am. J. Hum. Genet.*, **49**, 773–785.

24. Dupuis, J. Brown, P. O., and Siegmund, D. (1995). Statistical methods for linkage analysis of complex traits from high-resolution maps of identity by descent. *Genetics,* **140**, 843–856.

25. Eaves, L. and Meyer, J. (1994). Locating human quantitative trait loci: guidelines for the selection of sibling pairs for genotyping. *Behav. Genet.*, **24**, 443–455.

26. Efron, B., Halloran, E., and Holmes, S. (1996). Bootstrap confidence levels for phylogenetic trees. *Proc. Natl. Acad. Sci. USA.*, **93**, 13,429–13,434.

27. Elston, R. C. (1992). *Designs for the global search of the human genome by linkage analysis.* Proc. XVIth Int. Biometric Conf., pp. 39–51.

28. Elston, R. C., Guo, X., and Williams, L. V. (1996). Two-stage global search designs for linkage analysis using pairs of affected relatives. *Genet. Epidemiol.*, **13**, 535–558.

29. Elston, R. C. and Sobel, E. (1979). Sampling considerations in the analysis of human pedigree data. *Am. J. Hum. Genet.*, **31**, 62–69.

30. Elston, R. C. and Stewart, J. (1971). A general model for the genetic analysis of pedigree data. *Hum. Hered.*, **21**, 523–542.

31. Ewens, W. J. and Schute, N. E. (1986). A resolution of the ascertainment sampling problem. 1. Theory. *Theor. Pop. Biol.*, **30**, 388–412.

32. Ewens, D. J. and Spielman, R. S. (1995). The transmission/disequilibrium test: history, subdivision, and admixture. *Am. J. Hum. Genet.*, **57**, 455–464.

33. Falconer, D. S. (1989). *Introduction to Quantitative Genetics*, 3rd ed. Longman, London.

34. Falk, C. T. and Rubinstein, P. (1987). Haplotype relative risks: An easy reliable way to construct a proper control sample for risk calculations. *Ann. Hum. Genet.*, **51**, 227–233.

35. Feingold, E., Brown, P. O., and Siegmund, D. (1993). Gaussian models for genetic linkage analysis using complete high resolution maps of identity by descent. *Am. J. Hum. Genet.*, **53**, 234–251.

36. Felsenstein, J. (1983). Statistical inference of phylogenies. *J. Roy Statist. Soc. A*, **142**, 246–272.

37. Felsenstein, J. (1985). Confidence limits on phylogenies: an approach using the bootstrap. *Evolution*, **39**, 783–791.

38. Fisher, R. A. (1918). The correlation between relatives on the supposition of Mendelian inheritance. *Trans. Roy Soc. Edinburgh*, **52**, 399–433.

39. Fulker, D. W. and Cardon, L. R. (1994). A sibpair approach to interval mapping of quantitative trait loci. *Am. J. Hum. Genet.*, **54**, 1092–1103.

40. Gauderman, W. J. and Faucett, C. L. (1997). Detection of gene–environment interactions in joint segregation and linkage analysis. *Am. J. Hum. Genet.*, **61**, 1189–1199.

41. Gauderman, W. J. and Thomas, D. C. (1994). Censored survival models for genetic epidemiology: a Gibbs sampling approach. *Genet. Epidemiol.*, **11**, 171–188.

42. Goldgar, D. E. (1990). Multipoint analysis of human quantitative trait variation. *Am. J. Hum. Genet.*, **47**, 957–967.

43. Griffiths, A. J. F., Miller, J. H., Suzuki, D. T., Lewontin, R. C., and Gelbert, W. M. (1996). *An Introduction to Genetic Analysis*, 6th ed. Freeman, New York.

44. Grove, J. S., Zhao, L. P., and Quiaoit, F. (1993). Correlation analysis of twin data with repeated measures based on generalized estimating equations. *Genet. Epidemiol.*, **10**, 539–544.

45. Guerra, R., Wang, J., Grundy, S. M., and Cohen, J. C. (1997). A hepatic lipase (LIPC) allele associated with high plasma concentrations of high density lipoprotein cholesterol. *Proc. Natl. Acad. Sci. (USA)*, **94**, 4,532–4,537.

46. Guo, S. W. (1994). Computation of identity-by-descent proportions shared by two siblings. *Am. J. Hum. Genet.*, **54**, 1104–1109.

47. Guyer, M. S. and Collins, F. S. (1995). How is the Human Genome Project doing, and what have we learned so far? *Proc. Natl. Acad. Sci. (USA)*, **92**, 10,841–10,848.

48. Hanis, C. L., Boerwinkle, E., Chakraborty, R., et al. (1996). A genome-wide search for human non-insulin-dependent (type 2) diabetes genes reveals a major susceptibility locus on chromosome 2. *Nat. Genet.*, **13**, 161–166.

49. Haseman, J. K. and Elston, R. C. (1972). The investigation of linkage between a quantitative trait and a marker locus. *Behav. Genet.*, **2**, 3–19.

50. Hasstedt, S. J. (1991). A variance components/major locus likelihood approximation to quantitative data. *Genet. Epidemiol.*, **8**, 113–125.

51. Hauser, E. R., Boehnke, M. Guo, S. W., and Risch, N. (1996). Affected-sib-pair interval mapping and exclusion for complex genetic traits: sampling considerations. *Genet Epidemiol.*, **13**, 117–137.

52. Heath, S. C., Snow, G. L., Thompson, E. A., Tseng, C., and Wijsman, E. M. (1997). MCMC segregation and linkage analysis. *Genet. Epidemiol.*, **14**, 1,011–1,016.

53. Holmans, P. (1993). Asymptotic properties of affected-sib-pair linkage analysis. *Am. J. Hum. Genet.*, **52**, 362–374.

54. Hopper, J. L. and Mathews, J. D. (1982). Extensions to multivariate normal models for pedigree analysis. *Ann. Hum. Genet.*, **46**, 373–383.

55. Kaplan, N. L., Hill, W. G., and Weir, B. S. (1995). Likelihood methods for locating disease genes in nonequilibrium populations. *Am. J. Hum. Genet.*, **56**, 18–32.

56. Kaplan, N. L., Martin, E. R., and Weir, B. S. (1997). Power studies for the transmission/disequilibrium tests with multiple alleles. *Am. J. Hum. Genet.*, **60**, 691–702.

57. Khoury, M. J., Beaty, T. H., and Cohen, B. H. (1993). *Fundamentals of Genetic Epidemiology*. Oxford University Press, New York.

58. Kong, A. and Cox, N. J. (1997). Allele-sharing models: LOD scores and accurate linkage tests. *Am. J. Hum. Genet.*, **61**, 1,179–1,188.

59. Kruglyak, L. (1996). Thresholds and sample sizes. *Nature Genet.*, **14**, 132–133.

60. Kruglyak, L., Daly, M. J., Reeve-Daly, M. P., and Lander, E. S. (1996). Parametric and nonparametric linkage analysis: A unified multipoint approach. *Am. J. Hum. Genet.*, **58**, 1347–1363.

61. Kruglyak, L. and Lander, E. S. (1995). Complete multipoint sib-pair analysis of qualitative and quantitative traits. *Am. J. Hum. Genet.*, **57**, 439–454.

62. Lalouel, J. M., Rao, D. C., Morton, N.E., and Elston, R. C. (1983). A unified model for complex segregation analysis with pointers. *Am. J. Hum. Genet.*, **35**, 816–826.

63. Lander, E. S. and Botstein, D. (1986). Strategies for studying heterogeneous genetic traits in humans by using a linkage map of restriction fragment length polymorphisms. *Proc. Natl. Acad. Sci. (USA)*, **83**, 7353–7357.

64. Lander, E. S. and Botstein, D. (1989). Mapping Mendelian factors underlying quantitative traits using RFLP linkage maps. *Genetics,* **121**, 185–199.

65. Lander, E. S. and Green, P. (1987). Construction of multilocus genetic linkage maps in humans. *Proc. Natl. Acad. Sci. (USA)*, **84**, 2363–2367.

66. Lander, E. S. and Kruglyak, L. (1995). Genetic dissection of complex traits: guidelines for interpreting and reporting linkage results. *Nature Genet.*, **11**, 241–247.

67. Lander, E. S. and Schork, N. J. (1994). Genetic dissection of complex traits. *Science,* **265**, 2037–2048.

68. Lander, E. S. and Waterman, M. S. (1988). Genomic mapping by fingerprinting random clones: a mathematical analysis. *Genomics,* **2**, 231–239.

69. Lander, E. S. and Waterman, M. S. (1995). *Calculating the Secrets of Life*. National Academy Press, Washington, DC.

70. Lange, K. (1997). *Mathematical and Statistical Methods for Genetic Analysis*. Springer-Verlag, New York.

71. Lange, K. and Elston, R. C. (1975). Extensions to pedigree analysis I. Likelihood calculations for simple and complex pedigrees. *Hum. Hered.*, **25**, 95–105.

72. Lathrop, G. M., Lalouel, J. M., Julier, C., and Ott, J. (1984). Strategies for multilocus linkage analysis in humans. *Proc. Natl. Acad. Sci. (USA)*, **81**, 3443–3446.

73. Lee, H. and Stram, D. O. (1996). Segregation analysis of continuous phenotypes by using higher sample moments. *Am. J. Hum. Genet.*, **58**, 213–224.

74. Li, W.-H. (1997). *Molecular Evolution*. Sinauer Associates, Sunderland, MA.

75. Liang, K.-Y. and Beaty, T. H. (1991). Measuring familial aggregation by using odds-ratio regression models. *Genet. Epidemiol.*, **8**, 361–370.

76. Liang, K.-Y., Rathouz, P. J., and Beaty, T. H. (1996). Determining linkage and mode of inheritance: mod scores and

other methods. *Genet. Epidemiol.*, **13**, 575–593.

77. Liang, K.-Y. and Self, S. G. (1996). On the asymptotic behaviour of the pseudolikelihood ratio test statistic. *J. Roy Statist. Soc. B,* **58**, 785–796.

78. Lin, S., Thompson, E., and Wijsman, E. (1996). An algorithm for Monte Carlo estimation of genotype probabilities on complex pedigrees. *Ann. J. Hum. Genet.*, **58**, 343–357.

79. Lunnetta, K. L., Boehnke, M., Lange, K., and Cox, D. R. (1996). Selected locus and multiple panel models for radiation hybrid mapping. *Amer. J. Hum. Genet.*, **59**, 717–725.

80. Mack, W., Lanholz, B., and Thomas, D. C. (1990). Survival models for familial aggregation of cancer. *Environ. Health Perspect.*, **87**, 27–35.

81. Martin, N., Boomsma, D., and Machin, G. (1997). A twin-pronged attack on complex traits. *Nat. Genet.*, **17**, 387–392.

82. Miki, Y., Swensen, J., Shattuck-Eidens, D., et al. (1994). A strong candidate for the breast and ovarian cancer susceptibility gene BRCA1. *Science,* **266**, 66–71.

83. Morton, N. E. (1997). Genetic epidemiology. *Amer. J. Hum. Genet.*, **67**, 1–13.

84. Mulcahy, B., Waldron-Lynch, F., McDermott, M. F., et al. (1996). Genetic variability in the tumor necrosis factor–lymphotoxin region influences susceptibility to rheumatoid arthritis. *Am. J. Hum. Genet.*, **59**, 676–683.

85. Neale, M. C. and Cardon, L. R. (1992). *Methodology for Genetic Studies of Twins and Families,* NATO ASI Series, Series D: Behavioral and Social Sciences, 67, Kluwer Academic, Boston.

86. Nelson, D. O. and Speed, T. P. (1994). Statistical issues in constructing high resolution physical maps. *Statist. Sci.,* **9**, 334–354.

87. O'Connell, J. R. and Weeks, D. E. (1995). The VITESSE algorithm for rapid exact multilocus linkage analysis via genotype set-recoding and fuzzy inheritance. *Nat. Genet.,* **11**, 402–408.

88. Olson, M. V. (1993). The human genome project. *Proc. Natl. Acad. Sci. (USA),* **90**, 4,338–4,344.

89. Olson, J. and Wijsman, E. (1993). Linkage between quantitative trait and marker loci: Methods using all relative pairs. *Genet. Epidemiol.*, **10**, 87–102.

90. Ott, J. (1974). Estimation of the recombination fraction in human pedigrees: Efficient computation of the likelihood for human linkage studies. *Am. J. Hum. Genet.*, **26**, 588–597.

91. Ott, J. (1977). Counting methods (EM algorithm) in human pedigree analysis: Linkage and segregation analysis. *Ann. Hum. Genet.*, **40**, 443–454.

92. Ott, J. (1979). Maximum likelihood estimation by counting methods under polygenic and mixed models in human pedigrees. *Am. J. Hum. Genet.*, **31**, 161–175.

93. Ott, J. (1989). Computer-simulation methods in human linkage analysis. *Proc. Natl. Acad. Sci. (USA),* **86**, 4175–4178.

94. Ott, J. (1991). *Analysis of Human Genetic Linkage*. Johns Hopkins University Press, Baltimore, MD.

95. Pericack-Vance, M. A. and Haines, J. L. (1995). Genetic susceptibility to Alzheimer disease. *Trends Genet.*, **11**, 504–508.

96. Ploughman, L. M. and Boehnke, M. (1989). Estimating the power of a proposed linkage study for a complex genetic trait. *Am. J. Hum. Genet.*, **44**, 543–551.

97. Prentice, R. L. and Zhao, L. P. (1991). Estimating equations for parameters in means and covariances of multivariate discrete and continuous responses. *Biometrics,* **41**, 825–839.

98. Province, M. A. and Rao, D. C. (1995). General purpose model and a computer program for combined segregation and path analysis (SEGPATH): Automatically creating computer programs from

symbolic language model specifications. *Genet. Epidemiol.*, **12**, 203–219.

99. Qaqish, B. F. and Liang, K. Y. (1992). Marginal models for correlated binary responses with multiple classes and multiple levels of nesting. *Biometrics*, **48**, 939–950.

100. Rao, D. C. (1985). Applications of path analysis in human genetics. In *Multivariate Analysis B IV*, P. R. Krishnaiah, ed., pp. 467–484. Elsevier, New York.

101. Risch, N. (1990). Linkage strategies for genetically complex traits. I. Multilocus models. *Am. J. Hum. Genet.*, **46**, 222–228.

102. Risch, N. (1990). Linkage strategies for genetically complex traits. II. The power of affected relative pairs. *Am. J. Hum. Genet.*, **46**, 229–241.

103. Risch, N. (1990). Linkage strategies for genetically complex traits. III. The effect of marker polymorphism on analysis of affected relative pairs. *Am. J. Hum. Genet.*, **46**, 242–253. Erratum: **51**, 673–675 (1992).

104. Risch, N. and Zhang, H. (1995). Extreme discordant sib pairs for mapping quantitative trait loci in humans. *Science*, **268**, 1584–1589.

105. Risch, N. and Zhang, H. (1996). Mapping quantitative trait loci with extreme discordant sib pairs: Sampling considerations. *Am. J. Hum. Genet.*, **58**, 836–843.

106. Roeder, K. (1994). DNA fingerprinting: A review of the controversy. *Statist. Sci.*, **9**, 222–278. (With discussion.)

107. Sawcer, S., Jones, H. B., Judge, D., et al. (1997). Empirical genomewide significance levels established by whole genome simulations. *Genet. Epidemiol.*, **14**, 223–230.

108. Schaid, D. J. (1996). General score tests for associations of genetic markers with disease using cases and their parents. *Genet. Epidemiol.*, **13**, 423–449.

109. Schork, N. J., Boehnke, M., Terwilliger, J. D., and Ott, J. (1993). Two-trait-locus linkage analysis: A powerful strategy for mapping complex genetic traits. *Am. J. Hum. Genet.*, **53**, 1127–1136.

110. Schork, N. J. and Schork, M. A. (1989). Testing separate families of segregation hypotheses: Bootstrap methods. *Am. J. Hum. Genet.*, **45**, 803–813.

111. Singer, S. (1985). *Human Genetics: An Introduction to the Principles of Heredity*. Freeman, New York.

112. Sobel, E. and Lange, K. (1993). Metropolis sampling in pedigree analysis. *Statist. Methods Med. Res.*, **2**, 263–282.

113. Speed, T. P. and Waterman, M. S. (1996). *Genetic Mapping and DNA Sequencing*. Springer-Verlag, New York.

114. Spielman, R. S. and Ewens, W. J. (1996). The TDT and other family-based tests for linkage disequilibrium and association. *Am. J. Hum. Genet.*, **59**, 983–989.

115. Stewart, E. A., McKusick, K. B., Aggarwal, A., et al. (1997). An STS-based radiation hybrid map of the human genome. *Genome Res.*, **7**, 422–433.

116. Thompson, E. (1986). *Pedigree Analysis of Human Genetics*. Johns Hopkins University Press, Baltimore, MD.

117. Thompson, E. A. (1986). Genetic epidemiology: A review of the statistical basis. *Statist. Med.*, **5**, 291–302.

118. Thompson, E. A. (1994). Monte Carlo likelihood in genetic mapping. *Statist. Sci.*, **9**, 355–366.

119. Thompson, M. W., McInnes, R. R., and Willard, H. F. (1991). *Genetics in Medicine*, 5th ed. Saunders, Philadelphia.

120. Todorov, A. A. and Rao, D. C. (1997). Tradeoff between false positives and false negatives in the linkage analysis of complex traits. *Genet. Epidemiol.*, **14**, 453–464.

121. Tregouet, D.-A., Herbeth, B., Juhan-Vague, I., et al. (1998). Bivariate familial correlation analysis of quantitative traits, by use of estimating equations: application to a familial analysis of the insulin resistance syndrome. *Genet. Epidemiol.*, **16**, 69–83.

122. Valdes, A. M. and Thomson, G. (1997). Detecting disease-predisposing variants: the haplotype method. *Am. J. Hum. Genet.*, **60**, 703–716.

123. Wan, Y., Cohen, J. C., and Guerra, R. (1997). A permutation test for the robust sib-pair linkage method. *Ann. Hum. Genet.*, **61**, 79–87.

124. Weeks, D. E., Lehner, T., Squires-Wheeler, E., Kaufmann, C., and Ott, J. (1990). Measuring the inflation of the lod score due to its maximization over model parameter values in human linkage analysis. *Genet. Epidemiol.*, **7**, 237–243.

125. Weir, B. S. (1996). *Genetic Data Analysis II*. Sinauer Associates, Sunderland, MA.

126. Weiss, K. M. (1995). *Genetic Variation and Human Disease: Principles and Evolutionary Approaches*. Cambridge University Press.

127. Whittemore, A. S. and Gong, G. (1994). Segregation analysis of case-control data using generalized estimating equations. *Biometrics*, **50**, 1073–1087.

128. Whittemore, A. S., Gong, G., and Itnyre, J. (1997). Prevalence and contribution of BRCA1 mutations in breast cancer and ovarian cancer: results from three U.S. population-based case-control studies of ovarian cancer. *Am. J. Hum. Genet.*, **60**, 496–504.

129. Whittemore, A. S. and Halpern, J. (1997). Multistage sampling in genetic epidemiology. *Statist. Med.*, **16**, 153–167.

130. Wijsman, E. and Amos, C. I. (1997). Genetic analysis of simulated oligogenic traits in nuclear and extended pedigrees: Summary of GAW10 contributions. *Genet. Epidemiol.*, **14**, 719–736.

131. Williamson, J. A. and Amos, C. I. (1990). On the asymptotic behavior of the estimate of the recombination fraction under the null hypothesis of no linkage when the model is misspecified. *Genet. Epidemiol.*, **7**, 309–318.

132. Xiong, M. and Guo, S. W. (1997). Fine-scale genetic mapping based on linkage disequilibrium: Theory and applications. *Am. J. Hum. Genet.*, **60**, 1513–1531.

133. Xiong, M. and Guo, S. W. (1997). Fine-scale mapping of quantitative trait loci using historical recombinations. *Genetics*, **145**, 1201–1218.

134. Xu, S. and Atchley, W. R. (1995). A random model approach to interval mapping of quantitative trait loci. *Genetics*, **141**, 1189–1197.

135. Zhao, L. P. and Grove, J. S. (1995). Identifiability of segregation parameters using estimating equations. *Hum. Hered.*, **45**, 286–300.

136. Zhao, L. P., and Le Marchand, L. (1992). An analytical method for assessing patterns of familial aggregation in case-control studies. *Genet. Epidemiol.*, **8**, 141–154.

81 Statistics in Medical Diagnosis

D. M. Titterington

81.1 Introduction

Medical diagnosis is one area in which the application of statistical analysis has made a rather enigmatic impact. The associated problems can be abstracted neatly as well-developed statistical ones, and real-life applications are reported in ever-increasing numbers, but acceptance of statistical methods has not yet been widespread for reasons that will be delineated below.

81.2 The Diagnosis Problem

Formally, it is assumed that patients belong each to one of d disease classes, $\Pi_1, \ldots, \Pi_d$, say, but it is not known to which. The objective is to use all information available from "past experience" and from data gathered from a particular patient to make some inference or decision about the unknown class membership of that patient. The data from a patient usually take the form of records of the presence or absence of various symptoms or the results of relevant clinical tests or measurements. These variables may be grouped together under the term *indicants*. In a very simple case, with $d = 2$, Π_1 represents patients with appendicitis, for which the appropriate treatment is surgery, and Π_2 subsumes patients suffering from abdominal pain resulting from causes that require less drastic treatment. Indicants are used from which, statistically or otherwise, the members of Π_1 can be identified without having to operate on everyone.

In some applications, the $\{\Pi_i\}$ may represent *prognostic* categories [13].

Even at this level of generality there may be practical difficulties. Is the term "disease" appropriate for a particular condition? Can all possible disease classes be identified? Is the presence of a symptom well defined? (Often patients are asked to describe the degree of some pain or other, which is likely to lead to very subjective, nonstandardized responses.) Is even a particular clinical test properly standardized?

Such very real difficulties beset both clinical and statistical diagnosis. They have helped to maintain antipathy toward statistical methods that depend on well-defined structure and have promoted some exploration of "fuzzy" methods; see Reference 14.

Table 1: $\{f_i(\mathbf{x})\}$ for presence or absence of disease; x = diastolic blood pressure.

	Low $x=1$	Moderate $x=2$	High $x=3$	p_i
Presence Π_1	0.2	0.4	0.4	0.3
Absence Π_2	0.7	0.2	0.1	0.7

81.3 Relevant Statistical Methods

81.3.1 Discriminant Analysis

If the indicants measured on a given patient are represented by a $k \times 1$ vector $\mathbf{x}$, then of direct interest are the conditional probabilities $\{p(\Pi_i|\mathbf{x}), i = 1, \ldots, d\}$, or ratios such as $p(\Pi_1|\mathbf{x})/p(\Pi_2|\mathbf{x})$, sometimes called *odds ratios*. On the basis of these, a patient may be assigned to a particular disease class [that corresponding to the largest of the $\{p(\Pi_i|\mathbf{x})\}$, for instance] and treated accordingly. If the consequences of the various misclassifications are not equally costly, then some modification of the above based on decision theory can be used, provided relative costs can be assessed—another point of dispute in the medical context.

Statistical modeling of the $\{p(\Pi_i|\mathbf{x})\}$, or of some valid discriminant rule, is required, with subsequent estimation using "past experience," often in the form of datasets from patients already identified as belonging to the various disease classes. Modeling may be of the $\{p(\Pi_i|\mathbf{x})\}$ or their ratios directly, using logistic models, for instance. Alternatively, modeling may be based on the components of the right-hand side of the following expression of Bayes' theorem:

$$p(\Pi_i|\mathbf{x}) \propto p(\Pi_i, \mathbf{x}) = p_i f_i(\mathbf{x}).$$

Here $f_i(\cdot)$ denotes the sampling density of $\mathbf{x}$ within the ith class and $\{p_i\}$ the prior probabilities or incidence rates of the disease classes. This latter formulation is by far the more prevalent in the medical literature, although Dawid [2] argues that, particularly for medical problems, direct modeling of the $\{p(\Pi_i|\mathbf{x})\}$ is more sensible in that the $\{p(\Pi_i|\mathbf{x})\}$ are (1) less influenced by any bias incurred by the method of selecting patients; (2) often more stable, over time, for instance, than the joint probabilities $\{p(\Pi_i, x)\}$.

The Bayes theorem approach can be illustrated by the analysis in Tables 1 and 2. The two "disease categories" indicate presence (Π_1) or absence (Π_2) of a certain disease the indicant is diastolic blood pressure, coded low ($x = 1$), moderate ($x = 2$), or high ($x = 3$). Table 1 gives the $\{f_i(\mathbf{x})\}$ and the $\{p_i\}$, with the resulting $\{p(\Pi_i|\mathbf{x})\}$ and odds ratios $\{p(\Pi_1|\mathbf{x})/p(\Pi_2|\mathbf{x})\}$ in Table 2.

The information in Table 1 can be used to illustrate the use of blood pressure in a *diagnostic screening test*. One sensible diagnostic screening test is derived by specifying a cutoff value for blood pressure above which we decide to treat a patient as diseased (*positive* diagnosis) and below which we treat as unaffected (*negative*). Table 3 shows the proportions of positives classified correctly as positives and falsely as negatives, and the corresponding proportions of negatives, for two such screening procedures. In Table 3(a) a +, that is, Π_1, is predicted if $x = 2$ or 3, and in Table 3B only if $x = 3$.

The proportion of actual positives correctly predicted as such is called the *sensi-*

Table 2: $\{p(\Pi_i|\mathbf{x})\}$ and odds ratios for the model of Table 1.

	$x=1$	$x=2$	$x=3$
Π_1	6/55	6/13	12/19
Π_2	49/55	7/13	7/19
Odds ratios $(\Pi_1 : \Pi_2)$	6/49	6/7	12/7

Table 3: Proportions of true/false positives and negatives for two screening procedures.[a]

		(a) Predicted		
		$-(x=1)$	$+(x=2,3)$	Total
True	$+(\Pi_1)$	0.2	0.8	1
	$-(\Pi_2)$	0.7	0.3	1
		(b) Predicted		
		$-(x=1,2)$	$+(x=3)$	Total
True	$+$	0.6	0.4	1
	$-$	0.9	0.1	1

[a] Based on the model in Table 1.

tivity of the test; the proportion of actual negatives correctly predicted is the *specificity*. Note that the sensitivity is higher in Table 3(a) than in Table 3(b) (0.8 vs. 0.4) but that the corresponding specificities show, in compensation, an opposite trend (0.7 vs. 0.9). Which of the tables represents the more acceptable compromise depends critically on the relative costs of the two types of error.

In the general structure all the special methods and considerations of discriminant analysis are relevant. This is reflected in the medical literature, at least some of which is right up to date with developments in discriminant analysis and pattern recognition. It is largely the result of direct involvement in medical diagnosis of appropriate experts in the methodology [8].

Two points have particular importance to the treatment of the medical problems.

The Use of Independence Models. Many early applications of Bayes' theorem incorporated independence models for the class-conditional densities $\{f_i(\mathbf{x})\}$; that is, it was assumed that

$$f_i(\mathbf{x}) \equiv \prod_{j=1}^{k} f_{ij}(x_j) \equiv f_i^I(\mathbf{x}),$$

say, where the $\{f_{ij}(\cdot)\}$ are all univariate densities. This occurred mainly because indicants often are categorical and, in particular, binary (symptom present or absent). The independence model is estimated easily and the only elementary alternative model, the multinomial, which for large k has many cells, can seldom be well estimated. (A typical dataset might initially contain about 300 patients in 6 disease classes with up to 50 binary indicants.) Such was the popularity of the independence model that some medical writers still seem to be under the misapprehension that the "Bayes" method has as one of its precepts that the indicants be independent, given the disease class [5]! Some empirical studies have shown that the independence model can perform quite well in diagnostic terms even when it is wrong. Indeed, odds ratios can be calculated correctly using the independence model [4] even when the more general assumption is made that

$$f_i(\mathbf{x}) = c(\mathbf{x}) f_i^I(\mathbf{x}), \quad \text{for each } \mathbf{x} \text{ and } i.$$

More recently, models intermediate between the independence and multinomial models have also been used [13]. General multivariate normal assumptions have also been made, leading to linear and quadratic discriminant functions.

Variable Selection and Created Indicants. In medical data there are often very many indicants, as exemplified earlier. In contrast, seldom are more than a very few variables required to extract the best out of a dataset from the perspective of discriminant analysis.

In the head injury study reported by Titterington et al. [13], for instance, over 200 variables could be reduced to six and yet still provide a good discrimination procedure. Variable selection is therefore a vital part of the statistical exercise, which may then be a stepwise discriminant analysis, for instance. Reduction of dimensionality also allows us to explore beyond the independence models toward more highly parameterized and realistic structures. Variable selection has an appealing parallel in clinical practice, where the experienced diagnostician often is distinguished by asking a few pertinent questions instead of accumulating a mass of possibly irrelevant information [3]. Another means of reducing dimensionality is to create new indicants, each as a function of several original variables. This might be to cope with missing data (provided at least one of the original variables is available, the created indicant may be evaluated) or to produce a set of indicants for which the conditional independence assumption is more reasonable.

81.3.2 Diagnostic Keys and Sequential Testing

In an ideal situation there would be enough information in any patient's $\mathbf{x}$ to identify the disease class uniquely, whatever $\mathbf{x}$ is. In particular, with data about k binary symptoms, the sample space of 2^k possible $\mathbf{x}$'s could be partitioned into d subsets corresponding to the various disease classes. If k is large, it is advantageous to streamline the testing procedure into a sequential pattern in which the symptoms are investigated one by one, the order depending on the results of previous tests. Such a rule is called a *diagnostic key* [9] and it can be represented diagrammatically by a binary tree. An optimal key would minimize, say, the expected number of tests required to complete a diagnosis. Construction of such a key is, combinatorially, a difficult problem, for large k and d.

In medical problems perfect discrimination is seldom possible and a further requirement of a good key is that, when testing is complete, the discrimination achieved is as good as possible [12]. If $\{\theta_i\}$ denote probabilities of the disease classes for a given patient after testing, a good key should lead to the $\{\theta_i\}$ being close to the correct degenerate probability measure as frequently as possible. This closeness can be measured by expected entropy. Furthermore, expected change in entropy can be used as a criterion for sequential selection of tests and can be played off against the cost of a test to decide when to end the testing procedure.

81.4 Application and Acceptance in Practice

Performance of statistical techniques often is assessed by simulation, so that properties can be discovered or checked under known conditions. In medical diagnosis, because of the variety of datasets that arise, empirical comparison of various methods on the same real datasets is also of great importance. Some comparisons (e.g., [1,13]) have shown that the choice of technique is not as important as, for instance, variable selection and data preparation. Refinement of technique is far less of a priority to Croft [1] than the following:

1. Improved standardization of tests and symptom definition

2. Generation of large datasets from which rules can be derived that can also be reliably transferred for use in other establishments

3. Increased acceptance by the medical profession

Physicians have been wary of the anonymous power of computers, which conflicts with the fact that clinical diagnosis follows a variety of patterns, depending on the individual physician, who blends hard data with subjective experience, opinions, and even emotion; see Mitchell [6]. The decision trees associated with diagnostic keys and sequential testing are far more acceptable attempts to model this logical emotional process than the probabilistic perfection of Bayes' theorem and discriminant analysis. Gradually, however, the latter methods are finding more favor. For instance, the use of Bayes' theorem can produce more accurate diagnoses in practice than those of clinicians; see Stern et al. [10]. Taylor et al. [11] show that clinicians often do not choose the most efficient tests and do not then process the results appropriately, according to Bayes' theorem, in particular being very conservative about the resulting probabilities or odds ratios. They are also liable to ignore the prior probabilities of disease classes in their mental assessments. If we add to this the difficulty of interpreting multivariate data intuitively, particularly the correlations therein, the advantages of computer-assisted diagnosis seem great.

A final problem is that a procedure, once developed, seldom lives up to expectations. Error rates claimed in the development stage often are not met in later application, although this may be because the program has been tested on the training set, which gives a falsely optimistic indication of performance. Variable selection may also lead to bias in that the "best subset" as judged by one dataset may well not be the best for another. This happens even in artificially generated situations in which the datasets have identical statistical properties [7].

References

1. Croft, D. J. (1972). *Comput. Biomed. Res.*, **5**, 351–367.

2. Dawid, A. P. (1976). *Biometrics*, **32**, 647–648.

3. De Dombal, F. T. (1978). *Methods Inform. Med.*, **17**, 28–35.

4. Hilden, J. (1982). *Personal communication.*

5. Kember, N. F. (1982). *An Introduction to Computer Applications in Medicine*, Chap. 10. Arnold, London.

6. Mitchell, J. H. (1970). *Int. J. Biomed. Comput.*, **1**, 157–166.

7. Murray, G. D. (1977). *Appl. Statist.*, **26**, 246–250.

8. Patrick, E. A., Stelmack, F. P., and Shen, L. Y. L. (1974). *IEEE Trans. Syst. Man. Cybern.*, **SMC-4**, 1–16. (Includes well-annotated references to applications.)

9. Payne, R. W. and Preece, D. A. (1980). *J. Roy Statist. Soc. A*, **143**, 253–292. (Good reference list.)

10. Stern, R. B., Knill-Jones, R. P., and Williams, R. (1975). *Br. Med. J.*, **2**, 659–662.

11. Taylor, T. R., Aitchison, J., and McGirr, E. M. (1971). *Br. Med. J.*, **3**, 35–40.

12. Teather, D. (1974). *J. Roy Statist. Soc. A.*, **137**, 231–244.

13. Titterington, D. M., Murray, G. D., Murray, L. S., Spiegelhalter, D. J., Skene, A. M., Habbema, J. D. F., and Gelpke, G. J. (1981). *J. Roy Statist. Soc. A*, **144**, 145–175. (Includes discussion.)

14. Wechsler, H. (1976). *Int. J. Biomed. Comput.*, **7**, 191–203.

82 Statistics in Medicine

Vern T. Farewell and Stephen J. Senn

82.1 Introduction

Until the mid-1990s, the history of statistics in medicine (or medical statistics) was the history of statistics itself. It would be both difficult and superfluous to relate this general history, and we limit ourselves to a very few selective remarks on matters that affect statistics in medicine. The reader who is interested in finding out more should consult Hald [14] and Stigler [28] for the history of statistics in general and Gehan and Lemak [13] and Armitage [1] for medical statistics. Lancaster's account [19] is also highly relevant, and he makes a vigorous, if controversial, case that all the most important quantitative progress in biology (including medicine) was made by life scientists and not by mathematicians.

However, some key figures in the development of statistics itself were qualified both as mathematicians and as physicians. Arbuthnot (1667–1753) was a teacher of mathematics before becoming a fashionable physician, and Daniel Bernoulli (1700–1782) was a physician before becoming a mathematician. It is curious that these two scientists provide the two most commonly cited early applications of the significance test (Arbuthnot in 1712 to the sex ratio of infants, and Daniel Bernoulli in 1734 to the coplanarity of planetary orbits.) Perhaps the judgment that this is an unnatural device that has been foisted on the medical profession by frequentist statisticians is a little unfair. The reverse may be the case.

Of course, the medical statistician is proud of statistical techniques and the associated theory. It is this that is often meant nowadays by "statistics" rather than "data." Nevertheless, data are crucial to the subject; perhaps one of the most important of all advances was the realization that collecting records of vital events and (later) clinical outcomes would make an important contribution to medical understanding. Pioneers in this respect were William Petty (1623–1687), who proposed the setting up of a central statistical office for England some 150 years before the general registry was established, and his friend John Graunt (1620–1674), who carried out a remarkable early investigation of the London bills of mortality. It was William Farr (1807–1883), however, who, at the General Registry Office of England and Wales, presided over the implementation of the first thorough system for recording vital statistics. This system, and the data it yielded, were essential elements in the eventual triumph of the sanitary reforms of the late Victorian era. Farr was a physician by training and no outstanding mathematician. However, if greatness is measured in terms of effects on health, he has a claim to be one of the greatest of all medical statisticians. Also important is Florence Nightingale. Statistics were a fundamental weapon in her campaign to persuade the British Army to carry out necessary health reforms, and she was an enthusiastic devotee of the growing science.

An important early application for statistics was the construction of life tables and the study of population dynamics. The Dutch politician Jan De Witt (1625–1672), the English scientist Edmund Halley (1656–1742), the French–English mathematician Abraham de Moivre (1667–1754), the Dutch mathematician Nicolaas Struyck (1687–1769), the Swiss mathematician Leonhard Euler (1707–1783), the German theologian Johann Süssmilch (1707–1767), and the English mathematician John Simpson (1710–1761) all made important contributions to this field. Life tables, of course, are a form of survival analysis, an extremely important topic of medical statistics today. However, these early investigations were developed as much for their economic relevance (for determining annuities and so forth) as for any medical purpose.

A major field of application for medical statistics is in the design and analysis of clinical trials. These have acquired considerable prestige as the perfect paradigm of the application of the experimental method in medicine. Gehan and Lemak [13] trace the idea back to the Islamic–Persian scholar Avicenna (980–1037). Early examples include the famous controlled trial carried out in 1747 by James Lind (1716–1794) of the effects of citrus fruit on scurvy, and the investigations (in 1825) of the effectiveness of bloodletting on phthisis by Pierre Louis (1787–1872). Louis's students spread his ideas on medical evidence to Britain and the USA. The appreciation of the value of controls made ground slowly.

A famous example is the within-patient trial conducted by Cushny and Peebles [12] and used for illustration by Student [30]. Possibly the first modern randomized clinical trial (RCT), however, is the Medical Research Council's investigation of the effect of streptomycin in tuberculosis, published in 1948. The key figure here was the medical statistician Bradford Hill (1897–1991).

82.2 Key Features

Four general points can be made about statistics in medicine. These were identified by Peter Armitage in a talk that he gave in 1979 in Belgium, entitled "Statistics in medical research."

The statistician who applies statistical thinking to medical research generally finds that this must be a collaborative activity. This collaboration reflects the multidisciplinary aspects of medical research. And the need for collaboration across many disciplines is becoming increasingly important in the evaluation of medical data. Many medical statisticians find this one of the most interesting features of their work.

A second distinctive feature is the predominant fact that most medical research is carried out on people. Therefore the design of studies and the interpretation and use of analysis results must be done within the confines of ethical constraints. As part of a collaborative team, statisticians cannot divorce their role from this necessity. This was stressed by Hill [15], who wrote that the statistician involved in a clinical trial "cannot sit in an armchair, remote and Olympian, comfortably divesting himself of all ethical responsibility."

Third, statistics in medicine has a very wide scope. Laboratory studies, animal experiments, clinical research, epidemiological research, and health service research all fall within the purview of medical statistics. Each has distinctive features, but there is considerable overlap in the statistical methodology useful to them.

Finally, a key and recurring theme in the methodology used in medical research is that of comparisons between groups with adjustment for group differences. Historically, this is seen in early efforts at standardization of vital statistics data. Now, the extensive use of regression models reflects the continuing importance of the topic.

82.3 More Recent History

While the history of medical statistics, as briefly discussed earlier, goes back a considerable period, the 1970s saw a burst of activity that has given the area a prominence in the general statistical community in succeeding years. For example, Cox [10] has indicated that one might consider three periods in the development of modern statistical methodology for experiments. The first would derive from applications in agriculture, the second from industrial applications, and the last from medical research.

In the statistical literature, much of the methodological development in medical statistics during this period related to survival data and to categorical data. After a number of years of focus on parametric models for survival data, often paralleling work done in industrial life testing, Cox's 1972 paper [9], "Regression models in life testing," presented an elegant semiparametric alternative that allowed the use of regression techniques for explanatory variables without the need to completely specify an underlying survival distribution. This was a very liberating development for medical research. Building on developments in logistic regression, various regression models were also developed for categorical data, a notable contribution being the consideration of models for ordered categorical data by McCullagh [22]. Generally applicable tools were thus made available for the fundamental problem of group comparisons with adjustment for other explanatory variables.

The growth of medical statistics was nurtured by a variety of individuals in government agencies and academia in the UK and the USA. An important factor in the USA was the extensive federal funding directed toward public health. Research groups were spawned that combined new methodological development with substantive applied work. Now medical statistics, or often simply biostatistics in North America, is recognized as a discipline in its own right with an increasing number of departments, independent of any formal link with more general departments of statistics. As well, many statisticians work primarily in medical research establishments, some in statistical research groups, others as part of research groups in particular medical specialties.

82.4 Observational Plans

Data to be dealt with by the medical statistician can derive from a variety of studies. Here we highlight three observational plans and, for illustration of the types of questions with which a medical statistician might be concerned, comment briefly on some selected statistical issues relevant to each.

82.5 Clinical Trials

As indicated earlier, the design and analysis of clinical trials, in particular the RCT, is a major topic in medical statistics. The RCT has not been without its critics both on ethical and (more rarely) on scientific grounds. Its importance cannot be ignored, however, and it seems likely that whatever the future holds in store for clinical statistical investigation, the RCT will continue to provide a standard by which other devices will be judged. While the basic principles were ably laid down by Hill, there are, nevertheless, ongoing developments and issues.

Randomization is seen by many to be the key feature of clinical trials. In the evaluation of treatments, randomization provides the most direct route to statements of causality. The criticism has been raised, however, that randomization sacrifices ethical behavior towards the individual patient, and that alternative methods could be used to acquire the needed treatment information. This criticism has been countered, but discussion continues. In drug regulation, randomization remains a dominant requirement for trials. However, here, especially in so-called phase I studies (to establish tolerability and basic pharmacokinetics) and phase II studies (preliminary studies to find potentially efficacious doses), there is increasing interest in alternatives to the RCT: for example the so-called continuous reassessment method for phase I dose finding in cytotoxic compounds [23] and dose-escalation studies combined with pharmacokinetic–pharmacodynamic modeling for dose finding for chronic diseases. However, for phase III studies (attempts to establish efficacy definitively and assess safety in target populations) the RCT has remained the design of choice.

The ongoing monitoring of data as they are acquired in a clinical trial is often ethically necessary. If endpoint information is evaluated repeatedly, then techniques for this sequential analysis need to be considered. Much early statistical work dealt with the problem of continuous monitoring of a trial, usually assuming that outcome measures were available for individual patients soon after trial entry. The influence of this work on medical research appears to have been minimal. More recently, designs that allow for the periodic appraisal of study results have been developed under the label of *group sequential designs*. These have been much more readily accepted by the medical profession (although their analysis has been controversial within the statistical profession, highlighting, as it does, a major difference between frequentist statisticians and Bayesians) and are currently very widely used, particularly in publicly funded clinical trials. This, among other factors, has helped to give statisticians an increasingly visible role in data and safety monitoring committees which oversee clinical trials. On such committees, statisticians bring their expertise to be combined with that of physicians, ethicists, community representatives, and others to ensure the ethical conduct of this type of medical research.

After randomization to treatment in a clinical trial, a patient may or may not continue in the treatment as described in the trial protocol. In the analysis of the trial, the comparison of treatment groups defined by their randomized treatment, irrespective of what treatment was actually received, is referred to as an *intent-to-treat* analysis. The motivation for this is that what most trials should be comparing is treatment strategies, and it is to be expected that in some cases an initially prescribed treatment will have to be altered. It is the comparison of the intent to treat in one manner with the intent to treat in another that is of clinical relevance in a comparative trial. It is recognized that there are relevant secondary questions that might be examined through different comparisons, but that it is prudent in most cases to view the intent-to-treat analysis as primary.

There are many other areas of specific interest in clinical trials. Special techniques need to be employed in bioequivalence trials, where the aim is to establish that treatments are equally bioavailable, and in active control equivalence studies, where the object is to show that two treatments are therapeutically equivalent or at least that a new treatment is not inferior to a standard, rather than to show the superiority of experimental treatment. When a considerable period must elapse after treatment before a definitive clinical endpoint can be observed, there is interest in studying the value of intermediate endpoints, which may serve as surrogates for the actual endpoint of interest and allow earlier determination of trial results. There has also been increasing interest in developing methods for dealing with dropouts in clinical trials and for adjusting for compliance as well as for using internal pilot studies for sample size determination. Specialized designs such as crossover trials (changeover designs) also continue to attract much research interest.

82.5.1 Case–Control Studies

In epidemiological investigations of the relationship between the occurrence of a specific disease D and exposure to a possible risk factor E, the so-called case–control study has been very useful. In this type of study, separate samples are taken of diseased and nondiseased individuals, with great care taken to ensure that they are otherwise comparable groups. Observations are then made on the exposure E in both groups, as well as recording additional information on other factors that may be related to disease or the disease-exposure relationship. During the 1970s and early 1980s, logistic regression emerged as the preeminent tool in the analysis of such studies. However, logistic regression models focus on $\Pr(D|E)$, but in a case–control study, sampling is from the density $f(E|D)$. For a binary exposure, Cornfield [8] is credited with demonstrating the equivalence of the odds ratio relating D and E for a binary exposure variable under prospective (i.e., $D|E$) and retrospective (i.e., $E|D$) sampling. A comparable equivalence for logistic regression models was suspected [7]. In a series of papers, culminating with that by Prentice and Pyke [25], it was ultimately demonstrated that a prospective logistic regression model for $\Pr(D|E, z)$, where z is a vector of explanatory variables, can be applied in the usual manner to data from a case–control study, and that valid estimates of odds-ratio parameters can be obtained.

The use of other statistical procedures for such data have also been investigated. More recently, there has been interest in developing modified sampling plans, such as case–cohort studies [24] and two-stage case–control studies [4], to overcome some of the limitations of the traditional case–control study. A comprehensive discussion of the history of statistical work related to case–control studies is given by Breslow [3].

82.5.2 Longitudinal Studies

Other data in medical research derive from long-term observation of a group of individuals. In epidemiological research, this is seen primarily in the many cohort studies where individuals are identified and followed through time to observe disease incidence. Alternatively, equivalent information is collected in a historical fashion for a group of individuals who can be defined at some point in the past. In clinical research, there is use of longitudinal databases, in which information on patients with specified conditions is collected routinely over time. Information on prognostic factors, treatment, disease progression, and so forth is then analyzed to learn something about the course of the disease of interest.

Many of the general techniques for survival data and categorical data can be applied to such data. For example, most of the traditional epidemiological measures used in long-term cohort studies, such as standardized mortality ratios, can be seen to be special cases of more general methodology such as Poisson regression or Cox's [9] regression model, sometimes referred to as *relative-risk regression*. Breslow and Day [5] provide a comprehensive discussion.

When repeat observations are made on an endpoint of interest, more specialized techniques are needed. Two important developments have been the adaptation of random-effects models for such data, as in the work of Laird and Ware [18], and the adaptation of generalized linear models, for example through the work of Liang and Zeger [20], the latter development being closely linked with the general and increasing interest, in the statistical community, in estimating equations. Another topic recently investigated is that of multistate models, which, for example, have been used extensively in characterizing the course of HIV disease [21].

82.6 Stochastic Processes

The application of statistics in medicine frequently, as in the case of multistate models, needs to make use of the general theory of stochastic processes. There is also a long history of disease modeling that makes use of this theory. While much of medical statistics still focuses on quantities such as relative risks, with the associated calculation of significance levels, estimates, and confidence intervals, there is increasing communication between those whose primary research interest has been disease modeling and the more "classical" medical statisticians.

A particular impetus to this communication has been work on the AIDS epidemic where both population forecasting and the understanding of disease progression can benefit from the range of quantitative methodologies available. A more recent reference is Isham and Medley [16].

82.7 Meta-Analysis

Many clinical studies have been too small to permit definitive statements about the treatments investigated. For many diseases it is also the case that numerous such small studies have been carried out. There has been an explosion of interest more recently in formal overviews, so-called meta analyses, of such studies. Methods for pooling trials have been developed, with an accompanying debate as to whether fixed- or random-effects models for treatment are more appropriate. More important, however, has been that an increasing number of such meta-analyses are being carried out, often with very interesting results. The rapid progress in computing together with the establishment of relevant databases has made this possible, and organizations, such as the Cochrane Collaboration, have been set up to promote good meta-analyses. The increasing importance being attached to meta-analysis is a very significant recent development in medical statistics.

82.8 Bayesian Medical Statistics

Since the mid-1980s there has been an explosion of activity related to the use of Bayesian statistical methodology in medical research. Much of this has been possible through the enormous increase in computing power and the development of statistical algorithms such as Gibbs sampling to exploit it. This has moved Bayesian statistics on from being primarily a means to think about data to being a technique for their analysis. Perhaps the most ideologically Bayesian work is in the area of clinical trials, where research, for example, is ongoing in the areas of dose finding in phase I [23] and phase II [31] trials, sample size determination in phase II trials, and the interpretation of phase III trials [26]. General consideration of the potential use of Bayesian methods in medical research is given in Breslow [2]. Many workers regard multiplicity problems, the evaluation of many treatments, subgroup analyses, etc., to be particularly amenable to Bayesian methods. Some drug development programs have also begun to use Bayesian methods with explicit introduction of subjective probabilities.

So-called hierarchical Bayes modeling is being applied to a wide range of problems in medicine. The role of subjective probabilities is much less central in this work than in clinical trials. The formulation of these models is convenient, however, for the introduction of random effects and smoothing techniques, and this has been seen as very useful. Areas of application for these models include population pharmacokinetics (especially where measurement is sparse) [30], geographic epidemiology, genetic epidemiology, longitudinal studies, and performance indicators in health services research.

82.9 Current Research Interests

As in the past, advances and new trends in medical research will spawn new research questions, the answers to which will often require careful statistical thought. Topics that have been of interest include missing data (particularly the handling of informative dropout from medical studies), correlated response data, and quality-of-life data. A major area of future work will relate to the analysis of DNA sequencing data.

82.10 The Practice of Statistics in Medicine

Because medical statistics is more than just a collection of topics, some aspects of the current professional life of the subject (as of 1996) will also be covered.

Whereas, until relatively recently, there was no specific outlet for methodological research in medical statistics, the subject now has a number of journals largely or entirely devoted to it. This reflects the increasing specialization of the subject. The journal space given to medical statistics has seen a phenomenal growth. For example, *Statistics in Medicine*, although started only in 1981, now has 24 issues and nearly 3000 pages a year, and a further journal devoted entirely to this topic, *Statistical Methods in Medical Research*, was started in 1992.

We are also beginning to see further specialization of the subject, so that medical statistics as applied to drug development is now sometimes referred to as "biopharmaceutical statistics" and has a journal (*Journal of Biopharmaceutical Statistics*, started in 1991). Other specialization has been by topic. An enormous amount of research has taken place in the field of survival analysis since Cox's seminal paper [9], and this too now has a journal, *Lifetime Data Analysis*, which, while not exclusively oriented toward medicine in terms of applications, will no doubt provide another important outlet for papers on this topic.

In understanding current trends in medical statistics, another interesting journal to consider is *Controlled Clinical Trials*. Such trials are far from being the exclusive preserve of the medical statistician (as physicians would no doubt be very ready to point out); nevertheless, the majority of the papers in *Controlled Clinical Trials* have a medical statistics flavor to them, an indication of the importance that the subject has acquired in this field. The *Drug Information Journal* also carries many articles with statistical content. *Biometrics*, for many years a major outlet for medical statistics, continues to include papers on medical statistics, and *Biometrika* is a well-respected outlet for methodological developments. General statistics journals also publish many relevant papers.

Almost anything one writes about the Internet will become almost instantly out of date, but a number of bulletin boards and lists currently provide forums for discussion in statistics and related topics such as evidence-based medicine. Krause's paper [17], itself available on the Internet, gives a useful review.

The explosion of journal activity has been matched by other professional activity. Medical statistics has now emerged as a discipline for a master's degree that may be studied at a number of universities throughout the world. (It will be interesting to see whether the general development of statistics itself, which moved from a postgraduate discipline only, prior to the late 1960s, to being an undergraduate discipline in many universities by the late 1970s, is also followed.) In terms of geographic spread, while (as in statistics generally) English-speaking countries remain predominant, the discipline has now become extremely important in a number of others. The contribution of statisticians from Scandinavia and the Low Countries, for example, has been particularly notable.

A major source of employment of medical statisticians is in drug development, not only with pharmaceutical sponsors but also with regulators and, increasingly, with third parties such as contract research organizations. It will be interesting to see whether the rise in the importance of medical purchasing authorities provides another important employment outlet. The essential role of medical statistics in drug development has now been formally recognized in the European Union, for example, by the official requirement that medical statisticians shall be employed in all stages of designing, supervising, and analyzing clinical trials and by the issuing of statistical guidelines [11]. The Food and Drug Administration of the USA has long been an important influence on the practice of medical statistics in the field. Publicly funded research organizations such as, for example, the National Institutes of Health in the USA and the Medical Research Council in the UK have also been important employers of medical statisticians, as, of course, have universities, national health services, and the World Health Organization. It is interesting to contrast the research interests of statisticians working in the public sector and those working in or around the pharmaceutical industry. The former are more likely to work on survival analysis, sequential analysis, meta-analysis, and binary outcomes. The latter are more likely to work on bioequivalence [27], crossover trials or change over designs, continuous outcomes, and the pharmacokinetic–pharmacodynamic modeling.

A number of societies serve medical statistics exclusively. The International Society for Clinical Biostatistics was founded in 1978. Statisticians in the Pharmaceutical Industry (PSI), a largely UK-based society, dates from the same time, and there is a European federation of similar national societies. Older established societies such as the American Statistical Association and the Royal Statistical Society have specialist sections devoted to medical statistics.

If "statistics in medicine" is the collection and analysis of medical data, however, it would be wrong to regard this as the exclusive preserve of the medical statistician. There are a number of reasons. First, in methodological research, many advances have been made in the subject by others, and there continue to be other disciplines whose interests often coincide with that of medical statistics. Epidemiology is a major area of application for statistics, where much of the necessary statistical methodological development has been worked on by epidemiologists as well as statisticians. An interesting consequence is that statisticians and epidemiologists sometimes use different terms for the same concept and sometimes similar but not identical approaches to the same topic; for example, the epidemiologist is more likely to speak of relative risk and the statistician, to be content with log-odds ratios. Genetics is another case in point. R. A. Fisher himself was a master in both genetics and statistics, but despite the fact that genetics has an important quantitative side to it, the average medical statistician is unlikely to be very familiar with the material. Survival analysis was for years the preserve of the actuary and demographer; these are two professions that continue to contribute to medical statistics in its broadest sense. Another interesting example is given by pharmacokinetic and pharmacodynamic modeling. Despite a certain amount of interest in the 1950s and 1960s by various statisticians and the fact that this topic is a natural outlet for complex mathematical modeling, until relatively recently the key work had been done by pharmacologists with statisticians arriving rather later on the scene.

Second, when it comes to actual analysis, statistics is such a fundamental, inevitable, and hence common part of any scientific approach to medicine that the sheer volume of work means that much analysis will be carried out by physicians themselves. A consequence is that there is often a considerable disparity between theoretical advances and practical applications; the former tend to percolate through slowly and imperfectly to the latter. This has been the subject of much discussion. It would be wrong to blame physicians alone for this state of affairs. The medical profession itself is taking important steps to remedy it, and a number of leading medical journals, such as, for example, the *British Medical Journal* and *The Lancet*, have instituted statistical review. On the research side a distinction needs to be drawn between investigations of a purely theoretical nature, pursued because of their intellectual interest, which may or may not have eventual practical application but that contribute to general understanding, and development of techniques of immediate utility. Medical statisticians need the necessary practical insight to distinguish between the two in order to successfully communicate with physicians. Even a journal such as *Statistics in Medicine*, which has as a goal to "... enhance communication between statisticians, clinicians and medical researchers ...," is likely to be daunting reading for the average physician, and the continuing importance of review articles in medical journals cannot be overstressed.

Third, *statistics*, singular, is the subject, but *statistics*, plural, are data, and these are collected by all sorts of healthcare professionals. Gaunt, Petty, Louis, Farr, and Nightingale are all important medical statisticians in this sense, but they could hardly have dreamed of the sorts of tools we now have available. The potential influence of well-structured databases on medical investigations is enormous. For example, the people of Saskatchewan are among the most studied on earth, not because they are regarded as typical, but because of the comprehensive approach to logging health care data employed by that province [29].

In short, the first of Armitage's key features, that medical statistics is collaborative in nature, has applied and seems likely to continue to apply, not only to the practice of medical statistics but also to its general development.

References

1. Armitage, P. (1983). Trials and errors: The emergence of clinical statistics.

J. Roy. Statist. Soc. A, **146**, 321–334.

2. Breslow, N. (1990). Biostatistics and Bayes. *Statist. Sci.*, **5**, 269–284.

3. Breslow, N. E. (1996). Statistics in epidemiology: The case control study, *J. Am. Statist. Assoc.*, **91**, 14–28.

4. Breslow, N. E. and Cain, K. C. (1988). Logistic regression for two-stage case-control data. *Biometrika*, **75**, 11–20.

5. Breslow, N. E. and Day, N. E. (1980). The analysis of case-control studies. In *Statistical Methods in Cancer Research*, Vol. I. IARC, Lyon.

6. Breslow, N. E. and Day, N. E. (1987). The design and analysis of cohort studies. In *Statistical Methods in Cancer Research*, Vol. II. IARC, Lyon.

7. Breslow, N. E. and Powers, W. (1978). Are there two logistic regressions for retrospective studies? *Biometrics*, **34**, 100–105.

8. Cornfield, J. (1951). A method of estimating comparative rates from clinical data. Applications to cancer of the lung, breast and cervix. *J. Natl. Cancer Inst.*, **11**, 1269–1275.

9. Cox, D. R. (1972). Regression models and life tables (with discussion). *J. Roy Statist. Soc. B*, **34**, 187–220.

10. Cox, D. R. (1984). Present position and potential developments: Some personal views; design of experiments and regression. *J. Roy Statist. Soc. A*, **147**, 306–315.

11. CPMP Working Party on Efficacy of Medicinal Products (1995). Biostatistical methodology in clinical trials in applications for marketing authorizations for medicinal purposes. *Statist. Med.*, **14**, 1659–1682.

12. Cushny, A. R. and Peebles, A. R. (1905). The action of optical isomers. II. Hyocines. *J. Physiol.*, **32**, 501–510.

13. Gehan, E. A. and Lemak, N. A. (1994). *Statistics in Medical Research*. Plenum, New York.

14. Hald, A. (1990). *A History of Probability and Statistics and Their Applications before 1750*. Wiley, New York.

15. Hill, A. B. (1963). Medical ethics and controlled trials. *Br. Med. J.*, **3**, 1043–1049.

16. Isham, V. and Medley, G., eds. (1995). *Models for Infectious Diseases*. Cambridge University Press, Cambridge, UK.

17. Krause, A. (1995). *Electronic Services in Statistics* (http://www.med.uni-muenchen.de/gmds/ag/sta/serv/ess/internet.html).

18. Laird, N. M. and Ware, J. H. (1982). Random-effects models for longitudinal data. *Biometrics*, **38**, 963–974.

19. Lancaster, H. O. (1994). *Quantitative Methods in Biological and Medical Sciences: A Historical Essay*. Springer, New York.

20. Liang, K.-Y. and Zeger, S. (1986). Longitudinal data analysis using generalized linear models. *Biometrika*, **73**, 13–22.

21. Longini, I. M., Clark, W. S., Byers, R. H., Ward, J. W., Darrow, W. W., Lemp, G. F., and Hethcote, H. W. (1989). Statistical analysis of the stages of HIV infection using a Markov model. *Statist. Med.*, **8**, 831–843.

22. McCullagh, P. (1980). Regression models for ordinal data (with discussion). *J. Roy Statist. Soc. B*, **42**, 109–142.

23. O'Quigley, J., Pepe, M., and Fisher, L. (1990). Continual reassessment: a practical design for phase I clinical trials. *Biometrics*, **46**, 33–48.

24. Prentice, R. L. (1986). A case-cohort design for epidemiologic cohort studies and disease prevention trials. *Biometrika*, **73**, 1–12.

25. Prentice, R. L. and Pyke, R. (1979). Logistic disease incidence models and case-control studies, *Biometrika*, **66**, 403–411.

26. Spiegelhalter, D. J., Freedman, L. S., and Parmar, M. K. B. (1994). Bayesian approaches to randomised trials. *J. Roy Statist. Soc. A*, **157**, 357–416.

27. Steinijans, V. W., Hauschke, D., and Schall, R. (1995). International harmonization of regulatory requirements for average bioequivalence and current issues

in individual bioequivalence. *Drug Inform. J.* **29**, 1055–1062.

28. Stigler, S. (1986). *The History of Statistics*, Belknap, Harvard.

29. Strand, L. M. and Downey, W. (1994). Health databases in Saskatchewan. In *Pharmacoepidemiology*, B. L. Strom, ed. Wiley, Chichester, UK.

30. Student (1908). The probable error of a mean. *Biometrika*, **6**, 1–25.

31. Wakefield, J. and Racine-Poon, A. (1995). An application of Bayesian population pharmacokinetic/pharmacodynamic models to dose recommendation. *Statist. Med.*, **14**, 971–986.

83 Statistics in the Pharmaceutical Industry

P. A. Young

83.1 Introduction

Pharmaceuticals are the medicines used in humans and animals for the treatment or prevention of disease. Many are "ethical" products in that throughout their manufacture and marketing, standards are maintained consistent with the ethics of clinical and veterinary practice. In spite of the great variety of substances involved, with regard to both their origin and the formulation in which they are presented, they possess two types of property in common: (1) the dependence of their effect on the dose at which they are administered and (2) the liability to induce harmful toxic effects at higher dose levels. Thus the dose–response relationship is of fundamental importance to the industry.

83.2 The Clinical Response to Treatment

When a treatment is prescribed it is with the intention of achieving one or more specific clinical objectives. If these be couched in purely qualitative terms (allowing time for the treatment to act), the outcome might be recorded as success or failure with respect to a particular objective. Replicated observations of this all-or-none type may provide a so-called quantal measure of response by noting the relative frequency of successes.

Commonly the clinical objectives will be capable of expression in quantitative terms and the degree of success achieved by the treatment may then be assessed by appropriate measurement. This may involve some physical or chemical determination, or it may be a subjective rating by the clinician or even self-rating by the patient.

Clinical experiments have shown that even a dummy treatment may at times induce effects simulating drug action—the "placebo" response.

If the medicine be given at too low a dose, either no response will be seen or it will be insufficient to meet the required criterion. Too high a dose may induce too profound an effect with its attendant dangers, or it may lead to other undesirable responses which reflect the toxicity of the substance. The various types of response that are associated with the complete dose range make up an "activity profile" for the preparation, from a knowledge of which it is possible to judge the margin of safety attaching to its use.

83.2.1 The Aims of Drug Research

Just as one drug may elicit several types of response, depending on its dosage, so is it that any required response may be produced by a number of quite distinct chemical compounds. Research effort is therefore deployed in attempts to discover substances that will produce the desired effect but with a greater freedom from side effects and a greater margin of safety. Another area for research is concerned with the duration of action of substances—in some cases to produce longer or, in others, shorter action. There will obviously also be a call to devise remedies for conditions where no satisfactory treatment at present exists.

Although the principles of drug design have been the subject of many recent publications, the medicinal chemist cannot predict the activity of a novel compound with anything approaching the level of assurance required for it to be tested in humans. The only option that permits that step to be taken is to use experimental animals for preclinical testing.

Much preliminary work can be carried out in vitro on isolated tissues, their extracts, or on cultures of cells or parasites. Among the advantages of this approach is the humane aspect of reducing the numbers of conscious animals used.

Among mammalian species there is a broad similarity in the pattern of responses at different dose levels that provides the basis for preliminary studies in humans. Unpredictable species differences occur, however, which call for exteme caution. Dosage is commonly adjusted on the basis of body weight, but in certain cases the two-thirds power of body weight appears preferable.

General theoretical arguments, based on the bioavailability of active constituent(s) of a drug, lead to the following formula for response (R)

$$R = \frac{EQ \text{ dose}}{1 + Q \text{ dose}} \qquad (1)$$

in which E is the maximum response or "efficacy" and Q is a potency factor. Since the equation does not indicate the direction of E relative to control, this must be specified by terms such as "stimulant" or "depressant." A similar, but more physiological approach yields as the equation for the observed variable

$$V = \frac{C + LQ \text{ dose}}{1 + Q \text{ dose}} \qquad (2)$$

where C is the control value, L is the asymptotic value, and Q the potency factor. In both equations the effect will be at the midpoint of its range when dose $= 1/Q$. This dose may conveniently be described as the $D50$ but care should be taken to distinguish it from the $ED50$ relating to quantal response data.

Rearranging the terms of (1) and (2), taking logarithms, and inserting a slope constant b gives the logit of effect:

$$\begin{aligned} Y &= \ln\left(\frac{R}{E-R}\right) = \ln\left(\frac{V-C}{L-V}\right) \\ &= \ln(Q) + b\ln(\text{dose}). \end{aligned} \qquad (3)$$

This is the equation of the Hill plot [2]. For data that conform to the simple model, the Hill plot will be linear with unit slope ($b = 1$).

For many practical purposes it is unnecessary to go beyond the simple linear relationship between the measurement of effect and the logarithm of dose. Over the full range of effect the regression will be sigmoid, but between about 20% and 80% of the range of deviations from linearity can usually be ignored. In this usage, effect can be scaled in any way to suit the circumstances transformed as necessary to achieve homoscedasticity.

83.3 Quantal Responses

The initial response to treatment will be in the form of quantitative changes in one or more of the response systems. A qualitative criterion of effect might be expected to correspond to some required threshold of response being achieved in these systems. Such a mechanistic approach to modeling the quantal response situation has not yet been fully developed, and empirical statistical procedures dominate the scene.

In a very few instances it is possible to use drug infusions to measure directly the amount of drug required to produce a given effect. This would be defined as the threshold dose, or tolerance, for the effect. Tolerances so measured appear to be distributed either in a normal fashion or, more commonly, lognormally. The quantal response, being the proportionate incidence of effect at a given dose, is therefore seen as corresponding to an estimate on the cumulative distribution of tolerances. Probit analysis permits the estimation of the parameters of the tolerance (or log-tolerance) distribution, but by convention the median rather than the mean is quoted. This is symbolized as the $ED50$.

The logit transformation of proportionate response, $\ln[p/(1-p)]$, provides a very different approach since it differs from the simple models for quantitative drug action solely by the slope parameter b, given in (3).

For practical purposes there is little to choose between the results of probit and logit analyses.

Results of titrations of infective materials such as bacteria or viruses may be treated as ordinary quantal responses, but Peto [3] advanced the theoretical relationship

$$\log S = -pn,$$

where S is the proportion of test subjects failing to develop symptoms, n is the dose in infective units, and p the probability of symptoms developing from each such unit.

83.4 Drug Interactions

Various types of interaction between drugs can be identified. In the simplest case of the additive effect, one drug will substitute for another when the two are given together, the substitution being in inverse proportion to their potencies. Drugs that act in the same mode, but with different efficacies, will display antagonism of one another over part or all of the response range. When they act in different modes, the result may be one of either noncompetitive antagonism or potentiation.

The nature of the interaction will determine the form of analysis to be employed, but for a visual assessment the isobologram can be useful. An *isobole* is a contour line for a given level of effect drawn on a graph with axes corresponding to the scales of dose for the two drugs. The shape of the isobole, particularly with respect to its intercepts (if any), assists in the interpretation of the data [1].

83.4.1 Screening Drugs for Activity

Of the very large numbers of novel compounds synthesized, very few will qualify for eventual clinical study. The tests to which they are subjected are organized as a sequential screening process, some being designed to investigate the desirable properties of the compounds, others to assess side effects and toxicity. In the earlier stages of the screening, the tests are designed to be capable of a high throughput at minimal cost, usually with preset rules for the rejection of unsuitable compounds. Thereafter, as the numbers passing through diminish, so the nature of the testing becomes more detailed and specific. The emphasis shifts from merely detecting quantitative activity toward identifying the mechanism by which it is attained.

On the few compounds that pass through the whole screening process a large amount of information will have been assembled concerning their actions and acute toxicity. It is then a matter of economic necessity to select only one or two for the final stage of preclinical testing—the toxicity study. The selection process will often be extremely difficult since the advantages and disadvantages of each of the contenders must be considered from the multidimensional data, with the relative clinical importance of each action being taken into account.

The formal study of toxicity involves daily administration of the drug at two or more dose levels to groups of animals of both sexes drawn from two, three, or more species. At the higher levels the purpose is to identify the nature and incidence of pathological changes associated with near-lethal dosage, while at the lowest level, chosen to represent a possible clinical dosage, it is to test for freedom from such changes. The duration of the study may range from 1 month to 6 months, to 1 year or even longer, depending on the intended clinical use of the drug.

Even more sophisticated tests may be required to examine for embryotoxicity or for mutagenic effects that might adversely influence reproduction.

83.5 Biological Standardization (Bioassay)

Chemical and physical analytical methods may be inadequate to estimate the biological activity of pharmaceutical products of natural origin. The standardization of such products must then be achieved by biological testing. Since any unit of activity based on response to a given dose will vary with the sensitivity of the animals employed, standard preparations are maintained against which samples from each batch of product may be directly compared. Potency is then quoted in standard units instead of animal units.

The statistical methods developed for the analysis of bioassays lend themselves to research applications where variation of animal sensitivity may also confuse the interpretation of various tests of activity. The place of the standard is taken by a reference compound, when possible selected from established drugs whose general profiles of activity are widely recognized. Since the substances to be compared in this type of test are not chemically identical, Finney proposed the term *comparative assay* to distinguish it from the analytical assay on which the statistical methods were originally based.

83.6 Stability of Products

Many medicines, particularly the synthetic chemical compounds, remain substantially unchanged over long periods of storage provided extreme conditions are avoided. Others, however, will deteriorate on keeping, and for them it is necessary to specify an expiry date together with the conditions of storage which will ensure that prior to that date the change of activity will be of little clinical significance. For therapeutic drugs a loss of about 10% may be tolerated, while for vaccines losses of up to 50% may be allowed. To establish the conditions of storage and the shelf life will, in the first instance, require predictive testing. This is achieved by estimating the rate of loss at a series of elevated temperatures. The temperature coefficient of decay then allows, by extrapolation, the rates to be estimated for lower temperatures, which then become the basis for prediction.

Vaccines, and possibly other products of natural origin, may lose activity in a multistage process. This factor, combined with the larger errors of biological assay by which their potency must be tested, leads to data poorly represented by the simpler models of chemical decay. It is customary therefore to make retrospective checks between the predictions made and the potency found after prolonged storage of samples at lower temperatures. The shelf life of further batches of the same material can then be adjusted if necessary.

83.7 Quality Control

The pharmaceutical industry is not alone in seeking to ensure the quality of its products by tests carried out both during manufacture and on the finished products. In spite of the esoteric nature of many of the tests involved, the statistical problems that they present are common to many sampling procedures. These concern the establishment of rules governing the type of test, the size of a sample (often related to batch size), and of criteria leading to acceptance or rejection of the batch or to its retesting.

To avoid intolerable expense, these rules and criteria have been established by the common consent of the industry and the statutory bodies charged with licensing its products, taking into account the accumulated data from past experience. If the precious element of trust that this implies is to be preserved, constant vigilance is required to guard against exceptional circumstances failing to be recognized.

83.8 Experiment Design

The importance of design derives from a desire (1) to maximize the information from a given set of resources, and (2) to minimize the costs of achieving requisite information; costs in terms of materials, labor, time and, where appropriate, animals.

The choice of design will often be determined by how much is known about the drug to be tested, how well understood is the method of testing, and on the nature of the test system. In clinical work crossover designs are commonly used, but obviously would be pointless if the effect of a treatment were to bring about a cure of the diseased state. In analytical assays, crossover designs, including Latin squares, would often be the most economical. In preclinical studies the type of design is determined more by the test system. When intact animals are used, parallel groups are by far the most common. These will sometimes be controlled by a group receiving dummy treatment, sometimes by pretreatment measurements, and occasionally both. Isolated tissues and animals under anaesthetic will commonly be subjected to sequential testing, although sometimes a form of crossover might be attempted.

For obvious reasons the incomplete block design is least common in such work and is limited to those areas where the maximum of information must be extracted from a system constrained in some way.

83.9 Nonparametric Methods

The traditional objective of drug research is to characterize the dose–response relationship for as many response systems as may be relevant. With so much emphasis on estimation, nonparametric methods have little to offer. They are welcomed, however, when experimental results do not lend themselves to valid analysis by parametric methods and there are numerous types of localized study in which evaluation is achieved by significance testing based on nonparametric techniques. These can be regarded as contributing qualitative statements to points in a broadly quantitative structure. The major problem is how to reconcile statistical significance with clinical importance in a highly multivariate system.

One must reserve a special place among the nonparametric methods for the double-dichotomy test. Its unpretentious form gives it a justifiable appeal to biologists and clinicians alike through all stages of drug investigation.

References

1. De Jongh, S. E. (1961). In *Quantitative Methods in Pharmacology*, H. De Jonge, ed. North-Holland, Amsterdam.
2. Hill, A. V. (1910). *J. Physiol.*, **40**, 4–7.
3. Peto, S. (1953). *Biometrics*, **9**, 320–335.

84 Statistics in Spatial Epidemiology

Andrew B. Lawson

84.1 Introduction

Spatial epidemiology is a wide subject and encompasses a range of topics where the geographic or spatial distribution of disease is of importance. Statistical methods employed in this area are also diverse in their range, and, besides basic exploratory and descriptive methodology common to many subject areas, there is a need to employ particular *spatial* statistical methods that are designed for such data. The basic characteristic of data encountered in this application area is their *discrete* nature, whether in the form of spatial locations of cases of disease, or counts of disease within defined geographic regions. Hence methods developed for continuous spatial processes, such as kriging, are not directly applicable or are only approximately valid.

Often hypotheses of interest in spatial epidemiology focus on whether the residential address of cases of disease yields insight into etiology of the disease or, in a public health application, whether adverse environmental health hazards exist locally within a region (as exemplified by local increases in disease risk. For example, in a study of the relationship between malaria endemicity and diabetes in Sardinia a strong negative relationship was found [1; 13, Chap. 10]. This relation had a spatial expression and the geographic distribution of malaria was important in generating explanatory models for the relation. In public health practice it is of considerable importance to be able to assess whether localized areas that have larger-than-expected numbers of cases of disease are related to any underlying environmental cause. Here spatial evidence of a link between cases and a source is fundamental in the analysis. Evidence such as a decline in risk with distance from the *putative* source of hazard or elevation of risk in a preferred direction is important in this regard.

There are four main areas where statistical methods have seen development in spatial epidemiology: Disease Mapping, Disease Clustering, Ecological Analysis, and Disease Map Surveillance. Before looking in detail at each of these areas, it is appropriate to consider some common themes or issues which arise in all areas of the subject.

84.2 Basic Definitions and Models

A fundamental feature of data available for analysis in spatial epidemiology is that they are usually discrete (in the form of either a point process or counting process), and the cases of concern arise from within a local human population that varies in spatial density and in susceptibility to the disease of interest. Hence any model or test procedure must make allowance for this background (nuisance) population effect. The background population effect can be allowed for in a variety of ways. For count data it is commonplace to obtain *expected* rates for the disease of interest based on the age–sex structure of the local population (see, e.g., Ref.6, Chap. 3), and some crude estimates of local relative risk are often computed from the ratio of observed to expected counts [e.g., standardized mortality/incidence ratios SMRs. For case event data, expected rates are not available at the resolution of the case locations and the use of the spatial distribution of a control disease has been advocated. In that case the spatial variation in the case disease is compared to the spatial variation in the control disease. A major issue in this approach is the correct choice of control disease. It is important to choose a control that is matched to the age–sex structure of the case disease but is unaffected by the feature of interest. For example, in the analysis of cases around a putative health hazard, a control disease should not be affected by the health hazard. Counts of control disease cases could also be used instead of expected rates when analyzing count data. Figures 1 and 2 display case event and control data maps for the a region of the UK for a fixed time period. Figure 3 displays a typical count data example.

84.2.1 Case Event Data

Case event locations often represent residential addresses of cases and the cases arise from a heterogeneous population that varies both in spatial density and in susceptibility to disease. A heterogeneous Poisson process model is often assumed as a starting point for further analysis. Define the first-order intensity function of the case event process as $\lambda(\mathbf{x})$, representing the mean number of events per unit area in the neighbourhood of location $\mathbf{x}$. This intensity may be parameterized as

$$\lambda(\mathbf{x}) = \rho.g(\mathbf{x}).f(\mathbf{x};\theta),$$

where ρ is the overall rate of the process, $g(\mathbf{x})$ is the "background" intensity of the population at risk at $\mathbf{x}$, and $f(\mathbf{x};\theta)$ is a parameterized function of risk. The focus of interest for making inferences regarding parameters describing excess risk lies in $f(\mathbf{x};\theta)$, treating $g(\mathbf{x})$ as a nuisance function.

It is possible that population or environmental heterogeneity may be unobserved in the dataset. This could be because either the population background hazard is not directly available or the disease displays a tendency to cluster (perhaps due to unmeasured covariates). The heterogeneity could be spatially correlated, or it could lack correlation, in which case it could be regarded as a type of "overdispersion". One can include such unobserved heterogeneity within the framework of conventional models as a random effect.

This approach can lead to maximum a posteriori estimators similar to those found for universal kriging in geostatistics [10]. This approach can also be implemented in a fully Bayesian setting (see, e.g., Refs. 11 and 14).

84.2.2 Count Data

A considerable literature has developed concerning the analysis of count data in

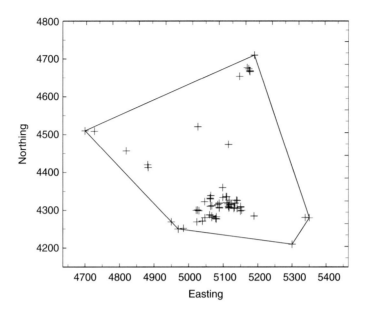

Figure 1: Distribution of cases of childhood lymphoma and leukemia in Humberside UK (1974–1986). (*Source*: Cuzick and Edwards [5].)

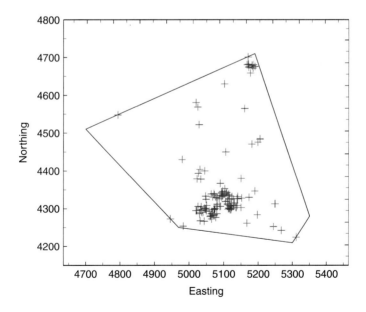

Figure 2: Control distribution: distribution of a sample of live births from the birth register in Humberside UK (1974–1986). (*Source*: Cuzick and Edwards [5].)

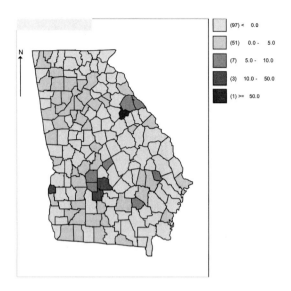

Figure 3: Distribution of the standardized incidence ratio (SMR) for oral cancer within the counties of Georgia USA 1990. (*Sources*: OASIS online health access system, Division of Public Health State of Georgia USA (http://oasis.state.ga.us/); and Lawson [14].)

spatial epidemiology (e.g., see reviews in Refs. 13, 6, and 14).

The usual model is adopted for the analysis of region counts $\{y_i, i = 1, \ldots, p\}$ to be independent Poisson random variables with parameters $\{\lambda_i, i = 1, \ldots, p\}$. Here

$$\lambda_i = \int_{W_i} \lambda(\mathbf{x}) d\mathbf{x}, i = 1, \ldots, p,$$

where $\lambda(\mathbf{x})$ is the first-order intensity of the underlying cases and W_i is the ith subregion. Often the λ_is are assumed to be constant within areas. Usually the expected count is modeled as

$$E(y_i) = \lambda_i = e_i \theta_i, \qquad i = 1, \ldots, p.$$

This model may be extended to include unobserved heterogeneity between regions by introducing a prior distribution for the log relative risks ($\log \theta_i$, $i = 1, \ldots, p$), or by modeling the log of the risk as a linear combination of terms. For example, a common additive specification is $\log \theta_i = \alpha_0 + u_i + v_i$, which includes an intercept (α_0), and a *convolution* of two (random effect) terms that have prior distributions. The first term (u_i) is often assumed to describe spatial correlation in the risk, while the second term (v_i) is an uncorrelated random noise term. Often the first term is assumed to have a conditional autoregressive (CAR) prior distribution, while the second term has an i.i.d. Gaussian distribution with zero mean and given variance. Incorporation of such heterogeneity has become a common approach, and the Besag–York–Mollié (BYM) model is now a standard model [2, 19, 26, and 15]. It has been shown that this model is very robust in estimating true risk under a range of different true risk specifications. A full Bayesian analysis using this model is available on WinBUGS [16, 4, 22].

84.3 Special Application Areas

84.3.1 Disease Mapping

In this area, focus is on the processing of the disease map to take out random noise. Often applications in health services research require the production of an 'accurate' map of relative risks. Models for relative risk range from simple SMRs to posterior expected estimates from Bayesian models (see, e.g., Ref. 13, Chap. 8 for a more recent review).

84.3.2 Disease Clustering

In this area, the focus is not on reduction of noise, per se, but the assessment of the clustering tendency of the map and in particular the assessment of which areas of a map display clustering. Here, clustering could be around a known putative source of hazard (*focused* clustering) or have no known locations of clustering (*nonfocused* clustering). A variety of testing methods are available for cluster detection (9,21), while model-based approaches have been receiving greater attention (11,3,8,18,7).

84.3.3 Ecological Analysis

In this area, the relation between disease incidence and explanatory variables is the focus, and this is usually carried out at an aggregate level, such as with counts in small areas. The ecological fallacy concerns the issue of the appropriateness of making inferences from aggregate data to individuals in the population (see, e.g., Refs. 27, 25, 20, and 26).

84.3.4 Disease Map Surveillance

In this area, the focus is the construction of methods which, usually, examine the spatiotemporal variation of disease. However, this is done in real- or near-real time. Certain optimal methods are available for the detection of changes in a disease incidence and clustering in space-time (see, e.g., Refs. 12, 17, 24, and 23).

References

1. Bernardinelli, L., Clayton, D. G., Pascutto, C., Montomoli, C., Ghislandi, M., and Songini, M. (1995). Bayesian analysis of space-time variation in disease risk. *Statist. Med.*, **14**, 2433–2443.

2. Besag, J., York, J., and Mollié, A. (1991). Bayesian image restoration with two applications in spatial statistics. *Ann. Inst. Stat. Math.*, **43**, 1–59.

3. Clark, A. B. and Lawson, A. B. (2002). Spatio-temporal cluster modelling of small area health data. In *Spatial Cluster Modelling*, A. B. Lawson and D. Denison, eds., Chap. 14, pp. 235–258. CRC Press, New York.

4. Cowles, M. K. (2004). Review of WinBUGS 1.4. *Am. Statist.*, **58**, 330–336.

5. Cuzick J. and Edwards, R. (1990). Spatial clustering for inhomogeneous populations (with discussion). *J. Roy. Stat. Soc. B*, **52**, 73–104.

6. Elliott, P., Wakefield, J. C., Best, N. G., and Briggs, D. J., eds., (2000). *Spatial Epidemiology: Methods and Applications*. Oxford University Press, London.

7. Gangnon, R. and Clayton, M. (2000). Bayesian detection and modeling of spatial disease clustering. *Biometrics*, **56**, 922–935.

8. Hossain, M. and Lawson, A. B. (2005). Local likelihood disease clustering: Development and evaluation. *Environ. Ecol. Statist.*, **12**, 259–273.

9. Kulldorff, M., Tango, T., and Park, P. J. (2003). Power comparisons for disease clustering tests. *Comput. Statist. Data Anal.*, **42**, 665–684.

10. Lawson, A. B. (1994). On using spatial Gaussian priors to model heterogeneity

in environmental epidemiology. *Statistician*, **43**, 69–76.

11. Lawson, A. B. (2000). Cluster modelling of disease incidence via rjmcmc methods: A comparative evaluation. *Statist. Med.*, **19**, 2361–2376.

12. Lawson, A. B. (2001). Population health surveillance. In *Encyclopedia of Environmetrics*. Wiley Chichester, UK.

13. Lawson, A. B. (2006). *Statistical Methods in Spatial Epidemiology*, 2nd ed., Wiley, Hoboken, NJ.

14. Lawson, A. B. (2009). *Bayesian Disease Mapping: Hierarchical Modeling in Spatial Epidemiology*. CRC Press, New York.

15. Lawson, A. B. and Banerjee, S. (2008). Bayesian spatial analysis. In *Handbook of Spatial Analysis*, S. Fotheringham and P. Rogerson, eds., Chap. 9. Sage, New York.

16. Lawson, A. B., Browne, W. J., and Vidal-Rodiero, C. L. (2003). *Disease Mapping with WinBUGS and MLwiN*. Wiley, Hoboken, NJ.

17. Lawson, A. B. and Kleinman, K. eds., (2005). *Spatial and Syndromic Surveillance for Public Health*. Wiley, Hoboken, NJ.

18. Lawson, A. B., Kulldorff, M., Simeon, S., Biggeri, A., and Magnani, C. (2007). Line and point cluster models for small area health data. *Comput. Statist. Data Anal.* in press.

19. Mollié, A. (1999). Bayesian and empirical Bayes approaches to disease mapping. In *Disease Mapping and Risk Assessment for Public Health*, A. Lawson, A. Biggeri, D. Boehning, E. Lesaffre, J.-F. Viel, and R. Bertollini, eds., Chap. 2, pp. 15–29. Wiley, Chichester, UK.

20. Salway, R. and Wakefield, J. (2005). Sources of bias in ecological studies of non-rare events. *Environ. Ecol. Statist.*, **12**, 321–347.

21. Song, C. and Kulldorff, M. (2003). Power evaluation of disease clustering tests. *Int. J. Health Geogr.*, **29**, 1-8.

22. Spiegelhalter, D., Thomas, A., Best, N., and Lunn, D. (2007). *WinBUGS Manual*, version 1.4.3. MRC Biostatistics Unit, Cambridge, UK.

23. Vidal-Rodeiro, C. and Lawson, A. B. (2006). Monitoring changes in spatiotemporal maps of disease. *Biometr. J.*, **48**, 1-18.

24. Vidal-Rodeiro, C. and Lawson, A. B. (2006). Online updating of space-time disease surveillance models via particle filters. *Statist. Methods Med. Res.*, **15**, 423–444.

25. Wakefield, J. (2004). Prior and likelihood choices in the analysis of ecological data. In *Ecological Inference: New Methodological Strategies*, G. King, O. Rosen, and M. Tanner, eds., Chap. 1, pp. 13–50. Cambridge University Press.

26. Wakefield, J. and Shaddick, G. (2006). Health-exposure modeling and the ecological fallacy. *Biostatistics*, **7**, 438–455.

27. Wakefield, J. C. and Salway, R. (2001). A statistical framework for ecological and aggregate studies. *J. Roy. Statist. Soc.*, **164**, 119–137.

85. Stochastic Compartment Models

M. A. Woodbury and K. G. Manton

85.1 Introduction

A discussion of stochastic compartment models is fraught with the difficulty of providing a precise definition. This is a difficult task because of the wide variety of problems to which compartment models have been applied and their even broader range of potential applications. Indeed, in conceptual terms, the discussion of compartment models is nearly as general as the discussion of systems. As a consequence, we shall view compartment models as a general analytic strategy rather than as a specific set of analytic procedures. The crux of the analytic strategy of compartment modeling is the explicit consideration of the mathematical model of the time-directed behavior of a given system. Once the mathematical structure of the system behavior is described, statistical issues may be addressed. In order to evaluate the compartment model strategy in greater depth, in the remainder of this chapter, we discuss (1) the origin of compartment analysis in the biological sciences, (2) the formal aspects of the general linear compartment system, and (3) issues of parameter identifiability and estimability. As will be seen, identifiability and estimability are two crucial concerns in the development of specific compartment models.

85.2 Origin of Compartment Models in the Biological Sciences

Compartment models were originally developed for the analysis of complex biological systems—a problem that was not amenable to analysis by classical statistical procedures. The crucial feature in the analysis of complex biological systems is that certain parameters in the system could not be experimentally manipulated without distorting the behavior of the system. Indeed, much of the system behavior was unobservable as well as unmanipulatable. As a result, analysis of the behavior of the system was confined largely to the study of the inputs and outputs of the system. The only practical approach to analysis under such constraints was to posit a mathematical model, based on ancillary information, to link inputs and outputs. Inference in this case is restricted to determining if the mathematical structure consistently linked the temporal schedule of inputs and outputs.

Biologists, in developing such compartment models, found that certain classes of system structure facilitated computation and were consistent with a number of theoretical precepts about biological systems. In particular, biologists found it convenient to model complex systems as a series of discrete states or compartments linked by flows of particles governed by partially observed transition rates that were functions of continuous time. This basic type of compartment system could be described alternatively as a discrete-state, continuous-time stochastic process with partially observed transition parameters. Because the intercompartment transfers are observed only partially, a critical feature of such analyses was the generation of sufficient restrictions on parameters to achieve structural identifiability of the equations relating compartment interchanges to time. Another important feature of this particular formulation of the compartment model is that it had a dual nature, being described either probabilistically, in terms of the waiting-time distributions of particles within compartments, or in terms of the transition rate functions themselves. For biologists the most familiar description was in terms of the transition-rate functions, because they led naturally to solutions for linear ordinary differential equation systems [6].

85.3 Linear Compartment Systems

It is a relatively recent development that statisticians have realized that there are broad classes of problems that could be approached only by some "modeling" strategy such as compartment analysis. From this realization there have emerged a series of attempts to develop general analytic strategies for the analysis of compartment systems. One such attempt involves the development of strategies appropriate to the analysis of the linear compartment systems, or

$$\dot{x}(t) = A(t)x(t) + B(t), \quad (1)$$

where $\dot{x}(t)$ represents the change in the number of particles in each of n compartments, $A(t)$ is an $n \times n$ matrix of time-dependent transition rates, $x(t)$ is the vector of the number of particles in each of n compartments, and $B(t)$ is the vector of inputs to each compartment. Often, these equations are simplified by assuming $A(t)$ and $B(t)$ to be time-invariant. Bellman [1] has derived a solution using a matrix exponential form for the time-invariant form of (1). When the eigenvalues of A are real and distinct, a solution is available using the "sums of exponential" model [5].

Matis and Wehrly [9] have pointed out that the linear compartment system represented in (1) is often applied deterministically. They propose that this model may be usefully generalized by introducing several different types of stochasticity. They identify two types of stochasticity that they feel are frequently present in applying the linear compartment model. The first type of stochasticity is associated with individual units or particles. Particle "stochasticity" is further divided into that due to sampling from a random process and that due to particles having random rate coefficients. The second type of stochasticity they identify is associated with a replication of the entire experiment. "Replicate" stochasticity can also be divided into two types. The first is replicate stochasticity due to the initial number of particles being random. The second type of replicate stochasticity is due to the variability of rate coefficients across replicates. Naturally, in any given situation, one, or a combination, of these four types of stochasticity may be present.

The basic rationale for dealing with the complications produced by including consideration of the four types of stochasticity in the linear compartment system is that a deterministic system may not adequately represent the behavior of individuals within the system. For example, the age-specific mortality probabilities from a cohort life table are often used to describe the trajectory of individuals through different "age" compartments. However, if individuals within a cohort have different "susceptibility" to death, individuals will be systematically selected by mortality and the cohort life table will no longer be a valid model for the age trajectory of mortality risks for individuals. Instead, it will only be descriptive of the age-specific mean risk of death among survivors to a given age [7,10].

85.4 Estimation and Identification of Stochastic Compartment Model Parameters

Although it is clear that appropriately representing various types of stochasticity in compartment models will greatly improve their theoretical and analytic worth, it is also clear that estimation of the parameters in such models will be far more difficult. For example, estimation of the parameters of a stochastic compartment system often runs into identifiability problems because a consideration of only the means is insufficient information to identify the parameters for individual transitions. Two basic approaches are employed to deal with these issues in the analysis of stochastic compartment systems. The first approach involves the development of basic compartment models with standard computational procedures that will be applicable to broad classes of analytic problems. In this approach statistical information, say, as contained in the error covariance structure, is used to achieve identifiability. Representative of such approaches is the nonlinear least-squares strategy proposed by Matis and Hartley [8]. The second basic approach is to develop compartment models specific to individual problems. In this approach it is explicitly recognized that each compartment model is composed of a substantively determined mathematical structure and a statistical model and that each needs to be developed for the specific problem at hand. Thus identification is achieved by imposing restrictions derived from substantive theory or ancillary data on the parameter space.

In either approach one must deal with the central analytic issue of parameter identification, i.e., that each parameter be observationally independent of the set of other parameters. Estimability is a related concept suggesting that the available data contain sufficient statistical information so that precise estimates of parameters can be made. In general, identifiability must be achieved with information on certain measurable outputs (i.e., the rate of exit of particles into one of a set of external, observable compartments) that we can identify as the vector y, certain (possibly manipulatable) inputs (rate of entry of particles to the system from controllable external compartment) that we can identify as the vector u, and the temporal relations of u and y. From these observed quantities, one hopes to be able to determine the number of particles in each theoretically specified internal compartment, which we will represent as a vector x, and the matrix F, of transfer coefficients governing the flow of particles between compartments. As can be seen, we are restricting our discussion to linear, time-invariant systems. Let us assume further that the transfer of inputs to internal compartments is governed by an observed matrix B, and that another transfer matrix C determines the measured output vector y, or

$$\dot{x} = Bu + Fx \quad (2)$$
$$y = Cx \quad (3)$$

as functions of time.

By taking the Laplace transform on both sides of (2) and (3), and under the assumption that the system is empty at time $t = 0$, we obtain

$$sX = BU + FX, \quad (4)$$
$$Y = CX, \quad (5)$$

where U, X, and Y are the Laplace transforms of u, x, and y and s is the transformation-domain variable. Equations (4) and (5) can be solved to determine the relation between U and Y, or

$$Y = C(sI - F)^{-1}BU. \quad (6)$$

In (6), U and Y are vectors, so that, to consider a single transfer coefficient, T_{ij}, we have to consider the ith element of Y and the jth element of U, as in

$$T_{ij} = e_i^T C(sI - F)^{-1} Be_j. \quad (7)$$

Here T_{ij} represents the transfer coefficient between input state j and output state i and where e_i and e_j are unit vectors with a 1 in the appropriate ith and jth positions and zeros elsewhere. In (7), U is, in effect, replaced by the identity matrix and the vector Y is replaced by the elements of the matrix of transfer coefficients, T. This situation corresponds to the analysis of cases where the time-domain input of particles to the system is in the form of a single "pulse." It can be seen that the relation between F and T is nonlinear. In this case identifiability is achieved only if it is possible to generate empirical estimates of the T_{ij} from temporal measurements of output i of the system resulting from a "pulse" of particles directed through input j by computing its Laplace transform. To consider this in greater detail, let us write the expression that relates a change in a T_{ij} to a "small" change, df_{kl}, of a given transfer coefficient, f_{kl} (for k, l such that $f_{kl} \neq 0$), or

$$\frac{\partial T_{ij}}{\partial f_{kl}} = e_i^T C(sI - F)^{-1}$$
$$\times e_k e_l^T (sI - F)^{-1} Be_j. \quad (8)$$

If f_{kl} is not identifiable, then, for all i and j, $(\partial T_{ij}/\partial f_{kl})$ is identically zero in s. This implies that f_{kl} is not identifiable if there does not exist an i such that $e_i^T C(sI - F)^{-1} e_k \neq 0$ and a j such that $e_l^T(sI-F)^{-1} Be_j \neq 0$. To understand what this implies for a structure of a compartment system, we should note that, for sufficiently large s

$$(sI - F)^{-1} = \frac{I}{s} + \frac{F}{s^2} + \frac{F^2}{s^3} + \frac{F^3}{s^4} + \cdots, \quad (9)$$

so that a given entry $(sI - F)^{-1}_{kl}$ is not identically zero in s if there is some power F^P of F that has $F^P_{kl} \neq 0$. It can be seen that this is so if there is a "path" from k to l in F^P, i.e., if there exist $k = k_1, k_2, k_3, \ldots, k_p = l$ such that $f_{k_1,k_2} \neq 0, f_{k_2,k_3} \neq 0, \ldots, f_{k_{p-1},k_p} \neq 0$. This can be determined readily from the "flowchart" of the compartment system [2–4].

We now introduce the notion of input-connectable and output-connectable systems. A compartment l is input-connectable if there is a path from an input to the compartment l. This is equivalent to the "nonvanishing" of $e_l^T(sI - F)^{-1}B$. Similarly, the nonvanishing of $C(sI - F)^{-1}e_k$ is equivalent to the output connectability of compartment k. Hence a necessary condition for the identifiability of a compartmental system is that for every $f_{kl} \neq 0$, k is output-connectable and l is input-connectable [3].

Structural identifiability for a given compartment system can be achieved either by introducing statistically imposed constraints (e.g., by developing constraints on the error covariance structure) or by deriving parameter restrictions from ancillary data and theory. Estimability will, practically, be the more difficult condition to achieve because of the near collinearity of parameters in the nonlinear functional forms often employed.

85.5 Model Building in Statistical Analysis

The foregoing discussion is meant to sensitize the statistician to the fact that compartment analysis represents an effort to formalize a series of mathematical model-building strategies for analytic problems where direct information is not available to estimate all parameters of a model. As a consequence of such problems, it is necessary to assess explicitly the detailed mathematical form of the response model assumed in the analysis. As such, this type of analysis represents an interesting synthesis of mathematical modeling, statistical inference, and substantive theory in an attempt to model behavior that is only partially observable. It also requires careful attention to aspects of computational theory and numerical analysis. It represents a very flexible tool for the statistician faced with a problem whose complexity makes a more traditional approach impossible. As the statistician becomes involved with a variety of important policy questions involving the behavior of complex human systems, the need to employ a modeling strategy of this type increases.

References

1. Bellman, R. (1960). *Introduction to Matrix Analysis.*, McGraw-Hill, New York. (A general reference work on the mathematical aspects of matrix applications.)
2. Cobelli, C. and Jacur, G. R. (1976). *Math. Biosci.* **30**, 139–151. (An analysis of the conditions for establishing the identifiability of the parameters of a compartment model.)
3. Cobelli, C., Lepschy, A., and Jacur, G. R. (1979). *Math Biosci.*, **44**, 1–18. (Extends their earlier 1976 results and responds to questions raised by Delforge.)
4. Delforge, J. (1977). *Math. Biosci.*, **36**, 119–125. (Illustrates that Cobelli and Romanin Jacur had provided only necessary and not sufficient conditions for identifiability.)
5. Hearon, J. Z. (1963). *Ann. NY Acad. Sci.*, **108**, 36–68. (Provides some useful results for the analysis of compartment systems.)
6. Jacquez, J. A. (1972). *Compartmental Analysis in Biology and Medicine*. Elsevier, New York. (A broad overview of the practice and theory of the applications of compartment models in the biological sciences.)
7. Manton, K. G. and Stallard, E. (1980). *Theor. Popul. Biol.*, **18**, 57–75. (An analytic model of disease dependence represented via a compartment system with a heterogeneous population.)
8. Matis, J. H. and Hartley, H. O. (1971). *Biometrics*, **27**, 77–102. (A basic methodological exposition of the application of nonlinear least squares to the estimation of the parameters of a compartment system.)
9. Matis, J. H. and Wehrly, T. E. (1979). *Biometrics*, **35**, 199–220. (A useful and comprehensive review article on the application of linear compartment systems and of various types of stochastic formulations of such systems.)
10. Vaupel, J. W., Manton, K. G., and Stallard, E. (1979). *Demography*, **16**, 439–454. (A presentation of a model based on explicit mathematical assumptions to represent the effects of heterogeneity on the parameters of a basic human survival model.)

86. Surrogate Markers

Geert Molenberghs

86.1 Introduction

Surrogate endpoints are defined as endpoints that can replace or supplement other endpoints in the evaluation of experimental treatments or other interventions. For example, surrogate endpoints are useful when they can be measured earlier, more conveniently, or more frequently than the endpoints of interest, which are referred to as the "true" endpoints [10].

Prentice [18] proposed a formal definition of surrogate endpoints and outlined how potential surrogate endpoints could be validated. Much debate ensued, for the criteria set out by Prentice are not straightforward to verify [13]. In addition, Prentice's criteria are only equivalent to his definition in the case of binary endpoints [4]. Freedman et al. [14] supplemented Prentice's approach by introducing the *proportion explained* (PE), which is the proportion of the treatment effect mediated by the surrogate. Buyse and Molenberghs [4] proposed replacing it by two new measures. The first one, defined at the population level and termed *relative effect* (RE), is the ratio of the overall treatment effect on the true endpoint over that on the surrogate endpoint. The second one is the individual-level association between both endpoints, after accounting for the effect of treatment, and referred to as *adjusted association*.

In turn, a drawback of the RE is that, when calculated from a single trial, its use depends on strong unverifiable assumptions, the main one being that it should be constant across a class of trials. A way out of this problem is the combination of information from several groups of patients (multicenter trials or meta-analyses). Such an approach was suggested by Albert et al. [1] and was implemented by Daniels and Hughes [9] and by Buyse et al. [5]. Gail et al. [15] contrasts the work by Daniels and Hughes [9] and Buyse et al. [5], and address several important issues. The latter extended the adjusted association and the RE to an individual-level measure of association and a trial-level measure of association, respectively. They suggest using these or similar measures as an alternative way to assess the usefulness of a surrogate endpoint. An important aspect of such measures is that they allow one to quantify the quality of a surrogate. Thus, one is not confined to an "all or nothing" situation where a candidate endpoint is either perfect or no surrogate at all.

A question that then arises naturally is whether, in addition to these new measures, single-trial-based quantities such as the PE or the RE still convey useful information. Several authors have argued that they can be misleading. Nevertheless, they still enjoy a great deal of support. Arguably, it will take a while before the controversy is satisfactorily resolved.

We start with the case of a single unit and then we briefly review Prentice's definition and criteria. The proportion explained is introduced next, followed by the relative effect and adjusted association. Thereafter, a multiunit framework is introduced. The notions of trial-level and individual-level surrogacy are based thereupon.

86.2 Data from a Single Unit

We will first discuss the single unit setting (e.g., a single trial). The notation and modeling concepts introduced are useful in presenting and critically discussing the key ingredients of the Prentice–Freedman framework. Therefore, this section should not be seen as setting the scene for the rest of the chapter. This is reserved for the multiunit case.

Throughout, we will adopt the following notation: T and S are random variables that denote the true and surrogate endpoints, respectively, and Z is an indicator variable for treatment. For ease of exposition, we will assume that S and T are normally distributed. The effect of treatment on S and T can be modeled as follows:

$$S_j = \mu_S + \alpha Z_j + \varepsilon_{Sj}, \quad (1)$$
$$T_j = \mu_T + \beta Z_j + \varepsilon_{Tj}, \quad (2)$$

where $j = 1, \ldots, n$ indicates patients, and the error terms have a joint zero-mean normal distribution with covariance matrix:

$$\Sigma = \begin{pmatrix} \sigma_{SS} & \sigma_{ST} \\ & \sigma_{TT} \end{pmatrix}. \quad (3)$$

In addition, the relationship between S and T can be described by a regression of the form

$$T_j = \mu + \gamma S_j + \varepsilon_j. \quad (4)$$

Note that this model is introduced because it is a component of the Prentice–Freedman framework. Given that the fourth criterion will involve a dependence on the treatment as well, as in Equation (5), it is of legitimate concern to doubt whether Equations (4) and (5) are simultaneously plausible. Also, the introduction of Equation (4) should *not* be seen as an implicit of explicit assumption about the absence of treatment effect in the regression relationship, but rather as a model that can be used when the uncorrected association between both endpoints is of interest.

We will assume later that the n patients come from N different experimental units, but for now the simple situation of a single experiment will suffice to explore some fundamental difficulties with the validation of surrogate endpoints.

86.2.1 Definition and Criteria

Prentice [18] proposed the definition a surrogate endpoint as "a response variable for which a test of the null hypothesis of no relationship to the treatment groups under comparison is also a valid test of the corresponding null hypothesis based on the true endpoint" 18, (p. 432). In terms of our simple model (1)–(2), the definition states that for S to be a valid surrogate for T, parameters α and β must simultaneously be equal to, or different from, zero. This definition is not consistent with the availability of a single experiment only, since it requires a large number of experiments to be available, each with tests of hypothesis on both the surrogate and true endpoints. An important drawback is also that evidence from trials with nonsignificant treatment effects cannot be used, even though such trials may be consistent with a desirable relationship between both endpoints. Prentice derived operational criteria that are equivalent to his definition. These criteria require that

1. Treatment has a significant impact on the surrogate endpoint [parameter α differs significantly from zero in Eq. (1)].

2. Treatment has a significant impact on the true endpoint [parameter β differs significantly from zero in Eq. (2)].

3. The surrogate endpoint has a significant impact on the true endpoint [parameter γ differs significantly from zero in Eq. (4)].

4. The full effect of treatment on the true endpoint is captured by the surrogate.

The last criterion is verified through the conditional distribution of the true endpoint, given treatment *and* surrogate endpoint, derived from (1)–(2):

$$T_j = \tilde{\mu}_T + \beta_S Z_j + \gamma_Z S_j + \tilde{\varepsilon}_{Tj}, \quad (5)$$

where the treatment effect (corrected for the surrogate S) β_S and the surrogate effect (corrected for treatment Z) γ_Z are

$$\beta_S = \beta - \sigma_{TS}\sigma_{SS}^{-1}\alpha, \quad (6)$$
$$\gamma_Z = \sigma_{TS}\sigma_{SS}^{-1}, \quad (7)$$

and the variance of $\tilde{\varepsilon}_{Tj}$ is given by

$$\sigma_{TT} - \sigma_{TS}^2 \sigma_{SS}^{-1}. \quad (8)$$

It is usually stated that the fourth criterion requires that the parameter β_S be equal to zero. Essentially, this last criterion states that the true endpoint T is completely determined by knowledge of the surrogate endpoint S. Buyse and Molenberghs [4] showed that the last two criteria are necessary and sufficient for binary responses, but not in general. Several authors, including Prentice, pointed out that the criteria are too stringent to be fulfilled in real situations [18,12].

In spite of these criticisms, the spirit of the fourth criterion is very appealing. This is especially true if it can be considered in the light of an underlying biological mechanism. For example, it is interesting to explore whether the surrogate is part of the causal chain leading from treatment exposure to the final endpoint. While this issue is beyond the scope of the current chapter, the connection between statistical validation (with emphasis on association) and biological relevance (with emphasis on causation) deserves further reflection. A detailed study of the criticism is given in Reference 17.

86.2.2 The Proportion Explained

Freedman et al. [14] argued that the last Prentice criterion raises a conceptual difficulty since it requires the statistical test for treatment effect on the true endpoint to be *non*-significant after adjustment for the surrogate. The nonsignificance of this test does not prove that the effect of treatment on the true endpoint is *fully* captured by the surrogate, and therefore Freedman et al. [14] proposed calculating the proportion of the treatment effect mediated by the surrogate:

$$\mathrm{PE} = \frac{\beta - \beta_S}{\beta},$$

with β_S and β obtained respectively from Equations (5) and (2). In this paradigm, a valid surrogate would be one for which the proportion explained (PE) is equal to one. In practice, a surrogate would be deemed acceptable if the lower limit of its confidence interval of PE was "sufficiently" large.

Some difficulties surrounding the PE have been described in the literature [4,9,20,7,16,11]. PE will tend to be unstable when β is close to zero, a situation that is likely to occur in practice. As Freedman et al. [14] themselves acknowledged, the confidence limits of PE will tend to be rather wide (and sometimes even unbounded if Fieller confidence intervals are used), unless large-sample sizes are available or a very strong effect of treatment on the true endpoint is observed. Note that large-sample sizes are typically available in epidemiological studies or in meta-analyses of clinical trials. Another complication arises when Equation (4) is not the correct conditional model, and an interaction term between Z_i and S_i needs to be included. In that case, defining the PE becomes problematic.

86.2.3 The Relative Effect

Buyse and Molenberghs [4] suggested to calculate another quantity for the validation of a surrogate endpoint: the relative effect (RE), which is the ratio of the effects of treatment on the final and the surrogate endpoint. Formally; this is

$$\text{RE} = \frac{\beta}{\alpha}, \quad (9)$$

They also considered the treatment-adjusted association between the surrogate and the true endpoint, ρ_Z:

$$\rho_Z = \frac{\sigma_{ST}}{\sqrt{\sigma_{SS}\sigma_{TT}}}. \quad (10)$$

Now, a simple relationship can be derived between PE, RE, and ρ_Z. Let us define $\lambda^2 = \sigma_{TT}\sigma_{SS}^{-1}$. It follows that $\lambda\rho_Z = \sigma_{ST}\sigma_{SS}^{-1}$ and, from Equation (6), $\beta_S = \beta - \rho_Z\lambda\alpha$. As a result, we obtain

$$\text{PE} = \lambda\rho_Z\frac{\alpha}{\beta} = \lambda\rho_Z\frac{1}{\text{RE}}. \quad (11)$$

A similar relationship was derived by Buyse and Molenberghs [4] and by Begg and Leung [2] for standardized surrogate and true endpoints. Next, let us introduce a multiunit framework

86.2.4 Data from Several Units

Buyse et al. [5] extended the setting and notation by supposing that we have data from $i = 1, \ldots, N$ units (e.g., centers, investigators, trials), in the ith of which $j = 1, \ldots, n_i$ subjects are enrolled. We now denote the true and surrogate endpoints by T_{ij} and S_{ij}, respectively, and by Z_{ij} the indicator variable for treatment. (See also Ref. 9.) Several authors have considered other than normally distributed outcomes, but we will develop the ideas in the normal context and provide some references to other types of outcomes at the end.

The linear models (1) and (2) can be rewritten as

$$S_{ij} = \mu_{Si} + \alpha_i Z_{ij} + \varepsilon_{Sij}, \quad (12)$$
$$T_{ij} = \mu_{Ti} + \beta_i Z_{ij} + \varepsilon_{Tij}, \quad (13)$$

where μ_{Si} and μ_{Ti} are trial-specific intercepts, α_i and β_i are trial-specific effects of treatment Z_{ij} on the endpoints in trial i, and ε_{Si} and ε_{Ti} are correlated error terms, assumed to be zero-mean normally distributed with covariance matrix (3), as before. Because of the replication at the trial level, we can impose a distribution on the trial-specific parameters

$$\begin{pmatrix} \mu_{Si} \\ \mu_{Ti} \\ \alpha_i \\ \beta_i \end{pmatrix} = \begin{pmatrix} \mu_S \\ \mu_T \\ \alpha \\ \beta \end{pmatrix} + \begin{pmatrix} m_{Si} \\ m_{Ti} \\ a_i \\ b_i \end{pmatrix}, \quad (14)$$

where the second term on the right-hand side of Equation (14) is assumed to follow a zero-mean normal distribution with covariance matrix

$$D = \begin{pmatrix} d_{SS} & d_{ST} & d_{Sa} & d_{Sb} \\ & d_{TT} & d_{Ta} & d_{Tb} \\ & & d_{aa} & d_{ab} \\ & & & d_{bb} \end{pmatrix}. \quad (15)$$

This setting lends itself naturally to introducing the concept of surrogacy at both the trial level and the individual level. We discuss them in turn.

86.2.5 Trial-Level Surrogacy

As indicated before, the key motivation for validating a surrogate endpoint is to be able to predict the effect of treatment on the true endpoint on the basis of the observed effect of treatment on the surrogate endpoint *at the trial level*. It is essential, therefore, to explore the quality of the prediction of the treatment effect on the true endpoint in trial i by (1) information obtained in the validation process based on trials $i = 1, \ldots, N$ and (2) estimation of the effect of Z on S in a new trial $i = 0$. Fitting model (12)–(13) to data from a meta-analysis provides estimates for the parameters and the variance components. Suppose then that the new trial $i = 0$ is considered for which data are available on the surrogate endpoint but not on the true endpoint. We then fit the following linear model to the surrogate outcomes S_{0j}:

$$S_{0j} = \mu_{S0} + \alpha_0 Z_{0j} + \varepsilon_{S0j}. \quad (16)$$

Estimates for m_{S0} and a_0 are

$$\widehat{m}_{S0} = \widehat{\mu}_{S0} - \widehat{\mu}_S, \quad (17)$$
$$\widehat{a}_0 = \widehat{\alpha}_0 - \widehat{\alpha}. \quad (18)$$

Note that such an approach is closely related to leave-one-out regression diagnostics [8,6].

We are interested in the estimated effect of Z on T, given the effect of Z on S. To this end, observe that $(\beta + b_0 | m_{S0}, a_0)$ follows a normal distribution with mean and variance:

$$E(\beta + b_0 | m_{S0}, a_0)$$
$$= \beta + \begin{pmatrix} d_{Sb} \\ d_{ab} \end{pmatrix}^T$$
$$\times \begin{pmatrix} d_{SS} & d_{Sa} \\ d_{Sa} & d_{aa} \end{pmatrix}^{-1}$$
$$\times \begin{pmatrix} \mu_{S0} - \mu_S \\ \alpha_0 - \alpha \end{pmatrix}, \quad (19)$$

$$\text{Var}(\beta + b_0 | m_{S0}, a_0)$$
$$= d_{bb} - \begin{pmatrix} d_{Sb} \\ d_{ab} \end{pmatrix}^T$$
$$\times \begin{pmatrix} d_{SS} & d_{Sa} \\ d_{Sa} & d_{aa} \end{pmatrix}^{-1} \begin{pmatrix} d_{Sb} \\ d_{ab} \end{pmatrix}. \quad (20)$$

In practice, these equations can be used as follows. Using Equations (17) and (18), a prediction can be made using Equation (19), with prediction variance (20). Of course, one has to properly acknowledge the uncertainty resulting from the fact that the parameters in (17)–(18) are not known but merely estimated. This follows from a straightforward application of the iterated expectation law.

A surrogate could thus be called *perfect at the trial level* if the conditional variance (20) were equal to zero. A measure to assess the quality of the surrogate at the trial level is the coefficient of determination

$$R^2_{\text{trial}}$$
$$= R^2_{b_i | m_{Si}, a_i}$$
$$= \frac{\begin{pmatrix} d_{Sb} \\ d_{ab} \end{pmatrix}^T \begin{pmatrix} d_{SS} & d_{Sa} \\ d_{Sa} & d_{aa} \end{pmatrix}^{-1} \begin{pmatrix} d_{Sb} \\ d_{ab} \end{pmatrix}}{d_{bb}}. \quad (21)$$

Similar to the logic in Equations (19) and (20), the conditional model for β_i given μ_{Si} and α_i can be written

$$\beta_i = \theta_0 + \theta_a \alpha_i + \theta_m \mu_{Si} + \varepsilon_i, \quad (22)$$

where expressions for the coefficient $(\theta_0, \theta_a, \theta_m)$ follow from Equations (14) and (15). In case the surrogate is perfect at the trial level ($R^2_{\text{trial}} = 1$), the error term in Equation (22) vanishes and the linear relationship becomes deterministic, implying that β_i equals the systematic component of Equation (22).

This approach avoids problems surrounding the RE, since the relationship between β_i and α_i is studied across a family of units rather than in a single unit. Even if the posited linear relationships do not hold, it is possible to consider alternative regression functions, although one has to be aware of a potentially low power to discriminate between candidate regression functions. By virtue of replication, it is possible to *check* the stated relationships for the treatment effects. Moreover, the use of a measure of association to assess surrogacy is more in line with the adjusted association suggested in the single-trial case.

A key issue when using the proposed meta-analytic framework, and in particular its prediction facility (19), is the coding of the treatment indicators Z_{ij}. While the framework is invariant to coding reversal of *all* treatment indicators at the same time, more caution is needed when the coding of a single trial is considered, such as in Equation (16). In such a case, invariance is obtained only when the fixed effects in Equations (12) and (13) are equal to zero. This issue is intimately linked to the question as to how broad the class of units, to be included in a validation study, can be. Clearly, the issue disappears when the same or similar treatments are considered across units (e.g., in multicenter or multi-investigator studies, or when data are used from a family of related study such as in a single drug development line). In a more loosely connected, meta-analytic, setting, it is important to ensure that treatment assignments are logically consistent. This is possible, for example, when the same standard treatment is compared to members of a class of experimental therapies.

Next, we will show that the adjusted association carries over naturally to the multiunit setting as well.

86.2.6 Individual-Level Surrogacy

We now return to the association between the surrogate and the final endpoints after adjustment for treatment. As before, we need to construct the conditional distribution of T, given S and Z. From (12)–(13) we derive

$$T_{ij}|Z_{ij}, S_{ij} \sim N\{\mu_{Ti} - \sigma_{TS}\sigma_{SS}^{-1}\mu_{Si} \\ + (\beta_i - \sigma_{TS}\sigma_{SS}^{-1}\alpha_i)Z_{ij} \\ + \sigma_{TS}\sigma_{SS}^{-1}S_{ij}; \sigma_{TT} - \sigma_{TS}^2\sigma_{SS}^{-1}\}, \quad (23)$$

which is an extension of (5). Note that

$$\beta_{Si} = \beta_i - \sigma_{TS}\sigma_{SS}^{-1}\alpha_i \quad (24)$$

The association between both endpoints after adjustment for the treatment effect is captured by

$$R^2_{\text{indiv}} = R^2_{\varepsilon_{Ti}|\varepsilon_{Si}} = \frac{\sigma^2_{ST}}{\sigma_{SS}\sigma_{TT}},$$

the squared correlation between S and T after adjustment for both the trial effects and the treatment effect. R^2_{indiv} generalizes ρ^2_Z by adjusting the association both for treatment and for trial. We call a surrogate *perfect at the individual level* if $R^2_{\text{indiv}} = \rho^2_Z = 1$.

Taken together, the R^2 measures allow one to quantify the properties of a candidate surrogate endpoint. In addition, by using a hierarchical model such as (12)–(13), measurement error in the surrogate is automatically taken into account. When a two-stage approximation (i.e., fitting a separate model to each unit in the first stage and fitting a regression on the resulting treatment-effect parameters in the second stage) is used for such a model [5], this is no longer true. Burzykowski et al. [3] illustrate how measurement error can be incorporated in such a context.

A methodological issue is that the choice of an individual-level measure of agreement, such as the R^2, is not universal. We have concentrated on the situation where the true and surrogate endpoints are both normally distributed (in which case the individual-level R^2 follows naturally as the coefficient of determination of the adjusted regression). In practice, endpoints will often be binary, time-dependent, or repeatedly measured over time, and so different association measures will have to be used depending on the problem at hand. Fortunately, in most settings, it is possible to retain an R^2 measure for the trial-level surrogacy. For the individual-level surrogacy, it depends on the type of joint model for the surrogate and true outcome that is used. A bivariate probit model for binary data [19] would produce a tetrachoric correlation, while a Dale model produces odds ratios [19]. For survival endpoints [3], copula-based models have been used, of which the natural association parameters may be quite difficult to interpret. Fortunately, they can often be transformed into Kendall's tau or Spearman's rank correlation.

Acknowledgments. We gratefully acknowledge support from Belgian IUAP/PAI network "Statistical Techniques and Modeling for Complex Substantive Questions with Complex Data."

References

1. Albert, J. M., Ioannidis, J. P. A., Reichelderfer, P., Conway, B., Coombs, R. W., Crane, L., Demasi, R., Dixon, D. O., Flandre, P., Hughes, M. D., Kalish, L. A., Larntz, K., Lin, D., Marschner, I. C., Muñoz, A., Murray, J., Neaton, J., Pettinelli, C., Rida, W., Taylor, J. M. G., and Welles, S. L. (1998). Statistical issues for HIV surrogate endpoints: Point/counterpoint. *Statist. Med.*, **17**, 2435–2462.

2. Begg, C. B. and Leung, D. H. Y. (2000). On the use of surrogate endpoints in randomized trials (with discussion). *J. Roy. Statist. Soc. A*, **163**, 15–28.

3. Burzykowski, T., Molenberghs, G., Buyse, M., Geys, H., and Renard D. (2001). Validation of surrogate endpoints in multiple randomized clinical trials with failure-time endpoints. *Appl. Statist.*, **50**, 405–422.

4. Buyse, M. and Molenberghs, G. (1998). The validation of surrogate endpoints in randomized experiments. *Biometrics*, **54**, 1014–1029.

5. Buyse, M., Molenberghs, G., Burzykowski, T., Renard, D., and Geys, H. (2000). The validation of surrogate endpoints in meta-analyses of randomized experiments. *Biostatistics*, **1**, 49–67.

6. Chatterjee, S. and Hadi, A. S. (1988). *Sensitivity Analysis in Linear Regression*. Wiley, New York.

7. Choi, S., Lagakos, S., Schooley, R. T., and Volberding, P. A. (1993). CD4+ lymphocytes are an incomplete surrogate marker for clinical progression in persons with asymptomatic HIV infection taking zidovudine. *Ann. Intern. Med.*, **118**, 674–680.

8. Cook, R. D. and Weisberg, S. (1982). *Residuals and Influence in Regression*. Chapman & Hall, London.

9. Daniels, M. J. and Hughes, M. D. (1997). Meta-analysis for the evaluation of potential surrogate markers. *Statist. Med.*, **16**, 1515–1527.

10. Ellenberg, S. S. and Hamilton, J. M. (1989). Surrogate endpoints in clinical trials: cancer. *Statist. Med.*, **8**, 405–413.

11. Flandre, P. and Saidi, Y. (1999). Letters to the editor: estimating the proportion

of treatment effect explained by a surrogate marker. *Statist. Med.*, **18**, 107–115.

12. Fleming, T. R. and DeMets, D. L. (1996). Surrogate endpoints in clinical trials: Are we being misled? *Ann. Intern. Med.*, **125**, 605–613.

13. Fleming, T. R., Prentice, R. L., Pepe, M. S., and Glidden, D. (1994). Surrogate and auxiliary endpoints in clinical trials, with potential applications in cancer and AIDS research. *Statist. Med.*, **13**, 955–968.

14. Freedman, L. S., Graubard, B. I., and Schatzkin, A. (1992). Statistical validation of intermediate endpoints for chronic diseases. *Statist. Med.*, **11**, 167–178.

15. Gail, M. H., Pfeiffer, R., van Houwelingen, H. C., and Carroll, R. J. (2000). On meta-analytic assessment of surrogate outcomes. *Biostatistics*, **1**, 231–246.

16. Lin, D. Y., Fleming, T. R., and DeGruttola, V. (1997). Estimating the proportion of treatment effect explained by a surrogate marker. *Statist. Med.*, **16**, 1515–1527.

17. Molenberghs, G., Buyse, M., Geys, H., Renard, D., and Burzykowski, T. (2002). Statistical challenges in the evaluation of surrogate endpoints in randomized trials. *Controlled Clin. Trials*, **23**, 607–625.

18. Prentice, R. L. (1989). Surrogate endpoints in clinical trials: definitions and operational criteria. *Statist. Med.*, **8**, 431–440.

19. Renard, D., Geys, H., Molenberghs, G., Burzykowski, T., and Buyse, M. (2002). Validation of surrogate endpoints in multiple randomized clinical trials with discrete outcomes. *Biometr. J.*, **44**, 1–15.

20. Volberding, P. A., Lagakos, S. W., Koch, M. A., Pettinelli, C., Myers, M. W., Booth, D. K., Balfour, H. H., Reichman, R. C., Bartlett, J. A., and Hirsch, M. S. (1990). Zidovudine in asymptomtic human immunodeficiency virus infection: A controlled trial in persons with fewer than 500 CD4-positive cells per cubic millimeter. *New Engl. J. Med.*, **322**, 941–949.

87 Survival Analysis

Per Kragh Andersen and Michael Væth

87.1 Survival Data

Survival analysis is concerned with statistical models and methods for analyzing data representing life times, waiting times, or more generally times to the occurrence of some specified event. Such data, denoted as *survival data*, can arise in various scientific fields including medicine, engineering, and demography. In a clinical trial the object of the study is the comparison of survival times with different treatments in some chronic disease; in an engineering reliability experiment, a number of items could be put on test simultaneously and observed until failure; a demographer could be interested in studying the length of time that a group of workers stay in a particular job. Thus, survival data are basically nothing but realizations of nonnegative random variables but the statistical inference is usually complicated by presence of incomplete observations. The most common form of incomplete survival data is right censoring which reflects that limitations in time and other restrictions on data collection prevent the experimenter from observing the event in question for every individual or item under study. Rather, for some individuals only partial information will be available about their survival time, namely, that it exceeds some observed censoring time, whereas the actual survival time is not observed or cannot be observed. The survival analysis methodology aims at estimating aspects of the distribution of the complete data from observation of the incomplete data. The mechanisms generating the incompleteness of the data must satisfy certain conditions of "independent censoring" to ensure that such inference is feasible. The presentation here will mainly consider incomplete data formed by right censoring and an adequate model of the censoring mechanism will depend on aspects of the particular problem to be described and the design of the study. A well-established way of formulating conditions for independent censoring is based on the following two assumptions [7, p. 194]:

1. Given all that has happened up to time t, the failure mechanisms for different individuals act independently over the interval $[t, t + dt)$.

2. For an individual alive and *uncensored* at t, the conditional probability of failing in $[t, t + dt)$ given all that has happened up to time t coincides with the conditional probability of failing in $[t, t + dt)$ given survival up to time t.

(For alternative formulations of independent censoring, see e.g. [3, p. 139]). An implication of independent censoring (assumption 2) is the exclusion of censoring mechanisms withdrawing individuals from risk when they appear to have particularly high or low risk of failure. A simple censoring model, often applicable in biomedical contexts and fulfilling both assumptions, is random censorship in which survival times and censoring times are stochastically independent.

The object of a survival analysis is to draw inferences about the distribution of the survival times T. This distribution can be characterised by the survival distribution function (sdf),

$$S(t) = \Pr[T > t], \qquad t \geq 0,$$

or equivalently (provided that S is differentiable) by the hazard rate function (hrf)

$$\lambda(t) = -d(\log S(t))/dt = f(t)/S(t), \qquad t \geq 0, \quad (1)$$

where $f(t)$ is the probability density function (pdf), or the cumulative hazard rate function (chrf)

$$\Lambda(t) = \int_0^t \lambda(u)du, \qquad t \geq 0, \quad (2)$$

Statistical models for continuous survival data can be specified via any of these quantities but it has become customary to formulate these models in terms of the hazard rate function. In particular, the influence of covariates on the survival time is often specified via the hrf.

Throughout it will be assumed that the censoring mechanism is independent and that the available data are of the form of times of observation t_1, ..., t_n and indicators d_1, ..., d_n. The latter are used to distinguish between uncensored and censored observations, i.e. $d_i = 1$ if t_i is an actual survival time and $d_i = 0$ if t_i is a censored observation. For methods where $S(t)$ is discontinuous, corresponding to survival distributions with discrete components, see [7, p.46]. Methods for dealing with grouped survival data do exist; the classical actuarial life table is the main example. Such methods are applicable in situations where the sample size n is so large that it is not feasible to record the exact survival and censoring times but only to which of a number of prespecified time intervals they belong. They also apply in situations where individuals are only observed at prespecified follow-up times so that it is only known in which interval between successive follow-up times that the event under study has occurred. A somewhat extreme example would be to analyze only the status alive/dead after a single time interval. This is the situation in the analysis of quantal response data in bioassay.

87.2 Nonparametric Survival Models

Presence of censored observations implies that classical nonparametric methods based on ranks are not directly applicable. In particular, standard graphical procedures such as an empirical cdf or a histogram cannot be used with censored data. To present the required modifications, it is convenient to consider the following quantities defined from the basic observations $(t_i, d_i), i = 1, \ldots, n$:

$$N(t) = \#\{i = 1, \ldots, n : t_i \leq t, d_i = 1\}$$

and

$$Y(t) = \#R(t),$$

where

$$R(t) = \{i = 1, \ldots, n : t_i \geq t\}$$

is the *risk set* at t. Thus, $N(t)$ is the *counting process* counting the number of failures before or at t and $Y(t)$ is the number of individuals *at risk* at $t-$ (that is, just before t). The nonparametric methods to be discussed in this section have all been reviewed in [3, Ch. 4-5].

Estimation. The sdf can be estimated by the Kaplan-Meier or product limit estimate [8]

$$\hat{S}(t) = \prod_{t_i \leq t}\left[1 - \frac{\Delta N(t_i)}{Y(t_i)}\right],$$

where $\Delta N(t) = N(t) - N(t-)$ is the number of failures at t. Similarly, the chrf defined in (2) can be estimated by the Nelson-Aalen estimate [1]

$$\hat{\Lambda}(t) = \sum_{t_i \leq t} \frac{\Delta N(t_i)}{Y(t_i)}.$$

Conditions (including independent censoring) can be found under which $\hat{S}(t)$ and $\hat{\Lambda}(t)$ behave asymptotically ($n \to \infty$) as normal processes, and approximate standard errors can be calculated as

$$\hat{\sigma}(\hat{\Lambda}(t))$$
$$= \left[\sum_{t_i \leq t} \frac{\Delta N(t_i)}{Y(t_i)(Y(t_i) - \Delta N(t_i) + 1)}\right]^{1/2},$$

$$\hat{\sigma}(\hat{S}(t)) = \hat{S}(t)\hat{\sigma}(\hat{\Lambda}(t)).$$

Estimates $\hat{\lambda}(t)$ of the hrf can be obtained, e.g., by smoothing $\hat{\Lambda}(t)$ using some kernel function.

Hypothesis Tests. Nonparametric comparison of the survival distributions in $k = 2$ groups of individuals can be based on test statistics of the form

$$Z = \sum_{i=1}^{n}\left[K(t_i)\left\{\frac{\Delta N_2(t_i)}{Y_2(t_i)} - \frac{\Delta N_1(t_i)}{Y_1(t_i)}\right\}\right],$$

where $K(t)$ is a stochastic "weight" process. Under suitable conditions $ZV^{-1/2}$, where

$$V = \sum_{i=1}^{n}\left\{K^2(t_i)\frac{\Delta N_1(t_i) + \Delta N_2(t_i)}{Y_1(t_i)Y_2(t_i)}\right\},$$

has an asymptotic standard normal distribution (as $n \to \infty$). Different choices of K lead to test statistics discussed in the survival data literature. For example, the choice $K = Y_1Y_2/Y$ yields the logrank test where Z reduces to the difference between the observed $O_j = N_j(\infty)$ and the "expected" number of failures in group $j(= 1$ or $2)$:

$$E_j = \sum_{i=1}^{n}\left[\frac{Y_j(t_i)}{Y(t_i)}\Delta N(t_i)\right].$$

Sometimes it is of interest to compare an observed survival distribution with an sdf $\exp(-\Lambda^*(t))$ that is known, for example, from the life tables for some reference population. Conditions can be found where a test statistic of the form

$$\sum_{i=1}^{n} K(t_i)\frac{\Delta N(t_i)}{Y(t_i)} - \int_0^\infty K(u)\,d\Lambda^*(u)$$

has an asymptotic standard normal distribution (as $n \to \infty$) when properly normalized by an approximate standard deviation. As for the two sample situation, $K(t)$ is a stochastic "weight" process and special choices of K yield various previously discussed test statistics.

Nonparametric comparison of the survival distributions in $k \geq 2$ groups of individuals can be based on test statistics of the form

$$Z_j = \sum_{i=1}^{n}\bigg[K(t_i)Y_j(t_i)$$
$$\times \left\{\frac{\Delta N_j(t_i)}{Y_j(t_i)} - \frac{\Delta N(t_i)}{Y(t_i)}\right\}\bigg],$$
$$j = 1,\ldots, k.$$

Defining the $k \times k$ matrix $\mathbf{V} = (V_{jl})$ by

$$V_{jl} = \sum_{i=1}^{n}\bigg[K^2(t_i)\frac{Y_j(t_i)}{Y(t_i)}$$
$$\times \left\{\delta_{jl} - \frac{Y_l(t_i)}{Y(t_i)}\right\}\Delta N(t_i)\bigg],$$

where δ_{jl} is the Krönecker delta, conditions can be found under which $\mathbf{Z}'\mathbf{V}^-\mathbf{Z}$ has an asymptotic chi-squared distribution with $k-1$ degrees of freedom (as $n \to \infty$). Here $\mathbf{Z} = (Z_1, \ldots, Z_k)'$ and $\mathbf{V}^-$ is a generalized inverse of $\mathbf{V}$.

87.3 Parametric Survival Models

Survival data are often skewed so the normal distribution plays a less prominent role in the analysis of such data. Popular parametric survival models include the exponential distribution and the Weibull distribution. The hrf has a simple form for these distributions and parameter estimation is simplified because a closed form expression for the survival function is available. The basic properties of these distributions are reviewed below, the emphasis being on the aspects of the distributions that are important in survival analysis. Moreover, survival distributions with piecewise-constant hazard rate and the log-normal distribution will be briefly discussed. For further reading, see [9].

The simplest lifetime distribution, the exponential distribution, is characterized by a constant hrf $\lambda(t) = \lambda$ for all $t \geq 0$, and the sdf has the form $S(t) = \exp(-\lambda t)$. Although the assumption of a constant hazard rate is restrictive, the exponential distribution was the first survival model to become widely used, partly due to its computationally attractive features. The appropriateness of the exponential distribution for a given set of survival data may be checked by plotting $\hat{\Lambda}(t)$ or equivalently $-\log \hat{S}(t)$ versus t. Such a plot should approximate a straight line through the origin. With censored data from an exponential distribution maximum likelihood estimation is particularly simple and the maximum likelihood estimate becomes the rate $\hat{\lambda} = D/\sum t_i$, where $D = N(\max(t_i)) = \sum_i d_i$ is the total number of events. Improved flexibility without sacrificing the simplicity of the computations can be obtained by assuming that the hazard rate is piecewise constant on a number of pre-specified time intervals, e.g. 1 or 5 year intervals. With censored data from this distribution the maximum likelihood estimate of the hazard rate λ_j on the jth interval becomes $\hat{\lambda}_j = D_j/T_j$, where D_j is the number of events in the jth interval and T_j is the sum of the time periods each individual spends in the jth interval.

The Weibull distribution is probably the most widely used parametric survival model in technical as well as biomedical applications. The Weibull model provides a fairly flexible class of distributions and includes the exponential distribution as a special case. The hrf and the sdf have the form

$$\lambda(t) = \lambda \rho (\lambda t)^{\rho-1}, \quad t \geq 0,$$

and

$$S(t) = \exp(-(\lambda t)^\rho), \quad t \geq 0, \quad (3)$$

where $\lambda > 0$ is an inverse scale parameter and $\rho > 0$ is a shape parameter. The hrf is monotonically increasing if $\rho > 1$, monotone decreasing if $0 < \rho < 1$, and constant if $\rho = 1$. The Weibull distribution appears as one of the asymptotic distributions of the smallest extreme and this fact motivates its use in certain applications. From (3) it follows that

$$\log \Lambda(t) = \rho \log \lambda + \rho \log t,$$

and the appropriateness of the Weibull model can therefore be checked by plotting $\log \hat{\Lambda}(t)$ versus $\log t$.

The lognormal distribution has also been used in survival analysis, although no closed form expressions are available for $S(t)$ and $\lambda(t)$. The distribution is most easily specified through $\log T$ having a normal distribution with mean μ and variance σ^2, say, when T is a lognormal variate. With this parametrization the lognormal sdf is

$$S(t) = 1 - \Phi\left(\frac{\log t - \mu}{\sigma}\right),$$

where Φ is the cdf of the standard normal distribution. The lognormal hrf has the value 0 at $t = 0$, increases to a maximum, and then decreases with a limiting value of 0 as t tends to infinity. This behavior may be unattractive in certain applications.

Inference. For the parametric survival models, including the Weibull model and the lognormal model, maximum likelihood estimation and large-sample likelihood methods are the inference procedures generally used, as the presence of censoring makes exact distributional results complicated in most situations. In the case of no censoring or so-called type II censoring, several alternative procedures, including methods giving exact distributional results for the parameter estimates, are available for the exponential and the Weibull models. A review of these methods is given in [9].

For the class of independent censoring schemes the likelihood function becomes (apart from a constant of proportionality)

$$L(\boldsymbol{\theta}) = \prod_{i=1}^{n} \lambda(t_i; \boldsymbol{\theta})^{d_i} S(t_i; \boldsymbol{\theta}), \quad (4)$$

where $\boldsymbol{\theta}$ is the vector of unknown parameters to be estimated. From (4) it is apparent that survival models admitting a closed form expression for the sdf are computationally attractive and in general an iterative procedure is required to solve the likelihood equations. In most cases, the standard errors of the parameter estimates have to be estimated from the observed information matrix

$$-\left(\left(\frac{\partial^2 \log L(\boldsymbol{\theta})}{\partial \theta_i \partial \theta_j}\bigg|_{\boldsymbol{\theta}=\hat{\boldsymbol{\theta}}}\right)\right),$$

since calculation of the expected information requires detailed knowledge of the distribution of the censoring times.

The theoretical justification of the asymptotic normality of the maximum likelihood estimate and the limiting χ^2 distribution of the likelihood ratio statistic with censored data has been established in many special cases. A unified approach to the asymptotic theory of maximum likelihood estimation for parametric survival models with independent censoring has been given in [3, Ch. 6].

87.4 Regression Models

Application of survival analysis usually involves evaluation of the dependence of the survival time on one or several concomitant variables ("covariates") measured on each individual. Regression modeling is the standard approach to identify and evaluate such prognostic variables. Several regression models, which allow for right censoring, have been developed, the most important being Cox's proportional hazards regression model. In most of these regression models (including Cox's regression model) the influence a prognostic variable on the survival time is modeled by letting the hrf depend on the prognostic variables.

The Cox Regression Model. Cox's regression model [5], is a semi-parametric model in which the dependence on time is described by an unspecified reference hazard function and the influence of a prognostic variable is modeled by modifying this hazard function by a multiplier independent of time. The hrf for a subject with covariates $\mathbf{x} = (x_1, \ldots, x_p)$ can be written as

$$\lambda(t, \mathbf{x}) = \lambda_0(t) \exp(\boldsymbol{\beta}'\mathbf{x}), \quad (5)$$

where $\boldsymbol{\beta}$ is a p-dimensional vector of unknown regression coefficients reflecting the effects of $\mathbf{x}$ on survival and $\lambda_0(t)$, the baseline hazard rate function, is an unspecified function of time. The basic features of the model (5) are the assumption of proportional hazards, that is, $\lambda(t, \mathbf{x}_1)/\lambda(t, \mathbf{x}_2)$ does not depend on t, and the assumption of log linearity in $\mathbf{x}$ of the hrf. The parameters of the model are most easily interpreted as log hazard ratios. In particular, for a prognostic factor x_1 with two categories exposed ($x_1 = 1$) and unexposed ($x_1 = 0$), $\exp(\beta_1)$ becomes the hazard ratio between an exposed individual and an unexposed individual, who is otherwise identical to the exposed individual. The statistical analysis of Cox's regression model is based on the partial likelihood function which in the case of no ties between the survival times is given by

$$L(\boldsymbol{\beta}) = \prod_{i=1}^{n} \left(\frac{\exp(\boldsymbol{\beta}'\mathbf{x}_i)}{\sum_{j \in R(t_i)} \exp(\boldsymbol{\beta}'\mathbf{x}_j)} \right)^{d_i}. \quad (6)$$

Here, $\mathbf{x}_i$ is the covariate vector of the individual with observation time t_i and $R(t_i)$ is the risk set at t_i. Estimates of the parameters $\boldsymbol{\beta}$ are obtained by maximizing $L(\boldsymbol{\beta})$ and the usual type of large-sample likelihood methods also apply to partial likelihoods when censoring is independent and certain regularity assumptions are satisfied; see, e.g. [3, Ch. 7]. The chrf $\Lambda_0(t)$ corresponding to $\lambda_0(t)$ can be estimated by

$$\hat{\Lambda}_0(t) = \sum_{t_i \leq t} \frac{d_i}{\sum_{j \in R(t_i)} \exp(\hat{\boldsymbol{\beta}}'\mathbf{x}_j)}$$

in the case of no tied survival times. In applications, the parameter estimates are usually presented as hazard ratios with 95% confidence intervals. These are obtained as the exponential function of $\hat{\beta}_j$ and of the 95% confidence limits for β_j. Regression diagnostics for Cox's regression model also include methods for checking the validity of the proportional hazards assumption. If a particular prognostic variable influences the survival time in a way that can not be described by proportional hazards one may consider a stratified Cox's regression model. This model includes separate, not necessarily proportional, baseline hazard functions for each category of a prognostic variable, whereas the dependence on the remaining prognostic variables is described by proportional hazards.

Cox's regression model can be generalized to allow for time-dependent covariates. This is an important extension that increases the flexibility of the model considerably. Three broad classes of time-dependent covariates can be identified. As an alternative to a Cox regression model stratified on a covariate, x, one may add a known function of time, e.g. $x \cdot \log(t)$, as a covariate in order to describe an influence of x acting on survival in a non-proportional way. Another type of time-dependent covariate is used to model the influence of prognostic variables for which information is updated during follow-up. Rather than using the value available at entry, one may want to include the most recent value in the model. Finally, one may want the model to reflect events that may be observed during follow-up. This can be done by introducing a time-dependent binary covariate that takes the value 0 until the event occurs and the value 1 thereafter. Comprehensive reviews of the model are given by [7, Ch. 4-6] and [3, Ch. 7]. It should be noted that if random time dependent covariates (e.g., of the second and third type just described) are included in the model for the hrf then the relation $S(t) = \exp(-\Lambda(t))$ implied by (1) no longer holds. This is because the cumulative survival probability will also depend on the random future development of the time dependent covariate. As a consequence, effects of time dependent covariates should be interpreted with caution.

Parametric Regression Models. Parametric proportional hazards models are obtained by replacing the arbitrary function $\lambda_0(t)$ by a function belonging to some parameterized family and then statistical inference is usually based on a likelihood function analogous to (4). The simplest examples arise when $\lambda_0(t)$ is replaced by a constant (exponential regression) or by a power function of time (Weibull regression). Using a piecewise constant hrf, the likelihood function becomes proportional to a likelihood function obtained by formally treating the number of events in each interval as independent Poisson variates with mean equal to the hazard rate multiplied by the time at risk. This model can therefore be analyzed with software for generalized linear models and the model is usually called the Poisson regression model. If all covariates are categorical, the data can be aggregated to a multi-way table of counts and person-years at risk. This may be a useful approach for analyzing data from large cohort studies.

Additive Hazard Models. As an alternative to the multiplicative structure in Cox's regression model, additive hazard rate models have been developed, the most flexible one being *Aalen's additive hazard rate model*, e.g. [3, Ch. 7], [2, Ch. 4] and [10, Ch. 5]. In this model, the hrf for a subject with covariates $\mathbf{x} = (x_1, \ldots, x_p)$ (which may be time dependent) is given by

$$\lambda(t, \mathbf{x}) = \beta_0(t) + \boldsymbol{\beta}(t)'\mathbf{x}.$$

Here, $\beta_0(t)$ is an unspecified baseline hrf and $\beta_j(t), j = 1, \ldots, p$, are unspecified regression functions quantifying the influence of the covariates. Thus, for a binary covariate, x_j, taking values 0 or 1, the value at time t of the corresponding regression function, $\beta_j(t)$, describes the hazard difference at t between subjects with $x_j = 1$ and $x_j = 0$ and all other covariates identical. Both the cumulative baseline hazard and the cumulative regression functions, $B_j(t) = \int_0^t \beta_j(u)du, j = 0, 1, \ldots, p$ can be estimated non-parametrically by "generalized Nelson-Aalen estimators" whose increments at a failure time t_i is obtained by solving a least squares equation at that time. Asymptotic inference including standard errors and test statistics for linear hypotheses of the form $\mathbf{c}'\boldsymbol{\beta}(t) = 0$ are also available [2, Ch. 4]. A simple special case is when all regression functions are constant:

$$\lambda(t, \mathbf{x}) = \beta_0(t) + \boldsymbol{\beta}'\mathbf{x}$$

but where the baseline hazard is still unspecified. Also semi-parametric versions where some regression functions are constant and others are left unspecified have been studied [10, Ch. 5].

The Accelerated Failure Time Model. The *accelerated failure time regression model* is conveniently introduced via the logarithm of the survival time T. The model specifies a linear relationship

$$\log T = \boldsymbol{\gamma}'\mathbf{x} + \sigma W, \qquad (7)$$

where σ is a scale parameter and W is a random variable giving the error. Various choices of the error distribution lead to regression versions of the parametric survival models already discussed. Specifically, if W has an extreme value distribution (Gumbel distribution), a Weibull regression model is obtained, the exponential regression being a special case corresponding to $\sigma = 1$. A log normal regression model is obtained if W is a standard normal variate. Parametric statistical inference for the model (7) based on a likelihood function analogous to (4) has been described by [7, Ch. 3] and [9, Ch. 5]. Nonparametric analysis of the model (7) has also been developed [3, Ch. 7] and [10, Ch. 8].

The accelerated failure time regression model can alternatively be formulated via the HRF, viz.,

$$\lambda(t, \mathbf{x}) = \lambda_u[t\exp(-\boldsymbol{\gamma}'\mathbf{x})]\exp(-\boldsymbol{\gamma}'\mathbf{x}),$$

where $\lambda_u(\cdot)$ denotes the HRF of the random variable $U = \exp(\sigma W)$. The only such models that are also proportional hazards models are the exponential and the Weibull regression models.

87.5 Some Extensions

Delayed Entry. In a clinical trial, patients are typically followed from some well-defined event, e.g. diagnosis, treatment or randomization, until the endpoint occurs or the patient is censored alive. The relevant time scale is time since entry and all patients are therefore followed from time 0. This is, however, not always the case with other types of study designs. Individuals with a chronic disease may be identified on a given date and then followed forward in time until death. The relevant time scale could here be time since diagnosis, but patients are not followed from the date of diagnosis. To be included in the sample a patient must survive until the date that the sample is identified. This type of incomplete observation is denoted left truncation or delayed entry, because entry occurs after time 0 and patients are excluded if they experience the event before entry. The statistical methods described above can be modified to allow for both left truncation and right censoring. The formulas for the non-parametric methods and the partial likelihood function used in Cox's regression analysis are still valid, when the risk set $R(t)$ is defined as

$$R(t) = \{i = 1, \ldots, n : v_i < t \leq t_i\},$$

where v_i and t_i are the entry time and the exit time of patient i. For the parametric methods the likelihood function becomes

$$L(\boldsymbol{\theta}) = \prod_{i=1}^{n} \lambda(t_i; \boldsymbol{\theta})^{d_i} S(t_i; \boldsymbol{\theta}) / S(v_i; \boldsymbol{\theta}).$$

Some studies may involve several time variables, e.g. age at diagnosis, current age, time since first diagnosis, time since last episode, and identification of the most important time scale may be part of the purpose of the analysis. Statistical methods in survival analysis often require a primary time scale and other time scales may then be introduced as time dependent covariates. However, with Poisson regression several time variables can be introduced and studied simultaneously without selection of a primary time scale.

Interval Censoring. Incomplete observation may also have the form of *interval censoring*. The survival time is here only known to lie in some interval. Grouped survival data are a special case of interval censoring data for which the possible intervals are the same for all individuals. In general, interval censoring requires special methodology, however, parametric inference based on likelihood functions is easily developed for interval censored data. If the survival time of an individual is known to lie in the interval from L_i to R_i, this person contributes with a factor of $S(L_i) - S(R_i)$ to the likelihood function. Non-parametric inference for interval censored data has also been developed, e.g. [11].

Multi-State Models, Competing Risks. The models mentioned so far can be thought of as describing transitions from one state "alive" to another state "death", the transition rate being the HRF $\lambda(t)$. Many of the methods described can be extended to situations where more than one type of transition can occur, so-called "multi-state models". The simplest such extension is the *competing risks* model where the state "death" is split into, say, a states: "death from cause 1," ..., "death from cause a." In this model, there are a transition intensities, so-called "cause-specific HRFs", $\lambda_j(t), j = 1, \ldots, a$ corresponding to the different causes of death, each of which may be analysed using, e.g. hazard regression models [7, Ch. 8]. In the competing risks model there is still one SDF,

$$S(t) = P(T > t)$$
$$= \exp\left(-\sum_{j=1}^{a} \int_0^t \lambda_j(u) du\right)$$

giving the probability of surviving beyond time t. The CDF, $F(t) = 1 - S(t)$, however, is split into a sum of a "cumulative incidences", $F(t) = \sum_{j=1}^{a} F_j(t)$, where

$$F_j(t) = P(T \leq t, D = j)$$
$$= \int_0^t S(u-) \lambda_j(u) du.$$

Here, D is the cause of death indicator. The cumulative incidence for cause $j, j = 1, \ldots, a$ is the transition probability from the initial state "alive" to the final state "death from cause j", that is, it gives the probability of failure from cause j before time t. Note that $F_j(t)$ (via $S(u)$) depends on the cause-specific hazards for *all* causes and it may be estimated non-parametrically by plugging-in Nelson-Aalen estimators for the cumulative cause-specific HRFs and the Kaplan-Meier estimator for the SDF [3, Ch. 4]. Direct regression models for the cumulative incidences are also available [10, Ch. 10]. In other multi-state models the transition probabiliities are more complicated functions of the transition intensities [3, Ch. 4].

Multivariate Survival Data, Recurrent Events. In the survival data models described in previous sections, an implicit assumption has been *independence* between the n observations. However, *clustering* may appear both in the form of "serial" and "parallel" survival data. Serial data, or recurrent events, corresponds to the situation where, for the same subject, the event of interest could occur repeatedly, e.g. episodes of a non-chronic disease. Examples of parallel data could be survival times observed in families, e.g. twins, or in medical centers.

In both situations, independence within clusters is questionable while independence between clusters may still be a reasonable assumption and the survival data inference needs to address this potential intra-cluster dependence. As for other types of outcome variables, e.g. quantitative or binary data, there are basially two types of approach for addressing the dependence. One, *random effects models*, explicitly describes the dependence by letting the outcome for subjects from the same cluster share common random effects while the other, *marginal models*, treats the intra-cluster association as a nuisance parameter and aims at estimating model parameters (and standard errors) in a way which is robust to the lack of independence. Both types of approach have been studied for both serial and parallel survival data.

In random effects models for survival data the HRFs for subjects, $j = 1, \ldots, n_i$ from the same cluster, i, are typically related by sharing a common, unobservable random cluster-specific "frailty" Z_i. Such models, known as frailty models, specify the HRF for individual j in cluster i by

$$\lambda_{ij}(t) = z_i \lambda_j(t), \qquad (8)$$

when $Z_i = z_i$ and $\lambda_j(t)$ is the HRF for an individual with unit frailty [6, Ch. 7-10]. Models of the form (8) are analogous to variance components models for quantitative outcomes and inference typically assumes that $Z_1, \ldots, Z_m$, where m is the number of clusters, are independent and identically distributed with a known type of distribution. For mathematical convenience, the gamma distribution has played a prominent role for frailty models.

In marginal models for clustered, parallel survival data, on the other hand, a model for the HRF of the marginal distribution for the survival time of subject j in cluster i is typically specified using one of the hazard based models described above, such as the Cox regression model. Parameters are then estimated using generalized estimating equations, ("GEE"), and standard errors, robust to lack of independence within clusters, are obtained using "sandwich-type" formulas [7, Ch. 10]. Marginal models for serial data [4, Ch. 3] are typically specified for the mean number of events, $E(N_i(t))$, for subject i, where $N_i(t)$ is the process counting the number of recurrent events for i before time t.

87.6 An Example

In the period 1964-1973, 205 patients with malignant melanoma had radical surgery performed at the Department of Plastic Surgery, University Hospital of Odense, Denmark [3, Ch. 1]. At the end of the follow-up period (1 January 1978), 57 of the patients had died from the disease and 14 patients had died from other causes. Figure 1 shows a stacked plot of the cumulative incidences for these two causes of death. The plot shows that after five years the estimated risk of dying from malignant melanoma is 22.4%, the risk of dying from other causes is estimated to 4.4%, and the chance of being alive is estimated to 73.2%. Among the variables registered in the study were the sex of the patient and the age at entry, tumor thickness, and ulceration (absent or present).

The first analysis considers mortality from all causes with survival time measured from the date of operation. Figures 2-3 show the Kaplan-Meier estimates of the survival functions in groups defined by the sex of the patients and by ulceration. Table 1 shows the results of two logrank analyses comparing the survival distributions in the groups. It is seen that, when considered separately, both sex and ulceration have a significant influence on survival. Cox regression analyses of each covariate separately confirm these findings. The estimated hazard ratios with 95% confidence intervals for sex (male relative to female), and ulceration (present relative to absent) were 1.94 (1.15; 3.26) and 4.36 (2.44; 7.77), respectively. Moreover, the estimated hazard ratios from single factor Cox regression analyses of age (per 10 years) and tumor thickness included as ln(thickness in mm) were 1.21 (1.02; 1.44) and 2.02 (1.55;2.64).

The four variables were included simultaneously in a Cox regression model

$$\lambda(t) = \lambda_0(t) \times \exp(\beta_1 x_1 + \beta_2 x_2 + \beta_3 x_3 + \beta_4 x_4).$$

The estimates in this model are presented in Table 2. The top panel shows the estimated hazard ratios, obtained as $\exp(\hat{\beta}_j)$ and the corresponding 95% confidence interval. A p-value for the test of the hypothesis of no association between the covariate and survival, i.e. $\beta_j = 0$, is also shown. The estimates in Table 2 are mutually adjusted, and a comparison of these estimates with those obtained from separate analyses of each variable shows that the adjusted hazard ratio for males is considerably smaller and no longer statistically significant. This reflects that the simple analysis does not adjust for the fact that male patients have thicker tumors than females. The excess risk associated with ulceration is also reduced considerably, but still statistically significant.

Cox regression analyses of the cause-specific hazards, either for death from the disease or for death from other causes, provide further insight in the association between the four predictors and the mortality of the patients. The center panel of Table 2 shows the result of a Cox regression analysis of mortality from malignant melanoma. Compared to the hazard ratios obtained from the analysis of total mortality (top panel), the estimates for tumor thickness

Table 1: Single-factor logrank analyses of total mortality.

Variable	Number of patients	Deaths Observed	Deaths Expected	logrank test (p value)
Sex of patient				
Female	126	35	46.27	7.90
Male	79	36	24.73	(0.005)
Ulceration				
Absent	115	23	44.46	27.87
Present	90	48	26.54	(<0.0001)

Table 2: Hazard ratios (HR) with 95% confidence intervals estimated by Cox regression analysis of total mortality (top), mortality from disease (center), and mortality from other causes (bottom).

	All causes		
Variable	HR	95% CI	p value
Male	1.45	0.90;2.33	0.123
Age (per 10 years)	1.24	1.06;1.44	0.006
ln(thickness in mm)	1.54	1.13;2.09	0.006
Ulceration	2.23	1.28;3.87	0.004
	Malignant melanoma		
Variable	HR	95% CI	p value
Male	1.44	0.85;2.44	0.179
Age (per 10 years)	1.12	0.95;1.32	0.164
ln(thickness in mm)	1.74	1.23;2.48	0.002
Ulceration	2.57	1.37;4.83	0.003
	Other causes		
Variable	HR	95% CI	p value
Male	1.41	0.48;4.15	0.529
Age (per 10 years)	2.12	1.37;3.27	0.001
ln(thickness in mm)	0.96	0.52;1.80	0.908
Ulceration	1.28	0.37;4.36	0.697

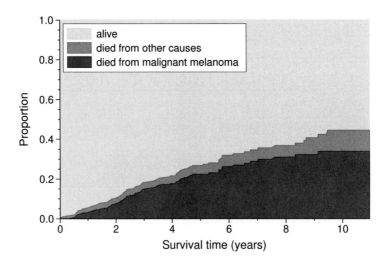

Figure 1: Stacked plot of the cumulative incidences for death from malignant melanoma and death from other causes.

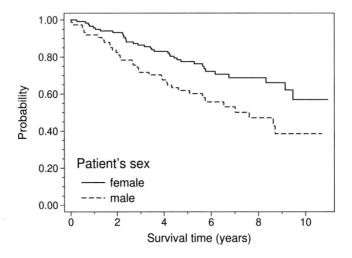

Figure 2: Kaplan–Meier estimates for male and female patients with malignant melanoma.

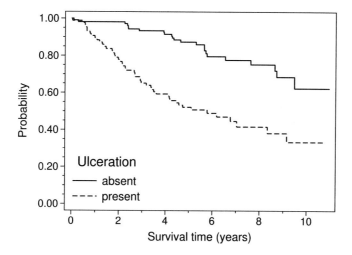

Figure 3: Kaplan–Meier estimates for malignant melanoma patients with and without ulceration.

and ulceration are increased, whereas the estimate for sex is essentially unchanged and the estimate for age is reduced and no longer statistically significant. The bottom panel of Table 2 presents the results of a Cox regression analysis of the mortality from other causes. These results should be cautiously interpreted, as the analysis is based on only 14 deaths. In general it is not advisable to study the simultaneous effect of four variables based on so little data. In the analysis of mortality from other causes the disease-specific covariates (thickness and ulceration) seem not important and the hazard ratio for sex is similar to the previous estimates. Age, on the other hand, is significantly associated with mortality from other causes.

The estimates from the analysis of mortality from other causes are asymptotically independent of the estimates from the analysis of mortality from malignant melanoma. A Wald test based on the difference between the estimated regression coefficients can be used to compare the effect of a given covariate on each cause of death. For age the hazard ratios from the two cause-specific analyses are significantly different ($p = 0.007$), but for the other three covariates this is not the case.

Acknowledgments. This research was supported by grant R01-54706-12 from the National Cancer Institute.

References

1. Aalen, O. O. (1978). *Ann. Statist.*, **6**, 701–726.
2. Aalen, O. O., Borgan, Ø., Gjessing, H.K. (2008). *Survival and Event History Analysis. A Process Point of View*. Springer, New York.
3. Andersen, P. K., Borgan, Ø., Gill, R. D., and Keiding, N. (1993). *Statistical Models Based on Counting Processes*. Springer, New York.
4. Cook, R. J. and Lawless, J. F. (2007). *The Statistical Analysis of Recurrent Events*. Springer, New York.
5. Cox, D. R. (1972). *J. Roy. Statist. Soc. B*, **34**, 187–220.
6. Hougaard, P. (2000). *Analysis of Multivariate Survival Data*. Springer, New York.

7. Kalbfleisch, J. D. and Prentice, R.L. (2002). *The Statistical Analysis of Failure Time Data (2nd ed.)* Wiley, New York.

8. Kaplan, E. L. and Meier, P. (1958). *J. Am. Statist. Assoc.*, **53**, 457–481.

9. Lawless, J. F. (2002). *Statistical Models and Methods for Lifetime Data (2nd ed.)* Wiley, New York.

10. Martinussen, T. and Scheike, T. H. (2006). *Dynamic Regression Models for Survival Data.* Springer, New York.

11. Sun, J. (2006). *The Statistical Analysis of Interval-Censored Failure Time Data.* Springer, New York.

Index

A-optimality, 788
Aalen plots, 2
Aalen's nonparametric linear regression model, 564
absolute risk, 184, 750
accelerated failure time model, 1, 457
accelerated failure time regression model, 959
accelerated life model, 323
acceptance sampling plans, 735
activity profile, 927
actuarial hazard function, 393
actuarial survival rates, 393
adaptive group sequential tests, 384
adaptive learning, 574
adaptive weights smoothing procedure, 98
additive genetic variance, 874
additive logistic distribution, 774
additive model, 233
additive neutrality, 773
additive risk model, 1, 4, 752, 756
additive–multiplicative hazards model, 5
adjusted p value, 549
adjusted association, 945
adjusted chi-square statistics, 154
adverse drug reaction, 451
affinal model, 263
age incidence curves, 288
age-dependent incidence rates, 517
age-related macular degeneration, 91
age-specific fertility rate, 649
aggregate measures, 157
aggregation, 9, 10
Agresti–Coull confidence interval, 723
AIDS, 15
AIDS projections, 275
AIDS surveillance, 274
Akaike information, 852
Akaike's information criterion, 896

all-or-none compliance, 37
allele frequency matching, 84
allele frequency-matching probability, 87
allele-sharing methods, 897
alleles, 893
allelic disequilibrium, 804
analog-to-digital converter (ADC), 398
analyses, 776
analysis of variance, 219, 243
ANOVA-based tests, 797
ANOVA-type methodology, 472
approximate Bayesian bootstrap, 430
archimedean copula, 350
array, 795
artificial neural networks, 405
ascertainment sampling, 43
ascertainment sampling procedures, 46
assessment bias, 47
assessment of harms, 49
associate studies, 900
asymptotic minimax, 493
attributable risk, 303
autocorrelation function (ACF), 410
autocovariance, 12
autoregressive–moving average (ARMA) models, 11
available-case missing-value restrictions (ACMV), 541
average bioavailability, 54
average mean squared error, 636
average response, 245

backprojection, 420
backward elimination procedure, 622
backward reliability curve, 840, 842, 849
Balaam's design, 55
balanced treatment incomplete block (BTIB) designs, 173
balanced uniform designs, 218

band shifting, 264
bandit problems, 573
bandlimited, 398
Barrett–Marshall–Schwartz approach, 834
basic reproduction number, 272
Bayes' formula, 619
Bayes' theorem, 400, 574, 881, 910, 912
Bayesian approach, 640
Bayesian bootstrap, 430
Bayesian discriminant analysis, 405
Bayesian estimation, 641
Bayesian HMM approach, 99
Bayesian image processing, 401
Bayesian iterative simulation, 431
Bayesian iterative simulation methods, 431
Bayesian network, 742, 843
Bayesian paradigm, 644
Bayesian restoration models, 417
Bayesian scan statistics, 742
Bayesian segmentation modeling, 99
Bayesian spatial scan statistics, 742
behavioral factors, 649
bending energy analysis, 507
bending energy matrix, 506
Bernoulli bandits, 573
Bernoulli distribution, 659
Bernoulli model, 738
Bernoulli processes, 734
Bernoulli random variables, 736
Bernoulli sequence, 736
Besag–York–Mollié (BYM) model, 936
best linear unbiased predictor (BLUP), 636
beta distribution, 85, 656
beta geometric distribution, 656
beta geometric model, 818
between-cluster variation, 143
bias and trial quality, 687
bias due to differential history, 607
bias due to differential maturation, 607
bias-adjusted estimate, 371
bias-corrected test, 83
binary clinical endpoint, 721
binary response, 726
Bingham distribution, 508
binning, 265
binomial distribution, 270, 721
binomial random variables, 740
bioassay, 118, 807
bioavailability, 53, 339
bioavailability studies, 338
bioequivalence, 922

bioequivalence measures, 54
bioequivalence studies, 339
bioequivalent, 53
biological factors, 649
biological grids, 505
biological landmark, 501
biological standardization (bioassay), 930
biomarker, 833
biomedical studies, 368
biomedical trials, 577
biometric functions, 496
biometrical genetics, 801
biometrical method, 894
BioSense program, 869
birth–death process, 61, 65, 72
bivariate exponential model, 184
bivariate extreme-value distribution, 350
bivariate frailty model, 350
bivariate lifetime distribution, 496
bivariate mixed model prediction method, 630
bivariate noncentral t distribution, 57
bivariate normal distribution, 507
bivariate normal model, 184
Bland–Altman plot, 463
blinded randomized trials, 147
blinding, 47
block diagonal matrix, 252
blurring, 415
body mass index, 302
Bonferroni critical values, 172
Bonferroni inequality, 100
Bonferroni method, 236
Bonferroni test, 550
Bonferroni-type inequalities, 743
Bongaarts model, 654
Bookstein coordinate system, 504
bootstrap, 797
bootstrap distribution, 485
bootstrap resampling, 550
box filter, 417
bracketing, 785
bracketing design, 785
Bradley–Blackwood test, 462
Brownian bridge, 495
Brownian motion, 366, 899
Brownian motion processes, 486
Brownian path, 368

χ^2 goodness-of-fit tests, 325
χ^2 tests, 325
c-statistic, 628

calibration, 846
cancer incidence, 62
cancer stochastic models, 61
candidate regions, 899
canonical discriminant, 776
carcinogenesis, 62
carcinogenicity, 332
carryover effect, 55, 216
carryover effects, 230
case k interval censoring, 452
case event data, 934
case fatality rates, 147
case–cohort data, 185
case–cohort sampling, 559
case–control, 367, 569
case–control data, 283
case–control method, 714
case–control study, 81, 189, 305, 366, 919
case–referent study, 714
categorical data, 311
categorical data methods, 232
Cauchy–Schwartz inequality, 843
causal graphics, 843
causal modeling, 39
causality and association, 119
cause–effect relationship, 567, 570
cause-specific cumulative incidence function, 751
cause-specific hazard function, 179, 750, 751
cDNA array, 795
censored, 312
censored counting process, 194
censored data, 194
censored times, 481
censored-data rank, 199
censoring, 122, 392
census, 159
center–satellite models, 12
center–satellite process, 10
centered landmarks matrix, 503
central-limit theorem, 172
centralized genomic control, 83
chain binomial distribution, 27
chain multinomial models, 19
chance agreement, 466
chance error, 306
change over design studies, 225
change over designs, 229, 922
change point methods, 95
changeover trials, 215
charge-coupled devices (CCDs), 398

CHARM (chromosomal aberration region miner), 98
Chernoff's conjecture, 576
chi-square distribution, 238, 394, 724, 802, 957
chi-square method, 872
chi-squared test statistic, 83, 804
child mortality, 154
circular binary segmentation, 101
circular-cone test, 174
circularity, 240
classification, 405
classification stopping rule, 618
clinical response to treatment, 927
clinical trials, 121, 131, 209, 304, 579, 917
clonal expansion models, 66
closed family of null hypotheses, 551
closed testing, 550
closure method, 171
cluster randomization, 143
Cochran–Armitage test statistic, 727
codominant, 801
coefficient of individual agreement (CIA), 470, 471
coefficient of interobserver variability, 470
cofactors, 275
cognate model, 263
cohort, 281
cohort analysis, 157
cohort analysis of fertility, 162
cohort sampling techniques, 559
cohort study, 305, 677
cohort table, 161
combination designs, 232
combinatorial and geometric probability methods, 743
combining evidence, 882
community diagnosis, 122
comparative assay, 930
comparative bioavailability study, 53
comparative clinical trials, 383
comparative designs, 606
comparative trial, 606
comparisons with a control, 167
comparisonwise error rate, 548
compartment analysis, 939
compensator, 194
competing risk models, 193
competing risks, 2, 49, 179, 394
competing-risk data, 749
complete case analysis (CCA), 535

complete factorial design, 782
complete isometry, 772
complete neutrality, 772
completely randomized, 145
complex Bingham distribution, 507, 508
complex multivariate normal distribution, 508
complex segregation analysis, 804
composite outcomes, 49
compound geometric distribution, 655
compound Poisson approximation, 741, 743
compound symmetry, 240, 848
compound symmetry structure, 852
computed tomography, 415
computer vision algorithms, 399
computer vision system, 400
computer vision technique, 412
computerized tomography, 404
computing software, 373
conception model, 829
concomitant variables, 233
concordance correlation coefficient (CCC), 465
concurrent follow-up, 343
conditional approach, 507
conditional autoregressive (CAR) prior distribution, 936
conditional hazard function, 1
conditional hazard rates, 350
conditional likelihood, 221
conditional maximum likelihood method, 845
conditional maximum likelihood procedures, 355
conditional power, 137
conditional survival probability, 4
conditionally, 843
confidence bands, 486
confidence sequences, 383
confidentiality, 121
confounding, 90, 227
confounding bias, 307
confounding variables, 120, 569
conjugate beta distribution, 818
CONSORT checklist, 690
CONSORT extension, 690
CONSORT Group, 689
CONSORT statement, 689
contingency table, 124, 150, 309, 392, 536
continuous clinical endpoint, 720
continuous-time Markov process, 185

continuous-time martingale, 2
controlled clinical trials, 147
controlled randomized clinical trials, 215
controlled randomized experiment, 570
convergence in law, 196
convergence in mean, 493
convergence in probability, 196
convolution, 936
convolution integral, 416
correlated frailty, 200
correlation matrix, 151, 169
countermatched sampling, 189, 564
counting process, 2, 193, 586, 755, 756
counting process decomposition, 324
covariance analysis, 121
covariance function, 493
covariance matrix, 2
covariate processes, 199, 586
coverage probability, 371
coverage probability (CP), 463
Cox proportional hazards model, 797
Cox regression, 189, 559, 751
Cox regression model, 199, 452, 758, 818, 920
Cox's partial likelihood approach, 320
Cox's proportional hazards, 957
Cronbach alpha coefficient, 840
cross-classification, 227
cross-correlation, 12
cross-ratio function, 350
cross-sectional and longitudinal data, 282
cross-sectional area, 776
cross-sectional effects, 695
cross-sectional sampling, 496
cross-sectional study, 281, 305, 452, 695
cross-validation, 421
crossover, 803
crossover design, 55, 215
crossover events, 359
crossover studies, 308
crossover trials, 922
cryptic relatedness, 81
cumulative hazard rate function, 954
cumulative incidence function, 184, 750
cumulative risk, 184
cumulative sum (CUSUM) type statistics, 742
cumulative type I error, 370
current status data, 451
curse of dimensionality, 366
cycle viability, 648, 659, 815

δ methodology, 198
D-efficiency, 790
D-optimality, 788
data, 312
data augmentation, 431
data imputation and augmentation procedures, 535
data structure, 392
day-specific conception probabilities, 821
day-specific probability of conception, 659
decision processes, 137
decision rules, 54
decision-tree classification, 405
deconvolution, 416
definitive analysis, 536
deformation, 505
degradation rates, 791
delta centralization, 81
demography, 119, 279, 528
dendrogram, 805
dentistry, 871
dependent observation process, 598
derived landmarks matrix, 503
derived model, 637
design comparison, 788
design moment, 790
design parameters, 368
design weights, 444
detectable difference, 611
deviance, 313
diagnostic keys, 913
diagnostic medicine, 366
diagnostic methods, 131
diagnostic screening test, 910
Diana, 796
differential equations, 185
differentially expressed genes, 797
differentiation, 64
differentiation rate, 61
digital image processing, 397
direct determinants, 652
Dirichlet distribution, 773
Dirichlet prior, 811
Dirichlet process, 811
disability models, 193
discount sequence, 573
discrete boundary values, 379
discrete Fourier transform (DFT), 403
discrete scan statistic, 735, 737
discrete survival approach, 834
discrete-time stochastic process, 19

discrete-time survival models, 657
discriminant analysis, 448, 910
discriminating power, 880
disease clustering, 937
disease map surveillance, 937
disease mapping, 937
disease occurrence, 1
disease surveillance, 743
disease-screening trials, 131
disease-specific mortality, 48
distance-based methods, 502
distributions over eyes, 580
distributions over individuals, 580
DNA fingerprinting, 259
DNA markers, 260
DNA microarray data, 795
DNA molecules, 260
DNA profiling, 259, 879
DNA repair system, 61
DNA sequence, 359, 893
DNA typing, 259
Doléans–Dadé exponential formula, 491
dominance variance, 874
dominant allele, 801
dosage–response relationship, 123
dosage–response relation, 673
dose ranging trials, 551
dose–response curve, 66, 725
dose–response experiments, 808
dose–response modeling, 867
dose–response relationship, 551, 931
dose-escalation studies, 918
double-blind, 306
double-blind techniques, 119
double-blinding, 47
double-censoring, 454
double-interval censoring, 454
double-sampling scheme, 331
doubly censored model, 496
drift parameter, 368
drug absorption, 53
drug efficacy study implementation, 337
drug interactions, 929
drug research, 928
drug surveillance studies, 339
Dunnett test, 550
duration dependence, 322
dynamic allocation index, 575
dynamic programming, 98, 575
dynamic representation, 505
dynamics of human reproduction, 647

Early Pregnancy Study, 816
early stopping, 365
ecological analysis, 937
edge pixel, 406
effects function, 670, 673
effects of aging, 164
elimination, 273
elimination half-life, 54
Elston–Stewart algorithm, 803, 897
EM (expectation–maximization) algorithm, 802
EM algorithm, 200, 275, 324, 351, 356, 420, 421, 441, 591, 660, 846, 897
embedded Markov chain, 739
empirical Bayes, 797
empirical Bayes estimate, 324, 643, 699
empirical cumulative hazard function, 494
empirical distribution function, 481, 482, 496
empirical studies, 303
empirical subsurvival functions, 484
endemic, 269
endpoint, 133, 451
environment factors, 649
epidemic, 269
epidemiology, 279, 299
eradication, 273
error spending, 370, 381
error spent, 379
error-spending function, 366
estimators, 482
Euler number, 410
event-history analysis, 193, 319
event history data, 523, 585
evidentiary sample, 260
evolutionary genetics, 873
evolutionary trees, 901
Ewens–Karlin–McGregor–Kingman–Watterson sampling formula, 556
ex post facto, 569
exit probability, 379
expectation–maximization (EM), 616
expectation–maximization (EM) algorithm, 447
expected information, 957
expected sample size, 380
experimental contamination, 146
experimental method in medicine, 916
experimental units, 226
experimentwise error rate, 548

experimentwise type I error rate, 724
expiration dating period, 781
explanatory variables, 392
exponential distribution, 349, 446, 956
exponential failure distributions, 135
exponential family, 844
exponential random variables, 776
exponential regression, 959
exponential variables, 770
exposure time, 658
extended multievent model, 63
extended Risch's method, 898
extended two-stage model, 65
extreme value distribution, 959
eye-specific basis, 580

F distribution, 84, 237
F statistic, 175
F test, 699, 700
factor analysis, 441
factorial analysis, 842
factorial designs, 132, 442
factorial linear model, 219
factoring the likelihood, 446
failure rate, 491
fallacies, 882
false discovery rate (FDR), 102, 549, 743, 797
false negative, 119, 366
false positive, 119, 366
familial aggregation analysis, 895
family-based association tests, 362
family-based designs, 185
family-wise error rate, 101, 548
feature extraction, 409
fecundability, 815
fecundability of the woman, 647
fertile window, 821
fertility, 161
fertility control, 651
fertility rate, 709
field trial, 304
Fieller confidence intervals, 948
Fieller's theorem, 809
filtered backprojection (FBP), 420
filtering, 98
finite Gaussian mixture model, 97
finite Markov chain embedding in, 743
finite memory, 574
finite-mixture models, 441
Finsen Institute data, 520
first-order interactions, 124

Fisher density, 507
Fisher distribution, 848
Fisher scoring, 820
Fisher scoring methods, 802
Fisher's exact test, 136
fixed effect model, 138, 533, 636
fixed effects, 699
fixed-bin method, 265
fixed-size procedure, 365, 367
floating-bin, 265
focused clustering, 937
follow-up, 343
follow-up studies, 481
force of mortality, 179, 491
force of transition, 179, 185
forensic science, 879
forensic statistics, 879
forward-looking studies, 284
four-sequence design, 55
Fourier analysis, 416
Fourier deconvolution methods, 417
Fourier descriptors, 410
Fourier transform, 398, 416
fraction of missing information, 428
fractional factorial design, 784
frailty, 657
frailty distribution, 350
frailty models, 200, 320, 323, 349
Framingham heart study, 345
Framingham study, 353, 523
full design, 782
fully conditional specification in multivariate imputation, 431
fully sequential tests, 372
functional delta method, 757
futures initiative, 865
fuzzy-logic systems, 411

G-efficiency, 790
gametic phase disequilibrium, 804
gamma distribution, 349, 591, 657, 661, 820, 823
gamma frailty distribution, 200
Gauss–Hermite quadrature, 846
Gaussian convolution, 407
Gaussian data, 697
Gaussian distribution, 416, 456, 751
Gaussian martingales, 196
Gaussian process, 183, 485, 493, 563, 757, 899
Gaussian stationary processes, 444
GEE approach, 472

general linear compartment system, 939
general linear model, 124, 776
generalized Bayesian method, 31
generalized birthday probability, 739
generalized estimating equation, 151, 846
generalized estimating equation approach, 586
generalized estimating equations, 154, 315, 700, 701
generalized gamma distribution, 323, 773
generalized inverse, 756
generalized least squares estimator, 640
generalized least-squares function, 590
generalized likelihood ratio type statistics, 742
generalized linear mixed effects model, 824
generalized linear mixed models (GLMM), 658, 700
generalized linear model, 320, 656, 822, 844
generalized mixed models, 644
generalized Nelson–Aalen estimators, 200
generalized Wilcoxon test, 199
genes within genotypes, 875
genetic epidemiology, 742, 893
genetic map distance, 360
genetic markers, 361
genetic variation, 801
genetics, 801
genome mismatch scanning, 363
genome-wide association studies, 84
genomic cloning, 106
Genomic control, 81
Genomic control (GC)-corrected test, 83
genomics, 893
genotypes, 801
geometric discounting, 575
geometric distribution, 817
geometric random variable, 654
Gibbs distribution, 418
Gibbs image models, 418
Gibbs sampler, 324, 431
Gibbs sampler algorithm, 661
Gibbs sampling, 25, 29, 431, 661, 920
Gibbs sampling algorithm, 824
Gibbsian form, 401
Gittins index, 575
gold-standard correlation, 468
goodness of fit, 5, 208, 324, 394, 533
goodness-of-fit test, 758, 852
graphical latent variable models, 853
graphical modeling, 842

graphical Rasch model, 845
gray-level concurrence matrix, 410
Greenwood estimated variance, 485
group randomization, 143
group sequential designs, 918
group sequential methods, 137
group sequential stopping rule, 383
group sequential tests, 366
group-randomized trial, 605, 612, 635, 636
group-sequential experimental designs, 378
group-sequential methods, 365
group-sequential procedure, 367
group-sequential tests, 377
grouped data, 183, 391
grouped survival data, 391
growth curves, 101, 523
Gumbel distribution, 959

Hájek—Le Cam convolution theorem, 493
Hagerstown morbidity study, 516
HapMap data, 82
Hardy–Weinberg equilibrium, 81, 261, 803, 895
Haseman–Elston linkage method, 898
Haseman–Elston method, 897
hazard function, 68, 203, 314, 349, 855
hazard rate, 818
hazard rate function, 189, 954
hazard ratio, 135
health promotion studies, 304
health related quality of life (HrQoL), 839
Henderson's approach, 640
Henderson's mixed-model equations, 640
heredity, 801
heritability, 894
heterogeneity, 349
heterogeneous Poisson process, 934
heterozygous, 261, 801
hidden Markov model, 23, 897
hierarchical Bayes modeling, 921
hierarchical Bayesian models, 510
hierarchical likelihood, 222
hierarchical model, 641
hierarchical parameter interpretation, 704
hierarchy, 885
high-dimensional Markov process, 71
high-pass filters, 404
higher-order crossover design, 55
higher-order moments, 700
highest posterior density (HPD) interval, 57
Hill plot, 928
historical cohort study, 713

historical follow-up, 345
HIV incidence, 275
HIV incubation period, 18
HIV infection, 19
Holm procedure, 551
homogeneous covariance structure, 237
homogeneous Markov chain, 521
homogeneous Poisson process, 741
homozygosity, 261
homozygous, 801
hot deck imputation, 444
hot-deck procedure, 430
Hotelling's T^2 statistic, 237
Hough transform, 407
HPD intervals, 829
human fecundity, 815
human fertility, 647, 815
human genetics, 893
human genome, 893
Human Genome Project, 361
human health risks, 332
human immunodeficiency virus, 15
human surveys, 121
Hunter upper bound, 737
hypergeometric distributions, 736

identical by descent, 263, 897
identical by state, 264
identifiability, 350, 491, 492, 661
identifiability and estimability, 939
identification, 879
identity-by descent, 803
identity-by-state, 803
ignorable models, 427
ignorance, 537
image, 397
image algebra, 403
image analysis, 399
image enhancement, 404
image processing, 397, 399
image restoration, 402, 415
imaging system, 398
imbalance functions, 135
implication relationships, 551
importance sampling, 736
imprecision, 537
imputation, 425
incidence, 269
incidence density ratio, 679
incidence function, 68
incidence of a disease, 302
inclusion–exclusion criteria, 153

incomplete block design, 55
incomplete data, 441
incomplete longitudinal data, 536
independent increments, 367
independent models, 912
index numbers, 9
index of dispersion, 9
indicants, 909
individual-level analysis, 150
individual-level surrogacy, 950
individualization, 879
infection incidence, 18
infectivity, 275
infinite-type process, 556
information matrix, 788
inhomogeneous Markov chains, 520
instrument, 845
instrumental variable, 39
integrated hazard, 197
integrated hazard function, 349
integrated intensity, 196
integrated regression functions, 2, 200
integrated transition rates, 520
integrated weighted difference, 592
intensity process, 2, 193, 320
intent-to-treat, 918
interaction among events, 515
intercourse pattern analysis, 824
interim analyses, 137, 365, 368, 377
interim looks, 365
interim monitoring, 137, 365
intermediate events, 210
interrater agreement, 461
intersection-union principle, 475
intersection–union multiple testing, 170
interval crossing, 451
interval mapping, 362
interval-censored covariates, 454
intervening variable, 300
intervention, 609
intraclass coefficient, 853
intraclass correlation, 228
intraclass correlation coefficient (ICC), 467
intraclass correlation model, 580
intracluster correlation, 148
intrarater agreement, 468
inversion formula, 491
island problem, 884
isobole, 929
isometry, 771
isotone regression method, 811

isotonic regression estimator, 587
isotropic distribution, 504
item response theory models, 845
iterated conditional modes (ICM), 409
iterated logarithm, 493
iterative convex minorant algorithm, 590

jackknife, 421
joint analysis, 855
joint correlation matrix, 847
joint covariance matrix, 847
joint coverage probability, 169
joint segregation, 896
joint survival distribution, 181

k-armed Bernoulli bandits, 577
K-means, 796
k-means clustering, 98
k-nearest-neighbor classification, 405
k-ratio method, 175
k-sample comparison, 593
Kalman filter model, 25, 63
Kalman filtering, 640
Kaplan–Meier, 955
Kaplan–Meier curves, 204
Kaplan–Meier estimator, 4, 452, 481, 489, 658, 961
Karhunen–Loeve transforms, 411
Kendall's correlation coefficient, 872
Kendall's tau, 351, 951
kernel function, 4
kernel smoothing, 197
Kolmogorov forward and backward differential equations, 520
Kolmogorov forward differential equations, 198
Kolmogorov forward equation, 68, 69
Kolmogorov–Smirnov-type tests, 4
Krönecker product, 625
kriging, 12, 505, 640, 933
kriging interpolator, 505, 506
Kruskal–Wallis statistic, 235
Kullback–Leibler information, 843
Kullback–Leibler measure, 840
Kullback–Leibler measure of association, 843
Kullback–Leibler measure of conditional association, 843

L test, 107
labeled vertices, 778
labor economics, 528

laboratory tests, 451
Lan–DeMets method, 381
Lander–Green algorithm, 803
landmark data, 501
landmark free methods, 510
landmarks, 501
Laplace transform, 349, 942
large deviation, 495, 743
large deviation approximation, 735
large deviation theory, 899
last observation carried forward (LOCF), 535
latent distribution, 846
latent parameters, 846
latent regression, 854
latent unidimensionality, 843
latent value, 639
latent-failure time model, 181, 185
Latin square, 230
laws of independent assortment, 801, 803
LD50, 674, 809
least squares, 125
least-squares filter, 417
least-squares methods, 242
Lebesgue–Stieltjes integrals, 490
Lenglart's inequality, 196
Lexis diagram, 157
life table, 396, 516, 916
likelihood ratio, 262
likelihood ratio statistic, 321, 699, 957
likelihood ratio test, 174, 207, 802
likelihood ratio test statistic, 100, 102
likelihood-based linkage analysis, 803
limit of life-table, 482
limits of agreement (LOA), 462
linear and quadratic discriminant functions, 912
linear compartment systems, 940
linear drift, 505
linear hazard model, 1
linear mixed model, 697
linear mixed-effects regression model, 616
linear model, 533
linear probability model, 671
linear programming, 743
linear regression model, 97
linear transformation matrix, 247
linkage, 803
linkage analysis, 359, 896
linkage disequilibrium, 804
linkage disequilibrium mapping, 899

linkage equilibrium, 261
linked cross-sectional, 229
local independence, 845
local influence, 538
local region, 894
local sufficiency, 845
location score, 360
lod score, 360, 803
log normal regression model, 959
log–log plots, 211
log-rank statistic, 393
logistic distribution, 809
logistic dose–response model, 727
logistic link, 845
logistic model, 672
logistic multiple regression, 355
logistic regression, 145, 151, 313, 309, 538, 563, 659, 900
logistic regression model, 221, 702, 919
logistic–normal model, 151
logistic-normal regression models, 154
logit function, 123
logit transformation, 672
loglinear analysis, 221
loglinear model, 125, 321, 395
loglinear size variables, 774
lognormal distribution, 773, 956
logrank analyses, 963
logrank test, 207, 313
long-term testing schedule, 781
longitudinal data, 319, 587
longitudinal data analysis, 515
longitudinal effects, 695
longitudinal population studies, 355
longitudinal study, 225, 280, 695, 919
lowpass filters, 404
lurking variables, 232

m-order spacings, 741
magnetic resonance imaging (MRI), 398, 419
main effects, 124
malignant melanoma data, 758
Mann–Whitney U, 872
Mantel–Haenszel chi-square statistic, 680
Mantel–Haenszel methods, 150
Mantel–Haenszel odds-ratio estimator, 384
Mantel–Haenszel statistic, 151, 393
Mantel–Haenszel test, 155, 621
Mantel–Haenszel test statistic, 145
Mantel–Haenszel–Cochran test, 313
map function, 803

Mardia–Dryden distributions, 507
marginal approach, 507
marginal covariance, 848
marginal hazards, 181
marginal maximum-likelihood method, 845
marginal models, 220
marginal survival distributions, 181
marked point process, 564
markers, 893
Markov approach, 71
Markov chain, 23, 804, 899
Markov chain embedding method, 741
Markov chain modeling, 743
Markov chain Monte Carlo (MCMC), 363, 661
Markov chain Monte Carlo algorithms, 418
Markov chain Monte Carlo methods, 894, 901
Markov dependence, 736
Markov models, 319, 322
Markov process, 185
Markov property, 519, 855
Markov random fields, 401, 418
martingale central limit theorem, 757
martingale central limit theory, 196
martingale methods, 495
martingale residual processes, 5
martingale theory, 320
martingales, 193
masked cause of failure, 754
match probability, 262
matched case–control data, 560
matched case–control studies, 221
matched controls, 282
matched-pair, 145
matched pair design, 147, 308
matching, 308
mathematical landmarks, 501
mathematical model, 123
mathematical morphology, 403
matrix product–integral, 198
matrixing, 785
matrixing design, 785
max–min formula, 588
maximal deviation, 199
maximum (partial)-likelihood estimates, 205
maximum a posteriori (MAP) estimation, 418
maximum a posteriori (MAP) estimator, 401

maximum backward one step Cronbach alpha coefficient, 850
maximum concentration, 54
maximum likelihood, 125
maximum-likelihood estimates, 180
maximum-likelihood estimators, 482
MCMC sampling, 824
McNemar's test, 901
mean change point model, 103
mean imputation, 444
mean integrated squared error, 197
mean shape, 503
mean-square error, 240
mean-squared deviation (MSD), 465
measurement model, 845
measures of intrarater agreement, 469
median filter, 402
median survival time, 312
median-filtered image, 403
medical diagnosis, 909
medical epidemiology, 528
medical follow-up study, 489
medical history status, 392
medical imaging, 398
Mendelian genetics, 801
Mendelian inheritance, 873, 893
Mendelian system, 873
meta-analysis, 138, 143, 154, 311, 531, 687, 920, 922
Metropolis algorithm, 896
Metropolis–Hastings, 661
microarray chips, 795
microsatellite markers, 361
migration, 164
minimum χ^{2*}, 125, 810
minimum prediction squared error, 640
missing at random (MAR), 441, 535
missing completely at random (MCAR, 535
missing data, 441
missing data in time series, 444
missing-data problem, 425
missing-information principle, 356, 447
missing non-future-dependent (MNFD), 541
missing not at random (MNAR), 535
missing-plot problem, 442
mixed longitudinal design, 229
mixed model, 63, 67, 138, 533, 635, 639
mixed parametric Poisson processes, 586
mixed Poisson processes, 586
mixed Rasch model, 845
mixed-case interval censoring, 452

mixed-effects model, 471
mixed-effects regression model, 613, 616
mixed-model analysis of variance, 441
mixture of distributions, 30
mixture of geometric distributions, 656, 818
mixtures of Markov models, 516
model checking, 209
model evaluation, 324
model fitting, 2
model-based clustering, 796
model-based inference, 642
model-based procedures, 445
model-free methods, 897
modeling, 209
models with heterogeneity, 660
modified inverse filters, 417
modified likelihood ratio test, 102
modified randomized block design, 55
Mokken model, 845
molecular genetics, 893
molecular technology, 259
monotonicity, 845, 853
Monte Carlo EM algorithm, 99
Monte Carlo procedures, 899
Monte Carlo sequential procedure, 29
Monte Carlo studies, 20
Monte Carlo testing, 743
Moore–Penrose generalized inverse, 507
morbidity, 344
morphometry, 777
mortalcity, 1
mortality, 162, 344
mortality studies, 164
multiallelic markers, 901
multicenter clinical trials, 138
multiclinic trial, 137, 339
multidimensional models, 842
multidimensional scaling (MDS) method, 509
multievent model, 63
multilocus genotype, 261
multinomial distribution, 16, 72, 125, 675
multinomial likelihood, 394
multipath change point problem, 101
multiple comparison procedure, 167
multiple comparisons, 121, 338, 547
multiple comparisons methods, 236
multiple correlation coefficient predicting, 727
multiple imputation, 425
multiple imputation by chained equations, 431
multiple loci, 901
multiple logistic regreswsion analysis, 728
multiple primary endpoints, 547
multiple regression analysis, 727
multiple regression methods, 242
multiple regression models, 150
multiple scan statistics, 735
multiple variance change points, 100
multiple-arm dose–response trial problem, 725
multiple-contrast tests, 174, 175
multiple-decrement life table, 180
multiple-decrement life tables, 186
multiple-imputation methods, 536
multiple-looks bias, 380
multiple-response variables, 119
multiple-sample problem, 723
multiple-scan statistics, 741
multiplicative intensity model, 194, 196, 320
multiplicative logistic distribution, 774
multipoint linkage analysis, 362, 896
multirandomization group clinical trials, 133
multistage model, 62
multistage procedure, 377
multistate models, 209, 321
multivariate t distribution, 168, 169
multivariate analysis of covariance model, 338
multivariate counting process, 2, 193, 494
multivariate incomplete data, 443
multivariate linear model, 219, 699
multivariate logistic distribution, 672
multivariate logistic regression model, 728
multivariate lognormal distribution, 773
multivariate lognormal family, 773
multivariate mixed-effects regression model (MLME), 623
multivariate normal assumptions, 912
multivariate normal distribution, 237, 394, 507, 524, 593, 625, 773
multivariate prediction model, 623
multivariate profile, 229
multivariate rank test, 220
multivariate stochastic models, 63
multivariate tolerance distributions, 672
multiway contingency table, 125
mutation, 555
mutation processes, 72, 555
mutation rate, 63
mutator phenotype pathway, 66

myelamatosis data, 3
myopic strategies, 574

n-way cross-classification, 123
National Death Index, 346
natural age specific fertility rate, 651
natural fertility, 651
natural heterogeneity, 614
nearest-neighbor distances, 9
neighboring-case missing values (NCMV), 541
Nelson–Aalen estimator, 195, 321, 324, 587
Nelson-Aalen estimate, 955
nested case–control sampling, 189
nested case-control sampling, 559
nested models, 321
neutral model, 876
neutrality, 771
New York angler cohort study, 824
new-drug applications, 335
Newton–Raphson algorithm, 595
Newton–Raphson procedures, 395, 616
Newton–Raphson method, 675
noise, 416
noise removal, 402
noisy image, 400
non-future dependent missing-value (NFMV) restrictions, 541
nonabsorbing states, 186
noncausal association, 900
noncompliance, 37, 135
nondifferential item functioning, 845, 853
nonfocused clustering, 937
nonhomogeneous Markov processes, 198
nonhomogeneous Poisson processes, 590, 591
nonidentifiability, 63, 182
nonignorable models, 427
nonlinear least-squares strategy, 941
nonlinear mixed model, 846
nonparametric approach, 472
nonparametric joint estimation, 331
nonparametric maximum pseudolikelihood estimator, 588
nonparametric maximum-likelihood estimator (NPMLE), 195, 456
nonparametric rank methods, 232
nonparametric survival models, 954
nonproportional hazards, 210
nonrandomized studies, 147
nonresponse process, 535
nontherapeutic interventions, 146

nonuniform background, 737
normal response, 725
number of holes, 410
numerical integration, 895
nuptiality, 164

O'Brien–Fleming test, 380
observation model, 27, 73, 75
observational plans, 917
observational studies, 567
observed at random, 441
occurrence dependence, 322
odds, 303
odds ratio, 120, 145, 729, 910
oligonucleotide, 795
oligonucleotide (affymetrix) chips, 795
one-armed bandit, 576
one-sample problem, 720
one-sided intervals, 169
one-way analysis of variance (ANOVA) model, 723
ophthalmology, 579
optimal classification error, 621
optimal diagnostic criteria, 630
optimal strategies, 577
optimum minimum distance estimators, 496
optional variation process, 195
optional-sampling bias, 380
order statistics, 122
order statistics and spacings, 743
order-restricted inference, 174
ordered null hypotheses, 551
ordinal variables, 301
ordinally scaled categorical data, 235
ordinary least squares, 2
orthogonal contrast test, 175
orthogonal polynomial coefficients, 524
orthogonal polynomials, 524
orthonormal contrasts, 248
orthonormal functions, 245
outbreak data, 272
outlier detection, 5
overdispersion, 349, 934
overinterpretation, 383
ovulation, 648, 815, 831

P-scan method, 737
pair matching, 569
paired t test, 462, 97, 871
paired availability design, 38
pairwise t tests, 97
pandemic, 269

panel analysis, 517
panel count data, 585
panel design, 229
panel studies, 319
parallel model, 841
parallel-group, 215
parallel-line assay, 808, 812
parametric approach, 472
parametric single-step tests, 550
Pareto distribution, 349
partial credit model, 845
partial least squares, 797
partial least-squares regression, 797
partial likelihood, 189, 204, 321, 323, 559, 751
partial likelihood approach, 754
partial likelihood function, 206, 958
partial likelihood techniques, 5
partial quasi likelihood methods, 846
partial response, 133
partial warp, 507
partially balanced design, 218
partition testing, 552
path diagram, 826
patient-reported outcome (PRO), 839
pattern–mixture modeling (PMM), 535, 539
pearl index, 658
pearl rate, 658
Pearson chi-square test, 150
pedigrees, 895
penetrance, 895
Penrose's partition, 777
percentage of item responses, 853
perfect at the trial level, 949
permutation techniques, 797
permutation test, 40
persistence, 515
person-specific basis, 580
pharmaceuticals, 927
pharmacokinetic–pharmacodynamic modeling, 918, 922
phase I, 920
phase I trial, 304
phase II trial, 304, 920
phase III trial, 304
phase IV trials, 304
phase 2 trials, 132
phase 3 trial, 132
phenotypes, 801, 893
physical mapping, 894
piecewise exponential distribution, 394

piecewise exponential likelihood function, 394
piecewise exponential model, 321
Pitman–Morgan test, 462
pixels, 397
plausibility of paternity, 885
point process, 193, 320
point spread function, 416
Poisson, 734
Poisson approximation, 493
Poisson distribution, 9, 72, 73, 402, 420, 734
Poisson process, 9, 61, 322, 360, 586, 734, 770, 776
Poisson random variable, 650, 740
Poisson regression, 144, 314, 395
Poisson–Dirichlet distribution, 557
Poisson-type approximations, 741
polygenes, 895
polymerase chain reaction, 260
polynomial models, 523
polytomous logistic regression, 313
pool adjacent violators, 588
pool adjacent violators algorithm, 588
population averaged models, 817
population genetics, 801
population stratification, 81
population substructure, 262
positive control trials, 136
positive stable distribution, 200, 350
positron emission tomography (PET), 420
posterior probability, 619
posterior probability distribution, 400
potential confounding, 608
power, 135, 137, 310, 611
power of the test, 368
preassessment model, 885
preclinical disease, 613
predictable variation process, 194
predicted longitudinal trends, 627
prediction, 208
prediction classification method, 628
predictive encoding, 411
pregnancy rate, 658
preshape landmarks, 503
prevalence, 302
prevalence of disease, 119
prevention trials, 131
primary sampling unit (PSU), 638
principal components, 90, 504
principal warps, 506
principal-component analysis, 81, 777, 842,

848
privacy, 121
probability integral transformation, 405
probability of conception, 661, 817
probability-generating function, 68
proband, 43
probit, 671
probit analysis, 669
probit model, 671
process control and monitoring, 742
procrustes analysis, 507
procrustes mean shape, 503
product limit estimate, 183, 955
product-limit estimator, 481
product–integral, 193, 197
profile likelihood, 200, 351
prognostic categories, 909
proliferation rate, 61
proportion explained, 945
proportional excess hazards model, 5
proportional hazard alternatives, 135
proportional hazards assumption, 313
proportional hazards model, 1, 203, 319, 349, 395, 457, 492, 598, 752, 758, 958
proportional hazards regression, 314
proportional hazards structure, 395
proportional mean model, 586
proportional odds, 313
proportional odds model, 457
prospective studies, 284, 677
proximate determinants of fertility, 652
pseudolandmarks, 501
purely random, 649
pyrogen test, 339

QR decomposition, 508
quadrat sampling, 12
qualitative interactions, 138
qualitative response models, 670
quality, 683
quality control, 931
quality control tests, 735
quality of life, 839
quality-adjusted time without symptoms or toxicity, 862
quality-control measures, 612
quality-of-life data, 921
quantal measure, 927
quantal models, 669
quantal response analysis, 669
quantal response curves, 810

quantal response models, 809
quantal response studies, 813
quantal responses, 807
quantile ratio, 657
quantitative, 801
quantitative traits, 804
quasi-experiment, 568
quasi-experimental comparisons, 146
quasi-experimental designs, 146
quasi-likelihood modeling, 895
quasisymmetry, 251

radioimmunoassay, 808
radon transformation, 404
Ramlau-Hansen's kernel estimator, 4
random assignment, 145
random censorship, 183, 485
random effects, 635
random permutation model approach, 640
random permutation model predictors, 642
random permutation models, 643
random-effect models, 533, 639, 820, 817
randomization, 131, 134, 203, 918
randomization-based nonparametric methods, 242
randomized clinical trial, 121, 695, 916
randomized consent design, 38
randomized controlled trials (RCTs), 683
randomized designs, 38
randomized response, 122
randomized trial receives, 37
randomized trials, 143
randomly stopped sum distribution, 9
rankits, 672
Rasch model, 840, 844
rate constant, 54
ratio vector, 770
real Bingham distribution, 508
realized group, 639
receiver operating curve (ROC) analysis, 132
recessive, 801
recombinant DNA, 893
recombination parameter, 359
record linkage, 293
recurrent event, 320, 585
recursive box filter, 417
redistribute-to-the-right estimator, 484
reduced design, 784
reference population, 260
reflection size and shape, 508
region of ignorance, 538

region of uncertainty, 538
regression analysis, 594
regression imputation, 444
regression on the mean, 20
regression techniques, 797
regression to the mean, 873
regression with grouped or censored data, 448
regressive models, 895
regular, 771
regular discounting, 576
related pair, 771
relative effect, 945, 948
relative potency, 808, 812
relative risk, 302
relative warp analysis, 507
relative-risk regression, 920
reliability, 469
reliability function, 482, 489
reliability of linear systems, 736
reliability studies, 585
renewal process, 323
repeat fecundability, 820
repeat victimization, 525
repeatability coefficient, 470
repeated confidence intervals, 137
Repeated confidence intervals (RCIs), 382
repeated cross-sectional design, 152
repeated cross-sectional surveys, 152
repeated imputations, 429
repeated measurement designs, 528
repeated measurements, 695
repeated measures analysis of variance, 232, 240
repeated measures studies, 225
repeated significance test, 380
replacement rates, 710
replication, 131, 145
reproduction rates, 709
resampling-based single-step tests, 550
resampling-based testing procedures, 550
research design, 606
residual effects, 230
restricted maximum likelihood (REML), 219, 616
restricted maximum likelihood (REML) estimation, 699
restriction enzymes, 260
retrospective, 282
retrospective case–control studies, 119
retrospective cohort study, 714

retrospective studies, 713
retrospective surveys, 284, 319
reversed supermartingales, 492
Ricatti equations, 70
right-censored, 481, 489
right-censoring-left-truncation model, 496
risk difference, 303
risk ratio, 679
Robbins–Monro process, 812
robust HMM approach, 99
robust modeling schemes, 896
robust multiarray average, 796
robust variance estimators, 151
ROC curve located, 619
Roessler's adjustment, 168
rotating groups, 229
run-in period, 216

σ-algebras, 193
saddle point approximations, 743
sample size, 611
sample size determination, 57, 310, 719
sample survey data, 443
sampling variability, 307
scale mixtures normal distribution, 824
scaled original landmarks, 503
scales, 685
scan statistics, 733
scanning arc, 739
scatterplot, 107
Scheffé method, 238
Schwarz information criterion, 105
score test, 207
second-order interaction, 124
secondary sampling unit (SSU), 638
segmentation, 405
selection bias, 152, 607
selection effect, 37
selection modeling, 538
selection models, 538
selection procedure, 167
selective mixing pattern, 24
self-consistency criterion, 490
self-consistent estimator, 484
self-organizing maps, 796
semi-Markov models, 323
semiparametric additive model, 760
semiparametric analysis, 199
semiparametric approach, 472
semiparametric estimation, 321
semiparametric models, 751

sensitivity, 912
sensitivity analysis, 536
separability, 852
sequence of Bernoulli trials, 655
sequential analysis, 922
sequential decision problems, 574
sequential designs for quantal responses, 808
sequential probability ratio test (SPRT), 365
sequential procedures, 365
sequential regression multivariate imputation, 431
sequential testing, 913
sequentially rejective algorithm, 551
seroprevalence data, 272
severe acute respiratory syndrome (SARS), 867
shape comparisons, 505
shape distributions, 507
shape variability, 504
shape vector, 770
shared frailty, 200
shelf life, 781
shelf life estimation, 789, 791
shift-invariant, 416
short tandem repeats, 260
shrinkage, 636
Šidák-adjusted p value, 549
sign test, 235
signal-to-noise ratio (SNR), 399
significance level, 137
significance probability, 880
significance testing, 720
Simes test, 549
simple logistic regression, 313
simple response error model, 643
simple tree ordering, 174
simulated annealing, 402
simultaneous confidence bands, 4
simultaneous confidence interval, 168
simultaneous confidence limits, 2
simultaneous inference, 167
single-blind, 306
single-institution trials, 137
single-locus genotype, 261
single-marker linkage analysis, 896
single-nucleotide polymorphism, 82
single-photon emission computed tomography (SPECT), 398, 419
single-step methods, 169
single-step procedure, 170
single-step tests, 549

size and shape, 501
size and shape analysis, 769
size variable, 770
slope-ratio assay, 808
software, 764
sources of variability studies, 225
spatial aggregation, 12
spatial clustering, 315
spatial detection, 743
spatial epidemiology, 933, 936
spatial resolution, 399
spatial sampling, 398
spatial scan statistics, 741
spatial-temporal cases, 738
spatially invariant, 416
Spearman's rank correlation, 951
Spearman–Brown formula, 847
Spearman–Kärber estimator, 811
specificity, 852, 912
spherical refractive error, 583
sphericity, 240
spline curve, 797
split-block experiment, 227
split-plot experiments, 225, 226
split-split-plot experiment, 227
sporadic risk, 895
square integrable martingales, 757
squared integrated deviation, 199
stability, 781
stability of multivariate relationships, 515
stability of products, 930
stability protocol, 783, 791
stability study designs, 781
stability testing, 339
stable population, 709
stage space models, 63
stagewise ordering, 372
standard crossover design, 55
state space models, 25, 73
statistical analysis of reproduction, 647
statistical clustering, 796
statistical decision theory, 573
statistical evidence, 330
statistical genetics, 118, 801
statistical guidelines, 337
statistical ignorance, 537
statistical imprecision, 537
statistical power, 790
statistical principle of parsimony, 445
statistical quality of life, 839
statistical uncertainty, 537

stay with winner rule, 574
stepdown Bonferroni method, 172
stepdown method, 170
stepdown procedures, 797
stepup method, 171
stepwise regression analysis, 124
stepwise testing procedures, 170
stochastic compartment model, 518, 939
stochastic curtailment, 137, 378
stochastic difference equations, 72
stochastic differential equations, 71
stochastic integrals, 193
stochastic proliferation, 64
stochastic system model, 74
stochastic transmission models, 15
stopping boundaries, 365
stopping rule, 368, 377
stopping time, 365
stratification, 134, 145
stratification variable, 311
stratified, 146
stratified design, 146
stratified proportional hazards model, 204
stratified randomization, 134
stratified sampling, 85
strong law, 493
strongly balanced uniform designs, 218
structure of individual time paths, 515
structured association, 81
student's t distribution, 169
student's t test, 168
subcompositional independence, 774
subdistribution hazard function, 753
subject-specific models, 221
subject-specific regression parameters, 699
subset analysis, 137
subset pivotality condition, 550
sum of item responses, 853
superpopulation, 641
superpopulation model approach, 640
superpopulation model predictors, 641
superpopulation models, 643
superregular, 577
surrogate effect, 947
surrogate endpoints, 945
surrogate measure of exposure, 189
surveillance data, 271
survival data, 953
symbolic computing, 743
symmetry, 251
syntactic grammars, 412

systematic error, 306

2×2 contingency table, 119
t statistic, 170, 175
t test, 219, 245, 699
t-test statistics, 591
tangent space, 504
temporal pattern, 716
temporal point process, 741
temporal surveillance for bioterrorism, 737
teratogenicity, 332
terminal decision rule, 377
thematic mapper, 398
therapeutic equivalence trial, 136
thin-plate splines, 505
three-point distribution, 818
threshold analysis, 97
tied data, 205
time axis split, 211
time curve, 54
time reversibility, 555
time to achieve maximum concentration, 54
time to pregnancy, 648, 817
time-dependent covariate, 5, 209, 212, 657, 958
time-homogeneous Markov model, 522
time-inhomogeneous Markov chain, 522
time-series analysis, 640
time-to-event analysis, 451
time-to-event data, 204
time-to-event variable, 203
time-to-pregnancy, 660
time-to-pregnancy models, 656
time-varying covariates, 528
time-varying covariate histories, 564
time-varying effects, 324
timing of outcomes, 49
tobits, 673
tolerance components, 670
tolerance distribution, 671, 809
tolerance interval, 463
tolerance interval (TI), 463
total agreement, 469
total deviation index (TDI), 463
total fertility rate, 649, 653
total mortality, 47
tracking, 523
transfer evidence, 879
transform coding, 411
transform features, 411
transition probabilities, 198
transmission disequilibrium test, 900

transmission models, 276
treatment effect, 37, 947
tree-based scan statistic, 742
trellis plot, 474
trend test statistic, 83, 727
trial-level surrogacy, 949
triangular boundary, 366
triangular density, 818
truncated distribution, 443
tuberculosis mortality, 163
Tukey's T procedure, 724
tumor-incidence data, 452
Turnbull's algorithm, 456
two sample $t$tests, 797
two-armed bandits, 576
two-armed Bernoulli bandit, 574
two-dimensional discrete scan statistic, 737, 740
two-dimensional histogram, 411
two-dimensional scan statistics, 740
two-parameter class of tolerance distributions, 672
two-period crossover design, 133
two-period design, 217
two-sample $t$test, 697
two-sample problem, 721
two-sequence dual design, 55
two-stage acceptance sampling plan, 377
two-stage hierarchical models, 742
two-stage regression approach, 150
two-way contingency table, 444
two-way mixed model, 467
type I error, 85, 137, 143, 167, 310, 547
type II error, 310

U statistic, 107
U-shaped fecundability distributions, 657
unbalanced design, 580
unbiased measurement, 131
unblinded randomized studies, 147
uncensored times, 481
uncertainty, 790
unified theory, 367
uniform consistency, 196
uniform distribution, 87
uniform matrix designs, 790
uniformly minimum variance unbiased estimate (UMVUE), 372
uniformly minimum-variance unbiased estimator (UMVUE), 382
uniformly strongly consistent, 492
union–intersection (UI), 549

union–intersection multiple testing, 169
union–intersection testing, 549
unitary transforms, 404
univariate counting process, 193
unlabeled vertices, 778
unsupervised hidden Markov models, 98
untestable assumptions, 538
up-and-down algorithms, 588
up-and-down method, 812

vaccinations, 273
vaccine efficacy, 273
value of the evidence, 881
variable number of tandem repeats, 260
variable-size windows, 737
variance components model, 471
variance–covariance matrix, 207, 625
variance-balanced design, 218
variance-covariance structure, 524
variation, 801
vital statistics, 117
von Mises density, 507

Wald confidence interval, 721
Wald goodness-of-fit statistic, 254
Wald statistic, 235, 251, 394
Wald test statistic, 729
Wald tests, 321, 699
Wald's test, 207
Wald-type test, 702
warping, 505
washout period, 216
weakest link, 671
Weibull distribution, 323, 956
Weibull regression, 959
Weibull survival model, 457
weight of evidence, 881
weight process, 955
weighted least squares, 125, 810
weighted least squares methods, 254, 396
weighted least-squares (WLS) estimator, 4
weighted sum of item responses, 853
weighting procedures, 444
Wiener filter, 417
Wilcoxon signed rank test, 235, 238
Wilcoxon's signed rank, 872
Williams designs, 218
WinBUGS, 936
within-cluster dependences, 143
within-patient trial, 916
within-subject functions, 233
within-subject linear functions, 237

within-subject responses, 233
Wright–Fisher model, 875
Wright's measure of degree, 83